WITHDRAWN
0 2 MAR 2022

Handbuch der experimentellen Pharmakologie
Handbook of Experimental Pharmacology

Heffter-Heubner New Series

XXXVI

Add: HODGE, H. C. and others editors

Uranium · Plutonium Transplutonic Elements

Contributors

W. J. Bair · J. E. Ballou · B. Bleaney · G. W. Dolphin · P. W. Durbin
A. Ghiorso · J. E. Gindler · J. W. Healy · H. C. Hodge
D. A. Holaday · J. B. Hursh · W. H. Langham · A. de G. Low-Beer
J. F. Park · H. M. Parker · C. L. Sanders · L. M. Scott
G. T. Seaborg · N. L. Spoor · J. N. Stannard · D. M. Taylor
J. Vaughan · C. L. Yuile

Editors

H. C. Hodge · J. N. Stannard · J. B. Hursh

With 243 Figures

Springer-Verlag Berlin · Heidelberg · New York 1973

Harold C. Hodge, Department of Pharmacology, University of California, San Francisco, California 94122, USA

John B. Hursh, University of Rochester School of Medicine and Dentistry, Department of Radiation Biology and Biophysics, Rochester, New York 14642, USA

J. Newell Stannard, University of Rochester School of Medicine and Dentistry, Department Radiation Biology and Biophysics, Rochester, New York 14642, USA

ISBN 3-540-06168-1 Springer-Verlag Berlin · Heidelberg · New York
ISBN 0-387-06168-1 Springer-Verlag New York · Heidelberg · Berlin

Type setting, printing and binding: Universitätsdruckerei H. Stürtz AG Würzburg

Table of Contents

Chapter 2

Physical and Chemical Properties of Uranium. James E. Gindler. With 8 Figures

Chapter 3

Animal Experiments. C. L. Yuile. With 2 Figures

Chapter 4

Data on Man. J. B. Hursh and N. L. Spoor. With 8 Figures

Chapter 5

Protection Criteria. N. L. Spoor, and J. B. Hursh. With 1 Figure

Chapter 6

Environmental Monitoring and Personnel Protection in Uranium Processing. L. M. Scott. With 14 Figures

Chapter 7

Uranium Mining Hazards. D. A. Holaday

Plutonium

Chapter 11

Plutonium in Soft Tissues with Emphasis on the Respiratory Tract. W. J. Bair, J. E. Ballou, J. F. Park, and C. L. Sanders. With 43 Figures

Chapter 12

Maximum Permissible Body Burdens and Concentrations of Plutonium: Biological Basis and History of Development. W. H. Langham and J. W. Healy

Chapter 13

Bioassay of Plutonium. Anne de G. Low-Beer. With 2 Figures

Chapter 14

Plutonium. Industrial Hygiene, Health Physics, and Related Aspects. H. M. Parker. With 14 Figures

Chapter 15

Plutonium in the Environment. J. N. Stannard

Transplutonic Elements

Chapter 16

A History of the Transplutonic Elements. A. Ghiorso. With 11 Figures

Chapter 17

Chemical and Physical Properties of the Transplutonium Elements. DAVID M. TAYLOR. With 2 Figures

Chapter 18

Metabolism and Biological Effects of the Transplutonium Elements. PATRICIA W. DURBIN. With 40 Figures

Chapter 19

Maximum Permissible Concentrations and Maximum Permissible Body Burdens for Transplutonic Elements. G. W. Dolphin

Chapter 20

Bioassay of Transplutonium Elements. Anne de G. Low-Beer.

Chapter 21

Contributors

W. J. Bair, Biology Department, Battelle Pacific Northwest Laboratories, Battelle Boulevard, Richland, WA 99352 / USA

J. E. Ballou, Biology Department, Battelle Pacific Northwest Laboratories, Battelle Boulevard, Richland, WA 99352 / USA

Betty Bleaney, Garford House, Garford Road, Oxford / Great Britain

Geoffrey W. Dolphin, National Radiological Protection Board, Biology Department, HQ & Southern Ctr., Bldg. 565 T, Harwell, Didcot, Berkshire / Great Britain

Patricia W. Durbin, Donner Laboratory and the Division of Biology and Medicine, Lawrence Berkeley Laboratory, University of California, Berkeley, CA 94720 / USA

Albert Ghiorso, Director, Heavy Ion Linear Accelerator, Building 71, 268 Room, LRL, University of California, Berkeley, CA 94720 / USA

James E. Gindler, Chemistry Division, Argonne National Laboratory, 9700 South Cass Avenue, Argonne, IL 60439 / USA

J. W. Healy, Los Alamos Scientific Laboratory, P.O. Box 1663, Los Alamos, NM 87544 USA

Harold C. Hodge, Department of Pharmacology, University of California, San Francisco, CA 94122 / USA

Duncan A. Holaday, 7 Carlton Road, Wellesley, MA 02181 / USA

John B. Hursh, University of Rochester School of Medicine and Dentistry, Dept. Radiation Biology and Biophysics, Rochester, New York 14642 / USA

Wright H. Langham†, Group Leader, Biomedical Research, Los Alamos Scientific Laboratory, P.O. Box 1663, Los Alamos, NM 87544 / USA

Anne de G. Low-Beer, Lawrence Radiation Laboratory, University of California, Berkeley, CA 94720 / USA

J. F. Park, Biology Department, Battelle Pacific Northwest Laboratories, Battelle Boulevard, Richland, WA 99352 / USA

Herbert M. Parker, Senior Staff Consultant, Battelle-Northwest Lab., Battelle Boulevard, Richland, WA 99352 / USA

C. L. Sanders, Biology Department, Battelle Pacific Northwest Laboratories, Battelle Boulevard, Richland, WA 99352 / USA

L. M. Scott, Union Carbide Corporation, Nuclear Division, Y-12 Plant, Bldg. 9711-1, P.O. Box Y, Oak Ridge, TN 37830 / USA

Glenn T. Seaborg, Lawrence Berkeley Laboratory, University of California, Berkeley, CA 94729 / USA

Norman L. Spoor, National Radiological Protection Board, United Kingdom Atomic Energy Authority, Health and Safety Branch, Harwell, Didcot, Berkshire / Great Britain

J. N. Stannard, University of Rochester School of Medicine and Dentistry, Dept. Radiation Biology and Biophysics, Rochester, NY 14642 / USA

David M. Taylor, Royal Cancer Hospital, Department of Biophysics, Clifton Avenue, Belmont, Sutton, Surrey / Great Britain

Janet Vaughan, 1 Fairhaven End, First Turn, Wolvercote, Oxford, OX2 8 AR / Great Britain

Charles L. Yuile, Dept. of Radiation Biology and Biophysics, University of Rochester, Medical Center, Rochester, NY 14642/ USA

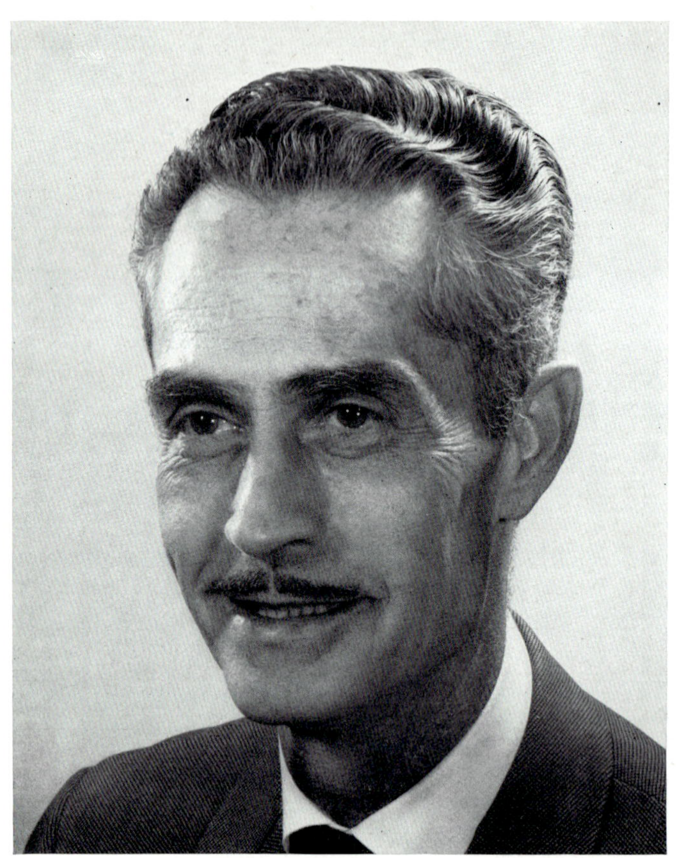

Wright Haskell Langham
1911—1972

Dedication to Wright Haskell Langham

This volume is dedicated by its scientific editors to the memory of WRIGHT HASKELL LANGHAM, Ph.D., whose scientific career was abruptly and prematurely ended in May 1972. Dr. LANGHAM devoted a major portion of his research and scientific administrative years to understanding the biomedical aspects of the actinide elements, particularly plutonium. Indeed, on the biomedical front, WRIGHT LANGHAM deserves fully the title, "Mr. Plutonium."

Soon after he joined the staff of the old "Metallurgical Laboratory" at Chicago during World War II he began work on the new actinides and their metabolism and effects in animals and man. Carrying on at the Los Alamos Scientific Laboratory from the summer of 1944 until his death, he early developed bioassay procedures for estimation of body burdens of plutonium in man and animals and pioneered studies of its distribution and excretion in man, both workers and hospital patients, as described in Chapters 8 and 10 herein. From these he constructed the universally used "Langham equation", which describes the kinetics of plutonium excretion. Indeed, he was active in stimulating and correlating much of the toxicological work on plutonium and other actinides, not only at Los Alamos but also at the "Chicago Toxicity Laboratory" and the University of Rochester during the years of the Manhattan Project and later in the development of programs with the bone-seeking radionuclides at the University of Utah and other laboratories.

With Doctors AUSTIN BRUES, ROBLEY EVANS, KARL MORGAN, HERBERT PARKER and SHIELDS WARREN as the principal other participants, Dr. LANGHAM took an active part in developing the logic for and stimulating the decisions concerning setting the maximum permissible body burden for plutonium and derived figures for permissible air and water concentrations in installations in the U.S.A. and later, with Canadian and United Kingdom participation, the internationally recommended values were established that remain essentially unchanged today. The saga of these developments, fortunately, is assembled for posterity in Chap. 12 of this volume by W. LANGHAM and J. HEALY.

Dr. LANGHAM combined to an extraordinary degree the abilities of a talented scientific researcher with those of the practical "operations man". These talents led him to participate in many of the field operations of the United States Atomic Energy Commission, the Department of Defense, and of various international groups. Thus, he was almost immediately summoned to take major responsibilities in the Broken Arrow incidents at Thule, Greenland and Palomeras, Spain. These activities are described in part in Chap. 15 of this book. Earlier he took an active part in biomedical aspects of the nuclear weapons testing program and what is now described as the "nuclear weapons fall-out controversy".

While we identify WRIGHT LANGHAM primarily with plutonium toxicology, he did many other things. With E. PINSON and others, for example, he pioneered the study of the metabolism of tritium in man. In addition, he contributed significantly to an intensive program on the relative biological effectiveness of different types of radiation, particularly neutrons *versus* X and gamma radiation, to our understanding of the radiobiological problems of manned space flight and to an examination of the potential for radiotherapy of various high energy particles such as mesons.

President of the Health Physics Society in 1968–1969, member of numerous national and international committees, chairman of several, advisor to both the scientific establishment and to governments, Dr. WRIGHT H. LANGHAM exerted a most significant influence on the field encompassed by this book. Accordingly, we dedicate it to him with our respect for his abilities and character, with gratitude for his contributions, and with affectionate memories of his friendship.

Uranium

Preface

More than a year ago the three editors sat down at a table and worked out a set of six chapter headings which they believed might serve, in turn, for each of the three sections of this handbook. (The reader will note a similarity in order of presentation and in emphasis.) However, as our editorial plans progressed it became apparent that for each element and for the element group, there were one or two special topics appropiate for that section alone.

Accordingly, in the section on uranium the common pattern holds for Chaps. 1 through 6 which include: an introduction (Chap. 1), a discussion of the physical and chemical properties (Chap. 2), experimental data on animals (Chap. 3), experimental data on man (Chap. 4), the rationale and development of air concentration limits to control industrial worker exposure (Chap. 5), and the practical problems of applying such limits in the uranium industry (Chap. 6). Chap. 7 entitled "Uranium Mining Hazards" is the subject category which is special for uranium; the chapter brings up to date the account of an important occupational hazard which was first noted by GEORGIUS AGRICOLA (1490–1555).

It will be readily appreciated that the approach throughout this handbook focuses on the control of these hazardous elements in the industrial situation and in the total environment. We believe this to be a proper matter of concern for the toxicologist. As a professional person it is imperative that he be informed in this area of increasing importance as the power-hungry world depends more and more on reactors to satisfy its needs. As a private individual living in that world he has a special duty to bring an alert balanced judgement to bear on problems of contamination of the environment by these elements and their by-products.

JOHN B. HURSH

Chapter 1

A History of Uranium Poisoning (1824–1942)

HAROLD C. HODGE

With 5 Figures

I. Early History. Homeopathic Uses

CHRISTIAN GOTTLOB GMELIN, professor of chemistry at the University of Tübingen, at the age of 32 published a treatise[1] (1824) describing his studies of the effects of salts of 18 metals, including uranium, on animals. These are the first known observations of the biological effects of uranium. A brief summary (GMELIN, 1825) in the Journal für Chemie and Physik (Halle) the following year presents his conclusions, and a more detailed critical condensation by an unknown editor appeared in the Edinburgh Medical and Surgical Journal in 1826. Gmelin's treatise "intented to supply the deficiencies existing in our knowledge of the poisonous properties of the metals, and so to inform the physician what are deleterious and what are not, and to enable the toxicologist to complete the classification of the metallic poisonings". GMELIN characterized uranyl salts given by mouth as "a feeble poison" with "keine sehr bedeutende Einwirkung", whereas injected intravenously "bewirken sie schnell den Tod" by destroying the irritability of the heart and coagulating the blood (Fig. 1.1–1.3).

GMELIN prepared pure uranium oxide from pitchblend by Arfvedson's method and then made the uranyl nitrate, sulfate and chloride. He noted that at least one sample of uranium nitrate had a perceptible excess of acid, and that the sulfate was only slightly soluble. Two dogs received each a dose of a uranium salt mixed with meat, a dog and a rabbit were given larger doses by stomach tube, and 2 dogs were given sizable doses (180 and 600 mg, resp.) intravenously in aqueous solution.

The dogs treated per os showed no illness from doses of 300 mg of the sulfate and 900 mg of the nitrate; emesis repeated once was observed in a third dog given 4 g of Uranyl nitrate (UNO_3) in 30 cc of water. In contrast, the rabbit given about 2 g of the chloride by gavage died without convulsions 52 hours later (paralysis of the extremities preceded death). An autopsy examination showed inflamed loci on the pleura and the lung. The heart was normal. The lining of the stomach, mottled and gangrenous, showed inflammation from the cardia to the pylorus, with many areas of extravasated blood. The upper intestine was sound.

Injecting 600 mg of UNO_3 or 180 mg of the chloride into the jugular vein killed the dogs within a minute. In the dog given UNO_3, immediate examination showed the right heart full of black, fluid blood. In the dog given the chloride, the right heart was filled with clotted blood that also could be pulled in red

1 In examining a copy of GMELIN's treatise in the rare book collection of the Countway Medical Library of the Harvard Medical School, I was astonished to find the pages uncut.

Versuche

über die

Wirkungen

des

Baryts, Strontians, Chroms, Molybdäns, Wolframs, Tellurs, Titans, Osmiums, Platins, Iridiums, Rhodiums, Palladiums, Nickels, Kobalts, Urans, Ceriums, Eisens und Mangans

auf den

thierischen Organismus.

Von

C. G. Gmelin,

Doctor der Medicin

und Professor der Chemie an der Universität zu Tübingen.

Tübingen

in der H. Laupp'schen Buchhandlung

1824.

(Druk von Hopfer de l'Orme.)

Fig. 1.1. Reproduction of the frontis page from the treatise of C. G. Gmelin, published in Tübingen by the Laupp'schen Buchhandlung in 1824. Courtesy of the Boston Medical Library in The Francis A. Countway Library of Medicine

strings from the great vessels. Considerable fluid was found in the pericardial space. The left heart was empty but a clot filled the beginning of the aorta. Gmelin wondered whether the astonishing difference in coagulating effect depended on the nature of the acids. He contrasted the feeble character of the heart beat with the still vigorous contraction of the diaphragm elicited by touching the phrenic nerve. Gmelin pointed out that the salts of only 3 of the many metals tested coagulated blood, namely, those of barium, uranium, and palladium, which drew his comment that the 3 have very different chemical properties. He concluded on a broader scope, having considered his studies of many elements, that "no relation whatever subsists between their action on the economy, and their physical and commoner chemical properties".

Nearly 30 years passed before Leconte (1853, 1854) discovered the remarkable ability of uranium acetate to induce glycosuria along with oliguria and anuria, the effect which ultimately made uranium one of the agents of choice in the production of experimental, acute or chronic, nephritis. Leconte (1853) "always found sugar in the urine of dogs slowly poisoned by small doses of nitrate of uranium". His brief statement (1853, 1854) included a) an estimate of a lethal dose range for small animals (0.5 to 1 g), b) a description of the local irritant

(78)

Aus diesen Versuchen ergibt sich, dafs das Kobalt dem Nickel ziemlich ähnlich wirkt, nur zeigt sich der Unterschied, dafs es auch, wenn man es in das Zellgewebe unter die Haut bringt, Erbrechen bewirkt, was bei dem Nickel nicht der Fall ist.

Versuche mit Uran. *)

Erster Versuch.

Einem Pudel gab man 5 Grane schwefelsaures Uranoxyd mit etwas Fleisch. (Das Salz war in Wasser wenig auflöslich und schien basisch zu seyn.) Der Hund befand sich, nachdem er das Metallsalz verschlukt hatte, fortwährend wohl.

Zweiter Versuch.

Ein Hund verschlukte 15 Grane salpetersaures Uran mit etwas Fleisch. Es erfolgte nicht die geringste Störung in dem Wohlbefinden des Thiers, es war fortwährend munter, erbrach sich nicht.

Dritter Versuch.

Einem Hunde mittlerer Grösse sprizte man in den Magen 1 Drachme salpetersaures Uranoxyd in $1\frac{1}{2}$ Unzen Wasser gelöst. Nach $1\frac{1}{4}$ Stunde erfolgte Erbrechen, das

*) Bei der Darstellung des reinen Uranoxyds aus der Pechblende bediente ich mich der Methode von Arfvedson.

(79)

sich einigemale wiederholte; am andern Tag aber war das Thier schon wieder vollkommen hergestellt, war munter, frafs begierig.

Vierter Versuch.

Einem starken Kaninchen sprizte man 34 Grane salzsaures Uranoxyd in 1 Unze Wasser gelöst in den Magen. Das Thier starb 52 Stunden nach der Injection ohne Convulsionen, nachdem $\frac{1}{2}$ Stunde vor dem Tode Lähmung der Extremitäten eingetreten war. Am ersten Tag nach der Injection hatte es noch gefressen.

Bei der einige Stunden nach dem Tode vorgenommenen Section fanden sich einige entzündete Flecken an der Pleura und an den Lungen, das Herz im normalen Zustand. Von den Eingeweiden des Abdomens war nur der Magen, aber dieser heftig entzündet, schon von aussen sah er blauroth und brandig aus; im Innern erstrekte sich die Entzündung von der Cardia über den ganzen Fundus bis an den Pylorus; an mehreren Stellen fand sich ausgetretenes Blut. Die übrigen Eingeweide waren alle gesund, nur die Contenta des Darmkanals ganz flüssig.

Fünfter Versuch.

Einem sehr kleinen jungen Hunde sprizte man in die äussere Jugularvene 10 Grane salpetersaures Uranoxyd *)

*) Das Salz hatte einen merkbaren Ueberschufs von Salpetersäure.

Figs. 1.2 and 1.3. Reproductions of the pages from the Gmelin treatise presenting the first report of the biological effects of uranium. Courtesy of the Boston Medical Library in The Francis A. Countway Library of Medicine

action in the stomach and the gut after which despite repeated vomiting the toxic action persisted because UNO_3 easily coagulated albuminoids and therefore penetrated the mucus, c) a description of large fecal masses in rabbit intestine and of constipation in dogs attributed to muscular paralysis, and d) several suggestions of possible causes of death. He postulated that the *liver* undoubtedly contracted and diminished the circulation into the hepatic veins despite his observation that changes in the liver were hard to see. He proposed an action of uranium in the *heart* and pulmonary *blood vessels* reducing the size of the capillaries, nearly blocking the flow and accounting for clotting and increased blood pressure which, if excessive, ruptured vessel walls, permitting internal hemorrhages and death if the vessels were weak, but, if the vessels were robust, led to death from asphyxia. He shrewdly guessed that there is also some direct action of UNO_3 on the *kidney* although his preferred explanation of the cause of the anuria lay in a greatly diminished renal circulation and an accumulation of blood in the veins. He commented on the usual delay (up to 2 weeks) in death but ruled out inanition as a cause of death citing data from Chossut, Magendie, and Bouchardat on the survival of dogs for 22–24 days without food, and of rabbits for 12 days. Leconte attributed the glycosuria he observed in dogs and rabbits to an impaired sugar usage (respiration is profoundly altered) which allowed excess sugar for urinary excretion; in his opinion, normal sugar production remained until the liver of the severely poisoned animal no longer formed sugar.

(80)

in 2 Drachmen Wasser gelöst. Das Thier athmete noch etwa eine Minute, aber der Herzschlag war sogleich nicht mehr zu fühlen. Eine Minute, nachdem die Flüssigkeit injicirt worden war, öffnete man den Thorax. Das Herz zuckte nur schwach; die linke Hälfte desselben enthielt hellrothes, flüssiges Blut; die rechte Abtheilung war voll von schwarzem flüssigen Blut.

Sechster Versuch.

Einem kleinen kräftigen Hunde sprizte man 3 Grane neutrales salzsaures Uranoxyd in 40 Granen Wasser aufgelöst in die Vena jugularis externa. Das Thier schrie ein paar mal auf, und starb noch vor einer Minute.

Bei der 3 Minuten nach dem Tode vorgenommenen Section fand sich folgendes: die willkührlichen Muskel zogen sich auf das lebhafteste zusammen, namentlich zog sich das Zwerchfell sehr lebhaft zusammen, wenn man den Nervus phrenicus berührte; die Contraction des Herzens war kaum mehr merkbar; die rechte Hälfte desselben war voll von coagulirtem Blut, welches sich aus den grossen Gefässen, wie Arter. pulmonalis, Cava superior als rothe Fäden ausziehen liefs. Diese Coagulation des Bluts war von der sehr verschieden, welche salzsaures Palladiumoxyd hervorbringt; die coagulirte Masse war nämlich zusammenhängend, wie in der Placenta sanguinis, bildete keine einzelne Klümpchen. Zwischen dem Herzbeutel und dem Herzen fand sich ziemlich viel wässe-

(81)

rige Flüssigkeit. Die linke Hälfte des Herzens war fast ganz blutleer, jedoch war der Anfang der Aorta mit geronnenem Blut erfüllt.

Es ergibt sich aus diesen Versuchen, dafs Uranoxydsalze vom Magen aus keine sehr bedeutende Einwirkung zeigen, nur in grossen Gaben Brechen erregen, oder, wenn dieses, wie bei Kaninchen, nicht möglich ist, eine Magen-Entzündung herbeiführen, an welcher das Thier stirbt. — In das Gefässsystem injicirt bewirken sie schnell den Tod, durch Zerstörung der Irritabilität des Herzens und Coagulation der Blutmasse.*)

Versuche mit Cerium.**)

Erster Versuch.

Einem Hunde mittlerer Grösse sprizte man 20 Grane salzsaures Ceriumoxydul in 1 Unze Wasser gelöst in den Magen. Nach 10 Minuten erfolgte Erbrechen, welches sich innerhalb 2 Stunden 5 bis 6 mal wiederholte. Nachher war das Thier sehr munter und gefräfsig.

Zweiter Versuch.

Einem Hunde mittlerer Grösse sprizte man 1 Drachme salzsaures Ceriumoxydul in 1 Unze Wasser gelöst in

*) Auffallend ist es, dafs salzsaures Uranoxyd, nicht aber salpetersaures Uranoxyd Coagulation des Bluts bewirkte. Hängt wohl die Verschiedenheit in der Wirkung von der Natur der Säure ab?

**) Zu der Darstellung des reinen Ceriumoxyds aus dem Cerit bediente ich mich der Methode von Laugier.

6

Fig. 1.3

It is easy to imagine that Leconte's report (1853, 1854) of glycosuria in dogs and rabbits following doses of uranium nitrate was greeted by an intense interest that sooner or later would lead to its use by homeopathic physicians of that day. In the words of Hughes (1876), "this fact, curious only in the eyes of an ordinary reader, was to a homeopathist pregnant with suggestiveness". The dictum, similia similibus curantur, stimulated the treatment of patients suffering from "saccharine urine" with this potent glycosuric agent and Blake (1868) phrased his introduction on this concept. "With the view of supplementing and elaborating the researches of Leconte, I purpose giving the details of experiments, made on the human subject, and on some of the lower animals, which I trust will contribute towards our very scanty knowledge of the pathogenesis of Nitrate of Uranium, and will assist us in defining accurately those cases in which its use would be appropriate in obedience to the great law of similars."

The first trials in man were reported by Bradford (1860): "It occurred to me that this Nitrate of Uranium might prove a valuable homeopathic remedy in the treatment of diabetes in the human subject. Accordingly I had it prepared in trituration, from the first to the third, and, although I have had, as yet, but few opportunities of administering it in cases of diabetes-mellitus, I feel warranted, from its satisfactory effect in those few cases, in recommending those who have patients suffering from this disease to make a trial of the remedy... I am fully persuaded that it merits a careful and scientific proving."

What pharmacopeia first listed uranium cannot be stated. The first edition of HUGHES's "A Manual of Pharmacodynamics" (1867) gives a one page citation, and the third edition (1876) refers to the proving of UNO_3 on "3 human subjects and 19 animals" (dogs, cats, rabbits) by BLAKE (1868), which results "have been wrought by him into a monograph upon the drug which constitutes the second part of the *Hahnemann Materia Medica.*" HERING (1891) refers to this monograph and attributes it to part 2 of the 1871 edition of Hahnemann's classic.

BLAKE (1868) gave repeated doses of UNO_3 to one man and one woman and also to 8 rabbits, 2 cats, 6 kittens and 2 pups. The 25-year old man who enjoyed average health was subjected to a 2-month-long treatment with increasing doses of various UNO_3 solutions. As a homeopathic physician, BLAKE started with a few drops of the second decimal dilution "from a saturated aqueous solution". HALE (1875) specified that UNO_3 was soluble in 0.5 parts of water, which corresponds to a little over one gram per ml of solution. If this rough calculation can provide an estimate of the doses, the patient received about 0.6 mg UNO_3 by mouth initially increasing to 4 mg, 8 mg, 10 mg, 13 mg, 14 mg, 35 mg and 64 mg at weekly intervals. Thereafter the dose increased more rapidly: the dose rose to about 350 mg daily for 5 days. Finally, on the last day of the study a solution containing 8 g was injected into the rectum. The "solution forcibly returned in 20 minutes, causing sharp colic and tenesmus, with raw feeling in rectum". Earlier, soft stools and flatulence had been noted; urine was acid, increased in volume, chlorides, and phosphates. No sugar could be detected in samples of urine collected at several times "either by Moore's or by Fehling's tests".

The young female subject suffered from chronic albuminuria. Daily treatments p.o. with 30 to 60 milligrams of UNO_3 in solution on 4 successive days followed by a single dose of 600 milligrams produced polyuria without change in specific gravity, a trace of albumin, copious chlorides but no sugar nor lithates.

The first rabbit received about 2 g of UNO_3 into the stomach on 5 successive days and was found dead on the 6th day. A hemorrhagic spot in the stomach was the main gross change; the kidneys were normal histologically. Two other rabbits were treated with daily oral doses of 30 milligrams for 4 days; 1 died on the 4th, 1 on the 5th days. Both had stomach ulcers and both had discolored kidney tissue. The remaining 5 rabbits, given subcutaneously single doses of 30 to 600 milligrams died in 2 to 5 days. One had a stomach ulcer, one a yellow patch and one congestion. Three exhibited ascites, and one also had pleural and pericardial fluid. One rabbit showed albuminuria without sugar.

Two cats were treated subcutaneously with repeated doses; one cat received increasing doses from 6 milligrams to over a gram in an irregular pattern and died on the 18th day. This cat showed sugar in the urine with copious chlorides and albumin on the 16th day. The other cat was given a total of 12 mg and died on the 9th day. Other urine samples from both cats were negative for sugar. The cat given 6 mg on the 1st and 3rd days had "a great deal of coagulated fluid in the peritoneal cavity." Three kittens that received UNO_3 solutions subcutaneously and 3 that were treated by rectum died on the 6th to the 15th days (1 was sacrificed). Only one showed a stomach ulceration. A few urine samples showed albumin but no sugar. Kidneys were reported to be healthy. Some kittens received as little as 3 to 8 mg UNO_3.

Two pups given UNO_3 solutions subcutaneously showed albumin but no sugar in the urine. One pup died on the 2nd day; the other was sacrificed on the 45th day. The kidney was slightly anemic. Fecal analyses showed increased albumin, perhaps a decreased phosphate, and traces of glucose.

BLAKE was disappointed not to have succeeded in establishing a distinct diabetes. He had discovered instead "a specific action on the circulation of the stomach and duodenum... usually ulcerative... to the most marked degree in the neighborhood of the pylorus." BLAKE commented in the monograph he prepared for Hahnemann's Materia Medica (as quoted by HALE, 1875) that "It seems singular that it should have cured veritable cases of "sugar diabetes", which it undoubtedly has done." With homeopathic logic, therefore, he attempted to find some effect of uranium similar to the pathology of diabetes. He hit upon "the homeopathic rapport which exists between the pathogenesis of Nitrate of Uranium and the digestive symptoms so commonly seen in the diabetic". BLAKE concluded that "Probably any cause that will arrest lung action, or materially disturb the harmony of the assimilative functions, will cause the urine to become saccharine. ...It appears to me that diabetes, a disease in which we can find no distinct pathological lesion, is essentially amenable to pure symptomatic treatment, and to that, perhaps, we shall always return from our "butterfly-hunts" after reputed specifics which ever elude our grasp."

HUGHES (1867) mentioned case reports (mostly the treatment of diabetics) by HALE, LOWDER, CURIE, JOUSSET, BÄHR, DRYSDALE, JOHN BLAKE, CORNELL, HUGHES, ZWINGENBERG, FISCHER, MAGDEBURG and CAREY.

RABUTEAU's monograph, Elements de Toxicologie et de Medicine Legale (1873) devoted 3 pages to Uranium although he never heard of a human uranium poisoning despite the use of several salts in industry and in the arts. RABUTEAU gave a dog increasing oral doses of uranyl acetate over a period of 4 days (total 3.75 g), and found neither albumin nor sugar in the urine early during polydipsia or later during oliguria. Attempts to detect uranium in the urine failed; traces were found in the bile obtained at autopsy.

The Handbuch der Pharmaceutischen Praxis authored by HERMANN HAGER in 1880 includes a matter of fact sentence attributing to KENNEDY[2] the indication of UNO_3 as an agent against diabetes mellitus and specifying the usual dose as 10–20 mg t.i.d., not to exceed 100 mg as a single dose (dilution uncertain). In the ensuing years, up to about 1900, reports kept appearing of this use of uranium (see examples in Table 1.1). HERING (1891) gives under the heading, "clinical authorities", references to treatment of ophthalmia, diabetes, albuminuria, diuresis, nocturnal enuresis, and incontinence of urine. WEST (1895), for example, gave detailed histories of patients under treatment with a) UNO_3, and b) the double chloride of uranium and quinine, between which he could see no difference in action. "These cases, taken together with several others not systematically observed, all point to the conclusion that we have in uranium a drug which has a powerful effect upon diabetes." Messers Oppenheimer, Sons and Co. (1896) offered practitioners a "convenient form" of medication, UNO_3 Platinoids. Use of uranium in diabetes continued into the 20th century. BLACKLEY (1912) in a review of "Diabetes Mellitus" for british homeopathic physicians recommended as the two principal therapeutic drugs: phosphoric acid and uranium. In his opinion, the cases cured comprised largely intermittent or symptomatic glycosuria, although in his experience these drugs frequently only held the inevitable denouement of genuine diabetes at bay for some time. WILCOX (1917) supported the use of UNO_3; thirst, polyuria and glycosuria were reduced in 46 of 54 diabetic patients by doses of $^1/_2$ to 3 grains without untoward symptoms despite periods

2 In The Lancet, i, 835–836 (1874), one of the items in the section called "A Mirror of Hospital Practice", a case from the "West Ham, Stratford, and South Essex Dispensary" under the care of Mr. KENNEDY is presented, giving the details furnished by Mr. R. J. CAREY, house surgeon.

Table 1.1. Selected reports of human administration (usually UNO_3)

Reference	Dose	Comment
Bradford (1860)	gr. 2 or 3[a]	less urine; after continued use, less sugar
Carey (1874)	gr. 1/3 to 1/6 t.i.d.	single doses not over gr. 1–1/2, daily doses not over gr. 8
West (1895, 1896)	gr. 5 to 15, t.i.d.	slow acting; dyspepsia in 1 at gr. 5, t.i.d.; loose bowels
Burton (1896)	gr. 9 to 12, t.i.d.	diarrhea at gr. 12, not at gr. 9
Angermeyer (1897)	gr. 5 to 10	toxic hazard
Soundby (1897)	—	not effective in diabetes
Tyson (1897)	—	not effective in diabetes; loose bowels
Duncan (1897)	gr. 20 t.i.d. gr. 3 to 11	improved health of diabetics sometimes continued for 3 months
Bond (1897, 1898)	gr. 30 t.i.d.	lessened glycosuria; no symptoms in diabetic mental patients; no improvement of mental symptoms
Wilson (1903a and b)	—	increased uric acid excretion
Tylecote (1904)	gr. 1 t.i.d. initially	weight gain in phthisis; constipation
Kobert (1906)	—	U contraindicated in diabetes; total dose of 16 g. (?) UNO_3 in 100 days dangerous
Wilcox (1917)	gr. 1/2 to 3	reduced thirst, polyuria, glycosuria in 46 of 54 diabetic patients; no untoward symptoms; treatment extended sometimes for months or years

[a] Third trituration. This dilution was frequently specified although apparently dilutions as great as the sixth attenuation or more were used on occasion. Hughes (1886) stated that only the first and second dilutions "have hitherto been used".

of treatment sometimes extending for months or years. This paper apparently marked the end of an era; no later reports have been found. Nevertheless, in the 1930 edition of Hager's Handbuch, for the treatment of diabetes, a preparation is described called *Vin Urané* attributed to Pesqin of Paris which had the following formula: UNO_3-1.0, glycerol-50, and red wine-1000. Presumably the publication in the first decades of the 1900s of a wealth of experimental animal studies clearly delineating the renal pathology from relatively small doses of uranium salts and revealing the origin of the glycosuria halted their misuse in diabetes. The lack of evidence of human injury may be traceable, at least in part, to the relatively small homeopathic doses that became popular in higher and higher "potency", i.e., greater and greater dilution.

Therapeutic failures and warnings of toxic hazards had been voiced. Some patients seemed to respond, others did not. Tyson (1897) and Soundby (1897) referred to unsuccessful attempts to treat diabetic patients in about the year, 1880. Side effects, especially dyspepsia and loose bowels, were mentioned (Table 1.1). Carey (1874) gave upper limits for dosage size. Angermeyer (1897) pointed out the toxic potentiality, and Kobert (1906) flatly concluded that uranium is contraindicated in diabetes.

Tests of the efficacy of uranium in other diseases included the recommendation (cited by Bond, 1898) of effervescing UNO_3 for "throat affection connected with a saccharine urine" from a book written (pre-1868) by Sir George Duncan Gibb. Wilson (1903) reduced hemorrhage from a cancer of the larynx by a topical UNO_3 spray. He treated another patient suffering from an "inoperable cancer" with Bourroughs Welcome preparations of uranium salicylate and uranium acetate tablets. Tylecote (1904) observed a weight gain in phthisis under U therapy. Drysdale (1880) has cured a stomach ulcer, according to Wilson. Cook (1886) used UNO_3 in a case of incontinence.

II. Toxicity of Uranium Compounds

"The nitrate of uranium is a feeble poison". So began the brief, translated summary of the section on uranium in Gmelin's 1824 treatise. A teaspoonful given orally to a dog produced only vomiting and half a teaspoonful killed a rabbit after a lapse of two days. In contrast, 3 grains intravenously "caused instant death". The characteristic delayed death and the importance of the route of administration were thus recognized from the first.

Le Conte (1851), who "always found sugar in the urine of dogs slowly poisoned by small doses of nitrate or uranium", set the stage for two quite different uses of the effects of uranium, 1) as a homeopathic remedy for diabetes mellitus, and 2) as an agent to induce experimental nephritis. The attempts to treat diabetes at least demonstrated clearly in man that small doses of UNO_3 can be tolerated for extended periods. The rich literature growing from the pathophysiologic research on nephritis established many important biological properties of uranium.

a) Renal changes, structural and functional, are the primary effects of uranium. With small doses, injury is restricted to the terminal third of the proximal convoluted tubules.

b) The source of the glycosuria can not be laid to liver injury.

c) The minor glomerular changes (vascular effects) following relatively large doses fail to reproduce the pathological sequence of glomerulonephritis.

d) Flaccid paralysis in dying animals does not arise from specific neurological effects of uranium.

e) Uranium shares with many metals the property of coagulating proteins.

The preoccupation with experimental nephritis led investigators into extensive histopathologic work. Investigators gave widely divergent doses (Table 1.2) of a few uranium compounds, but mostly of uranium nitrate, by several routes (usually subcutaneously) to nearly a dozen species of animals (especially to the rabbit and the dog). During these studies, toxicologic facts accumulated, more or less as by-products.

a) The extreme toxicity of uranium given parenterally. Over and over the statement was made that uranium is highly toxic; in fact the most toxic metal, for example, by Kobert (1906) and Autenreith and Warren (1928).

b) The typical delay in death was described by the earliest investigators and frequently commented on by those who followed.

c) The dose-related distinction between acute and chronic poisoning.

d) The remarkable tolerance and its basis in the atypical regeneration of tubular epithelium. A few references may be mentioned. Garnier et al. (1928, 1928b), Mauriac (1928, 1928b, 1932), MacNider (1929b), Garnier and Marek (1930, 1931a, b, c, 1932), Gil y Gil (1923–1924), Dominguez (1928), Larson (1936), and Mentzer (1934).

e) The marked individual variation in response, especially in rabbits.

Table 1.2. Selected data on the toxicity of uranium compounds

Species	Compound	Dose	Route	Effect	Reference
Dog	UNO_3	1 g	p.o.	nil	Gmelin (1825)
	UNO_3	4 g	p.o.	emesis	
Rabbit	UNO_3	1.4 g	p.o.	fatal	
	UNO_3	0.18 g	i.v.	fatal	
Dog, rabbit	UAc[c]	0.5–1 g	p.o.	fatal	LeConte (1854)
Man	UNO_3	1.8 mg ?	p.o.	in diabetes	Bradford (1860)
Man	UNO_3	0.6–350 mg repeated	oral	loose bowels, flatulence	Blake (1868)
Rabbit, cat, pup	UNO_3	3 mg–2 g (dilution ?)	oral	fatal, glycosuria in 1 cat	
Dog	UAc	0.25–2 g	p.o.	fatal	Rabuteau (1873)
Man	UNO_3	≦20 mg ?	p.o., t.i.d.	nil ?	Carey (1874)
Rabbit, dog	UNO_3	3.5–4400 mg	subc	fatal	Chittenden (1888)
Dog	UNO_3	50–450 mg repeated	p.o.	LD	Chittenden and Lambert (1889)
Dog, cat, rabbit, goat, goose, dove, crow, chicken, frog	UNO_3 NaU tartrate[a]	U mg/kg dog 1.66 goat 1.66 rabbit 0.83 cat 0.41 rat 0.41	p.o.	MLD	Woroschilsky (1889)
Man	UNO_3	?	p.o.	dyspepsia, loose bowels	West (1895)
	UNO_3	0.3 g ? (dilution ?)	p.o., t.i.d.	nil	West (1896)
Man	UNO_3	0.7 g ? (dilution ?)	p.o., t.i.d.	diarrhea	Burton (1896)
	UNO_3	0.5 ? (dilution ?)	p.o., t.i.d.	nil	
Man	UNO_3	≦1.8 g ? (dilution ?)	p.o.	nil	Bond (1898)
Dog	UNO_3	120 mg	subc	glycosuria	Lepine (1903)
Rabbit	UNO_3	7.5 mg	subc	LD	Richter (1904, 1905)
Rabbit	UNO_3	0.5–7.5 mg	subc	LD	Blanck (1906)

Table 1.2. (continued)

Species	Compound	Dose	Route	Effect	Reference
Rabbit Rabbit	UNO_3 UNO_3	5–500 mg 5	p.o. subc	nephritis nephritis	Fleckseder (1906, 1907)
Rabbit	UNO_3	5 mg	subc	LD	Georgopulos (1906)
Dog, cat	UNO_3, UAc NaU tartrate[a]	0.5–2 mg/kg	subc	LD	Kobert (1906)
Rabbit	UNO_3	7.5–10 mg repeated	subc	LD	Heineke and Meyerstein (1907)
Dog	UNO_3	10–15 mg, twice	subc	nephritis	Siegel (1907)
Rabbit	UNO_3	1–5 mg, single or repeated	—	nephritis	Christian (1908)
Rabbit	UNO_3	5 mg repeated	p.o., subc	edema	Shirokauer (1908)
Rabbit	UNO_3	1–5 mg repeated	subc	nephritis	Smith (1908)
Rabbit	UNO_3	6.3 to 39 mg/kg	—	nephritis	Takayasu (1908)
Rabbit	UNO_3	5 mg daily	—	nephritis	Bence (1909)
Guinea pig	UNO_3	0.25–5 mg, single or repeated	subc	nephritis	Dickson (1909)
Dog, rabbit	UNO_3	1.8–37.5 mg	subc	nephritis	Pearce (1909)
Guinea pig	UNO_3	0.5–15 mg, repeated	subc	some survival	Dickson (1910)
Dog Rabbit	UNO_3, UAc UNO_3	5–20 mg 2.5–20 mg	subc —	nephritis nephritis	Theohari and Giurea (1910)
Rabbit Guinea pig	UNO_3 UNO_3	5 mg/kg max. >0.5 mg/kg	subc subc	often survival usually fatal	Christian et al. (1911)
Rabbit Rabbit	UNO_3 UNO_3	2 mg/kg 1 mg/kg	subc subc	MLD never LD	Boycott and Ryffle (1913)

Dogfish (*Mustelis canis*)	UNO_3	100 mg/kg 150 mg/kg	subc	5 day survival dead in 1–3 days	Denis (1913/1914)
Rabbit	UNO_3	0.35 mg	subc	delayed LD	Dünner and Siegfried (1920)
Rabbit	UNO_3	1 mg	subc	LD	Gil y Gil (1923–1924)
Man	U salts	industrial exposure	inhalation, skin	anemia	DeLaet and Meurice (1925)
Dog Rabbit	UNO_3 UNO_3	2 mg/kg 3 mg/kg or 10 mg/kg	subc subc	LD $^3/_4$ died	Garnier et al. (1928)
Rabbit	UNO_3	1–4 mg/kg	subc	4/13–4/5 dead	Garnier and Marek (1929)
Rabbit	UNO_3, UAc USO_4[b]	0.5–1.3 mg	injected	renal death	Messini (1929)
Rabbit	UNO_3	0.5 mg/kg 1 mg/kg 2–10 mg/kg $>$ 10 mg/kg	subc subc subc subc	male / female 1/2 / 1/1 dead 12/17 / 6/29 dead 4/5 to 6/7 dead all / all dead	Garnier and Marek (1932)
Rabbit	UAc	0.05 mg U	intracisternal ?	MLD	Pentschew and Kassowitz (1932)
Rabbit	UNO_3	1 mg 0.83 mg 0.5 mg 0.35 mg	subc subc subc subc	LD-Mikawa LD-Sollmann LD-Angermayer LD-Pohl	Sato (1933)
Rabbit	UNO_3	5–30 mg/kg	subc	min LD to max LD	Vetri (1933)
Dog, kitten, mouse, cat, guinea pig	"Colloidal uranium"	1–15 mg/kg/day	subc	LD	Maloney and Burton (1938)

[a] NaU tartrate = Sodium uranium tartrate.
[b] USO_4 = Uranium sulfate.
[c] UAc = Uranium acetate.

f) The species variation, especially between rabbits, dogs and guinea pigs. See WOROSCHILSKY (1889), DICKSON (1910).

g) The susceptibility of adult and old dogs contrasted to the resistance of puppies; studied in detail by MACNIDER and others. MACNIDER (1927) also found that pregnant dogs were considerably more susceptible than non-pregnant females.

h) The importance of the route of administration. PATRASSI and ROGERS (1931) contrasted the effects of subcutaneous doses with those given intravenously or intra-arterially. A number of early workers gave uranium compounds by mouth. PENTSCHEW and KASSOWITZ (1932) injected UAc[3] into the occipital region of rabbit brain.

i) The comparative toxicities of certain uranium compounds. Water solubility or "insolubility" gave two classes, but even among soluble compounds differences were reported. MESSINI (1929) found uranium acetate to be less toxic than uranium sulfate which in turn was less toxic than uranium nitrate.

j) The influence of other substances given during the course of uranium poisoning. DÜNNER (1914–1915) found that calcium intravenously increased the toxic effect of uranium, sodium chloride did not. SACHS (1934) observed that ephedrine increased uranium toxicity, whereas HESS and WIESEL (1913) had reported a beneficial action of adrenaline on uranium-poisoned animals. THEOHARI and GIUREA (1910) fed meat to dogs chronically poisoned with uranium and showed that they immediately developed an acute uremia and died. HOLMAN and MEBANE (1940) discovered that when the plasma proteins of dogs were depleted, the dogs were more susceptible to uranium toxicity than were normal dogs.

k) The relative insignificance of radiation as contrasted with the chemical toxicity.

l) Loss of body weight usually accompanied acute poisoning from voluntary inanition. Organs also lost weight, e.g., MEYNER (1898) found the following percentage weight losses in mice: liver-36, spleen-29, heart-20, stomach-19, lung-18 and kidney-18.

Few systematic toxicity evaluations of uranium compounds were reported, even for estimates of minimal lethal doses. Most investigators were satisfied with effective doses, i.e., those providing frank evidence of renal injury.

III. Experimental Nephritis; Kidney

Human kidney disease, of widespread and common occurrence, devastating in its course, gave impetus to the study of the biological effects of uranium during the ninety year span from 1850 to 1940. PEARCE (1910) traced the unfolding knowledge of nephritis through 3 phases.

Phase 1. Clinical observations. The classical work of RICHARD BRIGHT (1827) "demonstrated that albuminuria and dropsy had an intimate relation to certain pathological changes in the kidney".

Phase 2. Pathologic anatomy. FRERICHS (1851) proposed a landmark hypothesis: nephritis is a single process beginning with an acute illness and injury and progressing through successive stages ultimately to the small granular kidney.

Phase 3. Experimental pathology. WEIGERT (1879) demonstrated that nephritis cannot be viewed as a process of successive stages but comprises a complex group of diseases differing in cause, in gross and histological changes, in functional derangements, and in clinical correlates, "representing the varied

3 UAc — uranium acetate.

reactions of kidney tissue". For example, he described two forms of chronic Bright's Disease: in one, the kidney is large and white (histologically altered in glomeruli, tubules, and interstitial tissues), in the other, the kidney is contracted, small and granular.

The development of experimental pathology set the stage for testing CLAUDE BERNARD's (1857) dictum that "poisons are the reagents of life". LE CONTE's (1853) discovery that uranium injures the kidney, placed uranium among the kidney poisons available as tools to induce experimental nephritis. The classical pattern of uranium poisoning—polyuria, albuminuria, casts, glycosuria, oliguria, sometimes anuria—ending either in uremic death or in recovery was not delineated immediately. LE CONTE (1853) described the glycosuria and the terminal anuria. BLAKE (1868) observed polyuria in man (and animals), albuminuria in animals, copious urinary chlorides and phosphates, and the presence of pleural, peritoneal and pericardial fluid.

CHITTENDEN and colleagues (1888) observed polyuria from small doses, the albuminuria preceding glycosuria, and the broken tubule cells. WOROSCHILSKY (1889) followed the entire sequence including the development in some animals of edema and ascites.

Because "the glomerular contracted kidney is, as a matter of fact, really the predominating type in human pathology, the most important problem was to find a means of producing a glomerulonephritis in animals" (ASCHOFF, 1912). Major attention in many early studies was therefore devoted to agents producing glomerular as well as tubular lesions. Glomerular and other vascular reactions were carefully described anatomically and the chemical sequelae began to be identified. Functional measurements initially were largely limited to effects of diuretics and to changes in kidney volume and urine production. Renal disease in experimental animals was roughly classified into effects following glomerular injury (held to be vascular effects) and those following tubular injury (held to be epithelial effects). Kidney poisons were therefore divided into three groups: those primarily affecting the tubules (e.g., uranium, chromium, mercury); those primarily affecting the glomeruli (e.g., arsenic, cantharidin, and certain snake venoms); and those producing diffuse injuries, (e.g., diptheria toxin). All caused albuminuria, glycosuria, and casts. As time went on, the variations in responses were observed a) when various doses were given, b) to various species, c) and in various pattern of dosing, single and repeated. Acute poisoning was identified, processes of repair and regeneration were discovered. Studies of functional lesions became more and more sophisticated and illuminating.

The interest of many investigators centered on uranium, partly because in early studies (e.g., those of CHRISTIAN, and colleagues, 1908 and later) only in uranium nitrate poisoning were observed "certain changes in the glomerular loops which would seem to indicate a possible vulnerability of the glomeruli in regard to uranium" (ASCHOFF, 1912). The task of discriminating glomular from tubular effects occupied a great deal of attention (see below). No other investigator rivals the place MACNIDER made for himself in his studies of the pathology of uranium poisoning.

Among the early studies of function were those which attempted to identify vascular reactions on the assumption that these would distinguish glomerular nephritis from tubular nephritis. Stimuli to contraction of blood vessels included blowing tobacco smoke in the nose of the animal, momentary suffocation, or the administration of adrenalin, e.g., to decrease kidney volume and increase blood pressure. Vessel dilation was induced by caffeine or by concentrated sodium cloride solution and the evidence of change was sought in records of urine volume.

Uranium was used in animal studies (up to 1910) by RABUTEAU (1873), CHITTENDEN et al. (1888, 1889), WOROSCHILSKY (1889), SCHNEIDER (1903), RICHTER (1904), Lepine; in 1906 by GEORGOPULOS, LEOPOLD, FLECKSEDER, BLANCK; in 1907 by HEINEKE and MEYERSTEIN, SIEGEL, SCHLAYER and colleagues, SCHUR and WIESEL, OPHÜLS, SCHIROKAUER; in 1908 by CHRISTIAN, PEARCE, TAKAYASU, SMITH; in 1909 by DICKSON, BENCE; in 1910 by TRIA.

Despite a considerable and growing effort to learn from animals the nature of human nephritis, CHRISTIAN (1916) concluded that "however close the resemblance, it cannot be said that in animals a condition identical with chronic nephritis in man has been produced as yet by any experimenter". The ensuing quarter century did not alter this generality. Nevertheless the accumulation of knowledge of kidney function and structure in health and disease benefitted from studies of uranium nephritis (as well as that from other poisons). To toxicologists, defining the biological effects of uranium is inherently important.

A. Pathology

1. Glomerular Injury vs. Tubular Injury

The importance attached to this distinction can be judged by PEARCE'S (1910) statement. "Here I may at once call your attention to the fundamental problem of experimental nephritis, that is, the influence of the glomerulus as contrasted with the influence of the tubule. This enters into all phases of renal pathology, in some partially elucidated, but in most still a matter of doubt and speculation." A continuing scientific argument ran on for years supported by such comments as BAEHR (1912) made to the effect that to reproduce human nephritis, an injury

Table 1.3. Tubular injury

Species	Cmpd	Dose (Subc)	References
Dog, cat, rabbit, rat, goat, goose, dove, crow, chicken, frog	UNO_3, tartrate	0.5 mg to 1 g	WOROSCHILSKY (1889)
Rabbit	UNO_3	7.5 mg	RICHTER (1905)
Rabbit	UNO_3	1 to 5 mg	SMITH (1908)
Dog	UNO_3	1.5 mg/kg	PEARCE et al. (1910)
Rabbit	UNO_3?	8–12 mg/kg	SCHLAYER and TAKAYASU (1910)
Rabbit	UNO_3	2.5 to 20 mg	GROSS (1911)
Dog, cat, rabbit	UNO_3	0.35 to 1.4 mg	POHL (1911/1912)
Rabbit, guinea pig, rat, mouse	UNO_3	—	SUZUKI (1912)
Rabbit	UNO_3	2 mg/kg	BOYCOTT and RYFFLE (1913)
Rabbit	UNO_3	10 mg	GHOREYEB (1913)
Dog	UNO_3	8.5 to 30 mg	KRAUS (1913)
Dog	UNO_3	3–12 mg/kg	O'HARE (1913)
Dog	UNO_3	6.7 mg/kg	MACNIDER (1912–1913, 1914, 1915, 1919)
Rabbit, guinea pig, rat	UNO_3	5 mg	OLIVER (1915)
Guinea pig	UNO_3	100 mg/kg	SUZUKI (1926)
Rabbit	UNO_3	7 mg	AURIAT (1927)
Rabbit	UNO_3	6 mg, intrap.	BINET and MAREK (1933)
Rabbit	UNO_3	0.5 mg/kg	MENTZER (1934)

Table 1.4. Glomerular injury (minimal to severe)

Species	Cmpd	Dose (Subc)	References
Dog, rabbit	UNO_3	3.5 to 440 mg po 2.5 to 200 mg	CHITTENDEN and LAMBERT (1888, 1889)
Rabbit	UNO_3	7.5 to 10 mg	HEINEKE and MEYERSTEIN (1907)
Rabbit	UNO_3	5 mg	SHIROKAUER (1908)
Rabbit	?	6 to 39 mg	TAKAYASU (1908)
Guinea pig	UNO_3	0.25–5 mg	DICKSON (1909)
Rabbit	UNO_3	1 to 5 mg	CHRISTIAN et al. (1911)
Rabbit	UNO_3	0.43 to 0.7 mg intra-renal artery	BAEHR (1912, 1912–1913)
Cat	UNO_3	0.8 to 2 mg	FOLIN et al. (1912)
Rabbit	UNO_3	3 to 5 mg/kg	CHRISTIAN and O'HARE (1913)
Rabbit	UNO_3	12.5 to 25 mg intrap	HESS and WIESEL (1913)
Dog	UNO_3	6.7 mg/kg	MACNIDER (1912, 1913, 1920)
Rabbit	UNO_3	3–12 mg/kg	O'HARE (1913)
Rabbit	?	?	FAHR (1914)
Rabbit	UNO_3	3 to 9 mg	FITZ (1914)
Dog	UNO_3	100 mg	MOSENTHAL (1914)
Rabbit	UNO_3	25 to 50 mg, intrap.	WIESEL and HESS (1915)
Rabbit	UNO_3	0.35 0.013 to 0.042 mg intra-renal artery	DÜNNER and SIEGFRIED (1920)
Rabbit	UNO_3	25 mg, intrap.	ROTH and BLOSS (1922)
Cat	UNO_3	8 to 15 mg	KELLAWAY et al. (1925)
Rabbit	UNO_3	3.5 to 5 mg	SUZUKI (1926)
Rabbit	UNO_3	0.4 to 452 mg	DOMINGUEZ (1928)
Dog	UNO_3	15 mg/Kg	MILHORAT and DEUEL (1928)
Toad	UNO_3	6 mg	GÉRARD and CORDIER (1932)
Rabbit	UNO_3	0.5 to 303 mg	HUNTER and ROBERTS (1932)
Rabbit	UNO_3	10 mg/kg	GARNIER and MAREK (1931 b)
Rabbit	UNO_3	i.v. and intrarenal artery	PATROSSI and ROGERS (1931)
Dog	UNO_3	0.2 to 3.6 mg/kg 1.1 to 6.3 mg/kg	SACHS (1934)
Guinea pig	UNO_3	8 to 100 mg/kg	
Rabbit	UNO_3	2.5 to 10 mg/kg	

primarily of the glomerulus must be obtained. In the opinion of some investigators, uranium injury was limited to the tubule (Table 1.3), others found evidence of glomerular pathology varying from minimal, transient changes to severe and in some cases permanent injury (Table 1.4). Some of the factors contributing to the recurrent differences in observations may be enumerated. a) The most important factor is the size of the dose; with small doses, uranium is "ein rein tubulares Gift" (BAEHR, 1912). b) Species differences are important: the dog, for example, is resistant to glomerular injury, the rabbit relatively susceptible (O'HARE, 1913). c) The route of uranium administration alters the response if only by changing the concentration of uranium reaching the kidney. Intrarenal arterial injections (BAEHR, 1912; PATRASSI and ROGERS, 1931), for example, produced small infarcts under the capsule, hemorrhagic glomeruli, and other abnormalities in about half of the glomeruli. d) The time elapsed between the dose and the histological observations, and e) whether a single dose or repeated doses were given are factors (Austin and Eisenbrey, 1911; Wiesel and Hess, 1915). f) Stresses imposed concomitantly, such as feeding rabbits cholesterol (Weltmann and Biach, 1913), giving ephedrine (Sachs, 1934), or E. coli (O'Hare, 1913) alter the response in uranium poisoning.

MacNider (1919) succinctly described the renal changes in the uranium-poisoned "old" dog following a moderate dose of uranium nitrate. The kidneys showed no characteristic *gross* pathological change. The surface was smooth and the capsule not adherent. On section the cortex was pale, the tissue at the cortico-medullary junction showed a large amount of fat, which extended into the cortex in streaks. Histological study revealed a degree of degeneration which depended for its severity and extent upon the interval between uranium treatment and pathological examination.

2. Glomerular Lesions

Christian (1908b) first noted fine hyaline granules or droplets in the wall of the glomerular tuft of uranium-treated rabbits. Christian and O'Hare (1913b) described the degenerative processes which led to the development of fibrin thrombi, hemorrhage, and a dilation of capsular space with granular material. A proliferation of capillary and of capsular epithelium represented the final stage. Baehr (1912) described the evolution of the glomerular lesions following intra-arterial injections. Four stages of severe glomerular injury occurred with small doses of uranium: 1. coagulation necrosis, 2. swelling and proliferation of the glomeruli and the capsular epithelium, 3. capillaries and vasa efferentia refilled with blood, and 4. hyaline degeneration. Baehr likened the appearance to that of human focal nephritis. This experimental glomerulonephritis developed with only slight changes in the tubular epithelium. Nevertheless, Christian (1916) wrote "Still there remained a definite discrepancy between functional disturbance and anatomical change in the kidney, and our histological studies did not throw much light on the nephritis problem in man whether looked at from a functional or from a structural view-point".

MacNider (1920) confirmed O'Hare's (1913) observation that the dog glomerulus is resistant. "When uranium is given subcutaneously to the dog in small doses, properly spaced, the injury induced in the kidney is characterized by an acute tubular injury with an acute engorgement of the renal vessels, occasionally a small amount of exudate into the subcapsular space, but without degeneration in the vascular tissue" (MacNider, 1920). During the recovery process, a chronic diffuse type of nephropathy appeared, the glomeruli showed a capsular and an intercapillary injury. "Functional studies... show these kidneys to resemble very closely the chronic diffuse nephropathy of human material" (MacNider, 1920).

3. Tubular Lesions

In a definitive study, Suzuki (1912) and Aschoff and Suzuki (1912) considered the proximal convoluted tubule as comprising three portions and showed that, in effective doses, uranium injury was confined to the distal third of the proximal convoluted tubule. This site was also that primarily injured by mercury. In contrast, the renal injury produced by chromium was shown to involve the proximal portion of the proximal convoluted tubule (Fig. 1.4).

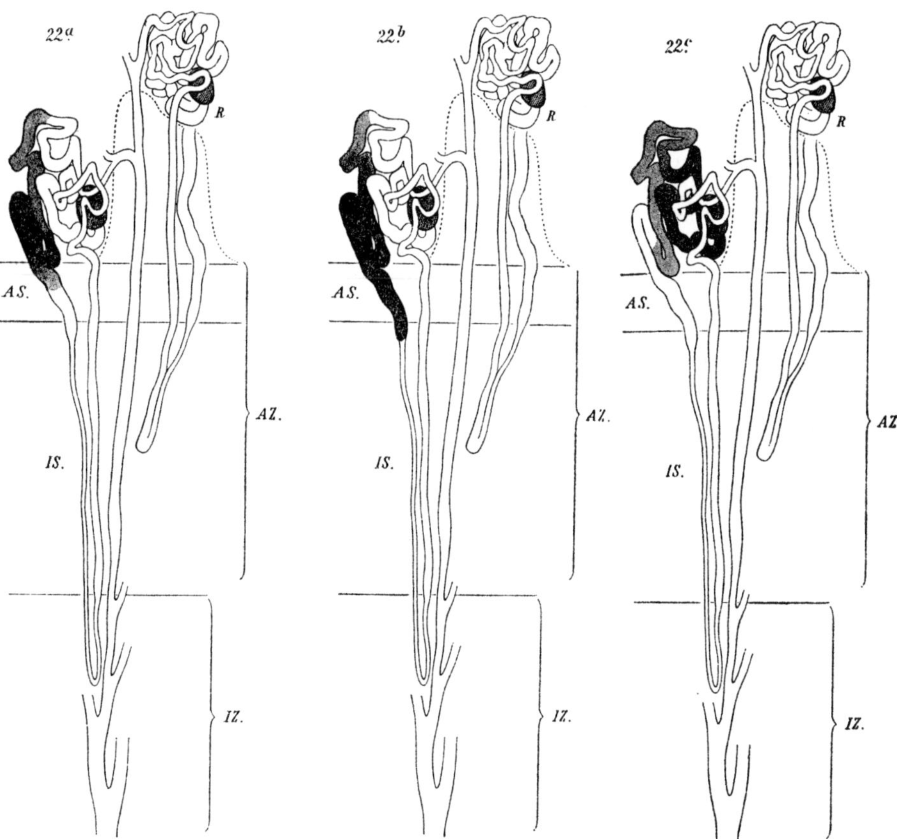

Fig. 1.4. An illustration showing the localisation of the action of uranium in the proximal convoluted tubule of the kidney. 22a. Lokalisation der Uranwirkung. 22b. Cantharidin oder Sublimat. 22c. Chromwirkung. Taken from the monograph by TATZUO SUZUKI, „Zur Morphologie der Nierensekretion", by permission of the publisher, Gustav Fischer Verlag, Jena 1912. Courtesy of the Boston Medical Library in The Francis A. Countway Library of Medicine

MACNIDER, in an extended series of studies from 1912 to 1937, repeatedly described in detail the histopathology of uranium poisoning. In the early days (up to the 7th to the 10th day) following a moderate dose (4 mg), the injury in the dog is largely limited to the tubular epithelium, and to the ascending limb of the loop of Henle. These changes consisted in an acute swelling and vacuolation, which first appeared in the third portion of the convoluted tubule, and were followed by pyknosis of the nuclei and necrosis of the epithelium. The lumen of some tubules was occluded by the edematous cells. Even at this early stage of the intoxication the ascending limb of Henle's loop contained a large amount of stainable fat in the form of large droplets. By the 6th to the 10th days of the intoxication, the epithelial injury in the "old" dog has become more diffuse. All the proximal convoluted tubules have participated in the degeneration, most cells are necrotic and regenerative changes are absent. The epithelium of the distal convoluted tubules and junctional tubules have shown evidence of injury. Glomerular vessels are engorged, thrombi are present, fibrin can be seen in the subcapsular space. Oliver (1915) described similar changes in guinea pigs, rabbits, and rats, differing mainly in susceptibility. "Necrosis of practically the entire proximal convoluted tubule" occurred, filling the tubules "with dense masses of granular detritus". Collecting tubules containing many casts. "A fact of much importance is the persistence of a few cells with normal nuclei and only slightly disturbed cytoplasm" (Oliver, 1915).

In the young dog given the same dose, beginning repair and regeneration of the renal epithelium is evident by the 14th to the 18th day.

Among the large number of investigators who examined the histopathology of uranium nephritis (e.g., see Tables 1.3 and 1.4) may be mentioned the serial studies of Feyel (1939) on the rat, and the attempt by Gérard and Cordier (1932) to relate probable uranium concentrations with site of injury in the toad. Karsner et al. (1918–1919) showed that the injury demonstrated histologically occurred when uranium was being excreted by the kidney. See also Schneider (1903), Schlayer and Hedinger (1907), Pearce (1908), Smith (1908), Theohari (1911), Dickson (1912), Aschoff (1913), Heilig (1923), Kellaway et al. (1925), Garnier and Marek (1929), Frola (1930), Esser (1937), and a number of other papers by MacNider from 1914–1915 to 1937.

MacNider (1920) assembled a group of 45 dogs suffering from naturally acquired chronic nephropathy. This disease is characterized by the vascular changes of chronic glomerulonephritis; the injury to the glomerulus is "out of proportion to the severity of the changes in the renal epithelium..., like certain nephropathies in human material" (MacNider, 1920). Some of these dogs received uranium nitrate thereby superimposing "an acute tubular injury (early edema and necrosis of the tubular epithelium) without any acute injury having been superimposed on the chronic glomerular pathology" (MacNider, 1920). MacNider was struck by the severe functional disturbances (rapid reduction in PSP elimination, retention of urea and creatinine, acute anuria, death in coma or convulsions), emphasizing the importance of the tubular epithelium.

4. Repair of Tubular Injury

Suzuki (1912) and Aschoff and Suzuki (1912) described the processes of repair and the regeneration of the tubular epithelium in animals surviving uranium intoxication. If doses were not too large, the regenerated epithelium appeared much like that of the normal kidney. Oliver (1915) examined the process in great detail in 3 species, finding only minor differences in the severity of injury and in the time and rate of repair, but basically a common pattern of regeneration. The essential steps include a) the appearance of "comparatively enormous nuclei" along with smaller, intensely staining nuclei (the latter presumably original renal cells) and evidence of mitoses (mitotic figures in various stages), b) "the gradual development of syncytiumlike masses and giant cells" projecting into the lumen, c) a replacement of the lost cell layer frequently with excess cells in disorderly arrangement, and d) a contemporaneous connective tissue proliferation. Some of the regenerated cells "have become indistinguishable from the normal cells around them". Other epithelium "surrounded by connective tissue still differs widely from the normal renal cells" (Oliver, 1915).

The origin of regeneration was described by Aschoff (1912) based on his work with Suzuki. "We were able to observe the extraordinarily rapid epithelial regeneration in the damaged Hauptstucke (proximal convoluted tubule) in all animals, and could readily determine that this epithelialization had its point of origin primarily in the narrow limb of Henle's loop and in the remains of the lower transitional division of the Hauptstuck." MacNider (1920) agreed that regeneration began in the epithelium of the terminal portion of the proximal convoluted tubule and the upper portion of descending limb of Henle's loop.

MacNider (1929b) showed that this kind of epithelium was just as susceptible to uranium injury as the original normal epithelium had been.

5. Tolerance

With larger doses and with repeated doses another type of regenerated epithelium appears which exhibits resistance to many effects of uranium (tolerance).

The existence of a chronic type of uranium nephritis, proposed by Siegel (1907), was investigated in detail by Dickson (1909, 1912) who gave repeated doses during extended periods, e.g., guinea pigs received 4 mg 86 times during 696 days, rabbits 5 mg 5 times in a period of 111 days, and dogs 4 doses of 10 mg each and later 3 doses of 20 mg each in a period of 45 weeks. Dickson's studies were made doubly valuable by his description of changes in the type of tubular epithelium, of the presence of masses of connective tissue, and of glomerular changes. Chronic uranium nephritis was inherently interesting to those searching for a model of human nephritis. Investigators promptly took up the study of chronic uranium poisoning, a topic which has proved to hold continuing interest, for example, Smith (1911), Gil y Gil (1924), and Hunter (1928) in the rabbit; O'Hare (1913), and MacNider in a lengthy series of reports in the dog; Frothingham et al. (1913) and Oliver (1915–1921) in the rat, guinea pig and rabbit; Karsner et al. (1918–1919) in the cat.

Some lengthy periods of observation have been reported. Dominguez (1928) treated rabbits repeatedly with doses of 0.2 to 5 mg over periods up to 579 days. Mauriac (1928) gave rabbits daily doses of 7 mg of uranium nitrate from October 1926 to June 1928. The longest follow-up period was that of MacNider (1929) who observed one group of dogs for five years and four months after uranium treatment.

Remarkably large doses have been given. Garnier and Marek (1931b) gave a total of 3.58 g of uranium nitrate in 153 injections to one rabbit. The same investigators in 1932 treated rabbits at intervals of 12 days in a program of increasing doses which on the 7th dose equalled 140 mg/kg. Goto (1917) in a period of 45 days increased doses given periodically to a dog to a maximum of 45 mg/kg.

The secret of the amazing tolerance can be found in atypical cells of the regenerated tubular epithelium. Dickson (1909) had described atypical epithelial cells and Suzuki (1912) first recognized the resistance to uranium of this kind of regenerated epithelium. MacNider (1916a) demonstrated that the resistance included resistance to such a toxic substance as chloroform. MacNider (1929) in several large scale studies on dogs explored thoroughly the importance of the atypical cell. For example, in a dog that had been given 4 mg of uranium nitrate per kg and maintained for 9 months and 15 days befor sacrifice, he found on histological examination that "the epithelium of the convoluted tubules showed two entirely different types of repair". In the first type, "the epithelium of the proximal convoluted tubules has shown an advanced stage of repair in terms of regenerating epithelial cells, not of an atypical character, but cells normal for this location in the tubule". "The second form of tubular repair is an entirely different type of cell. This type of repair takes place in those tubules in which the severest epithelial destruction has occurred and develops an ingrowth from the lower end of the proximal convoluted tubule or from the upper end of the descending limb of Henle's loop of a flattened, at first non-specialized cytoplasmic layer in which cell differentiation is frequently imperfect. Later cell differentiation develops; these cells stain uniformly and intensely." MacNider (1929b) described the responses of groups of dogs to a second injection of uranium given either before the processes of repair had taken place or after complete functional recovery. MacNider's conclusions were clearcut. If the degenerated tubular epithelium was repaired by convoluted tubule cells not too severely injured this type of epithelium had no resistance to uranium. If the regeneration had occurred by flattened imperfectly differentiated cell layers, the animal was resistant to the second injury from uranium.

Another form of renal tissue response following uranium poisoning, the so-called "sclerotic areas", was described by Dickson (1912) who believed that these areas developed from a blockade of regeneration by necrotic masses or by a collapse of a necrotic tubule and thereafter in either case an inactive atrophy. He thought this process would lead to a tubular contracted kidney.

Other studies of tolerance may be cited: MacNider (1920, 1931), Gil y Gil (1923–1924), Hunter (1928), Mauriac (1928, 1928b), Garnier et al. (1928), Garnier and Marek (1930, 1931, 1931c), Patrassi and Rogers (1931), Mauriac (1932), Carson (1936), Donnelly and Holman (1942).

B. Age

Age is an important factor in the response of dogs to subcutaneous doses of UNO_3. MacNider (1912–1913) described the relative resistance of young dogs (1.5 years old), polyuria, glycosuria and acetonuria were less severe than in old dogs (1.5 to 6 years). MacNider (1913, 1913–1914, 1916) continued to explore the differences in histopathological and biochemical responses of puppies (4.5 months) and of adults (4 to 5 years). Patterns of histologic changes, while similar, differed quantitatively, e.g., in the degree of fatty change in kidney and liver. Differences were demonstrated in the structural and functional changes of kidneys in puppies and in adult dogs subjected to anesthesia a) by morphine, chloroform and alcohol or b) by morphine and ether. Chloroform exacerbated the U damage found in tubular epithelium in both but the injury was more marked in adult dogs; adults became anuric, puppies did not. Ether produced similar but less severe changes; puppies were less affected. MacNider (1917a and b, 1919a and b) tested dogs of 3 age groups, 1. puppies 5 to 11 months old, 2. adults 1 to 4 years old, and 3. old dogs 6 to $10^1/_2$ years old. Acidosis was more severe in older dogs. Tubular epithelium changed in different patterns; cells shrank in young dogs, became swollen in old dogs. Puppies survived longer and exhibited more effective epithelial repair, Karsner et al. (1918–1919 IV) confirmed the differences in histological responses between older dogs and puppies and also confirmed the greater severity of acid intoxication in older dogs.

C. Anuria

Students of uranium nephritis beginning with Le Conte (1853), Chittenden and Lambert (1888), and Woroschilsky (1889) regularly reported anuria in certain animals after moderate to large doses. Dünner (1914–1915) injected uranium into both renal arteries and reproduced the effects following subcutaneous administration showing that other organs are not important. The controlling role of dose and the influence of NaCl were demonstrated by DAKE (1926). Large doses caused decreases in urine volume sometimes to zero; moderate doses produced polyuria. A small dose which failed to alter urine volume in $^2/_3$ rabbits, caused anuria in rabbits previously treated with sodium chloride. CHRISTIAN (1908) showed that NaCl given intravenously to uranium-poisoned rabbits led promptly to the onset of anuria. GROSS (1911) postulated that anuria came not from an inability to excrete water but from a failure of access to NaCl from body water. MACNIDER (1911–1912) proposed a mechanical blockade of tubules by masses of necrotic cells based on his findings that diuretics had little or no effect. He later (MACNIDER, 1914–1915) observed that renal vessels in uranium nephritis were dilated by caffeine and constricted by epinephrine; anuria therefore was not secondary to a loss of vascular function. Anuria was precipitated by anesthetizing poisoned dogs (MACNIDER, 1912–1913) and sometimes by giving a diuretic.

D. Casts

The presence of casts in the urine of uranium-poisoned animals was recorded routinely, with or without comment, so often that attempts at documentation would be futile. SMITH (1908) provided what may be the most detailed study of the stages of cast production. 1. Swelling of cells, loss of boundaries, convoluted tubule lumen obscured. Less change in straight tubules. 2. Degeneration marked, presence of necrotic masses. Changes in cell nuclei, chromatin in a dark speck. 3. Early—*granular* detritus filling segments of tubules. 4. Tubules farther from cortex exhibit less necrosis but granular material in lumen without cell remnants. Later—*hyaline*. 5. Uniformly granular casts in the collecting tubules. In a survivor (26th day), dilation of tubules; necrosis and granular material in non-dilated. Few casts in collecting tubules.

PEARCE (1908) gave serum from a uranium-treated dog to an untreated recipient and discovered casts in that dog's urine. The response was not found in rabbits. GROSS (1911) stained casts in vivo with toluidine blue. NUZUM and ROTHSCHILD (1923) showed that casts are not related to blood chemistries.

E. Edema

The discovery that uranium stands unique among kidney poisons in its ability to produce edena and ascites (LE CONTE, 1853; BLAKE, 1868; CHITTENDEN et al., 1887, 1888, 1889; RICHTER, 1904) and therefore that uranium nephritis offers a closer parallel to human disease may account for the predominant choice of uranium in the study of experimental nephritis over a period of a half century. The role of vascular injury was explored (RICHTER, 1904, 1905, and others later) and an extended series of investigations developed around the mechanism of edema formation.

RICHTER (1904) fed uranium-poisoned rabbits wet food (beets) or dry food (oats); the former developed edema, the latter did not. Little fluid accumulated in fasted rabbits; those fed salt and given a little water by mouth showed some tissue edema; and those given more water formed fluid in pleural and peritoneal cavities. Adding sodium phosphate or magnesium sulfate to the diets had no such consistent effects. Varying degrees of edema developed in rabbits not given extra water (e.g., CHRISTIAN, 1908, 1909, 1911). Many investigators confirmed Richter's results under a variety of salt and water treatments (see Table 1.5.)

Table 1.5. Selected reports of experimental edema in the rabbits with uranium nephritis

Treatment	Uranium dose	Effect	Investigator
Sodium chloride water, Beets vs. Oats	UNO_3, 5–7.5 mg, subc. S, R[a]	edema, pleural and peritoneal fluid	Richter (1904, 1905)
Serum from uranium nephritic		edema in recipient	Heineke (1905)
Serum from uranium nephritic	UNO_3, 5–7.5 mg, subc	inconstant incidence of edema in recipient	Blanck (1906)
Nephrectomy then serum from U nephritic	UNO_3, 5 mg, subc	edema	Georgopulos (1906)
No extra water		grades of edema	
Nephrectomy	then, UNO_3, 5 mg, intrap.	edema	Fleckseder (1906, 1907)
Repeated chromate and sodium chloride then a) control or b) U nephritic	UNO_3, 7.5 mg, subc, S, R	edema, slightly more often in U nephritic	Heineke and Meyerstein (1907)
Sodium chloride effective only on water-rich diet	UNO_3, 5 mg, subc, S, R	edema	Christian et al. (1908, 1909, 1911)
Some given water p.o.	UNO_3, 7.5 mg, subc.	edema and fluid only when water given	Pearce (1909)
Fasted then A. Thirsty + U	UNO_3, 5 mg, subc ?, R	no edema, loss of body weight, rbc ↑ or ↓, serum refractive index (RI) ↑ or ↓	Bence (1909)
B. Nephrectomized + water p.o. + U		edema, less loss of body weight, rbc ↑, RI little change	
C. Water p.o. + U		edema, weight loss at times only 4–5%, rbc ↑, RI ↑	
D. Nephrectomized + water excess + U		usually edema, body weight usually ↑, rbc ↑, RI ↑ In each, blood volume ↑ most near death	
Acute U nephritis		enormous edema	Théohari and Giurea (1910)
i.v. infusion of hypotonic sodium chloride 4 days after U	?	no edema, sometimes ascites, ↓ Hgb, ↑ blood sodium chloride	Schmidt and Schlayer (1911)
Acute U nephritis	UNO_3, 1–5 mg, subc, S, R	1/16 rabbits with fluid in serous cavities, no subcutaneous edema. No heart failure	Dickson (1912)
Subcutaneous infusion of sodium chloride then U	UNO_3, 0.3–20 mg/kg, subc	peritoneal and pleural fluid	Dake (1926)

Table 1.5 (continued)

Treatment	Uranium dose	Effect	Investigator
After U, 0.8 g NaCl in 20 ml H_2O	UNO_3, 7 mg, subc	edema, whether Cl, SO_4 acetate, PO_4, citrate, etc. also given. Not same as tieing ureters or nephrectomy	MAURIAC and AUBEL (1926)
Rapid vs. slow U poisoning NaCl p.o.	UNO_3, 5 mg, subc, R	slow: ↓ Na/Cl in blood and urine, more Na and less H_2O in muscle after NaCl p.o. rapid: ↑ NaCl in muscle with normal Na/Cl, K in blood and muscle	AUBEL et al. (1926)
U nephritis	UNO_3, injected	striated muscle shows histological changes secondary to edema and not a uranium effect	AURIAT (1927a)
Sodium given to some	UNO_3, 7 mg, subc, R	some without edema, some with slight edema, abdominal, lumbar	AURIAT (1927b)
Acute U nephritis fed after 2nd day, given 0.8 g NaCl	UNO_3, 7 mg, subc, R	marked edema with serous effusion	
Subacute, 3 weeks	3 weekly doses	degenerative lesions	
Subacute, total 2 months	3 weekly doses, then daily doses	loss of body weight, no edema	
Chronic U nephritis up to 6 months	UNO_3, 7 mg, subc, R	in edema, fluids are imbibed into tissues, because of a mineral complex disequilibrium	AUBEL and MAURIAC (1927)
U nephritis	UNO_3, 10 mg/kg, subc	edema about hilus of liver and root of mesentery	WHITNEY (1928)
Subcutaneous 0.1 cm^3 of isotonic NaCl in ear	UNO_3, 3–10 mg/kg, subc	more rapid disappearance in a nephritic then in control. Extreme thirst	SCHULMANN and MAREK (1928)
Physiological NaCl or H_2O, p.o.	UNO_3, 7 mg, subc	less urine with physiological saline, more exudate with water	GOVAERTS (1928)
U nephritis, moderate or severe	UNO_3, 10, 15 mg, subc	little difference in blood or in plasma volume in mild poisoning. When severe, ↑ in blood and plasma vol.	LITZNER (1930)
U nephritis	UNO_3, subc	edema formation not accompanied by other changes (e.g., in serine to globulin ratio). Hypoproteinemia no greater than in "dry" nephritic	TRAISSAC (1932)

Table 1.5 (continued)

Treatment	Uranium dose	Effect	Investigator
Guinea pig given repeated doses, water not forced	UNO_3, 0.5–1 mg, subc, S, R	marked ascites in some, no edema in others	DICKSON (1910, 1912)
Dog	UNO_3, 3–10 mg, subc, S, R	edema	
Dog given serum from renal vein	UNO_3	sometimes ↓ edema	TRIA (1910)
Guinea pig given NaCl and water	UNO_3, 0.5–5 mg, S, R	edema in some	CHRISTIAN et al. (1911)
Dog, U nephritis	U	no edema, marked increase in blood	PLESCH (1922)
Dog	UNO_3, 1–4 mg/kg subc, S, R	no edema in dog although fluid in rabbit peritoneally, pleurally and pericardially	GARNIER et al. (1928)

[a] S = Single; R = Repeated.

Species variability in response was recognized: rabbits are susceptible (WOROSCHILSKY, 1889), so are guinea pigs (DICKSON, 1910), but dogs are resistant (PLESCH, 1922).

Strenuous attempts to apply the findings from animal studies to the elucidation of human renal edema were not successful. PEARCE (1909) concluded that three factors interact in uranium nephritis, a) uranium plethoric hydremia ("increased water in the circulating fluid without decrease in proteins"), b) vascular injury, and c) nephritis; all 3 must co-exist to cause edema. GOVAERTS (1928) echoed these conclusions and described the edema of uranium poisoning as an exudate due to a) toxic changes in the endothelium of blood vessels, b) decreased osmotic pressure of the blood perhaps involving alterations in protein metabolism, and c) water retention from the kidney injury. The role of sodium chloride has been accepted since RICHTER'S work; as MAURIAC and AUBEL (1926) commented, water alone rarely works. A running argument over the contribution of liver disease, essential or incidental, will be discussed elsewhere. Many authors accepted DICKSON'S (1910) insight that uranium nephritis is accompanied by a true renal edema.

The composition of the edema fluid claimed the attention of several investigators. RICHTER (1904) showed by measuring the freezing point depression that the concentration of electrolytes might equal but did not exceed that of plasma. An increase in water content of muscle and liver with little change in ash or phosphate percentage recorded by SHIROKAUER (1907) was confirmed by CHISHOLM (1913–1914) who found that spleen, heart, lungs, alimentary canal and the "rest of the body" were especially affected, liver maybe. Edematous muscle tended to be high in sodium, low in potassium (AUBEL, MAURIAC and BOUTIRON, 1926; AUBEL and MAURIAC, 1927). Edema fluid coagulated spontaneously (GARNIER et al., 1928c); one 15 cm^3 sample of ascites contained 13 cm^3 of water, 2.6% albumin, and 0.3% urea. GOVAERTS (1928) found the edema fluid to contain about the same total protein and NPN and to have the some osmotic pressure as blood.

F. Diuretics

Interest in the effects of diuretics in uranium-poisoned animals seemed to center on the advisability of using a diuretic in a human nephritic patient. Demonstrated benefits were doubtful and limited to mild nephritis (FITZ, 1914); in acute poisoning diuretics were ineffective (SCHLAYER et al., 1907; THEOHARI and GIUREA, 1910; BOYCOTT and RYFFLE, 1913; MOSENTHAL and SCHLAYER, 1913), and in severe poisoning survival time was shortened (CHRISTIAN, 1913; WALKER and DAWSON, 1913). In chronic poisoning, although sodium sulfate produced no diuresis, caffeine did (THEOHARI and GIUREA, 1910).

MACNIDER (1916, 1917, 1920) found characteristic differences in the response to diuretics in anesthetized[4] uranium-poisoned dogs depending on whether the anesthetic was ether or chloroform. Before anesthesia, all the dogs were diuresing. Under morphine-ether anesthesia, urine flow continued and the dogs responded to diuretics, such as urea, glucose, caffeine, and theobromine. Under chloroform anesthesia, the dogs became acutely anuric and showed no response to diuretics. Interestingly enough, there was an increased kidney volume with the diuretic reaction. Histological studies showed a well-preserved epithelium with slight necrotic loci in dogs anesthetized with ether, whereas those anesthetized with chloroform showed swelling of the tubular epithelium and a marked necrosis. Several investigators used diuretics to gain information about the role of vascular injury in experimental nephritis. Renal vessels were hypo- or hyper-reactive to diuretics depending on the stage of uranium poisoning (SCHLAYER and HEDINGER, 1907; SCHLAYER and TAKAYASU, 1910). When responsive, vessels dilated (SCHLAYER et al., 1907, BOYCOTT and RYFFLE, 1913), although to a lesser than normal extent (WEBER, 1911). If "secretory" duiretics, e.g., caffeine, failed to increase urine flow, "mechanical" diuretics, e.g., NaCl solution, were also ineffective (BOYCOTT and RYFFLE, 1913). In early uranium nephritis, doses of sodium chloride reduced urine volume and chloride excretion and vessels constricted; caffeine increased urine volume and dilated the vessels (MOSENTHAL and SCHLAYER, 1913). THEOHARI and GIUREA (1910) had shown that in acute uranium poisoning, diuretics increased kidney volume (in dogs and in rabbits) despite an absence of urine secretion but less than in normal animals.

The composition of the urine was altered by the uranium poisoning. In the oliguric phase, chloride (BOYCOTT and RYFFLE, 1913), nitrogen and PSP excretion were reduced. WEBER (1911) reported a decrease in NaCl excretion following sodium sulfate and theocin, and BOYCOTT and RYFFLE (1913) found that the rate of urine excretion was reduced by several diuretics (NaCl, dextrose, caffeine, theocin, or urea). In contrast, FITZ (1914) reported that theobromin and theocin increased the output of water, chloride and nitrogen, although with repeated doses the kidney response diminished more easily than in normal rabbits.

G. Kidney Composition

A few observations on the alterations in kidney composition produced by uranium poisoning will be mentioned here.

Titratable acid, attributed to lipase activity (autolysis) was reduced (QUINAN, 1915–1916) in kidney samples from rabbits repeatedly injected with UNO_3. Both cortex and medulla gave values about half the normal values. Increases on in-

4 The effects of anesthetics in uranium nephritis had first been described by PEARCE (1910); the responses, diuretic or anuric, occurred while the vessels were still responsive. MACNIDER (1911–1912, 1912–1913, 1913, 1914–1915) confirmed and extended PEARCE's observations.

cubation were likewise reduced. Adding ethyl butyrate and incubating gave higher values but less than that from normal kidney. Liver and muscle samples showed similar reductions.

Brain and Kay (1927) showed a reduction in percentage dry weight of rabbit kidneys following subcutaneous or intravenous doses of 1.3 to 4 mg/kg of UNO_3 (control: 22.4–24.0%; U-treated: 15.6–17.6%). Kidney phosphatase activity was markedly depressed (control: 8.1–12.6 units/g; U-treated: 2.2–3.7 u/g).

Sanada (1936) gave single doses of 25 mg/kg of UNO_3 to rabbits and 40 hours later exsanguinated them. Kidney brei extracts were analyzed for total nitrogen, rest-N, albumin-N, and colloid osmotic pressure. Osmotic pressures were decreased from a control value of 72 to 57. The average nitrogen analyses, expressed as mg per dl were as follows:

	Total-N	Rest-N	Albumin-N
Normal	447	74	373
U-treated	382	109	274

Total nitrogen and albumin-N decreased; rest-N rose.

Rathery et al. (1953) found kidney chlorides always higher than normal.

IV. Experimental Nephritis; Urine

A. Glycosuria and Albuminuria

Following a usually toxic dose of uranium, the appearance and intensity of glycosuric and albuminuric responses were neither uniform nor dependable. The tolerant rabbit surviving a lethal dose for a non-tolerant rabbit continued to excrete sugar and albumin (Garnier and Marek, 1932). During long periods of repeated dosing, e.g., 132 days, the glycosuric response gradually increased; some albumin still appeared (Garnier and Marek, 1931b). In a tolerant dog given 2 doses per week for 5 months, 24 hour urine samples contained 2 to 8 g glucose. Decreases in urinary sugar in diabetic patients (e.g., West, 1895) perhaps were related to an augmentation of glycolysis (Lepine, 1903).

Albuminuria often preceded glycosuria; Milhorat and Deuel (1928, 1937), however, could find no relation in dogs given large doses (15 mg/kg).

The origin of the urinary glucose and the means by which glucose excretion occurred in uranium poisoning claimed the interest of a number of investigators. An increased blood glucose was often suggested and often denied. Le Conte (1853) argued that uranium decreased sugar usage (presumably making sugar available for urinary excretion). Glycogen was thought to be a likely source (Chittenden and Hutchinson, 1887) and the delayed appearance secondary to a liver effect. MacNider (1916) early looked either to liver or kidney as the source. Meyner (1898) reported a glycemia averaging 180 mg-% and Fleckseder (1906–1907) added plausibility claiming that urinary sugar concentrations rose and fell with blood sugar levels. Fleckseder (1906–1907), Wallace and Myers (1913–1914) and Weekers (1935b) lowered the blood sugar in uranium-poisoned dogs by giving insulin and observed decrements in glycosuria without abolishing it. Phloridizin thereafter always increased the glycosuria. Lepine (1903) found hypoglycemica early in U poisoning in a dog but contradicted this observation (Lepine and Bouled, 1904).

A direct action on the kidney was suggested by LECONTE (1853). CARTIER (1891) believed that the nature of the poison not the existence of glycosuria produced the histological lesions. POLLACI (1909) cut the splanchnic nerve and showed that the glycosuric response remained. Further evidence of the renal origin was provided by POLLACK (1911) in the demonstration that adrenalin raised blood sugar without glycosuria whereas uranium still induced glycosuria but without raising blood sugar levels. The latter effect was immediately confirmed by WALLACE and MYER (1913–1914). FRANK (1913) terminated the glycosuria by fasting dogs and rabbits and showed that it promptly reappeared on resumption of feeding. LUZZATTO (1914) held that glycogen was not important since starving rabbits showed glycosuria although of reduced intensity and delayed. Dietary alterations in dogs were not reflected in urinary sugar excretion (GENAUD, 1930). The most persuasive data came from crossed circulation experiments. BRULL and FANIELLE (1931) perfused the kidneys of a normal dog with blood from a uranium-poisoned donor currently glycosuric and from a normal donor. Neither blood evoked a glycosuric response in the recipient. (See also BRULL and FANIELLE, 1932). BRULL and ROERSCH (1933) confirmed these results. WEEKERS (1934, 1934b) transplanted a kidney from a uranium-poisoned dog into the neck of a normal dog and showed that the nephritic kidney continued to produce urine with an elevated glucose concentration. WEEKERS (1936) concluded that the glycosuria must have a renal origin and that liver changes in chronic uranium intoxication are accessory only.

Beginning with the studies of CHITTENDEN and colleagues (1887, 1888), many observations of albuminuria can be found. Urinary concentrations as high as 8% have been found (HESS and WIESEL, 1913) and 24 hour excretions of 2 to 3 g GARNIER et al., 1928). MARTIN and SCICLOUNOFF (1935) notably diminished the albuminuria in poisoned rabbits by giving glucose solutions subcutaneously but the rabbits died quicker.

The origin of the urinary protein claimed attention. CHITTENDEN (1888) thought that the magnitude of the albuminuria showed how greatly blood vessels are involved. PEARCE (1908) proposed a toxic factor in the serum of uranium-poisoned dogs capable of inducing albuminuria in recipients; no such effect could be shown in rabbits. GARNIER and MAREK (1931) echoed this and asked if an albumin might be the cause of the nephritis and not its consequence. GARNIER and SCHULMANN (1925) had concluded earlier that the albuminuria from egg white must have a different mechanism of production than that from parenteral U compounds. Uranium-poisoned dogs given a meat diet put out more urinary albumin (THEOHARI, 1911). BARTFELD (1922) held that some extra action must be involved. NUZUM and ROTHSCHILD (1923) gave data showing that albuminuria was not related to changes in blood chemistry. KELLAWAY et al. (1925) using immunological tests provided evidence that kidney protein probably is the source of the urinary proteins although urinary albumin in cats may have been derived from plasma. WEEKERS (1934) transplanted a nephritic kidney into the neck of a normal dog. The existing albuminuria was not affected by changing the blood supply to normal blood or to blood from a uranium-poisoned dog.

B. Urinary Nitrogen

The frequent observations of albuminuria called attention to the desirability of quantitative urinary nitrogen analyses. The results (Table 1.6) whether for total-N, uric acid, creatinine, urea or ammonia never gave a simple concordant picture probably because various investigators gave small to large doses and the

Table 1.6. Selected reports of urinary nitrogen excretion in uranium poisoning

Total N	Uric acid	Crea-tinine	Urea	NH_3	U dose	Species	References
↓	—	—	—	—	15 mg	dog	PEARCE (1910)
—	↑	↓	—	—	7–60 mg, R[a]	dog	KRAUS (1913)
↑	—	—	—	—	0.35–1.4 mg, R	dog, cat, rabbit	POHL (1911)
↑	—	—	—	↑	2–4 mg, R	rabbit	NUZUM and ROTHSCHILD (1923)
↑	—	—	—	—	0.35 mg	rabbit	DÜNNER (1914—15)
—	—	—	—	↓	38–110 mg	dog	HENDRIX and BODANSKY (1924)
—	S1. Δ[b]	S1. Δ	—	↑	5 mg/kg, R	dog	MACNIDER (1916)
—	—	↓	↓	—	4 mg/kg	dog	MACNIDER (1920)
—	—	—	—	↑	—	rabbit	PRIBYL (1929)
—	no Δ	↑	↑	—	— R	dog	GENAUD (1930)
↓	—	—	—	↓	6 mg/kg	dog	ROERSCH (1933)
—	—	—	—	S1. Δ ↑ ↓	3 mg/kg 3 mg/kg 6 mg/kg	rat rabbit dog	DEVESE (1934)
S1. ↑	—	—	—	↑	6 mg/kg	dog	ROERSCH (1934a)
↓	—	—	—	—	0.05–1 mg/kg	dog	QUINBY and FITZ (1915)
↓	—	↓	—	—	10 mg/kg	rabbit	UNDERHILL and KAPSINOW (1922)
↓	—	—	—	—	0.3–20 mg/kg	rabbit	DAKE (1926)
—	—	—	↓	↓	4 mg/kg	dog	POLONOVSKI et al. (1935)
—	—	↑	↑	—	1 mg/kg	rabbit	DONINI (1939)
—	↓	—	—	—	?	man	DELAET and MEURICE (1925)
—	—	—	↑	—	10–20 mg, R	dog	LEONE (1930)

[a] R = repeated doses.
[b] S1. Δ = slight change.

samples analyzed were collected early or late in acute poisoning or during the chronic phase with repeated doses. A few attempts to correlate stage of poisoning, dose size and urinary nitrogen elimination gave more coherent results. In early acute poisoning with small doses, total N excretion was elevated, with larger doses, total N excretion diminished (AUSTIN and EISENBREY, 1911; MOSENTHAL, 1914; DÜNNER, 1914–1915; WATANABE, 1917). QUINBY and FITZ (1915) showed that decreasing total N excretion accompanied increasingly severe pathological changes. After relatively small doses, urea decreased early (MACNIDER, 1920; POLONOVSKI et al., 1935) as did creatinine (MACNIDER, 1920), uric acid (MOSENTHAL, 1914), rest-N (DAKE, 1926), and NH_3-N (POLONOVSKI et al., 1935). Large doses sometimes gave little or no change in rest-N (DAKE, 1926). In late poisoning, NH_3-N was decreased (HENDRIX and BODANSKY, 1924). Indican excretion, decreased initially, was increased later (KISHI, 1928).

The urinary concentrations in the acute phase of total-N and NH_3-N were reported by ROERSCH (1933) for the dog as follows:

	Pre-UNO_3	Post-UNO_3 (6 mg/kg)
Total-N	28–51	5–14 (g/l)
NH_3-N	0.8–1.9	0.2–0.3 (g/l)

MAURIAC (1928b) found in the 24-hour urine samples of a tolerant dog under treatment for 5 months, 0.2–0.8 g of albumin, 4–21 g of urea, and 0.2–0.7 g of ammonia. Urinary creatinine excretion in uranium poisoning was reduced by giving creatinine (KRAUS, 1913; MAJOR, 1924).

Correlations with renal histological changes were best for urinary to blood urea concentrations (WATANABE et al., 1918).

Injecting renal vein serum from a uranium nephritic dog into a normal recipient increased the urinary volume but depressed the urea-N content (TRIA, 1910).

The source of the urinary nitrogen was held by CHITTENDEN (1888) to be "proteids", and by MOSENTHAL (1914) to be increased protein catabolism. POHL (1911) ascribed it to "irritation". GENAUD (1930) found a dependence on the diet of dogs in certain responses, e.g., the increase in creatinuria during uranium poisoning was diminished by milk or by meat (creatinine excretion was not altered).

C. Kidney Function

In contrast with the variable excretion of nitrogen fractions, all investigators agreed on the marked reduction in PSP excretion. MACNIDER (1920) proposed it as the earliest evidence of renal injury from small doses of uranium and the reappearance of PSP in the urine as the first evidence of repair and improvement in renal functions. Decreases to zero PSP excretion were often reported in acute poisoning (e.g., MACNIDER, 1927, 1929a) and sometimes values as low as 5% were found after the first hour (MENTZER, 1934). The rabbit given subcutaneously a dose of 1.25 to 3 mg exhibited a frequently described injury pattern, sharp and continued decrease in PSP excretion, progressive elevation in NPN and in urea-N, followed by beginning recovery as shown by the reversal of these changes (e.g., Fig. 1.5 from FROTHINGHAM et al., 1913). See also EISENBREY (1911), AUSTIN and EISENBREY (1911), FROTHINGHAM et al. (1913), CHRISTIAN (1914), CHRISTIAN et al. (1911), QUINBY and FITZ (1915), MACNIDER (1916, 1919), TRAISSAC (1932).

LARSON (1936) examined the time course of xylose and galactose tolerance tests in rabbits given 0.2 to 1 mg/kg of UAc. Both substances were retained abnormally in acute poisoning. In chronic poisoning, little effect on xylose tolerance was found.

RICHARDS et al. (1936) examined the excretion (and reabsorption) of inulin and of creatinine given subcutaneously or intravenously to dogs given UNO_3 by MACNIDER (in 1932, 1933). Normally, the inulin : creatinine ratio is 1.02, i.e., neither is resorbed. In one of two dogs that had once received uranium, the ratio was 1.0; in the other, the ratio was 1.1, i.e., 13% of the creatinine had been resorbed. Diffusion *in* had occurred as a result of some change in permeability. Further tests showed no evidence of glomerular alteration but a high urea resorption.

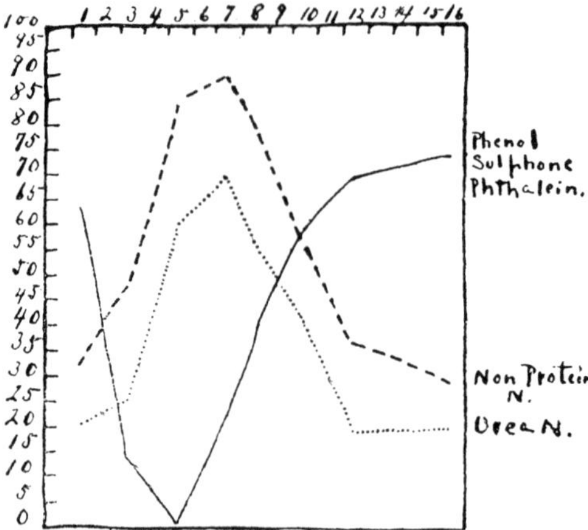

Fig. 1.5. "Curves showing phenolsulphonephthalein excretion, non-protein nitrogen and urea nitrogen in severe nephritis, Experiment 16". Taken from the work of C. FROTHINGHAM, JR., R. FITZ, OTTO FOLIN and W. DAVIS (1913). By permission of the publisher, Archives of Internal Medicine

Tests on a toad kidney by SHOJI (1936) after the toad received uranium also showed evidence of altered permeability. Blood albumin introduced into the ureter passed back and was detected in the renal vein blood. Very little if any could be found in a toad with a normal kidney. Similar studies with creatinine (SHOJI and TAKEDA, 1935) had given similar data.

D. Urinary Enzymes

QUINBY and FITZ (1915) injected UNO_3 in doses of 0.05 to 1 mg into the left kidney artery of a dog and compared the composition of the urine from the kidney with that of urine from the untreated kidney hour by hour for 3 hours. Diastase activity proved to be a valuable index of the degree of injury found later on histological examination except when there were marked differences in urine volumes from the two kidneys. FITZ (1915) discovered that considerable variation exists in "normal" urinary amylase in rabbits. In uranium poisoning, amylase activity increased in 6 of 10 rabbits in amounts not accounted for by increased urine volume in every case. In severe nephritis, activity decreased. FITZ thought that the amylase test is similar to but less quantitative than the PSP test.

E. Urinary Acetone, Acetoacetic Acid, Phenols, Organic Acids

Urinary acetone and diacetic acid roughly mirrored the intensity of uranium poisoning (MACNIDER, 1912–1913, 1916, 1920, 1927, 1929a; NUZUM and ROTHSCHILD, 1935). Older dogs excreted more acetone than younger ones. Normal dog urine was free from acetone. On day 1, after treatment with 4 mg/kg of UNO_3, urinary acetone levels were 0, 89, 89 and 86 mg/l in 4 dogs. On day 2, the concentrations were 16, 38, 54, and 289; on day 6 — 0, 30, 56 and 51 mg/l (MACNIDER, 1920). When the kidney was severely injured, the output of acetone and diacetic acid decreased. MACNIDER (1920) concluded that the quantities of acetone and diacetic acid in the urine do not reflect the disturbance in acid-base balance in the blood.

BARAC (1935) showed that substances giving a diazo reaction (phenols and imidazoles) increase during uranium poisoning despite polyuria and renal injury.

Increases in organic acid excretion after doses of uranium have long been noted. CHITTENDEN (1888) and CHITTENDEN and LAMBERT (1889) described large amounts of calcium oxalate crystals invariably present. MACNIDER (1916) proposed that the degree of uranium poisoning could be assessed from the production of organic acids and acetone bodies. KARSNER et al. (1918–1919 IV) stated "that as far as our methods are dependable there is conclusive evidence that in uranyl nitrate poisoning the primary renal irritation is not dependent on an acid intoxication as revealed by the determination of reserve alkali of the blood." Referring to this work of KARSNER et al., MACNIDER (1920) refuted the interpretation. "In a certain number of his (KARSNER'S) animals the albuminuria developed prior to the reduction of the alkali reserve of the blood. This had not occurred in our studies with a large number of animals." NUZUM and ROTHSCHILD (1923) showed that organic acids and ammonia go together with N in the urine. DEVESE (1934) found that the rat, rabbit and dog were similar in many ways in excreting organic acids during uranium poisoning.

Urine	Rat	Rabbit	Dog
Total organic acids	↑ or ↓	↑	no △
Volatile organic acids	↑	↑	no △
Lactic acid	↑ ↑	↑ ↑	↑ ↑ ↑
Hippuric acid	↓	↓	↓

DE LAET and MEURICE (1925) reported an increase in the acidity of the urine in workmen with occupational exposures to uranium compounds.

F. Urinary Sodium, Chloride, Potassium, Phosphate

BLAKE (1868) reported an increased specific gravity of urine, the excess either in urates or chlorides. With few exceptions, investigators since then have found that urinary chlorides increased in concentration and in amount, especially early in not too severe uranium poisoning. Later and especially with larger doses, chloride excretion fell off. See AUSTIN and EISENBREY (1911), GROSS (1911), POHL (1911–1912), MACNIDER (1912), KRAUS (1913), DÜNNER (1914–1915), DAKE (1922), MILHORAT and DEUEL (1937). BOYCOTT and RYFFLE (1913) and LEONE (1930) reported lower chloride concentrations in nephritic urines. GENAUD (1930) could not detect an influence of diet on the chloride excretion. Two authors treated nephritic animals with sodium chloride solutions, one subcutaneously, the other by mouth (DAKE, 1926; AUSTIN and JONAS, 1918). In both cases, urinary chloride excretion decreased. TRIA (1910) found that chloride excretion fell after a dose of renal vein serum from a poisoned animal.

DE LAET and MEURICE (1925) showed increased urinary chlorides in exposed workmen.

Sodium and potassium excretion tended to follow chlorides (POHL, 1911–1912; MOSENTHAL, 1914—sodium only).

Contrary to the report by LEONE (1930) that not much change occurred in phosphate excretion, SHIBUYA (1935) perfused kidneys with phosphate esters. The slight decrease in rotatory power was less than in the normal. SHIBUYA attributed the difference to the lessening of urine secretion with an attendant decrease in the urinary secretion of phosphate esters and of kidney phosphatase.

An early transient excretion of phosphates (Hendrix and Bodansky, 1924; Weekers, 1936, 1936b) was described which was followed by a decrease. Chittenden and Lambert (1888) reported increased phosphate excretion after 9 days. Weekers (1936) showed that the phosphaturia was not due to an increased blood phosphate nor to a change in blood phosphates but to a decreased kidney threshold. When blood from a uranium-treated dog was perfused through a transplanted kidney placed in a recipient dog's neck, phosphate excretion increased.

G. Acidosis

The acidosis of uranium poisoning demonstrated in blood analyses produced profound changes in urine composition. Brull and Roersch (1934) gave extensive data on urinary changes in uranium-treated dogs, some fed on a carbohydrate diet, some on a meat diet. Urinary volume increased on the meat diet,

	Urine volume	N	NH_3	Fixed base	Cl	P	S	Total inorganic acid	Organic acid
Meat diet	↑	↓	↓	++	±	↑	↑ or ↓	++	+, no △
Carbo-hydrate diet	↓	↓	less ↓	+	±	Sl. ↓	±	±	Sl. ↓

decreased on the carbohydrate diet; total N and ammonia decreased on both; more fixed base, inorganic acid, and phosphate appeared on the meat diet; chloride was unaltered, sulfate inconstant, organic acids increased or remained unchanged on the meat diet, but decreased slightly on the carbohydrate diet. Roersch (1934b) transplanted a normal kidney into the neck of a dog and perfused it with normal blood or with blood from a nephritic dog. The urine composition showed some remarkable differences (the data are given in milliequivalents per hour).

Perfusion	Organic Acids	Cl	P	SO_4	Total acid	NH_3-N	Fixed base	Total base	Acid base
Normal blood	0.6	3.3	0.17	0.07	4.1	0.1	4.4	4.5	0.9
U-poisoning	4.3	—	0.70	4.3	9.8	0.5	5.2	5.7	1.6

The 7-fold increase in organic acids, the 6-fold increase in sulfates, the 4-fold increase in phosphate stand out.

Brull et al. (1934) showed that along with the intense polyuria, N excretion dropped 50%, ammonia 57% and total base 9.6%, while fixed base elimination increased 28%. Mineral acid excretion rose 40%, organic acid 52%. The acid to base ratio in the poisoned dog was 2.8 (normal 1.8). Increased acid production, not renal changes must have been involved.

Roersch (1935) found a decrease in NH_3-N excretion beginning on the day UNO_3 was injected despite polyuria, presumably sequelae of cell injury. When a normal kidney transplanted into the neck was perfused by blood from a nephritic donor, however, NH_3-N excretion increased (see Texttable) along with the volume of urine, i.e., the kidney secretory activity was not lost. The additional NH_3-N came from increased splitting of ammonia complexes (Polonovski et al., 1935). Analyses of the urines collected from the transplanted kidney showed the effects of perfusion with blood from a nephritic donor (data in milliequivalents per liter of urine).

Perfusion	Organic acid	Cl	SO_4	Total acidity	NH_3-N	Fixed base	Total base
Normal blood	34	101	33	170	3	219	222
U-poisoning	48	82	6	136	12	131	144

ROERSCH commented on the variability of mineral acid excretion and pointed out that the acidosis was not the result of retention of acids.

V. Effects of Uranium on the Cardiovascular System

The retention of fluids in certain kidney diseases and the mysteries of the complex mechanism of fluid retention and edema production kept the question of a uranium effect on the blood vessels as one of the focal points in the attention of those investigating experimental nephritis. This property of uranium, in fact, made it the first choice of many investigators who hoped to draw conclusions from uranium poisoning that would be applicable to human patients suffering from nephritis.

A. Capillary Permeability

The keenest attention was devoted to the matter of capillary permeability. KOBERT (1906) described a degeneration of the intima and the subsequent outward diffusion of blood. SCHLAYER, HEDINGER and TAKAYASU (1907) postulated an increased permeability of cutaneous vessels on the basis of evidence that PIERCE (1909) found inconclusive. Tissue fluids of uranium-treated rabbits contained more proteids and more chlorides than normal (BOYCOTT and DOUGLAS, 1914–1915). WALLACE and PELLINI (1921) stated without data that uranium causes widespread capillary damage. BARTFELD (1922) believed that edema came in part from capillary injury. WHITNEY (1928) concluded that a generalized capillary damage would account for the immunological evidence of slowed penetration of lymph taken from liver or from the forelimb of uranium-poisoned animals following the intravenous administration of typhoid-immune rabbit serum. Editorial comment, regretting Dr. WHITNEY's death, offered several alternate explanations, e.g., less muscular movement. Other workers provided conflicting data. WEBER (1911) could find no evidence of changes in vessels walls in uranium poisoning. Blood volume restoration was slowed following an intravenous dose of gelatin (BOYCOTT, 1913–1914b). Intravenously infused Ringer's solution passed out of the circulation at a normal rate (CHISHOLM, 1914–1915) ruling out a direct action of uranium on the capillary wall.

This question was not rapidly settled; "One of the questions still under discussion is the relation of capillary permeability to edema" (WHITNEY, 1928).

Perfusion experiments led WOROSCHILSKY (1889) to assume vessels dilation in an ox kidney accounted for an increased perfusion rate when uranium was added to the blood. GHOREYEB (1913) found a decrease in perfusion rate of a uranium-poisoned rabbit kidney.

Studies of blood volume were pertinent. BOYCOTT (1913–1914a, 1913–1914b) showed delays in the restoration of normal blood volume after giving a saline solution or a gelatin solution intravenously to nephritic rabbits. Dogs showed only slight changes from normal (BOGERT et al., 1916a, b, c).

Blood volume restoration following hemorrhage was slowed but attained normal serum protein and chloride concentrations in uranium-treated rabbits (ITO, 1929–1930). In uranium poisoning, TAKAHASHI (1935a) showed an apparent diminution in circulating blood volume. Saline solution intravenously decreased the blood volume; gum arabic solution given intravenously increased the blood volume more than in controls (TAKAHASHI, 1935b).

B. Vessel Tone

LECONTE (1853) postulated constriction of vessels and thereby accounted for clots. TYLECOTE (1904) measured the outflow from the vena cava of pithed frogs perfused with solution of UNO_3 or uranyl chloride. Solutions containing less than 1:10000 had no effect; with increasing concentration, contraction was observed. BRULL and ROERSCH (1933) demonstrated vasodilation in the kidney of a recipient dog when perfused with blood from a uranium-poisoned donor. Blood flow was slightly increased (TRIBE et al., 1915–1916) through the poisoned kidney, the average was 2.4 cm^3/g/min (normal — 2.0).

C. Blood Pressure

An increase in blood pressure in the dog, dose dependent and transient, was mentioned by WOROSCHILSKY (1889). PEARCE (1909a) could not confirm such an effect with a dose of 7.5 mg but did detect an increase after giving 37.5 mg. The rabbit showed only a protracted fall of about 10 mm Hg. THEOHARI and GIUREA (1910) found the blood pressure changes following the administration of diuretics to be like those in untreated control animals. Blood pressure increase was reported in 1 of 2 dogs (MOSENTHAL, 1914) and in rabbits (WILCOX, 1917; SAKAMOTO, 1933). DOMINGUEZ (1928) in studies of 52 rabbits described vessel pathology without change in blood pressure. ERICKSON (1926) observed a slight increase when 5 cm^3 of a 2% uranium acetate solution were added to the calves' blood perfusing a dog hind leg preparation. In studies of rabbits, TRIBE et al. (1915–1916) reported a tendency to lower blood pressures: 74–120 mm Hg in uranium poisoning vs. 94 to 154 mm Hg in normals.

D. Cardiac Hypertrophy

Hypertrophy in guinea pigs, rabbits and dogs (SIEGEL, 1907; SMITH, 1911; DICKSON, 1912) was detected only in half of a group of rabbits (CHRISTIAN et al., 1911). TRAISSAC (1932) found it difficult to confirm cardiac hypertrophy (in rabbits). Workmen occupationally exposed showed no abnormalities in heart or vessels (DE LAET and MEURICE, 1925).

E. Arterial Lesions

About 6% of some 3500 control rabbits exhibited arteriosclerotic lesions in the aorta and certain peripheral vessels. In contrast, 23 of 52 uranium-treated rabbits had calcified rings or plaques (DOMINGUEZ, 1926, 1928). In dogs in which plasma proteins were increased by repeated injections of dog plasma, or decreased by plasmapheresis, subsequent uranium treatment produced necrotizing arteritis (HOLMAN, 1941).

The numerous studies of the histopathology of the glomerulus in uranium-treated animals were reviewed above.

F. Heart Rate

No significant changes were observed in the heart rate of a dog given uranium intravenously. Subcutaneous doses in a frog ultimately led to cardiac standstill. A rabbit heart perfused with defibrinated blood showed a transient fall in rate and force when uranium was added but recovered and kept beating for 5 hours (WOROSCHILSKY, 1889). Solutions of UNO_3 and uranyl chloride of 1:15000 to 1:10000 did not alter the rate of a perfused heart; size of contractions were diminished by concentrations over 1:10000; the heart was arrested by solutions of 1:2500. Stimulation of the vagus still slowed and stopped the heart after uranium solutions were applied to the vagus nerve.

G. Hemoglobin

Adding uranium solutions to blood produced a brown-black precipitate (WOROSCHILSKY, 1889), or a "muddier, browner red" to reflected light (TYLECOTE, 1904) without change in oxyhemoglobin bands by transmitted light. No methemoglobin was detected when a solution of 1:500 was applied. Red blood cell counts were reduced (TRAISSAC, 1932; SANADA, 1936; YASUDA, 1937), or elevated (HEILIG, 1923). Oxygenation of venous blood increased (TRIBE et al., 1915–1916) as it emerged from the kidney; average percent saturation was 76% in uranium-poisoned rabbits, 59% in normals.

H. Viscosity

BAEHR (1912–1913) found the blood viscosity to be increased in uranium poisoning.

I. Coagulation

LECONTE (1853) described clots in uranium-poisoned animals. Clotting time was increased (TYLECOTE, 1904), i.e., blood was less coagulable (WILCOX, 1917).

J. Hemolysis

A solution of 1:500 of uranium failed to hemolyze cells (TYLECOTE, 1904). LEOPOLD (1906) suspended washed, rabbit red blood cells in a diluted solution of urine from a uranium-poisoned rabbit and observed marked hemolysis, whereas the urine of normal rabbits was only weakly hemolytic. Ascites from uranium-treated rabbits was also strongly hemolytic.

K. Toxic Serum Factor

That the serum of an animal suffering from uranium nephritis might contain some toxic material led to tests by several investigators. SCHUR and WIESEL (1907) found that serum from a poisoned rabbit would dilate a frog pupil; human nephritic serum gave this response and also showed a ferric chloride reaction for adrenalin. The basis was thought to be an influence of uranium on chromaffin tissue. HESS and WIESEL (1913) gave poisoned rabbits adrenalin with what seemed to be benefit. SACHS (1934) found in guinea pigs, rabbits and dogs that ephedrine worsened uranium toxicity. Serum from a nephritic rabbit increased the blood pressure in recipient rabbits in contrast to the reaction of dogs in which blood pressure decreased if the donor had received a moderate dose but increased after a large dose (PEARCE, 1908). PEARCE (1909) produced transient (seconds) blood pressure decrease in recipient dogs and increases or more frequently decreases in rabbits.

VI. Effects of Uranium on the Blood Chemistry

A. Nitrogen

Nitrogen retention, increasing with time and under certain conditions with dose, uniformly characterizes uranium poisoning. The *total protein* of blood changed little (BRULL and FANIELLE, 1932); serum albumin often decreased (RATHERY et al., 1932; SANADA, 1936; FUKUHARA and HIRATSUKA, 1937; YASUDA, 1937) but could show little change (BRULL and FANIELLE, 1932); globulin sometimes was elevated, sometimes decreased, usually unchanged (BRULL and FANIELLE, 1932; RATHERY et al., 1933).

1. Non-protein Nitrogen

NPN consistently increased, sometimes to remarkably high concentrations. From normal levels of about 40 mg-% in dogs, MACNIDER (1927) observed levels as high as 159 mg-%, CONNELLY and HOLMAN (1942) as high as 200 mg-%; in rabbits (normal about (20 mg-%), FROTHINGHAM et al. (1913) observed peak values of 266 mg-%; in cats (normal about 40%), KARSNER and DENIS (1914) observed peak values of 228 mg-%. See also: FOLIN et al. (1912), HEILIG (1923), MACNIDER (1916, 1929a), PRIBYL (1929), FITZ (1915), UNDERHILL and KAPSINOW (1922), DAKE (1922), FUKUHARA and HIRATSUKA (1937), MILHORAT and DEUEL (1937). The time course of NPN changes was observed by MACNIDER (1916, 1920) in the dog and by FEYEL (1939) in the rat. The NPN values in chronic uranium poisoning were reported by MACNIDER (1916) and by TRAISSAC (1932).

HOLMAN and MEBANE (1940) gave data on the marked elevation in NPN and also reported simultaneous analytical values for urea-N, creatine and creatinine-N, and undetermined N (Table 1.7).

Table 1.7. Fractionation of non-protein nitrogen before and after administration of uranium nitrate to standard hypoproteinemic dogs[a]

Dog. no.	N.P.N.	Before	1 week after	2 weeks after
38–25	total	32.3	145.0	332.0
	urea N	18.7 (58)	70.3 (48)	126.5 (38)
	Creatine and creatinine N	1.7 (5)	6.6 (5)	13.8 (4)
	undetermined N	11.9 (37)	68.1 (47)	191.7 (58)
39–27	total	23.0	50.5	39.6
	urea N	11.6 (51)	26.1 (52)	18.1 (46)
	creatinine and creatine N	1.5 (6)	1.9 (4)	1.6 (4)
	undetermined N	9.9 (43)	22.5 (44)	19.9 (50)

[a] Table I, p. 302, HOLMAN and MEBANE (1940). With permission of the Journal of Experimental Medicine.

2. Rest-N

Rest-N values of remarkable size have been recorded. FRANKE (1928) reported a peak value of 386 mg-% and SACHS (1934) of 500 mg-% in the dog, and JACOBY (1935) of 385 mg-% in the rabbit. See also MIKAWA (1925). DONINI (1939) showed that rabbits on a milk diet developed higher rest-N values.

3. Urea-N

Urea-N concentrations increased remarkably also. In the dog, RATHERY et al. (1933) observed a peak value of 2.4 g/l, HAGI-PARASCHIV and CIMINO-BÉRANGER (1934) of 4 g/l; in the rabbit, GARNIER et al. (1928) of 4.4 g/l; and in the sheep, DELAGE (1938) of 6 to 7 g/l. DENIS (1913–1914) saw no change in the blood urea values in dogs fed meat. MAURIAC et al. (1932) confirmed this observation in rabbits on a milk diet and reduced the levels by insulin treatment. SATO (1933 b) reduced blood urea levels by treatment with YAKRITON, a liver detoxification "hormone" of A. SATO. For other observations of urea N, see FOLIN et al. (1912), KARSNER and DENIS (1914), MACNIDER (1916, 1919, 1920, 1927, 1929 a), BALIFF et al. (1932), SATO (1933), FROTHINGHAM et al. (1913), WATANABE et al. (1918), LARSON (1936), MILHORAT and DEUEL (1937), FEYEL (1939), JACOBY (1935), MENTZER (1934), GOTO (1917).

4. Azotemia

Azotemia was quantified by GARNIER et al. (1928), MAURIAC (1928 b), BINET and HALBERSZ-MARYNOWSKA (1933), SACHS (1934), CHRISTIAN (1913), SMILLIE (1915); SANADA (1936), FUKUHARA and HIRATSUKA (1937). BRULL and FANIELLE (1930) found a marked increase near death despite starvation self-imposed. BARTFELD (1922) concluded that the increase in blood N could not be explained by a decreased excretion.

5. Uric Acid

Uric acid concentration was reported to increase early (WILCOX, 1917). GENAUD (1930 b) found that altering the diet of dogs did not influence uric acid levels. MARTIN and SCICLOUNOFF (1935) treated uranium-poisoned rabbits with glucose solutions subcutaneously and found a shortened survival with lower blood uric acid values.

6. Creatinine

Creatinine concentrations increased (MACNIDER, 1916, 1920, 1929 a; UNDERHILL and KAPSINOW, 1922), e.g., to values of 3.8 mg-% in pregnant dogs (compare Table 1.7 from HOLMAN and MEBANE, 1940). Little or no change in creatinine levels was reported by GENAUD (1930 b) and HAGI-PARASCHIV and CIMINO-BÉRANGER (1934) in the dog.

7. Ammonia

Ammonia concentrations increased (MACNIDER, 1916; GENAUD, 1930 b; KOPROWSKI and UNINSKI, 1941) by as much as 300% (POLONOVSKI et al., 1935). The kidney cannot be the chief source of blood NH_3 in the opinion of KOPROWSKI and UNINSKI (1941); pathological changes in the circulation of the blood and convulsions probably account for the marked increase. JACOBY (1935) proposed that the liver contributed ammonia, although the kidney must be the primary site of origin.

8. Serine

Serine concentrations were rarely increased, sometimes decreased in the dog (Rathery et al., 1933).

9. Xanthoproteic Reaction

The *Xanthoproteic* reaction was increased (Franke, 1928; Plaszek, 1928) on occasion; Sachs (1934) reported no change.

10. Indican

The *Indican* content of blood increased in early uranium poisoning, however, in the later stages, with recovery the indican levels decreased. The over-all indican formation lessened in U poisoning (Kishi, 1928).

11. Diazo-reacting Substances

Diazo-reacting substances increased in acute uranium poisoning not as a result of retention but from excess production (Barac, 1935, 1935b).

12. Index of Pathological Change

Watanabe et al. (1918) chose the ratio of urinary urea to blood urea as the best correlate of the severity of kidney injury. MacNider (1919, 1920) preferred the percentage retention of urea and creatinine which increased "with the severity of injury to the tubular epithelium" (MacNider, 1920). Franke (1928) concluded that less histological change went along with increased residual N, but that the correlation was not a good one.

B. Acidosis

An increased urinary excretion of carbonate was recognized by Chittenden (1888). The degree of "acid intoxication" was correlated with the changes in renal epithelium by (MacNider, 1912–1913). Later, acidosis as revealed by lowered blood CO_2 (Goto, 1917) and by lowered CO_2 in blood and in alveolar air (MacNider, 1917, 1920) was recognized as the earliest chemical sign of intoxication. Lowered blood pH and reserve alkali always appeared prior to or simultaneously with albuminuria (Karsner et al., 1918). Urinary ketone bodies appeared still later (Nuzum and Rothschild, 1923). The acidosis on occasion was profound; blood pH decreased sharply and reserve pH values were recorded of 7.7 (normal — about 8) along with alveolar carbon dioxide tensions of 30, 20, even 12 mm Hg (MacNider, 1920). MacNider (1916) argued that uranium poisoning really was due to acid because giving UNO_3 and at the same time a soluble carbonate produced a uranium intoxication identical with that from UNO_3 alone. Acidosis was more severe in older dogs which condition paralleled the greater histological injury (MacNider, 1917).

Wallace and Pellini (1921) showed that the kidneys need not be intact for CO_2 loss to occur. CO_2 bound as HCO_3 fell to a value of 24 (normal — 58) in the intact dog; following a double nephrectomy, a value of 30 was recorded (preoperative — 50).

Other evidence. Reduction in CO_2: Nuzum and Rothschild (1923), Roersch (1935). Fall in pH: Pribyl (1929), Baliff and Gherscovici (1932), Roersch (1934). Decreased reserve alkali: MacNider (1919, 1929a), Nuzum and Rothschild (1923), Ptaszek (1928), Garnier et al. (1928), Baliff and Gherscovici

Table 1.8. The influence of a disturbance in acid-base equilibrium of the blood on renal function and pathology (MacNider, 1925–1926)

Dog	Treatment	Alkaline reserve	Urinary		PSP excretion	Urine volume	Pathology
			albumin	casts			
No uranium dose							
Normal	i.v. HCl	↓	+	+	↓↓	↓↓ or anuria	no vascular change; tubular epithelium: edema, vacuolization, necrosis, lipid
Normal	i.v. HCl then i.v. $NaHCO_3$	↑	+	+	↓↓ or 0	usually anuria; alkaline	except for no lipid, the same
Spontaneous nephropathy	i.v. $NaHCO_3$ then i.v. HCl	↓	— or trace	— or trace	↓↓ or 0	↓	glomerular nephropathy; tubular changes
Uranium poisoning							
Normal, young	U then $NaHCO_3$	no Δ	— or trace	— or trace	little Δ	little Δ	less change
Normal, old	U then $NaHCO_3$	no Δ	+ or trace	+	↓	little Δ	some dogs show changes
Spontaneous nephropathy	U	↓↓	↑	↑	↓↓ or 0	↓	glomerular nephropathy; tubular edema, necrosis, lipoid masses
Spontaneous nephropathy	U then $NaHCO_3$	↓	+	+	↓	↑ then ↓	similar except with no acute injury

(1932), RATHERY et al. (1933), BINET and HALBERSZ-MARYNOWSKA (1933). Acidosis: HENDRIX and BODANSKY (1924), MACNIDER (1927).

Evidence that acid itself given intravenously produced nephritis gave rise to the hypothesis that nephritis, spontaneous or experimental from whatever cause, might depend on an intermediate acidosis and the toxic effects of "acid bodies". The preventive or beneficial effects of alkalis strengthened this concept. MACNIDER (1916, 1917) demonstrated the beneficial action of carbonates in uranium poisoning, a finding confirmed by GOTO (1917) and extended by KARSNER et al. (1918–1919). The kinds of evidence developed to support the acidosis vs. carbonate mechanism are shown in the table (Table 1.8) which summarizes the work of MACNIDER (1925–1926) on "the influence of a disturbance in acid-base equilibrium of the blood on renal function and pathology". Intravenous acid produced renal injury with little evidence of alteration when bicarbonate was given to the normal dog but less injury in spontaneously nephropathic dogs. In uranium poisoning, bicarbonate was protective both in young and in old dogs and markedly reduced the acute injury in the highly susceptible old dogs already suffering from spontaneous nephropathies.

BRULL and ROERSCH (1934) sought the origin of the acidosis by giving uranium to dogs on diets high in carbohydrate or in raw meat. Blood analyses pre- and post-UNO_3 (6 mg/kg) showed no change in fixed base, no change or reduced chlorides, phosphate increased or unchanged, reduced bicarbonate, reduced base bound to serum protein, lowered pH, and increased acids of undetermined nature. Since the acidosis was not associated with increased phosphates, chlorides or sulfates, they reasoned that the origin of the acidosis must be accumulating organic acids. Their suspicion was confirmed by an experiment in which a normal kidney, transferred to a nephritic dog, markedly increased the urinary output of organic acids.

ROERSCH (1935) reported lowered bicarbonate and chloride, but increased blood phosphates in uranium poisoning. BRULL (1935) argued that the acidosis could not be attributed to an inability to metabolize organic acid. He "reversed" the circulation by anastomasing liver and muscle vessels, and he implanted four normal kidneys in the circulation of an acidotic, nephritic dog and failed to show an increase in the alkaline reserve. BRULL and HAIRS (1935) also failed to influence the degree of acidosis by insulin treatment. ROERSCH (1935) found that the urinary acid to base ratio pre-U treatment averaged 1.8, contrasted with an average of 2.8 after treatment. The acidosis thus cannot be laid to an accumulation of acid. BRULL et al. (1934) had clearly shown that an increase in acid production, reflected in the enormous increase in urinary acid excretion (40% in mineral acids, 52% in organic acids), could not be a renal effect.

C. Other Blood Constituents

1. Chloride

Chloride concentrations in uranium-poisoned animals were reported to be lowered (GEORGOPULOS, 1907; CHRISTIAN, 1908; HEILIG, 1923) or raised (GROSS, 1911; AUSTIN and JONAS, 1918), or sometimes higher and sometimes lower (DAKE, 1926; BALIFF and GHERSCOVICI, 1932; BINET and HALBERSZ-MARYNOWSKA, 1933; RATHERY et al., 1933) or without meaningful change (DENIS, 1923). In the rat, FEYEL (1939) showed elevated plasma chlorides at 24-hours which fell to the 7th day and had returned to normal in survivors by the 11th to 20th days. HEILIG (1923) reversed the depressed chloride levels by giving NaCl solution intravenously. GENAUD (1930b) obtained no evidence of alteration in chloride by dietary change.

2. Sodium

Sodium concentration were lower (Heilig, 1923), or unchanged (Denis, 1923), or variable (Rathery et al., 1933).

3. Potassium

Potassium concentrations were unchanged (Denis, 1923) or variable (Rathery et al., 1933). K levels in red blood cells were lowered.

4. Magnesium

Magnesium concentrations were not altered (Denis, 1923).

5. Calcium

Calcium concentrations were not changed (Denis, 1923) but Rathery et al. (1933) found lower values frequently.

6. Sulfate

Sulfate concentrations were increased (Shirokauer, 1908; Denis, 1923). In a later study, however, Denis and Reed (1927) found lower inorganic and ethereal sulfate levels after small uranium doses which were further reduced after moderate doses. Larger doses were followed by transient decreases and then marked increases in sulfates.

7. Phosphate

Phosphate concentrations were unaltered (Denis, 1923; Rathery et al., 1933) or increased (Delage, 1928).

8. Cholesterol

Cholesterol concentrations were increased (Genoud, 1930b; Mauriac, 1931; Mauriac et al., 1932a and b; Mauriac and Servantie, 1933) and could be reduced by insulin treatment (Mauriac et al., 1932). Politzer (1936) quantified the increase in free cholesterol (2.3 fold) despite a 50% reduction in cholesterol esters.

9. Lipids

Blood lipids usually were higher than normal in rabbits injected with doses of 1 mg of uranyl acetate in blood samples collected 4 to 9 days later (death ensued after 5 to 11 days). *Free fatty acids* increased 5-fold, neutral fat over 3-fold, *phospholipids* by 70% and *total fatty acids* by 70% (Politzer et al., 1936). In sheep, lipids sometimes increased, sometimes were not changed (Delage, 1928).

10. Phenols

Phenols were elevated (Marenzi, 1931).

11. Sugar

Sugar concentrations increased to 180 mg-% on the average (Meyner, 1898) in rabbits given 1 mg/kg of uranyl tartrate subcutaneously but were not maintained at that level, in fact by the 4th day blood sugar levels were 140 mg-%. Urinary glucose levels rose and fell independently. Later workers found less change in blood sugar; levels were normal (Mauriac, 1928b; Massaat, 1933). Milhorat and Deuel (1928, 1937) detected transient small decreases in blood sugar during glycosuria but later when anuric, the blood sugar levels were elevated.

12. Freezing Point

The *freezing point* rose by 0.05° on the 4th day then fell by 0.108° on the 9th day and returned to normal in about 2 weeks (MOSONYI, 1924).

13. Refractive Index

The *refractive index* increased (HEILIG, 1923).

VII. Effects of Uranium on the Liver

LECONTE (1853) wrote that the effects of uranium in the liver were difficult to define; the disappearance of liver sugar was laid to a cessation of formation. CHITTENDEN and HUTCHINSON (1886) thought glycogenolysis was increased and CHITTENDEN and LAMBERT (1888) related the liver effects to an insufficiency in nephritis. Hepatic lesions and an ephemeral diabetes were linked by CARTIER (1891) who described the persistence of the liver changes even though the urine soon ceased to be diabetic. CARTIER contrasted the ability of phloridzin to cause urinary sugar excretions as great as 50 g in 24 hours without killing the dog with the much lower urinary sugar output ($<1\%$) from a dose of uranium which proved to be promptly fatal. MEYNER (1898) reported both a glycogen loss and a fat loss in poisoned rabbits which he attributed to starvation. Thus began a debate on whether the pathological changes and functional derangements of the liver in uranium poisoning were an essential or a minor part of uranium nephritis. Undoubtedly the relatively large doses of uranium used in many of the studies (and not just in the earlier ones, e.g., MACNIDER (1937) gave dogs 4 mg/kg of UNO_3), confused the issue and prevented a ready evaluation of the relative importance of the liver changes.

A. Pathology

Histological descriptions, first of degeneration (KOBERT, 1906), and then of dilated, congested sinusoids (CHRISTIAN et al., 1911) preceded the classical work of MACNIDER (1919). Beginning with the earliest appearance of dust-like fat particles in the cells around the central vein, MACNIDER traced in sequence the granular degeneration, the accumulation of large fat masses, the cloudy swelling, and finally the edema and necrosis. Neither the rapidity of change nor the intensity of liver destruction paralleled the size of the uranium dose. Older dogs developed more severe reactions with greater quantities of stainable fat. The liver injury in pregnant dogs was out of proportion to the renal injury (MACNIDER, 1927). Air hunger preceded coma, although some died in convulsions. Recovery was "associated with processes of repair in liver and kidney". MACNIDER's interest in the liver effects of uranium brought him back on several occasions to further research (see below).

Fatty degeneration and alterations in the liver cell nuclei were reported repeatedly, e.g., in the rabbit (MACNIDER, 1929c; in chronic poisoning, TRAISSAC, 1932 and 1933); in the dog (MARRAS, 1934). According to TRAISSAC (1932), the liver pathology in uranium poisoning resembled hepatonephritic disease. MARRAS (1934) called the chronic response in the rabbit, a typical, diffuse, periportal hepatitis; he was struck by the contrast in the chronic response in the guinea pig i.e., a disintegration of the periphery of the lobule with an increase of intra-trabecular connective tissue. VERNE (1931) used the term, sclerosis, to describe the loci of hyperplastic, disorganized hepatic cells in the chronically poisoned rabbit.

SATO (1933 a) referred to liver studies by J. WATANABE, Shakai Igaku Zasshi, *34*, No. 516 (1930), M. HARA, Tokyo Igakkai Zasshi, *35*, 869 (1921), und K. TANAGUCHI, Nihon Byori Gakkai Kaishi, *14*, 579 (1924). SATO stated that S. INABA, Fukuoka-Idadaigaku Zasshi, *11*, 437 (1918) and Y. MIKAWA, idem., *18*, 819 (1925) had observed remarkable increases of liver autolysis in uranium nephritis (Translations unavailable).

MACNIDER conducted over more than a decade a series of experiments on "the resistance of fixed tissue cells morphologically altered through processes of repair". The initial liver injury was usually induced in dogs by UNO_3 (MACNIDER, 1932-1933 a and b, 1934, 1934–1935, 1935, 1936 a and b, 1937) although on occasion by another chemical. The processes of repair were examined both by tests of liver function using phenoltetrachlorphthalein (PClP), and by histological examinations of liver biopsies; these studies were made after 2 weeks and again after one month. A second uranium dose (or a dose of some other agent, e.g., the inhalation of chloroform) given at intervals from a few days up to more than 4 years established the susceptibility or the resistance of the regenerated cells. If the first uranium dose (e.g., 2 mg/kg) induced a diffuse injury with cloudy swelling, edema, vacuolation, stainable lipoid material, and some, not extensive necrosis, and, if repair restored the liver "to its normal histological structure" and normal function as shown by the rapid removal of PClP from the plasma, a second uranium dose, given, for example, 4 weeks later would again call forth a typical acute uranium poisoning with no evidence of resistance, or "tolerance". On the other hand, if the first uranium dose (e.g., 4 mg/kg) produced a more severe injury which then was followed by an atypical histological repair, and an abnormal, prolonged PClP retention in the blood, the second dose of UNO_3 would not induce acute poisoning; the marked resistance was demonstrable histologically and functionally. The resistant (atypical) liver cells were of a "very flattened type, with large deeply staining nuclei... in narrow, cord-like structures which radiated in an irregular course from the central vein to the periphery of the lobule". The degree of resistance was correlated with the extent of repair by the flattened cells.

In one dog, retested 4 years and 7 months after the initial uranium intoxication, no characteristic centrolobular necrosis developed, only small foci of epithelial change, i.e., the acquired resistance had persisted.

B. Biochemistry

Biochemical lesions of the liver in uranium poisoning have claimed the attention of a few investigators. The loss of glycogen was mentioned above. The ability of liver enzymes to split ethyl butyrate was markedly reduced (QUINAN, 1915/1916). Rest-N was greater than normal after 24 hours autolysis either in liver from poisoned rabbits or in rabbit or mouse liver treated in vitro with UNO_3 or UAc (MIKAWA, 1924). SATO (1933 a) established a correlation between the ammonia detoxifying power and survival and on this basis (SATO, 1933 b) treated uranium poisoning with the ammonia detoxicating hormone (YAKRITON) of A. SATO. Daily treatments of YAKRITON lowered mortality and decreased blood urea concentrations. MAURIAC and TRAISSAC (1927 b) found in the rabbit, briefer, higher glucose tolerance curves with coefficients of Maillard increased severalfold, a finding confirmed and extended to the dog by GARNIER and MAURIAC (1930). Extracts of liver brei from rabbits given large uranium doses (25 mg/kg) showed no change in total N, albumin-N, or osmotic pressure, but a doubled rest-N (Sanada, 1936). Nakayama (1940 and 1940 b) examined proteolytic enzyme activity in rabbit liver brei preparations 24 hours after uranium treatment (5 mg/kg, repeated). Liver dipeptidase activity fell and rose again. Casein and gelatin splitting slowed with some activation when cysteine was added. Nakayama quoted Nakai [Jikken Shokakibyo Gaku, *9*, 1211, 1424 (1934)] who found an increase in reduced glutathione during the 24 hours after uranium treatment with a gradual decrease later.

C. Biliary Excretion

Uranium excretion in bile was demonstrated in the dog (Karsner, Reimann and Brooks, 1918–1919 IV) but could not be found in the cat (Karsner, Reimann and Brooks, 1918–1919 III).

D. Fecal Analysis

Little attention has been given to possible changes in fecal composition in uranium poisoning. Mosenthal (1914) found *fecal N* to be unchanged unless a diarrhea developed which would greatly increase N excretion. A negative calcium balance in rabbits given 0.2 mg/kg of UNO_3 subcutaneously, was ascribed by Pribyl (1929) to increased *fecal calcium* without a change in urinary calcium. Brull and Roersch (1934) analyzed feces pre- and post-uranium treatment.

	Dry weight of feces (g)	N (g)	Milliequivalents			
			Ca	Mg	Cl	P
Normal	8.8	0.16	22	6.8	0.003	11.2
Uranium treated	13.8	0.22	25	6.2	0.004	22

The average excess of calcium plus magnesium over phosphate was considerably reduced: 9.1 after treatment, 17.6 before. In the opinion of Brull and Roersch, this was a matter of secondary import.

VIII. Effects of Uranium on Muscle

The effect of uranium on the contractions of isolated frog muscle were likened in many respects to the effect of barium or veratrine (Tylecote, 1904). In low concentrations, uranium did not decrease the amount of work, in fact the height and frequency of contractions were like those in saline solution. A series of equal stimuli produced contractions of remarkable uniformity. Only in higher concentrations, e.g., 1:500, did the characteristic contractures develop.

In a preparation of frog heart in Ringer's solution, uranium can replace potassium (Zwaardemaker, 1919–1920); for winter frogs the uranium concentration must be double that for summer frogs. Calcium and uranium interact; if calcium is increased, uranium must be increased. Responses were obtained in winter frogs to 200 mg calcium chloride plus 25 mg of uranium, in summer frogs to 250 mg of $CaCl_2$ plus 0.6 to 6 mg of uranium nitrate. Adding fluorescein permitted the potassium concentrations to be decreased and uranium concentrations to be markedly decreased. In one preparation, 40 mg of KCl and 10 mg of UNO_3 were added to potassium-free Ringer's solution. Weak β-radiation from mesothorium in a glass bulb initiated a heart beat after 13 minutes which was terminated by adding more uranium.

Uteri of mouse, guinea pig or rabbit suspended in potassium-free Tyrode's or Ringer's solution started beating with UNO_3 concentrations of 1:100000; contractions were less powerful than with potassium (De Road, 1914). Soref (1920–1921) produced submaximal contractions in the guinea pig uterus with lower concentrations of uranium salts than of potassium in the Locke-Ringer's solution. When a uranium solution was given intravenously to a cat, the blood pressure rose and the uterus and intestinal muscle responded as they do to doses of adrenalin. Uranium could not replace potassium in a potassium-free Tyrode's solution bathing a preparation of pig uterus (Maki, 1924); in one test, however, regular contractions were observed for a time.

Maloney and Burton (1938) gave dogs repeated intravenous doses of "colloidal" uranium. No blood pressure response occurred until the 5th or 6th dose when a marked stimulation of respiration occurred probably the result of an "allergic-like" response to colloid, not a response to uranium per se.

IX. Effects of Uranium on the Nervous System

Only a few investigators remarked on and fewer still studied the effects of uranium on the nervous system. Prostration, hind leg paralysis, loss of sight, and loss of coordination sometimes occurred in the rabbit near death from uranium poisoning (Chittenden, 1888; Chittenden and Lambert, 1888; Garnier and Marek, 1931 b; Verne, 1931; Esser, 1937). Such nervous symptoms are not unique, most metals including uranium can induce convulsions (Esser, 1937). Dilated pupils, almost unreactive to the light, and optic nerve atrophy was reported in animals during the final months of a 132 day period of repeated doses (Garnier and Marek, 1931 b). Tylecote (1904) showed that nerve stimuli were just as effective in a frog nerve-muscle preparation bathed in a uranium solution as in the control saline preparation.

Verne (1931) examined the tissues of rabbits repeatedly dosed with uranium over periods of three to five months (total doses of 1.7 to 3.6 g of UNO_3). Nerve changes came late. The animals became incapable of normal posture and lay with the rear legs extended. Histological changes in the central nervous system involved the cerebral and cerebellular cortices and were essentially localized in pyramidal cells and in Purkinje cells. Hemolysis, more or less marked in Nissl bodies, often took the form of perinuclear lamellae. Diffuse chromatic substance gave a uniform color to the cytoplasm. Some pyramidal cells were retracted with one or more satellite cells having the character of macrophages. Purkinje cells had retracted dendrites. Some nerve fibers had degenerated. Purjesz et al. (1930) gave large enough doses of uranium subcutaneously, intravenously, and intraarterially to produce death in a few minutes. Nevertheless renal tubular degeneration was obvious. The choroid plexus was hyperemic and the epithelium degenerated. In other dogs large repeated subcutaneous doses gave rise to massive glomerulonephritis. A serous exudate in the choroid plexus accompanied severe epithelial degeneration. The brain in all dogs examined showed hyperemia and capillary hemorrhages. Purjesz et al. (1930) proposed 1. that poisons injuring kidney tubules also injure the choroid plexus epithelium, 2. that poisons producing glomerular changes will alter the villi, and 3. that the elaboration of cerebrospinal fluid probably involves mechanisms comparable to those of urine secretion.

After doses of 25 mg/kg, extracts of brain brei showed a decrease in total nitrogen, a two-fold increase in rest-N, a decrease in albumin-N to one-third and in colloid osmotic pressure to one-third normal (Sanada, 1936).

In the only published observation on industrially-exposed workmen, neurological examinations including sense organs gave normal findings (DE LAET and MEURICE, 1925).

X. Effects of Uranium on Enzymes

The paucity of studies of the effects of uranium compounds on enzyme systems perhaps reflects the failure to recognize any characteristic pattern of enzyme inhibition. The protein precipitating property of uranium in solution was recognized at least by the late-1880s (KOWALSKY, 1885; CHITTENDEN and HUTCHINSON, 1886). Some of the purported enzyme effects may really have been coagulation; concentrations in parts per thousand or higher were sometimes used.

The inhibition of amylase activity (on starch) was demonstrated at high dilutions of several uranium salts, e.g., about 50% inhibition at concentrations ranging from 0.002 to 0.005% of UNO_3, UAc, NH_4 or $NaUSO_4$[5], $KUClO_3$, and NH_4Ucitrate. Low concentrations sometimes stimulated the starch conversion, e.g., 0.0003 to 0.0005% of $NaUSO_4$ (CHITTENDEN and HUTCHINSON, 1886). The action of pepsin on fibrin was partially inhibited by 1% solutions of UNO_3 and UAc; USO_4, NH_4USO_4, NH_4Ucitrate, $NaUSO_4$, and $KUClO_3$ also were inhibitory. $KUClO_3$ stimulated pepsin at high dilutions. The proteolysis of fibrin by pancreatic trypsin was inhibited by the same salts. The effects on amylase and protease were confirmed (CHITTENDEN, 1888).

Glycolysis in normal defibrinated blood was stimulated by a small amount of uranyl acetate (LEPINE, 1903).

TYLECOTE (1904) confirmed the inhibitory action on pepsin and trypsin and showed also that salivary digestion and amylopsin were blocked. In contrast, uranium increased the clotting rate of rennet on milk.

HIRATA (1910) reported increased diastase activity in blood but sharply diminished in urine.

AGULHON and SAZERAC (1912) found UAc to be much less antiseptic than UNO_3. At a concentration of 1:1000, uranyl acetate stimulated growth and acetic acid production by mycoderma aceti; a 1% solution blocked growth, perhaps by precipitating phosphate. Glycerol splitting by *bacteria du sorbose* also was increased. Alcohol production by fermentation (yeast) was prevented by a concentration of 1:10000, however, volatile acid production increased in a solution containing 1:50000. Diastase activity was inhibited.

KAYSER (1912) showed that uranium nitrate, and certain other salts depressed the ability of yeast to form alcohol. In other experiments, various concentrations of uranium, e.g., 50 to 200 mg/l were added and then the medium was sterilized. Precipitated salts were filtered off and yeast was added. Under these conditions with reduced concentrations of uranium a stimulation in alcohol production of 8 to 18% was observed. Sucrase was more sensitive then zymase to uranium. Yeast raised in a uranium solution of 100 mg/l for up to 9 generations became tolerant and was thereafter less sensitive to uranium.

Growth of the tuberculosis bacterium was blocked by 1:100 UNO_3. The bacteria showed degenerated forms (BECQUEREL, 1913).

The observation of NAKAI (1934) that the liver content of reduced glutathione increased during the 24 hours after giving uranium and gradually decreased

5 $NaUSO_4$	Sodium uranium sulfate
NH_4USO_4	Ammonium uranium sulfate
$KUClO_3$	Potassium uranium chlorate
NH_4U Citrate	Ammonium uranium citrate.

thereafter might be taken as a hint that SH groups are important (quoted by NAKAYAMA, 1940).

Kidney phosphatase activity was decreased (SHIBUYA, 1935).

WEEKERS (1936, 1936d) showed that uranium slows glycolytic enzymes in blood without blocking them. Uranium did not alter the reduction in blood phosphate following insulin administration.

The kidney brei prepared from uranium-poisoned rabbits exhibited reduced cathepsin activity which was restored a little by cysteine addition (NAKAYAMA, 1940). Peptonase activity also appeared to be somewhat less than normal. Almost no change was found in dipeptidase activity. The acetone precipitate from the enzyme solution and the total N content were each reduced to 75 or 80% normal.

DOZZI (1941) showed an increase in urinary amylase in uranium poisoning without any change in blood amylase activity. Blood lipase activity was unaltered.

XI. Selected Metabolic Effects of Uranium

Metabolic derangements secondary to the renal injuries of uranium were described above in the section devoted to the kidney. In the following paragraphs, a few reports of other metabolic alterations and their sequelae are presented.

Starvation is not the cause of death in uranium poisoning, although poisoned animals frequently refuse food (LE CONTE, 1853). Initially *protein metabolism* (CHITTENDEN and LAMBERT, 1888) was not deranged; after a few days evidence of abnormal protein breakdown appeared in the increased urinary nitrogen, sulfate and phosphate. BARTFELD (1922) attributed these changes to extrarenal uranium effects. Plasma protein formation proceeded normally in uranium poisoning even when NPN values rose to more than 10 times normal levels (HOLMAN and MCBANE, 1940). (See other effects under Kidney, Urine, Blood Chemistry.)

Disordered "respiration" was blamed for the glycosuria (LE CONTE, 1853; WOROSCHILSKY, 1889; FRANKE, 1913). CHAPHEAU (1934) showed from data on blood alcohol levels that the rate of alcohol metabolism in uranium poisoning was not significantly altered. Elevated blood sugar levels were rapidly reduced (in 3 or 4 hours) by insulin treatment (MAURIAC, 1931). WEEKERS (1936a, c and d) showed in the dog and rat that the liver glycogen content was lowered, although glycogenesis was not impaired. Blood sugar levels in the dog decreased on insulin treatment the same as in normals, however, glycosuria continued. Blood phosphorus decreased in a parallel course to the sugar. The ratio of plasma sugar to red blood cell sugar was unaltered (WEEKERS, 1936a).

Blood lipids were determined by POLITZER (1936) in rabbits 4 to 11 days after uranyl acetate injections. Total lipids, free fatty acids, free cholesterol and phospholipids were increased, neutral fatty acids dramatically so. Total fatty acids decreased as did total cholesterol and chloresterol esters. These changes were attributed to liver lesions. DELAGE (1938) found irregular changes in blood lipids in sheep, sometimes no change, sometimes two-fold increases. Phospholipids decreased although blood phosphorus values increased. Cholesterolemia typically develops in uranium poisoning (GENAUD, 1930b; MAURIAC, 1931; MAURIAC et al., 1932a, b). Alterations in the diet of treated dogs failed to change the plasma cholesterol levels (GENAUD, 1930b). Insulin sometimes decreased the blood levels (MAURIAC et al.,), for example, daily insulin treatments of rabbits chronically poisoned over a period of 8 months, slowly reduced the hypercholesterolemia. Rabbits maintained on a milk diet, but not on a carbohydrate diet, developed abnormally high cholesterol levels (Traissac, 1932). A general dislocation of metabolic state (including N metabolism) was held responsible.

Alterations in cellular processes decreasing the normal acid production were demonstrated in muscle samples. The titratable acidity in normal tissues averaged 2.2 cm^3 of N/20 NaOH, in tissues from uranium-treated muscle—1.1 cm^3. After a 24 hour autolysis, normal values averaged 0.9 cm^3, poisoned—0.4 cm^3 (Quinan, 1915–1916).

Oxygen use by the kidney decreased, venous blood showed a higher percentage oxygen saturation of hemoglobin, and blood flow was slightly increased (Tribe et al., 1915–1916) in uranium poisoning.

The oxidation rate of reduced methylene blue was unaltered in liver, kidney or muscle from poisoned rabbits (Aubertin et al., 1928). Reduced glutathione decreased in kidney (Vetri, 1933), in liver, and in muscle (Yamamoto, 1940). In liver, oxidized glutathione increased moderately, whereas a slight decrease was found in muscle.

XII. Treatment of Uranium Poisoning

Rabuteau (1873) recommended emesis, gastric lavage with tannin or albumin solutions, oil purgative or sodium phosphate (to precipitate dissolved uranium). The remarkable recuperative power of the kidney was cited by Christian (1909); if the uranium poisoning was not lethal, animals recovered. Tria (1910) offered the first of several attempts to provide antidotal assistance. The administration of *renal vein serum* sometimes reduced edema, increased urine volume, decreased albuminuria, casts, urea-N and chloride excretion. Diuretin reduced survival (Christian and O'Hare, 1913). Creatinine administration reduced polyuria, uric acid and creatinine excretion (Kraus, 1913). Doses of adrenalin given rabbits intravenously increased urine volume, diminished albuminuria, the animals seemed improved with perhaps increased survival; no improvement could be shown histologically, nor could anuria be relieved (Hess and Wiesel, 1913). Sachs (1934) showed that ephedrine worsened uranium poisoning in guinea pigs, rabbits and dogs. Dünner (1914–1915) injected uranium into one renal artery and found that the removal of that kidney a few hours later would allow the rabbit to survive. Twice daily treatment with calcium lactate solution intravenously for 8 and for 19 days worsened the poisoning; NaCl solutions were ineffective.

Effective antidotal therapy came from the demonstration by MacNider (1916) that the "acid intoxication" in dogs could be offset by sodium bicarbonate, and that this treatment not only reduced organic acids but also lessened the loss of reserve alkali and diminished the histologically demonstrable renal injury. Goto (1917) confirmed these findings in dogs given bicarbonate by mouth. (See also Wilcox, 1917.) MacNider (1917) examined the action of diuretics in poisoned dogs treated with bicarbonate. Examples of the evidence for diminished uranium toxicity in young dogs (11 months), adult dogs (3 years old), and dogs with spontaneous nephropathy are given in Table 1.9 (MacNider, 1915–1926). Evidence from histological examinations correlated with the tabulated data. Other investigators have confirmed the effectiveness of bicarbonate in the dog (Rathery et al., 1933; Binet and Halbersz-Maryowska, 1933) and in the rabbit (Binet and Marek, 1933). Donnelly and Holman (1942) may have achieved much the same end by giving dogs daily intravenous doses of 230 mg/kg of sodium citrate for 5 days. Twelve of 13 control (untreated) dogs died as contrasted with 1 of

Table 1.9. Protection by sodium bicarbonate against kidney injury from uranium nitrate (MacNider, 1925–1926)

Dog	Day of experiment	(g/kg bid) $NaHCO_3$	(mg/kg) UNO_3	Albumin and casts	Acetone (mg-%)	PSP (%)	Reserve pH	Blood urea (mg-%)
Normal, young		0	0	0	0	76	8.05	16
	1	0	3	+	0	63	7.95	14
	10	0	0	+	4.3	0	7.75	78
Normal, young		1	0	0	0	77	8.05	14
	1	1	3	0	0	72	8.0	15
	10	1	0	trace	0	74	8.05	12
Normal, adult		0	0	0	0	74	8.1	17
	1	0	3	trace	0	50	7.95	17
	10	0	0	+	trace	trace	7.6	69
Normal, adult		0	0	0	0	72	8.05	20
	1	1	3	0	0	73	8.0	18
	10	1	0	+	trace	60	8.0	20
Spontaneous		0	0	trace	0	67	8.05	18
nephropathy	1	0	3	+	6.2	18	7.9	20
	10	0	0	+	1.4	0	7.6	104
Spontaneous		1	0	+	0	58	8.05	16
nephropathy	1	1	3	+	3.8	51	7.95	22
	10	1	0	+	2.0	10	7.85	24

14 citrate-treated dogs. The CO_2-combining power (alkali reserve) fell in the control dogs from an average of 48 to 28 and remained low; in the citrate-treated dogs similar decrements were recorded in the first week, thereafter values increased to about 40.

Pribyl (1931) demonstrated improved survival with a lowering of blood rest-N, an increase in PSP excretion and a lessening of histological injury when magnesium hydroxide was given intravenously.

An alkalinizing action may also have played a role in the lifesaving effects in rabbits of a diet of beets and greens, cauliflower stalk or juice (fresh, or heated and filtered, or alcohol-treated) reported by Eisner (1931). Jacoby and Eisner (1934) carried these studies further but were unable to identify the nature of the protective substance(s). Yeast extracts were also effective. The survival of rabbits and dogs was improved by oral doses of an extract of sarsparilla root (Schneider, 1932).

Attempts to demonstrate a protective action of the kidney or of kidney extracts began with the work of Tria (1910). Injecting renal vein serum sometimes reduced edema, increased urine volume and decreased albuminuria, casts and urea-N. Single or repeated doses of uranium were followed by doses of a renal extract (Serio and Sofia, 1936) which decreased azotemia, glycemia, plasma chloride, the glucose tolerance curve, urine volume, and hemoglobinuria. Donini (1939) could not confirm this work; kidney extracts given subcutaneously or sometimes intravenously did not alter decisively the course of uranium poisoning. Vacirca (1940) repeated Donini's experiments and showed increased survival from daily injections of kidney extracts. If, following uranium treatment, the animal was thyroidectomized, the renal extract was ineffective; protection again became evident if a thyroid extract was given along with the renal extract.

The liver hormone, Yakriton, of A. Sato decreased mortality (T. Sato, 1933b). Jacoby (1935) proposed that the experimental therapy attack should begin with the liver because the increase in the urea-N to rest-N ratio involves the liver. Serio and Sofia (1936) found a liver extract ineffective.

XIII. Tissue Contents of Uranium in Uranium Poisoning

Rabuteau (1873) reported that traces of uranium could be found in the bile of poisoned animals but none in urine. Human urine from patients under homeopathic treatment gave negative tests (West, 1895) even when the patient was given by mouth a large daily dose. The limit of detection by the method used was 1:2000. The urine probably contained less than 10% of the daily dose.

Schneider (1903) obtained a positive uranium test in the tubular epithelial cells of petromyzon fluviatilis when a uranyl acetate-sodium bicarbonate solution was injected into the back muscle and subcutaneously. Potassium ferrocyanide and picric acid gave a yellow brown pigment at the injection site and in the canals of the mesonephros. Pigment could also be seen in the cytoplasm and in the nucleus of some epithelial cells.

Only traces of uranium could be detected in the urine (Kobert, 1906). Baehr (1912–1913) concluded that uranium *in* the kidney must be responsible for the injury seen histologically in the kidney treated by intraarterial injection because no changes developed on the contralateral side.

Karsner and Reimann (1918–1919 I) could find uranium in the bile of 2 dogs. None, however, was found in cat liver, spleen, blood or urine 48 hours after an intraperitoneal dose. Karsner et al. (1918–1919 III) gave intraperitoneal doses or in some cases used whole body perfusion to administer 50 to 70 mg/kg of UNO_3 to cats. Only in the kidney (16 organs, tissues or fluids were tested) was uranium detected (amount estimated as about 20% of the dose); bile was negative. In preliminary tests with minced tissues mixed with a uranium solution containing 0.125 g per liter of UNO_3, qualitatively positive reactions were uniformly obtained. Dogs given 0.5 mg/kg excreted 12–14% of the dose in bile samples of 14 to 32 ml.

Neither phosphate precipitation nor ferrocyanide color developed in kidney sections of rabbits that had been given 1 to 50 mg of UNO_3 subcutaneously or intravenously (Gil y Gil, 1923–1924). Attempts to obtain radioautographs were negative. A "biological" method was devised. Two or 3 cm^3 of urine from a treated rabbit sterilized and injected intraperitoneally into a mouse, killed the mouse in 4 or 5 days with typical renal tubular lesions. A procedure of forcing fluid also worked. A rabbit given 80 mg UNO_3 subcutaneously plus 110 cm^3 of water by mouth excreted 14 cm^3 of urine containing 2 mg of uranium. When 10 mg uranium was given intravenously plus 80 cm^3 of NaCl solution, 25 cm^3 of urine contained 5 mg of uranium.

Auriat (1927) described a green color (attributed to uranium) when cochineal stain was applied to kidney sections; the color developed in the areas of most marked destruction.

The limit of detection of uranium with ferrocyanide is about 25 mg/kg whereas the yellow-green fluorescence under ultraviolet light can be seen at 1:500000 or a ten-fold greater dilution (Eitel, 1928). Guinea pigs and rabbits were given 7.5 mg UNO_3 subcutaneously. Kidney samples ashed and treated with citric acid showed fluorescence. Urine collected 2 hours after treatment was positive. Using

Table 1.10. Results of analysis of blood, urine, liver, spleen, and kidney from rabbits by use of the magneto-optic method (JONES and GOSLIN, 1933). (Results are expressed in percent of uranium injected)

Duration of experiments (hours)	U injected mgm	Blood	Liver	Spleen	Kidney	Urine and urinary system	Total
Blank	0.0	0.0	0.0	0.0	0.0	0.0	0.0
1	0.53	89.0	1.0	1.0	6.2	1.3	98.50
2	0.30	31.0	44.0	0.31	15.0	0.66	90.97
3	0.38	27.0	50.0	0.34	23.0	0.34	100.68
4	0.32	16.0	26.0	0.10	17.0	25.0	84.10
5	0.99	13.0	22.0	0.084	22.0	23.0	80.084
6	0.36	5.6	16.0	0.072	23.0	33.0	77.672
7	0.48	4.0	13.0	0.0014	17.0	45.0	79.0014
8	0.33	0.15	15.0	0.0017	18.0	47.0	80.1517
9	0.37	0.16	23.0	0.06	18.0	35.0	76.22
10	0.34	0.39	20.0	0.22	28.0	33.0	81.61
11	0.43	0.79	13.0	0.21	30.0	39.0	83.0
12	0.31	2.1	15.0	0.19	35.0	28.0	80.29
13	0.41	1.6	13.0	0.20	32.0	43.0	89.81
14	0.36	1.7	15.0	0.26	26.0	32.0	74.96
18	0.36	0.60	12.0	0.025	35.0	43.0	89.625
20	0.42	4.0	13.0	0.039	40.0	29.0	86.039
24	0.39	0.34	14.0	0.086	34.0	38.0	86.426

Reproduced by permission of the American Journal of Physiology and of Dr. ROY N. GOSLIN.

another procedure, not then quantified, tissue sections ashed with borax gave a glass. The kidney contour was preserved. Only the cortex fluoresced. Other tissues (heart, liver, lung, small intestine, muscle and skin) were negative. These results were confirmed by STRAUB (1928) using the same borax fluorescence method.

JONES and GOSLIN (1933) applied the magneto-optic method with a sensitivity of 0.7×10^{-12} g U/cm³, error about 10%, to tissues of rabbits given 0.5 mg/kg of UNO_3 subcutaneously. Tissue samples were taken in the ensuing 24 hours. Their data are given in Table 1.10. The blood, initially containing 89% of the dose, fell to less than 1%. The liver at 3 hours held 50% of the dose and from 6 hours on kept fairly constant with about 15% of the dose. The spleen never contained more than 1%. The kidney content rose to 15–20% from 2 to 10 hours and rose again to values of 25–40%. Urine samples contained small amounts in the first 3 hours and thereafter jumped to 25% at 4 hours, about 30% at 12 hours and about 40% at 24 hours. Total recoveries ranged from 74 to 100%.

ESSER (1937) attempted to find uranium by a spectographic method in the brain and spinal cord of treated animals. Even with huge doses (190–697 mg–kg) results were negative in 5 of 7 animals and questionable in 2. Uranium could be detected in vitro when solutions contained 0.01%, 0.001% was doubtful, and 0.0001% was undetectable.

Not until the fluorescence of uranium in a sodium fluoride glass was used in a quantitative procedure was it possible to analyze normal animal tissues for uranium (HOFFMANN, 1941, 1942). Beef blood contained minute but measurable traces (Table 1.11). Beef muscle and tooth substance, and normal human urine had in the order of 10^{-5} µg/g or ml, resp. A sample of fresh beef bone, mostly of

Table 1.11. Uranium content in normal tissues (HOFFMANN, 1941, 1942)

Sample	Analyzed uranium content (μg U/g or ml)
Beef blood	1.6×10^{-7} (probably)
Beef muscle, fresh	4.0×10^{-5}
Beef muscle, ash	9.0×10^{-3}
Beef tooth, fresh	6.1×10^{-5}
Beef tooth, ash	2.75×10^{-3}
Beef bone, mostly joint, fresh	1.3×10^{-2}
Beef bone, mostly joint, ash	7.0
Human bone, ash	1.1×10^{-2} (by fluorescence)
	$< 10^{-3}$ (uranyl band)
Human urine	1.2×10^{-5}

articular bone, contained 1.3×10^{-2} μg/g on the average; the ashed sample —7 μg/g. Human bone ash had less uranium: 1.1×10^{-2} μg/g by fluorescence; less than 10^{-3} μg/g by estimation using the uranyl band.

References

AGULHON, H., SAZERAC, R.: Activation de certains processus d'oxydation microbiens par les sels d'urane. C. R. Acad. Sci. (Paris) **155**, 1186–1188 (1912).

AGULHON, H., SAZERAC, R.: De l'action de l'uranium sur certains microorganismes. Bull. Soc. chim. Fr., Ser. 4, **2**, 868–872 (1912).

ANGERMAYER: Toxicity of uranium salts. Pharm. J. p. 378, Oct. 30 (1897).

ASCHOFF, L.: The pathogenesis of the contracted kidney. Arch. intern. Med. **12**, 723–738 (1913).

AUBEL, E., MAURIAC, P.: Étude de rôle de quelques ions et des variations de leur repartition dans la pathogénie des oedèmes. Congrès Francais de Médecine, 19th Session, Paris 1927, Rapports II, Physiopathologie des oedèmes.

AUBEL, E., MAURIAC, P., BOUTIRON: Rôle de l'équilibre minéral dans les oedèmes expérimentaux. C. R. Soc. Biol. (Paris) **95**, 1557–1559 (1926).

AUBERTIN, E., MAURIAC, P., AUBEL, E.: Sur la vitesse des processus d'oxydoréduction chez des lapins nephrétique. C. R. Soc. Biol. (Paris) **99**, 321 (1928).

AURIAT, G.: Modifications des muscles striés dans l'oedème expérimental chez le lapin. C. R. Soc. Biol. (Paris) **97**, 73–75 (1927).

AURIAT, G.: Anatomie pathologique des néphrites expérimentales à l'urane avec et sans oedème. C. R. Soc. Biol. (Paris) **96**, 111–114 (1927).

AUSTIN, J. H., EISENBREY, A. B.: Experimental acute nephritis: The elimination of nitrogen and chlorides as compared with that of phenolsulphonephthalein. J. exp. Med. **14**, 366–376 (1911).

AUSTIN, J. H., JONAS, L.: Effects of diet on the plasma chlorides and chloride excretion in the dog. J. biol. Chem. **33**, 91–101 (1918).

AUTENRIETH, W., WARREN, W. H.: Laboratory manual for the detection of poisons and powerful drugs. Philadelphia: P. Blakiston's Son & Co. 6th Amer. edition from the 5th German edition. 1928.

BAEHR, G.: Über experimentelle Glomerulonephritis. Beitr. path. Anat. (Jena) **55**, 545–574 (1912).

BAEHR, G.: Über die Polyurie bei subakuter Nephritis. Dtsch. Arch. klin. Med. **109**, 417–432 (1912–1913).

BALLIF, L., GHERSCOVICI, I.: L'équilibre acido-basique dans les intoxications expérimentales par le nitrate d'urane chez le chien. C. R. Soc. Biol. (Paris) **109**, 229–231 (1932).

BARAC, G.: Nouvelles recherches sur la diazo-réaction en milieu alcalin. Rev. belge Sci. méd. **7**, 669–703 (1935).

Barac, G.: Les diazo-valeurs de sang et de l'urine au cours de la néphrite aiguë à l'urane chez le chien. C. R. Soc. Biol. (Paris) **118**, 1250–1251 (1935b).

BARTFELD, B.: Über die subacute Uranvergiftung der Kaninchen. Biochem. Z. **129**, 534–548 (1922).

BENCE, J.: Experimentelle Beiträge zur Entstehung der nephritischen Oedeme. Z. klin. Med. **67**, 69–111 (1909).

Binet, L., Haulbersz-Marynowska, H.: La néphrite aiguë expérimentale. Son syndrome humoral. Bull. Soc. méd. Hôp. Paris, 583–585, Séance de 28 Avril (1933).
Binet, L., Marek, J.: La néphrite aiguë expérimentale. Son traitement. Bull. Soc. méd. Hôp. Paris, 586–587, Séance du 28 Avril (1933).
Binet, L., Marek, J.: Recherches thérapeutiques sur les néphrites aiguës expérimentales. Presse méd. **41**, 1217—1218 (1933).
Binet, L., Marek, J.: Les injections sous-cutanées de gaz carbonique comme traitemente des néphrites expérimentales à l'urane. R. C. Soc. Biol. (Paris) **116**, 911–913 (1934).
Black, F.: Notes on diabetes. Brit. J. Homeopathy **37**, 113–132, 346–371 (1879).
Blackley, J. G.: Diabetes mellitus. Brit. homeopath. J. **2**, 433–450 (1912).
Blake, E. T.: Provings of nitrate of uranium. Brit. J. Homeopathy **26**, 1–12, 585–594 (1868).
Blanck, P.: Experimentelle Beiträge zur Pathogenese der Nierenwassersucht. Z. klin. Med. **60**, 472–479 (1906).
Bogert, L. J., Underhill, F. P., Mendel, L. B.: The regulation of the blood volume after injections of saline solutions. Studies of the permeability of cellular membranes. I. Amer. J. Physiol. **41**, 189–218 (1916).
Bogert, L. J., Underhill, F. P., Mendel, L. B.: The action of saline-colloidal solutions upon the regulation of blood volume. Studies of the permeability of cellular membranes. II. Amer. J. Physiol. **41**, 219–228 (1916).
Bogert, L. J., Underhill, F. P., Mendel, L. B.: The influence of alkaline-saline solutions upon regulation of blood volume. Amer. J. Physiol. **41**, 229–233 (1916).
Bond, C. H.: Further points in the relation of diabetes, including glycosuria, to insanity. J. ment. Sci. **43**, 291–313 (1897).
Bond, C. H.: Remarks upon the value of uranium nitrate in the control of glycosuria. Practitioner **61**, 257–264 (1898).
Bonis, V. de.: Experimentelle Untersuchungen über die Nierenfunktionen. Arch. Anat. Physiol., Physiol. Abt. **30**, 271–296 (1906).
Boycott, A. E.: On the regulation of the blood volume in normal and nephritic animals. I. J. Path. Bact. **18**, 11–31 (1913–1914).
Boycott, A. E.: The regulation of the blood volume in normal and nephritic animals. II. J. Path. Bact. **18**, 498–512 (1913–1914).
Boycott, A. E., Douglas, J. S. C.: On the regulation of the blood volume in normal and nephritic animals. III. Experiments with tissue fluid. J. Path. Bact. **19**, 221–225 (1914–1915).
Boycott, A. E., Douglas, J. S. C.: On an attempt to discover the composition of tissue fluid in normal and nephritic animals. J. Path. Bact. **19**, 528–548 (1914–1915).
Boycott, A. E., Ryffel, J. H.: The action of diuretics in experimental nephritis. J. Path. Bact. **17**, 458–501 (1913).
Bradford, F. S.: Case from practice. N. Amer. homeopath. J. **8**, 502–503 (1860).
Bradford, J. R.: Bright's disease and its varieties. Lancet **1904 II**, 191–194.
Brain, R. T., Kay, H. D.: Kidney phosphatase. II. The enzyme in disease. Biochem. J. **21**, 1104–1108 (1927).
Brull, L.: Études expérimentales du mécanisme des modifications urinaires et sanguines de la néphrite au nitrate d'urane. I. La polyurie. Arch. Mal. Reins **8**, 569–578 (1934).
Brull, L.: Étude expérimentale du mécanisme des modifications urinaires et sanguines de la néphrite au nitrate d'urane. II. La glycosurie. III. L'albuminurie. Arch. Mal. Reins **9**, 83—89 (1935).
Brull, L.: Réversibilité de l'acidose de la néphrite aiguë au nitrate d'urane. C. R. Soc. Biol. (Paris) **118**, 811–812 (1935).
Brull, L., Fanielle, G.: Origine rénale de la glycosurie au nitrate d'urane. C. R. Soc. Biol. (Paris) **108**, 1163–1165 (1931).
Brull, L., Fanielle, G.: Étude expérimentale de la néphrite. Mécanisme des modifications urinaires de la néphrite uranique. Arch. int. Pharmacodyn. **42**, 1–38 (1932).
Brull, L., Hairs, E.: Néphrite expérimentale à l'urane: Action de l'insuline sur l'acidose. C. R. Soc. Biol. (Paris) **118**, 1632–1634 (1935).
Brull, L., Lambert, J., Roersch, C.: L'acidose de la néphrite expérimentale. Excrétion rénale des bases et des acides. C. R. Soc. Biol. **115**, 182–184 (1934).
Brull, L., Roersch, C.: Néphrite à l'urane; polyurie et vasodilatation des reins entre deux donneurs. C. R. Soc. Biol. (Paris) **114**, 919–920 (1933).
Brull, L., Roersch, C.: On the mechanism of acidosis in experimental uranium nephritis. Biochem. J. **28**, 1513–1515 (1934).
Burton, A.: Case of diabetes mellitus treated by uranium nitrate. Brit. med. J. **1896 II**, 847.
Cannava, A.: Note sperimentali sull'avvelenamento da sali di uranio. Boll. Soc. ital. Biol. sper. **8**, 1551–1555 (1933).
Carey, R. J.: Case of diabetes mellitus. Lancet **1874 I**, 835–836.

Cartier, F.: Toxic glycosurias, particularly that produced by uranium nitrate. Ther. Gaz. **15**, 776–777 (1891).
Chapheau, M.: Combusion de l'alcool éthylique chez le lapin au cours de quelques intoxications (intoxication par le nitrate d'urane et par le phosphore). C. R. Soc. Biol. (Paris) **116**, 889–890 (1934).
Chisholm, R. A.: The water content of the tissues in experimental nephritis. J. Path. Bact. **18**, 404–413 (1913–1914).
Chisholm, R. A.: The regulation of the blood volume in experimental nephritis. (II.) J. Path. Bact. **19**, 265–275 (1914–1915).
Chittenden, R. H.: The physiological action of uranium salts. Ther. Gaz. **12**, 698–699 (1888).
Chittenden, R. H., Hutchinson, M. T.: Influence of uranium salts on the amylolytic action of saliva and the proteolytic action of pepsin and trypsin. Trans. Connecticut Acad. Arts Sci. **7** (Nov. 1886), 261–273 (1885–1888).
Chittenden, R. H., Lambert, A.: Some experiments on the physiological action of uranium salts. Trans. Connecticut Acad. Arts Sci. **8**, 1–18 (1888).
Chittenden, R. H., Lambert, A.: Untersuchungen über die physiologische Wirkung der Uransalze. Z. Biol. **25**, 513–532 (1889).
Christian, H. A.: Experimental nephritis. Boston med. surg. J. **158**, 416–420, 452–457 (1908).
Christian, H. A.: A glomerular lesion of experimental nephritis. Boston med. Surg. J. **159**, 8–9 (1908).
Christian, H. A.: On the study of renal function: the relation of functional tests to pathological diagnosis. Trans. Congr. Amer. Phycns Surg. **9**, 1–13 (1913).
Christian, H. A.: Further studies of experimental nephritis; some effects of diuretics. Trans. Congr. Amer. Phycns Surg. **28**, 198–201 (1913).
Christian, H. A.: The effect of thoebromin sodium salicylate in acute experimental nephritis, as measured by the excretion of phenolsulphonephthalein. Arch. intern. Med. **14**, 827–843 (1914).
Christian, H. A.: Some phases of the nephritis problem (Harvey lecture). American J. med. Sci. **151**, 625–642 (1916).
Christian, H. A., O'Hare, J. P.: A study of the therapeutic value of a diuretic (thoebromin sodium salicylate or diuretin) in acute experimental nephritis. (Study XVI.) Arch. intern. Med. **11**, 517–522 (1913).
Christian, H. A., O'Hare, J. P.: Glomerular lesions in acute experimental (uranium) nephritis in the rabbit. J. med. Res. **28**, 227–234 (1913).
Christian, H. A., Smith, R. M., Walker, I. C.: Experimental cardiorenal disease. Arch. intern. Med. **8**, 468–551 (1911).
Cook, G.: Nitrate of uranium in incontinence of urine. Brit. J. Homeopathy **24**, 331–332 (1866).
Cushny, A. R.: The secretion of the urine, 1st ed., chap. XV. New York: Longmans, Green and Company 1917.
Dake, Toshi: Beiträge zur Kenntnis der akuten Nephropathien durch Bakterientoxine. IV. Mitteilung. Nephropathie durch Uranylnitrat. Schlußbetrachtungen aus den gesamten Versuchen. Mitt. allg. Path. (Sendai) **2**, 519–556 (1926).
De Laet, M., Meurice, C.: Etude sur la pathologie professionnelle de l'uranium. Ingénier Chimiste (Brussels) **9**, 257—252 (1925).
Delage, B.: Le système lipoprotéidique du sérum an cours de la néphrite expérimentale au nitrate d'urane. C. R. Soc. Biol. (Paris) **128**, 735–736 (1938).
Denis, W.: Note on the tolerance shown by elasmobranch fish towards certain nephrotoxic agents. J. biol. Chem. **16**, 395–398 (1913/14).
Denis, W.: A study of the inorganic constituents of the blood in experimental nephritis. J. Biol. Chem. **56**, 473–481 (1923).
Denis, W., Reed, L.: A study of the influence of kidney function on the concentration of certain non-protein sulfur compounds in the blood. J. biol. Chem. **73**, 41–50 (1927).
Devèze, R.: L'aminoacidurie et l'ammoniurie au cours de la néphrite uranique aiguë chez le chien, le lapin et le rat. C. R. Soc. Biol. (Paris) **117**, 1111–1112 (1934).
Devèze, R.: L'Organoacidurie au cours de la néphrite aiguë à l'urane chez le chien, le lapin et le rat. C. R. Soc. Biol. (Paris) **117**, 1113–1114 (1934).
Dickson, E. C.: A report on the experimental production of chronic nephritis in animals by the use of uranium nitrate. Arch. intern. Med. **3**, 375–410 (1909).
Dickson, E. C.: Edema formation in guinea pigs in chronic experimental uranium nephritis. Proc. Soc. exp. Biol. (N.Y.) **8**, 46–48 (1910).
Dickson, E. C.: A further report on the production of experimental chronic nephritis in animals by the administration of uranium nitrate. Arch. intern. Med. **9**, 557–591 (1912).

DOMINGUEZ, R.: Note on arteriosclerosis in rabbits caused by some samples of uranium nitrate. Science **64**, 407–408 (1926).
DOMINGUEZ, R.: Effect on the blood pressure of the rabbit of arteriosclerosis and nephritis caused by uranium. Arch. Path. **5**, 577–606 (1928).
DONINI, P.: Opoterapia renale e nefrosi sperimentale da nitrato di uranio. Rass. Clin. Ter. **38**, 265–282 (1939).
DONNELLY, G. L., HOLMAN, R. L.: The stimulating influence of sodium citrate on cellular regeneration and repair in the kidney injured by uranium nitrate. J. Pharmacol. exp. Ther. **75**, 11–17 (1942).
DOZZI, D. L.: Origin of blood amylase and blood lipase in the dog. Relation between blood amylase and urinary amylase following induction of uranium nephritis. Arch. intern. Med. **68**, 232–240 (1941).
DRYSDALE: The true place of specifics in pharmacodynamics. Brit. J. Homeopathy **27**, 253–308 (1869).
DÜNNER, L.: Über das Wesen der experimentellen Ausschwemmungsnephritis (POHL) nach Uranvergiftung. Z. klin. Med. **81**, 355–376 (1914–1915).
DÜNNER, L., SIEGFRIED, K.: Experimentelle und pathologisch-anatomische Untersuchungen an den Nieren bei Vergiftung mit kleinen Gaben Uran. Z. exp. Path. Ther. **21**, 380–392 (1920).
DUNCAN, E.: The treatment of diabetes mellitus by nitrate of uranium. Brit. med. J. **1897 II**, 1044–1047.
EISENBREY, A. B.: A study of the elimination of phenolsulphonphthalein in various experimental lesions of the kidney. J. exp. Med. **14**, 462–475 (1911).
EISNER, G.: Über die lebensrettende Wirkung von Pflanzenteilen und daraus isolierten Säften bei der tödlich verlaufenden, subakuten Uranvergiftung. Biochem. Z. **232**, 218–228 (1931).
EITEL, H.: Über eine empfindliche Methode des Urannachweises und die Lokalisation des Urans im tierischen Organismus bei der Uranvergiftung. Naunyn-Schmiedebergs Arch. exp. Path. Pharmak. **135**, 188–193 (1928).
Erichson, K.: Über die Gefäßwirkungen einiger Substanzen an der mit Kalbsblut durchströmten Hundeextremität. Naunyn-Schmiedebergs Arch. exp. Path. Pharmak. **114**, 375–383 (1926).
ESSER, A.: Klinischanatomische und spektographische Untersuchungen des Zentralnervensystems bei akuten Metallvergiftungen unter besonderer Berücksichtigung ihrer Bedeutung für gerichtliche Medizin und Gewerbepathologie. III. Teil: Epikrise. Dtsch. Z. ges. gerichtl. Med. **27**, 253–289 (1937).
FAHR, T.: Zur Frage der sogenannten hyalintropfigen Zelldegeneration. Verh. dtsch. path. Ges. **17**, 119–123 (1914).
FAHR, T.: Pathologie des Morbus Brightii. Ergebn. allg. Path. path. Anat. **19**, 1–116 (1919).
FEYEL, P.: La néphrite expérimentale aiguë au nitrate d'urane. Ann. Anat. path. **16**, 561–596 (1939).
FITZ, R.: The immediate effect of repeated doses of theobromin sodium salicylate and theocin on renal function in acute experimental nephritis. Arch. intern. Med. **13**, 945–956 (1914).
FITZ, R.: The relation between amylase retention and excretion and non-protein-nitrogen retention in experimental uranium nephritis. Arch. intern. Med. **15**, 524–542 (1915).
FLECKSEDER, R.: Über Hydrops und Glykosurie bei Uranvergiftung. Naunyn-Schmiedebergs Arch. exp. Path. Pharmak. **56**, 54–67 (1906–1907).
FOLIN, O., KARSNER, H. T., DENIS, W.: Nitrogen retention in the blood in experimental acute nephritis of the cat. J. exp. Med. **16**, 789–796 (1912).
FRANK, E.: Über experimentelle und klinische Glykosurien renalen Ursprungs. Naunyn-Schmiedebergs Arch. exp. Path. Pharmak. **72**, 387–443 (1913).
FRANKE, M.: Sur la retention de l'azote residuel et des corps aromatiques du sang et sur sa localisation anatomique dans l'insuffisance rénale expérimentale. C. R. Soc. Biol. (Paris) **98**, 1053–1057 (1928).
FROLA, E.: Sulla nefrite tossica sperimentale nella idronefrosi. Pathologica **21**, 109–115 (1929).
FROTHINGHAM, C., FITZ, R., FOLIN, O., DENIS, W.: The relation between non-protein nitrogen retention and phenolsulphonephthalein excretion in experimental uranium nephritis. Arch. inter. Med. **12**, 245–258 (1913).
FROUIN, A.: Influence des sels d'uranium et de thorium sur le développement du bacille tuberculeux. C. R. Soc. Biol. (Paris) **74**, 282–284 (1913).
FUKUHARA, R., HIRATSUKA, G.: Beeinflussung des Bluteiweißes und dessen kolloid-osmotischen Drucks durch Infusion von Gummilösung bei experimenteller Nieren- sowie Leberschädigung. Tohoku J. exp. Med. **31**, 38–59 (1937).
GARNIER, M., MAREK, J.: Variations de la toxicité du nitrate d'urane en injection souscutanée chez le lapin. C. R. Soc. Biol. (Paris) **102**, 978–981 (1929).

Garnier, M., Marek, J.: Les limites de l'accoutumance au nitrate d'urane en injection sous-cutanée chez le lapin. C. R. Soc. Biol. (Paris) **103**, 1077–1080 (1930).

Garnier, J., Marek, J.: Sur albuminurie déterminée par le nitrate d'urane en injection sous-cutanée. J. Physiol. Path. gén. **29**, 752–766 (1931).

Garnier, M., Marek, J.: Des effets de la répétition d'une même dose de nitrate d'urane en injection sous-cutanée chez le lapin. C. R. Soc. Biol. (Paris) **107**, 99–102 (1931).

Garnier, M., Marek, J.: L'intoxication chronique par le nitrate d'urane en injection quotidienne chez le lapin. C. R. Soc. Biol. (Paris) **107**, 938–940 (1931).

Garnier, M., Marek, J.: Effets des doses croissantes de nitrate d'urane en injection sous-cutanée chez le lapin. C. R. Soc. Biol. (Paris) **108**, 651–654 (1931).

Garnier, J., Marek, J.: D'un phénomène d'accoutumance dans l'intoxication expérimentale par le nitrate d'urane. Presse méd. **40**, 829–932 (1932).

Garnier, M., Schulmann, E.: Sur l'albuminurie déterminée par l'injection de blanc d'œuf au lapin. C. R. Soc. Biol. (Paris) **93**, 600–602 (1925).

Garnier, M., Schulmann, E., Marek, J.: Toxicité du nitrate d'urane en injection sous-cutanée chez le lapin. C. R. Soc. Biol. (Paris) **99**, 269–271 (1928).

Garnier, M., Schulmann, E., Marek, J.: L'accoutumance au nitrate d'urane. C. R. Soc. Biol. (Paris) **99**, 707–709 (1928).

Garnier, M., Schulmann, E., Marek, J.: Evolution de l'albuminurie dans les néphrites expérimentales par nitrate d'urane chez le chien. C. R. Soc. Biol. (Paris) **98**, 285–287 (1928).

Genaud, P.: Etude de l'excrétion de quelques constituants urinaires chez le chien normal et le chien néphrétique en fonction de divers régimes alimentaires. C. R. Soc. Biol. (Paris) **104**, 548–549 (1930).

Genaud, P.: Etude des variations quantitatives de quelques constituants chimiques du sang chez le chien normal et le chien néphrétique en fonction de divers régimes alimentaire. C. R. Soc. Biol. (Paris) **104**, 550 (1930).

Georgopulos: Experimentelle Beiträge zur Frage der Nierenwassersucht. Z. klin. Med. **60**, 411–471 (1906).

Gérard, P., Cordier, R.: Sur le méchanisme de production des néphrites à l'urane chez le crapaud. C. R. Soc. Biol. (Paris) **109**, 59–60 (1932).

Ghoreyeb, A. A.: A study of the mechanical obstruction to the circulation of the kidney produced by experimental acute toxic nephropathy. J. exp. Med. **18**, 29–49 (1913).

Gil y Gil, C.: Die Immunität im Nierenepithelgewebe. Beitr. path. Anat. **72**, 621–653 (1923–1924).

Gmelin, C. G.: Versuche über die Wirkungen des Baryts, Strontians, Chroms, Molybdäns, Wolframs, Tellurs, Titans, Osmiums, Platins, Iridiums, Rhodiums, Palladiums, Nickels, Kobalts, Urans, Ceriums, Eisens und Mangans auf den thierischen Organismus. Tübingen: Laupp'schen Buchhandlung **1824**.

Gmelin, C. G.: Versuche über die Wirkungen, etc. (Same as 1824). J. Chemie u. Physik (Halle) **43**, 110–115 (1825).

Gmelin, C. G.: Experiments on the effects of baryta, strontia, chrome, molybdenum, tungstene, tellurium, titanium, osmium, platinum, iridium, rhodium, palladium, nickel, cobalt, uranium, cerium, iron and manganese, on the animal system. Edinb. med. surg. J. **26**, 131–139 (1826).

Goto, K.: A study of the acidosis, blood urea, and plasma chlorides in uranium nephritis, and of the protective action of sodium bicarbonate. J. exp. Med. **25**, 693–719 (1917).

Govaerts, M. P.: Pathogénie de l'oedème au cours de l'intoxication aiguë par l'urane. Bull. Acad. roy. Méd. Belg. 8, 33–46 (1928).

Gross, W.: Experimentelle Untersuchungen über den Zusammenhang zwischen histologischen Veränderungen und Funktionsstörungen der Nieren. Beitr. path. Anat. (Jena) **51**, 528–575 (1911).

Hager, H., Hermann, J.: Handbuch der Pharmazeutischen Praxis, Teil 2, S. 1168–1171. Berlin: Springer 1880.

Hagi-Paraschiv, A., Cimino-Béranger, E.: Recherches sur la créatininémie dans les néphrites expérimentales chez les chiens. C. R. Soc. Biol. (Paris) **117**, 124–126 (1934).

Hale, E. M.: Materia medica and special therapeutics of the new remedies, 4th ed., p. 458–463. New York and Philadelphia: Boericke and Tafel 1875.

Heilig, R.: Über Urandiurese. Klin. Wschr. **2**, 1029 (1923).

Heilig, R.: Über Urandiurese. Z. ges. exp. Med. **37**, 163–174 (1923).

Heineke, A., Meyerstein, W.: Experimentelle Untersuchungen über den Hydrops bei Nierenkrankheiten. Dtsch. Arch. klin. Med. **90**, 101–131 (1907).

Hendrix, B. M., Bodansky, M.: The relation of acidosis and hyperglucemia to the excretion of acids, bases and sugar in uranium nephritis. J. biol. Chem. **60**, 657–676 (1924).

HERING, C.: The guiding symptoms of our materia medica, vol. 10, p. 357–361. Philadelphia: Globe Printing House 1891.

HESS, L., WIESEL, J.: Über die Wirkung von Adrenalin bei akuten experimentellen Nephropathien. Wien. klin. Wschr. **26**, 317–319 (1913).

HIRATA, G.: Beitrag zum Verhalten der Diastase in Blut und in Urin beim Kaninchen. Biochem. Z. **28**, 23–28 (1910).

HOFFMANN, J.: Uran im tierischen Organismus. Wien. tierärztl. Mschr. **24**, 561–566 (1941).

HOFFMANN, J.: Fluoreszenzen an menschlichen Knochenaschen. Biochem. Z. **311**, 247–251 (1942).

HOLMAN, R. L.: Acute necrotizing arteritis, aortitis, and auriculitis following uranium nitrate injury in dogs with altered plasma proteins. Amer. J. Path. **17**, 359–376 (1941).

HOLMAN, R. L., MEBANE, J. G.: The influence of nitrogen retention upon the regeneration of plasma proteins. J. exp. Med. **71**, 299–304 (1940).

HUGHES, R.: On the nature and treatment of diabetes. Brit. J. Homeopathy **24**, 253–269 (1866).

HUGHES, R.: A manual of pharmacodynamics. 1867, 1st edit., p. 537; 1876, 3rd. ed., p. 752–755; 1886, 5th ed., p. 866–868. London: Henry Turner & Co.

HUGHES, R.: A few chronic cases. Brit. J. Homeopathy **31**, 367–373 (1873).

HUNTER, W. C.: Experimental study of acquired resistance of the rabbit's renal epithelium to uranyl nitrate. Ann. intern. Med. **1**, 747–789 (1927–1928).

HUNTER, W. C., ROBERTS, J. M.: Glomerular changes in the kidneys of rabbits and monkeys induced by uranium nitrate, mercuric chloride and potassium biochromate. Amer. J. Path. **8**, 665–688 (1932).

ITO, W.: Über die Veränderung des Stoffaustausches zwischen Blut und Gewebe nach Aderlaß. Tohoku J. exp. Med. **14**, 236–253 (1929–1930).

JACKSON, D. E., MANN, F. C.: On the pharmacological action of uranium. Amer. J. Physiol. **26**, 381–395 (1910).

JACOBY, M.: Über die Zusammensetzung des Reststickstoffs bei der experimentellen Uranvergiftung. Biochem. Z. **281**, 198–199 (1935).

JACOBY, M., EISNER, G.: Über die bei der Uranvergiftung wirksamen Substanzen aus höheren Pflanzen und Hefe. Biochem. Z. **268**, 322-325 (1934).

JONES, H. D., GOSLIN, R.: Some quantitative studies of the localization of uranium in the principal organs of rabbits during the course of uranium intoxication by the use of the magneto-optic method. Amer. J. Physiol. **105**, 693–696 (1933).

KARSNER, H. T., DENIS, W.: A note on nitrogen retention following repeated injections of nephrotoxic agents. J. exp. Med. **19**, 270–276 (1914).

KARSNER, H. T., REIMANN, S. P.: Studies of uranium poisoning. I. The toxicity of certain water-insoluble salts of uranium. J. med. Res. **39**, 157–161 (1918–1919).

KARSNER, H. T., REIMANN, S. P., BROOKS, S. C.: Studies of uranium poisoning. II. The solubility of uranium oxide in artificial and human gastric juice. J. med. Res. **39**, 163–168 (1918–1919).

KARSNER, H. T., REIMANN, S. P., BROOKS, S. C.: Studies of uranium poisoning. III. The question of renal tissue affinity for uranium. J. med. Res. **39**, 157–161 (1918–1919).

KARSNER, H. T., REIMANN, S. P., BROOKS, S. C.: Studies of uranium poisoning. IV. The relation of acid intoxication to nephritis. J. med. Res. **39**, 177–187 (1918–1919).

KARSNER, H. T., SHEN, T. C., WAHL, S. A.: Studies of uranium poisoning. V. The influence of light on uranium poisoning in guinea pigs. J. med. Res. **43**, 1–19 (1922).

KAYSER, E.: Influence des sels d'urane sur les ferments alcooliques. C. R. Acad. Sci. (Paris) **155**, 246–248 (1912).

KELLAWAY, C. H., DAVIES, G. F. S., WILLIAMS, F. E.: The source of the protein in the albuminuria of experimental nephritis. Aust. J. exp. Biol. med. Sci. **2**, 139–149 (1925).

KIMURA, K.: Sauerstoffzehrung des Bluts bei Nierenschädigung. Tohoku J. exp. Med. **15**, 267–284 (1930).

KISHI, Y.: Experimental studies on the indican formation. Tohoku J. exp. Med. **11**, 504–543 (1928).

KISHI, Y.: Experimental studies on the indican formation. II. Tohoku J. exp. Med. **12**, 75–80 (1928).

KOBERT, R.: Lehrbuch der Intoxikationen, Bd. II, p. 321–323. Stuttgart: Ferdinand Enke 1906.

KOPROWSKI, H., UNINSKI, H.: Taxa de amônia do sangue de cães com nefrite urânica e após nefrectomia bi-lateral. Rev. bras. Biol. **1**, 253–262 (1941).

KOWALEWSKY, N.: Essigsaures Uranoxyd, ein Reagens auf Albuminstoffe. Z. analyt. Chem. **24**, 551–556 (1885).

Kraus, W. M.: The effect of uranium nephritis on the excretion of creatinin, uric acid and chloride, and the effect of creatinin injections during uranium nephritis. Arch. intern. Med. **11**, 613–629 (1913).
Larson, H. W.: Xylose tolerance of rabbits with uranium nephritis. J. Lab. clin. Med. **22**, 117–125 (1936).
Le Conte, C.: Résumé des expériences sur l'azotate d'uranium. C. R. Soc. Biol. (Paris) **5**, 171–173 (1853).
Le Conte: Expériences sur l'azote d'uranium. La Lancette Française. Gaz. Hôp. (Paris) **27**, (40) 157 (4 Avril 1854).
Le Conte: Résumé des expériences sur l'azotate d'uranium. Gaz. méd. Paris **9**, 196 (1854).
Le Conte: Résumé des expériences sur l'azotate d'uranium. Jber. Fortschr. ges. Med. in allen Landen **5**, 108–109 (1854).
Le Conte: Wirkung des salpetersauren Urans. Schmidt's Jb. in- und ausl. ges. Med. **83**, 22 (1854). Abstract by Julius Clarus.
Leone, G.: Il tasso di colesterina e quello di N nel sangue nella nefrosi sperimentale da acetato di uranio. Rass. Ter. Pat. Clin. **2**, 1–14 (1930).
Leopold, E. J.: Über die Hämolyse bei Nephritis. Z. klin. Med. **60**, 480–489 (1906).
Lépine, R.: Les glycosuries toxiques. Revue critique. Arch. Méd. exp. **15**, 129–161 (1903).
Lépine, R., Boulud: Sur l'absence l'hyperglycémie dans la glycosurie uranique. Rev. Méd. **24**, 1–3 (1904).
Lévy, J.: Sur l'accoutumance expérimentale à quelque poisons. Bull. Soc. Chim. biol. (Paris) **16**, 631–709 (1934).
Litzner, S.: Experimentelle und klinische Untersuchungen über das Verhalten der Blutmenge bei Nierenerkrankungen. Z. klin. Med. **112**, 93–123 (1930).
Luzzatto, R.: Die Glykosurie bei experimentellen Nephritiden. Z. exp. Path. Ther. **16**, 18–67 (1914).
MacNider, W. de B.: A study of the action of various diuretics in uranium nephritis. J. Pharm. exp. Ther. **3**, 423–439 (1911–1912).
MacNider, W. de B.: A study of the renal epithelium in various types of acute experimental nephritis and of the relation which exists between the epithelial changes and the total output of urine. J. med. Res. **26**, 79–126 (1912).
MacNider, W. de B.: A study of the action of various diuretics in uranium nephritis with special reference to the part played by anesthetic in determining the efficiency of the diuretic. J. Pharmacol. exp. Ther. **4**, 491–516 (1912–1913).
MacNider, W. de B.: The effect of different anesthetics on the pathology of the kidney in acute uranium nephritis. J. med. Res. **28**, 403–422 (1913).
MacNider, W. de B.: On the difference in the response of animals of different ages to a constant quantity of uranium nitrate. Proc. Soc. exp. Biol. (N.Y.) **11**, 159–162 (1913–1914.
MacNider, W. de B.: The vascular response of kidney in acute uranium nephritis. The influence of the vascular response of diuresis. J. Pharmacol. exp. Ther. **6**, 123–146 (1914–1015).
MacNider, W. de B.: A pathological and physiological study of the naturally nephropathic kidney of the dog, rendered acutely nephropathic by uranium or by an anesthetic. J. med. Res. **34**, 199–230 (1916a).
MacNider, W. de B.: The inhibition of the toxicity of uranium nitrate by sodium carbonate, and the protection of the kidney acutely nephropathic from uranium from the toxic action of an anesthetic by sodium carbonate. J. exp. Med. **23**, 171–187 (1916b).
MacNider, W. de B.: A consideration of the relative toxicity of uranium nitrate for animals of different ages. I. J. exp. Med. **26**, 1–17 (1917a).
MacNider, W. de B.: The efficiency of various diuretics in the acutely nephropathic kidney, protected and unprotected by sodium carbonate. II. J. exp. Med. **26**, 19–35 (1917b).
MacNider, W. de B.: On the occurrence of degenerative changes in the liver in animals intoxicated by mercuric chloride and by uranium nitrate. Proc. Soc. exp. Biol. (N.Y.) **16**, 82–84 (1919a).
MacNider, W. de B.: A functional and pathological study of the chronic nephropathy induced in the dog by uranium nitrate. J. exp. Med. **29**, 513–529 (1919b).
MacNider, W. de B.: A study of renal function and the associated disturbance in the acid-base equilibrium of the blood in certain experimental and naturally acquired nephropathies. Arch. intern. Med. **26**, 1–37 (1920).
MacNider, W. de B.: A review of acute experimental nephritis. Physiol. Rev. **4**, 595–638 (1924).
MacNider, W. de B.: Studies concerning the influence of a disturbance in the acid-base equilibrium of the blood on renal function and pathology. J. metab. Res. **7–8**, 1–28 (1925–1926).

MacNider, W. de B.: The development of the chronic nephritis induced in the dog by uranium nitrate. A functional and pathological study with observations on the formation of urine by the altered kidneys. J. exp. Med. **49**, 387–409 (1929a).

MacNider, W. de B.: The functional and pathological response of the kidney in dogs subjected to a second subcutaneous injection of uranium nitrate. J. exp. Med. **49**, 411–434 (1929b).

MacNider, W. de B.: Urine formation during the acute and chronic nephritis induced by uranium nitrate. A consideration of the functional value of the proximal convoluted tubule. Amer. J. med. Sci. **178**, 449–469 (1929c).

MacNider, W. de B.: The morphological basis for certain tissue resistances. Science **73**, 103–105 (1931).

MacNider, W. de B.: Concerning the naturally acquired resistance of the livers of certain senile dogs to alcohol and to chloroform. Proc. Soc. exp. Biol. (N.Y.) **30**, 237–238 (1932–1933a).

MacNider, W. de B.: Acquired resistance of altered type of liver epithelium to uranium nitrate and to chloroform. Proc. Soc. exp. Biol. (N.Y.) **30**, 238–241 (1932–1933b).

MacNider, W. de B.: The resistance of fixed tissue cells morphologically altered through processes of repair. Trans. Ass. Amer. Phycns **49**, 14–22 (1934).

MacNider, W. de B.: Acquired resistance of liver cells to the toxic action of uranium nitrate. Proc. Soc. exp. Biol. (N.Y.) **32**, 791–793 (1934–1935).

MacNider, W. de B.: The resistance of fixed tissue cells to the toxic action of certain chemical substances. Science **81**, 601–605 (1935).

MacNider, W. de B.: A study of the acquired resistance of fixed tissue cells morphologically altered through processes of repair. I. The liver injury induced by uranium nitrate. A consideration of the type of epithelial repair which imparts to the liver resistance against subsequent uranium intoxications. J. Pharmacol. exp. Ther. **56**, 359–372 (1936a).

MacNider, W. de B.: A study of the acquired resistance, etc. (same as 1936a). II. The resistance of liver epithelium altered morphologically as the result of an injury from uranium, followed by repair, to the hepatotoxic action of chloroform. J. Pharmacol. exp. Ther. **56**, 373–381 (1936b).

MacNider, W. de B.: A study of the acquired resistance, etc. (same as 1936a). IV. Concerning the persistence of an acquired type of atypical liver cell with observations on the resistance of such cells to the toxic action of chloroform. J. Pharmacol. exp. Ther. **59**, 393–398 (1937).

MacNider, W. de B., Helms, S. T., Helms, S. C.: The course of uranium nitrate intoxications in pregnant dogs. Bull. Johns Hopk. Hosp. **40**, 145–159 (1927).

Maddock, A. G.: The toxicology of uranium. National Research Council, Montreal Laboratory, 28 Sept. 1943.

Magdeburg, W.: Diabetes mellitus. Brit, J. Homeopathy **34**, 67–77 (1876). Translated from Z. homeopath. Klin. **20**, No 14.

Major, R. H.: The excretion of creatinine in uranium nephritis. J. Lab. clinical Med. **9**, 701–704 (1924).

Maki, S.: Über die Wirkung radioaktiver Substanzen auf den isolierten Uterus. Biochem. Z. **152**, 211–227 (1924).

Maloney, A. H., Burton, A. F.: Colloidal uranium. J. Pharmacol. exp. Ther. **63**, 58–64 (1938).

Marenzi, A.-D.: Rôle du rein dans l'élimination des phénols du sang. C. R. Soc. Biol. (Paris) **107**, 741–742 (1931).

Marras, S.: Ricerche sulla intossicazione de nitrato di uranio. I. Modificazioni del fegato. Boll. Soc. ital. Biol. sper. **9**, 820–823 (1934).

Martin, E., Sciclounoff, F.: L'action du sérum glucose sur l'évolution de la néphrite uranique expérimental. C. R. Soc. Biol. (Paris) **118**, 752–754 (1935).

Massaut, C.: Sur le mode de formation du liquide céphalo-rachidien (action du violet de méthyle, du nitrate d'urane et du citrate de soude). Arch. int. Physiol. **37**, 310–316 (1933).

Massaut, C.: Action du violet de méthyle, du nitrate d'urane et du citrate de soude sur la formation du liquide céphalo-rachidien. C. R. Soc. Biol. (Paris) **114**, 921–922 (1933).

Mauriac, P.: Contribution à l'étude de la néphrite chronique expérimentale chez le lapin. C. R. Soc. Biol. (Paris) **99**, 322–323 (1928).

Mauriac, P.: Contribution à l'étude de la néphrite expérimentale chez le chien; de l'influence des régimes chlorurés et déchlorurés sur le sang. C. R. Soc. Biol. (Paris) **99**, 1733–1734 (1928).

Mauriac, P.: L'atteinte de foie dans la néphrite chronique recherches expérimentales. J. Méd. Fr. **19**, 351–353 (1930).

Mauriac, P.: La néphrite expérimentale par les sels d'urane chez le lapin et chez le chien. Arch. int. Pharmacodyn. **39**, 345–370 (1930).

Mauriac, P.: Action de l'insuline sur l'hypercholestérinémie des lapins néphrétiques. C. R. Soc. Biol. (Paris) **108**, 54–56 (1931).
Mauriac, P.: A propos d'un article de M. Garnier et Marek sur "un phénomène d'accoutumance dans l'intoxication expérimentale par le nitrate d'urane". Presse méd. **40**, 921–922 (1932).
Mauriac, P., Aubel, E.: Reproduction expérimentale de la néphrite hydropigène. C. R. Soc. Biol. (Paris) **95**, 593–594 (1926).
Mauriac, P., Broustet, P., Dubarry: Action de l'insuline sur les malades atteints de néphrite chronique. Presse méd. **40**, 1844 (1932).
Mauriac, P., Broustet, P., Traissac: Action de l'insuline sur l'azotémie et la cholesterolémie des lapins atteints de néphrite chronique à l'urane. Presse méd. **40**, 1844 (1932).
Mauriac, P., Broustet, P., Traissac, J.-F.: Action de l'insuline sur l'azotémie et la cholesterolémie des lapins atteints de néphrite chronique à l'urane. Bull. Acad. nat. Méd. (Paris) **108**, 1443–1445 (1932).
Mauriac, P., Servantie, L.: Azotémie expérimentale et régime lacté chez le lapin atteint de néphrite chronique a l'urane. C. R. Soc. Biol. (Paris) **105**, 921–923 (1930).
Mauriac, P., Servantie, L.: Production expérimentale d'hypercholestérole chez le lapin en état de néphrite chronique à l'urane. C. R. Soc. Biol. (Paris) **114**, 1105–1108 (1933).
Mauriac, P., Traissac, F.-J.: Etude sur l'élimination de la phénolsulfonephthaléine dans la néphrite expérimentale à l'urane. C. R. Soc. Biol. (Paris) **97**, 79–81 (1927).
Mauriac, P., Traissac, F.-J.: Contribution à l'étude des troubles fonctionnels de foie au cours de la néphrite expérimentale à l'urane. C. R. Soc. Biol. (Paris) **97**, 81–84 (1927).
Mentzer, J. H.: The effect of diuretics on rabbits during the recovery stage from acute uranium nephritis. J. Pharmacol. exp. Ther. **52**, 246–258 (1934).
Messini, M.: Sulla tossicità di alcuni sali di uranile. Biochim. Terap. sper. **16**, 120–124 (1929).
Meyner, H.: Der Kohlehydratverbrauch bei Uranvergiftungen. (Thesis) Universität Würzburg (1898), 26 pp.
Mikawa, Y.: Über die chemischen Veränderungen des Organstoffwechsels bei Nierenerkrankungen mit besonderer Berücksichtigung der Uranvergiftung. I. Biochem. Z. **146**, 545–561 (1924).
Milhorat, A. T., Deuel, H. J., Jr.: The mechanism of the glycosuria of acute uranium nephritis. Proc. Soc. exp. Biol. (N.Y.) **25**, 294–295 (1927–1928).
Milhorat, A. T., Deuel, H. J., Jr.: Acute uranium nephrosis. The mechanism of the glycosuria. Arch. intern. Med. **60**, 77–87 (1937).
Mosenthal, H., Schlayer, C.: Experimentelle Untersuchungen über die Ermüdbarkeit der Niere. Dtsch. Arch. klin. Med. **111**, 217–251 (1913).
Mosenthal, H. O.: Nitrogen metabolism and the significance of the non-protein nitrogen of the blood in experimental uranium nephritis. Arch. intern. Med. **14**, 844–868 (1914).
Mosonyi, J.: Chemische und physikalisch-chemische Veränderungen im Blute bei experimentellen Nephritiden. Z. klin. Med. **99**, 500–505 (1924).
Nakayama, M.: Über die Nierenproteolyse. II. Über die Nierenproteolyse bei der Cantharidin- und Uranylnitratvergiftung. Tohoku J. exp. Med. **38**, 493–502 (1940a).
Nakayama, M.: Über die Leberproteolyse bei der Cantharidin- und Uranylnitratvergiftung. Tohoku J. exp. Med. **38**, 545–553 (1940b).
Nuzum, F. R., Rothschild, L. L.: Experimental uranium nephritis. Arch. intern. Med. **31**, 894–909 (1923).
O'Hare, J. P.: Acute renal lesions produced by uranium nitrate in the dog in comparison with the rabbit. Arch. intern. Med. **12**, 61–63 (1913).
O'Hare, J. P.: Experimental chronic nephritis produced by the combination of chemical (uranium nitrate) and bacteria (B. coli communis). Arch. intern. Med. **12**, 49–60 (1913).
Oliver, J.: The histogenesis of chronic uranium nephritis with especial reference to epithelial regeneration. J. exp. Med. **21**, 425–451 (1915).
Opie, E. L., Alford, L. B.: The influence of diet upon necrosis caused by hepatic and renal poisons. II. Diet and the nephritis caused by potassium chromate, uranium nitrate, or chloroform. J. exp. Med. **21**, 21–37 (1915).
Patrassi, G., Rogers, N. W.: Alterazioni renali e manifestazioni immunitarie nell'intossicazione sperimentale da uranio. Sperimentale **85**, 259–287 (1931).
Pearce, R. M.: The theory of chemical correlation as applied to the pathology of the kidney. Arch. intern. Med. **2**, 77–106 (1908).
Pearce, R. M.: An experimental study of the influence of kidney extracts and of the serum of animals with renal lesions upon the blood pressure. J. exp. Med. **11**, 430–443 (1909a).
Pearce, R. M.: The production of edema. An experimental study of the relative etiologic importance of renal injury, vascular injury and plethoric hydremia. Arch. intern. Med. **3**, 422–437 (1909b).

PEARCE, R. M.: The problems of experimental nephritis. Arch. intern. Med. **5**, 133–167 (1910).

PEARCE, R. M.: Concerning the depressor substance of dog's urine and its disappearance in certain forms of experimental acute nephritis. J. exp. Med. **12**, 128–138 (1910).

PEARCE, R. M., HILL, M. C., EISENBREY, A. B.: Experimental acute nephritis: The vascular reactions and the elimination of nitrogen. J. exp. Med. **12**, 196–226 (1910).

PEARCE, R. M., SAWYER, H. P.: Concerning the presence of nephrotoxic substances in the serum of animals with experimental nephritis. J. med. Res. **19**, 269–280 (1908).

PENTSCHEW, A., KASSOWITZ, H.: Vergleichende Untersuchungen über die Wirkung verschiedener Metallsalze auf das Zentralnervensystem von Kaninchen. Naunyn-Schmiedebergs Arch. exp. Path. Pharmak. **164**, 667–684 (1932).

PLESCH, J.: Untersuchungen über die Physiologie und Pathologie der Blutmenge. Z. klin. Med. **93**, 241–284 (1922).

POHL, J.: Über subakute Nephritis. Naunyn-Schmiedebergs Arch. exp. Path. Pharmak. **67**, 233–246 (1911–1912).

POLITZER, M.: Il ricambio lipidico nella nefrite sperimentale da uranio. Arch. Farmacol. sper. **62**, 70–76 (1936).

POLLAK, L.: Kritisches und Experimentelles zur Klassifikation der Glykosurien. Naunyn-Schmiedebergs Arch. exp. Path. Pharmak. **61**, 376–386 (1909).

POLLAK, L.: Über renale Glykosurie. Naunyn-Schmiedebergs Arch. exp. Path. Pharmak. **64**, 415–426 (1911).

POLONOVSKI, M., BIZARD, G., GOULANGER, P.: L'ammoniophanérèse au cours des néphrites expérimentales uraniques chez le chien. C. R. Soc. Biol. (Paris) **118**, 989–990 (1935).

PŘIBYL, E.: L'anisotonie dans le néphrite provoquée par le nitrate d'urane. C. R. Soc. Biol. (Paris) **102**, 402–404 (1929).

PŘIBYL, E.: Chemotherapeutische Beeinflussung der durch Gifte experimentell verursachten Leber- und Nierenschädigungen mittels kolloiden Magnesiumhydroxyds. Naunyn-Schmiedebergs Arch. exp. Path. Pharmak. **160**, 255–268 (1931).

PTASZEK, L.: Réserve alcaline et corps aromatique du sang dans l'insuffisance rénale expérimentale chez le chien. C. R. Soc. Biol. (Paris) **98**, 150–152 (1928).

PURJESZ, B., DANCZ, M., HORVATH, K.: Die Rolle des Plexus chorioideus bei der Ausscheidung des Liquor cerebrospinalis. Mschr. Psychiat. Neurol. **77**, 319–347 (1930).

QUINAN, C.: Lipase studies. III. On acid production in the tissues with especial reference to uranium nephritis. J. med. Res. **33**, 345–377 (1915–1916).

QUINBY, W. C., FITZ, R.: Observations on renal function in acute experimental unilateral nephritis. Arch. intern. Med. **15**, 303–328 (1915).

RABUTEAU, A.: Éléments de toxicologie et de médecine légale appliqués à l'empoisonnement, p. 839. Paris: Librairie Lauwereyns 1873.

RAAD, H. DE: Uterusbewegungen en Radioactiviteit. Thesis, Utrecht, 1922; Physiol. Abstr. **8**, 145 (1923).

RATHERY, F., DEBIENNE, BINET DU JASSONEIX: Néphrite uranique expérimentale. Étude de certains troubles humoraux. C. R. Soc. Biol. (Paris) **112**, 1652–1654 (1933).

RICHARDS, A. N., WESTFALL, B. B., BOTT, P. A.: Inulin and creatinine clearances in dogs, with notes on some late effects of uranium poisoning. J. biol. Chem. **116**, 749–755 (1936).

RICHTER, P. F.: Festschrift Herrn Geheimrat Professor Senator für Feier seines Siebenzigsten Geburtstages gewidmet. Die Experimentelle Erzeugung von Hydrops bei Nephritis. Beitr. klin. Med. 283–293 (1904).

RICHTER, P. F.: Experimentelles über die Nierenwassersucht. Berliner Klin. Wschr. **42**, 384–387 (1905).

ROERSCH, C.: Excrétion d'ammoniaque chez le chien néphretique par l'urane. C. R. Soc. Biol. (Paris) **114**, 1356–1357 (1933).

ROERSCH, C.: Néphrite expérimentale à l'urane. L'excrétion ammoniacale etudiée sur des reins sains au cou d'animaux néphrétiques. C. R. Soc. Biol. (Paris) **117**, 79–81 (1934).

ROERSCH, C.: L'acidose de la néphrite expérimentale. Excrétion des bases et des acides etudiée sur les reins intercalés au cou de deux donneurs. C. R. Soc. Biol. (Paris) **117**, 81–83 (1934).

ROERSCH, C.: Contribution à l'etude de la néphrite expérimentale: Méchanisme de l'acidose. Arch. int. Physiol. **40**, 329–356 (1935).

ROTH, W., BLOSS, K.: Über die experimentelle Nephritis (Glomerulonephritis). Virchows Arch. path. Anat. **238**, 325–358 (1922).

SACHS, I.: Die Wirkung des Ephedrins auf den Ablauf der Urannephritis. Naunyn-Schmiedebergs Arch. exp. Path. Pharmak. **176**, 248–254 (1934).

SAKAMOTO, H.: Über die Blutdrucksteigerung bei experimenteller Nephritis. Jap. J. med. Sci. IV. Pharmakology **7**, 118 (1933).

Sanada, Y.: Studien über den kolloid-osmotischen Druck des Gewebseiweißes. II. Gewebseiweiß und dessen kolliod-osmotischer Druck bei Nierenschädigung. Tohoku J. exp. Med. **29**, 156–168 (1936).

Saundby: See discussion, Duncan (1897).

Sato, T.: Studies on the detoxicating hormone of the liver (Yakriton), 38th Report. Uranyl nephritis in individuals with different liver power of detoxication. An evidence of close relation between the liver and the kidney. Tohoku J. exp. Med. **20**, 399–407 (1933).

Sato, T.: Studies on the detoxicating hormone of the liver (Yakriton), 39th Report. Effect of yakriton upon uranyl nephritis. Tohoku J. exp. Med. **20**, 408–419 (1933).

Schirokauer, H.: Über den Salzstoffwechsel bei experimenteller Nierenwassersucht. Z. klin. Med. **64**, 329–358 (1907).

Shirokauer, H.: Weitere Beiträge zum Salzstoffwechsel bei experimenteller Nephritis. Z. klin. Med. **66**, 169–182 (1908).

Schlayer: Neuere klinische Anschauungen über Nephritis. Med. Klin. **8**, B.-H. 9 (1912).

Schlayer, Hedinger: Experimentelle Studien über toxische Nephritis. Dtsch. Arch. klin. Med. **90**, 1–51 (1907).

Schlayer, Hedingr, Takayasu: Über nephritisches Ödem. Dtsch. Arch. klin. Med. **91**, 59–97 (1907).

Schlayer, Takayasu, R.: Untersuchungen über die Funktion kranker Nieren. Dtsch. Arch. klin. Med. **98**, 17–92 (1910).

Schmid, P., Schlayer: Über nephritisches Ödem. Dtsch. Arch. klin. Med. **104**, 44–93 (1911).

Schneider, F.: Die Beeinflussung der experimentellen Urannephritis durch Renotrat. Naunyn-Schmiedebergs Arch. exp. Path. Pharmak. **166**, 56–61 (1932).

Schneider, G.: Ein Beitrag zur Physiologie der Niere niederer Wirbelthiere. Skand. Arch. Physiol. **14**, 383–389 (1903).

Schulmann, E., Marek, J.: Recherches sur l'oedème local provoqué. Action expérimentale de quelques substances néphrotoxiques. C. R. Soc. Biol. (Paris) **99**, 272–273 (1928).

Schur, H., Wiesel, J.: Beiträge zur Physiologie und Pathologie des chromaffinen Gewebes. Verh. der Dtsch. Path. Ges., 11th Session, 16–19 Sept. 1907, in Dresden, S. 175–182 (1907).

Serio, F., Sofia, F.: Sull'azione biologica di alcuni estratti renali. Arch. Sci. med. **62**, 553–565 (1936).

Shibuya, K.: Ein Durchblutungsversuch der Niere mittels Phosphorsäureester. Arb. 3. Abt. anat. Inst. Kais. Univ. Kyoto, Ser. C, **5**, 207–214 (1935).

Shoji, T., Takeda, K.: Durchlässigkeit der Nierenepithelien für Kreatinin bei gesunder und pathologischer Krötenniere. Tohoku J. exp. Med. **26**, 592–602 (1935).

Shoji, T.: Durchlässigkeit der Nierenepithelien für Eiweiß bei den gesunden und den pathologischen Krötennieren. Tohoku J. exp. Med. **29**, 1–7 (1936).

Shoji, T.: Über den Einfluß des Novasurols auf den Durchtritt des Kreatinins durch uranvergiftete Nierenpithelien. Tohoku J. exp. Med. **29**, 8–16 (1936).

Siegel, W.: Ein Stoffwechselversuch bei Urannephritis am Hunde. Z. exp. Path. Ther. **4**, 561–575 (1907).

Siegel, W.: Mikroskopische Präparate zur Erkältungsnephritis. Dtsch. med. Wschr. **34**, 411 (1908).

Smillie, W. G.: Potassium poisoning in nephritis. Arch. intern. Med. **16**, 330–339 (1915).

Smith, R. M.: The origin of urinary casts; An experimental study. Boston med. surg. J. **158**, 696–701 (1908).

Sollmann, T.: Drugs irritant to the kidneys and hence to be avoided in impaired kidney function. J. Amer. med. Ass. **43**, 1591–1597 (1904).

Soref, E.: Radioactivity and smooth muscle. J. Physiol. (Lond.) **54**, lxxxiii (1920–1921).

Straub, W.: Urannephritis und Uranverteilung. Naunyn-Schmiedebergs Arch. exp. Path. Pharmak. **138**, 134–135 (1928).

Suzuki, T.: Zur Morphologie der Nierensekretion unter physiologischen und pathologischen Bedingungen. Jena: Gustav Fischer 1912.

Suzuki, T.: Über Urannephritis. Tohoku Univ. (Sendai, Miyagi Prefecture) Arb. Anat. Inst. Kais. **12**, 169–181 (1926).

Takahashi, S.: Über den Einfluß der Volumveränderungen der Blutflüssigkeit auf die zirkulierende Blutmenge im normalen und pathologischen Zustand. I. Die Veränderungen der zirkulierenden Blutmenge durch Kochsalzinfusion bei gesunden sowie Kantharidin- und Urantieren. Tohoku J. exp. Med. **25**, 531–549 (1935).

Takahashi, S.: Über den Einfluß der Volumveränderungen der Blutflüssigkeit auf die zirkulierende Blutmenge im normalen und pathologischen Zustand. II. Die Veränderungen der zirkulierenden Blutmenge durch die Infusion von Gummilösung bei gesunden sowie Kantharidin- und Urantieren. Tohoku J. exp. Med. **25**, 550–563 (1935).

TAKAYASU, R.: Über die Beziehungen zwischen anatomischen Glomerulusveränderungen und Nierenfunktion bei experimentellen Nephritiden. Dtsch. Arch. klin. Med. **92**, 127–153 (1908).

THÉOHARI, A.: L'alimentation carnée dans certains néphrites expérimentales. J. Physiol. Path. gén. **13**, 916–927 (1911).

THÉOHARI, A., GIUREA, C.-N.: Etude sur l'action des diurétiques dans les néphrites expérimentales. J. Physiol. Path. gén. **12**, 538—552 (1910).

TRAISSAC, F.-J.: Contribution à l'étude de la pathogénie des néphrites. L'intoxication expérimentale par le nitrate d'urane. Thèse, Univ. Bordeaux, No 57 (1932).

TRAISSAC, F.-J.: Les lésions du foie dans l'intoxication expérimentale du lapin par le nitrate d'urane. C. R. Soc. Biol. (Paris) **112**, 875–876 (1933).

TRIA, P.: Il siero di teissier nelle nefriti sperimentale. Rif. med. **26**, 617–620 (1910).

TRIBE, E. M., HOPKINS, F. G., BARCROFT, J.: Vascular and metabolic conditions in kidneys of rabbits injected with uranium acetate. J. Physiol. **50**, Proc. Physiol. Soc. 22 Jan. 1916, xi–xii.

TYLECOTE, F. E.: The pharmacology and therapeutics of uranium. Med. Chron., Ser. **4**, **7**, 379–390 (1904).

TYSON: See discussion, Duncan (1897).

UNDERHILL, F. P., KAPSINOW, R.: The relationship of blood concentration to nitrogen retention in experimental nephritis. J. Urol. (Baltimore) 8, 307–315 (1922).

VACIRCA, F.: Osservazioni sul lavoro del Dott. P. Donini, „Opoterapia renale e nefrosi sperimentale da nitrato di uranio". Nuovi dati sul decorso e sulla evoluzione della nefrite sperimentale da nitrato di urano nel coniglio. Arch. ital. Sci. farmacol. **9**, 273–279 (1940).

VERNE, J.: Lésions histologiques des centres nerveux supérieurs chez les lapins soumis a l'intoxication chronique par l'urane. Ann. Anat. path. 8, 757–758 (1931).

VETRI, M.: Il glutatione renale nella nefrite tossica da sali di uranio. Boll. Soc. ital Biol. sper. 8, 1542–1545 (1933).

WALTER, C., DAWSON, R. P.: The effect of diuretic drugs on the life of animals with severe acute nephritis. Arch. intern. Med. **12**, 171–177 (1913).

WALLACE, G. B., MYERS, H. B.: Uranium glycosuria. J. Pharm. exp. Ther. **5**, 511 (1913–1914).

WALLACE, B. G., PELLINI, E. J.: Acidosis from capillary poisons. Proc. Soc. exp. Biol. Med. **18**, 115–117 (1921).

WATANABE, C. K.: A comparative study of the rate of excretion of the nitrogenous waste products to their blood concentration in experimental uranium nephritis. J. Urol. (Baltimore) **1**, 485–494 (1917).

WATANABE, C. K., OLIVER, J., ADDIS, T.: Determination of the quantity of secreting tissue in the living kidney. J. exp. Med. **28**, 359–376 (1918).

WEBER, S.: Untersuchungen über die Permeabilität der Gefäßwand. Naunyn-Schmiedebergs Arch. exp. Path. Pharmak. **65**, 389–400 (1911).

WEEKERS, R.: Origine rénale de la glycosurie dans la néphrite expérimentale au nitrate d'urane. C. R. Soc. Biol. (Paris) **115**, 1393–1395 (1934).

WEEKERS, R.: Origine rénale de l'albuminurie dans la néphrite experimentale au nitrate d'urane. C. R. Soc. Biol. (Paris) **117**, 817–818 (1934).

WEEKERS, R.: Glycosurie uranique, ses modifications au cours de l'hypoglycémie expérimentale par l'insuline. C. R. Soc. Biol. (Paris) **118**, 1252–1253 (1935).

WEEKERS, R.: Métabolisme du sucre au cours de la néphrite expérimentale par le nitrate d'urane. C. R. Soc. Biol. (Paris) **118**, 1254–1255 (1935).

WEEKERS, R.: Intoxication expérimentale au nitrate d'urane. Arch. int. Pharmacodyn. **54**, 423–476 (1936).

WEEKERS, R.: Excrétion urinaire du phosphore au cours de l'intoxication expérimentale par le nitrate d'urane. C. R. Soc. Biol. (Paris) **121**, 866–868 (1936).

WEEKERS, R.: Rôle du foie dans le méchanisme de la glycosurie uranique. C. R. Soc. Biol. (Paris) **121**, 868–870 (1936).

WEEKERS, R.: Action du nitrate d'urane sur l'hypoglycemie insulinique et sur la glycosyle sanguine. C. R. Soc. Biol. (Paris) **122**, 78–80 (1936).

WELTMAN, O., BIACH, P.: Zur Frage der experimentellen Cholesteatose. Z. exp. Path. Ther. **14**, 367–378 (1913).

WEST, S.: The treatment of diabetes mellitus by uranium nitrate. Brit. med. J. **1895 II**, 467–472.

WEST, S.: Further contribution on the treatment of diabetes mellitus by uranium nitrate. Brit. med. J. **1896 II**, 729–730.

WHITNEY, C.: Capillary permeability in acute uranium nephritis. J. Path. Bact. **31**, 699–704 (1928).

WIESEL, J., HESS, L.: Über experimentellen Morbus Brightii. Z. exp. Path. Ther. **17**, 74–87 (1915).

Wilcox, R. W.: The therapeutics of uranium nitrate. Med. Rec. **92**, 361–364 (1917).
Wilson, A. C.: Uranium salts in cancer: A suggestion. Lancet **1903** !, 476.
Wilson, A. C.: Uranium salts in cancer. Lancet **1903 I**, 1126.
Woroschilsky: Wirkung des Urans. Arb. pharmak. Inst. **5**, 1–41 (1890).
Woroschilsky, J.: Thesis, Univ. Dorpat, 1889, Wirkung des Urans, p. 86.
Yamamoto, K.: Veränderung des Gluthathionsgehaltes in der Leber und dem Muskel der Ratte nach der Injection von Nierengiften. Kyoto-Ikadaigaku-Zasshi **29**, 817–822 (1940).
Yanagihara, D.: Über die Beziehung zwischen dem Phenol und der Autolyse verschiedener Organe bei experimentell hervorgerufener Vergiftung mit Kantharidin, Sublimat und Uran. Chosen Med. Ass. J. **29**, 489 (1939). German Abstract, p. 163.
Yasuda, H.: Studien über die bluteiweiß- und kolloidosmoregulierende Tätigkeit der Leber. II. Über die Veränderungen des Eiweißes und des kolloid-osmotischen Drucks des zu- und abströmenden Blutes der Leber bei Nierenschädigung (Kantharidin- und Uranvergiftung. Tohoku J. exp. Med. **31**, 456–468 (1937).
Zwaardemaker, H.: On physiological radio-activity. J. Physiol. (Lond.) **53**, 273–289 (1919–1920).

Chapter 2

Physical and Chemical Properties of Uranium

JAMES E. GINDLER

With 8 Figures

I. Physical Properties of Atomic Uranium

A. Extra-nuclear Properties

Uranium is the ninety-second element of the periodic system. The atomic ground state is designated $^5L_6^0$ in spectroscopic notation and has a $1s^2\,2s^2\,2p^6\,3s^2\,3p^6\,3d^{10}\,4s^2\,4p^6\,4d^{10}\,4f^{14}\,5s^2\,5p^6\,5d^{10}\,5f^3\,6s^2\,6p^6\,6d\,7s^2$ electronic configuration (LAUN, 1966; STEINHAUS et al., 1966). The valence electrons are those of the $5f^3\,6d\,7s^2$ shells. Binding energies of electrons in the various shells are given in Table 2.1 (SIEGBAHN et al., 1967). The tabulated binding energies are given with zero binding energy at the Fermi level. The total binding energy is therefore larger by an amount equal to the work function. Recent measurements of the uranium work function and the methods of measurement are given in Table 2.2.

The first three ionization potentials for uranium have been calculated by KURDIN and MAZEEV (1967) on the basis of the statistical model considering the exchange interactions and correlation of electrons. These potentials are: $I_1 = 5.65$ eV, $I_2 = 14.36$ eV and $I_3 = 25.13$ eV. Measured values of I_1 only are 6.25 ± 0.02 eV (WERNING, 1958), 6.08 ± 0.08 eV (BAKULINA and IONOV, 1959) and 6.22 ± 0.06 eV (HERTEL, 1967).

Table 2.1. Electron binding energies of uranium[a]

Shell		Binding energy (Electron-volts)	Shell		Binding energy (Electron-volts)
$1s_{1/2}$	K	115606	$4d_{3/2}$	N_{IV}	780
$2s_{1/2}$	L_I	21758	$4d_{5/2}$	N_V	738
$2p_{1/2}$	L_{II}	20948	$4f_{5/2}$	N_{VI}	392
$2p_{3/2}$	L_{III}	17168	$4f_{7/2}$	N_{VII}	381
$3s_{1/2}$	M_I	5548	$5s_{1/2}$	O_I	324
$3p_{1/2}$	M_{II}	5181	$5p_{1/2}$	O_{II}	260
$3p_{3/2}$	M_{III}	4304	$5p_{3/2}$	O_{III}	195
$3d_{3/2}$	M_{IV}	3728	$5d_{3/2}$	O_{IV}	105
$3d_{5/2}$	M_V	3552	$5d_{5/2}$	O_V	96
$4s_{1/2}$	N_I	1442	$6s_{1/2}$	P_I	71
$4p_{1/2}$	N_{II}	1273	$6p_{1/2}$	P_{II}	43
$4p_{3/2}$	N_{III}	1045	$6p_{3/2}$	P_{III}	33
			$6d_{3/2}$	P_{IV}	4
			$6d_{5/2}$	P_V	

[a] SIEGBAHN et al. (1967).
The M_V level has been used as a reference level.
The tabulated binding energies are given with zero binding energy at the Fermi level.
The total binding energy is larger by an amount equal to the work function (Table 2.2).

Table 2.2. Recent measurements of the uranium work function

Method[a]	Work function (Electron-volts)	Reference
TE	3.47 ± 0.03	RAUH and THORN (1959)
PE	3.47 ± 0.01 (α)[b]	FRY and CARDWELL (1962)
PE	3.52 ± 0.01 (β)[b]	FRY and CARDWELL (1962)
PE	3.39 ± 0.01 (γ)[b]	FRY and CARDWELL (1962)
KCPD	3.23 ± 0.013	RIVIERE (1962)
TE	2.97[c]	HAAS and JENSEN (1963)
KCPD	3.63 ± 0.01	HOPKINS and SARGOOD (1967)
KCPD	3.73 ± 0.02 (100)[d]	LEA and MEE (1968)
KCPD	3.90 ± 0.03 (110)[d]	LEA and MEE (1968)
KCPD	3.67 ± 0.03 (113)[d]	LEA and MEE (1968)
PE	3.73 ± 0.02 (100)[d]	LEA and MEE (1968)
PE	3.90 ± 0.03 (110)[d]	LEA and MEE (1968)
PE	3.66 ± 0.03 (113)[d]	LEA and MEE (1968)
EBCPD	3.78 ± 0.03 (100)[d]	LEA and MEE (1968)
EBCPD	4.00 ± 0.04 (110)[d]	LEA and MEE (1968)
EBCPD	3.73 ± 0.04 (113)[d]	LEA and MEE (1968)

[a] TE = thermionic emission; PE = photoelectric emission; KCPD = Kelvin contact potential difference; EBCPD = electron beam contact potential difference. [b] Uranium crystal structure: α = orthorhombic, β = tetragonal, γ = body-centered cubic. [c] Effective work function calculated for 300° K. [d] Tungsten surface on which uranium was deposited is given in parentheses.

The most important K X-ray transitions are designated in SIEGBAHN notation as

$$
\begin{aligned}
K_{\alpha 1} &= K - L_{III} \\
K_{\alpha 2} &= K - L_{II} \\
K_{\beta 1} &= K - M_{III} \\
K_{\beta 2} &= K - N_{III} \\
K_{\beta 3} &= K - M_{II} \\
K_{\beta 4} &= K - N_{II} \\
K_{\beta 5} &= K - M_{IV}.
\end{aligned}
$$

Energies of the uranium K X-rays may be derived from the data in Table 2.1 by taking the differences between the binding energies of electrons in the appropriate shells. For example, the energy of the $K_{\alpha 1}$ X-ray is 115606 eV minus 17168 eV or 98438 eV. Unresolved $K_{\beta 1}+K_{\beta 3}+K_{\beta 5}$ and $K_{\beta 2}+K_{\beta 4}$ X-rays may appear as $K_{\beta 1}'$ and $K_{\beta 2}'$ transitions. The intensities of the $K_{\alpha 2}$, $K_{\beta 1}'$, and $K_{\beta 2}'$ X-rays per 1000 $K_{\alpha 1}$ X-rays emitted are 565, 382 and 138, respectively (WAPSTRA et al., 1959). The energy of the $K_{\beta 1}'$ transition is 111×10^3 eV and that of the $K_{\beta 2}'$ transition is 114.5×10^3 eV (LEDERER et al., 1967). The fluorescent yields ω_K and ω_L of the uranium K and L shells are, respectively, 0.963 (WAPSTRA et al., 1959) and 0.450 (MARTIN and BLICHERT-TOFT, 1970). These yields represent the probabilities that a vacancy in those shells will result in the emission of an X-ray. The number n_{KL} of L-shell vacancies created by a K-shell vacancy in uranium is 0.78 (MARTIN and BLICHERT-TOFT, 1970).

B. Nuclear Properties

1. Isotopes

Decay characteristics have been reported for uranium isotopes having mass numbers $A = 227$ through $A = 240$. Of these, only ^{235}U and ^{238}U have half-lives sufficiently long to be found as primordial substances in nature. Uranium-234

Table 2.3. Nuclear properties of uranium isotopes[a]

Mass No.	Neutron No.	Δ[b] (MeV)	Means of production	Half-life	Ground state $I\pi$	Q[c] (MeV)	Type of decay	Principal radiations: Approximate energies (MeV) intensities[d]
227	135	29	^{232}Th(α, 9n) ^{231}Pa(p, 5n)[e]	1.3±0.3 m 1.1±0.1 m[e]		α = (7.1) EC = (2.2)	α	α: 6.8±0.1 α[e]: 6.87±0.02
228	136	29.23	^{232}Th(α, 8n) ^{231}Pa(p, 4n)[e]	9.2 m	0+	α = 6.803 EC = 0.361	$\alpha \geqq 95\%$ EC $\leqq 5\%$	α: 6.684 (70 R), 6.59 (29 R) γ: Th X-rays, 0.152 (0.2%), 0.187 (0.3%), 0.246 (0.4%) daughter radiations from ^{224}Th, etc
229	137	31.20	^{232}Th(α, 7n) ^{231}Pa(p,3n)[e]	58 m	(3/2+)	EC = 1.32 α = 6.473	EC $\simeq$ 80% $\alpha \simeq$ 20%	γ: Pa X-rays α: 6.362 (64 R), 6.334 (20 R), 6.229 (11 R), daugther radiation from ^{225}Th, etc
230	138	31.60	daughter of ^{230}Pa	20.8 d	0+	α = 5.991	α	α: 5.887 (67.5%), 5.816 (31.9%) γ: Th L X-rays, 0.0721 (0.54%), 0.1543 (0.21%), 0.1585 (0.13%), 0.2304 (0.18%) e^-: 0.054, 0.068 daughter radiations from ^{226}Th, etc
231	139	33.8	^{230}Th(α, 3n) ^{231}Pa(d, 2n) ^{232}Th(α, 5n)	4.3 d	(5/2−)	EC = 0.36 α = 5.55	EC > 99% α = 0.0055%	γ: Pa X-rays, 0.0256 (12%), 0.0842 (7%), 0.220 (1%) e^-: 0.040, 0.054, 0.063 α: 5.46
232	140	34.60	daughter of ^{232}Pa ^{232}Th(α, 4n)	72 y	0+	α = 5.414	α	α: 5.320 (68%), 5.263 (31.7%) γ: Th L X-rays, 0.0575 (0.21%), 0.1291 0.082%), 0.270 (0.0038%) 0.3275 (0.0034%) e^-: 0.040, 0.054 daughter radiations from ^{228}Th, etc

Table 2.3 (continued)

Mass No.	Neutron No.	Δ[b] (MeV)	Means of production	Half-life	Ground state $I\pi$	Q[c] (MeV)	Type of decay	Principal radiations: Approximate energies (MeV) intensities[d]
233	141	36.94	$^{232}Th(n, \gamma)^{233}Th$ $(\beta^-)^{233}Pa$ (β^-)	1.62×10^5 y $(1.6210\pm0.0032)\times10^5$ y[f] $(1.540\pm0.001)\times10^5$ y[g] $(1.553\pm0.010)\times10^5$ y[h] $(1.583\pm0.007)^5\times10$ y[i]	$5/2^+$	$\alpha=4.909$	α	α[j]: 4.824 (84.4%), 4.783 (13.23%), 4.729 (1.61%) γ: Th X-rays, 0.0290 (60 R), 0.042 (310 R), 0.0546 (68 R), 0.097 (100 R), many others e^-: 0.023, 0.038 daughter radiations from ^{229}Th, etc
234	142	38.16	daughter of ^{238}Pu daughter of ^{234}Pa	2.48×10^5 y	0^+	$\alpha=4.856$	α	α: 4.773 (73%), 4.723 (27.5%) 4.603 (0.3%) γ: Th L X-rays, 0.0531 (100 R), 0.121 (34 R), 0.456, 0.508, 0.585 (1.2×10^{-5}%) daughter radiations from ^{230}Th, etc
235	143	40.93	found in nature	7.1×10^8 y $(7.038\pm0.005)\times10^8$ y[k]	$7/2^-$	$\alpha=4.681$	α	α: 4.597 (4.6%), 4.556 (3.7)%, 4.502 (1.2%), 4.415 (4%), 4.396 (57%), 4.366 (18%) 4.323 (3%), 4.216 (5.7%), others γ: Th X-rays, 0.110 (2.5%), 0.143 (11%), 0,163 (5%), 0.185 (54%), 0.204 (5%) daughter radiations from ^{231}Th, etc
235 m	143	40.93	daughter of ^{239}Pu	26 m[l]	$1/2^+$		IT	e^-: <0.00008
236	144	42.46	$^{235}U(n, \gamma)$	2.40×10^7 y $(2.342\pm0.002)\times10^7$ y[m]	0^+	$\alpha=4.572$	α	α: 4.493 (74%), 4.443 (26%), 4.331 (0.26%) γ: [Th L X-rays] e^-: 0.032, 0.045

237	145	45.41	$^{238}U(n, 2n)$ $^{236}U(n, \gamma)$ daughter of ^{241}Pu	6.75 d	1/2+	$\beta^- = 0.517$ $\alpha = 4.228$	β^-	β^-: 0.248 max γ: 0.02635 (2.4%), 0.0594 (36%), 0.1646 (2.0%), 0.208 (23%), 0.2675 (0.76%), 0.3324 (1.3%), others e^-: 0.008, 0.011, 0.038, 0.089, 0.186
238	146	47.33	found in nature	4.49×10^9 y $(4.468 \pm 0.0034) \times 10^9$ y[k]	0+	$\alpha = 4.267$	α	α: 4.195 (77%), $\simeq$ 4.145 (23%), γ: [Th L X-rays] e^-: 0.030, 0.043 daughter radiations from ^{234}Th, ^{234m}Pa
239	147	50.60	$^{238}U(n, \gamma)$	23.54 m	(5/2+)	$\beta^- = 1.28$	β^-	β^-: 1.29 (20%), 1.21 (80%) γ: Np L X-rays, 0.0435 ($\simeq$ 1.2%), 0.0747 (62%), many others[n] e^-: 0.011, 0.023, 0.052, 0.069 daughter radiations from ^{230}Np
240	148	52.74	$^{238}U(n, \gamma)$ $^{239}U(n, \gamma)$	14.1 h	0+	$\beta^- = 0.52$ $\alpha = (4)$	β^-	β^-: 0.36 max γ: Np L X-rays, 0.0441 (1.69%) e^-: 0.022, 0.038 daughter radiations from ^{240m}Np.

[a] Data tabulated from the "Table of Isotopes" (Lederer et al., 1967) and/or "Nuclear Data Sheets" (Wapstra and Gove, 1966; Ellis, 1970; Schmorak, 1970) unless specific reference is made to other, generally more recent work. [b] Δ = (mass) − (mass number) = $M - A$. 1 MeV = 1.07356×10^{-3} atomic mass unit. [c] Total decay energy between ground states of parent and daughter systems whether for alpha (α) decay, negatron (β^-) decay, or orbital electron capture (EC). Values in parentheses are estimated from theoretical considerations of alpha or beta systematics. [d] Intensities of radiations expressed as percentages refer to percentages of the total decay events. A number followed by an R indicates the relative intensity compared to other radiations of the same type. [e] Hahn et al. (1969). [f] Ihle et al. (1967). [g] Oetting (1968). [h] Keith (1968). [i] Durham (1969). [j] Baranov et al. (1967). [k] Jaffey et al. (1971a). [l] The half-life is observed to be influenced by the environment of the sample. Mazaki and Shimizu (1966); Neve de Mevergnies (1970). [m] Jaffey et al. (1971b). [n] Mackenzie and Connor (1968).

occurs naturally as one of the decay products of ^{238}U. The relative abundances of ^{234}U, ^{235}U and ^{238}U are 0.0055%, 0.720% and 99.2741%, respectively (ELLIS, 1970). The atomic weight of natural uranium is 238.03 atomic mass units on the ^{12}C scale.

Pertinent data on the nuclear properties of the uranium isotopes are summarized in Table 2.3.

2. Nuclear Fission

When a nucleus is excited with sufficient energy to pass over a so-called fission barrier, there is a certain probability, depending on the nature of the nucleus and its excitation energy, that the nucleus will fission. The excitation energy is supplied by an incident particle. Fission has been induced in a variety of nuclei by bombarding them with photons, electrons, mesons, neutrons, or charged particles such as protons, deuterons, etc. Occasionally the nucleus may not pass over the fission barrier but penetrates through it to fission. This process is known as spontaneous fission.

In fissioning a nucleus separates into at least two large fragments of roughly half the size of the fissioning nucleus. Neutrons may be emitted at the time of fission and, less frequently, a light mass, charged particle, such as an alpha particle or trition, is emitted. Even less frequently, the nucleus may split into three or more large fragments. The fission fragments are fast moving, highly excited nuclei. They have kinetic energies that range from approximately 70 to 100 MeV and they may emit several neutrons and photons before reaching their ground states. Even then these nuclei are generally radioactive and decay by the emission of beta-particles, gamma-rays, X-rays and occasionally a neutron. The total energy released in the fission of a uranium nucleus is about 200 MeV.

In the following sections, the various aspects of fission described above are examined more closely for the uranium isotopes. More complete descriptions of the fission process are given in review articles by HYDE (1964), FRASER and MILTON (1966), and GINDLER and HUIZENGA (1968).

a) The Probability of Fission

The probability for a nucleus to fission spontaneously is given by the ratio of its total half-life to its partial spontaneous fission half-life. Spontaneous fission half-lives for the uranium isotopes are listed in Table 2.4. Comparison of these half-lives with the total half-lives listed in Table 2.3 shows that spontaneous fission is not a very probably mode of decay for the uranium isotopes.

Table 2.4. Spontaneous fission half-lives for uranium isotopes[a]

Mass number	T_{SF}	Reference
232	$(8 \pm 6) \times 10^{13}$ y	
233	$(1.2 \pm 0.3) \times 10^{17}$ y	ALEKSANDROV et al. (1966)
234	$(2 \pm 1) \times 10^{16}$ y	
235	$(3.5 \pm 0.9) \times 10^{17}$ y	ALEKSANDROV et al. (1966)
235	$(1.9 \pm 1.0) \times 10^{17}$y	SEGRE (1952)
236	2×10^{16} y	
238	$(9.86 \pm 0.15) \times 10^{15}$ y	
238	$(7.191 \pm 0.036) \times 10^{15}$ y	ISHIMORI et al. (1967)
238	$(9.50 \pm 0.21) \times 10^{15}$ y	LEME et al. (1971)

[a] Half-lives with no reference given are from "Nuclear Data Sheets" (ELLIS, 1970; SCHMORAK, 1970). References to the original literature may be found there.

Table 2.5. Neutron induced fission and activation cross sections in barns for uranium isotopes[a]

Mass number	Thermal		Neutron energy		
	σ_γ	σ_f	1 MeV σ_f	3 MeV σ_f	14 MeV σ_F
230		25 ± 10			
231		400 ± 300			
232	78 ± 4	77 ± 10			
233	49 ± 6	524.5 ± 1.9	1.95	1.93	2.12
233	47.0 ± 0.9[b]	530.6 ± 1.9[b]			
233	44.6 ± 0.9[c]	537.9 ± 1.9[c]			
234	95 ± 7	—	1.12	1.52	
235	101 ± 2	577.1 ± 0.9	1.25	1.30	2.20
235	98.3 ± 1.1[b]	580.2 ± 1.8[b]			
235	97.3 ± 1.1[c]	585.7 ± 1.8[c]			
236	6 ± 1	—	0.36	0.93	
238	2.73 ± 0.04	—	0.017	0.54	1.13
239	22 ± 5[d]	14 ± 3[d]			

[a] Data, unless indicated otherwise, are taken from "Neutron Cross Sections" by Hughes and Schwartz (1958) and Stehn et al. (1965). [b] Hanna et al. (1969). [c] de Volpi (1971). [d] Pile neutrons.

The probability for inducing fission is related to its fission cross section. Neutron induced cross sections have been the subject of extensive research because of their importance to reactor technology, especially the fission and absorption cross sections of ^{233}U, ^{235}U and ^{238}U. Cross sections for the (n, γ) reaction between the various uranium isotopes and thermal neutrons are listed in Table 2.5. Fission cross sections are also given for neutrons having different energies.

b) *The Energy Released in Fission*

The energy released in fission is equal to the sum of (1) the total kinetic energy E_K of the fission fragments after neutron emission, (2) the energy E_{np} of the neutrons emitted by the fragments, (3) the energy $E_{\gamma p}$ of the photons emitted in the de-excitation of the fragments, and the energies of the (4) beta particles E_β, (5) photons $E_{\gamma d}$, (6) anti-neutrinos $E\bar{\nu}$, and (7) delayed neutrons E_{nd} emitted in the radioactive decay of the fission products. For a statistically large number of fission events the average total energy Q_f released per fission is the sum of the component energies.

$$Q_f = \bar{E}_K + \bar{E}_{np} + \bar{E}_{\gamma p} + \bar{E}_\beta + \bar{E}_{nd} + \bar{E}_{\gamma d} + \bar{E\bar{\nu}}, \tag{1}$$

where $\bar{E}$ signifies an average energy.

Thermodynamically Q_f is equal to the sum of the energy-equivalent masses of the reactants minus the sum of the energy-equivalent masses of the products,

$$Q_f = M(A, Z) + M_n + E_i - \Sigma_i Y_i M(A_i, Z_i) - \bar{\nu}_T M_n. \tag{2}$$

where $M(A, Z)$ is the atomic mass in MeV of the uranium nucleus with mass number A and nuclear charge $Z = 92$, M_n is the energy-equivalent mass of the neutron and E_i its incident energy (both in MeV), Y_i is the yield of the stable fission product (A_i, Z_i), and $\bar{\nu}_T$ is the average total number of neutrons emitted per fission.

Unik and Gindler (1971), in a review of published, experimental data on the energy released in fission, have determined the energy Q_f' which is simply Q_f minus E_i. Their results for Q_f' are summarized in Table 2.6 for ^{233}U, ^{235}U,

Table 2.6. The energy (MeV) released in the fission of uranium isotopes by neutrons of different energy[a]

Uranium mass no.	Neutron energy	Q'_f [b,c]	$\overline{E}_K$	$\overline{E}_n$ [d]	$\overline{E}_{\gamma p}$	$\overline{E}_{\gamma d}$	$\overline{E}_\beta$	$\overline{E}_{\bar{\nu}}$	$\overline{E}_i$	Q'_f [b,e]
233	thermal	197.99±0.21	167.5±2.0	4.90±0.15	7.6 ±1.5	5.9±1.3	5.6±0.3	7.8±0.5	0	199.3±2.9
235	thermal	202.74±0.21	167.9±2.0	4.69±0.11	7.64±0.75	7.1±1.3	7.4±0.4	10.3±0.6	0	205.0±2.6
235	fission spectrum	201.92±0.26								
235	14 MeV	187.15±0.83								
238	fission spectrum	205.89±0.45	165.9±2.0	5.60±0.16	7.6±1.5	9.0±1.5	10.5±0.6	14.7±0.9	2.8	210.5±3.1
238	14 MeV	194.21±0.67								

[a] UNIK and GINDLER (1971). [b] $Q'_f = Q_f - E_i$ (see Text). [c] Calculated using Eq. (2). [d] $\overline{E}_n = \overline{E}_{n\gamma} + \overline{E}_{nd}$ (see text). [e] Calculated using Eq. (1).

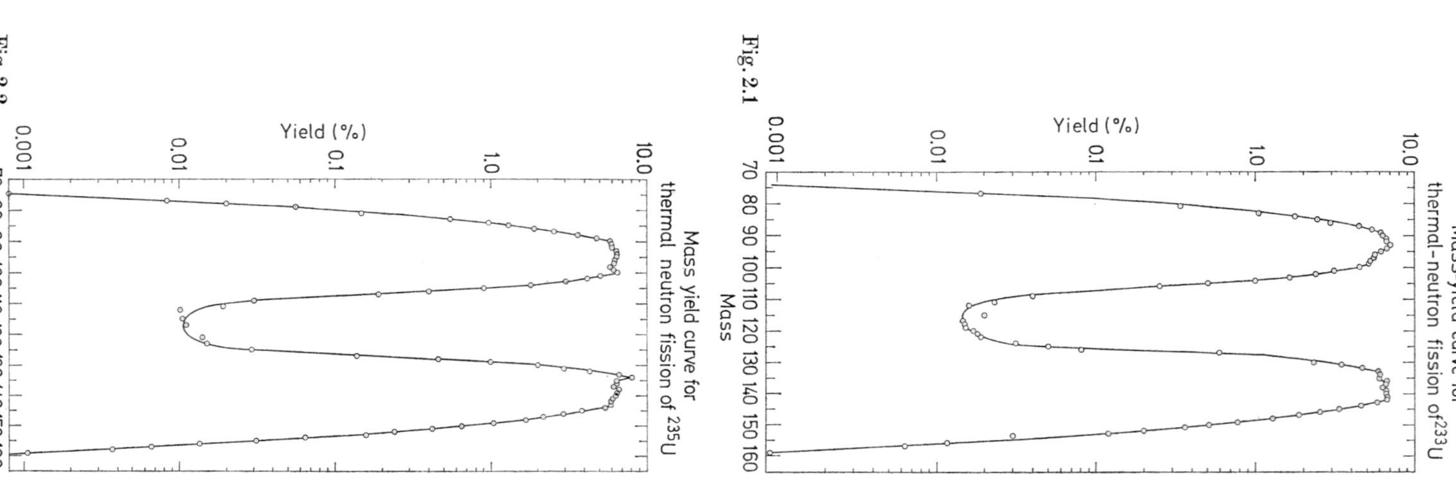

Fig. 2.1

Fig. 2.2

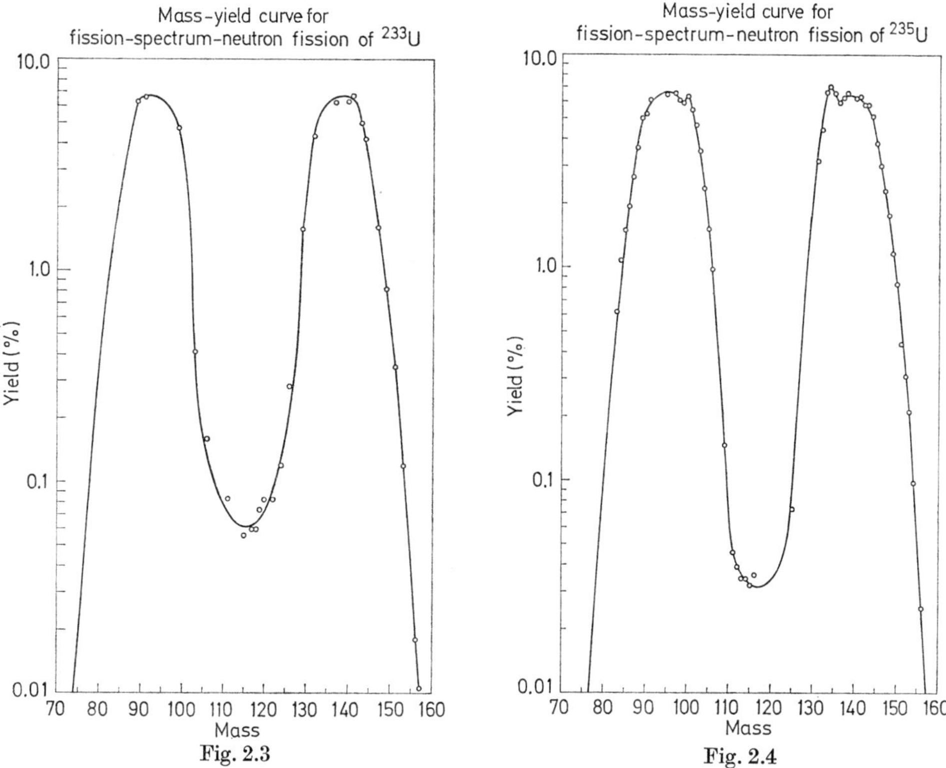

Fig. 2.3

Fig. 2.4

Fig. 2.3. Fission product mass distribution for the fission of ^{233}U by fission spectrum neutrons. (Flynn and Glendenin, 1970)

Fig. 2.4. Fission product mass distribution for the fission of ^{235}U by fission spectrum neutrons. (Flynn and Glendenin, 1970)

and ^{238}U. Their results for the component energies E_K, $\bar{E}_n$, etc. are also given in the table.

c) *Fission Product Mass Distribution*

It was stated in the beginning of this section on fission that there are usually two fragments per fission event. For a large number of such events there is a distribution of fragments having different masses. The shape and character of these mass distributions depend both on the mass of the uranium isotope and the energy of the incident neutron. Fission product mass distributions for ^{233}U, ^{235}U and ^{238}U excited by thermal, fission spectrum, and 14-MeV neutrons are shown in Fig. 2.1–2.7. These curves show the percent yield of a given mass as a function of final mass number, i.e. the mass number after neutron emission. Mass distributions of the initial fission fragments are displaced to slight higher mass numbers.

Fig. 2.1. Fission product mass distribution for thermal neutron induced fission of ^{233}U. (Flynn and Glendenin, 1970)

Fig. 2.2. Fission product mass distribution for thermal neutron induced fission of ^{235}U. (Flynn and Glendenin, 1970)

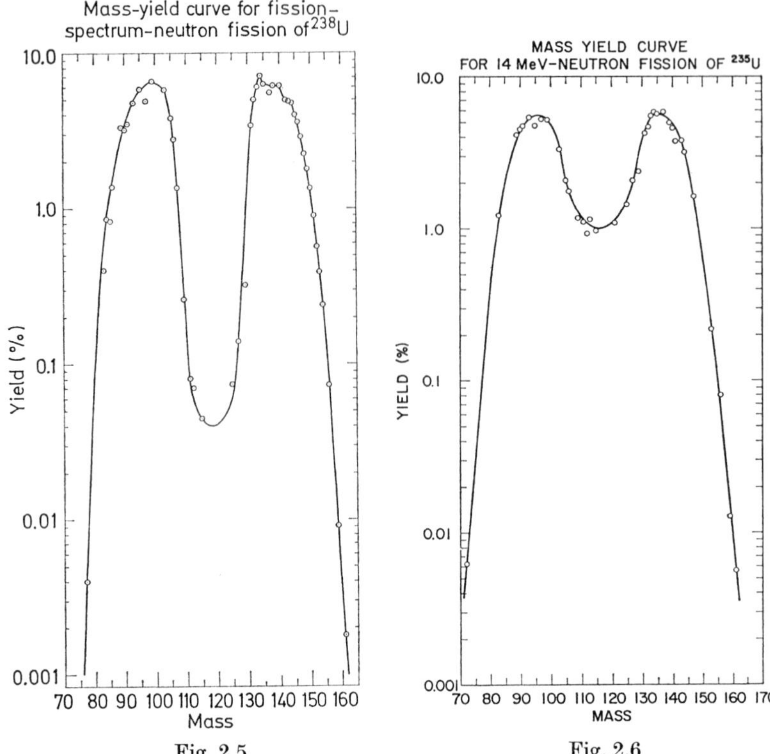

Fig. 2.5 Fig. 2.6

Fig. 2.5. Fission product mass distribution for the fission of ^{238}U by fission spectrum neutrons. (FLYNN and GLENDENIN, 1970)

Fig. 2.6. Fission product mass distribution for the fission of ^{235}U by 14-MeV neutrons. (FLYNN and GLENDENIN, 1970)

d) Charge Distribution

The mass yields plotted in Fig. 2.1–2.7 are cumulative yields for a particular chain of isobars. For example, the isobaric chain of mass number 95 includes:

$$\underset{(?)}{95_{Br}} \xrightarrow{\beta^-} \underset{(0.8\,s)}{95_{Kr}} \xrightarrow{\beta^-} \underset{(0.36\,s)}{95_{Rb}} \xrightarrow{\beta^-} \underset{(26\,s)}{95_{Sr}} \xrightarrow{\beta^-} \underset{(10.9\,m)}{95_{Y}} \xrightarrow{\beta^-} \underset{(65.5\,d)}{95_{Zr}} \xrightarrow{\beta^-} \underset{(35\,d)}{95_{Nb}} \xrightarrow{\beta^-} \underset{(stable)}{95_{Mo}}$$

The half-life of each of the isobars is given in parentheses. Each of these isobars may be formed as one of the primary fission products (the fission fragment after de-excitation by the emission of prompt neutron and gamma rays). Radiochemical experiments on mass chains in which the independent fission yields of several isobars have been measured have shown that these yields may be represented by a Gaussian distribution,

$$P(Z) = \frac{1}{\sqrt{C\pi}} \exp\,[-(Z-Z_p)^2/C], \tag{3}$$

where $P(Z)$ is the fractional independent yield of the isobar and C and Z_p are constants which describe the isobaric distribution. WAHL *et al.* (1969) have made an extensive study of these constants. They concluded that in thermal neutron

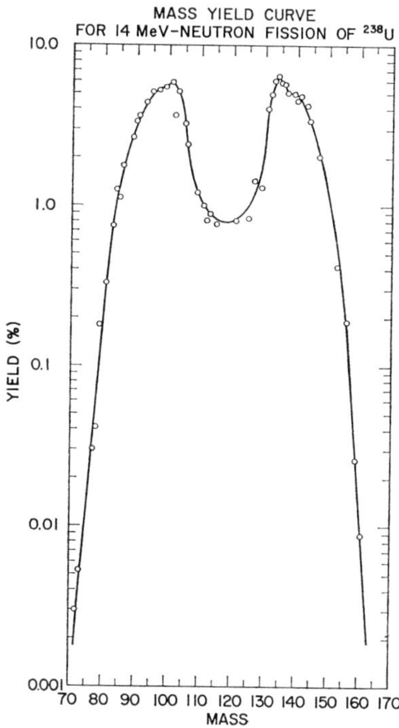

Fig. 2.7. Fission product mass distribution for the fission of ^{238}U by 14-MeV neutrons. (FLYNN and GLENDENIN, 1970)

induced fission of ^{235}U, a normal distribution of nuclear charge for a given isobaric chain is characterized by a width constant $C = 0.80 \pm 0.14$. They have also determined the most probable charge Z_p as a function of mass number A.

e) Fission Neutrons

Neutrons are emitted during three stages of the fission process: at the time of fission, during the de-excitation of the fission fragments, and in the radioactive decay process. Neutrons emitted in the first two stages are usually referred to as *prompt* neutrons. Those neutrons emitted at the time of fission may be further distinguished as *central* or *scission* neutrons. The neutrons emitted during radioactive decay are described as *delayed* neutrons.

The average number of neutrons $\bar{\nu}$ emitted per fission is a function of the energy of the incident neutrons. Table 2.7 lists values of $\bar{\nu}$ for thermal neutron induced fission of some uranium isotopes. The subscripts d, p, and T refer to delayed, prompt and total (delayed plus prompt) neutrons, respectively. The apparently small differences in $\bar{\nu}$ values listed in Table 2.7 are important in the design of nuclear reactors. SKARSVÅG and BERGHEIM (1963) indicate that approximately 15% of the prompt neutrons emitted in thermal neutron induced fission of ^{235}U are scission neutrons.

One of the striking features of prompt neutron emission is the manner in which neutrons are emitted by the various fragment masses. This is illustrated in Fig. 2.8 in which the average number of neutrons emitted per fragment $\bar{\nu}_A$

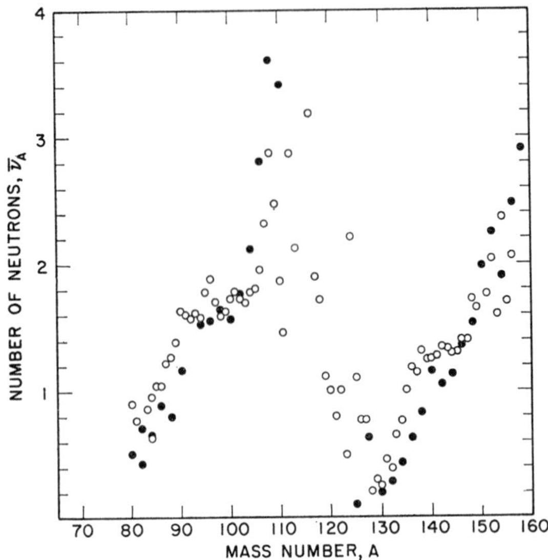

Fig. 2.8. Average number of neutrons emitted as a function of mass number in thermal neutron induced fission of ^{233}U (Closed circles) and ^{235}U (open circles). (Adapted from MILTON and FRASER, 1965)

is plotted as a function of mass number A for thermal neutron induced fission of ^{233}U and ^{235}U. The saw-tooth type distribution is correlated with the ease of deformation of the nuclear masses.

Delayed neutron emission occurs when a parent nucleus beta decays to a nuclear level in the daughter nucleus which has an energy greater than the neutron binding energy in the daughter nucleus. Delayed neutron emission therefore occurs with a half-life which is characteristic of the beta half-life of the parent nucleus (or precursor). The apparent decay of the neutrons is analyzed mathematically into an optimum six periods of approximately 55, 21, 5, 2, 0.5 and 0.2 sec (KEEPIN *et al.*, 1957; EAST and KEEPIN, 1969).

Table 2.7. Average number of neutrons $\bar{\nu}$ emitted per thermal neutron induced fission[a]

Uranium mass No.	$\bar{\nu}_d$	$\bar{\nu}_p$	$\bar{\nu}_T$
232		3.04 ± 0.05[b]	
233	0.0066 ± 0.003[c]	2.497 ± 0.008[b]	2.4866 ± 0.0069[d] 2.453 ± 0.007[e]
235	0.0158 ± 0.0005[c]	2.426 ± 0.006[b]	2.4229 ± 0.0066[d] 2.400 ± 0.007[e]

[a] The subscript d, p, and T refer, respectively, to delayed, prompt, and total neutrons.
[b] STEHN et al. (1965).
[c] KEEPIN et al. (1957).
[d] HANNA et al. (1969).
[e] DE VOLPI (1971).

II. The Chemistry of Uranuim

The chemistry of uranium has been and continues to be the subject of extensive research. "Gmelins Handbuch der Anorganischen Chemie" (Meyer and Pietsch, 1936) reviews the chemistry of uranium and its compounds as it was known until 1936. Later comprehensive reviews include those of Katz and Rabinowitz (1951), Katz and Seaborg (1957), parts 1, 2, and 4, Volume 15 of "Nouveau Traité de Chemie Minérale" (Caillat and Elston, 1960, 1961; Elston and Caillat, 1967), and Cordfunke (1969).

Prior to the discovery of the transuranium elements, uranium was considered to be a transition element belonging to group VIB of the periodic system—a homolog of Cr, Mo, and W. Although uranium does resemble these elements in many of its chemical properties, it is now generally accepted that uranium is the fourth member of the actinide family of elements. This includes elements with atomic numbers 89 through 103. A divergent view is that uranium is the first member of the uranide family which includes elements through americium.

A. Metallic Uranium

Katz and Rabinowitz (1951) have described several methods by which uranium metal can be prepared: the reduction of uranium oxides with carbon; the reduction of uranium oxides with aluminium, calcium, or magnesium; the reduction of uranium halides with alkali or alkaline earth metals; the electrolytic reduction of uranium halides; and the thermal decomposition of uranium halides. The thermal decomposition of uranium tetraiodide UI_4 on a hot filament is a means of preparing uranium of high purity, but only on a small scale. The most important method technologically is the reduction of uranium halides by magnesium or calcium. Uranium tetrafluoride UF_4 is the halide most commonly used, although the tetrachloride UCl_4 can be used.

The processing of uranium from the ore to the metal has been described by many authors including Vdovenko (1960), Wilkinson (1962), Gittus (1963), Galkin and Sudarikov (1964), Cordfunke (1969), in "Nouveau Traité de Chemie Minérale" volume 15, and in the proceedings of the three conferences on "Peaceful Uses of Atomic Energy" held in Geneva in 1955, 1958, and 1964. The reprocessing of irradiated uranium fuels is also described in most of these references and in the volume edited by Chiotti (1969).

Uranium is a dense, silvery white metal that may exist in one of three allotropic forms. The crystal parameters of the allotropes are given in Table 2.8 together with a number of physical properties of the metal. The accuracy to which these properties are representative of uranium depends on the purity of the samples measured. In uranium, the physical properties may also depend on the microstructure, orientation, and metallurgical history of the sample.

Pile irradiation produces dimensional changes in both single and polycrystals of uranium. Both anisotropic and isotropic growth occur at temperatures $< 400°$ C. The latter growth is associated with the accumulation of fission products. At temperatures $> 500°$ C, anisotropic growth apparently does not occur; whereas, isotropic growth does. The magnitude of the latter is enhanced by the accumulation of fission-product gases within pockets in the metal. Other irradiation effects include surface roughening and pimpling, warping, high hardness, extreme brittleness, cracks and porosity, decreased thermal and electrical conductivity, and broadened X-ray diffraction lines. Thermal cycling in the alpha temperature range produces dimensional changes in textured, polycrystalline samples

Table 2.8. Physical properties of uranium metal[a]

I. Crystal structure[b]

Allotropic form	Crystal structure	Lattice parameter (Å) a	b	c	Temperature[c] (°C)	Density[d] (g/cm³)
α	orthorhombic	2.854	5.869	4.955	room	19.05
β	tetragonal	10.754	$a=b$	5.6525	700	18.13
γ	body-centered cubic	3.534	$a=b=c$		800	17.91

II. Thermal properties

	Temperature (°C)[e]	Latent heat, ΔH (kcal/mole)[f]	Entropy of transformation, ΔS (cal/deg mole)[b]
α→β transformation	667.7 ± 1.3	0.75	0.72
β→γ transformation	774.8 ± 1.6	1.15	1.1
Melting point	1132.3 ± 0.8	2.5–2.9	
Boiling point	4690[g]	119.9 ± 2.5 (2000° K)[g] 115.5 ± 1.7 (2200° K)[h]	
Sublimation		129.0 ± 2.0 (0° K)[g] 126.3 ± 1.0 (298° K)[h]	

Vapor pressure

$$\log p\ (\text{atm}) = -\frac{(26210 \pm 270)}{T} + (5.920 \pm 0.135)^{g}\ (1720\text{–}2340°\ K)$$

$$\log p\ (\text{atm}) = -\frac{(25230 \pm 370)}{T} + (5.71 \pm 0.17)^{h}\ (1980\text{–}2420°\ K)$$

Specific heat, C_p (cal/deg mole)[e,f]

α $2.61 + 8.95 \times 10^{-3}\,T + 1.74 \times 10^{-8}\,T^2$ (298–941° K)
β 10.38
γ 9.1
liquid $-13.15 + 16.17 \times 10^{-3}\,T$

Entropy, S° (cal/mole)[e] 12.00 ± 0.02 (25° C)

Enthalpy, H° (cal/mole)[e] 1521 ± 3 (25°)

Thermal conductivity (cal/cm sec deg)[e]
high purity U (70° C) 0.0714 ± 0.003
reactor grade U (100° C) 0.0668

Thermal expansion[b]

α (25–650° C)

$$L_{[100]t} = L_{[100]0°C}\,(1 + 23.53 \times 10^{-6}\,t + 13.74 \times 10^{-9}\,t^2 + 9.94 \times 10^{-12}\,t^3)$$
$$L_{[010]t} = L_{[010]0°C}\,(1 + 1.16 \times 10^{-6}\,t - 9.43 \times 10^{-9}\,t^2 - 11.79 \times 10^{-12}\,t^3)$$
$$L_{[001]t} = L_{[001]0°C}\,(1 + 19.38 \times 10^{-6}\,t + 21.58 \times 10^{-9}\,t^2 + 3.32 \times 10^{-12}\,t^3)$$

β [100] and [010] directions $(23.4 \pm 1.5) \times 10^{-6}$/deg
23.0×10^{-6}/deg (20–700° C)
[001] direction $(6.0 \pm 2) \times 10^{-6}$/deg
4.6×10^{-6}/deg (20–700° C)

γ $(22.5 \pm 1.3) \times 10^{-6}$/deg

Table 2.8 (continued)

III. Electrical properties

Resistivity of α-uranium at 25° C[f]				Wiedemann-Franz ratio ([volts/deg]^2)[e]		
Direction	[100]	[010]	[001]	Theoretical value		2.45×10^{-8}
ϱ (μ ohm cm)	36.4	25.1	32.1	α-U	20° K	2.34×10^{-8}
Superconductivity critical temperature, T_c					60° K	2.46×10^{-8}
					100° K	2.75×10^{-8}
α-U	0.21–0.25° K[i]				300° K	2.67×10^{-8}
β-U	~0.8° K[j]			β-U		2.29×10^{-8}
Photoelectric and thermionic emission work functions given in Table 2.2.				γ-U		2.15×10^{-8}

IV. Magnetic properties

Magnetic susceptibility (e.m.u./g)[e]		
α-U	25° C	1.673×10^{-6}
	200° C	1.747×10^{-6}
	637° C	1.914×10^{-6}
β-U	679° C	1.934×10^{-6}
	766° C	2.113×10^{-6}
γ-U	785° C	2.116×10^{-6}
	992° C	2.193×10^{-6}

Hall coefficient, R_H (cm^3/coulomb)		
T (°K)	Reactor grade U[k]	High purity U[k]
4.2	$+1.45 \times 10^{-5}$	-0.31×10^{-5}
77	$+3.9 \times 10^{-5}$	$+4.75 \times 10^{-5}$
273		$+3.93 \times 10^{-5}$
300	$+4.1 \times 10^{-5}$	
$R_H = (3.8 \pm 10\%) \times 10^{-5}$ (298–1523° K)[l]		

[a] References to original literature are given in [b, e, f].
[b] Data taken from the review by GITTUS (1963).
[c] Temperature corresponding to measured lattice parameters and density calculation.
[d] Calculated from the lattice parameters according to the relationship $\varrho = \frac{1}{abc}\frac{Mn}{A}$, where M is the atomic weight of uranium, n is the number of atoms per unit cell, and A is Avogadro's number.
[e] Data taken from "Nouveau Traité de Chemie Minérale" (ELSTON and CAILLAT, 1967).
[f] Data taken from the review by CORDFUNKE (1969).
[g] PATTORET et al. (1969).
[h] ACKERMAN and RAUH (1969).
[i] GEBALLE et al. (1966).
[j] MATTHIAS et al. (1966).
[k] BERLINCOURT (1959).
[l] BUSCH et al. (1970).

but not in single crystals of alpha uranium. Uranium heated through the alpha-beta phase change increases in volume by about one percent. Alpha-beta cycling causes distortion, reduction in density, and porosity. The uranium surface may also wrinkle and split.

A more complete description of the physical properties of metallic uranium and the effects of irradiation and thermal cycling is given by HOLDEN (1958), GRAINGER (1958), GITTUS (1963), and in "Nouveau Traité de Chemie Minérale" (ELSTON and CAILLAT, 1967).

Chemically uranium metal is a strong reducing agent. A potential of $+1.80$ eV for the half-cell reaction, $U \rightarrow U^{+3} + 3e^-$, places uranium between beryllium and aluminium in the electromotive force series (LATIMER, 1952). It reacts with most non-metallic elements and with compounds of these elements. The rate of reaction and the temperature at which the reaction occurs depend on the degree of subdivision of the uranium. These reactions have been described by KATZ and RABINOWITZ (1951) and are summarized in Table 2.9.

Table 2.9. Chemical reactions of uranium metal with non-metallic elements and compounds[a]

Reactant	Temperature of reaction (°C)		Product
	Massive	Powder	
H_2	250	25	UH_3
C	1800–2400	800–1200	UC, U_2C_3, UC_2
N_2	700	500	UN, $UN_{1.75}$, UN_2
P		600–1000	U_3P_4
O_2	150–350	Pyrophoric	UO_2, U_3O_8
S	500		US_2
F_2	25		UF_6
Cl_2	500–600	150–180	UCl_4, UCl_5, UCl_6
Br_2	650	210	UBr_4
I_2	350	260	UI_3, UI_4
H_2O	100	25	UO_2
HF		200–400	UF_4
HCl		250–300	UCl_3
NH_3	700	400	$UN_{1.75}$
H_2S		500	US, U_2S_3, US_2
NO	400		U_3O_8
N_2O_4		25	$UO_2(NO_3)_2 \cdot 2NO_2$
CH_4		635–900	UC
CO	750		UO_2+UC
CO_2	750		UO_2+UC

[a] Adapted from HOEKSTRA and KATZ (1954) and KATZ and SEABORG (1957).

The pyrophoricity of uranium is described in detail by WILKINSON (1962). At room temperature finely divided uranium may ignite spontaneously in air, oxygen, and even water. At temperatures of ~200 to 400° C uranium powders may self-ignite in atmospheres of carbon dioxide or nitrogen. When air has access to a piece of uranium surrounded by other pieces, the interior piece tends to ignite at a lower ambient temperature than would an isolated piece. If the piece is a powder particle with a limited size range, there is a critical height of a column of such powder, in a uncovered container of any diameter, beyond which the ignition temperature exhibits a constant value. The relationship between particle size range, specific area, ignition temperature, and critical height is given in Table 2.10 for roughly spherical particles (SCHNIZLEIN et al., 1960). Pyrophoricity is associated with large surface areas per unit volume and heat build-up in micropores. Materials such as oxygen and moisture from the air, some organic reagents such as acetone, and hydrogen resulting from reactions with water and acids absorb on uranium and react with it. Displacement of the absorbed reactive

Table 2.10. Pyrophoricity of uranium powders[a]

Mesh	Specific area of particle (cm^2/g)	Critical height (mm)	Ignition temperature (°C)
−200+230	51.2	1.5	255
−120+140	30.2	2.4	260
−80+100	21.3	3.0	270
−40+45	9.06	4.3	295
−18+20	3.78	7.0	310
−12+14	2.25	8.8	320

[a] SCHNIZLEIN et al. (1960).

Table 2.11. Reactions of massive uranium metal with aqueous acid solutions[a]

Acid	Reaction rate	Reaction products	Remarks
HF	slow	U(IV) fluoride	UF_4 forms an insoluble coating
HCl	rapid	U(III), U(IV), black residue	Complex reaction. Black residue contains complex uranium hydrides
HNO_3	moderate	Uranyl nitrate	Finely divided uranium may react with explosive violence
H_2SO_4	slow	U(IV) acid sulfate	$<6N$ acid does not attack uranium
H_3PO_4	slow	U(IV) acid phosphate	Heating increases the reaction rate
$HClO_4$	rapid		Uranium shavings react violently with $>60\%$ $HClO_4$

[a] Adapted from HOEKSTRA and KATZ (1954).

materials with inert materials or materials which later volatilize reduces the likelihood of pyrophoricity.

The oxidation of uranium under suitable conditions may release sufficient energy to cause an explosion. The lower explosive limit for dust clouds of uranium is 55 mg/liter. Powders of metals that may be mixed with uranium, such as aluminium and zirconium, may be pyrophoric and potentially explosive.

The effect of aqueous acids on massive uranium is summarized in Table 2.11. The black residue formed by the dissolution of uranium in hydrochloric acid may be brought into solution by the addition of oxidizing agents such as hydrogen peroxide, nitric acid, etc. The presence of fluosilicic acid permits complete solution of uranium metal in concentrated hydrochloric acid. Hydrobromic acid is similar to hydrochloric acid in its effect on uranium, except the reaction is slower. Hydroiodic acid reacts even more slowly (KATZ and RABINOWITZ, 1951). Finely divided uranium should be added in small portions to nitric acid since nitric acid vapors may react violently with uranium. Dilute sulfuric acid in conjunction with oxidizing agents (H_2O_2 or HNO_3) will dissolve uranium. Dilute perchloric acid in conjunction with oxidizing agents dissolves uranium smoothly. The use of concentrated perchloric acid with large quantities of uranium metal is not recommended (KATZ and RABINOWITZ, 1951). Organic acids (formic, acetic, propionic or butyric), dilute or anhydrous, do not react with metallic uranium. However, violent reactions occur in the presence of hydrogen chloride with the formation of the corresponding uranium (IV) salts. Uranium (IV) benzoate is formed by the reaction of uranium with benzoic acid in ether solution. Uranium reacts with acetic anhydride or acetyl chloride to form uranium acetate.

Solutions of alkali metal hydroxides have little effect on metallic uranium. The addition of hydrogen peroxide causes uranium to dissolve with the formation of soluble peruranates (KATZ and RABINOWITZ, 1951).

Metals with lower oxidation potentials than uranium may be displaced from solutions of their salts when treated with metallic uranium. This has been done in solutions of $Hg(NO_3)_2$, $AgNO_3$, $CuSO_4$, $SnCl_2$, $PtCl_4$, and $AuCl_3$ (KATZ and RABINOWITZ, 1951).

Intermetallic compounds are formed between uranium and a wide variety of metals. Many of these compounds are listed in Table 2.12 according to groups of the periodic system. Solid solutions are formed with Nb, Hf, Zr, Mo and Ti.

Table 2.12. Intermetallic binary compounds of uranium[a]

Group	Element	Compounds
Group IB	Cu:	UCu_5
	Au:	U_2Au_3, UAu_3
Group IIA	Be:	UBe_{13}
Group IIB	Zn:	U_2Zn_{17}
	Cd:	UCd_{11}
	Hg:	UHg_2, UHg_3, UHg_4
Group IIIA	Al:	UAl_2, UAl_3, UAl_4
	Ga:	UGa, UGa_2, UGa_3
	In:	UIn_3
	Tl:	UTl_3
Group IVA	Ge:	U_5Ge_3, UGe, U_3Ge_5, UGe_2, UGe_3
	Sn:	U_3Sn_2, U_3Sn_5, USn_3
	Pb:	UPb, UPb_3
Group IVB	Ti:	U_2Ti
	Zr:	UZr_2
Group VA	Sb:	USb, U_3Sb_4, USb_2
	Bi:	UBi, U_3Bi_4, UBi_2
Group VIIB	Mn:	U_6Mn, UMn_2
	Re:	U_2Re, URe_2
Group VIII	Fe:	U_6Fe, UFe_2
	Co:	U_6Co, UCo, UCo_2, UCo_3, UCo_4, U_2Co_{11}
	Ni:	U_6Ni, U_7Ni_9, U_5Ni_7, UNi_2, UNi_5
	Ru:	U_2Ru, URu, U_3Ru_4, U_3Ru_5, URu_3
	Rh:[b]	U_4Rh_3, U_3Rh_4, U_3Rh_5, URh_3
	Pd:	UPd_3, UPd_4, UPd_5, U_2Pd_{11}, U_2Pd_{17} metastable – UPd, U_5Pd_6
	Os:	U_3Os, U_2Os, U_5Os_4, UOs_2
	Ir:[c]	U_3Ir, U_3Ir_2, UIr, UIr_2, UIr_3
	Pt:	UPt, UPt_2, UPt_3, UPt_5

[a] Compiled from GITTUS (1963), "Nouveau Traité de Chemie Minérale" (CAILLLAT and ELSTON, 1960; ELSTON and CAILLIAT, 1967) and CORDFUNKE (1969).
[b] PARK (1968).
[c] PARK and MULLEN (1968).

B. Compounds of Uranium and Non-Metallic Elements

Compounds of uranium and non-metallic elements are of interest as possible nuclear fuels or, as in the case of uranium hydride, as an intermediate in the preparation of other uranium compounds. The chemical preparation of primarily binary uranium compounds and their chemical reactions are reviewed. Some uranium mixed halide, oxyhalide, and nitrogen halide compounds are also included. The order of review is by groups of the periodic system.

1. Uranium Hydride

The compound UH_3 may be prepared by heating uranium in a hydrogen atmosphere. At 250° C the reaction is very rapid and results in the complete disintegration of the metal. At temperatures $>300°$ C the rate of hydrogenation decreases rapidly. The hydrogen pressure of the reaction

$$UH_3 \rightleftharpoons U + \frac{3}{2} H_2$$

has been measured between 450° C and 650° C (LIBOWITZ and GIBB, 1957) to be

$$\log p\mathrm{H}_2\,(\mathrm{atm}) = \frac{-4410}{\mathrm{T}} + 6.26.$$

The hydride is also formed by the reaction between uranium and boiling water or steam. The uranium reacts with the hydrogen evolved.

Uranium hydride is very reactive. It reacts with N_2, O_2, Cl_2, Br_2 and I_2 to give uranium nitrides, oxides, and tetravalent halides, respectively. Finely divided UH_3 may ignite spontaneously in air or oxygen at $-76°$ C (Wilkinson, 1962). It may react violently with water depending on the relative amounts of the reactants and the rates of contact. The hydride also reacts with gaseous H_2S, NH_3, PH_3, HF, HCl, HBr, HI, CO, CO_2, C_2H_4, HCN, $COCl_2$, CH_3I and CCl_4. The reaction with halogenated organic solvents is dangerous. Non-halogenated organic solvents do not dissolve UH_3 or react with it (Katz and Rabinowitz, 1951). Hydrochloric acid reacts more slowly with UH_3 than with uranium metal. Dilute perchloric acid dissolves the hydride very slowly. Concentrated perchloric acid reacts with the hydride upon heating. With large amounts of UH_3, the reaction with concentrated perchloric acid is potentially dangerous. Concentrated phosphoric acid and hot concentrated sulfuric acid react with UH_3. Dilute sulfuric acid reacts very slowly. Nitric acid attacks UH_3 with the liberation of NO_2. Solutions of hydrogen peroxide, silver fluoride, or silver nitrate may react vigorously with UH_3.

2. Group IIIA—Uranium Borides

The phase diagram of the uranium-boron system has been given by Howlett (1959). Three compounds have been identified. UB_2, UB_4 and UB_{12}. They may be prepared by heating calculated amounts of boron and finely divided uranium in the form of cold pressed compacts in molybdenum crucibles under an argon atmosphere (Cordfunke, 1969). The reaction between the two elements begins between 1000° and 1250° and is exothermic. The temperature is then raised to 1400 to 2000° C, depending on the sample. The compounds UB_4 and UB_{12} can be prepared by electrolysis of a U_3O_8—B_2O_3 melt containing alkaline earth oxides and fluorides at 1000 to 1100° C (Andrieux and Blum, 1949). The tetraboride can be prepared also by the reactions (Cordfunke, 1969)

$$3UF_4 + 16B \xrightarrow{1600°\,C} 3UB_4 + 4BF_3$$

and

$$UO_2 + 6B \xrightarrow{2000°\,C} UB_4 + (BO)_2(g).$$

The diboride is insoluble in aqueous alkalis and acids with the exception of nitric acid and hydrofluoric acid. It dissolves in melted alkali with the liberation of hydrogen (Katz and Rabinowitz, 1951). The tetraboride dissolves in cold hydrofluoric or hydrochloric acid and in a mixture of nitric acid and concentrated hydrogen peroxide. Sulfuric acid is reduced by UB_4 forming SO_2 and S. The tetraboride is rapidly decomposed by alkaline fusion or in the presence of carbonate and bisulfate and reacts vigorously with peroxides (Caillat and Elston, 1961). The compound UB_{12} is not attacked by hydrofluoric or hydrochloric acid which permits the separation of UB_4 from UB_{12}. Boiling, concentrated sulfuric acid slowly dissolves UB_{12}. The latter is soluble in a mixture of nitric acid and concentrated hydrogen peroxide (Caillat and Elston, 1961).

3. Group IVA—Uranium Carbides and Silicides

a) Uranium Carbides

Three uranium carbides are known, UC, U_2C_3 and UC_2. The uranium-carbon system has been studied extensively. Phase diagrams have been prepared by Wilson (1961) and others. The carbides can be prepared by direct combination of the elements with control of the elemental proportions and the temperature. However, the purity of the product is difficult to control by this method. The monocarbide may be prepared by heating between 625 and 900° C finely divided uranium powder with methane. The dicarbide is formed in a similar reaction at temperatures >950° C. The monocarbide and dicarbide may be prepared by heating uranium dioxide with carbon. The sesquicarbide is prepared by heating a mixture of UC and UC_2 at temperatures between 1300 and 1800° C. Above 1800° C U_2C_3 decomposes to UC and UC_2.

Uranium dicarbide is a pyrophoric substance that may ignite on being ground. In air it decomposes slowly, probably by reaction with water vapor. At temperatures near 400° C, the carbide ignites in either air or oxygen forming U_3O_8 and CO_2. Both UC and UC_2 react slowly with water at ordinary temperatures. Heating causes a rapid reaction with the formation of UO_2, CH_4 and higher hydrocarbons. Hydrogen reduces UC_2 to UC and CH_4. The dicarbide reacts explosively when heated slightly with fluorine and ignites when heated with chlorine or bromine at temperatures of 350° C or 390° C, respectively. Iodine reacts with UC_2 and U_2C_3 to give uranium iodide. The carbides react with nitrogen at temperatures of ~1100° C to form nitrides. UC and UN form a continuous series of solid solutions depending on the pressure of the nitrogen and the temperature. UC_2 burns in sulfur vapor to give uranium sulfide and carbon disulfide. A similar reaction occurs with selenium. Carbon monoxide and CO_2 react with UC and UC_2 to give UO_2. Hydrochloric, nitric and sulfuric acids attack UC and UC_2 with the formation of hydrogen, hydrocarbons and uranyl salts. Concentrated acids react slowly at ordinary temperatures but vigorously when heated. UC_2 is decomposed by alkalis. The interactions between uranium carbides and other metals and their alloys have been studied extensively (Elston and Caillat, 1967; Cordfunke, 1969).

b) Uranium Silicides

Six uranium silicides are identified in the phase diagram presented by Kaufmann (1957). They are U_3Si, U_3Si_2, USi, U_2Si_3, USi_2, and USi_3. The silicides are generally prepared by synthesis from the elements present in appropriate proportions. The reactions between uranium and silicon begin about 450° C. They are exothermic and temperature dependent. Several phases are generally present after such preparation (Caillat and Elston, 1961). The silicide U_3Si is not formed except by heating uranium and U_3Si_2 at about 800° C. At 930° C, U_3Si decomposes into γ-U and U_3Si_2 (Cordfunke, 1969).

Uranium silicides are generally resistant to corrosion by water but not by oxygen, air, or nitrogen. Katz and Rabinowitz (1951) report that USi_2 is insoluble in cold or hot concentrated hydrochloric, nitric, or sulfuric acid or aqua regia but soluble in concentrated hydrofluoric acid; that it reacts with molten alkalis or alkali carbonates but reacts slowly with molten $KHSO_4$; and that it burns in oxygen at 800° C and reacts with chlorine at 500° C. Dubuisson (Caillat and Elston, 1961) reports a somewhat different effect of acids on USi_2. At concentrations $>1\,N$, sulfuric, nitric and hydrochloric acids react with USi_2. The amount of USi_2 dissolved in a 7 hour period increases with acid concentration to about $5\,N$ for sulfuric acid and then decreases to zero near $15\,N$. In nitric acid, the amount of USi_2 dissolved increases with acid concentration to $\sim 8\,N$ and then decreases slightly. In hydrochloric acid, the amount of USi_2 dissolved increases continuously with acid concentration to about $8\,N$ with no indication of a subsequent decrease.

4. Group VA — Uranium Nitrides, Phosphides, and Arsenides

a) Uranium Nitrides

Uranium nitrides include the compounds UN, U_2N_3, and UN_2. A mixture of the nitrides is formed by heating uranium and nitrogen at temperatures greater than 400° C. Above 1200° C only UN is stable at a nitrogen pressure of one atmosphere. The mononitride may therefore be prepared by heating U_2N_3 in vacuum or in an inert atmosphere at 1250–1500° C. At 1700° C, UN is formed by the reaction

$$UO_2 + \frac{1}{2} N_2 + 2C \xrightarrow{1700^\circ C} UN + 2CO.$$

However UN forms a solid solution with UO and UC. Therefore UN prepared by this method contains impurities.

The sesquinitride U_2N_3 exists in two allotropic forms. Cubic U_2N_3 is prepared by the reaction of uranium and nitrogen in the presence of hydrogen. A product $UN_{1.75}$ is obtained. In four hours at 500° C the composition $UN_{1.52}$ is obtained (Caillat and Elston, 1961). Hexagonal U_2N_3 is formed by heating cubic U_2N_3 at 0.01 mm pressure for one hour at 1100° C (Price and Warren, 1965) and by the reaction between UN and nitrogen at 0.4 atm pressure at 1315° C (Benz and Bowman, 1966).

Uranium dinitride UN_2 has been recognized through X-ray data. A composition with the limit $UN_{1.84}$ has been prepared (Cordfunke, 1969). Cordfunke (1969) gives a partial phase diagram for the uranium nitrogen system.

Uranium nitrides are oxidized readily. UN powder prepared below 1200° C may ignite in air at room temperature. With water below 300° C, UN reacts slowly forming a protective coating of UO_2. The mononitride dissolves in nitric acid, concentrated perchloric acid and hot phosphoric acid, but not in hot or cold hydrochloric or sulfuric acid or in sodium hydroxide solutions. It does react rapidly with molten alkali (Katz and Rabinowitz, 1951; Cordfunke, 1969). UN reacts with a number of metals through the formation of alloys or nitridation (Elston and Caillat, 1967; Cordfunke, 1969).

b) Uranium Phosphides

Three uranium phosphides are known: UP, U_3P_4, and UP_2. The diphosphide UP_2 is prepared by the reaction of finely divided uranium and phosphine carried in a stream of argon at 400° C (Baskin and Shalek, 1964). Above 500° C, UP_2 decomposes to U_3P_4 which decomposes to UP above 1100° C. UP_2 can be prepared also by heating the proper proportions of uranium and red phosphorus at 400° C in vacuum (Tróc et al., 1966). UP formed by direct reaction between the elements suffers from impurity contamination, mainly UO_2.

UP heated in oxygen at 1000° C forms $(UO_2)_2\ P_2O_7$. It dissolves slowly in mineral acids with the evolution of phosphine. Sintered UP is inert to boiling water (Cordfunke, 1969). U_3P_4 burns in air and is attacked slowly by water if air is present. It decomposes instantly in concentrated nitric acid, aqua regia or molten NaOH (Katz and Rabinowitz, 1951).

c) Uranium Arsenides

Uranium and arsenic combine to form UAs, U_3As_4, and UAs_2. UAs_2 and U_3As_4 are prepared by heating the proper proportions of the powdered elements in evacuated, sealed tubes at 800° C and 950° C, respectively. UAs is prepared by the decomposition of the higher arsenides (Cordfunke, 1969). UAs has also been prepared by the reaction of finely divided uranium with arsine at 300° C followed by heating the resulting mixture in vacuum between 1200 and 1400° C (BASKIN, 1967).

UAs oxidizes slowly in air at room temperature but ignites at 290° C (BASKIN, 1965). All uranium arsenides are soluble in concentrated nitric acid (ELSTON and CAILLAT, 1967).

5. Group VIA—Uranium Oxides, Sulfides, Selenides, and Tellurides

a) Uranium Oxides

The uranium-oxygen system is complex. Four uranium oxides have been established: UO_2, U_4O_9, U_3O_8, and UO_3. There is also evidence for U_3O_7 and U_2O_5 phases. Each of the higher oxides, U_4O_9, U_3O_8, and UO_3, has several allotropic modifications. In addition, a large number of non-stoichiometric uranium-oxygen combinations occur (CAILLAT and ELSTON, 1961; ELSTON and CAILLAT, 1967; CORDFUNKE, 1969).

The trioxide UO_3 may be prepared in a number of ways. Orange-colored amorphous UO_3 is formed by heating uranium compounds such as hydrated uranium peroxide, uranyl oxalate or ammonium uranyl tricarbonate in air at 400° C. α-UO_3 is brown-colored and may be prepared by heating hydrated uranium peroxide, containing some nitrate ions, at 450° C or by heating amorphous UO_3 between 470 and 500° C under an oxygen pressure of 40 atmospheres. Orange-colored β-UO_3 can be made by heating ammonium diuranate slowly to 450° C or by heating uranyl nitrate rapidly to the same temperature. Another method is to heat U_3O_8 between 500 to 550° C under an oxygen pressure of 40 atmospheres. Yellow γ-UO_3 is prepared by heating uranyl nitrate in air to 200° C, where it is homogenized, and then to 500° C. It may also be prepared by heating amorphous UO_3 with the other forms of UO_3 at 650° C under an oxygen pressure of 40 atmospheres. δ-UO_3 is dark red and is prepared by heating β-$UO_3 \cdot H_2O$ in air at 375° C for more than 24 hours. Brick-red ε-UO_3 is prepared by passing NO_2 over U_3O_8 at 250° to 375° C. Another form η-UO_3 has been prepared at a pressure of 30 kilobars and temperature of 1100° C (ELSTON and CAILLAT, 1967; CORDFUNKE, 1969).

U_3O_8 may be prepared by heating amorphous and other forms of UO_3 in air at a temperature near 750° C. U_3O_8 is also prepared by heating uranyl salts, such as the fluoride, chloride, bromide, nitrate, sulfate, or organic salts to a temperature between 750 and 850° C. Ammonium diuranate and uranium peroxide heated in air at 700° C also form U_3O_8. The product obtained by heating the lower oxides UO_{2+x} in air at 800° is U_3O_8 (CAILLAT and ELSTON, 1961; ELSTON and CAILLAT, 1967).

UO_2 may be prepared by the thermal decomposition of a uranyl salt in the presence of a reducing agent, such as dry hydrogen, carbon or metallic sodium, magnesium or aluminium. Organic salts and the halide salts UO_2Br_2 and UO_2Cl_2 give UO_2 when thermally decomposed in the absence of air or under vacuum. The dioxide is also prepared by electrolysis of a uranyl chloride solution in a molten salt bath. UO_2 is deposited at the cathode (CAILLAT and ELSTON, 1961; ELSTON and CAILLAT, 1967). Finely divided UO_2 may be pyrophoric. Pyrophoric UO_2 oxidizes in air at room temperature to give U_3O_8.

The chemical reactions of uranium oxides have been studied extensively. They are described by KATZ and RABINOWITZ (1951) and in "Nouveau Traite de Chemie Minerale" (CAILLAT and ELSTON, 1961; ELSTON and CAILLAT, 1967). Some of these reactions are listed in Table 2.13. The rate of dissolution of uranium oxides in acids depends on the particle size, acid concentration and temperature. Hydro-

Table 2.13. Reactions of UO_2, U_3O_8, and UO_3 with various reagents[a]

Reagent	Uranium products	Reagent	Uranium products
UO_2			
HF(g)(550), $NH_4F \cdot HF$, SF_4(500)		Cl_2 (red heat)	UO_2Cl_2
fluorinated hydrocarbons	UF_4	Br_2(g)+C	UBr_4
Hot fuming HCl(aq)(slow),		CS_2(900)	$UO_2 \cdot 2\,US_2$
S_2Cl_2, $COCl_2$, CCl_4, C_3Cl_6, Cl_2+C	UCl_4	C	UC, UC_2
$SOCl_2$	$UOCl_2$		
U_3O_8			
HF(aq)	UF_4, UO_2F_2	Br_2(g), HBr(g)	UO_2Br_2 +lower oxides
F_2+C(300), F_2(g)(370), CoF_3, BrF_3, SF_4(400)	UF_6(g)		
Cl_2(900)	UO_2Cl_2	CS_2	$UO_2 \cdot 2\,US_2$
S_2Cl_2(230–250), CCl_4(360), $COCl_2$(500)	UCl_4	CS_2+CO_2 (red heat)	UO_2
UO_3			
HF(aq), SeF_4	UO_2F_2	CCl_4(400), S_2Cl_2, $SOCl_2$(350)	UCl_4
$NH_4F \cdot HF$, SF_4(300)	UF_6	$SiCl_4$(370–380)	UCl_4, UO_2Cl_2
CCl_4	UO_2Cl_2		
CCl_4(115–185)	UCl_5	NH_4NO_3 (molten)	$UO_2(NO_3)_2$

[a] Compiled from Katz Rabinowitz (1951), Caillat and Elston (1961), and Elston and Caillat (1967). Temperature of the reaction in °C are given in parentheses. The rate of the reaction, eg. slow, may also be given in parentheses.

chloric acid has little effect on UO_2. The concentrated acid reacts with U_3O_8 and, with oxidizing agents present, dissolves U_3O_8 to give UO_2Cl_2. UO_3 dissolves in hydrochloric acid to give UO_2Cl_2. All three oxides dissolve in nitric acid. Dilute sulfuric has little effect on UO_2, but prolonged boiling gives a mixture of U(IV) and U(VI) sulfates.

Water reacts with UO_3 to form hydrates which are thermally unstable and liberate water when heated between 50 and 300° C. UO_2 and U_3O_8 do not react directly with water although hydrates of these oxides may be formed by hydrolysis or photolysis of uranyl salt solutions.

Strong bases do not react with UO_2 or U_3O_8 but form water-insoluble uranates with UO_3. In the presence of peroxide, water-soluble peruranates are formed. In solution of alkaline carbonates or bicarbonates, the oxides dissolve, forming double carbonate salts of uranium and the alkali metal. Alkaline earth salts of this type have also been prepared.

Uranium oxides may react with metallic oxides at high temperatures to form the metal uranates. Solid solutions may also be formed. These reactions are important from the standpoint of reactor technology.

b) Uranium Sulfides

Five uranium sulfides are known: US, U_2S_3, U_3S_5, US_2 and US_3. There are three allotropic forms of US_2. The monosulfide US may be prepared by heating finely divided uranium with H_2S at 500° C. The product is homogenized by heating in vacuum or in an argon atmosphere at 2000° C. US may also be prepared by the dissociation of U_2S_3 under vacuum at 1800° C, by the reaction between stoichiometric amounts of US_2 and UH_3 at successive temperatures of 300, 400–600, and ~2000° C, and by the reduction of UOS with Al or C. The latter reaction may result in a mixture of uranium sulfides and in the formation of uranium carbides. The sesquisulfide U_2S_3 is obtained by the reduction of US_2 with powdered Al between 1200 and 1350° C. Excess Al and Al_2S_3 formed in the reaction are removed with boiling acetic acid. U_2S_3 is formed in the reaction between finely divided uranium and H_2S or US_2 in stoichiometric amounts. The resulting product is homogenized by heating between 1000 and 1200 °C. U_3S_5 is prepared by the dissociation of US_2 in vacuum at 1500° C or by the reduction of α-US_2 with hydrogen at 1500° C. α-US_2 is obtained by the action of H_2S on U_3O_8. Relatively pure US_2 is formed around 1150° C. β-US_2 is formed by heating H_2S with uranium or lower uranium sulfides at 800 to 900° C or with UCl_4 at 600° C. β-US_2 may also be prepared by the reaction between gaseous UI_4 and sulfur at 950° (Smith and Cathey, 1967). γ-US_2 is prepared by the action of H_2S on uranium or U_3S_5 but at 400° C. The polysulfide US_3 is obtained by the reaction of sulfur with US_2 in a sealed tube between 500 and 800° C (Cordfunke, 1969; Elston and Caillat, 1961).

US is fairly resistant to oxidation up to 300° C. At 400° C, the main oxidation product is UO_2; at 500° C, it is U_3O_8. US is not attacked by boiling water, dissolves slowly in mineral acids, and is stable towards acetic acid and solutions of strong alkalis and ammonia. Picon and Flahaut (1955a) report that U_2S_3 and U_3S_5 are similar in their chemical properties to US_2. U_2S_3 and U_3S_5 are attacked by mineral acids and aqueous solutions of oxidizing agents, such as iodine-water, H_2O_2 and $KMnO_4$. They are not attacked by aqueous solutions of $K_2Cr_2O_7$ or strong alkalis. Acetic acid diluted by half does not attack these sulfides even with boiling. US_2, however, reacts with acetic acid diluted ten-fold and with aqueous solutions of $K_2Cr_2O_7$ (Picon and Flahaut, 1953b, 1955b). β-US_2 appears to be less reactive than the α or γ forms. The latter three sulfides are reasonably stable in air to 300° C. US_3, on the other hand, is oxidized in air at room temperature giving free sulfur and uranyl sulfate.

Uranium oxysulfide UOS is formed by passing H_2S over UO_2 or U_3O_8 at 1000° C. It is also formed by heating a mixture of US_2 and UO_2 in vacuum at 1350° C. UOS is reduced by carbon or aluminium to give US. UOS is more resistant to acids than are uranium sulfides. Acetic acid has no action on UOS nor has dilute hydrochloric or sulfuric acid. Concentrated nitric acid does attack UOS fairly rapidly. Aqueous solutions of alkalis do not dissolve UOS but solutions of oxidizing agents such as bromine-water, H_2O_2, or $KMnO_4$ convert UOS to uranyl sulfate (Picon and Flahaut, 1953a).

c) Uranium Selenides

Six uranium selenides, USe_3, USe_2, U_3Se_5, U_2Se_3, U_3Se_4, and USe, have been reported as well as three allotropic modifications of USe_2 (Khodadad, 1961). The general method for preparing uranium selenides is to treat USe_3 under vacuum but at different temperatures between 500 and 1500° C; alternatively, a stream of hydrogen may be passed over USe_3 at different temperatures. Uranium polyselenide USe_3 is obtained by the action of selenium vapor on UCl_4 in the presence of hydrogen at 620° C. It may also be obtained by the passage of gaseous H_2Se over metallic uranium turnings at 750° C or by the reaction between H_2Se and USe_2 below 600° C. USe_3 dissociates under vacuum (10^{-3} mm Hg) at temperatures of $\sim$600, 1000, 1350, 1450 and 1500° C to form the respective uranium selenides USe_2, U_3Se_5, U_2Se_3, U_3Se_4, and USe. Relatively pure compounds may be obtained by maintaining the proper temperature for several hours. Khodadad (1957a) has reported the formation of α-USe_2 by treating USe_3 in a current of hydrogen for 10 hours at 700° C; β-USe_2 was obtained by heating USe_3 under vacuum for 3 hours at 760° C or by treating the double salt $UCl_4 \cdot 2NaCl$ with H_2Se at 900° C; and γ-USe_2 was prepared by heating USe_3 under vacuum for 6 hours between 570 and 580° C. U_2Se_3 may be prepared also by reducing UOSe or USe_2 with aluminium under vacuum at 1100° C (KHODADAD, 1959). USe has been prepared by sealing a powdered mixture of uranium metal and selenium in a quartz ampoule under vacuum and then heating (FERRO, 1954).

The chemical properties of uranium selenides have been reported by KHODADAD (1961). The selenides more rich in selenium are markedly more sensitive toward water. USe_3 decomposes easily in water at room temperature to give UO_2, H_2Se, and elemental selenium. The decomposition of U_3Se_5 by water is very slow at room temperature. At 100° C, decomposition becomes appreciable. However, the reaction products are UO_2, H_2Se, and hydrogen rather than elemental selenium. USe is practically insensitive to water up to 100° C. The hydrolysis product UO_2, dried under vacuum, is susceptible to ignition in air. Pure oxygen does not alter uranium selenides at room temperature. However a noticeable reaction occurs around 200° C which gives UO_2, Se, and SeO_2. Above 350° the selenides ignite forming U_3O_8, Se and SeO_2. Sulfur readily displaces selenium in uranium selenides. The displacement can be made either in a stream of hydrogen or with sulfur alone in a sealed tube starting at 600° C. A series of mixed intermediate structures are formed.

Of the halogens, fluorine attacks the selenides most vigorously and iodine the least. At room temperature fluorine diluted in an inert gas transforms uranium selenides to UF_4 and SeF_4 with a small amount of SeF_6 being formed. If fluorine alone is used, combustion occurs with the formation of UF_6 and SeF_6. At room temperature chlorine reacts strongly to give UCl_4 and Se_2Cl_2. The latter may be converted to $SeCl_4$. Bromine vapor reacts in a similar manner giving UBr_4, Se_2Br_2 and $SeBr_4$. Iodine reacts with uranium selenides around 400° C to form UI_4 and selenium. The selenides react readily with acids. With even dilute hydrochloric and acetic acids, U(IV) ions and H_2Se gas are formed. The polyselenide USe_3 forms elemental selenium as a third reaction product; whereas the lower selenides going from U_3Se_5 to USe form molecular hydrogen. Nitric acid attacks the selenides and give uranyl ions and selenious acid. With alkaline solutions in the absence of oxygen selenides form $U(OH)_4$ and alkaline selenides or polyselenide. Hydrogen also is released with the selenides from U_3Se_5 through USe. USe_3 stirred with mercury in the absence of air forms γ-USe_2 and HgSe.

The oxyselenide UOSe may be prepared in a number of ways. FERRO (1954) reports the formation of UOSe in the fusion of U_3O_8, KCN, and Se at 1000° C and by the action of gaseous H_2Se on $UCl_4 \cdot 2NaCl$ between 600 and 700° C. The first reaction probably results in a mixture of uranium products. KHODADAD (1957b) has prepared UOSe by heating at 500° C an equimolecular mixture of UO_2 and USe_2 in a sealed tube, by the action of H_2Se on uranium oxides heated to 1050–1080° C in the presence of graphite, and by heating USe_3 or USe_2 to 600° C in a stream of H_2Se saturated with water vapor. If uranium oxalate is used in the second reaction rather than the oxide, no graphite is required.

UOSe does not react with water at room temperature. At 350° C it is attacked giving UO_2 and H_2Se. Concentrated acetic and hydrochloric acids do not attack UOSe even with boiling. Concentrated nitric acid attacks UOSe immediately forming selenious acid and uranyl ions. Solutions of sodium or potassium hydroxide do not react with UOSe. However, a vigorous reaction occurs in the fusion of UOSe with sodium or potassium hydroxide pellets. By treating the products of this reaction with water, a solution of the alkaline polyselenide and UO_2 is obtained (KHODADAD, 1957b).

d) Uranium Tellurides

The uranium-tellurium system is similar to the uranium-selenium system with the compounds UTe_3, UTe_2, U_2Te_3, U_3Te_4, and UTe having been identified (Cordfunke, 1969). Ferro (1954) reports the formation of UTe_2, U_2Te_3, U_3Te_4, and UTe by sealing a powdered mixture of uranium metal and tellurium in a quartz ampoule under vacuum and then heating. Matson et al. (1963) have prepared these compounds in a similar manner. The latter report that the tellurides react radily with both acids and bases. With acids, H_2Te is generated. However, UTe_2 does not appear to react with concentrated hydrochloric acid. Ferro (1954) reports that the tellurides are pyrophoric and flammable in their reaction with concentrated nitric acid.

Murasik and Niemiec (1965) report the formation of uranium oxytelluride from pure components of U, Te, and UO_2.

6. Group VIIA—Uranium Fluorides, Chlorides, Bromides, and Iodides

Uranium halide compounds have been studied extensively. Their preparation, chemical and physical properties are well-described in the reviews by Katz and Rabinowitz (1951), Caillat and Elston (1961), Elston and Caillat (1967), and Cordfunke (1969). Table 2.14 lists some of the physical properties of these compounds.

Table 2.14. Some physical properties of the uranium halides[a]

Halide compound	Color	Crystal structure	Heat of formation $-\Delta H_{298}$ (kcal/mole)	Melting point (°C)
UF_3	dark violet to black	hexagonal	345	1425
UF_4	green	monoclinic[d]	450	1036; 1003±2[g]
U_4F_{17}	black	—	—	—
U_2F_9	black	body-centered cubic	—	—
UF_5	α-white[b]; β-white[b]	α-tetragonal; β-tetragonal[e]	380.8 (g)	346–348
UF_6	white	rhombic	522.6 (s) 510.8 (g)	56.4 (sublimes)
UCl_3	olive-green	hexagonal	213.5	842
UCl_4	dark green	tetragonal[f]	251.3	590; 564±1[g]
UCl_5	red-brown	monoclinic	261.5	—
UCl_6	dark green	hexagonal	270.7	177.5±2.5
UBr_3	dark brown	hexagonal	172.3	730–755
UBr_4	brown	monoclinic	197.5	519
UBr_5	black[c]	—	—	—
UI_3	black	orthorhombic	114.2	766
UI_4	black	—	126.5	506

[a] Compiled from Cordfunke (1969) and Caillat and Elston (1961).
[b] Tinted by impurities.
[c] Lux (1969).
[d] Phase transition reported at 837±3°C (Khripin et al., 1965).
[e] α form exists at temperatures >150°C; β form at <135°C.
[f] Phase transition reported at 547±3°C (Khripin et al., 1965).
[g] Khripin et al., 1965.)

a) Uranium Fluorides

Uranium fluoride compounds UF_3, UF_4, U_4F_{17}, U_2F_9, UF_5 and UF_6 have been identified. UF_4 is an important compound in the preparation of uranium metal. UF_6 sublimes at 56.4° C (Table 2.14). It is used in the separation of uranium isotopes by gaseous diffusion.

UF_3 is prepared by the reduction of UF_4 with hydrogen, aluminium or uranium metal. With hydrogen the reduction is carried out under vacuum at 1000° C. Great precautions must be taken to remove all traces of oxygen or water. The reduction of UF_4 by aluminium may be performed under vacuum in a graphite crucible at 900° C. The UF_3 obtained is slightly contaminated with UO_2 attributable to traces of moisture in the UF_4 starting material. The reduction of UF_4 by uranium metal is carried out in an inert atmosphere at 1050° C. Small amounts of UO_2 and UO_2F_2 may also be formed.

UF_3 is relatively stable in air at room temperature. It is not very hygroscopic. At 900° C, UF_3 is converted to U_3O_8. In vacuum UF_3 disproportionates above 1000° to UF_4 and uranium metal. The trifluoride is slowly oxidized in cold water to UF_4. At 100° C oxidation is more rapid with the formation of uranyl ions. In water vapor at 800° C, U_3O_8 and HF are formed. UF_3 is not reduced by hydrogen at 250° C but is reduced with calcium. The halogens react with UF_3 around 250 to 300° C to give the corresponding uranium halotrifluoride. UF_3 is insoluble in ammonium oxalate. Oxidizing acids convert UF_3 to uranyl salts. Cold hydrochloric, sulfuric and nitric acids attack it slowly. Hot nitric acid dissolves UF_3 fairly rapidly with the formation of nitrogen oxides. Hot sulfuric acid attacks UF_3 less rapidly. At 400° C gaseous HCl and UF_3 form a mixture of fluorides and chlorides. The addition of boric acid to another acid facilitates dissolution by the formation of the fluoroborate complex.

UF_4 can be prepared using either dry or wet processes. A dry process is preferred in industry. It involves the reaction between UO_2 and HF at temperatures from 500 to 700° C. The degree of conversion to UF_4 critically depends on the surface area, particle size and preparative history of the UO_2 as well as the temperature. UO_2 prepared from ammonium diuranate reacts to a greater degree than UO_2 prepared from uranyl nitrate. UF_4 is also obtained (1) by the action of freons (for example, CCl_2F_2) on UO_3 at 400° C, (2) by the reaction between a mixture of ammonia, hydrogen fluoride and UO_3 at 500 to 750° C or between NH_4F and UO_2 at 250° C to 300° C, (3) by the simultaneous reaction between hydrazine fluoride and ammonium bifluoride on a variety of uranium starting compounds such as UO_3, U_3O_8, $(NH_4)_2U_2O_7$ or $(UO_2)_2C_2O_4$, (4) by the action of HF on UH_3 at 200° C, (5) by the reaction between UCl_3 and anhydrous HF at 450°, and (6) by the reduction of UF_6. In reactions (2) and (3) the compound NH_4UF_5 (or $NH_4F \cdot UF_4$) is formed. This decomposes at 450° C to give UF_4 and NH_4F.

The general procedure for preparing UF_4 by a wet process involves the reduction of uranyl ions to U(IV) ions and treatment of the latter with fluoride ions. The hydrated UF_4 which is precipitated is dried and dehydrated at 400° C. The reduction of U(VI) to U(IV) may be done electrolytically or with stannous (Sn^{2+}) ion. Reduction has also been done with SO_2 in the presence of formic acid, with sodium hydrosulfite ($Na_2S_2O_4$), and by the action of sun-light (photoreduction). UF_4 may also be prepared by treating UO_2 with hydrofluoric acid or with ammonium fluoride and sulfuric acid. The reaction between UO_2 and cold sulfuric acid is weak and the compound NH_4UF_5 is obtained. The latter is decomposed to give UF_4.

Table 2.15. Conditions for the preparation of intermediate uranium fluorides by the reaction between UF_4 and UF_6[a]

Fluoride	Temperature (°C)	Pressure (mm Hg)
U_4F_{17}	320	18
U_2F_9	255	18
	270	50
α-UF_5	160	12–14
β-UF_5	115	12–14

[a] Compiled from Caillat and Elston (1961), Elston and Caillat (1967).

UF_4 is relatively stable when heated in air to 200° C. Calcination in air gives U_3O_8. The solubility of UF_4 in water at room temperature is very slight. At temperatures greater than 100° C, hydrolysis becomes appreciable. In the presence of oxygen, the hydrolysis product is UO_2F_2. With oxygen or air at 700 to 850° C, UF_4 reacts to form both UF_6 and UO_2F_2. Fluorine and CoF_3 convert UF_4 to UF_6. UF_6 reacts with UF_4 to give the intermediate uranium fluorides, U_4F_{17}, U_2F_9 and UF_5. Chlorine has little effect on UF_4 at 500 to 675° C. Aluminium, uranium metal, and pure hydrogen reduce UF_4 to UF_3. Alkali and alkaline earth metals with the exception of beryllium reduce UF_4 to metallic uranium. Fuming perchloric acid dissolves UF_4 readily to give a solution of uranyl ions. Nitric acid attacks the fluoride slowly. The addition of boric acid facilitates dissolution through the formation of the BF_4^- complex. A mixture of dilute sulfuric acid and silica dissolves UF_4 with the formation of $U(SO_4)_2$ and fluosilicic acid. Hot phosphoric acid dissolves UF_4 with the formation of U(IV) phosphate. UF_4 is moderately soluble in ammonium oxalate solution. It reacts vigorously with alkali peroxides and ammonia-hydrogen peroxide mixtures to form soluble peroxyuranates. With metal fluorides, UF_4 forms a series of double salts. Hydrates having 0.5 to 2.5 water molecule per UF_4 molecules have been identified. The hydrated form precipitated by the addition of HF to a U(IV) ion solution below 90° C is $UF_4 \cdot 2^1/_2 H_2O$.

The fluorides U_4F_{17}, U_2F_9 and UF_5 are produced by the reaction between UF_4 and UF_6. The conditions of temperature and pressure for forming the different compounds are given in Table 2.15. U_2F_9 may also be prepared by the reaction between U_4F_{17} and UF_6, the thermal dissociation of UF_5 at 175° C, the hydrolysis of UF_5, and the action of hydrogen on UF_6 between 400 and 600° C. UF_5 is formed by the action of HF on UCl_5 or UCl_6 and by the action of fluorine on UF_4 between 150 and 250° C.

The intermediate uranium fluorides disproportionate to UF_6 and the next lowest fluoride. With water they form UO_2F_2, UF_4, and HF. However, the rate of reaction varies considerably, UF_5 reacting most rapidly. Pentavalent uranium forms a number of complex fluorides generally of the form MUF_6, where M usually represents a monovalent metal. However, heptafluoro, octofluoro, and dodecafluorouranium(V) salts have also been reported (Selbin and Ortego, 1969).

The preparation of UF_6 has been studied extensively. Two of the most important methods are the fluorination of UF_4 by fluorine at 300° C and the oxidation of UF_4 by oxygen at 800° C. The first reaction is complicated since intermediate compounds may be formed at temperatures between 250 and 450° C. The yield is also subject to impurities such as uranium oxides and traces of water. The second reaction is the basis for the Fluorox process for preparing UF_6.

Table 2.16. Methods for preparing UF_6 [a]

Starting uranium compound	Reagent	Temperature	Products other than UF_6
U	F_2	(violent reaction)	
U	BrF_3	50–125	Br_2
U	BrF_5	50–75	BrF_3
U	ClF_3	25–75	ClF
UC_2	F_2	350	CF_4
UO_2	F_2	500	O_2
UO_2	BrF_3		Br_2, O_2
U_3O_8	F_2	360	O_2
U_3O_8	F_2, C	300	CO_2
U_3O_8	SF_4	400	UF_4, SOF_2
U_3O_8	BrF_3		Br_2, O_2
UO_3	F_2	400	O_2
UO_3	SF_4	300	SOF_2
UO_3	BrF_3		Br_2, O_2
$Na_2U_2O_7$	F_2		NaF, O_2
CaU_2O_7	F_2		CaF_2, O_2
UO_2F_2	F_2	270	O_2
UO_2F_2	SF_4	300	SOF_2
UO_2F_2	BrF_3		Br_2
UF_4	F_2	340	
UF_4	BrF_3	70	Br_2
UF_4	ClF_3	<100	ClF
UF_4	CoF_3	250	CoF_2
UF_4	HgF_2	400	HgF
UF_4	ClO_3F		UO_2F_2, Cl_2
UF_4	O_2	600–800	UO_2F_2
U_4F_{17}	F_2		
U_4F_{17}	—	270–350	UF_4
U_2F_9	F_2	300	
U_2F_9	—	>200	UF_4
UF_5	F_2	270	
UF_5	—	175	U_2F_9
$NaUF_5$	F_2	340	NaF
UCl_5	F_2	−20 to −30	UF_4, Cl_2
UCl_6	HF	60–150	HCl
UO_2HPO_4	F_2	370	PF_3, PF_5
$U(SO_4)_2$	ClF_3	>300	Residue of UO_2SO_4+ UO_2F_2

[a] Compiled from CAILLAT and ELSTON (1961) and ELSTON and CAILLAT (1967).

UO_2F_2 is formed as a byproduct which is recycled through UO_2 to give UF_4. Table 2.16 lists various methods for the preparation of UF_6.

The vapor pressure of solid UF_6 has been measured by OLIVER et al. (1953) to be

$$\log_{10} p(\text{mm Hg}) = 6.38363 + 0.0075377t - 942.76/(t + 183.416),$$

where t is in °C. Pressure data on liquid UF_6 from 64 to 116° C are represented by the equation

$$\log_{10} p(\text{mm Hg}) = 6.99464 - 1126.288/(t + 221.963).$$

Above 116° C, the pressure is given by

$$\log_{10} p(\text{mm Hg}) = 7.69069 - 1683.165/(t + 302.148).$$

UF_6 reacts violently with water with the evolution of considerable heat. The reaction products are primarily UO_2F_2 and HF. With bases, such as NaOH, the (di)uranate is formed. UF_6 is reduced to UF_4 by hydrogen at 250° C, by HCl from 200 to 300° C, by HBr at 100° C, by NH_3 at 900° C, and by non-metallic elements such as carbon, silicon, phosphorus, and arsenic. It is also reduced by H_2S, CS_2 and many organic compounds. It reacts with various metals and common materials of construction. Pure fluorocarbons do not react with UF_6 and stable solutions of the latter may be formed. Addition compounds are formed between UF_6 and NOF, NO_2F, and the metal fluorides, NaF, KF, RbF, AgF, TlF.

b) Uranium Chlorides

UCl_3 can be prepared by the reduction of UCl_4. Hydrogen at 550° C, zinc at 450°, uranium metal, and hydrogen iodide at 300 to 350° C have been used successfully as reducing agents. Ammonia has also been used at elevated temperatures. However, the resulting product contains nitrogen. UCl_3 is also prepared by the action of HCl on finely divided uranium or uranium hydride at 250 to 300° C and by the reaction between finely divided uranium metal and $PbCl_2$ at 600 to 700° C. Solutions of UCl_3 have been prepared by the electrolytic reduction of UO_3 in concentrated hydrochloric acid and by the reduction of UO_2Cl_2 with zinc, also in hydrochloric acid.

UCl_3 decomposes thermally. The decomposition products formed depend on the temperature. UCl_3 is hygroscopic and dissolves in water to give an initial purple color which soon turns green indicative of U(IV) ions. UCl_3 dissolves more rapidly in dilute hydrochloric acid than in more concentrated acid. Anhydrous liquid HF reacts with UCl_3 to give HCl and presumably UF_3. UCl_3 dissolves in glacial acetic acid and reacts with methanol or formamide. It is insoluble in non-polar solvents, CCl_4, $CHCl_3$, acetone and pyridine. Ammonolysis occurs with liquid ammonia. UCl_3 reacts with pure oxygen at 150° to give UO_2Cl_2. With chlorine, it reacts at 250° C to form UCl_4. Bromine and iodine react with UCl_3 at 500° C to give, respectively, UCl_3Br and UCl_3I. The latter reaction is used to purify UCl_3 since UCl_3I decomposes to UCl_3 and iodine on cooling.

UCl_4 can be prepared by several methods. At 180° C finely divided uranium and chlorine gas react violently to form UCl_4. Chlorine also reacts with UH_3 starting at 200° C to form UCl_4. However, the reaction is difficult to control. A better process is the conversion of UH_3 to UCl_3 by HCl and the subsequent conversion of UCl_3 to UCl_4 by chlorine at 250° C. UF_4 can be converted to UCl_4 in the vapor phase by its reaction with $AlCl_3$ or BCl_3. UF_4 can also be fused with $BeCl_2$, $MgCl_2$, $CaCl_2$, $BaCl_2$, or $NaAlCl_4$ at 700 to 800° C to give UCl_4. UCl_4 can be obtained by heating UCl_5 under vacuum or in an atmosphere of nitrogen or carbon dioxide at 250°. Uranium oxides may be converted to UCl_4 by a number of chlorinating agents: Cl_2 in the presence of carbon, CCl_4, hexachloropropene, hexachlorocyclopentadiene, benzotrichloride, phosgene ($COCl_2$), thionyl chloride ($SOCl_2$), and sulfur chloride (S_2Cl_2). Depending on the starting uranium oxide, reagent and temperature, variable amounts of UCl_5 and other reaction products may be formed. An aqueous solution of UCl_4 can be prepared by the reduction of uranyl chloride solutions. Reduction can be effected electrolytically, photochemically or by amalgams of silver, cadmium or bismuth, by metallic copper, or by sodium hydrosulfite ($Na_2S_2O_4$) (Katz and Rabinowitz, 1951; Caillat and Elston, 1961).

UCl_4 can be sublimed or distilled without decomposition. It is very sensitive to water. In humid air, it deliquesces with decomposition to $UOCl_2$ and HCl. In water, it dissolves with formation of a green U(IV) solution. With steam at 600° C, U_3O_8 is formed. UCl_4 is generally insoluble in non-polar solvents (hydrocarbons, benzene, chloroform, and ether) and soluble in polar solvents. Ammonolysis occurs in liquid ammonia. In anhydrous liquid HF, UCl_4 forms a precipitate $UF_4 \cdot HF$ at room temperature. Heating to 600° C yields UF_4. UCl_4 is reduced by HBr to UBr_3 between 300 and 350° C, by HI to UCl_3 at 350 to 420° C, and by H_2 to UCl_3 at 550° C. The metals Li, Na, Ca, Mg, and Al reduce UCl_4 to uranium metal. Oxygen converts UCl_4 to UO_2Cl_2 at 230° C. The latter is converted to U_3O_8 at 250° C. Fluorine may react violently with UCl_4 to give UF_6. Chlorine reacts with the tetrachloride at high temperatures to form a mixture of UCl_5

and UCl_6. H_2S and UCl_4 react at 550° C to form US_2 and HCl. A similar reaction occurs with H_2Se. Double chloride compounds are formed of the type M_2UCl_6 with alkali metals, pyridine, or quinoline or of the type $MUCl_6$ with alkaline earth metals. UCl_4 has been used as the initial reactant in the preparation of such organic compounds of U(IV) as aluminium alkoxides $U[Al(OR)_4]_4$ (ALBERS et al., 1952), dicarbonyls $U(RCOCHCOR')_4$ (GILMAN et al., 1956), amides $U(NR_2)_4$, alkoxides $U(OR)_4$, and mercaptides $(U(SR)_4$ (JONES et al., 1956a), cyclopentadienyls UCp_4, UCp_3BH_4, and UCp_3Cl, where $Cp = C_5H_5$ (REYNOLDS and WILKINSON, 1956; ANDERSON and CRISLER, 1969), and bis(cyclooctatetraenyl) uranium or uranocene $U(C_8H_8)_2$ (STREITWIESER and MÜLLER-WESTERHOFF, 1968).

UCl_5 can be prepared by the chlorination of UCl_4 with chlorine gas at 550° C., liquid chlorine at elevated temperature and pressure, or CCl_4 vapor contained in a stream of air at 550° C. Uranium oxides can be converted to UCl_5 by CCl_4 in the liquid phase under pressure of 5 to 8 atmospheres and temperature of 80 to 200° C. The reaction is autocatalytic and facilitated by the addition of a small amount of UCl_5 to the starting reactants. UO_3 can be converted to UCl_5 at the boiling point of CCl_4 after many hours provided a large excess of CCl_4 is used, considerable UCl_5 has been added, and chlorine is bubbled continuously through the mixture.

UCl_5 decomposes to UCl_4 and Cl_2 when heated. Some decomposition occurs below 100° C. At 250° C, decomposition is rapid. Under vacuum UCl_5 disproportionates into UCl_4 and UCl_6 at 100 to 175° C. UCl_5 is very hygroscopic. In water it decomposes immediately to give UCl_4, UO_2Cl_2, and HCl. UCl_5 reacts vigorously with dimethyl ether, reacts immediately with diethyl ether, acetone, ethyl acetate, formamide, dioxane, and alcohol, and reacts slowly with chloroform, acetophenone, and isopropyl ether. It dissolves in CCl_4, CS_2, and $SOCl_2$. Oxygen reacts with UCl_5 to give a mixture of uranium oxyhalides. Fluorine converts UCl_5 to UF_6 and UF_4 at −40° C. Chlorine reacts with UCl_5 to give UCl_6. The reaction is reversible and depends on temperature and pressure. UCl_5 is reduced by sodium metal to metallic uranium. It is reduced to UCl_4 by hydrogen at 300° C and by $CHCl_3$ in an autoclave at 140° C. Anhydrous liquid HF converts UCl_5 to UF_5. A number of hexachloruranate(V) salts have been prepared which are generally more stable than UCl_5 itself. An octachloro compound has also been isolated and there is evidence of a heptachloro species (SELBIN and ORTEGO, 1969). Addition compounds such as $UCl_5 \cdot PCl_5$ and $UCl_5 \cdot SOCl_2$ are known. However, these compounds are formed by the reaction between PCl_5 or $SOCl_2$ and UO_3. UCl_5 reacts with alcohols or sodium alkoxides to form uranium(V) alkoxides (JONES et al., 1956b, c). These compounds may also be prepared by the reaction between UCl_4 and the appropriate alcohol or sodium alkoxide followed by oxidation to give the uranium(V) compound.

UCl_6 is the only hexavalent uranium compound other than UF_6 that does not contain oxygen. It is prepared by the disproportionation of UCl_5 at 80 to 180° C under vacuum (10^{-6} to 10^{-3} mm Hg). The hexachloride sublimes and is collected on a cold finger of the apparatus. It can also be prepared by the action of chlorine gas on UCl_4 above 350° C.

UCl_6 is stable in dry air, nitrogen, and helium at room temperature. Decomposition occurs at temperatures of 120 to 150° C. UCl_6 is very hygroscopic and reacts violently with water to form UO_2Cl_2. It is soluble in CCl_4, slightly soluble in C_7F_{16} and isobutyl bromide, and reacts with tetrachloroethylene and naphthenic hydrocarbons. UCl_6 reacts with anhydrous liquid HF at room temperature to give UF_5, HCl, and chlorine.

c) *Uranium Bromides*

UBr_3 can be prepared by the reaction between UH_3 and HBr at 300° C, by direct synthesis with metallic uranium and bromine vapor at 570° C, or by the reduction of UBr_4 by metallic uranium (550° C) or hydrogen (550° C). A solution of UBr_3 is obtained by electrolytic reduction of a solution prepared by dissolution of UO_3 in hydrobromic acid.

UBr_3 is very hygroscopic, even more so than UCl_3. It dissolves readily in water with the evolution of hydrogen. It is almost insoluble in non-polar solvents. Its dissolution in polar solvents, such as formamide, acetamide, and ethanol, is usually accompanied by a reaction. UBr_3 is reduced to metallic uranium by sodium and calcium. It is oxidized to UBr_4 by bromine and reacts with oxygen at room temperature. UBr_3 is very corrosive. It attacks most metals and reacts with quartz.

UBr_4 can be prepared in a number of ways. Most of them involve the reaction of uranium metal or a uranium compound and bromine. The bromination of UBr_3 at 300° C is an excellent method. UBr_4 may be purified by vacuum distillation or sublimation in a stream of nitrogen or helium containing some bromine vapor. Solutions of UBr_4 can be made by the reduction of uranyl bromide. Reduction can be effected electrochemically or chemically with uranium metal. Solutions can also be prepared by photochemical reduction of UO_3 in an alcoholic solution of HBr or by the dissolution of $UO_2 \cdot 2H_2O$ in hydrobromic acid.

UBr_4 is partially decomposed to UBr_3 and bromine by distillation in a stream of nitrogen. Its vapors are reduced to metallic uranium on a hot filament in an evacuated tube. UBr_4 dissolves readily in H_2O giving a green solution which is strongly hydrolyzed. It does not dissolve in non-polar liquids but does in polar liquids. Acetic acid, methanol, ethanol, phenol and aniline react with UBr_4 to evolve HBr.

Nitrobenzene, nitromethane, and benzaldehyde oxidize UBr_4 to UO_2Br. Dioxane, acetone, diethyl ether, ethyl acetate, and amyl acetate form stable solvates with UBr_4. Stable solutions are formed with formamide at room temperature and with molten acetamide. UBr_4 is converted by oxygen to UO_2Br_2 and by chlorine to UCl_4. It reacts with UO_2 to give $UOBr_2$. UBr_4 is reduced by hydrogen between 470 and 700° C to UBr_3 and by a number of metals, particularly magnesium and calcium, to metallic uranium.

UBr_5 has been reported by PRIGENT (1954a, b) who prepared it by the reaction between UO_3 and CBr_4 in a sealed tube at 126° C. The preparation can be effected between UO_3 and $COBr_2$ at a lower temperature. LUX (1969) has reported the formation of UBr_5 by the reaction between UBr_4 or uranium metal and boiling bromine. The compounds are extremely sensitive to moisture. UBr_5 is stabilized by formation of 1:1 addition compounds or by formation of the hexabromouranate(V) anion (SELBIN and ORTEGA, 1969).

d) *Uranium Iodides*

UI_3 is prepared by distilling the stoichiometric amount of iodine into an evacuated tube containing finely divided uranium at 350° C. The tube is then sealed and heated first to 130° C for several hours and then to 570° C for 15 to 20 hours. The reaction between UH_3 and methyl iodide at 275 to 300° C also gives UI_3.

UI_3 is very hygroscopic and dissolves vigorously in water. Freshly prepared UI_3 may react violently with water. The triiodide can be reduced to uranium metal by alkali or alkaline earth metals. It can also be reduced electrochemically in a solution of molten SrI_2 at 540° C using a tungsten anode and a molybdenum cathode. UI_3 is moderately stable in oxygen-free solutions. Dry air or oxygen converts UI_3 to UO_2I_2 at room temperature and to U_3O_8 at elevated temperatures. UI_3 is corrosive at higher temperatures, attacking glass and porcelain.

UI_4 is prepared in the same manner as UI_3. However, it is necessary to maintain the partial pressure of iodine sufficiently high ($\sim$200 mm) to prevent dissociation of the UI_4. Iodination of UI_3 is another method for preparing UI_4.

As indicated above, UI_4 dissociates to UI_3 and iodine. At sufficiently high temperatures UI_4 decomposes to uranium metal and iodine. This is the basis of the *hot-wire* method for preparing uranium metal. UI_4 is hygroscopic and dissolves readily in water. The solution is strongly hydrolyzed. Hydrogen reduces UI_4 to UI_3 and HI at moderately elevated temperatures. Oxygen or dry air converts UI_4 to UO_2I_2 at room temperature and to U_3O_8 at raised temperatures. Chlorine reacts with UI_4 at room temperature.

e) Uranium Mixed Halides and Uranium(IV) Borohydride

Most combinations of mixed halide compounds of uranium (III) and uranium (IV) are known (Katz and Rabinowitz, 1951; Caillat and Caillat and Elston, 1961). Methods for the preparation of these compounds are indicated in Table 2.17.

Table 2.17. Methods for the preparation of uranium (III) and uranium (IV) mixed halides[a]

Uranium mixed halide	Reactants and temperature (°C)	Other reaction or residual products of uranium
	U(III) compounds	
$UBrCl_2$	2 UCl_3+UBr_3 (850)	
$UBrCl_2$	$UBrCl_3$ + H_2	UCl_3 (removed by iodination)
UBr_2Cl	UBr_3Cl+H_2	UBr_3 (removed by iodination)
$UICl_2$	UI_2Cl_2 (400, vacuum)	
$UIBr_2$	UI_2Br_2 (>375, vacuum)	
UI_2Br	UI_3Br (heat, vacuum)	
	U(IV) compounds	
$UClF_3$	UF_3+$^1/_2$ Cl_2 (315)	
$UClF_3$	UF_4+UCl_4 (600, helium atmosphere)	
$UClF_3$	UF_4+CCl_4 (420)	UCl_4, UF_4
$UClF_3$	UO_2F_2+CCl_4 (450)	UCl_4
$UClF_3$	UO_2F_2+CCl_2=$CClCCl_3$ (175)	UCl_4
$UBrF_3$	UF_3+$^1/_2$$Br_2$ (250)	
UIF_3	UF_3+$^1/_2$ I_2 (180)	
$UBrCl_3$	UCl_3+$^1/_2$$Br_2$ (500)	
$UBrCl_3$	UBr_4+3 UCl_4 (550)	UCl_4
$UBrCl_3$	UCl_3+UBr_4 (550)	UBr_3, UBr_4
UBr_2Cl_2	UCl_4+UBr_4 (590)	
UBr_3Cl	UCl_4+3 UBr_4 (590)	
$UICl_3$	UCl_3+$^1/_2$ I_2 (500)	
UI_2Cl_2	UCl_4+UI_4	
UI_3Cl	UCl_4+3 UI_4	
UI_3Cl	UI_3+UCl_4 (600)	UCl_3
$UIBr_3$	UBr_3+$^1/_2$ I_2 (500)	
UI_2Br_2	UBr_4+UI_4 (520, helium atmosphere)	
UI_3Br	UBr_4+3 UI_4	
$UIBrCl_2$	$2UBr_2Cl_2+H_2 \rightarrow UBrCl_2+UBr_2Cl+HBr+HCl$	
$UIBr_2Cl$	$UBrCl_2+UBr_2Cl+I_2 \rightarrow UIBrCl_2+UIBr_2Cl$	

[a] Compiled from Katz and Rabinowitz (1951) and Caillat and Elston (1961).

The mixed halides generally are hygroscopic and dissolve in water. The mixed fluorides decompose in water and form an insoluble phase.

Uranium(IV) borohydride $U(BH_4)_4$ is treated by KATZ and RABINOWITZ (1951) as a *pseudo-halide*. It is formed in the reaction under vacuum between UF_4 and $Al(BH_4)_3$ in which considerable heat is liberated. $U(BH_4)_4$ sublimes without melting and melts at 126° C with decomposition. It reacts vigoriously with water and alcohol, is somewhat soluble in diethyl ether, and is slightly soluble in non-polar solvents.

f) Uranium Oxyhalides

The oxyhalides of uranium(VI) are the uranyl salts UO_2F_2, UO_2Cl_2, UO_2Br_2, and UO_2I_2. They are discussed in Section II C. Other U(VI) oxyhalides, $UOCl_4$ (see CAILLAT and ELSTON, 1961) and UOF_4 (see ELSTON and CAILLAT, 1967), have been proposed but have not been isolated.

Pentavalent uranium oxyhalides reported are $UOCl_3$ (see KATZ and RABINOWITZ, 1951; BUDAYEV and VOLSKY, 1958), $UOBr_3$ (PRIGENT, 1953), and UO_2Br (LEVET, 1965). Spectral evidence for the existence of UO_2Cl has also been reported (ADAMS et al., 1963). $UOCl_3$ has been identified as an intermediate compound in the chlorination of UO_2, U_3O_8, and UO_2Cl_2 by CCl_4 at 400 to 500° C and of UO_2 by hexachloroethane at 450 to 500° C (BUDAYEV and VOLSKY, 1958). It has since been prepared by the reaction between stoichiometric quantities of UO_2Cl_2 and UCl_4 in vacuum at 370 to 390° C (SHCHUKAREV et al., 1958) and by the reaction between UO_3 and $MoCl_5$ in vacuum at 200 to 220° C (GLUKHOV et al., 1968). $UOCl_3$ is hygroscopic and dissolves in water, acetone, methanol and ethanol but not in carbon tetrachloride or benzene. $UOBr_3$ is prepared by the action of CBr_4 on UO_3 at 110° C (PRIGENT, 1953). It decomposes into $UOBr_2$ and Br_2 when heated in a stream of nitrogen. Decomposition is nearly complete at 300° C. It is oxidized to UO_2Br_2 by oxygen at 140° C. $UOBr_3$ is very soluble in water. UO_2Br has been prepared by the reaction between amorphous UO_3 and HBr at 250° C (LEVET, 1965). The bromide is not soluble in water but is soluble in $2\,N$ H_2SO_4.

The tetravalent uranium oxyhalides, $UOCl_2$ and $UOBr_2$, have been known for sometime. UOF_2 has been reported, but its preparation is questionable (see KATZ and RABINOWITZ, 1951)[1]. $UOCl_2$ can be prepared by dissolving UO_2 in an excess of molten UCl_4 at 600° C. After cooling, the excess UCl_4 is removed under vacuum at 450° C. $UOCl_2$ dissolves slowly to give a green solution. Liquid HF reacts with $UOCl_2$ to give UF_4; chlorine reacts at 170° C to give UCl_4. $UOBr_2$ can be prepared by the reaction between $UO_2 \cdot 2US_2$ and bromine at 600° C, the action of UBr_4 on UO_2 at 500° C, and by the dissociation of $UOBr_3$ at 300° C. $UOBr_2$ decomposes at 800° C to UBr_4 and UO_2. It dissolves in water to give a green solution which gradually darkens and gives a slight precipitate after several hours. $UOBr_2$ is relatively stable in air at room temperature. At high temperature it is oxidized to U_3O_8.

g) Uranium Nitrogen Halides

UNF can be prepared by the reaction of UF_4 with silicon in a nitrogen atmosphere at 900° C (YOSHIHARA et al., 1969). The nitrogen fluoride reacts readily with nitric acid but not with sulfuric or hydrochloric acid. When heated in a nitrogen atmosphere at 1100° C, UNF is converted to U_2N_3; heated in an argon atmosphere, it is gradually converted to UN. The conversion to UN is facilita-

1 A compound with the composition UOF_2 and its crystal hydrate $UOF_2 \cdot H_2O$ have been prepared by VDOVENKO et al. (1967).

ted in the presence of a reducing agent such as silicon. UNF is gradually oxidized in air at room temperature.

UNCl, UNBr, and UNI can be prepared by heating the respective tetrahalides in a current of ammonia and by the reaction of the corresponding tetrahalides with UN (JUZA and MEYER, 1969). These nitrogen halides hydrolyze in moist air with the formation of β-UO_2 and the corresponding ammonium halide. Untempered UNI is pyrophoric. Heated to red heat in air, the nitrogen halides are converted to U_3O_8. They are soluble in dilute mineral acids, especially with heating.

C. Uranium Salts, Hydrated Oxides, Uranates, and Peruranates

The chemical preparation and properties of a few uranium(IV) and uranium(VI) salts as well as those properties of uranium hydrated oxides, uranates, and peruranates, important to uranium technology are reviewed in the present section. For more detailed descriptions of these and other uranium compounds, the reader is referred to one of the several reviews mentioned at the beginning of Sect. II or to CHERNYAEV's (1964, 1966) review on "Complex Compounds of Uranium".

1. Uranium(IV) Salts

The preparation and chemical properties of uranium(IV) halides are described in Sect. IIB6.

a) Carbonates

Uranium(IV) compounds, including those which are only slightly soluble, such as the dioxalates, tetrafluorides, phosphates, and freshly precipitated hydroxides are dissolved in an excess of concentrated carbonate solutions (CHERNYAEV, 1964). Compounds with the following types of U(IV)—CO_3^{2-}—OH^- combinations have been prepared: $[U(CO_3)_2(OH)_2]^{2-}$, $[U_2(CO_3)_5(OH)_4]^{6-}$, $[U(CO_3)_3]^{2-}$, $[U(CO_3)_4]^{4-}$, and $[U(CO_3)_5]^{6-}$. The U(IV) pentacarbonate complex is both readily oxidized and hydrolyzed. Oxidation occurs according to the reaction:

$$[U(CO_3)_5]^{6-} + O_2 \rightarrow [UO_2(CO_3)_3]^{4-} + 2\,CO_2.$$

b) Sulfates

Uranium(IV) sulfate $U(SO_4)_2$ may be prepared by the reaction between uranium dioxide UO_2 and sulfuric acid, by heating UCl_4 with sulfuric acid or by the reduction of UO_2SO_4. Reduction may be effected either electrolytically, photolytically in the presence of alcohol, or chemically with iron, zinc, or zinc amalgam. The uranium sulfate separated from aqueous solution is hydrated. The anhydrous salt may be obtained by prolonged heating of the hydrate in concentrated sulfuric acid.

Uranium(IV) sulfate is soluble in dilute acid. Strongly acidic solutions are relatively stable and are only slightly oxidized in the presence of air. Oxidizing agents such as chlorine, nitric acid, nitrates, and MnO_2 convert the uranium to the hexavalent state.

Tetravalent uranium sulfate is not saturated in its coordination sphere. A large number of complex compounds have been prepared in which the $U^{4+}:SO_4^{2-}$ ratio varies 1:0.8, 1:1, 1:1.25, 1:1.75, 1:2, 1:2.5, 1:2.75, 1:3, 1:3.5, 1:4, 1:5, and 1:6. Hydrated uranium(IV) oxysulfate $UOSO_4$ is formed by the photochemical reduction of uranyl sulfate in aqueous solution with formic acid and ethanol present. The anhydrous oxysulfate is obtained by heating the hydrate at 150–160° C under vacuum for several hours.

c) *Oxalates*

Uranium(IV) oxalate may be prepared by the addition of oxalic acid to a solution of uranium(IV) ions or to a suspension of uranium hydroxide. The uranium(IV) solution is prepared by dissolving a salt, such as UCl_4, or reduction of a uranyl salt in solution. Reduction may be effected photochemically, electrolytically, or chemically with copper, zinc, or hydrosulfite. The salt $U(C_2O_4)_2 \cdot 6H_2O$ is only slightly soluble in water and dilute acids. In water at 25° C, its solubility is 0.05 g per 1000 g of solution. In 1 *N* solutions of acetic, perchloric, hydrochloric, nitric, and sulfuric acids the solubility is approximately 0.04, 0.07, 0.08, 0.1, and 0.18 g per 1000 g of solution, respectively (ELSTON and CAILLAT, 1967). Thermal decomposition under vacuum of $U(C_2O_4)_2 \cdot 6H_2O$ permits isolation of the di- and mono-hydrates as well as the anhydrous salt. Numerous complex compounds containing U(IV) and oxalate ion have been prepared. The ratio of U(IV) to oxalate in these compounds varies 1:2, 1:2.5, 1:3, 1:3.5, and 1:4.

2. Uranium(VI) Salts

a) *Carbonates*

Anhydrous UO_2CO_3 may be prepared by the reaction between a saturated alcoholic solution of uranyl nitrate and CO_2 under pressure at 120° C. By using a 30% aqueous uranyl nitrate solution in a similar reaction the monohydrate $UO_2CO_3 \cdot H_2O$ is obtained (CHERNYAEV, 1964). ČEJKA (1959) has prepared the carbonate by first reducing uranyl nitrate to $UO_{2.84} \cdot xH_2O$ photochemically in an alcoholic solution followed by passage of an air and CO_2 stream through the suspended uranium oxide.

UO_2CO_3 is thermally stable to ~500° C where it decomposes to UO_3 and CO_2. The UO_3 formed is converted to U_3O_8. UO_2CO_3 is only slightly soluble in water. At 30° C its solubility is 0.028 g per 1000 g of solution (CAILLAT and ELSTON, 1961). The carbonate is decomposed by acids and is soluble in solutions of alkali metal and ammonium carbonate.

A large number of complex compounds have been prepared in which the $UO_2^{2+}:CO_3^{2-}$ ratio is 1:1, 1:1.5, 1:2, 1:2.5, or 1:3. Of particular importance in uranium technology are the 1:3 compounds formed in reactions of the following type (GALKIN and SUDARIKOV, 1964):

$$UO_2(OH)_2 + 3M_2CO_3 = M_4[UO_2(CO_3)_3] + 2MOH;$$

$$M_2U_2O_7 + 6M_2CO_3 + 3H_2O = 2M_4[UO_2(CO_3)_3] + 6MOH;$$

$$UO_2SO_4 + 3M_2CO_3 = M_4[UO_2(CO_3)_3] + M_2SO_4,$$

where M is an ammonium or alkali metal ion. Ammonium tricarbonatouranylate $(NH_4)_4[UO_2(CO_3)_3]$ is thermally unstable. Decomposition begins at 80° C and is complete at 150° C. Dry crystals of the salt are stable in air, but moist crystals decompose on standing. The compound hydrolyzes rapidly in water which does not contain an excess of carbonate or bicarbonate ions. The pH of a 0.01 *M* solution is 8.0. The salt is decomposed by acids. Strong alkalis also decompose the salt to form hydroxy compounds (CHERNYAEV, 1964).

Sodium tricarbonatouranylate is a thermally stable compound to 300° C. Above this temperature decompositions occurs and is complete at 350° C. This salt also hydrolyzes rapidly in an aqueous solution not containing an excess of carbonate or bicarbonate. The pH of a 0.01 *M* solution is 9.2. The salt is stable in dry air. It is decomposed by acids and strong alkalis (CHERNYAEV, 1964).

b) *Nitrates*

Evidence for uranyl nitrate hydrates containing 0, 1, 2, 3, or 6 water molecules has been obtained (CORDFUNKE, 1969). However, the most important compound from the technological point of view is the hexahydrate $UO_2(NO_3)_2 \cdot 6H_2O$. It is prepared by dissolving UO_3 in nitric acid. The hexahydrate separates from such a solution at room temperature. The trihydrate forms above 60° C and the dihydrate above 120° C. The monohydrate has been recognized by X-ray and infrared measurements after heating hydrated $UO_2(NO_3)_2$ under a water pressure of 0.1 to 10 mm at temperatures between 100 and 165° C (CHOTTARD, 1968). Anhydrous uranyl nitrate is prepared by carefully heating the product $UO_2(NO_3)_2 \cdot 2NO_2$ under vacuum at 163 to 165° C.

Anhydrous $UO_2(NO_3)_2$ dissolves readily in water. It reacts explosively with aniline at room temperature. It also reacts vigorously with ether releasing nitrogen oxides. The anhydrous nitrate is soluble in alcohols, ketones, esters, acetonitrile, nitromethane, and to a limited degree in nitrobenzene.

The hydrates of uranyl nitrate are readily soluble in water. The solubility of the salt decreases with increasing nitric acid or nitrate concentration. Uranium is precipitated from an aqueous uranyl nitrate solution as a uranate by the addition of ammonium, alkaline metal, or alkaline earth hydroxide or as the hydrated peroxide by the addition of hydrogen peroxide. Uranyl nitrate hydrates are also soluble in a number of organic solvents, such as ethers, alcohols, aldehydes and ketones.

Uranyl nitrate is insoluble in non-polar organic solvents, such as carbon tetrachloride, benzene, and toluene.

A number of complex nitratouranylate compounds have been prepared. Their composition may be represented by the general formulas

$$M[UO_2(NO_3)_3] \text{ or } M_2[UO_2(NO_3)_4].$$

c) *Phosphates*

Uranyl phosphates may be prepared by the addition of phosphoric acid to a solution of uranyl nitrate. The nature of the resulting precipitate depends on the molecular ratio of P_2O_5 to UO_2 (DOMINE-BERGES, 1953). For a uranyl nitrate solution 0.2 *M* in UO_2^{2+} and 0.02 *M* in acid, no precipitate is formed with a $P_2O_5:UO_2$ ratio $<1/3$; with the ratio >2, the acid phosphate UO_2HPO_4 precipitates; with the ratio >2 but <3, a mixture of the acid phosphate and the normal phosphate $(UO_2)_3(PO_4)_2$ is obtained; and with a ratio >3, the normal phosphate is formed. The normal phosphate contains four water molecules at ordinary temperatures. The dihydrate is obtained by heating the tetrahydrate in the presence of P_2O_5. The monohydrate is obtained under vacuum at 100° C and the anhydrous salt at 140 to 250° C. SCHREYER and BAES (1954) have studied the solubility of uranium (VI) orthophosphates in 0.001 to 15 *M* phosphoric acid. Three compounds were identified: $(UO_2)_3(PO_4)_2 \cdot 6H_2O$ at <0.014 *M* total phosphate, $UO_2HPO_4 \cdot 4H_2O$ between 0.014 and 6.1 *M* total phosphate, and $UO_2(H_2PO_4)_2 \cdot 3H_2O$ at >6.1 *M* total phosphate. The maximum solubility was observed at ~ 6 *M* total phosphate where the total uranium concentration in solution was about 2 *M*. Both $(UO_2)_3(PO_4)_2 \cdot 4H_2O$ and $UO_2HPO_4 \cdot 4H_2O$ are insoluble in water.

A number of complex compounds having the $UO_2^{2+}:PO_4^{3-}$ ratio 1:1, 2:2, and 5:2 have been prepared.

d) Sulfates

The salt obtained on crystallization from a saturated uranyl sulfate solution has, until recently, been designated as a trihydrate, $UO_2SO_4 \cdot 3H_2O$. However, CORDFUNKE (1969b) has shown that all such samples below 125° C have the same x-ray pattern and that this corresponds to $UO_2SO_4 \cdot 2.5H_2O$. At 75° C the hydrate begins to loose its water and only above 250° C is the anhydrous salt formed. X-ray analysis gives no indication of an intermediary monohydrate. $UO_2SO_4 \cdot 2.5H_2O$ was also prepared by placing anhydrous UO_2SO_4 over a saturated solution of sodium chloride. The monohydrate $UO_2SO_4 \cdot H_2O$ may be prepared by heating a mixture of the 2.5 hydrate and anhydrous UO_2SO_4 in the ratio $H_2O/UO_2SO_4 = 1.0$ in a sealed quartz tube at 200° C for 36 hours. At room temperature in a dry atmosphere the monohydrate gradually changes into the original mixture of 2.5 hydrate and anhydrous salt. CORDFUNKE (1969b) has determined the solubility of the 2.5 hydrate in water to be 60.83 weight percent at 26° C. At this temperature there are two stable phases, $UO_2SO_4 \cdot 2.5H_2O$ and the acid salt $UO_2SO_4 \cdot H_2SO_4 \cdot 2.5H_2O$. A metastable phase with the composition $UO_2SO_4 \cdot 2H_2O$ has also been found.

Sulfatouranylate compounds with UO_2^{2+}: SO_4^{2-} ratios of 1:0.5, 1:1, 1:1.5, 1:2, 1:2.5, and 1:3 have been reported.

e) Uranyl Halides

Uranyl fluoride can be prepared by either dry or wet processes (CAILLAT and ELSTON, 1961). Dry processes are represented by the following reactions:

$$U_3O_8 + 8HF \longrightarrow UF_4 + 2UO_2F_2 + H_2O;$$

$$UO_3 + 2HF \xrightarrow{300\text{–}550°\,C} UO_2F_2 + H_2O;$$

$$UO_2HPO_4 \cdot xH_2O + 2HF \xrightarrow{350\text{–}500°\,C} UO_2F_2 + H_3PO_4 + xH_2O;$$

$$2UF_4 + O_2 \xrightarrow{800°\,C} UF_6 + UO_2F_2;$$

$$UO_3 + 2F_2 \xrightarrow{300°\,C} UO_2F_2 + OF_2(?);$$

$$U_3O_8 + 5F_2 \xrightarrow{300°\,C} 3UO_2F_2 + 2OF_2(?);$$

$$UO_2 + F_2 \xrightarrow{300°\,C} UO_2F_2.$$

Uranyl fluoride can be prepared in aqueous solution by the action of hydrofluoric acid on uranium trioxide or uranium peroxide. Careful adjustment of the amount of HF is necessary to give the proper end-product. Uranyl fluoride is also prepared by treating a uranyl salt, such as the acetate, with hydrofluoric acid. The solution is evaporated repeatedly on a hot water bath.

Anhydrous uranyl fluoride is very soluble in water. It is also soluble in ethyl alcohol and acetone but not in ether and isoamyl alcohol. Its solubility in hydrofluoric acid decreases with an increase in acid concentration.

Uranyl fluoride decomposes to uranium oxides and uranium hexafluoride when heated in air to temperatures of 700° C or more. UO_2F_2 is reduced when heated to 700 to 850° C with hydrogen or to 500 to 600° C with sulfur.

In a study of the HF—UO_3—H_2O system at 20° C, BUSLAEV et al. (1963) observed the formation of the compound $UO_2(OH)F \cdot 0.5H_2O$ which probably exists as the dimer $U_2O_5F_2 \cdot 2H_2O$. The subsequent solid phase was the dihydrate $UO_2F_2 \cdot 2H_2O$ which was found to exist up to a concentration of 22.8% HF. The compound $UO_2F_2 \cdot 2HF \cdot 4H_2O$ was formed when the HF concentration was increased to 24.11%. The region of crystallization of this latter compound was traced to an HF concentration of 91.40%. In addition to these compounds, complex compounds have been identified in which the $UO_2^{2+}:F^-$ ratio is 1:1, 1:1.5, 1:2, 1:2.5, 1:3, 1:3.5, 1:4, 1:4.5, 1:5, and 1:6.

Uranyl chloride (CAILLAT and ELSTON, 1961) can be prepared by the action of HCl on UO_3. The reaction with anhydrous UO_3 is very slow. If the hydrated oxide is used, $UO_2Cl_2 \cdot H_2O$ is obtained. Anhydrous UO_2Cl_2 can be obtained by heating the monohydrate in a stream of HCl gas at 3(0° C. Another method is to oxidize UCl_4 in a stream of oxygen at 300 to 350° C If a chlorine pressure of about 0.5 atm is maintained, oxidation to U_3O_8 does not occur.

Uranyl chloride can be crystallized from solutions in which (1) UCl_4 has been oxidized by nitric acid, (2) $UO_3 \cdot H_2O$ has been dissolved in hydrochloric acid, or (3) uranyl sulfate has been treated with barium chloride. The trihydrate $UO_2Cl_2 \cdot 3H_2O$ is crystallized by slow evaporation. The monohydrate is prepared by evaporation at 120° C or by treating the evaporated residue with thionyl chloride.

Anhydrous UO_2Cl_2 is very hygroscopic. In moist air it successively forms the monohydrate, the trihydrate and finally, a solution. Its solubility in water at 18° C is 320 g per 100 ml. The solubility of the trihydrate is 746 g per 100 g of water. Anhydrous UO_2Cl_2 is also soluble in acetone, alcohols, acetophenone, pyridine and dioxan. It is not soluble in CCl_4, benzene, or xylene.

Acid $UO_2Cl_2 \cdot HCl \cdot 2H_2O$ and basic $UO_2(OH)Cl \cdot 2H_2O$ salts are formed as well as a rather larger number of complex compounds with ammonia, alkali metals, and organic compounds (CAILLAT and ELSTON, 1961; ELSTON and CAILLAT, 1967).

Uranyl bromide is prepared by passing bromine vapor over a heated mixture of UO_2 and charcoal. The UO_2Br_2 formed is extracted with ether. Evaporation of this solution yields the dissolvate $UO_2Br_2 \cdot 2(C_2H_5)_2O$. The anhydrous bromide is obtained under vacuum dessication. An alternative process for preparing anhydrous UO_2Br_2 is the oxidation of UBr_4 with oxygen at 150 to 160° C. Hydrated UO_2Br_2 is obtained by evaporation of a uranyl bromide solution prepared by heating UO_2 or a uranyl salt such as the acetate with hydrobromic acid. PETERSON (1961) has shown that the trihydrate is formed by this method rather than the heptahydrate as reported earlier (see KATZ and RABINOWITZ, 1951).

The uranyl bromides are deliquescent and very soluble in water. They are also soluble in alcohol and ether. UO_2Br_2 is thermally unstable. At room temperature the salt slowly loses bromine. At 350° C it decomposes completely within two days.

The basic salt $UO_2(OH)Br \cdot 2H_2O$ is obtained from acid-deficient solutions of uranyl bromide (PETERSON, 1961). Complex compounds with organic molecules and ammonium or alkali metal bromides are formed (CHERNYAEV, 1964).

Uranyl iodide solutions can be prepared by a double exchange reaction. The reactants are chosen such that one of the products is uranyl iodide and the other is an insoluble salt in the solvent used (KATZ and RABINOWITZ, 1951). An example of this is the addition of an equivalent amount of barium iodide to a uranyl sulfate solution. It is doubtful whether uranyl iodide has been obtained in the solid state since iodine tends to separate out during concentration of this aqueous solution.

Uranyl iodide solutions are also prepared by the double exchange reaction in organic solvents such as alcohols, acetone, diethyl ether, pyridine, and methyl and ethyl acetates. Solvated uranyl iodide compounds may be prepared from such solutions.

f) Acetates

Uranyl acetate separates as the dihydrate, $UO_2(CH_3COO)_2 \cdot 2H_2O$, from a solution of UO_3 dissolved in acetic acid. The anhydrous salt is obtained by heating the dihydrate at 115° C or by the action of acetic anhydride on uranyl nitrate.

Uranyl acetate is soluble in water (7.7 g/100 g H_2O at 15° C) and is slightly soluble in methylamine and ethylamine. Basic salts and acid salts of uranyl acetate have been identified in addition to a number of complex acetatouranylate compounds in which the $UO_2^{2+}:CH_3COO^-$ ratio is 1:1, 1:2, 1:2.5, and 1:3. Sodium, potassium, and guanidine acetates form complexes $M[UO_2(CH_3COO)_3]$ of low solubility with the uranyl ion.

g) Oxalates

The trihydrate $UO_2C_2O_4 \cdot 3H_2O$ is obtained when oxalic acid is added to a uranyl salt solution. Lower hydrates are formed by heating the trihydrate. The anhydrous salt is formed at 250° C. Decomposition occurs above 300° C.

Uranyl oxalate is only slightly soluble in water (0.45 g per 100 g H_2O at 20° C). Its solubility increases with temperature and in the presence of mineral acids, oxalic acid, and alkali oxalates. It is slightly soluble in ethyl and methyl alcohol.

A large number of complex oxalatouranylate compound have been identified as well as mixed oxalate-sulfate, oxalate-carbonate, oxalate-fluoride, oxalate-thiocyanate, and oxalate-acetate compounds of uranium.

3. Uranium Oxide Hydrates

The hydrate obtained when UO_3 reacts with water is the dihydrate $UO_3 \cdot 2H_2O$. The monohydrate $UO_3 \cdot H_2O$ can be prepared by heating the dihydrate in air to temperatures above 80°. This results in the α-modification of $UO_3 \cdot H_2O$. Two others modifications, the β- and ε-forms, are known. A third hydrate $UO_3 \cdot 0.5H_2O$ has been reported but its existence is dubious (CORDFUNKE, 1969). The hydrates have been referred to in several contexts: as hydrated oxides, as uranic acids, or as uranium hydroxides. The equivalent formulas denoting the various functions are:

Hydrates	Acids	Hydroxides
$UO_3 \cdot H_2O$	H_2UO_4	$UO_2(OH)_2$
$UO_3 \cdot 2H_2O$	H_4UO_5	$UO_2(OH)_2 \cdot H_2O$
$UO_3 \cdot 0.5H_2O$	$H_2U_2O_7$	$U_2O_5(OH)_2$

Infrared absorption and nuclear magnetic resonance spectra of the hydrates indicate the existence of the compounds as the hydroxides (DEANE, 1962 and SCHWARZMAN and GLEMSER, 1962).

CORDFUNKE et al. (1968) have prepared two non-stoichiometric hydrates $UO_{2.86} \cdot 1.5H_2O$ and $UO_{2.86} \cdot 0.5H_2O$ by irradiating a solution containing uranyl acetate, ethanol, and water with two 150 W electric light bulbs. Analyses showed that both U(IV) and U(VI) were present in an amorphous form.

Uranium dioxide dihydrate is reportedly prepared by precipitating tetravalent uranium hydroxide from solution (see KATZ and RABINOWITZ, 1951). This compound is very reactive. It oxidizes rapidly in air and dissolves readily in acids.

Uranium peroxide $UO_4 \cdot 4H_2O$ is formed by the addition of hydrogen peroxide to a uranyl salt solution at room temperature. The tetrahydrate is converted to the dihydrate in air at temperatures between 80 and 100° C. The dihydrate decomposes around 200° C into UO_3, oxygen, and water. Uranium peroxide is decomposed also by concentrated acids forming solutions of uranyl salts. Precipitation of the peroxide is hindered in the presence of alkaline or alkaline earth metals by formation of the corresponding peruranates.

4. Uranates and Peruranates

The formation of *uranates* by the reaction of uranium oxides with strong bases or with metal oxides at high temperatures was mentioned in Sect. IIB5a. Uranates or rather polyuranates are obtained also by the addition of a base to a uranyl salt solution. These polyuranates were long assumed to be diuranates. It is now established that the composition of the precipitate depends on the conditions of precipitation. These polyuranates are insoluble in water and alkaline solutions, but are soluble in acids and alkaline or ammonium carbonate solutions.

The term "uranate" is really a misnomer since the compounds were considered salts of uranic or polyuranic acid. In reality the "uranates", prepared by alkaline precipitation are complex aquohydroxy compounds where the ligands are water or other neutral substituents and hydroxy groups (Chernyaev, 1964).

Peruranates or peroxyuranates are formed by the action of a peroxide, notably H_2O_2, on a uranate or polyuranate. The peruranates are soluble in water. Crystals are obtained by evaporation of the solution or by precipitation with the addition of alcohol.

The peruranates have been regarded as double peroxides of uranium and an alkali metal, as salts of peracids, and as derivatives of the peruranic acid ion UO_8^{4-}. However, the peroxide group O_2^{2-} should be considered as a typical ligand, capable of forming complexes with the uranyl ion in a manner analogous to the carbonate and similar ions (Chernyaev, 1964).

D. Uranium in Solution

1. Aqueous Solution

a) Oxidation States

Uranium is known to exist in oxidation states of +3, +4, +5, and +6. These states represent the loss of electrons from the $5f^3 6d^1 7s^2$ shells of the ground-state atom and correspond, respectively, to electron configurations of $5f^3$, $5f^2$, $5f^1$, and $5f^0$ (Cordfunke, 1969). In acid solution, uranium ions in these four oxidation states are represented as U^{3+}, U^{4+}, UO_2^+, and UO_2^{2+}. The potentials which relate the various oxidation states are given in the following diagrams (Latimer, 1952).

Acid solution:

$$U \xrightarrow{+1.80\text{ eV}} U^{3+} \xrightarrow{+0.61\text{ eV}} U^{4+} \xrightarrow{-0.62\text{ eV}} UO_2^+ \xrightarrow{-0.05\text{ eV}} UO_2^{2+}$$

$$U^{4+} \xrightarrow{-0.334\text{ eV}} UO_2^{2+}$$

Basic solution:

$$U \xrightarrow{+2.17\text{ eV}} U(OH)_3 \xrightarrow{+2.14\text{ eV}} U(OH)_4 \xrightarrow{+0.62\text{ eV}} UO_2(OH)_2$$

Thermodynamic data for the formation of the aqueous ions and for oxidation-reduction reactions of the ions in solution are given in Table 2.18.

Evidence for the existence of U^{3+} ions in solution comes from the reversibility of the U(III)/U(IV) couple. The electron configuration of the U^{3+} ion has been

Table 2.18. Thermodynamic data for uranium ions in aqueous solution[a]

Ion formation

Valence state	Ion in solution	Heat of formation H_{298} (kcal/mole)	Free energy of formation G_{298} (kcal/mole)	Standard entropy S_{298} (cal/deg mole)
III	U^{3+}	-112.6 ± 3.0		-35 ± 6
IV	U^{4+}	-146.3 ± 3.0		-81 ± 7
V	UO_2^+		-237.3[b]	
VI	UO_2^{2+}	-250.0 ± 2.0		-20 ± 5

Change in oxidation state

Valence change	Reaction	ΔG_{298} (kcal/mole)	ΔH_{298} (kcal/mole)	ΔS_{298} (cal/deg mole)
III–IV	$[U^{3+} \cdot aq] + [H^+ \cdot aq] \rightarrow [U^{4+} \cdot aq] + {}^1/_2(H_2)$	-14.6	-23.7	-30.5
IV–VI	$[U^{4+} \cdot aq] + 2\{H_2O\} \rightarrow [UO_2^{2+} \cdot aq] + 2[H^+ \cdot aq] + (H_2)$	$+15.4$	$+32.9$	$+59$
V–VI	$[UO_2^+ \cdot aq] + [H^+ \cdot aq] \rightarrow [UO_2^{2+} \cdot aq] + {}^1/_2(H_2)$	$+1.5$		
IV–V	$[U^{4+} \cdot aq] + 2\{H_2O\} \rightarrow [UO_2^+ \cdot aq] + 3[H^+ \cdot aq] + {}^1/_2(H_2)$	$+13.9$[c]		

[a] Compiled from RAND and KUBASCHEWSKI (1963). Meanings of the various brackets are: {} liquid state, () gaseous state, [] dissolved (suffix denoting medium).
[b] Calculated using the reaction for the IV–V valence change.
[c] Calculated from the reactions given for the valence changes IV–VI and V–VI.

reported as $5f^3$ (JØRGENSEN, 1956) and $5f^2 6d^1$ (JEZOWSKA-TRZEBIATOWSKA, 1963). SOMEYA (1927) reports the color of U^{3+} solutions to be deep red in concentrated hydrochloric acid and greyish green in less acidic solutions. Solutions of tervalent uranium can be prepared by dissolution of a uranium trihalide (UCl_3, UBr_3 or UI_3) or by reduction of a U(IV) or U(VI) solution. Reduction can be effected electrolytically at a mercury cathode (PERETRUKHIN et al., 1966) or chemically with zinc amalgam (SATO, 1967) or stannous chloride (PETIT, 1969). Uranium(III) is a strong reducing agent. It reacts with water to form hydrogen and U^{4+} ions:

$$2U^{3+} + 2H_2O \rightarrow 2U^{4+} + H_2 + 2OH^-.$$

Solutions of U^{3+} in 0.5 N acid are reasonably stable in an inert atmosphere. The stability of the solution depends on both the acid concentration and the nature of the acid. At acid concentrations $> 0.5\ N$, the stability decreases. Solutions of hydrochloric acid are more stable than those of perchloric acid or sulfuric acid. In the presence of air, U^{3+} ions are not very stable in either perchloric acid or sulfuric acid at any concentration. However, in hydrochloric acid the stability tends to increase with acid concentration (SATO, 1967).

Confirmation for the existence of U^{4+} ions in solution has been obtained by measuring the amount of acid liberated in the dissolution of UCl_4 (KRAUS and NELSON, 1950) and by solvent extraction studies of U(IV) (BETTS and LEIGH, 1950; RYDBERG and RYDBERG, 1956). The $5f^2$ electron configuration of the ion appears to be well-established (JØRGENSON, 1955; JEZOWSKA-TRZEBIATOWSKA, 1963). Solutions of tetravalent uranium are green in color. They can be prepared by dissolution of a tetravalent uranium salt such as UCl_4 or $U(SO_4)_2$, by the dissolution of uranium metal in sulfuric or phosphoric acid, or by reduction of a uranyl solution. Reduction can be effected either photochemically, electrochemically, or chemically by lead or silver amalgam in hydrochloric acid (PETIT, 1969), sodium hydrosulfite, or stannous chloride (CAILLAT and ELSTON, 1960). The solutions are stable in an inert atmosphere but oxidize slowly in the presence of air according to the reaction

$$2U^{4+} + 2H_2O + O_2 \rightarrow 2UO_2^{2+} + 4H^+.$$

The presence of cations such as Hg^{2+} and Cu^{2+} catalyze the reaction; whereas Ag^+, Fe^{2+}, and Cl^- act as inhibitors (HALPERN and SMITH, 1956). Uranium(IV) is readily oxidized in the presence of oxidizing agents (ELSTON and CAILLAT, 1967).

The presence of pentavalent uranium in solution as the UO_2^+ species is strongly supported by the reversible nature of the U(V)/U(VI) couple (HERASYMENKO, 1928; HARRIS and KOLTHOFF, 1945; HEAL 1949) and the existence of U(VI) in solution as the UO_2^{2+} ion. Further support of the UO_2^+ species comes from the analogy with Np(V) and Am(V) ions which have been given formulas NpO_2^+ and AmO_2^+ from both infrared (JONES and PENNEMAN, 1953) and crystallographic (NIGON et al., 1954; ZACHARIASEN, 1954) studies. Solutions of UO_2^+ can be prepared by dissolving UCl_5 (KRAUS and NELSON, 1949) or by reducing UO_2^{2+} either electrolytically or with U(IV) ions, hydrogen or zinc amalgam (KRAUS et al., 1949). The UO_2^+ ion is most stable in the pH range of 2 to 4 where disproportionation into U^{4+} and UO_2^{2+} is slow. The equilibrium constant for the disproportionation reaction

$$2UO_2^+ + 4H_3O^+ \rightarrow UO_2^{2+} + U^{4+} + 6H_2O$$

has been determined to be $(1.7 \pm 0.3) \times 10^6$ (NELSON and KRAUS, 1951). The rate of disproportionation increases with increasing hydrogen ion concentration (provided no organic acid is present), increasing chloride concentration, increasing pH of organic acid present, increasing organic acid concentration (the rate being second order in U(V) and first order in organic anion), and decreasing pK of the organic acid. The rate decreases slightly in the presence of hydrazine (SELBIN and ORTEGO, 1969). These results indicate that anions but not undissociated organic acid molecules participate in the disproportionation reaction by UO_2^+ complexing of the anions. The UO_2^+ ion is oxidized to UO_2^{2+} by oxygen, ferric ion and ceric ion (KRAUS et al., 1949).

The hexavalent state of uranium is present in solution as the UO_2^{2+} species. This has been established by a number of physical-chemical measurements (see HOEKSTRA and KATZ, 1954, or KATZ and SEABORG, 1957). The UO_2^{2+} ion is the most stable uranium species found in solution. Uranyl solutions are readily prepared by dissolving in water such uranyl salts as the halide, nitrate, sulfate or acetate. Uranyl solution can be prepared also by dissolving a U(VI) compound such as UO_3 or UCl_6 in an appropriate solvent, by dissolving a lesser-valent uranium compound in an oxidizing medium, or by oxidizing a lesser-valent uranium species already in solution. Uranyl solutions are yellow in color. As indicated in the preceding paragraphs, U(VI) can be reduced in solution by chemical, electrochemical, or photochemical methods.

b) Hydrolysis

Hydrolysis is but a special case of complex ion formation in which the hydroxyl ion is the ligand L and water is the protonated ligand HL. In the more general case of complex ion formation, m central atoms or ions M react with n ligands L to form the complex $M_m L_n$. Various equilibrium constants may be measured by a variety of methods. These equilibria are described in Table 2.19 following the notation of SILLÉN and MARTELL (1964).

Table 2.19. Notation used in Tables 2.20–2.25

The notation used in Tables 2.20–2.25 on hydrolysis and complex ion formation is patterned after that used by SILLÉN and MARTELL (1964). This reference should be consulted for further details concerning notation.

"Method" refers to the method by which the constants were measured. The abbreviations used are:

aix	anion exchange	pol	polarography
cfu	centrifuge or ultracentrifuge	pre	preparative work
cix	cation exchange	qh	quinhydrone electrode
col	colorimetry	red	emf with redox electrode
con	conductivity	sol	solubility
dis	distribution between two phases	sp	spectrophotometry
E	emf, not specified	tp	electrical migration or transference number
fp	freezing point		
gl	glass electrode	X	X-ray diffraction
iE	current-voltage studies	ΔG	combination of thermodynamic data
ix	ion exchange	Pt, Ag	emf measurement with electrode of metal stated
mag	magnetic susceptibility		
nmr	nuclear magnetic resonance	$Ag_2C_2O_4$	emf measurement with $Ag_2C_2O_4$ electrode
pH_2O, pL	partial pressure of substance indicated	?	method not known to compilers
pH	pH method, not specified		

"Temp." gives the temperature in °C, "RT" indicates room temperature, and "?" may be used if the temperature is unknown to the compilers.

"Medium" denotes the nature of the medium to which the equilibrium constants refer. The concentrations given in terms of moles per liter or moles per kilogram are not distinguished. Water is the solvent unless otherwise stated. Symbols used are:

$\rightarrow 0$	constants extrapolated to zero ionic strength
0 corr	constants corrected to zero ionic strength by application of some theoretical or empirical formula
0.1	an ionic strength of 0.1 mole per liter
1 $NaClO_4$	constant concentration of the substance stated (1 mole per liter $NaClO_4$)
1 ($NaClO_4$)	ionic strength held constant at value stated (1 mole per liter) by addition of the inert salt indicated.
I($NaClO_4$)	measurements made at series of ionic strengths (I) with $NaClO_4$ as the inert salt
1 (Na)ClO_4	concentration of the anion (ClO_4^-) held constant at the value stated (1 mole per liter) with the ion shown in parenthesis as the inert cation
C(ClO_4^-)	measurement made at a series of perchlorate concentrations
dil	dilute solution, concentration usually not more than 0.01 mole per liter
var	ionic medium varied, and in some cases no special attempt was made to control the ionic strength
KCl var	the medium was mainly aqueous KCl at various concentrations
EtOH	ethanol as solvent
org	various organic solvents
50% MeOH	50% methanol-water as solvent

"Log of equilibrium constant, remarks." The following conventions are used: "$K_1 6$" means "$\log_{10} K_1 = 6$"; "$K_2 > 3$" means "$\log_{10} K_2 > 3$"; i.e., the "log" and "=" symbols are omitted. Concentrations [ ] are usually expressed as moles per liter, but is not distinguished from moles per kilogram. Pressures p are in atmospheres.

The equilibrium constants refer to the follwing types of reactions:

1. Consecutive or step-wise constants: K

   a) Addition of ligand (L)

$$ML_{n-1} + L \rightleftharpoons ML_n \qquad K_n = \frac{[ML_n]}{[ML_{n-1}][L]}$$

b) Addition of protonated ligand (HL) with elimination of proton

$$ML_{n-1}+HL \rightleftharpoons ML_n+H^+ \qquad {}^*K_n=\frac{[ML_n][H^+]}{[ML_{n-1}][HL]}$$

c) Addition of protonated ligand (H_pL)

$$M(H_pL)_{n-1}+H_pL \rightleftharpoons M(H_pL)_n \qquad H_pL:K_n=\frac{[M(H_pL)_n]}{[M(H_pL)_{n-1}][H_pL]}$$

d) Addition of central atom (M)

$$M_{n-1}L+M \rightleftharpoons M_nL \qquad K_{1n}=\frac{[M_nL]}{[M_{n-1}L][M]}$$

2. Cumulative or gross constants: β

In β_{nm} and $^*\beta_{nm}$ the subscripts n and m denote the composition of the complex M_mL_n formed. When $m=1$, the second subscript is omitted.

a) Addition of central atoms (M) and ligands (L)

$$mM+nL \rightleftharpoons M_mL_n \qquad \beta_{nm}=\frac{[M_mL_n]}{[M]^m[L]^n}$$

b) Addition of central atoms (M) and protonated ligands (HL) with elimination of protons

$$mM+nHL \rightleftharpoons M_mL_n+nH^+ \qquad {}^*\beta_{nm}=\frac{[M_mL_n][H^+]^n}{[M]^m[HL]^n}$$

3. Solubility constants: K_s

a) Solid M_aL_b in equilibrium with free ions in solution

$$M_aL_b(s) \rightleftharpoons aM+bL \qquad K_{s0}=[M]^a[L]^b$$

b) Solid M_aL_b in equilibrium with complex M_mL_n and ligand L in solution

$$\frac{m}{a}M_aL_b(s) \rightleftharpoons M_mL_n+\left(\frac{mb}{a}-n\right)L \qquad K_{s_{nm}}=[M_mL_n][L]^{\left(\frac{mb}{a}-n\right)}.$$

In $K_{s_{nm}}$ the subscripts n and m denote the composition of the complex M_mL_n formed in solution. When $m=1$, the second subscript is omitted.

c) Protonated ligand reacts with the elimination of proton

$$\frac{m}{a}M_aL_b(s)+\left(\frac{mb}{a}-n\right)H^+ \rightleftharpoons M_mL_n+\left(\frac{mb}{a}-n\right)HL$$

$${}^*K_{s_{mn}}=\frac{[M_mL_n][HL]^{\left(\frac{mb}{a}-n\right)}}{[H^+]^{\left(\frac{mb}{a}-n\right)}}$$

4. Acidic and basic constants:

a) When L is hydroxide (OH^-), HL is water and $^*K_n$ is the nth acid dissociation constant for the hydrolysis of a metallic ion.

b) The use of H^+ as the central atom to represent protolytic constants is illustrated by 1.d) above.

c) Other acidic constants are denoted by K_a, followed by parentheses enclosing the formula of the species donating the proton.

d) Basic constants are denoted by K_b followed, if necessary, by parentheses enclosing the formula of the species accepting the proton.

5. Special constants:

a) K (equation)

The equation defines the reaction to which K refers.

b) †K, †$K_{s_{nm}}$

The corresponding reaction is given in parentheses after the constant when the latter is first used for a particular ligand or central atom, or the reaction is given immediately below the equilibrium constants for a particular ligand or central atom.

c) β (formula)

The formula gives the composition of the complex in terms of the species from which it is formed. Species with negative subscripts are eliminated in the formation of the complex.

d) K_s (formula)

The formula gives the composition of the solid phase in terms of the species with which it is in equilibrium in solution. Species with negative subscripts are eliminated in the formation of the solid.

Enthalpy and entropy changes are included in the "remarks" column. ΔH is usually written in kilocalories and ΔS in calories per degree. They are related to the corresponding cumulative equilibrium constants as follows: β_n, $\Delta H_{\beta n}$; β_{nm}, $\Delta H_{\beta nm}$; $^*\beta_{nm}$, $\Delta^* H_{\beta nm}$. Where the symbol K is used for the equilibrium constant, H or S is given the same superscript or subscript as the corresponding K: e.g., K_n, ΔH_n; K_{s0}, ΔH_{s0}; $^+K_{nm}$, $\Delta^+ H_{nm}$; etc.

Other abbreviations used in the "remarks" column are:

alt	alternative	unch	uncharged
ev M_mL_n	evidence for the existence of the complex M_mL_n	polyn	polynuclear
cpx	complex	?	authors' doubt expressed in reference given
cat	cationic	(?)	compilers' doubt
ani	anionic		

"References" without asterisks list references found in the compilation by SILLÉN and MARTELL (1964). References with single asterisks are given at the end of this chapter. References with double asterisks list those found in the review by GINDLER (1962). A reference d+c indicates conclusions of d from data in d and c; d+, conclusions of d from data in d and in several or unspecified other references; c/o, conclusions of o from data in c; f=h, conclusions and data of f are identical with those of h.

In hydrolysis, a relatively small ion with a large electric charge is transformed into a larger ion which decreases the density of electric charge. Hydrolysis is not simply hydration but consists of an interaction sufficiently great to remove a proton from the water undergoing reaction with formation of a metal-oxygen bond (KATZ and SEABORG, 1957). In uranium ions, the order of increasing hydrolysis is $U^{4+} > UO_2^{2+} > U^{3+}$.

It is difficult to investigate the hydrolysis of U^{3+} in solution because of the instability of the ion. Evidence based on pH measurements of such solutions indicates that the hydrolytic behavior of U^{3+} is similar to that of rare earth 3+ ions.

Little is known also of the hydrolytic behavior of UO_2^+ because of the limited pH range (2 to 4) in which the ion is relatively stable. KRAUS et al. (1949) have estimated the pK for the hydrolytic reaction

$$UO_2^+ + 2H_2O \rightarrow UO_2OH + H_3O^+$$

to be about 8 and, that at pH somewhat above 8, negative $UO_2(OH)_2^-$ species would be formed.

Uranium (IV) undergoes extensive hydrolysis with evidence in the early stages for the formation of the monomeric species according to the reaction

$$U^{4+} + H_2O \rightleftharpoons UOH^{3+} + H^+.$$

The equilibrium for this reaction has been determined by a number of investigators. Their results $^*K_1$ are summarized in Table 2.20, together with thermodynamic data for the reaction. The mechanism of hydrolysis is more complex than the simple reaction given above. KRAUS (1956) has recognized that at least two other types of hydrolysis products may exist: low-molecular-weight polymers in equilibrium with each other and U^{4+} and high-molecular-weight polymers not in equilibrium with the monomer. HIETANEN (1956) has assumed that in addition

Table 2.20. Hydrolysis of U(IV)[a]

The central ion M in this table is U^{4+}. The ligand L is OH^-; the protonated ligand HL is water, H_2O.

Method	Temp.	Medium	Log of equilibrium constant; remarks	Reference
mag	20	var	$*\beta_2$ −2.30	b
sp	25	C($NaClO_4$)	$*K_1$ −1.63(C=2), −1.56(C=1), −1.50(C=0.5)	c
sp	10–43	0.5(Na)ClO_4	$*K_1$ −1.90(10°), −1.47(25°), −1.00(43°)	d
sp	10–43	→0	$*K_1$ −1.12(10°), −0.68(25°), −0.18(43°), $\Delta *H_1$=11.7, $\Delta *S_1$=36(25°)	d+c
sp	15–25	0.19($HClO_4$)	$*K_1$ −1.38(15.2°), −1.12(24.7°)	e
sp	25	0 corr	$\Delta *H_1$=10.7, $\Delta *S_1$=33, ΔS_1=52	
ΔG	25	0 corr	$*K_{S0}$ 3.80, $*K_{S1}$ 2.60 [UO_2(s)]; $*K_{S0}$ 9.95, $*K_{S1}$ 8.78 [U $(OH)_4$ (s)]	f
gl, red	25	3(Na)ClO_4	$*K_1$ −2.0, $*\beta_{3n,n+1}$ −1.2–3.4n	g
sol	25	0 corr	K_{S5} −3.77 ["U$(OH)_4$(s)"]	h
gl, sp	25	2($NaClO_4$)	$*K_1$ −1.68 (in H_2O), −1.74 (in D_2O)	i
sol, gl			K_{S0} −51.96	j
H	25	UF_4 var	apparent $*\beta_n$ given. F^- complexing neglected, $*K_1$ −2.48, $*\beta_2$ −6.21, $*\beta_3$ −10.38, $*\beta_4$ −16.37, K_1 11.35	k
H	25 ?	UCl_4 var	$*K_1$ 3.05, $*K_2$ −1.95, $*K_2$ −3.13 (1.5% EtOH)	l
nmr		var	K_1 12.5 (K_w −14 ?)	m
sp	20	1(H+Li) 0.98 ClO_4^- 0.07 NO_3^-	$*K_1$ −1.57	n*
sp	25	0 corr	$*K_1$ −1.11	c/o*

$*K_1$: $U^{4+}+H_2O \rightleftharpoons UOH^{3+}+H^+$; $*\beta_2$: $U^{4+}+2H_2O \rightleftharpoons U(OH)_2^{2+}+2H^+$; $*\beta_{3n,n+1}$: $(n+1)U^{4+}+3nH_2O \rightleftharpoons U[(OH)_3U]_n^{n+4}+3nH^+$; K_1: $U^{4+}+OH^- \rightleftharpoons UOH^{3+}$

[a] Most of the data in this table has been taken from the compilation by SILLÉN and MARTELL (1964). References in the last column without asterisks may be found in the aforementioned compilation. References with asterisks represent later information and are given at the end of this chapter. The notation of SILLÉN and MARTELL (1964) is used in the table and is explained in Table 2.19. [b] LAWRENCE (1934). [c] KRAUS and NELSON (1950). [d] KRAUS and NELSON (1955). [e] BETTS (1955). [f] DELTOMBE et al. (1956). [g] HIETANEN (1956). [h] GAYER and LEIDER (1957). [i] SULLIVAN and HINDMAN (1959). [j] STEPANOV and GALKIN (1960). [k] NIKOLAEV and LUK'YANCHEV (1961). [l] ROACH and AMIS (1962). [m] VDOVENKO et al. (1963). [n*] MCKAY and WOODHEAD (1964). [o*] SOLOVKIN et al. (1967).

to the monomer, the hydrolysis of U^{4+} may be explained by the formation of polymers according to the "core-links" reaction

$$(n+1)U^{4+}+3nH_2O \rightleftharpoons U[(OH)_3U]_n^{4+n}+3nH^+.$$

The equlibrium constant $*\beta_{3n,n+1}$ for this reaction is also given in Table 2.20.

The hydrolysis of the uranyl ion has been studied extensively by many investigators (Table 2.21). Recent reviews of the subject have been made by HERMANS (1964, 1970) and CORDFUNKE (1969). Evidence for a number of complex hydrolytic species has been reported and is summarized in Table 2.21.

c) Complex Ion Formation

In the general case of complex ion formation, a metal ion M will interact with anions to a degree depending on the strength of M as an acid. The higher

Table 2.21. Hydrolysis of U(VI)[a]

The central ion M in this table is UO_2^{2+}. The ligand L is OH^-; the protonated ligand HL is water, H_2O.

Method	Temp.	Medium	Log of equilibrium constant; remarks	References
qh	25–26	0 corr	$^*K_1 - 4.3$	b
gl	20–25	var	$^*\beta_{22} - 5.87$	c
col	?	var	$^*K_1 - 4.50$, $^*\beta_{22} - 4.95$, ev UO_2NO_3OH, $UO_2(NO_3)_3OH^{2-}$ (?)	d
qh	25	0 corr	$^*K_1 - 4.09$	e
gl	15	$C[Ba(NO_3)_2]$	$^*\beta_{22} - 5.97$ (C=0.6), -5.72 (C=0.06)	g+f=h
qh, gl	20	$1(Na)ClO_4$	$^*K_1 - 4.70$ (?, see ref. j), ev polyn cpx	i
	20	$1(Na)ClO_4$	$^*\beta_{22} - 6.05$, $^*\beta_{2n,\,n+1}\,0.30–6.35\,n$	i/j
	20	$1(Na)ClO_4$	$^*\beta_{22} - 6.10$, $^*\beta_{2n,\,n+1}\,0.30–6.40\,n$ ($^*K_1 - 4.77$, $^*\beta_{22} - 6.10$, $^*\beta_{53} - 16.74$)	i/k
gl, fp, sp	25 ?	0.15 $(NaClO_4)$	$^*\beta_{22} - 5.99$, $^*\beta_{43} - 13.29$, $K_a[U_3O_8(OH)_n^{2-n}]$: -3.55 (n=0), -6.5 (n=1), -7.4 (n=2), -11.0 (n=3), -11.4 (n=4)	l
gl, fp, sp	25 ?	0.1 ClO_4^-	$^*\beta_{22} - 5.94$, $^*\beta_{43} - 12.90$	
pH_2O	25	2 $UO_2(ClO_4)_2$ $+UO_3$	ev$(UO_2OH)_2^{2+}$, not UO_2OH^+	m
sol	25	0 corr	$^*K_{S0}\,6.04$, $^*K_{S1}\,1.90$, $K_{S3} - 3.60$ $K_{S4} - 3.77$, $^*K_1 - 4.14$, ev polyn cpx	n
sp	?	var	$^*K_1 - 4.19$	o
dis	25	$0.1(Na)ClO_4$	$^*K_1 - 4.2$, $^*K_2 - 5.20$	p
ΔG	25	0 corr	$^*K_{S0}\,14.74[UO_3(s)]$, $4.97[UO_2(OH)_2(s)]$, $5.60[UO_2(OH)_2H_2O(s)]$	q
gl	25	0 corr	$^*\beta_{22} - 5.06$, $^*\beta_{24} - 1.26$	r
gl	25	$0.347[Ba(ClO_4)_2]$	$^*K_1 - 5.40$, $^*\beta_{22} - 5.82$	s
gl	25–40	$0.0347[Ba(ClO_4)_2]$	$^*K_1 - 5.82$ (25°), -5.10 (40°), $\Delta^*H_1 = 20.8$, $^*\beta_{22} - 6.15$ (25°), -5.92 (40°), $\Delta^*H_{\beta 22} = 6.7$	
	25	$0.0347[Ba(ClO_4)_2]$	$\Delta^*S_1 = 43$, $\Delta^*S_{\beta 22} = -6$	
sol	20	0 corr	$K_{S0} - 17.22$ (?), $K_{S1} - 10.89$ (?)	t
gl	25	0.16 (NaCl)	$\beta_{2n,\,n+1}\,0.30–6.42\,n$	u
gl, qh	25	0.4 UO_2, $1(Na)ClO_4$	$^*\beta_{12} - 3.66$, $^*\beta_{22} - 6.02$	v
gl, qh	25	1.4 UO_2, 3 $(Na)ClO_4$	$^*\beta_{12} - 3.68$, $^*\beta_{22} - 6.31$, $^*\beta_{43} - 12.6$?	
sol, gl	RT ?	0.2 NH_4NO_3	$K_{S0} - 21.74$	w
qh, gl	20	$1(Na)ClO_4$	$^*\beta_{22} - 5.96$, $^*\beta_{53} - 16.74$ or $^*K_1 - 4.77$, $^*\beta_{22} - 6.10$, $^*\beta_{53} - 16.74$	i/x
gl	25	$1(Na)Cl$	Apparent $^*\beta_{nm}$ given. Cl^- complexing neglected, $^*\beta_{22} - 6.17$, $^*\beta_{43} - 12.33$, $^*\beta_{53} - 17.00$ (also cfu)	w
sol, gl	20	0 corr	$K_{S0} - 17.22$, $K_{S1}(UO_2OH^+ OH^-) - 11.89$ $K_{S2} - 5.98$	y
gl	25	0.1 KNO_3	$^*K_1 - 6.10$ and $^*\beta_{22} - 5.84$; or $^*\beta_{22} - 5.83$ and $^*\beta_{64} - 17.6$	z
gl, qh	25	$3(Na)ClO_4$	$^*\beta_{22} - 6.03$, $^*\beta_{43} - 13.20$, $^*\beta_{53} - 16.55$, $^*\beta_{64} - 19.42$	aa

Table 2.21 (continued)

Method	Temp.	Medium	Log of equilibrium constant; remarks	References
dis		var	K_{S0} −23.74	bb
dis, gl	20	0.1 ($NaClO_4$)	K_1 9.2, β_2 17.2, β_3 25.5 (H^+: K_1 14.2)	cc
gl	25	1.5 (Na)SO_4	$*\beta_{22}$ −8.17, $*\beta_{43}$ −16.20, $*\beta_{64}$ −24.51, $*\beta_{85}$ −32.14 or unlimited series, $*\beta_{2n,n+1}$ −(7.66 n + log $n!$ + 0.52)	dd
cal	25	3 (Na)ClO_4	$\Delta{*H_{\beta22}}$ 9.44, $\Delta{*H_{\beta43}} \sim 18$, $\Delta{*H_{\beta53}} = 25$, $\Delta{*H_{\beta64}} \simeq 24$, $\Delta{*S_{\beta22}} = 4.09$, $\Delta{*S_{\beta53}} \simeq 8$, equil. constants from ref. z	ee + aa
sp, gl	25	1 ($NaClO_4$)	$*\beta_{22}$ −5.94 (sp, gl), −5.91 (gl), $*\beta_{53}$ −16.41 (sp, gl), −16.43 (gl)	ff
cal	25	3 (Na) ClO_4	$\Delta{*H_{\beta22}} = 9.55$, $\Delta{*H_{\beta43}} \simeq 14.5$, $\Delta{*H_{\beta53}} = 25.7$	ee/gg
gl, sp		1 (Ba,NO_3)	$*\beta_{22}$ −6.1 (gl), −6.3 (sp)	hh
cix		0.5 ($NaNO_3$)	$*\beta_{53}$ −16	
sol	25	dil	K_{S0} −19.82, $UO_2(OH)_2H_2O$ (s)	ii
gl	25	0.5 (K)NO_3	$*K_1$ −5.7, $*\beta_{22}$ −5.92, $*\beta_{53}$ −16.22	jj
	94.4	0.5 (K)NO_3	$*K_1$ −4.19, $*\beta_{22}$ −4.51, $*\beta_{53}$ −12.74	
	25	0.5 (K)NO_3	$\Delta{*H_1} = 11$, $\Delta{*H_{\beta22}} = 10.2$, $\Delta{*H_{\beta53}} = 25.1$, $\Delta{*S_1} = 11$, $\Delta{*S_{\beta22}} = 7.1$, $\Delta{*S_{\beta53}} = 10$	
gl, qh	25	3 (Na) ClO_4	$*\beta_{22}$ −6.04, $*\beta_{53}$ −16.53, $*\beta_{43} \simeq -13.6$, $*\beta_{64} < -19.2$, $*K_1 < -5.9$ (range with up to 0.1 M UO_2^{2+})	kk = ll
	25	3 (Mg)ClO_4	$*\beta_{12}$ −3.81, $*\beta_{22}$ −6.25, $*\beta_{53}$ −17.18, $*\beta_{43} \simeq -13.3$, $*\beta_{64} \simeq -20.2$	
		3 (Ca) ClO_4	$*\beta_{12}$ −3.96, $*\beta_{22}$ −6.20, $*\beta_{53}$ −16.91, $*\beta_{43} \simeq -13.4$	
gl, qh	25	3 (Na) Cl	$*\beta_{22}$ −6.64, $*\beta_{53}$ −18.07, $*\beta_{43}$ −12.54, $*\beta_{64}$ −20.0, $*\beta_{74}$ −24.9	mm = ll
gl, qh	25	1 (K)NO_3	$*\beta_{12}$ −4.2, $*\beta_{22}$ −5.96, $*\beta_{53}$ −16.21, $*\beta_{43}$ −12.8	nn = ll
gl	16–27	$UO_2(NO_3)_2$ var	$*K_1$ −5.05 (16°), −4.59 (27°), ev polyn cpx	oo
gl, sp	RT	0.1	$*\beta_{22}$ −6.09 (gl), −6.28 (sp)	pp*
		30% EtOH	$*\beta_{22}$ −6.2 (sp)	
		50% EtOH	$*\beta_{22}$ −4.8 (sp), $*\beta_{12}$ −2.5 (gl), −1.9 (sp)	
pH	?	var	β_{22} 23.20 $K_A[(UO_2)_2(OH)_2^{2+} + 2\,OH^- \rightleftharpoons 2\,UO_2(OH)_2]$ 15.81, $K_B[(UO_2)_4(OH)_6^{2+} + 2\,OH^- \rightleftharpoons 4\,UO_2(OH)_2]$ 8.80, $K_C[UO_2(OH)_2 + OH^- \rightleftharpoons UO_2OOH^- + H_2O]$ 6.34, $K_D[2\,UO_2(OH)_2 + 2\,OH^- \rightleftharpoons U_2O_7^{2-} + 3\,H_2O]$ −10.92	qq*
temperature jump	25	0.5 (KNO_3)	log forward rate constant for $*\beta_{22}$ 2.06 (indicator present)	rr* + ff, jj
	25	0.5 ($NaClO_4$)	log forward rate constant for $*\beta_{22}$ 2.02 (indicator present), 2.06 (no indicator) (rate constant in terms of M^{-1} sec^{-1})	

Table 2.21 (continued)

Method	Temp.	Medium	Log of equilibrium constant; remarks	References
con	100–200	→0	$^*K_1$ −4.0 (100°), −3.5 (150°), −3.4 (200°), $^*\beta_{22}$ −4.3 (100°), −3.8 (150°), −3.7 (200°)	ss*
pressure jump	25	0.06	log reverse rate constant for $^*K_1$ > 6.43 log forward rate constant for $^*\beta_{22}$ 1.80	tt*+jj
gl	25	3(Mg)NO_3 5(Mg)NO_3	$^*K_1$ −5.38, $^*\beta_{22}$ −6.34, $^*\beta_{53}$ −17.37 $^*K_1$ −5.53, $^*\beta_{22}$ −6.52, $^*\beta_{53}$ −17.76	uu*
cal	25	3(Na)ClO_4	$\Delta^*H_{\beta 22}$ = 9.5, $\Delta^*H_{\beta 53}$ = 24.4, $\Delta^*S_{\beta 22}$ = 4.3, $\Delta^*S_{\beta 53}$ = 6	vv*+kk
pH	?	var	$^*\beta_{22}$ −4.8, $^*\beta_{64}$ −14.8, $^*\beta_{21}$ −8.5, $^*\beta_{31}$ −30.1, $^*\beta_{62}$ −55.9	ww*+qq*
gl	25	0.1 $NaClO_4$	$^*K_1$ −4.39, $^*\beta_{12}$ −2.22, $^*\beta_{22}$ −6.09, $^*\beta_{53}$ −15.64, $^*\beta_{73}$ −24.03	ww*
gl	25	3(Na)ClO_4	In D_2O: $^*\beta_{22}$ −6.80, $^*\beta_{43}$ −14.00, $^*\beta_{53}$ −18.63	xx*

$^*\beta_{pq}$: $qUO_2^{2+} + pH_2O \rightleftharpoons (UO_2)_q(OH)_p{}^{(2q-p)} + pH^+$.

[a] See footnote [a] to Table 2.20. [b] HEIDT (1942). [c] MACINNES and LONGSWORTH (1942). [d] GUITER (1947). [e] HARRIS and KOLTHOFF (1947). [f] SCHAAL and FAUCHERRE (1947). [g] FAUCHERRE (1948). [h] FAUCHERRE (1954). [i] AHRLAND (1949). [j] AHRLAND et al. (1954). [k] ROSOTTI et al. (1956). [l] SUTTON (1949). [m] ROBINSON and LI (1951). [n] GAYER and LEIDER (1955). [o] KOMAR and TRETYAK (1955). [p] RYDBERG (1955). [q] DELTOMBE (1956). [r] ORBAN et al. (1956). [s] HEARNE and WHITE (1957). [t] BRUSILOVSKII (1958). [u] LI et al. (1958). [v] HIETANEN and SILLÉN (1959). [w] BABKO and KODENSKAYA (1960). [x] RUSH et al. (1962). [y] BRUSILOVSKII (1960). [z] GUSTAFSON et al. (1960). [aa] HIETANEN (1960). [bb] OOSTING (1960). [cc] STARÝ (1960). [dd] PETERSON (1960). [ee] SCHLYTER (1962). [ff] RUSH and JOHNSON (1963). [gg] SCHLYTER (1963). [hh] NIKOLAEVA et al. (1962). [ii] PEREZ-BUSTAMANTE et al. (1962). [jj] BAES and MEYER (1962). [kk] HIETANEN et al. (1963). [ll] DUNSMORE et al. (1963). [mm] DUNSMORE and SILLÉN (1963). [nn] DUNSMORE (1963). [oo] POZHARSKII et al. (1963). [pp*] BARTUSEK and SOMMER (1964). [qq*] ISRAELI (1965). [rr*] WHITTAKER et al. (1965). [ss*] RYZHENKO et al. (1967). [tt*] HURWITZ and ATKINSON (1967). [uu*] SCHEDIN and FRYDMAN (1968). [vv*] ARNEK and SCHLYTER (1968). [ww*] TSYMBAL (1969). [xx*] MAEDA and KAKIHANA (1970).

the positive charge on M, the stronger will be its acid properties and the greater will be its association with anions as complex ions. For a given metal ion, the amount of complex ion formation will be roughly related to the basicity of the anion L which can be inferred from the degree of dissociation of the acid H_pL. An acid which is largely dissociated has an anion which has little affinity for a proton H^+ or for positively charged cations. The anion of a highly dissociated acid is therefore a weak base and is weakly complexing. Conversely the anion of a weak acid is a strong base and is strongly complexing (KATZ and SEABORG, 1957).

Little information is available concerning complex ion formation of U^{3+}. The red color of U(III) in concentrated hydrochloric acid has been attributed to the formation of a chloro complex of U(III) by JØRGENSEN (1956). MARCUS et al. (1965) have reported the formation of UCl^{2+} and UBr^{2+} in concentrated solutions of LiCl and LiBr, respectively, with formation constants of $10^{-2.85}$ and $10^{-3.96}$. They have also reported the formation of a sulfate complex and a very unstable thiocyanate complex. RULFS and ELVING (1955) have indicated the formation of a U(III) cupferrate complex.

Complex ion formation by U(IV) has been reviewed recently by PETIT (1969) as part of a general review of the chemistry of U(III) and U(IV) in aqueous, non-aqueous, and molten salt solutions. Complexes formed in solution with inorganic ligands are given in Table 2.22. The ability of uranium (IV) to form complexes with inorganic anions has been used in its separation from other elements as well as from uranium (VI). The general order of increasing stability of the anionic complexes is (PETIT, 1969):

$$ClO_4^- < I^- \sim NO_3^- < Br^- < Cl^- < SCN^- < SO_4^{2-} < F^-.$$

Complexes formed between uranium (IV) and organic ligands are listed in Table 2.23. Particularly stable U(IV) complexes ($\log K_1 > 20$) are formed with ethylenediamine tetraacetic acid (EDTA), cyclohexanediamine tetraacetic acid (CDTA), diethylenetriamine pentaacetic acid (DTPA) and hexamethylenediamine tetracetic acid. The U(IV)-arsenazo (III) complex is used in the photometric determination of uranium (NEMODRUK and PALEI, 1963).

Table 2.22. U(IV) Complex ion formation. Inorganic ligands[a]

Complexing agent	Method	Temp.	Medium	Log of equilibrium constant; remarks	Reference
Thiocyanate, SCN^-	red	20	1 $(NaClO_4)$ 0.6 H^+	K_1 1.49, K_2 0.46, K_3 0.22	b
	dis	10	2 $(Na)ClO_4$, 1 H^+	K_1 1.78, K_2 0.52	c
	dis	25	2 $(Na)ClO_4$, 1 H^+	K_1 1.49, K_2 0.62, $\Delta H_1 = -5.7$, $\Delta H_2 = -1.8$, $\Delta S_1 = -10$, $\Delta S_2 = 9.7$	
	dis	40	2 $(Na)ClO_4$, 1 H^+	K_1 1.30, K_2 0.68	
	pre, con	25	MeOH	ev UL_8–in MeOH	d
Carbonate, CO_3^{2-}	pre, sol		var, solid	ev UL_5^{6-}, $UL_3(OH)_3^{5-}$	e
	sol, pol	25	var	ev cpx	f
	sp			ev cpx, UL_5^{6-} ?	g
	pol	25	Na_2L, NaHL var	ev cpx, probably $UL_4(OH)_2^{6-}$ UL_5^{6-}	h
	pre		var	ev UL_4^{4-}, UL_5^{6-}	i
Phosphate, PO_4^{3-}	sol	35	var	ev $UAl_6(HL)_8^{6+}$ or $UAl_8(HL)_8^{12+}$	j
	red	25	H_3L var	ev cpx, stronger than for UO_2^{2+}	k
	?	25 ?		$K_s(U^{4+}HL_2^{2-}) - 27.5$	l
	sol	20	?	ev cpx with 1–4 P/U. $K_s(U^{4+}HL_2^{2-}) - 27.74$, $K_s(UHL^{2+}HL^{2-}) - 15.92$	m
	dis		H_3L var	ev $U(HL)_3^{2-}$	n*
	sol		0.35	HL^{2-}: K_1 12.0, K_2 10.0 K_3 8.6, K_4 8.0, β_4 38.6 $[K_{s0}(U^{4+}HL_2^{2-}) - 27.80]$ †$K_1 - 7.34$, †$K_2 - 5.74$, †$K_3 - 5.60$, †$K_4 - 5.96$ †$K_n = K_{s0}(HL^{2-}) \cdot \beta_n(HL^{2-})$ $[K_a(H_3L)\,K_a(H_2L^-)]^{n-2}$: $U(HL)_2(s) + (n-2)H_3L \rightleftharpoons$ $U(HL)_n^{(2n-4)} + (2n-4)H^+$	o*
Diphosphate, pyrophosphate, $P_2O_7^{4-}$	sol, pre		var	ev UL_2^{4-}	e
Sulfite, SO_3^{2-}	pre, sol		var, solid	ev ani cpx, UL_4^{4-}, $U_2L_7^{6-}$	e

Table 2.22 (continued)

Complexing agent	Method	Temp.	Medium	Log of equilibrium constant; remarks	Reference
Sulfate, SO_4^{2-}	dis	25	$2(HClO_4)$	$^*K_1$ 2.53, $^*K_2$ −0.13	p
		25	$2(HClO_4)$	$^*K_1$ 2.41, $^*K_2$ 1.32	p/q
		25	$2(HClO_4)$	K_1 3.24, K_2 2.18 [$K_1(H^+)$ 1.125] (?)	p/q/r
	dis	10	$2(Na)ClO_4$, $1H^+$	$^*K_1$ 2.63, $^*K_2$ 1.34	s
	dis	25	$2(Na)ClO_4$, $1H^+$	$^*K_1$ 2.52, $^*K_2$ 1.35, $\Delta H_1 = -3.2$, $\Delta H_2 = 0.9$, $\Delta S_1 = 0.7$, $\Delta S_2 = 9.3$	
	dis	40	$2(Na)ClO_4$, $1H^+$	$^*K_1$ 2.38, $^*K_2$ 1.38	
	aix		$(NH_4)_2L$ var	ev ani cpx	t
	pre		var	ev UL_2, UL_3^{2-}, UL_4^{4-}, UL_5^{6-}, UL_5^{8-}, $U_2L_5^{2-}$	u
	con	25	var	ev UL_3^{2-}	u*
	nmr		2 $HClO_4$	K_1 1.70	v*
	red	25	$1(H_2L)$	K_1 2.62	w
	gl		0 corr	$K_s(U^{4+}OH^-{}_2L^{2-})$ −31.17, [$^*K_1$ 2.41, $^*\beta_2$ 3.73, $K_1(H^+)$ 1.92, $K_a(U^{4+})$ −1.14, $K_a(UOH^{3+})$ −1.14]	x
	dis	25	$2(HClO_4)$	$^*K_1$ 2.32, $^*K_2$ 1.25	p/q/y*
	dis	?	$3.8(NaClO_4)$ $1H^+$, $1NO_3^-$ $0.01U^{4+}$	$^*K_1$ 2.27, $^*K_2$ 1.59	z*
Fluoride, F^-	X		solid	ev UF_7^{3-} in $K_3UF_7(s)$	aa
	dis	25	$2(NaClO_4)$, $1H^+$	$^*K_1 \geqq 6$, $^*\beta_2 \geqq 8$	s
	sol	25	$0.12(HClO_4)$	K_{S2} 2.30, K_{S3} 1.59, K_{S4} −4.23, K_{S5} −4.27, K_{S6} −3.96 (H^+: K_1 2.98, K_2 0.59)	bb
	sol	25	HF var	K_1 4.7 (?)	cc
	sol		var	K_{S0} −21.24 [$UF_4(H_2O)_{2.5}$], $\beta(Al^{3+}F_2^-)$ −9.06	dd
	nmr			K_1 7.15, β_2 12.4, β_3 17.7	ee*
	sp, nmr	20	var	K_1 8.78, K_2 5.84 or 5.60	ff*
Chloride, Cl^-	sp	25	$0.5(NaClO_4)$	K_1 −0.20	gg
		25	0 corr	K_1 0.85	
	X		solid	ev UCl_6^{2-} in $(Me_4N)_2UCl_6(s)$	hh
	red	20	$1(Na)ClO_4$, $0.6H^+$	K_1 0.30	b
	dis	10	$2(NaClO_4)$, $1H^+$	K_1 0.52	s
		25	$2(NaClO_4)$, $1H^+$	K_1 0.26; alt K_1 0.08, K_2 −0.02	
		40	$2(NaClO_4)$, $1H^+$	K_1 0.18; alt K_1 −0.04, K_2 −0.06	
	aix	25	HCl var	ev ani cpx in > 5.5 *M*-HCl	ii
	aix		HCl var	ev ani cpx in ≧ 12 *M*-HCl	jj
	sp	160	$PyH^+Cl^-(1)$	ev UCl_6^{2-} in $PyH^+Cl^-(1)$	kk
	sp	400	(Li, K)Cl(1)	ev UCl_8^{4-} in (Li, K)Cl(1, eutect)	
	red	25	→0	K_1 0.8 (also for some other I values)	w
	nmr		2 $HClO_4$	K_1 0.78	v*
Perchlorate, ClO_4^-	nmr		2 $HClO_4$	K_1 −0.92	v*
Bromide, Br^-	red	20	$1(Na)ClO_4$, $0.6H^+$	K_1 0.18	b
	aix		HBr var	ev ani cpx in > 4 *M*-HBr	ll
	nmr		2 $HClO_4$	K_1 0.30	v*
Iodide, I^-	nmr		2 $HClO_4$	K_1 0.18	v*
Halogenides: mixed	sp		$MeNO_2$	$^†K_1$ −0.5, $^†K_2$ −1.3 in $MeNO_2$ (Bu_4N^+ var), $^†K_n$: $UCl_{7-n}Br_{n-1}^{2-} + Br^- \rightleftharpoons UCl_{6-n}Br_n^{2-} + Cl^-$	mm

Table 2.22 (continued)

Complexing agent	Method	Temp.	Medium	Log of equilibrium constant; remarks	Reference
Nitrate, NO_3^-	dis, sp	25	HL var	ev UL_6^{2-}	nn
	sp	26.5	$3.5\,H(ClO_4)$	$K_1\,0.36$, $\beta_2\,0.47$, $\beta_3\,0.42$, $\beta_4\,0.18$	oo
			$3\,H(ClO_4)$	$K_1\,0.28$, $\beta_2\,0.31$, $\beta_3\,0.17$, $\beta_4 -0.15$	
			$2.5\,H(ClO_4)$	$K_1\,0.21$, $\beta_2\,0.19$, $\beta_3\,0.03$, $\beta_4 -0.31$	
			$2H(ClO_4)$	$K_1\,0.20$, $\beta_2\,0.17$, $\beta_3 -0.02$, $\beta_4 -0.46$	
	sp	20	$4(H, Li, ClO_4)$	$K_1\,0.18$, $\beta_2\,0.78$, $\beta_3 < -1$, β_4 small	pp*
			$3(H, Li, ClO_4)$	$K_1\,0.20$, $\beta_2\,0.30$, $\beta_3 < -1$, β_4 small	
			$2(H, Li, ClO_4)$	$K_1\,0.06$, $\beta_2\,0.00$, $\beta_3 < -1$, β_4 small	
			$1(H, Li, ClO_4)$	$K_1\,0.04$, $\beta_2 -0.30$, $\beta_3 < -1$, β_4 small	
	dis	?	$3.8(NaClO_4)$, $1\,H^+$, $0.01\,U^{4+}$	$K_1\,0.1$ (TTA)	z*
	dis	20	$8(HClO_4)$	$K_1 -0.08$, $\beta_2\,0.58$, $\beta_3 -0.43$, $\beta_4 -0.30$	qq*
	sp	26.5	$IH(ClO_4)$	no ev $UL_n^{(4-n)+}$cpx	oo/rr*
	sp	20	$I(H, Li, ClO_4)$	no ev $UL_n^{(4-n)+}$cpx	pp*/rr*

[a] See footnote [a] to Table 2.20. [b] Ahrland and Larsson (1954). [c] Day et al. (1955). [d] Markov and Traggeim (1961). [e] Rosenheim and Kelmy (1932). [f] Harris and Kolthoff (1947). [g] McClaine et al. (1955). [h] Stabrovskii (1960). [i] Golovnya et al. (1960). [j] Strandell (1957). [k] Marcus (1958). [l] Denotkina et al. (1960). [m] Markov et al. (1960). [n*] Tserkovnitskaya and Bykhovtseva (1967). [o*] Moskvin et al. (1967). [p] Betts and Leigh (1950). [q] Sullivan and Hindman (1952). [r] Whiteker and Davidson (1953). [s] Day et al. (1955). [t] Saito and Sekine (1957). [u] Golobnya et al. (1960). [u*] Goldinov and Stabrovskii (1963). [v*] Vdovenko et al. (1963b). [w] Sobkowski (1961). [x] Stepanov and Galkin (1962). [y*] Varga (1969). [z*] Rao and Pai (1969). [aa] Zachariasen (1954). [bb] Savage and Brown (1960). [cc] Luk'yanychev and Nikolaev (1962). [dd] Luk'yanychev and Nikolaev (1963). [ee*] Vdovenko et al. (1963a). [ff*] Vdovenko et al. (1966). [gg] Kraus and Nelson (1950). [hh] Staritzky and Singer (1952). [ii] Kraus et al. (1955). [jj] Katz and Seaborg (1957). [kk] Gruen and McBeth (1959). [ll] Andersen and Knutsen (1962). [mm] Jørgensen (1963). [nn] Wilson and Keder (1961). [oo] Ermolaev and Krot (1962). [pp*] McKay and Woodhead (1964). [qq*] Lahr and Knoch (1970). [rr*] Solovkin (1969).

Little information is available on the complexing of uranium (V) in solution. Evidence for UO_2^+ complex ions in aqueous solutions comes mainly from the polarographic reduction of uranyl ions (e.g., Stabrovskii (1960) Branica and Kuta (1966), and Lai and Chen (1967)). Elving and Krivis (1959) have reported a weak U(V)-sulfate complex with $\log K_1 \approx -0.9$ if UO_2SO_4 is the most stable uranyl complex present. Hála (1964) has determined $\log K_1$ for the UO_2^+-sulfosalicylic acid complex to be 5.14.

Complex ion formation with uranyl ion as the central ion has been and continues to be studied extensively. Much of the quantitative information available on complexing with inorganic ligands is given in Table 2.24. The displacement series of ligands for uranyl ion is (Chernyaev, 1964):

$$CO_3^{2-} \geqq O_2^{2-} > OH^- > F^- > SO_4^{2-} > Cl^- > NO_3^- > Br^- > I^-.$$

Complex ion formation between uranyl ion and organic ligands is summarized in Table 2.25. Table 2.25 is extensive; however, not exhaustive. No information has been included in the table on complex ion formation with mixed central ions. For example, Adin et al. (1969) have investigated reactions of the type,

$$In(III)L_2 + 1/2\,[U(VI)L]_2 \overset{K}{\rightleftharpoons} In(III)U(VI)L_2 + L,$$

where L represents the citrate, malate, or tartrate ligand. Average equilibrium constants K determined for the above reaction over a wide range of metal and ligand concentrations are, for the respective ligands, 720, 30 and 31.

Table 2.23. U(IV) complex ion formation. Organic ligands[a]

Complexing agent: formula, name, ligand L definition.

Method	Temp.	Medium	Log of equilibrium constant; remarks	Reference
$C_2H_2O_4$, Oxalic acid, H_2L				
sol		var	$K_1 9$, $K_2 7$, $K_3 6$, $K_4 4$	c*+b*
sol		var	$K_1 8.60$, $K_2 8.25$, $K_3 5.92$, $K_4 4.47$	d*
		0 corr	$K_1 10.9$, $K_2 9.8$, $K_3 5.7$, $K_4 3.4$	
$C_5H_8O_2$, Pentane-2,4-dione (acetylacetone), HL				
dis	20	$0.1 (ClO_4^-)$	$K_1 8.6$, $K_2 8.4$, $K_3 6.4$, $K_4 6.1$	e, f, g
$C_6H_8O_7$, Citric acid, H_4L				
sp	20	$0.1 NaClO_4$	HL^{3-}: $K_1 13.5$	h*
gl			HL^{3-}: $K_1 11.53$, $K_2 7.93$	i*
$C_6H_9O_6N$, Nitrilotriacetic acid (NTA), H_3L				
pH	25	0.1 (KCl)	ev UL_2	j*
NTA — mixed ligands:				
NTA + 8-hydroxy-5-quinolinesulfonic acid (HQS), H_2A				
pH	25	0.1 (KCl)	ev ULA^-, ULA_2^{3-}	k*
NTA + disodium 1,2-dihydroxybenzene-3,5-disulfonate (Tiron), H_2A^{2-}				
pH	25	0.1 (KCl)	ev UL_2^{2-}, $U_2A_3^{4-}$	k*
$C_6H_{11}O_5N$, N-Hydroxyethyliminodiacetic acid (HIMDA), H_3L				
pH	25	0.1 (KCl)	$K[U(HL)_2(OH)^- + H^+ \rightleftharpoons U(HL)_2] 3.67$	j*
$C_7H_7O_2N$, Benzohydroxamic acid, HL				
gl	25	$0.01 HClO_4$, $0.001 U^{4+}$	$K_1 9.89$, $K_2 8.11$, $K_3 8.32$, $K_4 6.62$	l*
$C_8H_5O_2SF_3$, 1,1,1-Trifluoro-3,2′-thenoylacetone (TTA), HL				
sp	25		$K_1 7.2$	m
$C_{10}H_{16}O_8N_2$, Ethylenediamine-NNN′N′-tetracetic acid (EDTA), H_4L				
sp			$K(UOH^{3+} + L^{4-} \rightleftharpoons UOHL^-) 17.32$	n
col		$0.02 H^+$, $0.0375 SO_4^{2-}$	$K_1 25.6$	o*
sp		var	pH = 0.5–1.0: $U(H_2O)n^{4+} + H_4L \rightarrow UL + 4H^+ + nH_2O$ $U(OH)(H_2O)_{n-1}^{3+} + H_4L \rightarrow U(OH)L^- + 4H^+ + (n-1)H_2O$; $U(H_2O)_n^{4+} \rightarrow UHL^+ + 3H^+ + nH_2O$ pH = 2.0: $U(H_2O)_n^{4+} + H_4L \rightarrow UL + 4H^+ + nH_2O$ $U(H_2O)_n^{4+} + H_3L^- \rightarrow UL + 3H^+ + nH_2O$ $U(OH)(H_2O)_{n-1}^{3+} + H_4L \rightarrow U(OH)L^- + 4H^+ + (n-1)_2O$ $U(OH)(H_2O)_{n-1}^{3+} + H_3L^- \rightarrow U(OH)L^- + 3H^+ + (n-1)H_2O$ weakly acid medium: UL_2^{4-} alkaline medium: unstable cpx pH > 10: $U(OH)_4$ precipitated	p*
sp	25	$I(NaClO_4)$	$K(U^{4+} + H_5L^+ \rightleftharpoons UL + 5H^+)$ K 2.28 (I = 1.0), 1.98 (I = 1.5), 1.90 (I = 2.0), 1.65 (I = 2.5), 1.61 (I = 3.0) K 2.67 (I = 0.5, extrapolated) K 3.23 (I = 0.1, corr) $K_1 25.83$	q*

Table 2.23 (continued)

Method	Temp.	Medium	Log of equilibrium constant; remarks	Reference
sp,gl pre	25	I($NaClO_4$)	$K_A[UL + OH^- \rightleftharpoons ULOH^-]$ $K_B[2\,ULOH^- \rightleftharpoons (ULOH)_2^{2-}]$ $K_C[ULOH^- + OH^- \rightleftharpoons UL(OH)_2^{2-}]$	r*

I	0	0.01	0.1	0.25	0.5	1.0
K_A	8.95	9.0	9.07	9.08	9.17	9.13
K_B	2.78	2.84	2.75	2.79	2.84	2.86
K_C		5.91	6.29	6.41	6.49	6.87

$K_D\ [2\,UL(H_2O)_2 + H_2L^{2-} \rightleftharpoons U_2H_2L_3^{2-} + 4\,H_2O]$
$K_E\ [2\,UL(H_2O)_2 + H_2L^{2-} \rightleftharpoons U_2L_2HL^{3-} + H^+ + 4\,H_2O]$

I	0.04	0.1	0.5	2.0
K_D	3.64	3.57	2.85	3.30
K_E	0.46	0.79	0.82	1.07

$K_F[U_2HL_3^{3-} \rightleftharpoons U_2L_3^{4-} + H^+]$ −4.3 (I = 0.1),
$K_G[U_2(OH)_2L_3^{6-} + H_2O \rightleftharpoons (ULOH)_2^{2-} + HL^{3-} + OH^-]$ −9 (I = 0.1),
$K_H[U_2(OH)_2L_3^{6-} \rightleftharpoons U_2L_3^{4-} + 2\,OH^-]$ −12.11 (I = 0.1)

Method	Temp.	Medium	Log of equilibrium constant; remarks	Reference
pH	20	0.1 (KCl)	K_1 25.8	j*
	25	0.1 (KCl)	$K_A[ULOH^- + H^+ \rightleftharpoons UL + H_2O]$ 4.72 $K_B[(ULOH)_2^{2-} + 2\,H^+ \rightleftharpoons 2\,UL + 2\,H_2O]$ 6.53 $K_C[2\,ULOH^- \rightleftharpoons (ULOH)_2^{2-}]$ 2.9	
		1 (KCl)	K_A 4.58, K_B 6.68, K_C 2.5	
sp		var	K_1 17.00	s*
EDTA — mixed ligands:				
EDTA + fluoride, F^-				
sp		0.1 $NaClO_4$	$K[UF^{3+} + L^{4-} \rightleftharpoons UFL^-]$ 17.50	t
EDTA + iminodiacetic acid (IMDA), H_2A				
pH	25	0.1 (KCl)	$K[UL + A^{2-} \rightleftharpoons ULA^{2-}]$ 8.2	k*
EDTA + pyrocatechol (PY), H_2A				
pH	25	0.1 (KCl)	$K[UL + A^{2-} \rightleftharpoons ULA^{2-}]$ 14.16	k*
EDTA + disodium 1,2-dihydroxybenzene-3,5-disulfonate (Tiron), H_2A^{2-}				
pH	25	0.1 (KCl)	$K[UL + A^{4-} \rightleftharpoons ULA^{4-}]$ 15.61	k*
EDTA + salicylic acid (SA), H_2A				
pH	25	0.1 (KCl)	ev ULA^{2-}	k*
EDTA + 5-sulfosalicylic acid (SSA), H_3A				
pH	25	0.1 (KCl)	$K[UL + A^{3-} \rightleftharpoons ULA^{3-}]$ 11.08	k*
EDTA + potassium acid phthalate (Ph), HA^-				
pH	25	0.1 (KCl)	$K[UL + A^{2-} \rightleftharpoons ULA^{2-}]$ 4.2	k*
EDTA + 8-hydroxy-5-quinoline sulfonic acid (HQS), H_2A				
pH	25	0.1 (KCl)	$K[UL + A^{2-} \rightleftharpoons ULA^{2-}]$ 9.72 $K[ULA(OH)^{3-} + H^+ \rightleftharpoons ULA^{2-} + H_2O]$ 7.14	k*
EDTA + disodium 1,8-dihydroxynaphthalene-3,6-sulfonate (CS), H_2A^{2-}				
pH	25	0.1 (KCl)	$K[UL + A^{4-} \rightleftharpoons ULA^{4-}]$ 16.22	k*
$C_{10}H_{18}O_7N_2$, N-Hydroxyethyethylenediaminetetraacetic acid (HEDTA), H_4L				
pH	25	0.1 (KCl)	ev UHL^+	j*

Table 2.23 (continued)

Method	Temp.	Medium	Log of equilibrium constant; remarks	Reference
$C_{12}H_{20}O_9N_2$, 2,2′Oxy *bis* [ethyliminodi(acetic acid)], H_4L				
sp		var	K_1 18.80	s*
$C_{14}H_{16}O_8N_2$, 1,2-Cyclohexanediaminetetraacetic acid (CDTA), H_4L				
pH	20	0.1 (KCl)	K_1 26.9	j*
	25	0.1 (KCl)	$K[ULOH^- + H^+ \rightleftharpoons UL + H_2O]$ 4.85 $K[(ULOH)_2 + 2H^+ \rightleftharpoons 2UL + 2H_2O]$ 6.24 $K[2ULOH^- \rightleftharpoons (ULOH)_2^{2-}]$ 3.5	
$C_{14}H_{23}O_{10}N_3$, Diethylenetriaminepentaacetic acid (DPTA), H_5L				
pH	25	0.1 (KCl)	ev UL^- $K[ULOH^{2-} + H^+ \rightleftharpoons UL^- + H_2O]$ 7.69	j*
sp		var	K_1 21.60	s*
$C_{14}H_{24}O_8N_2$, Hexamethylenediaminetetraacetic acid, H_4L				
sol	23	0.1 (H, Na) ClO_4	K_1 24.64	x*
$C_{16}H_{13}O_{11}N_2S_2As_2$, Benzene-2-arsonic acid-(1-azo-7)-1,8-dihydroxynaphthalene-3,6-disulfonic acid (Arsenazo), H_6L				
sp		var	ev UL at pH 2.0–2.5, ev UL_2 at pH > 2.5 K_1 (?) 16.22	v* + u*
$C_{18}H_{30}O_{12}N_4$, Triethylenetetraminehexacetic acid (TTHA), H_6L				
pH	25	0.1 (KCl)	$K[UL^{2-} + H^+ \rightleftharpoons UHL^-]$ 2.28	j*
sp		var	K_1 20.70	s*
$C_{22}H_{18}O_{14}N_4S_2As_2$, 1,8 Dihydroxynaphthalene-3,6-disulfonic acid-2,7-bis [(azo-2)-phenylarsonic acid] (Arsenazo III), H_8L				
sp		6 HCl	ev UL_1, UL_2, UL_3	w*
		0.1 HCl	ev UL_1, UL_2	
$C_{22}H_{16}O_{14}N_4Cl_2S_2P_2$, 2,7-Bis(2′-phosphono-4′-chlorophenylazo)chromotropic acid (Chlorophosphonazo III), H_8L				
sp	25	0.2 (KNO_3)	β_{12} (?) 11.09	y*
$C_{24}H_{20}O_{14}N_4Cl_2S_2P_2$, 2,7-Bis(2′-phosphono-4′-chloro-5-methyl-phenylazo)chromotropic acid, H_8L				
sp	25	0.2 (KNO_3)	β_{12} (?) 10.48	y*

[a] References with asterisks may be found at the end of this chapter. References without asterisks are given in the compilation of SILLÉN and MARTELL (1964). Table notation is explained in Table 2.19. When mixed organic ligands are indicated, the primary ligand is designated as L; the secondary ligand as A, eg. NTA and HQS are respectively designated as H_3L and H_2A. [b*] GRINBERG et al. (1958). [c*] GRINBERG et al. (1960). [d*] ZAKHAROVA and MOSKVIN (1960). [e] RYDBERG (1950). [f] RYDBERG (1953). [g] RYDBERG (1955). [h*] ADAMS and SMITH (1960). [i*] NEBEL and URBAN (1966). [j*] CAREY and MARTELL (1968). [k*] CAREY and MARTELL (1967). [l*] BARCOAS et al. (1966). [m] CALVIN and ZEBROSKI [n] SMITH (1957) quoted in t. [o*] KLYGIN et al. (1959). [p*] PALEI and HSU (1961). [q*] KROT et al. (1962). [r*] ERMOLAEV and KROT (1963). [s*] HAFEZ (1968). [t] SMITH (1959). [u*] KUTEINIKOV (1958). [v*] KUTEINIKOV (1962). [w*] NEMODRUK and PALEI (1963). [x] MERKUSHEVA et al. (1969). [y*] BUDESINSKY et al. (1967).

In comparing the ability of organic ligands to complex the uranyl ion with that of inorganic ligands, the acetate and oxalate ions are found between the fluoride and sulfate ions in the ligand displacement series given above. Certain amines, such as acridine, β-naphtho-quinoline, 2,2′-dipyridyl, and 1,10-phenanthroline, determine their position between the nitrate and bromide ion (CHERNYAEV, 1964).

Table 2.24. Uranium (VI) complex ion formation. Inorganic ligands.[a]

Complexing agent	Method	Temp.	Medium	Log of equilibrium constant; remarks	Reference
Vanadate, $L^{2-}=$ $VO_2(OH)_3^{2-}$ $=HVO_4^{2-}$	aix, sp		var	ev ani cpx, UO_2LOH^-	b
	pre, sol		var	ev UO_2LOH^-, $UO_2HL_2^-$, $UO_2H_3L_3^-$, K_s (?) for $NH_4UO_2L_3(s)$, etc.	c
	sol	25	dil	$K_s[K_2^+(UO_2^{2+})_2HL_2^-H_{-4}^+(H_2O)_3]$ -13.7 (carnotite)	d
Cyano-ferrate (II), $Fe(CN)_6^{4-}$	sol	25	var	K_{s0} -13.15	e
Cyanate, OCN^-	aix		var	ev ani cpx	f
Thio-cyanate, SCN^-	sp	20	1 ($NaClO_4$)	K_1 0.76, K_2 -0.02, K_3 0.44	g
	sp	25	→0	K_1 0.93	h
	sp	15–45	20% MeOH	K_1 0.99 (15°), 1.00 (25°), 1.05 (35°), 1.07 (45°) $\Delta H_1=1.15$, $\Delta S_1=8.5$ (25°) in 20% MeOH	i
	con	25	MeOH	ev $UO_2L_3^-$, $UO_2L_4^{2-}$, $UO_2L_5^{3-}$ in MeOH	j
	cix	32	1 ($NaClO_4$)	K_1 -1.3, β_2 1.05, β_3 1.08	k
	X		solid	ev $UO_2L_5^{3-}$ in $CsUO_2(NCS)_5(s)$	l
Carbonate, CO_3^{2-}	pre		solid	ev $UO_2L_3^{4-}$	m
	sol, sp ?	25 ?	0 corr	K_3 3.78	n
	sol, sp	25 ?	0 corr	$K[UO_2(OH)_2H_2O(s)+CO_2(g)\rightleftharpoons UO_2CO_3(s)$ $+2H_2O]$ 4 $K_{s2}/K_{p0}K_1^2(H^+)$ 1.42, $K_3/K_{p0}K_1^2(H^+)$ 1.81	o
	$K_{s2}/K_{p0}K_1^2(H^+)$: $UO_2CO_3(s)+2HCO_3^-\rightleftharpoons UO_2(CO_3)_2^{2-}+CO_2(g)+H_2O$ $K_3/K_{p0}K_1^2(H^+)$: $UO_2(CO_3)_2(s)+2HCO_3^-\rightleftharpoons UO_2(CO_3)_3^{4-}+CO_2(g)+H_2O$				
	ΔG	25	0 corr	β_2 14.6, β_3 18.3	
	aix		var	K_3 7.0	p
	sp, sol	26	~2	$K_3\sim 3.5$ ev $(UO_2)_2(OH)_3L^-$	q
	sol	26	→0	$K_s[Na_4^+(UO_2L_3)^{4-}]$ -2.8 to -2.0	
	var		var	ev $UO_2L_2^{2-}$, $UO_2L_3^{4-}$, ev $(UO_2)_2(OH)_2L_2^{2-}$ and $(UO_2)_2(OH)_2L$ or $(UO_2)_2(OH)_3L^-$, no ev cpx HL^-—UO_2^{2+} (sp, sol, gl, tp)	r
	sol, gl	25	1 (NH_4Cl)	β_3 22.8	s
	sol, gl	RT	0.2 (NH_4NO_3)	β_2 15.57, β_3 20.70 (H^+: K_1 10.30, K_{12} 6.46, $K_{s0}(UO_2^{2+}(OH^-)_2)$ -21.74)	t
		?		K_{s0} -10.73 (should be -11.73) (K_{s2} from ref. [v]) $K[UO_2(OH)_2(s)+H_2L\rightleftharpoons UO_2L(s)+2H_2O]$ 0.22 (should be 1.22)	t+o
	pol	25	Na_2L, NaHL, var	ev cpx, probably $UO_2L_3^{4-}$, $UO_2L_2(OH)_2^{4-}$	u
	gl	18–20	var	K_3 5.5	v
	aix		0.5 ($NaNO_3$)	K_3 7.0, $\beta_3\simeq 23$	w
	var		var	ev polyn cpx with $\leqq 1L^{2-}$ per UO_2^{2+} (sol, sp, aix, cix)	x
	gl	25	0.1 $NaClO_4$	$\beta_{1pq}[qUO_2^{2+}+pH_2O+1CO_3^{2-}\rightleftharpoons$ $((UO_2)_q(OH)_p(CO_3)_1)^{+2q-p-21}+pH$ $\beta_{3,0,1}$ 21.57, $\beta_{2,0,1}$ 16.16, $\beta_{1,1,1}$ 4.10, $\beta_{0,7,3}$ -24.20	y*
Hydroxyl-amine: $H_2N\cdot OH$	var		var	ev $UO_2L_2OOH^-$ (con, pre, tp)	z
	sol, gl	22	2 (KCl)	$L'=NH_2O^-=LOH^-$ $K'_{s2}[UO_2L'_2(H_2O)_3(s)\rightleftharpoons UO_2L'_2+3H_2O]$ -3.39 $K(UO_2^{2+}+L\rightleftharpoons UO_2L'+H^+)$ -1.15 $K(UO_2L'_2+L\rightleftharpoons UO_2L'^-_3+H^+)$ -7.38 $K_a(HUO_2L'_3)$ -7.47 [$K_b(L)$ -7.97, K_w ?]	aa

Table 2.24 (continued)

Complexing agent	Method	Temp.	Medium	Log of equilibrium constant; remarks	Reference
Azide, N_3^-	sp	25 ?	0.3 ($NaClO_4$)	K_1 3.50 [$K_1(H^+)$ 4.72]	bb
	sp	25	var	K_1 2.31 ev UO_2L_2	cc
	con, tp		var	ev ani cpx ($UO_2L_3^-$?) at 0.5 *M* L^-	
	sp		var	ev UO_2L^+, UO_2L_2, $UO_2L_3^-$	dd
	sp	RT	var	K_1 2.64	ee
Nitrate, NO_3^-	sp	25	I ($NaClO_4$), 2H^+	K_1 −0.57 (I=7), −0.68 (I=5.38)	ff
	X		solid	ev $UO_2L_3^-$ in $RbUO_2L_2$(s)	gg=hh
	qh	20	1 ($NaClO_4$)	K_1 −0.3	ii
	dis	25	0 corr	K_{d2} [$UO_2^{2+}+2L^- \rightleftharpoons UO_2L_2$(org)] −3.20 (org=$Bu_2O$), −2.52 ($Pr^i_2O$), −1.82 (i-$C_5H_{11}OCOCH_3$), 0.73 ($Bu^iCOMe$), 0.87 (cyclohexone)	jj/kk
	dis	25	0 corr	K_{d2} −0.94 (org=Et_2O)	ll+jj/kk
	dis	10–40	2 ($NaClO_4$)	K_1 −0.52 (10°), −0.62 (25°), −0.77 (40°)	mm
	aix		HL var	ev ani cpx in > 8 *M* HL	nn
	dis	25	0 corr	K_{d2} 1.80 [20% $(BuO)_3PO$, 80% kerosene]	oo/kk
	sp		Me_2CO	K_3 3.6 in Me_2CO	pp
	dis	25 ?	0 corr	K_d [$UO_2^{2+}+2L^-+2T(org) \rightleftharpoons UO_2L_2T_2$(org)] 1.71, T=TBP, org=AMSCO 125-90W	qq
	dis	25	HL var	K_d 1.08, T=TBP, org=kersone	
	aix, sp		L^- var	ev HUO_2L_3, $UO_2L_3^-$, no higher cpx	ss
	con	25	EtOH	K_2 3.15, K_3 (?) 1.39 in EtOH, →0	tt
		25	Me_2CO	K_2 3.96, K_3 (?) 2.46 in Me_2CO	
	dis	0–25	1HL	K_d 1.35 (25°), ΔH_d=6.3, ΔS_d=−15, T=TBP, org=n-dodecane, also K_d, ΔH_d, ΔS_d for T=twenty other compounds R_3PO_4 and R_3PO_3	uu
	dis	19	0 corr	K_d 3.40, T=(Isopentyloxy)$_2$(CH_3)PO, org=kerosene	vv
	dis	25	0.72 H(ClO_4)	K'_d [$UO_2^{2+}+2L^-+2T(in\ CCl_4) \rightleftharpoons UO_2L_2T_2$(in CCl_4)] ≃ 1 [T=$(BuO)_3PO$], ≃ 2 [T=$(BuO)_2BuPO$], ≃ 6 (T=Bu_3PO), K_1 −0.2	ww
	dis	25	0 corr	K''_d [$UO_2^{2+}+2L^-+hH_2O \rightleftharpoons UO_2L_2(H_2O)_h$(org)] −1.0 ($Et_2O$, *h*=4), −2.37 ($Pr^i_2O$, *h*=4), −3.22 ($Bu_2O$, *h*=2.5), −4.06 (Isopentyl ether, *h*=2.2), −3.84 [$(ClC_2H_4)_2O$, *h*=3.8], −1.80 (Me_2phthalate, *h*=4), −3.72 (Isopentyl benzoate, *h*=2.5), −0.48 [$(BuOC_2H_4)_2O$, *h*=4], also K''_d for 9 other esters	xx
	X		solid	ev $UO_2L_2(H_2O)_2$ units in $UO_2L_2(H_2O)_6$(s)	yy
	dis	25	0 corr	K_d 1.43, ΔH_d=−4.43, ΔS_d=−7.7, T=TBP, org=CCl_4 or kerosene	zz
	sp		BuOH	K ($UO_2L_2+T \rightleftharpoons UO_2L_2T$) 0.8 in BuOH, T=TBP	a1
	dis	25	0.71H(ClO_4)	K_1 ≃ −0.2, K'_d 2.87 [T=$(BuO)_2BuPO$], 4.47 (T=$BuOBu_2PO$), ev mixed ClO_4^--L^- cpx with $(BuO)_3PO$	b1
	dis		HL var	K [BHL(org)+$UO_2^{2+}+2L^- \rightleftharpoons$ $(BH)_2UO_2L_3$(org)] 0.31 (org=CCl_4), 0.46 (org=o-xylene), B=$(C_8H_{17})_3N$	c1
	cix	32	1 ($NaClO_4$)	K_1 −1.4, K_2 0, K_3 0.9	k
	con	25	org	K_2 3.3 to 3.5 in MeOH, 3.3 to 3.9 in EtOH 6.1 to 6.7 in Me COEt, also K_2 and K_3 values for other solvents.	d1
	dis	25	0 corr	K_d 4.75 (org=kerosene), 4.40 (org=dodecane), T=Bu_3PO_4 [K_1 −0.7]	/e1
	sp		$MeNO_2$	K_4 0.67 in $MeNO_2$	f1

Table 2.24 (continued)

Complexing agent	Method	Temp.	Medium	Log of equilibrium constant; remarks	Reference
	dis	?	0 corr	$K_d'''[UO_2^{2+}+2L^-+tT(\text{in } C_6H_6)+hH_2O \rightleftharpoons UO_2L_2T_t(H_2O)_h]$ −4.74 ($t=h=2$), −3.82 ($t=h=3$) ($T=Bu_2O$); −3.55 ($t=h=3$, $T=Pr_2O$; −3.04 ($t=h=3$, $T=PrOEt$); −3.37 ($t=3$, $h=2$); −3.05 ($t=4$, $h=3$) ($T=PhCH_2OMe$)	g1
	sp, tp	20–60	HL var	ev ani cpx in > 5 M HL (20°), K_1 and K_2 rise which T	h1
	con	25	I ($NaClO_4$)	K_1 −0.43 (I=0.54), −0.70 (I=0.82), −0.72 (I=1.06).	i1*
	tp ?		L^- var ?	ev $UO_2(H_2O)_2L_2$	j1*
	dis	20	8 ($HClO_4$)	K_1 0.47, β_2 −1.52	k1*
	$K_d[UO_2^{2+}+2L^-+2T(org) \rightleftharpoons UO_2L_2T_2(org)]$ $K_d'[UO_2^{2+}+2L^-+2T(\text{in } CCl_4) \rightleftharpoons UO_2L_2T_2(\text{in } CCl_4)]$ $K_d''[UO_2^{2+}+2L^-+hH_2O \rightleftharpoons UO_2L_2(H_2O)_h(org)]$				
Phosphate, PO_4^{3-}	sp	25	1 $HClO_4$	$K[UO_2^{2+}+H_3L \rightleftharpoons UO_2H_{1+x}L^{x+}+(2-x)H^+]$ 1.58	l1
	sol	25	1 $HClO_4$	$K[UO_2^{2+}+H_3L \rightleftharpoons UO_2H_{1+x}L^{x+}+(2-x)H^+]$ 1.57 $x=1$ or 2	m1+n1
			1 $HClO_4$?	$K[UO_2^{2+}+2H_3L \rightleftharpoons UO_2(H_2L)_2+2H^+]$ 1.18	
			1 $HClO_4$?	$K[UO_2^{2+}+3H_3L \rightleftharpoons UO_2(H_2L)_2H_3L+2H^+]$ 2.30	
			1 $HClO_4$	$K_s[UO_2HL(s)+2H^+ \rightleftharpoons UO_2^{2+}+H_3L]$ −2.85	
			1 $HClO_4$	$K_s[UO_2HL(s)+xH^+ \rightleftharpoons UO_2H_{1+x}L^{x+}]$ −1.29 $x=1$ or 2	
			H_3L var	$K_s[UO_2HL(s)+H_3L \rightleftharpoons UO_2(H_2L)_2]$ −1.7	
			H_3L var	$K_s[UO_2HL(s)+2H_3L \rightleftharpoons UO_2(H_2L)_2H_3L]$ −0.55	
			H_3L dil	$K_s[(UO_2)_3L_2(s)+6H^+ \rightleftharpoons 3UO_2^{2+}+2H_3L]$ −6.15	
			H_3L dil	$K_s[(UO_2)_3L_2(s)+H_3L+3H^+ \rightleftharpoons 3UO_2(H_2L)_2]$ −2.40	
			H_3L dil	$K_s[(UO_2)_3L_2(s)+4H_3L \rightleftharpoons 3UO_2(H_2L)_2]$ −2.89	
			H_3L dil	$K_s[(UO_2)_3L_2(s)+7H_3L \rightleftharpoons 3UO_2(H_2L)_2H_3L]$ 0.53	
			solid: "$UO_2HL(s)$" $= UO_2HPO_4(H_2O)_4(s)$		
	sol	25	var	ev $UO_2H_2L^+$, $UO_2H_3L^{2+}$, $UO_2(H_2L)_2$	o1≃p1
	aix, gl	20	0 corr	H_2L^-: K_1 3.0, K_2 2.5, K_3 1.9; H_3L: $K_1<1.8$, β_2 3.9, K_3 1.4 $K_a[UO_2(H_3L)_2^{2+}]$ −0.5, −2.3, −1.4; $K_a[UO_2(H_3L)_2^{2+}]$ −1.5, −1.1 (H^+: K_{13} 2.1)	q1/r1
	sp	25	var	ev $UO_2H_2L^+$, $UO_2H_3L^{2+}$, $UO_2(H_2L)_2$, $UO_2H_5L_3^+$	s1
	sp	25	1 ($NaClO_4$)	$\beta[UO_2^{2+}(H_3L)H_{-1}^+]$ 0.72, $\beta(UO_2^{2+}H_3L)$ 0.76 $\beta[UO_2^{2+}(H_3L)_2H_{-2}^+]$ 0.41, $\beta[UO_2^{2+}(H^3L)_2H_{-1}^+]$ 1.33 (H^+: K_{13} 1.68)	s1/t1
	sol	19–20	var	$K_s(UO_2^{2+}NH_4^+L^{3-})$ −26.36 $K_s(UO_2^{2+}K^+L^{3-})$ −23.11 $K_s(UO_2^{2+}HL^{2-})$ −10.67, (H^+: K_1 12.44, K_{12} 6.71, K_{13} 1.96)	u1
	sp	25	1.07 $NaClO_4$)	$\beta(UO_2^{2+}H_{-1}^+H_3L)$ 1.19	v1=w1
	dis	25	1.07 ($NaClO_4$)	$\beta(UO_2^{2+}H_{-2}^+(H_3L)_2)$ 1.34 $\beta(UO_2^{2+}H_{-2}^+(H_3L)_3)$ 1.01	
	aix	25	H_3L var	ev ani cpx	x1
	sol	19–20	dil	$K_s[(UO_2^{2+})_3L_2^{3-}]$ −49.1	y1
	sol, gl	25	var	$K_s[(UO_2^{2+})_3L_2^{3-}]$ −46.68 (K_1 12.44, K_{12} 6.71, K_{13} 1.96)	z1
	sol	25	dil	$K_s[NH_4^+UO_2^{2+}L^{3-}(H_2O)_3]$ −25.44 (H^+: K_1 12.32, K_{12} 7.21, K_{13} 2.12)	a2
Triphosphate, $P_3O_{10}^{5-}$	con		var	ev cpx $2L/UO_2^{2+}$	b2

Table 2.24 (continued)

Complexing agent	Method	Temp.	Medium	Log of equilibrium constant; remarks	Reference
Polyphosphate $(P_nO_{3n+1})^{(n+2)-}$	pH	25	dil	$n \approx 5$: K_1 3.0, or β_2 6.0, or β_3 9.5	c2
Arsenate, AsO_4^{3-}	sol, gl	20	var	$K_s(UO_2^{2+}HL^{2-})$ −10.50, $K_s(UO_2^{2+}Li^+L^{3-})$ −18.82, $K_s(UO_2^{2+}Na^+L^{3-})$ −21.87 $K_s(UO_2^{2+}K^+L^{3-})$ −22.60, $K_s(UO_2^{2+}NH_4^+L^{3-})$ −23.77, (H^+: K_1 11.53, K_{12} 6.77, K_{13} 2.25)	d2
Peroxide, O_2^{2-}	sol, gl	25	0 corr	$K[UO_2L(s) + H^+ \rightleftharpoons UO_2LH^+]$ −1.44 $K[UO_2L(s) + 2H^+ \rightleftharpoons UOL^{2+} + H_2O]$ 0.18 $K[UO_2L(s) + OH^- \rightleftharpoons UO_2LOH^-]$ −1.96 $K[UO_2L(s) + 2OH^- \rightleftharpoons UO_3L^{2-} + H_2O]$ −0.05	e2
	sp, gl		var	$K[2UO_2^{2+} + 2H_2L + H_2O \rightleftharpoons H_2U_2O_5L_2 + 4H^+)$ −2.7 $K_a(H_2U_2O_5L_2) \sim -7$ $K_a(HU_2O_5L_2^-) \sim -10$	f2=g2
	sp		Na_2CO_3 var	apparent K_1 4.71 (in 0.05–1.5 M Na_2CO_3)	h2
	sol	25	var	ev $UO_2L_3^{4-}$, $UO_2L_2^{2-}$	i2
	gl, sol	25	0 corr	$K_s[UO_2L(s) + H^+ \rightleftharpoons UO_2LH^+]$ −1.44 ?, $K_s[UO_2L(s) + OH^- \rightleftharpoons UO_2LOH^-]$ −1.96 ?, polyn. cpx ?, "$UO_2L(s)$" $\rightleftharpoons UO_2L(H_2O)_{2 \text{ or } x}$	j2
	var		var	$K_a(HUO_2L_3^{3-})$ −12.3 (Bi, gl, sp)	g2
	gl		0.85	ev $HUO_2L_3^{3-}$	k2+g2
	fp	−1	$KClO_3$ sat	ev $UO_2L_3^{4-}$	k2
	fp	−19	$NaNO_3$ sat	ev $(UO_2)_2L_4^{4-}$	
	sp		0.4 (KNO_3) →0	$K[UO_2(CO_3)_3^{4-} + H_2L \rightleftharpoons UO_2(CO_3)_2HL^{2-} + HCO_3^-)$ 2.0 K 2.2	l2
	var		var	ev $H_2(UO_2)_2OL_2$, $UO_2(CO_3)_2HL^{3-}$ $UO_2(CO_3)_2L^{4-}$, $K_a(UO_2(CO_3)_2HL^{3-})$ −10.6 (sp, fp, gl)	m2
	sol	78–114	var	$K_{s1}[UO_2L(s) \rightleftharpoons UO_2L]$ −4.0 (78°) $^*K_{s0}[UO_2L(s) + 2H^+ \rightleftharpoons UO_2^{2+} + H_2L]$ −1.44 (78–114°)	n2
	gl, sp		var	ev $UO_2L_2^{2-}$, $UO_2L_3H^{3-}$, $UO_2L_3^{4-}$ $(UO_2)_2L_2OH_2$, $(UO_2)_2L_2OH^-$, $(UO_2)_2L_2O^{2-}$, $(UO_2)_2L_4OH^{5-}$, $(UO_2)_4L_5O_2H^{5-}$, $(UO_2)_4L_4O_3^{6-}$; $K_a(UO_2L_3H^{3-}) \simeq -12.5$	o2
	sol	20	?	$^*K_s[UO_2L(H_2O)_4(s) + 2H^+ = UO_2^{2+} + H_2L + 4H_2O)]$ −2.86	p2
see also F^--O_2^{2-}, mixed ligands (this table) and Table 2.25: $C_6H_8O_7$—O_2^{2-} and $C_{10}H_{16}O_8N_2$—O_2^{2-}					
Thiosulfate, $S_2O_3^{2-}$	dis			ev cpx	f
	sol	25	var	K_{s0} −3.4, no ev cpx [$K_a(UO_2^{2+})$ −4.2]	q2
Sulfite, SO_3^{2-}	sp		var	ev $UO_2L_2^{2-}$, strong cpx	r2
	sol	25	var	K_{s0} −8.59, β_2 7.10 (H^+: K_1 7.00, K_{12} 1.77)	s2
Sulfate, SO_4^{2-}	sp	25	2.65 ($NaClO_4$) 2H+	$^*K_1$ 0.70	ff
		25	3.5 (?), 2H+	K_1 1.83 [$K_1(H^+)$ 1.125]	ff/t2
	qh	20	1 ($NaClO_4$)	K_1 1.70, K_2 0.84, K_3 0.86 $\beta(UO_2^{2+}L^{2-}Ac^-)$ 3.78, $\beta(UO_2^{2+}L^{2-}{}_2Ac^-)$ 4.60	ii
	sp	20	1 ($NaClO_4$)	K_1 1.75, K_2 0.90	
	qh	20	0 corr	K_1 2.93, K_2 0.85, K_3 −0.38	ii/u2
	sp	20	0 corr	K_1 2.98, K_2 0.90, $K_3 \ll -0.5$	
	con	25	0 corr	K_1 3.23 ?	v2
	dis	10	2 ($NaClO_4$)	K_1 1.80, K_2 0.96 [$K_1(H^+)$ 1.01]	mm
	dis	25	2 ($NaClO_4$)	K_1 1.88, K_2 0.97 [$K_1(H^+)$ 1.08] $\Delta H_1 = 2.3$, $\Delta H_2 = -0.9$, $\Delta S_1 = 16$, $\Delta S_2 = 2$	
	dis	40	2 ($NaClO_4$)	K_1 1.98, K_2 0.93 [$K_1(H^+)$ 1.17]	
	aix		var	ev $UO_2L_3^{4-}$, $U_2O_5L_3^{4-}$	w2
	aix	25	var	ev UO_2L, $UO_2L_2^{2-}$, $UO_2L_3^{4-}$, $U_2O_5L_3^{4-}$	x2
	sp	25	→0	K_1 2.96, $K_2 \sim 1$	h

Table 2.24 (continued)

Complexing agent	Method	Temp.	Medium	Log of equilibrium constant; remarks	Reference
	sol	250	0 corr	$\beta(UO_2^{2+}L^{2-}{}_2Ba^{2+})$ 9.3, no ev polyn cpx $UO_2^{2+}-L^{2-}$	y2/z2
	sol	39	0 corr	$\beta(UO_2^{2+}L^{2-}{}_2Ag^+{}_2)$ 6.18 [$K_{s0}(Ag^+{}_2L^{2-})$ − 4.66]	
	sol	150-250	0 corr	K [0.5 La_2L_3(s) + 0.5L^{2-} + $UO_2^{2+} \rightleftharpoons UO_2L_2La^+$] 0–1	
	dis	25	1	K_1 1.53, K_2 0.78, $\beta_3 < 2.1$	u2
		25	0 corr	K_1 2.76, K_2 0.78, $\beta_3 < 2.1$	
	sol	25–200	0 corr	K_1 2.72 + [0.02939 (t − 25) + 3.230 × 10^{-4} (t − 25)^2] log e K_2 1.48 + [5.388 × 10^{-3} (t − 25) − 9.392 × 10^{-7} (t − 25)^2] loge ΔH_1 = 5 (25°), 63 (200°), ΔS_1 = 30 (25°), 176 (200°), ΔH_2 = 0.9 (25°), 2.5 (200°), ΔS_2 = 10 (25°), 14 (200°)	a3
	sp	25	1 ($NaClO_4$)	K_1 1.81, K_2 0.48	b3
	aix	32	1 ($NaClO_4$)	K_1 1.63, K_2 2.15	k
	sp	25	20% MeOH	K_1 3.88, K_2 ca 1.6 in 20% MeOH	c3
	gl	25	UO_2L var	K_1 3.85, $\beta(UO_2OH^+L^{2-})$ 3.32 [$K_a(UO_2^{2+})$ − 4.59]	d3
	con	25	var	ev $UO_2L_2{}^{2-}$	e3*
	tp	25	var	K_1 1.7, K_2 0.85	k4*
Selenite, SeO_3^{2-}	sol, gl	20	var	K_{s0} − 10.42 (H^+: K_1 8.0, K_{12} 2.4	f3
Fluoride, F^-	aix	25	HCl var	* K_1 1.18	g3
				K_1 4.32	g3/h3
	sp		var	K_1 5.5, $\beta_4 \sim 8$	h3
	fp	~0	UO_2F_2 var	† K_{42} [2$UO_2F_2 \rightleftharpoons (UO_2F_2)_2$] 0.18	i3
	qh	20	1 ($NaClO_4$)	K_1 4.59, K_2 3.34, K_3 2.56, K_4 1.36 [$K_1(H^+)$ 2.94]	j3
	con	25	0 corr	$K_2 \sim 4.4$ ev other cpx	k3
	dis	10–40	2$NaClO_4$)	* K_1 1.74 (10°), 1.42 (25°), 1.32 (40°)	mm
		25	2 ($NaClO_4$)	$\Delta^* H_1$ = − 5.4, $\Delta^* S_1$ = − 12	
	dis	25	C ($NaClO_4$)	* K_1 1.42 (C=2), 1.43 (C=1), 1.38 (C=0.5), 1.57 (C=0.25), 1.71 (C=0.05)	
	cfu	0–30	UO_2F_2 var	† K_{42} 0.48 (0°), 0.85 (30°)	l3
	X		solid	ev $UO_2F_3^{3-}$ in $K_3UO_2F_5$(s)	m3
	qh	20	1 (Na)ClO_4	K_1 4.54, K_2 3.34, K_3 2.57, K_4 1.34 [$K_1(H^+)$ 2.93], no ev polyn cpx for < 0.1 M UO_2^{2+}	n3+j3
	aix		var	ev $UO_2F_3^-$	o3
	aix	25	1 HF, HCl var	ev ani cpx	p3=q3
	con	−5	HF	K [UO_2F_2(+ 4HF) $\rightleftharpoons$ UF_6 + 2H_2O] − 3.95 in HF(1), m units	r3
	sp	25 ?	0.65 ($NaClO_4$)	* K_1 1.18	s3
	sp	20	dil	K_1 4.77 [$K_1(H^+)$ 3.10]	t3
	sol	25	var	$K_1 \simeq 4.6$, ev UO_2F_2	u3
	pre, con		var	ev $UO_2F_2SO_4^{2-}$ and other mixed cpx	v3
	con	25	var	ev $UO_2L_3^-$	e3
	cix	25	I $HClO_4$	* K_1 1.31 (I=2.10), 1.52 (I=1.04), 1.54 (I=0.51), 1.57 (I=0.20)	w3*
F^-—O_2^{2-}, mixed ligands	sp		var	K [2$(UO_2F_4)^{2-}$ + H_2O_2 = $((UO_2)_2(OO)F_x)^{(x-2)}$ + 2H^+ + (8 − x) F^-] − 8.8 (x=5) K [2$(UO_2F_4)^{2-}$ + 2$H_2O_2 \rightleftharpoons ((UO_2)_2(OO)_2F_5)^{5-}$ + 4H^+ + 3F^-] − 20.5 K [$((UO_2)_2(OO)_2F_5)^{5-}$ + $H_2O_2 \rightleftharpoons ((UO_2)_2(OO)_3F_2^{4-})$ + 2H^+ + 3F^-] − 14.3 K [$((UO_2)_2(OO)_3F_2)^{4-}$ + $H_2O \rightleftharpoons ((UO_2)_2(OO)_3(OH)F_2)^{5-}$ + H^+] − 8.1 K [$(UO_2)_2(OO)_3(OH)F_2)^{5-}$ + 3$H_2O_2 \rightleftharpoons$ 2$(UO_2)_2(OO)_2(OOH)(H_2O))^{3-}$ + 3H^+ + 2F^-] − 30.3 K [$(UO_2F_4)^{2-}$ + 3H_2O_2 = $(UO_2(OO)_2(OOH))^{3-}$ + 5H^+ + 4F^-] − 36.6 K [$(UO_2F_4)^{2-}$ + 3$H_2O_2 \rightleftharpoons (UO_2(OO)_3)^{4-}$ + 6H^+ + 4F^-] − 49	x3*

Table 2.24 (continued)

Complexing agent	Method	Temp.	Medium	Log of equilibrium constant; remarks	Reference
				$K[UO_2^{2+}+3H_2O_2 \rightleftharpoons (UO_2(OO)_3)^{4-}+6H^+]-37.2$	
				$K[UO_2^{2+}+3O_2^{2-} \rightleftharpoons UO_2(OO)_3^{4-}]73.0$ est	
				$K[2UO_2^{2+}+5F^-+O_2^{2-} \rightleftharpoons (UO_2)_2(OO)F_5^{3-}]51.6$ est	
				$K[2UO_2^{2+}+5F^-+2O_2^{2-} \rightleftharpoons (UO_2)_2(OO)_2F^{5-}]76.4$ est	
				$K[2UO_2^{2+}+2F^-+3O_2^{2-} \rightleftharpoons (UO_2)_2(OO)_3F_2^{4-}]99$ est	
				$K[(UO_2)_2(OO)_2(H_2O)_8+5F^- \rightleftharpoons (UO_2)_2(OO)_2F_5^{5-}]5.87$	
Chloride, Cl^-	qh	20	1 $(NaClO_4)$	$K_1-0.10$	ii
	sp	20	1 $(NaClO_4)$	$K_1-0.30$?	
	var	25	0 corr	$K_1 0.38$ (gl, pol, sp)	y3
	X		solid	ev $UO_2Cl_4^{2-}$ in $(Me_4N)_2UO_2Cl_4$(s)	z3
	dis	10–40	2 $(NaClO_4)$	$K_1-0.24(10°), -0.06(25°), 0.06(40°)$	mm
		25	2 $(NaClO_4)$	$\Delta H_1=3.8, \Delta S_1=12$	
	aix	25	HCl var	ev ani cpx in >0.5 M HCl	a4
		25	0 corr	$K_1-0.1, K_2-0.82, K_3-1.70$	a4/b4
	sp	25	0 corr	$K_1 0.22$	i
	sp	25	→0	$K_1 0.21$	h
	aix		HCl var	ev ani cpx in ≧6 M HCl	nn
	dis, sp	15 ?	var	$K_d[UO_2^{2+}+2Cl^-+2BHCl(org)=(BH)_2UOCl_4(org)]1.5$, org$=CCl_4$, B$=(C_8H_{17})_3N$	c4
	con	25	→0	K_2(?)1.3 to 1.84; also plots of K_2(?) for EtOH$-H_2O$ mixtures of 25–45°	d4
	sp		Me_2CO	ev UO_2Cl^+, UO_2Cl_2, $UO_2Cl_3^-$ in Me_2CO	e4
	sp	25	0–90% EtOH	$K_1 1.64(0\%), 0.78(30\%), 0.29(60\%)$, in p% EtOH, 1.24$(NaClO_4)$, ev UO_2Cl_2, $UO_2Cl_3^-$	f4
	sp	25	90% EtOH	$K_1 2.83$ in 90% EtOH, 0.08 $NaClO_4$	
	cix	32	1 $(NaClO_4)$	$K_1 0.3$	k
	sp	25	50% EtOH	$K_1 1.24$ in 50% EtOH (0 corr)	c3
	fp	−3	KNO_3 sat	$K_1 0.26$	g4
	sp		HL, Me_2CO, var	ev $UO_2L_4^{4-}$	h4*
	con	25	I $(NaClO_4)$	$K_1 0.38(I=0.54), 0.15(I=0.82), -0.05(I=1.06)$	i1*
			→0	$K_1 1.18$	
	tp ?		L^- var ?	ev $UO_2(H_2O)_3L^+\cdot L^-$ at>1M L^-, $UO_2(H_2O)_3L^+\cdot 2L^-$ at>7M L^-	j1*
	dis	20	8 $(HClO_4)$	$K_1-0.09, \beta_2-1.30$	k1*
Perchlorate, ClO_4^-	sp	25	2–6 ClO_4^-	no ev cpx	ff
	sp	25	HL var	ev cpx in >80% (vol) HL	i4
	tp ?		L^- var ?	no ev UO_2L_2 cpx	j1*
Bromide, Br^-	qh	20	1 $(NaClO_4)$	$K_1-0.30$	ii
	sp	25	→0	$K_1-0.20$	h
	sp		HL, MeCO, var	ev $UO_2L_4^{2-}$	h4*
Iodate, IO_3^-	sol	25	0.2 (NH_4Cl)	$K_{s0}-7.01, K_{s2}-4.28, K_{s3}-3.34, \beta_2 2.73, K_3 0.94$	j4
	sol	60	0.2 (NH_4Cl)	$K_{s0}-6.65, K_{s2}-3.91, K_{s3}-3.32, \beta_2 2.74, K_3 0.69$	

[a] References with asterisks may be found at the end of this chapter. References without asterisks are given in the compilation of SILLÉN and MARTELL (1964). Table notation is explained in Table 2.19. [b] MORACHEVSKII and BELYAEVA (1956). [c] MORACHEVSKII et al. (1958). [d] HOSTELTLER and GARRELS (1962). [e] TANANAEV et al. (1956). [f] ZIEGLER (1959). [g] AHRLAND (1949). [h] DAVIES and MONK (1957). [i] BALE et al. (1957). [j] MARKOV and TRAGGEIM (1960). [k] BANERJEA and TRIPATHI (1961). [l] ARUTYUNYAN and PORAI-KOSHITS (1963). [m] HASTINGS and SENDROY (1925). [n] BULLWINKEL (1954). [o] MCCLAINE et al. (1955). [p] PARAMONOVA (1955). [q] BLAKE et al. (1956). [r] PORTER (1957). [s] KLYGIN and SMIRNOVA (1959). [t] BABKO and KODENSKAYA (1960). [u] STABROVSKII (1960). [v] PARAMONOVA et al. (1962). [w] PARAMONOVA and NIKOLAEVA (1962). [x] PEREZ-BUSTAMANTE et al.

(1962). [y*] Tsymbal (1969). [z] Zvyaginstsev and Kuznetsov (1959). [aa] Kozlov (1961). [bb] Nair et al. (1961). [cc] Sheriff and Awad (1961). [dd] Sheriff and Awad (1962a). [ee] Sheriff and Awad (1962b). [ff] Betts and Michels (1949). [gg] Hoard and Stroupe (1949). [hh] Hoard and Stroupe (1958). [ii] Ahrland (1951). [jj] Glueckauf et al. (1951). [kk] Rozen (1957). [ll] Karpacheva et al. (1957). [mm] Day and Powers (1954). [nn] Katz and Seaborg (1951). [oo] Rozen and Khorkhorina (1957). [pp] Vdovenko et al. (1957). [qq] Codding (1958). [rr] Iwase et al. (1959). [ss] Foreman et al. (1959). [tt] Jézowska-Trzebiatowska et al. (1958). [uu] Siddall (1959). [vv] Shevchenko et al. (1959). [ww] Voden et al. (1959). [yy] Fleming and Lynton (1960). [zz] Naito (1960). [a1] Minc and Libús (1960). [b1] Pushlenkov et al. (1960). [c1] Shevchenko et al. (1960). [d1] Jézowska-Trzebiatowska and Chmielowska (1961). [e1] Marcus (1961). [f1] Ryan (1961). [g1] Maslova and Fomin (1962). [h1] Pozharskii (1962). [i1*] Ohashi and Morozumi (1967). [j1*] Celeda et al. (1968) [k1*] Lahr and Knoch (1970). [l1] Baes et al. (1953). [m1] Baes and Schreyer (1953). [n1] Schreyer and Baes (1953). [o1] Schreyer and Baes (1954). [p1] Schreyer and Baes (1955). [q1] Marcus (1955). [r1] Marcus (1968). [s1] Baes (1956). [t1] Baes (1958). [u1] Chukhlantsev and Stepanov (1956). [v1] Thamer (1956). [w1] Thamer (1957). [x1] Freiling (1959). [y1] Chukhlantsev and Alyamovskaya (1961). [z1] Karpov (1961). [a2] Klygin et al. (1961). [b2] Giesbrecht and Vincenti (1960). [c2] Vanwaser and Campanella (1950). [d2] Chukhlantsev and Sharova (1956). [e2] Thompson (1955). [f2] Gurevich et al. (1957). [g2] Gurevich and Preobrazhenskaya (1958). [h2] Martin-Frère (1957). [i2] Ratner et al. (1957). [j2] Gayer and Thompson (1958). [k2] Gurevich and Komarov (1959). [l2] Komarov (1959). [m2] Komarov et al. (1959). [n2] Gilpatrick et al. (1959). [o2] Gurevich et al. (1960). [p2] Markov et al. (1960). [q2] Klygin and Kolyada (1960). [r2] Nyiri (1942). [s2] Klygin and Kolyada (1959). [t2] Whiteker and Davidson (1953). [u2] Allen (1958). [v2] Brown et al. (1954a). [w2] Arden and Wood (1956). [x2] Arden and Rowley (1957). [y2] Jones et al. (1957). [z2] Glueckauf (1957). [a3] Lietzke and Stoughton (1960). [b3] Matuso (1960). [c3] Morgans and Monk (1961). [d3] Pozharskii et al. (1963). [e3*] Gol'dinov and Stabrovskii (1963). [f3] Krylov and Chukhlantsev (1957). [g3] Moore and Kraus (1950). [h3] Blake et al. (1951). [i3] Johnson and Kraus (1952). [j3] Ahrland and Larsson (1954). [k3] Brown et al. (1954b). [l3] Johnson et al. (1954). [m3] Zachariasen (1954). [n3] Ahrland et al. (1956). [o3] Bhat and Gokhale (1959). [p3] Nelson et al. (1960). [q3] Faris (1960). [r3] Nikolaev et al. (1960). [s3] Connick and Paul (1961). [t3] Kuteinikov (1961). [u3] Tananaev and Deichman (1961). [v3] Chernyaev et al. (1963). [w3*] Krylov et al. (1968). [x3*] Gurevich and Susorova (1968). [y3] Nelson and Kraus (1951). [z3] Staritzky and Singer (1952). [a4] Kraus et al. (1955). [b4] Moore et al. (1956). [c4] Bizot and Trémillon (1959). [d4] Godlenberg and Amis (1959). [e4] Vdovenko et al. (1959). [f4] Hefley and Amis (1960). [g4] Faucherre and Crego (1962). [h4*] Prigent and Padiou (1963). [i4] Silverman and Moudy (1954). [j4] Klygin et al. (1959). [k4*] Carpentier (1969).

Table 2.25. Uranium (VI) complex ion formation. Organic ligands[a]
Complexing agent: formula, name, ligand L definition.

Method	Temp.	Medium	Log of equilibrium constant; remarks	Reference
CH_2O_2, Formic acid, HL				
gl	31	0.1 ($NaClO_4$)	K_1 2.61	b* = c*
$C_2H_2O_2$, Oxalic acid, H_2L				
Pt	25	0 corr	K_1 5.82, K_2 4.74, H_2L: K_1 2.57	d**
cix	25	0.16 $HClO_4$	HL^-: K_1 3.40, K_2 2.56	e**
cix	25	1 $HClO_4$	HL^-: K_1 2.83, $K_2 \sim 1.85$	e**
cix	25	2 $HClO_4$	HL^-: K_1 2.89, $K_2 \sim 1.85$	e**
$Ag_2C_2O_4$	25	0.069	†K $[(UO_2)_2L_3^{2-} + 2L^{2-} \rightleftharpoons (UO_2)_2L_5^{6-}]$ 4.42	f**
$Ag_2C_2O_4$	25	0.022	†K 4.60, Δ K $[(UO_2)_2L_3^{2-} + 2L^{2-} \rightleftharpoons 2(UO_2L_2)^{2-}]$ 1.32	f**
			K $[2(UO_2L_2)^{2-} + L^{2-} \rightleftharpoons (UO_2)_2L_5^{6-}]$ 3.28	
$Ag_2C_2O_4$	25	0.008	Δ K −0.48	f**
sol	20	var	K_1 6.77, K_2 5.23, β_2 12	g**
			K $[UO_2^{2+} + H_2L \rightleftharpoons UO_2L + 2H^+]$ 1.60	
			K $[UO_2^{2+} + 2H_2L \rightleftharpoons UO_2L_2^{2-} + 4H^+]$ 1.68	
sol	20	$HClO_4$, 1.5 H^+	K_s $[UO_2^{2+}L^{2-}(H_2O)_3]$ −8.66	g**
sol	20	HNO_3, 1.5 H^+	K_s $[UO_2^{2+}L^{2-}(H_2O)_3]$ −8.52	g**

Table 2.25 (continued)

Method	Temp.	Medium	Log of equilibrium constant; remarks	Reference
cix	25	1 $HClO_4$	K_1 6.58 graphically, 6.40 analytically, β_2 10.74	e**/g**
cix	25	2 $HClO_4$	K_1 6.92	e**/g**
sol	25	$\rightarrow 0$	K_1 6.00, K_2 5.08, 4.44 (preferred value)	h**
sp, gl		0.312	K_3 2.1	i**
sp, gl		0.05	$K[2UO_2L+H_2O_2 \rightleftharpoons (UO_2L)_2(00)^{2-}+2H^+]$ −1.62	i**
sp, gl		0.312	$K[2UO_2L_2^{2-}+H_2O_2 \rightleftharpoons (UO_2)_2(00)L_4^{6-}+2H^+]$ −3.66	i**
gl	25	1.0 (KNO_3)	β_2 9.1	j*
$C_2H_4O_2$, Acetic acid, HL				
pol		NaL-HL buffer	K_1 2.63, K_2 2.03, K_3 1.60	k**
E, sp	20	1 $NaClO_4$	K_1 2.38, K_2 1.98, K_3 1.98	l
cix	25	0.16 (NaCl)	K_1 2.38, β_3 6.38	m**
ix	RT	0.5 ($NaNO_3$)	K_3 1.22, β_3 5.89	n**
pH	25	0.2 (KNO_3)	K_1 2.70	o**
tp			β_3 5.61	p**
$C_2H_4O_3$, Glycolic acid, HL				
gl	20	1 $NaClO_4$	K_1 2.42, K_2 1.54, K_3 1.24	q**
cix	25	0.16 (NaCl)	K_1 2.78, K_2 1.30	e**
pH	25	0.16–0.19	K_1 2.75, K_2 1.52	e**
gl	30	0.1 KCl	K_1 2.97, K_2 2.40	r*
dis	20	0.1 $NaClO_4$	K_1 2.71, β_2 4.08, β_3 5.5	s*
gl	25	0.1 ($NaClO_4$)	K_1 2.57	t*
$C_2H_3O_2Cl$ Chloroacetic, HL				
E	20	1 $NaClO_4$	K_1 1.44, K_2 0.85, K_3 0.51	u*
sp	20	1 $NaClO_4$	K_1 1.38, K_2 0.80, K_3 0.37	u*
E	20	1 $NaClO_4$	K_1 1.45, β_2 2.26, β_3 2.82	u*/v*
$C_2H_4O_2S$, Thioglycolic acid (Mercaptoacetic acid), H_2L				
gl	30	0.1 KCl	K_1 2.88, K_2 2.40	r*
$C_2H_5O_2N$, Glycine, HL				
dis	25	0.45 (NaCl) pH 1.90	K_1 1.43	m**
gl	30	0.1 KCl	K_1 7.53, K_2 7.15	r*
$C_2H_6ON_2$ Glycine amide, L				
gl	25	0.15–0.25	K_1 5.15	w
$C_3H_4O_3$ Pyruvic acid, HL				
gl	31	0.1 ($NaClO_4$)	K_1 2.15, K_2 0.59	b*=c*
$C_3H_4O_4$ Malonic acid, H_2L				
gl	25	1.0 (KNO_3)	K_1 5.66, K_2 4.00	j*
gl	31	0.1 ($NaClO_4$)	K_1 5.28, K_2 4.01	b*=c*
$C_3H_6O_2$ Propionic acid, HL				
gl	31	0.1 ($NaClO_4$)	K_1 3.03	b*=c*
$C_3H_6O_3$ β-Hydroxypropionic acid, HL				
pH	31	0.1 $NaClO_4$	K_1 2.74, K_2 2.2, K_3 2	x*
gl	30	0.1 KCl	K_1 3.25, K_2 2.88	r*
$C_3H_6O_3$ Lactic acid, HL				
sp	RT		ev UO_2^{2+}/L cpx in ratio 1/1 pH 3.5	y**
sp, pH	RT		ev UO_2^{2+}/L cpx in ratio 1/1 pH 6; presupposes dimerization	z**
dis	20	0.1 $NaClO_4$	K_1 2.81, β_2 4.56, β_3 5.46	s*
pH	31	0.1 $NaClO_4$	K_1 3.36, K_2 2.2, K_3 2	x*
pH	20	1.0 ($NaClO_4$)	K_1 2.45, K_2 1.66, K_3 1.17	aa*
gl	25	0.1 ($NaClO_4$)	K_1 2.67, β_2 4.66	t*

Table 2.25 (continued)

Method	Temp.	Medium	Log of equilibrium constant; remarks	Reference
$C_3H_6O_2S$ Thiopropionic acid (β-Mercaptopropionic acid), H_2L				
pH	30		K_1 3.7, K_2 7.4	bb*
gl	20	0.1 KNO_3	K_1 7.59, K_2 7.13	cc*
	30	0.1 KNO_3	K_1 7.83, K_2 7.23, $H_{\beta 2} = -10.1$, $S_{\beta 2} = 35.5$	
	40	0.1 KNO_3	K_1 8.02, K_2 7.28	
$C_3H_7O_2N$ β-Aniline, HL				
gl	30	0.1 KCl	K_1 7.78, K_2 7.53	r*
$C_3H_7O_3N$ Serine, HL				
dis	25	0.45 (NaCl) pH 2.05	K_1 0.87	e**
$C_4H_2O_4$ Diketocyclobutenediol (Squaric acid), H_2L				
sp	25	0.5 $NaClO_4$	K_1 3.08	dd*
$C_4H_4O_4$ Maleic acid (*cis*-Butenedioic acid), H_2L				
gl	25	1.0 (KNO_3)	K_1 4.46	j*
pol	30	0.1 $NaClO_4$	ev $UO_2L_2^{2-}$ cpx only	ee*
gl	31	0.1 ($NaClO_4$)	K_1 5.15	b* = c*
$C_4H_4O_4$ Fumaric acid (*trans*-Butenedioic acid), H_2L				
pol	30	0.1 $NaClO_4$	ev $UO_2(HL)_2$ and $UO_2L_2^{2-}$ cpx	ee*
gl	31	0.1 ($NaClO_4$)	K_1 3.05	b* = c*
$C_4H_6O_2$ Crotonic acid (2-Butenoic acid (trans)), HL				
gl	31	0.1 ($NaClO_4$	K_1 2.74	b* = c*
$C_4H_6O_4$ Succinic acid (Butanedioic acid), H_2L				
pH	25	0.2 (KNO_3)	HL^-: K_1 2.62	o*
gl	25	1.0 (KNO_3)	K_1 3.68	j*
gl	31	0.1 ($NaClO_4$)	K_1 4.48	b* = c*
$C_4H_6O_5$ Diglycolic acid (Oxydiacetic acid), H_2L				
gl	31	0.1 ($NaClO_4$)	K_1 4.90	b* = c*
$C_4H_6O_5$ Malic acid, H_2L				
sp	RT		ev UO_2^{2+}/malate cpx in ratio 1/1 and 2/1 pH 3.5	y**
sp, pH	RT		ev UO_2^{2+}/malate cpx in ratio 1/1 as dimer acid solution; 3/3 and 3/2 pH ~ 8	z**
gl	25	I (KNO_3)	$K[2UO_2^{2+} + 2HA^{2-} + 2H_2O \rightleftharpoons (UO_2)_2(HA)_2(OH)_2^{2-} + 2H^+]$ 7.76 (I = 0.136), 9.08 (I → O)	ff*
gl	25	0.20 (KNO_3)	$K[3/2(UO_2)_2(HA)_2(OH)_2^{2-} + 2H_2O \rightleftharpoons (UO_2)_3(HA)_3(OH)_5^{5-} + 2H^+]$ −10.56	ff*
gl	25	1.0 (KNO_3)	$K[UO_2^{2+} + HA^{2-} \rightleftharpoons UO_2A^- + H^+]$ 1.66 $K[UO_2^{2+} + H_3A \rightleftharpoons UO_2A^- + 3H^+]$ −5.55, $K[2UO_2^{2+} + 2H_3A \rightleftharpoons (UO_2)_2A_2^{2-} + 6H^+]$ −7.75, $K[2UO_2A^- \rightleftharpoons (UO_2)_2A_2^{2-}]$ 3.35, $K[(UO_2)_2A_2^{2-} \rightleftharpoons (UO_2)_2A_2(OH)^{3-} + H^+]$ −6.07, $K[3(UO_2)_2A_2^{2-} \rightleftharpoons (UO_2)_3A_3(OH)_2^{5-} + 4H^+]$ −19.35	gg**
$C_4H_6O_6$ Tartaric acid, H_2L				
sp	RT		ev UO_2^{2+}/tartrate cpx in ratio 1/1, 2/1 ?, and 3/1 pH 3.5, 4.6	y**
sp, pH	RT		ev UO_2^{2+}/tartrate cpx in ratio 1/1as dimer acid solution; 3/3 and 3/2 pH ~ 8	z**
gl	25	I (KNO_3)	$K[2UO_2^{2+} + 2H_2A^{2-} + 2H_2O \rightleftharpoons (UO_2)_2(H_2A)_2(OH)_2^{2-} + 2H^+]$ 5.54 (I = 0.136), 6.76 (I → O)	ff*
gl	25	0.20 (KNO_3)	$K[3/2(UO_2)_2(H_2A)_2(OH)_2^{2-} + 2H_2O \rightleftharpoons (UO_2)_3(H_2A)_3(OH)_5^{5-} + 2H^+]$ −9.7	ff*

Table 2.25 (continued)

Method	Temp.	Medium	Log of equilibrium constant; remarks	Reference
gl	25	1.0 (KNO_3)	$K[UO_2^{2+} + H_2A^{2-} \rightleftharpoons UO_2HA^- + H^+]$ 0.75, $K[UO_2^{2+} + H_4A \rightleftharpoons UO_2HA + 3H^+]$ −5.62, $K[UO_2^{2+} + 2H_4A \rightleftharpoons (UO_2HA)_2^{2-} + 6H^+]$ −8.00, $K[2UO_2HA^- \rightleftharpoons (UO_2)_2(HA)_2^{2-}]$ 3.24, $K[(UO_2)_2(HA)_2^{2-} \rightleftharpoons (UO_2)_2(HA)_2(OH)^{3-}]$ −5.26 $K[(UO_2)_2(HA)_2 \rightleftharpoons 2(UO_2)_3(HA)_3(OH)_2^{5-}]$ −17.91	gg*
$C_4H_8O_2$ n-Butyric acid, HL				
gl	31	0.1 $(NaClO_4)$	K_1 2.91, K_2 1.62	hh*
$C_4H_8O_2$ Isobutyric acid, HL				
gl	31	0.1 $(NaClO_4)$	K_1 3.40, K_2 2.43	hh*
$C_4H_8O_3$ α-Hydroxybutyric acid, HL				
pH	31	0.1 $NaClO_4$	K_1 3.29, K_2 1.7	hh*
gl	25	0.1 $(NaClO_4)$	K_1 2.72, β_2 4.67	t*
$C_4H_8O_3$ β-Hydroxybutyric acid, HL				
pH	31	0.1 $NaClO_4$	K_1 2.70, K_2 1.4	x*
$C_4H_8O_3$ α-Hydroxyisobutyric acid, HL				
dis	20	0.1 $NaClO_4$	K_1 3.58, β_2 5.30, β_3 7.01	s*
$C_4H_6O_4S$ Thiomalic acid (Mercaptosuccinic acid), H_3L				
gl	30	0.1 KCl	K_1 3.56, K_2 3.42	r*
sp	25	var	K_1 3.0	ii*
gl	31	0.1 $(NaClO_4)$	K_1 3.75, 3.82	jj*
gl	45	0.1 $(NaClO_4)$	K_1 3.91	jj*
gl	31	0.1 $(NaClO_4)$	K_1 3.71	b* = c*
$C_4H_7O_4N$ Aspartic acid, H_2L				
gl	30	0.1 KCl	K_1 8.00	r*
pH	25	0.2 (KNO_3)	HL^-: K_1 2.62	o*
$C_4H_7O_4N$ Iminodiacetic acid, H_2L				
gl	25	I (KNO_3)	K_1 8.93 (I = 0.1), 8.73 (I = 1.0)	kk*
pol	30	0.15 $NaClO_4$	HL^-: β_2 3.92, ev $UO_2L_2^{2-}$ and $UO_2(OH)L_2^{3-}$ cpx	ll*
$C_4H_{11}O_4P$ Diethylphosphoric acid, HL				
cix	20	1 $NaClO_4$	K_1 2.32	mm*
sp	20	1 $NaClO_4$	K_1 2.46, β_2 4.16	mm*
$C_5H_6O_4$ Itaconic acid (Methylenesuccinic acid), H_2L				
gl	28, 31	0.1 $(NaClO_4)$	K_1 4.85 (28°), 4.72 (31°)	jj*
gl	31	0.1 $(NaClO_4)$	K_1 4.86	b* = c*
$C_5H_8O_2$ Pentane-2,4-dione (acetylacetone), HL				
gl	30	50% dioxan	K_1 9.32, K_2 7.60	nn
gl	10–40	0 corr	K_1 7.94 (10°), 7.66 (20°), 7.74 (30°) 7.42 (40°), K_2 6.53 (10°), 6.49 (20°), 6.43 (30°), 6.26 (40°), $\Delta H_2 = -40$, $\Delta S_2 = 16$	oo, pp
dis	25	0.1 $NaClO_4$ $(H_2O—CHCl_3)$	K_1 6.8, K_2 6.3 $K[UO_2^{2+} + OH^- + L^- \rightleftharpoons UO_2LOH] \leq 15.6$ $K[UO_2^{2+} + L^- + HL \rightleftharpoons UO_2LHL^+]$ 8.7 $K[UO_2^{2+} + 2L^- + HL \rightleftharpoons UO_2L_2HL]$ 14.8 $K[UO_2^{2+} + HL \rightleftharpoons UO_2HL^{2+}] < 0$ $K[UO_2^{2+} + 2HL \rightleftharpoons UO_2(HL)_2^{2+}] < 1$ $K[UO_2^{2+} + 3HL \rightleftharpoons UO_2(HL)_3^{2+}] < 2$	qq
sp	25	EtOH	HL: K_1 2.43	rr*
pol		1 $(NaClO_4)$	K_1 7.5, K_2 4.8	ss*
pol		1 $(NaClO_4)$(?)	β_2 12.5	tt*

Table 2.25 (continued)

Method	Temp.	Medium	Log of equilibrium constant; remarks	Reference
$C_5H_8O_7$ Trihydroxyglutaric acid (2,3,4-Trihydroxypentanedioic acid), H_2L				
sp	20	2(H^+, $NaClO_4$)	†K_n $[UO_2^{2+} + n(H_3A) \rightleftharpoons UO_2(H_2A)_n^{2-n} + nH^+]$ where $H_3A \rightleftharpoons H_2L$ †K_1 0.30, †K_2 0.46	uu*
$C_5H_{10}O_3$ α-Hydroxy-n-valeric acid, HL				
gl	25	0.1 ($NaClO_4$)	K_1 2.80, β_2 4.45	t*
$C_5H_9O_4N$ Glutamic acid (2-Aminopentanedioic acid), H_2L				
pH	25	0.2 (KNO_3)	HL^-: K_1 2.66	o*
$C_5H_9O_4N$ Methyliminodiacetic acid, H_2L				
gl	25	0.1 (KNO_3)	K_1 9.70	vv*
gl	25	1.0 (KNO_3)	$K[UO_2(OH)L^- + H^+ \rightleftharpoons UO_2(H_2O)L]$ 5.92, $K[2UO_2(OH)L^- \rightleftharpoons (UO_2)_2(OH)_2L_2^{2-}]$ 3.41 $K[UO_2(H_2O)L \rightleftharpoons (UO_2)_2(OH)_2L_2^{2-} + 2H^+]$ −8.43	vv*
$C_5H_{11}NS_2$ Diethyldithiocarbamic acid, HL				
sp	25	1 ($NaClO_4$)	β_4 17.82–17.16	ww
$C_6H_6O_2$ Catechol(1,2-Dihydroxybenzene), H_2L				
pH, sp	30	0.1 KNO_3	K_1 14.59 $K[UO_2^{2+} + H_2L \rightleftharpoons UO_2L + 2H^+]$ −6.54 $K[2UO_2^{2+} + 2H_2L \rightleftharpoons (UO_2L)_2 + 4H^+]$ −10 $K[(UO_2L)_2 + H_2O \rightleftharpoons (UO_2L)_2(OH)^- + H^+]$ −5.48 ev $(UO_2L)_3(OH)_2^{2-}$ or $(UO_2L)_6(OH)_4^{4-}$	xx*
$C_6H_6O_3$ 3-Hydroxy-2-methyl-4-pyrone (Maltol), HL				
sp	25	0.1 $NaClO_4$	*K_1 −0.34, −0.31 *K_2 −1.94, −1.89 *K_3 −5.35 K_1 8.3, K_2 6.7, K_3 3.26	yy*
$C_6H_6O_3$—CO_3^{2-} Maltol (HL)-carbonate mixed ligands				
sp	25	0.1 $NaClO_4$	$K[UO_2L_3^- + HCO_3^- \rightleftharpoons UO_2L_2CO_3^{2-} + HL]$ 0.2 $K[UO_2L_2CO_3^{2-} + 2HCO_3^- \rightleftharpoons UO_2(CO_3)_3^{4-} + 2HL]$ −0.9 $K[UO_2(CO_3)_3^{4-} + HL \rightleftharpoons UO_2(CO_3)_2L^{3-} + HCO_3^-]$ 0.9 $K[UO_2(CO_3)_2L^{3-} + 2HL \rightleftharpoons UO_2L_3^- + 2HCO_3^-]$ 0.7	yy*
$C_6H_6O_4$ Kojic acid, HL				
gl	30	75 vol % dioxan	K_2 10.23	zz
gl	30	50% dioxan	K_1 10.1, K_2 7.4	a1
$C_6H_8O_6$ Ascorbic acid, (H_2L)				
sp, pH	20–21	0.1 ($NaClO_4$) pH 2–3	HL^-: K_1 2.48	b1**
$C_6H_8O_7$ Citric acid, H_4L				
sp	RT	(NaCl)	ev UO_2^{2+}/citrate cpx in ratio 1/1 and 2/1 pH 2–6	c1**
pol	30		ev UO_2^{2+}/citrate cpx in ratio 1/1 present as dimer pH 4.6	d1**
sp,pH	RT		ev UO_2^{2+}/citrate cpx in ratio 3/3 and 3/2 pH 8	z**
sp, pH con	?	var	HL^{3-}: K_1 3.165 pH 4–7 ev $UO_2HL^- \rightleftharpoons UO_2L_2^{2-} + H^+$ pH 4.6 ev UO_2^{2+}/citrate cpx in ratio 2/3 pH 7–9	e1**
gl	25	I (KNO_3)	$K[2UO_2^{2+} + 2H_2L^{2-} \rightleftharpoons (UO_2)_2(H_2L)_2(OH)_2^{2-} + 2H^+]$ 7.68 (I = 0.136), 9.04 (I → 0)	ff*
gl	25	I (KNO_3)	$K[UO_2^{2+} + HL^{3-} \rightleftharpoons UO_2HL^-]$ 7.40 (I = 0.1), 6.87 (I = 1.0) $K[2UO_2^{2+} + 2HL^{3-} \rightleftharpoons (UO_2HL)_2^{2-}]$ 18.87 (I = 0.1), 17.70 (I = 1.0) $K[2UO_2HL^- \rightleftharpoons (UO_2HL)_2^{2-}]$ 4.07 (I = 0.1), 3.96 (I = 1.0)	f1*

Table 2.25 (continued)

Method	Temp.	Medium	Log of equilibrium constant; remarks	Reference
$C_6H_8O_7$—O_2^{2-} Citric acid (H_4L)-peroxide mixed ligands				
sp		0.1 $NaClO_4$	HL^{3-}: K_1 8.52, 8.57, 8.63 $K[2UO_2HL^- + 2H_2O_2 + 8H_2O \rightleftharpoons (UO_2)_2(00)_2(H_2O)_8 + 4H^+ + 2HL^{3-}]$ −19.72 $K[UO_2HL^- + 2H_2O_2 \rightleftharpoons (UO_2)_2(00)_2HL^{3-} + 4H^+ + HL^{3-}]$ −15.62 $K[2UO_2^{2+} + H_2O_2 + HL^{3-} \rightleftharpoons (UO_2)_2(00)HL^- + 2H^+]$ 9.26 $K[(UO_2)_2(00)HL^- + H_2O_2 \rightleftharpoons (UO_2)_2(00)_2HL^{3-} + 2H^+]$ −7.74 $K[(UO_2)_2(00)_2HL^{3-} + H_2O \rightleftharpoons (UO_2)_2(00)_2HL(OH)^{4-} + H^+]$ −6.3	g1*
$C_6H_{10}O_4$ Adipic acid, H_2L				
gl	31	0.1 ($NaClO_4$)	K_1 4.08	b* = c*
$C_6H_{12}O_3$ α-Hydroxycaproic acid, HL				
gl	25	0.1 ($NaClO_4$)	K_1 2.90	t*
$C_6H_6O_2N_2$ Nicotinohydroxamic acid, HL				
pH	30	0.1 ($NaClO_4$)	K_1 7.50, K_2 7.15	h1*
$C_6H_6O_3N_2$ 4-Nitro-2-aminophenol, HL				
pH	30	50% dioxan, 0.1 ($NaClO_4$)	K_1 7.59, K_2 7.13	i1*
pH	30	50% dioxan, 0.1 ($NaClO_4$)	K_1 7.50, K_2 7.00	j1*
C_6H_7ON o-Aminophenol, HL				
pH	30	50% dioxan, 0.1 ($NaClO_4$)	K_1 8.88, K_2 7.69	j1*
$C_6H_9O_2N_3$ Histidine, HL				
gl	25	0.15–0.25	K_1 7.71	w
$C_6H_9O_6N$ Nitrilotriacetic acid, H_3L				
dis		0.1 $NaClO_4$	K_1 9.56	k1*
cix		0.1 $NaClO_4$	K_1 9.41	k1*
tp			K_1 7.88	l1*
$C_6H_{11}O_2N$ N, N-Dimethylacetoacetamide, HL				
gl	20	50% dioxan, 0.2 $NaClO_4$	K_1 12.10, K_2 10.04	m1*
$C_6H_{11}O_5N$ N-2-Hydroxyethyliminodiacetic acid, H_2L				
gl	25	0.10 (KNO_3)	K_1 8.32 $K[H^+ + UO_2(OH)L^- \rightleftharpoons UO_2L]$ 5.92 $K[2UO_2(OH)L^- \rightleftharpoons (UO_2(OH)L)_2^{2-}]$ 3.50	kk*
		1.0 (KNO_3)	K_1 7.99 $K[H^+ + UO_2(OH)L^- \rightleftharpoons UO_2L]$ 5.87 $K[2UO_2(OH)L^- \rightleftharpoons (UO_2(OH)L)_2^{2-}]$ 3.65	
gl	25	0.1 (KNO_3)	K_1 8.34	vv*
gl	25	1.0 (KNO_3)	$K[UO_2(OH)L^- + H^+ \rightleftharpoons UO_2(H_2O)L]$ 5.86 $K[2UO_2(OH)L^- \rightleftharpoons (UO_2)_2(OH)_2L_2^{2-}]$ 3.40 $K[2UO_2(H_2O)L \rightleftharpoons (UO_2)_2(OH)_2L_2^{2-} + 2H^+]$ −8.32	
$C_6H_{12}O_4N_2$ Ethylenediamine-NN′-diacetic acid, H_2L				
gl	25	0.1 (KNO_3)	K_1 11.34 $K[UO_2(OH)L^- + H^+ \rightleftharpoons UO_2(H_2O)L]$ 5.96	vv*
$C_6H_5O_4Cl$ 3-Chloro-5-hydroxy-2-hydroxymethyl-4-pyrone (chlorokojic acid), HL				
gl	30	75 vol% dioxan	K_2 9.93	n1

Table 2.25 (continued)

Method	Temp.	Medium	Log of equilibrium constant; remarks	Reference
$C_6H_6O_8S_2$ 4,5-Dihydroxybenzene-1,3-disulfonic acid (Tiron), H_4L				
gl	25	0.1 (KNO_3)	K_1 15.90	o1
C_6H_6ONCl 4-Chloro-*o*-aminophenol, HL				
pH	30	50% dioxan, 0.1 ($NaClO_4$)	K_1 8.40, K_2 7.63	j1*
$C_6H_7O_4NS$ 4-Sulfo-*o*-aminophenol, H_2L				
pH	30	50% dioxan, 0.1 ($NaClO_4$)	K_1 8.43, K_2 7.10	j1*
$C_7H_6O_2$ Benzoic acid, HL				
gl	31	0.1 ($NaClO_4$)	K_1 2.59	jj*
gl	31	0.1 ($NaClO_4$)	K_1 2.57	b*=c*
$C_7H_6O_2$ Salicylaldehyde, HL				
sp	?	50% EtOH	K_1 1.81 uranyl acetate, acetic acid present, pH 3	p1**
		var	β_2 2.63 uranyl nitrate, pH 5	
gl		50% C_2H_5OH	β_2 12.83	q1
$C_7H_6O_3$ 2,4-Dihydroxobenzaldehyde (β-Resorcylaldehyde), H_2L				
gl	30	50% dioxan, 0.1 ($NaClO_4$)	K_1 6.60, K_2 5.20	r1*
$C_7H_6O_3$ 2,5-Dihydroxybenzaldehyde (Gentisinaldehyde), H_2L				
gl	30	50% dioxan, 0.1 ($NaClO_4$)	K_1 8.20, K_2 6.85	r1*
$C_7H_6O_3$ Salicylic acid, H_2L				
sp			K_1 13.4	s1**
dis	25	0.1 ($NaClO_4$)	$K[UO_2^{2+} + HL^- + H^+ \rightleftharpoons UO_2H_2L^{2+}]$ −0.62	t1
			$K[UO_2^{2+} + HL^- \rightleftharpoons UO_2HL^+]$ 2.2	
			$K[UO_2^{2+} + HL^- + 2H^+ \rightleftharpoons UO_2H_3L^{3+}]$ −4.5	
			$K[UO_2^{2+} + L^{2-} + OH^- \rightleftharpoons UO_2L(OH)^-]$ 12.1	
sp	35		K_1 4.91	u1
sp		0.5 (KNO_3)	K_1 11.82, β_2 21.4, β_3 29.7	v1*
ix		0.5 (KNO_3)	K_1 11.30, β_2 20.68	w1*
$C_7H_6O_4$ 2,4-Dihydroxybenzoic acid (β-Resorcylic acid), H_3L				
gl	25	0.2 ($NaClO_4$)	H_2L^-: K_1 2.10	x1*
			$K[UO_2^{2+} + H_2L^- \rightleftharpoons UO_2HL + H^+]$ −0.66	
			$K[2UO_2^{2+} + 2H_2L^- \rightleftharpoons (UO_2)_2HL_2 + 3H^+]$ −4.17	
$C_7H_6O_4$ 2,5-Dihydroxybenzoic acid (Gentisic acid), H_3L				
gl	25	0.2 ($NaClO_4$)	H_2L^-: K_1 1.51	x1*
			$K[UO_2^{2+} + H_2L^- \rightleftharpoons UO_2HL + H^+]$ −1.00	
$C_7H_6O_4$ 3,5-Dihydroxybenzoic acid (α-Resorcylic acid), H_3L				
gl	25	0.2 ($NaClO_4$)	H_2L^-: K_1 2.13	x1*
			$K[UO_2^{2+} + H_2L^- + H_2O \rightleftharpoons UO_2(H_2L)(OH) + H^+]$ −2.02	
$C_7H_6O_5$ Gallic acid (3,4,5-Trihydroxybenzoic acid), H_4L				
sp			K_1 5.16	y1*
$C_7H_4O_7N_2$ 3,5-Dinitrosalicylic acid, H_2L				
gl	35	var	K_1 7.0, K_2 5.5	z1*
$C_7H_7O_2N$ Benzohydroxamic acid, HL				
gl	25	0.1 $HClO_4$	K_1 8.72, β_2 16.77	a2*
pH	30	0.1 ($NaClO_4$)	K_1 9.03, β_2 17.94	h1*

Table 2.25 (continued)

Method	Temp.	Medium	Log of equilibrium constant; remarks	Reference
$C_7H_7O_2N$ Salicylamide, HL				
sp			K_1 6.40, K_1 4.97	b2**
$C_7H_7O_3N$ Salicylhydroxamic acid, H_2L				
pH	30	0.1($NaClO_4$)	HL^-: K_1 7.71 HL^-: K_2 6.80	h1*
$C_7H_6O_6S$ 5-Sulfosalicylic acid, H_3L				
gl	25	0.1($NaClO_4$)	K_1 11.14, K_2 8.06	c2
sp	25	~0.015	$K[UO_2^{2+} + HL^{2-} \rightleftharpoons UO_2HL]$ 3.89	d2
gl	25	I(KNO_3)	K_1 10.62 (I=0.10), 10.44 (I=1.0)	kk*
pol	18–22	1.0($NaNO_3$)	K_1 11.65, β_2 17.64	e2*
$C_7H_{11}O_2N_3$ Histidine methyl ester, L				
gl	25	0.15–0.25	K_1 5.76	w
$C_8H_6O_4$ Phthalic acid, H_2L				
gl	25	1.0 (KNO_3)	K_1 4.38	j*
$C_8H_8O_2$ β-Methyltropolone, HL				
sp		50% EtOH, 0.5(KNO_3)	K_1 9.62, K_2 6.98	f2*
pH		50% EtOH, 0.5(KNO_3)	K_2 6.93	f2*
$C_8H_8O_2$ Phenylacetic acid, HL				
gl	31	0.1($NaClO_4$)	K_1 3.25	jj*
gl	31	0.1 ($NaClO_4$	K_1 3.21	b*=c*
$C_8H_8O_3$ Phenoxyacetic acid, HL				
hl	31	0.1 (NaClO)	K_1 2.41	jj*
gl	31	0.1 ($NaClO_4$)	K_1 2.59	b*=c*
$C_8H_8O_3$ m-Methoxybenzoic acid, HL				
dis	25	0.1 ($NaClO_4$)	$K[UO_2^{2+} + L^- + OH^- \rightarrow UO_2LOH]$ 11.9 $K[UO_2^{2+} + HL + 2H^+ \rightarrow UO_2HLH_2^{4+}] - 5.85$	g2
$C_8H_6ON_2$ 8-Hydroxycinnoline, HL				
gl	20	50% dioxan, 0.3 $NaClO_4$	K_1 8.68, K_2 7.16	h2
$C_8H_6ON_2$ 5-Hydroxyquinoxaline, HL				
gl	20	50% dioxan, 0.3 $NaClO_4$	K_1 8.40, K_2 7.51	h2
$C_8H_6ON_2$ 8-Hydroxyquinazoline, HL				
gl	20	50% dioxan, 0.3 $NaClO_4$	K_1 8.99, K_2 7.70	h2
$C_8H_{15}O_2N$ N,N-Diethylacetoacetamide, HL				
gl	20	50% dioxan, 0.2 $NaClO_4$	K_1 1 2.29, K_2 10.16	m1*
$C_8H_8O_2S$ 3-Thenoylacetone, HL				
gl	30	75% dioxan, →0	K_1 12.25, K_2 10.32	i2*
C_8H_9ONS o-Mercaptoacetanilide				
gl	30	75% dioxan,	K_1 10.14, K_2 8.96	j2
$C_8H_5O_2SF_3$ 2-Thenoyltrifluoracetone				
gl	30	75% dioxan, →0	K_1 8.71, K_2 7.92	i2*

Table 2.25 (continued)

Method	Temp.	Medium	Log of equilibrium constant; remarks	Reference
$C_9H_8O_4$ α-Carboxy-β-methyltropolone, H_2L				
sp		0.5 (KNO_3)	K_1 9.22, K_2 6.75	f2*
pH		0.5 (KNO_3)	K_2 6.80	f2*
sp		0.2 $(NaClO_4)$	K_1 9.60	k2*
pH		0.2 $(NaClO_4)$	K_1 9.72, K_2 6.78	k2*
$C_9H_{10}O_2$ 2′-Hydroxy-5′-methylacetophenone, HL				
sp, con	30	var	β_2 7.17	l2*
C_9H_7ON 8-Hydroxyquinoline (oxine), HL				
dis	25	0.1	β_3 23.76	m2**
gl	20	50% dioxan, 0.3 $NaClO_4$	K_1 11.25, K_2 9.64	h2
dis	20	0.1 $(NaClO_4)$	$K[UO_2^{2+} + 3L^- + H^+ \rightleftharpoons UO_2L_2HL(org)]$ 36.18 $K[UO_2L_2OH^- + 2H^+ + L^- \rightleftharpoons UO_2L_2HL(org)]$ 25.40, where org = $CHCl_3$, no ev $UO_2L_3^-$ cpx	n2*
$C_9H_8ON_2$ 8-Hydroxy-4-methylcinnoline, HL				
gl	20	50% dioxan, 0.3 $NaClO_4$	K_1 9.00, K]7.30	h2
$C_9H_{10}O_2N_2$ N-(1-Pyridyl)-acetoacetamide, HL				
pH	25	0.1 KNO_3	K_1 7.26, K_2 6.63	o2*
$C_9H_{15}O_2N$ Acetoacetipiperidine				
gl	20	50% dioxan, 0.2 $NaClO_4$	K_1 12.98, K_2 11.28	m1*
$C_9H_7O_4NS$ 8-Hydroxyquinoline-5-sulfonic acid, H_2L				
gl	25	0.1 (KNO_3)	K_1 8.52, K_2 7.16 $K[UO_2L_2OH^{3-} + H^+ \rightleftharpoons UO_2L_2^{2-}]$ 6.68 $K[(UO_2L_2OH)_2^{6-} + 2H^+ \rightleftharpoons 2UO_2L_2^{2-}]$ 11.7	p2
$C_9H_{11}ONS$ N-(2-Mercapto-4-methylphenyl)acetamide (N-acetyl-2-mercapto-p-toluidine), HL				
gl	30	75% dioxan	K_1 10.18, K_2 9.03	j2
$C_{10}H_6O_3$ 2-Hydroxy-1,4-naphthaquinone, HL				
gl	30	75% dioxan	K_1 7.27, K_2 6.07	n1
$C_{10}H_6O_3$ 5-Hydroxy-1,4-naphthaquinone, HL				
gl	30	75% dioxan	K_1 11.84, K_2 10.53	n1
$C_{10}H_{10}O_2$ Benzoylacetone, HL				
gl	30	75% dioxan	K_1 12.15, K_2 11.12	q2
$C_{10}H_{10}O_3$ o-Hydroxybenzoylacetone, HL				
gl	30	75% dioxan	K_1 10.97, K_2 10.23	q2
$C_{10}H_{12}O_2$ β-Isopropyltropolene, HL				
sp		50% EtOH, 0.5 (KNO_3)	K_1 9.62, K_2 6.92	f2*
pH		50% EtOH, 0.5 (KNO_3)	K_2 6.88	f2*
$C_{10}H_9ON$ 8-Hydroxy-2-methylquinoline, HL				
gl	20	50% dioxan, 0.2 $NaClO_4$	$K_1 \sim 9.4$, $K_2 \sim 8$	h2
$C_{10}H_9ON$ 8-Hydroxy-5-methylquinoline, HL				
gl	20	50% dioxan, 0.3 $NaClO_4$	K_1 11.25, K_2 9.52	h2

Table 2.25 (continued)

Method	Temp.	Medium	Log of equilibrium constant; remarks	Reference
$C_{10}H_9ON$ 8-Hydroxy-6-methylquinoline, HL				
gl	20	50% dioxan, 0.3 $NaClO_4$	K_1 10.89, K_2 9.26	h2
$C_{10}H_9ON$ 8-Hydroxy-7-methylquinoline, HL				
gl	20	50% dioxan, 0.3 $NaClO_4$	K_1 11.28, K_2 9.78	h2
$C_{10}H_{10}ON_2$ 8-Hydroxy-2,4-dimethylquinazoline, HL				
gl	20	50% dioxan, 0.3 $NaClO_4$	K_1 8.77, K_2 7.33	h2
$C_{10}H_{10}O_4N_2$ 3-Nitroacetoacetanilide				
gl	20	50% dioxan, 0.2 $NaClO_4$	K_1 8.99, K_2 7.51	m1*
$C_{10}H_{10}O_4N_2$ 4-Nitroacetoacetanilide				
gl	20	50% dioxan, 0.2 $NaClO_4$	K_1 9.39, K_2 7.66	m1*
$C_{10}H_{11}O_2N$ Acetoacetanilide				
gl	20	50% dioxan, 0.2 $NaClO_4$	K_1 9.94, K_2 8.08	m1*
$C_{10}H_{16}O_8N_2$ Ethylenediamine-NNN'N'-tetraacetic acid (EDTA), H_4L				
dis		0.1 $NaClO_4$	HL^{3-}: K_1 7.32	k1*
ix		0.1 $NaClO_4$	HL^{3-}: K_1 7.13	k1*
sp		0.1 (NH_4Cl)	K_{12} $[UO_2^{2+} + UO_2L^{2-} \rightleftharpoons (UO_2)_2L]$ 10.4 β_{12} $[2UO_2^{2+} + L^{4-} \rightleftharpoons (UO_2)_2L]$ 15.2	r2*
sp	25	0.15 ($NaClO_4$)	HL^{3-}: K_1 7.90, 7.79, 7.96 β_{12} 17.85, 17.69, 18.01 $K[UO_2(OH)HL^{2-} + H^+ \rightleftharpoons UO_2(H_2O)HL^-]$ 5.59, 5.66 $K[(UO_2OH)_2L^{2-} + 2H^+ \rightleftharpoons (UO_2 \cdot H_2O)_2L]$ 11.02, 11.18	s2*
gl	25	0.1 (KNO_3)	HL^{3-}: K_1 7.40, β_{12} 17.87	t2=vv*
gl	25	1.0 (KNO_3)	HL^{3-}: K_1 7.35, β_{12} 17.77 $K[UO_2(OH)L^{3-} + H^+ \rightleftharpoons UO_2(OH)HL^{2-}]$ 6.30 $K[UO_2(OH)L^{3-} + 2H^+ \rightleftharpoons UO_2(H_2O)HL^-]$ 11.92 $K[2UO_2(OH)L^{3-} + 2H^+ \rightleftharpoons (UO_2)_2(OH)_2H_2L_2^{4-}]$ 15.87	t2*
gl	25	1.0 (KNO_3)	$K[UO_2(OH)HL^{2-} + H^+ \rightleftharpoons UO_2(H_2O)HL^-]$ 5.62 $K[2UO_2(OH)HL^{2-} \rightleftharpoons (UO_2)_2(OH)_2H_2L_2^{4-}]$ 3.27 $K[2(UO_2)(H_2O)HL^- \rightleftharpoons (UO_2)_2(OH)_2H_2L_2^{4-} + 2H^+]$ −7.97	vv*
sp		var	K_1 13.40	u2*
tp	20	0.1 (KCl)	HL^{3-}: K_1 6.9, 6.8 H_2L^{2-}: K_1 3.60	v2*
$C_{10}H_{16}O_8N_2 - O_2^{2-}$ EDTA (H_4L)-peroxide mixed ligands				
sp		var	$K[(UO_2)_2L + H_2O_2 \rightleftharpoons (UO_2)_2(OO)L^{2-} + 2H^+]$ −3.77 $K[2UO_2L^{2-} + H_2O_2 \rightleftharpoons (UO_2)_2(OO)L_2^{6-} + 2H^+]$ −1.15 $K[(UO_2)_2(OO)L^{2-} + H_2O \rightleftharpoons (UO_2)_2(OO)L(OH)^{3-} + H^+]$ −8 $K[(UO_2)_2L + O_2^{2-} \rightleftharpoons (UO_2)_2(OO)L^{2-}]$ 32 $K[2UO_2L^{2-} + O_2^{2-} \rightleftharpoons (UO_2)_2(OO)L_2^{6-}]$ 35	w2*
$C_{10}H_{18}O_7N_2$ N-(Hydroxyethyl)-ethylenediaminetriacetic acid, H_3L				
pol	30	0.5 ($NaClO_4$)	β_2 8.61, 9.81 $\beta[UO_2^{2+} + 2HL^{2-} \rightleftharpoons UO_2(HL)_2^{2-}]$ 5.73, 7.13 $\beta[UO_2^{2+} + 2H_2L^- \rightleftharpoons UO_2(H_2L)_2]$ 5.57	x2*
$C_{10}H_{19}O_2N$ N-Dipropylacetoacetamide, HL				
gl	20	50% dioxan, 0.2 $NaClO_4$	K_1 12.31, K_2 10.99	m1*

Table 2.25 (continued)

Method	Temp.	Medium	Log of equilibrium constant; remarks	Reference
$C_{10}H_8O_8S_2$ 1,8-Dihydroxynaphthalene-3,6-disulfonic acid (chromotropic acid), H_4L				
gl	25	0.06	K_1 3.20	y2
$C_{10}H_9O_2NCl_2$ 2,5-Dichloroacetoacetanilide, HL				
gl	20	50% dioxan, 0.2 $NaClO_4$	K_1 8.66, K_2 7.02	m1*
$C_{10}H_{10}O_2NCl$ 2-Chloroacetoacetanilide, HL				
gl	20	50% dioxan, 0.2 $NaClO_4$	K_1 8.97, K_2 7.28	m1*
$C_{10}H_{10}O_2NCl$ 3-Chloroacetoacetanilide, HL				
gl	20	50% dioxan, 0.2 $NaClO_4$	K_1 9.69, K_2 7.85	m1*
$C_{10}H_{10}O_2NCl$ 4-Chloroacetoacetanilide, HL				
gl	20	50% dioxan, 0.2 $NaClO_4$	K_1 9.78, K_2 8.02	m1*
$C_{10}H_9O_2N_3S$ 2-(2-Thiazolylazo)-4-methoxyphenol, HL				
sp		30% EtOH, 0.1	$^*K_1$ 0.3–0.4, K_1 8.8	z2*
$C_{10}H_9O_2N_3S$ 2-(2-Thiazolylazo)-5-methoxyphenol, HL				
sp		30% EtOH, 0.1	$^*K_1$ 0.4–0.5, K_1 8.1	z2*
$C_{10}H_{13}ONS$ α-Mercapto-N-(2,6-dimethylphenyl)acetamide (N-(mercaptoacetyl)-2,6-xylidine), HL				
gl	30	75% dioxan,	K_1 10.30, K_2 9.17	j2
$C_{10}H_{13}ONS$ α-Mercapto-N-phenylbutyramide (α-mercaptobutyranilide), HL				
gl	30	75% dioxan,	K_1 10.78, K_2 9.68	j2
$C_{10}H_{13}O_3NS$ α-Mercapto-N-(2,5-dimethoxyphenyl)acetamide (N-(mercaptoacetyl)-2,5-dimethoxyaniline), HL				
gl	30	75% dioxan,	K_1 10.21, K_2 9.10	j2
$C_{10}H_6O_7S_2Na_2$ Disodium 2-naphthol-3,6-disulfonate (R-salt,), HL^{2-}				
gl	25	0.1 $NaClO_4$	K_1 7.42, K_2 5.70	a3*
$C_{10}H_8O_7NS_2K$ Monopotassium 7-amino-1-naphthol-3,6-disulfonic acid (2R-salt), H_2L^-				
gl	25	0.1 $NaClO_4$	K_1 6.19, K_2 4.87	a3*
$C_{11}H_8O_3$ 2-Hydroxy-3-methyl-1,4-naphthaquinone, HL				
gl	30	75% dioxan,	K_1 8.39, K_2 7.20	n1
$C_{11}H_9O_3N$ N-Furoyl-N-phenylhydroxylamine, HL				
pH	30	0.1 ($NaClO_4$)	K_1 8.14, K_2 7.91	h1*
$C_{11}H_{12}O_3N_2$ 4-(p-Nitrophenylimino)pentan-2-one, HL				
gl	30	50% dioxan,	K_1 10.56, K_2 10.52	b3
$C_{11}H_{13}ON$ 4-(Phenylimino)pentan-2-one, HL				
gl	30	50% dioxan,	K_1 10.97, K_2 10.00	b3
$C_{11}H_{13}O_2N$ 2-Methylacetoacetanilide, HL				
gl	20	50% dioxan, 0.2 $NaClO_4$	K_1 9.35, K_2 7.68	m1*
$C_{11}H_{13}O_2N$ 3-Methylacetoacetanilide, HL				
gl	20	50% dioxan, 0.2 $NaClO_4$	K_1 10.14, K_2 8.23	m1*

Table 2.25 (continued)

Method	Temp.	Medium	Log of equilibrium constant; remarks	Reference
$C_{11}H_{13}O_2N$ 4-Methylacetoacetanilide, HL				
gl	20	50% dioxan, 0.2 $NaClO_4$	K_1 10.24, K_2 8.34	m1*
$C_{11}H_{13}O_3N$ 2-Methoxyacetoacetanilide, HL				
gl	20	50% dioxan, 0.2 $NaClO_4$	K_1 9.70, K_2 8.06	m1*
$C_{11}H_{13}O_3N$ 3-Methoxyacetoacetanilide, HL				
gl	20	50% dioxan, 0.2 $NaClO_4$	K_1 10.00, K_2 8.18	m1*
$C_{11}H_{13}O_3N$ 4-Methoxyacetoacetanilide, HL				
gl	20	50% dioxan, 0.2 $NaClO_4$	K_1 10.28, K_2 8.42	m1*
$C_{11}H_{18}O_8N_2$ 1,3-Diaminopropanetetraacetic acid, H_4L				
gl	25	0.1(KNO_3)	HL^{3-}: K_1 8.94	t2*
$C_{11}H_{12}ONCl$ 4-(p-Chlorophenylimino)pentan-2-one, HL				
gl	30	50% dioxan,	K_1 11.67, K_2 10.63	b3
$C_{11}H_{12}O_2NCl$ 2-Methyl-3-chloroacetoacetanilide				
gl	20	50% dioxan, 0.2 $NaClO_4$	K_1 9.32, K_2 7.75	m1*
$C_{11}H_{12}O_2NCl$ 2-Methyl-5-chloroacetoacetanilide				
gl	20	50% dioxan, 0.2 $NaClO_4$	K_1 9.26, K_2 7.63	m1*
$C_{12}H_{12}O_3$ 1-Phenylhexane-1,3,5-trione, HL				
gl	30	75% dioxan	K_2 10.64	zz
$C_{12}H_{11}ON$ 2-Amino-4-phenylphenol, HL				
pH	30	50% dioxan, 0.1 ($NaClO_4$)	K_1 8.94, K_2 7.75	j1*
$C_{12}H_{15}O_2N$ 2,4-Dimethylacetoacetanilide, HL				
gl	20	50% dioxan, 0.2 $NaClO_4$	K_1 9.84, K_2 8.14	m1*
$C_{12}H_{15}O_2N$ 3,5-Dimethylacetoacetanilide, HL				
gl	20	50% dioxan, 0.2 $NaClO_4$	K_1 9.83, K_2 8.10	m1*
$C_{12}H_{15}O_3N$ 2,4-Dimethoxyacetoacetanilide, HL				
gl	20	50% dioxan, 0.2 $NaClO_4$	K_1 10.05, K_2 8.31	m1*
$C_{12}H_{15}O_3N$ 2,5-Dimethoxyacetoacetanilide, HL				
gl	20	50% dioxan, 0.2 $NaClO_4$	K_1 9.47, K_2 7.81	m1*
$C_{12}H_{20}O_9N_2$ 2,2′-Oxybis(ethyliminodi(acetic acid))				
sp		var	K_1 14.60, β_{12} 9.70	u2*
$C_{12}H_{17}ONS$ N-(2,6-Diethylphenyl)mercaptoacetamide (2,6-diethyl-N-(mercaptoacetyl)anal-line), HL				
gl	30	75% dioxan,	K_1 10.42, K_2 9.26	j2
$C_{13}H_{11}O_2N$ N-Benzoyl-N-phenylhydroxylamine, HL				
pH	30	0.1 ($NaClO_4$)	K_1 8.77, K_2 8.21	h1*

Table 2.25 (continued)

Method	Temp.	Medium	Log of equilibrium constant; remarks	Reference
$C_{13}H_{12}ON_2$ Benzanilidoxime, HL				
gl	25	50% dioxan, 0.1 $NaClO_4$	K_1 10.22, K_2 9.83	c3*
$C_{13}H_{13}ON$ 1,2,3,4-Tetrahydro-9-hydroxyacridine, HL				
gl	20	50% dioxan, 0.3 $NaClO_4$	K_1 10.10, K_2 8.20	h2
$C_{13}H_{17}ON$ 4-(2,6-Dimethylphenylimino)pentan-2-one, HL				
gl	30	50% dioxan	K_1 11.44, K_2 9.74	b3
$C_{14}H_8O_4$ 1,8-Dihydroxyanthraquinone, H_2L				
gl	30	75% dioxan	K_1 12.13, K_2 11.03	n1
$C_{14}H_{11}O_2N_5$ N,N'-Di-(2-hydroxyphenyl)-C-cyanoformazan				
sp			K_1 8 (pH 5.7)	d3*
$C_{14}H_{13}O_2N$ N-Benzoyl-N-*o*-tolylhydroxylamine, HL				
pH	30	0.1($NaClO_4$)	K_1 8.64, K_2 8.43	h1*
$C_{14}H_{13}O_2N$ N-Benzoyl-N-*p*-tolylhydroxylamine, HL				
pH	30	0.1($NaClO_4$)	K_1 8.90, K_2 8.67	h1*
$C_{14}H_{13}O_2N$ N-*p*-Toluyl-N-phenylhydroxylamine, HL				
pH	30	0.1($NaClO_4$)	K_1 8.80, K_2 8.53	h1*
$C_{14}H_{13}O_3N$ N-*p*-Methoxybenzoyl-N-phenylhydroxylamine, HL				
pH	30	0.1($NaClO_4$)	K_1 8.68, K_2 8.35	h1*
$C_{14}H_{13}O_3N$ 3-Methoxysalicylaldehydeaminophenol, H_2L				
dis		0.3 (KNO_3)	K_1 4.34	e3*
$C_{14}H_{22}O_8N_2$ 1,2-Diaminecyclohexanetetraacetic acid, H_4L				
dis		0.1 $NaClO_4$	HL^{3-}: K_1 5.27	k1*
ix		0.1 $NaClO_4$	HL^{3-}: K_1 5.27	k1*
$C_{14}H_{23}O_{10}N_3$ Diethylenetriaminepentaacetic acid (DTPA), H_5L				
sp		var	K_1 15.5, β_{12} 13	u2*
$C_{14}H_{24}O_8N_2$ 1,6-Diaminohexanetetraacetic acid, H_4L				
gl	25	0.1(KNO_3)	HL^{3-}: K_1 9.96	t2*
$C_{14}H_{24}O_{10}N_2$ Ethyleneglycol-bis (aminoethyl ether) tetraacetic acid, H_4L				
gl	25	0.1 (KNO_3)	HL^{3-}: K_1 7.40, β_{12} 17.87	vv*
gl	25	1.0 (KNO_3)	$K[UO_2(OH)HL^{2-} + H^+ \rightleftharpoons UO_2(H_2O)HL^-]$ 5.98 $K[2UO_2(OH)HL^{2-} \rightleftharpoons (UO_2)_2(OH)_2H_2L_2^{4-}]$ 3.48 $K[2UO_2(H_2O)HL^- \rightleftharpoons (UO_2)_2(OH)_2H_2L_2^{4-} + 2H^+]$ −8.48	vv*
$C_{14}H_8O_7S$ Alizarin red, H_3L				
	30	0.1($NaClO_4$)	K_1 4.56	f3
sp	25	0.15($NaClO_4$	K_1 4.22	f3
$C_{15}H_{10}O_6$ 3,4',5,7-Tetrahydroxyflavone (Kaempferol), H_4L				
sp		var	K_1 6.90(pH=6.0)	g3*
$C_{15}H_{12}O_3$ o-Hydroxydibenzoylmethane, HL				
gl	30	75% dioxan	K_1 11.40, K_2 11.03	q2
$C_{15}H_{14}O_3$ 2-Hydroxy-3-(3-methylbut-2-enyl)-1,4-naphthaquinone, HL				
gl	30	75% dioxan	K_1 8.73, K_2 7.30	zz
$C_{15}H_{12}ON_2$ 4-Methyl-2-phenylquinazolin-8-ol, HL				
gl	20	50% dioxan, 0.3 $NaClO_4$	K_1 8.53, K_2 7.85	h2

Table 2.25 (continued)

Method	Temp.	Medium	Log of equilibrium constant; remarks	Reference
$C_{16}H_{14}O_3$ o-Methoxydibenzoylmethane, HL				
gl	30	75% dioxan	K_1 13.30, K_2 11.06	q2
$C_{16}H_{15}ON$ 3-Phenylimino-1-phenylbutan-1-one, HL				
gl	30	50% dioxan,	K_1 11.32, K_2 10.42	y2
$C_{16}H_{16}O_2N_2$ Bis-salicylaldehyde-ethylene-diimine, H_2L				
dis		0.3	K_1 24.35	h3*
$C_{16}H_{13}O_{10}N_2S_2As$ 4-(p-Arsonophenylazo)-3-hydroxynaphthalene-2,7-disulfonic acid, H_5L				
gl	30		K_1 15	i3
$C_{16}H_{13}O_{11}N_2S_2As_2$ Benzene-2-arsonic acid-(1-azo-7)-1,8-dihydroxy-naphthalene-3,6-disulfonic acid (Arsenazo), H_6L				
sp		var	K_1(?) 11.60	j3*+k3*
$C_{17}H_{16}O_4$ Di-o-anisoylmethane, HL				
gl	30	75% dioxan,	K_1 13.28, K_2 11.52	q2
$C_{18}H_{14}O_4N_2$ o-(2-Hydroxy-1-naphthylazo)phenoxyacetic acid, H_2L				
gl	30	75% dioxan,	K_1 15.01	l3
$C_{18}H_{16}O_4N_4$ o-(4,5-Dihydro-3-methyl-5-oxo-1-phenyl-1H-pyrazol-4-ylazo)phenoxyacetic acid, H_2L				
gl	30	75% dioxan,	K_1 11.93	m3
$C_{18}H_{20}O_4N_2$ Bis-3-methoxy-salicylaldehyde-ethylenediimine, H_2L				
dis		0.3	K_1 19.6	h3*
$C_{18}H_{30}O_{12}N_4$ Triethylenetetraminehexaacetic acid, H_6L				
sp		var	K_1 15.00, β_{12} 11.80	u2*
$C_{18}H_{14}O_{11}N_2S_2$ 2-(1,8-Dihydroxy-3,6-disulfo-2-naphthylazo)-phenoxyacetic acid, H_5L				
sp	25	0.1(KNO_3)	HL^{4-}: K_1 10.10	n3*
$C_{19}H_{18}O_4N_4$ *o*-(4,5-Dihydro-3-methyl-5-oxo-1-phenyl-1*H*-pyrazol-4-ylazo)-3-phenoxypropanoic acid, H_2L				
gl	30	75% dioxan,	K_1 12.08	o3*
$C_{19}H_{12}O_7SBr_2$ Dibromopyrogallol sulfonphthalein, H_4L				
sp	25	var	HL^{3-} or H_2L^{2-}: K_1 4.3, 5.7	p3*
$C_{20}H_{16}O_2N_2$ Bis-salicylaldehyde-phenylene-diimine, H_2L				
dis		0.3	K_1 20.9	h3*
$C_{21}H_{17}ON$ 1-(Phenylimino)-1,3-diphenylpropan-3-one, HL				
gl	30	50% dioxan,	K_1 11.59, K_2 10.49	b3
$C_{22}H_{14}O_9$ Aurintricarboxylic acid, H_5L				
sp	25	0.01	K_1 4.77	q3
gl	25	0.1 $NaClO_4$	K_1 7.40, K_2 2.95, K_3 2.73	r3*
$C_{22}H_{18}O_{14}N_4S_2As_2$ 1,8-Dihydroxynaphthalene-3,6-disulfonic acid-2,7-bis [(azo-2)-phenylarsonic acid] (Arsenazo III), H_8L				
sp		var	K_1 4.02(0.5HNO_3), 3.69(1.0HNO_3), 3.81(5.8HNO_3), 3.51(11.5HNO_3), 3.88(1.12$HClO_4$), 5.01(5.5$HClO_4$)	s3*
$C_{22}H_{18}O_{14}N_4S_2P_2$ 2,7-Bis(2′-phosphonophenylazo)chromotropic acid (Phosphonazo III), H_8L				
sp	25	0.2(KNO_3)	β_{12}(?) 14.24	t3*
$C_{22}H_{16}O_{14}N_4Cl_2S_2P_2$ 2,7-Bis (2′-phosphono-4′-chlorophenylazo)chromotropic acid (Chlorophosphonazo III), H_8L				
sp	25	0.2(KNO_3)	β_{12}(?) 11.96	t3*

Table 2.25 (continued)

Method	Temp.	Medium	Log of equilibrium constant; remarks	Reference
$C_{22}H_{17}O_{14}N_4ClS_2P_2$ 2-(2′-Phosphono-4′-chlorophenylazo)-7-(2″-phosphonophenylazo) chromotropic acid, H_8L				
sp	25	0.2 (KNO_3)	β_{12}(?) 11.32	t3*
$C_{23}H_{15}O_9SCl_2$ Sulfodichlorohydroxydimethylfuchsindicarboxylic acid (Chrome azurol S), H_4L				
sp	30	0.2 (KCl)	K_1 4.47	f3
sp	30	0.2 (KCl)	K_1 4.94	f3
sp	30	0.2 (KCl)	K_1 4.5, 4.9	u3
$C_{24}H_{20}O_{14}N_4Cl_2S_2P_2$ 2,7-Bis(2′-phosphono-4′-chloro-5-methylphenylazo)chromotropic acid, H_8L				
sp	25	0.2 (KNO_3)	β_{12}(?) 11.75	t3*
$C_{26}H_{51}O_{10}N_6S$ Desferrioxaminemethanesulfonate (Desferal)				
sp	RT	var	$\beta_{nm}[mUO_2^{2+} + nH_{x-p}L^{-p} \rightleftharpoons UO_2^{(2m-nx)} + n(x-p)H^+]$ β_{11} 5.10 (pH=4.5), β_{32} 11.34 (pH=7.5)	v3*
$C_{37}H_{44}O_{13}N_2$ 3′,3″-Bis{[N,N-bis(carboxymethyl)amino]methyl}thymolsulfone phthalein (methylthymol blue), H_6L				
sp	30	0.1 ($NaClO_4$)	H_4L^{2-}: K_1(?) 5.6	w3*

[a] References without asterisks may be found in the compilation by Sillén and Martell (1964). References with a single asterisk are given at the end of this chapter. References with double asterisks may be found in the review by Gindler (1962). The notation used in this table is explained in Table 2.19. [b*] Ramamoorthy and Santappa (1968b). [c*] Ramamoorthy and Santappa (1969). [d**] Heidt (1942). [e**] Li et al. (1957). [f**] Tekster et al. (1959). [g**] Moskvin and Zakharova (1959). [h**] Ptitsyn and Tekster (1959). [i**] Komarov and Gurevich (1959). [j*] Rajan and Martell (1967). [k**] Tishkoff (1949). [l] Ahrland (1951). [m**] Li et al. (1958). [n**] Nikolsky and Paramonova (1958). [o*] Feldmann and Koval (1963). [p*] Marcu and Botar (1967). [q**] Ahrland (1953). [r*] Cefola et al. (1962). [s*] Starý and Balek (1962). [t*] Schaefer (1968). [u*] Ahrland (1949). [v*] Varga (1969). [w] Li et al. (1957). [x*] Crutchfield et al. (1962). [y**] Feldmann and Havill (1954). [z**] Feldman et al. (1954). [aa*] Sobkowska (1967). [bb*] Mishra and Nigam (1968). [cc*] Saxena and Gupta (1969). [dd*] Tedeco and Walton (1969). [ee*] Lai and Chen (1968). [ff*] Feldman et al. (1960). [gg*] Rajan and Martell (1964b). [hh*] Ramamoorthy et al. (1969). [ii*] Mathur et al. (1963). [jj*] Ramamoorthy and Santappa (1968a). [kk*] Rajan and Martell (1964a). [ll*] Lai and Chen (1967). [mm*] Ul'yanov et al (1969). [nn] Bryant (1954). [oo] Izatt et al. (1954). [pp] Izatt et al. (1955). [qq] Rydberg (1955). [rr**] Jezowska-Trzebiatowska et al. (1958). [ss*] Jeftic and Brànica (1963a). [tt*] Jeftic and Brànica (1963b). [uu*] Vorob'ev et al. (1969). [vv*] Fraústo da Silva and Simoes (1970). [ww] Zingaro (1956). [xx*] Verma and Agarwal (1968). [yy*] Chiacchierini et al. (1968). [zz] Kido and Fernelius (1960). [a1] Bryant and Fernelius (1954). [b1**] Gal (1956). [c1**] Feldman and Neuman (1951). [d1**] Neumann et al. (1951). [e1**] Heitner and Bobtelsky (1954). [f1*] Rajan and Martell (1965). [g1*] Gurevich and Osicheva (1968). [h1*] Dutt and Seshadri (1969). [i1*] Vartak and Menon (1966). [j1*] Vartak and Menon (1967). [k1*] Starý and Prašilová (1961). [l1*] Marcu et al. (1968). [m1*] Ketrup and Specker (1969). [n1] Kido et al. (1960). [o1] Gustafson et al. (1958). [p1**] Bobtelsky (1952). [q1] Hertnee and Shamir (1957). [r1*] Vartak and Menon (1969). [s1**] Babko (1949). [t1] Hok-Bernstrom (1956b). [u1] Dutt and Goswami (1959). [v1*] Paramonova et al. (1964a). [w1*] Paramonova et al. (1964b). [x1*] Ostacoli et al. (1968). [y1*] Chou and Ts'ui (1965). [z1*] Dube and Dhindsa (1970). [a2*] Barnocelli and Grossi (1965). [b2**] Chakraburtty et al. (1953). [c2] Banks and Singh (1960). [d2] Foley and Anderson (1949). [e2*] Hála (1964). [f2*] Dutt and Singh (1965). [g2] Hok-Bernstrom (1956a) [h2] Irving and Rossotti (1954). [i2*] Rosenstreich and Goldberg (1965). [j2] Martin (1961). [k2*] Gupta et al. (1967). [l2*] Gupta and Malik (1969). [m2**] Dyrssen and Dahlberg (1953). [n2*] Oki (1969). [o2*] Harries (1967). [p2] Richard et al. (1959). [q2] Holst (1955). [r2*] Kozlov and Krot (1963). [s2*] Bhat and Krishnamurthy (1964). [t2*] Fraústo da Silva and Simoes (1962). [u2*] Hafez (1968). [v2*] Stepanov and Makarova (1969). [w2*] Gurevich et al. (1968). [x2*] Lai and Lee (1969). [y2] Jantti

(1957). [z2*] SOMMER et al. (1968). [a3*] BANERJEE and DEY (1968a). [b3] MARTIN et al. (1961). [c3*] MANOUSSAKIS and KOUIMTZIS (1969). [d3*] SERGOVSKAYA and SOBOLEVA (1966). [e3*] ZIELINSKI and STROŃSKI (1968). [f3] SRIVASTAVA and DEY (1960). [g3*] GARG et al. (1968). [h3*] STROŃSKI et al. (1966). [i3] SNAVELY and QUINN (unpub.). [j3*] KUTEINIKOV (1958). [k3*] KUTEINIKOV (1962). [l3] SNAVELY and KRECKER (unpub.) [m3] SNAVELY and CRAVER (1962). [n3*] SHIMOISHI (1969). [o3*] SNAVELY et al. (1965). [p3*] PANDE and SANGAL (1968). [q3] MUKHERJI and DEY (1958). [r3*] BANERJEE and DEY (1968b). [s3*] BORÁK et al. (1970). [t3*] BUDĚŠÍNSKÝ et al. (1967). [u3] SRIVASTAVA and DEY (1963). [v3*] HAFEZ and PATTI (1969). [w3*] SRIVASTAVA and BANERJI (1969).

2. Physiological Systems

a) Complexing Agents

One of the principal differences between simple aqueous systems and physiological systems, as far as the chemistry of uranium is concerned, is the existence in the latter of a number of complexing agents that compete for any uranium present. The chemical properties of the uranium complex ion are different from those of the uranium ion (specifically UO_2^{2+} or U^{4+}) alone. An example is the precipitation of an aqueous 0.5 per cent solution of uranyl acetate and uranyl nitrate by the addition of 0.1 N NaOH. The acetate ion (Table 2.25) forms a much stronger complex with uranyl ion than does the nitrate ion (Table 2.24). If NaOH is added carefully the pH of the acetate solution can be brought to a value of 8 or more without the appearance of a precipitate. On the other hand, precipitation begins in the nitrate solution at pH 4 (DOUNCE et al., 1949).

The complexes formed between uranium and bicarbonate ions are of great physiological importance. DOUNCE et al. (1949) have shown that uranium compounds relatively insoluble in water have limited solubility rates in serum through which CO_2 has been passed. This can be attributed to complex ion formation between the uranium and bicarbonate ions present (see Tables 2.22 and 2.24). Furthermore, the U(VI)-bicarbonate complex is ultrafilterable (DOUNCE et al., 1949; MUNTZ and BARRON, 1951). It is this complex that is probably responsible to a large extent for the transport of uranium to the blood stream from deposits on or within the body and from the blood stream to the tissues and organs where the uranium is redeposited.

Uranyl ion is also complexed by protein. This complex is not ultrafilterable. However, once uranium has entered the blood, a rapid equilibrium is established between the uranium-bicarbonate complex and the uranium-protein complex. Early work indicated that approximately 40% of the uranium (VI) in the blood was protein bound and 60% was bicarbonate bound (NEUMAN, 1950; HODGE, 1956). Later work by CHEVARI and LIKHNER (1968) indicates that 47% of U(VI) in the blood is associated with the inorganic part of the plasma, 32% with protein, and 20% with erythrocytes. CHEVARI and LIKHNER (1968) have also determined stability constants for complexes between U(VI) and bicarbonate, plasma albumin, and lipoprotein of erythrocytic membrane. Their results are given in Table 2.26.

Protein can be precipitated by both U(IV) and U(VI). This is best accomplished at pH values less than 7 since at greater pH values complexes are formed. Precipitation of protein by either U(IV) or U(VI) has little or no tendency to cause denaturation provided that the pH and other conditions are kept within the limits normally tolerated by the protein (DOUNCE et al., 1949).

The complex formed between U(IV) and bicarbonate is not as strong as the one formed between U(VI) and bicarbonate, as determined from the ability of bicarbonate to remove U(IV) and U(VI) from denatured egg albumin (DOUNCE et al., 1949). Also, U(IV) is only slightly ultrafilterable from serum. An increase

Table 2.26. Stability constants β for complexes formed between U(VI) and some complexing agent found in blood[a]

Complexing agent	n	$\log \beta$
Bicarbonate	1.7	18.04
Plasma albumin	0.6	10.48
Lipoprotein of erythrocytic membrane	0.46	8.90

[a] Stability constants for the reaction $UO_2^{2+} + nL^{p-} \rightleftharpoons UO_2L_n^{2-np}$ CHEVARI and LIKHNER (1968).

in bicarbonate concentration tends to increase its ultrafilterability only slightly (DOUNCE et al., 1949).

An important difference between U(IV) and U(VI) is that following their precipitation by the addition of NaOH, U(VI) can be redissolved by the addition of strong complexing agents, such as citrate or bicarbonate, while U(IV) will not redissolve. This implies that if some oxide were formed upon injection of uranium into an animal, the uranium would be brought into solution if the uranium were hexavalent. If the uranium were tetravalent, it would remain insoluble until oxidation might occur or a colloidal uranium oxide might be formed in the plasma (DOUNCE et al., 1949).

Studies of the oxidation-reduction potentials of U(IV)-U(VI) in various complexing media indicate that U(VI) could be reduced in the body providing the uranium should penetrate cells (DOUNCE et al., 1949; DOUNCE, 1949). Reduction in the plasma is improbable. *In vitro* oxidation of U(IV) to U(VI) by molecular oxygen is retarded by protein and by cysteine and glutathione. The rate of oxidation is markedly reduced by cysteine plus ascorbic acid. Ascorbic acid alone has little or no effect on the rate of oxidation.

b) Prevention and Therapy of Uranium Poisoning

The chief deposition sites of U(VI) are the bone and kidney. Uranium (IV) accumulates in the liver and spleen as well as in the bone and kidney. However, the uranium in the soft tissues decreases fairly rapidly, whereas the concentration in the bone decreases very slowly (DOUNCE, 1949). Rats fed on a fox chow diet excreted in the urine an average of about 60 per cent of the injected U(VI) within the first three days after an intravenous injection. Only 10 per cent of an injected dose of UCl_4 partly neutralized with sodium acetate was detected in the urine during the first three days after injection. However, appreciable fecal excretion of U(IV) began after the second day and continued for some time. The urinary excretion of U(IV) and U(VI) after the first three days was very low (DOUNCE, 1949).

NEUMAN and co-workers (1953) have shown that when chemically detectable amounts of uranium (VI) are administered to animals, the bone mineral is the principal site of deposition. Experiments with radioactive isotopes of calcium and phosphorus have shown that for every uranyl ion deposited on the crystal surface of the bone, two calcium ions are no longer available for exchange with calcium ions in solution and two phosphate ions are no longer available for exchange with phosphate ions in solution. These results may be explained by assuming that each uranyl ion is complexed by two phosphate ions located adjacent to each other on the crystal surface of the bone. At the same time two calcium ions, previously associated with the phosphate ions, are released from the bone.

Uranium (VI) exerts a toxic effect on living cells by blocking the carbohydrate metabolism, apparently by blocking an enzyme system (ROTHSTEIN, 1953). Experimental results indicate that uranium specifically inhibits hexokinase in the surface of a cell. HODGE (1956) has described the mechanism of inhibition as follows. In the absence of uranium, glucose tends to enter the cell membrane through an appropriate portal assumed to bear a negative charge. Glucose is adsorbed on the surface of the hexokinase, a surface in which there is a characteristic binding of adenosine-triphosphate (ATP) through magnesium to hexokinase. The adsorption permits a phosphate from ATP to be taken by the sixth carbon atom of glucose. Glucose-6-phosphate thus formed bears a negative charge and cannot return to the medium through the negatively charged portal. In the presence of uranium, the uranyl ion competes with and replaces the magnesium ion. Thus, ATP is bound through uranyl ion to hexokinase. When glucose penetrates and is adsorbed, the ATP-uranyl-hexokinase complex will not release a phosphate to the glucose. In this manner the first step in glucose utilization cannot be taken and the glucose remains unchanged in the medium.

From *purely chemical considerations* of the above mechanisms of uranyl ion adsorption by bone and uranyl ion inhibition of hexokinase, the effects of uranium poisoning may be combated in at least three ways. (1) Assuming the adsorption of uranyl ion by the phosphate groups to be reversible, an excess of calcium ion and magnesium ion may be added to the system. According to the law of mass action the reaction should be shifted to reduce the adsorption of uranyl ion. (2) Phosphate or polyphosphate ions may be added to the system. These ions should compete with those located on the bone or hexokinase surface in complexing the uranium. (3) A complexing agent that forms an extremely stable complex with the uranyl ion may be added to the system. Such an agent should not only compete with the phosphate groups located at the adsorption sites for the uranium, but could supposedly remove uranium already adsorbed by the fixed phosphate groups.

In the use of complexing agents to effect the removal of uranium from a body, consideration must be given not only to the affinity between the complexing ligand and the uranium ion but between the complexing ligand and metal ions, such as calcium, which are found in serum and which may react strongly with the ligand. Chelating agents, such as ethylenediaminetetraacetic acid may be administered as the calcium sodium salt or in combination with an equivalent amount of calcium since the chelating agent is less toxic in this manner. (A chelating agent is a type of complexing agent having a polydentate ligand which forms a hetrocyclic ring structure with the metal.) The equilibrium between the calcium complex CaL and uranium ion U may be written

$$\mathrm{CaL} + \mathrm{U} \rightleftharpoons \mathrm{UL} + \mathrm{Ca},$$

where L represents the chelating or complexing ligand and U represents a uranium ion, U^{4+} or UO_2^{2+}. The ionic charges have been omitted for simplicity. The equilibrium constant K for the reaction is then

$$\mathrm{K} = \frac{[\mathrm{UL}][\mathrm{Ca}]}{[\mathrm{CaL}][\mathrm{U}]} = \frac{\mathrm{K_{UL}}}{\mathrm{K_{CaL}}},$$

where K_{UL} and K_{CaL} are, respectively, the stability constants for the uranium and calcium complexes. In logarithmic form

$$\log \mathrm{K} = \log \mathrm{K_{UL}} - \log \mathrm{K_{CaL}}.$$

For the complexing agent to be effective the value of K_{UL} should be larger than K_{CaL}. A measure of the effectiveness of the complexing agent is given by the concentration of complexed uranium to the concentration of the free uranium ion,

$$\frac{[UL]}{[U]} = \frac{K_{UL}}{K_{CaL}} \frac{[CaL]}{[Ca]}.$$

This ratio may be re-written (SCHUBERT, 1955) approximately as

$$R_{UL} = \frac{[UL]}{[U]} = \frac{K_{UL}}{K_{CaL}} \frac{[L_{total}]}{[Ca]}.$$

A more exact equation for the ratio (CATSCH, 1964) is given by

$$R_{UL} = \frac{K_{UL} \cdot [L_{total}]}{\alpha + K_{UL} \cdot [U] + K_{CaL} \cdot [Ca]},$$

where $\alpha = 1 + K_{HL} \cdot [H] + K_{HL} \cdot K_{H_2L} \cdot [H]^2 + K_{HL} \cdot K_{H_2L} \cdot K_{H_3L} \cdot [H]^3 + \cdots$ and K_{H_xL} are basicity constants defined as

$$K_{H_xL} = \frac{[H_xL]}{[H][H_{x-1}L]}.$$

In therapy, one attempts to have R_{UL} as large as possible.

The above equations for R_{UL} neglect the hydrolysis of uranium, reactions between hydrolyzed forms of uranium and the complexing agent, and reactions between body constituents such as protein (SCHUBERT, 1955). These reactions may be written in simplified form as

$$CaL + UOH \rightleftharpoons UL + Ca + OH,$$

$$CaL + UOH \rightleftharpoons UOHL + Ca,$$

and

$$CaL + UP \rightleftharpoons UL + CaP,$$

where P represents the binding by substances found in the body. Considering these reactions the approximate value of R_{UL} (SCHUBERT, 1955) becomes

$$R_{UL} = \frac{[UL]^3[UOHL]}{[U][UOH]^2[UP]} = \frac{K_{UL}{}^3 \cdot K_{UOHL}}{K_{CaL}{}^4 \cdot K_{UOH} \cdot K_{UL} \cdot K_W} \cdot \frac{[L_{total}]^4}{[Ca]^4} \cdot \frac{[H]}{[P]},$$

where $K_W = [H^+][OH^-] = 10^{-14}$.

In addition to these chemical considerations it is desirable that the complexing agent form complexes that are water soluble and diffusible; does not react with normal body constitutents, especially calcium; be of low toxicity; possess high adsorption when taken orally; and not be metabolized (SCHUBERT, 1955).

A number of complexing agents have been tested *in vivo* to determine their effectiveness in preventing or treating uranium poisoning. Among the first therapeutic agents used were the sodium salts of bicarbonate, citrate, malate, lactate, fumarate, and succinate (NEUMAN et al., 1949). These agents were administered by intraperitoneal injection or by forced feeding. None exhibited any better preventive or therapeutic action in uranium (VI) poisoning than the bicarbonate. Presumably these complexing agents, with the exception of sodium bicarbonate, are metabolized in the body leaving an alkaline residue which takes the form of sodium bicarbonate. The disodium salt of *d*-malate, which is not readily metabolized by the body, was found to be no more effective than the *l*-malate in preventing uranium poisoning.

It was also found that the condition of rats poisoned with uranium (VI) could be improved by treating with bicarbonate as late as 12 hours after administration of the uranium (NEUMAN et al., 1949). The best effects were obtained by pretreating the animal two or three days prior to the injection of uranium. The mortality rate was reduced from 80 per cent to 0 per cent under these circumstances by using sufficient bicarbonate to produce a moderate alkalosis.

SINGER et al. (1951) have investigated the therapeutic effectiveness of citrate and hydroxyaspartate anions in rabbits injected with lethal doses of uranyl nitrate. Hydroxyaspartic acid is a simple hydroxyaminoacid synthetically produced. The experimental results indicated that hydroxyaspartate was at least as effective as the citrate; and with respect to survival and regeneration of renal tissue, the aminoacid appeared to be more effective than citrate.

LUSKY and BRAUN (1950) have reported some protection against uranium poisoning by sodium catechol disulfonate.

HODGE et al. (1951) have found that calcium salts of polyphosphates administered intraperitoneally reduced the mortality in rats poisoned with uranium (VI). However, the polyphosphates are hydrolyzed *in vivo* which causes a metabolic acidosis. The polyphosphates also complex the serum calcium causing hypocalcemic effects (DAGIRMANJIAN et al., 1956).

The chelating agent ethylendiaminetetraacetic acid (EDTA) has been one of the most effective in treating animals poisoned by metals (SCHUBERT, 1955; CATSCH, 1964). Its effectiveness in treating uranium poisoning in rats and in mice has been demonstrated by DAGIRMANJIAN et al. (1956) and CATSCH (1959), respectively. The toxicity of calcium disodium ethylendiaminetetraacetate (CaEDTA) in young adult female albino rats was determined by DAGIRMANJIAN et al. (1956) to be 3.8 g/kg. Intraperitoneal injections of 50 mg CaEDTA per rat at 2,4, and 6 hours during a 24 hour feeding period with diets containing 4 or 8 percent uranyl nitrate hexahydrate reduced the mortality. Two injections of CaEDTA at 2 and 6 hours or single injections at 4 or 24 hours were ineffective. The 30 day LD_{50} for rats given a solution of uranyl nitrate intraperitoneally was 3.2 mg U/kg. The 30 day LD_{50} for rats treated intraperitoneally with CaEDTA doses at 2,4, and 6 hours was 9.3 mg U/kg. CaEDTA increased the 24 hour urinary excretion of parenterally administered uranium by about 50 per cent whether the CaEDTA was administered simultaneously with the uranium as one injection or as three injections at 0, 4, and 6 hours. The excretion of uranium in both control and treated rats was negligibly small during the second and third 24 hour periods. CaEDTA did not mobilize uranium deposited in the bone. Femurs of CaEDTA treated and control rats contained equal amounts of uranium.

CATSCH (1959) used white mice to determine the effectiveness of the calcium sodium salts of EDTA, diethylenetriaminepentaacetic acid (DTPA), triethylenetetraaminehexaacetic acid (TTHA), tetraethylenepentaamineheptaacetic acid (TPHA), and diaminodiethylethertetraacetic acid (DDETA) in treating uranium poisoning. The 28 day LD_{50} for control mice injected intraperitoneally with uranyl nitrate was determined to be 6.7 mg U/kg. The injection of EDTA $^1/_2$ hour after the injection of uranyl nitrate increased the LD_{50} to 12.4 mg U/kg. With DTPA the LD_{50} was increased to 16.2 mg U/kg. The other polyamino acids were not found to be effective. In fact, with the injection of TTHA or TPHA into poisoned animals there was a rapid deterioration of the animals' general condition, not observed after the injection of TTHA or TPHA into unpoisoned animals. The greater effectiveness of DTPA in treating poisoned animals compared to that of EDTA is a general pattern observed for metals other than uranium (CATSCH, 1964).

IVANNIKOV et al. (1964) have reported the greater biological effectiveness of the thioether, diaminodiethylthioethertetraacetic acid, compared to DDETA in increasing the elimination of uranium from the organisms of the rat and in reducing the build-up of uranium in the organs.

E. References to the Analytical Chemistry of Uranium

It is beyond the scope of this section to give a description of the analytical chemistry of uranium. Rather a number of references are given in which pertinent information may be found.

A number of review articles have been written on this particular subject. Among the more recent ones are those by BOOMAN and REIN (1962), PALEI (1962, 1963), RODDEN (1964), and the summaries given in "Nouveau Traité de Chemie Minérale" (CAILLAT and ELSTON, 1960; ELSTON and CAILLAT, 1967). GINDLER (1962) has made a survey of methods for the separation of uranium and for its radiochemical determination. Detailed procedures for the separation and/or determination of uranium are given in the references by BOOMAN and REIN (1962), GINDLER (1962), PALEI (1962, 1963), and RODDEN (1964). The most recent review is an excellent one by KORKISCH, J., and HECHT, F. with SORANTIN, H. Handbuch der Analytischen Chemie. FRESENIUS, W. (Ed.). Bd. VI bβ. Berlin-Heidelberg-New York: Springer-Verlag 1972.

In addition to the review papers specifically devoted to uranium, there are a number of surveys that are applicable to uranium analysis. The use of long-chain amines as solvent extractants has been reviewed by MOORE (1960) and the use of tri-n-octyl phosphine oxide has been reviewed by WHITE and ROSS (1961). ISHIMORI and co-workers (1963, 1964, 1966) have determined the distribution coefficients for a large number of elements in various acid-extractant-diluent systems. The acids used were HCl, HNO_3, or H_2SO_4. The extractants investigated were tri-n-butyl phosphate, tri-n-octyl phosine oxide, tri-n-butyl phosphine oxide, tetrabutyl methylene diphosphonate, tetrabutyl ethylene diphosphonate, di-(2-ethylhexyl) phosphoric acid, a number of amines, hexone, diisopropyl ketone, isopropyl ether, diethyl cellosolve, and pentaether.

A number of surveys have been made on the adsorption of elements with anion exchange resins from various acidic media. The review by GINDLER (1962) contains the results of such surveys for HCl, HF, HNO_3, H_2SO_4, $(NH_4)_2CO_3$, and H_3PO_4 aqueous phase systems. Similar surveys have been made by ICHIKAWA et al. (1961) for the Dowex 1 anion exchange resin-nitric acid system and by ANDERSON and KNUTSEN (1962) for the Dowex 1 resin-hydrobromic acid system. MAJUMDAR and MITRA (1965) have separated a number of elements using Amberlite IRA-400 resin and aqueous thiocyanate solutions. KORKISCH (1966) has reviewed much of the work done with both anion and cation exchange resins in mixed and non-aqueous solutions.

NELSON et al. (1964) have determined the distribution of a large number of elements between Dowex 50 cation exchange resin and HCl media, as well as $HClO_4$ media. STRELOW et al. (1965) have made similar determinations using Bio-Rad AG 50W cation exchange resin and nitric acid or sulfuric acid media. Inorganic cation exchangers have been investigated by MAECK et al. (1963). They have reported distribution coefficients for many elements between hydrous zirconium oxide, zirconium molybdate, zirconium phosphate, and zirconium tungstate ion exchangers and nitric acid solutions. BIGLIOCCA et al. (1967) have measured the adsorption of sixty elements on MnO_2 from nitric acid media. Uranium is not adsorbed.

The use of cation exchange resins in the separation of a large number of elements including uranium have been reported. FRITZ and UMBREIT (1958) have used Dowex 50 resin and EDTA to effect some separations. The EDTA complexes the metallic ions at different pH values. The complexed ion passes through the exchange column while the metallic cation is adsorbed. FRITZ and co-workers (1961, 1962) have also used Dowex 50 resin to separate a number of elements with hydrofluoric acid and with hydrobromic acid. MAJUMDAR and MITRA (1965) have separated some elements with Amberlite IR-120 resin and thiocyanate solutions. QURESHI et al. (1971) have separated uranyl ions from numerous metallic ions with a stannous arsenate column and aqueous nitric acid.

In the field of paper chromatography, R_F values for uranyl ions and a number of other cations have been determined for several systems (LEDERER and OSSICINI, 1964; LEDERER and MOSCATELLI, 1964; O'LAUGHLIN and BANKS, 1964; PRZESLAKOWSKI, 1966; PRZESLAKOWSKI and SOCZEWINSKI, 1967; ELBEIH and ABOU ELNAGA, 1967). A review of paper chromatographic and electromigration techniques has been written by BAILEY (1962).

Volatilization as a means of separation has been reviewed by DE VOE (1962).

O'KELLEY (1962) has reviewed the methods of detecting and measuring nuclear radiation.

References

ACKERMAN, R. J., RAUH, E. G.: The vapor pressure of liquid uranium; effects of dissolved tantalum, phosphorous, sulfur, carbon, and oxygen. J. phys. Chem. **73**, 769–778 (1969).

ADAMS, A., SMITH, T. D.: The formation and photochemical oxidation of uranium (IV) citrate complexes. J. chem. Soc. 4846–4850 (1960).

ADAMS, M. D., WENZ, D. A., STEUNENBERG, R. K.: Observation of uranium (V) species in molten chloride salt solution. J. phys. Chem. **67**, 1939–1941 (1963).

ADIN, A., KLOTZ, P., NEWMAN, L.: Mixed-metal complexes. In: Brookhaven National Laboratory Report BNL 50502 (S-72) (1969).

AHRLAND, S.: On the complex chemistry of the uranyl ion. II. The complexity of the uranyl monochloroacetate. A comparative potentiometric and extinctiometric investigation. Acta chem. scand. **3**, 783–808 (1949).

ALBERS, H., DEUTSCH, M., KRASTINAT, W., VAN OSTEN, H.: Über flüchtige organische Uranverbindungen. Chem. Ber. **85**, 267–278 (1952).

ALEKSANDROV, B. M., KRIVOKHATSKII, L. S., MALKIN, L. Z., PETRZHAK, K. A.: Determination of the probabilities of spontaneous fission in ^{233}U, ^{235}U, and ^{243}Am. Atom. Energ. **20**, 315–317 (1966); Soviet Atom. Energ. **20**, 352–353 (1966).

ANDERSON, M. L., CRISLER, L. R.: Spectra, structures, and bonding of some uranium (IV) cyclopentadienyls. J. organomet. Chem. **17**, 345–348 (1969).

ANDERSON, T., KNUTSEN, A. B.: Anion-exchange study. I. Adsorption of some elements in HBr-solutions. Acta chem. scand. **16**, 849–854 (1962).

ANDRIEUX, J. L., BLUM, P.: Sur la préparation électrolytique et les propriétés des borures d'uranium. C. R. Acad. Sci. (Paris) **229**, 210–212 (1949).

BAILEY, R. A.: Paper chromatographic and electromigration techniques in radiochemistry. NAS-NS 3106 (1962).

BAKULINA, I., IONOV, I.: Determination of the ionization potential by a surface ionization method. Zhur. Ekesper i Teoret. Fiz. **36**, 1001–1005 (1959); Soviet Phys. JETP **9**, 709–712 (1959).

BANERJEE, A., DEY, A. K.: Stepwise formation of uranium (VI) compounds with sulphonated phenolic compounds. Anal. chim. Acta **42**, 473–479 (1968a).

BANERJEE, A., DEY, A. K.: Stepwise formation of metal chelates of uranium (VI) and thorium (IV) with ammonium aurintricarboxylate. J. Inorg. Nucl. Chem. **30**, 3134–3139 (1968b).

BARANOV, S. A., GADZHIEV, M. K., KULAKOV, V. M., SHATINSKII, V. M.: New data on the α decay of ^{233}U. Yadern. Fiz. **5**, 518–526 (1967); Soviet J. Nucl. Phys. **5**, 365–372 (1967).

BARNOCELLI, F., GROSSI, G.: The complexing power of hydroxamic acids and its effect on the behavior of organic extractants in the reprocessing of irradiated fuels. I. The complexes between hydroxamic acid and zirconium, iron (III) and uranium (VI). J. Inorg. Nucl. Chem. **27**, 1085–1092 (1965).

BAROCAS, A., BARONCELLI, F., BIONDI, G. B., GROSSI, G.: The complexing power of hydroxamic acids and its effects on behavior of organic extractants in the reprocessing of irradiated fuels. II. J. Inorg. Nucl. Chem. **28**, 2961–2967 (1966).

BARTUŠEK, M., SOMMER, L.: Über die Hydrolyse von UO_2^{2+} in verdünnten Lösungen. Z. phys. Chem. **226**, 309–332 (1964).

BASKIN, Y.: Oxidation behavior of uranium monophosphide. J. Amer. ceram. Soc. **48**, 153–156 (1965).

BASKIN, Y.: Synthesis of uranium monoarsenide (UAs). J. Inorg. Nucl. Chem. **29**, 2480–2482 (1967).

BASKIN, Y., SHALEK, P. D.: Synthesis of uranium monophosphide by phosphine reaction. J. Inorg. Nucl. Chem. **26**, 1679–1684 (1964).

BATZAR, K., GOLDBERG, D. E., NEWMAN, L.: Effect of β-diketone structure on the synergistic extraction of uranyl ion by tributylphosphate. J. Inorg. Chem. **29**, 1511–1518 (1967).

BELYAEV, YU. I., VDOVENKO, V. M., SKOBLO, A. I., SUGLOBOV, D. N.: Bromide anionic complexes of uranyl in nonaqueous solutions. Radiokhim. **9**, 720–723 (1967); Soviet Radiochem. **9**, 680–682 (1967).

BENZ, R., BOWMAN, M. G.: Some phase equilibria in the uranium nitrogen system. J. Amer. chem. Soc. **88**, 264–268 (1966).

BERLINCOURT, T. G.: Hall effect, resistivity, and magnetoresistivity of Th, U, Zr, Ti, and Nb. Phys. Rev. **114**, 969–977 (1959).

BETTS, R. H., LEIGH, R. M.: Ionic species of tetravalent uranium in perchloric and sulfuric acids. Canad. J. Res. B **28**, 514–525 (1950).

BHAT, T. R., KRISHNAMURTHY, M.: Studies on EDTA complexes. II. Uranyl-EDTA system. J. Inorg. Nucl. Chem. **26**, 587–594 (1964).

BIGLIOCCA, C., GIRARDI, F., PAULI, J., SABBIONI, E., MELONI, S., PROVASOLI, A.: Radiochemical separations by adsorption on manganese dioxide. Analyt. Chem. **39**, 1634–1639 (1967).

BOOMAN, G. L., REIN, J. E.: Uranium. In: Treatise on analytical chemistry. Part II, vol. 9. New York: Interscience Publishers 1962.

BORÁK, J., SLOVÁK, Z., FISCHER, J.: Use of moderately dissociated complexes in spectrophotometric determinations. II. Reactions of Arsenazo III with uranyl and thorium (IV) ions. Talanta **17**, 215–229 (1970).

BRANICA, M., KUTA, J.: Polarographic study of reduction and dismutation of uranium in aqueous solutions of acetylacetone. Collect. czech. chem. Commun. **31**, 2833–2840 (1966).

BUDAYEV, I. V., VOLSKY, A. N.: The chlorination of uranium dioxide and plutonium dioxide by carbon tetrachloride. Proceed. Second. U. N. Internat. Conf. Peaceful Uses Atom. Energ. **28**, 316–330 (1958).

BUDĚŠÍNSKÝ, B., HAAS, K., BEZDĚKOVÁ, A.: Spektralphotometrische Untersuchung der Reaktion von Phosphonazo III und seiner Analoga mit Metallionen. Collect. czech. chem. Commun. **32**, 1528–1540 (1967).

BURGER, L. I.: Uranium and plutonium extraction by organophosphorus compounds. J. phys. Chem. **62**, 590–593 (1958).

BUSALEV, YU. A., NIKOLAEV, N. S., TANANAEV, I. V.: The solubilities and compositions of the solid phases in the system HF–UO_3–H_2O. Dokl. Akad. Nauk USSR **148**, 832–834 (1963); Doklady Chem. **148**, 90–92 (1963).

BUSCH, G., GÜNTHERODT, H.-J., KÜNZI, H. U.: Hall coefficient and electrical resistivity of liquid uranium. Phys. Letters **32** A, 376–377 (1970).

BYKHOVTSEV, V. L., MELIKHOVA, G. N.: Peculiarities of synergism in the extraction of uranium from solutions of nitric with an equimolar mixture of D2EHPA–TBP in kerosene. Radiokhim. **11**, 619–620 (1969); Soviet Radiochem. **11**, 608–609 (1969).

CAILLAT, R., ELSTON, J.: Nouveau traité de chimie minérale, vol. 15, part 1. Paris: Masson 1960.

CAILLAT, R., ELSTON, J.: Nouveau traité de chimie minérale, vol. 15, part 2. Paris: Masson 1961.

CAREY, G. H., MARTELL, A. E.: Mixed ligand chelates of uranium (IV). J. Amer. chem. Soc. **89**, 2859–2865 (1967).

CAREY, G. H., MARTELL, A. E.: Formation, hydrolysis, and olation of uranium (IV) chelates. J. Amer. chem. Soc. **90**, 32–38 (1968).

CARPENTIER, J. M.: Étude par électrophorèse de la complexation de l'uranium (VI) de l'étain (II) et du bismuth (III). Bull. Soc. chim. Fr, 3851–3855 (1969).

CATSCH, A.: Die Wirkung einiger Chelatibildner auf die akute Toxicität von Uranylnitrat. Klin. Wschr. **37**, 657–650 (1959).

Catsch, A.: Radioactive metal mobilization in medicine. Translated by B. Kawain. Springfield, Ill.: Ch. C. Thomas, publ. 1964.
Cefola, M., Taylor, R. C., Gentile, P. S., Celiano, A.: Coordination compounds. III. Chelate compounds of the uranyl ion with hydroxy, mercapto and amino acids. J. phys. Chem. **66**, 790–791 (1962).
Čejka, J.: Herstellung von Uranylcarbonat. Collect. czech. chem. Commun. **24**, 3180–3181 (1959).
Celeda, J., Jedinakova, V., Smirous, F.: Uranyl ion complexes in concentrated chloride, perchlorate, and nitrate solutions. Chem. Zvesti **22**, 93–101 (1968); Chem. Abstr. **49**, 54828u (1968).
Chernyaev, I. I. (Ed.): Complex compounds of uranium. Moscow: Izdatel'stvo „Nauka" 1964; New York: Daniel Davey and Co. Inc. 1966.
Chevari, S., Likhner, D.: Complex formation of natural uranium in the blood. Med. Radiol. **13** (8), 53–57 (1968).
Chiacchierini, E., Havel, J., Sommer, L.: UO_2 complexes with phenolic ligands. XI. On the reaction of uranium (VI) with maltol. Collect. czech. chem. Commun. **33**, 4251–4247 (1968).
Chiotti, P. (Ed.): Reprocessing of Nuclear Fuels. Conf-690801 (1969).
Chottard, G.: Isolement et caractérisation du nitrate d'uranyle monohydraté. C. R. Acad. Sci. (Paris) **267**, C-147–148 (1968).
Chou, M., Ts'ui, C.: Gallic acid complexes. II. Spectrophotometric study of gallic acid-uranyl complexes in aqueous solutions. Acta Sci. Natur. Univ. Jilinensis, No 2, 115–121 (1965); Nucl. Sci. Abstr. **23**, 15866 (1969).
Cordfunke, E. H. P.: The chemistry of uranium. Amsterdam-London-New York: Elsevier Publ. Co. 1969.
Cordfunke, E. H. P.: The system uranyl sulphate-water. I. Preparation and characterization of the phases in the system. J. Inorg. Nucl. Chem. **31**, 1327–1335 (1969b).
Cordfunke, E. H. P., Prins, G., Vlaanderen, P. van: Preparation and properties of the violet "U_3O_8 hydrate". J. Inorg. Nucl. Chem. **30**, 1745–1750 (1968).
Crutchfield, C. A., Jr., McNabb, W. M., Hazel, J. F.: Complexes of uranyl ion with some simple organic acids. J. Inorg. Nucl. Chem. **24**, 291–298 (1962).
Dagirmanjian, R., Maynard, E. A., Hodge, H. C.: The effects of calcium disodium ethylenediamine tetraacetate on uranium poisoning in rats. J. Pharm. exp. Ther. **117**, 20–28 (1956).
Deane, A. M.: The infrared spectra and structures of some hydrated uranium trioxides and ammonium diuranates. J. Inorg. Nucl. Chem. **21**, 238–252 (1961).
De Voe, J. R.: Application of distillation techniques to radiochemical separations. NAS-NS 3108 (1962).
De Volpi, A.: Discrepancies and possible adjustments in the 2200 m/s fission parameters. Argonne National Laboratory ANL 7830 (1971).
Domine-Berges, M.: Conditions de formation et caractérisation de phosphates d'uranium hexavalent. C. R. Acad. Sci. (Paris) **236**, 2242–2244 (1953).
Dounce, A. L.: The mechanism of action of uranium compounds in the animal body. In: Voegtlin and Hodge, p. 951–976 (1949).
Dounce, A. L., Tishkoff, G. H., Fanta, P., Ho Lan, T.: Chemistry of uranium compounds with particular reference to biochemical and toxicological studies. In: Voegtlin and Hodge, p. 61–113 (1949).
Dube, S. S., Dhindsa, S. S.: Potentiometric study of the composition and stability of uranyl complexes of 3,5-dinitrosalicylic acid. J. Inorg. Nucl. Chem. **32**, 1041–1042 (1970).
Durham, R. W.: Half-life of ^{233}U. In: Chemistry and materials division progress report. Atomic Energy of Canada Ltd., AECL-3477, 35 (1969).
Dutt, N. K., Seshadri, T.: Organic reagents used in inorganic analysis. V. Determination of formation constants of uranium (VI) chelates with several hydroxamic and N-arylhydroxamic acids. J. Inorg. Nucl. Chem. **31**, 2153–2157 (1969).
Dutt, Y., Singh, R. P.: Studies in complex formation of metal ions with tropolones. II. Dissociation constants of some tropolones and composition and stability of uranyl complexes. J. Indian chem. Soc. **42**, 767–772 (1965).
East, L. V., Keepin, G. R.: Fundamental fission signatures and their applications to nuclear safeguards. In: Proceedings of the second IAEA symposium on the physics and chemistry of fission, Vienna (1969), p. 647–666. Vienna: IAEA 1969.
Elbeih, I. I. M., Abou-Elnage, M. A.: Chromatographic separation of iron (III) and its photometric determination. Chemist-Analyst **56**, 99–100 (1967).
Ellis, Y. A.: Nuclear data sheets. Nucl. Data B **4**, No. 6 (1970).
Elston, J., Caillat, R. (directors): Nouveau traité de chimie minérale, vol. 15, part 4. Paris: Masson 1967.

ELVING, P. J., KRIVIS, A. F.: Polarographic reduction of uranium (VI) under complexing conditions and noncomplexing conditions. Nature of the uranium (V) sulphate complex. J. Inorg. Nucl. Chem. **11**, 234–241 (1959).

ERMOLAEV, N. P., KROT, N. N.: Complex formation between uranium (IV) and ethylenediaminetetraacetic acid. Zh. Neorg. Khim. 8, 2447–2461 (1963); Russ. J. Inorg. Chem. 8, 1282–1289 (1963).

FELDMAN, I., KOVAL, L.: Reactions of the uranyl ion with amino acids. Bidentate carboxylate chelation. Inorg. Chem. **2**, 145–150 (1963).

FELDMAN, I., NORTH, C. A., HUNTER, H. B.: Equilibrium constants for the formation of polynuclear tridentate 1:1 chelates in uranyl-malate, -citrate and -tartrate systems. J. phys. Chem. **64**, 1224–1230 (1960).

FERRO, R.: Über einige Selen- und Tellurverbindungen des Urans. Z. anorg. allg. Chem. **275**, 320–326 (1954).

FLYNN, K. F., GLENDENIN, L. E.: Yields of fission products for several fissionable nuclides at various incident neutron energies. ANL-7749 (1970).

FRASER, J. S., MILTON, J. C. D.: Nuclear fission. Ann. Rev. Nucl. Sci. **16**, 379–444 (1966).

FRAÚSTO DA SILVA, J. J. R., SIMÕES, M. L. S.: Studies on uranyl complexes. III. Uranyl complexes of EDTA. Talanta **15**, 609–622 (1968).

FRAÚSTO DA SILVA, J. J. R., SIMÕES, M. L. S.: Studies on uranyl complexes. IV. Simple and polynuclear uranyl complexes of some polyamino carboxylic acids. J. Inorg. Nucl. Chem. **32**, 1313–1322 (1970).

FRITZ, J. S., GARRALDA, B. B.: Cation exchange separation of metal ions with hydrobromic acid. Analyt. Chem. **34**, 102–106 (1962).

FRITZ, J. S., GARRALDA, B. B., KARRAKER, S. K.: Cation exchange separation of metal ions by elution with hydrofluoric acid. Analyt. Chem. **33**, 882–886 (1961).

FRITZ, J. S., UMBREIT, G. R.: Ion exchange separation of metal ions. Analyt. chim. Acta **19**, 509–516 (1958).

FRY, R. K., CARDWELL, A. B.: Photoelectric properties of natural uranium and changes occurring at crystallographic transformations. Phys. Rev. **125**, 471–474 (1962).

GALKIN, N. P., SUDARIKOV, B. N. (Ed.): Technology of uranium. Moscow: Atomizdat (1964); AEC-tr-6638 (1966).

GARG, B. S., TRIKHA, K. C., KATYAL, M., SINGH, R. P.: Spectrophotometric studies of colour reactions of UO_2^{2+}, Th^{4+} and MoO_4^{2+} with kaempferol and robinetin. Indian J. Chem. **6**, 334–336 (1968).

GEBALLE, T. H., MATTHIAS, B. T., ANDRES, K., FISCHER, E. S., SMITH, T. F.: Superconductivity of alpha-uranium and the role of 5f electrons. Science **152**, 755–757 (1966).

GILMAN, H., JONES, R. G., BUNDSCHADLER, E., BLUME, D., KARMAS, G., MARTIN, G. A., Jr., NOBIS, J. F., THIRTLE, J. R., YALE, H. L., YOEMAN, F. A.: Organic compounds of uranium. I. 1,3-dicarbonyl chelates. J. Amer. chem. Soc. **78**, 2790–2792 (1956).

GINDLER, J. E.: The radiochemistry of uranium. NAS-NS 3050 (1962).

GINDLER, J. E., HUIZENGA, J. R.: Nuclear fission. In: Nuclear chemistry, vol. II, p. 1–183. New York-London: Academic Press 1968.

GITTUS, J. H.: Uranium. Washington: Butterworths 1963.

GLUKHOV, I. A., ELISEEV, S. S., VOZHDAEVA, E. E.: Production of uranium (V) oxychloride. Zh. Neorg. Khim. **13**, 919–921 (1968).

GOLDINOV, A. L., STABROVSKII, A. I.: Determination of the composition of some U(IV) and U(VI) complexes in aqueous solutions. Zh. Neorg. Khim. 8, 1612–1616 (1963); Russ. J. Inorg. Chem. 8, 840–842 (1963).

GRAINGER, L.: Uranium and thorium. New York-Toronto-London: Pitman Publ. Corp. 1958.

GRINBERG, A. A., PETRZHAK, G. I., EVTEEV, L. I.: Investigations on the chemistry of uranium and thorium oxalates. Zh. Neorg. Khim. **3**, 204–211 (1958); Russ. J. Inorg. Chem. **3**, 315–326 (1958).

GRINBERG, A. A., PETRZHAK, G. I., EVTEEV, L. I.: The problem of instability constants of oxalate complexes of uranium. Radiokhim. **2**, 505–507 (1960); Soviet Radiochem. **2**, 246–247 (1960).

GUPTA, B. P., DUTT, Y., SINGH, R. P.: Stability of bivalent metal complexes with α-carboxy-β-methyltropolone. Indian J. Chem. **5**, 322–323 (1967).

GUPTA, B. D., MALIK, W. U.: Complexes of uranium (VI) and thorium (IV) with 2-hydroxy-5-methylacetophenone. J. Less-Common Metals **17**, 271–275 (1969).

GUREVICH, A. M., OSICHEVA, N. P.: Investigation of equilibria in the uranyl-peroxo-citrate system UO_2^{2+}—H_2O_2—$C_6H_5O_7^-$—H_2O. Radiokhim. **10**, 202–210 (1968); Soviet Radiochem. **10**, 192–198 (1968).

Gurevich, A. M., Polozhenskaya, L. P., Solntseva, L. F.: Complex formation of the system UO_2^{2+}—H_2O_2—EDTA—H_2O. II. Investigation of equilibria in solution. Radiokhim. **10**, 195–202 (1968); Soviet Radiochem. **10**, 186–191 (1968).

Gurevich, A. M., Susorova, N. A.: Peroxofluoride complexes of uranyl. II. Investigation of equilibria in solution. Radiokhim. **10**, 211–221 (1968); Soviet Radiochem. **10**, 199–207 (1968).

Haas, G. A., Jensen, J. T., Jr.: Thermionic studies of various uranium compounds. J. appl. Phys. **34**, 3451–3457 (1963).

Hafez, M. B.: Contribution à l'étude de la complexation du cerium et des uranides par l'acide diethylentriaminepentaacetic acid. Centre d'Etudes Nucléaires de Fontenay-aux-Roses. CEA-R-3521 (1968).

Hafez, M., Patti, F.: Characterization of the complexes of uranium (VI) with desferrioxaminemethanesulfonate (desferral) by absorption spectrophotometry. Bull. Soc. chim. Fr. 1419–1423 (1969).

Hahn, H. T., Vander Wall, E. M.: Complex formation in the dilute uranyl nitrate-nitric acid-dibutyl phosphoric acid-tributyl phosphate-AMSCO system. J. Inorg. Nucl. Chem. **26**, 191–202 (1964).

Hahn, R. L., Roche, M. F., Toth, K. S.: Alpha decay of ^{227}U. Phys. Rev. **182**, 1329–1331 (1969).

Hála, J.: Polarographische Studie der Komplexe des Urans mit Sulfosalicylsäure. Collect. czech. chem. Commun. **29**, 905–914 (1964).

Halpern, J., Smith, J. G.: Kinetics of the oxidation of uranium (IV) by molecular oxygen in aqueous perchloric acid solution. Canad. J. Chem. **34**, 1419–1427 (1956).

Hanna, G. C., Westcott, C. H., Lemmel, H. D., Leonard, B. R., Jr., Story, J. S., Attree, P. M.: Revision of values for the 2200 m/s neutron constants for four fissile nuclides. Atom. Energ. Rev. **7**, 3–92 (1969).

Harries, H. J.: Complex formation of N-substituted acetoacetamides. II. Stabilities of divalent metal complexes of N-(1-pyridyl)-acetoacetamide. J. Inorg. Nucl. Chem. **29**, 2484–2486 (1967).

Harris, W. E., Kolthoff, I. M.: The polarography of uranium. I. Reduction in moderately acid solutions. Polarographic determination of uranium. J. Amer. chem. Soc. **67**, 1484–1490 (1945).

Heal, H. G.: Some observations on the electrochemistry of uranium. Trans. Faraday Soc. **45**, 1–11 (1949).

Healy, T. V., McKay, H. A. C.: The extraction of nitrates by tri-n-butyl phosphates (TBP). Part 2. The nature of the TBP phase. Trans. Faraday Soc. **52**, 633–642 (1956).

Herasymenko, P.: Electroreduction of uranyl salts by means of the mercury dropping cathode. Trans. Faraday Soc. **24**, 272–279 (1928).

Hermans, M. E. A.: The urea process for UO_2 production. Thesis, Delft, Netherlands. Technische Hogeschool. NP-14078 (1964).

Hermans, M. E. A.: Hydrolytic phenomena in U(VI)-precipitation. In: Physical chemical aspects of adsorbents and catalysts (B. G. Linsen, Ed.), p. 373–424. London-New York: Academic Press 1970.

Hertel, G. R.: Surface ionization. II. The first ionization potential of uranium. J. chem. Phys. **47**, 335–336 (1967).

Hietanen, S.: Studies on the hydrolysis of metal ions. 17. The hydrolysis of uranium (IV) ion, U^{4+}. Acta chem. scand. **10**, 1531–1546 (1956).

Hodge, H. C.: Mechanism of uranium poisoning. Proceed. Internat. Conf. Peaceful Uses Atom. Energ., Geneva (1955) **13**, 229–232, P/73. New York: United Nations 1956.

Hodge, H. C., Maynard, E. A., Downs, W. L.: Antidotal action of polyphosphates in uranium poisoning. J. Pharmacol. exp. Ther. **101**, 17–18 (1951).

Hoekstra, H. R., Katz, J. J.: The chemistry of uranium. In: The actinide elements (G. T. Seaborg and J. J. Katz, Ed.). New York: McGraw-Hill Book Co. 1954.

Holden, A. N.: Physical metallurgy of uranium. Reading, Mass.: Addison-Wesley Publ. Co., Inc. 1958.

Hopkins, B. J., Sargood, A. J.: Some properties of vapor-deposited uranium films in ultra-high vacuum and in hydrogen. Nuovo Cim., Suppl. (1) **5**, 459–465 (1967).

Howlett, B. W.: A note on the uranium-boron alloy system. J. Inst. Met. **88**, 91–92 (1959).

Hughes, D. J., Schwartz, R. B.: Neutron cross sections. BNL 325, 2nd ed. (1958).

Hurwitz, P. A., Atkinson, G.: Pressure-jump study of the kinetics of uranyl ion hydrolysis and dimerization. J. phys. Chem. **71**, 4142–4144 (1967).

Hyde, E. K.: The nuclear properties of the heavy elements. III. Fission phenomena. Englewood Cliffs, New Jersey: Prentice-Hall, Inc. 1964.

Ichikawa, F., Uruno, S., Imai, H.: Distribution of various elements between nitric acid and anion exchange resin. Bull. chem. Soc. Japan **34**, 952–955 (1961).

IHLE, H. R., LANGENSCHEIDT, E., MURRENHOFF, A. P.: Bestimmung der spezifischen α-Aktivität des ^{233}U mit Hilfe einer Flüssigszintillations-Zählmethode. Kernforschungsanlage, Julich, JUL-491-P (1967).

ISHIMORI, T., AKATSU, E., CHENG, W., TSUKEUCHI, K., OSAKABE, T.: Data of inorganic solvent extraction (2). JAERI 1062 (1964).

ISHIMORI, T., AKATSU, E., TSUKEUCHI, K., KOBUNE, T., USUBA, Y., KIMURA, K., ONAWA, G., UCHIYAMA, H.: Data of inorganic solvent extraction (3). JAERI 1106 (1966).

ICHIMORI, T., NAKAMURA, E.: Data of inorganic solvent extraction (1). JAERI 1047 (1963).

ISHIMORI, T., UENO, K., KIMURA, K., AKATSU, E., KOBAYASHI, Y., AKATSU, J., ONO, R., HOSHI, M.: The spontaneous fission of ^{238}U. Radiochem. Acta **7**, 95–102 (1967).

ISRAELI, Y. J.: Mechanism of hydrolysis of the uranyl ion. Bull. Soc. chim. Fr. 193–196 (1965).

IVANNIKOV, A. T., et al., reported by BALABUKHA, V. S., TIKHONOVA, L. I., RAZBITNAYA, L. M., SMOLIN, D. D., RAZUMOVSKII, N. O., TORCHINSKAYA, O. L.: Physiochemical approaches to the selection of organic compounds designed to eliminate radioactive substances from the organism. Translated from p. 462–470 of Raspredelenie i biologicheskoe desistvie radioaktivnykh izotopov. Moscow: Atomizdat 1966. AEC-tr-6944, 581–591 (1964).

JAFFEY, A. H., FLYNN, K., BENTLEY, W. C., ESSLING, A.: Precision measurement of half-life and specific activity of ^{236}U. To be published in J. Inorg. Nucl. Chem. (1971b).

JAFFEY, A. H., FLYNN, K., GLENDENIN, L., BENTLEY, W. C., ESSLING, A.: Precision measurement of half-lives and specific activities of ^{235}U and ^{238}U. Phys. Rev. C **4**, 1889–1906 (1971a).

JEFTIC, L., BRANICA, M.: Square-wave polarograph of uranium (VI). I. Aqueous solutions of acetylacetone. Croat. chem. Acta **35**, 203–210 (1963a); Chem. Abstr. **60**, 2549h (1964).

JEFTIC, L., BRANICA, M.: Square-wave polarography of uranium (VI). II. The influence of surface-active agents. Croat. chem. Acta **35**, 211–215 (1963b); Chem. Abstr. **60**, 2550b (1964).

JEZOWSKA-TRZEBIATOWSKA, B.: Structure électronique de l'uranium dans ses composés tri-et-tetra-valents. J. chim. Phys. **60**, 843–848 (1963).

JONES, L. H., PENNEMAN, R.: Infrared spectra and structure of uranyl and transuranium (V) and (VI) ions in aqueous perchloric acid solution. J. chem. Phys. **21**, 542–544 (1953).

JONES, R. G., BINDSCHADLER, E., KARMAS, G., MARTIN, G. A., JR., THIRTLE, J. R., YOEMAN, F. A., GILMAN, H.: Organic compounds of uranium. IV. Uranium (V) alkoxides. J. Amer. chem. Soc. **78**, 4289–4290 (1956c).

JONES, R. G., BINDSCHADLER, E., KARMAS, G., YOEMAN, F. A., GILMAN, H.: Organic compounds of uranium. III. Uranium (V) ethoxide. J. Amer. chem. Soc. **78**, 4287–4288 (1956b).

JONES, R. G., KARMAS, G., MARTIN, G. A., JR., GILMAN, H.: Organic compounds of uranium. II. Uranium (IV) amides, alkoxides, and mercaptides. J. Amer. chem. Soc. **78**, 4285–4286 (1956a).

JØRGENSEN, C. KLIXBÜLL: Studies of absorption spectra. VI. Actinide ions with two 5f-electrons. Kgl. Danske Videnskat. Selakab. Mat-Fys. Medd. **29**, 1–21 (1955).

JØRGENSEN, C. KLIXBÜLL: Absorption spectra of red uranium (III) chloro complexes in strong hydrochloric acid. Acta chem. scand. **10**, 1503–1505 (1956).

JUZA, R., MEYER, W.: Über Uran-Nitrad-Chlorid-Bromid und -Jodid. Z. anorg. allg. Chem. **366**, 43–50 (1969).

KATZ, J. J., RABINOWITZ, E.: The chemistry of uranium. New York: McGraw-Hill Book Co. Inc. 1951.

KATZ, J. J., SEABORG, G. T.: The chemistry of the actinide elements. London: Methuen and Co. Ltd. 1957.

KAUFMANN, A., CULLITY, B., BITSIANES, G.: Uranium-silicon alloys. J. Metals **9**, AIME Trans. **209**, 23–27 (1957).

KEEPIN, G. R., WIMETT, T. F., ZIEGLER, R. K.: Delayed neutrons from fissionable isotopes of uranium, plutonium and thorium. Phys. Rev. **107**, 1044–1049 (1957).

KEITH, R. L. G.: The half-life of ^{233}U. J. Nucl. Energy **22**, 471–475 (1968).

KETRUP, A., SPECKER, H.: Stabilitätsuntersuchungen an Uranylkomplexen substituierter Acetessigsäureamide. Z. analyt. Chem. **246**, 108–111 (1969).

KHRIPIN, L. A., GAGARINSKII, YU. V., LUK'YANOVA, L. A.: Phase transitions of uranium tetrafluoride and tetrachloride. Izv. Sibirsk. Otd. Akad. Nauk SSSR, No 3, Ser. Khim. Nauk, No 1, 13–19 (1965).

KHODADAD, P.: Sur les séléniures de l'uranium tétravalent. C. R. Acad. Sci. (Paris) **245**, 934–936 (1957a).

KHODADAD, P.: Sur l'oxyséléniure d'uranium. OSeU. C. R. Acad. Sci. (Paris) **245**, 2286–2288 (1957b).

KHODADAD, P.: Sur le séléniure d'uranium U_2Se_3. C. R. Acad. Sci. (Paris) **249**, 694–696 (1959).

KHODADAD, P.: Composés du système uranium-sélénium. Préparations. Propriétés physiques et chimiques. Bull. chim. Soc. Fr. 133–136 (1961).

KLINE, R. J., KERSHNER, C. J.: Oxidation of uranium (IV) acetate by silver acetate in liquid ammonia. Inorg. Chem. **5**, 932–934 (1966).

KLYGIN, A. E., SMIRNOVA, I. D., NIKOL'SKAYA, N. A.: The solubility of ethylenediaminetetraacetic acid in ammonia and hydrochloric acid and its reaction with uranium (IV) and plutonium (IV). Zh. Neorg. Khim. **4**, 2766–2771 (1959); Russ. J. Inorg. Chem. **4**, 1279–1282 (1959).

KORKISCH, J.: Ion exchange in mixed and non-aqueous media. In: Progress in nuclear energy. Series IX. Analytical chemistry, vol. 6. Oxford: Pergamon Press 1966.

KOZLOV, A. G., KROT, N. N.: Spectrophotometric study of complex formation between uranyl ion and ethylenediaminetetraacetic acid. Zh. Neorg. Khim. **5**, 1959–1963 (1963); Russ. J. Inorg. Chem. **5**, 954–957 (1963).

KRAUS, K. A.: Hydrolytic behavior of the heavy elements. Proceedings Internat. Conf. Peaceful Uses Atom. Energ., Geneva (1955) **7**, P/731, 245–257. New York: U.N. 1956.

KRAUS, K. A., NELSON, F.: Chemistry of aqueous uranium (V) solutions. II. Reaction of uranium pentachloride with water. Thermodynamic stability of UO_2^+. Potential of U(IV)/(V), U(IV)/U(VI) and U(V)/U(VI) couples. J. Amer. chem. Soc. **71**, 2517–2522 (1949).

KRAUS, K. A., NELSON, F.: Hydrolytic behavior of metal ions. I. The acid constants of uranium (IV) and plutonium (IV). J. Amer. chem. Soc. **72**, 3901–3906 (1950).

KRAUS, K. A., NELSON, F., JOHNSON, G. L.: Chemistry of aqueous uranium (V) solutions. I. Preparation and properties. Analogy between uranium (V), neptunium (V) and plutonium (V). J. Amer. chem. Soc. **71**, 2510–2517 (1949).

KROT, N. N., ERMOLAEV, N. P., GEL'MAN, A. D.: The behavior of ethylenediaminetetraacetic acid in acid solutions and its reactions with uranium (IV). Zh. Neorg. Khim. **7**, 2054–2060 (1962); Russ. J. Inorg. Chem. **7**, 1062–1066 (1962).

KRYLOV, V. N., KOMAROV, E. V., PUSHLENKOV. M. F.: Complexing of U(VI) with fluoride ions. Radiokhim. **10**, 723–725 (1968); Soviet Radiochem. **10**, 708–710 (1968).

KURDIN, L. P., MAZEEV, M. YA.: Potential of the uranium atom and calculation of the ionization energy. At. Energ. (USSR) **22**, 83–85 (1967); Soviet At. Energ. **22**, 85–90 (1967).

KUTEINIKOV, A. F.: The use of arsenazo reagent for the determination of rare elements. Zavodsk. Lab. **24**, 1050–1052 (1958); Industrial Lab. **24**, 1171–1173 (1958).

KUTEINIKOV, A. F.: The stability of complex arsenazo derivatives. Zavodsk. Lab. **28**, 1179–1182 (1962); Industrial Lab. **28**, 1255–1258 (1962).

LAHR, H., KNOCH, W.: Bestimmung von Stabilitätskonstanten einiger Aktinidenkomplexe. II. Nitrat- und Chloridkomplexe von Uran, Neptunium, Plutonium und Americium. Radiochim. Acta **13**, 1–5 (1970).

LAI, T., CHEN, T.: Polarography of uranyl-iminodiacetate complex. J. Inorg. Nucl. Chem. **29**, 2975–2981 (1967).

LAI, T., CHEN, T.: Polarographic studies of uranyl complexes with trans- and cis-butenedioicacids. Analyt. chim. Acta **43**, 63–70 (1968).

LAI, T., LEE, S.: Polarography of uranium (VI) complexes with N-hydroxyethylethylenediamine-triacetic acid. Analyt. Chem. **41**, 1316–1319 (1969).

LATIMER, W. M.: Oxidation potentials, 2nd ed. New York: Prentice-Hall 1952.

LAUN, D. D.: Self-reversal in the spectral lines of uranium. J. Res. nat. Bur. Stand. **70** A, 323–324 (1966).

LEA, C., MEE, C. H. B.: Work-function measurements on monolayer films of uranium on (100), (110), and (113) oriented faces of tungsten single crystals by photoelectric and contact potential difference techniques. J. appl. Phys. **39**, 5890–5896 (1968).

LEDERER, C. M., HOLLANDER, J. M., PERLMAN, I.: Table of isotopes, 6th ed. New York: John Wiley and Sons, Inc. 1967.

LEDERER, M., MOSCATELLI, V.: Chromatography on ion exchange paper X. J. Chromatog. **13**, 194–198 (1964).

LEDERER, M., OSSICINI, L.: Chromatography on paper impregnated with ion exchange resins. IX. J. Chromatog. **13**, 188–193 (1964).

LEME, M. P. T., RENNER, C., CATTANI, M.: Determination of the decay constant for spontaneous fission of ^{238}U. Nucl. Instr. Methods **91**, 577–579 (1971).

LEVET, J.-C.: Sur un nouvel oxybromure d'uranium pentavalent. C. R. Acad. Sci. (Paris) **260**, 4775–4776 (1965).

LIBOWITZ, G. G., GIBB, T. R. P., JR.: High pressure dissociation studies of uranium hydrogen system. J. phys. Chem. **61**, 793–795 (1957).

LUSKY, L. M., BRAUN, H. A.: Sodium catechol disulphonate protection in experimental uranium nitrate poisoning. Fed. Proc. **9**, 297 (1950).

LUX, F.: Uranpentabromid. Angew. Chem. **81**, Nachr. Chem. Techn. **17**, 129–130 (1969).
MACKENZIE, D. R., CONNOR, R. D.: The decay of ^{239}U. Nucl. Phys. A **108**, 81–93 (1968).
MAECK, W. J., KUSSY, M. E., REIN, J.: Adsorption of the elements on inorganic ion exchangers from nitrate media. Analyt. Chem. **35**, 2086–2090 (1963).
MAEDA, M., KAKIHANA, H.: Hydrolysis of the uranyl ion in heavy water. Bull. chem. Soc. Japan **43**, 1097–1100 (1970).
MAJUMDAR, A. K., MITRA, B. K.: Ion exchange separation of metal ions with thiocyanate. Z. analyt. Chem. **208**, 1–7 (1965).
MANOUSSAKIS, G., KOUIMTZIS, T.: Study of the complexes of Fe^{3+}, Cu^{2+}, and UO_2^{2+} with benzanilidoxime. J. Inorg. Nucl. Chem. **31**, 3851–3854 (1969).
MARCU, G., BOTAR, A.: Determination of the stability constants of UO_2^{2+}, Ce^{3+}, Y^{3+}, Ni^{2+}, and Sr^{2+} compounds with acetic acid by paper electrophoresis. Stud. Univ. Babes-Bolyai, Ser. Chem. **12**, (2) 11–17 (1967); Chem. Abstr. **69**, 30703y (1968).
MARCU, G., TOMUS, M., SOLEA, M.: Paper-electrophoretic study of the formation of complex combinations of Ce(III), Eu(III), Sr(II), and uranyl ions in nitrilotriacetic acid. Stud. Univ. Babes-Bolyai, Ser. Chem. **13** (2), 15–19 (1968); Chem. Abstr. **71**, 16391e (1971).
MARCUS, Y., GIVON, M., SHILON, M.: The chemistry of the trivalent actinides in aqueous solution and their recovery. Proc. Third Internat. Conf. Peaceful Uses Atom. Energ., Geneva (1964) **10**, P/819, 588–595. New York: U.N. 1965.
MARTIN, M. J., BLICHERT-TOFT, P. H.: Radioaktive atoms. Auger-electron, α-, β,- γ-, and X-ray data. Nucl. Data A 8, 1–198 (1970).
MATHUR, V. K., NIGAM, H. L., SRIVASTAVA, S. C.: Studies on the complex formation between uranium and thiomalic acid. Bull. chem. Soc. Japan **36**, 1658–1661 (1963).
MATSON, L. K., MOODY, J. W., HIMES, R. C.: Preparation and properties of some uranium selenides. J. Inorg. Nucl. Chem. **25**, 795–800 (1963).
MATTHIAS, B. T., GEBALLE, T. H., CORENZWIT, E., HULL, G. W., JR., HO, J. C., PHILLIPS, N. E., WOHLLEBEN, D. K.: Superconductivity of beta-uranium. Science **151**, 985–986 (1966).
MAZAKI, H., SHIMIZU, S.: Effect of chemical state in the decay constant of U^{235m}. Phys. Rev. **148**, 1161–1167 (1966).
MCKAY, H. A. C., WOODHEAD, J. L.: A spectrophotometric study of the nitrate complexes of uranium (IV). J. chem. Soc. 717–723 (1964).
MERKUSHEVA, S. A., SKORIK, N. N., SEREBRENNIKOV, V. V.: Uranium (IV) hexamethylenediaminetetraacetate. Radiokhim. **11**, 600–601 (1969); Soviet Radiochem. **11**, 583–584 (1969).
MEYER, R. J., PIETSCH, E.: Gmelins Handbuch der anorganischen Chemie, 8th ed. System No 55. Berlin: Chemie 1936.
MILTON, J. C. D., FRASER, J. S.: The energies, angular distribution and yields of the prompt neutrons from individual fragments in the thermal-neutron fission of ^{233}U and ^{235}U. In: Proceedings of the symposium on physics and chemistry of fission, Salzburg (1965). Vol. II, p. 39–55. Vienna: IAEA 1965.
MISHRA, M. B., NIGAM, H. L.: Complex formation between uranyl (VI) and thiopropionic acid. Acta chim. Acad. Sci. hung. **57**, 1–4 (1968); Nucl. Sci. Abstr. **23**, 146 (1969).
MOORE, F. L.: Liquid-liquid extraction with high-molecular-weight amines NAS-NS 3101 (1960).
MOSKVIN, A. I., ESSEN, L. N., BUKHTIYAROVA, T. N.: Complex formation of tetravalent thorium and uranium in phosphate solutions. Zh. Neorg. Khim. **12**, 3390–3392 (1967); Russ. J. Inorg. Chem. **12**, 1794–1795 (1967).
MUNTZ, J. A., GUZMAN BARRON, E. S.: The transport of uranium to the tissues. In: TANNENBAUM, p. 182–198 (1951).
MURASIK, A., NIEMIEC, J.: Neutron diffraction study of uranium oxytelluride (UOTe). Bull. Acad. pol., Ser. Sci. Chim. **13**, 291–296 (1965).
NEBEL, D., URBAN, G.: Potentiometrische Untersuchungen zur Komplex-Bildung von Ce(III), Ce(IV), Th(IV), U(IV) und Citrat in wäßriger Lösung. Z. phys. Chem. **233**, 73–84 (1966).
NELSON, F., KRAUS, K. A.: Chemistry of aqueous uranium (V) solutions. III. The uranium (IV)-(V)-(VI) equilibrium in perchlorate and chloride solutions. J. Amer. chem. Soc. **73**, 2157–2161 (1951).
NELSON, F., MURASE, T., KRAUS, K. A.: Ion exchange procedures. I. J. Chromatog. **13**, 503–535 (1964).
NEMODRUK, A. A., PALEI, P. N.: A photometric study of the interaction of tetravalent uranium with arsenazo III. Zh. Anal. Khim. **18**, 480–485 (1963); J. analyt. Chem. USSR **18**, 416–420 (1963).
NEUMAN, W., HAVEN, F., DOUNCE, A. L., HO LAN, T., ROBERTS, E.: Attempted prevention and therapy of uranium poisoning. In: VOEGTLIN and HODGE, p. 976–980 (1949).

Neuman, W. F.: Urinary uranium as a measure of exposure hazard. Industr. Med. Surg. **19**, 185–191 (1950).

Neuman, W. F.: Deposition of uranium in bone. In: Voegtlin and Hodge, p. 1911–1991 (1953).

Neve de Mevergnies, M.: Energy of the isomeric transition in ^{235}U. Phys. Lett. **32** B, 482–484 (1970).

Nigon, J. P., Penneman, R. A., Staritzky, E., Keenan, T. K., Asprey, L. B.: Alkali carbonates of Np(V), Pu(V), and Am(V). J. phys. Chem. **58**, 403–404 (1954).

Oetting, F. L.: Recent calorimetric determinations of the specific powers and half-lives for certain long-lived alpha-emitting radionuclides. In: Proceedings of a symposium on the thermodynamics of nuclear materials with emphasis on solution systems, p. 55–66. Vienna: IAEA 1968.

Ohashi, H., Morozumi, T.: Electrometric determination of stability constants of uranyl-chloride and uranyl-nitrate complexes with pCl-stat. Nippon Genshiryoku Gakkaishi **9**, 65–71 (1967); Chem. Abstr. **67**, 111876t (1967).

O'Kelley, G. D.: Detection and measurement of nuclear radiation. NAS-NS 3105 (1962)

Oki, S.: Complexes formed in the chloroform extraction of uranium (VI) with oxine. Talanta **16**, 1153–1158 (1969).

O'Laughlin, J. W., Banks, C. V.: Separation of various cations by reversed-phase partition chromatography using neutral organophosphorus compounds. Analyt. Chem. **36**, 1222-1229 (1964).

Oliver, G. D., Milton, H. T., Grisard, J. W.: The vapor pressure and critical constants of uranium hexafluoride. J. Amer. chem. Soc. **75**, 2827–2829 (1953).

Ostacoli, G., Campi, E., Gennaro, M. C.: Complexes of some dihydroxybenzoic acids with uranyl ion in aqueous solution. Gazz. chim. ital. **98**, 301–315 (1968).

Palei, P. N. (compiler): Analytical chemistry of uranium. Moscow: Izdatel'stvo Akad. Nauk SSSR 1962: New York: Daniel Davey & Co. Inc. 1963.

Palei, P. N., Hsu, L. Y.: Complex formation between uranium (IV) and complexone III. Zh. Neorg. Khim. **6**, 2649–2653 (1961); Russ. J. Inorg. Chem. **6**, 1337–1340 (1961).

Pande, S. C., Sangal, S. P.: Composition and stability constants of thorium (IV) and uranium (VI) chelates with dibromopyrogallol red in aqueous solution. Microchem. J. **13**, 674–684 (1968).

Paramonova, V. I., Platunova, N. B., Baklanovskii, E. D.: Complex formation of the uranyl ion with salicylic acid. II. Study of complex formation in solution by the ion exchange method. Radiokhim. **6**, 513–518 (1964b); Soviet Radiochem. **6**, 495–500 (1964b)

Paramonova, V. I., Platunova, N. B., Dubrovin, V. S.: Complexing of uranyl ion with salicylic acid. I. Spectrophotometric studies of complexing solutions. Radiokhim. **6**, 505–513 (1964a); Soviet Radiochem. **6**, 487–494 (1964a).

Park, J. J.: Reactions of uranium and the platinide elements. II. The uranium-rhodium system. J. Res. nat. Bur. Stand. **72** A, 11–17 (1968).

Park, J. J., Mullen, L. R.: Reactions of uranium and the platinide elements. III. The uranium-iridium system. J. Res. nat. Bur. Stand. **72** A, 19–25 (1968).

Parker, J. R., Banks, C. V.: Bis-(disubstituted phosphinyl)-alkanes. III. Crystalline complexes of bis(di-2-ethylbutylphosphinyl)-methane and bis(dicyclohexylphosphinyl)-methane with uranium (VI) nitrate and with thorium nitrate. J. Inorg. Nucl. Chem. **27**, 583–587 (1965).

Pattoret, A., Drowart, J., Smoes, S.: Mass spectrometric determination of heat of sublimation of uranium. Trans. Faraday Soc. **65**, 98–112 (1969).

Peretrukhim, V. F., Krot, N. N., Gel'man, A. D.: Preparations of aqueous solutions of trivalent uranium. Radiokhim. 8, 670–673 (1966); Soviet Radiochem. **8**, 610–612 (1966).

Peterson, S.: Uranyl bromides obtained from aqueous solution. J. Inorg. Nucl. Chem. **17**, 135–137 (1961).

Petit, N.: Propriétiés de l'uranium (III) et de l'uranium (IV) en solution. Centre d'Etudes Nucléaires de Fountenay-aux-Roses. CEA-BIB-141 (1969).

Picon, M., Flahaut, J.: Sur l'oxysulfure d'uranium. C. R. Acad. Sci. (Paris) **236**, 816–818 (1953a).

Picon, M., Flahaut, J.: Propriétés des sulfures d'uranium S_2U α et β. C. R. Acad. Sci. (Paris) **237**, 1160–1162 (1953b).

Picon, M., Flahaut, J.: Sur les propriétés des sulfures d'uranium S_5U_3 et S_3U_2. C. R. Acad. Sci. (Paris) **240**, 784–785 (1955a).

Picon, M., Flahaut, J.: Sur une nouvelle forme cristalline γ du sulfure d'uranium, S_2U. C. R. Acad. Sci. (Paris) **240**, 2150–2151 (1955b).

Price, C. E., Warren, I. H.: Some observations on uranium-nitrogen compounds. Inorg. Chem. **4**, 115–116 (1965).

PRIGENT, J.: Un novel oxybromure d'uranium pentavalent. C. R. Acad. Sci. (Paris) **236**, 710–711 (1953).
PRIGENT, J.: Préparations simples de l'oxybromure d'uranium tetravalent, du tetrabromure d'uranium et du bromure d'uranyle anhydres. Existence du pentabromure d'uranium. C. R. Acad. Sci. (Paris) **238**, 102–104 (1954a).
PRIGENT, J.: Préparation du pentabromure d'uranium. C. R. Acad. Sci. (Paris) **239**, 424–426 (1954b).
PRIGENT, J., PADIOU, J.: Sur de nouveaux halogéno-complexes de l'ion uranyle. Bull. Soc. chim. Fr. 1136 (1963).
PRZESLAKOWSKI, S.: Chromatography of anionic thiocyanate complexes on paper impregnated with higher tertiary amines. Chem. Anal. (Warsaw) **12**, 57–66 (1967).
PRZESLAKOWSKI, S., SOCZEWINSKI, E.: Chromatography of anionic bromide and iodide complexes of metals on paper impregnated with tri-n-octylamine. Chem. Anal. (Warsaw) **11**, 895–903 (1966).
QURESHI, M., RATHORE, H. S., KUMAR, R.: Influence of temperature on the ion-exchange properties of stannic arsenate. J. Chromatog. **54**, 269–276 (1971).
RAJAN, K. S., MARTELL, A. E.: Equilibrium studies of uranyl complexes. I. Interaction of uranyl ion with some hydroxycarboxylic and aminocarboxylic acids. J. Inorg. Nucl. Chem. **26**, 789–798 (1964a).
RAJAN, K. S., MARTELL, A. E.: Equilibrium studies of uranyl complexes. II. Interaction of uranyl ion with tartaric and malic acids. J. Inorg. Nucl. Chem. **26**, 1927–1944 (1964b).
RAJAN, K. S., MARTELL, A. E.: Equilibrium studies of uranyl complexes. III. Interaction of uranyl ion with citric acid. Inorg. Chem. **4**, 462–469 (1965).
RAJAN, K. S., MARTELL, A. E.: Equilibrium studies of uranyl complexes. IV. Reactions with carboxylic acids. J. Inorg. Nucl. Chem. **29**, 523–529 (1967).
RAMAMOORTHY, S., RAGHAVAN, A., SANTAPPA, M.: Complexes of uranyl ion with butyric and isobutyric acids. J. Inorg. Nucl. Chem. **31**, 1765–1769 (1969).
RAMAMOORTHY, S., SANTAPPA, M.: Stability constants of some uranyl complexes. Bull. chem. Soc. Japan **41**, 1330–1333 (1968a).
RAMAMOORTHY, S., SANTAPPA, M.: Complexes of uranyl ion with various carboxylic acids. Curr. Sci. (India) **37**, 403–404 (1968b).
RAMAMOORTHY, S., SANTAPPA, M.: Stability constants of some uranyl complexes. II. Bull. chem. Soc. Japan **42**, 411–416 (1969).
RAND, M. H., KUBASCHEWSKI, O.: The thermochemical porperties of uranium compounds. Edinburgh and London: Oliver and Boyd 1963.
RAO, C. L., PAI, S. A.: Study of nitrate and sulphate complexes of uranium (IV). Radiochim. Acta **12**, 135–140 (1969).
RAUH, E. G., THORN, R. J.: Thermionic properties of uranium. J. chem. Phys. **31**, 1481–1485 (1959).
REYNOLDS, L.T., WILKINSON, G.: π-Cyclopentadienyl compounds of uranium-IV and thorium-IV. J. Inorg. Nucl. Chem. **2**, 246–253 (1956).
RIVIERE, J. C.: The work function of uranium. Proc. phys. Soc. (London) **80**, 116–123 (1962).
ROACH, D., AMIS, E. S.: The equivalent conductance of electrolytes in mixed solvents. IX. Uranium (IV) chloride in the water ethanol system. Z. phys. Chem. **35**, 274–288 (1962).
RODDEN, C. J.: Uranium. In: Analysis of essential nuclear reactor materials. Washington, D.C.: U.S. Government Print. Off. 1964.
ROSENSTREICH, J. L., GOLDBERG, D. E.: Formation constants of complexes of 2-thenoyltrifluoroacetone and 3-thenoylacetone. Inorg. Chem. **4**, 909–910 (1965).
ROTHSTEIN, A.: Studies in cell metabolism. In: VOEGTLIN and HODGE, p. 1992–2103 (1953).
RULFS, C. L., ELVING, P. J.: Uranium (III) cupferrate. J. Amer. chem. Soc. **77**, 5502–5503 (1955).
RYDBERG, J., RYDBERG, B.: Studies on the extraction of metal complexes. XIII. The complex formation between U(IV) and acetylacetone. Arkiv. Kemi **9**, 81–94 (1956).
RYZHENKO, B. N., NAUMOV, G. B., GOGLEV, V. S.: Hydrolysis of uranium ions at elevated temperatures. Geokhim. No 4, 413–417 (1967); Geochem. Int. **4**, 363–367 (1967).
SATO, A.: Studies of the behavior of trivalent uranium in an aqueous solution. I. Its reduction and stability in various acids solutions. Bull. chem. Soc. Japan **40**, 2107–2110 (1967).
SAXENA, R. S., GUPTA, K. C.: Composition and the stability of the complexes of uranyl ion with β-mercaptopropionic acid. J. indian chem. Soc. **46**, 303–307 (1969).
SCHAEFER, J. B.: Complex formation of five- and six-valent actinide elements with α-hydroxy carbonic acids. Kernforschungszentrum, Karlsruhe, Germany, KFK-765 (1968).
SCHEDIN, U., FRYDMAN, M.: Studies on the hydrolysis of metal ions. 59. The uranyl ion in magnesium nitrate medium. Acta chem. scand. **22**, 115–127 (1968).
SCHMORAK, M. R.: Nuclear data sheets. Nucl. Data B **4**, No. 6 (1970).

SCHNIZLEIN, J. G., BAKER, L., VOGEL, R. C.: Metal oxidation and ignition kinetics and metal water reactions. Argonne National Laboratory ANL-6231 (1960).

SCHREYER, J. M., BAES, C. F., JR.: The solubility of uranium (VI) orthophosphates in phosphoric acid solutions. J. Amer. chem. Soc. **76**, 354–357 (1954).

SCHUBERT, J.: Removal of radioelements from the mammalian body. Ann. Rev. Nucl. Sci. **5**, 369–412 (1955).

SCHWARZMANN, E., GLEMSER, O.: Die Art der Wasserbindung im System Urantrioxid-Wasser. Z. anorg. allg. Chem. **315**, 305–308 (1962).

SEGRÈ, E.: Spontaneous fission. Phys. Rev. **86**, 21–28 (1952).

SELBIN, J., ORTEGO, J. D.: The chemistry of uranium (V). Chem. Rev. **69**, 657–671 (1969).

SERGOVSKAYA, V. V., SOBOLEVA, T. A.: Spectrophotometric study of uranium (VI) reaction with N,N'-bis(2-hydroxyphenyl)-C-cyanoformazan. Tr. Ural. Politekh. Inst. **148**, 117–123 (1966); Chem. Abstr. **68**, 45972p (1968).

SHCHUKAREV, S. A., VASIL'KOVA, I. V., MARTYNOVA, N. S., MAL'TSEV, YU. G.: Heats of formation of uranyl chloride and uranium monoxy-trichloride. Zh. Neorg. Khim. **3**, 2647–2650 (1958); Russ. J. Inorg. Chem. **3**, 70–74 (1958).

SHIMOISHI, Y.: Spectrophotometric study of the uranyl chelate of 2-(1,8-dihydroxy-3,6-disulfo-2-naphthylazo)-phenoxyacetic acid. Bull. chem. Soc. Japan **42**, 690–693 (1969).

SIEGBAHN, K., NORDLING, C., FAHLMAN, A., NORDBERG, R., HAMRIN, K., HEDMAN, J., JOHANSSON, G., BERGMARK, T., KARLSSON, S., LINGREN, I., LINDBERG, B.: ESCA atomic molecular and solid state structure studied by means of electron spectroscopy. Nova Acta Reg. Soc. Sci. Upsal., Ser. IV **20**(1967).

SILLÉN, L. G., MARTELL, A. E.: Stability constants of metal-ion complexes. London: The Chemical Society, BurlingtonHouse, W. 1 (1964).

SINGER, T. P., MUNTZ, J. A., MEYER, J., GASVODA, B., GUZMAN BARRON, E. S.: The reversible inhibition of enzymes by uranium. In: TANNENBAUM, p. 208–245 (1951).

SKARSVÅG, K., BERGHEIM, K.: Energy and angular distributions of prompt neutrons from slow neutron fission of ^{235}U. Nucl. Phys. **45**, 72–97 (1963).

SMITH, P. K., CATHEY, L.: The preparation and electrical resistance of single crystals of β-US_2. J. electrochem. Soc. **114**, 973–975 (1967).

SNAVELY, F. A., MAGEN, W., KOZART, D.: Metal derivatives of arylazopyrazolone compounds. VII. Molarity quotients of azopyrazolone compounds containing *o*-carboxyethoxy and *o*-carboxythioethoxy groups. J. Inorg. Nucl. Chem. **27**, 679–681 (1965).

SOBKOWSKA, A.: Equilibrium of reactions between uranyl ions and lactic and tartaric acids. Rocz. Chem. **41**, 1673–1679 (1967).

SOLOVKIN, A. S.: Absence of complex formation between U^{4+} and NO_3^- ion in aqueous solution. Zh. Neorg. Khim. **14**, 1124–1126 (1969); Russ. J. Inorg. Chem. **14**, 589–591 (1969).

SOLOVKIN, A. S., TSVETKOVA, Z. N., IVANTSOV, A. I.: Thermodynamic constants for U^{4+} complexing with OH^- ions and Zr^{4+} with OH^- and NO_3^- ions. Zh. Neorg. Khim. **12**, 626–632 (1967); Russ. J. Inorg. Chem. **12**, 326–330 (1967).

SOMEYA, K.: Über die Frabe der Chromo-, Vanado-, und dreiwertigen Uranionen. Z. anorg. allg. Chem. **161**, 46–50 (1927).

SOMMER, L., ŠEPEL, T., IVANOV, V. M.: Complexation of uranium by 2-(2-thiazolylazo)-4-methoxyphenol and 2-(2-thiazolylazo)-5-methoxyphenol. Talanta **15**, 949–961 (1968).

SRIVASTAVA, K. D., BANERJI, S. K.: Uranium (VI)-methylthymol blue (pentasodium salt) chelate: A spectrophotometric study. J. prakt. Chem. **311**, 769–774 (1969).

STABROVSKII, A. I.: Polarography of uranium compounds in carbonate and bicarbonate solutions. Zh. Neorg. Khim. **5**, 811–820 (1960); Russ. J. Inorg. Chem. **5**, 389–394 (1960).

STARÝ, J., PRAŠILOVÁ, J. L.: Extraction and ion exchange investigation of uranium (VI) chelates. J. Inorg. Nucl. Chem. **17**, 361–365 (1961).

STARÝ, J., BALEK, V.: Untersuchung der Uran (VI)-Komplexe mit α-Hydroxysäuren durch Extraktionsmethode. Collect. czech. chem. Commun. **27**, 809–815 (1962).

STEHN, J. R., GOLDBERG, M. D., WIENER-CHASMAN, R., MUGHABGHAB, S. F., MAGURNO, B. A., MAY, V. N.: Neutron cross sections. BNL 325, 2nd ed., Suppl. 2 (1965).

STEINHAUS, D. W., BLAISE, J., DIRINGER, M.: Present status of the analysis of the arc spectrum of uranium (UI). Los Alamos Scientific Lab. Report LA-3475 (1966).

STEPANOV, A. V., MAKAROVA, T. P.: Electromigration method for studying complex formation of U(VI) with EDTA. Radiokhim. **11**, 290–295 (1969); Soviet Radiochem. **11**, 286–289 (1969).

STREITWIESER, A., JR., MÜLLER-WESTERHOFF, U.: Bis(cyclooctatetraenyl) uranium (Uranocene). A new class of sandwich complexes that utilize atomic f orbitals. J. Amer. chem. Soc. **90**, 7364 (1968).

STRELOW, F. W. E., RETHEMEYER, R., BOTHMA, C. J. C.: Ion exchange selectivity scales for cations in nitric acid and sulfuric acid media with a sulfonated polystyrene resin. Analyt. Chem. **37**, 106–110 (1965).

STRÓNSKI, I., ZIELÍNSKI, A., SAMOTUS, A., STASICKA, Z., BUDĚŠINSKY, B.: The determination of distribution coefficients and stability constants of complexes of nickel and uranium with quadridentate Schiff bases. Z. analyt. Chem. **222**, 14–22 (1966).

TANNENBAUM, A.: Toxicology of uranium. New York-Toronto-London: McGraw-Hill 1951.

TEDESCO, P. H., WALTON, H. E.: Metal complexes of "squaric acid" (diketocyclobutenediol) in aqueous solution. Inorg. Chem. **8**, 932–937 (1969).

TRÓC, R., LECIEJEWICZ, J., CISZEWSKI, R.: Antiferromagnetic structure of uranium diphosphide. Phys. Status Solidi **15**, 515–519 (1966).

TSERKOVNITSKAYA, I. A., BYKHOVTSEVA, T. T.: Determination of the valence forms of uranium in mixtures of oxides containing iron. Zh. Analit. Khim. **22**, 96–100 (1967); J. analyt. Chem. USSR **22**, 79–82 (1967).

TSYMBAL, C.: Contribution à la chimie de l'uranium (VI) en solution. Centr d'Etudes Nucléaires de Grenoble. CEA-R-3476 (1969).

UL'YANVO, V. S., SVIRIDOVA, R. A., LASKORIN, B. N.: Extraction of uranium (VI) from aqueous perchlorate solutions of di(2-ethylhexyl)-phosphoric acid in n-octane. Radiokhim. **8**, 416–420 (1966); Soviet Radiochem. **8**, 386–389 (1966).

UL'YANOV, V. S., SVIRIDOVA, R. A., ZARUBIN, A. I.: Determination of the association constants of uranyl ions with diethyl ester of phosphoric acid. Radiokhim. **11**, 13–18 (1969); Soviet Radiochem. **11**, 11–15 (1969).

UNIK, J. P., GINDLER, J. E.: A critical review of the energy released in nuclear fission. ANL-7748 (1971).

VARGA, L. P.: An objective computer-oriented method for calculation of stability constants from the formation function. Analyt. Chem. **41**, 323–330 (1969).

VARTAK, D. G., MENON, N. G.: Solution stability constants of complexes of 4-nitro-2-aminophenol with some divalent metal ions. J. Inorg. Nucl. Chem. **28**, 2911–2917 (1966).

VARTAK, D. G., MENON, N. G.: Complex equilibria of UO_2^+ with ligands having oxygen and nitrogen donor atoms. Proceedings of the Nuclear and Radiation Chemistry Symposium. Poona, March 6–9, 1967, p. 466-468. Bombay: Department of Atomic Energy 1967.

VARTAK, D. G., MENON, K. R.: Proton ligand and metal ligand stability constants of substituted salicylaldehydes. J. Inorg. Nucl. Chem. **31**, 3141–3147 (1969).

VDOVENKO, V. M.: The chemistry of uranium and transuranium elements. Moscow-Leningrad: Publ. House of the Acad. Sci. USSR (1960); AEC-tr-642 (1964).

VDOVENKO, V. M., ROMANOV, G. A., SHCHERBAKOV, V. A.: Study of the complex formation of uranium (IV) with fluoride ion by the proton resonance method. Radiokhim. **5**, 581–585 (1963a); Soviet Radiochem. **5**, 538–541 (1963a).

VDOVENKO, V. M., ROMANOV, G. A., SHCHERBAKOV, V. A.: Proton resonance study of complex formation of U(IV) with halide, sulfate, and perchlorate anions. Radiokhim. **5**, 664–668 (1963b); Soviet Radiochem. **5**, 624–627 (1963b).

VDOVENKO, V. M., ROMANOV, G. A., SHCHERBAKOV, V. A.: A study of fluoro-complexes of uranium (IV) in aluminum salt solutions. Zh. Neorg. Khim. **11**, 252–255 (1966); Russ. J. Inorg. Chem **11**, 139–141 (1966).

VDOVENKO, V. M., ROMANOV, G. A., SOLNTSEVA, L. V.: On the question of the existence of UOF_2. Radiokhim. **9**, 727–729 (1967); Soviet Radiochem. **9**, 688–690 (1967).

VERMA, S. K., AGARWAL, R. P.: Stability of uranyl catechol chelates. J. prakt. Chem. **38**, 280–288 (1968).

VOEGTLIN, C., HODGE, H. C.: Pharmacology and toxicology of uranium compounds. New York-Toronto-London: McGraw-Hill, Parts I, II (1949); Parts III, IV (1953).

VOROBEV, S. P., DAVYDOV, I. P., SHILIN, I. V.: Complexes of uranium (VI) and molybdenum (VI) with trihydroxyglutaric acid. Zh. Neorg. Khim. **14**, 1029–1033 (1969); Russ. J. Inorg. Chem. **14**, 536–539 (1969).

WAHL, A. C., NORRIS, A. E., ROUSE, R. A., WILLIAMS, J. C.: Products from thermal-neutron-induced fission of ^{235}U: a correlation of radiochemical charge and mass distribution data. In: Proceedings of the second IAEA symposium on the physics and chemistry of fission, Vienna (1969), p. 813–848. Vienna: IAEA 1969.

WAPSTRA, A. H., GOVE, N. B.: Nuclear data sheets. Nucl. Data B **1**, No 5 (1966).

WAPSTRA, A. H., NIJGH, G. J., VAN LIESHOUT, R.: Nuclear spectroscopy tables. Amsterdam: North-Holland Publ. Co. 1959.

WERNING, J. R.: Thermal ionization at hot surfaces. Thesis. UCRL-8455 (1958).

WHITE, J. C., ROSS, W. J.: Separations by solvent extraction with tri-n-octylphosphine oxide. NAS-NS 3102 (1961).

WHITTAKER, M. P., EYRING, E. M., DIBBLE, E.: Kinetics of uranyl ion hydrolysis and polymerization. J. phys. Chem. **69**, 2319–2323 (1965).

WILKINSON, W. D.: Uranium metallurgy. Vol. I: Uranium process metallurgy. New York-London: Interscience Publ. 1962.

Wilson, W. B.: High-temperature x-ray diffraction investigation of the uranium-carbon system. J. Amer. ceram. Soc. **43**, 77–81 (1961).

Yoshihara, K., Kanno, M., Mukaibo, T.: A new compound-UNF. J. Inorg. Nucl. Chem. **31**, 985–988 (1969).

Zachariasen, W. H.: Crystal chemical studies of the 5f-series of elements. XXIII. On the crystal chemistry of uranyl compounds and of related compounds of transuranic elements. Acta cryst. **7**, 795–799 (1954).

Zakharova, F. A., Moskvin, A. I.: The solubility product of uranium (IV) oxalate and the composition and dissociation constants of oxalato-uranium (IV) complexes in aqueous solution. Zh. Neorg. Khim. **5**, 1228–1233 (1960); Russ. J. Inorg. Chem. **5**, 592–595 (1960).

Zielinski, A., Stronski, I.: Radiotracer studies of the extraction of metal ions. XIII. The determination of distribution coefficients and stability constants of chelates of Ni^{2+} and UO_2^+ with some Schiff bases. Nukleonika **13**, 765–773 (1968).

Chapter 3

Animal Experiments

C. L. YUILE

With 2 Figures

I. Introduction

Since the first demonstration of the acute nephrotoxic action of uranium salts by CHITTENDEN and LAMBERT (1889), very numerous animal experimental studies have been reported. Most of the earlier work reviewed by MCNIDER (1924) and HUNTER (1928) was concerned with the use of uranium poisoning as a model system for the study of experimental nephritis. The morphological effect of uranium action on the kidney was early described as predominantly a degenerative or necrotizing tubular change, involving the proximal convoluted tubule. Many publications, notably those of DICKSON (1909, 1912), CHRISTIAN et al. (1911) and SUSUKI (1912), were concerned with what, in retrospect, appears to have been a rather futile search for an experimental lesion, analogous to chronic nephritis seen in humans, which could be used for studies in pathogenesis, diagnostic functional testing and therapy. The moderate degrees of focal interstitial inflamation and fibrosis, coupled with nephron atrophy and glomerular scarring, were described enthusiastically by some authors as chronic nephritis but as relatively insignificant and not progressive by others. Regeneration of damaged tubular epithelium in non-fatal cases was briefly alluded to in some of the very early reports and OLIVER (1915) and MCNIDER (1919) gave good descriptions of this process. An interesting observation made early in the history of uranium experiments concerned the apparent tolerance which developed as a result of repeated doses, whereby much larger amounts were subsequently required to produce the initial toxic effect of a small dose. The regenerated epithelium was considered to be of a metaplastic or different type, incapable of responding to the toxic material as did the normal cells. The mechanism of such tolerance was never fully clarified although HUNTER (1928) suggested that uranium failed to unite with protein of the regenerated cell. Most of the investigators during the early decades of the century turned their attention to study of the effects of these renal lesions on function of the kidney—chiefly in the acute stages of uranium nephritis. In Paper 13 by SCHWARTZ and KATZ in TANNENBAUM (1951), the literature prior to 1947, dealing with biochemical effects of uranium poisoning, was reviewed, with emphasis on tests for nitrogen retention, acidosis, albuminuria and altered renal clearance. A comprehensive bibliography of publications in this field is also included. While the nephrotoxic action of uranium was of major interest in this period, there were a few studies concerned with damage to the liver, GARNIER and MAREK (1931a and b) and VERNE (1931), the central nervous system, VERNE (1931), and to blood coagulation, JACKSON (1910). There were also some indications of differences in species susceptibility, as well as differences in sensitivity to the material related to age, sex, and animal strain, reported by the authors already cited.

The project for the development of atomic energy during World War II was a tremendous stimulus for biological studies designed to characterize uranium poisoning and determine relative toxicities of the various compounds of this element. These studies, begun in 1943, were to serve as a guide for the safety of the numerous persons handling uranium in laboratories and plants. The results have been described by Carl Voegtlin as "the most comprehensive experimental investigation of an industrial poison ever carried out by any group of scientific workers in such a short time", Voegtlin and Hodge (1949).

Much of the animal experimental work was carried out for the Manhattan Project either at the University of Rochester or by the Metallurgical Laboratory of the University of Chicago. Detailed descriptions of the activities of the Project at Rochester are contained in a four-volume publication entitled "The Pharmacology and Toxicology of Uranium Compounds", the first two volumes of which were published in 1949 and the second two volumes in 1953, Voegtlin and Hodge (1949, 1953). A volume entitled "Toxicology of Uranium", detailing the activities of the program carried out in Chicago, was published under the editorship of Albert Tannenbaum, Tannenbaum (1951). It was early decided that the major industrial hazard probably would result from inhalation of uranium dust, and much of the work of the University of Rochester Atomic Energy Project was, and has been, directly or indirectly concerned with this aspect, while the Chicago group concerned themselves largely with studies of injection and ingestion in the mouse. Apart from the obviously practical aspect of the experimental work carried out during and for a short time after the war years, many basic scientific studies were also a part of the program.

In the remainder of this chapter, a brief outline of the important animal experimental work described in these volumes, as well as studies reported in the literature since their publication, will be made. It seems appropriate to mention at this time that an extensive bibliography of the biological effects of uranium contained in the open literature from the beginnings of such experimental work until approximately 1950 is contained in Volume IV of "The Pharmacology and Toxicology of Uranium Compounds, p. 2270–2289, Voegtlin and Hodge (1953).

II. Criteria of Uranium Toxicity in Experimental Animals

In the experiments reported in the literature prior to 1943, as well as in experiments to be described below, numerous methods for estimating uranium toxicity have been used. These are based on different general and local manifestations of the action of uranium on the mammalian body with much of the focus being upon the alteration in renal structure and function. Among the measures of toxicity are mortality, either absolute without relation to time, or mortalities at different time intervals, such as 24 hours or 2 to 3 weeks. Doses causing various percent mortalities are referred to as, for example, the LD_{50} (lethal dose causing an approximate 50% fatality). Other general effects such as survival time, life-span shortening, and diminution of growth rate have also been utilized. In some experiments the evidence for toxicity has been based entirely on the presence or absence of histological renal lesions while in others biochemical effects reflecting alterations in renal function, have been the basis for toxicity evaluation. Examples of the latter are measurements of increased urinary protein, sugar or amino acid axcretion, elevations in blood non-protein nitrogen, and increased levels of various enzymes in the urine. Excretion of the dye phenolsulfonephthalein and measurements of renal clearance have also been employed to detect early and subtle alterations in renal function produced by the administration of uranium. Among other methods employed have been changes in urine volume, the urinary excretion of uranium and its deposition in various tissues of the body. Local irritative effects when the material has been applied to such areas as the skin or the eye, and finally, in certain long term experiments, radiation effects have been described. The above criteria have been used alone or in various combinations, the age, sex, species and strain of animal has not always been controlled, and the effects have varied considerably depending on the method of administration. It is, therefore, readily apparent that actual quantitative comparisons between experiments are not possible and it should be borne in mind that the data to be described can be compared only in relative terms.

III. Parenteral Administration

Because of the need to obtain at least semi-quantitative data relating to the effects of different uranium compounds in different species, extensive studies of acute parenteral toxicity were initially undertaken in the Medical Division of the Manhattan Project at the University of Rochester. Most of these, which were in reality studies of mortality, were done in the rat, with three different compounds: uranyl nitrate, $UO_2(NO_3)_2 \cdot 6H_2O$; uranyl fluoride, UO_2F_2; and uranyl tetrachloride, UCl_4 given intraperitoneally. In addition, a few mice were given intraperitoneal injections of uranyl nitrate, and groups of rabbits, guinea pigs, rats and mice were given intravenous injections of this same material in a preliminary study. A considerable amount of additional information on the subject of parenteral toxicity will be referred to in subsequent sections of this chapter. The data were gathered in studies devoted to metabolic and functional changes, a search for therapeutic procedures against uranium poisoning, studies of sensitive methods for detection of the earliest signs of poisoning and studies of tolerance. In much of this work toxicity measurements were based on criteria other than mortality. However, in general, when mortality was used, the data were in line with those to be reported in this section.

Although in many of the studies reported in the early literature, uranium was administered parenterally, usually as uranium nitrate, only a few reports provided information relating dose to mortality. The studies of McNider (1929, 1936) indicated that the fatal dose of uranyl nitrate given subcutaneously to dogs was somewhat greater that 2 mg U/kg, while from the work of Hunter (1928) it would appear that about 0.4 mg of uranium intravenously and 0.7 mg of uranium subcutaneously would kill the average rabbit probably weighing from 1 to 3 kg.

The data from many experiments performed at the University of Rochester reported by Haven and Hodge in chapter 6, Voegtlin and Hodge (1949), and dealing with intraperitoneally injected uranium compounds are summarized in Table 3.1. In this table, comparisons are made between the mortalities occurring at 24 hours and those occurring between 14 and 21 days, as well as between male and female and between young and older rats. Conclusions drawn from these studies included a striking increase in toxicity with increasing age in the rat based on the 24 hour LD_{50}. However, there was little evidence of dependence of mortality on age for the 2 to 3 week period. Uranyl fluoride was found to be more toxic than uranyl nitrate or uranium tetrachloride, and on the basis of 14–21 day mortalities, a sex difference was noted, with males being more resistant than females.

Table 3.1. Comparative parenteral toxicity of uranium compounds after intraperitoneal injection

Animal	Uranium compound — mg U/kg					
	Uranyl nitrate		Uranyl fluoride		Uranium chloride with sodium acetate buffer	
	LD_{50} 24 hr	LD_{50} 14–21 days	LD_{50} 24 hr	LD_{50} 14–21 days	LD_{50} 24 hr[a]	LD_{90} 14–21 days
Young male rats	305	2.0	78	2.5	615 (335)	—
Young female rats	—	1.0	78	1.0	615 (335)	—
Adult male rats	204	2.5	87	2.5	434	25[b]
Adult female rats	135	1.0	87	1.0	359 (190)	25[b]
Old male rats	128	—	40	—	297	—

[a] Values in parentheses are for non-buffered 10% aqueous UCl_4.
[b] 90 percent mortality.

The LD_{50} of intraperitoneal uranyl nitrate determined by the Rochester group in two strains of mice, varied from 6–8 mg U/kg in the albino strain and from 20–25 mg U/kg in the C—3H strain. In 4 different strains of mice, Tannenbaum (1951) found that the subcutaneous injection of 1.0 mg of uranium as the nitrate resulted in mortalities ranging from 0 to 100%. Tannenbaum (1951) also noted that male mice were more resistant than females but did not find an age differential in this species.

The relative susceptibilities of four species of animals, based on approximate lethal doses of uranyl nitrate hexahydrate administered intravenously as reported by Orcutt in chap. 6, Voegtlin and Hodge (1949), were rabbits, 0.1 mg U/kg; guinea pigs, 0.3 mg U/kg; rats, 1.0 mg U/kg; mice, 10 to 20 mg U/kg.

IV. Oral Toxicity of Uranium

In the early days of the Manhattan Project considerable emphasis was placed on oral toxicity experiments because of the feeling that the most important and insiduous exposure to uranium from an industrial standpoint was inhalation. Since these exposure are not simple, uncomplicated pulmonary problems, because the subjects also receive toxic materials through the gastrointestinal tract, it was important to obtain information on the entry of uranium into the body by the latter route. The report of Karsner and Reiman (1918) outlines the only related study found in the early literature. These authors recognized the possible industrial hazard of inhalation of uranium dust and that associated with it, considerable amounts of material would, in all probability, be swallowed. They fed three uranium compounds, U_3O_8, sodium uranate and uranyl nitrate to guinea pigs, using qualitative tests for urinary albumin and sugar, as well as mortality to estimate toxicity. They found that all the materials used were toxic orally but subsequent studies have not confirmed some of the details of their experiments.

A. Thirty Day Feeding Experiments

The first studies begun in early 1943 involved a large number of experiments carried out for periods of 30 days, 1 year und 2 years. In the 30 day experiments, 9 different uranium compounds were tested while in the longer feeding experiments the compounds were limited to four. Large numbers of rats and mice were used by the workers in Rochester, chap. 7, VOEGTLIN and HODGE (1949), and Chicago, TANNENBAUM (1951), respectively, and the former also performed feeding experiments on smaller numbers of dogs and rabbits. Relative susceptibilities of these various species were determined, as well as the effects of age and sex on the toxicity of the material utilized. Dosages varied considerably, being based, for the smaller animals, on the percentage of uranium compound in the diet, while for dogs the number of grams per kilogram fed daily was used. The presence or absence of toxicity was based on the criteria of mortality, if this occurred, diminution in growth rate, and histopathological changes. In some of the longer feeding experiments, tests for elevation of blood non-protein nitrogen, urinary protein and sugar, and alterations in the hematologic picture were also done but were found to be essentially normal.

In the 30 day feeding experiments with rats, the relatively insoluble UO_2, U_3O_8, and UF_4 were found to be non-toxic while the remaining 6 compounds UO_3, UO_2Ac_2, UO_4, UCl_4, UO_2F_2 and $UO_2(NO_3)_2 \cdot 6H_2O$, all of which were water-soluble to a considerable degree, were found to be toxic. Diets containing up to 20% of any of the non-toxic compounds caused no mortality and little, if any, interference with growth. One hundred percent mortality resulted when from 2 to 10% of any of the toxic compounds was added to the diet, and 0.1 to 1.0% of any of these compounds in the diet caused growth depression. Mature rats were found to be more susceptible to poisoning by uranyl nitrate in the diet than were weanling rats. The frequency and degree of renal changes seen histopathologically in animals fed the toxic group of compounds varied considerably, depending on the dose and compound. No significant histopathological changes were seen following 30 day feeding of the three insoluble non-toxic uranium compounds.

A small number of rabbits were fed uranyl nitrate hexahydrate at levels of 0.02, 0.1 and 0.5% in the diet for a period of 30 days. Body weight losses were noted at all levels but were minimal in the rabbits receiving 0.02%. All rabbits receiving 0.5% uranyl nitrate in the diet died but none receiving 0.02% did so. Histological renal damage was observed at all three levels. On the basis of somewhat limited data, it was estimated on the one hand by relating mortality to amount fed, and on the other hand, by calculating the minimal amount of uranyl nitrate in the diet which caused detectable weight depression, that the rabbit was considerably more susceptible to the feeding of uranyl nitrate than was the dog, which, in turn, was considerably more susceptible than the rat.

B. One- and Two-year Feeding Experiments

Large groups of male and female rats were fed diets containing various levels of 2 of the toxic compounds, namely UO_2F_2 and uranyl nitrate, as well as 2 insoluble, non-toxic compounds, UF_4 and UO_2, for a period of 1 year, chap. 20, VOEGTLIN and HODGE (1953). The minimum percentages in the diet which produced growth depression were 0.1 for UO_2F_2 and 0.5 for uranyl nitrate. Some growth depression was produced by approximately 20% of UF_4 in the diet, while with this percentage of UO_2, no growth depression was noted. Except for a slight increase noted during the 1 year period when 0.5% UO_2F_2 and 3% uranyl nitrate were added, most of the diets failed to produce any change in mortality over that observed in control rats. Some idea of the relative toxicity of these compounds is obtained by comparing the dietary percentages that produced equal body weight depressions. These are as follows: UO_2F_2, 0.25%; uranyl nitrate, 1.0%; UF_4, 20%. Twenty percent UO_2 in the diet failed to influence growth. Some pathological findings were noted in rats fed 0.1% UO_2F_2, while with uranyl nitrate, except for slight transient injury and nearly complete recovery with time at a level of 0.5%, the findings were restricted to the animals receiving 2% in the diet. These showed moderately severe initial injury followed by regeneration and progressive atrophic changes in the renal tubules. At the 2% level of uranyl nitrate some degenerative changes in the testes of some of the male rats were seen after 9 months. These consisted of focal atrophy and degeneration in the seminiferous tubular epithelium.

The same 4 uranium compounds fed in the 1 year study were continued for a second year, in the diet, of groups of 15–25 male and female rats. In general, the growth tendencies in the latter part of the first year were continued throughout the second year. There were no further significant changes in the mortality rate following ingestion of any of the compounds and pathological changes observed were essentially similar to those seen at the end of the first year.

Uranium analyses of the kidneys and femurs of rats receiving these four uranium compounds were made. There was no evidence of an increase in the uranium content in either tissue when UO_2 was fed. However, there was some indication of a slight increase in uranium centent following the ingestion of increasing amounts of UF_4. The uranium content of the kidney increased to values of between 20 and 30 μg/g of fresh tissue when the highest levels of UO_2F_2 and uranyl nitrate were fed. Uranium values in femur ash ranged from 1 μg/g for rats on a 20% diet of UO_2, to about 28 μg/g at the highest percentage of UF_4. Much larger values for the uranium content in bone were found in rats fed the soluble materials. Rats fed 0.5% UO_2F_2 in the diet showed levels of approximately 60 μg/g at both 12 and 24 months, while the levels after feeding uranyl nitrate at the 2% level reached values between 150 and 200 μg/g of ash. These values are of considerable interest from a radiological hazard standpoint in view of the accepted tolerance figure of 25 μg/g wet weight. However, no long-term feeding studies in long-lived species are available and continued exposure would be required to maintain or increase these levels in the face of the continuous excretion from the bone which occurs following cessation of ingestion. It has been estimated by MAYNARD et al., in sec. 4, chap. 20 of VOEGTLIN and HODGE (1953) that, in rats on a diet containing 0.5 percent uranyl nitrate, the amount of uranium absorbed and deposited in bone is balanced by the amount mobilized from bone and excreted. About 2 percent of the total uranium in the skeleton, or 2.4 μg, would thus be excreted daily.

By comparing the mortality, body-weight depresion and histopathological abnormalities in groups of rats given daily intraperitoneal injections of uranyl nitrate with groups fed various percentages of this material in the diet, it has been estimated that a diet containing 1% uranyl nitrate results in the absorption of an amount equivalent to the daily injection of between 0.2 and 0.7 mg/kg, or that about 0.1% of this soluble uranium compound is absorbed.

Thirty dogs were fed various daily doses of five uranium compounds, UO_2, UF_4, $UO_2(NO_3)_2 \cdot 6H_2O$, UCl_4, UO_2F_2, for a period of one year. The doses ranged from 0.0002 g/kg of body weight of soluble UO_2F_2 to 10 g/kg of body weight of insoluble UO_2. An adverse effect on growth was noted only in those dogs which received uranyl nitrate at a daily dose of 0.2 g/kg and a slight, questionable effect, possibly attributed to fluorine, was seen at a UF_4 level of 5 g/kg daily.

A few animals receiving the highest dose of all five uranium compounds showed some renal cortical tubule changes with little evidence of either necrosis or regeneration evident at the end of the one year period. The occasional alterations in urinary protein and sugar excretion which were found were of questionable significance. A few scattered uranium analyses were made on dogs fed 0.01 and 0.05 g/kg/day of uranium tetrachloride and 0.1 and 0.2 g/kg/day of uranyl nitrate. Kidney values following ingestion of either of these materials for 1 year were under 3 μg/g. With uranium tetrachloride, bone values ranged from 1 to 5 μg/g without much evidence of dependence on dose, while a dog fed 0.2 g/kg/day of uranyl nitrate had deposited about 14 μg U per gram of femur. This compares with approximate values of 20 μg U/g and 200 μg U/g in the femur of rats fed diets containing 0.5 and 2.0 percent uranyl nitrate respectively.

V. Uranium Compounds Applied to the Skin

Because of possible direct dermal exposure of industrial workers to salts of uranium and because of the importance of evaluating inadvertent percutaneous intoxication during studies involving other routes of administration, a rather extensive series of experiments dealing with the toxicology of compounds of uranium following their application to the skin, was carried out. These were reported in some detail by Orcutt in chap. 8 of The Pharmacology and Toxicology of Uranium Compounds, Voegtlin and Hodge (1949).

The findings may be briefly summarized as follows. Soluble uranium compounds, including uranyl nitrate, uranyl fluoride, uranium pentachloride, uranium trioxide, sodium diuranate and ammonium diuranate were found to be absorbed through the skin of experimental animals in sufficient amounts to cause severe uranium poisoning and death. In rats, following dermal exposure to uranyl nitrate, significant quantities of uranium were found in the blood stream. While UCl_4 also produced poisoning, it was believed that oxidation to the hexavalent form was necessary before significant absorption occurred. The insoluble oxides of uranium, UO_2, UO_4, and U_3O_8, as well as the tetrafluoride, UF_4, were not found to cause significant evidence of poisoning following their application to the skin. A marked species difference noted for uranyl nitrate, was presumed to apply to other toxic compounds as well. Rabbits, rats, guinea pigs and mice, in order of decreasing susceptibility, were poisoned, with more than a hundred-fold difference in the LD_{50} of rabbits and mice. Although the size of the dose required to produce poisoning varied as a result of this species difference, the course of the response to the toxic material was essentially similar in all species. Toxocological manifestations consisted of increased urinary protein, elevation of blood non-protein nitrogen (determination made only in rabbits), body weight depression and an increase in mortality. In survivors, these manifestations began to decrease rapidly following the tenth day and were essentially normal by approximately the end of the third week. Changes in urinary protein, body weight and mortality were essentially the same in all species tested. The renal damage, histologically, was characteristic of that seen in uranium poisoning by other means of administration and in those animals that recovered there was evidence of a remarkable ability for repair. Repeated exposures by skin application resulted in a tolerance to cumulative doses that would have been quite lethal if applied as a single exposure initially. As far as uranyl nitrate, uranium pentachloride and uranium tetrachloride were concerned, a mild to moderate local skin irritation occurred.

VI. Uranium Compounds Applied to the Eye

Because of the possible hazard to workers in uranium plants and laboratories from ocular exposure to uranium compounds and because, in experimental animals in dust chambers, the eye represents a possible extraneous route of absorption in toxicological study of pulmonary exposure, a series of investigations was undertaken between 1943 and 1945 at the University of Rochester and reported by Orcutt in chap. 9, Voegtlin and Hodge (1949), on the toxicity of a number of compounds following instillation into the conjunctival sac in experimental animals. The animals used were rabbits, rats and guinea pigs and the compounds consisted of uranyl nitrate, uranyl fluoride, uranium trioxide, sodium diuranate, ammonium diuranate, uranium tetrachloride, uranium pentachloride, uranium tetrafluoride, uranium dioxide, uranium peroxide and uranium tritaoctoxide. Both acute and chronic exposures were undertaken. However, since the data were assembled from various isolated experiments or groups of experiments, not all designed for the same purpose, the conditions were not identical in all cases with respect to vehicle and concentration. In the studies of acute poisoning, a single large dose of each compound was instilled into the conjunctival sac in the form of either a dry powder, an aqueous solution or a suspension in lanolin. Local reactions ranged from mild conjunctival reddening through edema, exudation and ulceration. Evidence was also sought of corneal corrosion or vascularization. Degrees of local damage were recorded on a scale of 1+ to 4+. On the basis of the findings, the various compounds listed above were classified according to relative local damage and systemic toxicity in rabbits. Uranyl nitrate, uranyl fluoride, uranium tetrachloride, uranium pentachloride, uranium trioxide, sodium diuranate, ammonium diuranate and uranium tetrafluoride were absorbed in sufficient quantities to produce severe systemic poisoning and death in some animals. The most severe local conjunctival reactions resulted following the placing of dry uranium pentachloride in the conjunctiva. Of the 4 animals used, death occurred in 2, while all 4 showed necrosis, separation of the conjunctiva, and marked changes in the periorbital tissues. The severe local reaction was attributed to the caustic action of a high local concentration of hydrogen chloride liberated on hydrolysis of this compound. Uranyl nitrate, uranyl fluoride, and sodium diuranate all caused local damage of varying degrees and all were absorbed in sufficient quantities to cause systemic poisoning. Uranium tetrafluoride and uranium diuranate caused relatively less local irritation, but both caused systemic effects. Uranium trioxide was unique in that there were no significant local effects but the absorption caused sufficient systemic reaction to produce death in all 4 animals studied when the material was given as a dry powder. On the other hand, its supension in lanolin prevented any evidence of systemic poisoning. Uranium dioxide, uranium peroxide, and uranium tritaoctoxide produced no systemic effects and only mild local irritiation, possibly of a mechanical nature. A few chronic studies were done with the use of uranium tetrafluoride as a dry powder only. Rabbits, as well as guinea pigs and rats, were used in this study. Local and systemic effects were observed only in rabbits.

VII. Respiratory Tract Exposure to Uranium Dusts

A. Studies by Means of Tracheal Insufflation

In the experimental study of the entrance of dust or aerosols into the respiratory system of animals, two major techniques may be employed. In the first, measured amounts of the material are placed directly into the upper respiratory passages, usually the trachea, by injection through a cannula or other device by the technique known as tracheal insufflation. The other method is the exposure in chambers to an atmosphere containing the dispersed dust or aerosol, which reaches the respiratory tract by spontaneous inhalation on the part of the experimental animal. The main advantage of the former technique is the ability to introduce measured quantities of a test substances in a relatively short space of time. This technique, therefore, has much in common with other methods of administration such as intravenous, intramuscular, subcutaneous or intraperitoneal injection. Two series of experiments using the insufflation method were carried out at the University of Rochester during the active period in the early days of the Manhattan Project. The first of these, described by C. W. LaBelle in chap. 21, Part E of Voegtlin and Hodge (1953), involved rats, and was designed to study the relationship of particle size to toxicity of the relatively insoluble uranium dioxide. Single doses of seven fractions differing in particle size over the range 11 to 0.35 microns, were given to anesthetized rats intratracheally. The resulting toxic response was estimated in 6 days by measurements of change in body weight, accumulation of non-protein nitrogen in the blood, and the ability of the kidney to excrete phenolsulfonephthalein. The quantitity of uranium was also measured in lungs, kidneys and femurs. The 2% suspensions utilized were carefully chosen to provide a dose yielding approximately 30 mg of uranium dioxide per kilogram of body weight. Instillation of this quantity of uranium into the lungs of rats led to a typical response in the kidneys characteristic of uranium poisoning. The intensity of response was more or less inversely proportional to the particle size, with a greater response being associated with particles smaller than 1 micron and there was no indication that a limit had been reached at particles sizes of 0.35 microns, the smallest size tested. Apparently, the variation in particle size reflected a corresponding variation in the quantity of uranium gaining access to the systemic circulation, since the intensity of the renal reaction was proportional to the quantity of uranium measured in the kidney at a given time. The uranium content of bone was cumulative and tended to be highest in animals showing the greatest toxic effects, whereas kidney uranium showed less relation to overall toxicity and did not increase significantly in 6 days. This latter finding suggested that uranium was passing through rather than accumulating in the kidney. In the second group of experiments, under the direction of H. E. Thompson and L. T. Steadman, chap. 21, Voegtlin and Hodge (1953), the animal used was the rabbit, and two uranium compounds—the nitrate, $UO_2(NO_3)_2 \cdot 6H_2O$, which is water soluble, and the oxide U_3O_8, which is relatively insoluble, were studied. Toxicity was evaluated mainly by measurements of weight change, survival time, and the occurrence of albumin and sugar in the urine. A principal object was the study of the absorption, distribution, and excretion of uranium following its administration by tracheal insufflation and this will be considered in a later section when the general aspects of these problems are discussed in more detail. Relatively large doses of 100, 10 and 5 mg were administered. With uranyl nitrate, at the highest dose, all animals died within 7 days. Death was preceded by a reduction in the volume of urinary output, an increase in albumin excretion and the presence of significant amounts of sugar in about 20% of the urine samples. At the 10 and 5 mg levels, animals survived for a longer time, but more than 90% eventually died of uranium poisoning within 14 days. Effects on urinary excretion of albumin and sugar were similar to, but less than, those seen at the 100 mg level. The administration of U_3O_8 by this route was found to be considerably less toxic to rabbits than was the nitrate. Only 2 out of 33 animals receiving the 100 mg dose died within 20 days and the remaining animals continued to live until sacrified at intervals up to 28 days. No toxic effects were found following insufflation of 10 mg of U_3O_8. In a comparative study using single intramuscular injections of 100 mg of uranyl nitrate, it was determined that toxicity was similar to that of a similar dose given intratracheally but that absorption from muscle tissue was slower than from the lung.

B. Toxicity Following Inhalation

1. 30-day Studies

A very high priority was given to the medical section of the Manhattan District Project for studies of inhalation toxicology. This was designed to develop safety standards for the control of possible health hazards resulting from contamination of factory air with uranium dusts. Initially, information on chemical toxicity alone was sought, since it was felt that any radiation hazard would be extremely remote because of the low levels of activity in the materials being used. At the time the inhalation studies were being planned, no record of exposures of animals to uranium dust could be found in the literature. A number of reports of inhalation of relatively non-toxic dust, such as coal, shale, granite, asbestos, and lead were found. However, the purpose of most of these investigations was primarily in the animal response and the experiments were semi-quantitative at best. An exception was the work which had recently been done by FAIRHALL and SAYERS (1940). While these workers had been mainly interested in the toxic effects of heavy metals, they had devised and described important advances in the techniques of dust exposures chiefly through the use of improved methods for the production of aerosols and their accurate estimation. FAIRHALL's method for the estimation of dust concentration was used almost exclusively, with slight modifications, in the work reported from Rochester. Details of the techniques which were devised and used in the inhalation program which comprised a large percentage of the continuing toxicological work on uranium during the war years in the 1940's and subsequently, are described in some detail in chap. 10 of VOEGTLIN and HODGE (1949) and by LEACH et al. (1959). These consisted of development of exposure chambers, dust feed mechanisms and aerosol particle size measurements. The experiments were divided into relatively short-term studies, lasting approximately 30 days and chronic studies with exposures of from 1 to 5 years.

The importance of particle size in relation to inhalation has been shown by WILSON et al. (1948, 1952, 1955). They found with exposures to aerosols of UO_2 having particle sizes of 0.5 μ and 2.5 μ, that about 10 times as much of the finer than of the coarser material was deposited in the lungs of animals killed after 20 hours. The ratio was considerably increased at 2 to 3 months, due to the more rapid clearance of larger particles demonstrated by STOKINGER et al. (1951).

In the 30-day inhalation studies eleven uranium compounds, of which most were of a high degree of purity, and one high-grade ore of industry, were used. These materials were four oxides, UO_2, UO_3, $UO_4 \cdot 3\,H_2O$ and U_3O_8; three fluorides, UF_6, UF_4, and UO_2F_2; sodium and ammonium diuranate; uranium tetrachloride, UCl_4, and uranyl nitrate hexahydrate. Because of the known species differences in susceptibility to uranium poisoning, six different animals were used. These were rats, dogs, rabbits, guinea pigs, mice, and hamsters. Because of the known relationship between toxicity and particle size, which had been shown by intratracheal insufflation studies, the size distribution of dusts in the inhalation experiments was kept as small as possible. Criteria of toxicity were similar to those used in parenteral and oral studies and consisted of loss of, or failure to gain, weight, mortality, increased levels of blood non-protein nitrogen and blood sugar, histological examination of target organs, notably kidney, liver, and lung, changes in the hematological picture, and, in most instances, buildup of uranium content of tissue. It was also shown by biochemical testing that increases in urinary catalase, protein, and in the amino acid nitrogen-creatinine ratio were early, sensitive indicators of uranium poisoning. In 46 short-term inhalation exposures, each lasting 30 days, the approximate concentrations in air of the 11 uranium-containing dusts in the exposure chambers ranged from 20 to 0.05 mg/m^3. From a consideration of the results of the numerous exposure experiments with these materials, it is possible to define definite toxic levels, levels producing mild, transitory changes, and those which are followed by no detectable damage in otherwise moderately susceptible species. The relatively soluble compounds, UF_6, UO_2F_2, UCl_4, $UO_2NO_3 \cdot 6H_2O$ are definitely toxic, being fatal to some species and causing the occurrence of considerable renal damage, as well as some pulmonary changes, at dust levels of 2.5 and 20 mg/m^3. They are slightly toxic to occasional animals at concentrations of 0.2 mg/m^3 and non-toxic at lower levels. The less soluble dusts, UF_4, high-grade ore and UO_2 are rarely fatal even at 20 mg/m^3 and produce little or no renal damage at the 2.5 mg/m^3 level. Of the compounds tested which are of lesser importance from the point of view of industrial health, the most toxic was uranium trioxide. The peroxide and diuranates of ammonium and sodium had an intermediate effect while the tritaoctoxide (U_3O_8) was almost without effect. In preliminary studies FISH (1961) reported on dogs, mice and rats subjected to single and multiple inhalation exposures to U_3O_8 fume. The main purpose of these experiments was to simulate conditions in several accidental human exposures to this material and to relate lung burden to urinary excretion. Long term retention of U_3O_8 in lungs and pulmonary lymph nodes was indicated but no toxic effects were noted. The characteristic time lag between exposure and response in other types of uranium intoxication was also noted with inhalation exposure and varied with the solubility and concentration of uranium dusts used. The sequence of events in uranium poisoning following inhalation is not unlike that following administration by other routes and presumably depends on the maximum levels in the blood stream attained at any given time by solution and absorption of uranium.

Obviously it is not possible, other than by analogy, to estimate actual dose levels as they affect the animal in this way following inhalation. While the relative degree of susceptibility of different animal species can be estimated with reasonable accuracy for a single uranium dust, there is no satisfactory method for determining the order of species susceptibility to uranium dusts in general. Many of the test materials are complicated by the presence of a potentially toxic anion of an element other than uranium to which different animals may show different susceptibilities. For example, the mouse is relatively highly susceptible to UF_6 and UCl_4 and the rat resistant to UO_2F_2. Toxic responses, such as mortality or renal damage, may also differ between species, the dog and rabbit being about equally susceptible to renal injury, while under some similar conditions of exposure, the dog will survive, whereas the rabbit will die.

Based on these short-term experiments and with the recognition of these differences, an approximate order of susceptibility following inhalation would read rabbit and cat, as most susceptible; dog and mouse, intermediate susceptibility, and guinea pig and rat, least susceptible. With respect to relative inhalation toxicity of the 11 uranium dusts under consideration, based on the criteria of mortality, weight response, biochemical and histological changes, the most soluble compounds were the most toxic. As for other routes of administration, only small differences existed between UF_6, UO_2F_2, UCl_4 and uranyl nitrate. The relatively great toxicity of high-grade ore was thought to relate to its content of numerous materials, such as lead, copper, nickel, and other heavy metals which accompany uranium oxides. UO_2 and U_3O_8 were far below the usual range of toxicity of uranium compounds. The effects of oral intake on the toxicity following inhalation are assumed to be significant with the more soluble dusts but are probably negligible with the more insoluble materials. Percutaneous and ocular absorption probably contribute very little to the over-all toxicity except for certain of the highly soluble compounds.

2. Studies of Toxicity Following Inhalation of Uranium Compounds for One Year

In the short-term 30-day inhalation studies, emphasis was placed on uranium levels producing demonstrable injury, while in the longer-term studies, the emphasis was on the assignment of dust concentrations which would either be safe for an exposure of 1 year or more, or produce minimal or borderline injury in more susceptible species. In the 1-year studies, the number of test compounds was restricted to five, namely uranium tetrachloride, uranium hexafluoride, and uranyl nitrate representing highly soluble and toxic uranium compounds; uranium dioxide and uranium tetrafluoride representing insoluble and decidedly less toxic substances. With the exception of uranyl nitrate, which throughout these experiments was considered as a sort of standard, 2 exposure concentrations of each dust were used. The experimentation was largely confined to two species, the dog and the rat, although rabbits and guinea pigs were used for part of the period at 2 exposure levels for uranyl nitrate and one level for each of the other compounds.

a) Uranyl Nitrate Hexahydrate

Large groups of dogs and rats were exposed to each of the following four concentrations in the air: 0.04, 0.25, 0.4 and 2 mg U/m³. Groups of rabbits and guinea pigs were added about four months after the beginning of the exposure and were exposed to levels of 2 and 0.25 mg/m³. Mortalities were low in all groups except rabbits, high and low doses killing 60 and 50 percent of these respectively. Pathological changes were noted in the kidneys of all dogs, rats and rabbits exposed to the highest level. These were mild to moderate in degree. Guinea pigs failed to show pathological evidence of uranium poisoning even at the highest dose level of uranyl nitrate. Occasional microscopic renal changes were noted at the 0.25 mg/m³ level in a few of the rats and dogs and in a few of the rabbits. Biochemical studies were, in general, negative, as far as protein and carbohydrate were concerned. The urinary catalase activity of rabbits exposed to 0.25 mg/m³ showed unusually high values in about one-third of the cases. Renal clearance studies, with the use of diodrast and inulin, showed some indication of renal dysfunction in dogs at the higher dose levels. Extensive hematological examination revealed no significant abnormalities. Initial losses in body weight, following the 2 mg/m³ dose, were regained by dogs but not by rabbits, and guinea pigs failed to gain as much weight as the control groups throughout the experiment.

The uranium content of lung, kidney and bone was determined at intervals in rats, at the end of 1 year in dogs and after $6^1/_2$ months in rabbits and guinea pigs. Only small amounts of residual uranium (less than 1.5 µg/g) were found in the lungs of any species even after inhaling the highest level (2 mg/m³) of uranyl nitrate for up to 1 year. Kidney values were also low, except in the rat which showed a mean concentration of 5.6 μg/g at 1 year. Higher uranium concentrations occurred in bone ash the mean values for which were dog 7, rabbit 5.6, rat 4.2 and guinea pig 1.5 μg/g. A tendency for bone deposition to increase with time was observed in rats. The accumulation of uranium in the bones of animals breathing uranyl nitrate dust became appreciable only at air concentrations higher than 0.25 mg U/m^3.

b) Uranium Hexafluoride

This compound was given a high priority because of its inherent toxicity and because, presumably, numerous persons might suffer an exposure from it during the wartime development of the atomic bomb. The hexafluoride decomposes instantly on contact with moist air so that rather than exposure to UF_6, it should be described as exposure to UO_2F_2 fumes plus HF. During the entire exposure period of approximately 1 year for dogs and rats, and 8 months for rabbits and guinea pigs at two levels, namely, 0.05 and 0.2 mg U/m^3, no deaths were attributed to uranium poisoning. Biochemical studies revealed no significant abnormalities and pathological examination of the kidneys revealed a fairly consistent but mild injury in all species at the higher level. Hematological studies were unrevealing. Uranium deposition in tissues, while showing trends similar to those following uranyl nitrate inhalation, never reached levels of more than 1 μg/g.

c) Uranium Tetrachloride

This material was used for exposures to aerosols containing 0.05 and 0.2 mg/m^3. The results of examination of all four species exposed to the higher level gave evidence of only borderline uranium toxicity and, following the lower exposure, the degree of toxicity was considerably less. Very little uranium was present in the tissues of any animals and there was no evidence of accumulation with time.

d) Uranium Dioxide

Year-long inhalation exposures were conducted with UO_2 at levels of 10 mg U/m^3 and 1 mg U/m^3. It was demonstrated that no chemical injury occurred in any animals exposed to 1 mg U/m^3. However, a few dogs and rabbits showed slight, inconsistent evidence of renal injury when atmospheres containing 10 mg/m^3 were breathed for this protracted period. Large amounts of uranium were deposited in lungs and pulmonary lymph nodes but insignificant amounts were detected in other tissues, including bones and kidneys.

e) Uranium Tetrafluoride

Groups of the same four animal species used in other inhalation experiments of this group were exposed to atmospheres containing 3.0 and 0.5 mg U/m^3. Except for the histological finding of mild to border-line renal changes in some of the animals at the higher level, none of the criteria used for estimating toxicity were positive following exposure to this material. Accumulation of uranium in tissues was much lower than with UO_2.

3. Studies of Toxicity Following Inhalation of Uranium Compounds for Two Years

During a second year following these exposure to the three relatively soluble and two relatively insoluble dusts of uranium, 23 dogs and 70 rats, selected from the higher levels of the first year exposures, were subjected to further exposure, this time only to uranyl nitrate hexahydrate at a concentration of about 2 mg U/m^3. This material was chosen because of the greater amount of chemical and toxicological data available for it, because it represented a typical soluble uranium compound, and because more animals were at hand from the first year exposures to it. While a prime objective of this second year study was to investigate deposition and accumulation of uranium in tissues and its elimination therefrom (to be discussed in a subsequent section of this chapter) data were also collected on the toxic response based on mortality, weight depression, blood non-protein nitrogen, urinary protein and bromsulfalein removal, as well as histological and hematological changes. The conditions of exposure were much like those previously used in the highest level test on uranyl nitrate hexahydrate. The mass median particle size was 3 microns with a range of 1.2 to 7.8. In the dog, the mortality was negligible, with only two dogs dying, both of which had previously been exposed to UO_2. There was very little change in body weight in the rats, while dog weights were maintained at the initial levels. There were no biochemical changes of significance except for a slight increase in non-protein nitrogen in the dogs previously exposed to the insoluble dusts UO_2 and UCl_4. Except for these groups of dogs, the pathological changes were essentially similar to those at the end of the first year. There was extensive acute renal tubular injury observed in the dogs exposed to UO_2. This was of particular interest since animals in this group appeared to be the only ones which did not develop significant renal damage or some degree of tolerance during the first year. These was some gradual decline in the red blood cell count in dogs previously exposed to UO_2, UF_4 and UCl_4. The degrees of tubular atrophy noted at the end of one year, and in most of the animals also at the end of two years, may indicate some chronic injury. However, the changes appear to be rather static and may merely represent the end result of a reparative process.

In general, following a second year of exposure to uranyl nitrate, irrespective of which compound had been inhaled during the first year, further accumulation of uranium was confined to bone. The amount retained by this tissue was double or tripled to maximum levels of about 20 μg/g. More than 90% of the relatively large amount of insoluble UO_2 and all the UF_4 retained in lungs and pulmonary lymph nodes at the end of 1 year was eliminated during a second year of exposure to uranyl nitrate inhalation.

4. Inhalation of UO_2 for Five Years

Because of the fact that rat and dog lungs, following 1 year of exposure to UO_2 dust levels of 10 mg U/m^3, retained large amounts of uranium, ranging from 400 to 600 μg/g in the rat, and 400 to 1200 μg/g in the dog, attention at Rochester was shifted from an interest in the renal effects of uranium following UO_2 inhalation to the possibility that storage of insoluble uranium in the lung might constitute a radiological hazard. As a result, a five-year inhalation experiment using UO_2 was started in 1955. Rats, monkeys and dogs were exposed for 6 hours daily, 5 days per week, to atmospheres containing UO_2 dust in a concentration of approximately 5 mg/m^3. The mass median diameter of the particles averaged about 1 micron. Details of this experiment, up to the end of the five year inhalation period, have been reported recently by LEACH et al. (1970). Serial sacrifices of rats over a one-year period, and of dogs and monkeys over a five-year period were carried out. At no time during this study were evidences of acute uranium toxicity detected by changes in body weight, mortality, increased levels of NPN in the blood or the hematologic picture and no renal abnormalities were seen histologically. Fluorimetric analyses for uranium were carried out on major tissues of the body, and it was found that the only significant concentrations were present in the lungs and the lymph nodes draining them. These areas contained sufficient uranium to create a potential radiologic hazard, although at the end of the five year period, few changes were noted that could be attributed to this effect. Fibrotic changes suggestive of radiation injury were seen occasionally in the tracheobronchial lymph nodes of monkeys and dogs, and in the lungs of monkeys after exposure periods of longer than three years, when estimated alpha doses, approaching levels of other types of irradiation which produce pulmonary damage, had accumulated.

Some of the dogs from the 5-year inhalation study were removed after 1 year and the clearance rate of natural uranium dioxide was compared with the clearance rate of uranium dioxide from the lungs of dogs subjected to a single, 1 hour, inhalation exposure to an aerosol of $^{235}UO_2$. Measurement of the clearance rate following the multiple exposures was accomplished by *in vivo* counting of the 186 keV gamma rays associated with ^{235}U in natural uranium. A marked difference in the clearance of the lungs was found, with approximate biological half-times being 180 and 340 days for single and multiple inhalation exposures, respectively. This was attributed to movement of the dust into a compartment which is cleared more slowly than other pulmonary regions following multiple exposures.

Following the 5-year exposures, groups of animals were held for postexposure study. The results are being prepared for publication by LEACH et al. (1971). A number of rats were held for one year after the one year exposure, while monkeys and dogs were sacrified at intervals up to $6^1/_2$ years following one, two or five years of exposure. During the postexposure period, despite some clearance of uranium from the lungs and lymph nodes, the estimated radiologic doses to these tissues continued to increase to maximum levels in lung and tracheobronchial lymph nodes of 660 and 16000 rads respectively for the dog and 1280 and 24800 rads for the monkeys (see Figs. 3.1 and 3.2).

There was no evidence of chemical toxicity in any of the animals and the only pathological changes occurred in the lungs and in the tracheobronchial lymph nodes. Granular black to brown pigment, which had appeared to increase considerably in both lungs and the draining lymph nodes during the exposure periods, tended to remain constant as far as subjective evaluation was concerned, or to decline very slightly during the post-exposure years. This pigment was indistinguishable by routine light microscopy from pigment present to a considerably smaller degree in control animals so that its relationship to inhaled UO_2 is not entirely clear. Attempts to identify it by autoradiography were unsuccessful due to the excessive exposure time required to detect such low levels of alpha radiation. However, considerable quantities of uranium could be identified in the region of these pigment deposits by neutron activation. It was concluded that while the uranium was deposited in association with other inhaled particulate pigments, it could not be specifically identified histologically.

Pulmonary fibrosis comparable to that seen after inhalation of $^{238}PuO_2$ and $^{239}PuO_2$, CLARKE and BAIR (1964), PARK et al. (1964), YUILE et al. (1970), as well as after 2000 R of thoracic X-irradiation, SCHREINER et al. (1968) was noted focally in the lungs of monkeys exposed to uranium dust inhalation for approximately three and a half years. This interstitial fibrosis became progressively more marked during the remainder of the inhalation period and also after the cessation of exposure. Widespread hyaline fibrotic changes were seen in all lobes of monkeys whose lungs had accumulated approximately 1000 rads of alpha irradiation.

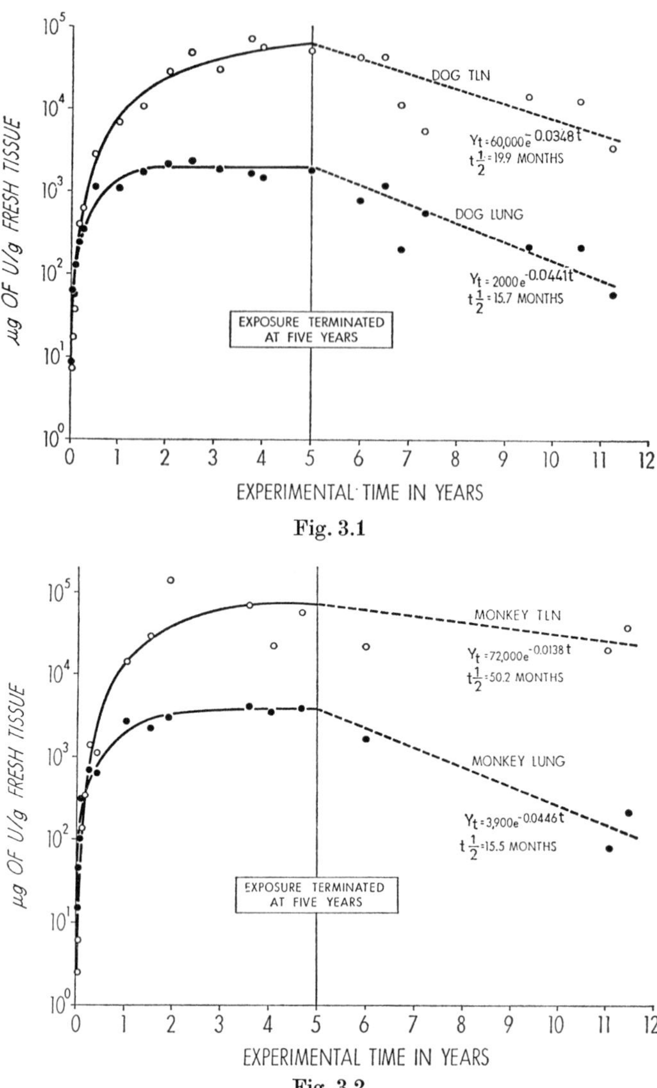

Fig. 3.1

Fig. 3.2

Figs. 3.1 and 3.2. Graphs indicating buildup and clearance of uranium from dog (*1*) and monkey (*2*) lungs and lymph nodes, during and after 5 year exposure to a UO_2 dust concentration of 5.8 mg/m^3 in air. Clearance rate, $t_{1/2}$, determined by equation $Y_t = Ae^{-Kt}$, where Y is tissue concentration of U in μg, t is post-exposure time, and A is tissue concentration at time $t = 0$

While no tests were carried out in these animals, the changes would appear to be of sufficient degree to cause considerable impairment of pulmonary function. These findings strongly suggest dose dependency for such fibrosis and this thought is further substantiated by the fact that in dogs, in these experiments, only moderate degrees of focal fibrosis were noted even at the end of the long post-exposure periods. At this time the estimated accumulated irradiation in dog lungs amounted to a maximum of only 660 rads. Lymph node changes, consisting of combinations of fibrosis and necrosis in the tracheobronchial chain were seen either focally or diffusely in both dogs and monkeys during the post-exposure interval after a 1 to 5 year period of uranium dioxide inhalation. The changes were similar to those seen after inhalation of $^{238}PuO_2$ and $^{239}PuO_2$ (see references at the beginning of this paragraph) and are considered to be effects of irradiation. Widespread changes of this type were noted with accumulated lymph node doses of 10000 rads or more, which, again, is comparable to the plutonium studies.

A markedly increased incidence of pulmonary neoplasms and of foci of epithelial proliferation and metaplasia, which probably represent early or preneoplastic lesions, were seen in a high percentage of the dogs exposed to UO_2 for 5 years and held for post-exposure periods of up to $6^1/_2$ years. This is of particular interest since spontaneous neoplasms in dog lungs are rarely encountered. The reports of MONLUX (1952) and NIELSON and HORAVA (1960) revealed less than 0.5%. In the present study, 20% of the dogs surviving a 1 year exposure period to UO_2, 33% of a small number of dogs exposed for 2 years and 45% of 13 dogs exposed for 5 years showed small tumors or tumor-like lesions. In the same group of 13 five-year dogs, 4 frank neoplasms were encountered, 2 of which were relatively localized adenomas, without evidence of frank malignancy, while the other 2 were adenocarcinomas. One of these occurred in multiple pulmonary locations, had invaded perivascular, peribronchial and pleural lymphatics and had metastasized to tracheobronchial lymph nodes. It is noteworthy that the commonest spontaneous pulmonary tumors in dogs are adenomas and adenocarcinomas of broncheolar origin. Thus it would seem that the build up of the insoluble uranium compound, UO_2, due to inhalation, to values far exceeding the calculated tissue tolerance, can and does produce pathological changes of significance. It is of interest that, while the fibrosis which develops is probably similar in different species with the same radiation dose, there appears to be a marked species difference in the incidence of tumors, since these, occurring only in the dog, did so at calculated radioactive levels much lower than those achieved in the monkey. It is possible that other unknown carcinogenic factors were operative in these dogs in addition to radiation.

5. Recent Inhalation Studies of Uranium Trioxide

Early UO_3 inhalation studies, briefly alluded to in section VII.B.1. of this chapter, had indicated that this uranium oxide behaved more like soluble uranyl nitrate than like the so-called insoluble oxides UO_2 and U_3O_8. A recent inhalation study with UO_3 enriched with ^{235}U, in dogs, by MORROW et al. (1971) confirmed the earlier observations and provided additional information concerning retention and excretion following a single exposure. In these experiments most of the UO_3 was removed rapidly from the lungs, with a biological halftime of 4.7 days, and systemic absorption amounted to more than 20 percent of the exposure burden. Measurement of UO_3 levels in the lungs by external gamma counting were found to be in error because of significant deposits of ^{235}U in extrapulmonary structures of the thorax.

VIII. Distribution and Excretion of Uranium

As pointed out by TANNENBAUM (1951), it is important to distinguish uranium toxicity from exposure to, absorption of and deposition of uranium in the tissues. For example, as shown above, very large exposures to some uranium compounds by the gastrointestinal or pulmonary route result in very low toxicity, while large amounts of uranium may gradually accumulate in some tissues following the repeated absorption of low levels over long periods of time without evidence of toxicity. It would appear to be a fair statement that the degree of acute uranium poisoning depends on the amount initially absorbed and the resulting level in the blood stream, irrespective of the route of administration for a given species or strain of animal. This is supported by most, if not all, experimental work which has already been described or cited in this chapter. Most of the studies involving distribution and excretion of uranium in the body have been based on introduction of material by intravenous or intraperitoneal injection. In a number early publications dealing with the subject of uranium, widely divergent opinions were expressed and different conclusions were reached, probably due in large measure to lack of a sufficiently sensitive method for measuring uranium in biological material. Accurate, quantitative and time-related determinations of the amount of uranium in different body tissues became a reality when Neuman and Bloor perfected the procedure of Hoffman (1942), utilizing the fluorescent light emitted when sodium fluoride, fused with a small amount of uranium, is exposed to ultraviolet light. With this method, one-two billionth of a gram or 0.5 nanograms of uranium in a biological sample can be measured with an error of $\pm 5\%$, by determining the intensity of emitted light, Neuman et al. (1948). Attempts have also been made to determine, at least semi-quantitatively, the distribution in various organs and tissues as well as in the whole animal body, radiochemically and by autoradiography, when radioactive isotopes of uranium have been incorporated in the test material.

Based on the extensive work done under the auspices of the Atomic Energy Commission between 1943 and 1950, briefly summarized by Hodge (1956), it was concluded that uranium exists in the mammalian body in the hexavalent form, as the uranyl ion. After reaching the blood stream by one of the several absorptive routes already alluded to, this uranyl ion reacts either with proteins in the plasma or with bicarbonate to form complexes. On the basis of studies involving intravenous injections of varying doses of uranium compounds, it was shown that uranium leaves the blood stream very rapidly, Neuman et al. (1948). At 40 minutes after injection, about 50% had been excreted in the urine, nearly one-quarter to one-third had been deposited in the skeleton, and of the balance, approximately one-half was in the kidney and one-half in the other soft tissues. At this early time, only a trace (approximately 1%) remained in the blood. With time, the uranium deposited in the kidney tended to move into the urine, while the uranium deposited in the other soft tissues and skeleton was excreted via kidneys. The skeletal elimination was slower than that in any other area so that at 40 days after administration of an intravenous dose, 20 to 25% was still retained in the bones, while practically the entire balance had been excreted from the body in the urine. In the longer studies previously described, continued feeding of uranium compound-containing diets for one to two years, or repeated inhalation exposures to relatively soluble uranium compounds in the form of aerosols, resulted in low rates of absorption into the blood stream. For example, based on a number of assumptions, only about 0.1% of uranium in the diet is ultimately incorporated into the body. This would mean that of 2500 μg of uranium in the daily diet, approximately 2.4 μg U per day would be absorbed. Following these long-term exposures of one type or another, there was a continued deposition in the bones of the skeleton and the teeth to levels which were not considered to constitute a radiological hazard, Hodge (1956), however, the experiments were too short to exclude such a possibility in a long-lived species. In a series of studies described by Neuman et al., in chap. 24 of Voegtlin and Hodge (1953) and in Neuman et al. (1948), it was shown that the deposition in bone is principally on the surfaces of extremely minute hydroxyapatite crystals. Uranium competes with calcium for sites of deposition and it has been shown, with the use of radioactive isotopes of calcium and phorphorous, that uranium-treated bone has less exchangeable calcium than "normal" bone. It would appear a reasonable assumption, on the basis of the acute distribution experiments, that the build-up of uranium in the kidneys, bones, teeth and other tissues during repeated exposures would tend to diminish to near zero levels with time after discontinuation of the exposure.

A number of autoradiographic studies complementing or extending those described above, dealing with distribution and excretion of uranium, have been done, either alone or in conjunction with fluorophotometric or radiochemical analysis. The first of these described by NEUMAN and WILLS in chap. 11 of VOEGTLIN and HODGE (1949), was designed to obtain autoradiographs of uranium distribution in rat kidney tissue following a large dose (5 mg/kg) of uranyl nitrate hexahydrate containing ^{232}U, equivalent to a radioactive dose of 2 microcuries per kg. Animals were killed at intervals of from 1 to 6 hours after injection and autoradiographs prepared. Techniques at this time had not been perfected sufficiently to give a close identification of the relationship of alpha tracks to tissue structures. However, tentative conclusions were drawn, indicating that although some of the uranium was distributed generally throughout the kidney, most of it tended to be concentrated in localized areas in the cortex. Studies were also done of autoradiographic localization of uranium in bone in conjunction with other studies of the deposition of this metal in skeletal tissues of the body, NEUMAN and NEUMAN (1948). In four rats given sublethal doses of uranium (2.5 mg U/kg as a 0.1% aqueous solution of either uranyl nitrate or uranyl acetate) and containing about 1.5 microcuries of ^{233}U, autoradiograms were prepared following sacrifice after 4 hours, 2, 5, and 40 days. It was shown that the element was deposited only in the mineral portions of the bone and was concentrated particularly on surfaces adjacent to the circulation where calcification occurs. Once fixed, there was little redistribution of the uranium and as normal bone growth and calcification processes continued, new bone accumulated over the lines of deposition of the uranium isotope. Resorption of bone appeared to be inhibited to some extent in these experiments but unenriched uranium failed to inhibit the resorptive processes. In teeth, the pulpal portion of the dentine contained the radioactive material.

In paper 7 of the monograph edited by TANNENBAUM (1951), studies dealing with the distribution and excretion of uranium were described in which ^{233}U was used either as a tracer or alone to obtain autoradiographic and analytical data. Labeled uranyl nitrate was given by subcutaneous injection, to mice, rats and dogs. In general, the findings of NEUMAN et al. (1948) on the distribution and excretion of injected uranium were confirmed with the kidneys and bones being the principal sites of accumulation. A greater proportion, up to 40 percent, was retained at the end of one month following a toxic dose as compared to only 10 percent with non-toxic doses. Autoradiographs indicated that uranium in the bones was located chiefly in cancellous portions and endosteum, while in the kidney it was located in the cortex with some tendency for concentration at the corticomedullary junction.

KISIELESKI et al. (1952) gave pure ^{233}U, as the nitrate, by intravenous injection to mice in doses of 0.05 and 0.005 μCi per mg in studies of distribution and excretion. With both doses, 60 percent remained in the body at 24 hours and 15 percent after 120 days. Again uranium was quantitatively deposited chiefly in the bones and kidney. Autoradiographic studies of cellular distribution revealed alpha tracks in the kidney to be localized mainly in relation to proximal convoluted tubular cells and in casts while a few were also seen emanating from Kupffer cells in the liver and macrophages in the splenic pulp.

The next pertinent study involving autoradiography was that of JONES (1966). The experiments described were designed to study the distribution of uranium in rat kidneys following the intravenous injection of varying amounts of uranyl nitrate buffered with sodium acetate to which 10 μg ^{233}U were added. Total masses ranged from 10 to 1000 μg/kg with each dose containing the same amount of ^{233}U, except when a 1 mg/kg dose of ^{233}U alone was used. The data on distribution derived from alpha track counting of the autoradiographs was compared with that found by radiochemical analysis. The experiments indicated a significantly higher percent retention at the two higher dose levels than at the lower level. A concentration ratio indicating higher deposition in the cortex than in the medulla was also found and at the early post-injection times, uranium tracks were found mainly in proximal tubules. There was a tendency to shift to the distal tubules as post-injection times increased. With the 1000 μg U/kg dose, numerous aggregates of alpha tracks in the outer cortex were noted which is of interest in view of the finding of NEUMAN and WILLS reported in chap. 11 of VOEGTLIN and HODGE (1949), of large localized aggregates of alpha tracks in the kidneys of rats given lethal doses (5 mg/kg), and sacrificed within 6 hours after injection. The prolonged retention found by JONES (1966) with the largest dose was suggested as possibly being due to the effect of toxicity injuring the tubules and slowing the clearance from the body, but the large aggregates may also have some bearing on this finding.

Two interesting papers from Sweden by WALINDER et al. (1967a, b) described experiments involving the distribution of uranium following intravenous and intraperitoneal injections of uranyl nitrate and uranyl bicarbonate in mice, as well as distribution of uranium following administration by inhalation to mice, and through the skin of rabbits and piglets. These investigators used the technique of whole body autoradiography in conjunction with the use of the fluorimetric method to determine uranium concentrations in various organs and tissues, as well as in 24 hour urine samples. For autoradiography, the nuclide ^{233}U, as sodium uranyl tricarbonate, was injected intravenously in a dose of 42 μg of uranium, having the equivalent of 0.4 μCi. The first observations were made at 24 hours after injection so that any very rapidly equilibrating changes would have been missed. The highest uranium concentration in any organ at 24 hours was found in the renal cortex, particularly its inner part. The concentration fell more rapidly in this inner zone, however, than in the outer cortical zone, so that at 30 days the kidneys showed accumulation of radioactivity in a narrow peripheral zone. At 6 months, only a very slight activity remained in the kidney with no detectable localization. Accumulations of the isotope were also noted in the liver, spleen, adrenal cortex, bones and teeth. Concentrations in these areas gradually decreased with time. However, radioactivity was still detectable six months after the injection. It was of considerable interest that the spleen seemed to retain its activity longer than any other tissue. The uptake was localized principally in the marginal sinuses and by microautoradiography at six months, the uranium appeared to be extracellular. In bones, the uranium was localized in endosteum and periosteum and in teeth, it was concentrated by newly formed enamel and dentine. During the first two months after intravenous injection, the uranium content of the mouse liver, spleen and skeleton decreased by a factor of 10 while that of the kidney decreased by a factor of 100. Although in most previous studies no mention is made of significant concentrations of uranium in the spleen, it is of interest that following the long term inhalation of UO_2 dust by dogs and monkeys, LEACH et al. (1970), the spleen occupied a position many times below but next in line to the lungs and lymph nodes draining the lung. It is not known what factors are responsible for this but at least between dogs and monkeys, there appeared to be a marked species difference with the monkey retaining much higher quantities of the element. WALINDER et al. (1967b) also noted that following U_3O_8 inhalation with the formation of considerable depots in the mouse lung, only small amounts of uranium were absorbed but that bones and spleen showed much higher relative uranium concentrations than the kidney under these experimental conditions. The slow rate of incorporation compared to a rapid excretory rate probably is important in this connection.

IX. Mechanisms of Uranium Action

A. Entry of Uranium into the Body

At the beginning of the Manhattan Project investigations into uranium toxicity, a large amount of work was done on the formation of complexes by uranium [see chap. 1, VOEGTLIN and HODGE (1949)]. Of particular interest was the ability of the uranyl ion to form complexes readily with protein and bicarbonate. These complexes are probably of great importance relative to the absorption of uranium but little specific information is available. As far as absorption from the eye is concerned, it can be assumed that this occurs following the formation of a soluble complex such as bicarbonate in the conjunctival sac. It has been shown that uranium remains in the region of the eye for a considerable period of time after instillation so that insoluble complexes with proteins of the lining membrane undoubtedly also occur. The exact mechanism of the absorption of uranyl ion through the skin is also obscure but it is possible that lipid plays a role in this process in addition to protein and bicarbonate. As far as the lung is concerned, it is probable that bicarbonate complexes are the chief form in which uranium is absorbed. Little is known about the site or sites of uranium absorption from the gastrointestinal tract but it is known that only a small percentage of uranium when mixed with the diet enters the body. The remainder is probably combined with protein break-down products or perhaps with other food residues, and is excreted in the feces.

B. Transport of Uranium in Plasma

Uranium, in the body, tends to be converted into the stable hexavalent form of the uranyl ion. In the plasma, approximately 40% of the uranium is bound to plasma proteins as complexes which are non-diffusible. The remaining 60% is bound to bicarbonate, where it exists in the form of a diffusible anion. The distribution of uranium between protein and bicarbonate depends on the CO_2-combining power of the plasma and uranium binds to protein principally with the carboxyl groups in a non-denaturing precipitation. The addition of bicarbonate is followed by prompt solution of the protein complex so that, within the plasma, the higher the bicarbonate concentration the more uranium will be carried by bicarbonate and the less by protein.

C. Excretion of Uranium

As previously outlined in sec. VIII, the ultimate fate of all but a minute fraction of absorbed hexavalent uranium is storage in the bones or excretion in the urine so that 1 month after a single injection, 20% of the dose will remain in bone while the balance will have been removed almost entirely by urinary excretion. Such excretion, discussed by DOUNCE in chap. 15 of VOEGTLIN and HODGE (1949), comes about by diffusion of the bicarbonate complex from the glomerular capillaries into the glomerular filtrate. Within the proximal convoluted tubular segment, bicarbonate is reabsorbed into the venous blood, freeing uranyl ion into the tubular urine. The uranyl ion is presumably complexed by proteins of the lining epithelial cells, thus causing the renal injury to be discussed below. The likelihood of such combination is increased by acidification and the absorption of water in this portion of the tubule. The uranyl ion which is still complexed with bicarbonate, and probably a smaller amount combined with organic acids such as citric, will be excreted in the urine. The likelihood that such a mechanism is operative in uranyl excretion is substantiated by the fact that when an animal is flooded with bicarbonate, administered by infusion so that bicarbonate ion concentration in the tubules is high, there is very little accumulation of uranium in the kidney after injection. However, during the 4 to 8 hours following such an injection, a large amount of uranium is excreted in the urine. On the other hand, increasing the urinary pH without adding significant bicarbonate has much less effect in reducing the deposition of uranium in the kidney, while lowering the pH to 6.5 or less in the tubules causes marked increase in renal retention of uranium and a reduction in its excretion.

A considerable number of studies involving tetravalent uranium, chiefly UCl_4, have also been described. Although these are of relatively little interest toxicologically, presumably because only hexavalent uranium enters the blood stream unless the method of direct intravenous injection is employed, the studies will be described briefly. In contrast to hexavalent uranium compounds, of which about 60% of an injected dose is excreted in the urine during the first three days, only about 10% of an injected dose of UCl_4 appears in the urine during the same time interval. This is due to the fact that tetravalent uranium has less tendency to complex with bicarbonate and is carried in the blood largely in a non-filterable form. The uptake by the kidney of U_4 is less than of U_6 and since an appreciable fecal excretion of uranium begins about the second day after UCl_4 injection and continues for some time, it is presumed that there is some biliary excretion of this form of uranium that has accumulated in the liver.

X. Toxic Action of Uranium on Kidney Structure and Function

A. Morphological Studies

Since the earliest animal experiments with uranium, there have been numerous descriptions of the morphological expression of its renal toxicity based on experiments on animals of various species, age and sex. Different uranium compounds have been given in widely different doses and by various routes as either single or repeated administrations. Despite these differences, remarkable similarities are noted between the numerous pathological descriptions and interpretations found in the literature. The changes, which share many features with poisoning by other heavy metals such as mercury, are not specifically diagnostic of the effect of uranium. There is, in general, an inverse relationship between the time of onset and severity of the renal lesions and dose in any given susceptible species.

The gross abnormalities described consist of swelling and/or pallor initially. This has been noted as early as six hours in mice given a relatively high dose of uranyl nitrate (1 mg per 25–32 g mouse or approximately 40–50 mg/kg) administered subcutaneously. MUELLER and MASON (1956) and BENSCOME et al. (1960) noted pallor of the kidneys in rats two days after the subcutaneous administration of doses of about 35 mg/kg and 14.4 mg/kg in mouse and rat respectively. The absence of obvious gross changes, unless moderately severe renal damage was observed histologically, was stressed by BARNETT and METCALF in chap. 4 of VOEGTLIN and HODGE (1949). These authors were impressed with the occurrence of a zone of grayish-yellow color and loss of normal striated appearance in the corticomedullary junction region. With very high doses of uranium compounds, a number of descriptions mention a mottled red and yellow color on the cortical surface, while repeated injections or continuous long-term dietary exposures to relatively high levels of uranium compounds often resulted in a granular surface with frequently fine, pinpoint yellow areas, associated with scarring.

There are very numerous histological descriptions of the kidney following administration of uranium in the literature over the past 70 years, representative examples of which are referred to in the introduction to this chapter. Many of these were complicated by the different experimental conditions indicated at the beginning of this section and descriptions were based on routine light microscopy, with only occasional application of special stains. The reader is specifically referred to the papers of DAHL (1953) and BENSCOME et al. (1960), in which sequential changes following single subcutaneous injections of uranyl nitrate are described and in which the advantages of refinements in histological technique and histochemistry were fully utilized. The former author gave doses of 2, 10, 20, and 30 mg/kg to rats and showed that the changes were dose-dependent chiefly in so far as speed of evolution of the lesions was concerned. The latter authors produced a moderately severe injury in young rats with a dose of 14.4 mg/kg which was usually lethal within 5 to 7 days. During the first 6 days after injection, lesions of glomeruli and tubules at different levels of the nephron were graded on a scale of 0–4, on the basis of frequency or size, and correlated with changes in urine volume and degree of proteinuria. The following is a composite description based on the work of DAHL (1953), BENSCOME et al. (1960) and BARNETT and METCALF in chap. 4 of VOEGTLIN and HODGE (1949).

The proximal convoluted tubules show the earliest change consisting of vacuolization and the accumulation of hyalin droplets and other cytoplasmic bodies. There is a lag period of from about 6 hours to several days in the appearance of these changes, depending on the dose, and ultimately, with administration of sufficient amounts of the material, cell necrosis is seen. The necrosis appears initially in the lower portion of the convoluted tubule. With lower doses it is spotty in distribution and following higher doses, may extend as wedge-shaped areas into the outer cortex toward the capsule. Casts, either hyalin or containing shed necrotic cells, are present in considerable numbers after the first few days at all levels of the tubular system. After 2 or 3 days, evidence of regeneration of the lining of the convoluted tubules becomes apparent, depending, to some extent, on the severity of the initial injury. In non-fatal cases, after about 2 to 3 weeks, the lining epithelium is completely reformed but the individual cells are morphologically and functionally abnormal for considerably longer periods of time. Regenerating epithelium takes several forms. Nuclei tend to be larger and darker than normal, with mitotic figures being frequently seen in the early stages. Cytoplasm is basophilic and often flattened. As regeneration continues, and the kidney becomes more and more normal in appearance, following sublethal doses of uranium compounds, subtle changes are seen for a considerable period of time, so that it is difficult to distinguish between regenerated cells and cells which have been "activated" in some fashion. The presence of calcium deposits within tubular epithelial cells and tubular casts has been described by many authors, while others have made no mention of such a change. DAHL (1953) carried out careful histochemical investigations of this process and concluded that, at least in the rat, it was the first change observed, as fine, specifically-staining granules in the cytoplasm of proximal convoluted tubular cells with movement into luminal casts after cell rupture. The amount of calcium was dose-dependent. It is probable that larger intratubular deposits of calcium in casts are, in considerable measure, responsible for permanent blockage and the late changes of scarring and nephron atrophy described by many authors, particularly following continuous or repeated dosage.

Reports of ultrastructural changes corresponding to the above described light-microscopic alterations have not been numerous but have helped to shed further light on the pathogenesis of the lesions. MUELLER and MASON (1956), described changes at 6 hours in the mouse. About the only significant observation made by these authors was a progressive swelling, enlargement and disruption of mitochondria. They stated that basement membranes, cell borders, brush borders and background cytoplasm looked normal. STONE et al. (1961) made electron microscopic observations between 1 and 6 days after injection of a lethal dose of uranium nitrate and observed marked swelling of the apical portion of the convoluted tubular epithelial cells with apical vacuoles and marked enlargement of sub-basilar compartments. These changes were maximal at the time polyuria was developing and were interpreted as being related to disturbed fluid transport. They appear to correspond with earlier descriptions of pallor and vacuolization of these cells as seen with the light microscope. Early pre-necrotic changes seen at one day, or later, consisted of swelling of mitochondria, endoplasmic reticulum and Golgi systems, as well as loss of microvilli of the brush border. Various cytoplasmic vacuoles and bodies, including myelin figures and hyalin droplets, increased in number, size and variety as proteinuria developed, and were thought to be suggestive evidence of tubular reabsorption of material from the tubular lumens. PORTE et al. (1963) have also described the early degenerative and necrotizing changes seen under the electron microscope in mice. Their findings were in essential agreement with those of STONE et al. (1961), but they also included a study of regeneration, seen best at 6 days after administration of uranyl nitrate. At this time they observed viable cells scattered among the necrotic debris, and from these, thin cytoplasmic extensions were seen to progress along the basement membrane beneath necrotic cells. The regenerating cells were described as being "embryonic" in character, with large nucleoli, many free ribosomes, fine profiles of endoplasmic reticulum and a reduced number of mitochondria in which few cristae were seen. These cells showed no microvilli on their luminal surfaces and were joined together by "terminal bars" near their apices. There was frequent dilation of intracellular spaces. The findings of these authors would help to explain the light microscopic appearance of those regenerating cells which have markedly basophillic cytoplasm and are flat, without evidence of brush borders.

The preceding descriptions of the highly dramatic renal tubular changes resulting from uranium poisoning are undoubtedly of great importance, but have tended to overshadow the less marked and more subtle changes produced in the glomerulus. Despite the earlier reports of CHRISTIAN and O'HARE (1913) and SUSUKI (1912), both of whom described moderate glomerular alterations in the kidneys of rabbits in addition to tubular changes, most authors who have discussed renal pathology in connection with uranium, have either denied glomerular changes or considered them insignificant. However, HUNTER and ROBERTS (1932), using special stains, outlined glomerular changes in the kidneys of rabbits and monkeys, induced by uranium and other nephrotoxins which they interpreted as affecting largely the basement membranes of the glomerular capillaries, with development of some secondary fibrosis. Endothelial and epithelial degenerative and regenerative changes associated with formation of hyalin droplets were also noted. Finally, BENSCOME et al. (1960) and STONE et al. (1961) applied a variety of sophisticated special stains to 2-micron paraffin sections and confirmed their light microscope findings by electron microscopy and established the fact that striking glomerular changes did occur in rat kidneys damaged by uranium. A centrolobular lesion which developed after 3 days, contained fibers which stained like collagen and which subsequently were identified as such under the electron microscope, varying amounts of PAS-positive material were deposited in basement membranes adjacent to smaller lesions, there was vacuolization and hyalin droplet formation in epithelial cells and hyalin droplets and large globules of amorphous material were noted in the capsular spaces. The electron microscope revealed, in addition, a loss of foot processes, dense deposits between the basement membrane and the epithelium, as well as hyalin droplest, other cytoplasmic bodies, myelin figures and cytoplasmic vacuoles in epithelial cells. These changes are not unlike those seen in some spontaneous proteinuric states in humans and in some animals with experimentally induced proteinuria. It is possible, therefore, that some of the protein found in the urine following uranium administration or exposure, including plasma protein-bound uranium, may have escaped through the glomerulus. The nephron atrophy and scarring, seen on occasions in the late stages of severe or repeated uranium poisoning, is probably related in some manner to the glomerular collagen deposition.

B. Functional Studies

Numerous studies of disturbed renal function, associated with uranium toxicity, have been reported. These have been directed at elucidation of the locus and mechanism of action of the metal on the kidney, the apparent tolerance which develops following repeated or continuous dosage, as well as in a search for a means of early detection, quantitation and prevention or treatment of the renal injury.

1. Renal Clearance Studies

As has been mentioned above, the clearance of uranium from the body is decreased by acidification of urine and increased by alkalinization with bicarbonate. The work of WALKER et al. (1941) and PITTS and LOTSPEICH (1946) indicated that the major site of reabsorption of bicarbonate is in the proximal convoluted tubule. This finding, in conjunction with the fact that 60 to 80% of water from the glomerular filtrate is reabsorbed in the convoluted segment, strongly suggests that the uranyl ion is available in this location for combination with lining cells of the tubules. Presumably the deleterious action results from binding or combination of the uranyl ion with the cell proteins, possibly in the surface membrane or after absorption through it in membranes of such organelles as mitochondria. Sodium bicarbonate has been among the successful materials used to prevent or treat acute uranium poisoning in experimental animals. It was most effective when administered parenterally and presumably acted by keeping the filterable uranium that passed through the glomerulus in a complexed, and therefore a non-toxic, form. It is also known to increase the excretion of citrate in the urine and this citrate can also act keep the uranium in a complexed state.

Further evidence concerning the locus and, to some extent, the mechanism of uranium action on the kidney has been derived from studies of the renal clearance of various substances following uranium injection. BOBY et al. (1943) found all clearances decreased at the height of uranium injury in dogs, 5 days after subcutaneous injection of uranyl acetate. They interpreted the marked lowering of inulin clearance as due to back diffusion and of diodrast clearance as due to complete cessation of tubular secretion and pointed out that because of the severe nephron damage which occurs within a day or two after the administration of uranyl salts, only immediate effects on renal function would be valid in attempting to identify initial sites of uranium action. Consequently, renal clearances of 9 substances were measured within a period of about 2 hours after intravenous injection of uranyl acetate in rabbits. The findings were outlined by DOUNCE in chap. 15 of VOEGTLIN and HODGE (1949). The clearances of creatinine, inulin and xylose, which are considered to be measures of glomerular filtration, showed no significant change, indicating no early glomerular functional effect. Few significant changes were observed in the clearance of chloride and urea, both thought to be partially reabsorbed in the proximal convoluted tubular segment, and partially in more distal portions, of the nephron. Since it was considered likely that reduced reabsorption in some localized tubular region would be compensated for by increased reabsorption through some other portion, these clearances gave no information about the site of uranium action. Of the remaining substances studied, glucose and amino acids, which are normally reabsorbed in convoluted tubules, showed a significant increase in clearance, whereas the other two, phenol red and diodrast, believed to be secreted by epithelial cells into the proximal tubular lumen, showed a significant decrease in renal clearance. All the evidence derived from these clearance studies indicated that the major action of uranium is probably in the lower half or lower two-thirds of the proximal convoluted segment.

2. Proteinuria

An early and sensitive indication of uranium injury is the finding of protein in the urine. Of this protein, a sizeable fraction has been found, by electrophoresis, to be albumin. It is probable that a major part of the protein excreted in the urine results from failure of convoluted tubular epithelium to reabsorb the proteins which are found in very low concentrations in the large amount of glomerular filtrate which is normally concentrated about one hundred fold. Urinary protein is possibly enhanced to some degree by increased permeability of the glomerular membrane to plasma proteins, as suggested by the work of STONE et al. (1961) and also by other proteins derived from damaged tubular cells, either by leakage from or by sloughing of these cells into the tubular lumina.

3. Enzyme Studies

Alkaline phosphatase was found by BREEDIS et al. (1943) to be increased in the urine and decreased in kidney tubules when estimated histochemically. This suggested that other enzymes found in high concentration in the epithelium of kidney tubules might appear in the urine as a result of uranium injury. Studies which followed, reported by DOUNCE et al. in chap. 14 of VOEGTLIN and HODGE (1949), demonstrated that the measurement of urinary catalase was one of the most sensitive early tests for uranium poisoning in some animals, particularly the rabbit. It was shown that very small doses of uranium, in the range of 0.01 to 0.02 mg/kg intravenously, increased the urinary catalase within a period of 2 to 4 days, and that the test was reproducible after repeated exposures. Fortunately, in tests on some humans exposed to uranium, no evidence of increased catalase excretion was found.

Since alkaline phosphatase is known to be localized particularly in the brush border area of the proximal convoluted tubular epithelial cells, the appearance of this enzyme in the urine in uranium poisoning is further evidence of tubular damage in this segment. However, the tendency for it to be absent after further exposures and the finding of lowered phosphatase activity in the kidneys of animals which had survived single or repeated doses of uranium, even after several months, indicated that newly regenerated cells in the proximal tubules were different from cells of the original undamaged tubule. In this respect, the studies of WACHSTEIN (1957) on enzymatic staining reactions in regenerating tubular cells of rat kidneys following mercury poisoning, are of interest. Of 10 enzymes studied, all of which disappeared to a greater or lesser extent in the damaged kidney, there was a slow reappearance in the regenerating cells, in from 5 to 7 days after regeneration had begun, of the reaction for all enzymes except alkaline phosphatase. This enzyme, even after 28 days, could be detected only in trace amounts. The importance of alkaline phosphatase in uranium poisoning is also stressed by the findings of ALIBERT et al. (1963), who measured enzyme activities in kidney homogenates following intraperitoneal uranyl nitrate injection. Of 5 enzymes, the only change was noted in alkaline phosphatase, which fell to approximately 50% of the control value. This was observed at 24 hours after injection and continued throughout the course of the experiment which lasted 96 hours. While it is not known what role these changes in alkaline phosphatase play in the development and repair of renal tubular injuries, the findings appear to support the view that tolerance to repeated or continuous doses of uranium compounds is associated with biochemical as well as morphological alterations in the convoluted epithelial cell. While it has been established that uranium inactivated certain kidney enzymes, it is not known whether or not the immediate cause of the uranium damage to the kidney is enzyme blockage. Uranium is not a strong non-specific enzyme poison and much of the enzyme inhibition can probably be explained by the formation of uranium complexes with ionized acid groups of the protein component of the enzymes.

The uranyl ion reversibly inhibits a number of enzyme systems such as citrate and hydroxyapatite. However, of 28 enzyme systems investigated *in vitro*, only 4 were susceptible to low concentrations of uranium. Certain metabolic reactions are also inhibited, such as those occurring on the cell surfaces of yeasts, the oxidation of glucose by the kidney and such reactions as oxidation of pyruvate, succinate, and alanine. However, the amounts of uranium which inhibit these reactions are too small for testing for enzyme inhibition. The occurrence of a lag period from the administration of uranium to the appearance of renal damage, either functional or morphologic, is also more in favor of an indirect mechanism related to protein complexing in general than to a direct effect on enzymes, see chap. 13 of VOEGTLIN and HODGE (1949).

On the basis of the studies outlined in this section, it is probable that the mechanism of the toxic action of uranium on the kidney cannot be explained in any simple fashion. All the following factors probably play some role in the final outcome; non-specific protein complexing, enzyme inhibition, interference with mitochondrial function, and membrane and fluid transport, as well as with the excretory functions of the tubular epithelial cells.

C. Tolerance to Acute Uranium Poisoning

The development of a tolerance to uranium following repeated doses, by experimental animals, was suggested in several of the very early publications on this subject, while experiments designed to study tolerance, as such, were reported by Gil y Gil (1924), Hunter (1928), and McNider (1929). Subsequent studies were described by Haven in chap. 12 of Voegtlin and Hodge (1949) and by Gruschow and Mann in chap. 23 of Voegtlin and Hodge (1953). From the work of these investigators, and others referred to by them, the following general conclusions can be drawn. Tolerance may be defined as the resistance of an animal to an ordinarily lethal, or at least injurious, dose of uranium following one or more previous administrations of the poison. It has been developed after subcutaneous, intravenous, and intraperitoneal injections, as well as after feeding or inhalation. The repeated doses have been of the same or increasing size, and have been given continuously or intermittently. Tolerance has been tested at various time intervals after cessation of the procedure used for its development, it does not develop immediately and after discontinuance of repeated exposures to uranium, it may be lost. However, some degree of resistance may persist for a considerable period of time. For example, on the basis of mortality, about 50% of groups of rats made tolerant by repeated intraperitoneal injections of uranyl nitrate, tend to lose their acquired tolerance if administration of the test dose is delayed for from 21 to 45 days. It has also been noted by Barnett and Metcalf in chap. 4 of Voegtlin and Hodge (1949), on the basis of blood NPN levels and microscopic examination of the kidney, that the tolerance which developed in rats as a result of feeding 2% uranyl nitrate in the diet for 6 months, was completely lost during a subsequent $6^1/_2$ months on a uranium-free diet. It has been shown experimentally, for intraperitoneally injected uranyl nitrate, that the lower limit of dosage that produces tolerance lies somewhere between 0.1 and 0.2 mg/kg in rats.

There are good indications that kidney damage with subsequent regeneration of tubular epithelium, is a prerequisite to tolerance. In addition to the microscopic findings which have been described in many tolerant animals, an increased ratio of kidney weight to body weight, a marked reduction or absence of kidney phosphatase, both chemically and histochemically, high urinary volumes, and variable urinary pH have been found to occur even without the persistence of obvious histopathological abnormalities by light microscopy. It has been found in other studies that in tolerant rats less uranium is retained by the kidney following an exposure, that higher initial levels of urinary excretion, even though not necessarily of total excretion, are noted, as well as the continued production of increased amounts of citric acid which appears in the urine. On the basis of all the foregoing findings, it has been suggested that the mechanism responsible for the development of tolerance is related to a failure of uranium to complex with cellular proteins in the regenerated or regenerating tubular lining cells in the kidney. It has been further suggested that this is due to combination and subsequent excretion of uranium, dissociated from the bicarbonate complex, with the extra citric acid present in the tubular lumen. While the available evidence indicates that the presence of citric acid in the kidney tubule in concentrations above normal may account, in some way, for the acquired tolerance to the toxic action of the material, nothing is known about the mechanism by which such increased citric acid becomes available. Other factors involved in this phenomenon are also still incompletely understood.

XI. Attempted Prevention and Therapy of Uranium Poisining

Initial efforts to find an effective prophylactic or therapeutic agent against the injurious effects of uranium were concerned with administration of chemicals known to be complexers of hexavalent uranium. These were given with the hope of increasing excretion of the complexed metal, thus decreasing renal retention or washing out uranium already fixed in the kidney.

The most satisfactory results were obtained by intraperitoneal injections or forced feeding of sodium bicarbonate to rats, see chap. 15 of VOEGTLIN and HODGE (1949). Pretreatment for 2 or 3 days was most effective although poisoning could be ameliorated if bicarbonate treatment was started as late as 12 hours after uranium administration. Injection of citrate, which is somewhat more toxic than bicarbonate, was also found to be effective but this material is burned to a considerable extent by the body tissues, leaving an alkaline residue of sodium bicarbonate. Moreover, an increase in plasma bicarbonate, which tends to cause the excretion of citrate in the urine, will act secondarily to keep more uranium in a complexed form.

HOLMAN and DOUGLAS (1944) were able to protect 20 dogs from the lethal effect of 5.0 mg/kg or uranyl nitrate by the intravenous injection of large doses (0.23 g/kg) of sodium citrate given for from 5–10 days. Protection occurred even though administration of citrate was delayed for as long as 67 hours.

Studies with 2,3-dithiopropanol (BAL) has shown it to be an effective antidote for several types of heavy metal poisoning in which inhibition of SH enzyme systems is thought to be the mechanism responsible for the toxic effects. However, it was found by NEUMAN and ALLEN (1949) to be without effect against the toxic action of uranyl nitrate, uranyl fluoride and uranium tetrachloride administered parenterally to rats. The negative results were considered to be further evidence that uranium does not exert its toxic effects by combining with SH groups.

Further efforts to find a good antidote for acute uranium poisoning included studies of polyphosphates, HODGE et al. (1951). Polyphosphates, when given parenterally, are effective against uranium. However, they are hydrolyzed *in vivo*, and give rise to a metabolic acidosis, and by complexing serum calcium, cause hypocalcemia.

Subsequently, a study was undertaken by DAGIRMANJIAN et al. (1956) to determine the usefulness of calcium disodium ethylenediaminetetraacetate (Ca EDTA) in modifying the toxic effects of uranium. The toxicity of Ca EDTA alone was found to be low when injected intraperitoneally in doses of 50 mg. When given at intervals of 2, 4, and 6 hours, during a 24 hour feeding period, with diets containing 4 or 8% of uranyl nitrate respectively, there was a significant reduction of mortality. However, two injections of the chelating material at 2 and 6 hours, or single injections at 4 or 24 hours were ineffective. The 30 day LD_{50} for rats was increased approximately 3-fold by a 2, 4, and 6 hour schedule of EDTA injections and the material counteracted 3 lethal doses of uranium. Ca EDTA also increased the 24-hour urinary excretion of parenterally administered uranium by approximately 50% However, it did not mobilize uranium previously deposited in the skeleton. Thus, it would seem that while Ca EDTA is effective against some toxic effects of uranium under certain experimental conditions, its value as a practical antidote is questionable.

References

Alibert, A., Rigal, M., Bastide, P., Turchini, J., Dastuge, G.: Examen de quelques activités enzymatiques dans le foie et le rein de souris avant et après administration de tétrachlorure de carbone ou de nitrate d'uranyle. C. R. Soc. Biol. (Paris) **157**, 1439–1441 (1963).

Benscome, S. A., Stone, R. S., Latta, H., Madden, S. C.: Acute tubular and glomerular lesions in rat kidneys after uranium injury. Arch. Path. **69**, 470–476 (1960).

Boby, M. N., Longley, L. P., Dickes, R., Price, J. W., Hayman, J. M., Jr.: The effect of uranium poisoning on plasma diodrast clearance and renal plasma flow in the dog. Amer. J. Physiol. **139**, 155–160 (1943).

Breedis, C., Florey, C. M., Furth, J.: Alkaline phosphatase level in urine in relation to renal injury. Arch. Path. **36**, 402–412 (1943).

Chittenden, R. H., Lambert, A.: Untersuchungen über die physiologische Wirkung der Uransalze. Z. Biol. (N.S.) **25**, 513–532 (1889).

Christian, H. A., O'Hare, J. P.: Glomerular lesions in acute experimental uranium nephritis. J. med. Res. **28**, 227–234 (1913).

Christian, H. A., Smith, R. M., Walker, C.: Experimental cardiorenal disease. Arch. intern. Med. **8**, 468–551 (1911).

Clarke, N. J., Bair, W. J.: Plutonium inhalation studies. VI. Pathologic effects of inhaled plutonium particles in dogs. Hlth Physics **10**, 391–398 (1964).

Dagirmanjian, R., Maynard, E. A., Hodge, H. C.: The effect of calcium disodium ethylenediaminetetraacetate on uranium poisoning in rats. J. Pharmacol. exp. Ther. **117**, 20–28 (1957).

Dahl, L. K.: The stages in calcification of the rat kidney after the administration of uranium nitrate. J. exp. Med. **97**, 681–694 (1953).

Dickson, E. C.: The experimental production of chronic nephritis in animals by the use of uranium nitrate. Arch. intern. Med. **3**, 375–410 (1909).

Dickson, E. C.: A further report on the production of experimental chronic nephritis in animals by the administration of uranium nitrate. Arch. intern. Med. **9**, 557–591 (1912).

Fairhall, L. T., Sayers, R. R.: The relative toxicity of lead and some of its common compounds. U.S. publ. Hlth Bull. **253** (1940).

Fish, B. R.: Inhalation of uranium aerosols by mouse, rat, dog, and man. Inhaled particles and vapours. I., p. 209–218. London: Pergamon Press 1961.

Garnier, M., Marek, J.: Des effets de la répétition d'une même dose de nitrate d'urane en injection sous-cutanée chez le lapin. C. R. Soc. Biol. (Paris) **107**, 99–102 (1931).

Garnier, M., Marek, J.: L'intoxication chronique par le nitrate d'urane en injection quotidienne chez le lapin. C. R. Soc. Biol. (Paris) **107**, 938–940 (1931).

Gil y Gil, C.: Die Immunität im Nierenepithelgewebe. Beitr. path. Anat. **72**, 621–653 (1924).

Hodge, H. C.: Mechanism of uranium poisoning. Arch. industr. Hlth **14**, 43–47 (1956).

Hodge, H. C., Maynard, E. A., Downs, W. L.: Antidotal action of polyphosphates in uranium poisoning. J. Pharmacol. exp. Ther. **101**, 17–18 (1951).

Hoffman, J.: Fluoreszenzen an menschlichen Knochenaschen. Biochem. Z. **311**, 247–251 (1942).

Holman, R. L., Douglas, W. A.: Studies of heavy metal poisoning. II. Sodium citrate therapy effective 67 hours after uranium injury. Proc. Soc. exp. Biol. (N.Y.) **57**, 75–78 (1944).

Hunter, W. C.: Experimental studies of resistance of the rabbit renal epithelium to uranyl nitrate. Ann. intern. Med. **1**, 747–789 (1928).

Hunter, W. C., Roberts, J. M.: Glomerular changes in the kidneys of rabbits and monkeys induced by uranium nitrate, mercuric chloride and potassium dichromate. Amer. J. Path. **8**, 665–688 (1932).

Jackson, D. E.: On the pharmacological action of uranium. Amer. J. Physiol. **26**, 381–395 (1910).

Jones, E. S.: Microscopic and autoradiographic distribution of uranium in the rat kidney. Hlth Physics **12**, 1437–1451 (1966).

Karsner, H. T., Reiman, S. P.: Studies of uranium poisoning. I. The toxicity of certain water-insoluble salts of uranium. J. med. Res. **39**, 157–161 (1918).

Kisieleski, W. E., Faraghan, W. G., Norris, W. P., Arnold, J. S.: The metabolism of uranium^{233} in mice. J. Pharmacol. exp. Ther. **104**, 459–467 (1952).

Leach, L. J., Maynard, E. A., Hodge, H. C., Scott, J. K., Yuile, C. L., Sylvester, G. E., Wilson, H. B.: A five-year inhalation study with natural uranium dioxide (UO_2) dust. 1. Retention and biologic effects in the monkey, dog, and rat. Hlth Physics **18**, 599–612 (1970).

LEACH, L. J., SPIEGL, C. J., WILSON, R. H., SYLVESTER, G. E., LAUTERBACH, K. E.: A multiple chamber exposure unit designed for chronic inhalation studies. Amer. industr. Ass. J. **20**, 13–22 (1959).

LEACH, L. J., YUILE, C. L., HODGE, H. C., SYLVESTER, G. E., WILSON, H. B.: A five-year inhalation study with natural uranium dioxide (UO_2) dust. II. Postexposure retention and biologic effects in the monkey and dog. Hlth Physics, to be submitted (1971).

MCNIDER, W. DE B.: Functional and pathological study of the chronic nephropathy induced in the dog by uranium nitrate. J. exp. Med. **29**, 513–529 (1919).

MCNIDER, W. DE B.: A review of acute experimental nephritis. Physiol. Rev. **4**, 595–638 (1924).

MCNIDER, W. DE B.: Functional and pathological response of the kidney in dogs subjected to a second subcutaneous injection of uranium nitrate. J. exp. Med. **49**, 411–433 (1929).

MONLUX, W. S.: Primary pulmonary neoplasms in domestic animals. Southwest. Vet. **6**, 131–133 (1952).

MORROW, P. E., GIBB, F. R., BEITER, H.: Inhalation studies of uranium trioxide. Hlth Physics, in press (1971).

MORROW, P. E., GIBB, F. R., LEACH, L. J.: The clearance of uranium dioxide dust from the lungs following single and multiple inhalation exposures. Hlth Physics **12**, 1217–1223 (1969).

MUELLER, C. B., MASON, A. D., JR.: A study of the pathological anatomy of the proximal tubules of the mouse produced by various nephrotoxins. Surg. Forum **7**, 641–654 (1956).

NEUMAN, M. W., NEUMAN, W. F.: The deposition of uranium in bone. II. Radioautographic studies. J. biol. Chem. **175**, 711–715 (1948).

NEUMAN, W. F., ALLEN, R. P.: The failure of 2,3-dithiopropanol (BAL) to affect acute, systemic uranium poisoning. J. Pharmacol. exp. Ther. **96**, 95–98 (1949).

NEUMAN, W. F., FLEMING, R. W., DOUNCE, A. L., CARLSON, A. B., O'LEARY, J., MULRYAN, B. J.: The distribution and excretion of injected uranium. J. biol. Chem. **173**, 737–748 (1948).

NEUMAN, W. F., NEUMAN, M. W., MULRYAN, B. J.: The deposition of uranium in bone. I. Animal studies. J. biol. Chem. **175**, 705–709 (1948).

NIELSEN, S. W., HORAVA, A.: Primary pulmonary neoplasms of the dog. Report of sixteen cases. Amer. J. vet. Res. **21**, 813–830 (1960).

OLIVER, J.: The histogenesis of chronic uranium nephritis with especial reference to epithelial regeneration. J. exp. Med. **21**, 425–451 (1915).

PARK, J. F., CLARKE, W. J., BAIR, W. J.: Chronic effects of inhaled plutonium in dogs. Hlth Physics **10**, 1211–1217 (1964).

PITTS, R. F., LOTSPEICH, W. D.: The renal excretion and reabsorption of bicarbonate. Fed. Proc. **5**, 82 (1946).

PORTE, A., CUSSAC, Y., STOEBNER, P., ZAHND, J. P.: Sur la formation et l'ultrastructure des cellules de régénération dans les tubules rénaux de souris intoxiquées par le nitrate d'uranyle. C. R. Soc. Biol. (Paris) **157**, 2079–2081 (1963).

SCHREINER, B. F., MICHAELSON, S. M., YUILE, C. L.: The effects of thoracic irradiation upon cardiopulmonary function in the dog. Amer. Rev. resp. Dis. **99**, 205–218 (1968).

STOKINGER, H. E., STEADMAN, L. T., WILSON, H. B., SYLVESTER, G. E., DZINBA, S., LABELLE, C. W.: Lobar deposition and retention of inhaled insoluble particulates. Arch. industr. Hyg. **4**, 346–353 (1951).

STONE, R. S., BENSCOME, S. A., LATTA, H., MADDEN, S. C.: Renal tubular fine structure studied during reaction to acute uranium injury. Arch. Path. **71**, 160–174 (1961).

SUZUKI, T.: Zur Morphologie der Nierensekretion unter physiologischen und pathologischen Bedingungen. Jena: Gustav Fischer 1912.

TANNENBAUM, A.: Toxicology of uranium. New York-Toronto-London: McGraw-Hill 1951.

VERNE, J.: Lésions histologiques des centres nerveux supérieurs chez les lapins soumis à l'intoxication chronique par l'urane. Ann. anat. path. et anat. normale méd. chir. **8**, 757–758 (1931).

VOEGTLIN, C., HODGE, H. C.: Pharmacology and toxicology of uranium compounds, Vols. I, II. New York-Toronto-London: McGraw-Hill 1949.

VOEGTLIN, C., HODGE, H. C.: Pharmacology and toxicology of uranium compounds, Vols. III, IV. New York-Toronto-London: McGraws-Hill 1953.

WACHSTEIN, M.: Enzymatic staining reactions in regenerating tubular, cells of the rat kidney. J. Mt Sinai Hosp. **24**, 1316–1322 (1957).

WALINDER, G., FRIES, B., BILLAUDELLE, U.: Incorporation of uranium. II. Distribution of uranium absorbed through the lungs and the skin. Brit. J. industr. Med. **24**, 313–319 (1967).

WALINDER, G., HAMMARSTRÖM, L., BILLAUDELLE, U.: Incorporation of uranium. I. Distribution of intravenously and intraperitoneally injected uranium. Brit. J. industr. Med. **24**, 305–312 (1967).

WALKER, A. M., BETT, P. A., OLIVER, J., MACDOWELL, M. C.: The collection and analysis of fluid from single nephrons of the mammalian kidney. Amer. J. Physiol. **134**, 580–595 (1941).
WILSON, H. B., SYLVESTER, G. E., LASKIN, S., LABELLE, C. W., SCOTT, J. K., STOKINGER, H. E.: The relation or particle size of uranium dioxide dust to toxicity following inhalation by animals. II. Arch. industr. Hyg. **6**, 93–104 (1952).
WILSON, H. B., SYLVESTER, G. E., LASKIN, S., LABELLE, C. W., SCOTT, J. K., STOKINGER, H. E.: Relation of particle size of U_3O_8 dust to toxicity following inhalation by animals. Arch. industr. Hyg. **11**, 11–16 (1955).
WILSON, H. B., SYLVESTER, G. E., LASKIN, S., LABELLE, C. W., STOKINGER, H. E.: The relation of particle size of uranium dioxide dust to toxicity following inhalation by animals. J. industr. Hyg. **30**, 319–331 (1948).
YUILE, C. L., GIBB, F. R., MORROW, P. E.: Dose-related local and systemic effects on inhaled plutonium-238 and plutonium-239 dioxide in dogs. Radiat. Res. **44**, 821–834 (1970).

Chapter 4

Data on Man

J. B. Hursh* and N. L. Spoor

With 8 Figures

I. Introduction

Animals, not men, are the common experimental subjects in research designed to predict the behavior of radioactive nuclides in man. This is true for uranium quite as much as for other nuclides; witness the abundance of experimental animal data contained in the preceding chapter relative to the amount of information available for presentation in the following pages.

Human data become potentially available when man is exposed either by intent or by inadvertence. The intentional administrations of a radionuclide include the planned experiments designed to gain biological data and the use by physicians for treatment or diagnosis of disease. The inadvertent intakes are experienced either by reason of man's employment or because he lives in a slightly contaminated world. Some of the problems in the collection and use of data from these four sources will be discussed in the following paragraphs.

Planned experiments on man may, for some elements, utilize a short half-life isotope of the element of interest. Such a device enables the experimenter to keep his radiation dose commitment small and yet the amount of the nuclide introduced may be great enough to ensure that the samples collected can be measured at levels well above the sensitivity limits of the analytical methods available. Likewise, because the radiation dose may be made negligible, the experimenter need have no compunction about assembling a volunteer group of normal adult subjects in any convenient age range. The circumstance that a short half-life isotope has been chosen limits the period of observation. While uranium has no such short half-life isotope, the specific activity of natural uranium may be increased by the enrichment, i.e. by increasing the proportion, of ^{235}U (and ^{234}U). The existence of very sensitive analytical methods permits carefully supervised experiments on man. It has been judged prudent to conduct these experiments under hospital conditions with members of the subject group chosen by a physician to be sufficiently alert mentally so that their knowing consent may be obtained, and sufficiently normal physiologically so that the experimental results will be meaningful. Such a group will usually be small and the experimental duration short. The kinds of information that can be obtained are limited by these circumstances.

Several of the long-lived members of the naturally radioactive series have, in the past, been administered orally and intravenously to patients by physicians.

* This author was supported in part by funds from a contract granted to the University of Rochester by the U.S. Atomic Energy Commission and the number UR-3490-191 has been assigned to this chapter.

At one time or another, uranyl nitrate (between 1890 and 1920), radium bromide or chloride (between 1915 and 1930), and thorium dioxide (between 1930 and 1950) have been administered for a variety of ailments and occassionally with an enthusiasm (FIELD, 1916) which is now regarded by the medical profession as misguided. The quality and the kind of observations recorded by the attending physician should not be expected to satisfy our present quest for precise information and, generally speaking, they do not. In the case of the radium and thorotrast patients much needed information has been gained by remeasurement programs (AUB et al., 1952; NORRIS et al., 1955; SPIESS and MAYS, 1970; LOONEY et al., 1960) using modern detection techniques. No comparable re-study was performed for the uranium-treated patients and the physicians' records constitute the sole source of data.

Information that can be obtained from uranium plant exposures has its own shortcomings. The investigator exercises no control over the circumstances surrounding the introduction of uranium. Information concerning the time of the size and composition of the subject group is often incomplete and sometimes not available to the investigator. Indeed, these data are not only fragmentary, they are in retrospect usually difficult or impossible for the investigator to interpret with confidence.

The fourth source of information, i.e. the study of background uranium in man's food and water and in man himself, has one outstanding advantage: the size of the subject group is limited only by the investigator's time and resources. The obvious disadvantages are the very low concentrations of uranium in nature and the severe demands on measurement techniques. The kinds of information obtainable are restricted. The intakes are derivable in terms of average values rather than individual values. The body distribution data taken from autopsy material are usually interpreted as representing a steady state condition, and yield little or no information on the kinetics of the nuclide's behavior in the body. Since ingestion is the principal route of entry, the background data on uranium do not apply to the industrially important question of lung clearance rates.

It is, therefore, fair to ask "Why consider data on man if they are subject to all these disadvantages? Why not rely entirely on mammalian animal experiments to yield the data needed for control of occupational toxicological hazards?". A review of the basic biological data used by national and international bodies for the calculation of maximum permissible concentrations in air and water for radioactive nuclides (ICRP, Publication 2, 1959) shows that the large majority of listed values depend primarily on animal data. This, however, has not been a matter of choice but of necessity, and the aforesaid values have been listed with the full knowledge that unforeseen differences in the behavior of rats or dogs and man may introduce uncertainties in the calculated protection criteria. It is possible that in some future stage of the art these differences may be predictable and a system of precise extrapolation may be developed. Similarity principles for functional systems in a variety of mammalian species have been studied by interested biologists for many years (i.e. HUXLEY, 1932; ADOLPH, 1949; GÜNTHER and DE LA BARRA, 1966). The generalizations developed by the studies may, on occasion, permit us to "understand" an already identified difference in behavior of a given element in man and in, e.g., a rat, but they do not as yet provide an adequate prediction system. STARA, NELSON, DELLA ROSA, and BUSTAD (1971) addressed themselves to the question of the validity of extrapolating the results of animal research to man. In a review incorporating published data, supplemented by original work, involving a variety of radionuclides and several mammalian species including man, they conclude in part that "Careful evaluation of data from several species permit a cautious extrapolation to man. However, the data suggest species specific metabolic pathways for individual isotopes...". The foregoing comments refer to the usefulness of animal data in specifying tissue distribution and excretion for man. An area of equal importance to the radiation toxicologist is that of the dose-injury relationship. The sorts of injury that are considered in the development of occupational air and water limits are those that may develop after tens of years of exposure. If cancer is the injury endpoint, the longer lifespan of man affects the use of animal data in two ways: raising the general question as to whether the incidence and latency of the development of malignancies are related to lifespan and the more specific question as to whether in any given experiment an injury may have been produced but never became manifest because the animal's lifespan was too short.

On the same subject, but with a reservation of a different kind, PARKER (1969), writing on the use of animals to unravel the dose-effect relationship for the radon daughter hazard to the lungs of man, expresses the opinion "On the one side no matter how compelling the animal results, they would be suspect for application to cancer induction in humans. This arises from the growing belief that radiation carcinogenesis requires not only some radiation dose, but some as yet undefined environmental factor...". His further comments define areas of the general problem that require animal experimentation.

The International Commission on Radiological Protection Task Group (ICRP Publication 10, 1968) state in their Introduction that "New (since ICRP Publication 2) data derived from studies of human beings are presented for all these radionuclides except ^{132}Te, ^{144}Ce and ^{239}Np for which the new data pertain to animal experiments...".

There appears to be a consensus among radiation toxicologists that animal data alone are insufficient. Whenever available the use of human data is necessary, sparse and unsatisfactory though they may be, to afford an extra measure of assurance in calculating limits to apply to human exposure.

There is one additional compelling argument for the collection and analysis of data from industrial exposures. It is precisely the consequences of exposure to this aerosol, in this environment, inhaled by these workers, that we hope to control. The experiment in the laboratory may give precise answers given the conditions imposed. To what extent do the conditions reproduce the conditions under which the occupational exposure will occur? To gauge the vulnerability of the laboratory-collected data it must be compared with observations in the industrial situation if such observations exist.

The human data for uranium presented in the following pages appear in three sections: 1) planned experiments organized according to route of administration; 2) data from industrial exposure; and 3) data from background studies.

II. Planned Administrations of Uranium

A. Intravenous Injection

1. The Rochester Intravenous Injection Experiment[1]

(BASSETT, FRENKEL, CEDARS, VAN ALSTINE, WATERHOUSE and CUSSON, 1948)

At the time these experiments were planned the status of the uranium toxicity study was as follows. Animal experiments had shown that severe tubular nephritis

1 Since this paper has hitherto been available only as University of Rochester Report UR-37 with very limited circulation, a detailed account will be presented here.

could be produced by large intravenous doses of uranium (LECONTE, 1853; RICHTER, 1905; MACNIDER, 1924). Sensitive methods had been developed to detect early kidney responses by DOUNCE, ROBERTS and WILLS (VOEGTLIN and HODGE, 1949, chap. 14). It had been found that the intravenous injection of 6 μg of U per kg as uranyl acetate produced increased urinary catalase and protein in rabbits and dogs (VOEGTLIN and HODGE, 1949, chap. 14). The view had been developed by DOUNCE (VOEGTLIN and HODGE, 1949, chap. 15) that hexavalent uranium in the blood complexed with bicarbonate, citrate, phosphate and protein, that it was readily filtered by the kidney glomerulus as the bicarbonate and that a large fraction of the dose (60–80 percent) was excreted in the urine in the first 24 hours. A smaller fraction (10–20 percent) was taken up in bone from which it was released only very slowly. With this general picture of the behavior of uranium in animals in mind, BASSETT and others (1948) performed an experiment on human subjects with the following three objectives: 1) to find that dose of soluble uranium salt which, when injected as a single intravenous dose, would produce barely detectable renal injury, 2) to measure the rate at which soluble uranium compounds are excreted after having been introduced into the circulation, and 3) to observe the effect of procedures designed to increase or decrease the elimination rate of uranium.

Six subjects were selected from a large group of hospital patients and transferred to the Metabolic Ward. It is stated that "Criteria of importance in making the selection were reasonably good kidney function with urine free of protein and with a normal sediment on clinical examination. The probability that the patient would benefit from continued hospitalization and medical care was also a factor in the choice". The higher doses were administered to the older patients in order to minimize the remote possibility of late radiation effects. Further protection for patients 5 and 6 was afforded by reducing the specific activity of the enriched uranium by diluting with natural uranium.

To prepare the dose, a uranyl nitrate solution containing ^{238}U, enriched with ^{234}U and ^{235}U, was diluted with a 1.15 percent solution of sodium acetate and adjusted to pH 4.5. After autoclaving and verifying the uranium content, a prescribed volume was injected into the median basilic vein of the patient. Two additional doses each containing the designated volume and prepared in the same way were analyzed for uranium and the average value taken as the patient dose.

Total urine and fecal collection was made from all subjects. Urine was collected as individual voidings on the day of injection and thereafter in pools of 24 hours. Blood specimens were routinely taken for clinical evaluation. Aliquots of urine were wet digested, brought to neutral pH and transferred to an electroplating cell. Plating recoveries were determined to be 95 ± 5 percent. After plating the foil was counted for alpha activity in a low background 2π alpha counter. Fecal samples were analyzed in a similar way after digestion of 1 gram (dry ashed) aliquots of the sample. Blood samples were analyzed in the same way as fecal samples except that 25 ml aliquots were digested.

The uranium used for these experiments was enriched in ^{235}U so that its composition was by weight 29.34 percent ^{238}U, 70 percent ^{235}U and 0.66 percent ^{234}U. This yielded an alpha disintegration rate of about 93 disintegrations per minute per μg of uranium. The authors set their lower limit of measurement at about 0.05 μg which may be translated into from about 0.3 to 1.2 μg per total sample depending on the aliquot to sample ratio.

Renal function tests (VOEGTLIN and HODGE, 1949, chap. 14) included urinary catalase, protein amino nitrogen to creatinine nitrogen ratio, glomerular filtration

Table 4.1. Patient data and urinary results for the Rochester experiment (BASSETT, FRENKEL, CEDARS, VANALSTINE, WATERHOUSE and CUSSON, 1948)

Patien						Urinary excretion	
no.	sex	age yr.	wt. kg	condition	Dose µg/kg	percent 24 h	percent 5 days
1	m	32	60.5	arthritis-rheumatoid	6.3	82.3	86
2	f	40	75.5	cirrhosis of liver	6.3	84.7	87
3	f	24	37.0	chronic undernutrition	15.8	69.2	71.5
4	m	42	64.0	chronic alcoholism	30.0	66.6	74.3
5	m	51	65.3	unresolved pneumonia	42.0	75.3	76.3[a]
6	m	61	55.1	pulmonary fibrosis	70.9	77.6	87.6
6	m[b]	61	55.1	gastric ulcer	54.5	57.3	79.5

[a] Collected 4 days only.
[b] A second dose was given to this patient.

Table 4.2. Summary of renal function tests on Rochester patients. (Taken from BASSETT et al., 1948)

Patient no.	1	2	3	4	5	6
Urinary catalase (ml of H_2O)	no change[a]	no change	no change	no change	3.6 (3.7)[b]	9.5 (10.5)[c]
Urinary protein (mg albumin/5 ml urine)	0 (0)	0 (0)	0 (0)	0 (20)[d]	0 (<5)[e]	
Amino N/Creat. N	no change	0.49 (0.48)	0.26 (0.24)	no change	0.15 (0.11)	0.14 (0.16)
Glom. filtration (ml/min)	—	115 (118)	118 (117)	—	—	99 (93)
Renal plasma flow (ml/min)	—	463 (402)	593 (638)	—	—	482 (499)
Max. tub. excr. cap. (mg/min)	—	70 (79)	—	—	—	95 (83)
Urea clearance (percent normal)	102 (96)	—	43 (52)	—	77 (—)	—

[a] To be read "tests were performed and no change observed". Dash to be read, "test was not performed".
[b] Except as noted average pre-injection values are listed first with average post-injection values in parentheses.
[c] Questionably positive values on days 4 and 5.
[d] Not an average. Listed value occurred on day 4.
[e] Not an average. Traces, <5 mg, noted on days 5 and 6.

(mannitol), renal plasma flow, maximum tubular excretory capacity (para amino-hippuric acid) and urea clearance. Not all of the more complicated tests were performed on every patient; however, urinary catalase and protein were followed carefully in each case.

Size of Single Dose Versus Renal Function. Table 4.1 presents pertinent data on the subjects and the dose of hexavalent uranium administered. Table 4.2 presents the results of renal function tests. Urinary catalase and urinary protein tests were performed on all urine samples. These were uniformly negative for patients 1, 2, 3 and 4. Patient 5 (dose = 42 µg/kg) showed normal catalase but

Table 4.3. Collection time and urine sample uranium content in percent dose for Rochester patients. (Taken from Bassett et al., 1948)

Patient 1		Patient 2		Patient 3	
hours post-inj.	dose %	hours post-inj.	dose %	hours post-inj.	dose %
3	53.3	0.5	26.2	1.5	34.8
4	9.7	2.1	21.2	6.3	26.0
9	15.5	5.3	20.0	9.8	4.2
11	2.1	7.8	4.9	18.8	3.8
21	1.8	9.5	3.9	23	0.48
46	2.3	16.8	6.6	47	1.14
69	0.97	22.0	2.0	71	0.55
93	<0.20	46	1.8	95	0.30
117	<0.20	70	0.34	119	0.26
141	<0.07	94	0.14	143	0.13
		118	<0.06	167	0.12
		142	<0.06	191	<0.1
				215	<0.1

Patient 4		Patient 5		Patient 6 (1st inj.)	
hours post-inj.	dose %	hours post-inj.	dose %	hours post-inj.	dose %
3.3	23.4	6.2	52.5	0.4	1.2
5.1	12.4	7.8	9.3	2.5	21.8
9.9	17.5	11.8	4.3	5.4	17.2
13.2	5.0	14.0	2.8	7.8	15.3
16.2	4.6	15.8	1.6	9.8	7.7
21	2.8	19.3	1.6	12.1	5.7
45	3.1	22.3	1.06	14.7	3.7
69	1.1	46	1.75	17.9	2.6
93	0.37	70	0.14	21.1	1.56
117	0.21	94	0.04	22	0.86
140	0.19			46	5.24
165	0.12			70	1.81
189	0.12			94	1.81
213	<0.04			118	1.00
237	0.10			142	0.75
261	0.10			166	0.42
285	<0.06			190	0.46
309	<0.04			214	0.44
333	0.04			238	0.44
357	0.10			262	0.46
381	0.03			286	0.41
405	<0.03			310	0.27
				334	<0.17

a trace of protein appeared in the fourth day urine sample. The authors state "it is suspected that this was a chance observation and without significance as far as uranium is concerned". No follow-up was possible because the patient insisted on being discharged from the hospital. Patient 6 (dose = 70.9 μg/kg; 3.9 mg uranium total) showed slightly higher catalase and traces of protein in the fifth and sixth day urinary samples. The authors remark "while it cannot be stated that either test is unequivocally positive, it would seem advisable that any further increase in the dose should be made in small increments".

Urinary Excretion of Hexavalent Uranium. Table 4.3 lists the urinary uranium results for the six patients. Fig. 4.1 plots the administered dose taken as 100 per-

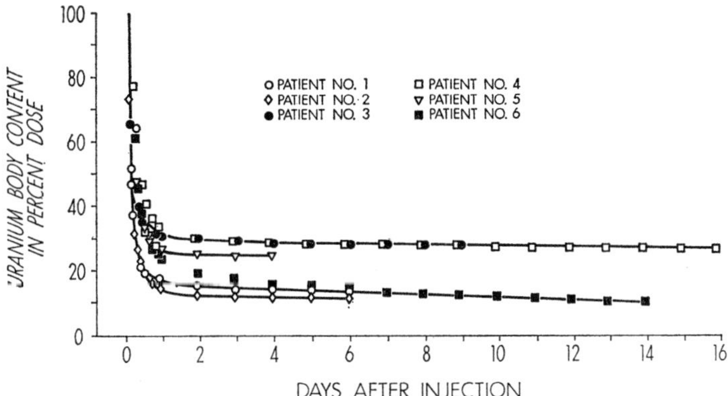

Fig. 4.1. Uranium body content after a single intravenous injection of six patients (Rochester experiment)

cent minus the cumulative urinary excretion in percent as a function of time after dose administration up to the termination of urinary collection. In the figure the ordinate is listed as percent body content because the fecal excretion of intravenous hexavalent uranium was found to be negligible compared to urinary excretion. For example, over the period of the experiment, Patient 1 excreted 86 percent via the urine and less than 0.87 percent via the feces. Consideration of Fig. 4.1 shows that the urinary excretion pattern is characterized by a very rapid early excretion so that 50 percent of the dose is excreted in from 3 to 10 hours and from 70 to 86 percent is excreted in 24 hours. The remaining 13 to 30 percent is excreted very slowly, so slowly in fact that the analytical sensitivity is frequently insufficient to measure the activity in the sample aliquot as reference to Table 4.3 indicates. For this reason, as well as the relatively short duration of the experimental periods, the loss rate from this slowly turning over compartment (assumed by BASSETT et al. to be bone) is poorly determined by these experiments.

Uranium in Fecal Samples and in Blood. As stated above urinary excretion was almost the exclusive route or uranium body loss. Fecal samples from the first day on contained insufficient amounts to be measured by the analytical technique used, i.e. were less than from 0.1 to 0.3 percent dose, the sample limit being a function of sample size and aliquot size. Two blood samples were analyzed for uranium. Twenty-eight ml of blood drawn from patient 4 eight minutes after administration of 1.92 mg U contained 1.2 μg U. Assuming a total blood volume of 5000 ml (0.078 times body weight), it may be calculated that at this time only 11 percent of the dose remained in the circulating blood. A blood sample of 26 ml was withdrawn from patient 6 at 16 minutes after administration of 3.91 mg U and found to contain 2.7 μg of U[2]. Assuming a total blood volume of 4000 ml (0.078 times body weight), it was again found that about 11 percent of the dose was in the circulating blood.

Procedures Designed to Alter the Excretion Rate. Three experiments were performed. It was estimated that patient 4, 12 days post-administration, retained 535 micrograms of uranium. Inasmuch as hexavalent uranium complexes readily with citrate an attempt was made to mobilize the residual uranium by giving 2000 ml of a solution containing 9.2 g sodium citrate and 2.2 g citric acid by

2 Additional data not included in UR-37, available to the authors through the courtesy of Dr. W. F. BALE.

intravenous drip. This procedure did not increase the rate of urinary excretion of uranium. The concentration of citrate in the urine increased very little suggesting that the citrate was metabolized rather than excreted.

Patient 6 made himself available for further studies after conclusion of the initial experimental period. At 12 days post-dose he retained about 10 percent of the injected uranium. At this time he was rendered acidotic by giving 8 g of ammonium chloride daily for 3 days, 12 g daily on the fourth and fifth days. On the fifth day he was given a second intravenous dose of 3.17 mg hexavalent uranium. The dose of ammonium chloride was reduced to 4 mg a day and discontinued after the ninth day. Measurements of CO_2 content of the blood serum on the day 5, 6, 9 and 13 showed a steady increase up to 70 volumes percent. This procedure had the effect of slowing down the initial rate of urinary uranium excretion but at 15 days post-injection the total amount of uranium excreted was the same for the normal and the acidotic condition.

At the conclusion of these two tests it was calculated that the total amount retained was about 1.17 mg uranium. In an attempt to mobilize this uranium dietary calcium was reduced to 0.2 g per day and gradually increasing doses of dihydrotachysterol (AT-10) were simultaneously administered. The maximum AT-10 dosage was 7.5 mg a day continued for 18 days. The urinary excretion of calcium was increased three times, but urinary phosphorus and urinary uranium were unchanged.

The Mechanism of Excretion of Hexavalent Uranium. The authors interpret their result in terms of the mechanism of excretion suggested by Dounce (Voegtlin and Hodge, 1949, chap. 15) and depending on animal results. It is proposed that in the blood where there is a large excess of uranium, about 60 percent of U(VI) forms a complex of the type $UO_2(CO_3)_2$. The diffusible bicarbonate complex is filtered at the glomerulus. In the tubule water is reabsorbed as well as base and acid carbonate. The resulting acidification leads to the partial breakdown of the uranium bicarbonate complex permitting either uranium combination with the cell protein of the tubules which in excess may produce injury, or forming a temporary binding complex formation with organic acids or phosphate in which case it is excreted directly. In subject 6 the administration of ammonium chloride created a highly acid urine, reducing the available bicarbonate and favoring the temporary binding of uranium by the tubular cell protein thus slowing up the rate of excretion.

2. The Boston Intravenous Injection Experiment

(Struxness, Luessenhop, Bernard and Gallimore, 1956; Luessenhop, Gallimore, Sweet, Struxness and Robinson, 1958; Bernard, Muir, and Royster, 1956)

A collaborative experiment involving the intravenous injection of uranium was performed by members of the Health Physics Division of the Oak Ridge National Laboratory and Dr. William H. Sweet of the Massachusetts General and the Veterans Administration Hospitals in Boston. Although the primary justification for these experimental injections of uranium in man was the possible value in neutron capture therapy (Sweet and Javid, 1952), the Oak Ridge group was interested in the study of the tissue distribution and excretion of uranium as it applied to calculation of protection criteria for uranium workers. The preparation of the dose and the analyses for uranium in excreta and tissues were done in Oak Ridge.

Table 4.4. Patient data and urinary excretion of intravenous uranium. (Data from LUESSENHOP et al., 1958; STRUXNESS et al., 1956)

Patient	Age yr	Sex	wt. kg	Dose[a] μg/kg	Excretion percent dose		Survival post-injection days
					24 h	total	
1	26	m	55.9	99	59.4	69	2.5
2	47	m	57.4	103	78.0	92	74
3	34	m	60.0	72	83.8	98	566
4	63	f	67.7	165	77.2	85	136
5	39	m	55.9	283	66.5	85	139
6	60	m	56.7[b]	907	49.1	63	18
7	—	—	71.8[b]	573[b]	20.0	68[b]	228[b]
8	—	m	63.2[b]	700[b]	16.9	57[b]	21

[a] These values differ slightly from those listed by LUESSENHOP et al. (1958). We have assumed that the body weights and total dose administered as listed by them are correct. If these numbers are used, the values listed above are obtained.
[b] These data were kindly provided by Dr. S. R. BERNARD.

The eight patients forming the subject group were in the terminal phase of severe central nervous system disease. The ages were 26 to 63 years. At the time of injection all patients except one were in coma and were receiving the usual hospital care consisting of frequent turning, skin care, gastric tube feedings, catheter drainage, and tracheal suction when necessary.

Uranyl nitrate solutions, enriched in ^{235}U and ^{234}U, were prepared and administered intravenously to the first six patients. Patients 7 and 8 were similarly injected with a solution of tetravalent uranium as UCl_4. The amounts received by each patient are listed in Table 4.4.

Collection and Analysis of Samples. One to 3 ml blood were taken by phlebotomy in the arm not used for the uranium injection at the rate of one sample each hour for the first 24 hours, one sample each 12 hours for several weeks, one sample each 24 hours until transfer from the hospital. Bone biopsy specimen to determine initial uptake were taken from the anterior tibia with a 0.5 in. trephine. Urine specimes were collected hourly from indwelling catheters for the first day post-injection, and at 12-hour intervals for 2 to 4 weeks. Thereafter 12-hour duration spot collections were made at 1 to 4 week intervals. Total collection of feces was made for the duration of the patients' hospital stay. Autopsy samples were obtained on patients 1, 2, 3, 5, 6 and 8. Twenty-one different organs or tissues were collected. Triplicate aliquots of each urine specimen were wet ashed, evaporated, made up to volume and electroplated for alpha particle counting. Blood samples were analyzed in generally the same way as urine samples. Tissue samples less than 2 g wet weight were dry ashed at 600° C for 24 hours, dissolved in 0.1 N nitric acid, and the entire sample electroplated and counted in an alpha particle counter. Feces, bone, and tissue samples greater than 2 g wet weight were dry ashed at 600° C for 24 hours. The resulting ash was weighed and uranium analysis was performed using the aluminum nitrate/diethyl ether extraction procedure with evaporation on an stainless steel planchet for alpha counting.

Urinary and Fecal Excretion. Table 4.4 presents data on the patients' administered dose, the percent dose excreted in 24 hours, and total urinary excretion. The largest amount measured in a single fecal sample was 0.03 percent dose

Table 4.5. Percent of injected dose per standard man organ or tissue estimated from autopsy results. (Taken from BERNARD and STRUXNESS, 1957: their Table IV abridged)

Patient no.	1	6	2	5	3	8[a]
Bone	10.0	4.9	1.4	0.6	1.3	14.4
Kidney	16.6	7.2	0.7	1.2	0.4	1.1
All other tissues and organs	8.4	5.9	1.9	1.2	0.3	15.7
Bone tibia biopsies 0–48 hours average % dose per 7000 g	9.1	6.4	6.5	3.9	4.2	1.3

[a] This patient was injected with uranium tetrachloride.

and the usual sample contained much less. The authors find the rates of urinary excretion of uranium are best represented as power functions and the best line through all the data describes the equation: $\%/\text{hr} = 34.3\, t^{-1.5}$. The power function fails at times shorter than 5 hours post-injection.

Loss from the Blood. The large number of blood samples collected in this experiment permitted the construction of detailed blood disappearance curves. It was found that at six minutes the blood contained about 33 percent of the injected dose and that 99 percent had left the blood at 20 hours.

The Minimum Single Dose Resulting in Kidney Damage. LUESSENHOP, the physician who reviewed the battery of clinical tests employed to evaluate kidney function in these patients, finds that the minimal dose to produce catalasuria and albuminuria is of the order of 0.1 mg per kg body weight (LUESSENHOP et al., 1958). When the patients came at autopsy the acute tubular damage was no longer visible and the histopathological findings were said to be comparable to those often seen in terminal patients.

Tissue Distribution of Uranium. The autopsy results on patients receiving hexavalent uranium are summarily listed in Table 4.5. The original data including all measurements show that apart from the bone and kidney there are no significant concentrations of uranium in any of the 21 tissues and organs sampled. The autopsy on subject 8 who received tetravalent uranium shows 9.2 percent in the liver and 5.6 percent in the spleen.

3. Experiments Using Uranium Injections to Evaluate Skeletal Metabolic Disorders

(TEREPKA et al., 1964)

TEREPKA et al. injected intravenously 30 micrograms hexavalent natural uranium per kg in three control patients without clinical, laboratory or x-ray evidence of bone disease and seven patients with a variety of bone disorders. The objective was to develop a means of evaluating the metabolic status of the skeleton. The rationale depended on earlier animal evidence that about 70 percent of a single injection of soluble uranium was rapidly cleared via urinary excretion and that of the amount retained "virtually all could be accounted for in bone. Both *in vivo* and *in vitro* studies indicated that the uranyl ion forms a stable linkage, probably with phosphate groups on the surface of hydroxyapatite crystals". The amounts of uranium excreted in the urine during the early post-injection period are listed in Table 4.6. The results are interpreted by TEREPKA et al. as showing that uranium retention is a function of the relative quantity of available or exchangeable bone. Experiments (not included in the table) which

Table 4.6. Urinary excretion after intravenous injection of uranyl nitrate. (Taken from TEREPKA et al., 1964)

Patient	age	sex	condition	Urinary excretion in % dose/24 h day 1	day 2	day 3	day 4	day 5	day 6
CM	53	m	normal	61.5	3.8	1.5	0.7	0.6	0.4
RW	53	m	normal	64.3	3.1	1.5	0.6	0.5	—
JP	72	m	normal	61.5	6.1	2.1	1.3	0.4	0.4
CW	53	m	Paget's disease	16.0	5.6	1.1	0.6	0.6	0.6
CW	53	m	Paget's disease[a]	18.6	7.2	1.5	—	—	—
CS	58	m	hyperparathyroid	31.0	—	—	—	—	—
AK	71	f	hyperparathyroid	63.3	6.9	—	—	—	—
MC	60	f	osteomalacia	34.2	5.0	2.6	2.1	1.3	1.1
JL	62	m	osteomalacia	32.6	7.1	—	—	—	—
MC	74	f	osteoporosis	59.5	6.3	1.9	1.2	0.9	0.5
AH	81	f	osteoporosis	60.1	6.1	2.1	0.9	0.8	0.6

[a] After treatment with Prednisone (5 mg twice daily).

unsuccessfully attempted to mobilize stored uranium by administration of parathyroid extract are believed to suggest that either (1) the uranium was deposited in areas of bone not affected by the resorbtive process or (2) that there is a cation specificity to the effects of parathyroid hormone. The authors held forth the possibility that "the marked difference between bone uptake and release of uranium and calcium may help to distinguish the relative contribution of bone accretion and bone resorption to the development of metabolic bone disease in man".

4. Interpretation of the Intravenous Experiments

Excretion. The three sets of experiments show excellent agreement where comparisons are possible. The 24 hour urinary excretion of hexavalent uranium yields an average of all values (except patient 6 in the Boston experiment) of 72.1 ± 8.3 percent dose. There is no apparent correlation with patient age or sex. The Boston data appear to indicate a reduced rate of excretion with increase in dose size, but no such relationship appears in the Rochester data. It is perhaps somewhat surprising that the two sets of data do not show greater differences. The Rochester patients were ambulatory in the metabolism ward and were not seriously ill, whereas the patients in the Boston study were with one exception in coma at the time of the injection.

The Minimal Dose to Produce Kidney Injury. Here again, the two sets of data are in reasonable agreement. The Rochester group found that transient injury was produced at their highest dose of 70.9 μg per kilogram and from the Boston experiment we learn that single doses of about 100 μg per kilogram produced catalasuria and proteinuria. From his consideration of the Boston data, and deriving a scaling factor from published findings on the rabbit, LUESSENHOP estimates (LUESSENHOP et al., 1958) the lethal intravenous dose for man at about 1 mg uranium per kilogram.

Tissue Distribution of Uranium. Although Terepka's data indicate that the metabolic status of the skeleton has a marked influence on the urinary excretion of uranium and by inference that the skeleton of normal man is an important repository for uranium, they do not include specific tissue distribution values. The Boston data on tissue distribution, in a quite general way, confirmed the

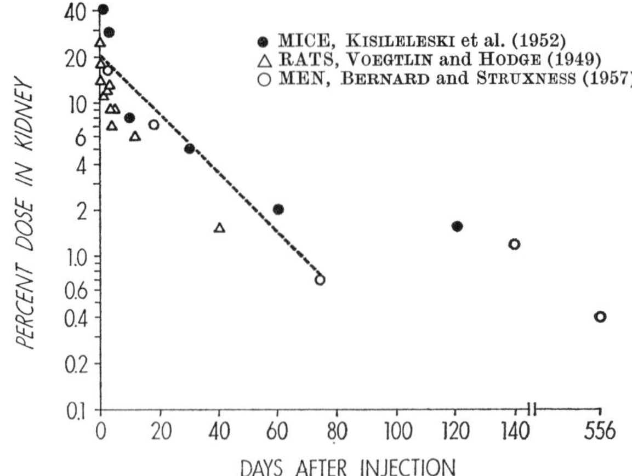

Fig. 4.2. Comparison of the uranium content of the kidney in several mammalian species after a single injection of uranium

pattern which might have been expected from the animal data: the intravenous hexavalent uranium which is not rapidly excreted deposits in the kidney and in the bone; when tetravalent uranium was introduced into the blood stream deposition in the kidney was reduced, and appreciable retention was observed in the liver and spleen. For the prediction of steady-state organ burdens after chronic occupational exposure by use of single exposure data, it is necessary to know the rate at which uranium is lost by the organ. Manipulation of the Boston data to this end introduces, in addition to the uncertainty caused by one subject per time point, uncertainties associated with the relatively rapid rate of introduction of massive doses of uranium and the parlous clinical condition of the subjects. It is well known that bed patients display a negative calcium balance and impaired circulation. Although these uncertainties are recognized, the Boston data are unique and afford the only opportunity to compare loss rates of uranium from the bone and kidney of man with similar animal data. BERNARD (1948) working with the Boston data, estimated biological half-times of 300 days for both bone and kidney. The value of 300 days for bone does not represent any change from that cited in N.B.S. Handbook 52 (1953) and based on animal data. On the other hand, acceptance of the 300 day half-life for the kidney would represent a substantial departure from the 15-day half-life currently in use by the NCRP and ICRP. From close consideration of the evidence it would appear that the existing data on animals and on man show no real differences and that the issue is one of interpretation rather than a conflict in the experimental findings. Fig. 4.2 is presented to support this view.

B. Oral Administration of Uranium

1. Use as a Therapeutic Agent

It is reported that oral administration of uranyl nitrate in small doses was used as early as 1851 to treat diabetes. Dr. SAMUEL WEST stimulated renewed medical interest in this form of treatment by two papers (1895, 1896) in which he described the alleviation of symptoms in 8 cases of diabetes mellitus treated with uranium. The motivation for using uranium calls to mind Mallory's explanation[3] of why he proposed to climb Mt. Everest. As West explains in his first paper "Some time ago my colleague at St. Bartholomew's Hospital, Dr. W. J. RUSSELL, the lecturer in chemistry, placed in my hand a double chloride of quinine and uranium, and suggested that I should try its action on disease". Early in his use of the drug West decided that uranyl nitrate was as effective as the double chloride of uranium and quinine. The regimen that he adopted was to begin by administering small doses, 30–60 mg U as uranyl nitrate in water, 2 or 3 times per day after meals, and as tolerance developed to increase this amount until beneficial results were observed. He notes that doses as high as 1.8 gram U per day were tolerated by the patient without gastric distress. He found in most cases that the sugar in the urine, the volume of urine, and the patients' thirst were reduced. When treatment was stopped the patients reverted to their pre-treatment status.

The findings of WEST were in general supported by DUNCAN (1897), BOND (1898), TYLECOTE (1904) and by WILCOX (1915). BOND's Case 9, a man of 62 years of age weighing 118 kg, is particularly notable because of the dose size. He was started at a dose of 185 mg U, 3 times per day and worked up to 925 mg U, 3 times per day, a dosage rate that was continued for one year. A sample of his urine taken during the latter part of the year was negative for uranium.

Experiments on animals had demonstrated that intravenously injected uranium in sufficiently high amounts could produce severe kidney injury. The physicians who made use of uranium as a therapeutic agent were aware of this work and their reports stress that the urine was examined for evidence of kidney injury and that none was found.

Medical opinion was not uniformly in support of this method of treatment [see, for example, the discussion following DUNCAN's paper (1897)] and although uranium was listed as a materia medica in the early 1900's, by the mid-thirties SOLLMANN (1936) was to advise that "The results are too indefinite to justify the further employment of so dangerous an agent". Quite apart from the merits of the treatment, the fact that these studies were made provides the interested toxicologist with data on a group of human subjects with a chronic exposure to oral uranium.

2. Experimental Ingestion by a Volunteer Subject

BUTTERWORTH (1955) published a paper examining the usefulness of analyzing the urine of workers for uranium as an exposure control measure. In this paper he included an account of a volunteer who ingested 1 gram uranyl nitrate (0.47 gram U) in 200 ml water in order to obtain data on uranium hazard evaluation. The results were dramatic. The person experienced rather violent vomiting, diarrhea, and slight albuminuria with a peak uranium urinary concentration of 8 mg U per liter (2 specimens of 30 ml). During the first week he excreted in his urine 2.5 mg U. Assuming a renal excretion of 66 percent of the uranium absorbed from the gut, it was estimated by the author that the subject absorbed at least 1 percent of the dose. Within 24 hours the subject's clinical recovery was complete. This experimental result created attention because the gut absorption factor for uranium then accepted by the ICRP and NCRP committees was 0.01 percent (based on rat data) and not 1 percent.

Regardless of the reservations which might properly be attached to an experimental result associated with such far-reaching gastrointestinal disturbances, it did, at that time, constitute the only uranium absorption data on man. As a

3 "Because it's there."

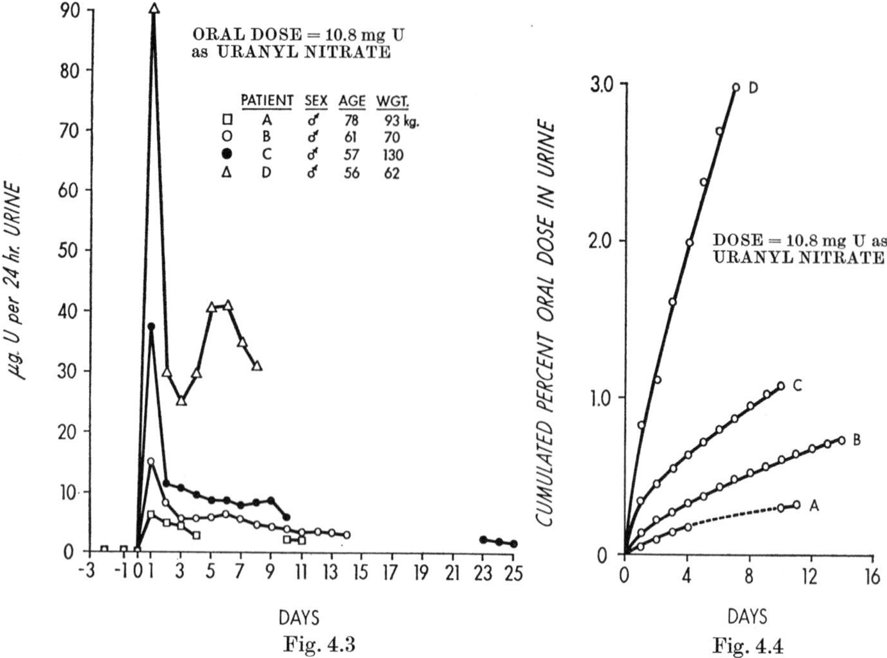

Fig. 4.3. The rate of urinary excretion after oral administration of uranyl nitrate to 4 human subjects

Fig. 4.4. The cumulative amount of uranium excreted in the urine after oral administration of uranyl nitrate to 4 human subjects

consequence of this finding, supported by experimental studies (Fish et al., 1960) on 10 dogs who received a single oral dose of UO_2F_2 in water and yielded an average absorption factor of 1.55 percent, the ICRP in Publication 6 (1964) increased the absorption factor from 1×10^{-4} to 1×10^{-2} for uranium.

3. Hospital Study on Oral Absorption of Uranyl Nitrate

Hursh et al. (1969) conducted an oral absorption study on four hospital patients. The patient selection criteria were (1) no known gastrointestinal disorder, (2) normal kidney function, and (3) availability for a minimum of 7 days. The selected subjects were transferred to the Clinical Research Unit for the testing period. Each subject received prior to his breakfast an oral dose of 10.8 mg U as uranyl nitrate hexahydrate dissolved in 100 ml of Coca-Cola. Urine and fecal samples were collected and analyzed for uranium. A pH measurement was made on the urine samples soon after collection and all samples were examined for protein. The patients reported no subjective symptoms and no protein was found in any of the urine samples. The results of the uranium determinations appear in Figs. 4.3 and 4.4. In the original experimental plan, only urine samples were to be collected[4]. The rationale behind this procedure was to rely on data from the

4 It might appear that the straightforward way of determining the absorbed fraction would be to collect all the feces and to subtract the total amount of uranium found therein from the administered dose. However, it must be remembered that the experimenters had good reason to anticipate that the absorbed fraction would be of the order of 1 percent or less and that they were aware of the variety of experimental and analytic errors which would conspire to defeat any attempt to determine the administered dose and the fecal uranium content to the precision necessary to determine such a small difference.

Table 4.7. Percent dose in feces on day as designated. (Taken from HURSH et al., 1969)

Patient	−2	−1	0	1	2	3	4	5	6	7	8	9	Cumulative
B[a]	0.002	0.004										0.22	
C			(—	—	86)	(—	12)	(—	—	—	—	1.7)	99.7
D[b]			(—	3.0)	(—	36.1)	(15.7)	(—	17.6)	(—	19.4)		91.8

[a] Spot collections only were made for patient B. Total collections were made for patients C and D; the collection period is indicated by brackets.
[b] One sample between 0 and 3 days missed.

intravenous experiments on man which indicated that a reasonably constant fraction, about 70 percent, of the absorbed uranium would be promptly excreted in the urine during the first 24 hours. Measurements of urinary uranium would therefore permit the ready calculation of the total amount absorbed. After the data from the first patient (patient A, Fig. 4.3) became available, it was apparent that, contrary to expectation, appreciable amounts of uranium continued to be excreted in the urine for a period of many days[5]. In an effort to assist in the interpretation of this finding, fecal samples were collected on the remaining three subjects. The results of the analyses of these samples are presented in Table 4.7.

It may be noted that the cumulative urinary loss had not reached a plateau for any subject at the end of the experimental period and that, therefore, the prospective total urinary loss can be estimated to be in excess of 0.3, 0.7, 1.1 and 3.0 percent dose for patients A to D respectively. Inasmuch as about one-third of the amount absorbed is taken up by the skeleton and the kidney, these numbers should be multiplied by 1.5 to yield minimum estimates of the uranium absorbed.

It is suggested by the authors that if the prolonged urinary excretion has the corollary of a prolonged period of uranium absorption, the radiological hazard to the gut cannot properly be calculated in terms of the more rapid passage of ingested food as is done in the current ICRP gastrointestinal model.

4. Interpretation of the Oral Experiments

The reported results of oral dosing of diabetic patients include three claims that need to be critically evaluated: that despite the large doses of uranium (1) gastrointestinal disturbances did not occur, (2) no uranium appeared in the urine, (3) no kidney injury occurred. None of these claims are supported by the scanty modern experimental data cited above in this section. Of the series of 22 cases total presented by WEST (1895, 1896), DUNCAN (1897), and BOND (1898), one patient received single doses of 0.91 g, two patients 0.61 g and three patients 0.46 g U repeated three times daily and continued for weeks, months, or as long as one year[6], whereas BUTTERWORTH's volunteer received a single dose of 0.47 g U and displayed severe gastrointestinal symptoms. Two explanations suggest themselves. The medical practice of beginning with small doses and increasing the dose until a beneficial effect was produced may have created a sort of habituation so that large doses could be tolerated with no gastric symptoms. Alternatively it may be supposed that the volunteer subject was susceptible to this

5 This pattern was not followed by Butterworth's subject. In that case 88% of the total urinary loss for the first week occurred on the first day. It may be assumed that the physiological processes of vomiting and diarrhea made a clean sweep of the gastrointestinal tract.
6 WILCOX's (1917) larger series of 84 treated patients are all of such low dosage as not to come into question.

manifestation of uranium toxicity and that if an equally large number of normal volunteers had been tested, individuals would be found displaying a tolerance comparable with that of the patient group. It should be noted that the "conditioned" patients were not uniformly free of gastrointestinal disturbances. Two of West's 9 cases (at 0.1 g and 0.15 g single dose) and one of Duncan's 5 cases (at 0.46 g single dose) reported dyspepsia and loose bowels.

Either explanation must take the conditions of administration into account. The patients received their nostrums directly after meals in the hopeful, trusting aura of the patient-physician relationship. The volunteer ingested his dose an hour or so after breakfast and probably appreciated he was drinking a toxic substance of no direct benefit to his health. Any choice between these two explanations must be intuitive rather than rational in the absence of adequate experimental data.

The second claim, i.e. that the urine contained no uranium, can be dealt with more satisfactorily. Modern data will be used to predict an upper limit for the daily uranium urinary excretion of Bond's Case 9 who received a total daily oral dose of 2.7 grams, as follows. Assuming a gut absorption factor of 0.05 and that 0.70 of the uranium passing into the circulation is excreted per day we might expect 95 mg uranium in one day's urine. Bond (1898) states that the daily urinary volume of this patient was about 76 ounces (2.2 liters). The concentration of uranium would therefore be about 4.3×10^{-5} grams per ml urine. It is apparent that an analytical method that could "readily detect 1 part in 2000" (Bond's comment on the chemical method for uranium as used at that time) would fail to detect the concentration of uranium predicted for Bond's exceptionally high-dose patient.

While the above calculation may serve to press the point that the failure to detect uranium in the urine was caused by insensitivity of the then current analytical methods it is almost certainly an overestimate.

The postulated 135 mg U per day (or $135/118 = 1.14$ mg/kg) passing into the circulation can be compared with the kidney injury threshold of about 0.1 mg/kg derived from the intravenous experimental studies and the assumed systemic intake of about 5 mg absorbed from the gut by Butterworth's volunteer subject.

Even if a more likely 1 percent absorption factor is assumed the contrast remains between the severe kidney injury to be expected based on data from the intravenous and oral experiments and the reported absence of albuminuria in the substantial number of medically treated patients who ingested amounts equal to or greater than 0.9 grams uranium per day for extended periods. This is yet more noteworthy in that, to the extent that the diabetic patients as a group had a chronically low blood bicarbonate reserve, the kidneys would be expected to fix a greater fraction of the available uranium and to suffer a more severe injury. It is plausible that the practice of gradually increasing the dose produced a tolerance in the medically treated subjects. This could be achieved either by a mechanism which reduced absorption from the gut or which in some way protected the kidney.

Both decreased absorption (Cloetta, 1906; Joachimoglu, 1916) and increased excretion (Hausmann, 1906) have been invoked to explain the habituation of the arsenic eaters of the Styrian mountains who are able to ingest once or twice weekly up to 0.3 g of arsenic trioxide, whereas 0.1 to 0.3 g is usually fatal. Perhaps more immediately to the point are the animal experiments demonstrating that a conditioning series of sublethal adminstrations of uranium will protect against the killing effect of a single large dose. Haven (Voegtlin and Hodge, 1949, chap. 12) finds that the kidneys of tolerant rats retain less uranium and concludes that elevated levels of citric acid in the kidney tubule may account for acquired uranium tolerance. Earlier experimenters (Hunter, 1928; MacNider, 1929) studied the tolerant kidney histologically and concluded that normal tubular epithelial cells were destroyed by the conditioning doses of uranium and replaced by a resistant epithelial cell type.

An important aspect of the animal experiments is that in order to produce tolerance, it was found necessary that the conditioning doses be high enough to produce some kidney injury. If the same rule holds for man, it may be inferred that the demonstration of tolerance in animals and the likelihood of its occurrence in man under some conditions does not warrant that it will be an ameliorating factor in the chronic exposure of uranium workers, where by intent the daily dose is regulated so that even transient kidney injury will not occur.

C. Inhalation of Uranium

1. Experimental Clearance of Uranium Dust from the Human Body

(Harris, 1961)

Examination of published reports have revealed only one experiment involving the planned inhalation of uranium compounds by man. A single experimental subject was exposed for 17 short periods to inhalation of first UO_3 (12 periods), and later UF_4 (5 periods) over a total elapsed time of 24 days. The design of the experiment and the interpretation of the results were closely related to the reason for performing the experiment. Harris starts from the position "Our experience in the uranium industry has been that, although workers appear to be exposed to dust concentrations which are very much higher than those which experiment had determined should produce illness, such illness has not been seen". He argues that a possible cause might be the characteristically larger average particle size of industrial dust as compared to that used for the animal experiments relating exposure and injury. Such a larger particle size would favor deposition in the naso-pharyngeal region and the other ciliated airways, and consequently a smaller fraction would be deposited in the parenchyma of the lung. The uranium dust deposited in the ciliated passages would be escalated, swallowed and passed into the gastrointestinal tract. Because even so-called "soluble" uranium is poorly absorbed from the gut only a few percent would find its way into the systemic circulation and the remainder would be excreted in the feces. Harris and his colleagues at the Health and Safety Laboratory, New York Operations Office of the AEC, pursued this interpretation by developing several dust sampling devices designed to separate dust into the two fractions, *viz.* that, which because of its inertial and aerodynamic properties, would deposit in the upper respiratory tract and that which would deposit in the lung parenchyma. In the report (Harris, 1961) it is stated that a preliminary check on several industrial processes in a uranium plant using one of these dust sampling devices indicated that 95–99 percent of the total dust concentration would be removed by the upper respiratory system. The objective of the experiment to be described was to compare the prediction of the device with human data in an actual inhalation experiment.

Experimental Design. The schematic drawing of the arrangement for inhalation exposure is reproduced from the original paper and appears as Fig. 4.5. The subject, using a valved face mask, inhaled UO_3 for 12 brief periods spaced in time as indicated in Table 4.8. About 7 days after the last UO_3 exposure another series was begun in which the subject inhaled UF_4 for five periods in the space of 3 days.

Table 4.8. Estimated amounts of uranium deposited in the lung parenchyma and in the upper respiratory tract (URT)

Time in days[a]	Lung μg U	URT μg U	Ratio URT / URT+lung
1.68	481 (0.20)	910 (0.38)	0.65
3.50	994	2086	0.68
5.30	607	1312	0.68
6.60	1702	2200	0.56
7.15	443	938	0.68
7.60	533	938	0.64
12.45	64	192	0.75
12.60	640	1280	0.67
12.70	1252	—	—
13.30	439	938	0.68
13.54	571	938	0.62
13.68	425	670	0.61
20.30	291	2662	0.90
20.60	117	1757	0.94
21.60	304	2068	0.87
22.50	461	1786	0.80
23.55	217	1998	0.90

[a] Read by present authors from Fig. 11. Harris (1961). To be construed as fractional days only; clock time was not given in the reference.

The device labelled "miniature cyclone" in Fig. 4.5 and the back-up fiberglass filter yielded estimates of the fractionation of the dust between the upper respiratory tract and the lung parenchyma, respectively, for each exposure period. The amount of uranium dust collected in the cyclone normalized to the subject's total air intake as measured by a pneumotachograph was taken as the amount deposited in the upper respiratory tract for the period under consideration. The amount of uranium dust collected on the filter, similarly normalized to the subject's air intake, minus the amount of uranium collected on the exhalation glass fiber filter was assigned to the deep lung. The results of these determinations transcribed from figures in the paper appear in Table 4.8.

Total urinary and fecal collections were made for a period of 32 days from the beginning of the experiment and all samples were analyzed for uranium content.

Results. Because of the temporal spacing of the exposure periods and the rapid clearance displayed in the urinary excretion rate data as reproduced in Fig. 4.6, Harris felt justified in combining data from several single and several more of less contemporaneous multiple exposures to estimate systemic uranium loss from a single exposure incident. The loss rates that he obtains are replotted in Fig. 4.7. The urinary excretion rate decreases with a half-time of about 7 hours for the period 3 to 20 hours post-exposure and then more slowly with a half-time of about 4 days for the period 20 to 200 hours postexposure. The relative excretion rates from one curve to the other are not significant inasmuch as they were derived from different sets of exposure data.

The summed urinary excretion from day 1 to day 33 was 3.7 mg uranium whereas the sum of the amounts assigned to the lower lung was 9.5 mg. The author assumes that the remaining 5.8 mg or 61 percent of the dose remains in the tissues.

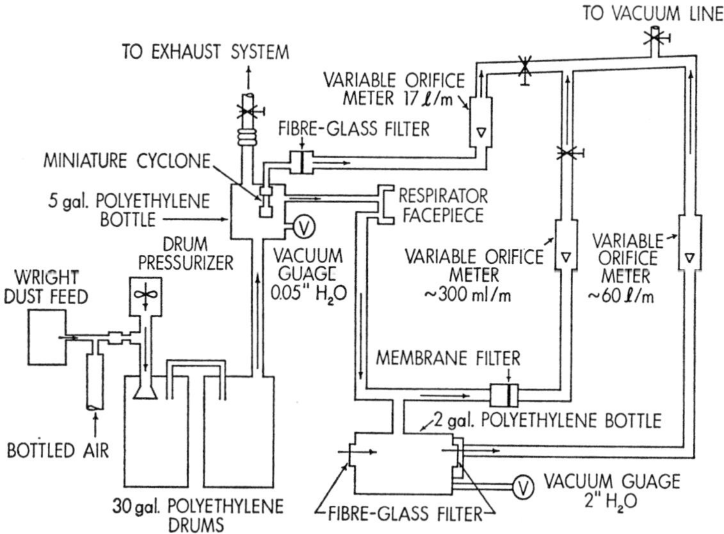

Fig. 4.5. Schematic drawing of the experimental arrangement used by HARRIS for inhalation exposures. (Taken from HARRIS, 1961)

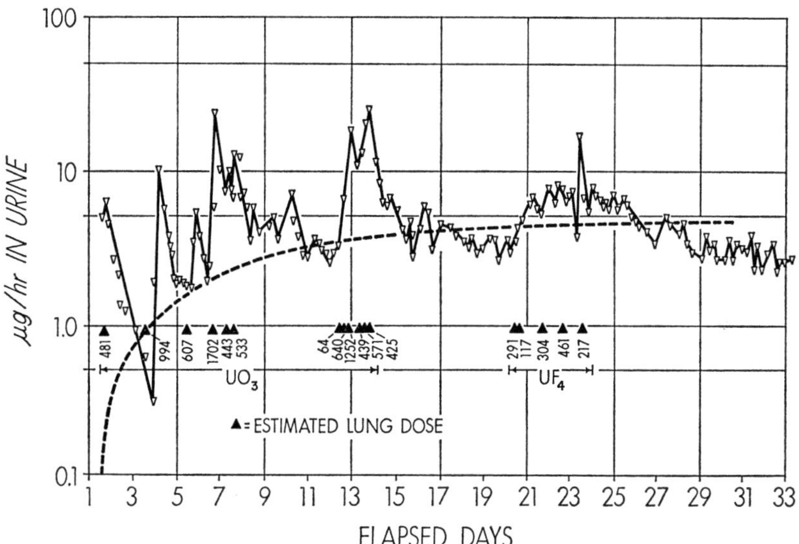

Fig. 4.6. The urinary excretion of a human subject after multiple inhalation exposures. The numbers associated with the triangles are the estimated pulmonary lung depositions in μg U

Similarly the summed fecal excretion for the period day 1 to day 30 was 8 mg, whereas the total amount assigned to the upper respiratory tract was 22.6 mg, indicating a "deficit" of 14.6 mg or 64 percent. The detail of the fecal excretion pattern shows no apparent holdup in the gut. For example, after the last exposure on day 23 the uranium content in the fecal samples decreased rapidly so that 7 days later it was down by a factor of 100. Because the deficit does not

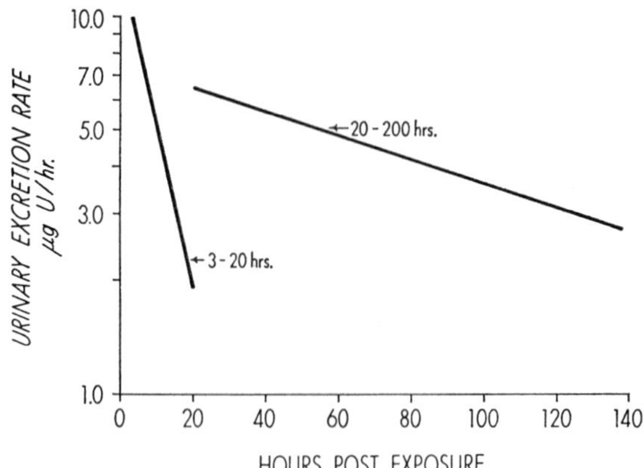

Fig. 4.7. Urinary excretion rates after a "single" inhalation exposure (see text). The half-time for the rapid decrease (applicable to 3–20 hours post-exposure) is about 7 hours. The half-time for the slower decrease (applicable to 20–200 hours post-exposure) is about 4 days. (Taken from Harris, 1961)

appear to be in the gut, the question as to its whereabouts is perplexing. The author considers this question but does not resolve it. He states that in the process of removing uranium from the cyclone some material on the outside of the device may have been included and that, therefore, the estimates of the URT deposit may be somewhat high. He characterizes the attempt to devise an air sampling device which would represent the upper and lower lung fractions as only moderately successful.

2. Interpretation of the Inhalation Experiment

It is regrettable that this rather complicated and potentially very useful experiment has not been repeated with other subjects and more extensive data gathering. The concept of respirable dust sampling is an old one and a variety of devices have been developed (Lippmann, 1970). The United States Atomic Energy Commission defined respirable dust (Lippmann, 1970) in terms of an acceptance curve which is plotted in Fig. 4.8 together with a similar curve adopted by the British Medical Research Council (MRC). The separation characteristics of the HASLcyclone used in the present experiment are also presented in Fig. 4.8 and show a reassuring correspondence.

Although the fractional separation in the air sampler is not given as such in Harris's account, the deposition in the upper respiratory tract and in the pulmonary region is furnished (see Table 4.8). The ratio of the URT deposited divided by the total deposited is entered in the last column. The average value of this ratio for the UO_3 exposures is 0.65 ± 0.06 and for the UF_4 exposures is 0.88 ± 0.07, indicating that the average UF_4 particle size was significantly greater than that for the UO_3 aerosol. On the basis of available knowledge from animal experiments, UO_3 would be regarded as a relatively soluble compound and UF_4 as relatively insoluble.

In the text the author treats the excretion data without regard to the differences between the two compounds inhaled. This position does not seem to be fully warranted by the data themselves. Without exception after exposure to UO_3, there is a prompt increase in urinary excretion of uranium, whereas this pattern is less striking after the UF_4 exposures. Taken together with the circum-

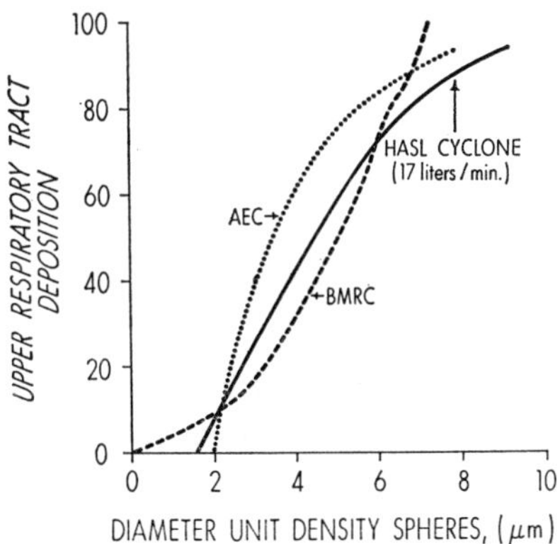

Fig. 4.8. The HASL respirable dust sambler performance is compared with the acceptance curves as defined by the USAEC and the MRC. (The curves are drawn from data in HARRIS, 1961 and LIPPMANN, 1970)

stance that 85 percent of the uranium deposited in the deep lung was in the form of UO_3 and that the early rapid phase of elimination depended entirely on UO_3 data, the urinary excretion rates determined by the author should perhaps be regarded as pertinent to UO_3 exposures and not necessarily appropriate for UF_4. The form in which the excretion data is presented and the multiple exposure format of the experiment defeat any attempt to assign different fractions of the body burden to the postulated two phases of urinary excretion.

Indeed attempts to interpret this experiment in terms of mechanisms are very frustrating. We are presented with plausible data for what went into the respiratory system and data for the excretory output. For radiation toxicity purposes, it is necessary to know the intermediate fate of the inhaled aerosol. Other information may be invoked to guess at the pattern of behavior. The data at hand show that UO_3 leaves the lung rapidly inasmuch as half-time components of 7 hours and 4 days have been measured for loss rates from the blood and loss from the lung can hardly be less rapid. If no slower undetected lung release rate exists, the 16 days lapse after the major exposures to uranium would ensure that most of it had been cleared by 32 days. Assuming that the uranium mobilized from the lung into the systemic circulation is either promptly excreted in the urine or deposited in the bone and the kidney, it might be anticipated that a complete accounting could be made in terms of the excretory[7] and tissue disposal routes. HARRIS performs such an accounting and assigns 61 percent of the pulmonary deposited uranium to the bone and kidney tissues. This respective distribution is almost the reverse of that found for intravenously injected uranium in man. A similar reversal from anticipated form is noted in the UO_3 inhalation experiments in dogs (MORROW et al., 1971) which show systemic uranium distributed in the ratio of 60 percent in tissue and 40 percent excreted in urine.

The deficit indicated by the sum of the URT deposited minus the uranium recovered in the feces may be partially explained as follows. Examination of deposition fraction versus particle size graphs such as that published by the ICRP Lung Task Group (1966) shows that the major site of deposition for the

7 A more detailed description would need to recognize that a part of the urinary loss would be accountable to a reduction in kidney burden. This may indeed be the principal component of urinary loss during the last days of the experiment.

URT is the Naso-Pharyngeal area. The unciliated nares and that ciliated portion of the passage which beats towards the nostrils require discharge of the material either by sniffing or by blowing out the nose. In an experiment such as that reported here an important fraction may be lost to the balance accounting.

The modest yield of information from this experiment will be supplemented in the section to follow by the data from industrial exposures, the major source of data on the inhalation of uranium by man.

III. Occupational Exposure to Uranium

A. Introductory Remarks

In all toxicological work, quantitative assessments based on occupational experience are weakened by the inaccuracies involved in the data, especially in the figures relating to personal occupational exposure. In general it is notoriously difficult to obtain a figure which represents a man's exposure history over several weeks or months. In the present context, however, this imprecision cannot be regarded as obviously greater than that necessarily involved in the interpretation of human injection experiments, like those conducted earlier by Bassett et al. (1948) and later by Bernard and Struxness (1957). In the second of these investigations, for example, it was assumed that six terminal patients would show physiological responses not unlike those of ordinary men under conditions of ordinary exposure. There was little point in questioning this assumption, as it could not be tested. In the event both the humans and the exposure showed marked deviations from the normal. The humans were all diseased; five were comatose and catheterized; three were given intravenous saline, and (incidentally) two were sixty years of age or over. The doses were between 6 and 70 percent of the estimated LD_{50}, but the dose rates were inevitably some thousands of times greater than those which might be encountered from inhalation under average industrial conditions. These considerations do not invalidate the results of the human injection experiments; they do, however, engender scepticism and point to the desirability of looking elsewhere for information concerning the reaction of the human organism to environmental uranium.

During the last 25 years large numbers of men have been occupationally exposed under medical supervision to various forms of natural uranium. Within the industry the consensus has emerged, and remained substantially unshaken, that natural uranium is much less toxic to humans than was formerly supposed. Admittedly, some transient injury has been sustained by a few workers exposed to acute levels under accidental conditions. There has been no evidence, however, among workers regularly exposed to any uranium compound of chronic toxicity which could reasonably be attributed to uranium. This view, which was tentatively expressed by Harris in 1958, was confirmed by Scott et al. in 1970. Harris based his assessment on a comparative study which included 34 employees who had been occupationally exposed to uranium. From the available evidence he was able to show that whereas employees who had been exposed to between 1 and 10 times the $(MPC)_a$ of lead or mercury showed symptoms of overexposure, those who had been exposed to between 40 and 100 times the $(MPC)_a$ of uranium did not. The recent report by Scott et al. was based on a study of 4500 employees with a total of about 14000 man-years of association with uranium processing. The enquiry showed that the mortality rate among these workers was more favorable than that among a population of non-uranium workers employed at the same facilities during the same period.

Occupational hygiene data are very relevant to the task of setting $(MPC)_a$ values. Both the International Commission of Radiological Protection (ICRP) and the Industrial Hygiene Foundation of America (IHFA) have emphasized the value of information based on studies on humans who have been exposed to a particular substance under working conditions (ICRP, 1959; IHFA, 1969; STOKINGER, 1969). It is significant that whereas the $(MPC)_a$ values used earlier were based on the results of a long series of animal experiments and on the results of some human experiments, acceptance of the $(MPC)_a$ value now in use rests largely on the accumulated experience from 25 years' industrial experience. The purpose of this section is to select some of the published data, to identify the most important variables and to extract any useful generalizations.

B. General Industrial Exposure

The information from industrial exposure to uranium includes data involving several chemical forms which are metabolized at different rates. The densities of these compounds lie within the range 2.8–10.9 g/cm³ and the solubilities in water at 25° C within the range 1–400000 mg U/l. Owing to the nature of the work in some plants, the air-borne dust is usually not a pure compound but is composed of a mixture of dusts from the various processes being carried out in or near the area concerned. The particle size distribution of the compounds varies over a wide range and fluctuates continually. To the complications stemming from these differences must be added those normally associated with occupational hygiene studies based on ordinary occupational exposure. The task of evaluating from environmental measurements the level of exposure sustained by an individual or group is one of familiar but formidable complexity. These considerations militate against the use of routine air sampling data and routine biological monitoring data for calculating parameters relating to the metabolism of natural uranium. Most of the data are retrospective and too sketchy for the purpose. The value of this information lies rather in the basis it provides for assessing the relationship, in any particular situation, between the level of atmospheric contamination on the one hand and on the other, medical histories and biological monitoring figures. Several investigators have published analytical data from postmortem tissues from uranium workers, and these results have provided useful information on the deposition of inhaled soluble uranium in bone and on the retention of inhaled insoluble uranium in the lung and tracheobronchial lymph nodes.

1. "Soluble" and "Insoluble"

Over many years ICRP and the American Conference of Governmental Industrial Hygienists (ACGIH) regarded it as inappropriate to recommend one $(MPC)_a$ value for natural uranium. They also regarded it as inappropriate to recommend separate $(MPC)_a$ values for each of the dozen or so uranium compounds widely used and handled on a large scale in the uranium industry. These organizations preferred to follow HODGE and STOKINGER (1949) in recommending separate $(MPC)_a$ values for "soluble" and "insoluble" uranium. This distinction was originally justified by the results of animal experiments which clearly indicated that whereas the limiting hazard from inhaled soluble compounds like $UO_2(NO_3)_2$ and UO_2F_2 was a chemical threat to the kidneys, the limiting hazard from inhaled "insoluble" compounds like UO_2 and U_3O_8 was a radiological threat to the lungs. Although this distinction between the hazard from soluble uranium and that from insoluble is still current, its practical significance has been eroded since 1970 when the ACGIH decided to fix the same figure (200 μg U/m³) as the $(MPC)_a$ for soluble and for insoluble uranium. This decision brought the ACGIH figure closer to those recommended by the ICRP in 1959 (210 μg U/m³ for soluble uranium and 180 μg/m³ for insoluble).

Although the relativity of the designation "insoluble" is obvious, misunderstandings may arise from reading statements out of context. For instance, uranium tetrafluoride which, according to two chemical dictionaries, is "insoluble in water" (Thorpe's Dictionary of Applied Chemistry, Vol. XI, 1954; Encyclopedia of Chemical Technology, Vol. 14, 1955) has, according to Seaborg and Katz (1954), a solubility of 100 mg/l at 25° C. The solubility of UO_2 in water is probably less than one percent of this figure, and the solubility of U_3O_8 lower still. The solubility of these and other "insoluble" compounds is affected by the particle size distribution and by thermal history; of all the compounds to which workers are normally exposed U_3O_8 is the least soluble in simulated plasma (Steckel and West, 1966).

2. Biological Monitoring

Intake of uranium at work is minimized by careful control of the environment, which is usually surveyed by routine measurements of air and surface contamination. Urine samples can be collected and analyzed for uranium to indicate whether or not the environmental control is satisfactory and, in case of a recognized accident, to indicate the degree of personal exposure. As a means of assessing the actual and potential body burden of the individual worker, routine urine analyses are of little value. In many plants the practice is to relate urine data in the first instance not to the individual's exposure, but to air contamination levels. From many years of operational experience the consensus is that in the control of personal exposures air sampling plays a vital but uncertain role. The accumulated data from air sampling, with or without data from personal air samplers (Sherwood and Greenhalgh, 1960), provide a rough assessment of the degree of air contamination but do not provide an adequate assessment of personal exposure. A tendency had inevitable arisen to rely on high uranium-in-urine figures rather than high uranium-in-air figures for detecting high personal exposures. This is generally true of workers exposed to soluble forms of uranium; it is less precisely true of those exposed to insoluble forms because in the case of insoluble uranium the ratio, amount of uranium excreted in urine (μg U/day) divided by the amount inhaled (μg U/day), is smaller and less predictable. This was emphasized by Neuman in 1950 when he published the first authoritative review concerning the value of urinary uranium figures as a measure of the personal exposure hazard. The subject was later discussed by several authors. Fish (1959) stressed the complexity of the relationship between the urinary figures and the uranium body burden; Lippmann (1959) reported on the capacity of certain groups of men to withstand high concentrations of various forms of uranium (as measured by air sampling and urine sampling) without discomfort or injury, and Jackson (1964) listed the various hazards associated with airborne uranium in different chemical forms and at different degrees of enrichment. These surveys and appraisals have demonstrated that the accumulated experience of many years in the management of urine sampling programs has not greatly facilitated the task of interpreting routine urine sampling data. The advantages which might be expected to accrue from the relatively simple metabolism of uranium are offset by the uncertainties associated with several variables: the physical and chemical form of the inhaled uranium, the shape of the biological monitoring program, and ordinary physiological idiosyncrasy.

Table 4.9. Details concerning some large accidental intakes of natural uranium[a]

Investigator	Nature of incident	Nature of intake	Chemical form	Duration of exposure min	Estimated uptake mg	Initial excretion rate mg/l
J. W. Howland (1949)	leak	inhalation	hexafluoride	5	3.0	0.50
	leak	inhalation	hexafluoride	5	0.7	0.15
	leak	inhalation	hexafluoride	5	1.5	0.15
M. Eisenbud and J. A. Quigley (1956)	explosion	inhalation[b]	oxide fume	—	>15.0	6.6
R. H. Wilson (1958)	fall into vat	ingestion and absorption	nitrate solution	1	2.0	0.8
	respiratory protection failure	inhalation and ingestion	trioxide	—	5.0	1.0
	fire	inhalation[b]	oxide fume	4	0.4	0.8
J. F. Wing, R. C. Heatherton and J. A. Quigley (1965)[c]	leak	inhalation	hexafluoride	7	3.0	2.6
	leak	inhalation	hexafluoride	7	2.5	2.1
	leak	inhalation	hexafluoride	7	1.5	1.3
	leak	inhalation	hexafluoride	0.5	1.5	1.2
	leak	inhalation	hexafluoride	<5	1.25	1.2
	leak	inhalation	hexafluoride	1.5	2.25	1.4
	leak	inhalation	hexafluoride	5	1.5	1.1
J. Chalabreysse (1970)	leak	inhalation	hexafluoride	—	2.0	—
	leak	inhalation	hexafluoride	—	1.0	—
	leak	inhalation	hexafluoride	—	0.94	0.6

[a] None of the workers involved in the incidents listed here sustained permanent injury resulting from internal exposure to uranium.
[b] The worker sustained thermal burns; at least part of the absorbed and excreted uranium may have been from uranium oxide deposited on the burned skin.
[c] The seven cases reported by Wing, Heatherton and Quigley in 1965 may include some of the five cases reported by Eisenbud and Quigley in 1956; although this seems likely, it cannot be regarded as certain.

Workers who have accidentally inhaled 1–2 mg uranium as uranyl fluoride within a few minutes have within a few hours excreted uranium in urine at a concentration of 1–2 mg/l (see Table 4.9). This concentration began to fall rapidly about 20 hours after the incident and continued to fall. In 1965 Wing et al. reported on several accidental inhalations of uranyl fluoride and from the urine data deduced that in these circumstances uranium was excreted from the body with a six-hour biological half-life. In 1970 Chalabreysse reported a similar incident and concluded that following accidental inhalation 80 percent of the absorbed uranium was excreted during the first 24 hours. As far as known accidental exposures are concerned, this relationship can be used to estimate from uranium-in-urine figures the approximate size of the uranium intake.

Lippmann et al. (1962) studied for several months the urinary excretion pattern of some overexposed individuals and on the basis of these data constructed an excretion pattern which could be applied to other cases of overexposure by inhalation. In a modified form Lippmann's equation was adopted by the ICRP (1968, 1971) who have recommended the following "fractional excretion equation":

$$Y_u = 0.8 \text{ (first day)}$$

$$Y_u(t) = 0.2\,t^{-1.5} \text{ (t > 1 day)},$$

where Y_u is the amount excreted in one day expressed as a fraction of the total urinary excretion.

Since soluble uranium is eliminated so rapidly by the body, only a urine sample donated during or immediately after exposure provides a reliable index of the extent of the exposure. Conceivably, for men exposed to a constant atmospheric concentration of soluble uranium, urine results might correlate closely with air concentration. For more typical industrial conditions where exposure levels vary with the operation performed and with time, the urine concentration at the end of the work day will depend on whether the peak concentration was early in the morning shift or late in the afternoon shift. It could even be influenced by a high exposure earlier in the week. It will almost certainly be influenced by the extent to which the individual is a "mouth-breather". A before-weekend urine concentration of 100 μg/l may be the result of an exposure of 500 μg/m³ two days before, or an exposure of 250 μg/m³ the day before, or an exposure of 50 μg/m³ on the day the sample was given. For this reason LIPPMANN (1959) failed to establish any simple correlation between the air concentration of soluble uranium compounds and before-weekend urine concentrations. HURSH (1959) concluded that, although the routine measurement of urinary uranium does not usually permit the calculation of the exact body dose, it provides the only simple method for testing air contamination with respect to the exposure of a particular individual on a particular job. HURSH also stressed the importance in certain situations of fecal sampling in evaluating personal exposure.

Discussions on the pulmonary deposition and retention of inhaled aerosols have generally included some reference to the proportion of inhaled dust which sooner or later enters the gastrointestinal tract (ICRP, 1959; HATCH and GROSS, 1964; Task Group on Lung Dynamics, 1966). Whenever an aerosol is inhaled some particles eventually find their way into the intestinal tract; the proportion of particles involved in this movement, the mode of entry, and the degree and speed of absorption from the gastrointestinal tract depend on the chemical and physical properties of the dust (FISCHOFF, 1965) and the nature of the inhalation. Elimination through the gastrointestinal tract occurs with both soluble aerosols and with insoluble. In the case of soluble aerosols a sizable fraction of the total excretion takes place via the kidneys and the urine data provide an estimate, which may be more or less reliable, of recent intake. In such a situation fecal data are of less interest. On this account and because they are less readily obtained they are often not requested. It is significant, however, that following an accidental inhalation of uranyl fluoride the amount of uranium excreted in the feces may be similar to that excreted in urine (CHALABREYSSE, 1970). In the case of insoluble aerosols a smaller proportion of the inhaled dust is excreted through the kidneys and the relationship between uranium-in-urine levels and uranium-in-air levels is always more elusive and sometimes wholly obscured. In such a situation, in which urine data are useful only as a possible indicator of overexposure, analytical data obtained from fecal samples may provide a more reliable guide to recent individual exposure.

In a special category are incidents in which a worker is exposed to a sudden overexposure during an accident which is recognized as such at the time. Two incidents have been reported in which a worker was exposed to U_3O_8 formed during the burning of uranium metal. The first, reported by EISENBUD and QUIGLEY (1956), involved an initial retention of U_3O_8 fume of at least 15 mg, and urine samples were analyzed for 70 days. The second, reported by WILSON (1959), involved an initial retention of U_3O_8 fume of about 0.4 mg, and urine samples were analyzed for 20 days. In both of these cases the employee at the time of the accident sustained severe thermal burns, and in both of these cases the first 24-hour urinary excretion contained between 60 and 70 percent of the total uranium excreted via the urine over the period of study. In incidents like these and generally only in incidents like these in which the overexposure is recognized as such at the time, biological monitoring can be initiated immediately. In the uranium industry moderate but significant overexposures are often not recognized until some days or weeks after the event. In such circumstances these rapid early excretion rates in urine and in feces, if they occur, remain unobserved.

3. Postmortem Data

The task of obtaining analytical data from the organs of persons who have been occupationally exposed to a particular compound requires a high standard of pathological and analytical skill; it also requires close collaboration between the medical officer, the pathologist and the analyst. There have been many situations in the uranium industry where these requirements have been met and usefully exploited, and the accumulated data from several investigations have made a significant contribution towards our understanding of the human metabolism of airborne uranium.

The most reliable analytical data are obtained when the tissues after excision from the cadaver are wrapped in a plastic bag and kept in a deep-freeze cabinet until required for analysis. Tissues which are kept in a solution of preservative (formalin, for example) may lose some uranium by leaching; alternatively, as uranium is an ubiquitous trace element, some uranium from the preservative may contaminate the tissues. QUIGLEY (1959) stressed the advisability of analyzing fresh postmortem tissue which has not been placed in a chemical preservative. Some of his data were, however, obtained from chemically-preserved tissues and on this account may be less reliable.

For the determination of uranium concentrations at these low levels (usually between 1.0 and 0.01 μg U/g wet tissues) in fresh postmortem samples weighing between 10 and 1000 g it is seldom necessary to achieve a standard error of less than 3 percent. There are indications that in some of the earlier investigations (BUTTERWORTH, 1959; QUIGLEY et al., 1959) the degree of precision was much less than this. Some inaccuracies were no doubt associated with the use of fluorophotometric method for estimating uranium (PRICE et al., 1953), which was inadequately adapted to the complexities of the analytical task in hand. In recent years, however, low levels of uranium have been estimated by counting the delayed neutron emissions after neutron bombardment of the sample (DYER et al., 1962; HAMILTON, 1966; GALE, 1967). As in this method the measured neutrons are produced by the fission of uranium-235, the estimation of total uranium rests not only on the number of neutrons counted but also on the isotopic ratio. In most cases which have so far been encountered in occupational hygiene the uranium handled has been natural uranium and no special estimation of isotopic ratio has been required. The neutron activation method is as sensitive and as specific as the fluorophotometric, but more amenable to standardization and less susceptible to interference from other cations. The availability of this tool has greatly improved the reliability of uranium data from postmortem tissues. DONOGHUE et al. (1972) employed this analytical method to measure the uranium content of the internal organs following the collapse and sudden death of an employee who had worked for ten years in a natural uranium workshop. The results are reproduced in Table 4.10. Relatively large organ samples were analyzed and the overall precision of the measurements was set at 2–3 per-

Table 4.10. Distribution of uranium in postmortem tissues. (DONOGHUE et al., 1971)

Tissue	Wet weight g	Uranium concentration μg U/g wet weight	Weight in tissues μg	(Likely) weight in whole organ μg
Lung lobes				
left upper	254	1.4	355.6	
left lower	252	1.1	278.6	
right upper	218	1.2	263.3	1250
right lower	245	1.1	263.8	
right middle	72	1.2	87.3	
Lymph nodes				
left	5.0	1.9	9.6	
right	7.0	1.8	12.3	35[a]
Sternum	114	0.09	10.1	800[b]
Kidney				
cortex	161	0.16	26.2	
medulla	56	0.09	5.3	30
Liver	177	0.02	3.8	35
Blood		(0.01)[c]	—	(50)

[a] In the tracheo-bronchial lymph nodes.
[b] In the entire skeleton.
[c] A conjectural figure, precariously based on data provided by Hamilton whose investigations on tissues from unexposed humans in the U.K. show that the concentration of uranium in blood is 0.84 ng U/g wet weight. The equivalent amount in total blood is 4.5 μg. See Nature **227**, No. 5257, 501–502 (1970).

cent. A careful retrospective study determined that the exposure was to a uranium aerosol composed of 85 percent U_3O_8 and 15 percent UO_2, and having Activity Median Aerodynamic Diameters in the range of 3.5 to 6.0 μm. The long-term average exposure level was estimated to be about 300 μg/m³. The authors state that the lung and lymph node uranium concentrations (between 1 and 2 μg U/g wet tissue) were less than one percent of the amounts calculated from the environmental data using the currently recommended lung model and biological parameters.

There are occasions (e.g., when a worker dies from natural causes in the course of his daily work) when postmortem data can be directly related to the conditions of exposure up to the time of death. In most of the 25 cases in which postmortem data have been obtained this was not the case; usually some days or weeks elapsed between the termination of exposure and the occasion of death. The extent to which these postexposure gaps limit the value of the data depends on the nature of the exposure and especially on the chemical and physical properties of the aerosol.

Three conclusions emerge from the postmortem data listed in Table 4.11, First, whereas there is no marked tendency for uranium to accumulate in the body, most of the uranium absorbed into the body from the lungs and the gastrointestinal tract is deposited on skeletal bone. DONOGHUE et al. (1972) estimated that in the body, excluding uranium retained in the lungs (which are metabolically outside the body), the uranium deposited on bone accounted for roughly 85 percent of that metabolized. This is the figure recommended by the ICRP (1959) and it is supported by analytical data obtained from tissues from unexposed persons (HAMILTON, 1971). Secondly, the amount of the uranium burden in the lungs of

Table 4.11. Analytical data on postmortem tissues from persons occupationally exposed to uranium and from some persons unexposed

Investigators	Case	Exposure			Uranium Concentration (μg U/g wet tissue)				
		years	soluble or insoluble	degree	lung	kidney	bone	liver	lymph modes
J. A. QUIGLEY and R. C. HEATHERTON (1959)	2	3	insoluble	moderate	0.42	0.071	0.018		
	3	5	mixed	low	1.02	0.038	0.048		
	4	5	mixed	low	0.06	0.008	0.035		
R. C. HEATHERTON, M. W. BOBACK and J. A. QUIGLEY (1963)	9[b]	5	mixed	low	0.05	0.03	0.03	0.01	0.03
	12[a,b]	7	mixed	moderate			0.21		
	13[a,b]	9	mixed	moderate	0.5				
	14[a,b]	9	mixed	high	5.4				2.1
	15	10	insoluble	moderate	1.7	0.05	0.07	0.01	
A. BUTTERWORTH and A. S. MCLEAN (1955)	D/46[a]	5	soluble	moderate	0.15	0.08		0.04	
A. BUTTERWORTH (1959)	D/75	3	soluble	low	0.05	0.04			
	D/87	9	mixed	low	0.20	0.04			
	D/91	11	mixed	low	0.02	0.02			
	D/97	3	mixed	low	0.05	0.04		0.02	
	D/99	5	mixed	low	0.05	0.06		0.04	0.12
F. W. MEICHEN (1962)	1	5	not known	not known	0.05	0.06		0.05	0.12
	2	3	not known	not known	0.05	0.05		0.02	
	3	10	not known	not known	0.06	0.03		0.01	
	4	10	not known	not known	0.03	0.02			
	5	9	not known	not known	0.12	0.06		0.02	
	6	8	not known	not known	0.20	0.04			
	7	9	not known	not known	1.6	0.12		0.04	0.32
	8	5	not known	not known	0.05	0.05			
	9	5	not known	not known	0.05	0.02		0.02	0.10
	10	6	not known	not known	0.03	0.02		0.02	
J. K. DONOGHUE, E. D. DYSON, J. S. HISLOP, A. M. LEACH and N. L. SPOOR (1972)	D.1[b,c]:	10	insoluble	high	1.2	0.14	0.09	0.02	1.8
J. A. QUIGLEY and R. C. HEATHERTON (1959)	5	0	—	—	0.089	0.026	0.028	0.092	
	6	0	—	—	0.006	0.020	0.004	0.008	
R. C. HEATHERTON, M. W. BOBACK and J.A. QUIGLEY (1963)	7	0	—	—	0.02	0.01	0.01		
	8	0	—	—	0.02	0.03	0.01	0.03	
	10	0	—	—	0.02	0.01	0.02	0.01	
A. BUTTERWORTH (1959)	D/95	0	—	—	0.06	0.02		0.01	
E. I. HAMILTON (1971)	d	0	—	—			0.007		

[a] The samples in the cases marked thus were from biopsy.
[b] In these cases the period between termination of exposure and (biopsy or) death was short.
[c] This investigation included analyses of two complete lungs and two complete kidneys.
[d] This figure for the concentration of uranium in bone (in the UK) is based on analyses of 63 samples.

those who have been exposed to insoluble uranium is considerably less—in several cases, less than one percent—than that calculated, using the usual parameters for inhalation from the evaluated air concentration data. This phenomenon was reported by two contributors to the 1958 Symposium (USAEC, 1959) and discussed by EISENBUD (1959); it has been confirmed by some other investigators (DONOGHUE et al., 1971) but not by the *in vivo* measurements reported by QUASTEL

Table 4.12. Human metabolism of insoluble uranium

Investigator	Place	Number of exposed persons	Chest burden mg	nCi	Biological half-life (days)	Substance
C. E. MILLER (1957, 1958) J. D. MCLENDON (1959) B. R. FISH (1958, 1961)	Oak Ridge	1	3	150[a]	21 and 121 (two mechanisms)	U_3O_8 (enriched)
L. M. SCOTT and C. M. WEST (1964, 1969)	Oak Ridge	4	0.5–2.0	20–50	range 550–1500	Uranium oxide, ceramic, or alloy[b] (enriched)
W. N. SAXBY, J. RUNDO et al. (1964)	Aldermaston Harwell	2	1[a]	60	11 and 360 (two mechanisms)	Uranium oxide or silicate[b] (enriched)
N. B. SCHULTZ (1966)	Oak Ridge	4	0.02–1.0	1.0–30	range, 120–250	U_3O_8 (enriched)
M. RONEN (1969)	Negev	1	207	140	100 and 1200; (two mechanisms)	Uranium oxides and (probably) other uranium compounds (natural)
M. R. QUASTEL H. R. TANIGUCHI T. R. OVERTON and J. D. ABBATT (1970)	Ottawa	15	10–100	6–60	no information	UO_2 (natural)

[a] Estimated value.
[b] Probably U_3O_8.

et al. (1970). Thirdly, the concentrations of uranium (expressed as μg U/g wet tissue) in the tracheobronchial lymph nodes are not greatly different from those in the lungs. These low figures have served to allay the fear, which was based mainly on earlier animal data, that inhaled insoluble uranium would accumulate more quickly in the lymph nodes than in the lung and present a critical radiological hazard (HODGE and THOMAS, 1959; THOMAS, 1965, 1968).

Any attempt to interpret these facts must fall far short of a satisfying explanation. It appears however that, on account of the small but appreciable solubility of the so-called "insoluble" compounds (STECKEL and WEST, 1966) the deposition of uranium in the lung following an accidental exposure may be followed by a brief but rapid mobilization of "insoluble" uranium from the lung to the systemic circulation. This may partly explain why the levels of uranium retained in the lung in such circumstances are so unexpectedly low; it may also indicate that under certain accident conditions transient kidney injury might be postulated to occur, even though the exposure were to "insoluble" uranium forms. Evidence for such relatively rapid removal after massive exposure to natural uranium fume is given by EISENBUD and QUIGLEY (1956), and after accidental exposure to enriched uranium fume by FISH (1961) and by SAXBY et al. (1964). See Table 4.12. Although no kidney injury was reported in any of these cases the short early half-time found (11 and 21 days) suggest that following a sufficiently high accidental intake of "insoluble" uranium transient albuminuria might be detectable. The possibility of this occurring would depend on the physical and chemical form of the inhaled material, its average particle size and its thermal history.

C. Uranium Hexafluoride

1. Normal Exposure

In 1949 HOWLAND reported on the medical condition of ten men who had been exposed to roughly 50 μg U/m³ for periods of between five and nine months. The number of employees was small and the exposure level admittedly low; the distinctive feature of this limited survey lay in the thoroughness of the physiological examinations. From the results of an impressively long series of tests HOWLAND concluded that at this level of exposure there were no discernible toxic effects that might have occurred as a result of exposure to uranium hexafluoride. When BUTTERWORTH later reported (1959) on the medical histories of some hundreds of workers who had been exposed for years to uranium hexafluoride, some of them to high concentrations, his observations were in close agreement with HOWLAND's. There had been many instances in which an individual exposed to high concentrations of uranium hexafluoride had shown symptoms of proteinurea. This was invariably a transient experience however; soon after removal from exposure the employee's condition returned to normal. The main conclusion reached by BUTTERWORTH (1955, 1959) from extensive operational experience was that there was no evidence to indicate that exposure to high air concentrations of uranium hexafluoride caused any permanent kidney damage. LIPPMANN (1959) reached the same conclusion from the results of a similar industrial survey.

2. Acute Overexposure

HOWLAND also reported (1949, 1953) an accident in an experimental laboratory in which large amounts of UF_6 and steam escaped simultaneously. This resulted in a very dense cloud of UF_6, UO_2F_2 and HF with excess steam. The accident resulted in the death of two employees, serious injury to three others, and slight injury to the remaining thirteen. The three seriously injured cases required 10 to 14 days of hospitalization before recovery was complete. The acute effects of exposure to high concentrations of UF_6 were shown to consist of corrosive changes in the skin, eyes, and respiratory mucosa, which were probably caused by the fluoride, and transient kidney changes related to the toxic action of the absorbed uranium. In this particular incident the more seriously injured individuals were unusually nervous and apprehensive for four to five days after the accident; the clinicians and other observers reported that the abnormality of mental reaction could not be explained only in terms of psychological reaction to trauma. In this accident the chemical insult was probably greatly increased by the presence of steam which immediately resulted in the liberation of so much hydrofluoric acid. This may account for the highly corrosive nature of the exposure, which killed two employees and seriously injured three others. In comparing these cases of HOWLAND with those reported later by EISENBUD and QUIGLEY (1965), it is significant that, although the seven cases reported more recently inhaled more uranium (between 5 and 10 mg in between two and ten minutes), they apparently sustained less transient injury. The six cases involved in the accident reported by BOBACK and HEATHERTON (1966) included one who inhaled 13 mg uranium as hexafluoride and who was sufficiently incapacitated to require several days' treatment in hospital. All these cases recovered quickly and none showed evidence of chest or kidney damage which was other than transient. CHALABREYSSE (1970) has reported in similar terms on an accident in which three workers inhaled 3.5, 4.0 and 7.5 mg uranium, respectively, within a few minutes.

These UF_6 exposure figures listed in Table 4.9 were calculated from the results of urine analyses, using constants derived partly from human experiments and partly from industrial experience. For example, in accordance with the recommended ICRP lung model (Publication 2, 1959) it was assumed that 25 percent of the inhaled UF_6 (or UO_2F_2) was deposited in the lungs and taken up into the body. In accordance with the results of the human experiments conducted first at Rochester and later at Boston, it was assumed that following a sudden assimilation of soluble uranium into the body roughly 75 percent of the mobile uranium is excreted in the urine during the first 24 hours.

D. Less-Soluble Uranium Compounds

1. Occupational Exposures

In 1955 Eisenbud and Quigley reported on the analyses of bone and lung tissues of men who had died from non-occupational causes after being exposed for some years to relatively high concentrations of UF_4 and UO_2. They concluded that natural uranium has a "low order of chemical toxicity in man", and they reckoned that the safety factor incorporated into the $(MPC)_a$ for insoluble compounds was unnecessarily conservative. In 1958 at the Symposium on Occupational Health Experience and Practices in the Uranium Industry, held in New York, Quigley et al. presented more information based on studies on tissues obtained from autopsies. The data were obtained from three men who had been exposed to levels of less-soluble uranium well below the $(MPC)_a$ for periods of between three and five years, and whose deaths were unrelated to their occupations. The amounts of uranium in the kidneys, which showed no signs of significant damage, were unexpectedly small. Much less uranium was found in the lung than expected, and this finding was regarded as more compatible with a biological half-life of 30 to 60 days than with the currently accepted (ICRP, 1954) biological half-life of 120 days. The value of Quigley's papers is limited by the inadequacy of the environmental data, which include no detailed information about chemical form or about particle size. At the time when Quigley's results were published, Eisenbud (1959) remarked that whereas the low lung retention figures were curious and reassuring, they required explanation. His own view was that the high density of the dust was responsible for the low alveolar deposition rate. Uranium oxides have a density of 9 or 10 g/cm^3, and 2 μm particles of this density may not have reached the alveoli at all.

The most interesting of the cases reported by Fish (1958, 1961) (see Table 4.12) concerned a machinist who had been exposed to uranium fume during an operation which consisted in deburring holes that had been drilled in pieces of uranium metal. The metal was highly enriched in ^{235}U (and therefore in ^{234}U) and in the course of normal operations the divided chips burned, producing a radioactive fume. This personal exposure was detected through the routine uranium-in-urine analysis program, and an initial estimate based on the urinary excretion placed the chest burden in the range of 1.3 to 13 mg U. As the data from the routine air sampling program provided inadequate information from which to calculate the individual's personal exposure, an operational simulation was arranged in which air samples were taken and analyzed for chemical composition, particle size distribution, and air radioactivity levels. Electron diffraction analysis confirmed the initial assumption that the compound was U_3O_8, and the mass median particle size 1.9 μm. Air activity levels varied greatly with location and with operating techniques, but from the data obtained it seemed possible for an individual to have been continually exposed to an average level of 300 μg U/m^3. Assuming a breathing rate of 10 m^3 per shift and an alveolar retention of 20 percent, it would have been possible to collect 6 mg in the alveoli during the known period of exposure. Subsequent estimates of the total chest burden, based on *in vivo* gamma spectrum analysis measurements, indicated that the initial mass was 3.0 mg. A sereies of *in vivo* measurements made over a period of 17 months indicated that there were at least two clearance mechanisms acting to remove retained U_3O_8 particles from the lung; these exhibited biological half-lives of 21 and 121 days. This investigation illustrates the value of the whole body monitor 8 in obtaining reliable estimates of chest burdens of retained uranium, especially enriched uranium, and especially when these measurements are related to the relevant environmental data and uranium-in-urine analytical data.

Scott and West (1964, 1969) reported on routine measurements on about 1 800 employees involved in the industrial-scale processing of uranium. Collectively the workers were exposed to uranium in a wide variety of chemical and physical forms. The compounds included UF_6, UF_4, $UO_2(NO_3)_2$, U_3O_8, UO_2, uranium metal and uranium alloys. The degree of uranium enrichment varied between zero and 90 percent and the mean particle size from less than 1.0 to 10 μm. Routine surveillance was achieved by means of uranium-in-urine analyses and by whole body monitor measurements. These investigators found that all personnel monitoring data of this kind can be put into one of four classes: (1) low urine analysis and low *in vivo* spectrum analysis, (2) high urine analysis and low *in vivo* spectrum analysis, (3) low urine analysis and high *in vivo* spectrum analysis, and (4) high urine analysis and high *in vivo* spectrum analysis. Class (1) accommodated most of those working at this particular plant. Of those workers showing elevated results, the majority were in Class (4). In class (3) four workers were found whose *in vivo* measurements showed that the biological half-lives of the uranium retained in the chest were 550, 1 040, 1 470 and 1 500 days, respectively. In cases like these a significant proportion of the deposited uranium is eventually excreted in the feces, and the task of evaluating the chest burden from excretion data alone is difficult if not impossible.

In vivo measurements of body radioactivity, together with urine and fecal analyses, also formed the basis of a study by Schultz (1966) of four male chemical operators who had accumulated significant chest burdens of the oxide U_3O_8. The biological half-lives for the dust retained in the chest region ranged from 120 to 150 days. The apparent decrease of uranium in the chest was greater than the total amounts excreted; this suggested that some uranium was transferred from the chest cavity to another organ. In three of these cases fecal excretion was found to be a significant mode of clearance, averaging 92 percent, 44 percent, and 183 percent, respectively, of the urinary uranium excretion. These variations may be explained by such factors as differences in particle size, effective solubility, physiological idiosyncrasy, and the chronological gap between the end of exposure and the beginning of fecal sampling.

Saxby et al. (1964) reported an investigation concerning an employee who, after 18 months' exposure to airborne insoluble uranium, had acquired a significant deposit of uranium in the chest. The uranium was known to be enriched and was assumed to be in the form of the oxide U_3O_8. The results of urine and fecal analyses were presented, together with measurements made in the whole-body monitor. The authors tentatively concluded that the apparent biological half-life of the dust in the chest was about 360 days; they also observed that for the first 500–600 days after inhalation the fecal excretion rate was higher than the urinary excretion rate by a factor of about three.

8 Use of the whole body gamma counter is described in chap. 6. Some of the advantages and disadvantages are also discussed at the end of this subsection. For a complete description of the equipment in use see IAEA (1964).

An unusual case of exposure to insoluble uranium reported by RONEN in 1969 concerned an employee who worked in a metallurgical laboratory; he had been engaged in a process for recovering uranium alloy scrap from graphite moulds. As part of the routine biological monitoring program he had provided several urine samples, most of which contained less than 5 μg uranium per litre. An incident involving hand contamination and the appearance of a urine sample containing a higher concentration of uranium provided the first indication that the worker had sustained a significant inhalation overexposure. The first abnormal urine sample contained 300 μg U/l, and subsequent samples between 30 and 4000 μg U/l. For several months the daily excretion rate remained high (*c.* 250 μg/day), but low concentrations in individual samples (*c.* 30 μg/l) were not uncommon. Measurements made by *in vivo* gamma spectrum analysis revealed that the employee had somehow acquired a chest burden of 207 mg natural uranium. The elimination of the uranium from the chest appeared to occur by a process involving at least two mechanisms, with half-lives of 100 and 1200 days. The task of interpreting this perplexing incident would be simplified if more detailed information were available concerning the nature of the inhaled dust. The report states only that the natural uranium dust involved in the contamination contained soluble, insoluble and moderately soluble particles of different size and different physical and chemical properties.

QUASTEL et al. (1970) reported an investigation in which 15 workers who had been chronically exposed to airborne insoluble uranium dust for long periods (1–20 years) were studied by means of *in vivo* gamma spectrum analysis together with urine and fecal analysis. The uranium was natural uranium and the dust predominantly UO_2. Individual thorax burdens varied from 10 to 100 mg and the average excretion rate in urine varied between 10 and 70 μg/day. Fecal concentrations of uranium decreased exponentially with a half-life of 6–30 hours and indicated that recent exposure to insoluble uranium is more reliably detected by analysis of feces than by analysis of urine.

The instrument used for obtaining the chest burden data listed in Table 4.12 is the whole body monitor, which can reliably detect *in vivo* 10 mg natural uranium or 1 mg enriched. In measuring the half-life of uranium in the chest this equipment possesses only two disadvantages: it cannot measure accurately the small chest deposits (1–5 mg) occasionally associated with occupational exposure to natural uranium, and it cannot distinguish between the uranium deposited in the lungs and that deposited in the tracheobronchial lymph nodes, the sternum and the ribs. This emphasizes the desirability of obtaining analytical data from postmortem tissues related to a known exposure history.

IV. Natural Background Levels of Natural Uranium

A. Uranium in Soil and Plants

According to KATZ and SEABORG (1957) uranium is widely distributed in nature. Although only a few ores contain a high concentration of uranium (i.e. 40–60 percent) about 100 mineral species contain 1 percent or more (VDOVENKO, 1960). The average concentration in the earth's crust is about 4×10^{-4} percent and the concentration in sea water is about 3.3 μg per litre (KEEN, 1968). For these reasons and because uranium is normally present in soil, in building materials (HAMILTON, 1971), in air (HAMILTON, 1970), and drinking water in measurable amounts, uranium is regarded as a ubiquitous trace element.

Uranium in soil is absorbed into plant tissues to an extent which depends on the species and on the depth of the root system. For many years the analysis of plants, especially leaves from deep-rooted trees, has been used as an aid to uranium prospecting (CANNON and KLEINHAMPL, 1955; MOISEENKO, 1959; and DEAN, 1966).

B. Uranium in Water and Foodstuffs

Several workers have reported on the levels of uranium in water and foodstuffs and related the average total rate of intake to the excretion rates in feces and urine.

Two Russian papers refer to situations in which relatively high levels (i.e. levels high in comparison with the levels normally encountered in the U.S.A. and the U.K.) of uranium were ingested in drinking water. NOVIKOV and REZANOV (1962) reported on the exposure of a local population to a water supply derived from "crack water" formed when atmospheric precipitation passes through areas where rocks of high uranium content are widely fissured. Measurements were made on the excreta from children aged 2 or 3 years who drank from a water supply containing 36 μg uranium per litre. As no information was provided concerning the amount of water consumed by the children, or the treatment the water received before consumption, or concerning the children's whole diet, the results are not amenable to precise interpretation. Roughly 0.4 percent of the uranium was excreted in the urine and 0.2 percent in the feces. During the short period of this investigation, therefore, about 99 percent of the ingested uranium was retained in the body. Data published by BERDNIKOVA (1964) of intake and excretion levels at two sites showed intake levels that differed by a factor of more than 25. In this study the water levels were the dominating factor, because the population was drinking spring water containing (according to the site) 200 or 7 μg uranium per litre. The concentrations of uranium in vegetables and in meat were measured and their contributions to the total were shown to be insignificant. In both locations the total uranium eliminated was reported to be roughly 10 percent of the daily intake, 90 percent of which was in the feces.

A more recent paper described the movement of uranium in the biological chain in an area where the average gross uranium content of the soil was 1.8×10^{-4} percent by weight, most of the uranium was in a virtually insoluble form and only 3 percent of this proportion was regarded as biologically available. In this survey PRISTER (1969) measured the concentrations of uranium in oats, wheat, millet and peas (between 10 and 100 ng U/g fresh seed) in potatoes, carrots and cabbage (15.0, 7.7 and 4.7 ng U/g fresh vegetable respectively). He also measured the concentrations in various pig tissues (43 ng U/g fresh skeleton for example, 23 ng U/g fresh kidney and 5 ng U/g liver). From these and other data PRISTER estimated the relative contributions of different foodstuffs and of drinking water to the total uranium intake in the adult human diet. See Table 4.13.

The figures obtained in a detailed and comprehensive American survey (WELFORD and BAIRD, 1967) showed that the concentration of uranium in the tap water in the New York City area was so low (32 ng U/l) that the contribution it made to the total human intake was negligible. These investigators analyzed for uranium a wide range of foodstuffs from New York, Chicago and San Francisco and concluded that in spite of significant differences in the uranium content of certain foods from these areas, the daily intake of uranium was virtually the same in each: 1.3 μg/day in New York, 1.4 μg/day in Chicago, and 1.3 μg/day in San Francisco. Potatoes, meat, fresh fish and bakery products contributed more than 70 percent to the annual uranium intake. See Table 4.14.

Table 4.13. Uranium content of food products and daily dietary intake by human adults[a]

Product	Concentration of uranium ng U/g raw product	Inclusion in dietary intake	
		μg	% of total
Bread and other baked products	12	6.0	19.6
Macaroni and flour	40	5.0	16.3
Potatoes and vegetables	18	5.4	17.7
Meat	20	4.0	13.0
Milk	2	1.0	3.3
Dairy products	35	3.5	11.4
Eggs	9.6	0.2	0.65
Water	2.5	5.5	18.0
Total	—	30.6	100

[a] After B. S. PRISTER, Moscow, 1969.

Table 4.14. Uranium content in foodstuffs and annual dietary intake by human adults[a]

Product	Intake (μg U/year)		
	New York	Chicago	San Francisco
Bakery and whole grain products	82	74	74
Potatoes	111	101	101
Meat and poultry	81.7	112.4	51.6
Flour, macaroni and rice	30	33.4	14.6
Vegetables (fresh and root)	37	34.4	56
Fresh fruit	76	76	42
Milk	16	31	62
Fish; shellfish	12.9	35.8	35.5
Eggs	3.4	3.4	3.4
Dried beans	4.5	11	11
Canned fruit; canned vegetables	8.2	8.4	8.0
Fruit juices	1.1	1.1	3.4
Total	463.8	521.9	462.5

[a] After G. A. WELFORD and R. BAIRD, New York, 1967.

In a similar survey conducted in the U.K., HAMILTON (1972) reported on the concentration of uranium in various items of raw foodstuffs and of prepared diet. The raw foodstuffs were bought directly from shops and the prepared diets were blended and cooked dishes produced, from eight major regions in the U.K., by colleges of domestic science. The results showed that for certain items in particular (cereals, for example, and vegetables) the uranium concentration in the raw foodstuff was significantly lower than that in the prepared dish. This difference was attributed to the relatively high level of uranium (40 ng U/g) in the salt—whether "table", "cooking" or "iodized"—used in cooking. HAMILTON concluded that the average intake of uranium from an adult diet from raw foodstuffs in the U.K. was about 1 μg/day. See Table 4.15.

Table 4.15. The concentration of uranium in various items of diet. With estimates of annual intake[a]

Foodstuff	Mean concentration ng U/g sample	Annual uranium intake μg	Fraction of annual intake %
Cereals: wheat flour, biscuits spaghetti, oats and rice	0.5	68	19
Vegetables: potatoes, parsnips, turnips and swedes[b]	1.0	138	38
Sugar	0.2	6	2
Vegetables and fruit: cabbages, sprouts, carrots and apples	0.8	92	26
Meat	0.4	16	4
Eggs	0.4	3	0.9
Fish	0.2	3	0.8
Milk	0.01	2	0.5
Fats and oils	2.0	32	9.0
Total		360	100

[a] After E. I. HAMILTON, Sutton, England, 1971.
[b] rutabage.

NOZAKI et al. (1970) have reported on the uranium analysis of drinking water (4.8–11.4 ng/l) and food (1.5 μg daily intake) for urban centers in Japan. They find no significant difference in the dietary intake for the two geographically separated cities where food sampling was done.

C. Uranium in Man

The figures published by WELFORD and BAIRD (1967) showed that uranium levels in human bone from the New York area were fairly uniformly distributed throughout the skeleton, with an average concentration of 0.02 μg U/g ash. The average concentration in lung tissue, from cadavers from the same area was 1 ng/g wet tissue, and the average concentration in blood 0.57 ng/g whole blood (WELFORD et al., 1970).

HAMILTON (1970) reported an average concentration of uranium in human blood for persons in the south of England of 0.84 ng/g whole blood. He has also reported an average concentration in bone of 0.024 μg U/g ash (equivalent to 6.96 ng U/g wet weight); this average is based on more than 60 analyses on a variety of bone tissues, which included sternum, femur, rib and vertebrae. On the basis of these and other data HAMILTON estimated that the skeleton and marrow taken together, which account for 14 percent of the body mass, contain about 70 μg uranium, and he tentatively suggested that the amount of uranium in a 70 kg man in England is between 100 and 125 μg. See Table 4.16.

NOZAKI et al. (1970) analyzed human bone samples of rib and sternum collected in Sapporo, Tokyo, Kyoto and Osaka. No geographical district, subject age or sex trends were found. They analyzed 47 samples and find an average concentration of 2.0 ng/g wet bone.

Some data have been published concerning the uranium content of body tissues from Indian adults with no occupational exposure. GANGULY (1968)

Table 4.16. Natural uranium in man and his diet

Description	U.S.A.[a]		U.K.	
	value	source[b]	value	source[b]
Tissue concentration (in ng/gram)				
Blood	0.57	1	0.84	4
Lung	0.7	1	—	
Liver	0.13	1	—	
Bone ash	20	2	24	5
Daily dietary intake (in μg)	1.3	2	1.2	6
Daily urinary excretion (in μg)[c]	0.154	2.3	0.380	7
Tap water concentration (in ng/l)	32	2	—	
Air concentration (in ng/m³)	0.4	2	0.02	7
Total estimated body content (in μg)	80	2	100	6

[a] All U.S.A. samples were drawn from the N.Y.C. area except those for dietary intake (see Table 4.14).

[b] Reference notation: 1) WELFORD et al., 1970; 2) WELFORD and BAIRD, 1967; 3) WELFORD et al., 1960; 4) HAMILTON, 1970; 5) HAMILTON, 1971; 6) HAMILTON, 1972; 7) DEAN, 1967.

[c] The values as listed in the references quoted are in μg/liter and have been converted into daily output by multiplying by 1.4.

analyzed bone and kidney samples collected during postmortem examinations of accidental deaths in Bombay. The uranium concentrations in bone were between 2.3 and 3.4 ng/g wet tissues (0.003–0.007 μg/g ash) and in kidney between 0.6 and 3.8 ng/g wet tissue (0.06–0.38 μg/g ash).

D. Use of Background Uranium Data to Estimate Metabolic Parameters

It is reasonable to suppose that adult "non-exposed" human subjects will be in a steady state with respect to the uranium in their environment. If this is true, data such as that presented in Table 4.16 may be used to estimate 1) absorption from the gastrointestinal tract and 2) the rate constant for loss of uranium from the total body.

1. Absorption

There are good experimental reasons to believe that urinary excretion of uranium is the only important route for loss of systemic uranium and that, therefore, the daily urinary output must be approximately equal to the uranium absorbed from the daily diet. Examination of the data in Table 4.16 verifies that uranium intake by the lungs or in the form of drinking water may be neglected by comparison with that in the food. The factor for absorption from the gastrointestinal tract may be estimated as 12 percent using WELFORD's data (U.S.A.) and as 31 percent using the data of HAMILTON and of DEAN (U.K.). These estimates are both higher than the value of 1 percent currently used in radiological protection and the values of 0.5 to 4.5 percent based on the oral administration of 10.8 mg U to hospital patients. The most evident differences are that, for the natural background "experiment", the size of the dose is smaller and it is necessarily always administered with food. On the technical side, the value for the daily dietary input is supported by the U.S.A., the U.K., and the Japanese data; the urinary output (U.S.A.) is based on samples from 37 subjects and the urinary output (U.K.) is based on 300 samples.

2. Rate Constant for Whole Body Loss

The rate of loss of systemic uranium (as distinct from that in the lung or gut) may be also estimated from these data. Grounded in the same central assumption that the urinary loss is the only important excretory route, the fraction of the whole body burden excreted per day becomes 0.154/80 (U.S.A. data) and 0.380/100 (U.K. data) or 1.9×10^{-3} day^{-1} and 3.8×10^{-3} day^{-1} as the respective loss constants. The corresponding half-times are 360 days and 180 days. The value listed in ICRP Publication 2 is 100 days. The body burdens entered in Table 4.16 and used here are estimates and depend primarily on the average uranium content of bone which is based on 63 sample measurements (U.K. data) and 8 sample measurements (U.S.A. data). It is to be hoped that the experimental interest in collecting data on the turnover of natural uranium will continue and that these estimates may be supported or modified in the future.

References

ADOLPH, E. F.: Quantitative relations in the physiological constitutions of mammals. Science **109**, 579–585 (1949).

AUB, J. C., EVANS, R. D., HEMPELMANN, L. H., MARTLAND, H. S.: The late effects of internally-deposited radioactive materials in man. Medicine (Baltimore) **31**, 221–329 (1952).

BASSETT, S. H., FRENKEL, A., CEDARS, N., VAN ALSTINE, H., WATERHOUSE, C., CUSSON, K.: The excretion of hexavalent uranium following intravenous administration. II. Studies on human subjects. USAEC Report UR-37 (1948).

BERDNIKOVA, A. V.: On the uranium content in the external environment and in human excreta. Vopr. Pitaniya **23**, 17–20 (1964).

BERNARD, S. R., MUIR, J. R., ROYSTER, G. W., JR.: The distribution and excretion of uranium in man. Proc. Hlth Phys. Soc., 33–48 (June 1956).

BERNARD, S. R., STRUXNESS, E. G.: A study of the distribution and excretion of uranium in man. An interim report. Oak Ridge National Laboratory Report ORNL-2304, June 18 (1957).

BOBACK, M. W., HEATHERTON, R. C.: Bioassay aspects of a UF_6 fume release. Proc. Twelfth Annual Bioassay and Analytical Chemistry Meeting, Gatlinburg, Tennessee (1966).

BOND, C. H.: Remarks upon the value of uranium nitrate in the control of glycosuria. Practitioner, 257–264, Sept. (1898).

BUTTERWORTH, A.: The significance and value of uranium in urine analysis. Trans. Ass. indstr. med. Offrs **5**, 36–43 (1955).

BUTTERWORTH, A.: Human data on uranium exposure. USAEC Report HASL-58 (1959).

BUTTERWORTH, A., MCLEAN, A. S.: Observations on the metabolism of soluble uranium in humans. UKAEA Report IGO-R/R. 8. (1955).

CANNON, H. L., KLEINHAMPL, F. J.: Botanical methods of prospecting for uranium. Proc. 1st Int. Conf. Peaceful Uses Atom. Energy, Geneva **6**, 801–805 (1955).

CHALABREYSSE, J.: Etudes et résultats d'examens effectués à la suite d'une inhalation de composés dits solubles d'uranium naturel Radioprotection **5**, 1–17 (1970).

CHALABREYSSE, J.: Etude et résultats d'examens effectués à la suite d'une inhalation de composés dits solubles d'uranium naturel. Radioprotection **5**, 305–310 (1970).

CLOETTA, M.: Über die Ursache der Angewöhnung an Arsenik. Naunyn-Schmiedebergs Arch. exp. Path. Pharmak. **54**, 196–205 (1906) (Cit. by SOLLMAN, 1936).

Committee on Threshold Limit Values: Documentation of threshold limit values. Amer. Conf. of Governmental Industrial Hygienists, 1014 Broadway, Cincinnati 2, Ohio (1962).

Committee on Threshold Limit Values: Documentation of threshold limit values. Amer. Conf. of Governmental Industrial Hygienists, Revised (1966).

DEAN, M. H.: A survey of the uranium content of vegetation in Great Britain. J. Ecol. **54**, 589–595 (1966).

DEAN, M. H.: Personal communication (1967).

DONOGHUE, J. K., DYSON, E. D.: HISLOP, J. S., LEACH, A. M., SPOOR, N. L.: Human exposure to natural uranium: a case history and analytical results from some postmortem tissues. Brit. J. industr. Med. **29**, No 1, 81–89 (1972).
DUNCAN, E.: The treatment of diabetes mellitus by nitrate of uranium. Brit. med. J. **(1897)**, **II**, 1044–1047.
DYER, F. F., EMERY, J. F., LEDDICOTTE, G. W.: A comprehensive study of the neutron activation analysis of uranium by delayed-neutron counting. USAEC Report ORNL-3342 (1962).
EISENBUD, M.: Discussion of apparent anomalies in lung retention of uranium. USEAC Report HASL-58, 212–213 (1959).
EISENBUD, M., QUIGLEY, J. A.: Industrial hygiene of uranium processing. Arch. industr. Hlth **14**, No 1, 12–22 (1956).
EISENBUD, M., QUIGLEY, J. A.: Industrial hygiene of uranium processing. Proceedings of the International Conference on the Peaceful Uses of Atomic Energy (8–20 August, 1955) **13**, 222–228 (1956).
Encyclopedia of Chemical Technology: Vol. 14. New York: Interscience Encyclopedia Inc. 1955.
FIELD, C. E.: Radium. Its physiochemical properties considered with relation to high blood pressure. Med. Rec. (N.Y.) **89**, 135–139 (1916).
FISCHOFF, R. L.: The relationship between and the importance of the dimensions of uranium particles dispersed in air and the excretion of uranium in the urine. Amer. industr. Hyg. Ass. J. **26**, No 1, 26–33 (1965).
FISH, B. R.: Single acute intake of U_3O_8 aerosol. Health Physics Annual Progress Report. USAEC Report ORNL-2590, 143 (1958).
FISH, B. R.: Urinalysis summary. USAEC Report HASL-58 (1959).
FISH, B. R.: Inhalation of uranium aerosols by mouse, rat, dog and man. In: Inhaled particles and vapours (Ed. C. N. DAVIES), 151–166. London: Pergamon Press 1961.
FISH, B. R., PAYNE, J. A., THOMPSON, J. L.: Ingestion of uranium compounds. Oak Ridge National Laboratory Report ORNL-2994, 269–279 (1960).
GALE, N. H.: Development of delayed neutron technique as rapid and precise method for determination of uranium and thorium at trace levels in rocks and minerals, with applications to isotope geo-chronology. In: Radioactive dating and methods of low-level counting. Vienna: Intern. Atom. Energy Agency 1967.
GANGULY, A. K.: Assessment of internal radioactive contamination: the present trends in the Indian programme. The detection and assessment of uranium and plutonium in the whole body and in the critical organs. Report 118. Vienna: IAEA 1970.
GÜNTHER, B., LEON DE LA BARRA, B.: Physiometry of the mammalian circulatory system. Acta physiol. lat.-amer. **16**, 32–43 (1966).
HAMILTON, E. I.: The determination of uranium in rocks and minerals by the delayed neutron method. Earth and Planetary Sci. Letters **1**, 77–81 (1966).
HAMILTON, E. I.: Uranium content of normal blood. Nature (Lond.) **227**, 501–502 (1970).
HAMILTON, E. I.: The concentration of uranium in air from contrasted natural environments. Hlth Phys. **19**, 511–520 (1970).
HAMILTON, E. I.: The relative radioactivity of building materials. Amer. industr. Hyg. Ass. J. **32**, 398–403 (1971).
HAMILTON, E. I.: The concentration and distribution of uranium in human skeletal tissues. Calc. Tiss. Res. **7**, 150–162 (1971).
HAMILTON, E. I.: The concentration of uranium in man and his diet. Hlth Phys. **22**, 2, 149–153 (1972).
HARRIS, W. B.: Introduction. USAEC Report HASL-58 (1959).
HARRIS, W. B.: The experimental clearance of uranium dust from the human body. In: Inhaled particles and vapours. I. (Ed. C. N. DAVIES), p. 209–215. London: Pergamon Press 1961.
HATCH, T. F., GROSS, P.: Pulmonary deposition and retention of inhaled aerosols. Prepared under the direction of the American Industrial Hygiene Association for the Division of Technical Information, USAEC, Academic Press (1964).
HAUSMANN: Zur Kenntnis der Arsengewöhnung. Pflügers Arch. ges. Physiol. **113**, 327–340 (1906) (Cit. by SOLLMAN, 1936).
HEATHERTON, R. C., BOBACK, M. W., QUIGLEY, J. A.: A continued program of analysis for uranium in human animal tissues. In: Proceedings of the ninth annual conference on bioassay and analytical chemistry. San Diego, California, 10–11 Oct. 1963, Report TID-7696 (1964).
HODGE, H. C., STOKINGER, H. E., NEUMAN, W. F.: Suggested maximal allowable concentration of soluble uranium compounds in air. USAEC Report AECD 2785 (1949).

HODGE, H. C., THOMAS, R. C.: The questions of health hazards from the inhalation of insoluble uranium and thorium oxides. In: Progress in nuclear energy. Series XII. Hlth Phys. **1** (1959).

HODGE, H. D., STOKINGER, H. E., NEUMAN, W. F., BALE, W. F., BRANDT, A. E.: Suggested maximum allowable concentration of insoluble uranium compounds in air. USAEC Report AECD 2784 (1949).

HOWLAND, J. W.: Studies in human exposures to uranium compounds, vol. 2, 993–1017. In: Pharmacology and toxicology of uranium compounds (Ed. VOEGTLIN, C., and H.C. HODGE). New York: McGraw-Hill 1949–1953.

HUNTER, W. C.: Experimental study of acquired resistance of rabbit's renal epithelium to uranyl nitrate. Ann. intern. Med. **1**, 747–789 (1928).

HURSH, J. B.: Urinary uranium as an indicator of dose to exposed personnel. USAEC Report HASL-58 136–138 (1959).

HURSH, J. B., NEUMAN, W. F., TORIBARA, T., WILSON, H., WATERHOUSE, C.: Oral ingestion of uranium by man. Hlth Phys. **17**, 619–621 (1969).

HUXLEY, J. S.: Problems of relative growth. London: Methuen & Co. Ltd. 1932.

Industrial Hygiene Foundation of America: Suggested principles and procedures for developing data for threshold limit values for air. Chemical-Toxicological Series, Bulletin No 8–69 (1969).

International Atomic Energy Agency: Directory of whole body radioactivity monitors. Vienna: IAEA 1964.

International Commission on Radiological Protection: Publication 2. Report of Committee II on Permissible Dose for Internal Radiation. London: Pergamon Press 1959.

International Commission on Radiological Protection: Publication 6. Recommendations as amended 1959 and revised 1962. London: Pergamon Press 1964.

International Commission on Radiological Protection: Publication 10. Report of Committee IV on evaluation of radiation doses to body tissues from internal contamination due to occupational exposure. London: Pergamon Press 1968.

International Commission on Radiological Protection: Publication 10A. A report of ICRP Committee IV. The assessment of internal contamination resulting from recurrent or prolonged uptakes. London: Pergamon Press 1971.

JACKSON, S.: The estimation of internal contamination with uranium from urine analysis results. In: Assessment of radioactivity in man, vol. II, 549–561. Vienna: International Atomic Energy Agency 1964.

JOACHIMOGLU, G.: Zur Frage der Gewöhnung an Arsenik. Naunyn-Schmiedebergs Arch. exp. Path. Pharmak. **79**, 419–442 (1916) (Cit. by SOLLMAN, 1936).

KATZ, J. J., SEABORG, G. T.: The chemistry of the actinide elements. London: Methuen 1957.

KEEN, N. J.: Studies on the extraction of uranium from sea water. J. Brit. Nucl. Energy Soc. **7**, 178–183 (1968).

LECONTE, CH.: Resumé des expériences sur l'azotate d'uranium, Paris, Soc. Biol. **5**, 171–173, Compt. Rendu (1853).

LIPPMANN, M.: Correlation of urine data and medical findings with environmental exposure to uranium compounds. USAEC Report HASL-58, 103–114 (1959).

LIPPMANN, M.: Environmental exposure to uranium compounds. Arch. industr. Hlth **20**, 211–226 (1959).

LIPPMANN, M., ONG, L. D. Y., HARRIS, W. B.: The significance of urine uranium excretion data. USAEC Report HASL-120 (1962).

LIPPMANN, M.: "Respirable" dust sampling. Amer. industr. Hyg. Ass. J. **31**, 138–159 (1970).

LOONEY, W. B., HURSH, J. B., COLODZIN, M., STEADMAN, L. T.: Acta Union Int. Contre le Cancer **16**, 435–447 (1960).

LUESSENHOP, A. J., GALLIMORE, J. C., SWEET, W. H., STRUXNESS, E. G., ROBINSON, J.: The toxicity in man of hexavalent uranium following intravenous administration. Amer. J. Roentgenol. **79**, 83–100 (1958).

MACNIDER, W. DE B.: A review of acute experimental nephritis. Physiol. Rev. **4**, 595–638 (1924).

MACNIDER, W. DE B.: Functional and pathological response of kidney in dogs subjected to second subcutaneous injection of uranium nitrate. J. exp. Med. **49**, 411–433 (1929).

MILLER, C. E.: Argonne National Laboratory Radiological Physics Division, Semi-Annual Progress Reports ANL-5829 (1957) and ANL-5919 (1958).

MOISEENKO, U.: Biochemical surveys in prospecting for uranium in marshy areas. Geochemistry (Wash.) **1**, 117–119 (1959).

MORROW, P. E., GIBB, F. R., BEITER, H. D.: Inhalation studies of uranium trioxide. Hlth Phys. **23**, 273–280 (1972).

MCLENDON, J. D.: A uranium inhalation exposure case history. Report AECU-4089 (1959).

National Bureau of Standards: Handbook 52. Maximum permissible amounts of radioisotopes in the human body and maximum permissible concentrations in air and water. U.S. Department of Commerce, March 20 (1953).

Neuman, W. F.: Urinary uranium as a measure of exposure hazard. Industr. Med. Surg. **19**, 185–191 (1950).

Norris, W. P., Speckman, T. W., Gustafson, P. F.: Studies of the metabolism of radium in man. Amer. J. Roentgenol. **73**, 785–802 (1955).

Novikov, Y. V., and Rezanov, I. I.: The content of uranium in drinking water of soil origin and its excretion in urine and feces. Gig. i Sanit. **27**, 103–105 (1962).

Nozaki, T., Ichikawa, M., Sasuga, T., Inarida, M.: Neutron activation analysis of uranium in human bone, drinking water, and diet. J. Radioanal. Chem. **6**, 33–40 (1970).

Parker, H. M.: The dilemma of lung dosimetry. Hlth Phys. **16**, 553–561 (1969).

Price, G. R., Ferretti, R. J., Schwartz, S.: Fluorophotometric determination of uranium. Analyt. Chem. **25**, 322–331 (1953).

Prister, B. S.: Behaviour of uranium in the biological chain. GKIAE report by Atomizdat, Moscow (1969); Canadian translation AEC/TR/7128 (1970); also USCEAR report A/AC. 82/G/L. 1298.

Quastel, M. R., Taniguchi, H., Overton, T. R., Abbatt, J. D.: Excretion and retention by humans of chronically inhaled uranium dioxide. Hlth Phys. **18**, 233–244 (1970).

Quigley, J. A., Heatherton, R. C., Ziegler, J. F.: Studies of human exposure to uranium. USAEC Report HASL-58, 34–40 (1959).

Richter, P. F.: Experimentelles über die Nierenwassersucht. Berl. klin. Wschr. **42**, 384–386 (1905) (Cit. by MacNider, 1924).

Ronen, M.: A case of insoluble natural uranium exposure: A two-year follow-up study. IAEA Paper SM-119/17, Vienna (1969).

Saxby, W. N., Taylor, N. A., Garland, J., Rundo, J., Newton, D.: A case of inhalation of enriched uranium dust. In: Assessment of radioactivity in man, vol. II, 535–547. Vienna: IAEA 1964.

Schultz, N. B.: Inhalation cases of enriched insoluble uranium oxides. Report K-C-822, prepared for presentation at the Technical Meeting of the First International Congress of the International Radiation Protection Association, Sept. 5–10 (1966).

Scott, L. M., Bahler, K. W., De la Garza, A., Lincoln, T. A.: Mortality experience of uranium and non-uranium workers. USAEC Report Y-1739 (1970).

Scott, L. M., West, C. M.: Detection and evaluation of uranium exposures. In: Assessment of radioactivity in man, vol. II, 523–533. Vienna: IAEA 1964.

Seaborg, G. T., Katz, J. J.: The actinide elements. Maidenhead: McGraw-Hill 1954.

Sherwood, R. J., Greenhalgh, D. M. S.: A personal air sampler. Ann. occup. Hyg. **2**, 127–132 (1960).

Sollmann, T.: A manual of pharmacology. London: W. B. Saunders Co. 1936.

Spiess, H., Mays, C. W.: Bone cancer induced by ^{224}Ra (ThX) in children and adults. Hlth Phys. **19**, 713–729 (1970).

Stara, J. F., Nelson, N. S., Della Rosa, R. J., Bustad, L. K.: Comparative metabolism of radionuclides in mammals: A review. Hlth Phys. **20**, 113–137 (1971).

Steckel, L. M., West, C. M.: Characterization of Y-12 uranium process materials correlated with *in vivo* experience. USAEC Report Y-1544-A (1966).

Stokinger, H. E.: Current problems of setting occupational exposure standards. Arch. environ. Hlth **19**, No 2, 277–281 (1969).

Struxness, E. G., Luessenhop, A. J., Bernard, S. R., Gallimore, J. C.: The distribution and excretion of hexavalent uranium in man. Proceedings International Conference on Peaceful Uses of Atomic Energy, vol. 10, 186–196 (1955). New York: United Nations 1956.

Sweet, W. H., Javid, M.: Possible use of neutron-capturing isotopes such as boron in treatment of neoplasms; intracranial tumors. J. Neurosurg. **9**, 200–209 (1952).

Task Group on Lung Dynamics: Deposition and retention models for internal dosimetry of the human respiratory tract. Prepared for Committee II of the International Radiological Protection Commission. Hlth Phys. **12**, 173–207 (1966).

Terepka, A. R., Toribara, T. Y., Neuman, W. F.: Skeletal retention of uranium in man. Presented at the 46th meeting of the Endocrine Society, San Francisco, June 18, 1964.

Thomas, R. G.: Transport of relatively insoluble materials from lung to lymph nodes. USAEC Report LF-21 (1965).

Thomas, R. G.: Transport of relatively insoluble materials from lung to lymph nodes. Hlth Phys. **14**, 111–117 (1968).

Thorpes Dictionary of Applied Chemistry: Vol. XI, Harlow: Longmans Green1954.

Tylecote, F. E.: The pharmacology and therapeutics of uranium. Med. Chron. Manchester **7**, 379–390 (1904).

United States Atomic Energy Commission: Symposium on occupational health experience and practices in the uranium industry. USAEC Report HASL-58 (1959).

Vdovenko, V. M.: Chemistry of uranium and transuranium elements. AEC-tr-6421. U.S. Atomic Energy Commission (1964). Translated from a publication of the Publishing House of the Academy of Sciences, USSR, Moscow, Leningrad (1960).

Voegtlin, C., Hodge, H. C.: Pharmacology and toxicology of uranium compounds, vol. 1, 2 (1949); vol. 3, 4 (1953). New York: McGraw-Hill.

Welford, G. A., Alercio, J. S., Morse, R. S.: Urinary uranium levels in non-exposed individuals. Amer. industr. Hyg. Ass. J. **21**, 68–70 (1960).

Welford, G. A., Baird, R.: Uranium levels in human diet and biological materials. Hlth Phys. **13**, 1321–1324 (1967).

Welford, G. A., Baird, R., Fisenne, I. M.: Preliminary results: uranium in man. The 16th Annual Bioassay and Analytical Chemistry Conf., Bethesda, Maryland, Oct. 8–9 (1970).

West, C. M., Scott, L. M.: Uranium cases showing long chest burden retention: an updating. Hlth Phys. **17**, 781–791 (1969).

West, S.: The treatment of diabetes mellitus by uranium nitrate. Brit. med. J. II, 467–472 (1895).

West, S.: Further observations on treatment or diabetes mellitus by uranium nitrate. Brit. med. J. **(1896) II**, 729–730.

Wilcox, R. W.: The therapeutics of uranium nitrate. Med. Rec. (N.Y.) **92**, 361–364 (1917).

Wilson, R. H.: The Hanford uranium bio-assay program. USAEC Report HASL-58 (1959).

Wing, J. F., Heatherton, R. C., Quigley, J. A.: Accidental acute inhalation exposure of humans to soluble uranium. Report NLCO-951 prepared for the Tenth Annual Meeting of the Health Physics Society, Los Angeles, California (1965).

Chapter 5

Protection Criteria

N. L. SPOOR and J. B. HURSH*

With 1 Figure

I. The History of Development of Prodection Criteria for Uranium in Air

A. Acceptable Limits to Human Exposure

Nowadays it is generally recognized that employers are under a moral and legal obligation to ensure that working conditions are safe. In areas where an industrial process involves the handling of a toxic substance personal exposures need to be kept as low as possible and always below limits which will not occasion personal discomfort or injury, whether immediate or long-term. The task of delineating for any particular toxic compound a safe limit for occupational exposure involves a detailed assessment of animal and human data, which may stem from a variety of sources and be of uneven quality. Occupational hygienists, health physicists, and medical officers assist in the task of marshalling and evaluating evidence, and their findings are interpreted from time to time by national and international organizations which accept responsibility for recommending exposure limits in air.

In most cases it is theoretically possible to reduce contamination to an extremely low level so that the air inhaled in the working environment is virtually pure. Where this degree of protection is necessary and feasible, it should be attained. In most situations, however, this is expensive and the benefits are not proportionate to the cost. Somewhere between the level of 100 percent safety and the higher level of anticipated risk there is a level at which the large majority of fit employees will experience no discomfort or injury. This level which provides adequate protection for the majority, under conditions of regular exposure, is the accepted standard which must be achieved and whenever possible surpassed. The underlying assumption is that with most toxic substances there is a threshold level below which, for the fit person, no discomfort or injury will be incurred. In radiological protection the safety limits are based on more hypothetical criteria and within a more elusive framework. Although the primary aim is to protect radiation workers against somatic or genetic injury, an ancillary consideration is the protection of the population as a whole against the genetic risks which might be associated with the contribution from this group to the cumulative population dose. The underlying assumption is that with radiation exposure there is a dose-rate below which the individual even after a lifetime's exposure is unlikely to experience discomfort, disease or life-shortening and below which

* This author was supported in part by funds from a contract granted to the University of Rochester by the U.S. Atomic Energy Commission and the number UR-3490-192 has been assigned to this chapter.

the population will not be exposed to any long-term risk which is not, to the best of our medical knowledge and enlightened imagination, offset by personal and social benefits.

B. The Meaning of "Maximum Permissible"

Even when the discussion is limited to absorption by inhalation, there are several ways of expressing safe exposure levels of an airbone material. There are basically five standards which have been used in certain cases and which may be useful to the occupational hygienist or health physicist who is trying to ensure that the environment in any particular process or laboratory is not potentially dangerous (Irish, 1965). The standards are:

a) The emergency exposure limit for a single event which is limited to a level and duration which will allow rescue with no irreversible injury;

b) An acceptable maximum for peaks above the acceptable ceiling for continued exposure;

c) An acceptable ceiling concentration, assumed to be safe for regular exposure of 8 hours a day;

d) An acceptable time-weighted average concentration, assumed to be safe for regular exposure of 8 hours a day; and

e) The minimum level of sensory detection.

In non-radiological protection against chemical toxins, whereas concepts (a), (c), and (d) all find occasional application, the most widely and frequently used is (d), the time-weighted average concentration.

As far as the hazards from non-radioactive substances are concerned, various interpretations of "maximum permissible" have been considered (Baier, 1965; Clayton, 1965 and Irish, 1965) and over the years the merits and demerits of the different kinds of standard have often been assessed (Cook, 1969; Smyth, 1962; Stokinger, 1962 and 1964). Although for the majority of substances in common use which present a non-radiological hazard the time-weighted average concentration provides the preferred maximum permissible value, the claims of the "ceiling" limit have long been too convincing to ignore and since 1963 the majority of specialists have recognized that for certain substances the best maximum permissible value is the "ceiling" limit (Stokinger, 1964). For a few other substances short-period emergency limits have been provisionally recommended (Smyth, 1966).

The organizations which set values for maximum permissible concentrations of substances which are chemically toxic emphasize that the recommended figures should be regarded not as fine lines between safe and dangerous conditions but as guides, chiefly for the benefit of specialists, in the control of occupational hazards. If for a particular substance the maximum permissible concentration for occupational exposure is x $\mu g/m^3$ air, it does not necessarily follow that a concentration of 1.5x $\mu g/m^3$ is dangerous or that one of 0.7x $\mu g/m^3$ is completely safe. Much depends on the physical and chemical properties of the material, and on the process, and on the persons exposed. There is a wide variation in individual susceptibility and consequently a small proportion of workers may experience discomfort or illness from exposures which to the majority are harmless and wholly acceptable (ACGIH, 1960).

The aim of (non-radiological) toxicological protection is to control exposure to an extent which prevents adverse immediate and chronic effects or reduces them to an acceptable level. In this context much useful information is often available from occupational exposure in industry (Stokinger, 1965). At exposure levels near the maximum permissible the effects of chemical toxins are—with a few exceptions like asbestos, beryllium and beta naphthylamine—immediate or early and usually identifiable. The aim of radiological protection is to control exposure to an extent which effectively prevents those delayed effects (e.g. cancer, life-shortening, and cataract) known to be related to the cumulative radiation dose. At low exposure levels these effects are presumed to be minutely small and statistically so elusive that their seriousness may be assessed only by theoretical estimates on large populations. In toxicological protection it is assumed that there is a "threshold": i.e. a level of exposure which can be accepted indefinitely without risk of discomfort or injury. In radiological protection it is assumed, in statistical calculations concerning large groups, that there is no threshold, i.e. there is no dose however low which may be assumed to involve no risk whatever to the individual or to the population. Also, in radiological protection there is usually—with the notable exceptions of radium and radon—little relevant information available from occupational exposures.

Radiation protection standards are set to limit the accumulated personal dose and the accumulated dose to the population. To this end limits are set to the quarterly dose to the radiation worker and to his annual dose. Within the quarterly and annual maxima the dose rate may fluctuate within wide limits. In calculating from the maximum permissible annual dose (mrem/y) the maximum permissible concentration in air for occupational exposure (μCi/ml) it is assumed that the occupational worker is regularly exposed to the limiting concentration for 8 hours each day and for 5 days each week. From this point of view it may be stated that the principles of radiological protection incorporate both a "ceiling" limit to the accumulated dose from 50 years' occupational exposure, and a time-weighted average for the quarterly and annual limits.

C. Towards an Occupational Air Standard for Natural Uranium

1. The Earliest Assessment at Oak Ridge (1947)

In 1947, three years after becoming Head of the Health Physics Division at the Clinton National Laboratory, Oak Ridge (now the Oak Ridge National Laboratory), MORGAN published a paper on tolerance concentrations of radioactive substances. The dose limits used—100 mr/24 h for X- and gamma radiation, and an RBE of ten for alpha particles—were those "set by the Clinton Laboratories and adopted by most of the laboratories associated with the Manhattan Project". MORGAN believed that the "safety factor is probably more than two or three". Using these basic criteria, he developed equations which, when solved for natural uranium (or uranium oxide), yielded an $(MPC)_a$ for continuous exposure of 211 μg uranium/m^3. For a 40-hour week the recommended figure was 886 μg/m^3.

This calculation rested on the assumption that the main hazard was from radiation and the "affected organ" was the lung, which retained on average 25 percent of the ingested uranium and contained 30 percent of the uranium in the body. The effective half-life in the lung was assumed to be two months. The $(MPC)_a$ values for the 16 radionuclides considered were offered tentatively by MORGAN with the following important qualification. "Perhaps the tolerance values given should be reduced by a factor of 10 to give added protection against the non-uniform distribution of the radioactivity in a given body organ, especially during the early stage of its fixation in the body." If it were decided to apply this qualification to the calculated $(MPC)_a$ for natural uranium, which MORGAN did not do, the 40-hour week value would be 89 μg/m^3.

The difference, implied by the reservation, resulted largely no doubt from the prevailing uncertainty among scientists about the precise meaning of "MPC" —or as it was then, MAC[1]. In the later paper Morgan (1948) gave 200 $\mu g/m^3$ as the "tolerance value" for natural uranium, "based on exposure for several years". The lung was assumed to be the critical organ and the main hazard was assumed to be radiological. The "daily tolerance value" for radiation was still 100 mr/day; the "permissible operating dose" was much lower than this, and some of the project leaders were considering reducing the maximum permissible exposure to 0.3 rem/week.

Morgan approached the problem of calculating an $(MPC)_a$ for uranium entirely as a radiobiological exercise. At the time he was responsible, on the operational side, for the control of the radiation exposures of laboratory workers, among whom the risks of overexposure were relatively small. During the same period (1944–1947) the team at the University of Rochester, New York, were approaching a similar problem from a different point of view. This team, which was composed of biochemists, pharmacologists, physicians and toxicologists, investigated the toxicity of several compounds of uranium—twelve in all—used in the commercial production of uranium from its ore.

2. The Emergency Guide at Rochester, N.Y. (1949)

In 1944, before any experimental work had been completed there was an urgent need, in the Medical Section of the Manhattan Project, for a rough toxicological guide for the safety of those handling uranium in laboratories and plants. As production under war-time conditions had to proceed without delay some working rule had to be adopted immediately. They decided to adopt as the $(MPC)_a$ for uranium for occupational exposure the current $(MPC)_a$ for lead, namely 150 $\mu g/m^3$. In the lead industry this limit had for a long time (since 1933, in fact) been regarded as an acceptable compromise between the demands of safety and those of engineering feasibility. In applying this figure to uranium there was no suggestion that lead and uranium were alike in their toxic effects or in the way they were handled by the body. It was chosen mainly because in the absence of any information to the contrary it was easy to assume that the toxicity of inhaled uranium, about which nothing was known, was comparable with that of lead, about which a great deal was known. As this air concentration had been attained in the lead industry, it seemed likely that it would be attainable in the uranium industry. This decision has been described by Hodge (1949) who pointed out that at this early stage the radioactive hazard from inhaled uranium seemed less important than the chemical toxicity.

3. Early Rochester Results (1949)

The early experiments at Rochester, N.Y., are summarized in two reports: the first by Hodge, Stokinger and Neuman (1949), dealing with soluble uranium compounds, and the second by Hodge, Stokinger, Neuman, Bale and Brandt (1949) dealing with insoluble uranium compounds.

1 In this context $(MPC)_a$ and MAC may be regarded as synonymous. The Maximum Permissible Concentration in Air—$(MPC)_a$—is the concept used by ICRP and defined in the Recommendations of that Organization (1955, 1959). The meaning of Maximum Acceptable Concentration (MAC), and of Threshold Limit Value (the TLV used by the ACGIH) have been discussed at length in the literature. See for example the papers by Smyth (1962), Stokinger (1962) and Clayton (1965).

The calculation of a $(MPC)_a$ for soluble uranium compounds was based on two established principles:

a) under exposures at or near the $(MPC)_a$ the amount of uranium present in the lung or in the kidney or the amount deposited on the bone will never, with natural uranium, constitute a radiological hazard,

b) the known typical uranium injury to the kidney is an extremely sensitive and reliable indicator of chemical toxicity, and an amount of soluble uranium compound in the atmosphere which fails to produce any kidney injury is a physiologically safe concentration.

From the results of experiments in which dogs, rats, rabbits and guinea pigs were exposed for one year to concentrations varying from 40 to 2000 μg uranium/m³, the air concentration of 50 μg/m³ was recommended as a tentative $(MPC)_a$ for soluble uranium. This figure was based on a considerable bulk of experimental data which was marshalled and assessed by the Rochester team.

The choice of a $(MPC)_a$ for insoluble uranium compounds was based on a calculation by Bale which assumed an RBE of 10 for alpha particles and showed that a concentration of 24 μg uranium in 1 g fresh tissue would deliver 0.045 rem/24 h, which corresponded almost exactly with the current tolerance dose rate. From the results of experiments in which dogs and rats were exposed for one year to concentrations varying from 1 to 10 mg uranium/m³ it was proposed, on the basis of the predicted uranium deposition in lung tissue and in pulmonary lymph nodes, that the $(MPC)_a$ for insoluble uranium for occupational exposure be tentatively fixed at 100 μg/m³.

The authors reckoned that this concentration, which was based on the radiation hazard, contained a safety factor of between two and five (Hodge et al., 1949) (AECD 2784, p. 5). However, that recommendation did not stand alone. On the following page of the report the statement appears "from the standpoint of chemical toxicity the suggested MAC range of 200–250 μg/m³ allows a reasonable margin of safety". Another account of the same work by the same authors (Voegtlin and Hodge, 1953, p. 2170) commented that "from the standpoint of chemical toxicity the suggested MAC of 100 μg/m³ allows a reasonable margin of safety". The occurrence of two MAC values in reports on the same work, by the same authors, in precisely similar contexts, inevitably caused some confusion.

The Rochester experiments were conducted on a scale that is probably unequalled in the history of toxicology. The aim of the project was not in the first instance to provide one or two $(MPC)_a$ values; it was rather to complete a comprehensive study, using different animal species, of the toxicology of several uranium compounds. The task of marshalling and interpreting all the data obtained presented a problem of considerable complexity. The results were published in detail, and the methods of interpretation explained in full. It is clear from the reports that no statistical methods were available, or conceivable, by which such a wide range of experimental data could be reduced to two $(MPC)_a$ values. The authors explained that "decisions were hard to make in some instances because it was difficult to fit large masses of data that were occasionally perplexing or even conflicting into the four arbitrary classifications of (1) safe, (2) MAC, (3) borderline, and (4) toxic for the response levels". One group of data, from which radiological considerations were omitted, were summarized in a form reproduced in Table 5.1. Another group, relating to the radiological hazard to the lung from insoluble uranium, were summarized in a form reproduced in Fig. 5.1. These data which are taken from two pages of the 2423 pages of the reports provide a rough indication of the kind of assessment involved. In each

Table 5.1. Effects on animals of one-year daily exposure to uranium dusts at various concentration levels. (From Stokinger, H. E., *in* Pharmacology and Toxicology of Uranium Compounds, Vol 3, p. 1660. Voegtlin, C. and Hodge, H. C. (Editors), McGraw-Hill, 1953)

Dust	Uranium in air ($\mu g/m^3$)			
	Non-injurious or safe level[a]	generally safe maximum allowable level[b]	level producing borderline injury[c]	toxic level[d]
UF_6		50	200	
UCl_4		50	200	
$UO_2(NO_3)_2 \cdot 6H_2O$	150[e]		250	2000
UF_4	50		3000	
UO_2	1000	10000		

[a] Non-injurious or safe; no animal showing evidence of renal injury. [b] Generally safe; an animal occasionally showing slight evidence of renal injury. [c] Borderline injury; a minority of the animals showing definite or moderate renal injury. [d] Toxic; a majority of the animals consistently showing renal injury. [e] Approximate level, as determined from the best possible estimates of dust concentration and distribution.

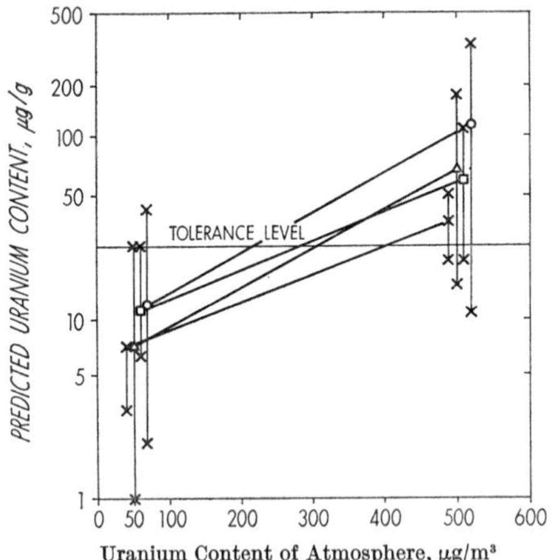

Fig. 5.1. Predicted uranium content of dog lung and pulmonary lymph nodes and of rat lung as a function of the uranium content of the atmosphere. The logarithm of the predicted uranium content (ordinate) is indicated for the highest, average, and lowest values in dog lung (△) and pulmonary lymph nodes (□ and ○) and in rat lung (×). ○, values as obtained for the dog pulmonary lymph nodes. □, values from the critically predicted data. The horizontal line across the figure at 25 µg/g represents the proposed tissue tolerance for uranium. [From Stokinger, H. E., and Neuman, W. F., in: Pharmacology and toxicology of uranium compounds, vol. 4. Voegtlin, C., and Hodge, H. C. (editors), McGraw-Hill, 1953]

case, when the task of interpolation was finished, a number was chosen which may at the time have been regarded as arbitrary, to represent the "margin of safety".

The tentativeness of the numerical recommendations made by the Rochester team stems almost inevitably from the thoroughness of their investigation. In 1947 when some of their results were being assessed, an added uncertainty was introduced by the comparative lack of information concerning the sensitiveness of humans to inhaled uranium compounds. Referring to the animal data, the authors observed: "It is quite important to state early in this account that such observations can be applied to man only in a relative fashion, inasmuch as the human being is anatomically, physiologically and biochemically somewhat different from the animal and his susceptibility to toxic substances may be different".

4. Recommendation of the Rochester Committee (1949)

A meeting held on 27th September, 1949 in Rochester, N.Y. was called by the University of Rochester Atomic Energy Project and members of the USAEC to discuss the toxicity data of uranium and to attempt to establish interim values for the $(MPC)_a$ for soluble and insoluble compounds of uranium. According to NBS Handbook 52 (1953) this meeting recommended that for both soluble and insoluble compounds the $(MPC)_a$ for continuous exposure be 1.7×10^{-11} μCi/ml (or 25 μg/m³).

5. Recommendation of the Chalk River Conference (1949)

A conference held on 29th and 30th September, 1949 at Chalk River, Canada, was attended by representatives of the radiation protection committees of the U.S.A., the U.K. and Canada[2]. These meetings initiated a discussion which continued by correspondence for some months, and the reports of the Conference also give part of the subsequent discussion. It is clear that at Chalk River the conference decided to raise the RBE for alpha particles from 10 to 20. It is probable, though less clear, that they decided to recommend 25 μg/m³ as the $(MPC)_a$ for uranium, both soluble and insoluble, for continuous exposure. In the NBS publication Handbook 52 of the U.S. Subcommittee on Internal Dose endorsed the "recommendation" of the Chalk River Conference. In fact, this publication quotes 1.7×10^{-11} μCi/ml as the $(MPC)_a$ for soluble and insoluble natural uranium and equated this value with 25 μg/m³. In this Handbook, published in 1953, it was assumed (p. 7) that 3 g natural uranium was equivalent to 7.5×10^4 alpha dis/sec, or 2 μCi. In some documents written between 1950 and 1954 there is evidence of confusion on the meaning of "curie" as applied to natural uranium. The position was clarified in the 1954 recommendations of the ICRP (published in 1955) in which the curie of natural uranium was defined as corresponding to the sum of the following activities: 3.7×10^{10} dis/sec from uranium 238, 3.7×10^{10} dis/sec from uranium 234, and 1.7×10^9 dis/sec from uranium 235[3]. According to this definition which was confirmed in 1959, one curie of natural uranium corresponds to 7.57×10^{10} alpha dis/sec, and 3 g natural uranium corresponds to 1 μCi.

6. The Tangled Reports of 1949

Between June 1949, when the results of the Rochester work were first published, and October 1949 when the first international consultation on the toxicity of uranium had taken place at Chalk River, the discussion had taken the following turns.

a) In June by all accounts the Rochester team had recommended 50 and 100 μg uranium/m³ respectively as the provisional $(MPC)_a$ values for soluble and

2 This was the first Tri-Partite Conference. The second was held in August 1950 at Buckland House, Berkshire, the third in Harriman, New York in March 1953.

3 The definition as stated here is correct. The contribution from ^{235}U was incorrectly given in the 1954 Recommendation of the ICRP (1955), the ICRP Publication 2 (1959), and the NCRP report HB 69 (1959). The error has been corrected in ICRP Publication 6 (1964) and Addendum 1 to HB 69 (1963).

insoluble uranium. (The period of exposure is not indicated but it is assumed that these concentrations are for a 40-hour week.)

b) On 27th September, 1949, apparently the Rochester team formally recommended 1.7×10^{-11} μCi/ml as the $(MPC)_a$ for both soluble and insoluble uranium. This level was associated with continuous exposure, and 1.7×10^{-11} μCi/ml was equated with 25 μg/m³.

c) On 29th–30th September, 1949, an international meeting of health physicists at Chalk River endorsed the recommendation of the Rochester meeting. Expressed in terms of curies, their recommendation was ambiguous. It is sufficiently clear, however, that they accepted 50 μg/m³ as 40-hour week $(MPC)_a$ and 25 μg/m³ as a continuous $(MPC)_a$.

BERNARD (1958) referred to the $(MPC)_a$ of 76 μg/m³ as the "NCRP recommendation". He presumably arrived at this figure by applying to 1.7×1^{-11} μCi/ml the early specific activity figure for natural uranium. This led to a continuous $(MPC)_a$ of 25 μg/m³ and a 40-hour week $(MPC)_a$ of 76 μg/m³. In giving this figure as the "NCRP recommendation", BERNARD proceeded to attribute the small factor of difference (1.5) between the NCRP value, 76 μg/m³, and the original Rochester value, 50 μg/m³, to the assumption by NCRP, that the Rochester recommendation was for an 8-hour day, 7-day week exposure rate. From a careful examination of the exposure periods of small animals he concluded that the Rochester animals were exposed to soluble uranium for an average of only 28 h/week, and that the 50 μg/m³ level corresponded therefore to an exposure rate of 28 h/week. This evidence could be used, and in fact was used, to support a values for the continuous $(MPC)_a$ of 9 μg/m³ (BERNARD, 1958).

7. Recommendation of the Harriman Conference (1953)

At the Tripartite Conference on Permissible Doses held at Harriman, N.Y., 30th–31st March and 1st April, 1953, reference was made to the relatively good health record of 26 men who had worked at least 10 years in a plant engaged in the extraction of thorium nitrate and thorium oxide from sand; EISENBUD recommended for insoluble thorium compounds for occupational exposure an $(MPC)_a$ of 500 μg/m³, and for uranium compounds 200 μg/m³. At one stage in this conference the suggestion was made that the RBE for alpha particles be reduced to five, but at a later stage in the proceedings it was decided to recommend an RBE of ten.

D. Recommendation of the ICRP Committee II (1954, 1958)

In December, 1954, the International Commission on Radiological Protection recommended that 3×10^{-11} μCi/ml (90 μg/m³) be accepted as $(MPC)_a$ for continuous exposure to soluble or insoluble natural uranium. In an appended note the Subcommittee made the following comment. "Values suggested by the Harriman, New York, Conference, April 1953. The original values were based on chemical toxicity. The present values are based on radiation exposure and are considered to be safe from the standpoint of chemical toxicity". The RBE for alpha particles remained unchanged at ten.

In 1958 the ICRP recommended 7×10^{-11} μCi/ml (210 μg/m³) as the $(MPC)_a$ for soluble uranium and 6×10^{-11} μCi/ml (180 μg/m³) for insoluble, both figures were for 40-hour week exposure. The $(MPC)_a$ value for soluble uranium was based on considerations of chemical toxicity. The value for insoluble uranium was based—as was HODGE's value of 1949—on the limiting dose-rate to the lungs of 15 rem/year. The ICRP confirmed these values in 1962.

E. Recommendation of the ACGIH (1957, 1968)

In 1957 the ACGIH recommended 50 and 250 $\mu g/m^3$ respectively as the $(MPC)_a$ values for soluble and insoluble uranium. These values were related explicitly to the results of the experiments conducted by HODGE et al. (1949) at Rochester, N.Y. In 1968 the ACGIH recommended that tho $(MPC)_a$ for uranium (soluble and insoluble) be set at 200 $\mu g/m^3$. The decision to raise the $(MPC)_a$ for soluble uranium was based on the lack of evidence, following 25 years' industrial experience, of any definite association between exposure to higher levels of airborne uranium and injury either to the kidney or to the blood.

F. Other Deliberating and Authorizing Bodies

The radiological standards applied in any particular establishment are those recommended for occupational exposure by the appropriate authority in the country concerned. The non-radiological toxicity standards recommended by the ACGIH and revised by them at regular intervals are endorsed by the responsible authorities in many countries. There is, however, no International Commission on Toxicological Protection and there are, strictly, no internationally-accepted $(MPC)_a$ figures for non-radiological hazards. There are two main sets of figures, one or other which is accepted in most countries. One is prepared by the ACGIH, the other is prepared by the health legislature of the Soviet Union. Most of the Russian levels are lower than the American, sometimes by a factor of 5 or 10 (ELKINS, 1961). Attempts have been made and still are being made by an ILO/WHO Committee on Occupational Health to examine and (if possible) reduce these differences. On the radiological side the ICRP has been active in establishing and keeping under review for many years a set of maximum permissible radiation doses and corresponding $(MPC)_a$ values for occupational and continuous exposure. These $(MPC)_a$ values have been endorsed by the IAEA (1962) and by responsible authorities in many countries.

Whereas the ACGIH values are based on exposures in an 8-hour working day, ICRP values are based primarily on exposures in a 40-hour working week.

Table 5.2. Recommended maximum permissible concentrations of natural uranium for occupational exposure

Authority	$\mu Ci/ml$	dis/min/m^3	$\mu g/m^3$
ACGIH (1969)[a]			
soluble compounds			200
insoluble compounds			200
ICRP (1959)[b]			
soluble compounds[c]	7×10^{-11}	315	210
insoluble compounds	6×10^{-11}	270	180

[a] These figures were adopted by, for example, the UK Department of Employment and Productivity (HM Factory Inspectorate, 1969). [b] These figures were adopted by the US National Committee on Radiation (US Department of Commerce, 1959) and by the International Atomic Energy Agency (1962). [c] This figure was adopted as a USSR safety standard (KULYAMIN, 1969).

II. Protection Criteria for Uranium in Drinking Water

A. History of Recommended $(MPC)_w$s

The control of uranium contamination in drinking water is a subject which has apparently received little attention from occupational hygienists and public health officers. The implication that the contamination of drinking water by uranium is an uncommon and relatively unimportant hazard is confirmed by the dearth of precise and unambiguous information in the literature. The WHO (1958, 1970) adopt the radiological standards of the ICRP, and refer to the ICRP as the appropriate source of information about the $(MPC)_w$ of uranium in drinking water.

In 1954 the ICRP recommended that the $(MPC)_w$ for continuous occupational exposure be 2×10^{-6} μCi/cm³, or 6 mg/l corresponding to a 40-hour week limit of 6×10^{-6} μCi/cm³ or 18 mg/l. This figure was based on the assumptions that the main risk was the radiological hazard to the gastrointestinal tract, and that f_1 (the fraction passing from the GI tract to the blood) was 5×10^{-4}. By 1959 the ICRP had raised the occupational figure (40-hour week) for the $(MPC)_w$ to 5×10^{-4} μCi/cm³, or 1500 mg/l. This increased limit was a consequence of a new set of assumptions about residence time, dilution, etc., in the various segments of the gastrointestinal tract. Again, this was an occupational standard which assumed that a worker consumed during each of the 8-hour days of the five-day working week 1100 cm³ water, or one-half the total water consumed. It was emphasized again that radiological toxicity was the limiting criterion (ICRP, 1959) and that the lower large intestine was the critical organ. When experimental data on man became available to show that f_1 was not 10^{-4} but more probably 10^{-2} or perhaps more (see Chap. 4, Sec. 2.B) the ICRP (Publication 6, 1964) recommended a 40-hour week $(MPC)_w$ of 2×10^{-5} μCi/cm³ or 60 mg/l, based on chemical toxicity and on the kidney as critical organ.

Table 5.3. The maximum permissible concentration (occupational) of natural uranium in water

Organization	Date	$(MPC)_w$		Parameters[a]	
		μCi/cm³	mg/l	f_1	f_w
NBS	1952	7×10^{-5} [b]	105	5×10^{-4}	2×10^{-4}
ICRP	1954	2×10^{-6} [c]	6	5×10^{-4}	—
NBS	1959	5×10^{-4} [d]	1500	—	—
ICRP	1959	5×10^{-4} [e]	1500	10^{-4}	—
IAEA	1962	5×10^{-4} [f]	1500	—	—
ICRP	1964	2×10^{-5}	60	10^{-2}	1.1×10^{-3}

[a] In accordance with the ICRP convention, "f_1" refers to the fraction going from the GI tract to the blood; "f_w" refers to the fraction taken orally that arrives in the critical organ. [b] In 1953 the NBS assumed that 3 g natural uranium were equivalent to 2 μCi. The $(MPC)_w$ recommended by them was for continuous exposure and based on chemical toxicity and the kidney as critical organ. [c] In 1954 the ICRP assumed that 3 g natural uranium were equivalent to 1 μCi. In calculating the $(MPC)_w$, for continuous exposure, they assumed that the GI tract was the critical organ. [d] In 1959 the NBS and the ICRP regarded the GI tract as the critical organ. These $(MPC)_w$ values are for occupational exposure (40-hour/week). [e] This value (for occupational exposure) assumes that the GI tract is the critical organ. [f] This value for occupational exposure assumes that the kidney is the critial organ.

B. The Limitation on Short-Term Intake

The Report of Committee 2 on Permissible Dose for Internal Radiation (ICRP, 1959) stated, in the section on Basic Standards of Maximum Permissible Internal Exposure (p. 4) that exceptionally "over a period of 13 weeks, the concentrations of the various radionuclides present in air or in water may be allowed to vary provided the total intake permitted by exposure at the constant levels indicated in sub-section 1 above". (Sub-section 1 refers to certain rules concerning the control of radiation dose-rates for occupational exposure.) There was no clear *a priori* justification for applying this principle, designed to control radiation exposure, to natural uranium, for which toxicity is related as much to chemical action as to radioactivity. When in Recommendations of the Commission (1959) the ICRP indicated that doses averaged over 13 weeks should be measured in rems, they were clearly not implying that uranium limits should be set by compliance with this principle, which would have permitted the ingestion of 100 g uranium in one event. Nevertheless, in order to remove grounds for possible ambiguity, the ICRP later made recommendations to restrict the permitted short-term intakes of soluble uranium to 150 mg by ingestion (averaged over two days). See p. 29 and 40 of Publication 6 (ICRP, 1964). These recommendations were accompanied by a reduction of the $(MPC)_w$ for natural uranium to 2×10^{-5} $\mu Ci/cm^3$, or 60 mg/l. This figure which is for occupational exposure (40-hour week), corresponds to an oral intake of about 70 mg/day.

No $(MPC)_w$ value for natural uranium has been recommended as a standard for the drinking water supplied to the general public. According to the WHO (1970), although it is recognized that the level of uranium in drinking water should be controlled, the information at present available is insufficient to form a sound basis for a recommendation.

III. The Calculation of the $(MPC)_a$ for Uranium According to Current ICRP-NCRP Practice

A. Guiding Principles and Dose Units for Radioactive Nuclides

The calculation of limits for radioactive nuclides in the working environment is based on the general premise that the production of injury is caused by and is directly proportional to the energy absorbed per gram tissue. Animal experimentation has clearly shown the need to modify this assumption to take account of the increased biological effectiveness of certain radionuclides operating as sources within the body. Generalizations to reflect this need for modification have been adopted by limit-setting bodies. For present purposes the two most important generalizations are: (1) that energy absorbed from alpha particle emissions is ten times as effective as energy absorbed from beta particle emissions, and (2) that for that class of nuclides that concentrate primarily in the skeleton an additional effectiveness factor of five is applied except for radium isotopes.

The unit of physical dose is the rad which is set equal to 100 ergs absorbed per gram tissue (or 6.24×10^7 Mev per g). The modified unit is the rem (rad equivalent man). According to dose-calculating convention:

1 rad = 1 rem (beta particles in soft tissue),
1 rad = 5 rems (beta particles in bone),
1 rad = 10 rems (alpha particles in soft tissue),
1 rad = 50 rems (alpha particles in bone except from radium),
1 rad = 10 rems (radium in bone).

A second premise is that dose rate limits may be set at a sufficiently low level so that the risk of late injury will be equal or less than the risk in other comparable occupations even though exposure is continued throughout a working lifetime. Such dose rate limits (ICRP, Publication 2, 1959; NCRP Report No. 39, 1971) are 5 rems per year for the gonads, the lenses of the eye, the red bone marrow and the whole body and 15 rems per year for other single body organs. Practice has been to set limits for bone seekers based on the effective dose delivered by 0.1 μg of ^{226}Ra which may be expressed as 28 rems per year.

B. Methods of Setting Limits where Chemical Toxicity is the Critical Endpoint

1. Alternative Approaches

As has been stated in Sec. I, the $(MPC)_a$ for soluble natural uranium[4] is determined by chemical toxicity with the kidney as the target organ. In this case dependence must be placed on the results of long-term animal experiments tempered by experience of occupational exposure to man where such experience can be interpreted. Permissible dust load limits are most simply set by selecting that exposure level which produced no injury in animals. More formally a permissible steady state concentration of the nuclide in the target organ may be set from experimental data on animals or man and the chronic exposure which will produce such a steady state concentration may be calculated.

2. Derivation of the (MPC_a) for Soluble Natural Uranium

The first direct method of setting $(MPC)_a$ is exemplified by the recommendations of Hodge et al. (Chap. 26, Voegtlin and Hodge, 1953). The more formal approach was used by NCRP Committee II and was developed as follows:

	Assumption	Source
1)	A maximum permissible concentration of 3 μg U per gram kidney	Animal experiment results (Voegtlin and Hodge, 1953; Committee judgment decision)
2)	An average kidney mass of 300 g	Standard Man (ICRP Publication 2, 1959)
3)	An effective half-life of 15 days for uranium in the kidney, i.e. $0.693/15 = 0.0462 \times$ the kidney burden is excreted per day	Animal experiments: (Voegtlin and Hodge, Chap. 11, 1949). Experiments on man: (Struxness et al., 1955; Butterworth and McLean, 1955)
4)	That 2.8 percent of the uranium inhaled was deposited in the kidney (f_a as denoted by ICRP-NCRP)	The lung model (ICRP Publication 2, 1959) specifies that the 25% deposited in the pulmonary lung is absorbed into the body for soluble compounds. The 50% deposited in the upper respiratory tract is transferred to the gut and because of the negligible absorption of uranium can be neglected. Of the systemic uranium, 78% is rapidly excreted and the remaining 22% is divided equally between bone and kidney. $f_a = 0.25 \times 0.11 = 0.028$
5)	That a worker breathes in an average of 6.9×10^6 cm^3 air per working day	Standard Man (ICRP Publication 2, 1959)

4 As well as for soluble ^{238}U and ^{235}U.

The calculation of the $(MPC)_a$ for soluble uranium then proceeds as follows:

1) The maximum permissible daily kidney input equals the daily rate of loss from the kidney when the burden is the maximum permissible $= 0.0462 \times 900 = 41.5\ \mu g$.

2) The corresponding lung daily input $= 41.5/0.028 = 1480\ \mu g$.

3) Therefore, the $(MPC)_a = 1480/6.9 \times 10^6 = 2.1 \times 10^{-4}\ \mu g/cm^3 = 210\ \mu g/m^3$.

As listed in NCRP Subcommittee 2 Report (NBS Handbook 69, 1959), ICRP Publication 2 (1959) and ICRP Publication 6 (1962), this limit has been expressed in terms of µCi per cm³ air using the special curie unit as used only for natural uranium and for natural thorium and as defined in this chapter, Sec. 5. Accordingly, 210 µg natural uranium per m³ air converts to $7 \times 10^{-11}\ \mu Ci$ per cm³ which is the value listed in the above references.

3. The Derivation of the $(MPC)_w$ for Natural Uranium

The calculation for this occupationally less important limit proceeds in exactly the same manner as for the air limit except that f_w, the fraction of the uranium taken by mouth which goes to the kidney is substituted for f_a and the average daily intake of water is substituted for the average daily intake of air. The values of these constants are as follows:

	Assumption	Source
1)	The fraction of the uranium taken in water which goes to the kidney is 0.0011	From oral intake experiments on man (see Chap. 4) a value of 1% absorption from gastrointestinal tract was selected 0.01×0.11 (see source material for air limit above) $= 0.0011$
2)	The average daily intake (occupational) of water is 750 cm³	Standard Man (ICRP Publication 2)

Therefore, if the average daily input to the kidney is set at 41.5 µg, the corresponding amount taken in by mouth is equal to 46 mg or 61 µg per cm³ water. Employing the natural uranium curie for conversion gives a limit of $2 \times 10^{-5}\ \mu Ci$ per cm³ water as listed in ICRP Publication 6 (1962).

Comment: Basis for Selection of 3 µg per Gram Kidney as Permissible

It should be noted that this value is not listed as such in the ICRP tables but is derived from three other listed values; namely, "q" or permissible whole body content listed as $5 \times 10^{-3}\ \mu Ci$, f_2 for kidney or the fraction of q, listed as 0.065, which is contained in the kidney and 300 g which is listed as the kidney mass for the Standard Man. Thus,

$$\frac{5 \times 10^{-3}}{0.33 \times 10^{-6}} \times \frac{0.065}{300} = 3.2\ \mu g/g.$$

The most extensive data on which choice of an acceptable concentration could be based is the one year inhalation study of soluble compounds of uranium reported by Voegtlin and Hodge (1953, 2174–2197). The data have been condensed and are reproduced in Table 5.4. It can be seen from this table that exposure of four mammalian species to concentrations of 200 µg U/m³ for somewhat less than 40 hours per week leads to average concentrations in the kidney of from 0.1 µg per gram to 2.7 varying with the uranium compound and

Table 5.4. Uranium concentration in the kidney after exposure of one year to inhalation of soluble uranium compounds[a]

Uranium dust concentration $\mu g/m^3$	Compound	Dogs		Rats		Rabbits[b]		Guinea Pigs[b]		Kidney injury
		no.	µCi/g	no.	µCi/g	no.	µCi/g	no.	µCi/g	
2000	$UO_2(NO_3)_2 \cdot 6H_2O$	5	1.7 (1.2–2.3)[c]	24	5.6 (1.9–11.3)	3	1.4 (0.8–2.2)	10	0.5 (0.3–0.8)	mild to moderate
250	$UO_2(NO_3)_2 \cdot 6H_2O$	15	1.0 (0.1–1.9)	23	1.6 (0.1–4.4)	5	0.9 (0.4–1.9)	10	0.1 (0.1–0.2)	occasional changes
200	UF_6	10	0.4 (0.0–0.7)	23	2.7 (0.0–5.8)	7	0.3 (0.0–0.8)	—	— —	mild
200	UCl_4	13	0.2 (0.0–0.5)	12	0.4 (0.1–1.9)	10	0.4 (0.2–0.6)	10	0.1 (0.0–0.3)	borderline
150	$UO_2(NO_3)_2 \cdot 6H_2O$	11	0.5 (0.2–1.0)	23	1.4 (0.6–3.3)	—	— —	—	— —	none
50	UF_6	12	0.3 (0.0–0.5)	26	0.9 (0.1–2.0)	—	— —	10	0.2 (0.1–1.1)	Extremely mild-very few animals
50	UCl_4	15	0.2 (0.9–0.5)	7	0.4 (0.1–0.9)	—	— —	—	— —	generally none-mild, few animals
40	$UO_2(NO_3)_2 \cdot 6H_2O$	17	0.4 (0.1–1.0)	25	0.4 (0.1–2.0)	—	— —	—	— —	none

[a] Data compiled from Voegtlin and Hodge (1953). [b] Exposure period was 7–9 months. [c] Range of values in parentheses.

with the mammalian species. Generally the histopathological examination of these kidneys showed mild tubular injury at sacrifice after one year of exposure.

Comparable results were obtained after one year's exposure of dogs, rats, rabbits, and guinea pigs to the relatively insoluble uranium compound, UF_4, at a dust load of 3000 µg uranium per cubic meter air. Average uranium concentrations in the kidney for dogs, rats, rabbits, and guinea pigs were respectively, 1.1, 2.0, 0.5, and 0.3 µg per g wet weight. Mild renal injury was detected in 47 percent of the dogs, 3 percent of the rats, 11 percent of the rabbits, and 10 percent of the guinea pigs.

Examination of these data make it clear that the ICRP and NCRP Committees responsible for setting the occupational limit for soluble uranium were less influenced in their choice of a safe kidney concentration by the animal data than by concern that the end result of the calculation reflect the experience of many years of occupational exposure. This experience (see Chap. 4, 3.A) shows no evi- of permanent kidney disfunction even though some workers were, in the past, exposed to many times the present $(MPC)_a$. In the two autopsy cases (Table 4.11) involving exposure to soluble uranium the concentrations of uranium found in the kidney are unexpectedly low.

C. Method of Setting Limits when Radiation Injury is the Critical Endpoint

1. Derivation of $(MPC)_a$ for Insoluble Natural Uranium

	Assumption	Source
1)	Lung is the critical organ	Animal experiments (VOEGTLIN and HODGE, 1953)
2)	Maximum permissible concentration in µCi per gram lung $= 8.4 \times 10^{-4}/\varepsilon$ where ε is equal to the effective energy in Mev (modified by biological effectiveness factor) per disintegration	Maximum permissible dose rate = 15 rem per year = 0.3 rem per week. $\mu Ci/g = [0.3 \times 6.24 \times 10^7]/[2.22 \times 10^6 \times 1440 \times 7 \times \varepsilon] = 8.4 \times 10^{-4}/\varepsilon$
3)	$\varepsilon = 94$ Mev (modified)/disintegration	The definition of the natural uranium special curie implies that for each ^{238}U disintegration (4.3 Mev) there occurs a ^{234}U disintegration (4.9 Mev) and 0.046 ^{235}U disintegration (4.6 Mev). Since all emissions are alpha disintegrations 1 rad = 10 rem
4)	τ, the effective half-life in the lung = 120 days	Lung model stipulation (ICRP Publication 2)
5)	f_a, the fraction of the inhaled nuclide which is deposited in the critical organ = 0.125	Lung model
6)	Mass of the lung = 1000 g	Standard Man (ICRP Publication 2)
7)	The average air intake is 6.9×10^6 cm³ per day	Standard Man

The calculation of $(MPC)_a$ is as follows:

1) $\mu Ci/g = 8.4 \times 10^{-4}/94 = 8.9 \times 10^{-6}$ µCi/g ($\simeq$26 µg/g).

2) Maximum permissible whole lung burden $= 8.9 \times 10^{-3}$ µCi (26 mg).

3) At a steady state the rate of loss from the lung will be $[0.693/120] \times 8.9 \times 10^{-3} = 5.1 \times 10^{-5}$ µCi per day.

4) The $(MPC)_a = 5.1 \times 10^{-5}/[6.9 \times 10^6 \times 0.125] = 6 \times 10^{-11}$ µCi/cm² ($\simeq$180 mg/m³).

2. Comment

a) The Critical Organ

It has been noted in the section on occupational exposure that even for so-called insoluble compounds of uranium the kidney may be the organ suffering the greatest injury after a single massive intake. The rationale for choosing the lung as the critical organ for chronic exposure rests entirely on animal experiments. The most recent experiment shows (LEACH et al., 1970) that after periods of up to 5 years' exposure to UO_2 the lungs of monkeys and dogs contain average concentrations more than two orders of magnitude greater than that in any other tissue except the pulmonary lymph nodes which contained substantially higher concentrations and the monkey spleen at about 1/10 the lung concentration. Moreover, recent data have shown (see Chap. 3) that dogs may develop lung neoplasia with high UO_2 burdens. In the face of these data it would appear prudent to continue to regard the lung as the critical organ despite the generally low lung burdens (exceptionally as high as about 20 percent of the maximum permissible burden and usually much lower [see Table 4.11]) found in postmortem tissues of workers, some of whom were exposed to many times the maximum permissible air concentration.

b) Further Consideration of the Differences in the Experimental Animal Results and Experience with Occupationally Exposed Human Subjects

Of all airborne toxic materials uranium has been the most thoroughly researched by animal experimentation. It might be reasonably expected that the results would permit a prediction of the steady state relation between air levels of insoluble uranium and the lung burden of man. Yet the indictment has been drawn and repeatedly confirmed that the lung burdens are in general substantially less than would be calculated.

Table 5.5 may serve to introduce the discussion of possible explanations of this discrepancy. The table reflects the Rochester data on long-term (one year or more) inhalation exposures of a variety of animals to UO_2 and UF_4 dust at a range of concentrations. For the purposes of the discussion it will be assumed

Table 5.5. A comparison of lung concentrations of uranium for $(MPC)_a$ (insoluble) calculated by direct scaling of data obtained at different dust concentrations

Mammalian species	Uranium compound	Dust conc. mg U/m³	Reference	Calculated[a] lung conc. µg/g wet
Dog	UO_2	1	VOEGTLIN and HODGE (1953), p. 2166	21
Rat	UO_2	1	VOEGTLIN and HODGE (1953), p. 2166	8
Dog	UO_2	10	VOEGTLIN and HODGE (1953), p. 2165	17
Rat	UO_2	10	VOEGTLIN and HODGE (1953), p. 2165	5
Rabbit	UO_2	10	VOEGTLIN and HODGE (1953), p. 2165	0.7
Guinea pig	UO_2	10	VOEGTLIN and HODGE (1953), p. 2165	3
Dog	UO_2	5	LEACH et al. (1970)	72
Monkey	UO_2	5	LEACH et al. (1970)	130
Rat	UO_2	5	LEACH et al. (1970)	29
Dog	UF_4	0.5	VOEGTLIN and HODGE (1953), p. 2168	2.1
Rat	UF_4	0.5	VOEGTLIN and HODGE (1953), p. 2168	1.3
Dog	UF_4	3.0	VOEGTLIN and HODGE (1953), p. 2168	0.7
Rat	UF_4	3.0	VOEGTLIN and HODGE (1953), p. 2167	0.5
Rabbit	UF_4	3.0	VOEGTLIN and HODGE (1953), p. 2167	0.6
Guinea pig	UF_4	3.0	VOEGTLIN and HODGE (1953), p. 2167	0.05

[a] The experimentally found concentration has been normalized to an exposure of 180 µg/m³ assuming direct proportionality between exposure level and steady state concentration.

that there is a direct proportion between the air concentration of uranium and the lung burden in the steady state after one year's exposure. The average concentration of uranium in the lung as reported is then normalized to an air level of 180 μg per m^3 [$(MPC)_a$] and listed in the last column. If the lung model used for man is applied to these data, the predicted lung concentration at a steady state would be 26 μg/g lung for 180 μg/m^3 exposure conditions. It may be seen that the calculated lung burdens in Table 5.5 differ by three orders of magnitude. Some patterns may be detected in the data: 1) the retained burden of UF_4 is consistently less than the retained burden of UO_2; 2) there is a clear species difference in the decreasing order: monkeys, dogs, rats, guinea pigs, and rabbits. There is a less compelling suggestion that higher air concentrations may be associated with lower lung burdens. Finally, the recent data from the study by LEACH et al. (1970) show substantially higher deposits indicating that there is an important variable which is not constant in the experiments yielding the data cited. It seems most likely that this variable is particle size. It is recognized that the particle size distribution determines the fractional deposition in the pulmonary lung (Task Group on Lung Dynamics, 1966) and the rate of dissolution from the lung (MERCER, 1967).

The lung model chosen by the responsible committee (ICR Publication 2, 1959) for the calculation of the expected lung burden in man was, of necessity, an approximation and, furthermore, prudence dictated that it be a conservative approximation, i.e. that the prediction overestimate rather than underestimate lung burdens. The parameters of the lung model pertinent to the present discussion are the division of inhaled compounds into the two categories of soluble and insoluble, the fraction deposited for the insoluble category, and the rate of removal from the lung. The model may in certain cases overestimate the fractional deposition of uranium in the pulmonary lung of occupationally exposed workers. The evidence for this is cited in Chap. 4, Sec. II.C, and consists of the statement (HARRIS, 1961) that dust measurements in some uranium plants showed that 95–99 percent of the dust would be removed in the upper respiratory tract. This may be compared with the 50 percent removal assumed by the lung model. The implication from HARRIS's statement is that only 1 to 5 percent of the dust inspired penetrates to the pulmonary region and that some fraction of these amounts is deposited. The lung model stipulates that 25 percent of that inhaled is deposited in the lower respiratory passages. The report of the ICRP Task Group on Lung Dynamics (1966) provides a new lung model with improved prediction capabilities. Using this model, fractional deposition in the nasopharyngeal, the tracheo-bronchial, and the pulmonary regions may be predicted if the particle size properties and ventilation states are known. The deposition is a sensitive function of the mass median diameter of the dust and, unfortunately, this information is lacking for the early exposure of uranium workers and, therefore, use of the new model does not provide the further needed insight.

The ICRP lung model divided all aerosols into soluble and insoluble compounds. For insoluble compounds half of that deposited in the lower respiratory tract (12.5%) was removed relatively rapidly and half was lost with a half-life of 120 days or at a rate of 0.0057 times the lung burden per day. The cumulative lung burden was referrable to the latter fraction. Application of this model to a specific industrial situation brings up the questions "how insoluble" and "insoluble in what". For example, UO_3, UF_4, UO_2, and U_3O_8 would all be generally classified as insoluble in water. Yet MORROW et al. (1971) have shown that inhaled UO_3 leaves the lungs of dogs with a 4.7 days half-time; the animal data of Table 5.4 show that exposure to UF_4 results in substantially lower lung burdens than an equivalent exposure to UO_2. SCOTT (see Chap. 6, Table 6.1) chooses to divide uranium compounds into three classifications: highly transportable, moderately transportable, and slightly transportable. In doing so he follows the recommendation of the ICRP Task Group on Lung Dynamics that three categories be set up based on solubility in tissue fluids.

The choice of 120 days half-time when applied to the more tissue fluid insoluble compounds such as UO_2 and U_3O_8 receives some limited support from animal and human data. The most extensive compilation of data for the dog is presented in Chap. 3, Table 3.1, and shows that, although the lung burden appears to be just short of a maximum after one year of exposure, the loss rate after termination of exposure at five years is shown as occurring with a half-time of 15.7 months. Human data presented in Chap. 4, Table 4.12, lists biological half-times which, if limited by lung loss rates, suggest a clearance described by two exponential terms with the long-term component yielding half-lives from 121 to 1500 days. It seems justifiable to conclude that the postulated lung model half-time is not overly long and thus cannot be held responsible for the overestimate of the lung burdens in man.

Having reviewed the grounds for questioning the predicted lung burdens, attention may be directed to the human exposure and postmortem data. At the outset it may be said that the force of these data resides not so much in a blithe confidence in the precision of the environmental and tissue analyses at any one installation as in the compelling fact that the data from all the installations point to the same conclusion. Consequently, any reservations should be general in their application. One such reservation hinges upon the nature of the human exposure. The model presumes, and the animals experience, a chronic uniform level of exposure year in and year out. This is not the usual pattern of human industrial experience, which is more likely to consist of a series of sporadic exposures. The effect of different levels of input loading of the lung on retention and clearance is not sufficiently understood to warrant a judgment as to whether this might be a minor perturbation or a major determinant of the steady state lung burden. A second potentially important aspect of the human data is the lack of specification of the chemical nature of the inhaled aerosol (see Chap. 4, Table 4.11). The exposure is in some cases classified as mixed and in other cases, even though classified as insoluble, may involve important fractions of compounds which are moderately transportable in the human body.

In summary, the differences between the prediction based on animal experiments and actual uranium lung burdens of workers exposed to insoluble uranium may be the result of a combination of factors including: 1) species differences; 2) an overestimate of the fractional deposition in the pulmonary lung; 3) exposure to industrial uranium aerosols in part not truly body-tissue-water "insoluble"; and 4) the difference in the pattern of animal inhalation exposure and that in the uranium industry. At the present time the relative importance of these factors is uncertain.

Attention should be directed to the possible future role of *in vivo* gamma measurement of insoluble uranium lung burdens as a means of resolving the discrepancy. The data reported on 15 workers by QUASTEL et al. (1970) lists a range of 0 to 4 times the maximum permissible lung burdens and an average value of 2 times. The Oak Ridge experience as represented in the two papers by WEST and SCOTT (1966, 1969) is based on *in vivo* gamma counting of 2500 subjects exposed to uranium. Their reports give details on only 5 cases, all of which displayed exceptionally long chest clearance half-times, presumably associated with highly insoluble forms of uranium. The *in vivo* chest burdens of these subjects were 1 to 1.5 times the maximum permissible lung burdens. SCHULZ (1968) reports on 80 employees exposed to enriched uranium oxides at the Oak Ridge Gaseous Diffusion Plant. He finds chest burdens of 6 to 50 nCi with only 2 subjects showing burdens greater than 17 nCi. He estimates the average air exposure level at 30 percent of the $(MPC)_a$ which might be expected to yield by calculation a steady state lung burden of about 17 nCi.

Although air exposure data are not cited in the reports by QUASTEL et al. and by WEST and SCOTT and, therefore, the measured chest burdens cannot be associated with calculated values, the *in vivo* measurements suggest that a minority, perhaps a small minority of active workers, have uranium lung burdens substantially higher than those found for the autopsied subjects. This tentative interpretation must be further qualified. The technique necessarily measures the chest burden and, as suggested by several authors (DOLPHIN, 1964; MORROW et al., 1971), may include uranium removed from the lung but remaining the survey field. The magnitude of the overestimate caused by this artifact will be reduced for the more sparingly soluble forms of uranium. The contention that such an error may have to be taken into account receives a degree of support from the single insoluble uranium exposure case reported by HEID and FUQUA (1971) who find a chest burden of 18 mg natural uranium measured by *in vivo* counting as contrasted with 6 mg determined as an upper estimate by analysis of postmortem lung samples. It is believed that the final resolution of the difference between the predicted and the real lung burdens will lead to improved protection practices and criteria. On the other hand, it needs to be emphasized that, in the light of present knowledge, the wisdom of choice of the $(MPC)_a$ equal to 180 $\mu g/m^3$ for insoluble uranium is not being brought into question. The industry has accepted this level and the long history of freedom from injury among uranium workers has confirmed its safety.

c) The Lymph Nodes as a Critical Organ

Monkeys and dogs exposed for five years to inhalation of UO_2 (LEACH et al., 1970) concentrate uranium in the tracheobronchial lymph nodes so that the steady state values are about 10 times higher than the concentration in the lung tissue. If the animal data were used for guidance and if the lymph nodes were designated as critical organ for man, the $(MPC)_a$ would be reduced to 20 μg per m^3. The responsible committees have not chosen to take this step. Quite apart from the problems that the industry might encounter in achieving this very low level it can be rejected on the grounds: 1) that loss of function of some of the tracheobronchial lymph nodes does not necessarily produce functional impairment for the individual; 2) that the postmortem analyses of uranium workers (see Chap. 4, Table 4.11) show uranium concentrations in lymph nodes not greatly different from the concentrations in lung tissue and both much lower than predicted; 3) that in dog experiments (see Chap. 3, Sec. VII, B.4) involving a five-year exposure to UO_2 and a five-year post-exposure study malignant lesions were detected in lung tissue, but only fibrotic changes were found in the lymph nodes.

IV. Modification of the Occupational MPCs for Application to the General Public

A. Lower Limits Recommended

For the development of basic dose limits both the ICRP and the NCRP consider two categories of public exposure; namely, the exposure of an individual in the population and the exposure of the population as a whole. It is argued that the basic dose limit for an individual member of the public should be lower than that for an occupationally exposed person inasmuch as (1) members of the general population include individuals more susceptible to injury, (2) medical and environmental supervision is focused for the worker and may be absent or diffuse for members of the public, and (3) that whereas the worker accepts the minimal risk of job-related injury together with his pay envelope, the individual in the public domain usually has little or no personal option and the benefit, if any, is usually indirect.

The advisability of specifying a lower average dose limit for the population as a whole (than for individuals) is based on genetic considerations and is intended to prevent the accumulation of an unacceptably large number of radiation-induced, recessive mutations in the genetic pool of that population. A more detailed discussion of background material for the recommendation that the dose limits for the general public be reduced from the occupational limits may be found in ICRP Publication 6 (1964) and in NCRP Report No. 39 (1971).

B. Basic Radiation Dose Limits and Related MPCs for the General Public

1. Individual

Exclusive of natural radiation and that incurred by practice of the healing arts (medical and dental X-rays) the ICRP recommends that an individual member of the population receive no more than an additional 0.5 rem per year to the whole body, gonads, or blood forming organs, no more than 3.0 rems per year to the skin, thyroid and bone, and no more than 1.5 rems per year to other organs. (The three limits are stipulated in order to include radiation from radionuclides which may concentrate in one or the other body organs with differing radiation sensitivity. For modified limits to apply to special cases, see ICRP Publication 9, 1966). The dose limits are set at one-tenth the corresponding occupational dose limits. The NCRP recommends an across-the-board dose limit of 0.5 rem per year, not because of any difference in interpretation of available data, but in the interests of simplification.

2. Population as a Whole

The ICRP (and NCRP) recommends that the average annual dose to the population be no more than 0.17 rem averaged over one year, excluding the dose from natural sources and from practice of the healing arts. Since the genetically significant dose is the ruling criterion, the critical target is considered to be the gonads.

3. The Derived MPCs

ICRP (Publication 6, paragraph 65, 1964) warns that the whole population suggested maximum dose of 5 rem ($\simeq 5/30 = 0.17$ rem per year) must not be used up by one single type of exposure, but goes on to concede that the proper apportionment will depend upon circumstances which may vary from country to country and, therefore, should be a matter for national, not international, decision. Nevertheless, because of the need of guidance in the planning of "the development of nuclear energy programs (with the associated waste disposal problems) and more extensive uses of radiation sources", an illustrative apportionment is provided in ICRP Publication 6, pp. 31–32, 1964 (see also ICRP Publication 9, p. 16, 1966, for a later de-emphasis). This apportionment assigns 1.5 rem total for the dose from internal emitters. The corresponding dose rate limit, 0.05 rem per year, would be satisfied by setting the MPC for the nuclide in question at 1/100 of the occupational MPC 168-hour week where the gonad is the critical target. For nuclides that concentrate in organs other than the gonads it is suggested that the MPCs for the population at large should not exceed one-thirtieth of the MPC 168-hour week values for occupational exposure.

There are no correpsonding specific examples of deriving MPC values for the individual member of the general public. If concern could be directed to a single nuclide and a single route of principal intake, and if there were no associated external source of radiation (other than natural sources and diagnostic X-rays), the dose limit to the individual would be satisfied by setting the maximum permissible concentration in air or water at 1/10 the occupational MPC 168-hour week for that nuclide.

C. At what Numerical Size Does a Group Become a Population?

The practical application of the recommendations raises the question: "How large must the group be before concern for somatic injury to the individual is replaced by concern for genetic injury to the social group?" ICRP (Publication 2, p. xxxi, 1959) avoids the issue: "In the case of, for example, average exposure, however, the "*whole*" *population* is involved and the responsibility for the interpretation of this concept will rest with national and international administrative authorities responsible for the legal application of protection requirements". The NCRP is more forthright. In paragraph 247 (Report No. 39, 1970) the dose limit is defined: "The dose equivalent to the gonads for the population of the United States as a whole from all sources of radiation other than natural radiation and radiation from the healing arts shall not exceed a yearly average of 0.17 rem (170 mrem) per person (see paragraph 162)". In the reference paragraph the position is qualified by the statement that there exist in the United States "several genetic pools and that currently there is relatively little mixing across certain racial or ethnic lines". They advise that the averaging should be done over these sub-population where they can be clearly defined.

In the event, the decision by the responsible administrative and legal authorities may present little difficulty. For example, if the size of the group drawing drinking water from a marginally contaminated water system is less than 5 percent of the population of the national unit, it would generally be true that the dose limit to the individual would be more restrictive and would constitute the proper choice. As the size of the affected group increased there would presumably occur some point, modified by assumptions about individual variability in uptake etc. and by considerations such as those mentioned above, where the average dose to the populations would be the appropriate criterion.

D. Limits for Uranium in Environmental Air and Water

1. Soluble Natural Uranium Compounds

Inasmuch as chronic exposure to soluble natural uranium compounds is judged to be limited by the chemo-toxic effect of uranium and because the limiting dose for whole populations is based on the radiation effect on the gonads, limits pertaining to the whole population will not be estimated here. On the other hand, there is some reason to believe that, in respect to exposure of the individuals in the population, the arguments for reducing the occupationally permissible concentrations (see this chap., 5.A) remain valid even though the postulated criterion is chemo-toxic and not radiation injury. Acceptance of this position would result in setting limits at one-tenth of the occupational MPCs yielding 9 $\mu g/m^3$ air ($\simeq 3 \times 10^{-12}$ μCi per cm^3) for soluble natural uranium. The corresponding limit for water would be 1.8 μg per cm^3 (6×10^{-7} μCi per cm^3).

2. Insoluble Natural Uranium Compounds

As has been discussed in an earlier part of this chapter, the occupational limit for airborne insoluble natural uranium is based on radiation and lung is regarded as the critical organ. The ICRP suggests that the MPC values for the whole population should not exceed one-thirtieth of the 168-hour week occupational MPC values. Following this suggestion the whole population upper limit for insoluble uranium compounds in air becomes 2 μg per m^3 ($7 \times 10-3$ μCi per cm^3). If circumstances be such that the dose to the individual in the population is the basis for decision, the limit in air becomes one-tenth the occupational value or 6 μg per m^3 (2×10^{-12} μCi per cm^3).

Although MPC values for "insoluble" natural uranium in water are included in ICRP Publication 2, they would appear to have little practical meaning in either occupational exposure control or protection of the general public. It is conceivable that food might be contaminated with insoluble uranium compounds and that the listed MPC values would facilitate calculation of maximum permissible oral daily intakes. However, further development of this subject is beyond the intended scope of the present discussion.

E. Comparison of Population MPCs and Present Environmental Levels

1. Uranium in the Environment

Environmental uranium does not presently, and is not anticipated in the future, to be a hazard to either individuals in the population or to the population as a whole. It would be expected that the rare reactor accident would present a fission product hazard rather than a uranium hazard. The possible exception might be a chemical explosion of fresh uranium. It is difficult to imagine circumstances in which this would occur and would lead to widespread hazardous contamination of the public domain with uranium. Routine industrial processing of uranium can be performed with relatively unimportant releases to the public environment. The relatively low specific activity of ^{238}U aids in its detection and control inasmuch as radioactively significant amounts involve gravimetrically appreciable amounts.

Table 5.6 lists the maximum permissible concentrations of natural uranium for the public and also the present environmental concentrations in air and water. It is seen that the environmental levels are lower than the MPCs by three to four orders of magnitude.

Table 5.6. Comparison of population MPCs for natural uranium and present environmental levels

	$MPC_{pop.}$	$MPC_{ind.}$	Environmental[a]
U in air μg per m^3	— 2 (insoluble)	9 (soluble) 6 (insoluble)	$0.06–0.4 \times 10^{-3}$
U in drinking water μg per l	—	1.8×10^3 (soluble)	$2.4–4.1 \times 10^{-4}$
Food μg per day			1.2–1.6

[a] See Table 4.16 for references.

2. Uranium Production Waste Products

Uranium processing does produce some waste products that may constitute an environmental hazard. The waste produced in the milling of uranium ore contains amounts of ^{226}Ra which depend on the percent uranium in the ore. Radium leached from these ore residues may find its way into waterways which supply drinking water for communities down-stream. WRUBLE et al. (1964) report on the contamination of the Colorado River basin by ^{226}Ra from ore tailings and the corrective measures that were instituted. Under certain conditions contamination of the air can result from uranium ore residues. In this case it is not radium but its radioactive daughter products that constitute the hazard. The situation is in every way similar to the occupational exposure of uranium miners (see Chap. 7) and may be brought about if the uranium ore residues are used as ground fill or are introduced as building materials for the communities in the neighborhood of the uranium mills.

V. Derived Occupational Limits for Personnel Monitoring

A. Urinary Excretion Limits

1. General Remarks

Although it is generally recognized that routine urinary assay for uranium is of limited value as a basis for calculation of body burdens, it continues to be a bioassay method widely used in the uranium handling industry. The reasons for its popularity and continued use are easily recognized. Collection of the sample involves no discomfort and little inconvenience to the worker. It is likely that the absence of such a program would be construed by the courts and by the labor unions as evidence of negligence on the part of the employer. In most cases the cost of the program can be charged to the customer which, for uranium, is the federal government. To these reasons of expediency may be added some reasons more responsive to the intent of the program. Some radiological health officers regard man as the best air sampler of the occupational environment and use urinary results to supplement mechanical air samplers for current control of air levels of uranium. Unreported accidental exposures may be detected by the routine urinary program and the worker burden assessed by special procedures. Inasmuch as the collected urine will accurately reflect the average blood plasma concentration over the collection period to the extent that this information relates to the critical organ content the procedure should give a valid estimate of the organ burden.

2. Urinary Levels as an Environmental Control Index

If the uranium aerosol is one of the more soluble compounds such as $UO_2(NO_3)_2$ or UO_2F_2 (from $UF_6 + H_2O$) because of the expected prompt appearance in the urine of a large fraction of the current intake, a urinary sampling program in which samples are collected during the work week will be a sensitive indicator of increases in the exposure of the working force. When urine sampling is used to this end, the interpretation of the urinary results is qualitative and increases are judged as significant by the local radiological protection officer on the basis of his experience.

If the exposure is to the more insoluble compounds of uranium such as UO_2 and U_3O_8, a urinary uranium program should not be expected to be a sensitive method of environmental control. For this purpose the collection and analysis of fecal samples would be advisable, or the alternative of collecting "nose-blows" (Lister, 1966) after a suspected release of insoluble airborne radioactive material.

3. Urinary Analysis as a Means of Detecting Unreported Accidents

The history of monitoring experience at all major installations includes examples of excessive individual exposures to uranium which were first detected by the urine sampling program. Little need be said to emphasize the importance of this role and the necessity to evaluate the dose incurred by the individual. It is supported by legal as well as ethical considerations.

Any sudden increases in the uranium content of the urine should receive special attention, particularly increases so that the amount excreted is 25 percent or more of the permissible urinary chronic outputs as defined below. Once an accidental exposure is detected, the usual procedures of removal from exposure, collection of 24-hour urine and fecal outputs, and an *in vivo* counting schedule should be instituted. Examples of such procedures may be found in the accounts by Scott and West (1964), by Saxby et al. (1964) and by Fish (1961).

4. The Use of Routine Urinary Uranium Data to Assess Individual Exposure

The uranium program shares the uncertainties inherent in any routine urinary bioassay program. The usefulness of the analytical data is limited by the precision, adequacy, and frequency of the urine collection procedure. Complete 24-hour collection samples are preferrable but in routine practice spot small volume samples have been more common. Samples collected in the working environment have an increased chance of unwanted contamination. Timing of the collection period in relation to the work week is particularly important for uranium. Even though these precautions are taken the body burdens inferred from urinary analysis for uranium have large uncertainties, as emphasized in Chap. 4 in the account of industrial experience.

B. Urinary Uranium Limits

Despite these reservations, current industrial dependence on the urinary sampling of uranium workers requires some general guidance in the form of calculated limits in urinary output so that the program may take its place in the overall protection scheme. Such a calculation was first published by Neuman (1950). Additional recommendations are contained in papers by Jackson (1964) and by Jackson and Dolphin (1966). Despite some changes in $(MPC)_a$ and in officially accepted biological parameters, the limits suggested 15 years ago by Neuman are not greatly different from those proposed in the more recent considerations.

1. Urinary Limit for Exposure to Soluble Forms of Uranium

A urinary limit for soluble uranium may be developed in the following way. Assume that the worker is and has been for some time exposed to the currently accepted occupational $(MPC)_a$ (soluble) so that as far as his metabolism of uranium is concerned he has reached a steady state; i.e. a state in which the rate of total uranium excretion is equal to the rate of uranium intake. Fecal excretion is considered to be negligible and, therefore, the total biological loss occurs by urinary excretion.

	Calculation	Assumptions
1)	*Working day* lung retention at the $(MPC)_a$ (soluble) $= 210 \times 10 \times 0.25 = 525$ μg/day	ICRP Publication 2 value of (MPC_a), 10 m^3 taken as the inspired working day air volume and 0.25 as the fraction retained by pulmonary compartment as per ICRP Publications 2 lung model
2)	Prompt urinary excretion $= 525 \times 0.78$ $= 410$ μg/day	78% of a single intravenous dose is excreted in 24 hours (Chap. 4). ICRP Publication 2 stipulated that 11% of lung retained soluble uranium goes to kidney and 11% to bone. The remainder is here assumed to be promptly excreted
3)	*Average* working day retention at $(MPC)_a$ $= 210 \times 6.9 \times 0.25 = 360$ μg/day	For purpose of calculation daily input to the kidney and bone because of their relatively long biological half-lives, the *average* daily input (i.e. averaged over the working year) is regarded as appropriate. See ICRP Publication 2, Standard Man
4)	Daily urinary contribution from the kidney = daily urinary contribution from skeleton $= 360 \times 0.11 = 40$ μg/day	In the steady state the average daily loss from the organ must equal the average daily input
5)	Therefore, the postulated maximum total urinary excretion would be $410 + 40 + 40$ $= 490$ μg/day	

The pattern of excretion will need to be kept in mind. A sample collected during the work week would be expected to conform to the total excretion as stated above if the standard man was in a steady state to an exposure of 210 μg/m^3. If the sample were collected on Monday morning, the promptly excreted component would be reduced greatly since two and one-half days absence from exposure have intervened. One would, therefore, set the corresponding maximum permissible urinary output of such a sample to include little more than the loss from the kidney and from the bone burden; namely, at 80 μg per 24-hour sample.

2. Urinary Limit for Exposure to Insoluble Uranium Compounds

A similar exercise performed on the ICRP Publication 2 model for insoluble uranium yields 155 μg per day as the urinary excretion for the individual worker in a steady state exposure to 180 μg/m^3 working environment. The usefulness of this urinary limit has been frequently brought into question and the consensus is that lung *in vivo* gamma counting should be used as a substitute for or an adjunct to the urinary bioassay program when the exposure is primarily to the sparingly soluble compounds of uranium.

C. The Urinary Limits Compared to Empirical Plant Findings

Jackson (1964) has calculated the mean of the ratios of average μg U per liter urine divided by average μg U per liter air using industrial data collected by Lippmann (1959). When the urine sample was taken before the weekend and the exposure was to soluble uranium, this ratio is 2.3×10^3. The comparable calculated value (see above) is 490 μg per day divided by 1.4 liters per day divided by 0.21 μg/liter $= 1.7 \times 10^3$. When the sample was taken after the weekend Jackson's ratio was 390 and the comparable value based on calculated urinary limits is 270.

Another set of values from industrial exposure experience is presented by QUASTEL et al. (1970). These investigators report an average of 26.1 ± 2.4 μg per day excreted by workers having a lung burden of 30 mg (i.e. the maximum permissible lung burden) as determined by *in vivo* lung gamma counting. The worker group studied numbered 15 and the exposure was to chronically inhaled uranium dioxide. The urinary excretion rate found by this empirical method is about one-sixth of that predicted (namely, 155 μg per day) by the calculation presented in B.2, this subsection.

The approximate agreement with JACKSON's ratios should not be interpreted as indicating that body burdens of an individual are predictable from a single routine urine sample. LIPPMANN (1959) and HURSH (1959) cite reasons to the contrary. However, the agreement does render some assurance within the limitation of the data that the lung model calculation is not wide of the mark for the average relationship between excretion and air levels of soluble uranium compounds.

VI. Practical Considerations Applicable to Industrial Conditions

The recommended $(MPC)_a$ values for natural uranium given in sec. 1 and 2 may appear to provide the health physicist with adequate standards on which to base any program of industrial radiological surveillance. In practice, however, this is not always the case. Because of the anomalous definition of the curie of natural uranium and the varying isotopic content of depleted uranium and enriched uranium[5] uncertainty occasionally arises about the interpretation of the ICRP $(MPC)_a$ values in terms of alpha dis/min/m³.

A. Selection of an Air Limit where Solubility of the Uranium Aerosol Cannot Be Stipulated

Two cases arise in practice. The first and most common occurs when it is not possible to state firmly that the material is either soluble or insoluble; the second arises when the uranium is known to be in soluble form. In the former case there is no option but to take the most restrictive $(MPC)_a$ value. For natural uranium this is 6×10^{-11} μCi/cm³ (ICRP Publication 6, 1964) and, taking into account the ICRP definition of the curie, this corresponds to 272 alpha dis/min/m³. For separated uranium-238 the figure is 7×10^{-11} μCi/cm³, which corresponds to 155 dpm/m³. The figure for uranium-238 is held to a lower figure than would be expected on pure radiation grounds because of the limiting effect of the chemical toxicity in the kidney. The figure for natural uranium is limited about equally by chemical toxicity to the kidney (soluble) and by the potential radiation dose to the lung (insoluble). For highly enriched uranium the alpha activity is due almost entirely to uranium-234 with an $(MPC)_a$ (insoluble) of 10^{-10} μCi/cm³, corresponding to 222 dis/min/m³. As these figures all lie within a narrow range there is clearly an advantage to be gained from using a single figure for all enrichments and a figure of 200 dis/min/m³ has been used in many establishment for many years.

5 An important aspect of uranium processing is the treatment of natural uranium by a gaseous diffusion process which increases or "enriches" the proportion of the thermal neutron fissionable ^{235}U and incidentally of ^{234}U in the isotope mixture. The corollary to the production of enriched uranium is the presence of the residual depleted uranium.

B. Selection of an Air Limit when Uranium Aerosol is Known to Be Soluble

In the case of materials known to be soluble, some relaxation is possible for enriched uranium because the $(MPC)_a$ for uranium-234 (6×10^{-10} μCi/cm^3) is higher than that for uranium-238 (7×10^{-11} μCi/cm^3), the former being limited by radiation toxicity and the latter by chemical toxicity. The critical organ for soluble uranium-238 is the kidney and for soluble uranium-234 the bone. The situation is complicated by the fact that the radiation dose from uranium-234 to the kidney is not insignificant, although it is not, in comparison to the bone dose, limiting.

C. Air Limits for Varying Degrees of Enrichment

In considering the effect of soluble uranium on the kidney it has been assumed in this assessment that the effects of chemical toxicity and radiological toxicity may be additive. This is probably a cautious assumption, but it makes only small changes in the $(MPC)_a$ values. Simplifying assumptions have also been made about the isotopic content of enriched uranium. It has been assumed that the mass ratio of uranium-234 to uranium-235 remains the same as in natural uranium for enrichments up to about 10 percent. The value of this ratio is 0.0076, and the $(MPC)_a$ values are not sensitive to small variations in the ratio. Table 5.7 shows the values of $(MPC)_a$ in alpha dis/min/m^3 for a range of enrichments. The values have been rounded off to two significant figures. Table 5.8 summarizes the recommended values rounded off to one significant figure for soluble and insoluble compounds. It must be emphasized that the figures for soluble compounds may be used only when it is certain that the material is in a soluble form.

It is indicated in Table 5.8, and stated in ICRP Publication 10 (1968), that when natural uranium has been enriched twelve-fold by the diffusion process the critical consideration ceases to be the chemical risk to the kidney and becomes the radiological risk to the bone. The choice of this enrichment factor as a transition-point in dose assessment is slightly arbitrary; it is based on a study of the contributing risks from uranium-238, uranium-235 and uranium-234 in natural uranium and in uranium of different degrees of enrichment.

For soluble compounds of natural uranium, irradiation of the bone, kidney or any other organ is a less vital consideration than the chemical toxicity to the kidney, and this is a function not of alpha activity but of mass. Natural uranium contains more than 99 percent by weight of uranium-238, with 0.71 percent of uranium-235. Also present in similar equilibrium with uranium-238 is uranium-234 (produced by the radioactive decay of uranium-238 via the shortlived nuclides

Table 5.7. Maximum permissible concentrations of soluble uranium compounds in air

^{235}U content (%)	Enrichment factor	$(MPC)_a$ (40-h week) dis/min/m^3	Critical organ
0	0	150	kidney
0.4	0.56	250	kidney
0.71	1.0	320	kidney
1.4	2.0	420	kidney
2.8	4.0	650	kidney
6.0	8.45	960	kidney
10.0	14.1	1200	bone

Table 5.8. Recommended values of $(MPC)_a$ for 40-hour week exposure of uranium workers

Material	$(MPC)_a$ (dis/min/m^3)	Critical organ
Insoluble Compounds		
Any enrichment or depletion	200	lung
Soluble Compounds		
Natural or depleted	200	kidney
Enriched up to 3%	400	kidney
Enriched 3% to 6%	700	kidney
Enriched 6% to 12%	1000	kidney
Enriched above 12%	1000	bone

Table 5.9. Alpha-emission of Naturally-Occurring Uranium Isotopes

Specific alpha activity of isotope (dis/min/g)	Isotope	Natural uranium	
		weight of isotope in 100 g natural uranium (g)	total alpha activity[a] (%)
7.4×10^5	^{238}U	99.283	49.07
4.7×10^6	^{235}U	0.712	1.86
1.4×10^{10}	^{234}U	0.0054	49.07

[a] The specific alpha activity of natural uranium is 1.5×10^6 dis/min/g

thorium-234 and protactinium-234). Although the mass concentration of uranium-234 in natural uranium is small, its specific alpha activity is by far the highest among the three uranium isotopes concerned (see Table 5.9).

When uranium is enriched in uranium-235 by the diffusion process, it is also enriched in uranium-234, which rapidly becomes the biggest contributor to the total alpha activity. Since the energy of alpha-emission of these three uranium isotopes is similar, the composition at which there is for soluble compounds a change in critical organ can be defined simply in terms of specific alpha-activity. If it be assumed that the proportion of uranium-234 to uranium-235 remains constant in the course of enrichment, it can be calculated that there is a change from kidney to bone as the critical organ at a uranium-235 concentration of 16 percent. If the alternative assumption be made that the proportion of uranium-234 to uranium-235 is that which would result from the diffusion of uranium hexafluoride according to the ideal gas laws, the calculated concentration for the change in critical organ is nearer 8.5 percent (Jackson, 1962). Both these values are hypothetical; the specific alpha-activity (8×10^6 dis/min/g) is the only means of stipulating without ambiguity the composition at which the transition occurs. This figure (8.5 percent uranium-235) like those in Table 5.7 is based on the assumption that chemical damage to the kidney and radiological damage to the kidney are complementary and additive. Although this assumption is cautious and plausible, its validity has not been demonstrated.

List of Acronyms

ACGIH	American Conference of Governmental Industrial Hygienists
IAEA	International Atomic Energy Agency
ILO	International Labour Organization
MAC	Maximum Allowable Concentration (in air)
$(MPC)_a$	Maximum Permissible Concentration (in air)
TLV	Threshold Limit Value
WHO	World Health Organization

References

American Conference of Governmental Industrial Hygienists: Threshold limit values for 1960. Arch. environm. Hlth **1**, 140–144 (1960).

BAIER, E. J.: Standards for exposure to airborne contaminants. Arch. environm. Hlth **11**, 846–849 (1965).

BERNARD, S. R.: Maximum permissible amounts of natural uranium in the body, air and drinking water based on human experimental data. Hlth Phys. **1**, 288–305 (1958).

CLAYTON, G. D.: Experience with non-radiation guides, standards and regulations. Hlth Phys. **11**, 855–858 (1965).

COOK, W. A.: Problems of setting occupational exposure standards—background. Arch. environm. Hlth **19**, 272–275 (1969).

ELKINS, H. B.: Maximum acceptable concentrations: A comparison in Russia and the United States. Arch. environm. Hlth **2**, 45–49 (1961).

European Standards for Drinking-Water: World Health Organization, Geneva, 1970.

FISH, B. R.: Inhalation of uranium aerosols by mouse, rat, dog, and man. In: Inhaled particles and vapours (edit. by C. N. DAVIES), 151–166. London: Pergamon Press 1961.

HARRIS, W. B.: The experimental clearance of uranium dust from the human body. In: Inhaled particles and vapours, vol. I (edit. by C. N. DAVIES), 209–215. London: Pergamon Press 1961.

HEID, K. R., FUQUA, P. A.: Review of uranium inhalation case. Personal communication, 1971.

HODGE, H. C., STOKINGER, H. E., NEUMAN, W. F.: Suggested maximum allowable concentration of soluble uranium compounds in air. USAEC Report AECD 2785, 1949.

HODGE, H. C., STOKINGER, H. C., NEUMAN, W. F., BALE, W. F., BRANDT, A. E.: Suggested maximum allowable concentrations of insoluble uranium compounds in air. USAEC Report AECD 2784, 1949.

HURSH, J. B.: Urinary uranium as an indicator of dose to exposed personnel. USAEC Report HASL-58, 136–138 1959.

International Atomic Energy Agency: Safe handling of radioisotopes. First edition with revised appendix 1. Safety series No. 1, Vienna, 1962.

International Standards for Drinking-Water: World Health Organization, Geneva, 1958.

IRISH, D. D.: Evolution of our concepts of standards. Arch. environm. Hlth **10**, 546–549 (1965).

JACKSON, S.: The interpretation of data on radionuclides in urine. Part I: Uranium. UKAEA Report AHSB(RP)R.15, 1962.

JACKSON, S.: The estimation of internal contamination with uranium from urine analysis results. In: Assessment of radioactivity in man, vol. II, 549–561. Vienna: International Atomic Energy Agency 1964.

JACKSON, S., DOLPHIN, G. W.: The estimation of internal radiation dose from metabolic and urinary excretion data for a number of important radionuclides. Hlth Phys. **12**, 481–500 (1966).

KULYAMIN, V. A. (editor): Radiation Safety Standards, NRB-69; Atomizdat, Moscow, 1970. English translation by J. E. BAKER, AERE Trans. 1134, HMSO, London, 1970.

LEACH, L. J., MAYNARD, E. A., HODGE, H. C., SCOTT, J. K., YUILE, C. L., SYLVESTER, G. E., WILSON, H. B.: A five-year inhalation study with natural uranium dioxide (UO_2) dust. I. Retention and biological effect in the monkey, dog and rat. Hlth Phys. **18**, 599–612 (1970).

LIPPMANN, M.: Environmental exposure to uranium compounds. Arch. ind. Hlth **20**, 211–226 (1959).

LISTER, B. A. J.: Early assessment of the seriousness of lung contamination by insoluble alpha and low energy beta emitting materials after inhalation exposure. Proc. First International Congress of Radiation Protection, Rome, Italy, Sept. 5–10, 1966, p. 1191–1198. Oxford: Pergamon Press 1968.

Maximum Permissible Amounts of Radioisotopes in the Human Body and Maximum Permissible Concentrations in Air and Water: National Bureau of Standards, U.S. Department of Commerce, Handbook 52, 1953.

Maximum Permissible Body Burdens and Maximum Permissible Concentrations of Radionuclides in Air and in Water for Occupational Exposure: National Bureau of Standards, U.S. Department of Commerce, Handbook 69, 1959.

MERCER, T. T.: On the role of particle size in the dissolution of lung burdens. Hlth Phys. **13**, 1211–1221 (1967).

Minutes of the Tripartite Conference on Permissible Doses: New York: Harriman 1953.

MORGAN, K. Z.: Tolerance concentrations of radioactive substances. J. phys. Colloid Chem. **51**, 984–1003 (1947).

Morgan, K. Z.: Protection against radiation hazards and maximum allowable exposure values. J. industr. Hyg. Toxicol. **30**, 286–293 (1948).
Morrow, P. E., Gibbs, F. R., Beiter, H. D.: Inhalation studies of uranium trioxide. Hlth Phys. **23**, 273–280 (1972).
N.B.S. Handbook 69: Maximum permissible body burdens and maximum permissible concentrations of radionuclides in air and in water for occupational exposure. Recommendations of the National Committee on Radiation Protection. Washington, D.C.: U.S.Government Printing Office 1959.
NCRP Report No 39: Basic radiation protection criteria. Recommendations of the National Committee on Radiation Protection and Measurements. NCRP Publications, Washington, D.C., 1971.
Neuman, W. F.: Urinary uranium as a measure of exposure hazard. Industr. Med. Surg. **19**, 185–191 (1950).
Quastel, M. R., Taniguchi, H., Overton, T. R., Abbatt, J. D.: Excretion and retention by humans of chronically inhaled uranium dioxide. Hlth Phys. **18**, 233–244 (1970).
Recommendations of the International Commission on Radiological Protection: British J. Radiol., Suppl. No 6 (1955).
Recommendations of the International Commission on Radiological Protection: ICRP Publication 2. Report of Committee II on Permissible Dose for Internal Radiation. London: Pergamon Press 1959.
Recommendations of the International Commission on Radiological Protection: ICRP Publication 6, 1964.
Recommendations of the International Commission on Radiological Protection: ICRP Publication 9. Oxford and New York: Pergamon Press 1966.
Safe Handling of Radioisotopes: First edition with revised appendix 1. Safety series No 1. Vienna: International Atomic Energy Agency 1962.
Saxby, W. N., Taylor, N. A., Garland, J., Rundo, J., Newton, D.: A case of inhalation of enriched uranium dust. In: Assessment of radioactivity in man, vol. II, 535–547: Vienna: IAEA 1964.
Schultz, N. B.: Inhalation cases of enriched insoluble uranium oxides. Report K-C-822, prepared for presentation at the Technical Meeting of the First International Congress of the International Radiation Protection Association, Sept. 5–10, Rome, Italy, 1966.
Scott, L. M., West, C. M.: Detection and evaluation of uranium exposures. In: Assessment of radioactivity in man, vol. II, 523–533. Vienna: IAEA 1964.
Smyth, H. F.: A toxicologist's view of threshold limits. Amer. industr. Hyg. Ass. J. **23**, 37–44 (1962).
Smyth, H. F.: Military and space short-term inhalation standards. Arch. environm. Hlth **12**, 488–490 (1966).
Stokinger, H. E.: Threshold limits and maximum acceptable concentrations. Arch. environm. Hlth **4**, 115–117 (1962).
Stokinger, H. E.: Modus operandi of threshold limits committee of ACGIH. Amer. industr. Hyg. Ass. J. **25**, 589–594 (1964).
Stokinger, H. E.: Industrial contribution to threshold limit values. Arch. environm. Hlth **10**, 609–611 (1965).
Task Group on Lung Dynamics: Deposition and retention models for internal dosimetry of the human respiratory tract. Prepared for Committee II of the International Radiological Protection Commission. Hlth Phys. **12**, 173–207 (1966).
Threshold Limit Values for 1957: Arch. industr. Hlth **16**, 261–265 (1957).
Threshold Limit Values for 1969: Technical Data Note 2/69. HM Factory Inspectorate; Department of Employment and Productivity, HMSO, London, 1970.
Voegtlin, C., Hodge, H. C.: Pharmacology and toxicology of uranium compounds, vol. 3, 4. New York: McGraw-Hill 1953.
West, C. M., Scott, L. M.: A comparison of uranium cases showing long chest burden retention. Hlth Phys. **12**, 1545–1555 (1966).
West, C. M., Scott, L. M.: Uranium cases showing long chest burden retention: an updating. Hlth Phys. **17**, 781–791 (1969).
Wruble, D. T., Shearer, S. D., Rushing, D. E., Sponagle, C. E.: Radioactivity in waters and sediments of the Colorado River Basin, 1950–1963. Radiol. Hlth Data **5**, 557–567 (1964).

Chapter 6

Environmental Monitoring and Personnel Protection in Uranium Processing

L. M. SCOTT

With 14 Figures

I. Introduction

Even before the discovery of radioactivity, uranium (pitchblende) was known to be toxic. Nineteenth century reports describe the use of uranium as a nephrotoxic agent in the study of kidney function. In the early days of the Manhattan Project, studies into both the chemotoxic and radiotoxic effects of uranium were undertaken. The original protection guides for uranium grew out of these studies. This chapter is limited to the problems of environmental and personnel protection. Of equal importance are those problems of criticality control which must be considered in the processing of uranium materials enriched in ^{235}U (American National Standards Institute Committee N16, 1970).

Since the Manhattan Project days, uranium processing has grown from small batch operations, similar to laboratory studies, to strictly industrial-type programs involving multikilogram quantities in a wide variety of chemical and physical forms. In addition, the isotopic distribution of the material varies widely. The chemical and physical forms plus the isotopic distribution govern the type and extent of the environmental and personnel monitoring programs needed to insure adequate protection.

II. Classification of Uranium

To aid in setting guidelines, some general type of material classification is necessary. The classification that follows is of a subjective nature and the classes tend to blend together. As far as this chapter is concerned, there are three levels of transportable material: highly transportable biological half-life of days, moderately transportable biological half-life of weeks to months, and slightly transportable biological half-life of months to years. "Transportability" refers to the rate that materials leave the critical organ without regard for the route or mode of movement. Table 6.1 lists some common uranium compounds falling into each of the three categories.

There are also three classes of material according to isotopic distribution. The first includes uranium which has not been in a reactor and contains less than 5–8 weight percent ^{235}U. For soluble materials, FORD (1964) and JACKSON (1964) have reported that for materials with less than 5–8 percent ^{235}U, the chemical toxicity to the kidney is the limiting value. The second class includes uranium which has been enriched in the ^{235}U isotope from greater than 5–8 percent to about 95–97 percent, but has not been in a reactor. The third class of uranium

Table 6.1. Classification of uranium compounds according to transportability[a]

Highly transportable	Moderately transportable	Slightly transportable
UF_6	UO_2[b]	UO_2[b]
UO_3[b]	U_3O_8[b]	U_3O_8[b]
$UO_2(NO_3)_2$	UO_3[b]	uranium oxides
UF_4[b]	UF_4[b]	uranium hydrides
Uranium sulfates	uranium nitrates	uranium carbides
Uranium carbonates		salvage ash

[a] Partially developed from data by STECKEL and WEST (1966), and Task Group on Lung Dynamics (1966). [b] Subjecting a particular uranium compound to higher temperatures tends to decrease rate of transportability.

includes material that has been in a reactor which resulted in an abnormal isotopic distribution and a specific activity of above 200 d/m/μg of uranium. Also included in the third class are all $^{233}U/^{232}U$ materials from thorium breeder reactors. The second and third class materials constitute the primary radiation hazards, and these limits are derived from basic radiation protection criteria.

In establishing protection systems, both classifications must be considered. For example, a highly transportable uranium with less than 5–8 percent ^{235}U could, at a given air concentration, constitute a serious toxicity problem for the kidney; whereas, the same level of a moderately or slightly transportable compound of less than 5–8 percent ^{235}U would not constitute a radiological problem to the lung and would also clear at a rate that is slow enough not to affect the kidneys.

In addition, in certain types of operations, less than 5–8 percent ^{235}U or the $^{233}U/^{232}U$ materials may constitute an external exposure potential of some magnitude, as explained in Sec. III.A. and III.C. to follow.

III. Process Equipment Design

The type of process equipment influences the quality of the environmental and personnel monitoring programs. In reality, if the process equipment is designed and operating properly, the only function that environmental and personnel monitoring programs serve is to give added assurance of the proper design and operation. Health Physics personnel should play an active part in the design of equipment and establishing operating procedures.

Since the design of equipment has been adequately covered in the literature, only a general discussion with the appropriate references will be given here.

A. Material Less than 5-8 Percent ^{235}U

Because of the low specific activity of these materials, they can usually be processed in areas with room ventilation. However, in the case of powders and extremely dusty operations, especially those with moderately and slightly transportable materials, open-face hoods with a face velocity of 125–200 ft^3/min should be considered. HARRIS and KINGSLEY (1959) and HARRINGTON and RUEHLE (1959) present detailed descriptions of process equipment for these types of material.

As previously stated, this material can present an external exposure problem. The first and second daughters of ^{238}U, ^{234}Th (UX_1) and $^{234}Pa^m$ (UX_2), are beta emitters with a 24-day controlling half life. In casting processes, these and other uranium daughters float to the surface of the molten uranium and may volatilize and plate out on the first surface which has a temperature below the melting point of uranium. Beta readings of several rad on the surface of the castings and furnace components such as crucibles, molds, liners, and lids can commonly occur. This problem becomes more serious as the size of the casting increases. Equipment should be such that these components can be handled without direct exposure to personnel. If close contact is required, a $^1/_4$-inch layer of aluminum may be used to adequately shield personnel. Individual protective measures such as heavy rubber aprons and gloves can also be used to minimize the exposure.

B. Material Greater than 5-8 Percent ^{235}U

These materials generally must be processed in open-faced hoods; however, in the case of dry powder and dusty operations, complete enclosures may be necessary. Such enclosures should have filtered exhaust ventilation to prevent the spread of contamination and to prevent loss of valuable material. Machining operations need not be enclosed but should have local ventilation to pull any material which becomes airborne away from the operator. PATTON et al. (1963) describe the types of equipment required for a wide variety of processes involving this material. Since there is less ^{238}U present in this material and the castings are smaller, due to criticality considerations, the external exposure potential from UX_1 and UX_2 is low and need not be considered except at levels around 5–8 percent ^{235}U.

C. Special Materials

These materials should always be processed in closed systems. The extent of the enclosure is governed primarily by the specific activity of the material and, in the case of ^{233}U/^{232}U, should follow the recommendations for plutonium (see Chap. 14).

It is not unusual for this material to pose an external hazard either from traces of fission products or, in the case of ^{233}U/^{232}U, the daughters of ^{232}U decay. The magnitude of this problem grows as the concentration of ^{232}U increases and the elapsed time since chemical purification increases. OWEN (1964) reports a growth in gamma levels by a factor of 55 between 4 and 60 days after purification. After two years, it has grown by a factor of 500. The growth of the beta and gamma emitting daughters of ^{232}U is controlled to the 1.9-year half life of the ^{228}Th daughter of ^{232}U. Consequently, remote handling should be considered for old materials.

IV. Environmental Monitoring

Monitoring of the environment is the first line of assurance that systems are operating and are being operated in a proper manner. An environmental monitoring program is not complete unless the following questions can be answered from the data gathered:

1. What is the general air concentration where employees are working (general air sampling)?
2. What is the air concentration associated with a particular process or operation (diagnostic and/or breathing-zone sampling)?
3. How much material is being released to the atmosphere (stack sampling, plant-effluent sampling: air, soil, and water)?
4. Is the material being adequately confined to hoods and process equipment (surface contamination sampling)?

5. Is the material being carried out of the area on shoes and clothing (surface contamination sampling)?

In addition to these types of monitoring, some groups have attempted to monitor the air breathed by the employee (personal air samplers).

A. General Air Sampling

All uranium processing plants should have some type of general air monitoring program. Such programs are generally required to assure regulatory authorities that specified air concentrations (Code of Federal Regulations, 1963) are not being exceeded.

There are no simple rules which can be followed in determining the number of samplers and their placement. As a general guide in areas of serious exposure potential, sampler density should be at least one per each 20-foot-square grid; placement on a 40–60-foot-square grid should be adequate for areas of minimal exposure potential. By pulling air through some filter medium, uranium particles in the respirable range (< 10 micron) are deposited on or in the filter. Table 6.2 gives a listing of some commonly used filter media with their general advantages and disadvantages. Fig. 6.1 shows an exploded view of a widely used sampling head; Fig. 6.2 is a view of a rather new sampling technique that undoubtedly will see wide use in the future (Sanders, 1967). As shown in the figure, the sample is taken on a piece of filter paper which is built into a computer card. These cards can be processed by standard card processing equipment (see Laboratory analysis).

Flow rate of the air through the sampling head should be relatively low (around 0.01 to 0.02 m^3/min) in order to catch the respirable particles. Frequency

Fig. 6.1. Exploded view of typical air sampler

Table 6.2. Filter media types[a]

Filter	Collection efficiency	Advantages	Disadvantages
Cellulose	40–98%	does not interfere with radiochemical analysis; inexpensive	particles can become embedded and affect counting
Cellulose-asbestos	96–99%	very durable; high collection efficiency	will interfere with radiochemical analysis
Glass-fiber	99+%	usable in high temperature areas	fragile; expensive
Membrane	99+%	does not interfere with radiochemical analysis	fragile

[a] Lockhart et al. (1964), Harley (1970), Atomic Industry Forum, Inc. (1969).

Fig. 6.2. Computer card sample and sample holder

of sample change out should be governed by experience; and, until such experience is gained, a daily sample change out should be followed.

Since general air samples are at fixed locations at levels above the breathing zone, it is not advisable to allow an area to approach the MPC (maximum permissible concentration). Table 6.3 presents the appropriate MPC for uranium. To assure that employees are not being overexposed, general air levels should be maintained below 50 percent of the appropriate MPC.

Table 6.3. Process area air limits[a]

	Transportability		
	high $\mu c/cm^3$	moderate[b] $\mu c/cm^3$	slight[b] $\mu c/cm^3$
$<$5–8% ^{235}U	7×10^{-11} kidney	10^{-10} lung	10^{-10} lung
$>$5–8%–97% ^{235}U	6×10^{-10} bone	10^{-10} lung	10^{-10} lung
$^{233}U/^{232}U$	5×10^{-10} bone	10^{-10} lung	10^{-10} lung

[a] National Bureau of Standards, 1959. International Commission on Radiological Protection, 1959. [b] Considered to be insoluble material.

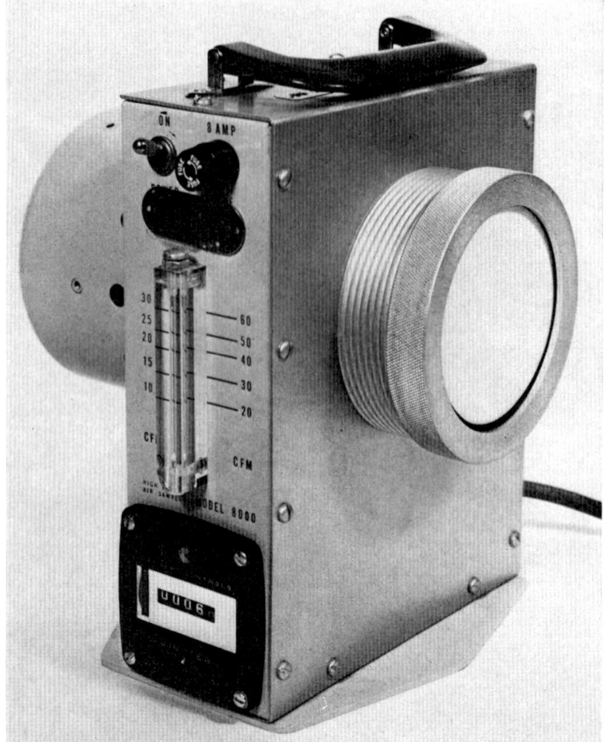

Fig. 6.3. Typical sampler used for breathing zone sampling

B. Diagnostic and Breathing-Zone Sampling

Diagnostic and breathing-zone samples are usually taken as short-term samples used to characterize the uranium air concentrations of a small area or operation. These samples are usually taken to pinpoint the source of a release of uranium to the environment which has caused high general air levels and/or personnel monitoring results. It is also advisable to sample around new process equipment to assure that proper operation has been achieved.

This sampling is done at a rather high air volume in order to collect sufficient amounts of material in a reasonable length of time. Fig. 6.3 shows a high-volume air sampler. This sampler can be set up and operated while an operation is being performed; or, if the operation is intermittently occurring at hours when Health Physics personnel are not available, the sampler can be equipped with a timer and actuated by electronic carpets, microswitches, or interruption of a light beam impinging on a photocell to take a sample any time the operation is performed. Sanders (1967) presents a detailed description of such uses. Fig. 6.4 depicts a typical installation.

Because of the proximity of the sampler to the operation and the high air flow, higher air concentrations occur than are experienced with general air samplers. Usually, levels at the MPC are acceptable, and concern should only be expressed when the MPC is consistently exceeded.

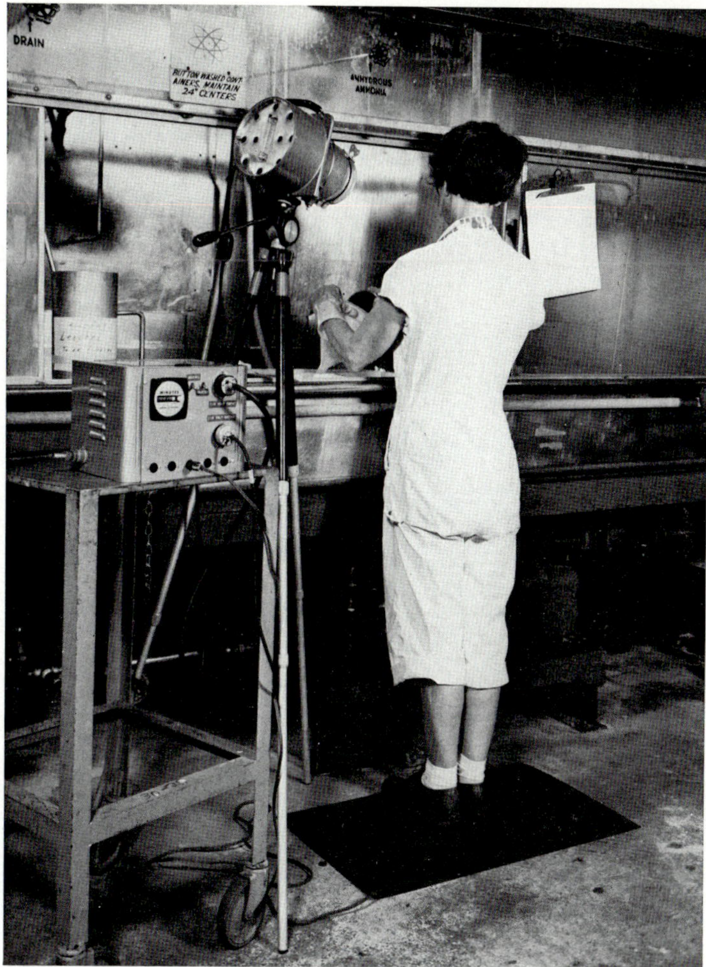

Fig. 6.4. Breathing zone sampling for automatic sampling using an electric carpet

C. Monitoring of Plant Releases

Monitoring for possible release of uranium from process areas should be conducted to assure that the plant environs are not being subjected to significant contamination. Additionally, in the case of uranium enriched in ^{235}U and $^{233}U/^{232}U$ materials, a strict material accountability is required due to the significant value of the material.

1. Stack Monitoring

Stack monitoring is used to determine the amount of material being discharged and to give assurance that minimal quantities are being released to the environment.

A representative sample from stack flow is not easy to collect. Attention must be given to the design of the sampler, sampler location, and the number of sampling heads. It is generally accepted that isokinetic sampling; i.e., sampling at the same

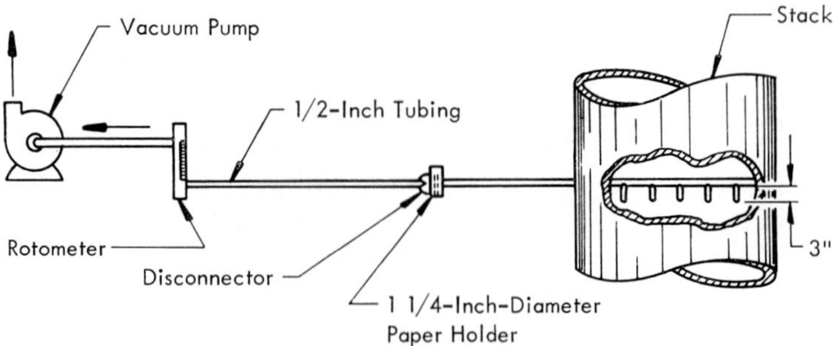

Fig. 6.5. Typical standardized stack sampler

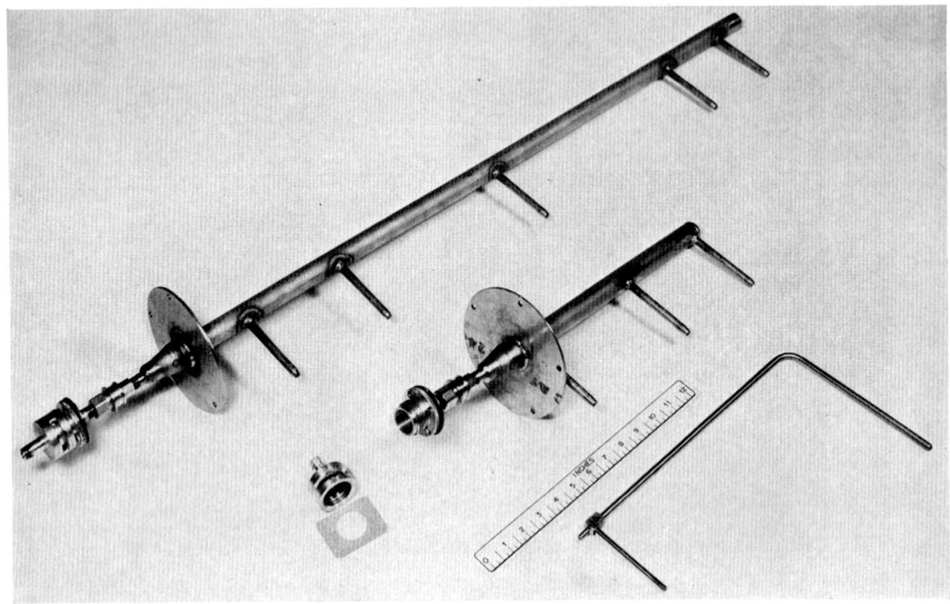

Fig. 6.6. Stack sampler assemblies

Table 6.4. Number of sample points for stack sampling[a]

Circular Stack		Rectangular Stack	
stack diameter (in.)	no. of points	stack area (sq. ft.)	no. of points
2–6	1	<0.5	1
8–12	2	1–2	4
14–18	3	2–25	6–12
20–28	4	>25	20
30–48	5		
>50	6		

[a] Atomic Industrial Forum, Inc., 1969.

flow rate as the stack being sampled, is the most appropriate sampling technique (Atomic Industrial Forum, Inc., 1969). The delivery tube from the sampling orifice to the collection point should be of minimum length to avoid the settling of large particles in the tube.

In placing the sampler in the stack, care must be taken to avoid bends and turns. Particles greater than 5 microns may be removed from the stream due to inertial effect in bends and turns. As a general rule, sampler locations should be downstream at least five stack diameters, and preferably ten diameters, from any bend or inlet in the stack. If the stack has both horizontal and vertical runs, sampling the vertical sections will avoid stratification due to gravity.

The number of sampling orifices required to obtain a representative sample is a function of the stack diameter. Table 6.4 can be used as a general guideline concerning the number of sampling orifices. Fig. 6.5 shows a typical stack sampling system, while Fig. 6.6 shows some typical sampler designs.

2. Plant Effluent Monitoring

Before startup of a plant, the effluent monitoring program should be put into force in order to gather background data for future comparisons.

Uranium concentrations in the air surrounding the plant should not exceed the appropriate value listed in Table 6.5. Monitoring is usually accomplished by

Table 6.5. Plant environs air limits[a]

	Transportability		
	high $\mu c/cm^3$	moderate[b] $\mu c/cm^3$	slight[b] $\mu c/cm^3$
$<$5–8% ^{235}U	3×10^{-12} kidney	5×10^{-12} lung	5×10^{-12} lung
$>$5–8%–97% ^{235}U	2×10^{-11} bone	4×10^{-12} lung	4×10^{-12} lung
$^{233}U/^{232}U$	2×10^{-11} bone	4×10^{-12} lung	4×10^{-12} lung

[a] International Commission on Radiological Protection, 1959 pq. 5. [b] Considered to be insoluble material.

pulling air through filter paper. Fig. 6.7 shows a sampling technique using an 8 by 10-inch piece of paper. Due to the low level of activity, samples need be changed only every two to four weeks. If electricity is not available, crude estimates of air concentrations can be made by collecting settled dust on a tray or trapping the material in water with a known surface area (Fig. 6.8). To aid in placing the samplers at the most advantageous locations, a study of meteorology data should be made to determine the prevailing wind direction. Obviously, the greater number of samplers should be located in the downwind direction of the prevailing wind.

Any creek or river into which plant effluent is dumped or one that receives plant runoff from rains should be sampled at least once a year. Uranium water concentration limits are given in Table 6.6. If the stream runs through the plant, sampling stations should be located at the plant perimeter and from three to five miles downstream. If there is no running stream, the stream bed that drains the plant runoff should be monitored.

Sampling of stream beds along with the plant atmosphere or dust fall will give an estimate for the levels in the soil. However, periodic soil sampling is

Fig. 6.7. Outside air sampler

Fig. 6.8. Dust-fall sampler

Table 6.6. Plant effluent water limits[a]

	Transportability		
	high $\mu c/cm^3$	moderate[b] $\mu c/cm^3$	slight[b] $\mu c/cm^3$
$<$5–8% ^{235}U	6×10^{-7}	4×10^{-5}	4×10^{-5}
$>$5–8%–97% ^{235}U	4×10^{-6}	3×10^{-5}	3×10^{-5}
$^{233}U/^{232}U$	4×10^{-6}	3×10^{-5}	3×10^{-5}

[a] International Commission on Radiological Protection, 1962. [b] Considered to be insoluble material.

advisable and can be very valuable in determining the spread of contamination due to such accidents as fires. Harley (1970) presents a procedure for collecting soil samples.

D. Surface Contamination Sampling

There are two techniques for measuring the degree of surface contamination: (1) smear or wipe sampling which measures the removable contamination and is indicative of the amount of material which might be resuspended in the air or constitutes a health hazard; (2) direct measurements with a survey instrument which, of course, measures the total contamination but may yield little or no information as to the potential health hazard. Smear sampling is the preferable method; and, since direct measurement is self explanatory, it will not be covered further. Even in situations where experience has shown that there is no exposure potential from the resuspension of material, it is a good practice to have a surface contamination sampling program to measure the degree of good housekeeping.

1. Process Area Smears

Smear values tend to vary and should be considered as only a relative indicator. Such programs do not need to be very extensive. A dry paper towel rubbed on the floor or equipment and then read with an alpha survey instrument will generally suffice. A more formalized program could be the use of filter paper or computer cards (Fig. 6.9) where a specified area is smeared (Sanders, 1964). Smears are of a subjective nature; and, partially for this reason, radiation protection guides have not been established. If a formal program of smearing a specified area is chosen, a control criterion of 4×10^{-4} $\mu c/100$ cm^2 will assure that housekeeping is adequate. Of more importance than an absolute value is the evaluation of changes in levels over extended periods of time. Care should be taken to be uniform in the frequency, number, and location of smears. A smear frequency of once a month is adequate unless a situation completely deteriorates.

2. Lunchroom and General Plant Area

To give assurance that uranium is not being carried out of process areas into lunchrooms and other noncontrolled areas, they should be smeared routinely. Concentration in these areas should be maintained below 2×10^{-5} $\mu c/100$ cm^2. If concentrations cannot be maintained at this level, consideration should be given to banning these areas from employees in work clothes.

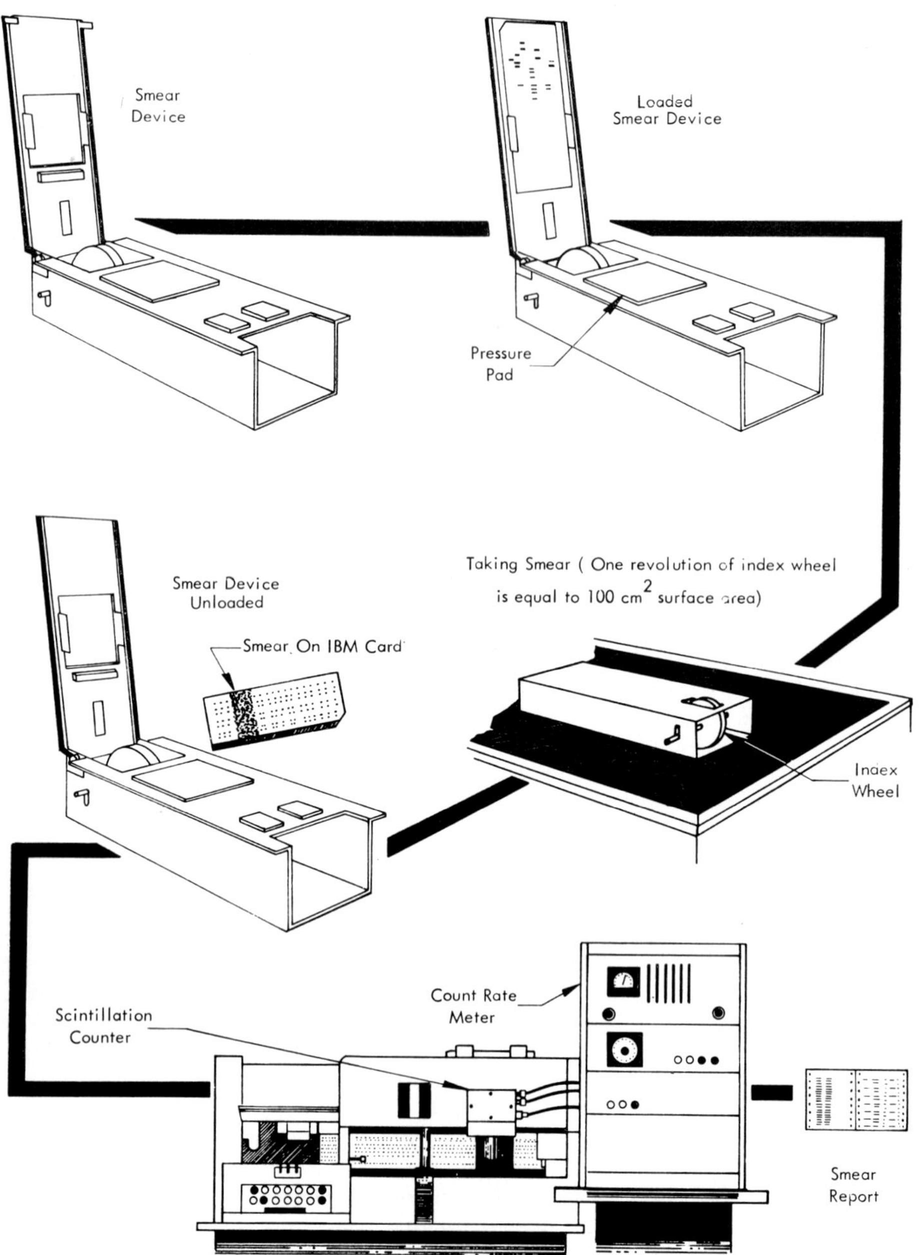

Fig. 6.9. Computer card smear sampling system

E. Personal Air Samplers

Several attempts have been made to sample the air a man breathes through the use of an air sampler carried on the individual employee. Fig. 6.10 shows a typical sampler.

Fig. 6.10. Personal air sampler

CALDWELL et al. (1968) present a rather convincing argument for their use through the evaluation of simulated releases of radioactive materials. The use of this type of sampler is rather widespread in plutonium processing plants in the United Kingdom (FRASER, 1967; BRUNSKILL, and HERMISTON, 1968).

These authors agree that personal air-sampler results will be higher than general air-sampler results by factors of up to 1000. CALDWELL et al. (1968) present similar results for uranium. The principal drawback to using personal air samplers is that in uranium processing areas certain levels of uranium contamination are common and acceptable. Personal air samplers used under these conditions are very susceptible to contamination; thus, rendering the results essentially meaningless. Personal air samplers may be of value in areas processing special high-specific-activity uranium materials where the presence of any material in the air may constitute an exposure potential.

F. Particle-Size Distribution in Processing Areas

The Task Group on Lung Dynamics (1966) proposed a revised model for the deposition and clearance of material from the respiratory tract. The model acknowledges three rates of pulmonary clearance and recognizes pulmonary deposition as partially a function of the particle size. If this model is adopted by the International Commission on Radiological Protection, it will have a major impact on radiological protection programs.

The adoption and implementation of the proposed model could result in one or both of the following effects:

Table 6.7. Airborne uranium particle size in process areas[a]

Operation	Mass median[b] diameter	Standard deviation
Foundry		
normal uranium	2.8	2.7
enriched uranium	3.3	2.2
enriched uranium[c]	2.1	2.0
Machining		
normal uranium		
machining with coolant	2.0	—
polishing wet and dry	2.6	—
milling dry	3.0	—
enriched uranium		
machining with coolant	1.9	—
machining with coolant[c]	1.6	—
Extruding normal uranium	3.2	2.7
Welding normal uranium	1.6	3.4
Reducing enriched UF_4 to metal	2.3	—
Conversion of enriched UF_6 to UF_4	1.1	2.0

[a] Partially developed from data by Hyatt et al., 1959. [b] Measurements made with a Cascade Impactor. [c] Independent measurements.

1. For slightly transportable material, the MPC in Table 6.3 would be reduced by a factor of 5. This would come about if, as proposed, the biological half-life in the lung model is changed from 120 days to 500 days and if the so-called standard aerosol AMAD of one micron is assumed in lieu of actual particle size measurement.
2. Particle size will have to be known for all processing areas. Table 6.7 presents some typical particle-size data from uranium processing areas.

V. Personnel Monitoring

Regardless of the quality of the process equipment and the levels of uranium air concentrations, some type of personnel monitoring program should be conducted for uranium workers. The kind and extent of such programs will depend upon the type of material and/or operation and past experience.

A. Urine Analysis

By far the most widely used technique for uranium personnel monitoring is the analysis of urine samples. Obviously, analysis of urine is indicative only of what has left the body and may or may not be an indicator of the amount of internal deposits. Therefore, in order to relate uranium in a urine sample to organ burden, some type of excretion and retention model must be assumed. Included in such a model must be a critical organ, pathways of elimination, and fraction eliminated by each pathway and biological half-life. Chap. 5 will aid in selecting the appropriate model.

1. Sample Type and Collection

The type of sample depends upon the type of uranium and the isotopic distribution. Table 6.8 shows the various types of samples and where each is generally appropriate. Usually it is easier to collect samples at the employee's convenience

Table 6.8. Types of urine samples

Type of sample	Description	Use
Spot	Sample of urine collected not necessarily a complete voiding. Must assume some standard excretion rate	Where low exposure potential to highly transportable material
Spot rate	Complete voiding of urine noting previous and present voiding times. Allow for individual variation in excretion rate in that a urine excretion rate can be computed for each employee	Where exposure potential is low to high for moderately and slightly transportable materials. For materials with specific activity less than ≈ 200 d/m/μg
24-hour voiding	Collection of complete voiding for a 24-hour period. Allow for individual variation in excretion rate. Provide large volume to increase total amount of uranium in sample	Where exposure potential is high to slightly transportable material. For materials with specific activity greater than 200 d/m/μg. Used for better evaluation of any exposure after such has been determined

during working hours; however, there are serious drawbacks to this type of sampling. For transportable-type materials where experience has shown that the possibility of employee exposure is low, sampling during the work period is adequate. However, for moderately and/or slightly transportable materials, sampling during the work period is not advisable due to biasing of results from highly transportable material, some of which is usually present with the other two types of material. Also, the possibility of contamination of the sample is always present with this type of sampling.

From the exposure evaluation standpoint, the best time to collect a sample is prior to changing into work clothes or entry into the work area. If moderately to slightly transportable materials are of concern, the sample should be collected after a work break of at least two days, and, preferably, longer. This time lapse allows for any highly transportable material to be eliminated.

2. Sampling Frequency

Sampling frequency must be governed by experience. In new operations, where there is no experience, a weekly to monthly sampling frequency should be followed, depending on the transportability and specific activity of the material. Periodically, all frequencies should be evaluated and adjusted as necessary. Factors to be considered in such evaluations are: group uranium excretion averages, individual variation, and number of individuals showing levels of concern. West and Reavis (1964) present a statistical approach which makes it possible to predict the degree of statistical confidence in results based on past experience. Generally, the evaluation of general air experience is of limited value in setting the urine sampling frequency. Schultz and Becher (1963) found little correlation between the urine excretion levels and general air levels in uranium processing areas.

The maximum allowable concentration of uranium in urine is covered in Chap. 5. However, there should be values below those at which resampling, requiring respirators and/or reassignment to areas of lower or no exposure potential, is desirable. The action dictated by the levels will depend on experience and knowl-

Table 6.9. Urine excretion action levels

	Transportability		
	high μc/day	moderate μc/day	slight μc/day
<5–8% ^{235}U	8×10^{-6a}	27×10^{-6}	14×10^{-6}
>5–8%–97% ^{235}U	5×10^{-5}	25×10^{-6}	13×10^{-6}
$^{233}U/^{232}U$	5×10^{-5}	25×10^{-6}	13×10^{-6}

[a] Based on chemical toxicity.

edge about such factors as transportability, specific activity of the exposure material, and freedom of action as to work assignment. Table 6.9 lists action points at which at least resampling should be required.

B. Relationship of Uranium Air Concentrations and Urinary Uranium Excretion

The present regulations (Code of Federal Regulations, 1963) governing the processing of uranium enriched in ^{235}U by private industry specify air concentration limits with no requirements as to the urinary excretion levels. Since, in the final analysis, the protection of the employee is the goal, a knowledge of the relationship between uranium air concentration and urinary uranium excretion would be of value.

Several studies have been conducted that have attempted to correlate air and urine levels. Heatherton and Huesing (1959) report: "...good correlation exists between average uranium concentration in urine and the air dust exposure." The urine samples were collected during the workday at a medical building. Exposures were to highly and moderately transportable materials. Fischoff (1963) reports results from the same plant when the air data were collected with a two-stage air sampler. As would be expected, the correlation was higher between the respirable fraction and urine than was the whole air sample. Lippman (1959) and Ross (1959) reported essentially no correlation for highly transportable or moderately transportable exposure materials. Schultz and Becher (1963) report a correlation between removable surface contamination and urine excretion, but no correlation between air concentration and urine levels. Both highly and moderately transportable materials were studied. This author has found little correlation between either surface contamination or air concentration and urine excretion levels. It is generally concluded that it is not possible to sample the same air that a man breathes; and, even if it were, it would be difficult with present techniques to relate to exposure levels because of various particle sizes and biological half lives.

C. In vivo Monitoring

During the 1960s, *in vivo* monitoring cames into routine use as a health physics monitoring tool. Although uranium is primarily an alpha emitter, ^{235}U emits a 186-keV gamma with about 55 percent of the alpha disintegrations. Routine monitoring of uranium by *in vivo* gamma spectrometry, where the lung is considered to be critical organ, has been in effect since 1961 and is spreading widely. Several reports (West and Scott, 1966, 1969; Scott and West, 1964; Schultz,

1968; SAXBY et al., 1964) have shown that, for moderately and slightly transportable materials, *in vivo* monitoring is superior to urinalysis and is the only method presently available to evaluate exposures to slightly transportable material.

Detailed descriptions of the technique of uranium *in vivo* monitoring have been reported by COFIELD (1960) and SCOTT and WEST (1968). The factor that sets uranium monitoring aside from other *in vivo* monitoring is the low energy of the uranium gamma. There is a Compton scatter peak from the body-stored ^{137}Cs and ^{40}K which is near its maximum at 186 keV. Some technique must be used to estimate the count rate in the uranium region due to Compton scatter so that the net count due to uranium can be determined. COFIELD (1960) and SCOTT and WEST (1968) used a linear predication with independent variables of count rate in the 250–300-keV region, ^{137}Cs region, ^{40}K region, and the subject's weight. The limit of sensitivity of this technique ranges from 120 micrograms, using one detector with a 20-minute count time, to 80 micrograms, using four detectors with a 20-minute counting time (COFIELD, 1960; SCOTT et al., 1969; SCOTT and WEST, 1968). HELGESON (1971) uses phoswich detectors developed by LAURER and EISENBUD (1968) and employs a mathematical fit of the energy spectrum above uranium and extrapolates the fit through the uranium region to determine the base count rate in the uranium region. The limit of sensitivity of this technique has not been published.

There is evidence that calibration factors may vary with the ^{235}U enrichment. This situation appears to be especially true with a ^{235}U content from 0.14 up to 2–3 percent. The author has observed that above approximately 10 percent ^{235}U, there is little variation in the calibration factors. The cause for this variation is not known, but is probably related to the inability of present detectors to resolve the 90-keV X-ray peak from ^{234}Th (first daughter of ^{238}U) and the 186-keV peak from ^{235}U independent of each other. COFIELD (1960) and SCOTT et al. (1969) have also reported techniques for determining ^{238}U utilizing the 90-keV ^{234}Th X-ray. The approach is the same as for ^{235}U. The limit of sensitivity of this technique is ± 6.5 mg ^{238}U. In utilizing this technique it should be realized that if the ^{234}Th daughter is enriched with respect to the ^{238}U parent, an overestimate can occur. Such enrichment of the daughter might occur in operation where the uranium is melted. An exposure to freshly purified uranium could go undetected by this technique until sufficient time had elapsed for the ^{234}Th daughter to grow back. It would require about six months to reach practical equilibrium.

There are two major problems which must be taken into consideration in the interpretation and use of *in vivo* monitoring data. First, a small amount of surface contamination can lead to spurious, high results. Material on the surface is detected with an efficiency six to ten times higher than lung-stored material (SCOTT and WEST, 1968). Most Health Physicists do not realize the high degree of sensitivity of *in vivo* monitors and are often misled by the failure to detect contamination with portable survey instruments. Clothing can also be contaminated, and new clothing is preferable. If clothing is to be reused, care should be taken to assure that it is not laundered with clothing from process areas. The second problem is the isotopic distribution of the exposure material. Knowing the amount of ^{235}U present is of only limited value since, in materials enriched in ^{235}U above a few percent, the majority of the exposure is from ^{234}U. Fig. 6.11 presents a general relationship between the ^{235}U and ^{234}U content which can be used to estimate the percent of ^{234}U and the specific activity of the exposure material. It is always best to determine the specific activity of the exposure material because the ^{234}U content varies, depending on the history of the material.

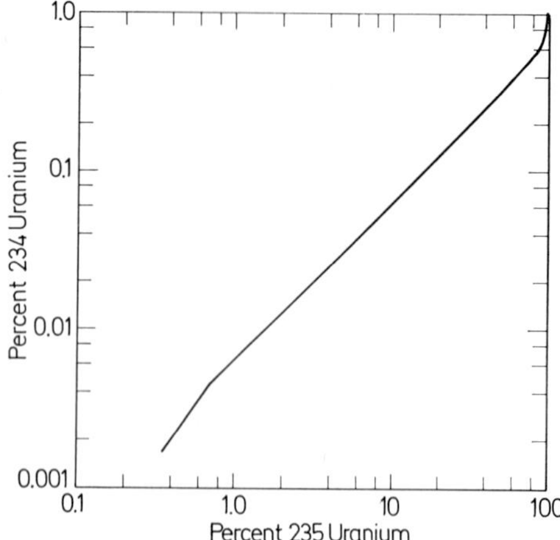

Fig. 6.11. $^{234}U/^{235}U$ relationship for uranium enriched in a gaseous diffusion plant

For example (as shown in Fig. 6.11) a 90 percent ^{235}U enriched material generally contains about 0.85 percent ^{234}U, but levels as low as 0.4 to 0.6 percent ^{234}U have been observed. Also, reactor fuels tend to increase in the percent of ^{234}U as the ^{235}U fissions away.

For the moderately to slightly transportable materials, the lung is considered to be the critical organ. The maximum permissible lung burden is 0.017 μc, regardless of the isotopic distribution, because the effective energies of ^{234}U, ^{235}U, ^{236}U, and ^{238}U are essentially the same.

As in the case of urinalysis, frequency of *in vivo* monitoring must be based on plant experience. However, the following guidelines should be generally followed until adequate experience is gained: Employees working with processes involving any powders of material enriched above a few percent in ^{235}U are usually monitored every three to six months; employees working with dry powder of lower enrichments or those in metal-forming operations of any enrichment are usually monitored every six to twelve months, while any employees who routinely may come in contact with moderately to slightly transportable materials are usually monitored at least once a year.

D. Correlation of Uranium Urine Excretion Rate and in vivo Measurement Results

Since *in vivo* measurements are of value for only moderately and slightly transportable materials where the lung is the critical organ, no correlation would exist for highly transportable materials. However, if it is assumed that urine excretion is measuring moderately to slightly transportable lung-stored materials, then they should correlate with *in vivo* measurements. Scott and West (1968) report that, on individual cases, correlation coefficients range from −1.00 to +1.00. This range is attributed to the low level of exposure, variability of the analytical techniques, and the fact that these employees were working in uranium processing areas and were excreting some highly transportable materials which had no relationship to lung-stored material. A correlation coefficient of 0.36 was found for a group of employees who had been restricted from uranium processing because of elevated *in vivo* or urinary uranium excretion results. Considering the errors in the analytical techniques, a correlation of 0.36 is quite good. When employees are grouped by work areas (exposure potential), the correlation coefficient increases as the exposure potential increases. It is felt that, even under the best of conditions, significant correlation cannot be expected between urine and *in vivo* data on individual cases.

E. Fecal Sampling

There is evidence that fecal excretion is a significant mode of eliminating moderately and slightly transportable materials (West and Scott, 1966; Schultz, 1968; Saxby et al., 1964; Quastel et al., 1970). However, due to the inconveniences of sample collection and analysis, routine fecal sampling is not recommended. Additionally, there is a rapid clearance of large particles from the upper respiratory tract which makes the interpretation of fecal data from employees still under exposure extremely difficult. The patterns of uranium excretion are not fully understood; and, to aid in this understanding, it is recommended that fecal samples be collected from all employees who receive exposure to uranium of such an extent that they are restricted from further uranium processing.

F. Personnel Practices

1. Respirators

In any operation involving uranium dust and/or mist to such a degree that the general air concentrations consistently exceed $10^{-10}\,\mu c/cm^3$, respirators should be worn. However, respirators should not be a replacement for proper equipment design and operating technique. It should never be an accepted practice to design an operation with the anticipation of its being a permanent respirator area.

Respirators are valuable pieces of equipment and should always be readily available for emergency or short-term operation use. It should be noted that respirators must be properly fitted to the face and that all employees should be instructed in their care, storage, and fitting. A thorough discussion of respirator usage has been presented by American Standards Association Committee Z2.1 (1960).

2. Protective Clothing

It is a good practice to provide work clothing to employees processing uranium. However, protective clothing is generally not required. In cases where equipment and/or ductwork must be entered for uranium recovery or equipment repair, clothing should be changed after the job and not used again until cleaned. Paper suits are good for such activities. Gloves should be provided for all employees who must handle uranium. Care should be taken to assure that the gloves are changed often enough so that they do not become highly contaminated and become a source of skin exposure to the hands and the fingers.

3. Personal Hygiene

As a general policy, all employees should be trained to work in such a manner that they and their clothing are kept as clean as possible. It is difficult to establish guidelines to govern smoking and eating in process areas. Usually, in areas processing material enriched to less than 5–8 percent ^{235}U, smoking and eating do not constitute a potential exposure condition. However, smoking and eating in areas processing higher enrichments, if allowed at all, should be confined to those operations where freshly machined metal is handled. Never allow smoking and eating in areas processing $^{233}U/^{232}U$ material or high-specific-activity materials (200 d/m/μg).

VI. Laboratory Analysis

Laboratory analysis of uranium has been covered in detail in the literature. The following sections present a general discussion of the more important points to be considered when establishing and conducting laboratory analyses for uranium.

A. Air and Smear Samples

These types of samples are analyzed by alpha counting. Usually either proportional or scintillation counters are used. Fig. 6.12 shows a typical proportional counter which is commonly used; however, individual handling of samples is

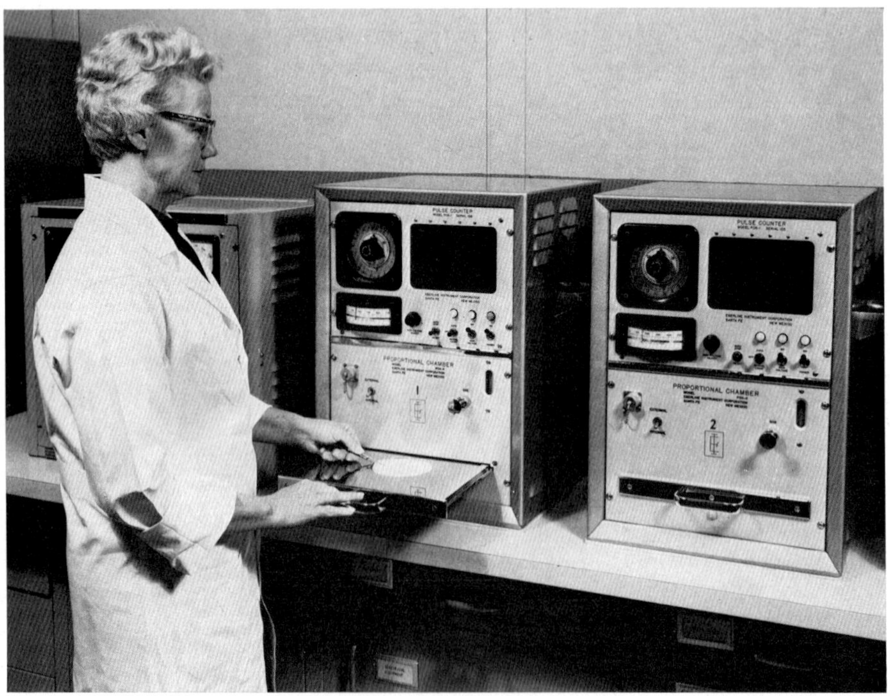

Fig. 6.12. Proportional counter

required. Fig. 6.13 shows a scintillation counter for counting general air and/or smear samples which are placed onto a planchet and automatically fed into the counting chamber. Note that the readout is punched into the computer cards. The counter used to count samples taken on computer cards is seen in Fig. 6.14; results are also automatically punched into the cards. These types of counters are marketed widely, and the selection of a particular counter is essentially one of individual preference.

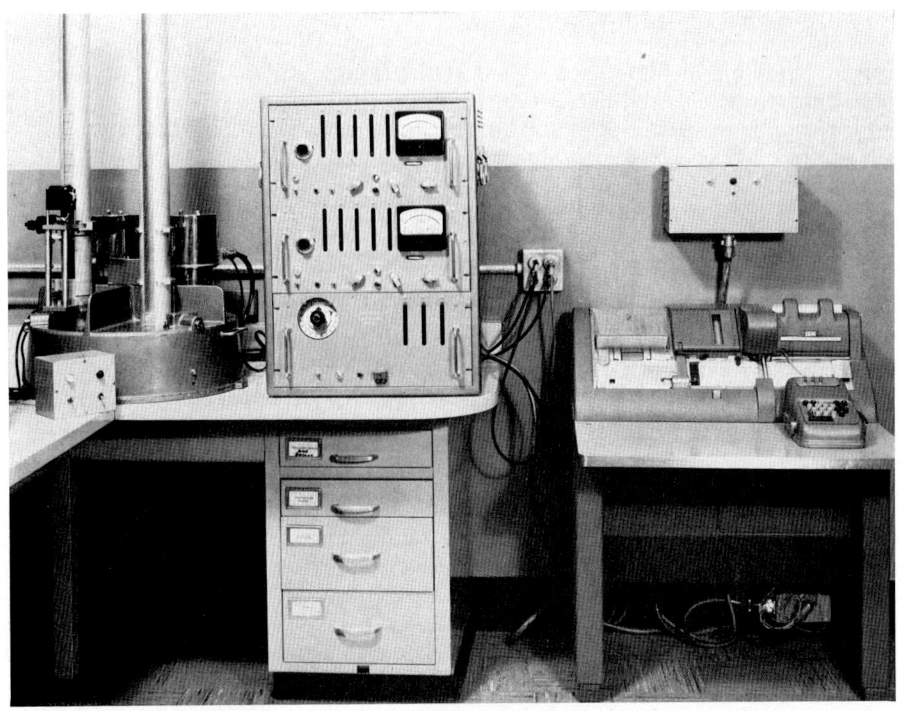

Fig. 6.13. Automatic counter and readout system

Fig. 6.14. Computer card counter and readout system

The counting efficiency can vary widely, depending on the counter and the type of sampling medium used to collect the sample. In setting up a system, the first thing to be checked out is the counting efficiency. One way to acquire this information is to chemically analyze fluorometrically (HARLEY, 1970) a group of air samples which have previously been counted to determine the weight of uranium present; then, by knowing the specific activity, efficiency can be determined by:

$$\text{Efficiency} = \frac{\text{Counts per minute}}{\mu\text{g U} \times \text{d/m}/\mu\text{g}} \times 100.$$

As a routine procedure, standard samples are usually counted to assure that the counter is operating properly. This operation will serve only as a check on the counter and not the overall efficiency since there is no way to set up standards that duplicate coverup and shielding which occur with actual samples.

Depending on atmospheric conditions, significant amounts of radon and/or thoron may be in the sampled air, and their daughter products can attach themselves to particulate matter in the air and be collected on the air sample. For general air samples it is customary to set the samples aside for 16–24 hours before counting to allow for the decay of these radon/thoron daughters. If this is not done, spurious, high results can occur. In diagnostic sampling, where quick results are sometimes needed, it is suggested that a sample be collected in a nonradioactive processing area and use it as a measure of the radon/thoron daughter activity. This radon/thoron value could then be subtracted from the sample of interest.

B. Analysis of Soil and Plant Water Discharge

Because of the low level of activity in such samples, very detailed laboratory preparation and analyses are required. SILL (1969) and SILL and WILLIAMS (1969) present discussions of an appropriate technique utilizing alpha spectrometry. HARLEY (1970) presents a uranium urine analysis technique which can be adopted for soil and/or water analysis.

C. Analysis of Urine Samples

Basically, there are two techniques for the analysis of urine samples for uranium. For materials with less than 5–8 percent ^{235}U, a fluorometric chemical analysis is used. For other materials, a radiochemical analysis is used. Table 6.10 lists some specific methods along with the limits of sensitivity and literature references. For counting samples for radiochemical analysis, proportional or

Table 6.10. Uranium urinalysis procedures

Uranium enrichment	Analysis technique	Reference	Sensitivity[a]
0.7% ^{235}U or less	fluorometric	HARLEY (1970)	—
	fluorometric	PRICE et al. (1953)	1.6 μg/liter
	neutron activation	KRAMER et al. (1967)	0.1 μg/liter
Greater than 0.7% ^{235}U	solvent extraction	TROJANOWSKI and BOUTON (1951)	5 d/m/liter
	ion exchange/precipitation	BONI (1960)	0.15 d/m/1500 ml
	neutron activation	KRAMER et al. (1967)	0.1 μg/liter
	solvent extraction	HURSH (1958)	—
	direct electroplating	PATTERSON (1959)	15 d/m/day

[a] Reported by laboratories using the technique and/or the authors.

scintillation counters are generally used. Usually a higher sensitivity can be achieved with a scintillation counter, but analysis cost will be higher than a proportional counter.

Independent of any internal control that the analytical laboratory may have on the technique, a control program should be run. Usually this step is accomplished by spiking urine samples with a known amount of uranium and submitting it to the laboratory in such a fashion that is not apparent to the laboratory that they are control samples.

References

American National Standards Institute: Nuclear criticality safety in operations with fissionable materials outside reactors. N16.1-1969, American Nuclear Society, Hinsdale, Illinois, 1970.

American Standards Association: American standard safety code for head, eye, and respiratory protection. Z2.1-1959, American Standards Association, Inc., New York, New York, 1960.

Atomic Industrial Forum, Inc.: American national standard guide to sampling airborne radioactive materials in nuclear facilities. ANSI N13.1-1969, American National Standards Institute, Inc., New York, New York, 1969.

Boni, A.: Urinalysis method for enriched uranium. Hlth Phys. **2**, **3**, 288–290 (1960).

Brunskill, R. T., Hermiston, S. T.: The detection and measurement of airborne contamination in major plutonium facilities. Proceedings of the First International Congress of Radiation Protection, part 2, p. 961–973. London, United Kingdom: Pergamon Press 1968.

Caldwell, R., Schnell, E., Potter, T.: Bioassay correlation with breathing-zone sampling. UCRL-18140, University of California, Berkeley, California, 1968.

Code of Federal Regulations: Atomic Energy Title 10, part 20. U.S.A. Office of the Federal Register, 1963.

Cofield, R. E.: *In vivo* gamma counting as a measurement of uranium in the human lung. Hlth Phys. **2**, 269–287 (1960).

Fischoff, L.: Correlation between two-stage air sampling data on the excretion of uranium in urine. NLCO-869, National Lead Company, Cincinnati, Ohio, April 1963.

Ford, M. R.: Comments on intake guides for various isotopes and isotopic mixtures of uranium. ORNL-3697, Union Carbide Corporation, Oak Ridge, Tennessee, October 1964.

Fraser, D. C.: Health physics problems associated with the production of experimental reactor fuels containing PuO_2. Hlth Phys. **13**, **10**, 1133–1143 (1967).

Harley, J. H.: Manual of standard procedures. NYO-4700, 3rd Ed., United States Atomic Energy Commission, New York, New York, Sept. 1970.

Harrington, C. D., Ruehle, A. E.: Uranium production technology. Princeton, New Jersey: D. van Nostrand Company, Inc. 1959.

Harris, W. B., Kingsley, I.: The industrial hygiene of uranium fabrication. Arch. industr. Hlth **19**, 540–565 (1959).

Heatherton, R. C., Huesing, J. A.: A uranium refinery and metal plant urine program and data. Symposium on Occupational Health Experience and Practices in the Uranium Industry, HASL-58, United States Atomic Energy Commission, New York, New York, 1959.

Helgeson, G. R.: Personal communication. Helgeson Nuclear Services, Inc., Pleasanton, California, 1971.

Hursh, J.: Bioassay procedures for uranium. Chemical Methods for Routine Bioassay, AECU-4024, University of Rochester, Rochester, New York, Nov. 1958.

Hyatt, E. C., Moss, W. D., Schulte, H. F.: Particle size studies on uranium aerosols from machining and metallurgy operations. Amer. industr. Hyg. Ass. J. **20**, 99–107 (1959).

International Commission on Radiological Protection: Report of committee II on permissible dose for internal radiation. London, United Kingdom: Pergamon Press 1959.

International Commission on Radiological Protection: Publication 6. London, United Kingdom: Pergamon Press 1962.

Jackson, S.: Estimation of uranium from urine analysis. Assessment of radioactivity in man, vol. II, p. 549–561. Vienna: International Atomic Energy Agency 1964.

Kramer, H. H., Malinski, V. J., Ness, H. W.: Urinalysis for uranium-235 and uranium-238 by neutron activation analysis. Hlth Phys. **13**, **1**, 27–30 (1967).

Laurer, G. R., Eisenbud, M.: *In vivo* measurement of nuclides emitting soft penetrating radiations. Diagnosis and treatment of deposited radionuclides, p. 189–207. Amsterdam, The Netherlands: Excerpta Medica Foundation 1968.

LIPPMAN, M.: Correlation of urine data and medical findings with environmental exposure to uranium compounds. Symposium on Occupational Health Experience and Practices in the Uranium Industry, HASL-58, United States Atomic Energy Commission, New York, New York, 1959.

LOCKHART, L. B., PATTERSON, R. J., JR., ANDERSON, W. L.: Characteristics of air filter media used for monitoring airborne radioactivity. NRL-6054, U.S. Naval Research Laboratory, Washington, D.C., March 1964.

National Bureau of Standards: Maximum permissible body burdens and maximum permissible concentrations of radionuclides in air and in water for occupational exposure. Handbook 69, U.S. Department of Commerce, Washington, D.C., June 1959.

OWEN, F. E.: Beta-gamma dose rates from ^{232}U and ^{233}U, HW-81964. General Electric Company, Richland, Washington, April 1964.

PATTERSON, G. R.: Evaluation and control of internal exposure from enriched uranium at Y-12. Symposium on Occupational Health Experience and Practices in the Uranium Industry, HASL-58, United States Atomic Energy Commission, New York, New York, 1959.

PATTON, F. S., GOOGIN, J. M., GRIFFITH, W. L.: Enriched uranium processing, international series of monographs on nuclear energy. Division IX. Chemical engineering, vol. 2. New York, N.Y.: The Macmillian Company 1963.

PRICE, G. R., FERRETTI, R. J., SCHWARTZ, S.: Fluorophotometric determination of uranium. Analyt. Chem. **25**, 322–331 (1953).

QUASTEL, M. R., TANIGUCHI, H., OVERTON, T. R., ABBOTT, J. D.: Excretion and retention by humans of chronically inhaled uranium dioxide. Hlth Phys. **18**, **3**, 233–244 (1970).

ROSS, J. E.: Correlation of urine data with environmental exposure to uranium. Symposium on Occupational Health Experience and Practices in the Uranium Industry, HASL-58, United States Atomic Energy Commission, New York, N.Y., 1959.

SANDERS, M.: Automated smears program for radioactive material. Hlth Phys. **10**, **5**, 341–344 (1964).

SANDERS, M.: Innovations in air-monitoring techniques for large-scale programs. Assessment of airborne radioactivity. Vienna: International Atomic Energy Agency 1967.

SAXBY, W. N., TAYLOR, N. A., GARLAND, J., RUNDO, J., NEWTON, D.: A case of inhalation of enriched uranium dust. Assessment of radioactivity in man, vol. II, p. 535–547. Vienna: International Atomic Energy Agency 1964.

SCHULTZ, N. B.: Inhalation cases of enriched insoluble uranium oxides. Proceedings of the first International Congress of Radiation Protection, part 2, p. 1205–1217. London, United Kingdom: Pergamon Press 1968.

SCHULTZ, N. B., BECHER, A. F.: Correlation of uranium alpha surface contamination, airborne concentrations, and urinary excretion rates. Hlth Phys. **9**, 901–909 (1963).

SCOTT, L. M., WEST, C. M.: Detection and evaluation of uranium exposures. Assessment of radioactivity in man, vol. II, p. 523–533. Vienna: International Atomic Energy Agency 1964.

SCOTT, L. M., WEST, C. M.: Health physics application of *in vivo* gamma spectrometry in a uranium processing plant. Diagnosis and treatment of deposited radionuclides, p. 543–552. Amsterdam, The Netherlands: Excerpta Medica Foundation 1968.

SCOTT, L. M., WEST, C. M., ABELE, H. M., BRYANT, E. H., CROMWELL, H. H.: Design and development of a mobile *in vivo* radiation monitoring laboratory. Amer. ind. Hyg. Ass. J. **30**, 165–169 (1969).

SILL, C. M.: Separation and radiochemical determination of uranium and the transuranium elements using barium sulfate. Hlth Phys. **17**, **1**, 89–107 (1969).

SILL, C. M., WILLIAMS, R. L.: Radiochemical determination of uranium and the transuranium elements in process solutions and environmental samples. Analyt. Chem. **41**, 1624–1632 (1969).

STECKEL, L. M., WEST, C. M.: Characterization of Y-12 uranium process materials correlated with *in vivo* experience. Y-1544-A, Union Carbide Corporation, Oak Ridge, Tennessee, July 1966.

Task Group on Lung Dynamics: Deposition and retention models for internal dosimetry of the human respiratory tract. Hlth Phys. **12**, **2**, 173–207 (1966).

TROJANOWSKI, M. M., BOUTON, R. Z.: A method of determining enriched uranium in urine. KAPL-667 General Electric Company, Schenectady, New York, 1951.

WEST, C. M., REAVIS, J. P.: Use of statistics in an applied health physics program. Hlth Phys. **10**, **5**, 345–351 (1964).

WEST, C. M., SCOTT, L. M.: A comparison of uranium cases showing long chest burden retentions. Hlth Phys. **12**, 1545–1555 (1966).

WEST, C. M., SCOTT, L. M.: Uranium cases showing long chest burden retention — an update. Hlth Phys. **17**, 781–791 (1969).

Chapter 7

Uranium Mining Hazards

D. A. Holaday

I. Introduction

Uranium mining involves a small fraction of those employed in the nuclear energy industry, but the work of these men is basic. Without the uranium which they produce there would be no power reactors and the Western world would be faced with extremely serious and difficult problems in attempting to meet power requirements from other sources. Uranium miners are confronted by health hazards which are complex and difficult to control and they comprise the largest group of industrial workers whose exposures to radiation have resulted in undeniable and excessive injuries. This chapter will discuss the problems inherent in producing uranium ores, methods of evaluation of hazards, and control procedures.

A. Distribution of Uranium Ore Deposits

Uranium is widely distributed in the earth's crust, usually in very low concentrations. Possible low grade source are black shales, phosphate beds or granites, none of which have produced any significant amounts of uranium or are now being exploited commercially. The most substantial reserves are all disseminated deposits in sedimentary or metamorphosed sedimentary rocks. The type of host rock has a marked influence on the radon emanation rates, and thus on the degree of potential hazard from radon and radon daughters.

The principal mining areas of the Western world are in the Colorado Plateau and the Tertiary Basin of Wyoming in the United States; the Beaverlodge and Blind River districts of Canada; The Rand mining area of South Africa; the Massif Central in France; Rum Jungle and Radium Hill in Australia; and in southcentral Spain.

B. Types of Mining Operations

Uranium ore deposits are exploited by essentially all of the methods used in mining other minerals. Approximately one-third of the uranium produced in the United States at present comes from open-pits and the remainder from underground operations. The tendency is to utilize open-pit production in all instances where this is economically possible. However, many of the new deposits which have been found are too deep to be worked as open-pits. In those cases where mills are located in close proximity to mines, the barren tailings are pumped back underground and used to fill worked out areas. This procedure reduces radon influx into active areas, prevents ground subsidence and reduces the size of tailings dumps.

C. Constituents of Uranium Ores

The average uranium content of ores mined in the United States and Canada ranges from about 0.1 percent to as high as 1.0 percent U_3O_8. Pockets of much higher grade materials are found and mined together with lower grade material. Thus the ranges given apply to average grade of ore shipped, not to extremes. South Africa ores are usually less than 0.1 percent and are produced in conjunction with gold production. Most uranium ores consist primarily of quartz with smaller amounts of other minerals. Table 7.1 lists the constituents of U.S. uranium ores.

Table 7.1. Constituents of Ores in U.S. Uranium Mines[a]

Constituents	Percent	Constituents	Percent
Free silica	30.0–40.0	Cobalt	0.0004–0.1
Aluminum	0.1–20.0	Arsenic	0.0004–0.05
Iron	0.1–6.0	Nickel	0.0002–0.04
Calcium carbonate	0.06–10.0	Chromium	0.0001–0.1
Uranium	0.01–0.5	Selenium	0.00003–0.4
Vanadium	0.0009–2.8		

[a] After U.S. Department of Interior, Geological Survey, Washington, D.C. All other metallic constituents individually are within the concentration range for selenium.

II. Health Hazards in Uranium Mining

The primary purpose of this section is to discuss radiation hazards in uranium mining. However, uranium miners are exposed to all the other hazards inherent to metal mining some of which may outweigh radiation hazards. In particular, injuries and deaths from mine accidents are still a very serious problem in uranium mining. Some of the non-radiation hazards are mentioned here so they will not be ignored by those responsible for protecting the health of miners.

A. External Radiation

Miners are exposed to gamma and beta radiation from the ore bodies and from radioactive dust suspended in the atmosphere or on contaminated clothing. Measurements of gamma radiation rates have been reported by Duhamel et al. (1964), Misawa (1964), Halvas et al. (1964), Iranzo and Liarte (1964), and Stewart (1964) for mines in France, Japan, Mexico, Spain, and Australia, respectively. In general the gamma radiation rates ranged from 0.02 to 4 millirem per hour (mrem/hr). Occasional readings as high as 20 mrem/hr were reported by Stewart. However, Stewart stated that even in mines with the highest grade ore the film badge and personal dosimeter readings showed that the average radiation doses were not excessive. The findings of external gamma radiation surveys in U.S. uranium mines are in accordance with those made in other countries. The most recent gamma radiation surveys were those reported by Breslin et al. (1969) who found radiation rates ranging from less than 0.1 to 2.6 mrem/hr in 9 mines. The mean radiation rates ranged from 0.20 to 0.70 mrem/hr. Such radiation rates cause relatively insignificant exposures.

B. Internal Radiation from Long-Lived Radioactive Elements

Radioactive ore dust can be inhaled and deposited in the lungs thus producing internal radiation exposure primarily from the long-lived alpha emitting elements ^{238}U, ^{234}U, ^{230}Th and ^{226}Ra. Fusamara and Misawa (1964) determined the atmospheric concentration of uranium in mines in Japan, and atmospheric concentrations of radium were measured in United States mines by the Occupational Health Field Station in the course of their environmental study. These studies showed that atmospheric concentrations of these elements were below the maximum permissible concentrations (ICRP values). Urinary uranium levels in miners found by the Occupational Healt Field Station were almost all below 10 micrograms per liter. The levels reported by Karajovic et al. (1964) in Yugoslavian miners were higher.

C. Radon and Radon Daughters

The primary radiation hazard has been exposure to the immediate daughters of radon, ^{218}Po, ^{214}Pb and ^{214}Po (RaA, RaB, RaC). These elements are formed by the radioactive decay of radon in mine atmospheres. As radon is continually emitted into open spaces in mines, their source is always present whether or not the area is being worked while dust is only produced during mining activities. The radon daughter atoms as formed are charged and will attach to the first surfaces which they encounter. These may be mine walls or particulate matter in the air. As either unattached atoms or atoms attached to particles are available for inhalation, the site of deposition in the respiratory tract will vary depending on the composition of mine aerosols. This subject is discussed further in Sec. VI.

D. Other Health Hazards

Crystalline free silica is present in almost all uranium ores which have been exploited and, therefore, the danger of silicosis must always be considered. Other usual mining hazards such as toxic gases produced by blasting and oil mists from drilling and loading ammonium nitrate-fuel oil mixtures into drill holes are always present. In many operations diesel powered equipment is used for loading broken ore and for trackless haulage. Where this equipment is used the miners are exposed to a wide variety of organic compounds in the combustion products which may contain carcinogens or co-carcinogens.

III. Experimental Exposures to Radon and Radon Daughters

Numerous investigations have been reported in which animals were exposed to radon and radon daughters. In early work the significant part played by radon daughters in delivering radiation to lung tissues was not recognized so atmospheric concentrations of these elements were not controlled or measured. Therefore, the results of these experiments are conflicting and cannot be compared with later studies.

Extensive and detailed animal experiments have been reported by Morken and Scott (1966) and Pohl (1965). Both of these reports contain references to other work which should also be consulted, particularly the reports of Rajewsky, Schraub, Aurand, Jacobi and other from the Max Planck Institute für Biophysik at Frankfort. Morken investigated the biological effects on mice of continued exposure throughout life to radon and radon daughters in an atmosphere containing 0.42 microcuries of radon per liter (μCi/l). The RaA, RaB and RaC·C' concentrations were 0.34, 0.20 and 0.10 μCi/l, respectively. The latent alpha energy content of the atmosphere was about 1400 Working Levels (WL). The daughters were attached to particles of a natural aerosol. Mortality, body weight, hematological and histological data were collected. Mice exposed for 24 hours were used to obtain data on distribution of radioactivity in tissue and dose estimates were made from these data. MORKEN estimated that average alpha radiation dose rates were: whole body, 5.5 rad per week; lung-trachea-bronchi, 280 rad per week; kidney, 18 rad per week; liver, 2 rad per week; gastrointestinal tract, stomach and contents, 60 rad per week. The primary biological effect noted was shortening of life span. The median value in the exposed animals was 35 weeks compared with 90 weeks for the control mice. No carcinomas were found and it was concluded that death was due to lung injury.

An extremely interesting findings from the distribution measurements was that much of the radioactivity was cleared from the respiratory tract with a physiological half-life of ten minutes. While a substantial part of the cleared radioactivity was in the gastrointestinal tract, sufficient was found in the systemic system to suggest that activity had dissolved from deposited dust particles and entered the blood. MORKEN concluded from these and other experiments that in these mice a rapid clearance phase with a half-life of about 30 minutes operated on much of any dust deposited in the lung.

POHL conducted investigations of the movement of radon-222 and its daughters from the lungs to other organs of the body. Mice and guinea pigs were exposed for at least 6 hours to atmospheres containing varying amounts of these gases and their daughter products. After exposure the animals were immediately killed, dissected and the organs counted. From the decay curves the radon content of each organ was calculated and also the excess RaB and RaC. The excess RaB and RaC was due to translocation of these elements from the lung to the blood and deposition in tissue. In some of the experimental animals the esophagus was ligated before exposure to eliminate swallowing of daughter products and subsequent absorption from the alimentary tract. These experiments showed that only an insignificant part of the daughter products reached the bloodstream via the alimentary route. POHL points out that the important finding of these experiments was that in mice 14 to 26 percent and in guinea pigs 27 percents of the accumulated decay products were found outside of the respiratory and alimentary systems. These results together with those reported by MORKEN show that in the species studied there is a clearance mechanism other than ciliary action which removes deposited radon duagthers from the lung with a biological half-time of a few minutes. The excess RaB and RaC concentrations were greatest in kidney, liver and blood.

Values for daughter concentrations in blood and urine of humans exposed to atmospheres containing radon and radon daughters were similar to those found in guinea pigs. Pohl then used the tissue distribution ratios found in guinea pigs to estimate organ doses in the test subjects. Assuming that the rapid lung clearance phase found in mice and guinea pigs operates in man, the estimated lung dose was decreased and that to the organs was increased somewhat. Studies of the removal of inhaled ^{212}Pb from the lung have showed that the rapid lung clearance phase found in small animals did not occur in men. HURSH et al. (1969, BOOKER et al. (1969) and HURSH and MERCER (1970) found lung clearance times ranging from 6.5 to 12 hours for inhaled ^{212}Pb adsorbed on natural aerosols. RaB (^{214}Pb) would be expected to behave similarly.

Several groups have reported investigations in which animals were exposed to radon and other substances. Such studies are important as uranium miners are almost invariably exposed to several agents, not radon and radon daughters alone. LAFUMA and MEDJEDOVIC (1964) and KUSHNEVA (1964) have reported experiments in which rats were exposed to silica dust and radon together while other test groups were exposed to radon or silica dust only. In both cases the animals exposed to both radon and silica dust showed greater fibrotic changes in the lungs than did those exposed to radon or silica dust separately. CHAMEAND and LAFUMA (1971) have described in a preliminary report induction of lung cancers in rats exposed either to radon alone or to radon and uranium ore dust. No lung cancers were found in a control group or in a group exposed to ore dust only. Work in progress at Battelle Northwest consists of exposing test animals to uranium ore dust and radon daughters, radon daughters plus cigarette smoke, and radon daughters plus diesel engine exhaust products. The results of these experiments will be interesting indeed. As stated by KUSCHNER et al. (1969): "The initiation of lung cancer undoubtedly depends on the interplay of a group of controlling or modifying factors in which the carcinogen may be essential but not sufficient in itself to complete the induction process".

IV. Human Exposures to Radon and Radon Daughters

As mentioned earlier, uranium is widely distributed in mineralized areas and therefore the atmosphere in many underground operations contains some radon in excess of that present in the general air whether or not the mining operation is called a "uranium mine". Thus there have been instances where miners were exposed to radon daughters in significant amounts in both uranium and non-uranium mines.

STEWART and SIMPSON (1964) reported a detailed historical review of European experience in the mines at Schneeburg and Jachmov in Germany and Czechoslovakia where mining was first started in the fifteenth century. This review has an extensive reference list and should be consulted. All of the medical studies of miners in these areas showed that incidence of lung cancer in the men was about 50 percent. EVANS and GOODMAN (1940) estimated that the average radon concentration to which the men were exposed was about 2.9×10^{-9} curie per liter (Ci/l). MITCHELL (1945) considered it possible that the average radon concentration was higher prior to the time that the hazard was recognized and estimated that it was about 1.5×10^{-8} Ci/l. While this value is about five times that estimated by EVANS and GOODMAN, it would only require a change in the effective ventilation rate in the mines by a factor of five to lower the radon levels by this amount. As there is essentially no information available on ventilation rates in the mines, all estimates of conditions prior to the period of environmental surveys are speculative. In any case, the levels of radon daughters to which the miners were exposed were high enough to produce an extremely elevated incidence of lung cancer and were many times higher than exist in controlled uranium mines. Other possibly carcinogenic elements such as cobalt, nickel and arsenic were present in the ores, so the European experience does not represent an example of uncomplicated exposures to radon and radon daughters.

The experience in Newfoundland fluorspar mines described by DE VILLERS and WINDISH (1964) is unique in that it does represent exposures to atmospheres containing radon daughters with no other known carcinogens. The history of the mines and the occupational records of the miners were precisely known, but the environmental conditions could only be estimated from surveys made after the hazards became apparent. A Royal Commission report (1969) describes the findings of its investigation in detail. The pertinent findings were that during the period 1933 to 1967, 142 deaths occurred in miners or ex-miners who had worked 2000 hours or more underground. Of these deaths, 51 were caused by lung cancer. The atmospheric concentrations of radon daughters to which the men were exposed were estimated by De Villers to range from 2.5–10 WL (Working Levels). The ores contained only traces of uranium and the source of radon contamination was ground water. Diesel engines were not a complicating factor as they were used on only one level of one mine.

The most detailed information on the biological effects of exposure to radon daughters is that obtained by the U.S. Public Health Service study of uranium miners in the Colorado Plateau. The findings of this study which was initiated in 1950 are reported by Lundin et al. (1971). In the study group of 3366 white and 780 non-white underground miners, a total of 437 deaths in the white and 72 in the non-white miners occurred during the period 1950 through September 1968, as compared with expected deaths of 296 and 90, respectively. The excess in the white miners was largely due to malignant neoplasms of the respiratory tract and violence. The deficit in non-white miners was primarily in cardio-vascular-renal disease. Almost all of the non-white miners were Navajo Indians who are known to have an extremely low death rate from these causes.

Estimates of total cumulative radiation exposures were made for each member of the study group in terms of working level months (WLM) and the men divided into exposure categories. A total of 70 deaths from lung cancer were observed versus 11.71 expected. No excess was observed in the less than 120 WLM category. In the 120–359, 359–839 and 840–1799 WLM categories a significant (approximately five-fold) excess of deaths from lung cancer was observed. In the 1800–3719 WLM category about a 15-fold excess was found and in the 1800 and over WLM category the lung cancer death rate was about 25 times that expected.

The environmental data on which the exposure estimates were made was not as complete as would be desired. Over 43000 air measurements of radon daughter concentrations were made, but about 2500 mines operated at various times during the period of observation. Many of the men worked in several mines and, before mechanical ventilation was widely used, the radon daughter levels could vary widely from place to place in the mines. However, the exposure records are probably at least as definite as those which were available for studies of the effects of other toxic substances.

V. Evaluation of Mine Atmospheres

A. Measurement of Radon Concentration

In some uranium-producing countries, the regulations for control are based on atmospheric concentrations of radon, so in these jurisdictions it is necessary to make radon measurements for regulatory purposes. In any case, mine ventilation officers often find that information on radon levels is of assistance in devising control programs and in undertanding the conditions in their mines. Two general procedures are available which are suitable for mine use. The most widely used in the scintillation flask method, one description of which is that by Billiard et al. (1964). The other procedure is the two-filter method developed by Thomas and Le Clare (1970). The advantage of the first method is that a large number of samples can be obtained rapidly be relatively untrained personnel. However, a competent laboratory is required to count the flasks and prepare the sampling equipment for use. The advantage of the two-filter method is that an answer can be obtained in a few minutes so it is more attractive to the ventilation officer who usually needs information as rapidly as possible, even though the results are not as accurate as could be obtained by other methods.

B. Measurements of Radon Daughters

Information on the atmospheric concentrations of the individual radon daughters is of assistance in estimating lung radiation doses and of some use to ventilation officers. The methods for measuring individual daughter concentrations are more time consuming and require more elaborate equipment than those for measuring total latent energy. However, a number of procedures have been developed and used in mine studies. Gusarov and Lyapidevskii (1964) discussed six methods and Lopez et al. (1970) investigated two. Both these groups of workers considered the advantages, disadvantages and errors in the procedures. Martz et al. (1969) developed an alpha spectroscopy method which is quite accurate and Raabe and Wren (1969) utilized a least squares technique to analyze counting data. All of these reports should be studied before a selection of methods is made, either for use in mines or in less restrictive situations.

C. Measurements of Latent Alpha Energy Content

A widely used index of atmospheric radon daughter concentrations is the latent alpha energy content of mine air. The unit, one Working Level (WL), is defined as that amount of any mixture of RaA, RaB, and RaC in one liter of air which will release 1.3×10^5 million electron volts of alpha energy on decay to RaD. The radiation dose delivered to various segments of the tracheo-bronchial tube by 1 WL will vary depending upon the proportion of each daughter present, the size distribution of particles present in the air, and the number of particles in each size range. However, it is a quantity which can be measured with field equipment and has been very useful for guiding control work.

Many different methods are available for measuring the working level. All depend upon drawing a known volume of air through a filter paper and counting the activity after the end of sampling. Details of the procedures will not be given here in the interest of brevity. The references should be consulted for specific information. Kusnetz (1956) described the original method which is still widely used. The primary disadvantage of this method is that a minimum waiting time of 40 minutes is required before counting. Rolle (1969) developed a variation with which a reading can be made in a few minutes. Reiter (1967) utilized a beta counting method for surveys of Bavarian mines. The most recent development is the Instant Working Level Meter devised and used by Schroder (1968) with which reading can be obtained in three minutes. Gusarov and Lyapidevskii (1964) also reported on comparisons of a number of methods of varying complexity and accuracy. The choice of a method will depend upon the speed with which a result is desired and the accuracy needed for the particular investigation.

D. Measurements of Condensation Nuclei

The composition of aerosols in uranium mines has been the subject of extensive investigations. The interest in this area is due to the fact that radon daughters are formed as individual, highly charged atoms by decay of radon in mine atmospheres. These charged atoms attach to the first surface which they encounter which may be mine walls or aerosol particles. In usual mine situations there are many more ultrafine particles than larger ones and studies have repeatedly shown that essentially all of the radon daughter activity is respirable. As the site of deposition of the daughters is largely dependent upon the size of the particles to which they are attached and the fraction existing as free atoms, characterization of mine aerosols is important to dosimetry studies. Studies of condensation nuclei in mines have been reported by Kirichenko et al. (1965) in Russian mines, Holleman (1968) in United States mines, and Cabrol (1970) in France. Additionally, the Health and Safety Laboratory of the U.S. Atomic Energy Commission has conducted a number of as yet unpublished mine studies. Cabrol's work is very detailed and shows the change in number concentration of condensation nuclei with changes in mine activity, residence time of air in the mine, and type of operation. In one mine the concentration of ultrafine particles varied from 5×10^3 per cubic centimeter during no activity to 2×10^5 with a diesel locomotive operating. The other investigators reported similar studies. All nuclei counters used were based on the principle of condensing water vapor from supersaturated air on particles contained in the sample of air. Commercially available equipment performs satisfactorily in mines.

E. Estimation of "Unattached Atoms"

Raabe (1969) described a theoretical procedure by which the fractions of radon daughter atoms existing as "unattached atoms" can be calculated. Knowledge of the concentrations of condensation nuclei and of the radon daughter atoms is required. Several different methods and types of equipment for measuring unattached atoms have been utilized for studying the fractions of unattached radon daughter atoms. Billiard et al. (1971), Craft et al. (1966), George and Breslin (1969) and others have described laboratory and field investigations. All the procedure are based on the high diffusion rate of atomic sized particles compared with the diffusion rates of even sub-micron sized particles. However, there probably is no sharp cut-off between the various size ranges, rather a continuous size spectrum in mine aerosols. Therefore, the fraction of unattached atoms reported in various studies are somewhat dependent upon the equipment used and so the results are not strictly comparable.

VI. Maximum Permissible Concentrations of Radon Daughters

The question of what constitutes a proper maximum permissible concentration for radon daughters has probably been the subject of more debate and more intensive study than has been devoted to any other radiation exposure. Stewart et al. (1964) recieved this subject in 1963 at the International Atomic Energy Authority Vienna Symposium. It was further reviewed in Federal Radiation Council Report Number 8 (Revised) (1967) and in the monograph by Lundin et al. (1969). The two possible approaches are to develop a level based on dosimetric calculations or to base a number on the results of studies of human experience. Dosimetry studies have been analyzed by Parker in Federal Radiation Council Report Number 8 and also in a paper presented at the Health Physics Society 1968 meeting (1969). The problem of lung dosimetry is complex, as many assumptions must be made in developing lung models. Furthermore, there is no precise information on what dose of alpha radiation is required for carcinogenesis. In these circumstances human experience has governed in recommendations of maximum permissible concentrations.

There are several reasons for the controversies which exist concerning interpretation of records of human experience. Foremost is probably the difficulty of maintaining atmospheric concentrations of radon daughters in uranium mines significantly below 1 WL. In addition, uranium miners are not only exposed to radon daughters, but also to other possible carcinogens. A very important factor is cigarette smoking which appears to be a powerful synergistic agent in this case. No reports are available of studies of men who worked in uranium mines for thirty or more years in concentrations of 1 WL or lower. Thus it is necessary to extrapolate from the findings of epidemiological studies of miners exposed to relatively high concentration of radon daughters to establish maximum acceptable concentrations for exposure for a period of thirty years. With all the uncertainties in the available information it is understandable that there are differences of opinion on what a maximum acceptable concentration of radon daughters should be. As is the case with other carcinogens, the goal should be to keep exposures at the lowest feasible level. The key word is "feasibility". It is probable that if there are advances in the technology of reducing radon influx into mines, in air-cleaning methods or in personal protective equipment, it will be feasible to reduce exposures of miners and maximum cencentration values will be lowered.

VII. Control of Radon Daughters Concentrations in Mines

Controlling atmospheric concentrations of radon daughters in mine atmospheres is an extremely difficult problem compared with control of other toxic substances which are usually encountered in mines. This situation is due to the facts that radon is continuously emanated into open areas of mines from ore or sub-ore and that extremely low concentrations of radon can be biologically significant. No attempt will be made to cover this subject thoroughly in this discussion, but general principles will be given and references made to standard treatises. As there are no typical uranium mines, the suggestions given will have to be adapted to fit actual conditions encountered. Several comprehensive manuals are available describing procedures for environmental control. Those by Rock and Walker (1970) and Beranek (1967) are very useful and should be studie. Mining and other technical journals now contain discussions of specific studies which are being reported frequently, so it is advisable to consult the current literature to be informed of recent developments.

A. Mine Planning

Proper mine planning is discussed by Rock and Walker and Beranek. These workers point out the great importance of planning mine ventilation and methods of ore extraction before operations are started. Development of a plan requires knowledge of the size and location of ore bodies in as great details as possible. Where deposits are irregular it may be difficult or impossible to obtain the necessary data while in other cases the ore bodies can be defined rather accurately.

The importance of utilizing retreat mining towards uncontaminated air is emphasized. Air courses should be driven in barren rock wherever possible and contaminated air kept downwind of working areas. The method of mine operation should be designed to minimize the number of working areas to keep open surfaces at a minimum. Storage of broken ore underground should be limited, transportation of ore minimized, and ore should be removed from stopes as completely as possible. Proper mine planning, then, will minimize radon influx and make it easier to remove contaminated air from working areas.

B. Reducing Radon Influx

It is highly important to limit influx of radon to the greatest extent possible. Mine ventilation is expensive and at present there is no way of removing radon from mine air except by dilution ventilation. In many mining districts it is necessary to heat ventilation air during the winter to prevent water from freezing. Even with warmed air, mine temperatures may often be as low as 34° F (1° C), so it is understandable that miners will shut off the ventilation in their stopes. In addition to minimizing the number of working areas, worked out and inactive stopes should be sealed off to prevent release of radon into mainstream air. Bulk heads and brattices will develop leaks from ground movement or shockwaves and need routine maintenance. In many situations it is possible to put the areas behind bulkheads under negative pressure with respect to the working areas thus preventing leakage into working areas. Materials such as polyurathane foams and polyvinyl chloride sheets are almost impervious to radon and have been used to seal off non-working areas. Techniques for installing barriers of these substances are still in the experimental stage, but use of sealants in selected circumstances is attractive.

Over-pressurization of mine areas to reduce radon influx has been studied by Schroder et al. (1966). Such procedure depend upon establishing a pressure gradient between mine areas and a "radon sink" which can be vented to the atmosphere. Again, further studies are required to evaluate the usefulness of over-pressurization and to develop criteria for selection of areas where it might be employed.

C. Removal of Radon and Radon Daughters by Ventilation

The general principles of mine ventilation are applicable to uranium mines. Rock et al. (1970) describe the factors to which particular attention must be given. These are:

1. Maintaining efficiency of air distribution through all connected mine openings.
2. Use of directional ventilation to keep contaminated air away from the miners.
3. Preventing recirculation of air.
4. Continued maintenance of seals and stoppings to prevent leakage of contaminated air into working areas.

Radon daughters grow in from radon rapidly so it is desirable to keep the residence time of air in the workings as short as possible. To do this requires that the volume ventilated be kept at a minimum, that recirculation be eliminated and that stagnation of air be avoided. These precepts may be difficult to follow in mature mines which were developed before the necessity of rigorous controls was realized. In new mines which are planned with the requirements for control in mind, many problems can be avoided. Even under favorable circumstances ventilation control in uranium mines is difficult and costly. For example, Frame (1969) in his report of the ventilation program in the Dennison mine in Canada states that to lower the mine average radon daughter levels from 1.08 WL to 0.64 WL required two and one-half years of work and the expenditure of several million dollars.

D. Removal of Radon Daughters by Air-Cleaning

In certain circumstances it is possible to remove radon daughters from ventilation air and use the cleaned air for further dilution ventilation. Where this can be done considerable conservation of ventilation air is possible. The factors which determine feasibility of air cleaning are the atmospheric radon concentration and the time required for use of the cleaned air. EVANS (1969) has described the basic principles and developed tables and graphs which can be used to calculate the possible usefulness of air cleaning in any particular situation. Both filters and electrostatic precipitators have been employed for removing radon daughters from mine air. Where diesel engines were operated, it was found that filter materials plugged rapidly, so electrostatic precipitators equipped with automatic washers were more satisfactory.

References

BERANEK, G.: Excerpts from problems of Occupational Hygiene in Underground Uuranium Mines, Bykovsky, A. V., Moscow, 1963 [in Russian]. Excerpts printed in Joint Committee on Atomic Energy Report Radiation Exposure of Uranium Miners, part. 2, 1967.

BILLIARD, F., MADELAINE, G. J., CHAPUIS, A., FONTAN, J., LOPEZ, A.: Contribution to the study of atmospheric pollution in uranium mines. Radioprotection **6**, 45 (1971).

BILLIARD, F., MIRABLE, J., MADELAINE, G., PRADEL, J.: Methodes de mesure de radon et de dosage dans les mines d'uranium. In: Radiological Health and Safety in Mining and Milling of Nuclear Materials, vol. I, p. 411. Vienna: IAEA 1964.

BOOKER, D. V., CHAMBERLAIN, A. C., NEWTON, D., STOTT, A. N. B.: Distribution of radioactive lead following inhalation and injection. Brit. J. Radiol. **42**, 457 (1969).

BRESLIN, A. J., GEORGE, A. C., WEINSTEIN, M. S.: Investigation of the radiological characteristics of uranium mine atmospheres. USAEC, HASL-220, December 1969.

CABROL, C.: Réalization d'un compteur de noyaux d'aitken automatique et portatif. These No. d'ordre 967, University of Toulouse, June 1970.

CHAMEAND, J., LAFUMA, J.: Lung cancers of different histological types induced in rats exposed to radon inhaled. Unpublished preliminary report, 1971.

CRAFT, B. F., OSER, J. L., NORRIS, F. W.: A method for determining the relative amounts of combined and uncombined radon daughters in underground uranium mines. Amer. industr. Hyg. Ass. J. **27**, 154 (1966).

DE VILLERS, A. J., WINDISH, J. P.: Lung cancer in a fluorspar mining community. I. Radiation, dust and mortality experience. Brit. J. industr. Med. **21**, 94 (1964).

DUHAMEL, F., BELAYGUE, M., PRADEL, J.: Organisation du contrôle radiologique dans les mines d'uranium françaises. In: Radiological Health and Safety in Mining and Milling of Nuclear Materials, vol. I, p. 59. Vienna: IAEA 1964.

EVANS, R. D.: Engineers guide to the elementary behavior of radon daughters. Hlth Phys. **17**, 229 (1969).

EVANS, R. D., GOODMAN, C.: Determination of the thoron content of air and its bearing on lung cancer hazards in industry. J. industr. Hyg. **22**, 89 (1940).

Federal Radiation Council: Report No 8, Revised, Sept. 1967.

FRAME, C. H.: Ventilation practices at Denison Mines, Ltd., Canadian Mining J., p. 48, Oct. 1969.

FUSAMURA, N., MISAWA, H.: Measurements of radioactive gas and dust as well as investigations into their prevention in Japanese uranium mines. In: Radiological Health and Safety in Mining and Milling of Nuclear Materials, vol. I, p. 391. Vienna: IAEA 1964.

GEORGE, A., BRESLIN, A. J.: Deposition of radon daughters in humans exposed to uranium mine atmospheres. Hlth Phys. **17**, 115 (1969).

GUSAROV, I. I., LYAPIDEVSKII, V. K.: Evaluation of the methods for determination of air pollution by the daughter products of radon. Gig i Sanit. **29**, 68 (1964).

HALVAC, J., TELICH, J., GONZALEZ, Y. R.: Condiciones de trabajo y niveles de radicion existenses en las mines que actualmente explota la cnen y en la planta piloto de la misma. In: Radiological Health and Safety in Mining and Milling of Nuclear Materials, vol. I, p. 97. Vienna: IAEA 1964.

HOLLEMAN, D. F.: Radiation dosimetry for respiratory tract of uranium miners. Special Report on USAEC Contract No. AT(11-1)-1500. Colorado State University, Dec. 1968.

HURSH, J. B., MERCER, T. T.: Measurement of ^{212}Pb loss rate from human lungs. J. appl. Physiol. **23**, 268 (1970).

HURSH, J. B., SCHRAUB, A., SATTLER, E. L., HOFFMAN, H. P.: Fate of ^{212}Pb inhaled by human subjects. Hlth Phys. **16**, 257 (1969).

IRANZO, E., LIARTE, J.: Control de los peligros de la radioactividad en las mines de uranio espanolos. In: Radiological Health and Safety in Mining and Milling of Nuclear Materials, vol. I, p. 111. Vienna: IAEA 1964.

Karajovic, D., Kilibarda, M., Panov, D., Djuric, D., Medjedovic, M., Raicevic, P., Delic, V.: Uranium in the urine of miners exposed to uranium compounds. In: Radiological Health andSafety in Mining and Milling of Nuclear Material, vol. II, p. 443. Vienna: IAEA 1964.

Kirichenko, V. N., Ogorodnikov, B. I., Isanov, V. D., Huish, A. A., Kachikin, V. I.: Submicroscopic aerosols of short-lived daughter products of radon in mine air. Gig. i Sanit. **30**, 308 (1965).

Kuschner, M., Laskin, S., Nelson, N.: Research in Environmental Health and Cancer, Fourth Annual Report, p. 72, Inst. Environ. Med., New York University (1969).

Kushneva, E. R.: Peculiarities of the course of experimental silicosis with inhalation of radon. In: Radiological Health and Safety in Mining and Milling of Nuclear Materials, vol. I, p. 317. Vienna: IAEA 1964.

Kusnetz, H. L.: Radon daughters in mine atmospheres: A field method for determining mine concentrations. Amer. industr. Hyg. Ass. Quart. **17**, 85 (1956).

Lafuna, J., Medjedovic, M.: Accroissement de l'activité de poumon dû à l'inhalation de radon. In: Radiological Health and Safety in Mining and Milling of Nuclear Materials, vol. I, p. 215. Vienna: IAEA 1964.

Lopez, A., Chapius, A., Fontan, J., Billiard, F., Madelaine, G. J.: Measurement of the state of equilibrium between radon and its daughters in the uranium mines. Aerosol Science **1**, 225 (1970).

Lundin, F. E., Wagoner, J. K., Archer, V. E.: Radon daughter exposures and respiratory cancer, quantitative and temporal aspects. National Institute for Occupational Safety and Health and National Institute of Environmental Health Service, Joint Monograph, Number 1, June 1971.

Martz, D. E., Holleman, D. F., McCurdy, D. E., Schiager, K. J.: Analyses of atmospheric concentrations of RaA, RaB, and RaC by alpha spectroscopy. Hlth Phys. **17**, 131 (1969).

Misawa, H.: Radiological health and safety in Japanese uranium mines. In: Radiological Health and Safety in Mining and Milling of Nuclear Materials, vol. I, p. 83. Vienna: IAEA 1964.

Mitchell, J. S.: Memorandum on some aspects of the biological action of radiation with special reference to tolerance problems. Montreal Laboratory Report HI-17, 20 Nov. 1945.

Morken, D. A., Scott, J. K.: The effects on mice of contimual exposure to radon and its decay products. University of Rochester Atomic Energy Project, UR-669, 1966.

Parker, H. M.: The dilemma of lung dosimetry. Hlth Phys. **16**, 553 (1969).

Pohl, E.: Biophysikalische Untersuchungen über die Inkorporation der natürlich radioaktiven Emanation und der Zerfallprodukte. S.-B. öst. Akad. wiss. Wien IIa, **174**, 309 (1965).

Raabe, O. G.: Concerning the interactions that occur between radon daughter products and aerosols. Hlth Phys. **17**, 177 (1969).

Raabe, O. G., Wren, McD. E.: Analyses of radon daughter activities by weighted least squares. Hlth Phys. **17**, 598 (1969).

Reiter, R.: Devices and procedures for protection against radon and its decay products in mine atmospheres. Nukleonik **9**, 367 (1967).

Report of Royal Commission Respecting Radiation, Compensation, and Safety at the Fluorspar Mines: St. Lawrence, Newfoundland, 1969.

Rock, R. L., Walker, D. K.: Controlling employee exposure to alpha radiation in underground uranium mines. U.S. Dept. of Interior, Bureau of Mines, Two volumes, 1970.

Rolle, R.: Improved radon daughter product monitoring procedures. Amer. industr. Hyg. Ass. J. **30**, 153 (1969).

Schroder, G. L.: Instrumentation for in-mine "instant" measuring of working level and radon concentrations. Mass. Inst. of Tech. Annual Progress Report, MIT-952, p. 346, 1968.

Schroder, G. L., Evans, R. D., Kraner, H. W.: Effects of applied pressure on the radon characteristics of an underground mine environment. AIMA Trans. **235** (1966).

Stewart, C. D., Simpson, S. D.: The hazards of inhaling radon 222 and its short-lived daughters: Consideration of proposed maximum permissible concentrations in air. In: Radiological Health and Safety in Mining and Milling of Nuclear Materials, vol. I, p. 333. Vienna: IAEA 1964.

Stewart, J. R.: Radiological health and safety in the Australian uranium mining and milling industry. In: Radiological Health and Safety in Mining and Milling of Nuclear Materials, vol. I, p. 161. Vienna: IAEA 1964.

Thomas, J. W., Le Clare, P. C.: A study of the two-filter method for radon-222. Hlth Phys. **18**, 113 (1970).

Plutonium

Preface

J. N. Stannard

The eight chapters in this section are intended to provide the pharmacologist-toxicologist with an overview of the development and present status of biomedical information on the isotopes of plutonium, the most used and quantitatively most important of the man-made actinide elements. While the coverage is far from complete the extent of developments over the last two decades, portents for the future and areas currently under intensive study should be clear. The intent is, as in the sections on uranium and the transplutonics, to provide pertinent historical perspective at a time when many of the pioneers in the field are approaching the time for completion of their most active involvement and to review in some depth the basic problems of behavior of the several isotopes, the mechanisms as currently understood along with some of the more applied problems of health physics and industrial medical management.

Fortunately there have recently been several extensive symposium volumes on plutonium. These have emphasized especially its radiobiology. They have been drawn upon extensively in some of the chapters herein. However their purpose was not coincident with that of this Handbook Series, and the amount of undesirable overlap is, we trust, minimal.

Two chapters *viz*: 8 and 12 are primarily historical in intent although Chap. 8 is also partly orientative. Chap. 9 provides the most salient features of the solution chemistry and biochemistry of plutonium isotopes and their bearing on biomedical problems. Chap. 10 is directed primarily at chemical, physical and biological aspects of the deposition of plutonium in bone and its effects after deposition. It includes a rather detailed consideration of the radiation dosimetry of alpha particles in bone, current ideas on the mechanisms behind the high efficiency of plutonium alpha particles in producing osteogenic sarcoma, a general review of its metabolism *en route* to bone and the current status of therapeutic measures. The chapter by Bair and colleagues (Chap. 11) while considering "Plutonium in Soft Tissue" lays special emphasis on inhalation problems. It presents a nearly complete summary of all work with plutonium isotopes entering the body by inhalation or reaching the lung or accessory respiratory structures by other means. The current status of our ideas and information concerning plutonium as a pulmonary carcinogen is especially emphasized.

As indicated in the dedication, Chap. 12 stands among the last of Dr. Langham's major works. We are indeed fortunate that he and Dr. Healy took seriously the need for an historical overview of the development of exposure standards for plutonium. While directly pertinent only to this nuclide, the chapter provides also some valuable insight into the procedures and philosophies needed to develop exposure standards for a new and potentially very dangerous agent when the quantities available were miniscule.

Chaps. 13 and 14 encompass matters more pertinent to the determination and control of plutonium in and around industrial-type nuclear facilities. Chap. 14 contains many details on the prevention and handling of plutonium contamination, on experience to date at installations such as the Hanford Works at Richland, Washington and a cogent reminder that external radiation and criticality potential assume more and more importance as the quantities of plutonium being handled increase. It also introduces the organization and purposes of the U.S. Transuranium Registry and a glimpse of some of the waste disposal and environmental problems to be expected in the decades ahead.

In the final chapter of the section, a survey is made of plutonium as an actual or potential environmental contaminant. Even though the amounts on a weight basis are small the great toxicity of plutonium and the high probability for addition of many more and different sources in the future have caused growing interest and concern for this aspect of the biology and medicine of plutonium. The chapter reviews briefly the major past work, which extends back to the construction of the first manufacturing facilities, the behavior of plutonium isotopes in comparison to the more commonly discussed nuclides in the environment such as isotopes of strontium and iodine, and the unclassified aspects of field tests and incidents including those at Thule, Greenland and Palomeres, Spain. Correlation of laboratory and field results is undertaken wherever possible.

The possible role of the pharmacologist-toxicologist in this field which has in the past been occupied more by representatives of other disciplines is emphasized where convient. As the uses of nuclear energy grow both in industry and in biology and medicine it is more and more likely that toxicologic aspects at least will require attention from the professional pharmacologist-toxicologist. Why should some of the most toxic substances known continue much longer to be outside the full purview of conventional toxicology?

The section editor wishes to express very special thanks to the several chapter authors many of whose manuscripts have "lain fallow" awaiting the completion of other chapters and other sections of the volume. Thanks are also expressed to my office staff, both full and part-time, for invaluable service in the course of both writing and editing.

Chapter 8[1]

Biomedical Aspects of Plutonium (Discovery, Development, Projections)

J. N. STANNARD

I. Foreword

There is probably no single element in the periodic table with a comparable saga of discovery, development, manufacture and use covering a span of only one generation. Uranium and its isotopes along with its daughter products such as radium and radon, have been on the scene for decades. Uranium was known in nature and used in industry long before its possession of radioactivity was known or became of interest and importance. By contrast plutonium was only a theoretical possibility in the late thirties in the minds of MEITNER, BOHR, IRENE JOLIET-CURIE and workers at the Cavendish Laboratory at Cambridge University until SEABORG and his colleagues isolated a tracer quantity of element $^{238}94$ on the night of February 23, 1941, and 0.5 microgram of $^{239}94$ on March 28 of the same year. (The name "plutonium" was not assigned until the following year.) The fact that this element could undergo fission with thermal neutrons was found essentially simultaneously with its discovery and thus began the tremendous effort to produce it in quantity for military purposes. This story is so familiar that nothing further need be added here except to point out that kilogram quantities were available by the summer of 1946 and we are now talking of thousands of kilograms.

If military uses were the end of the saga of plutonium this section of the Handbook would be of much less interest and much shorter. But plutonium is central to the development of electrical power by nuclear fission and to a growing number of other uses in industry and medicine. SEABORG (1969) estimates that the United States will be using 20000 kilograms of ^{239}Pu by 1980, 60000 kilograms in the decade 1980–1990 and 80000 kilograms in decade 1990–2000. Meanwhile the use of the shorter-lived isotope ^{238}Pu for power in space vehicles and other uses is estimated at 10–20 kilograms in 1970–1980, 100 kilograms in 1980–1990, while medical uses will grow from essentially zero now to 5 kilograms in 1980–1990 to some thousands of kilograms in 1990–2000. Even if these estimates are in error there is no question that we already have and will continue to have very sizable amounts of plutonium on hand. An understanding of its biomedical effects, ways to handle it and to avoid its harmful effects in either man or his environment are essential. Such knowledge is clearly within the purview of the modern pharmacologist-toxicologist, to whom this volume is directed, as well as to the radiation biologist and health physicist.

---

1 Work supported in part by Contract #AT(11-1) 3490 between the U.S. Atomic Energy Commission and the University of Rochester and has been assigned Report #UR-3490-116.

II. General Toxicology of Plutonium

It has now become traditional to refer to plutonium as "the most toxic element known to man". In a general sense this may indeed be true. Some of the trans-plutonic elements discussed in Part 3 of this Handbook may exceed plutonium in toxicity on the basis of potential effects per unit of mass or even per unit of activity. Other radioactive elements at the bottom of the periodic table will go plutonium a close second or even exceed plutonium in some aspects of effects per amount deposited. But none of these other elements are expected to be around in the quantities approaching those described above for plutonium. Thus if the presence of potential sources is taken into consideration as part of the potential hazard, plutonium occupies a central position among the radioactive heavy elements whether or not it should turn out to be *the* most toxic element known to man.

The biological effects of the plutonium isotopes of primary interest in biology and medicine, ^{238}Pu and ^{239}Pu (see Chaps. 9 and 10 for discussion of other isotopes of plutonium as well as details concerning 238 and 239), are due internally almost if not entirely to the alpha particles released in their decay. The mass of 1 Curie of Plutonium-239 is 16.28 grams, that of 1 Curie of Plutonium-238 is 57.5 milligrams. Acute effects occur with body burdens measured in microcuries. Long-term effects, our primary concern with plutonium, may occur with nanocuries per gram of tissue. Thus chemical effects of the type familiar to the pharmacologist-toxicologist seem quite unlikely to occur. Chemical toxicity of a type and magnitude per unit of mass wholly new would have to be present for the effect to be significantly a chemical one. The contrast to natural uranium, the subject of Part 1 of this Handbook, is clear. Yet the chemical properties of plutonium, like those of the other elements described in this volume, play a critical role in determining distribution of the element to tissues and organs and its microdistribution within tissues and within cells. These distributions in turn determine the ultimate effects to a degree quite unexpected when biological studies with plutonium and other trans-uranic elements were begun. Thus although the effects are to be considered primarily radiation effects[2] variation in such chemical parameters as valence, polymerism, presence or absence of complexers, etc. are critically important. The study of the effects of plutonium and of the other trans-uranics is thus far from a routine reiteration of well-established radiation effects predictable from the study of external radiation sources or from radioisotopes with very different chemical properties.

In the chapter immediately following this those chemical and physical properties of plutonium of direct bearing on biomedical effects are described in some detail. Those of particular pertinence to its deposition and retention in bone are considered further in Chap. 10.

It was found early on that plutonium injected intravenously deposited largely in and was retained avidly by bone. From this fact behavior analogous to that of radium in bone was anticipated. Grossly and in a qualitative sense this analogy has held. But as work progressed it became clear that the analogy was only a gross one and that plutonium, while a bone-seeker, represented a class of elements

2 This statement should not be taken to mean that *no* chemical effects occur beyond the role of chemistry in distribution and excretion. There is a growing number of facts which are leading investigators to consider that radiation dose alone may not be enough to explain all of the effects of deposited radionuclides. Nevertheless the statement made above that chemical effects if they do play a role would have to be of a type and magnitude not currently understood holds true. Also there are external radiation problems with large quantities of plutonium as detailed in Chap. 14.

which behaved very differently in bone than radium. While radium enters the mineral phase (hydroxyapatite) essentially in exchange for calcium, plutonium enters by a more complex route involving interaction with organic ligands and with essentially no simple exchange with calcium. The end result of deposit in bone for a long enough period is osteosarcoma just as occurs with radium. But the details of the process and the accompanying secondary phenomena are quite different and distinct. Because of this Chap. 10 deals extensively with plutonium "as a bone-seeker" both in terms of its behavior, i.e., metabolism, and its effects.

The contrast with uranium is also to be noted. Uranium exchanges for calcium in bone crystal as was made clear in the early work of NEUMAN and colleagues (NEUMAN and NEUMAN, 1948; NEUMAN et al., 1949a and b) and discussed in the first part of this Handbook. The exchange process may not be quite the same as for radium (FORMAN, 1971) but the contrast to plutonium is more marked.

It was also found quite early that plutonium is very poorly absorbed from the gastrointestinal tract. Accepted figures for the fraction going from the gastrointestinal tract to blood are 3×10^{-5} for plutonium isotopes, while a similar figure for radium is 0.3 (ICRP Publication II, 1959). Thus entry into the body by food or water ingestion is negligibly important, quite in contrast again to radium or uranium. Only puncture wounds and inhalation of plutonium in aerosol form are likely routes of occupational or environmental exposure with some exceptions to be noted in later sections.

Much effort has been expended on describing and understanding the metabolism of inhaled plutonium and its effects on lung, lymph nodes, bone and other tissues after inhalation. This work is summarized in Chap. 11 which concentrates on plutonium behavior and effects in soft tissue especially after inhalation.

It can be said with confidence that we have as yet recorded a case where clear-cut biological effects of plutonium deposited in the body have occurred in man. This reflects largely the foresight, engineering and fantastic effort in both resources and manpower devoted since its discovery to protecting workers and the environment from entry of plutonium into man or the biosphere. There have indeed been exposures, primarily from plutonium fires (it is quite pyrophoric), criticality accidents, "broken arrow" incidents (to be described subsequently), and the minute but inevitable introduction of this element into the environment of atomic energy workers and to a still lesser degree the biosphere itself particularly from atmospheric testing of nuclear explosives. But none of these exposures to date have produced body burdens of plutonium either acutely toxic or, by analogy to radium in man, expected to eventuate in detectable increases in the incidence of osteosarcoma.

It would be most unrealistic, however, to downgrade interest, research, and surveillance for plutonium because of these fortunate facts. The time needed to produce osteosarcoma and other forms of cancer in man is only now approaching. There is reason to believe that the body burdens are low enough not to produce cancer in the normal life span and we may escape seeing any cases at all from exposures to date. But even if we are so fortunate as to have no cases whatsoever appear from World War II exposures or those in the two decades since, the increase in quantities used and anticipated to be used, more non-military uses, and sheer probability tells us there will sooner or later be some demonstrated effects in man. Later chapters of this part summarize our experience to date and address the possibilities for the future (Chaps. 10, 11, 14).

Chap. 13 takes up ways to ascertain the body burden of plutonium in man ("Bioassay") while in Chap. 12 is described the fascinating story of how seemingly inadequate biological data, shrewd scientific intuition, and dedication to the protection of workers, population and environment led to development and adoption of protection standards which are closely comparable to those derived today from a much larger body of information.

Before closing this general introduction to the toxicology of plutonium the matter of dosimetry requires attention. The radiobiologist and health physicist use standardized measures of radiation dose such as the rad (absorbed energy equivalent to 100 ergs/gm), the roentgen (a measure of exposure dose for X or gamma radiation, 2.58×10^{-4} coulomb per kilogram of dry air), the rem (rad or roentgen equivalent in man or mammal), and standard nomenclature for radioactivity such as the Curie which is 3.7×10^{10} dis/sec, and its multiples and subunits[3].

Of perhaps greater importance to mention here is the contrast in attitude toward doses and dosimetry between the field in which we are working and more conventional pharmacology and toxicology. Whereas in classical pharmacology and toxicology investigators, indeed all concerned, are usually satisfied to know what amount of drug or chemical was given and by what route, in radiation toxicology this is only the start. The radiation dose is calculated wherever possible *at the probable site of action.* Frequently this means the dose to a "critical organ" or a "target organ". Sometimes it even implies, as described in the section on dosimetry in Chap. 10, the dose to a particular cellular structure or even subcellular structure. Also where there is more than one potentially "critical" organ several such calculations may be done. Such "microdosimetry" is only just beginning in most other areas of toxicology e.g. where the concentration of drug at some critical interface may be considered rather than the amount injected or ingested. While obviously desirable, and made possible by the sensitivity and localization possibilities of radiation measurement instrumentation the lament seen in some of the following chapters that some of these quantities are known imperfectly should be viewed in perspective by the reader of this Handbook. We are fortunate indeed that relatively so much can be said about the dose to the target structure despite some limitations in detail. The classical toxicologist should find this phenomenon both interesting and stimulating as he peruses these chapters.

III. Development of Biomedical Information on Plutonium

The several chapters to follow in this part of the Handbook will concern largely the *present status* of information and concepts regarding the metabolism and effects of this element. In many respects the history of the development of this information is unique (along with that for uranium in quantity and the other transuranics). It clearly is pertinent to the understanding of current work. Also the presentations in other chapters deal primarily with the findings and not the organization of the studies or their background. For these reasons a short summary of the development of biomedical information on plutonium is considered pertinent.

A. Early Beginnings

The recognition of potential biomedical hazards from the new radioactive materials brought into being by the atomic bomb project was essentially simultaneous with the decision to proceed with the project. But the accomplishment

3 The uninitiated pharmacologist-toxicologist can find these units defined in more detail in several places such as the Lapp and Andrews text (1971) or introductory radiology treatises.

of the first self-sustanining chain reaction on December 2, 1942 made the need for biomedical information urgent. Many details with dates are contained in the chapter herein by LANGHAM and HEALY (Chap. 12) on Maximum Allowable Concentrations and Maximum Permissible Body Burdens of Plutonium.

Much of the early work concerned primarily biological effects of *external* radiation sources. Particular emphasis was placed on those features which were "new" e.g. pile radiations (slow and fast neutrons plus gamma radiation), irradiation from beta particles, and the effects of doses of these in the "tolerance" range. This work had to be done to understand the effects of deposited radionuclides, particularly fission products, as well as in its own right. No attempt is made in this volume to summarize any of these studies since the literature is replete with reviews. (A good overview can be seen in the summary papers contained in a special issue of Radiology entitled "The Plutonium Project", issued in September 1947, and detailed in the bibliography.)

Because of the extremely small quantities of plutonium available in the early phases of the project and the high security classification of all such work the early information perforce involved only small animals and a small group of investigators. The first biological studies were carried out at the Radiation Laboratory of the University of California, Berkeley under the direction of Dr. JOSEPH HAMILTON and his associates. Early reports were "in-house" documents or letters, but the data can be seen in now declassified documents such as those by HAMILTON (1944), SCOTT et al. (1945), CROWLEY, LANZ, SCOTT, and HAMILTON (1946) and later in published papers from the same group (e.g. SCOTT et al., 1948, 1949; COPP, AXELROD and HAMILTON, 1947; and HAMILTON, 1947, 1948).

This work was carried out under the aegis of the Plutonium Project's Health Division which was in turn a part of the Metallurgical Laboratory Organization based at the University of Chicago. These early experiments showed quickly that intravenously injected plutonium went largely to bone, at least in the rat, and announced its potential similarity to radium.

B. The "Plutonium Project" Years

As larger quantities of plutonium became available, other laboratories began biomedical studies of its metabolism and effects. Various studies were undertaken at Chicago, Los Alamos, Rochester, and the Clinton Laboratories at Oak Ridge[4]. The fact that plutonium was indeed a "bone-seeker" was early and easily confirmed (LANGHAM, 1946; VAN MIDDLESWORTH, 1947; R. D. FINKLE, 1946). Details are contained in Chap. 10 and those especially pertinent to radiation protection in Chap. 12. Extension to many species and routes of administration more pertinent to potential exposure of man soon followed as described in some detail in Chap. 10. Comparison to other radionuclides particularly radium and strontium was contained in the Chicago reports (BRUES et al., 1946; LISCO et al., 1947) and to radium and polonium, an alpha emitter which does not seek bone, in the Rochester reports (BOYD et al., 1950a and b; FINK, 1950).

Considerations of relative energy and half-life predicted that plutonium would be less toxic on an energy basis than radium. A factor as high as 50 was proposed

4 The "Plutonium Project" officially encompassed work at the University of Chicago (Metallurgical Laboratory), University of California, Clinton Laboratory near Oak Ridge, Tennessee and the National Cancer Institute in Bethesda, Maryland. Work at other sites such as Rochester, Columbia, Los Alamos, etc. was under other divisions of the Manhattan District, but for the present discussion no distinction is made since we are considering the developments during the *years of* the Plutonium Project.

(see Langham and Healy, Chap. 12 herein for calculation). But the facts were very different. As a carcinogen for the production of osteosarcomas (as well as for many other end-points) plutonium turned out to be considerably *more* toxic than radium. The importance of this for understanding of mechanism is taken up in detail in the chapter on "Plutonium as a bone seeker" (Chap. 10) and its significance to radiation protection decisions in Chap. 12.

C. The Utah Project

Most of the more pertinent information on plutonium effects in the above reports involved small rodents, although a few dogs were used in some acute studies. Also the work naturally had some of the characteristics of the war-time pressures. For this reason, and because the need for work specifically focused on long-term effects was made especially evident by the Plutonium Project studies, a new approach was begun in the early 1950's under the sponsorship of the then new U.S. Atomic Commission. This was to organize a full-scale comparison of several bone-seeking radionuclides in a long-lived animal covering a wide range of doses. The project was domiciled in the Department of Anatomy at the University of Utah, Salt Lake City. The beagle dog was chosen as the test subject and the dosage a single intravenous injection. Because of the cost and organizational continuity required for such an experiment only the specific interest and financial commitment of a governmental agency could bring off such a program successfully. Yet the large national laboratories were off on other tacks, in part at least. Thus was born a new concept in the peace-time support of biomedical research. The proposal was to do the work entirely by contract to a University. This, I believe, was the first such contract for a *specific* long-term study using specific substances[5].

Another new feature was the introduction of a guiding group of experts from nearly all interested laboratories to *plan* the experiments. Known affectionately as "The Founding Fathers", this group planned every phase of the experiments to be done in collaboration with the responsible investigators at Utah. They have kept and been kept in close touch with the work, as exemplified by the proceedings of the 20th Anniversary symposium (Stover and Jee, 1972) where the "Founding Fathers" took an active part. The success of this particular mode of attack can be seen from the several key symposia sponsored by this group (e.g. Mays, Jee, and Lloyd, 1969; Stover and Jee, 1972) and the ubiquitous referencing of their work in the chapters to follow. The patience and dedication of all concerned deserve special mention since this type of research is expensive per unit of new information gained, is likely to lose "glamor" as the years go by, and is fraught with apprehension for the unexpected epidemic or mistake in planning. Yet there seems to be no substitute for the passage of time in unfolding the biological effects of low level exposures to radiation or other agents.

D. Inhalation Studies

While these animal studies at Utah using the intravenous route of entry gave and continue to provide a large body of information on mechanisms of effect and metabolism of plutonium once it had entered the blood stream it was clearly recognized as outlined in Sec. II, that inhalation was the most likely exposure

5 Parallels are seen in the long term inhalation study on uranium dioxide at Rochester described in Part 1, the external radiation and fission product study at Davis, California and several others begun more recently. However, the Utah project was the first one newly organized around so large and specific a task.

route of entry to man. In some of the Plutonium Project work (Berkeley and Chicago) small rodents received plutonium by inhalation or intratracheal injection and a very limited amount of information was obtained in man. The earliest report was that of SCOTT et al. (1945) followed by those of ABRAMS et al. (1946, 1947). Some of these were later published in the open literature as cited.

The work indicated long retention times in the lung of the dioxide and considerable dependence on valence state, compound, etc. as described in Chaps. 9 and 10 herein. But the inhalation route was clearly in need of much more thorough investigation, particularly since incidence of pulmonary cancer was reported from placement of radionuclides in the lung at about this time (WAGER et al., 1956). Just at the end of the Manhattan Project days a formal study of the fate and effects of inhaled plutonium was begun at Hanford using larger and longer-lived animals. This developed through the 1950's and 60's at Hanford (now Batelle-Northwest Laboratories) into an effort comparable in scope to that at Utah and it so continues (THOMPSON, 1967; TOTTER, 1972).

The inhalation work was not organized in the same way as the Utah project since much more range-finding work needed to be done before a long-term experiment at low doses could be mounted. Data of pertinence to this central problem are now ready and are reviewed and summarized for the first time in so complete a form in Chap. 11 of this volume.

The University of Rochester Atomic Energy Project was involved through the war years and thereafter in a major experiment with inhaled uranium as well as several on-going programs in other aspects of inhalation problems and had developed a strong group in fundamental aerosol physics. In the early 1950's they built a special facility for experiments with alpha emitters, particularly inhalation experiments. While a considerable fraction of their effort was devoted to ^{210}Po partly as a model for the effects of radon daughters, to long-term and short-term experiments with radon in mice and dogs, and to basic pulmonary mechanism and aerosol studies, a reasonably extensive plutonium inhalation experiment was completed. Dogs were exposed to both ^{239}Pu and ^{238}Pu. These results were described by YUILE, GIBB, and MORROW (1970), and are compared to other work in Chaps. 10 and 11.

These inhalation studies have proven beyond doubt that the metabolism of inhaled plutonium can be and usually is quite different from that of the systemically administered isotope and that the results could not have been predicted from studies using the parenteral route. The importance of the lymphatic system, particularly pulmonary lymph nodes, has been emphasized even though its role in effects is not yet entirely clear. Recently, strong evidence has been presented that plutonium deposited by inhalation can produce pulmonary cancer, in the dog at least, although there are differences between the major studies which need resolution. Thus bone may frequently *not* be the critical organ when certain poorly transportable (i.e. insoluble) compounds of plutonium are inhaled. The bearing of this on determination of maximum permissible air concentrations is obvious as discussed by LANGHAM and HEALY in Chap. 12.

E. Information from Experience with Man

In parallel with these experimental studies every effort was made to gather information on man. As described in Sec. II there is, to date, no documented case of plutonium *effects* in man. However, the efforts in control, bioassay, and tissue analyses of hospital patients and nuclear energy workers who received small amounts of plutonium has resulted in a modicum of metabolic data for man. The primary sources of information are Los Alamos and Hanford, although others contributed, particularly to the hospital patient studies. These show some evident differences between man and experimental animals, particularly in retention times in different organs and in excretion kinetics. The metabolic contrasts are summarized in Chap. 10 and from the point of view of occupational exposure in Chaps. 13 and 14 which take up respectively, bioassay (i.e. radiobioassay) and a detailed consideration of experience with plutonium under conditions of occupational exposure.

A review of the metabolic information obtained from man over the last twenty five years has been published recently by DURBIN both as a UCRL report with very full documentation (DURBIN, 1971) and as a chapter in the Utah anniversary volume cited earlier (STOVER and JEE, 1972). This reexamination and reconsideration of the accumulated information negates some of the older concepts of plutonium metabolism and emphasizes the role, in man, of complex formation with iron analogues such as transferrin.

It also reiterates and strengthens the views gained from animal work that not only is plutonium quite different from those nuclides which exchange for calcium in the way it deposits in bone but that it, along with the other actinides, has a component of concentration and effect in soft tissue of considerably greater importance in understanding its metabolism and effects than is the case with the "volume-seekers".

Of special interest is the work on excretion of plutonium by man. With minute body burdens and a substance with a very long half-life the measurements of amounts of plutonium in urine and feces require the ultimate in painstaking technique and sensitive instrumentation. The fact that excretion kinetics seem to follow a power function of time over a measurement period of five years was clear from the data summarized by LANGHAM (1959). Such information is most useful in attempting to estimate body burden from excreta contents even though it is subject to many restraints. But perhaps most notable is the magnitude of the effort in manpower, resources, and patience necessary and expended to produce this information.

F. Therapeutic Removal

Even though the number of exposures of man has been small the long term toxicity of plutonium is so great relative to many materials that considerable effort has been expended on possible means for therapeutic removal. Most of this work has been published since the middle 1950's and has emanated largely from laboratories with potential exposure situations or ongoing experimental programs. Although the first publications on therapeutic removal came from SCHUBERT (1947, 1955) then at Chicago, the Hanford group has probably been the most active among American laboratories in this area. A fairly considerable effort in this area has also occurred in the United Kingdom.

Although animal experimentation has progressed steadily, as reviewed in Chap. 10, the chances for *bona fide* trial of therapeutic removal methods in man have, fortunately, been scarce. Therefore the field has not been a rapidly progressing one although present practices are clearly based on firmer understanding and the wisdom of experience compared to a decade ago (Chap. 14).

Of the several approaches tried, only the chelating agents and such procedures as pulmonary lavage (McCLELLAND, 1971) now remain under serious study for possible use in accidental exposure cases in man. The present status of chelation therapy is summarized in this volume as much for its inherent interest to the pharmacologist-toxicologist as for demonstrated effectiveness in therapy. However, there are recent cases which indicate that a considerable fraction of the body burden may be removed if chelation therapy is begun early and handled correctly, particularly if combined with other measures.

G. Work Abroad

The above paragraphs have dealt entirely with work in the United States. Since the production of plutonium during the war years was entirely within the United States this is understandable. Unlike some other aspects of the development of nuclear energy for military purposes, it appears that no work was done with plutonium outside the United States until after World War II. However, every country with a reactor program now has an interest in the field, primarily for monitoring workers and the environment. Several quite extensive programs have developed over the last decade and a half.

The United Kingdom has developed a fluorishing bioassay program for plutonium, and interest in potential environmental contamination and, of course, a considerable research interest and capability in various fundamental aspects as evidenced by the authorship of two major chapters in this part of this Handbook by U. K. scientists.

Work in the Soviet Union on biomedical aspects of plutonium has been and is of considerable magnitude. Much of the Russian work is cited in Chaps. 10 and 11. A fine translation of the book "Problems of Plutonium Toxicology" by BULDAKOV, LYUBCHANSKI, MOSKALEV, and NIFATOV (1969) has been prepared by the Lovelace Foundation (translated by HORVATH, edited by THOMAS, 1970) and is recommended. Although the approach is somewhat different in the U.S.S.R. work, there is general agreement between the Soviet and U.S. work except for matters of detail. As seen in other aspects of radiobiology the Soviet work tends to stress physiological and biochemical, i.e. functional alterations, brought about by the agent rather more than does U.S. work. Also the Soviet work goes much further into detail concerning environmental contamination by plutonium especially from weapons tests, than does that from the U.S.A. at comparable times.

In continental Europe the "Euratom" group (European Atomic Energy Community) has issued regular reports on "the movement of certain isotopes in animals and in man" from the Centre d'Etudes Nucleaires de Fonteray-au-Roses, France (EUR 3918f, 1968; EUR 4527, 1971) and the installation at Saclay, France has an ongoing program with plutonium including an aerosol inhalation facility (LUTZ and ROUVROY, 1966). The Italian counterpart at Casaccia has a well developed laboratory for radiotoxicology (TESTA, 1969, 1970a) and special techniques for the determination of plutonium in biological materials (TESTA et al., 1970b). The primary thrust of many of these programs is not so much the Pharmacology and Toxiology of the element as its determination and control in and around installations where it is in use. However the development of well-equipped laboratories for inhalation studies presages interest and capabilities in these laboratories of importance to the subject of this volume.

Relatively recently Japan has developed a laboratory at the National Institute of Radiological Sciences in Chiba-City specifically for studies of the metabolism and effects of inhaled plutonium (SUZUKI et al., 1971; WATANABE et al., 1971).

The European Institute for Transuranium Elements was set up in 1960 at Karlsruhe, West Germany as part of a major nuclear research center already in operation. Centered around nuclear fuel fabrication and the physical and chemical properties of the various transuranium elements this Institute has 25 alpha laboratories, hot cells for various levels of activity, etc. VAANE (1969) working in this Institute presents a general summary of hazards connected with handling of transuranium elements and methods for personnel protection primarily, it appears, using this laboratory as a model. Again the thrust is primarily hazard control, measurements, and interpretation of the data from the working environment as they bear on good industrial medicine.

Recently a seminar on radiation protection problems relating to the transuranium elements was held under the auspices of the Commission of the European Communities and the European Nuclear Energy Agency. The proceedings (Commission of the European Communities, 1971) provide a good overview of work and views in the European laboratories and the relations between their work and that in other parts of the world. While titled "transuranics" the bulk of the discussions concern plutonium.

Thus capability for study and understanding and interest in the general radiotoxicology of plutonium are quite widespread. But the Centers for broad biomedical studies, particularly long-term effects, are not numerous abroad. This is not surprising in view of the cost in both time and patience of this type of research. Indeed we can feel fortunate that as many groups have entered this field as have been involved. The subsequent chapters will be drawing on nearly all of the major programs described.

IV. Newer Uses of Plutonium

Plutonium-239 as a fissile material of very long half-life will probably continue to be the most abundantly used isotope of this element on a mass basis. The increase in number and size of power reactors and particularly the almost certain addition of the breeder reactor to the nuclear power program brings plutonium from an element whose employment was limited to the military and its control primarily a problem for large governmental projects to an item of commerce. Thus a large increase is to be expected in amounts of this isotope available and in use. SEABORG (1969) predicts that power reactors alone will be producing approximately 10000 kilograms per year of plutonium and heavier isotopes by the year 1975 and that this may rise by a factor of three by 1980. Despite the increase in amounts, however, these developments do not presage any qualitatively new biomedical problems except that the Pharmacologist-Toxicologist is more likely to need some knowledge of the field.

Plutonium-238 with a half-life of about 90 years is now becoming an important isotope. With a significantly long but not too long half-life, alpha particle emission at 5.4 MeV with relatively little contaminating gamma or other "external" radiations the isotope is very useful in heat sources for specialized uses[6]. For example 1 gram of ^{238}Pu will liberate about 42000 watt-hours of energy in ten years while decaying by only 7 per cent in that time (SEABORG, 1969). Such sources are very useful in thermoelectric generators for use in the space program or in other situations of isolation. Many such devices (called SNAP devices for "Systems Nuclear Auxiliary Power") are powering satellites (weather and navigational) and also some components of the manned space program such as sources left on the moon to power long-term scientific experiments. The usefulness of such sources in vehicles for planetary exploration, or indeed for terrestrial power sources in isolated locations is obvious. Despite some problems with purity of isotope these sources are almost unique in requiring relatively little radiation

---

6 For details of external radiation hazards with plutonium in large quantities see Chap. 14.

shielding, e.g. when carried in or on a manned vehicle, because of the almost pure alpha particle emission produced.

Further Plutonium-238 is being considered seriously for new biomedical applications. It is a prime candidate for a radioisotope-powered artificial heart and as a miniaturized power supply for cardiac pacemakers thus eliminating the need for periodic replacement of batteries in such devices. Problems remain before such applications can be very general but projections for amounts of this isotope needed by the late twentieth century presume these problems will be overcome in another two decades at the most. The principal problem is the need for shielding against the small but inevitable amount of neutron and gamma radiation produced in the normal decay of ^{238}Pu and the presence as impurities of other radioisotopes with much larger contributions of external radiation. Thus development of sources of very high degrees of purity is important. The technology is essentially available to do this but the added cost of such purification is an obstacle to general use of such devices in medical practice or research outside of special governmental projects.

There are some potential uses of ^{238}Pu as a pure heat source in medical applications (i.e. not involving conversion to electrical power) but these are not very far along.

The development of these sources brings closer to the pharmacologist-toxicologist in a major medical center a need to know something of the radiobiological phenomena which are described in the subsequent chapters.

Since the specific activity of ^{238}Pu is much higher than that of the more commonly encountered ^{239}Pu the toxicity per unit of mass would be excepted to be higher. This is taken up in Chap. 10 and includes an indication that the 238 isotope may be metabolized somewhat differently from ^{239}Pu. This should be kept in mind in the unlikely but finite possibility of release of ^{238}Pu from a sealed source or environmental contamination with it.

At the present time none of the other isotopes of plutonium show sufficient potential for general use outside of very specialized installations to be considered here.

V. Summary[7]

The chapter is designed primarily to introduce the pharmacologist-toxicologist from another area to the primary features of plutonium toxicology, to guide the reader to the detailed discussions in other chapters, and to provide an historical and prospective overview. While many details are remanded to later chapters and other reviews, the chapter is intended to provide a condensed survey of the principal problems.

The development of biomedical knowledge regarding plutonium is very different from that of uranium, radium, and other naturally occurring radionuclides of long half-life. First produced in 1941 in less than microgram quantity, grams were available in a remarkably short time and this was followed within months by kilogram quantities. As for the future, about 20000 kilograms of the

7 Reference is made herein to many "in-house" documents important to historical perspective; most or all of these can be obtained through the Division of Technical Information Extension, U.S. Atomic Energy Commission by ordering photocopy from Microsurance, Inc., Oak Ridge Microreproduction Center, P.O. Box 3522, Oak Ridge, Tennessee 37830, or the National Technical Information Service, U.S. Department of Commerce, Springfield, Virginia, 22151. If difficulty is experienced the editors will attempt to aid interested readers in obtaining such documents for review or personal use.

isotope ^{239}Pu are expected to be needed by 1980, 60000 kilograms in the decade 1980–1990, and 80000 kilograms in the 1990–2000 decade. Amounts needed of the shorter half-life ^{238}Pu will be smaller by weight but very substantial if its shorter half-life and higher specific activity are taken into account. Further, instead of being largely limited to large governmental installations plutonium will almost certainly become in the future more an article of general commerce. Because of the increased quantities and this latter fact more general knowledge of its behavior and potential biological effects is needed among toxicologists and pharmacologists.

Plutonium has often been called "the most toxic element known to man". While not quite true it is indeed a most effective agent in producing long-term effects in the body. Its general toxicology is that of a "bone-seeker" superficially like radium, uranium, or strontium but with many significant differences. Because of its tendency to be a "surface seeker" rather than a "volume seeker" it is markedly more effective than radium on an activity basis in producing damage such as the induction of osteogenic sarcoma. Also the contrast in behavior between inhaled plutonium and that entering by other routes is so large that bone may not be a "critical organ" under conditions of inhalation exposure, and the component of deposition in and damage to soft tissue is much larger than with radium no matter what the route of administration.

The fact that the plutonium isotopes of most interest in biology and medicine are almost exclusively alpha emitters (in terms of internal contamination problems) puts them in a class with many other radionuclides in that effects occur at concentrations far below those associated with conventional "chemical" toxicity. Thus it is commonly assumed that the radiation dose is the prime if not the only cause of biological effect and that this is the more likely because of the high relative effectiveness (RBE) of alpha particles. Chemical properties affect primarily metabolism of the isotopes (absorption, distribution and excretion). Such properties include not only basic chemistry of the element but valence state, compound, complexes, etc. However, it cannot be proven beyond doubt that there is absolutely no chemical toxicity since there is no stable isotope for comparison and actually many inconsistencies persist.

The chapter considers primarily laboratory groups working in the United States both now and in the past. This occurs in part because plutonium was available only in the U.S.A. until the late 1940's and early 1950's. However, an abbreviated and undoubtedly incomplete survey of work and organizations in other countries is included in this chapter so that the technical reader not in the field of radiation toxicology can know where to look in his geographical area.

The chapter concludes with short review of prospective newer uses of plutonium isotopes in the space program and in medicine.

References

Abrams, R., Seibert, H. C., Forker, L., Greenberg, D., Lisco, H., Jacobson, L. O., Simmons, E. L.: Acute toxicity of intubated plutonium. USAEC Report, CH-3875, Metallurgical Lab. Univ. of Chicago (1946).

Abrams, R., Seibert, H. C., Potts, A. M., Forker, L. L., Greenberg, D., Postel, S., Lohr, W.: Metabolism and distribution of inhaled plutonium in rats. USAEC Report MDDC-677, Metallurgical Lab. Univ. of Chicago (1947).

Boyd, C. A., Silberstein, H. E., Fink, R. M., Frenkel, A., Minto, W. L., Metcalf, R. G., Casarett, G., Suter, C. M.: Pilot studies of the intravenous lethal dosage of polonium, plutonium and radium in rats. Chap. 7 in Biological studies with polonium, radium and plutonium, Ed. R. M. Fink, NNES, Div. VI, vol. 3, p. 211–294 (1950a) [also Report M-1878 (1946)].

BOYD, C. A., WILLIAMS, A., MINTO, W. L., TIEDEMAN, D. V., FINK, R. M., CASARETT, G., METCALF, R. G.: Simultaneous studies on the intravenous lethal dosage of polonium, plutonium and radium in rats. Chap. 8 in Biological studies with polonium, radium and plutonium, Ed. R. M. FINK, NNES, Div. VI, vol. 3, p. 295–404 (1950b) [also Report M-1902 (1946)].

BRUES, A. M., LISCO, H., FINKEL, M.: Carcinogenic action of some substances which may be a problem in future industries. USAEC Report No. MDDC-145, July (1946).

BULDAKOV, L. A., LYUBCHANSKII, E. R., MOSKALEV, Y. I., NIFATOV, A. P.: Problems of plutonium toxicology. Moscow: Atom Publications 1969. Translation by A. A. HORVATH, Edit. by R. G. THOMAS, LF-tr-41, UC-48 (1970).

Commission of the European Communities: Radiation protection problems relating to transuranium elements. Proceedings of a seminar sponsored by the Commission and the European Nuclear Energy Agency at Karlsruhe, West Germany, Sept. 1970. Published by Directorate General for Dissemination of Information — C.I.D. as report EUR-4612 d-j-3. Luxembourg, 1971.

COPP, D. H., AXELROD, D. J., HAMILTON, J. G.: The deposition of radioactive metals in bone as a potential health hazard. Amer. J. Roentgenol. 58, 10–16 (1947).

CROWLEY, J., LANZ, H., SCOTT, K., HAMILTON, J. G.: A comparison of the metabolism of plutonium in man and the rat. U.S.A.C.C. Report CH-3589 (1946).

DURBIN, P. W.: Plutonium in man: a Twenty-five year review. USAEC Report, UCRL-20850 (UC-41, TID-4500) June 1971, see also chap. VII.2 in STOVER and JEE, 1972, op. cit.

EURATOM: Movement de certains isotopes chez les animaux et chez l'homme. Annual Report of European Atomic Energy Community and Commissariat à l'Energie Atomique. Report EUR 3918f, Association No 049-64-3 B1AF, May (1968). Also Annual Report (1969), EUR 4527f (1971).

FINK, R. M. (Ed.): Biological studies with polonium, radium and plutonium. NNES Div. IV — vol. 3. New York: McGraw-Hill 1950.

FINKLE, R. D.: The toxicity and metabolism of plutonium in laboratory animals. USAEC, Report No CH-3783, August (1946).

FORMAN, H.: Translocation of drugs into bone. Handbook of experimental pharmacology, vol. XXVIII/I, Ed. by B. B. BRODIE and J. R. GILLETTE, chap. 13, p. 249–257 (1971).

HAMILTON, J. G.: Metabolism of product. Metallurgical Laboratory Report, No CN-2383, Nov. (1944) (see also MDDC-1018).

HAMILTON, J. G.: The metabolism of the fission products and the heaviest elements. Radiology 49, 325–343 (1947).

HAMILTON, J. G.: The metabolic properties of the fission products and the actinide elements. Rev. Mod. Physics 20, 718–728 (1948).

International Commission on Radiological Protection: Report of Committee II on permissible dose for internal radiation (1959). ICRP Publication 2, New York-London: Pergamon Press 1959. Also Hlth Phys. 3, 1–380 (1960).

LANGHAM, W. H.: Metabolism of plutonium in the rat. USAEC Report AECD-1914 (Los Alamos Report LADC-472), dated May (1946), declassified April (1948).

LANGHAM, W. H.: Physiology and toxicology of plutonium-239 and its industrial medical control. Hlth Phys. 2, 172 (1959).

LAPP, R. E., ANDREWS, H. L.: Nuclear radiation physics, fourth ed. Englewood Cliffs, New Jersey: Prentice-Hall 1972.

LISCO, H., FINKEL, M., BRUES, A.: Carcinogenic properties of radioactive fission products and plutonium. Radiology 49, 361–363 (1947).

LUTZ, M., ROUVROY, H.: Description d'un dispositif permet tant la contamination d'animaux de laboratoire par inhalation d'aérosols radioactifs. Report CEA-R 3086 Commissariat a l'Energie Atomique, Centre d'études nucléaires de Saclay, France (1966).

MAYS, C. M., JEE, W., LLOYD, R. D. (Eds.): Delayed effects of bone-seeking radionuclides. Proceedings of a symposium held at Sun Valley, Idaho. Salt Lake City: Univ. of Utah Press 1969.

MCCLELLAND, R.: Paper presented at Eleventh Hanford Biology Symposium, Richland, Washington, Sept. 1971 (to be published).

MIDDLESWORTH, L. VAN: Study of plutonium metabolism in bone. USAEC Report MDDC-1022, June (1947).

NEUMAN, W. F., NEUMAN, M. W.: The deposition of uranium in bone. II. Radioautographic studies. J. biol. Chem. 175, 711–714 (1948).

NEUMAN, W. F., NEUMAN, M. W., MAIN, E. R., MULRYAN, B. J.: The deposition of uranium in bone. IV. adsorption studies *in vitro*. J. biol. Chem. 179, 335–340 (1949a).

NEUMAN, W. F., NEUMAN, M. W., MAIN, E. R., MULRYAN, B. J.: VI. Ion competition studies. J. biol. Chem. 341 (1949b).

The Plutonium Project. Ten papers collected under this title in Radiology **49**, 269–365, Sept. (1947)[8].
SCHUBERT, J.: Treatment of plutonium poisoning by metal displacement. Science **105**, 389–390 (1947).
SCHUBERT, J.: Removal of radioelements from the mammalian body. Ann. Rev. Nuclear Sci. **5**, 369–412 (1955).
SCOTT, K. G., AXELROD, D. J., CROWLEY, J., HAMILTON, J. G.: Deposition and fate of plutonium, uranium and their fission products inhaled as aerosols by rats and man. Arch. Path. **48**, 31–54 (1949). Report No CH-3590, Dec. 1945, and MDDC-1276.
SCOTT, K. G., AXELROD, D. J., FISHER, H., CROWLEY, J., HAMILTON, J. G.: The metabolism of plutonium in rats following intramuscular injection. J. biol. Chem. **176**, 283–293 (1948).
SEABORG, G. T.: The synthetic actinides — from discovery to manufacture. Amer. Nuclear Soc., San Francisco, Calif., Dec. 2, 1969 (see also chapt 1.1 "Plutonium revisited" in STOVER and JEE, 1972, op. cit.).
STOVER, B. J., JEE, W.: Radiobiology of plutonium, p. 1–552. Salt Lake City: The J.W. Press, Univ. of Utah 1972.
SUZUKI, M., OKABAYASHI, O., WATANABE, S., HONGO, S., OHNO, S.: Studies on the inhalation of submicron plutonium nitrate aerosols in Wistar adult rats. Natl. Inst. Radiol. Sciences (Japan), Pub. 8, "Report of Project on Plutonium" (1971).
TESTA, C.: New radiotoxicological methods used at the Italian Nuclear Center of Casaccia. Comitato Naziconale Energia Nucleare (CNEN), Report CNEN-RT/PROT (69) 44 (1969).
TESTA, C.: Column reversed-phase partition chromatography for the isolation of some radionuclides from biological materials. Analyt. chim. Acta **50**, 447–455 (1970).
TESTA, C., MASI, G., MARCHIONNI, V.: Allestimento di un laboratorio per las determinazione di plutonio in campioni biologici: Messa a punta e confronto di vari methodi di analisi. Report CNEN-RT/PROT (7) 14, (1970).
THOMPSON, R.: Biological factors. Chap. 23 in Plutonium handbook. Edit. by O. J. WICK. New York: Gordon and Breach 1967.
TOTTER, J. R.: Biological research with plutonium, 1944–1984, chapt. I.2 in STOVER and JEE, 1972 (op. cit.).
VAANE, J. P.: Hazards connected with the handling of transuranium elements and methods used for the protection of personnel. Actinides Reviews **1**, 337–370 (1970).
WAGER, R. W., DOCKUM, N. L., TEMPLE, L. A., WILLARD, D. H.: Toxicity of radioactive particles IA. Intratracheal injection of radioactive suspensions. Hanfords Works Report HW-41500, pp. 61–72 (1956).
WATANABE, S., SUZUKI, M., HONGO, S., OHATA, T.: A study on the apparatus and method for inhalation experiment of plutonium aerosol. National Institute of Radiological Sciences (Japan). Pub. 8, Report of Project on Plutonium (1971).
YUILE, C. L., GIBB, F. R., MORROW, P. E.: Dose-related local and systemic effects of inhaled plutonium-238 and plutonium-239 dioxide in dogs. Radiation Res. **44**, 821–834 (1970).

8 Titles and authors are as follows: ZIRKLE: Introduction; Comments on the acute lethal action of slow neutrons; LORENZ et al: Biological studies in the tolerance range: JACOBSEN and MARKS: The hematological effects of ionizing radiation in the tolerance range; PROSSER et al.: The clinical sequence of physiological effects of ionizing radiation; RAPER: Effects of total surface beta irradiation; HAMILTON: The metabolism of the fission products and the heaviest elements; BLOOM: Histological changes following radiation exposures; HENSHAW et al.: The biologic effects of pile radiations; LISCO et al.: Carcinogenic properties of radioactive fission products and of plutonium; STONE: Editorial: The plutonium project.

Chapter 9

Chemical and Physical Properties of Plutonium

DAVID M. TAYLOR

With 3 Figures

I. Introduction

During the thirty years which have elapsed since the discovery of plutonium in 1940 (SEABORG et al., 1946) all aspects of its chemistry have been intensively studied and plutonium is now amongst the best understood elements in the periodic table.

In this chapter it is not proposed to present an extensive review of all the phases of plutonium chemistry but to concentrate on those aspects which are germane, or likely to be so, to our understanding of the biological behaviour and toxicology of the element. Thus the metallurgical chemistry of plutonium will receive only brief attention, whereas solution chemistry and complex formation will be discussed in some detail.

The general aspects of the chemistry of plutonium have been described by CLEVELAND (1970a) in his book "The Chemistry of Plutonium", and the chemistry and technology of plutonium have been reviewed in the two volumes of the "Plutonium Handbook" (WICK, 1967). More specialised aspects, such as complex ion formation, crystal chemistry and analytical chemistry have been discussed in the monographs by GELMAN et al. (1962), MAKAROV (1959), MOSES (1963), MILYUKOVA et al. (1967) and PASCAL (1970).

A. The Position of Plutonium in the Periodic Table

When the modern form of the Periodic Table, based on the concept of atomic number, was originally drawn up no elements of atomic number greater than 92 were seriously envisaged. The four heaviest elements actinium, thorium, protactinium and uranium were placed in Groups III to VI of the Sixth Period. This assignment was supported by their principal valency states and chemical properties, although there was general agreement that at some point the 5f electron shell would begin to fill up and, consequently, speculation about analogies to the rare earth or lanthanide series in which there is progressive filling of the 4f electron shell.

With the discovery of neptunium, plutonium and other transuranic elements the position changed, since even the earliest chemical studies, carried out on a few atoms of material, suggested that 5f electrons might be present. More detailed studies of chemical properties and of ultraviolet, visible and infrared spectra, magnetic susceptibility and paramagnetic resonance have confirmed the involvement of 5f electrons in the transuranic elements and we now firmly assign these elements to an *actinide series*, which starts with Element 89, actinium, and ends with Element 103, lawrencium. This series, which is illustrated in Table 9.1, is

Table 9.1 The actinide and lanthanide series

Lanthanide series (fifth period)															
Atomic number	57	58	59	60	61	62	63	64	65	66	67	68	69	70	71
Element	La	Ce	Pr	Nd	Pm	Sm	Eu	Gd	Tb	Dy	Ho	Er	Tm	Yb	Lu
Oxidation states						2	2								
	3	3	3	3	3	3	3	3	3	3	3	3	3	3	3
		4	4[a]						4[a]						

Actinide series (sixth period)															
Atomic number	89	90	91	92	93	94	95	96	97	98	99	100	101	102	103
Element	Ac	Th	Pa	U	Np	Pu	Am	Cm	Bk	Cf	Es	Fm	Md	No	Lr
Oxidation States						(2)	2[a]						2	*2*	
	3	3[a]	3[a]	3	3	3	*3*	*3*	*3*	*3*	*3*	*3*	*3*	3	*3*
		4	4	4	4	*4*	4[a]	4[a]	4						
			5	5	*5*	5	5								
				6	6	6	6								

[a] Indicates oxidation state unstable in solution. The figures in italics represent the most stable oxidation states. The figure in parenthesis indicates an unconfirmed oxidation state.

homologous with the lanthanide series although the analogies are not well marked with the earlier members of the series, and the actinides tend to exhibit multiple and higher valency states more frequently than the lanthanide elements. This is well illustrated by plutonium which exhibits all valency states from 2 to 7 in contrast to its homologue, samarium, which occurs only in the 2+ or 3+ states.

B. Electronic Structure

Information about the electronic configuration of elements and their compounds can be obtained from measurement of physical parameters such as magnetic susceptibility, paramagnetic resonance and ultraviolet, visible and infrared spectra. In the case of plutonium, such measurements have suggested that in the ground state, neutral gaseous atoms of plutonium have the following electronic configuration:

$$1s^2\,2s^2\,2p^6\,3s^2\,3p^6\,3d^{10}\,4s^2\,4p^6\,4d^{10}\,4f^{14}\,5s^2\,5p^6\,5d^{10}\,5f^6\,6s^2\,6p^6\,7s^2.$$

Similar data have also established that the six 5f electrons play important roles in bond formation in many plutonium compounds.

C. Isotopes of Plutonium

Since the discovery of ^{238}Pu by Seaborg et al. in 1940, 15 further isotopes of plutonium have been synthesised. These are listed in Table 9.2 which shows that all the isotopes are radioactive, most of them decaying by alpha particle emission, with half lives ranging from 0.18 seconds to 76 million years. Of these 16 isotopes only two, ^{239}Pu and ^{238}Pu, have found widespread application for peaceful or military purposes. Plutonium-239, the most important plutonium isotope, has a high cross-section for nuclear fission by thermal neutrons and it is this property which makes ^{239}Pu useful as both a nuclear fuel and a nuclear explosive. In addition to fission by neutrons ^{239}Pu may also capture a neutron to form ^{240}Pu,

Table 9.2. Plutonium isotopes

Isotope	Half-life	Mass per curie of α-radiation	Principal radiations (Energy in MeV)	Principal production reactions
^{232}Pu	36 min	2.22 μg	α 6.59 (2%); E.C.[a] 98%	$^{233}U(\alpha\ 5n)\ ^{232}Pu$; $^{235}U(\alpha\ 7n)\ ^{232}Pu$
^{233}Pu	20 min	2.48 μg	α 6.31 (0.1%); E.C.[a] 99%	$^{233}U(\alpha 4n)\ ^{232}Pu$
^{234}Pu	9.0 hrs	11.19 μg	α 6.20 (4%); 6.15 (1.9%); E.C.[a] 94%; γ (Np xrays)	$^{233}U(\alpha\ 3n)\ ^{234}Pu$; $^{235}U(\alpha\ 5n)\ ^{234}Pu$
^{235}Pu	26 min	1.1 mg	α 5.86 (0.003%); E.C.[a] 99 +%; γ (Np xrays)	$^{235}U(\alpha\ 4n)\ ^{235}Pu$; $^{233}U(\alpha\ 2n)\ ^{235}Pu$
^{236}Pu	2.85 yr (SF[b] 3.5×10^{9} yr)	1.88 mg	α 5.77 (69%); 5.72 (31%). γ UL xrays, 0.048 (0.31%), 0.109 (0.012%) e^- 0.028, 0.043	Daughter ^{236}Np $^{235}U(\alpha\ 3n)\ ^{235}Pu$
^{237}Pu	45.6 d	24.8 mg	α 5.66, 5.37 (0.0033%), γ Np xrays 0.060 (5%) e^- 0.026, 0.032, 0.038, 0.042, 0.056	$^{235}U(\alpha\ 2n)\ ^{237}Pu$ $^{237}Np(\alpha\ 2n)\ ^{237}Pu$
^{237m}Pu	0.18 s	—	γ Pu L xrays 0.145 (2%); e^- 0.125 (75%) 0.140 (23%)	Daughter ^{241}Cm
^{238}Pu	86.4 yr (SF[b] 4.9×10^{10} yr)	57.45 mg	α 5.50 (72%), 5.46 (28%); UL xrays 0.099 (8×10^{-3}%), 0.150 (10^{-3}%), 0.77 (5×10^{-5}%); e^- 0.024, 0.039	Daughter ^{238}Np Daughter ^{242}Cm
^{239}Pu	24,390 yr (SF[b] 5.5×10^{15})	16.28 g	α 5.16 (88%), 5.11 (11%); γ 0.039 (0.007%), 0.052 (0.020%), 0.125 (0.005%), 0.374 (0.0012%), 0.414 (0.0012%), 0.65 (8×10^{-5}%) 0.77 (2×10^{-5}%); e^- 0.008, 0.019, 0.033, 0.047	$^{238}U(n\gamma)\ ^{239}U(\beta^-)\ ^{239}Np$ $(\beta^-)\ ^{239}Pu$ $^{239}Pu(n\gamma)\ ^{240}Pu$
^{240}Pu	6580 yr (SF[b] 1.3×10^{11} yr)	4.41 g	α 5.17 (76%), 5.12 (24%); γ UL xrays 0.65 (2×10^{-5}%), e^-0.026, 0.040	
^{241}Pu	13.2 yr	—	β^- 0.021 max (99+%); α 4.90, 4.85 (2.3×10^{-3}%) γ UL xrays 0.145 (1.6×10^{-4}%)	$^{239}Pu(n\gamma)\ ^{240}Pu(n\gamma)\ ^{241}Pu$
^{242}Pu	3.79×10^{5} yr (SF[b] 7×10^{10} yr)	256.2 g	α 4.90 (76%), 4.86 (24%); γ UL xrays	$^{241}Pu(n\gamma)\ ^{242}Pu$
^{243}Pu	4.98 hr	—	β^- 0.58 max; e^- 0.019, 0.036; γ Am L xrays 0.084 (21%) 0.381 (0.7%)	$^{242}Pu(n\gamma)\ ^{243}Pu$
^{244}Pu	7.6×10^{7} yr (SF[b] 2.5×10^{10} yr)	51.81 kg	α 4.58	$^{243}Pu(n\gamma)\ ^{244}Pu$
^{245}Pu	10.1 hr	—	β^-	$^{244}Pu(n\gamma)\ ^{245}Pu$
^{246}Pu	10.85 hr	—	β^- 0.33 max (10%), 0.15 max; e^- 0.020, 0.038, 0.055, 0.156 γ Am L xrays 0.044 (30%), 0.180 (10%), 0.224 (25%)	$^{245}Pu(n\gamma)\ ^{246}Pu$

[a] E.C. = Electron capture. [b] S.F. = Spontaneous fission.
Compiled from LEDERER, HOLLANDER and PERLMAN (1967).

and higher plutonium isotopes may be formed by successive neutron capture.

$$^{239}\mathrm{Pu}\,(\mathrm{n}\gamma)\;^{240}\mathrm{Pu}\,(\mathrm{n}\gamma)\;^{241}\mathrm{Pu}\,(\mathrm{n}\gamma)\;^{242}\mathrm{Pu}\,(\mathrm{n}\gamma)\ldots\;^{245}\mathrm{Pu}.$$

Most of the studies of plutonium chemistry have been made with ^{239}Pu which is now produced in tonnage quantities. Serious difficulties in chemical studies with ^{239}Pu are posed by the intense alpha radiation, 1.36×10^{11} disintegrations per minute per gram. These range from chemical problems resulting from radiation decomposition of plutonium compounds, either directly by alpha particles or from the products of the radiolysis of water, to the exacting standards of laboratory design which are necessary to protect workers from contamination, since ^{239}Pu is one of the most potent carcinogenic agents known to man.

II. Plutonium and its Compounds

Plutonium is now produced on a large scale in nuclear reactors in various parts of the world. The nuclear reactions leading to the production of ^{239}Pu from ^{238}U are summarised in the following scheme:

$$^{238}\mathrm{U}\;(\mathrm{n}\gamma)\;^{239}\mathrm{U}\xrightarrow[23.5\,\mathrm{m}]{\beta}\;^{239}\mathrm{Np}\xrightarrow[2.33\,\mathrm{d}]{\beta}\;^{239}\mathrm{Pu}.$$

The plutonium is separated from the large mass of unchanged uranium and the accompanying, intensely radioactive fission products, by complex chemical processess generally involving oxidation-reduction reactions and solvent extraction. The technology of plutonium production lies outside the scope of this review; but the chemical aspects of plutonium production have been reviewed recently by Wick (1967) and Cleveland (1970a).

A. Plutonium Metal

Plutonium is a silvery white metal which melts at 639.5° C. It is a reactive metal and on warming oxidises readily in moist air. In finely divided form plutonium metal may be pyrophoric.

At elevated temperatures plutonium metal reacts with oxygen to form oxides; with the halogens to form halides; with nitrogen or ammonia to form nitrides; with hydrogen to form hydrides and with carbon monoxide to form carbides.

Plutonium metal is readily soluble in moderate to high concentrations of hydrochloric acid and hydrobromic acid but only slowly attacked by hydrofluoric acid. Similarly the metal is readily dissolved in concentrated perchloric or phosphoric acids. In contrast plutonium metal is insoluble in nitric acid at all concentrations, and it is only slowly attacked by sulphuric acid.

B. Oxidation States

It is now known that plutonium can exhibit all five oxidation states from three (III) to seven (VII), and the divalent state may exist in certain solid compounds. The four common oxidation states correspond to the ions Pu^{3+}, Pu^{4+}, PuO_2^+ and PuO_2^{2+}. Recently the heptavalent state has been prepared by oxidation of Pu(VI) by ozone in alkaline solution (Krot and Gel'man, 1967). This species is unstable in alkaline solution and it does not appear to exist in acid solutions. The composition of the Pu(VII) species has not yet been determined but is assumed to be the anionic, oxygenated species PuO_5^{3-}. This exhibition of multiple oxidation states introduces many complexities into the chemistry of plutonium.

The four lower oxidation states are stable in solution and, as will be discussed later, it is possible for all four oxidation states from Pu(III) to Pu(VI) to co-exist in the same solution.

C. Plutonium Compounds

A large number of plutonium compounds has now been prepared in the solid state. These have been reviewed extensively by CLEVELAND (1970a) and the present discussion will be limited to a few important compounds.

1. Oxides

Plutonium dioxide, PuO_2, is one of the most important of all plutonium compounds. It is formed when plutonium metal is burnt in oxygen or when oxygen containing compounds such as Pu(IV) oxalate or Pu(IV) peroxide are heated *in vacuo* at 1000° C. Plutonium dioxide is a highly refractory material, normally green in colour, which melts at temperatures between 2200 and 2400° C depending on the composition of the atmosphere in which it is melted. It is difficult to dissolve by normal methods but can be dissolved by heating in 85–100% phosphoric acid at 200° C. The refractory nature of PuO_2 tends to become more marked when high temperatures are used in its preparation. It is relatively resistant to self-irradiation by α-particles. Some other oxides of plutonium are known including the sesquioxide, Pu_2O_3; and, possibly a monoxide, PuO, representing the divalent state of plutonium.

Plutonium peroxide is formed by precipitation with hydrogen peroxide from aqueous solutions, and is of importance in the preparation of Pu metal.

2. Halogen Compounds

Plutonium reacts with halogens and halogen oxy-acids to form halides. The fluorides are the most important of the halides since they are used in plutonium processing.

The trifluoride, PuF_3, may be prepared by heating PuO_2 in hydrogen fluoride containing traces of hydrogen at 550° C. It is a purple crystalline solid containing less than one molecule of water of crystallisation.

Plutonium tetrafluoride, PuF_4, results from heating PuO_2 in hydrogen fluoride and oxygen at 550° C.

Both PuF_3 and PuF_4 are insoluble in water and acids, but they may be dissolved fairly easily in aqueous solutions of Zr(IV), Fe(III) or Al(III) or other ions which form stable complexes with fluoride ions.

The solubility products for PuF_3 and PuF_4 in nitric acid have been calculated to be 2.5×10^{-16} and 6×10^{-20} respectively (MANDELBERG et al., 1961).

Several double salts of PuF_3 and PuF_4, for example $KPuF_4$ and $LiPuF_5$, have been reported, and an oxyfluoride of Pu(III), PuOF, has been prepared. The double salt Cs_2PuCl_6 which is non-hygroscopic and stable towards alpharadiation, has been suggested as a primary analytical standard for plutonium. Several double salts of Pu(V) such as $CsPuF_6$ have been reported but the simple Pu(V) fluoride has not yet been prepared.

Plutonium hexafluoride, PuF_6, which results from the reaction of PuF_4 with fluorine at high temperature, is a volatile compound with a boiling point of 62.2° C. Plutonium hexafluoride is a thermodynamically unstable compound and is a powerful fluorinating agent. It reacts violently with water to yield PuF_4 and PuO_2, but under controlled conditions PuF_6 reacts with moist air to form, PuO_2F_2, plutonyl fluoride.

Radiation decomposition of PuF_6, to fluorine and lower plutonium fluorides, is caused by self-irradiation by alpha particles.

By comparison with the fluorides, other halogen compounds of plutonium have not been extensively studied. Plutonium(III) has been shown to form compounds with chlorine, bromine and iodine, the latter forming both iodide and iodate. However with Pu(IV) the only known compounds with the higher halogens are chloro complexes or double chloride salts, and there are no known halogen compounds of Pu(V) other than the fluoride double salts. Plutonyl chloride, $PuO_2Cl_2 \cdot 6H_2O$, is the only confirmed higher halide of Pu(VI) and this compound is unstable, being reduced to Pu(IV) by alpha-radiation.

3. Nitrates

Despite the fact that much of the chemical processing of plutonium is carried out in solutions in nitric acid, little is known concerning the plutonium nitrates. It seems probable that Pu(III) nitrate does not exist in view of the ease of oxidation of Pu(III) to Pu(IV) by nitric acid.

Crystalline Pu(IV) nitrate, $Pu(NO_3)_4 \cdot 5H_2O$, has been prepared. It is a greenish-black compound which is fairly stable in moist or dry air, but decomposes at temperatures above 40° C. The salt is readily soluble in water, but the resultant solution becomes extensively hydrolysed, see Sec. III B, unless the solution is made strongly acid.

A Pu(VI) nitrate, plutonyl nitrate $PuO_2(NO_3)_2 \cdot 6H_2O$, has also been prepared. This is a pink crystalline compound which turns reddish brown on exposure to air. It is stable to heating up to 176° C at which temperature the anhydrous salt is formed. On storage in air, plutonyl nitrate undergoes complete reduction to Pu(IV) and Pu(III).

4. Sulphates

Sulphates of Pu(III) and Pu(IV) have been prepared, but no Pu(VI) sulphate is known. These sulphates are of little technical importance but the double salt $K_4Pu(SO_4)_4 \cdot 2H_2O$ has been suggested as a possible primary standard for plutonium.

5. Phosphates

Phosphates are of importance in plutonium processing but relatively little is known about the plutonium phosphates. Phosphates of Pu(III), Pu(IV), Pu(VI) and probably Pu(V) have been reported.

6. Oxalates

Oxalates, or oxalate complexes, of plutonium are important in the preparation of plutonium metal. Stable oxalates of Pu(III), Pu(IV) and Pu(VI) have been prepared and characterised; a Pu(V) oxalate has also been reported. A number of oxalate complexes of Pu(III), Pu(IV) and Pu(V) are also known.

7. Hydride, Carbides, Nitride, Silicides and Sulphides

Plutonium hydride is of use in the preparation of plutonium metal powders, and the carbide, nitride, silicide and sulphide are of value, or potential value, as nuclear fuels.

Plutonium hydride which results from the direct reaction between plutonium metal and hydrogen at elevated temperatures has a slightly variable composition between PuH_2 and PuH_3. It is a hard black substance with solubility properties similar to Pu metal.

The reaction of PuO_2 with graphite at high temperatures, up to 2600° C, results in the production of three plutonium carbides; plutonium monocarbide, which has an average composition of $PuC_{0.85}$, a stoichiometric compound Pu_2C_3, the sesquicarbide, and a dicarbide PuC_2.

Plutonium monocarbide is relatively stable in air and cold water, but in hot water it decomposes to form $Pu(OH)_3$ plus hydrogen, methane and smaller amounts of other hydrocarbon gases. It is attacked by concentrated HNO_3, containing NaF, to yield a deposit of carbon.

The sesquioxide, Pu_2C_3, has similar properties to the monocarbide except that it is less stable towards atmospheric attack but more stable towards hydrolysis by hot water or acids. The hydrolysis products include higher, solid and liquid, hydrocarbons than those of the monocarbide.

Plutonium nitride, approximate formula PuN, is formed by reacting $PuH_{2.7}$ or PuO_2 with nitrogen at elevated temperatures. It is a hard black substance which decomposes fairly rapidly in moist air to produce PuO_2.

The reaction of plutonium metal, plutonium oxide or fluoride with silicon at high temperatures produces several silicides of plutonium. These are of uncertain composition but correspond approximately to $PuSi_2$ and Pu_3Si_5. These are hard, brittle, metal-like compounds, which are pyrophoric and decomposed rapidly by hot water.

A number of plutonium sulphides have been prepared by reaction of plutonium compounds with sulphur or sulphides, but only one, PuS, has been characterised. Like the nitrides, carbides and silicides, PuS has a high melting temperature, 2350° C, but unlike these compounds it is not decomposed by water or steam at temperatures up to 250° C.

8. Other Compounds

Most of the compounds discussed above have important technological uses, either as reactor fuels or in the chemical processing of plutonium, and these substances, together with a few complex compounds to be discussed in Sec. III C, represent the range of compounds which are most likely to be encountered in cases of accidental contamination of the human body by plutonium.

However, a number of other compounds of plutonium have been described, including arsenides and arsenates, antimonides, tellurides, alkoxides and various compounds formed with organic acids, or other organic compounds; these have been reviewed recently by CLEVELAND (1970a), but as they have few technological uses at the present time their chemistry has not been summarised here.

III. Solution Chemistry of Plutonium

A. Absorption Spectra of Plutonium

The various oxidation states of plutonium exhibit characteristic absorption spectra in the ultraviolet, visible and infrared regions. The spectra for the different oxidation states show sufficient variation from each other to make it possible to use spectrophotometry for the study of the oxidation states of plutonium in solution and for the investigation of hydrolysis and complex formation.

The absorption spectra for the various oxidation states of plutonium have been studied by CONNICK et al. (1949), COHEN (1961a and b) and KROT and GEL'MAN (1967) and some major absorption peaks are listed in Table 9.3. Complex formation or hydrolytic reactions may modify the spectra for any given oxidation state. For example spectrophotometric studies have shown that the sharp ab-

Table 9.3. Major absorption peaks of plutonium ions

Wavelength	Molar extinction coefficient				
(nm)	Pu (III)[a]	PU (IV)[a]	Pu (V)[b]	Pu (VI)[a]	Pu (VII)[c]
470	3	55	2	11	460
600	38	2	2	1	500
635	8	10	2	1	550
830	5	16	5	550	70
1130	16	13	22	0	0

[a] 1.0 M $HClO_4$ (25°C). [b] 0.2 M $HClO_4$ (10°C). [c] 0.5 M Na OH (25°C).

sorption peak of Pu(IV) which occurs at 470 nm is shifted to 496 nm when the Pu(IV) is complexed with 2 molecules of citrate (NEBEL, 1966a). Hydrolysis of Pu(IV) to form a polymeric Pu(IV) species is accompanied by a marked change in the absorption spectrum in the visible and near infrared regions (OCKENDEN and WELCH, 1956).

B. Hydrolytic Reactions

In aqueous solutions all positively charged ions tend to be associated with water to some extent. Thus even the simplest ion, the hydrogen ion, is present not as the free ion H^+ but in hydrated form as the hydronium ion H_3O^+. The tendency towards association with water, hydrolysis, depends on the electronic structure, and thus the general chemical properties, of the ion, but it is greatest for ions of high charge and relatively small ionic radius. Thus the ionic potential, defined as the ionic charge divided by the ionic radius (CARTLEDGE, 1928), affords a measure of the tendency of an ion to undergo hydrolysis.

The hydrolysis reaction sequence may be summarised by the following general equations:

$$M^{a+} + H_2O \rightleftharpoons M(OH)^{(a-1)+} + H^+,$$

$$M(OH)^{(a-1)+} + (n-1)H_2O \rightleftharpoons M(OH)^{(a-n)+} + (n-1)H^+,$$

$$M(OH)_n^{(a-n)+} + (a-n)H_2O \rightleftharpoons M(OH)_a + (a-n)H^+.$$

In addition to association with water, ions may also co-ordinate with other ligands to form complex ions. Complex formation is discussed in Sec. III C, but it must be remembered that hydrolysis and complex formation are closely related phenomena.

In view of their high ionic charge and relatively small radius, Table 9.4, plutonium ions would be expected to undergo extensive hydrolysis in aqueous solution, and in fact hydrolysis has been found to play an important role in the solution chemistry of plutonium. Hydrolytic reactions generally occur more readily at high pH, thus, together with complex formation, they are of very great importance in relation to the chemistry of plutonium in biological systems.

Table 9.4. Radius of plutonium ions[a]

Oxidation state	Pu (III)	Pu (IV)	Pu (V)	Pu (VI)	Pu (VII)
Ionic radius (nm)	0.100	0.090	0.087	0.081	?

[a] From CLEVELAND (1970).

Experimental studies of the hydrolysis of plutonium ions, reviewed by KRAUS (1956), have shown that the tendency towards hydrolysis decreases in the order:

$$Pu(IV) > Pu(VI) > Pu(III) > Pu(V).$$

1. Hydrolysis of Pu (IV)

The Pu^{4+} ion undergoes the most extensive hydrolysis of all the plutonium ions, and, since this is the most important oxidation state, the process has been extensively studied.

The first hydrolysis reaction of Pu^{4+} is

$$Pu^{4+} + H_2O \rightleftharpoons PuOH^{3+} + H^+$$

and $PuOH^{3+}$ is the only important species present in the early stages of the reaction.

As the hydrolysis proceeds the monomeric species aggregate irreversibly to form polymers of plutonium atoms linked together by hydroxide or oxide bridges. Finally as hydrolysis proceeds to completion precipitation of $Pu(OH)_4$ occurs. The reaction sequence may be visualised as

$$PuOH^{3+} + H_2O \rightleftharpoons Pu(OH)_2^{2+} + H^+,$$
$$Pu(OH)_2^{2+} + (n-1)Pu(OH)_2^{2+} \rightleftharpoons [Pu(OH)_2^{2+}]_n,$$
$$[Pu(OH)_2^{2+}]_n + 2nH_2O \rightleftharpoons nPu(OH)_4 + 2nH^+.$$

Pu(IV) hydroxide is a highly insoluble compound for which the solubility product has been calculated to be 7×10^{-56} (KATZ and SEABORG, 1957).

The Pu(IV) polymer has a characteristic absorption spectrum in the visible region, permitting it to be distinguished from monomeric Pu(IV), and if the concentration is high enough, it shows a bright green colour in solution. Polymer formation occurs fairly rapidly; it was found by OCKENDEN and WELCH (1956) that when solutions of Pu(IV) were diluted to contain 0.4 to 1.2×10^{-2} M Pu^{4+} in 0.1 M HNO_3, 40% of the plutonium polymerised in 30 minutes and 55% in 1 hour.

The hydrolysis of Pu^{4+} is influenced by pH, plutonium concentration, temperature and the presence of other ions, especially complexing anions.

In 10^{-2} M solutions of Pu^{4+} OCKENDEN and WELCH (1956) found that the polymeric species formed readily at a pH of about 1. At 2×10^{-5} M Pu^{4+} the polymeric species began to form at pH 1.5 (GREBENSHCHIKOVA and DAVYDOV, 1965). When the Pu^{4+} concentration was reduced to 6.8×10^{-8} M it was found that the Pu(IV) was present in anionic form up to pH 2.8; as a "pseudocolloid" between pH 2.8 and pH 7.5 and as a "true" colloid only between pH 7.5 and pH 12. (GREBENSCHIKOVA and DAVYDOV, 1962). Thus it appears that at very low concentrations colloid formation requires high values of pH. It is the behaviour at very low concentrations of Pu(IV) which is of most interest from the biological point of view, since 1 μg of plutonium, the presently accepted maximum permissible body burden (ICRP 1959), distributed throughout the extracellular fluids of the adult human body corresponds to a concentration of about 3×10^{-10} M.

The presence of complexing anions in sufficiently high concentration can completely prevent the formation of the Pu(IV) polymer. The effects of various anions on the pH at which 90% of the Pu(IV) could be separated from 10^{-5} M solutions by centrifuging have been studied by LUTZ et al. (1966) who found that 5×10^{-2} M acetate increased the pH of 90% removal from pH 2.8 to pH 5.7 whereas 5×10^{-4} M citrate, or 5×10^{-4} M diethylene-triaminepentaacetic acid (DTPA), completely prevented the formation of polymers of sufficient size to be separated by centrifugation. The polymerisation of Pu(IV) in dilute citrate solutions has also been studied by LINDENBAUM and WESTFALL (1965). These workers used ultrafiltration through specially prepared cellophane membranes in order to remove polymeric Pu(IV) of molecular weight greater than about 200000. Using a 1.96×10^{-5} M solution of Pu(IV) they found that at 3.4×10^{-2} M citrate about 92% of the Pu(IV) was ultrafilterable between pH 4 and pH 10. At equimolar citrate and Pu(IV) concentrations the ultrafilterability was only about 77% between pH 4 and pH 8.5 and fell sharply to 5% at pH 11. In the absence of citrate the ultrafilterability was 13% at pH 4, 9% at pH 8.5 and ~1% at pH 10–11. These authors also studied the resolubilisation of Pu(IV) colloid. A solution containing equimolar concentrations of Pu(IV) and citrate was adjusted to pH 11 and then within 1 hour the pH was reduced to pH 7.8 or pH 4.0 with HCl. At pH 7.8 the ultrafilterability increased from 5 to 29% in 1 hour and at pH 4.0 the increase was from 5 to 44%. Between 1 hour and 72 hours the increase in ultrafilterability was only about 1% per day at pH 7.8 and 4% per day at pH 4.0.

The depolymerisation of the Pu(IV) colloid has been found to be relatively slow, and also dependent on the age of the polymer. In 5 M HNO_3 at 25° C a freshly prepared Pu(IV) polymer was found to depolymerise with a half time of 20 hours whereas a polymer prepared under identical conditions and allowed to age had a depolymerisation half time of 320 hours at 25° C (COSTANZO and BIGGERS, 1963). The presence of complexing ions such as fluoride or sulphate, or oxidising agents such as permanganate, also increase the rate of depolymerisation of Pu(IV) polymer.

Hydrolysis and polymer formation can cause serious problems in chemical operations with plutonium. The polymer does not extract into many of the solvents required in the solvent extraction processes used in plutonium separation, it also readily adsorbs onto glass and metal surfaces with which it comes into contact. Thus considerable care must be exercised in preventing inadvertant hydrolysis of plutonium solutions.

The hydrolytic behaviour of plutonium must also be taken into consideration in biological studies with plutonium, since soluble and polymeric species show wide differences in their biological behaviour (LINDENBAUM and WESTFALL, 1965). Solutions of Pu(IV) in citrate have been widely used for the administration of plutonium in soluble form to animals, but the work of LINDENBAUM and WESTFALL (1965) indicates that even when great care is taken in the preparation of such solutions, they may still contain up to 10% of polymeric material. For animal studies TURNER and TAYLOR (1968a) have recommended the ultrafiltration of Pu(IV) citrate solutions through a 25 nm pore diameter filter immediately before use, in order to remove any large polymeric material.

2. Hydrolysis of Pu (VI)

The hydrolysis of the PuO_2^{2+} ion has been studied much less extensively than that of Pu^{4+}. Although PuO_2^{2+} could be regarded as a divalent ion the hydrolysis constants which have been measured are considerably higher than would be expected for a divalent ion and are closer to those of a tetravalent ion. Thus PuO_2^{2+} must have a strong positive charge on the metal and strong bonds between the metal and the two oxygen atoms.

3. Hydrolysis of Pu (III)

The first reaction in the hydrolysis of Pu(III) is

$$Pu^{3+} + H_2O \rightleftharpoons PuOH^{2+} + H^+$$

and continued hydrolysis results in the precipitation of $Pu(OH)_3$. Although still highly insoluble this hydroxide has a solubility product of 2×10^{-20} which is very much greater than that of $Pu(OH)_4$.

4. Hydrolysis of Pu (V)

The hydrolysis of Pu(V) in nitrate solutions is not apparent up to pH 5. At a concentration of about 10^{-2} M precipitation of $PuO_2(OH)$ occurs at about pH 6.8 (ZAITSEVA et al., 1968).

C. Plutonium Complexes

Complex formation and hydrolysis are closely related phenomena and the tendency of an ion to associate with co-ordinating groups, or ligands, to form complexes also depends on the ionic potential. Thus the small highly charged Pu^{4+} ion would be expected to form complexes readily, and, in fact, complex formation plays an extremely important part in the solution chemistry of Pu(IV). The other oxidation states of plutonium also form complexes, but less readily than Pu(IV). Complex forming ability decreases in the order:

$$\mathrm{Pu(IV)} > \mathrm{Pu(III)} > \mathrm{Pu(VI)} > \mathrm{Pu(V)}.$$

The chemistry of the plutonium complexes has recently been reviewed by CLEVELAND (1970a, 1970b). In this discussion the emphasis will be placed only on those complexes of plutonium which are of technological importance, and hence may possibly be involved in accidental contamination of the human body, or which are of biological interest.

In consideration of the chemistry of metal complexes it is necessary to know the stability of the complex. This is given by the stability constant, which is in fact the equilibrium constant for the overall reaction between the metal and the ligand. Stability constants can be measured by various techniques including ion exchange, potentiometric titration, and spectrophotometry; but their determination is frequently difficult, and in the case of readily hydrolysable metals such as plutonium further complications may arise because of the limited range of pH over which the metal can be maintained in ionic form.

The commonly accepted nomenclature for stability constants has been defined by SILLEN and MARTELL (1964). For the consecutive combination of ligands with metal ions

$$\mathrm{M} + \mathrm{L} \rightleftharpoons \mathrm{ML}$$

$$\mathrm{ML} + \mathrm{L} \rightleftharpoons \mathrm{ML_2}$$

which may be summarised by the general equation

$$\mathrm{ML_{n-1}} + \mathrm{L} \rightleftharpoons \mathrm{ML_n}.$$

The stepwise constant for each reaction stage is given by

$$\mathrm{K_n} = \frac{[\mathrm{ML_n}]}{[\mathrm{ML_{n-1}}]\,[\mathrm{L}]}.$$

The overall stability constant for the reaction of a metal ion M with n molecules of ligand is given by the product of the individual stepwise constants i.e.

$$\mathrm{K_1} \times \mathrm{K_2} \times \mathrm{K_3} \ldots \times \mathrm{K_n}.$$

The cumulative constant β_n for the overall reaction

$$M + nL \rightleftharpoons ML_n$$

is

$$\beta_n = \frac{[ML_n]}{[M]\cdot[L]^n}.$$

1. Complexes with Inorganic Anions

Plutonium forms complexes with many of the common inorganic anions, the complex forming ability being inversely related to the strength of the acid from which the anion is derived.

For Pu(III) the order of complexing with some common anions is

sulphate > nitrate > chloride > perchlorate.

The complexes formed between Pu(IV) and the common anions are of high stability. The order of complex stability with Pu(IV) is

fluoride > nitrate > chloride > perchlorate

for monovalent anions and

carbonate > sulphite > oxalate > sulphate

for divalent anions.

In concentrated HNO_3 Pu(IV) complexes to form all ions from $Pu(NO_3)^{3+}$ to $Pu(NO_3)_6^{2-}$ and these nitrate complexes are important in the chemical processing of plutonium. The fluoride, peroxide, oxalate and phosphate complexes have similar technological applications.

Perchlorate complexes only very weakly and studies of the absorption spectra of plutonium in $HClO_4$ are similar to those of solvated plutonium ions.

Plutonium(V) shows small tendency to form complexes with inorganic anions but chloride and oxalate complexes have been indicated.

Chloride anions have been found to form complexes with Pu(VI) more readily than nitrate ions but these complexes are much less stable than the corresponding Pu(IV) complexes.

2. Acetate Complexes

Complex formation between Pu(III), Pu(IV) and Pu(VI) and acetate ions has been reported.

Trivalent plutonium appears to form complexes containing up to five acetate molecules and the stability constant for the $Pu(C_2H_3O_2)_5^{2+}$ complex was estimated by SCHWABE and D. NEBEL (1962) to be $\beta_5 = 5 \times 10^{16}$. The complexing of Pu(IV) by acetate has been studied by E. NEBEL and SCHWABE (1963) who found that in solution below pH 4.8 complexes containing up to five acetate ions may be prepared. The stability constant, β_5, for the complex $Pu(C_2H_3O_2)_5^-$ at 25° C was found to be 8.06×10^{22} by spectrophotometry and 3.98×10^{22} by potentiometric titration.

The occurrence of mono-, di-, tri- and tetraacetate complexes of Pu(VI) have been inferred from solubility and spectrophotometric studies (GEL'MAN et al., 1962; SCHAEFER, 1968; RYAN and KEDER, 1967).

3. Lactate Complexes

No complexes between Pu(III) and the lactate anion are known. The lactate complexes of Pu(IV) have been studied by NEBEL (1966a and b) who found that the tetra-lactate $Pu(C_3H_5O_4)_4$ was the predominant species at concentrations of lactate greater than 5×10^{-4} M. From spectrophotometric and potentiometric studies he deduced a value for the overall stability constant β_4 of 1.5×10^{16}, at 25° C and I = 0.5 M. Attempts to detect the mono-, di- and tri-lactato complexes at lower lactate concentrations were frustated due to hydrolysis and disproportionation of the Pu(IV), and it was concluded that their stability is low (NEBEL, 1966b). Lactate complexes with Pu(VI) have been studied by SCHAEFER (1968) by spectrophotometry. Mono- and di-lactato complexes occur with stability constants of 6.94×10^2 and 1.212×10^3 respectively.

4. Citrate Complexes

Three complexes between Pu(III) and citrate, $Pu(C_6H_5O_7)$, $Pu(H_2C_6H_5O_7)_2^+$, $Pu(H_2C_6H_5O_7)_3$, have been reported by MOSKVIN et al. (1964).

Tetravalent plutonium forms very strong complexes with citrate. At citrate concentrations of only 10^{-15} M most of the Pu(IV) is complexed as the monocitrato complex, $Pu(C_6H_5O_7)^+$. But the di-citrato species $Pu(C_6H_5O_7)_2^{2-}$ is predominant between 10^{-15} and 10^{-13} M. At greater concentrations, $> 10^{-13}$ M, higher complexes appear to exist (NEBEL, 1966c). Spectrophotometric studies showed that, at the same citrate concentration, the extinction was independent of pH suggesting that the $(C_6H_5O_7)^{3-}$ anion is present in the complex. The stability constants for the mono- and di-citrato complexes with Pu(IV) were calculated from spectrophotometric data by two different methods and found to be

$$\beta_1 = 5.3 \times 10^{14} \quad \text{and} \quad 2.7 \times 10^{15} \quad \text{and}$$
$$\beta_2 = 1.6 \times 10^{30} \quad \text{and} \quad 6.9 \times 10^{29} \quad \text{respectively.}$$

(NEBEL, 1966c, d).

No studies on complex formation between Pu(V) or Pu(VI) and citrate ions have been reported.

5. Polyaminopolycarboxylic Acids

Complex formation between ethylenediaminetetraacetic acid (EDTA) and plutonium ions has been studied by FOREMAN and SMITH (1957) using an ion exchange method. Between pH 1 and pH 3 Pu(III) was found to form a 1:1 Pu(III) EDTA complex with a stability constant $K^1 = 1.3 \times 10^{18}$. At pH 3.3 Pu(IV) formed two complexes with Pu : EDTA ratios of 1:1 and 2:1 and stability constants of $K^1 = 4.5 \times 10^{12}$ and 1.6×10^{24} respectively. Hexavalent plutonium at the same pH probably forms a 1:1 Pu(VI) EDTA complex with a stability constant of $K^1 = 1.7 \times 10^{16}$. A 1:1 complex between Pu(V) and EDTA was observed by GEL'MAN et al. (1959) to have a stability constant $K^1 = 1.5 \times 10^{10}$.

Diethylenetriaminepentaacetic acid (DTPA) is of pharmacological importance since it is the most effective agent yet discovered for the treatment of plutonium poisoning in man. The complexes of plutonium with DTPA and other polyaminopolycarboxylic acids have been studied spectrophotometrically by HAFEZ (1968). Trivalent plutonium was found to form a complex which was stable at pH values greater than 8.2 and had a Pu(III):DTPA ratio of 2:3 and a stability constant $\log K^1 = 18.00$. Three Pu(IV) DTPA complexes were identified each existing in a well defined pH range. From pH 1 to 5.8 the 1:1 Pu(IV) DTPA complex was found to exist with a stability constant of $\log K^1 = 23.45$. Between pH 5.8 and 8.5, the pH range of greatest physiological interest, a 2:3 complex occurred with a much lower stability constant $\log K^1 = 18.00$. Above pH 8.5 an even less stable 1:2 complex was observed with $\log K^1 = 13.80$.

Table 9.5. Stability constants for Pu (IV) complexes with polyaminopolycarboxylic acids

Compound	log K [a]
Ethylenediaminetetraacetic acid (EDTA)	18.2
2,2 (Bis)-di-carboxymethylamino-diethylethertetraacetic acid (BAETA)	18.5
Diethylenetriaminepentaacetic acid (DTPA)	23.45
Triethylenetetraminehexaacetic acid (TTHA)	22.46

[a] From Hafez (1968).

Similar studies with Pu(VI) suggested that one molecule of PuO_2^{2+} was reduced by one molecule of DTPA and that Pu(IV) DTPA complexes were formed with metal-ligand ratios of 1:2 below pH 5.8 and 2:3 above pH 7.

The stability constants of the 1:1 Pu(IV) ligand complexes with four polyaminopolycarboxylic acids are listed in Table 9.5. These values were determined in acid solution in the pH range 3 to 4 and the stabilities of these 1:1 complexes are all probably very much greater than those of the complexes formed at physiological pH.

6. Complexes with 1 : 3 Diketones

Plutonium forms stable chelates with 1:2 diketones. The complexes of Pu(IV) with the simplest 1:3 diketone, acetyl acetone.

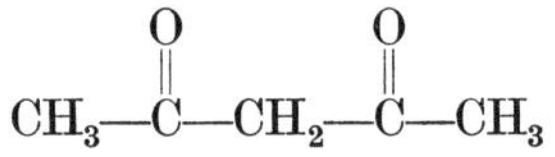

$$CH_3—\overset{\overset{\displaystyle O}{\|}}{C}—CH_2—\overset{\overset{\displaystyle O}{\|}}{C}—CH_3$$

have been studied by Rydberg (1956) who showed the existence of complexes containing up to four ligand molecules.

Thenoyltrifluoroacetone (TTA) is important in plutonium processing

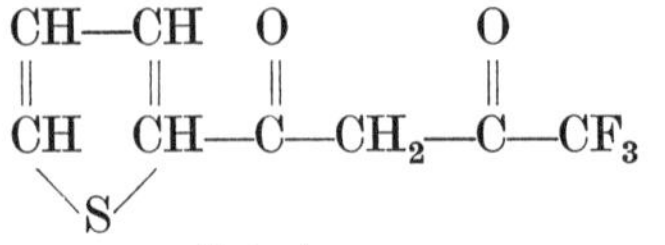

```
CH—CH  O       O            CH—CH  O       OH
||  ||  ||      ||            ||  ||  ||      ||
CH  CH—C—CH2—C—CF3         CH  CH—C—CH=C—CF3
  \S/                          \S/
     Keto form                     Enol form
```

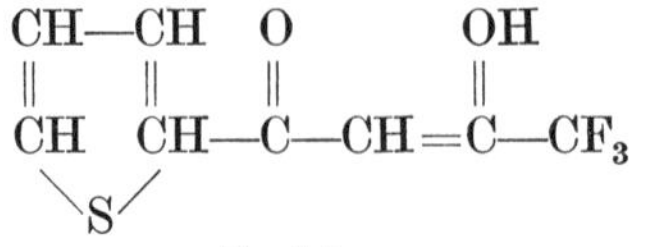

since the chelates with Pu(III) to Pu(VI) are soluble in organic solvents to varying extents, a property which has been employed in the solvent extraction process for separation of plutonium from uranium.

Chelate formation involves reaction of the acidic enol hydrogen with one equivalent of plutonium to form the chelates Pu(III) $(TTA)_3$, Pu(IV) $(TTA)_4$, Pu(VI) $(TTA)_2$. The Pu(IV) chelate is the most stable and also the most readily extracted into organic solvents.

7. Organophosphorus Complexes

Complex formation between plutonium and various types of organophosphorus compounds is important in the processing of plutonium by solvent extraction. Of these the most important is tri-*n*-butyl phosphate which reacts with Pu(IV) and Pu(VI) nitrates to form 1:2 metal: ligand complexes which are soluble in organic solvents (Healy and McKay, 1956).

Similar complexes with Pu(IV) and Pu(VI) nitrates are formed with tri-*n*-octyl phosphine oxide.

8. Other Complexing Agents

Plutonium forms complexes with a number of other compounds, for example salicylate, amino acids, sulphamates, but these have not been extensively studied.

Complexing of plutonium by desferrioxamine B has been shown to occur and this substance has been investigated as a potential therapeutic agent for the treatment of plutonium poisoning (TAYLOR, 1967). The formation of a complex between Pu(IV) and tetracycline has been detected by gel filtration, this complex appears to be less stable than the Pu(IV) citrate complex at pH 7.5 (TAYLOR et al., 1971).

D. Oxidation States in Solution

The tendency for ions to undergo oxidation or reduction can be summarised by the potential of the couple $M^{Reduced}/M^{Oxidised}$. Ideally this is expressed as the potential at infinite dilution, the *standard electrode potential*. However, standard electrode potentials are often difficult to measure, especially for metals, such as plutonium, which undergo extensive hydrolysis, and in such circumstances it is necessary to use the "*formal potentials*" measured under defined conditions of concentration and temperature.

The formal potentials for plutonium ions measured in 1 M perchloric acid, a medium in which negligible complexing occurs, and in neutral solution at pH 7 are summarised in Fig. 9.1.

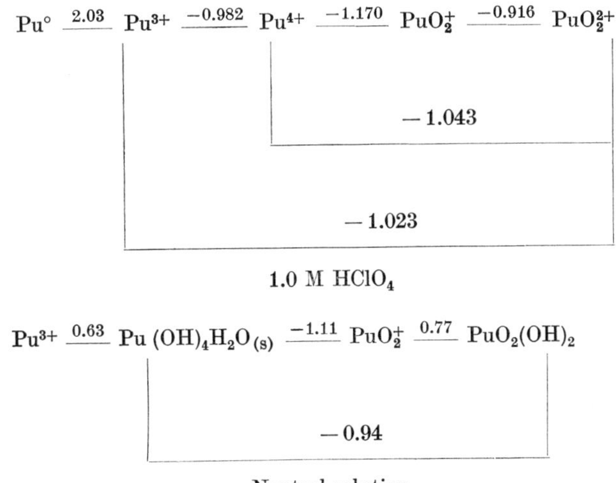

Fig. 9.1. Formal potentials of plutonium couples in acid and neutral solution at 25° C (from CLEVELAND 1970)

Oxidation-reduction reactions which involve the transfer of a single electron, e.g. $Pu^{3+} \rightleftharpoons Pu^{4+}$, tend to occur very rapidly but reactions involving transfer of two electrons, or the formation, or rupture, of an oxygen-metal bond e.g. $Pu^{4+} \rightleftharpoons PuO_2^+$ tend to be slow.

It has been mentioned earlier that in solution it is possible for Pu(III), Pu(IV), Pu(V) and Pu(VI) to co-exist in equilibrium. This highly unusual phenomenons predicted by the values of the formal potentials of the couples Pu^{3+}/Pu^{4+} and PuO_2^+/PuO_2^{2+} which are less negative than the Pu^{4+}/PuO_2^+ couple. This leads to the simultaneous oxidation and reduction of the Pu^{4+} and PuO_2^+ ions—the process called *disproportionation*.

Tetravalent plutonium disproportionates according to the following reaction:

$$3\,Pu^{4+} + 2\,H_2O \rightleftharpoons 2\,Pu^{3+} + PuO_2^{2+} + H^+.$$

The equilibrium constant for this reaction is proportional to the fourth power of the hydrogen ion concentration. It is also temperature dependent, disproportionation being more rapid at elevated temperatures (Rabideau and Cowan, 1955).

Pentavalent plutonium has been found to disproportionate below pH 1.5 or above pH 7 by the following slow reaction:

$$PuO_2^+ + PuO_2^+ \rightleftharpoons Pu^{4+} + PuO_2^{2+}.$$

If other Pu ions are present the reactions are:

$$PuO_2^+ + Pu_4^{3+} + 4\,H^+ \rightleftharpoons 2\,Pu^{4+} + 2\,H_2O$$

$$PuO_2^+ + Pu^{4+} \rightleftharpoons PuO_2^{2+} + Pu^{3+}$$

of which the latter is a fast reaction.

The overall reaction which describes the equilibrium of all four oxidation states in solution is

$$Pu^{4+} + PuO_2^+ \rightleftharpoons Pu^{3+} + PuO_2^{2+}.$$

In addition to hydrogen ion concentration and temperature, disproportionation reactions are influenced by the presence of complexing anions which may stabilise one particular valency state of plutonium by complex formation.

The plutonium alpha radiation may also cause the reduction of Pu(VI) to Pu(V) and Pu(III), by the products of the alpha particle-induced radiolysis of water. The usual product of alpha reduction is Pu(V); Pu(IV) and Pu(III) resulting primarily from disproportionation of Pu(V).

The oxidation-reduction reactions of plutonium in solution are important in the industrial processing of plutonium and the reactions have been widely studied. The reagents used in the oxidation and reduction of plutonium have been summarised by Cleveland (1970a) and only a few illustrations will be given here. Rapid oxidation of Pu(III) to Pu(IV) can be achieved at room temperatures by iodate or permanganate in dilute acid. Dilute nitric acid oxidises Pu(III) only slowly at room temperature. At neutral pH Pu(III) can be oxidised by water (Kraus, 1949). Reduction of Pu(IV) to Pu(III) by ascorbic acid is rapid in the presence of H_2SO_4, even in 4.7 M HNO_3 solutions. Oxidation of Pu(IV) to Pu(VI) occurs rapidly in the presence of Aq^{2+} at room temperature. Reduction of Pu(VI) to Pu(V) and Pu(IV) can be achieved with Fe^{2+}.

The oxidation-reduction reactions are generally fast in relation to the disproportionation reactions and the latter are generally most important at high plutonium concentrations and in the absence of oxidising agents.

IV. Interaction of Plutonium with Proteins and Other Substances of Biological Interest

It is evident from the discussion of hydrolytic reactions in Sec. III B that plutonium ions are extremely unlikely to exist at physiological pH. Consequently, hydrolysis and complex ion formation may be expected to play important roles in determining the biological behaviour of plutonium.

If plutonium in ionic form enters the mammalian body three main types of reaction may be expected to occur:

a) Hydrolysis to yield polymeric or colloidal species.

b) Complexing by proteins or other biological macromolecules.

c) Complexing, or chelation, with small molecular weight components of cells and tissues, for example citrate and other organic acid anions, amino acids etc.

If plutonium in ionic form enters the vascular compartment, then reaction "*b*" appear to predominate, since TURNER and TAYLOR (1968a) found that following the intravenous injection of plutonium citrate, in essentially monomeric form, into rats, 75 percent of the plutonium in the plasma was protein bound within one minute after injection. Direct injection of plutonium into the vascular compartment is unlikely to occur in the accidental contamination of the human body, but intravenous injection has been used extensively in the study of the behaviour of plutonium in experimental animals.

If plutonium ions enter directly into a tissue, such as muscle in the case of a contaminated wound, "*a*" and, to a lesser extent, "*b*" may be expected to be the predominant reactions. Thus, the plutonium would be expected to be deposited at the site of injection as an insoluble mass or as a complex with the structural proteins. Translocation of plutonium from the site of injection to other parts of the body would be expected to depend on attack by biological ligands of small molecular size to form stable, soluble plutonium complexes which could diffuse away from the site. Some material might also break away in particulate form and be removed by phagocytic action. In view of the known difficulties of solubilising polymeric plutonium, translocation would be expected to be a slow process. These predictions are substantiated by some of the earliest experiments on the metabolic behaviour of plutonium. SCOTT et al. (1948) showed that following intramuscular injection of solutions of Pu(IV), Pu(III) or Pu(VI) chlorides into rats, between 68 and 96 per cent of the injected material was present at the injection site at 4 days and considerable amounts still remained at the site 256 days after injection. The rate of clearance of the Pu(III) and Pu(VI) from the injection site was more rapid than the rate of translocation of Pu(IV) but the tissue distribution pattern were similar for all three oxidation states. It seems likely that *in vivo*, absorbed plutonium exists in a single oxidation state, probably Pu(IV).

In the gastro-intestinal tract it is probable that hydrolytic reactions and adsorption onto insoluble particles of food etc. are responsible for the very low absorption of plutonium following oral intake.

A. Complexing with Plasma Proteins

Interest in the proteins concerned in the transport of plutonium in the blood arose early in the study of the biological behaviour of plutonium. In 1947 MUNTZ and BARRON (1947) used moving boundary electrophoresis to study the distribution of plutonium amongst the serum proteins following injection of Pu(VI) and found that the largest fraction of the serum plutonium was associated with the β_1-globulin group. This observation was subsequently confirmed by paper electrophoresis (PALMER, 1955) and BELIAYEV (1959) showed that at least 90 per cent of the serum plutonium was protein bound, mainly to the globulin fraction.

The study of the interactions of heavy metals, such as plutonium, with proteins is complicated by hydrolytic reactions which may make it difficult to separate free and bound metal. In consequence, it is necessary to choose experimental methods very carefully so that hydrolytic reactions are minimised, and to obtain information by several different methods wherever possible.

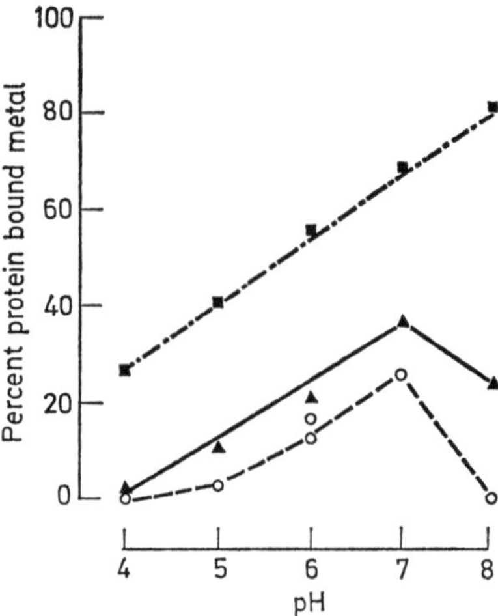

Fig. 9.2. The effect of pH on the binding of Fe(III) and Pu(IV) to transferrin and apotransferrin. ■ — — — — — ■ Fe(III)-transferrin; ▲———— ▲ Pu(IV)-apotransferrin; ○ — — — — ○ Pu(IV)-transferrin. (From data of CHIPPERFIELD and TAYLOR, 1970a)

In recent years the techniques of gel filtration, ion exchange and electrophoresis on cellulose acetate membranes have been used to study the interactions between Pu(IV) and serum proteins both *in vivo* and *in vitro*. These studies have established that the β_1-globulin, transferrin is the major plutonium-binding serum protein in man (STOVER et al., 1968), dog (STEVENS et al., 1968), horse (TURNER and TAYLOR, 1968b), rabbit (TAYLOR, 1969 and rat (BOOCOCK and POPPLEWELL, 1965; TURNER and TAYLOR, 1968a).

Transferrin is a glycoprotein with a molecular weight of about 90000 in which aspartic acid, glutamic acid, lysine, leucine and alanine are the most abundant amino acids. The transferrins from all the species studied bind two gram atoms of iron to form a complex which dissociates in acid solution but is stable up to pH 9 to 10 (FEENEY and KOMATSU, 1966). The effect of pH on the binding of Pu(IV) to transferrin has been studied by CHIPPERFIELD and TAYLOR (1970a) who showed that maximum binding occurred at pH 7, Fig. 9.2. The mechanism of Pu(IV) binding by transferrin is not yet understood. STOVER et al. (1968) and TURNER and TAYLOR (1968b) from studies on dog and horse sera concluded that bicarbonate ions were required for the binding of Pu(IV) to transferrin in a similar manner to their requirement in the binding of iron (FEENEY and KOMATSU, 1966).

As a result of more recent experiments CHIPPERFIELD and TAYLOR (1970a) have questioned this requirement for bicarbonate in the binding of Pu(IV) to transferrin. Plutonium can be displaced from its combination with transferrin by the addition of excess iron, and little plutonium is found to bind to iron-saturated transferrin (POPPLEWELL and BOOCOCK, 1968; STEVENS et al., 1968; TURNER and TAYLOR, 1968b; MASSEY and LAFUMA, 1968). These observations led POPPLEWELL and BOOCOCK (1968) to speculate that the iron- and plutonium-binding sites on transferrin might be identical. This hypothesis has been examined by CHIPPERFIELD and TAYLOR (1970a) who found that, under their experimental conditions, the binding of iron and plutonium to transferrin was similar in that both increased up to pH 7 (Fig. 9.2); above pH 7 the Pu(IV) transferrin complex was less stable than the Fe(III) complex, but this may be a reflection of the greater tendency of plutonium to undergo hydrolysis. The effect of pH on the binding of Pu(IV) to the iron-free protein, apotransferrin, was like that of transferrin which suggests that the same groups are involved in the binding of Pu(IV) to transferrin and apotransferrin. Further the recoveries were greater with apotransferrin suggesting that more sites were available for binding in the iron-free protein. However, studies in which iron or bicarbonate were added to the reaction medium showed no change in the binding of Pu(IV) to transferrin or apotransferrin. These results suggested that plutonium can bind at sites other than those which bind iron. On the present evidence it seems likely that Pu(IV) can bind at the iron-binding sites in transferrin but that it can also bind at other sites in the molecule.

It has not yet proved possible to measure the stability constant for the Pu(IV)-transferrin complex, STOVER et al. (1968) have concluded that the complex has a high stability but that it is less stable than the Fe(III) transferrin complex. In gel filtration experiments it has been shown that Pu(IV) can be displaced from the transferrin complex by citrate or DTPA (POPPLEWELL and BOOCOCK, 1968).

Although transferrin appears to be the major plutonium-binding protein in serum, other serum proteins can bind plutonium. Both electrophoretic and gel filtration analysis of sera labelled with plutonium *in vivo* or *in vitro* have indicated that a small proportion of the plutonium may be associated with albumin and with the γ- and other globulins (TURNER and TAYLOR, 1968a, b; STEVENS et al., 1968; POPPLEWELL and BOOCOCK, 1968). This association of plutonium with albumin and the γ- and other globulins in labelled sera could have arisen from some redistribution of the plutonium during the separation procedure and the biological significance of such binding is unknown. Studies with purified proteins labelled with plutonium *in vitro* have shown that plutonium binds to bovine γ-globulin less strongly than to transferrin. Human serum albumin also binds plutonium but the binding is very much weaker than the binding to transferrin (CHIPPERFIELD and TAYLOR, 1968).

B. Binding to Ferritin

An association between plutonium and haemosiderin in the macrophages of bone marrow was first described by ARNOLD and JEE (1957). Subsequent studies by BENO (1968) showed an increase in the number of iron containing cells and iron pigments in rabbit bone marrow 4 months after a single intravenous injection of $^{239}Pu(NO_3)_4$. A similar association of plutonium with iron-containing pigments was observed in the reticuloendothelial cells of canine liver at long times after injection of ^{239}Pu (TAYLOR et al., 1967).

Recent studies of the subcellular distribution of plutonium in rat liver after intravenous injection of ^{239}Pu citrate, in essentially monomeric form, have demonstrated the association of plutonium with ferritin in the soluble fraction of the liver cells at time intervals greater than 3 days after plutonium injection (BOOCOCK et al., 1970). Similar observations in canine liver have been reported by STOVER et al., (1970).

The binding of plutonium by ferritin has not been studied in any detail. BRUENGER et al. (1969) have studied the exchange of Pu(IV) between Pu(IV)-transferrin and ferritin *in vitro* using a gel filtration technique and have demonstrated that the Pu-ferritin complex is more stable than the Pu-transferrin complex.

In rat liver at 3 days, or longer, after injection the majority of the liver plutonium is associated with the lysosomes and less than 5 percent is associated with ferritin in the soluble fraction. The mode of fixation of plutonium in lysosomes is not yet understood, but binding by ferritin may be involved in this process.

C. Binding to Bone Proteins

Interest in the binding of plutonium by components of the organic matrix of bone was aroused by the observations that plutonium was deposited mainly on bone surfaces and especially on the endosteal surfaces (see Chap. 10, JANET VAUGHAN). The possible role of chelation in the fixation of plutonium on bone surfaces was first suggested by VAUGHAN and her colleagues (HERRING et al., 1962) and detailed studies of the proteins of the organic matrix of bone have been undertaken by HERRING (1969, 1970).

The plutonium binding properties of collagen and five glycoprotein fractions isolated from bovine cortical bone have been studied by CHIPPERFIELD and TAYLOR (1968, 1970a, b). These authors used a gel filtration procedure to separate protein-bound and non-protein bound plutonium. This technique is not an entirely satisfactory method for measuring protein-metal binding since the protein and metal do not remain in equlibrium. However the problems caused by the hydrolysis and precipitation of plutonium preclude the use of more conventional methods, such as equilibrium dialysis or potentiometric titration, to study plutonium-protein binding in the physiological pH range. Up to the present time the non-equilibrium gel filtration procedure has proved to be the most useful approach to the comparative study of the binding of plutonium by different proteins, and the recoveries of protein-bound metal are a useful, semi-quantitative, measure of the stabilities of the complexes. The relative binding of plutonium to bone proteins and to a number of other proteins or polypeptides at pH 7 is illustrated in Table 9.6 in which the data are expressed as the percentages of proteinbound metal recovered from the Sephadex gel columns.

The bone proteins studied were collagen, the main protein component of the organic matrix of bone, and five glycoprotein fractions originally isolated from bovine cortical bone by HERRING (1969, 1970). Only two of these glycoprotein fractions have been characterised; the first is bone sialoprotein (BSP) which has a molecular weight of 25000 ± 500 and contains 40 percent by weight of carbohydrate and unusually high contents of sialic, glutamic and aspartic acids (WILLIAMS and PEACOCKE, 1965; ANDREWS and HERRING, 1965). The second is the chondroitin sulphate protein complex which contains three components whose composition is consistent with their being either free chondroitin sulphate or one or two chondroitin sulphate chains attached to a protein core with similar properties to bone sialoprotein (HERRING, 1968). Glycoproteins I and II are mixtures of at least two components, both less acidic and containing less carbohydrate than

Table 9.6. Binding of Pu (IV) to proteins at pH 7.4. (Binding expressed as percentage of protein-bound plutonium recovered from Sephadex gel columns)

Protein	Percent of protein bound plutonum	Protein	Percent of protein bound plutonium
Bone sialoprotein (BSP)	54.7 ± 9.3	Human transferrin	18.9 ± 6.9
Chondroiton sulphate-protein complex (CSP)	49.2 ± 3.0	Human serum albumin	1.2 ± 0.8
		Bovine γ-globulins	13.8 ± 5.1
CPC-soluble glycoprotein	36.6 ± 5.1	Ovalbumin	18.9 ± 2.7
Glycoprotein I	30.0 ± 12.4	Chondroitin sulphate	13.2 ± 5.7
Glycoprotein II	50.3 ± 9.1	Poly-L-glutamic acid	68.7 ± 0.5
Soluble collagen	23.3 ± 4.3		

From CHIPPERFIELD and TAYLOR (1968, 1970b, 1972).

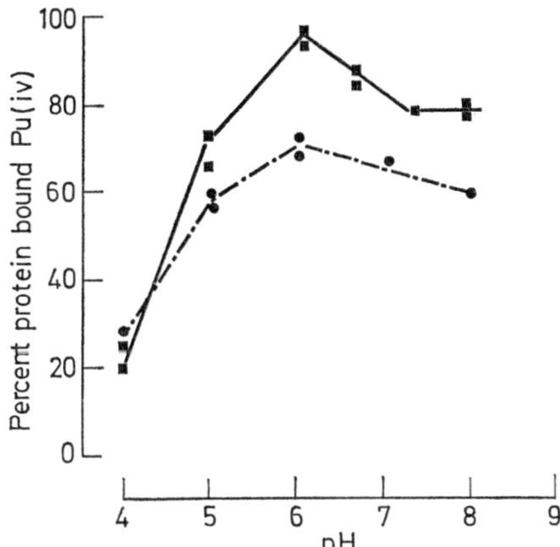

Fig. 9.3. The effect of pH on the binding of Pu(IV) to bone sialoprotein (■——■) and bone chondroitin-sulphate-protein complex (●—·—·—●). (Data taken from CHIPPERFIELD and TAYLOR, 1970b)

BSP (HERRING, 1969, 1970). The CPC-soluble glycoprotein has the analytical characteristics of a BSP deficient in sialic acid (HERRING, 1969, 1970). Glycoproteins I and II appear to be present in bone in much greater quantity than BSP or the chondroitin sulphate-protein complexes (HERRING, 1970a.)

The data in Table 9.6 shows that BSP, chondroitin sulphate-protein and glycoprotein II bind plutonium to an approximately equal extent and that this binding is greater than the binding to CPC-soluble glycoprotein, glycoprotein I, collagen or to transferrin.

The effects of pH on the binding of plutonium to BSP and chondroitin sulphate-protein complex are illustrated in Fig. 9.3 which shows that maximum binding occurs at about pH 6. These results have been interpreted to indicate that plutonium is bound by ionised carboxyl groups in the BSP and chondroitin sulphate protein molecules, a conclusion which is supported by the strong binding of plutonium to poly-L-glutamic acid which is suggested by the results in Table 6. The sialic acid groups appear to participate in the binding of plutonium to BSP since removal of the sialic acid reduced the binding of Pu(IV) to less than one third of that observed with intact BSP. The fifteen sialic acid groups in BSP occupy terminal positions and the reduced plutonium binding observed following removal of sialic acid is attributed to loss of carboxyl groups rather than to conformation changes in the BSP molecule (CHIPPERFIELD and TAYLOR, 1970a).

The possible biological significance of the binding of plutonium to proteins in the organic matrix of bone is discussed in Chap. 10 (JANET VAUGHAN).

D. Binding to Other Proteins

The data listed in Table 9.6 show that plutonium can bind to a number of other proteins *in vitro*, however, with the exception of bovine γ-globulin and ovalbumin, this binding is very much weaker than that observed with transferrin.

The studies of the subcellular distribution of plutonium in rat liver (BOOCOCK et al., 1970) have indicated that, during the first few hours after injection of essentially monomeric Pu(IV), there is an unidentified protein in liver which acts as the major binding species for plutonium in the soluble fraction. This protein has not yet been isolated but its chromatographic behaviour is similar to γ-globulin.

E. Binding to Other Substances

The binding of plutonium to non-protein ligands of biological interest has not been widely studied.

Complexes formed between amino acids and plutonium appear to be of low stability. The monoglycine complex with Pu(V) has been studied by Eberle and Wede (1968) who obtained a value of 3.04 ± 0.04 for log K_1. Chipperfield and Taylor (1970b) attempted unsuccessfully to study amino acid complexes of plutonium by gel filtration and concluded that the complexes formed, even with glutamic and aspartic acids were insufficiently stable to be measured by this method.

The complexes formed between lactic and citric acids have been discussed in Sec. III C. These complexes show high chemical stability but as the complexing ligand is readily metabolised they are probably of relatively low stability *in vivo*. This suggestion is supported by the fact that within 1 minute of the intravenous injection of an essentially monomeric solution of ^{239}Pu(IV) citrate more than 70 per cent of the injected radioactivity is bound to plasma proteins, mainly to transferrin (Turner and Taylor, 1968).

This difference between the chemical and biological stability of some plutonium complexes is important in relation to the choice of compounds for use in the treatment of plutonium poisoning. On the basis of the chemical stability of the plutonium complex which is likely to exist at physiological pH, citrate should be an effective agent for enhancing the rate of elimination of plutonium from the body. However studies in experimental animals show that citrate, which is readily metabolizable, is very much less effective than the non-metabolizable ligands EDTA or DTPA in hastening plutonium elimination (Smith and Chapman, 1969; Taylor and Sowby, 1962). A similar observation with another substance which is important in carbohydrate oxidation, α-lipoic acid, has been made by Dilley (1968). In leaching PuO_2 from a column of Sephadex gel, α-lipoic acid was found to be more effective than DTPA. However, *in vivo* α-lipoic acid did not increase the rate of excretion of Pu(IV) from rats although it did alter the tissue distribution pattern resulting in higher concentrations in the soft tissues.

In order to increase our understanding of the mechanisms governing the biological behaviour of plutonium a great deal more information is required on its interaction with the natural components of cells and tissues. For example the nature of the diffusible complexes which are concerned in the translocation of plutonium from a wound site is almost completely unknown; and little is known concerning possible interactions between plutonium and mucous secretions in the lung which might be important in the translocation of inhaled plutonium.

References

Andrews, A. T. de B., Herring, G. M.: Further studies on bone sialoprotein. Biochim. biophys. Acta (Amst.) **101**, 241–244 (1965).

Arnold, J. S., Jee, W. S. S.: Bone growth and osteoclastic activity as indicated by radioautographic distribution of plutonium. Amer. J. Anat. **101**, 367–418 (1957).

Beliayev, Yu. A.: Physicochemical state of plutonium (Pu^{239}) in blood after its intravenous administration. Med. Radiol. USSR **4**, 45–51 (1959).

Beno, M.: A study of "haemosiderin" in the marrow of the femur of normal young adult rabbits compared with that in rabbits 4 months after intravenous injection of $Pu(NO_3)_4$. Brit. J. Haemat. **15**, 487–493 (1968).

Boocock, G., Danpure, C. J., Popplewell, D. S., Taylor, D. M.: The subcellular distribution of plutonium in rat liver. Radiat. Res. **42**, 381–396 (1970).

Boocock, G., Popplewell, D.S.: Distribution of plutonium in serum proteins following intravenous injection into rats. Nature Lond.) **208**, 282—283 (1965).

BRUENGER, F. W., STOVER, B. J., STEVENS, W., ATHERTON, D. R.: Exchange of ^{239}Pu IV between transferrin and ferritin in vitro. Hlth Phys. **16**, 339–340 (1969).

CARTLEDGE, G. H.: Studies on the periodic system. I. The ionic potential as a periodic function. J. Amer. chem. Soc. **50**, 2856–2863 (1928).

CHIPPERFIELD, A. R., TAYLOR, D. M.: Binding of plutonium and americium to bone glycoproteins. Nature (Lond.) **219**, 609–610 (1968).

CHIPPERFIELD, A. R., TAYLOR, D. M.: Binding of plutonium to glycoproteins in vitro. Radiat. Res. **43**, 393–402, (1970a).

CHIPPERFIELD, A. R., TAYLOR, D. M.: The binding of americium and plutonium to bone glycoproteins. Europ. J. Biochem. **17**, 581–585 (1970b).

CHIPPERFIELD, A. R., TAYLOR, D. M.: The binding of thorium(IV), plutonium(IV) americium(III) and curium(III) to the constituents of bovine cortical bone *in vitro*. Radiat. Res. **51**, 15–30 (1972).

CLEVELAND, J. M.: The chemistry of plutonium. New York: Science Publishers, Gordon & Breach 1970a.

CLEVELAND, J. M.: Aqueous co-ordination complexes of plutonium. Cord. Chem. Rev. **5**, 101–137 (1970b).

COHEN, D.: Electrochemical studies of plutonium ions in perchloric acid solution. J. Inorg. Nucl. Chem. **18**, 207–210 (1961a).

COHEN, D.: The absorption spectra of plutonium ions in perchloric acid solution. J. Inorg. Nucl. Chem. **18**, 211–218 (1961b).

CONNICK, R. E., KASHA, M., MCVEY, W. H., SHELINE, G. E.: Spectrophotometric studies of plutonium in aqueous solution. In: The transuranium elements, vol. IV 14B. New York: McGraw Hill 1949.

COSTANZO, D. A., BIGGERS, R. E.: The polymerization, depolymerization and precipitation of quadrivalent plutonium as functions of temperature and acidity by spectrophotometric methods. A preliminary report. U.S.A.E.C. Report ORNL-TM-585-1963.

DILLEY, J. V.: In vivo and in vitro chelation of plutonium by α-lipoic acid and DTPA. Pacific NW Biol. Rept. 1967. BNWL-714 6.12–6.13. 1968.

EBERLE, S. H., WEDE, U.: Chelates of actinides with aminoacetic acid. Inorg. Nucl. Chem. Lett. **4**, 661–664 (1968).

FEENEY, R. E., KOMATSU, S. K.: The transferrins. Structure & Bonding **1**, 149–206 (1966).

FOREMAN, J. K., SMITH, T. D.: The nature and stability of the complex ions formed by ter-, quadri-, and sexa-valent plutonium ions with ethylene diaminetetraacetic acid. Part I. pH titrations and ion exchange studies. J. chem. Soc. 1752–1758 (1957).

GEL'MAN, A. D., ARTVYKHIN, P. I., MOSKVIN, A. I.: A study of complex formation by pentavalent plutonium in ethylenediaminetetraacetate solutions by means of ion exchange. Russ. J. Inorg. Chem. **4**, 599–602 (1959).

GEL'MAN, A. D., MOSKVIN, A. I., ZAITSEV, L. M., MEFOD'EVA, M. P.: Complex compounds of transuranium elements. New York: Consultants Bureau 1962.

GREBENSHCHIKOVA, V. I., DAVYDOV, YU. P.: Investigation of the state of Pu(IV) in dilute nitric acid. Soviet Radiochem. **3**, 51–60 (1962).

GREBENSHCHIKOVA, V. I., DAVYDOV, YU. P.: State of Pu(IV) in the region of pH = 1.0–12.0 at a plutonium concentration of 2.10^{-5} M. Soviet Radiochem. **7**, 190–193 (1965).

HAFEZ, M. B.: Contribution à l'étude de la complexation du cerium et des uranides par l'acide diethylenetriaminapentaacetique (DTPA). France: Commissariat à l'Energie Atomique CEA-R-3521 1968.

HEALY, T. V., MCKAY, H. A. C.: Complexes between tributyl phosphate and inorganic nitrates. Rec. Trav. Chim. Pays-Bas **75**, 730–736 (1956).

HERRING, G. M.: Mucosubstances and ion-binding in bone. Bibl. "Nutr. et Dieta" (Basel) **13**, 147–154 (1969).

HERRING, G. M.: A review of recent advances in the chemistry of calcifying cartilage and bone matrix. Calcif. Tiss. Res. **4** (Suppl.) 17–23 (1970a).

HERRING, G. M.: Private communication (1970b).

HERRING, G. M., VAUGHAN, JANET, WILLIAMSON, MARGARET: Preliminary report on the site of localization and possible binding agent for yttrium, americium and plutonium in cortical bone. Hlth Phys. **8**, 717–724 (1962).

ICRP (1959) — Recommendations of the International Commission on Radiological Protection: Report of Committee II on Permissible Dose for Internal Contamination. New York: Pergamon Press 1959.

KATZ, J. J., SEABORG, G. T.: The chemistry of the actinide elements. New York: Methuen 1957.

KRAUS, K. A.: Oxidation-reduction potentials of plutonium couples as a function of pH. Chap. 3.16. The transuranium elements NNES IV 14B. New York: McGraw Hill 1949.

Kraus, K. A.: Hydrolytic behaviour of the heavy elements. Proc. Intern. Conf. Peaceful Uses Atomic Energy, Geneva 1955, 7, 245–257 (1956).

Krot, N. N., Gel'man, A. D.: Preparation of neptunium(VII) and plutonium(VII). Dokl. Akad. Nauk SSSR **177** (1), 124–126 (1967).

Lederer, C. M., Hollander, J. M., Perlman, I.: Table of Isotopes, 6th ed. New York: J. Wiley & Sons 1967.

Lindenbaum, A., Westfall, W.: Colloidal properties of plutonium in dilute aqueous solution. Int. J. appl. Rad. Isotopes **16**, 545–553 (1965).

Lutz, M., Metnier, H., Langlois, Y.: Hydrolyse du plutonium tétravalent. Mise en évidence de la complexation par des anions d'intérêt biologique. CEA-R-3092 (1966).

Makarov, E. S.: Crystal chemistry of simple compounds of uranium, thorium, plutonium and neptunium, p. 145. New York: Consultants Bureau 1959.

Mandelberg, C. J., Francis, K. E., Smith, R.: The solubility of plutonium trifluoride, plutonium tetrafluoride, and plutonium(IV) oxalate in nitric acid mixtures. J. chem. Soc. 2464–2468 (1961).

Massey, P., Lafuma, J.: Fixation in vitro du plutonium sur la siderophiline humaine et réaction de compétition avec l'ion ferrique. CEA-R-3623 (CEA France) (1968).

Milyukova, M. S., Gusev, N. L., Sentyurin, I. G., Sklyarenko, I. S.: Analytical chemistry of plutonium. Analytical chemistry of elements series, p. 382. New York: Daniel Davey & Co. Inc. 1967.

Moses, A. J.: Analytical chemistry of the actinide elements, p. 187. New York: Pergamon Press 1963.

Moskvin, A. I., Zaitseva, V. P., Gel'man, A. D.: Complex formation of trivalent plutonium with acetate, citrate and tartrate anions by an ion exchange method. Soviet Radiochem. **6**, 206–222 (1964).

Muntz, J. A., Barron, E. S. G.: Combination of plutonium with plasma proteins. U.S. Atomic Energy Commission Report MDDC 1268 (1947).

Nebel, D.: Untersuchungen zum Gleichgewicht PuIV-laktat in wäßriger Lösung. Z. physik. Chem. **233**, 62–72 (1966a).

Nebel, D.: Über eine Untersuchung zur Komplexbildung von Plutonium in wäßriger Lösung. Z. analyt. Chem. **216**, 284–285 (1966b).

Nebel, D.: Spektralphotometrische Untersuchung des Gleichgewichtes PuIV-citrat in wäßriger Lösung. Z. physik. Chem. **232**, 161–175 (1966c).

Nebel, D.: Potentiometrische Untersuchungen des Gleichgewichtes PuIV-citrat in wäßriger Lösung. Z. physik. Chem. **232**, 368 (1966d).

Nebel, E., Schwabe, K.: Spektrophotometrische Untersuchungen von PuIV-acetatkomplexen in wäßriger Lösung. Z. physik. Chem. **224**, 29–50 (1963).

Ockenden, D. W., Welch, G. A.: The preparation and properties of some plutonium compounds. Part V. Colloidal quadrivalent plutonium. J. chem. Soc. 3358–3363 (1956).

Palmer, R. F.: Electrophoretic studies of the distribution of plutonium and ruthenium in the blood of rats. Biology Res. Ann. Rept. 1955. HW-35917. 68–69 (1956).

Pascal, P.: Nouveau traité de chimie minérale. Transuraniens, 5th edit., vol. XV. Paris: Masson 1970.

Popplewell, D. S., Boocock, G.: Distribution of some actinides in blood serum proteins. In: Diagnosis and treatment of deposited radionuclides, p. 45–55. Amsterdam: Excerpta Med. Found. 1968.

Rabideau, S. W., Cowan, H. T.: Chloride complexing and disproportionation of Pu(IV) in hydrochloric acid. J. Amer. chem. Soc. **77**, 6145–6248 (1955).

Ryan, J. L., Keder, W. E.: Anionic acetato complexes of the hexavalent actinides. Anion exchange and amine extraction of hexavalent actinide acetates. In: Lanthanide-Actinide chemistry, p. 335–351. Washington: Advan. Chem. Ser. 1967.

Rydberg, J.: Studies on the extraction of metal complexes. XXV. The complex formation of Pu(IV) with acetylacetone. Ark. Kem. **9**, 109–119 (1956).

Schaefer, J. B.: Complex formation of five and six-valent actinide elements with α-hydroxycarbonic acids. German Report KFK-765 (1968).

Scott, K. G., Axelrod, D. J., Fisher, H., Crowley, J. F., Hamilton, J. G.: The metabolism of plutonium in rats following intramuscular injection. J. biol. Chem. **176**, 283–293 (1948).

Schwabe, K., Nebel, D.: Potentiometrische Untersuchungen an Plutonium. Z. physik. Chem. **220**, 339–354 (1962).

Seaborg, G. T., McMillan, E. M., Kennedy, J. W., Wahl, A. C.: Radioactive element 94 from deuterons on uranium. Phys. Rev. **69**, 366 (1946).

Sillen, L. G., Martell, A. E. (Eds.): Stability Constants of metal-ion complexes. London: The Chemical Society 1964.

SMITH, H., CHAPMAN, I. V.: Use of citrate in mobilizing plutonium in rat. Nature (Lond.) **223**, 642–643 (1969).
STEVENS, W., BRUENGER, F. W., STOVER, B. J.: In vivo studies on the interactions of PuIV with blood constituents. Radiat. Res. **33**, 490–500 (1968).
STOVER, B. J., BRUENGER, F. W., STEVENS, W.: The reaction of PuIV with the iron transport system in human blood serum. Radiat. Res. **33**, 381–394 (1968).
STOVER, B. J., BRUENGER, F. W., STEVENS, W.: Association of americium with ferritin in canine liver. Radiat. Res. **43**, 173–186 (1970).
TAYLOR, D. M.: The effects of desferrioxamine on the retention of actinide elements in the rat. Hlth Phys. **13**, 135–140 (1967).
TAYLOR, D. M.: The metabolism of plutonium in adult rabbits. Brit. J. Radiol. **41**, 44–50 (1969).
TAYLOR, D. M., CHIPPERFIELD, A. R., JAMES, A .C.: The effects of tetracycline on the deposition of plutonium, and related elements, in rat bone. Hlth Phys., **21**, 197–204 (1971).
TAYLOR, D. M., SOWBY, F. D.: The removal of americium and plutonium from the rat by chelating agents. Phys. Med. Biol. **7**, 83–91 (1962).
TAYLOR, G. N., JEE, W. S. S., DOCKUM, N. L., HROMYK, E.: Translocation of ^{239}Pu and ^{241}Am in beagle livers. Radiat. Res. **31**, 554 (1967).
TURNER, G. A., TAYLOR, D. M.: The transport of plutonium, americium and curium in the blood of rats. Phys. Med. Biol. **13**, 535–546 (1968a).
TURNER, G. A., TAYLOR, D. M.: The binding of plutonium to serum proteins in vitro. Radiat. Res. **36**, 22–30 (1968b).
WICK, O. J. (Ed.): Plutonium handbook. A guide to the technology, vols. I and II. New York: Gordon & Breach Science Publisher 1967.
WILLIAMS, P. A., PEACOCKE, A. R.: The physical properties of a glycoprotein from bovine cortical bone (bone sialoprotein). Biochim. biophys. Acta (Amst.) **101**, 327–335 (1965).
ZAITSEVA, V. P., ALEKSEEVA, D. P., GEL'MAN, A. D.: Hydrolysis of nitrate solutions of Pu(V). Radiokhimiya **10**, 537–541 (1968).

Chapter 10

Distribution, Excretion and Effects of Plutonium as a Bone-Seeker

JANET VAUGHAN*, in collaboration with BETTY BLEANEY* and DAVID M. TAYLOR**, ***

With 81 Figures

I. Introduction

Sixteen isotopes of plutonium are known, (LEDERER et al., 1967). These are listed in Table 10.1. Only two of these are at present recognised as of biological importance. Plutonium-239 is used increasingly for industrial and military purposes and its biological effects have therefore been the subject of intense study. Plutonium-238 also has practical applications which have encouraged some analyses of its metabolic and biological effects. The shorter-lived ^{237}Pu, it has been suggested, may have value in metabolic studies.

The physical characteristics of these isotopes are shown in Table 10.2. There is a large difference in specific activities between ^{239}Pu and the other two isotopes. Plutonium-237 decays mainly by electron capture with the emission of x-rays but because of its relatively short half-life its specific activity for *alpha* emission is greater than that of ^{238}Pu. Table 10.2 gives the mass of the isotopes required to produce one microcurie of alpha activity for the three isotopes.

For each isotope the emission of a single alpha particle of roughly 5.4 Mev produces a long-lived isotope of uranium, so that one might expect the biological effects of all three isotopes to be the same. However, the high specific activities of ^{238}Pu and ^{237}Pu influence the transport of these isotopes round the body. ^{238}Pu appears to be more soluble in tissue fluids than ^{239}Pu and therefore more readily translocated (STUART, 1970). This makes the failure of authors to discriminate between the two isotopes, in some earlier experiments, unfortunate.

The metabolic behaviour of ^{239}Pu in relation to the skeleton will be described first, since this isotope has been the most extensively investigated. Any differences in the behaviour of ^{237}Pu and ^{238}Pu will be discussed in later sections. Further in order to discuss the biological behaviour of plutonium isotopes towards the skeleton, it will be necessary to cover some of the ground already discussed in the chapter dealing with the chemical and physical properties of this radionuclide since these condition its behaviour as a bone-seeker. In summary it may be said that the plutonium ion is readily hydrated giving rise to large polymeric particles. Furthermore, ions which undergo such extensive hydrolysis will also form complexes readily. Plutonium cannot exist as a simple ion at a physiological

* Oxford, England.
** Royal Cancer Hospital, Sutton, Surrey, England.
*** The authors and editors acknowledge with gratitude the contributions of Dr. C. W. MAYS, University of Utha in the preparation of and review of this chapter.

Table 10.1. Plutonium isotopes

Isotope	Half life	Decay			
		E.C.%	Alpha %	E(α)Mev	% alpha abundance
Alpha emitters					
232	36 min	98	2	6.58	
233	20 min	99	0.1	6.30	
234	9 hr	94	6	6.19	
235	26 min	99	3×10^{-3}	5.85	
236	2.85 yr	—	100	5.763 5.716	69 31
237	45.6 d	99	3.3×10^{-3}	5.65 5.36	21 79
238	86.4 yr	—	100	5.495 5.452	72 28
239	24360 yr	—	100	5.147 5.134 5.096	72.5 16.8 10.7
240	6580 yr	—	100	5.162 5.118	76 24
241	13.2±0.2 yr	Beta 99	2.4×10^{-3}	4.893 4.848	75 25
242	$3.79\cdot10^{5}$ yr	—	100	4.898 4.853	76 24
244	$76\cdot10^{7}$ yr	—	100	4.55 (calc.)	
Beta emitters				*E kev*	(% alpha abundance)
241	13.2±0.2 yr			20.8	
243	4.98 hr			579 490	62 38
245	10.1 hr	Q_β=1.2 Mev (calc.)			
246	10.85 d			330 150	27 73

E.C.=electron capture. Table adapted from Plutonium Handbook. A Guide to the Technology. Vol. 1. Ed. O. J. WICK. Publ. Gordon and Breach, London.

Table 10.2. Physical characteristics of ^{239}Pu, ^{238}Pu and ^{237} Pu

Pu Isotope	Half-life	Mass for 1 μCi of α activity
239	24360 yr	16 μg
238	86.4 yr	58 ng
237	45.6 d	25 ng

pH and a colloidal or complexed form will inevitably be presented to the tissues. Even if great care is taken to inject into the blood stream either a monomeric or a polymeric solution of known particle size, as in the experiments of LINDENBAUM and his colleagues (LINDENBAUM et al., 1968), the solution is likely to show polydyspersity (DOBSON et al., 1949).

Colloidal solutions are largely retained in the liver owing to the avidity with which colloidal particles are removed by reticulo-endothelial cells while monomeric solutions are retained to a greater extent in the skeleton (SCHUBERT et al., 1961; MARKLEY et al., 1964). Data reported in the literature for retention and excretion following different routes of entry and different forms of plutonium in different species are shown in Table 10.5–10.7, 10.10, 10.13, 10.16, 10.17. These data are discussed in detail under appropriate headings in the following pages.

The most extensive studies of plutonium metabolism have been made following intravenous injection of plutonium solutions. This clearly represents a rather simpler picture than that which would occur following the more usual routes of entry of plutonium into the body, i.e. by inhalation or wounding, when the plutonium will only reach the blood stream over a period of time after translocation from soft tissue deposits. Experiments using intravenous injection are therefore discussed before those employing the more usual accidental routes of entry, since the patterns of plutonium retention and excretion are not then complicated by the problems involved in translocation from different tissue sites.

The commonly encountered plutonium compounds used in industry are not readily soluble in body fluids and tend to move slowly from the site of entry to the blood stream or the lymphatics and thence to the skeleton (DOLPHIN et al., 1964). This retention of the plutonium at the site of entry, quite apart from the problems involved in the form in which the plutonium moves, makes it difficult to study its metabolism in relation to the skeleton. The precise physico chemical forms in which plutonium is translocated and the mechanisms of translocation are still far from clear. They may differ in different species and from different sites in the same species.

Apart from its distinctive metabolic behaviour plutonium owes its toxicity to the fact that it is an alpha emitter. It has been thought appropriate therefore to include a discussion of the problems and techniques of alpha dosimetry in this chapter since it is difficult to assess the value of the data available on radiation dose measurements to the skeleton without realizing both the physical and biological uncertainties involved. Sec. VI is a technical discussion of some of the problems of dosimetry. Sec. VII reviews what is known about the radiation dose received by osteogenic and bone marrow tissues from deposited plutonium.

II. Metabolism of ^{239}Pu with Special Reference to the Skeleton

JANET VAUGHAN

Different aspects of the metabolic behaviour of ^{239}Pu have been studied in *Mice* (SCHUBERT et al., 1961; LINDENBAUM et al., 1968, 1969; ROSENTHAL et al., 1968, 1969; ROSENTHAL and LINDENBAUM, 1969; ROSENTHAL et al., 1968, 1969, 1970; BULDAKOV et al., 1970); *Rats* (CARRITT et al., 1947; MIDDLESWORTH, 1947; KISIELESKI and WOODRUFF, 1947; SCOTT et al., 1948; SCHUBERT et al., 1950; WHITE and SCHUBERT, 1952; HAMILTON and SCOTT, 1953; LISCO and KISIELESKI, 1953; WEEKS and OAKLEY, 1953, 1954a, b, 1955; KATZ and WEEKS, 1954; FOREMAN et al., 1955; OAKLEY and THOMPSON, 1956; WEEKS et al., 1956; BALLOU and OAKLEY, 1957; ROSENTHAL and SCHUBERT, 1957; FRIED et al., 1959; TAYLOR et al., 1961; TAYLOR, 1962; MARKLEY et al., 1964; SMITH, 1964; KHODYREVA, 1965; BALLOU et al., 1967; TAYLOR, 1967; TURNER, 1967; BULDAKOV et al., 1970; STUART, 1970); *Rabbits* (LISCO and KISIELESKI, 1953; TAYLOR, 1969; BULDAKOV et al., 1970); *Pigs* (Weeks et al., 1956; Smith et al., 1961; Bustad et al., 1962; Cable et al., 1962; Buldakov et al., 1967; Buldakov et al., 1970); *Dogs* (Painter et al., 1946; Langham, 1959; Stover et al., 1959, 1962; Bair et al., 1962; Rysina and Erokhin, 1963; Morrow et al., 1967); *Sheep* (Buldakov et al., 1970); *Man* (Russell and Nickson, 1946; Crowley et al., 1946; Langham et al., 1950; Langham, 1959; Langham et al., 1962; Buldakov et al., 1970). The data on man are scanty and far from complete. Some information on excretion patterns and retention is available from a study of personnel who have been continually exposed to plutonium, largely by inhalation, or following acute industrial accidents (Foreman et al., 1954, 1960; Langham, 1957, 1959; Healy, 1957; Wilson, 1958; Hammond, 1959; Norwood et al., 1959; Wilson, 1959; Wilson and Silker, 1960; Norwood, 1960; 1962a, b, c; Sanders, 1961; Langham et al., 1962; Lister et al., 1963; Newton et al., 1967; Lagerquist et al., 1965, 1968; Larson et al., 1968; Schofield, 1969). Many of these reports of industrial accidents are largely concerned with the possible effects of agents that might serve to remove plutonium. They are far from complete since data on faecal excretion is rarely included (see Sec. II.H). Unfortunately also many of them are only described in special reports and are not available in the published literature. There are also some more recent studies of non-occupational exposure (Magno et al., 1969).

A. Species Differences

There are interesting and at present unexplained differences in the way both metallic plutonium and supposedly identical plutonium solutions are handled by different species. These differences must be kept in mind in examining Tables 10.5–10.7, 10.10, 10.13, 10.16 and 10.17. Factors that may be responsible for some of these differences are 1) differences in the proteins and other ligands available in the tissues of different species to bind plutonium presented in the blood stream; 2) differences in the degree of binding effected by different ligands; 3) differences in the character of binding to intracellular organelles; 4) differences in the tissue fluids that solubilize the plutonium and result in the complexes that translocate to the blood stream. Some of the recognised species differences in the handling of plutonium are discussed below. They indicate that it is unwise to extrapolate results from animal experiments to man except in broad terms. The results on the small rodents are particularly unsatisfactory as a pattern for human metabolism.

The first controlled experiment in which plutonium (given as ^{238}Pu) was administered to both rats and man by Hamilton and his colleagues in 1946 showed that retention in the skeleton of man was very much greater than in the skeleton of the rat (Crowley et al., 1946). In man only 5 percent of the injected dose was lost at the end of 296 days. In the rat the loss at the same time interval was 50 percent. In man excretion was largely by the urine, in the rat in the faeces. Later experiments using ^{239}Pu confirm these findings. ^{239}Pu elimination rate in the dog is comparable to that of man (Stover et al., 1959, 1962) and there are less complete observations suggesting that ^{239}Pu metabolism in the pig is approximately similar to that in the dog (Smith et al., 1961; Bustad et al., 1962) though Buldakov and his colleagues (1970) consider that plutonium is rapidly lost from the pig's skeleton. This may be due to the fact that they studied growing, rather than adult, pigs.

In Table 10.3 are shown figures for tissue retention of plutonium citrate given intravenously to beagle dogs and man showing that agreement for two species was quite good up to five months after injection (Langham et al., 1962).

Table 10.3. Distribution of intravenously administered plutonium in major tissues of the beagle and man approximately 5 months after injection (taken from LANGHAM et al., 1962, by courtesy of authors and publishers)

Tissue	Percent of injected dose Retained per tissue or organ	
	Beagle	Man
Skeleton	60	66
Liver	21	23
Spleen	0.2	0.4
Kidneys	0.2	0.4

There are even differences in metabolic behaviour between the small rodents. MARKLEY et al. (1964) found that both rats and mice deposited about one third of the monomeric plutonium and about two thirds of the polymeric form in the liver but the rat liver loses the polymeric plutonium faster than the mouse liver. This loss was not increased by DTPA therapy in the rat though it was in the mouse. The authors suggest that this difference may be explained by differences in the partitioning of polymeric plutonium in the liver among the Kupffer cells, the parenchymal cells and extracellular locations. Rats deposit approximately twice as much plutonium in bone as mice, 12 percent versus 6 percent of injected polymeric and about 60 percent versus 24 percent of monomeric plutonium (SCHUBERT et al., 1961). The reason for this difference is again not evident. MARKLEY and her co-workers (1964) suggest it may be due either to differences in the basic mechanisms of plutonium binding in bone or to a great relative increase in bone trabeculation in the rat compared with the mouse since plutonium is concentrated particularly on endosteal surfaces.

Following subcutaneous implantation of plutonium metal, LISCO and KISIELESKI (1953) noted considerable differences in the relative concentrations of plutonium in liver and skeleton in rabbits and rats. The rabbits appeared to absorb more subcutaneous plutonium than rats and to retain much of it in the liver. The rats concentrated, what they did absorb, in the skeleton rather than in the liver with the result that the only osteosarcoma observed occurred in the species with the lower total body burden, i.e. in the rat.

A difference has also been noted between the specific activity of the rat femur and tibia and that of the beagle dog femur and tibia following an intravenous injection of monomeric plutonium, the ratio was approximately one in the rat (MARKLEY et al., 1964) and approximately 2 in the dog (ATHERTON et al., 1958).

B. Age Differences

There is increasing evidence that the critical organ for plutonium may not be the same in the juvenile as the adult. This difference is dependent on many factors. Absorption from the gut is 100 times greater in the neonatal rat and dog than in the adult of either species (BALLOU, 1958; BULDAKOV et al., 1970; SIKOV and MAHLUM, 1972). These differences in the rat are illustrated in Fig. 10.1. Tissue retention following intravenous injection is greater in lambs than in sheep (BULDAKOV et al., 1970). The number of animals used is small but the figures recorded in Table 10.4 are suggestive. The authors emphasize the importance of recognizing that plutonium is more readily fixed initially in the young bone but

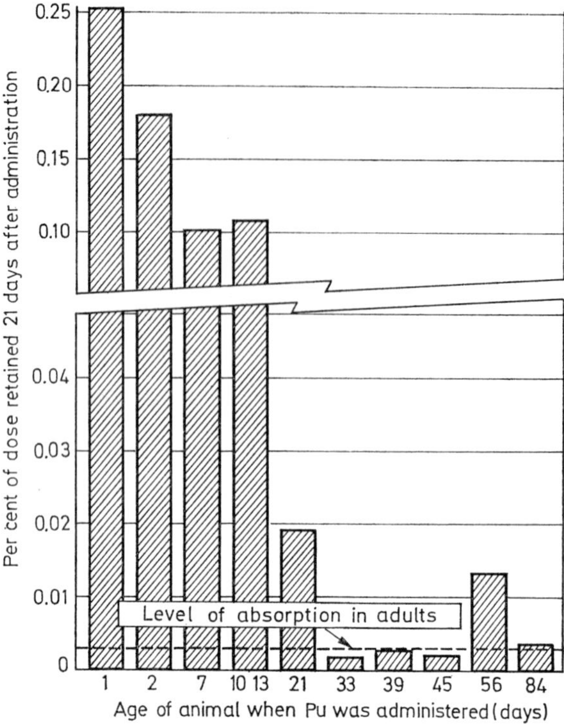

Fig. 10.1. Gastrointestinal absorption of plutonium by rats of different ages. (BALLOU, 1958, by courtesy of author and publishers)

also more rapidly lost from the skeleton. The initial distribution of plutonium from liver to bone following acute administration is more marked in the juvenile rat than in the adult. The size of the individual organ relative to that of the body changes with age, for example skeletal weight is 24 percent of body weight in the newborn rat but only 10 percent in the adult rat (SIKOV and MAHLUM, 1972). Such observations indicate that values obtained in the adult should not be accepted as correct for the young and immature.

Table 10.4. Content of ^{239}Pu citrate as percent of the administered dose in the organs of lambs and sheep at various times after intravenous administration. (Taken from BULDAKOV et al., 1970)

Time after administration of ^{239}Pu (days)	Age of animals[a] (days)	Number of animals	Skeleton	Liver	Spleen	Blood	Sum of internal organs
1	23	2	62.2	8.60	0.06	1.51	74.76
	180	1	36.00	12.85	0.43	12.40	69.35
8	25	2	45.50	9.30	0.06	0.47	66.70
	85	1	30.00	36.00	0.06	—	69.42
32	40	2	34.40	5.48	0.07	0.63	42.45
	210	1	43.40	12.85	—	0.46	58.84

[a] Age <40 days considered "lamb"; >40 days considered "sheep".

C. Routes of Entry with Special Reference to Skeletal Uptake

There are a variety of routes by which plutonium may reach the blood stream and hence the skeleton. Probably the most important from the point of view of radiation hazard is through some form of wound deeper than simple subcutaneous penetration. Translocation from lung and lymph glands as the result of inhalation may also lead to some significant skeletal deposition which may increase slowly with time (MORROW et al., 1967; CLARKE, PARK, PALOTAY, and BAIR, 1966; PARK, BAIR, and BUSCH, 1972).

Transport from the point of entry is thought to occur in a variety of ways: 1) by soluble complexes reaching the blood stream from the tissue fluids; 2) direct vessel injury; 3) by lymphatic drainage (ZNDANOV, 1952, CABLE et al., 1962; BULDAKOV et al, 1970).

1. Intravenous

Some of the available data on the behaviour of intravenously injected ^{239}Pu are shown in Table 10.5. This table is not a complete record of all the experimental data available but serves to illustrate certain points about the effects of the chemical and physical form in which plutonium reaches the blood stream, on its retention and excretion, and variations in different species. Some of the tabulated results, especially those of an early date, must be regarded with caution. Plutonium in solutions suitable for intravenous injection is frequently subject to hydrolysis and change in valence state. This may occur at such rates so that a solution that has stood for several hours before injection may have changed from its state when freshly prepared. The pattern of retention and excretion will then depend in part on the time elapsing between preparation and injection of the solution. Further, the elegant work of LINDENBAUM and his colleagues (SCHUBERT et al., 1961; LINDENBAUM et al., 1968; ROSENTHAL et al., 1968; ROSENTHAL and LINDENBAUM, 1969) illustrated in Table 10.5 has emphasised the extreme importance of the polymeric state and precise particle size of the plutonium in determining both its initial and subsequent distribution. Again this was not appreciated by the earlier workers.

The data in Table 10.5 show at once that in whatever form plutonium reaches the blood stream retention occurs primarily in the skeleton and liver, and that excretion, which occurs in both urine and faeces, though variable in different species, is extremely slow except when plutonium is given as a stable complex with a non metabolizable chelating agent such as TTPA (Sec. IX.C.2.c). In the rat, at least, it is then rapidly excreted, mainly in the urine. Further, when plutonium enters the blood in monomeric, or more strictly minimally polymeric form such as exists for example in citrate or 0.01 M nitric acid solutions filtered through 10 mμ pore diameter filters immediately prior to injection, the liver/skeleton uptake ratio is low, less than 0.4 and frequently less than 0.1. In contrast, when the injected solution contains a large proportion of polymeric material the initial liver/skeleton ratio approached 50 (LINDENBAUM et al., 1968). The results shown in the table from this group (LINDENBAUM et al., 1969) indicate that with a highly polymeric solution of narrow range of particle size, only 3 percent is found in the skeleton and 84 percent in the liver at six days after injection but that with time (350 days) plutonium is released from the liver and some of this apparently translocates to the skeleton. This is of course not the typical pattern as can be seen from Table 10.5. An earlier experiment using a monomeric solution that was 9 percent ultrafilterable gave 84 percent skeletal and 28 percent liver retention at 6 days. These authors also stress the importance of particle size. They found

Table 10.5. Distribution of ^{239}Pu after intravenous injection

Chemical form injected	Animal	Time after administration (days)	Percent dose retained in								Reference
			Skeleton	Liver	Spleen	Lungs	Kidneys	Excreta			
								Urine	Faeces	Total	
$PuCl_3$	Rat	4	45	23	0.7	—	2.2	0.9	17.1	18	CARRITT et al. (1947)
		16	42	7	1.0	—	0.7	1.4	37.6	39	
		32	52	5	0.7		0.5	1.2	33.8	35	
Pu(III) ascorbate	Rat	4	97	7	0.1	0.2	0.5				TAYLOR (1962)
$Pu(NO_3)_4$	Rat	4	29	40	1.2	—	1.4	0.9	15.1	16	CARRITT et al. (1947)
		16	31	26	1.4		0.7	1.7	26	28	
		48	32	22	1.4		0.4	2.2	39.6	42	
$Pu(NO_3)_4$	Rat	0.5	11	59	0.8						VAN MIDDLESWORTH (1947)
		10	26	28	0.9		0.7	1.1			
$Pu(NO_3)_4$ (0.02 N HNO_3)	Rat	4	46	3	0.2	0.1	0.6				TAYLOR (1962)
		21	47	6	0.6		0.5				
		28	55	3	0.3		0.6				
$Pu(NO_3)_4$ pH 1.5 100% filterable through 10 mμ filter	Rat	1	38	15	1.0	0.6	—				TURNER (1967)
pH 1.5 87% filterable through 10 mμ filter		1	43	25	1.2	0.6					
pH 2.0 42% filterable through 10 mμ filter		1	19	24	0.7	0.5					
$Pu(NO_3)_4$ neutralised with Na_2CO_3 just before injection	Rat	4	48	34	0.5	0.3	0.5	6	3	9	TAYLOR et al. (1961)
		21	55	12	0.5	0.2	0.4	7	11	18	
		62	58	9	0.5	0.4	0.4	8	29	37	
$Pu(NO_3)_4$ in 0.01 M HNO_3	Rabbit	1	39	11	0.3	0.4	0.4				TAYLOR (1969)
		8	34	10	0.1	0.1	0.2				
		112	35	18	0.1	0.06	0.2				
		365	22	7	0.01	0.03	0.1				

Table 10.5 (continued)

$Pu(NO_3)_4$ pH 2 four doses at 1 month intervals	Dog	~90	38	31	2.3	0.5	0.4				RYSINA and YEROKHIN (1963)
		~180	41	29	1.7	0.4	0.2				
		~640	39	32	0.5	0.4	0.2				
		~1650	29	30	0.4	0.2	0.1				
$Pu(NO_3)_4$ in 0.14 N HNO_3	Dog	30	6	80	7.5	0.3	0.1	0.46	0.74	1.2	BAIR et al. (1962)
$PuO_2(NO_3)_2$	Rat	2	56	8	0.2	—	3.1	10.7	2.6	13	VAN MIDDLESWORTH (1947)
$PuO_2(NO_3)_2$	Rat	4	56	9	0.5	—	1.9	7.9	5.7	13	CARRITT et al. (1947)
		16	58	3	0.5	—	1.1	9.2	12.9	22	
		32	57	3	0.3		0.7	9.3	20.5	30	
Pu(IV) Citrate	Mouse										LINDENBAUM et al. (1968)
<1% ultrafilterable (graded)[a]		5	16	80	1.9						
30% ultrafilterable (graded)[a]		5	68	28	0.6						
100% ultrafilterable (graded)[a]		5	65	22	0.08						
18% ultrafilterable (ungraded)[b]		5	25	80	2.1						
30% ultrafilterable (ungraded)[b]		5	39	47	1.8						
90% ultrafilterable (ungraded)[b]		5	85	28	0.2						
Pu (IV) Citrate in 1% Na citrate *monomeric* 93% ultrafilterable pH 6.2	Mouse	1	25	36				10.1	5.6	16	ROSENTHAL and LINDENBAUM 1967
		6	30	23				10.8	13.4	24	
		15	31	11.9				11.1	28.2	39	
		90	30								
		300	21								
Pu (IV) citrate in 0.02% Na citrate ungraded polymeric 14% ultrafilterable, pH~6	Mouse	5	20	39	1.5			10.3	3.8	14.1	BROWN et al. (1970)
		15	23	34	1.7			11.3	8.2	19.5	
		50	21	22	1.5			12.2	14.6	26.8	
Pu (IV) citrate in 1% Na citrate *monomeric* 92% ultrafilterable pH 5.4	Rat	3	53	31	0.3			1.5	4.6	6	MARKLEY et al. (1964)
		6	59	19					11.7		
		12	61	12				2.1	22.3	24	
Pu(IV) citrate in 0.01% Na citrate ungraded *polymeric* 15% ultrafilterable, pH 5.4	Rat	3	12	74	1.7			0.6	1.2	2	
		6	13	78					2.8		
		12	14	58				0.8	5.4	6	

Table 10.5 (continued)

Chemical form injected	Animal	Time after adminis-tration (days)	Percent dose retained in								Reference
			Skele-ton	Liver	Spleen	Lungs	Kid-neys	Excreta			
								Urine	Faeces	Total	
Pu(IV) citrate Monomeric 9% ultrafilterable	Mouse	6	84	28	0.2						
		90	56	2.5	0.3						
Polymeric 30% ultrafilterable	Mouse	6	49	46	1.7						ROSENTHAL et al. (1968)
		90	44	25	1.3						
Polymeric 18% ultrafilterable	Mouse	6	23	80	2.1						
		90	21	53	2.6						
Monomeric <0.01 μM particulate diameter	Mouse	6	45								LINDENBAUM et al. (1969)
polymeric graded narrow particle size	Mouse	6	3	84	6						
		350	15	56	6						
Pu(IV) citrate	Rat	4	70	10	0.3	2.1	0.5				TAYLOR (1962)
		7	74	2	0.1		0.3				
Pu(IV) citrate	Rat	8	60	174							BELYAEV (1962)
Pu(IV) citrate (filtered through 10 mμ filter)	Rat	7	80	2	0.1	0.4	—	—			TAYLOR (unpublished)
Pu(IV) citrate	Rat	4	57	9	0.7	—	1.6	1.3	10.9	12	CARRITT et al. (1947)
		16	60	4	0.6	—	0.8	2.1	20.3	22	
		48	60	3	0.6	—	0.4	2.8	29.7	32	
		1	54	21	0.3	—	1.4	0.9	1.2	2	ROSENTHAL and SCHUBERT (1957)
		1	45	24	—	—	—				SCHUBERT et al. (1950)
		1	42	34	0.6	—	2.2				WHITE and SCHUBERT (1952)
		8	50	10	0.2			3.5	11.7	15	
		15	55	9	0.6		1.4	6.2	20.9	27	HAMILTON and SCOTT (1953)

Table 10.5 (continued)

	Rat	16	56	10	—	—	—	2.4	26.3	29	Foreman et al. (1955)
		27	64	8	0.4		0.4				Fried et al. (1959)
		60	20				1.4	2.5	6.5	9	Smith (1964)
	Dog	29	62	22	0.3		0.2	3.9[c]	8.2[c]	12[c]	Stover et al. (1959), (1962)
		44	66	22	0.2		0.2				
		365	53	19	0.2		0.2				
		610	55	18	0.1		0.1				
		1324	44	13							
	Pig	7	57	14	0.2		0.2	13	15	3	Smith et al. (1961)
		480–690	34	32	—	—	—				Bustad et al. (1962)
	Growing pig	1	69	24	0.10	3.0	0.2				Buldakov (1968)
		330	23	23	0.15	0.2	0.08				
	Sheep	1	36	13	0.43	1.78	0.5				Buldakov et al. (1970)
		128	20	7	0.27	0.26	0.3				
	Man	150	66	23	0.4	1.0	0.4			5	Langham (1959)[d]
Ammonium plutonyl pentacarbonate	Rat	7	8	83							Belyaev and Lemberg (1964)
Pu-Desferrioxamine	Rat	4	6	1	0.15		1.4				Taylor (unpublished)
Pu-Diethylenetriamine-Pentaacetate (DTPA)		4	1	0.07	0.02		0.2				
Pu-Triethylenetetramine-Hexaacetic Acid (TTPA)		7	4	3	0.2	0.2	0.2				

[a] graded solutions contain a narrow range of particle size. [b] ungraded solutions contain a large variety of particle sizes. [c] Excreta at 22 days. [d] Mean value for 12 cases given $^{239}Pu_4$ citrate, 3 patients given hexavalent ^{239}Pu citrate (Russell and Nickson, 1946) and one patient given ^{238}Pu nitrate (Crowley et al., 1946).

that injection of graded solution, i.e. comprising a narrow range of particle size, containing 30 percent ultrafilterable plutonium from which both larger and smaller particles had been removed, resulted in a significantly different organ distribution of plutonium as compared with previously used ungraded solutions, i.e. containing a variety of particle sizes, of the same ultrafilterability, as shown in the table. In the case of the graded solution 68 percent of the retained dose was retained in the skeleton, and of the ungraded solution 39 percent was retained in the skeleton though both solutions were 30 percent ultrafilterable (LINDENBAUM et al., 1968). They conclude that the pattern of deposition in mouse tissues appears to depend on aggregation of plutonium-containing particulates to some critical particle size and conclude that this particle size is below 0.5 μ and well above 0.01 μ in diameter.

In the mouse, the factors influencing the relative retention of plutonium in skeleton and liver following intravenous injection are the proportion of colloidal plutonium and the particle size of the colloidal particles in the solution injected. Unfortunately no exactly comparable information is at present available on any animal other than CFI mice[1]. It would be of very great interest to know what would happen in dogs, pigs and man with solutions known to be extremely monomeric or extremely polymeric of defined particle size.

The classic report of the tissue distribution of an intravenous injection of plutonium in man is that made by LANGHAM and his colleagues in 1950 (LANGHAM et al., 1950) on 16 patients. This included 12 patients with short life-expectancy to whom they had given ^{239}Pu citrate intravenously, 3 patients previously reported by RUSSELL and NICKSON in 1946 who had received hexavalent ^{239}Pu citrate in 0.9 percent salt solution at pH 7 and a patient with gastric carcinoma already reported by CROWLEY and his colleagues (CROWLEY et al., 1946) who had received ^{238}Pu nitrate. The mean figures for these 16 patients as shown in Table 10.5 gave the highest retention in the skeleton, the figures being very similar to those for pigs and dogs also given a citrate solution.

These classical data on the results of intravenous injection in man have recently been reviewed by DURBIN (DURBIN, 1971). Her analysis makes little difference to the figures for total retention but results in a significant difference in organ retention. She estimates, that, in healthy humans, bone and liver will retain equal amounts of plutonium 15 years after exposure and that at later intervals the liver may gain further at the expense of the skeleton. This new assessment of the old data is dependent on a variety of factors, the most important of which is that allowances have been made for the fact that the patients investigated were extremely ill. Abnormalities in their circulatory system, in the capacity of their plasma proteins to bind plutonium, in their liver function and their dietary intake have been taken into account in making these new calculations. The same conclusion has been reached by a recent ICRP Task Group Report (International Commission on Radiological Protection Report, 1972) and by MAYS and his colleagues reviewing both human and animal data (MAYS et al., 1970).

2. Gastrointestinal

Early experiments on the gastrointestinal absorption of plutonium are somewhat confused as the authors do not always discriminate clearly between ^{238}Pu and ^{239}Pu (CARRITT et al., 1947; HAMILTON, 1947; SCOTT et al., 1948; KATZ and WEEKS, 1954; WEEKS et al., 1956). Where possible a distinction has been made in Table 10.6 though there appears to be no significant difference in the gastric ab-

---

1 See footnote page 455.

sorption of the two isotopes. WEEKS and his colleagues (1956) report that in chronic feeding experiments the concentration of ^{239}Pu in the solution fed had no effect on the fraction absorbed in rats over the range 10^{-5} μg/ml to 1 μg/ml though increased gastrointestinal absorption and therefore skeleton retention may occur when plutonium is ingested in an acidic solution or in the hexavalent state. The average absorption from a $Pu(NO_3)_4$ solution at pH 2 was 0.0028%. The presence of citrate (CARRITT et al., 1947) or TTHA (TAYLOR, 1970) also favours absorption in rats. The apparent difference in the absorption of $Pu(NO_3)_4$ in pigs recorded by WEEKS et al. (1956) and BUSTAD et al. (1962) shown in Table 10.6 is not thought by the workers concerned to be significant. It is attributed to the difficulties of measuring the very small amounts of plutonium involved. The retention in the skeleton in rats was approximately 90 percent of that absorbed (WEEKS et al., 1956) and did not change greatly between 4 and 8 days after administration. CARRITT and his fellow workers report that 0.3 percent of the ingested plutonium is absorbed when given in 5 percent sodium citrate solution to rats and that 74 percent of that absorbed was deposited in the skeleton and 7.5 percent in the liver. BALLOU (1958) emphasises the importance of age. He found that when $Pu(NO_3)_4$ pH 2 was fed to rats of varying age the percentage absorption was 0.25 percent in 1 day old rats; 0.1 percent at 7 days of age and 0.02 percent at 21 days. The detailed results are shown in Fig. 10.1. This higher absorption from the gut of young animals is also emphasised by BULDAKOV et al. (1970). As shown in Table 10.6 puppies had a higher skeletal retention than adult dogs. Absorption of plutonium applied directly to the tongue or fed in drinking water did not differ significantly from absorption following gavage feeding. Fasting, on the other hand, has been reported to increase the retention of ingested plutonium (FOREMAN et al., 1957).

In Fig. 10.2 are shown the results obtained by MOSKALEV and his colleagues (MOSKALEV et al., 1969) on tissue retention following continuous feeding of ^{239}Pu to rats from 3 months old till death. The ^{239}Pu was given daily, 0.5 μCi in citrate solution. The amount of ^{239}Pu in the skeleton was always 2–3 times higher than in the liver. The total amount (in percent of daily intake) in the skeleton, liver and kidneys after t days of intake is described by the following equations

$$\begin{aligned} R_t \text{ skeleton} &= 0.0478\ t^{0.878} \\ R_t \text{ liver} &= 0.0129\ t^{0.801} \\ R_t \text{ kidneys} &= 0.00245\ t^{0.835}. \end{aligned}$$

It is significant that 5 out of 68 rats developed osteosarcoma with an average bone dose of 57 rads calculated as a function of ^{239}Pu build up in the bones with time, representing the dose which would have resulted if the ^{239}Pu had been uniformly distributed throughout the bone.

It is clear from Table 10.6 that whatever the amount of plutonium absorbed most of it is retained in the skeleton though the liver/skeleton ratios showed considerable variation ranging from 0.025 (TAYLOR, 1962) to 0.46 (WEEKS et al., 1956) depending largely on the character of the plutonium solution. The length of time the plutonium was fed appears to be less important (WEEKS et al., 1956).

The *International Commission on Radiological Protection* (1960) has assumed that absorption from the gastrointestinal tract is 0.003 percent of the ingested dose. Such a figure may be reasonable for a contaminated water supply, in view of the fact that pigs absorb at most 0.0022 percent of pH 2 $Pu(NO_3)_4$ and rats reach the same following acute administration, but it is clear from Table 10.6 that hexavalent ^{239}Pu, ^{239}Pu in acid solutions or complexed with citrate are more readily absorbed than $Pu(NO_3)_4$ at a more physiological pH.

Table 10.6. Absorption and retention of ^{239}Pu in the rat, pig and dog after intragastric administration of various types of solutions

Chemical form ingested	Animal	Period studied (days)	Percent administered dose				Reference
			Bone	Liver	Soft tissue	Total	
$Pu(NO_3)_4$ pH 2	Rat	280	0.0025	0.0003		0.0028	KATZ and WEEKS (1954)[a]
Pu nitrate 97% Pu(IV) pH 4	Rat	4[b]	0.0011	0.0005	0.0001	0.0017	WEEKS et al. (1956)[b]
68% Pu(IV) pH 1	Rat	4[b]	0.17	0.09	0.015	0.275	WEEKS et al. (1956)
90% pH 2 Pu(III) 10% Pu(IV)	Rat	4[b]	0.0041	0.0016	0.0006	0.006	WEEKS et al. (1956)
100% Pu(VI) pH 1	Rat	4[b]	0.94	0.81	0.12	1.87	WEEKS et al. (1956)
7% Pu(III) 93% Pu(IV) pH 2	Rat	4[b]	0.0027	0.0015	0.0008	0.005	WEEKS et al. (1956)
96% Pu(IV) pH 2	Rat	4[b]	0.0007	0.0005	0.0001	0.0013	WEEKS et al. (1956)
Pu citrate 99% Pu(IV) pH 2	Rat	4[b]	0.017	0.009	0.03	0.029	WEEKS et al. (1956)
96% Pu(IV) 4% Pu(VI) pH 2	Rat	4[b]	0.18	0.084	0.023	0.29	WEEKS et al. (1956)
85% Pu(IV) 15% Pu(VI) pH 2	Rat	4[b]	0.26	0.14	0.012	0.41	WEEKS et al. (1956)
$Pu(NO_3)_4$ in 0.05 M HNO_3	Rat	6				0.07	TAYLOR unpublished
$Pu(NO_3)_4$ in 0.05 M HNO_3 +1 millimole TTHA orally 30 min after Pu	Rat	6				0.5	
$Pu(NO_3)_4$	Rat	10				0.01	CARRITT et al. (1947)
Pu(IV) citrate 5% citrate	Rat	10				0.30	
$PuCl_3$ in each of three valence states						0.007	SCOTT et al. (1948)
$Pu(NO_3)_4$ pH 2	Pig	9				0.0022	WEEKS et al. (1956)
$Pu(NO_3)_4$	Pig	664–694	0.0003	0.0002		0.0005	BUSTAD et al. (1962)
Pu(IV) citrate pH 6.5	Dog (adult)	300	0.0188	0.0082			BULDAKOV et al. (1970)
Pu(IV) citrate pH 6.5	Dog (3 days)	—	0.365	0.339			BULDAKOV et al. (1970)
Pu(IV) nitrate pH 2	Dog (3 days)	—	0.094	0.045			BULDAKOV et al. (1970)
Pu(IV) citrate	Pig	1	0.161	0.012			BULDAKOV (1968)

[a] KATZ and WEEKS (1954) – state one group only given ^{238}Pu, others ^{239}Pu. [b] The figures for retention after 80 days were not greatly different from the results at four days.

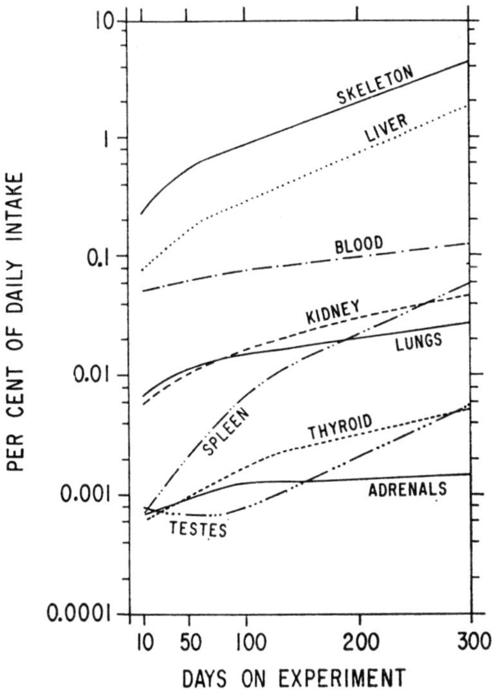

Fig. 10.2. Accumulation of ^{239}Pu in rat organs following continuous oral intake. (Moskalev et al., 1969, by courtesy of authors and publishers)

The significance of gastrointestinal absorption and skeletal retention cannot be ignored in considering radiation hazards since it is known that some of the ^{239}Pu that is inhaled reaches the gastrointestinal tract immediately, while continuous ingestion is known to have produced osteosarcoma in rats (Moskalev et al., 1969).

3. Unbroken Skin

It is disputed as to whether plutonium may gain access through the unbroken skin presumably by some process of diffusion. In some cases where it has been said to have occurred it may well be that there has been slight damage to the superficial layer of the skin perhaps by the acid in which the solution was made up or, as suggested by Khodyreva (1965), by a direct radiation effect. Such damage may well allow access to the underlying tissue. Reported figures for absorption through the unbroken skin of rats, rabbits, pigs and man are shown in Table 10.7. Early reports are confused as to whether the amount of plutonium absorbed is affected by the area of skin exposed or the concentration of plutonium applied (Weeks and Oakley, 1953, 1954a, 1955; Kornberg, 1954; Langham, 1959), but, in his authoritative review in 1967, Thompson concludes, that the area of skin exposed and the mass of plutonium applied, both affect the amount absorbed. Up to 9 percent of the plutonium in HNO_3 solutions is quickly absorbed if no immediate steps are taken to remove the contamination (Weeks et al., 1953). All workers are clear that increased acidity of the solution increases absorption (Weeks and Oakley, 1953, 1954; Langham, 1959; Thompson, 1967; Buldakov et al., 1970). Over a five day period about 0.3 percent of the plutonium applied

Table 10.7. Absorption and distribution of ^{239}Pu through unbroken skin

Chemical form	Animal	Time after application	Percent of Absorbed dose		Total absorbed % of applied	Reference
			Skeleton	Liver		
?	Pig		80	18		BULDAKOV et al. (1967)
$Pu(NO_3)_4$ in 10 N HNO_3	Rat	5 d	—	—	2.0	WEEKS and OAKLEY (1953)
$Pu(NO_3)_4$ in 0.1 N HNO_3	Rat	5 d	—	—	0.3	WEEKS and Oakley (1954)
Pu TBP complex	Rat	15 min	—	—	0.04	BALLOU and OAKLEY (1956)
$Pu(NO_3)_4$ in 10 N HNO_3	Rat	1 h	—	—	0.05	OAKLEY and THOMAS (1955)
$Pu(NO_3)_4$	Rabbit	14 d	77.5	7.5	0.15	KHODYREVA (1965)
Pu citrate pH 6.5	Piglets	3 h	21	4,6	0.105	BULDAKOV et al. (1967)
		6 h	45	2.4	0.185	
		1 d	44	10.0	0.261	
		3 d	72	7.5	0.187	
		6 d	68	8.7	0.347	
		10 d	72	5.2	0.254	
Pu in 9% HCl, EDTA and detergent	Man	?	—	—	0.01	LISTER et al. (1963)
Pu as TBP complex in CCl_4	Man	15 min	—	—	0.04	WILSON (1956)
$Pu(NO_3)_4$ in 0.4 N HNO_3	Man	1 h	—	—	0.002	LANGHAM (1959)

in 0.1 N nitric acid was absorbed, when the nitric acid was increased to 10 N the absorption was approximately 2 percent. The absorption of Pu-tributyl-phosphate (TBP) complex in carbon tetrachloride was also greatly increased (BALLOU and OAKLEY, 1957). Autoradiographic studies showed some evidence of plutonium migration down hair follicles (WEEKS and OAKLEY, 1955) and in the sweat and sebacious glands (BULDAKOV et al., 1970).

LANGHAM and his colleagues (LANGHAM et al., 1962) report that PuO_2 in fine particulate form applied to the skin of rabbits which had been rubbed with sand paper producing abrasions, did not translocate but was lost in the eschar which became detached. KHODYREVA (1965) applied ^{239}Pu nitrate pH 3.0 to an area of 50 cm^2 shaved skin of rabbits at a level of 0.4 or 4 $\mu Ci/cm^2$. Within three hours single alpha tracks could be seen in autoradiographs prepared from the blood. In blood studied 120 hours after application stars, indicating the presence of aggregates of plutonium, were found in the autoradiographs. The specific activity of the blood increased up to six days after the skin application to reach a maximum 2 to 3 times the initial value and then fell; as shown in Table 10.8 the level reached was related to the amount of ^{239}Pu in the solution applied. KHODYREVA attributes the rising blood level to skin damage from the radiation increasing the permeability. On the 14th day after application 77.5 percent of the dose absorbed was

Table 10.8. Radioactivity of the blood in rabbits after application of $^{239}Pu(NO_3)_4$ to unbroken skin. (Taken from KHODYREVA, 1965)

Time of taking blood after application	Specific radioactivity of blood $\times 10^{-8}$ μc/ml	
	Animals of group 1 (0.4 μc/cm²)	Animals of group 2 (4 μc/cm²)
1 h	3.37 ± 0.38	5.07 ± 0.54
6 h	6.1 ± 0.24	7.62 ± 0.18
1 d	$6.2 + 0.69$	7.93 ± 0.13
3 d	7.73 ± 0.81	9.27 ± 1.05
5 d	7.94 ± 0.89	13.3 ± 2.38
7 d	6.19 ± 1.62	7.41 ± 1.3
10 d	3.97 ± 0.43	3.74 ± 0.7

Table 10.9. Content of ^{239}Pu in organs and tissues of rabbits following a single application to the skin of $Pu(NO_3)_4$. (Taken from KHODYREVA, 1965)

Group of animals	Dose of ^{239}Pu applied to skin (in μc/cm²)	Radioactivity of 1 g of organ (in μc $\times 10^{-4}$			
		Bone	Liver	Spleen	Skin from experimental area
1	0.4	11.82	0.8	0.07	325.2
2	4.0	21.7	3.37	7.35	4340.83

in the skeleton and 7.5 percent in the liver. Of the applied dose 0.15 percent was in the body. These results are illustrated in Table 10.9. A rather higher value of absorption was obtained by her in rats. In the pig corresponding values for skeleton and liver were 80 and 18 percent respectively (BULDAKOV et al., 1967). The low liver/skeletal ratio suggests that the plutonium was absorbed in an essentially monomeric form, though the presence of stars in the blood autoradiographs indicates either that aggregates were present or that there was some form of cellular transport.

The skin of the rat and rabbit differs, however, from that of man. An experiment in man in which 10 μg of plutonium as $Pu(NO_3)_4$ in 0.4 N nitric acid was applied to the human palm gave an absorption rate of approximately 0.0002 percent per hour (LANGHAM, 1959). Some further information in man has been gained from industrial accidents. In one accident a man's hand was immersed for several minutes in a solution of plutonium in a TBP-CCl_4 solution. The total absorption was estimated from urine analysis alone, which may be unsatisfactory, (this Chap., Sec. II.H), to be 2×10^{-5} percent of the activity initially present on the hand (WILSON, 1956). Two other cases of supposedly unbroken human skin contamination have been studied. One man's hand was contaminated by a solution of plutonium in 9 percent HCl containing EDTA and a strong detergent. The other man cut his glove when handling a solution of mixed plutonium isotopes in aqua regia and nitric acid and his finger was not properly cleaned for $1^1/_2$ hours. The absorption based on both urine and faecal excretion was estimated to be about 10^{-2} percent (this chap., Sec. II.H). It is possible of course that the acid injured the skin rendering absorption easy (LISTER et al., 1963). The covering of contaminated skin with ointment or air tight dressings is also known to increase absorption (WEEKS and OAKLEY, 1955).

Table 10.10. Distribution of ^{239}Pu after intradermal or subcutaneous injection

Chemical form injected	Animal	Route of administration	Time after administration	Percent of administered dose retained in						Reference
				Skeleton	Liver	Spleen	Lungs	Kidneys	Excreta	
$Pu(NO_3)_4$ in 2N HNO_3	Pig	Intradermal	1 day	0.5	1.6	0.06	—	0.02	—	CABLE et al. (1962)
			7 d	3.1	4.1	0.25	—	0.05	—	
Ammonium plutonyl pentacarbonate pH 8	Piglet	?Subcutaneous	3 h	14.2	1.48	0.03				BULDAKOV et al. (1967)
			6 h	19.1	3.59	0.13				
			1 d	47.5	8.48	0.03				
			6 d	69.0	4.85	0.04				
			16 d	50.1	4.05	0.02				
			64 d	46.2	11.6	0.06				
			128 d	48.8	21.6	0.12				
Pu citrate pH 6.5	Piglet		3 h	25.8	4.47	0.14				BULDAKOV et al. (1967)
			6 h	22.7	6.9	0.05				
			1 d	47.3	5.5	0.05				
			6 d	40.7	8.5	0.06				
			16 d	39.6	4.9	0.05				
			64 d	24.8	6.4	0.06				
			128 d	25.0	15.3	0.07				
$Pu(NO_3)_4$ pH 2	Rat	Subcutaneous	6 d	29	1.9	0.11	0.13	0.3	43[a]	TAYLOR (unpublished)
Pu glutamate pH 6.0 (ultrafiltered)	Rat	Subcutaneous	6 d	31	2.9	0.09	0.23	0.9	55[a]	
Pu aspartate pH 6.6 (ultrafiltered)	Rat	Subcutaneous	6 d	32	1.6	0.17	0.11	0.17	44[a]	
Pu-Desferrioxamine pH 6	Rat	Subcutaneous	6 d	8	1.1	0.03	0.05	0.8	84[a]	
Pu-DTPA pH 6.0	Rat	Subcutaneous	6 d	2	0.1	0.01	0.09	0.1	97[a]	
Pu metal (0.6–1.8 mg)	Rat	Subcutaneous	see table 10.2							LISCO and KISIELESKI (1953)
	Rabbit	Subcutaneous	see table 10.2							

[a] Assumed from total body retention.

4. Intradermal

The results of intradermal injection are shown in Table 10.10. Intradermal injection in pigs of 0.04, 0.2, 1 and 5 μCi of plutonium nitrate in 0.01 ml of 0.2 N HNO_3 gives rise to deep ulceration (CABLE et al., 1962). Thirty to approximately ninety percent of the ^{239}Pu at the injection site was lost soon after ulcer formation, being sloughed off or later lost in the scab (HORSTMAN et al., 1961). Slightly less than 5 percent of the administered plutonium appeared to have translocated to the skeleton 168 hours after injection in the case of the animals given 1 μCi and rather less at the same time interval in those given 5 μCi. There was, however, an appreciable retention in the lymph nodes, increasing from 2 percent at 1 day to 11 percent at 7 days. This suggests that the plutonium is primarily translocated from the site of skin deposition via the lymph rather than the blood. Figures for retention in bones, liver and lymph nodes at the injection site are shown in Table 10.11. Since human skin is reported to be similar to pig skin these observations are important (MORITZ and HENRIQUES, 1947).

5. Subcutaneous

Early experiments (OAKLEY and THOMPSON, 1956) indicated that absorption increased with the depth of the subcutaneous cut. Liver deposition was about 50 times as great in the case of a subcutaneous cut as when the skin was intact. TAYLOR, as reference to Table 10.10 shows, has found considerable translocation to the skeleton from subcutaneous injection of $^{239}Pu(NO_3)_4$ at pH 2, ^{239}Pu glutamate at pH 6 and ^{239}Pu aspartate at pH 6.6, these solutions being largely monomeric; 5 days after injection about 30 percent of the injected dose was in the skeleton (TAYLOR, 1970). BULDAKOV et al., (1970) report even greater translocation to the skeleton following subcutaneous administration of ammonium plutonyl-pentacarbonate or plutonium citrate as shown in Table 10.10. In piglets as much as 50 percent of the administered dose had reached the skeleton 16 days later.

LISCO and KISIELESKI (1953) found some translocation from plutonium metal implanted subcutaneously into both rats and rabbits. The distribution in skeleton and liver of both animals is shown in Table 10.12. With amounts of plutonium expressed in percent of the total absorbed dose the following values were found: rabbit liver 55 percent and skeleton 44 percent, rat liver 6 percent and skeleton 85 percent. The concentration of ^{239}Pu in μCi/g wet weight in the liver of the rabbits exceeded that of the bones by a factor of almost six, whereas the reverse was the case in the rats in which the concentration in the bones was five times greater than that in the liver. On the same basis the skeleton of the rat contained on an average almost ten times more plutonium than that of the rabbit. It is therefore not surprising that the only bone tumour that occurred was an osteogenic sarcoma of the spine in the rat with the highest burden. The authors suggest that the different behaviour of the metal in the two species may be dependent upon differences in the character of the tissue fluids which result in ^{239}Pu reaching the blood stream in different forms.

Table 10.11. Distribution of plutonium 1 and 7 days after intradermal injection $^{239}Pu(NO_3)_4$ in pigs expressed as per cent of injected dose. (Taken from CABLE et al., 1962)

	Pig 1 (1 μc 1 day)	Pig 2 (1 μc 7 days)	Pig 3 (5 μc 1 day)	Pig 4 (5 μc 7 days)
Bones	0.9	4.8	0.2	1.5
Liver	2.3	6.6	0.9	1.6
Injection site	75.0	40.0	84.0	55.0
Lymph nodes	4.5	12.0	2.2	8.2

Table 10.12. Organ distribution of plutonium following subcutaneous implantation of the metal in rabbits and rats. (Taken from LISCO and KISIELESKI, 1955)

	Survival days	Skeleton		Liver	
		percent absorbed dose per organ	μg/gm	Percent absorbed dose per organ	μg/gm
Rabbits	260	37.06	0.002	62.58	0.010
	340	40.09	0.009	59.55	0.051
	516	62.31	0.003	36.74	0.008
	623	30.32	0.005	68.99	0.051
	710	38.77	0.018	59.47	0.125
	738	52.63	0.007	45.83	0.027
	1048	95.45	0.007	2.54	0.003
Rats	356	77.09	0.109	9.25	0.027
	484	78.39	0.044	1.80	0.004
	580	98.57	0.028	5.71	0.005

6. Intramuscular or Wound

Intramuscular injection of ^{239}Pu has been used experimentally to simulate what may happen when ^{239}Pu is introduced into the body by an accidental wound. The result of wound contamination will depend on the physicochemical form of the ^{239}Pu involved but also upon the nature of the wound, i.e. whether it is a tiny puncture or a major injury giving access to large vessels. Information about such contamination comes from animal experiment and the follow up of human accidents. Release from soft tissue deposition following a puncture wound and inhalation are probably the most likely routes by which appreciable amounts of plutonium will reach the blood stream, or possibly the lymphatics, and hence the skeleton (LANGHAM, 1959; LANGHAM et al., 1962). There is also a form of secondary release from tissues such as liver which have initially taken up large quantities from the blood stream and then released it over varying periods of time (BULDAKOV et al., 1970). This is known to occur in mice following the injection of a highly polymeric plutonium of known particle size (LINDENBAUM et al., 1969).

Plutonium deposited in a wound is likely to be in the form of an insoluble "mass" either because the original material was itself insoluble or became insoluble as a result of hydrolytic reactions occurring at the site of deposition. Part of this "mass" may be expected to react with constituents of the tissues and body fluids to form soluble complexes which will reach the blood stream and like other monomeric plutonium complexes will deposit largely in the skeleton. The rate and degree of solubilization will of course be variable. In addition, particles may break off from the original "mass", some of which may reach the blood stream from which they will be removed by the liver and other cells of the reticuloendothelial system but others will be engulphed by phagocytes which will migrate via the lymphatics to become concentrated again in the cells of the reticuloendothelial system, particularly adjacent lymph nodes. This plutonium may again slowly be converted into soluble forms which can then be transported from the reticuloendothelial system to other parts of the body, notably the skeleton. Such translocation has been studied under experimental conditions in animals. In Table 10.13 are tabulated the experimental results obtained in rats, rabbits

Table 10.13. Absorption and distribution of ^{239}Pu after intramuscular injection

Chemical administered	Animal	Time after administration (days)	Percent of injected dose retained at injection site	Percent of absorbed dose retained in								Reference
				Skeleton	Liver	Spleen	Lungs	Kidneys	Excreta cumulative			
									Urine	Faeces	Total	
PuO_2Cl_2	Rat	4	70	65	14.0	0.4	0.3	1.9	0.4	7.4	8	Scott et al. (1948)
		16	30	64	2.8	0.3	0.1	0.7	8.1	19.6	28	
		64	34	56	4.2	0.5	0.2	1.7	4.3	26.5	31	
		256	13	47	1.5	0.3	0.1	0.3	4.5	44.4	49	
$PuO_2(NO_3)_2$	Rat	4	70		10.0			1.6	7.3	3.3	11	Van Middlesworth (1947)
		10	51	52	2.4	0.2		0.7				Kisieleski and Woodruff (1948)
		64	40	51	2.0	0.2		0.2				
		406	20	38	1.1	0.2		0.2				
Citrate Pu(VI)	Rat	7	29	80	9.6	0.3		1.2				Kisieleski and Woodruff (1948)
		62	23	56	1.4	0.2		0.3				
		420	12	34	1.1	0.1		0.2				
	Dog	14	14		6.8							Painter et al. (1946)
		103	0.001	58	8.3	0.1	0.06				9	
		234	0.1	41	13.0	0.3	0.1	0.1			15	
$Pu(NO_3)_4$	Rat	4	96	79	4.4	0.3	0.2	1.7	0.6	6.1	7	Scott et al. (1948)
		16	88	69	6.0	0.3	0.2	2.5	1.3	13.6	15	
		64	68	63	2.7	0.1	0.1	0.5	1.9	22.9	25	
		256	67	50	1.8	0.2	0.1	0.3	1.9	41.2	43	
	Rat	4	84		13.0			0.6				Van Middlesworth (1947)
	Rat	6	56	71	9.0	0.3	—	—			—	Taylor (1967)
		7	72	86	3.4	—	—	—			—	Taylor and Sowby (1962)
	Rabbit	1	98	80	12.0	0.2	0.8	1.6			—	Taylor (1969)
		8	90	85	13.0	0.05	0.2	0.9			—	
		35	75	85	13.2	0.02	0.1	0.7			—	
		56	52	54	21.2	0.06	0.1	0.7			—	
		112	55	44	18.9	0.02	0.1	0.6			—	
		280	33	27	23.1	0.0	0.04	0.1			—	
		365	51	33	22.4	0.03	0.4	0.1			—	
Pu(IV) citrate	Rat	4	97	30	26.0	—	—	—			—	Foreman et al. (1955)
		7[a]	19	96	2.7	0.1	0.4	—			—	Taylor (unpublished)
$PuCl_3$	Rat	4	77	68	7.1	0.5	0.3	1.9	0.7	9.6	10	Scott et al. (1948)
		16	68	83	3.8	0.4	0.3	0.4	0.8	3.3	4	
		64	40	52	2.9	0.4	0.1	0.5	1.7	33.5	35	
		256	24	46	1.4	0.3	0.1	0.5	1.7	46.6	48	

[a] Solution filtered through 10 mμ filter before injection

Table 10.14. The retention of ^{239}Pu in adult rabbit tissues after intravenous injection of plutonium nitrate. (Taken from TAYLOR, 1969)

Time after injection (days)	Rabbit No.	Percentage of the administered dose retained in			
		Skeleton[a]	Liver	Spleen	Kidney
1	1620	51	9.1	—	0.51
	1634	27	12.7	0.28	0.33
8	1635	29	6.5	0.05	0.36
	1636	39	14.2	0.11	0.02
112	1621	31	18.9	0.02	0.16
	1637	35	27.0	0.16	0.24
	1678	32	13.1	0.07	0.14
	1683	24	10.0	0.08	0.31
	1684	45	21.5	0.11	0.18
	1688	45	16.3	0.37	0.12
365	1622	22	5.1	0.003	0.12
	1638[b]	22	9.1	0.02	0.18

[a] Estimated from plutonium concentration in certain bones. [b] Not included in published paper.

and dogs following intramuscular injection of a variety of ^{239}Pu solutions. Some more detailed figures are given in Table 10.14. It is at once apparent on looking at Table 10.13 and 10.14 that a high percentage of intramuscularly injected plutonium was retained for long periods at the site of the injection but that the greater part of what was absorbed finds its way to the skeleton, indicating that introduction of plutonium through a wound presents a severe skeletal hazard. Skeletal deposition was always higher than liver deposition suggesting of course that this plutonium was translocated in a largely monomeric form. The earliest experiments of SCOTT and his colleagues (SCOTT et al., 1948) indicated that translocation was a continuous relatively slow process which has indeed been confirmed in the more recent experiments (TAYLOR ,1969; BLEANEY and VAUGHAN, 1971). The early experiments on dogs suggest it may be more rapid in a larger animal but the doses given were lethal or subacutely lethal (PAINTER et al., 1946). Comparative studies of plutonium uptake in the skeleton of adult rabbits following intramuscular and intravenous injection of $^{239}Pu(NO_3)_4$ in 0.03 M nitric acid are tabulated in Tables 10.14 and 10.15 (TAYLOR, 1969; BLEANEY, 1969a, b). This solution was thought to contain some colloidal or polymeric ^{239}Pu since only about 70 percent of the total ^{239}Pu was filterable through a 10 mμ filter. At 112 days after injection the average skeletal retention following intravenous injection is 35 percent of the injected dose while after intramuscular injection the average skeletal retention is 19 percent of the injected dose, a figure which is reached much more slowly than in the case of intravenous injection. It fact after intramuscular administration the rate of migration from the site of injection is initially very slow. After about 28 days there appears to be a sharp increase in the amount of ^{239}Pu migrating to both liver and skeleton. The reason for this increased translocation is not clear but it was observed in two separate experiments. It is of interest to note that up to 49 days after injection the ratio of the percentage retention in the liver to that in the skeleton is less than 0.2 whereas from 56 days onwards, except for the single animal studied at 365 days it is greater than 0.5. This suggests that initially the plutonium is translocated from the injection site in some form of soluble complex or complexes but the subsequent increase in liver

Table 10.15. The retention of ^{239}Pu in adult rabbit tissues after intramuscular injection of plutonium nitrate (Taken from TAYLOR, 1969)

Time after injection (days)	Rabbit No.	Percentage of the administered dose retained				
		Skeleton [a]	Liver	Spleen	Kidney	Injection site
1	H332/A	4	0.6	0.008	0.08	98.0
8	1668	7	1.3	0.003	0.05	103.7
	H332/B	10	1.9	0.008	0.13	76.0
28	1673	6	0.9	0.006	0.13	105.2
35	H332/C	23	3.3	0.006	0.18	75.0
49	1669	22	1.3	0.025	0.11	59.4
56	H332/D	26	10.2	0.029	0.36	52.4
63	1671	18	9.3	0.020	0.16	—
112	1670	14	7.2	0.007	0.35	61.9
	H332/E	25	9.9	0.008	0.19	47.0
280	1672	18	14.4	0.01	0.09	33.3
288	H332/F	19	16.6	0.0066	0.033	29.3
365	1674	16	1.0	0.03	0.005	67.0
	H332/G	24	21.3	0.0085	0.096	34.8

[a] Estimated from plutonium concentration in certain bones.

uptake relative to the skeleton implies that the deposition is changing towards a more colloidal pattern which may mean that lymphatic drainage plays a part in the removal of ^{239}Pu from the site of injection at this stage. This hypothesis is indeed born out by a study of the marrow distribution of plutonium after intramuscular injection to the same rabbits (BLEANEY and VAUGHAN, 1971). For about 35 days after injection the distribution was largely diffuse in the marrow itself with few aggregates. Large aggregates of plutonium were however found in the marrow in rabbits killed 112 days after injection when about 90 percent of the marrow plutonium might be present in aggregates. This is suggestive of the presence of colloidal plutonium present in marrow macrophages.

Studies in men accidentally contaminated with plutonium from wounds have been largely confined to investigation of the effect of chelating agents. Such effects are discussed later (Sec. IX). SCHOFIELD (1969), who studied 3 similar cases of puncture wounds treated by early wound excision and intravenous administration of DTPA, emphasises from his experience that it is not possible to predict the fraction of the contaminating ^{239}Pu that will behave in a soluble manner and move away from the wound site. Therefore each wound, contaminated for instance with plutonium nitrate may lead to a different ultimate distribution of plutonium between the wound site and the rest of the body. In Fig. 10.3 is shown his model representing the transfer of plutonium from a wound. He concluded that the amount of plutonium transferred to the rest of the body is only a few percent of the original deposit at the wound site. This may have been in part due to the treatment given (see Table 10.48, page 479). It seems optimistic, in view of the experimental results already described in rabbits, when in one animal at least 50 percent of the intramuscular plutonium was found in skeleton and liver 280 days after administration.

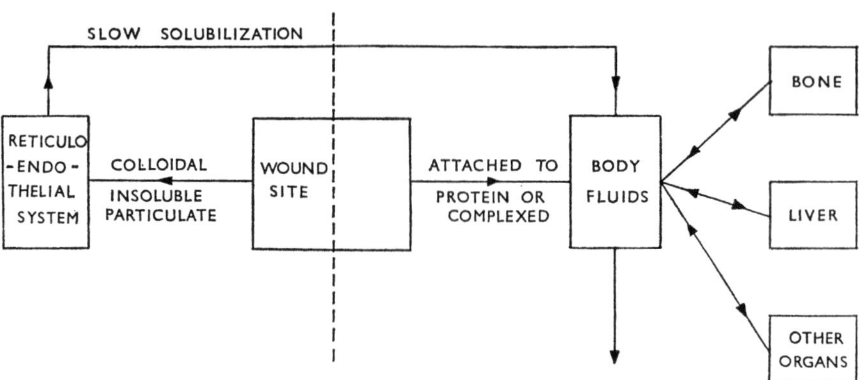

Fig. 10.3. A model representing the transfer of plutonium from a wound site. (SCHOFIELD, 1969, by courtesy of author and publishers)

7. Inhalation

The complex problems associated with inhalation of plutonium are discussed in considerable detail in Chap. 11. They are only briefly reviewed here in relation to translocation from the lung to the skeleton.

Since HAMILTON'S classic paper in 1947 it has been well recognised that plutonium may reach the blood stream from the inhalation of plutonium particles which are retained in the lung and adjacent lymph nodes. LANGHAM (1959) states that 10 percent of the original inhaled dose is rather rapidly absorbed into the "circulating" blood and deposited predominantly in the skeleton. It is difficult to tabulate the results of inhalation experiments as they affect skeletal retention in a meaningful way since to the already complex physicochemical characteristics of plutonium an additional complication is added, namely the variable size of the particles inhaled. Further, as BAIR and his colleagues (BAIR et al., 1961) warned in 1961, "extrapolation from acute studies to situations where plutonium may be inhaled in small quantities for long periods or to situations where late effects may occur following a single exposure, must be done with care because of the accumulation of plutonium in lymph nodes and the increased retention when small quantities are deposited". Variations in skeletal retention depending on particle size and filterability of a plutonium aerosol are shown in Table 10.16 (BAIR et al., 1962). Comparing skeletal retention, following the inhalation or intravenous injection of plutonium in 0.14 n HNO_3, they found 15 percent of the body burden in the skeleton following inhalation and only 6 percent following intravenous injection. Presumably the solution used for intravenous injection was extremely polymeric. They conclude that inhalation of aerosols with a small MMD (mass median diameter) resulted in greater translocation of plutonium than when large particle aerosols are used.

In 1966 CLARKE and his colleagues reported a study of beagle dogs given a single 10–30 minutes exposure to $^{239}PuO_2$ aerosols. The count median particle diameter of the aerosols ranged from 0.1 to 0.5 μ. Deaths occurred in some animals at 29 to 56 months post exposure. The terminal body burden and tissue distribution percent of body burden is shown in Table 10.17. The mean figure for liver retention in 10 dogs was 5.2. The mean figure for skeletal retention was 2.7 (CLARKE et al., 1966). A more recent report on animals from the same group of dogs that survived at least 4.5 years post exposure indicates that 10–15 percent of the plutonium was translocated from the lung to the liver and 5 percent to

Table 10.16. Tissue distribution of ^{239}Pu in dogs after inhalation of $^{239}PuO_2$ aerosols. (Taken from BAIR et al., 1962)

Group	Particle size CMD[a] (μ)	Particle size MMD[b] (μ)	Percent filterable	Dog no.	Percent of body burden one month after exposure: Skeleton	Liver	Spleen	Lung	Bronchial lymph nodes
I[c]	0.12	0.23	0.5	80	0.54	0.19	0.44	70	1.7
				89	0.56	0.71	0.15	70	3.4
				217	2.7	1.7	1.2	72	2.9
				187	0.51	0.10	0.0018	98	0.59
II[d]	0.32	2.3	0.2	84	0.50	0.24	0.02	96	1.4
				77	0.53	0.16	0.0078	98	0.25
				174	1.0	0.42	0.010	97	0.22
III[c]	0.43	3.3	0.2	208	0.53	0.21	0.0050	91	0.21
				198	0.38	0.22	0.0089	97	0.32
				214	0.32	2.3	0.022	95	0.42
IV[c]	0.24	6.1	2.0	242	1.7	0.89	0.0058	96	0.67
				241	20.0	2.1	0.021	76	0.58
				94	0.078	0.072	0.0016	96	1.8
V[c]	0.3	7.6	0.03	160	0.075	0.049	0.0035	98	0.70
				5	0.019	0.17	0.022	97	1.8
				181	0.050	0.040	0.00077	97	1.9
VI[e]	0.60	4.3	—	186	0.031	0.058	0.00074	97	1.6
				188	0.0049	0.049	0.001	98	0.79
				92	15.1	9.1	0.09	72	0.59
VII[f]	0.13	0.88	85	189	19.6	8.8	0.16	66	0.38
				190	13.4	17.0	0.26	63	0.44
				22	5.8	85.0	4.4	0.23	0.0056
VIII[f]	(Intravenous injection)			27	6.4	83.0	5.8	0.30	0.029
				243	6.9	78.0	9.6	0.31	0.0064

[a] CMD=count median diameter. [b] MMD=mass median diameter. [c] PuO_2. [d] Exposed to same plutonium as group III dogs except that the plutonium was suspended in distilled water instead of polypropyleneglycolethylene oxide polymer. [e] Exposed to dry PuO_2 dust. [f] Plutonium in 0.14 N HNO_3.

Table 10.17. Body burden and tissue distribution of plutonium (percent of body burden). (CLARKE et al., 1966, by courtesy of authors and publishers)

	Dog number: 182	184	215[a]	83[a]	173[a]	106[a]	183	180	76	213[a]
Lung	74	75	50	59	46	50	52	42	47	49
Bronchial and mediastinal lymph nodes	21	17	42	27	48	37	38	49	45	27
Liver	2	5	2	6	2	6	6	5	3	15
Bone	1	1	4	2	2	2	3	3	4	5
All other tissues	2.7	2	2	6	2	5	1	1	1	4
Final body burden (μc)	2.7	2.5	1.4	1.8	2.1	2.7	0.9	1.4	1.4	1.2
Time of death months post exposure	29	31	38	40	44	45	45	48	51	56

[a] Bronchio-alveolar tumours.

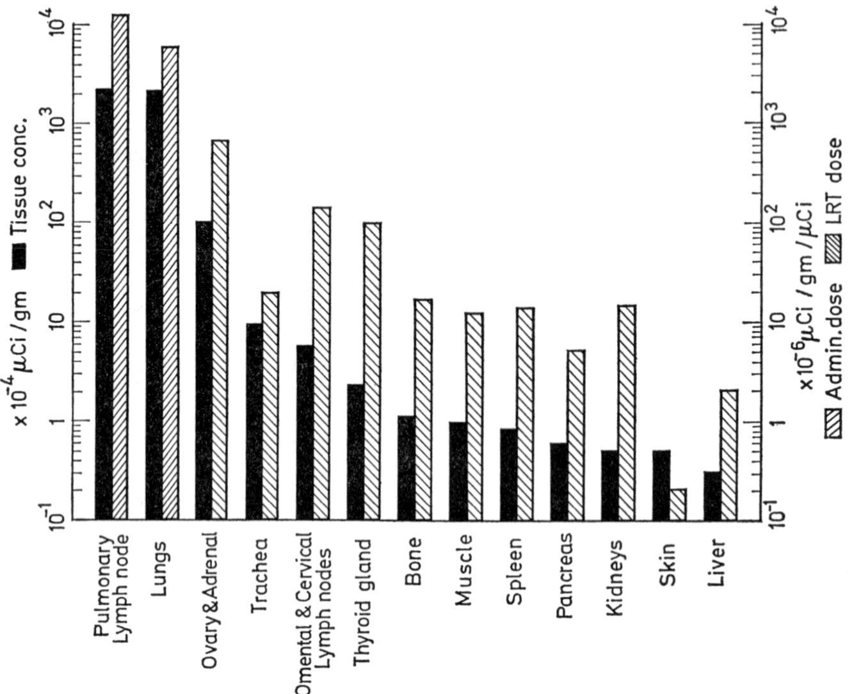

Fig. 10.4. Distribution of ^{239}Pu in tissues of the dog after inhalation. LRT = lower respiratory tract. (MORROW et al., 1967, by courtesy of authors and publishers)

the skeleton. Ten to eleven years post exposure the skeletal content was of the same order (PARK et al., 1972). These observations suggest that translocation from lung and lymph glands may continue over at least 4.5 years. The low skeletal retention, in the dogs described by MORROW and his colleagues, which are discussed below, may well be due to the fact, that the survival time of his dogs was only about eighteen months (MORROW et al., 1967).

MORROW and his colleagues (1967) who have made an extremely detailed study of the retention and fate of inhaled plutonium dioxide in dogs comment on the small retention found in the skeleton. "Neither the role of bone as a plutonium depot nor as a tissue with a particularly high concentration was manifest in this investigation." In MORROW's study (1967) the absorption of Pu averaged 3.5 percent of the PuO_2 deposited and retention in lungs and bronchial lymph nodes accounted for 95 percent of the body burden.

A summary of the mean plutonium concentrations in the lungs and twelve highest tissues is given in Fig. 10.4 from MORROW's data (MORROW et al., 1967). Also presented are the mean concentration values in terms of the administered dose (total Pu recovered) or in terms of the initial plutonium burden in the lower respiratory tract. The latter expression is applied only to the pulmonary tissues; the administered dose is used to normalize tissues in which the Pu level is probably due to both gastrointestinal and pulmonary absorption. The data in Fig. 10.4 were computed as though they were independent of time. The authors state, "All the bone specimens analysed contained their normal amount of marrow". There was possibly a factor of two in plutonium concentration in favour of spongy bone over compact bone which is also a feature of skeletal retention following

Table 10.18. Case of FOREMAN et al. (1960). Tissue concentration and extrapolated burden. (Intermittent exposure to Pu, primarily by inhalation, during the 12 yr prior to death). (From MAYS et al., 1970)

Tissue	Organ mass (g)	Av. conc. (dis/min/g)	Total content (dis/min)
Skeleton	10000	1.4	14000 (36%)
Liver	1950	9.9	19300 (49%)
Lungs	850	4.8	4080 (10%)
Bronchial lymph nodes	10	125.0	1250 (3%)
Muscle	30000	0.01	300 (0.8%)
Heart	400	0.06	24 (0.06%)
Spleen	116	0.18	21 (0.05%)
Kidney	270	0.05	14 (0.04%)
Balance	26400	0.01	264 (0.7%)
Total	70000		39253 (100%) [0.018 μCi]

intravenous injection (Sec. II.1). There appeared to be both a time and a dose effect expressed in the bone concentration of plutonium. Even dogs which were studied for less than 4 months but which received high initial doses showed appreciable skeletal concentration. Whether the early uptake of plutonium was based on the marrow is not certain. MORROW reports that marrow samples taken from dogs sacrificed at early times appear to account for at least one half of the total bone concentration. The significance of this finding is difficult to interpret unless it is known how the marrow was removed from the bone, since it is now recognised that some of the plutonium on bone surfaces may be within or on the osteogenic cells which are easily removed with the marrow (BLEANEY and VAUGHAN, 1971, Sec. IV.A).

BULDAKOV and his colleagues (1970) emphasize that the results of continuous exposure by inhalation differ from those of a single exposure. When they administered plutonium aerosols to rats over a long period they found the lungs and skeleton to be the organs of maximum retention. After the 80th day the skeletal retention was higher than that of the lungs, reflecting a slower rate of removal of plutonium from the skeleton. Even following a single exposure they indicate a relatively high skeletal retention. It is not altogether clear why their results are so different from those of MORROW and his colleagues (1967) already discussed. Many different factors such as species of animal, particle size and form of plutonium may be involved.

Reports are available on the tissue concentrations of plutonium found in two human cases exposed, largely through inhalation, for 12 and 9 years respectively. The skeleton and the liver were the major sites of deposition after the lymph nodes with only 10 percent of the total body burden remaining in the lung. When the skeletal and liver *contents* are comparable the liver *concentration* will be much higher due to the smaller mass of the liver (FOREMAN et al., 1960; LAGERQUIST et al., 1968). The figures for organ retention in these cases are shown in Tables 10.18 and 10.19. In the case described by FOREMAN et al. (1960) the skeletal retention may have been lower than 36 percent of the body burden since the skeletal samples from which total skeletal content was calculated consisted of pieces of rib, sternum and vertebrate which are likely to contain higher concentra-

Table 10.19. Case of LAGERQUIST et al. (1968). Tissue concentration and extrapolated burden. (Two contaminated puncture wounds and several inhalation exposures during the 9 yr prior to death). (From MAYS et al., 1970)

Tissue	Organ mass (g)	Av. conc. (dis/min/g)	Total content (dis/min)
Skeleton	10000	0.13	1300 (58%)
Liver	1649	0.32	326 (23%)
Lung	1015	0.21	214 (10%)
Bronchial lymph nodes	3	1.4	4 (0.2%)
Kidney	289	0.002	0.6 (0.03%)
Spleen	120	0.004	0.5 (0.02%)
Remaining tissue	69500	0.003	208 (9%)
Total	82576		2253 (100%) [0.001 μCi]

tions of Pu than the skeleton as whole. In the case of LAGERQUIST et al. (1968) the skeletal retention may have been less than 56 percent for the same reason (MAYS et al., 1970). In two other plutonium workers the vertebral concentrations were 5 and 19 percent respectively of that found in the liver (LANGHAM et al., 1962). The vertebral concentration in autopsy samples from non-occupationally exposed persons in 1969 averaged about 10 percent of that in the liver. This plutonium was presumably due to fall out (MAGNO et al., 1969). The combination of animal and human observation suggests, that, probably soon after a single inhalation, most of the body burden is in the lung, but that at long times, particularly after continuous inhalation at low levels, much of the residual plutonium is in liver and skeleton so that the average organ dose to the liver may exceed that to the lung over the long duration of the human life span (MAYS et al., 1970; International Commission on Radiological Protection, 1972).

8. Conclusions

In examining retention figures, following different routes of entry of plutonium into the body, it is important to realize that the results of a single exposure may differ from continuous exposure. This is particularly true of inhalation, continuous feeding and wound contamination. Further post mortem data, after continuous ingestion may only represent one point in time for a dynamic system.

This review of the experimental results covering deposition of plutonium in the skeleton together with scanty human data from exposed individuals suggests the following conclusions:

1) absorption from the intestinal tract into the blood stream is poor but cannot be ruled out as a skeletal hazard especially when absorption is continuous, or when those exposed are extremely young;

2) plutonium may be absorbed from apparently uninjured skin but the most likely route of entry to the blood stream in significant amounts is from a contaminated wound, or from inhalation;

3) plutonium reaching the blood stream by any route of entry, if in monomeric form, is deposited initially in the skeleton rather than in the liver; if in polymeric form, it is likely to be deposited in the liver rather than the skeleton though it may subsequently translocate to the skeleton;

4) little is known at present about the mechanisms and the form by which plutonium translocates from the site of entry to the blood stream, though experimental results suggest that it is initially translocated in the form of soluble complexes and may later take a more colloidal form. Initially, therefore, it can be expected to be deposited primarily in the skeleton rather than in the liver;

5) plutonium reaching the circulation from occupational and fall out exposure in man, after some years, has been assessed as redepositing roughly half in bone and half in liver;

6) age will affect both the absorption of plutonium into the blood stream and its partition between different organs. The critical organ for plutonium may not be the same for the juvenile as for the adult.

D. Plasma Transport

The plasma transport of plutonium, like everything else to do with this element, is complex. If it is administered intravenously as plutonium in a citric acid sodium citrate buffer at pH 3.5 (STEVENS et al., 1968) or as a freshly prepared solution of $Pu(NO_3)_4$ (BOOCOCK and POPPLEWELL, 1965, 1966, 1967), about 90 percent becomes bound to transferrin, the plasma protein that binds iron; 5 percent is associated with other macromolecules and a further 5 percent with low molecular weight substances like peptides, sugars, amino acids and citrate (BOOCOOCK and POPPLEWELL. 1965, 1966; STEVENS et al., 1968). A small amount of plutonium may also be bound to albumin (STEVENS et al., 1968). In normal plasma, transferrin is only 30 percent saturated, with respect to iron, leaving many binding sites available for plutonium though these can be blocked by saturation of transferrin with iron. As will be discussed in Sec. II.E the rate of removal of plutonium entering the blood as ^{239}Pu citrate or in soluble form from any source is relatively slow; it is in fact slower than the rate for removal of ^{59}Fe even though this probably forms a more stable complex with transferrin than does ^{239}Pu under physiological conditions (STEVENS et al., 1968).

It is known from KHODYREVA's experiments, already discussed that plutonium may also appear (KHODYREVA, 1965) in the blood in the form of aggregates of a size sufficient to give stars in an autoradiograph. Whether such aggregates are intracellular is not known. She demonstrated (Sec. II.C.3) that immediately following the application of plutonium nitrate to the shaved skin of rabbits plutonium was seen in the blood as single alpha tracks but within a few days aggregates were also present. Such aggregates are of course removed by the reticulo-endothelial system and do not reach the bone though they may reach the marrow.

It has been suggested (TURNER and TAYLOR, 1968; BULDAKOV et al., 1970) that plutonium that enters the body by gastric absorption, inhalation or tissue wounding may be transported from the site of entry as some type of freely diffusible complex with a low molecular weight component, or components, of the tissues or body fluids, to the blood stream. The nature of this "physiological plutonium complex" is still unknown but TURNER and TAYLOR consider that a similar type of diffusible complex is probably concerned in the transfer of plutonium from the transferrin of the blood to the sites of deposition in bone and other tissues. It is, of course, possible that some plutonium may be removed from the blood by diffusion of the plutonium transferrin complex but this is thought unlikely to be the major process (TURNER and TAYLOR, 1968).

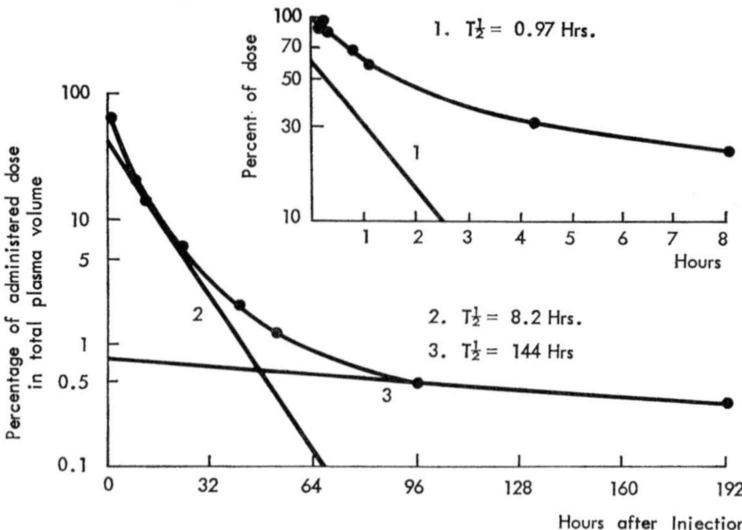

Fig. 10.5. The plasma clearance of plutonium citrate in the rat after intravenous injection. The experimental points represent the mean of 3 or more measurements. The curves shown were fitted to the experimental data by the method of least squares, assuming a multi-component exponential model. (TURNER and TAYLOR, 1968, by courtesy of authors and publishers)

E. Plasma Clearance

A wide range of rates of disappearance of intravenously injected plutonium from the plasma has been reported (SCHUBERT et al., 1950; ROSENTHAL and SCHUBERT, 1957; LANGHAM, 1959; STOVER et al., 1959; TURNER and TAYLOR, 1968); 15 to 50 percent of the injected material may disappear within a few minutes of injection (TURNER and TAYLOR, 1968). These differences are largely dependent upon the physicochemical form of the injected plutonium, as discussed in Chap. 2 (TURNER and TAYLOR, 1968). Removal of colloidal plutonium from the plasma is rapid as would be expected from other data on the removal of colloidal particles from the blood (DOBSON et al., 1949). Large colloidal particles are rapidly cleared by the liver. If polymeric species are removed from injection solutions before administration by filtration the rate of clearance is decreased (TURNER and TAYLOR, 1968).

Plasma clearance of plutonium following intravenous injection of ^{239}Pu citrate has been studied in rats (TURNER, 1967), dogs (STOVER et al., 1959) and man (LANGHAM, 1959). This of course presents an artificial situation since such citrate solutions do not gain access to the blood stream except possibly in rare accidents but the results are of importance in indicating the slow plasma clearance of plutonium in all species.

In Fig. 10.5 is shown the curve for rats, which, it has been suggested, affords a close approximation to the rate of removal of plutonium which enters the blood in soluble form from any source (TURNER and TAYLOR, 1968). STOVER and her colleagues have made an elaborate study of plutonium plasma clearance, excretion and retention over a period of 8 years in the Utah beagles (STOVER et al., 1959; STOVER et al., 1962). The solution used was tetravalent plutonium in citric acid with a sodium citrate buffer to pH 3.5. Careful analysis indicated that not more than 2 percent Pu(VI) was present and the solution was thought to be

Table 10.20. ^{239}Pu excretion and retention by dose level following intravenous injection of ^{239}Pu citrate to the beagle dog. (Taken from STOVER, ATHERTON, and KELLER, 1959)

Dose level	Injected dose (μc/kg)	Cumulative excretion 0–22 days (%)	Retention 22 days (%)	Cumulative excretion 22 days to 4 yrs (%)	Retention 4 years (%)
All	—	12.1	87.9	16.4	71.5
5	2.78	10.8	89.2	14.3	74.9
4	0.900	10.6	89.4	16.9	72.5
3	0.300	13.4	86.6	16.4	70.2
2	0.0955	13.3	86.7	16.5	70.2
1.7	0.0477	—		—	—
1	0.0159	14.2	85.8	—	—

monomeric. Data for excretion and retention in the first 4 years were largely obtained from dogs given a high initial dose as shown in Table 10.20. Many of these animals did not survive longer and the data in the later years came from animals given a lower dose. This may have been a factor in some of the slight differences observed which are discussed below.

The plasma concentration at 1 minute after injection was 0.193 percent per g plasma and it is calculated that about 85 percent of the dose is circulating at 1 minute. Between 1 minute and 10 hours the concentration drops only by a

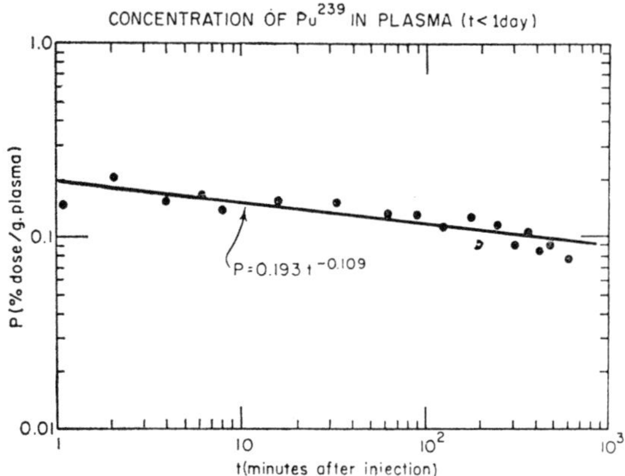

Fig. 10.6. Concentration of ^{239}Pu in beagle plasma during day of intravenous injection of plutonium citrate. (STOVER et al., 1959, by courtesy of authors and publishers)

factor of about 2. The plasma clearance for one day is shown in Fig. 10.6. From then until 3 weeks after injection the decrease is by a factor of 250. Then the decrease in concentration becomes slower and at 4 years the value is still one-fourth that at 3 weeks. The mathematical description of these results is best served by two power functions as shown in Fig. 10.7. The day of the injection was chosen as the first period for the plasma data and the calculated equation is

$$P = 0.193\, t^{-0.109}$$

where P is percent dose per gram of plasma and t is minutes. The second arbitrary time period chosen is 1 through 22 days after injection; the third is 22 days through

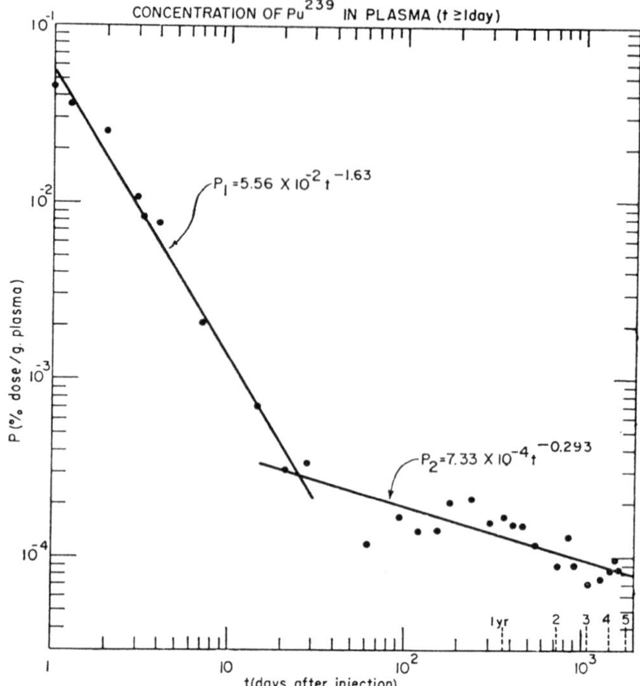

Fig. 10.7. Concentration of ^{239}Pu in plasma of the beagle dog from 1 day through 4 years after injection of a plutonium citrate solution. (STOVER et al., 1959, by courtesy of authors and publishers)

4 years. With t now expressed in days the calculated equations for these two periods are

$$P_1 = 5.56 \times 10^{-2}\, t^{-1.63} \quad (1 — t — 22 \text{ days}),$$

$$P_2 = 7.35 \times 10^{-4}\, t^{-0.293} \quad (t — 22 \text{ days}).$$

In the dogs at sacrifice between 4 and 8 years many of the measured terminal concentrations were considerably lower than the calculated values as shown in Table 10.21. The range of values of P observed/P calculated is from 0.05 to 0.94. This may have been due in part 1) to the moribund condition of the dogs, 2) the derivation of the equation from measurements made on higher dose level dogs or 3) invalid extrapolation of the equation. There was no correlation between the departure of P observed from P calculated with time after injection in these long period studies.

LANGHAM and his colleagues (LANGHAM et al., 1950) as already discussed, studied plasma clearance, excretion and retention of plutonium in a group of terminal male hospital patients over 45 years of age. Twelve patients were given ^{239}Pu citrate; 3 patients were given hexavalent ^{239}Pu (RUSSELL and NICKSON, 1946) and one patient was given ^{238}Pu nitrate (CROWLEY et al., 1946). The mean values observed for plasma clearance in these 16 patients are shown in Table 10.22. At four hours after injection 36 percent of the injected dose is still in the circulating blood falling to 3.4 percent 6 days later (LANGHAM, 1959).

The amount of ^{239}Pu remaining in the plasma at one day following intravenous injection of a monomeric solution in rabbits (TAYLOR, 1969) is in broad

Table 10.21. Examples of observed and calculated concentrations of ^{239}Pu in plasma (terminal values) following intravenous injection of ^{239}Pu citrate. (Taken from data in STOVER et al., 1962)

Dog	Days after injection	P_{obs} (% injected $^{239}Pu/g$) $\times 10^5$	P_{calc}[a] (% injected $^{239}Pu/g$) $\times 10^5$	$\frac{P_{obs}}{P_{calc}}$
M8P5	1192	1.2	9.2	0.12
F7P5	1491	2.7	8.6	0.31
M4P5	1562	0.41	8.5	0.05
F5P5	2059	0.90	7.8	0.12
M8P4	1157	2.4	9.3	0.26
F7P4	1198	3.6	9.2	0.39
F10P4	1241	0.78	9.1	0.09
M11P4	1288	3.5	9.0	0.39
F9P4	1343	3.7	8.9	0.42
F6P4	1357	0.39	8.9	0.04
M12P4	1462	2.0	8.7	0.23
M1P4	1724	6.1	8.3	0.74
M11P3	1198	0.63	9.2	0.07
F5P3	1504	5.0	8.6	0.58
F10P3	1547	6.4	8.5	0.74
F6P3	1617	5.2	8.4	0.62
F7P3	1627	1.7	8.4	0.20
M8P3	1771	1.6	8.2	0.19
F9P3	1895	7.6	8.0	0.94
F9P2	2014	3.2	7.9	0.40
F7P2	2093	3.2	7.8	0.41
F5P2	2423	1.0	7.5	0.14
F2P2	2780	1.0	7.2	0.14
M4P2	2948	1.7	7.1	0.24
M1P2	2985	1.7	7.0	0.24

[a] $P_{calc} = 7.33 \times 10^{-4}\, t^{-0.293(2)}$

Table 10.22. ^{239}Pu content of total blood volume[a] as a function of time after administration of ^{239}Pu citrate by intravenous injection to man. (Taken from LANGHAM, 1959)

Time after administration	^{239}Pu in blood (% administered dose)
4 hours	36
1 day	16
2 days	10
3 days	8.6
6 days	3.4
15 days	0.7
22 days	0.4

[a] Total blood volume taken as 7.7 percent of body weight.

agreement with results for dogs (STOVER et al., 1959), man (LANGHAM, 1959) and rats (TURNER and TAYLOR, 1968). Between 112 days and 365 days in rabbits the rate of removal of ^{239}Pu from the plasma is very slow and, if it is assumed to obey a monoexponential law, the half time is of the order of one year.

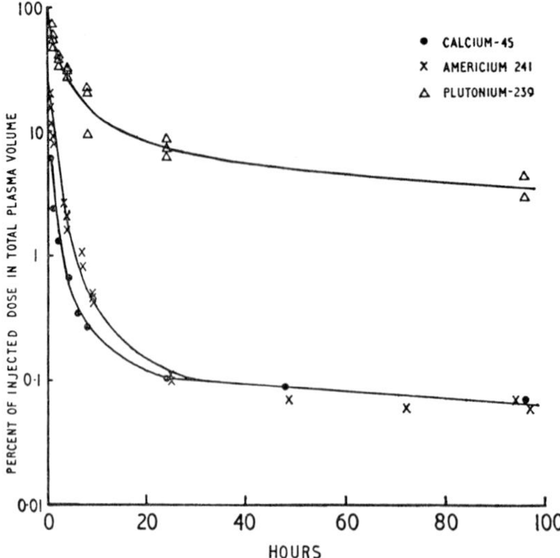

Fig. 10.8. Concentration of ^{45}Ca, ^{239}Pu and ^{241}Am in the plasma of the rat during the first four days after intravenous injection of the radionuclide in 0.02 M HNO_3. (TAYLOR, 1962, by courtesy of author and publisher)

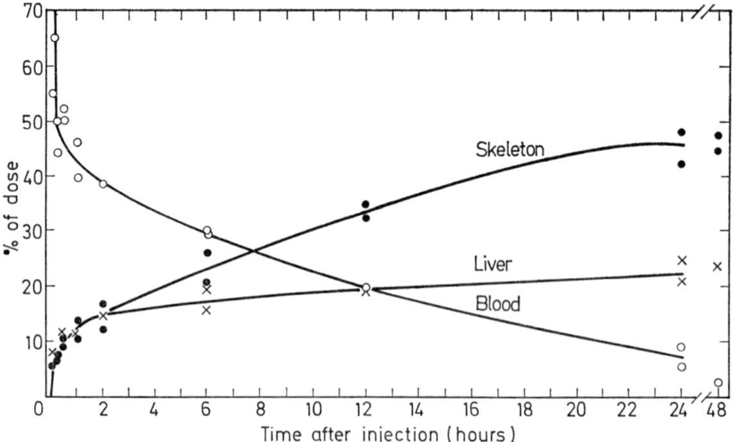

Fig. 10.9. Pu^{239} content of blood, liver and skeleton as a function of time after intravenous injection (PuIV citrate) in rats. (Solution injected completely diffusible. Acid solution of Pu + y diluted with excess of 2 percent citric acid, pH adjusted to 6.5 with NaOH, diluted to be isotonic with blood and then contained equivalent of 1 percent citric acid. Each animal received 0.4 cm^3 of solution containing 94.3 γ of Pu(+4) giving 1.7×10^6 counts per minute and 0.3 μCi of Y(+3) giving 4×10^3 counts per minute.) In cases in which only one point is shown for a time interval, the values for the two rats were almost identical. (SCHUBERT et al., 1950, by courtesy of authors and publishers)

This slow plasma clearance of ^{239}Pu is characteristic and differentiates this element from other bone seeking radionuclides, as can be illustrated by the following experimental results. In Fig. 10.8 (TAYLOR, 1962) is shown the concentration of ^{45}Ca, ^{239}Pu and ^{241}Am in the plasma of the rat during the first four days after intravenous injection. The plutonium was given in a solution of 0.02 M

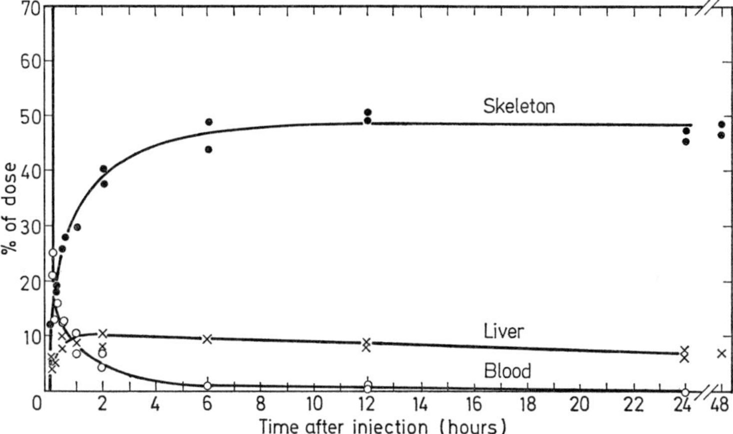

Fig. 10.10. ^{91}Y content of blood, liver and skeleton as a function of time after intravenous injection (Injection solution, see Fig. 10.9). In cases in which only one point is shown for a time interval, the values for the two rats were almost identical. (SCHUBERT et al., 1950, by courtesy of authors and publishers)

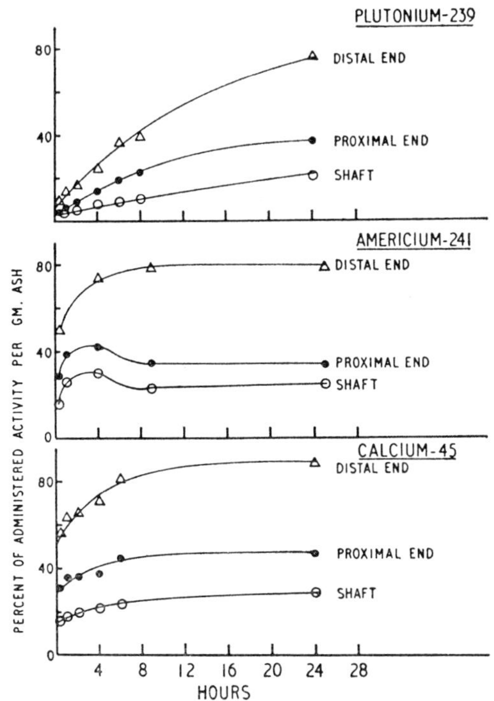

Fig. 10.11. Concentration of ^{45}Ca, ^{239}Pu and ^{241}Am in the proximal, shaft and distal regions of the rat femur during the first 24 hours after intravenous administration of the radionuclide in 0.02 M HNO_3. (TAYLOR, 1962, by courtesy of author and publishers)

HNO_3. The rate of removal of ^{239}Pu from the plasma was relatively slow while that of ^{241}Am was much more rapid and only a little slower than that of ^{45}Ca. The same difference has also been shown betwenn ^{239}Pu and ^{91}Y (SCHUBERT et al., 1950) as illustrated in Figs. 10.9 and 10.10.

The different rates of plasma clearance are reflected in the different rates of uptake of the nuclides into parts of the femur. In the case of ^{91}Y the comparative information as shown in Fig. 10.10 is for the whole skeleton; in the case of americium and calcium a differentiation has been made between the proximal, shaft and distal regions of the femur (Fig. 10.11). The overall picture is however the same, a much slower uptake of ^{239}Pu by the bone than of americium or calcium.

This slow removal of plutonium from the blood is considered by STOVER and her colleagues (1959) to result from the protein binding of the plutonium. As will be shown in Sec. V a similar but firmer binding of plutonium to certain glycoproteins of bone may well account for its long term skeletal retention (HERRING et al., 1962; TAYLOR and CHIPPERFIELD, 1970a, b).

F. Excretion

To obtain a further measure of the retention of plutonium following injection it is necessary to measure excretion. This was done by STOVER and her colleagues continuously through the first 22 days for beagles injected with the previously described ^{239}Pu citrate. Nineteen dogs representing all dose levels excreted 12.1 percent of the injected dose during this interval. Further measurements were made on fewer dogs through 36 days and twelve dogs receiving the higher doses were measured at 6 monthly intervals through 4 years. The daily total excretion X and the daily urinary excretion U are plotted in Fig. 10.12. Both sets of data have been divided into two time groups correpsonding to the second and third periods chosen for the plasma concentration data. For the first 22 days after injection the calculated equations for daily total and urinary excretion are respectively

$$X_1 = 6.98\, t^{-1.59} \quad (t \geq 22 \text{ days}),$$
$$U_1 = 1.44\, t^{-1.52}$$

where X_1 and U_1 are expressed in percent dose per day and t is in days. For the second period from 22 days through 4 years the corresponding equations with the same units are

$$X_2 = 0.223\, t^{-0.485} \quad (t \geq 22 \text{ days})$$
$$U_2 = 0.0605\, t^{-0.409}.$$

It is of some interest that, as illustrated in Fig. 10.13, the relative amounts excreted in the urine and faeces vary during the first 20–30 days and then remain approximately constant through 4 years. During the first 22 days the cumulative faecal excretion is 2.14 times the cumulative urinary excretion. From these results it can be calculated that the kidneys clear only 7.5 to 9 percent of the plasma plutonium in the course of a day. The results of the measurement of excretion in the 4–8 year period after injection show good agreement between observed and predicted excretion rates in most cases as do the observed urinary and faecal excretion rates. The average observed rates relative to the calculated rates for urinary excretion were 8 percent higher and the measured urinary plus faecal excretion rates relative to the calculated rates were 33 percent higher. STOVER and her colleagues discuss these discrepancies and think that the major error was probably due to contamination and that the effect on the retention equation was small (STOVER et al., 1959).

The urinary and urinary plus faecal excretion of plutonium in man are shown in Fig. 10.14 over a period of approximately five years expressed as a function

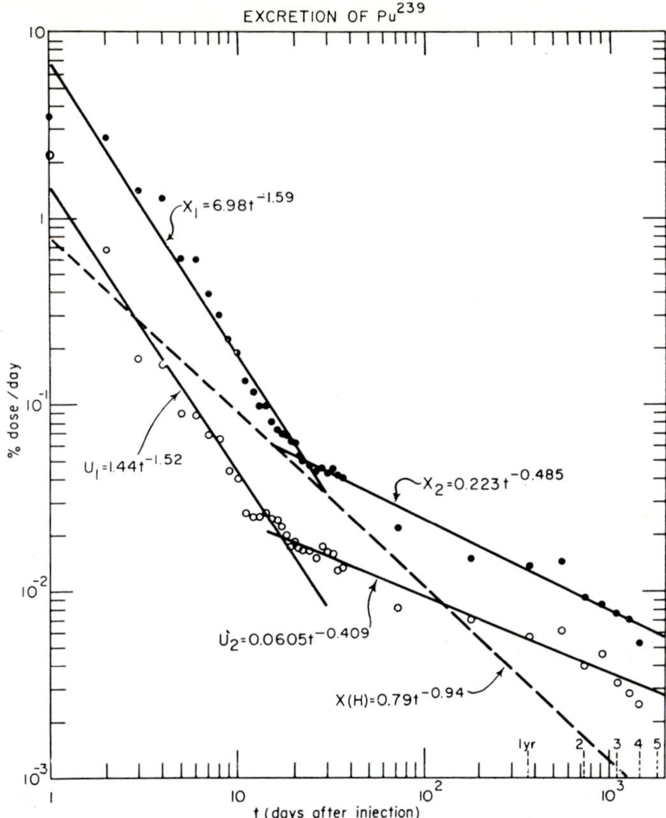

Fig. 10.12. Daily faecal plus urinary excretion (X) and daily urinary excretion U of ^{239}Pu by the beagle through 4 years. The equation for total excretion by man X(H) is also plotted where ($X(H) = 0.79\,t^{-0.94}$). (STOVER et al., 1959, by courtesy of authors and publishers)

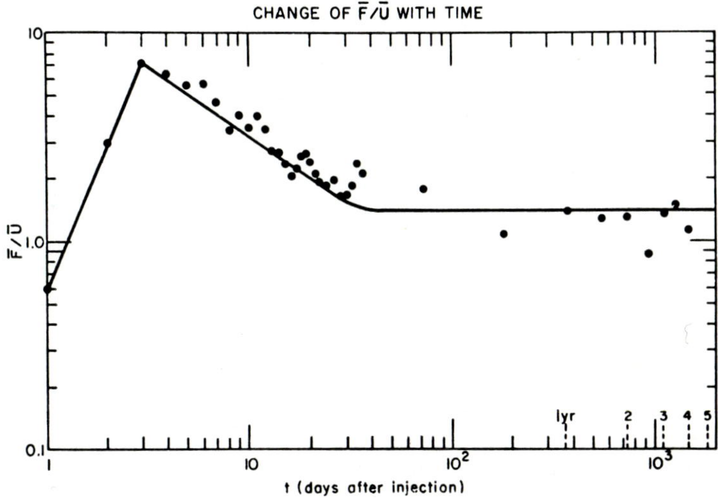

Fig. 10.13. The change with time of the ratio of average daily faecal excretion $\overline{F}$, to average daily urinary excretion, $\overline{U}$ of plutonium 239 in adult beagles. (STOVER et al., 1959, by courtesy of authors and publishers)

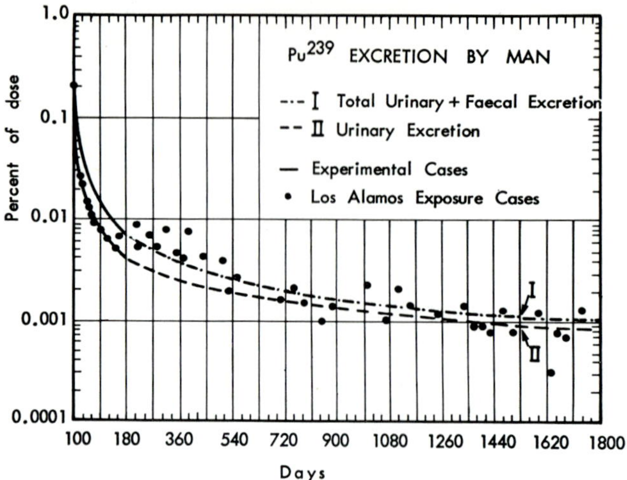

Fig. 10.14. Urinary and urinary and faecal excretion of plutonium by man over a period of approximately 5 years. (LANGHAM, 1959, by courtesy of author and publishers)

of time after administration (LANGHAM, 1959). Empirically the urinary excretion curve (curve I) fits the following power function:

$$Y_u = 0.20 \times t^{-0.74} \ (t > 1)$$

in which Y_u is the percentage of the administered dose excreted per day and t is the number of days between exposure and collection of the sample. From a survey of the early Los Alamos plutonium exposure cases LANGHAM considers that this expression holds reasonably well over a period of at least 12 years. The faecal excretion curve which is not shown in Fig. 10.14 is fitted by the expression:

$$Y_f = 0.63\, t^{-1.09} \qquad (t > 1)$$

in which Y_f is the percentage of the dose excreted per day in the faeces. The total urinary plus faecal excretion rate Y_{u+f} shown by curve I in Fig. 10.14 is represented by the sum of the two expressions.

These excretion rates in man show points of similarity with those already discussed in the dog: firstly, the excretion becomes slower as the period of fixation becomes longer and secondly, the faecal to urinary excretion rate is not constant. During the first 30 days the ratio is slightly greater than 1 after which it drops to less than 1 becoming progressively smaller with time. The rate of Pu excretion continues to decrease with time until only about 0.01 percent is being excreted per day at about 100 days after exposure. At approximately 5 years the rate has dropped to about 0.001 percent[2]. An interesting early comparison of excretion rates in man and rat following intravenous injection of Pu^{+4} nitrate is shown in Fig. 10.15 (CROWLEY et al., 1946).

It should be emphasised that these excretion measurements refer to intravenously injected plutonium. Excretion rates and the ratio of faecal to urinary excretion fractions may differ considerably from these when entry into the body is by other routes, particularly inhalation (see Chap. 11 — Editor).

2 It must be remembered that 12 of the patients on whom these data are based were given $^{239}Pu_4$ citrate, 1 was given ^{238}Pu nitrate and 3 were given hexavalent ^{239}Pu and they were all extremely ill.

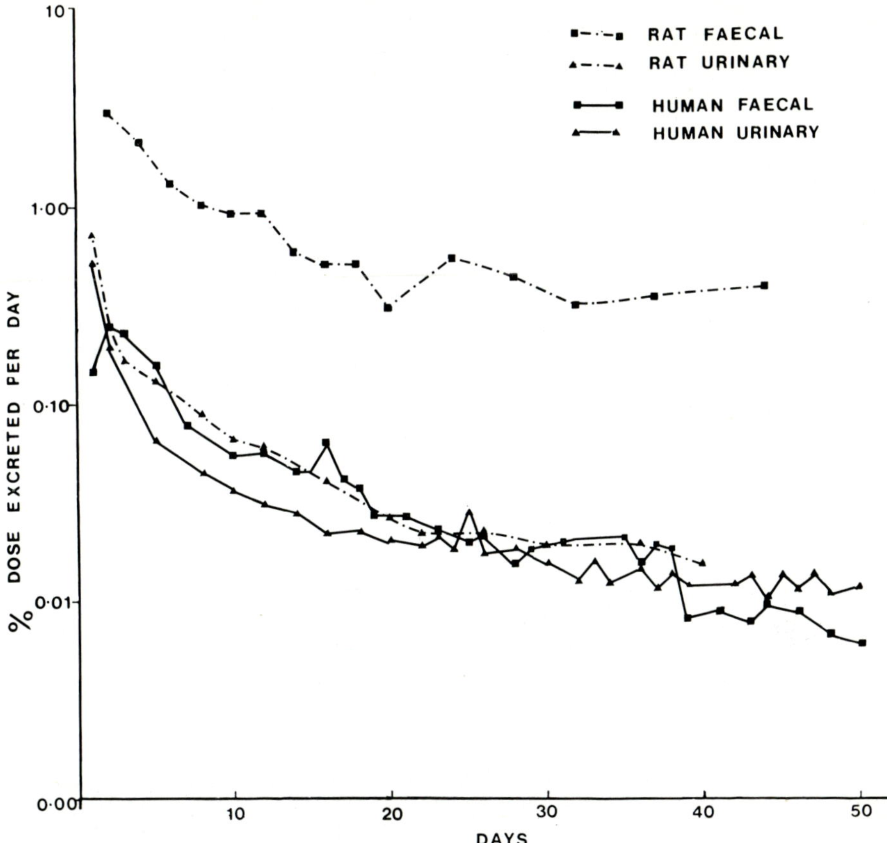

Fig. 10.15. Excretion patterns in man and rat following intravenous injection of Pu^{+4} citrate. (Taken from CARRITT et al., 1947)

G. Retention

Retention of plutonium in the Utah dogs was determined both by direct estimation of different tissues at different time intervals and by the use of the results of excretion studies. The agreement between the two was highly satisfactory. For times greater than 3 weeks after injection percent retention was calculated from excretion measurements using the following equation

$$R = 90.0 - 0.4346\, t^{0.515} \qquad (t \text{ in days}).$$

The retention in humerus and vertebrae of the beagle dog is shown in Fig. 10.16 and figures for both excretion and retention at 4 years in Table 10.20. Total retention at 4 years was about 72 percent of the injected dose. It is calculated that of the 16 percent of the injected dose excreted between 22 days and 4 years after injection six percent comes from the liver and about 10 percent from the skeleton. These values apply to what is described by the Utah workers as "our composite average dog" and agree well with early results obtained on mongrels (PAINTER et al., 1946). As already discussed in this section the small changes observed in the excretion between 4–8 years do not necessitate a change in this equation.

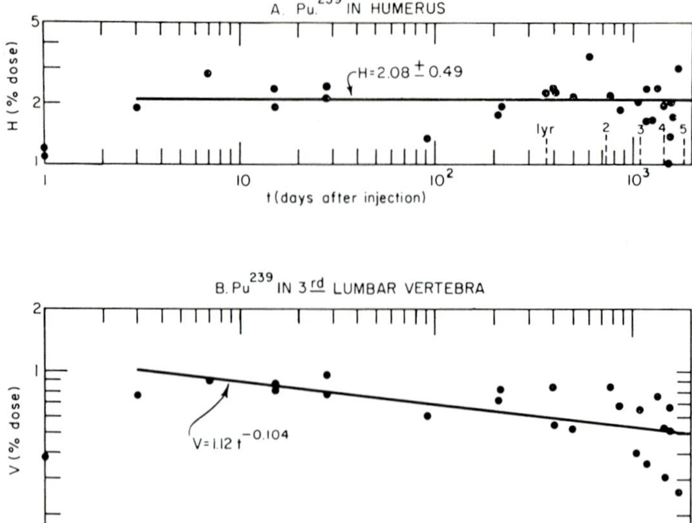

Fig. 10.16. A, amount of ^{239}Pu in one humerus. B, amount of ^{239}Pu in the third lumbar vertebrae of adult beagle after intravenous injection of plutonium citrate. (Stover et al., 1959, by courtesy of authors and publishers)

Total retention was found not to vary with dose level but liver retention was related to dose. It was therefore concluded that skeletal retention must vary with dose level. Skeletal retention appeared to be higher at high dose levels though absolute figures are not yet available.

In the case of man, Langham calculates that an individual may be expected to retain about 80 percent of his original plutonium body burden about 50 years after exposure. An interesting comparison with the excretion of radium is shown in Table 10.23. At 50 years after exposure the radium patient would in contrast retain only 0.3 percent of his original body burden. Approximately 200 years would be needed for man to eliminate half his plutonium body burden. Distribution of plutonium in different tissues of man about 5 months after injection are

Table 10.23. Accumulated urinary plus faecal excretion of ^{226}Ra and ^{239}Pu by man. (Taken from Langham, 1959)

Time after administration	Accumulated excretion (% administered dose)	
	^{226}Ra	^{239}Pu
1 day	46	0.5
10 days	84	2.6
100 days	95	4.7
1 years	97.5	6.3
5 years	98.9	8.7
20 years	99.5	12.2
50 years	99.7	17.6

Table 10.24. Distribution of ^{239}Pu in man following intravenous injection of ^{239}Pu citrate. (Taken from LANGHAM, 1959)

Organ or tissue	Weight of organ or tissue (g)	Pu % injected dose per organ or tissue
Skeleton including marrow	10000	66.0
Liver	1700	23.0
Spleen	300	0.4
Kidneys	700	0.4
Lungs	1000	1.0
Lymphoid tissue	700	0.5
Heart	300	0.1
Gastrointestinal tract	2000	0.5
Muscle and skin	36100	3.9
Blood	5400	0.2
Balance	11800	1.0
Excreted		5.0
	70000	102.0

shown in Table 10.24 (LANGHAM, 1959). The distribution in occupationally exposed individuals who received their ^{239}Pu largely from inhalation after longer time intervals has already been shown in Table 10.19 und 10.20.

H. Problems of Estimating Skeletal Burden

The formulae provided by LANGHAM'S data have been extensively used in estimating body burden in persons accidently exposed to ^{239}Pu contamination. There are however many theoretical difficulties in so doing. Firstly, contamination never takes the form of a single injection of ^{239}Pu citrate into the circulation; secondly, it usually occurs in a healthy young person and not in a hospital patient (LISTER et al., 1963); thirdly, the data from individual patients in the LANGHAM experiment show that the possible errors in applying an average expression to an individual whose metabolic and excretory behaviour may lie anywhere within a wide spectrum are considerable. Further, there are great daily variations in plutonium excretion in any one individual so that average figures over a period of time are required before making estimations of body burden (BEACH and DOLPHIN, 1964; BEACH et al., 1966). An ICRP committee set up to consider the use of power functions for describing excretion behaviour has suggested constants to be used for plutonium leading to the expression

$$Y_{u+f} = 0.0099\, t^{-1.01}.$$

Published cases of accidental internal contamination which have been studied all illustrate the wide variations possible in excretion behaviour. As well as differences in the rates of excretion there is variability in the relative amounts excreted in urine and faeces. This is of course not surprising when it is realized that translocation from different sites will vary in both rate and chemical form.

BEACH and DOLPHIN (1964) have calculated expected urinary excretion rates assuming that the plutonium is released slowly from the site of entry and the results of some of their calculations are shown in Fig. 10.17. Of the 3 curves one was obtained from LANGHAM'S original data (LANGHAM, 1957) and represents excretion following a plutonium citrate injection. The other two curves were calculated, one on the assumption of a hold-up at the site of entry with an ex-

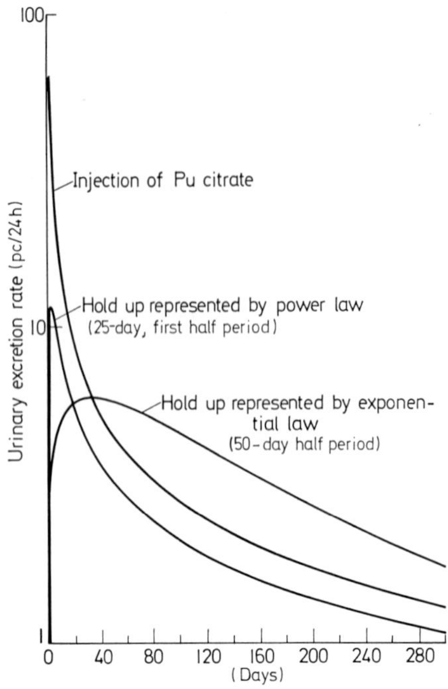

Fig. 10.17. Expected urinary excretion of plutonium in man following an intake of 0.04 μCi. (DOLPHIN et al., 1964, by courtesy of authors and publishers)

ponential release having a half period of 50 days, and the other on the assumption that the release follows a power law with a first half period of 25 days. These curves show that the initial excretion rate following an intake is very dependent on the rate of release from the site of entry. After about a year the excretion rate appears to be less dependent on the rate of release and these curves indicate that a more reliable estimate of the body content may be possible at longer times after the intake (DOLPHIN et al., 1964). These curves were based on urinary excretion data only. This is likely to be unsatisfactory as pointed out by LISTER and his colleagues (LISTER et al., 1963) who investigated two subjects who sustained high levels of contamination on the uncut skin of the hand from accidental contact with acid plutonium solutions. One man for whom the contaminant was mixed plutonium isotopes in aqua regia and nitric acid was followed for 150 days. The other, contaminated with a solution of plutonium in dilute hydrochloric acid containing EDTA and a detergent was followed for 110 days. The excretion patterns showed marked differences from LANGHAM's experimental data particularly in the high and variable amount of plutonium excreted in the faeces relative to urine. This is well illustrated in Fig. 10.18 showing the pattern of urinary and faecal excretion of plutonium. The plutonium content of the faeces varied from day to day by over a factor of 10 and the variations in the urine were greater.

An attempt was made to estimate body burden by the various available methods in the two cases and the results, expressed as fractions of the maximum permissible body burden for bone, as critical organ, are shown in Table 10.25. It is clear that the results vary considerably even when data on both faecal and

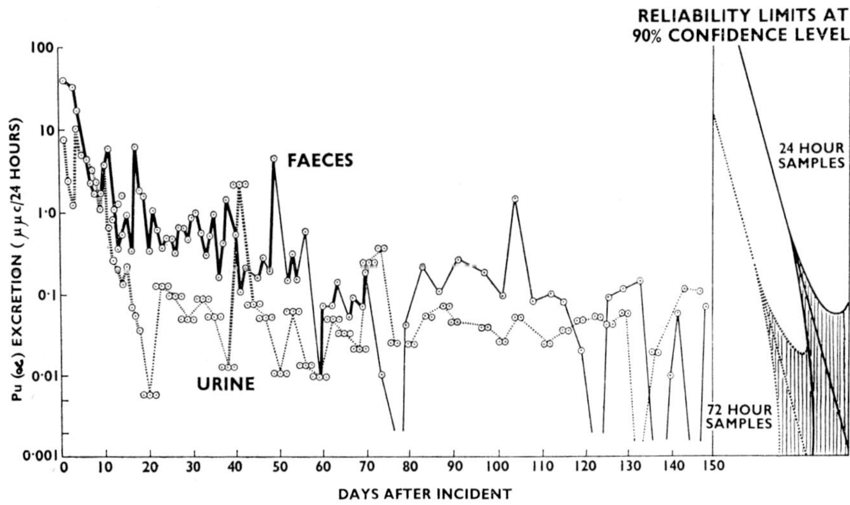

Fig. 10.18. Urinary and faecal excretion of plutonium by man after accidental contamination of the skin. (Adapted from LISTER et al., 1963)

urinary excretion are available and the sampling has continued over a reasonable period of time.

These observations on the plasma clearance, retention and excretion of ^{239}Pu in dogs and man suggest that plutonium reaching the blood stream, in other than colloidal form, is slowly cleared from the plasma where it is bound largely by transferrin, a plasma protein. The mechanism of transfer to the skeleton is not known but once it reaches the skeleton in the adult it is held tenaciously. Its subsequent slow excretion in urine and faeces appears to be variable, both in time and in different subjects, so that precise estimations of body burden are difficult.

Table 10.25. Body burden estimates as a fraction of maximum permissible body burden (with bone as critical organ) calculated from excretion data on 2 patients receiving skin contamination. (From LISTER et al., 1963)

Method of assessment	Biological material	Patient 1	Patient 2
Power function compared with LANGHAM expressions	Urine	~0.1	~0.01
	Faeces	~0.3	0.05–0.1
	U+F	~0.2	~0.05
Cumulative excretion compared with the mean of LANGHAM patients	Urine	~0.3	~0.003
	Faeces	~0.8	~0.13
	U+F	~0.6	~0.07
Power function compared with ICRP expression	U+F	~0.2	—
Spot blood samples compared with LANGHAM data	Blood	0–0.9	—
Integration of power function expression	U+F	0.02	—
Use of ICRP constant for $t=1$	U+F	0.9	0.025
Whole Body Monitoring		<12	

III. Other Plutonium Isotopes

David M. Taylor

Compared with ^{239}Pu the metabolic behaviour of ^{238}Pu and ^{237}Pu has been studied to a very limited extent. These latter two nuclides have a much greater specific activity than ^{239}Pu and this may result in apparent differences in metabolic behaviour when tissue distributions and retention of the three nuclides are compared on a microcurie for microcurie basis.

A. ^{238}Pu

In 1946 Hamilton and his colleagues published the results of administering ^{238}Pu to man and rats by intravenous injection (Crowley et al., 1946). This is the earliest record found of plutonium administration to man and as such is of some historical interest. They describe the solution used as $PuO(NO_3)_2$ in isotonic sodium chloride at pH 5.5. The man was a patient with gastric carcinoma and after 4 days a piece of rib was examined. The greater part of the plutonium was found on the endosteal surface of trabecular bone and in the marrow; little plutonium was present in the periosteum or cortical bone. They calculate that at this time 87 percent of the injected plutonium was in the skeleton. At the end of 296 days only 5 percent of the plutonium injected had been excreted largely in the urine while in the rat 50 percent had been eliminated largely in the faeces. The chart for excretion is shown in Fig. 10.15. Their final comment is that in man the half time of plutonium excretion is probably greater than 50 years.

^{238}Pu has been given to rats, rabbits, pigs and dogs as shown in Table 10.26 by intravenous, intragastric, intramuscular routes and by inhalation.

1. Intragastric

Early experiments feeding ^{238}Pu were carried out on rats by Katz and his colleagues at Hanford, the results being published in a series of reports (Katz, 1952; Katz, 1955; Katz and Weeks, 1954; Katz and Kornberg, 1955). It is a little difficult to be certain how far some of the results in these reports are included in a publication in 1955 in the open literature. The results are in close agreement. They concluded (Katz et al., 1954) that the concentration of the plutonium in the solution fed had no effect on absorption over the range 10^{-5} µg/ml to 1 µg/ml and that the average absorption was 0.0028 percent of the amount fed with 0.0025 percent of the amount fed being deposited in the skeleton. In 1955 Katz and his colleagues (Katz et al., 1955) published a series of experiments on the deposition of ^{238}Pu in bone and soft tissues after 517 low level feedings of $Pu(NO_3)_4$ at doses that varied from 0.04 to 1.2 times the currently accepted maximum permissible concentration of plutonium in drinking water, 10^{-6} to 3×10^{-5} µg/ml. The feeding lasted $9^1/_2$ months and the rats were given 1–4 doses a day. The dosage schedule is shown in Table 10.6. Within the 95 percent limits of confidence the mean percent total body, soft tissue and skeletal deposition was 0.00261 ± 0.00038, 0.00021 ± 0.00010 and 0.00234 ± 0.00028 respectively for rats receiving levels equivalent to 1.2 the maximum permissible concentration. No significant decrease in skeletal plutonium was observed over a period of 250 days following cessation of feeding.

2. Intravenous and Intramuscular Injection

When ^{238}Pu is given to rats, by intravenous injections as the citrate, at dose levels of 4 µCi or 78 µCi/kg 40 to 60 percent of the radioactivity is found in the

Table 10.26. Distribution of ^{238}Pu after administration by various routes

Route	Chemical form	Animal	Time days	Percent retained dose in					Reference
				Skeleton	Liver	Spleen	Lungs	Kidney	
Inhalation	$^{238}Pu_4O_2$	Rat	20	11	2.4	0.1	82	0.5	STUART (1970)
	$^{238}Pu_4O_2$		48.1	23	2.36	0.5	68	0.5	
	$^{238}Pu_4O_2$	Beagle	27	2	1	—	—	—	PARK et al (1970)
			180	13	5	—	—	—	
Intravenous	$^{238}PuO(NO_3)_3$ in iso-	Rat	296	+					CROWLEY et al. (1946)
	tonic sodium chloride	Man	4	87					
	$^{238}Pu(IV)$ citrate	Rat	1	52	14	0.3	—	1.9	BALLOU et al. (1967)
	90% ultrafilterable		21	52	3.2	0.4	—	0.8	
	4 μCi kg		1	58	13	0.1			
	~78 μCi/kg		64	60	1.8	0.2			
Intramuscular	$^{238}Pu(VI)$ citrate	Rat	15	64	6.5				HAMILTON and SCOTT (1953)
Intragastric	$^{238}Pu(NO_3)_4$ pH 2	Rat	250	0.00234[a]					KATZ and KORNBERG (1953)
	$^{238}Pu(NO_3)_4$		—	0.0025[a]					KATZ, WEEKS and THOMPSON (1954)
	$^{238}Pu(NO_3)_4$ pH 2		330	0.00234[a]					KATZ, KORNBERG and PARKER (1955)
	$^{238}PuO_2$	Pig	14	0.0001[a]					SMITH et al. (1966)

[a] Percent of dose absorbed.

skeleton at 1 to 3 days and 13–18 percent in the liver. This is similar to the distribution pattern observed following intravenous injection of 4 μCi ^{239}Pu/kg, but injection of 78 μCi ^{239}Pu/kg resulted in approximately equal distribution in liver and skeleton (BALLOU et al., 1967). In the case of intramuscular injection the tissue distribution pattern is essentially similar for ^{238}Pu and ^{239}Pu (HAMILTON and SCOTT, 1953).

3. Inhalation

Following inhalation of $^{238}PuO_2$ by both rats and dogs (STUART, 1970; PARK et al., 1970) translocation is again predominantly to the skeleton in the rats but to lymph nodes and liver in the dogs. The inhalation studies suggest that translocation from the lungs to other tissues would appear to be rather faster with ^{238}Pu than with ^{239}Pu, as reference to Tables 10.27 and 10.28 will show. In the dog the particle size of the ^{238}Pu which was inhaled was greater then that of the ^{239}Pu. However, this was not the case in the rat experiments yet the translocation of ^{238}Pu still appeared to be more rapid (STUART, 1970). Comparison of the rats which inhaled $^{238}PuO_2$ or $^{239}PuO_2$ of similar particle size and crystalline form showed that, at times greater than 1 year, between 23 and 48 percent of the body burden of ^{238}Pu was in the skeleton compared with less than 1 percent of the ^{239}Pu. The liver burden was also greater in the ^{238}Pu rats than in the ^{239}Pu-bearing animals (STUART, 1970). These observations suggest that ^{238}Pu is more soluble in tissue fluids and is more readily translocated from the lungs.

Table 10.27. Tissue distribution of ^{239}Pu in rats after inhalation of $^{239}PuO_2$ (percent of body burden at death). (Taken from STUART, 1970)

Rat No.	Days after exposure	Skeleton	Liver	Spleen	Lungs	Tracheobronchial lymph nodes
95	7	0.01	0	0	67	0.1
63	13	0.01	0.02	0	97	0.3
73	28	0.1	0.4	0.02	96	1.4
30	68	0.1	0.3	0.04	97	1.8
4	113	0.2	0.3	0	99	—

Table 10.28. Tissue distribution of ^{238}Pu in rats after inhalation of $^{238}PuO_2$ (percent of body burden at death). (Taken from STUART, 1970)

Rat No.	Days after exposure	Skeleton	Liver	Spleen	Lungs	Tracheobronchial lymph nodes
84	20	11	2.4	0.1	82	0.4
67	48	12	1.3	0.2	84	0.6
39	78	23	1.9	0.2	77	0.8
50	127	15	1.7	0.2	78	3.4
59	320	30	1.6	0.7	64	3.3
5E	465	48	3.6	0.6	40	0.1
4E	481	23	3.6	0.5	68	3.3

B. Relative Toxicities of ^{238}Pu and ^{239}Pu

Experiments to test the acute toxicity of ^{238}Pu in comparison with ^{239}Pu given intravenously on an equivalent microcurie basis were carried out by BALLOU and his colleagues on rats (BALLOU et al., 1967) using mortality data, weight loss,

Table 10.29. Comparative toxicity of ^{238}Pu and ^{239}Pu administered intravenously as the citrate (6 rats per group). (Taken from BALLOU et al., 1967)

Dose (μc/kg)		Survival time (average and range – days)	
^{238}Pu	^{239}Pu	^{238}Pu	^{239}Pu
—	79	—	23 (16–46)
98	98	80 (62–115)	16 (14–20)
120	110	59 (50–68)	12 (10–15)
130	140	62 (47–79)	10 (10–11)
160	—	43 (34–51)	

haematological observations and histological observations as indicators. When a high dose in excess of 70 μCi/kg was given as shown in the Table 10.29 ^{239}Pu appeared to be more toxic than ^{238}Pu. Unfortunately no data on toxicity at lower doses is given. The authors suggest that the difference in toxicity at high doses is due to differences in the pattern of distribution of the two isotopes which in turn they think may be explained by differences in the physical state of the isotopes resulting from the 270-fold greater mass associated with an equal microcurie quantity of ^{239}Pu. The differences in deposition are not striking, even at high doses, and when given in low doses i.e. 4 μCi/kg the differences in distribution between the two isotopes in rats would hardly appear significant. 4 μCi/kg of ^{238}Pu citrate or of ^{239}Pu both resulted in a skeletal retention at 21 days of 52 percent of the dose retained while liver retention was 3.4 and 3.2 percent respectively.

C. Metabolic Patterns of ^{237}Pu and ^{239}Pu

Studies to test the comparative metabolic behaviour of ^{237}Pu and ^{239}Pu have been carried out by BAIR and his colleagues (BAIR et al., 1970) on beagles. Both isotopes were given intravenously as the nitrate and the results are shown in Table 10.30. After 30 days an average of 46 percent of the body burden of ^{237}Pu was in the skeleton and 38 percent in the liver compared with an average of 6.4

Table 10.30. Comparative tissue distribution and excretion of ^{237}Pu and ^{239}Pu 30 days after intravenous injection of $Pu(NO_3)_4$ (percent of retained body burden). (After BAIR, NELSON, and WILLARD, 1970)

Dog. No.	^{237}Pu		^{239}Pu		
	322	300	22	27	243
Tissue					
Liver	33	44	85	83	78
Skeleton	50	43	5.8	6.4	6.9
Kidney	0.88	0.93	0.12	0.13	0.088
Spleen	0.75	0.82	4.4	5.8	9.6
Lung	0.46	0.60	0.23	0.30	0.31
Muscle	10	5.7	3.2	2.7	3.2
Pu excreted in urine[a]	19	7.2	0.48	0.43	0.47
Pu excreted in feces[a]	9	13	0.64	0.95	0.62
μCi injected	0.85	1.1	25	21	21
μg injected	0.0011[b]	0.0014[b]	400	340	340

[a] Percent of injected Pu. [b] Includes ^{239}Pu and ^{233}U contaminants.

percent of the ^{239}Pu in the skeleton and 82 percent in the liver. The excretion of ^{237}Pu in both urine and faeces was very much greater than of ^{239}Pu.

BAIR and his colleagues attribute these differences to the difference in mass of plutonium administered, about 350 μg in the case of ^{239}Pu and about 1 ng in the case of ^{237}Pu (including the mass of contaminants 420 pg ^{239}Pu and 500 pg ^{233}U). The small mass of ^{237}Pu would have provided little opportunity for colloid formation and as will be seen from earlier discussions of intravenous injection of ^{239}Pu the colloidal character of plutonium in the blood stream is relevant to the pattern of tissue distribution. $(Pu_4NO_3)_4$ is, as reference to Table 10.5 will show, unless filtered, largely in colloidal form. ^{237}Pu in the present experiment has behaved like a citrate complex of ^{239}Pu or one that has been filtered immediately before injection.

TODD and LOGAN (1966) have studied the distribution in rats of ^{237}Pu mixed with sufficient ^{239}Pu to give an injection mass of 0.05 μg Pu/kg body weight. The solution was deliberately prepared to contain an unknown proportion of colloidal material. In one rat killed 4 days after intraperitoneal injection 23 percent of the dose was found in the liver and 53 percent in the skeleton. This is similar to the distribution found by TAYLOR et al. (1961) at 4 days after intravenous injection of about 50 μg ^{239}Pu/kg body weight in a partially polymeric form, thus suggesting that within these limits the mass of material injected has little effect on the tissue distribution pattern.

D. Conclusions

The differences which have been observed in the metabolic behaviour of ^{237}Pu, ^{238}Pu and ^{239}Pu can be ascribed generally to mass effects resulting from the large differences in the specific activation of the isotopes. However, the apparently greater solubility *in vivo* of ^{238}Pu, as compared to ^{239}Pu, after inhalation of PuO_2 particles of the same particle size and similar crystalline form, is not easily explained in terms of a mass effect.

The radiation characteristics of ^{238}Pu and ^{239}Pu are very similar, thus because of the greater specific activity of ^{238}Pu the energy deposition in the region of a $^{238}PuO_2$ particle will be about 280 times greater per unit time than that deposited in the region of a similar sized particle of ^{239}Pu. This increased radiation dose could result in greater solubility of $^{238}PuO_2$ particles by either or both of two processes. Firstly the greater self-irradiation of the $^{238}PuO_2$ particle might result in some disturbance of the lattice structure which would render the $^{238}PuO_2$ particle more easily soluble in body fluids than a similar $^{239}PuO_2$ particle. Alternatively the higher radiation dose to the tissues around a $^{238}PuO_2$ particle might lead to increased production or secretion of metabolites which would lead to a more rapid solubilization of the particle.

One important point raised by the differences which have been observed between the metabolic behaviour of these three plutonium isotopes is the extent to which ^{237}Pu or ^{238}Pu may be used to predict the behaviour of ^{239}Pu and *vice versa*. The experiments by BALLOU et al. which compared the tissue distribution in rats of 0.24 μg ^{238}Pu/kg and 68 μg ^{239}Pu/kg after intravenous injection as citrate showed that no significant differences in tissue distribution occurred between 1 and 21 days after injection of this soluble form of plutonium. However, the scanty evidence which is available at present suggests that this may not be true for insoluble plutonium compounds. Until more information is available considerable caution must be exercised in using data obtained with one plutonium isotope to predict the biological behaviour of other plutonium isotopes.

IV. The Pattern of ^{239}Pu Distribution in the Skeleton

JANET VAUGHAN

The pattern of distribution of plutonium in the skeleton is unlike that of the alkaline earths, of yttrium or the other actinide elements. ^{228}Th, which is also characteristically deposited on bone surfaces, may follow the same pattern but the necessary experimental work to confirm this proposal has not yet been done. However, detailed studies are now available on certain aspects of plutonium distribution in both bone and bone marrow.

The characteristic distribution pattern of plutonium, americium and calcium is shown in three autoradiographs of longitudinal sections of the femurs of young rabbits injected respectively with americium chloride, plutonium citrate and calcium chloride (Fig. 10.19). There is no blackening in the marrow of the rabbits given either calcium or americium but a heavy blackening in the marrow in the section of the rabbit given plutonium citrate. There is an outlining of all bone surfaces in sections from the plutonium and americium rabbits but that on the endosteal surfaces is much heavier in the case of plutonium than of americium. There is heavy concentration in both bones beneath the epiphyseal plate. Both sections also show a spotty distribution throughout the cortex but this is more notable in the case of americium. In the case of calcium the picture is very different. There is a heavy concentration on the endosteal surface of the metaphysis and the periosteal surface of the diaphysis and beneath the epiphyseal plate while throughout the bone there is a faint diffuse reaction. Calcium is what MARSHALL (1969) has described as a "volume seeker" like the other alkaline earths. Both plutonium and americium are "surface seekers" but their patterns of surface deposition are different.

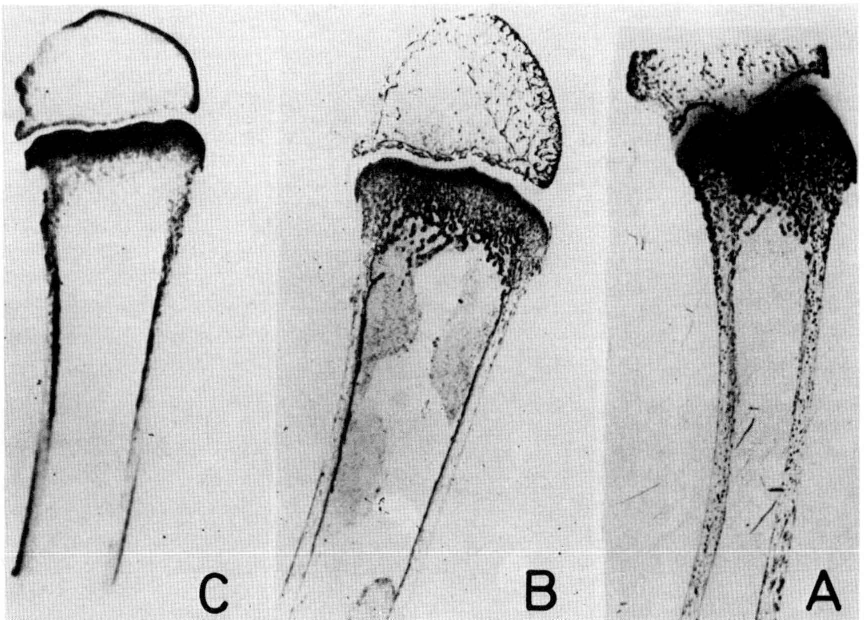

Fig. 10.19. Comparison of gross autoradiographs through femora of weanling rabbits injected with a) ^{241}Am, b) ^{239}Pu and c) ^{45}Ca and killed within 24 hours. Exposed on ILFORD ordinary plates. (WILLIAMSON, 1963, by courtesy of author)

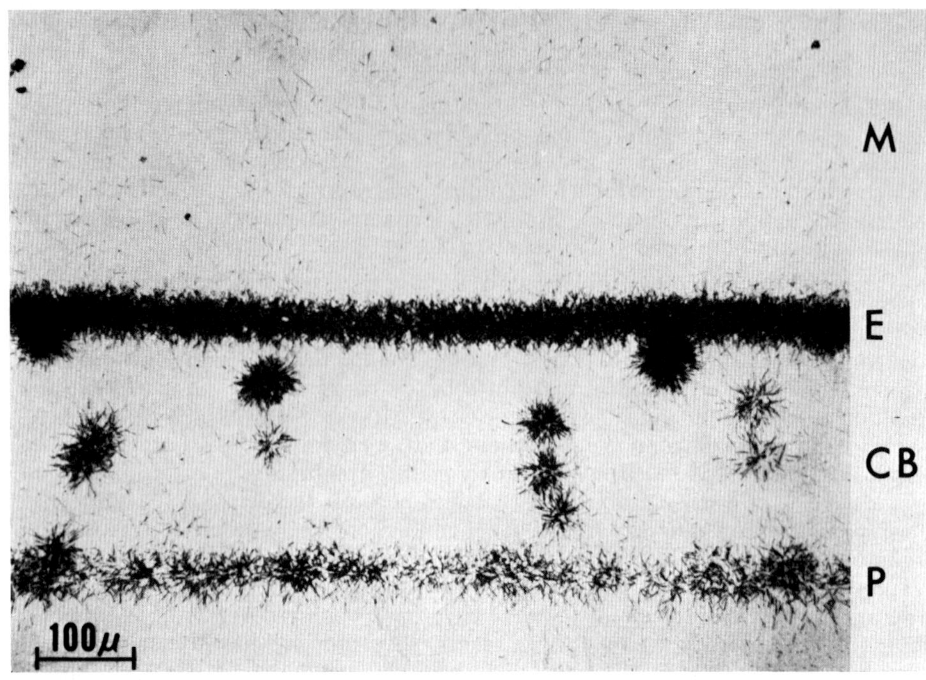

Fig. 10.20. Autoradiograph of mid shaft region of the tibia of a mouse sacrificed six days after intravenous injection of monomeric plutonium (90% ultrafilterable). The exposure time on NTA film was 3 months. The following symbols are used: *M* marrow, *E* endosteum, *CB* cortical bone, *P* periosteum. (ROSENTHAL et al., 1968, by courtesy of authors and publishers)

Table 10.31. Distribution of three forms of ^{239}Pu after intravenous injection of 0.1 μCi into mice[a] to show relative retention in bone and marrow. (Taken from ROSENTHAL et al., 1968)

	Mean percent of injected dose		
	Monomeric Pu (90% ultrafilterable)	Polymeric Pu (30% ultrafilterable) (mid-range)	Polymeric Pu (18% ultrafilterable)
	Six days after injection		
Liver	28	46	80
Spleen	0.2	1.7	2.1
Femurs	4.2	2.4	1.3
Tibial shaft			
bone	0.39	0.23	0.13
marrow	0.02	0.80	0.03
	90 days after injection		
Liver	2.5	25	53
Spleen	0.3	1.3	2.6
Femurs	2.8	2.2	1.0
Tibial shaft			
bone	0.41	0.27	0.14
marrow	0.05	0.08	0.03

[a] Values shown are the means of three determinations (four in group sacrificed 90 days after monomeric Pu). Each determination was based on samples pooled from two mice.

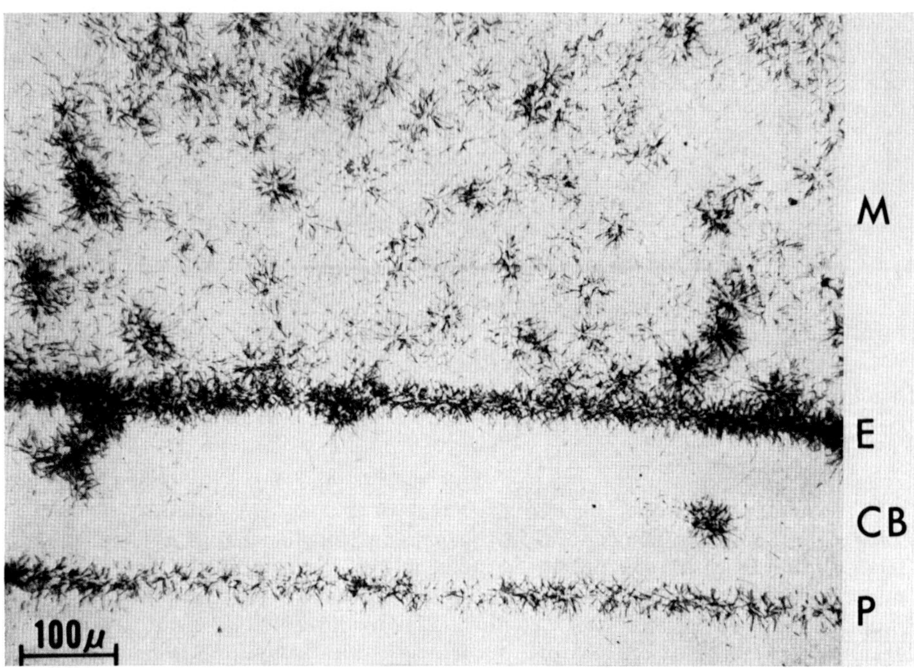

Fig. 10.21. Autoradiograph of the mid shaft region of the tibia of a mouse sacrificed six days after intravenous injection of mid range polymeric plutonium (30% ultrafilterable). The exposure time on NTA film was 3 months. (ROSENTHAL et al., 1968, by courtesy of authors and publishers). Symbols are as in Fig. 10.20

The relative amount of plutonium found in bone and marrow is now recognised to depend on the physicochemical nature of the solution injected. This was first shown in 1961 by SCHUBERT and his colleagues (SCHUBERT et al., 1961) in the mouse and confirmed in the rat in 1964 (MARKLEY et al., 1964).

BULDAKOV et al. (1970), quote, BELYAEV et al. on the relative retention of plutonium nitrate in the bone and marrow of rabbits after injection and draw attention to the high marrow retention for as long as 90 days. During the first three days these workers found the concentration of plutonium in the marrow was somewhat higher than in the spongy bone and from four to six times higher than in compact bone. Even at three months after its administration the concentration of plutonium in the bone marrow was higher than in the compact bone and only by the ninth month was it lower than in both spongy and compact bone (BELYAEV et al., 1970).

ROSENTHAL et al. (1968) have analysed the effect of the physicochemical nature of the solution on bone and marrow retention in a most elegant series of experiments. They prepared solutions of plutonium citrate, one of which was 90 percent, one 30 percent and the other 18 percent ultrafilterable, injected them into mice and estimated the retention of plutonium in both bone and marrow of the tibial shaft at 6 days and 90 days after injection. These results are shown in Table 10.31. At 6 days after injection of the monomeric solution the marrow contained 5 percent of the total activity present in the sample while injection of a medium polymeric solution gave a value of over 25 percent. After injection of an even more highly polymeric solution (^{239}Pu less than 1 percent ultrafilterable)

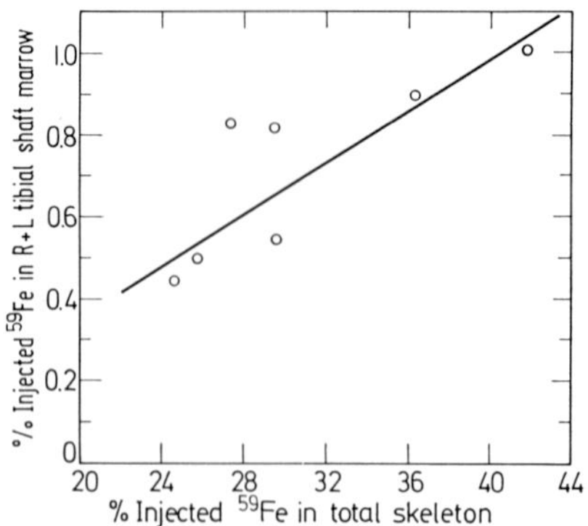

Fig. 10.22. Relationship of iron-59 uptake in mouse tibial shaft to total skeletal uptake at 5 hours after intravenous injection. (LINDENBAUM et al., 1969, by courtesy of authors and publishers)

60 percent of the isotope in tibial shaft was in the marrow (ROSENTHAL et al., 1969). The amount of plutonium in the marrow did not change between 6 and 90 days after injecting the polymeric solution but increased in the case of monomeric plutonium. These differences in broad outline are quite apparent in autoradiographs as shown in Figs. 10.20 and 10.21. In Fig. 10.20 is seen an autoradiograph of the mid shaft of the tibia of a mouse sacrificed 6 days after intravenous injection of monomeric plutonium. The greater number of the alpha tracks are on the bone surface and few appear in the marrow while in Fig. 10.21 an autoradiograph of the same area from a mouse given polymeric plutonium there is a heavy deposition of plutonium in the marrow and far less on the endosteal surface.

A question must be raised however about the validity of the answer obtained from radiochemical estimations in these experiments. It is known from the work of BLEANEY and VAUGHAN (1971) which is discussed in Sec. III. A that when marrow is extruded from the mid diaphysis of rabbit bones the osteogenic tissue is extruded with the marrow and that this may contain some of the plutonium retained on "the bone surface". It is possible that the same occurs when the mouse marrow is extruded so that all the marrow plutonium estimates reported are too high and contain some of the "bone surface" plutonium. Histological studies should serve to clarify this point. More recently ROSENTHAL and her colleagues (LINDENBAUM et al., 1969) have described another method of estimating skeletal marrow using ^{59}Fe uptake and propose to carry out double tracer experiments with both ^{59}Fe and ^{239}Pu. Total skeletal uptake of ^{59}Fe which they assume to be equivalent to marrow uptake however varies widely among individual mice (Fig. 10.22).

A. Bone Surfaces

HAMILTON (1947) first noted that plutonium was found characteristically on bone surfaces. In view of what is now known it is of interest to remember that SCOTT and his colleagues as early as 1948 (SCOTT et al., 1948) while commenting on the heavy deposition on the endosteal surface of the trabeculae published an autoradiograph showing "a heavy deposit of plutonium" in an area of periosteum which had become stripped from the bone surface as well as on the underlying mineral/matrix surface (Fig. 10.23).

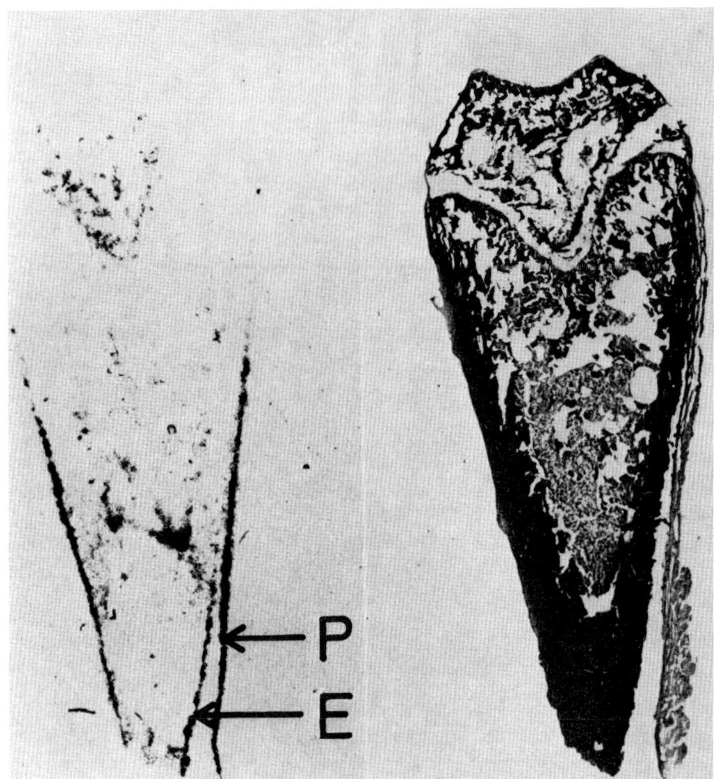

Fig. 10.23. Radioautograph (left) and section (right) of femur from a rat injected intramuscularly with 25 γ μCi of plutonium and sacrificed after 5 weeks. During the preparation of the sections the periosteum became separated from the shaft, and the autograph shows a heavy deposit of plutonium in this stripped layer. No calcium was detected in this particular area. *P* indicates periosteum, *E* indicates mineral in autoradiograph. Haematoxylin eosin and silver nitrate; × 10.4. (Adapted from SCOTT et al., 1948)

Further studies of this surface distribution were made by ARNOLD and JEE in both rats (1957) and dogs (1959, 1962). They indicated that directly after injection plutonium was found on all endosteal surfaces as seen in Fig. 10.24 which shows a contact autoradiograph of a costochondral junction demonstrating the heavy deposition. In Fig. 10.25 is shown a high power view of an autoradiograph of a trabecula from the vertebral body of a dog showing the detailed picture of the alpha tracks on the bone surface. Subsequently this surface deposition may be removed by osteoclastic resorption in the process of remodelling. In Fig. 10.26 is shown an autoradiograph from the lower end of the metaphysis of the femur of a young rabbit 8 days after injection of $^{239}Pu(NO_3)_4$. Osteoclasts and macrophages at the lower end of the metaphyseal trabeculae are seen heavily laden with plutonium aggregates (VAUGHAN, 1970a). In sites where new bone is laid down some at least of the endosteal concentration of plutonium becomes buried as shown in Fig. 10.27. This is an autoradiograph of a trabecula from a vertebra of a young dog injected 1200 days previously. The dense wavy line represents the deposition at the original endosteal border; the bone below formed after in-

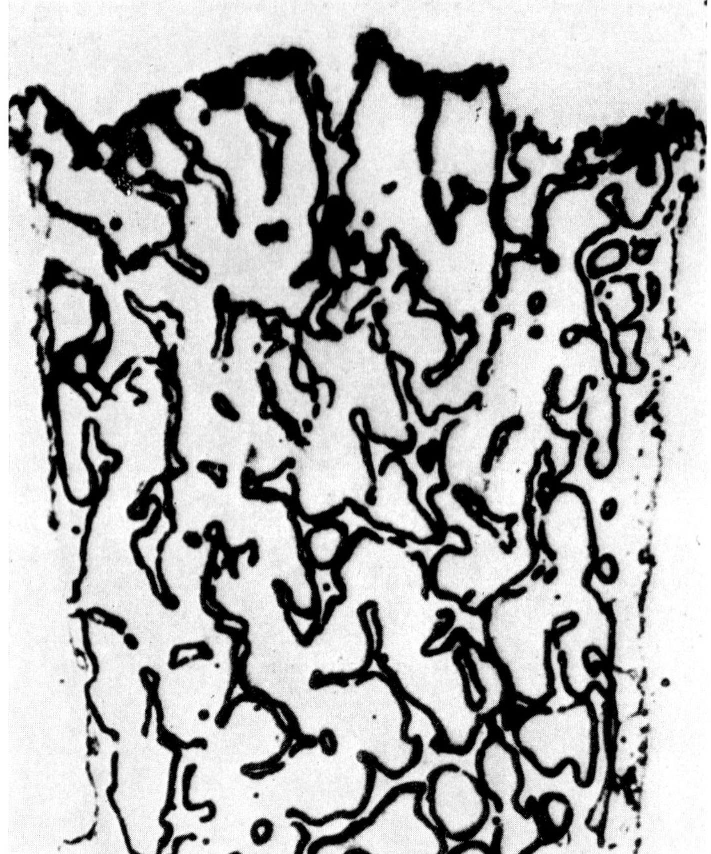

Fig. 10.24

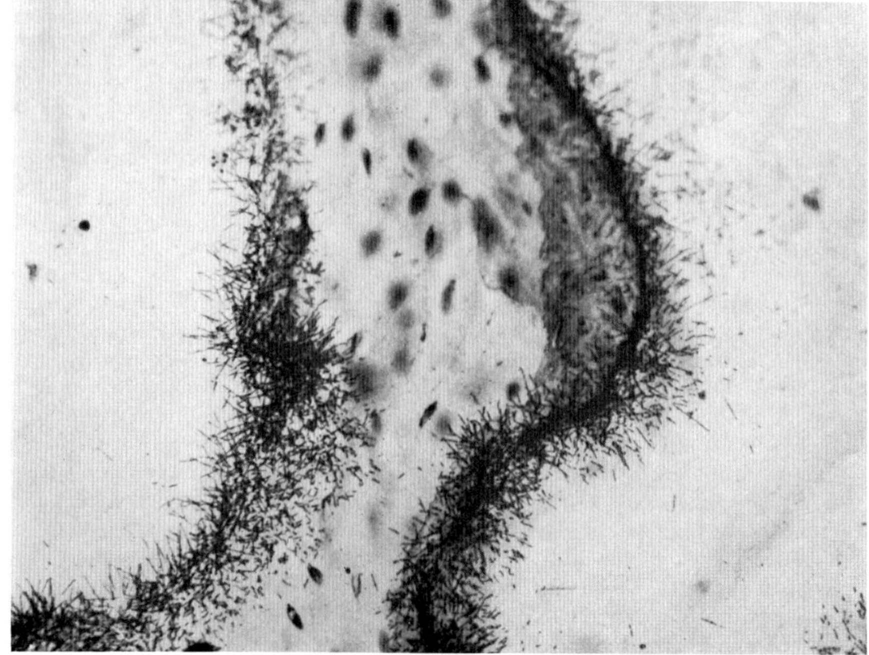

Fig. 10.25

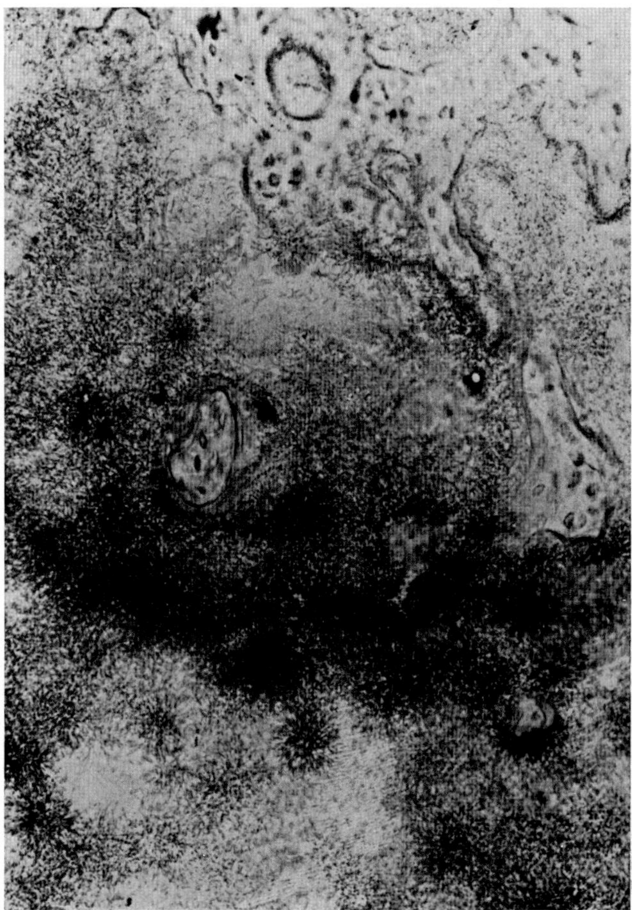

Fig. 10.26. Autoradiograph of thin section of the upper end of the femur at lower level of metaphyseal trabeculae of a weanling rabbit killed 8 days after an intravenous injection of 5 μCi/kg $Pu(NO_3)_4$. Kodak AR 10 film exposed 6 weeks. Note heavy concentration of Pu in the marrow at the end of the resorbing trabeculae. (VAUGHAN, 1970a)

Fig. 10.24. Contact radioautogram of costochondral junction demonstrating the heavy endosteal deposition and irregular lighter vascular and periosteal deposition in dog sacrificed 24 hours after intravenous ^{239}Pu injection. (ARNOLD and JEE, 1962, by courtesy of authors and publishers)

Fig. 10.25. High power view autoradiograph of a trabecula from vertebral body of a dog showing alpha tracks on the bone surface only. (WILLIAMSON, 1963, by courtesy of authors)

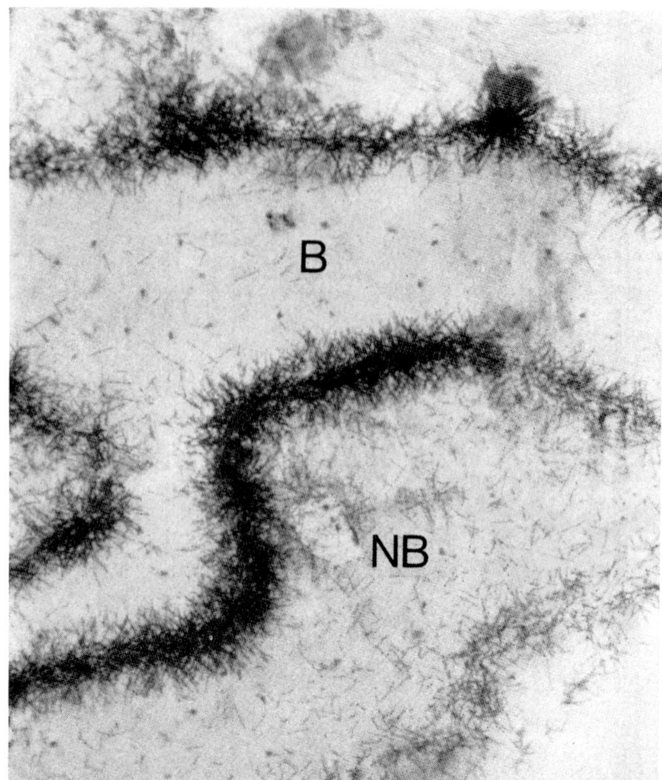

Fig. 10.27. Detailed radioautogram from the vertebra of a dog receiving 0.1 μCi/kg and sacrificed 1200 days after intravenous ^{239}Pu citrate injection. The dense wavy line represents the deposition at the original endosteal border; the bone below formed after injection contains a diffuse distribution of α activity, while the bone deposited before injection is free of activity. *B* indicates old bone. *NB* indicates post injection bone. (ARNOLD and JEE, 1962, by courtesy of authors and publishers)

jection contains a diffuse distribution of alpha activity, while the bone deposited before injection is free of activity.

Bone laid down before the plutonium is injected contains no alpha tracks in the mineralized matrix. A very few tracks are seen around HAVERSIAN systems which are actually laying down osteoid, as seen in Fig. 10.28, rather more around HAVERSIAN systems where active resorption is in progress, as in Fig. 10.29. This picture is again different from that seen in the case of americium. Americium is concentrated heavily in resorbing systems as shown in Fig. 10.30 and a little is seen in systems laying down bone as is shown in Fig. 10.31 (HERRING et al., 1962; WILLIAMSON and VAUGHAN, 1964). This heavy deposition in resorbing HAVERSIAN systems or building sites gives the picture so characteristic of americium uptake in autoradiographs of cross sections of cortical bone (see Fig. 10.32).

In the case of the heavy line of concentration in the region of the epiphyseal plate there are again differences between plutonium and americium. If autoradiographs are prepared at hourly intervals immediately after injection of both radionuclides the concentration of americium is rapid and can be seen within the half hour over the lowest hypertrophic cartilage cells as seen in Fig. 10.33. In the case

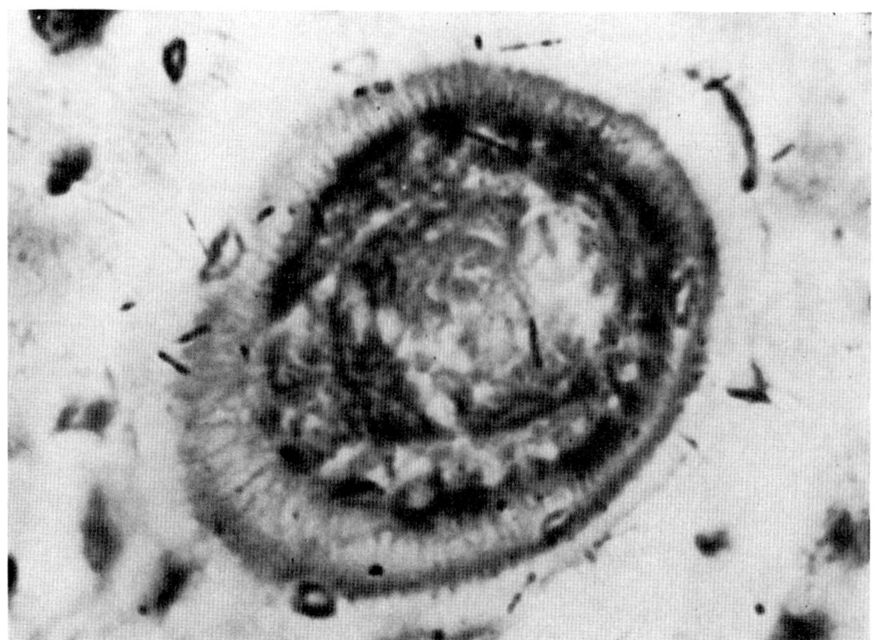

Fig. 10.28. Autoradiograph of a cross section of the mid diaphysis of the humerus of a dog 7 months old, injected intravenously with $^{239}PuCl_4$ (5 μCi/kg) and killed 4 hours later (Kodak AR 10, exposure 1 month) showing a few tracks behind the osteoid border and over the contents of an active Haversian system (× 780, reduced to $^6/_7$). (HERRING et al., 1962, by courtesy of authors and publishers)

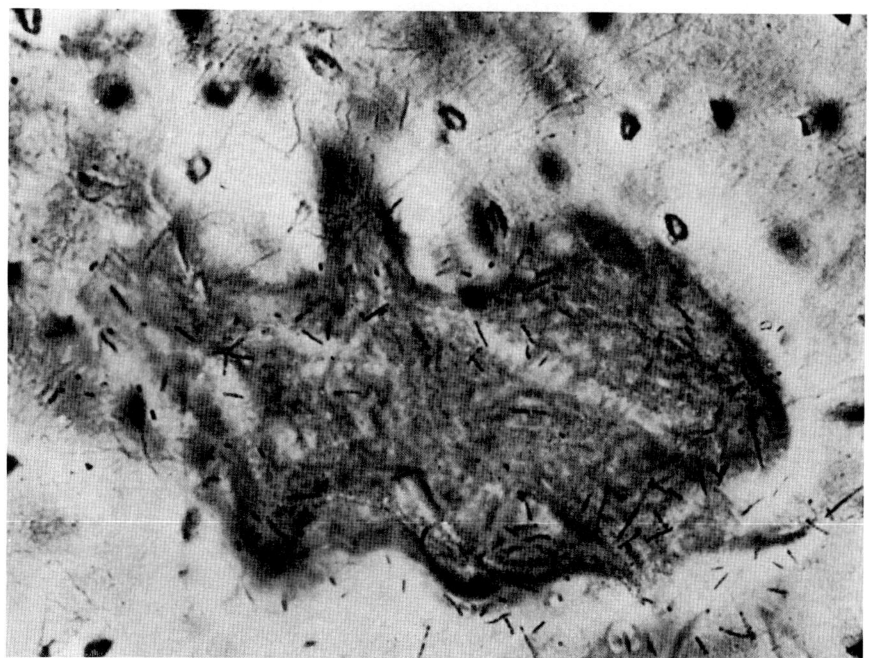

Fig. 10.29. Autoradiograph of a cross section of the mid diaphysis of the humerus of a dog 7 months old injected intravenously with $^{239}PuCl_4$ (5 μCi/kg) and killed 4 hours later. Kodak AR 10 exposure 1 month. Showing concentration of α tracks on irregular resorbing surface of an Haversian system (× 685, reduced to $^6/_7$). (HERRING et al., 1962, by courtesy of authors and publishers)

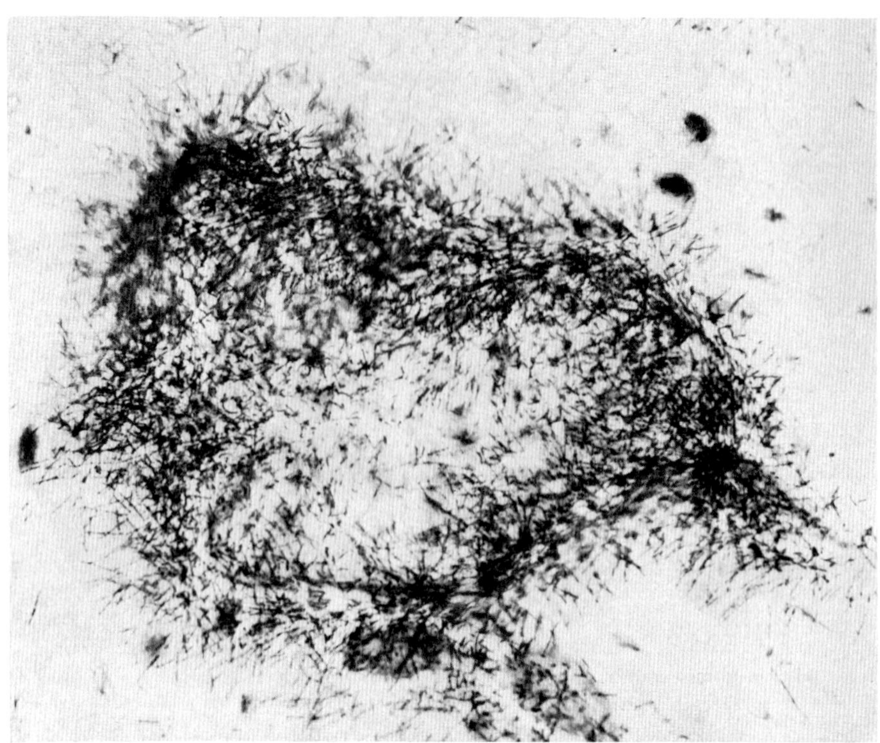

Fig. 10.30

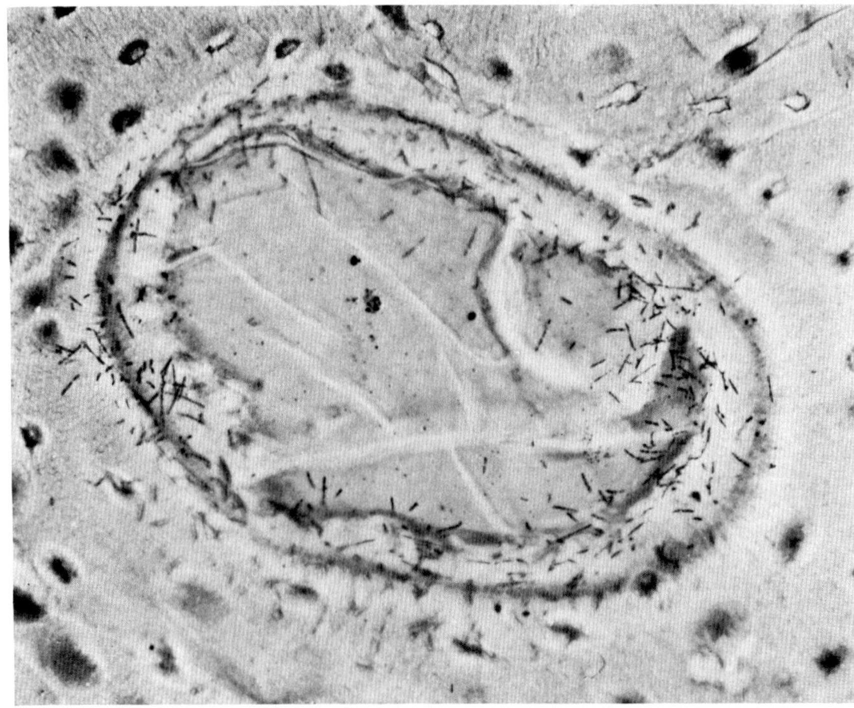

Fig. 10.31

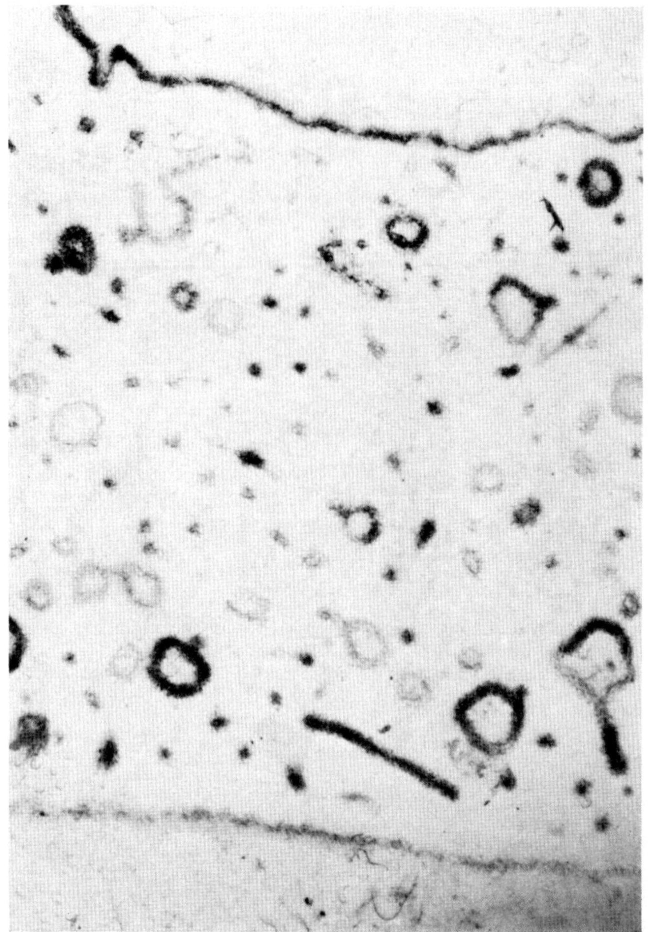

Fig. 10.32. Cross section, cortical bone, young dog killed 4 hours after injection intravenously 5 μCi $AmCl_3$ exposure 4 weeks. Note heavy periosteal uptake compared to endosteal and uptake round Haversian systems some of which is also heavy. (× 60, reduced to $^6/_7$)

Fig. 10.30. Autoradiograph of a cross section of the mid diaphysis of the humerus of a dog 7 months old injected intravenously with $^{241}AmCl_3$ (5 μCi/kg) and killed 4 hours later. (Kodak AR 10 exposure 5 weeks.) It shows heavy concentration of α tracks on irregular resorbing surface of an Haversian system (× 730, reduced to $^6/_7$). (WILLIAMSON, 1963)

Fig. 10.31. Autoradiograph of a cross section of the mid diaphysis of the humerus of a dog 7 months old, injected intravenously with $^{241}AmCl_3$ (5 μCi/kg) and killed 4 hours later (Kodak AR 10 exposure 5 weeks). Showing far lighter concentration of α tracks over and behind the osteoid border on active surface of part of a large Haversian system than in Fig. 10.30. (× 730, reduced to $^6/_7$) (WILLIAMSON, 1963)

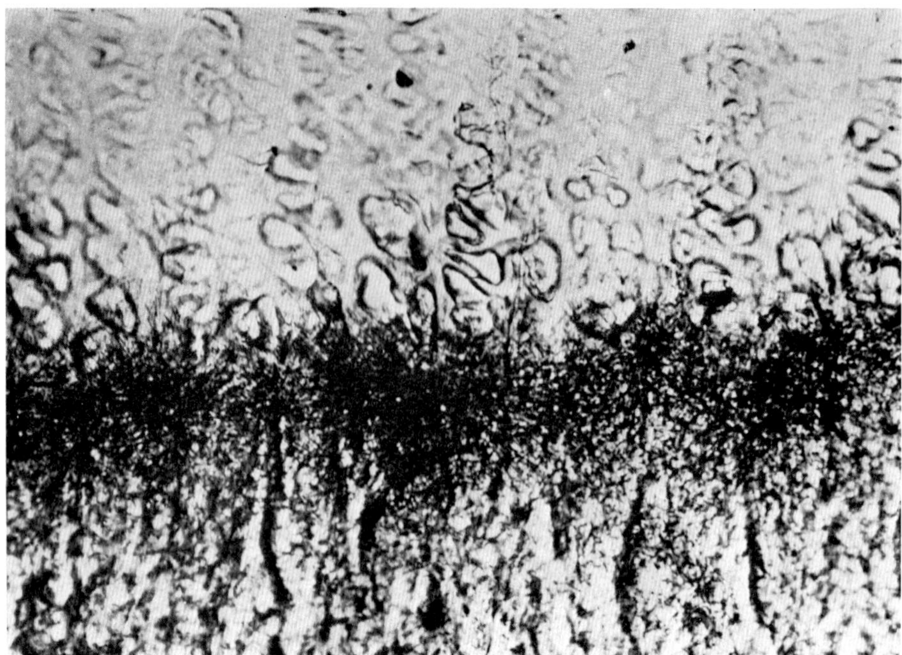

Fig. 10.33. Autoradiograph of longitudinal section through proximal cartilage plate of tibia of weanling rabbit killed 30 minutes after injection intravenously of 5 μCi/kg $AmCl_3$ to show activity over hypertrophic cells (AR 10 exposure 2 weeks) (× 350, reduced to $^6/_7$). (WILLIAMSON, 1963, by courtesy of authors)

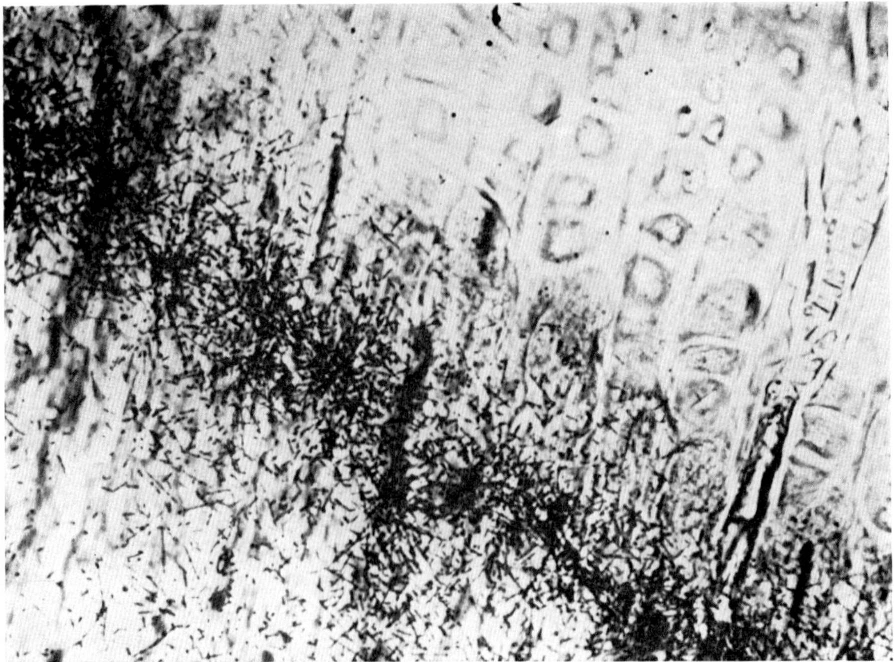

Fig. 10.34. Autoradiograph of longitudinal section through proximal cartilage plate of weanling rabbit injected intravenously with 5 μCi/kg $AmCl_3$ killed 4 hours after injection. Kodak AR 10 film $2^1/_2$ days exposure. × 280, reduced to $^6/_7$. (WILLIAMSON, 1963, by courtesy of authors)

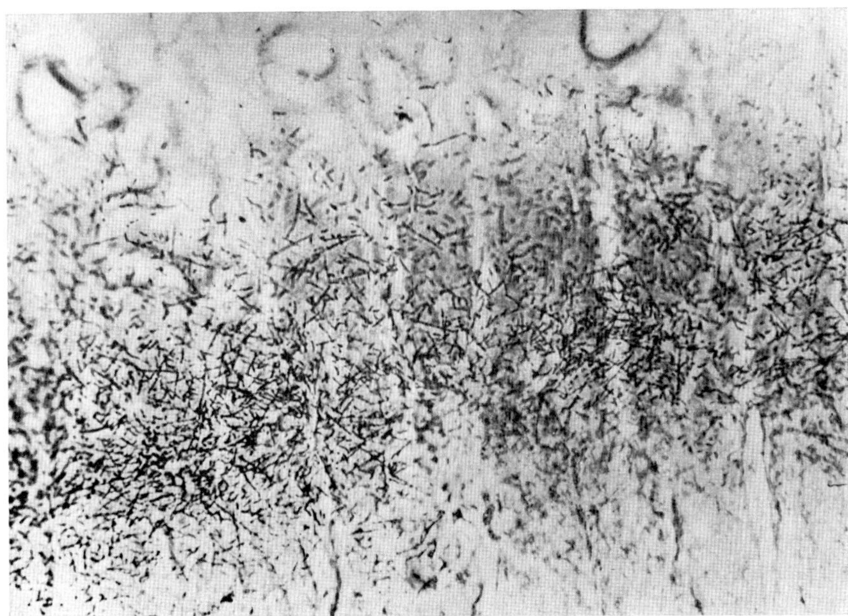

Fig. 10.35. Autoradiograph of longitudinal section of proximal cartilage plate of tibia of weanling rabbit killed 4 hours after intravenous injection of 5 μCi/kg ^{239}Pu citrate to show α tracks below hypertrophic cells. AR 10 3 weeks exposure. (WILLIAMSON, 1963, by courtesy of authors)

of plutonium alpha tracks are seen in the same site but in less concentration. At four hours after injection as shown in Figs. 10.34 and 10.35 the concentration of ^{241}Am is dense at some distance below the hypertrophic cells but the concentration of ^{239}Pu is very much less even and less dense. The two rabbits were each given 5 μCi/kg of the radionuclide but to obtain the ^{239}Pu autoradiograph 3 weeks exposure was needed compared with 4 days for ^{241}Am.

In association with these autoradiographic findings it is interesting to note that TAYLOR (1962), using radiochemical techniques, found that americium reached high levels of concentration in the distal and proximal ends of the rat bone more rapidly than plutonium. This has already been illustrated in the section on metabolism, Fig. 10.11.

Recently BLEANEY and VAUGHAN (1971) have made a detailed analysis of the precise distribution of plutonium on the endosteal surface of cortical bone in the rabbit. If marrow is extruded from the femoral shaft it takes with it the endosteal osteogenic tissue leaving the surface of the bone devoid of cells. In Fig. 10.36 is seen a longitudinal section of marrow extruded from the diaphysis of a long bone. On the edge in the adult rabbit is a layer of largely flattened spindle shaped cells which may be single or overlap one another forming a layer 2–3 cells deep. Osteoblasts and osteoclasts are rarely seen since in the adult most of the endosteal surface is neither resorbing nor actively laying down bone (SISSONS et al., 1967; JOWSEY et al., 1965). In Fig. 10.37 is seen a longitudinal section of the bone from which the marrow has been extruded. Its surface is devoid of cells. If several such sections are examined small groups of cells are sometimes seen suggesting that marrow extrusion had not been complete. Autoradiographs of the surface

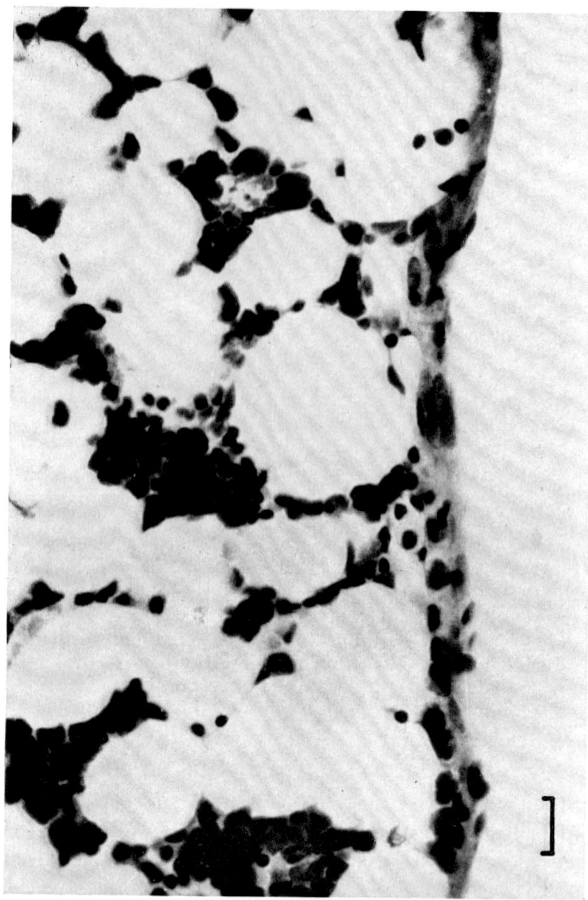

Fig. 10.36. A longitudinal section of extruded marrow from mid diaphysis of femur of adult rabbit stained with haematoxylin and eosin to show the osteogenic cells on surface. The marker line is 20 μ. (BLEANEY and VAUGHAN, 1971, by courtesy of authors and publishers)

of the extruded marrow indicate that it has an appreciable uptake of plutonium as seen in Fig. 10.38. Estimation of the radiation dose from alpha track counting on both the extruded marrow and the bone surface showed that the amount of plutonium in the endosteal cells was roughly proportional to that on the mineral/matrix surface, and that the same relative proportions were maintained though the overall figure for "bone surface" plutonium may change (BLEANEY and VAUGHAN, 1971). The calculations of radiation dose received from 1) the mineral/matrix surface, 2) the endosteal cells themselves and 3) from the two together based on alpha track counting after both an intravenous injection (1.5 μCi/kg) and an intramuscular injection (2.5 μCi/kg) of $^{239}Pu(NO_3)_4$ in animals 6 days post injection are shown in Table 10.32. About 40 percent of the "bone surface" plutonium appeared to be retained in the cells while the rest was present on the mineral/matrix. These observations have been made over a period of one year following both intramuscular and intravenous injections of a plutonium nitrate solution. TAYLOR (1969) describes the solution used for intravenous injection as

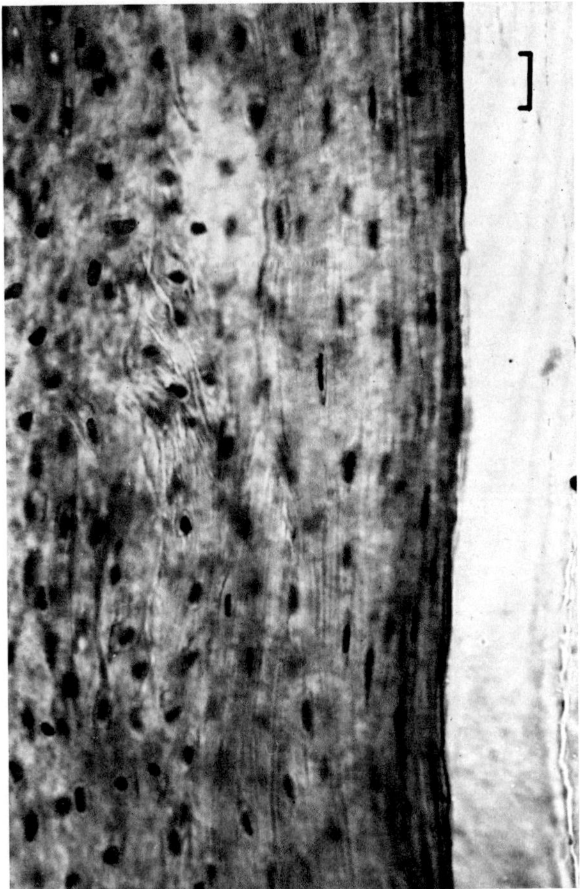

Fig. 10.37. Longitudinal section of diaphysis of femur of adult rabbit from which marrow has been extruded. The surface is devoid of endosteal cells. The marker line is 20 μ. (BLEANEY and VAUGHAN, 1971, by courtesy of authors and publishers)

containing "some colloidal or polymeric plutonium since only about 70 percent of the total ^{239}Pu was filterable through a 10 mμ filter. For intramuscular injection an appropriate volume of the stock solution of ^{239}Pu in 3 M nitric acid was neutralized with sodium hydroxide and diluted to a concentration of 12.5 or 25 μCi ^{239}Pu/cm^3". As discussed in Sec. II.C.6 he considers that the initial translocation of ^{239}Pu is in the form of soluble plutonium complexes and that later more polymeric forms may reach the bone and marrow perhaps through lymphatic drainage. This pattern of retention in both cells and on the mineral matrix surface has been questioned by JEE and his colleagues (JEE et al., 1969). They have studied the "residence time" of ^{239}Pu in the trabecular bones of beagles 1.5 months old given single intravenous injections of ^{239}Pu citrate 0.3 and 0.015 μCi per kg. They found that after 3 months most of the plutonium originally located upon trabecular bone surfaces had relocated, much of it being buried within the bone as shown in Fig. 10.27. They associate this with the known rate of remodelling of trabecular bone in dogs of approximately the age group studied (JEE et al., 1965). In older

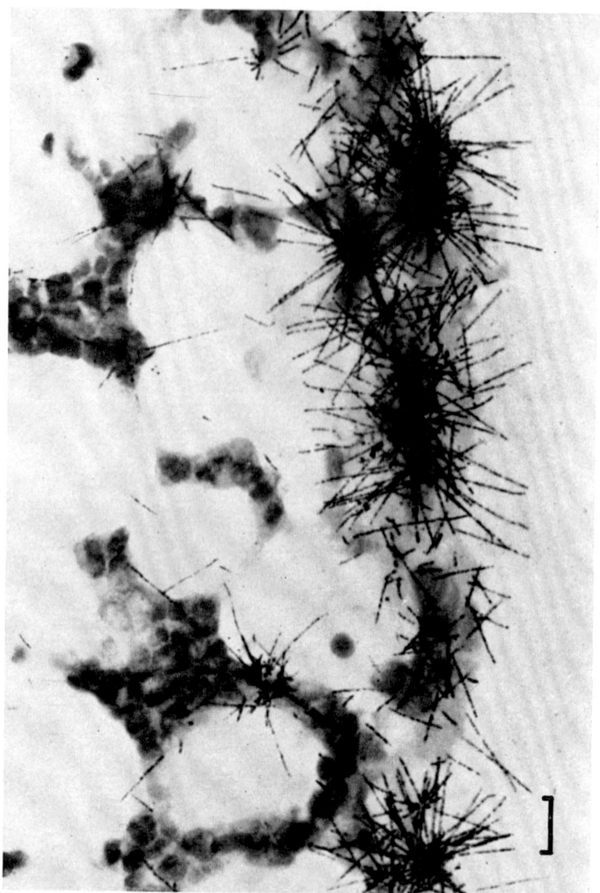

Fig. 10.38. Autoradiograph of a section of marrow extruded from the mid diaphysis of the femur of a rabbit injected intravenously with ^{239}Pu showing heavy concentration of plutonium in the osteogenic cells. The marker line is 20 μm. (BLEANEY and VAUGHAN, 1971, by courtesy of authors and publishers)

dogs the rate is much slower (AMPRINO and MAROTTI, 1964). JEE suggests that the surface retention in cells noted by BLEANEY and VAUGHAN (1971) is possibly due to damage to the osteogenic cells by the high dose of ^{239}Pu given. This may be true but long term surface retention was still found following what amounts to a lower dose, reaching the blood stream from an intramuscular injection, Table 10.14. Unfortunately the rate of remodelling of adult rabbit trabecular bone is not known, nor is that for human trabecular bone: the measurements both of SISSONS et al. (1965) and JOWSEY et al. (1967) indicate that in the trabecular bone of the ilium in man only 6–13 percent of the bone surface is resorbing at any point in time. This would suggest that the residence time of plutonium on adult human trabecular bone surfaces might be considerably longer than in young adult dogs.

Further it is difficult for technical reasons to make a really meaningful comparison of the data of JEE et al. (1969) and BLEANEY and VAUGHAN (1969). The thickness of the sections were different, and BLEANEY and VAUGHAN used solid state detection techniques which were equivalent to an exposure time for an

Table 10.32. The distribution of ^{239}Pu on the endosteal surface of bone of adult rabbits. Pu given as $^{239}Pu(NO_3)_4$ expressed in terms of dose-rate (rads/day/μCi/kg of injected dose). (Adapted from BLEANEY and VAUGHAN, 1971)

Rabbit	Dose (μCi/kg)	Dose rate (rad/day/μCi/kg of injected dose at 5 μm from surface)			Location of plutonium	P^a
		Min.	Max.	Average		
Intravenous injection	1.25	7	32	20	Bone+endosteal cells	44
		6	16	11	Bone only	
		1	31	10	Endosteal cells	
Intravenous injection	1.25	12	42	21	Bone+endosteal cells	30
		6	25	15	Bone only	
					Endosteal cells	
Intramuscular injection	2.5	2	15	7	Bone+endosteal cells	50
		1	7	4	Bone only	
		0	1	1	Endosteal cells	
Intramuscular Injection	2.5	3	14	7	Bone+endosteal cells	42
		1	10	4	Bone only	
		1	14	5	Endosteal cells	

[a] P is the percentage of the total plutonium on "the endosteal surface of bone" which is in endosteal cells. This is calculated as $100 \times \left(1 - \frac{D_a}{D_b}\right)$ where D_a is the average dose rate from Bone only, and D_b is the average dose rate from (Bone+endosteal cells) given in the previous column.

autoradiograph of about six times the exposure time used by JEE and his colleagues. Regions of low plutonium deposition that show no tracks in JEE's autoradiographs might show tracks with longer exposure. It will clearly be of very great importance to study the pattern of plutonium distribution in both cells and on the mineral/matrix of any trabecular bone that may become available from contaminated human beings. At present the observations of BLEANEY and VAUGHAN can only be taken as an example of the possible behaviour of plutonium reaching the marrow and the "bone surface". They indicate that plutonium, whether in monomeric or polymeric form is taken up and retained possibly both by the osteogenic cells and by the mineral/matrix surface. The possible dosimetric considerations arising from these observations are discussed in Sec. VII.B.1.c and the mechanism of plutonium binding to both cells and bone surfaces in Sec. V.

In summary it may be said that plutonium is concentrated particularly on endosteal surfaces where it is found possibly both in or on osteogenic cells and on the mineral/matrix bone border. The relative amounts of ^{239}Pu that are concentrated in marrow and on endosteal surfaces is conditioned by the physicochemical form of the plutonium gaining access to the blood stream.

At present no experimental evidence is available as to the distribution of americium between osteogenic cells and mineral/matrix surface.

B. Bone Marrow

In the earliest report of tissue retention in man, LANGHAM and his colleagues (1950) drew attention to the high level of plutonium in marrow. They gave a figure of 56 percent of the intravenously injected dose of ^{239}Pu citrate. MORROW

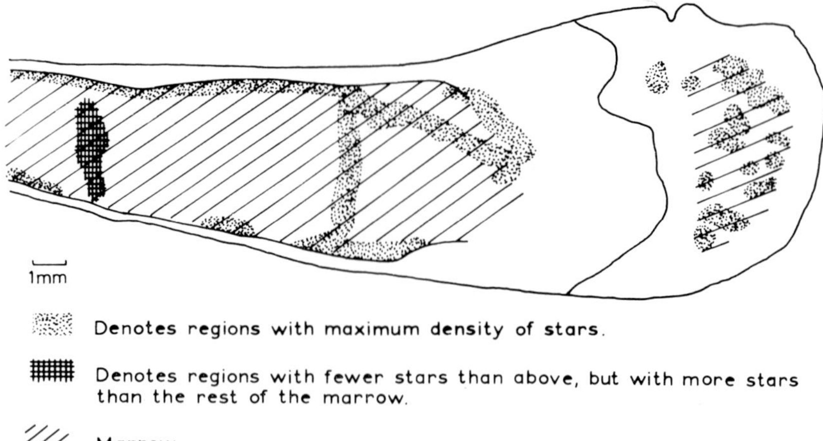

Fig. 10.39. "Star map" a diagram of longitudinal section of femur of rabbit killed 7 weeks after injection intravenously of 1.25 μCi/kg $^{239}Pu(NO_3)_4$ to show distribution of plutonium aggregates (stars). (VAUGHAN et al., 1967, by courtesy of authors and publishers)

and his colleagues (1967) later emphasised the high retention of inhaled ^{239}Pu by marrow and it is also mentioned by BULDAKOV et al. (1970) following all routes of administration.

Initial direct uptake of plutonium in the marrow of the dog 4 hours after injection of $PuCl_4$ was noted by HERRING et al. in 1962 (HERRING et al., 1962). It was described as both diffuse and aggregated. More recently detailed studies of both distribution and dosimetry of plutonium in the marrow have been made following intravenous and intramuscular injections of plutonium nitrate. Solutions which were probably at least 70 percent monomeric were used (BLEANEY, 1967; VAUGHAN et al., 1967; BENO, 1968; BLEANEY, 1969a, b; TAYLOR, 1969; BLEANEY and VAUGHAN, 1971). The importance of the colloidal character of the plutonium solution reaching the blood stream has been emphasised and illustrated by ROSENTHAL et al. (1968). As already discussed in Sec. IV these workers have shown that "the amount of plutonium in the marrow and its degree of aggregation are greater and the amount of plutonium on bone surfaces is smaller after injection of polymeric than monomeric plutonium". Whatever the solution used they consider, at least in the case of the mouse, the amount of plutonium in the marrow is less than that on "the bone surface". The following account therefore of plutonium distribution in the marrow can only be taken as applicable to the distribution observed following administration of the solutions used in the experiments quoted. The picture might well differ quantitatively though not qualitatively with different solutions administered by different routes.

Plutonium may reach the marrow as described by ARNOLD and JEE (1957, 1959, 1962) by resorption of bone containing plutonium by osteoclasts and its subsequent release and uptake by macrophages. This process has already been illustrated in Fig. 10.26. It appears probable that if this osteoclastic resorption occurs in young animals at the end of the metaphyseal trabeculae following a single injection of $^{239}Pu(NO_3)_4$ the heavy band of plutonium-containing macrophages may persist in the marrow as the bone grows in length. Such a band of macrophages is shown in Fig. 10.39. This is a diagram indicating, where, in a longitudinal section of the femur of a rabbit killed 7 weeks after injection of

1.25 μCi/kg $^{239}Pu(NO_3)_4$ (VAUGHAN et al., 1967) there are large aggregates of plutonium, often spoken of as stars, to distinguish them from the diffuse distribution of single tracks described in the next section.

The heavy band across the marrow shown in this diagram is due to the plutonium resorbed earlier from the metaphyseal trabeculae. Plutonium may also reach the marrow, directly from the blood stream, rather than by osteoclastic resorption as just described.

1. Distribution in Bone Marrow Following Intravenous Injection

The original studies of VAUGHAN and her colleagues (VAUGHAN et al., 1967; BLEANEY, 1969a, b) were made on marrow extruded from the diaphysis of the long bone of rabbits given $Pu(NO_3)_4$ by intravenous injection. A typical fission track[3] picture is shown in Fig. 10.40 taken from a longitudinal section of marrow from an adult rabbit injected intravenously with $^{239}Pu(NO_3)_4$ and killed 16 weeks later (BLEANEY, 1969a). The marrow contains both large aggregates of plutonium and diffuse single alpha tracks. It is now known, as discussed in Sec. IV.A, that the heavy band of stars which is a constant finding on the edge of all extruded marrow sections is not in the marrow but in or on the osteogenic cells of the endosteum which have been removed with the marrow. This must be remembered in looking at all the illustrations and tables in the papers published before 1970.

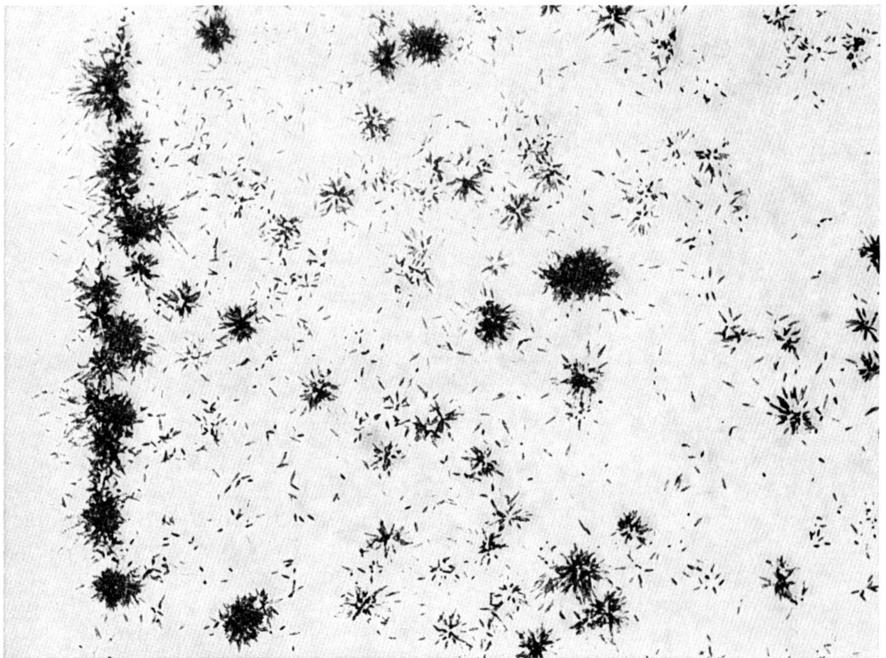

Fig. 10.40. Fission tracks in bone marrow which has been extruded from the shaft of the femur of rabbit injected intravenously with $^{239}Pu(NO_3)_4$ and killed 16 weeks after injection. The dense line of "stars" of tracks is in the osteogenic cells. (BLEANEY, 1969a, by courtesy of author and publishers). (Technique described in text V.A.4)

3 The technique of fission track autoradiography is discussed in Sec. V, A.4. The tracks appear shorter and thicker than they do in an ordinary autoradiographs made with stripping film.

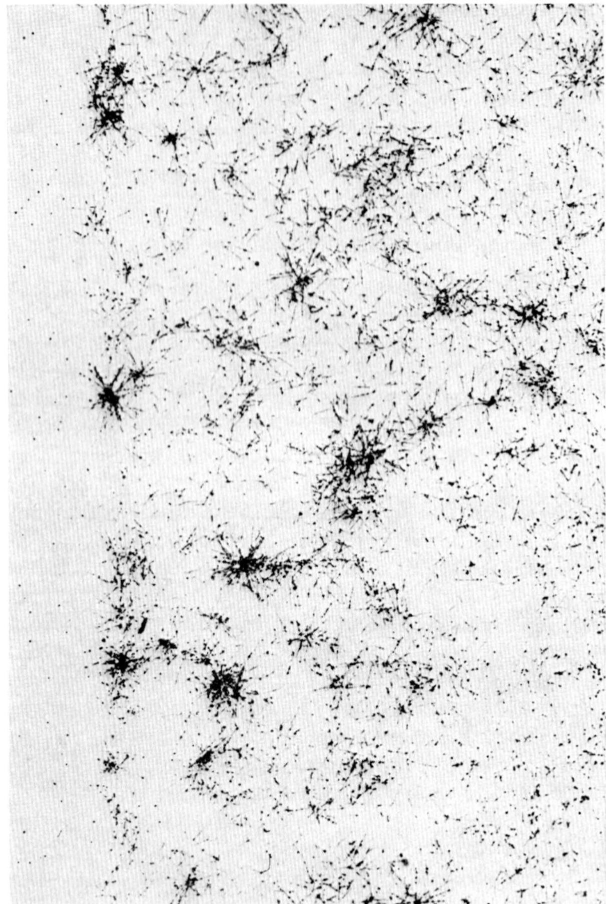

Fig. 10.41. Autoradiograph of thin longitudinal section (7.5 μ) of femur marrow of rabbit killed 24 hours after intravenous injection of 1.25 μCi/kg $^{239}Pu(NO_3)_4$ exposed 20 weeks (VAUGHAN et al., 1967, by courtesy of authors and publishers)

Typical autoradiographs of 7.5 μ sections of marrow from the femur of rabbits injected intravenously with 1.25 μCi/kg and killed 24 hours, 7 weeks and 7 months after injection are shown in Fig. 10.41, 10.42 and 10.43). These show that apart from the stars or aggregates at the edge of the marrow in osteogenic tissue the pattern of distribution throughout the marrow changes with time. Initially there are large stars throughout the marrow. These contain about 8×10^6 plutonium atoms and decrease in number with time after injection. Seven months after injection the aggregates are smaller containing about 2×10^6 plutonium atoms. These only show up on the autoradiograph after long exposure. There are also very small aggregates present 24 hours after injection but these do not give stars even at 20 weeks exposure and they decrease with time after injection. The diffuse distribution is largely due to these very small aggregates which appear as single tracks.

In an autoradiograph, the number of tracks observed depends on the amount of plutonium present and on the depth of the aggregates below the surface of the

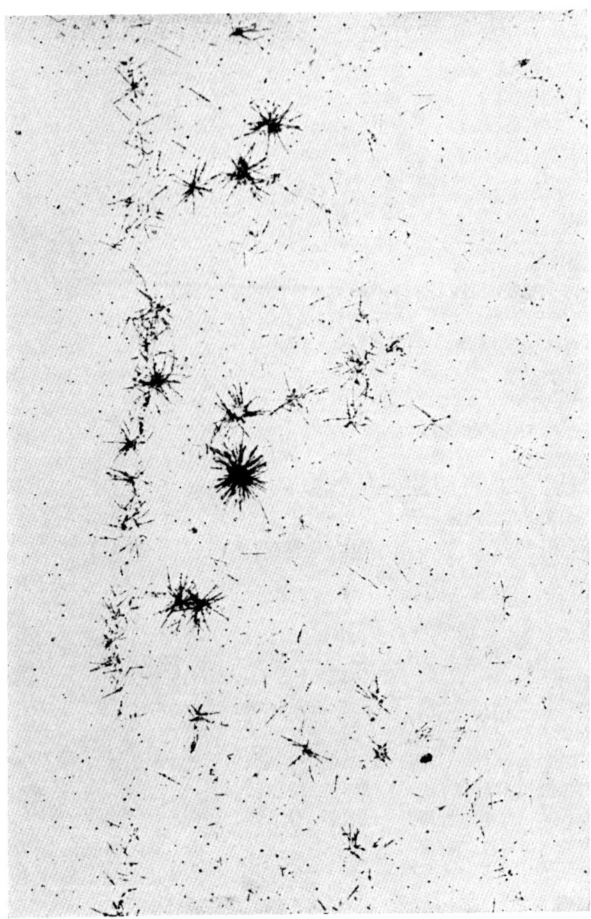

Fig. 10.42. Autoradiograph of thin longitudinal section (7.5μ) of femur marrow of weanling rabbit killed 7 weeks after intravenous injection 1.25 μCi/kg $^{239}Pu(NO_3)_4$ exposed 8 weeks. (VAUGHAN et al., 1967, by courtesy of authors and publishers)

film. In a thin section, 5 μm thick this effect is small, but in a 50 μm thick section alphas from aggregates at a depth greater than the range of the alpha particle in the marrow will not reach the film at all. The fraction of alphas emitted from the aggregates that reach the film decreases with the depth, see Sec. VI.C.3. The number of plutonium atoms is calculated by counting the number of tracks produced in a given time and using Eq. (3) in Sec. VI.C.3. For a given amount of plutonium in an aggregate the number of alphas emitted during the exposure time of the film is a variable quantitity, because the emission of an alpha particle by a plutonium nucleus is a random event. If the number of occurrences of a track count k are plotted against k, a Poisson distribution should be obtained as discussed in Sec. VI.C.6, provided that the aggregates are all the same size. (With a thick section, the distribution is still Poissonian because it is the sum of the Poisson distributions from all the horizontal layers from O to R, the alpha particle range.)

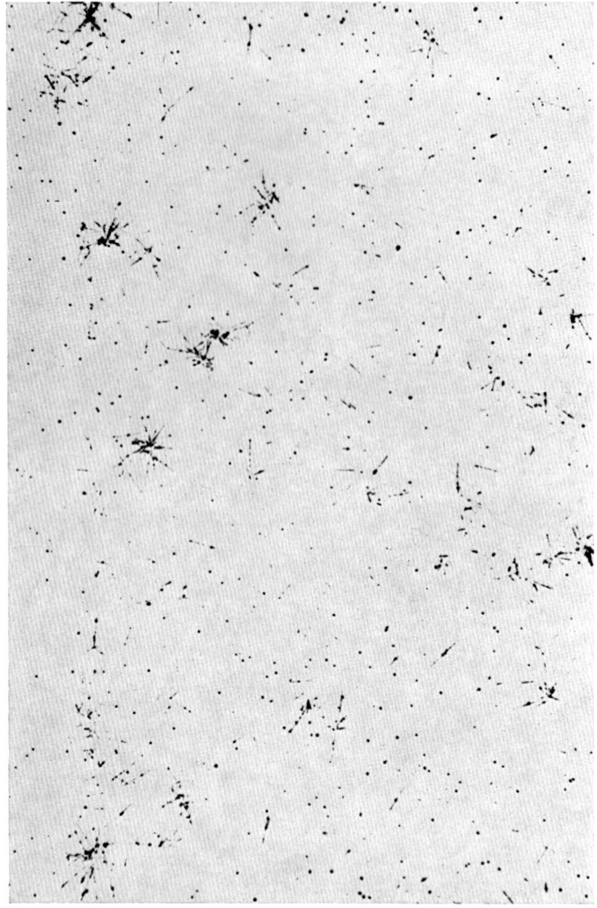

Fig. 10.43. Autoradiograph of thin longitudinal section (7.5 μ) marrow of rabbit killed 7 months after intravenous injection of 1.25 μCi/kg $^{239}Pu(NO_3)_4$ exposed 20 weeks. (VAUGHAN et al., 1967, by courtesy of authors and publishers)

The histogram in Fig. 10.44 shows the experimental data obtained by counting the number of tracks in stars on an autoradiograph. The data fit a Poisson distribution and therefore the aggregates in this section are all approximately the same size. This close fit to a Poisson distribution was also true for the medium sized aggregates, 10^6 atoms, provided that in all cases counts were taken in the central marrow.

2. Distribution in Bone Marrow Following Intramuscular Injection

Irrespective of the dose of ^{239}Pu the distribution of plutonium in the marrow of rabbits given an intramuscular injection was mainly diffuse shortly after injection (i.e. there were single tracks but few if any stars) and for the first 35 days after injection. Large aggregates were apparent in animals killed 112 days after injection. At this time about 90 percent of the plutonium may be in aggregates containing about 10^7 plutonium atoms. A few aggregates may contain double

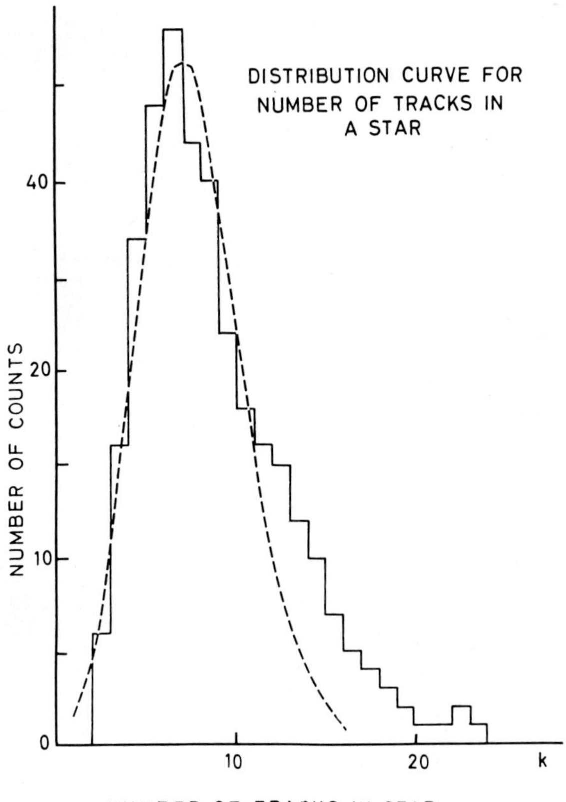

Fig. 10.44. Histogram of the number of tracks in stars counted in an autoradiograph of marrow from the femur of a rabbit given 5 μCi/kg $^{239}Pu(NO_3)_4$ intravenously and killed 24 hours later. (VAUGHAN et al., 1967, by courtesy of authors and publishers)

this number of atoms. The size of the aggregates appears to fall slightly to $\sim 3 \times 10^6$ eighteen months after injection. At 559 days after injection there is little plutonium in the centre of the marrow but large aggregates, 9×10^6 atoms, are found about 20 μ from the surface (BLEANEY and VAUGHAN, 1971). It should be noted that BLEANEY and VAUGHAN report data on two series of rabbits given an intramuscular injection of what was thought to be the same $Pu(NO_3)_4$ solution as regards valency and colloidal character and that for no obvious reason the level at the injection site fell more rapidly in one group than in another with a consequent difference in the marrow levels. This observation only serves to emphasise that the quantitative distribution of plutonium is extremely variable depending on many uncontrollable factors. The dosimetric interpretation of these results is discussed in Sec. VII.

Histological study of the marrow following both intravenous and intramuscular injection showed that aggregates or stars were always associated with yellow brown pigment contained in macrophages in sections stained with haemotoxylin and eosin. This pigment stained blue with PERL's reagent and was assumed to be haemosiderin (PEARSE, 1960; BENO, 1968). This association was noted as long ago as 1948 by BLOOM and discussed by ARNOLD in 1954. The precise nature

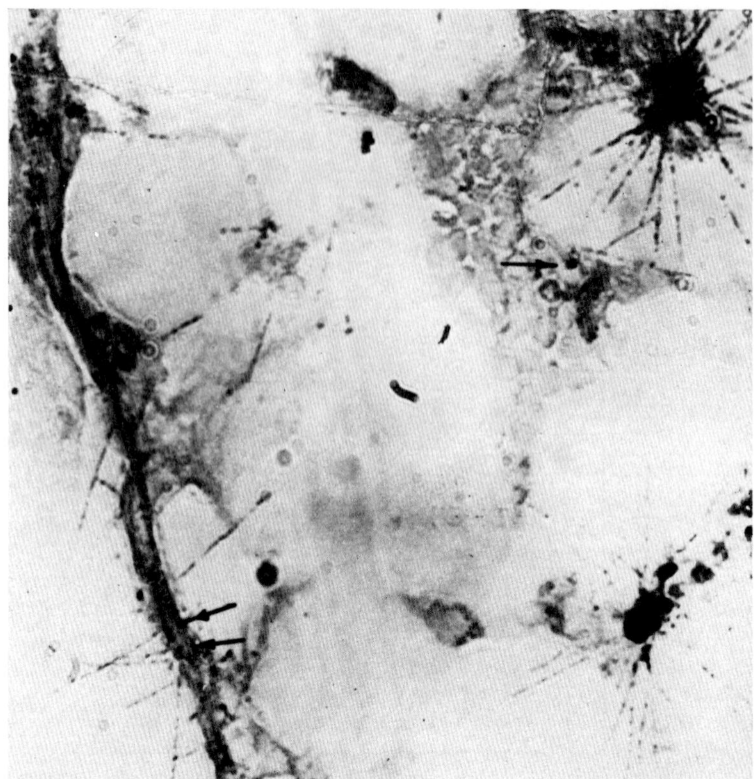

Fig. 10.45. Autoradiograph of rabbit bone marrow stained with Perl's reagent for iron prior to stripping film AR 10 application. No counterstain. The rabbit was injected intravenously with 1.15 μCi/kg of $^{239}Pu(NO_3)_4$ pH 4 and killed 16 weeks after injection. Note haemosiderin containing cells with ^{239}Pu aggregates. (BENO, 1968, by courtesy of author and publishers)

of haemosiderin is not known. BESSIS and BRETON GORIUS (1962) have said "it is erroneous to consider haemosiderin a uniquely defined compound" — it constitutes an extremely variable mixture of substances one of which is ferritin i.e. ferrihydroxyphosphate bound to a polysaccharide containing protein (PEARSE, 1960). The latter may be present in varying amounts and ferritin may also be present in other forms (BESSIS, 1962; BESSIS and BRETON GORIUS, 1959, 1962). Finch and his colleagues consider that in the present state of knowledge it is justifiable to regard the iron reserves i.e. ferritin and haemosiderin, in toto as a single functioning physiological unit rather than to attach any specific importance to their possible subdivision (FINCH et al., 1950; GALE et al., 1963). In Fig. 10.45 is shown a section of marrow stained with PERL's reagent with two macrophages containing both haemosiderin and plutonium aggregates. Aggregates on the surface of the marrow associated with osteogenic cells were not associated with haemosiderin. Haemosiderin-containing cells with ^{239}Pu were constantly found at a distance greater than 80 μ from the surface of the marrow cells. Further, a significant increase in the number of iron containing cells and in the amount of pigment the cells contained was noted by BENO (1968) in marrow from the femurs of rabbits four months after a single intravenous injection of $^{239}Pu(NO_3)_4$ compared with the normal controls.

Questions at once arise: What is the relationship between the ^{239}Pu and the haemosiderin? Are they both picked up by macrophages as they pursue their work as scavengers so that the association is purely one of chance or is the ^{239}Pu in some way associated with the haemosiderin? It is known that ^{239}Pu binds to transferrin in the plasma (STEVENS et al., 1968; BOOCOCK and POPPLEWELL, 1966, 1967) as already discussed (Sec. II.D) and to a variety of glycoproteins (TAYLOR and CHIPPERFIELD, 1970a, b, 1971). It is also known that ^{239}Pu appears to make a more stable complex with ferritin than with transferrin, so it may well be bound to haemosiderin (BRUENGER et al., 1969). It is perhaps also significant that BENO (1968) noted an increase in haemosiderin in rabbits treated with ^{239}Pu nitrate (1.25 μCi/kg) over normal controls. This may be associated with DOUGHERTY's observation (DOUGHERTY et al., 1960; DOUGHERTY, 1962) that ^{239}Pu given as the citrate (mean dose injected 2.7 μCi/kg) may cause some anaemia in dogs while ROSENTHAL has reported severe anaemia in mice given polymeric ^{239}Pu nitrate (7.8 μCi/kg) (ROSENTHAL et al., 1970). The relationship of plutonium to haemosiderin and iron metabolism requires further investigation. The very small aggregates of plutonium showing up single tracks in the marrow did not appear to be associated with haemosiderin.

In summary it may be said that plutonium is concentrated particularly on endosteal "bone surfaces" and in the marrow. There is some evidence that some of this plutonium is retained in the endosteal cells themselves. The extent to which plutonium on the mineral/matrix part of the surface becomes buried with time will depend on the rate of remodelling. This rate, in human trabecular bone is not known. The evidence available suggests that it is slow. The marrow distribution is in part diffuse and in part concentrated in macrophages associated with haemosiderin. It is not necessarily associated with osteoclastic resorption. The relative proportion in marrow and on "the bone surface" depends on the physicochemical character of the plutonium reaching the blood stream.

V. The Binding of Plutonium in Bone

DAVID M. TAYLOR

It was apparent from the earliest published autoradiographs that the microscopic distribution of ^{239}Pu in bone differed markedly from that of calcium (HAMILTON, 1947). These and subsequent studies showed that plutonium deposited on resting or resorbing areas on bone surfaces and that deposition on endosteal surfaces was much greater than that on periosteal surfaces (HAMILTON, 1947; HERRING et al., 1962).

The mechanisms responsible for this characteristic distribution pattern of plutonium on bone surfaces are not yet understood. As a result of their early work COPP et al. (1947) suggested that "plutonium was actively laid down in the uncalcified organic bone matrix below the epiphyseal cartilage and the endosteum and periosteum". This suggestion that plutonium was bound by the organic matrix of bone was apparently supported by the observations of TSEVELEVA (1960) who reported that, in rat and rabbit bone, 6 percent of the plutonium was associated with the bone mineral and 87.5 percent with the organic matrix. All the bone protein fractions were found to contain plutonium with the "albuminoid" and "mucoid" fractions showing the highest plutonium content per mg protein nitrogen. The carboxyl groups in the proteins and the sulphate groups of chondroitin sulphate were considered to be important in plutonium fixation. TSEVELEVA's results must however be treated with some caution as the chemical procedures used may well have caused redistribution of the plutonium during the separations, especially of plutonium which may originally have been fixed on bone mineral (Z. Z. Szot and D. M. Taylor, unpublished data).

Other workers have suggested that the bone mineral is the most important site of fixation of plutonium. Foreman (1962) studied the uptake of plutonium on bone preparations *in vitro* and showed that bone mineral took up greater quantities of plutonium and at a faster rate than the organic matrix. Complementary *in vivo* studies in which various bone fractions were implanted into the abdominal cavity of rats confirmed the preferential uptake of plutonium on bone mineral and Foreman (1962) concluded that plutonium was taken up by "adsorption on mineral surfaces along capillary routes". Autoradiographic studies by Jee and Arnold (1962) on the bones of normal and rachitic rats showed that "plutonium passed readily through the uncalcified osteoid and deposited on the underlying calcified bone surfaces". These authors concluded that plutonium is bound mainly to the mineral phase of bone but that some may be bound by the organic matrix.

In more recent studies Williamson and Vaughan (1964, 1967) have shown that plutonium and americium were deposited in areas of bone which gave a positive histochemical reaction with the Periodic acid—Schiff reagent and they suggested that mucosubstances might be important in the fixation of plutonium, and americium, in bone.

The mucosubstances in bovine bone have been studied extensively by Herring (1969) and five distinct glycoprotein fractions have been isolated. Only two of these have been studied in any detail. The first of these is bone sialoprotein which has a molecular weight of 25000 ± 500 and contains 40 percent by weight of carbohydrate. Bone sialoprotein is unusual in that it contains a high proportion of sialic, glutamic and aspartic acids (Williams and Peacocke, 1965; Andrews and Herring, 1965). The other is the bone chondroitin sulphate-protein complex, or rather the three chondroitin sulphate-protein complexes. These complexes probably consist of one or two chondroitin sulphate chains attached to a protein core with properties apparently similar to those of bone sialoprotein (Herring, 1968). The three other glycoprotein fractions, designated glycoproteins I and II and CPC-soluble glycoprotein, have not yet been characterised.

The metal binding properties of bone sialoprotein were first studied by Williams and Peacocke (1967) who showed, by potentiometric titration, that yttrium was bound much more strongly than calcium. Preliminary investigations with thorium suggested that this metal would be bound more strongly than yttrium. These authors suggested that the carboxyl groups of the bone sialoprotein were the binding sites for yttrium and thorium.

The binding of plutonium to the five glycoprotein fractions isolated from bovine bone has been studied by Taylor and Chipperfield (1971). These authors used a gel filtration procedure to study the metal binding. In this semi-quantitative method the percentage of protein bound plutonium which was eluted from the Sephadex gel was used as a measure of the relative stability of different plutonium protein complexes. The percentages of glycoprotein bound plutonium recovered when the plutonium-protein complexes with the five bone glycoprotein fractions were separated by gel filtration, on Sephadex G 50 in 0.15 M KCl — 0.01 M *tris*-HCl buffer at pH 7.5, are shown in Table 10.33. This table also includes data for transferrin, the protein which transports plutonium in the blood, collagen-free chondroitin sulphate and poly-L-glutamic acid. These data show that plutonium is bound by all five of the bone glycoproteins more strongly than to transferrin: collagen, the major protein of bone, or to free chondroitin sulphate.

Table 10.33. The binding of plutonium to bone glycoproteins and other substances, as determined by gel filtration. (Taken from CHIPPERFIELD and TAYLOR, 1971)

Protein	% recovery of protein-bound plutonium
Bone sialoprotein	54.7±9.3
Chondroitin sulphate	49.2±3.0
Protein complex	
Glycoprotein I	30.0±12.4
Glycoprotein II	50.3±9.1
CPC-soluble glycoprotein	36.6±5.1
Soluble collagen	23.3±4.3
Transferrin	18.9±6.9
Bone chondroitin sulphate	13.2±5.7
Poly-L-glutamic acid	68.7±0.5

Studies of the effect of pH on the binding of plutonium to bone sialoprotein and the chondroitin sulphateprotein complexes have shown that the maximum binding to both glycoproteins occurs at about pH 6. This has been interpreted to indicate that plutonium is bound to carboxyl groups on the glycoprotein molecules, a conclusion which is supported by the strong binding of plutonium to poly-L-glutamic acid (CHIPPERFIELD and TAYLOR, 1970a, 1970b).

Plutonium can also bind to mineral (FOREMAN, 1961) but the relative importance of the mineral and organic phases in the fixation of plutonium in bone is not understood. Studies of the binding of plutonium to bone mineral in competition with glycoprotein *in vitro* have been described by TAYLOR and CHIPPERFIELD (1971). In these experiments a thirty-fold excess of bovine cortical bone ash was mixed with a solution of glycoprotein and plutonium and after standing for 1 hour the percentage of soluble plutonium was measured. Under these conditions, in which the bone ash:glycoprotein ratio was similar to that present in living bone, the presence of bone sialoprotein increased the percentage of soluble plutonium in the system, from 1.5 ± 0.3 percent to 30.5 ± 1.6 percent but in the presence of bone chondroitin sulphate-protein complex the soluble plutonium increased to only 11.6 ± 1.9 percent. Thus both bone sialoprotein and the bone chondroitin sulphate complexes can bind significant amounts of plutonium even in the presence of a large excess of bone mineral.

At the present time it is difficult to assess the importance of glycoprotein-binding in the fixation of plutonium in bone *in vivo*. The functions of the bone glycoproteins and their detailed microscopic distribution in osseous tissue are unknown. If these substances are in fact present in those areas of the bone surface ar which plutonium is deposited, then the strong binding of plutonium to the glycoproteins discussed above, even in the presence of a large excess of bone mineral, would suggest that they could be important in the fixation of the metal in bone. Some indirect evidence in favour of the suggestion that glycoproteins may be important in the fixation of plutonium in bone may be inferred from studies with americium. Americium is bound by all the five bone glycoprotein fractions described above (CHIPPERFIELD and TAYLOR, 1970; TAYLOR and CHIPPERFIELD, 1971), even in the presence of thirty-fold excess of bone mineral, but the binding appears to be much weaker than that of ^{239}Pu. The binding of americium by bone mineral appears to be as strong, or even stronger than that of plutonium (TAYLOR and CHIPPERFIELD, 1971). Comparative studies of the long-term biological effects of ^{239}Pu and ^{241}Am have shown that, microcurie for microurie, ^{241}Am is much less efficient than ^{239}Pu in inducing bone tumours in rats, despite the fact that the radiation characteristics and gross distributions in bone are similar for the two nuclides (BENSTED et al., 1965; TAYLOR and BENSTED, 1969). It has been suggested that one possible explanation of this difference in the carcinogenicity of ^{239}Pu and ^{241}Am would be a preferential binding of ^{239}Pu by glycoproteins in, or on the surface of, the cells lining the endosteal surfaces of bone (TAYLOR and BENSTED, 1969). The presence of ^{239}Pu in such cells has been demonstrated by BLEANEY and VAUGHAN (1970) but the role of glycoprotein in this fixation remains to be elucidated.

The detailed description of the ways in which plutonium is fixed in living bone must await a clearer understanding of the biochemistry and physiology of bone, and especially of the bone surfaces. However the evidence which is at present available suggests that binding by glycoproteins may play as important a role as binding to bone mineral in explaining the characteristic distribution pattern of plutonium in bone.

VI. Theory and Technique of Alpha Dosimetry with Particular Reference to the Skeleton

BETTY BLEANEY

A. Experimental Techniques

1. Macroscopic and Microscopic Measurements

In order to study the metabolism of plutonium in animals and man it is necessary to measure both the macroscopic distribution in various tissues such as the liver, bone marrow or bone, and the microscopic distribution, in order to locate the plutonium with respect to particular cells, and regions within cells. The detection of ^{239}Pu in vivo is difficult since the alpha particles have a range of only 37 μm in soft tissue and the accompanying gamma rays are very weak. Plutonium in the lungs can be measured with sodium iodide crystals placed outside the body by determination of the low energy x-ray emission (Chap. 10 and 13).

Sections of bone or soft tissues can be ashed in a muffle furnace and the alpha activity of the section can be counted with a 2π proportional counter. Such a counter is described by TURNER et al. (1958). BRUENGER et al. (1963) developed a straightforward procedure for the preparation of the samples prior to counting. Sample preparation is rather time-consuming, and LINDENBAUM and LUND (1969) have used a liquid scintillation counter to count alpha particles in samples of bone, liver, spleen and other tissues. This method is not only much less laborious than proportional counting, it is also less sensitive to organic residues. LINDENBAUM et al. found agreement between proportional and scintillation counters, but the variations are consistently smaller with the latter method.

In order to locate the plutonium in the tissues with respect to the cells, in particular with respect to the radiosensitive cells referred to in Sec. III.A thin sections of tissue must be prepared. For soft tissue the sample can be embedded in a wax such as "paraplast", and cut on a microtome into 5 μm sections. For bone, the sections are embedded in methyl methacrylate (JOWSEY, 1955), or araldite (JAMES and KEMBER, 1970) and cut and ground to a suitable thickness. Owing to the different mechanical properties of bone and marrow, it is very difficult to obtain good sections of the bone-marrow interface; which is the region of great importance in the case of plutonium.

2. Autoradiography

The technique of autoradiography is well established and it has developed since 1945 into a standard laboratory procedure. Only a brief description will be given here since the details of the technique and its uses and limitations have been discussed fully in several monographs (ROGERS, 1967). In autoradiography the plutonium in the thin section on the slide is detected and measured by the alpha particle tracks produced in a layer of photographic emulsion placed over the slide for a suitable time and then developed.

The photographic emulsion can be in the form of stripping film. The film is supplied as a thin sheet on a glass plate. The film is cut to size (about 4 × 4 cm), and stripped off the glass and floated on water. In water the film swells and as it floats the section can be slid underneath. The film wraps round the slide and shrinks as it dries to make a good contact as shown in Fig. 10.46.

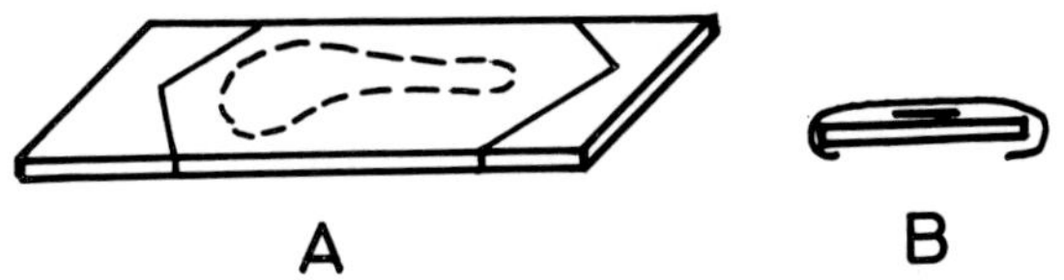

Fig. 10.46a and b. Figure illustrating an autoradiograph; the section is placed on a microscope slide and covered with film. After development the section is mounted in the normal way: a) view from above; b) end-on view

The alternative to stripping film is dipping film. This is a liquid emulsion into which the section already fixed on the slide is dipped and hung up to dry. Stripping film is more suitable for quantitative alpha particle work as the thickness of the film (3 μm) is constant over the section and from one section to another. Dipping film tends to be thicker at the lower end when it is hung up, but the speed of this technique is an advantage, and it is used a great deal for work with beta rays, for example in experiments using tritiated thymidine.

The alpha particles from ^{239}Pu that pass through the film during the exposure time produce straight black tracks with a maximum length of 22 μm in Kodak AR 10 or Ilford emulsion. Fig. 10.47 illustrates that the apparent length of the

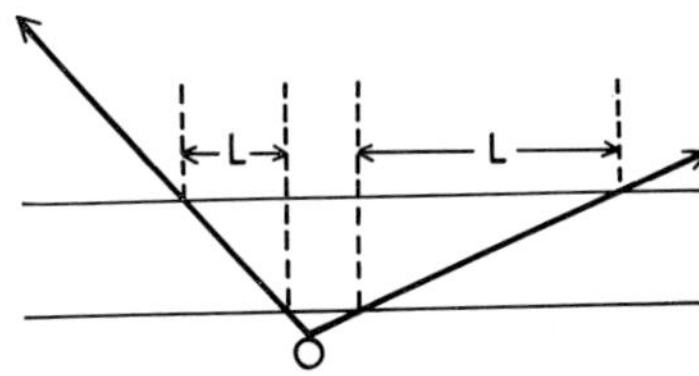

Fig. 10.47. Figure illustrating that the length of an alpha track L, as viewed through a microscope depends on the angle at which the alpha particle enters the film

track L, when viewed vertically through a microscope depends on the angle of the track to the vertical. A nearly vertical track appears to be very short. Since the developed film is transparent, so long as there are not too many tracks, the section can be viewed below the tracks. Aggregates of plutonium give rise to a star-shaped track pattern as in Fig. 10.41 radiating from a source of only a few μm diameter. If the tracks in a "star" are counted the approximate number of plutonium atoms in the aggregate can be calculated.

3. Microdensitometry

If the amount of plutonium present is large and the exposure time long, so that many tracks are produced, then at low magnification the tracks merge in to one another and the film has regions of more and less blackening, which is uniform enough for light absorption measurements. Then the relative distribution of tracks can be measured with a microdensitometer. This instrument shines a beam of light of standard intensity through a slit about 20 μm wide and through the film. One design has a split beam, where the second beam passes through an optical wedge. The wedge is moved until the light transmitted by both beams, (one through the wedge and the other through the film), are of equal intensity. The instrument can be automated and the data fed into a computer (JOWSEY et al., 1965; RIGGS et al., 1970). TWENTE and JEE (1961) used microdensitometry for measuring plutonium densities on bone surfaces. The resolution is not sufficient to locate the plutonium very precisely but for calcium 45 which has a beta ray of maximum energy 0.25 MeV, microdensitometry is a useful technique. The technique is also used for microradiographs, for example for measuring the amount of hydroxyapatite per unit volume of bone (JOWSEY et al., 1965).

4. Solid State Track Detectors

The technique of solid state track detection has been used in nuclear physics and in geophysics. The tracks are detected in a simple crystalline solid which is an electrical insulator, for example, mica or quartz. When a heavy charged particle (such as an alpha particle or a heavy charged nucleus from a nuclear reaction) passes through the solid, electrons are knocked off the atoms in the solid leaving positive ions, as illustrated in Fig. 10.48a. The electrostatic repulsive forces of these charged ions cause them to move apart by distances of about 5×10^{-9} m disturbing their neighbours in the process, as in Fig. 10.48b, and causing a path of dislocations along the track of the particle. If the solid is heated at this point the thermal agitation anneals these dislocations and the ions pick up stray electrons and return to their normal positions. At normal temperatures

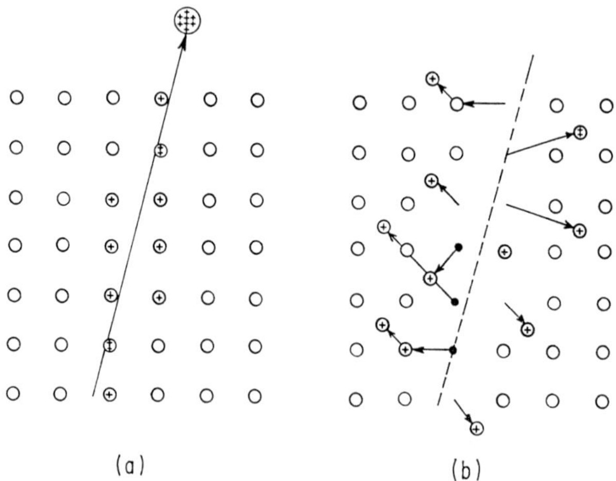

Fig. 10.48a and b. A two dimensional view of a simple crystal. a Passage of a heavy charged particle produces positive ions along its path. b The equilibrium in the lattice is disturbed and the ions move into interstitial positions leaving a gap along the particle track. (After FLEISCHER et al., 1965, by courtesy of authors and publishers)

the dislocations are too small to see under a microscope, but if the solid is immersed in a powerful solvent such as NaOH, any radiation damage trails are enlarged. The solvent attacks the whole surface, but the regions of dislocation are chemically more active than normal, and if the time of immersion in the solvent is suitable chosen, any trails crossing the surface are etched into tracks which can be viewed under a microscope. Various solids have been used, and a review of their properties and sensitivity to different charged particles is given in FLEISCHER et al. (1965). The length and width of the track are characteristic of the particle under constant etching conditions. It would be possible to detect the alpha particles from plutonium by this technique, and the exposure time would be the same as for an autoradiograph.

HAMILTON (1968) increased the sensitivity of detection by irradiating the section in contact with a polycarbonate foil in a beam of thermal neutrons in a nuclear reactor. The foil is the insulator in which the fission tracks are produced. Fission of the ^{239}Pu nucleus occurs, and the fission products are two heavy charged particles whose masses are between 91 and 149 atomic mass units. They are typically about 97 and 140 atomic mass units with energies of about 100 and 70 Mev respectively. Those that pass into the foil are recorded by the tracks that appear after etching. In an hour's irradiation in the reactor the numbers of fission tracks which are produced is equal to the number of alpha tracks that would be seen on an autoradiograph after several weeks exposure. The fission track method is a more rapid and more sensitive detector of small quantities of plutonium. This is because the half-life for the alpha particle decay of plutonium is very long, (24, 360 years), so that roughly one in a million plutonium nuclei emits an alpha particle in one week. In this time, in a flux of thermal neutrons of 10^{12} particles/cm^2/sec about 500 out of one million plutonium nuclei undergo fission, and each fission produces two particles, so that there are 1000 track-producing particles instead of one alpha particle. This gain in sensitivity of a factor of the order of a thousand means that often irradiations of the order of an hour are sufficient, but by irradiating for longer times very low concentrations of plutonium can be detected.

The procedure for fission track detection is as follows: Sections are prepared as for autoradiography but bone sections need not be thinner than 100 μm. A suitable foil is a polycarbonate such as the "Lexan" used by FLEISCHER et al. (1965) 200 μm thick which is cut to cover the section. About six sections can be mounted together in one can 7.5 cm high and 2.5 cm internal diameter for insertion in the reactor. Each section with its foil is separated from the next by an aluminium strip and uniform pressure and good contact is obtained between foil and section by means of aluminium clips, as shown in Fig. 10.49.

After irradiation the foils are removed from the section and etched 35 minutes in 6.25 N sodium hydroxide at 55° C. Since no tissue would stand this treatment

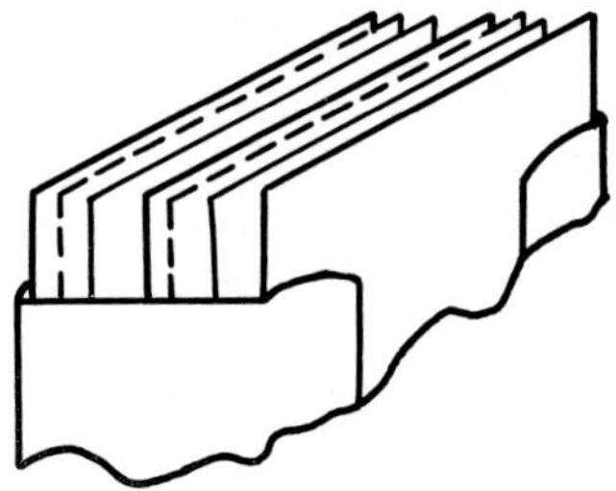

Fig. 10.49. Figure showing the assembly of sections for irradiation in a reactor. The thicker plates on the outside, and separating the sections, are aluminium. A section with its foil is sandwiched between two sheets of aluminium

the section cannot be viewed below the foil under the microscope as in the conventional autoradiograph. But if two irradiations are made of the same section (a procedure which cannot be carried out in autoradiography) one long one, giving many tracks which outline the bone surfaces, and one short one of about one hour for track counting, identification of the location of the plutonium in relation to the section is easy. An autoradiograph can then be made of the section and an adjacent section.

Relative intensity measurements can be made by direct comparison of track counts from sections in the same can, since they have had the same neutron flux. Absolute determinations demand a measurement of the neutron flux and comparison with a standard source of plutonium and an autoradiograph. Each irradiation of the standard source will produce some nuclei of ^{240}Pu and ^{241}Pu by neutron capture, as well as producing fission of ^{239}Pu nuclei. Therefore it is inadvisable to use the same standard source for every irradiation.

If the standard source of known activity P_1 is used with a (neutron flux $\times$ time) of Φ_1, then the number of tracks produced is $N_1 = A\,\Phi_1\,P_1$ where A is a constant. For another section of activity P_2 in a (neutron flux $\times$ time) of Φ_2 we have a number of tracks

$$N_2 = A\,\Phi_2\,P_2.$$

So that $P_2 = N_2/A\,\Phi_2 = \dfrac{N_2\,\Phi_1}{N_1\,\Phi_2} \cdot P_1$.

N_1 and N_2 are found by track counting; to find Φ_1/Φ_2 a cobalt 59 wire can be inserted in each can with the sections. The gamma ray count rate from the cobalt 60 produced is proportional to Φ, so that the ratio is equal to the ratio of the gamma ray count rates. The fraction of plutonium nuclei undergoing fission is very dependent on the neutron velocity spectrum. So long as this is constant the cobalt wire can be used to monitor the neutron flux, and the standard source is needed only for the initial calibration. An autoradiograph of the section provides an independent measurement. It is essential to keep the temperature in the reactor sufficiently low during the irradiation. Below 80° C there is no risk of the tracks annealing or the embedding wax in the soft tissue section melting.

Fission tracks are shorter than alpha tracks: the maximum length in polycarbonate foil is 11 µm (BLEANEY, 1969) instead of 22 µm for an alpha track in photographic film. A relation derived by BOHR (BOHR, 1941), predicts that the stopping power of various substances for fission particles varies in the same way as for alpha particles, so that the ratio of the ranges of fission particles in bone and bone marrow is equal to the ratio of the ranges of alpha particles in bone and bone marrow. Because fission tracks are shorter than alpha tracks they are closer to their point of origin. There are artefacts due to scratches on the foil, but these are easily distinguishable from tracks. The background is negligible.

Automatic counting techniques have been developed for fission tracks by JOHNSON et al. (1970). JEE (1970) has used this method for measuring the density of plutonium on the bone surfaces of vertebrae of beagle dogs. The etched foil is covered with a piece of alumunized Mylar. When a potential of 500 volts is applied across the combination, sparking causes evaporation of the aluminum where tracks occur, and these can be recorded with a quantitative television microscope. JEE also delayed the etching process for several months, and this resulted in a number of small dots appearing over the regions of bone but not over marrow, enabling the two regions to be easily distinguished. The origin of these dots is not yet known with certainty but is probably caused by induced beta activity in bone primarily from ^{32}phosphorus (JEE et al., 1971).

Electron microscope observations are needed to locate the plutonium more precisely with respect to particular regions in cells. This requires thin sections with a greater concentration of plutonium present. Electron microscopy combined with autoradiography of the beta emitter ^{241}Pu which has a half-life of 13.2 years appears at present to be the most promising method.

B. Plutonium Isotopes and Properties of Alpha Particle Irradiation

1. Plutonium Isotopes and the Decay of ^{239}Pu

Plutonium 239 is the commonest of the plutonium isotopes. It can be produced in a nuclear reactor by beta decay of uranium 239. Heavier isotopes are produced by successive neutron capture and lighter ones also exist. Isotopes 237 and 238 are discussed in sec. II, and 241 is a beta emitter, but because 239 is produced by nuclear power plants this is the isotope which is of most concern from the point of view of radiation protection. Table 10.1 lists the isotopes together with their half-lives and the particle energies. Plutonium 239 decays with a half-life of 24360 years, emits alpha particles of average energy 5.14 Mev, and very low energy gamma rays of about 38 kev. The daughter nucleus, uranium 235 has a half-life of 7.1×10^8 years, so that for purposes of radiation damage only the ^{239}Pu alpha particle need be considered. This alpha particle has a range of 25 μm in bone and 37 μm in soft tissue such as bone marrow, which means that the radiation damage is produced close to where the plutonium is located in vivo at the time of emission of the alpha particle.

2. The Linear Energy Transfer, LET, and Stopping Power

Alpha particles lose energy by collisions with the electrons in the atoms of the medium through which they pass. These collisions result in excitation and ionization producing a charged atom and a free electron. The free electrons, sometimes called delta rays, cause further ionization. For most of the range, when the alpha particle energy is above 1 Mev, the rate of loss of energy can be calculated. When the energy is less than 1 Mev and the alpha particle is moving more slowly, it captures electrons and loses them again. Also, whereas above 1 Mev excitation and ionization of the atoms of the nucleus is the dominant process whereby the alpha particle loses energy, below 1 Mev elastic collisions with the atoms, producing transfer of energy and momentum to the struck atom are important. At low energies the rate of loss of energy is much more difficult to calculate from theory, but some experimental information is available (Rotondi, 1968). The energy of the alpha particle falls from 5.14 Mev to zero at the end of its range. If at a distance x from its origin, the alpha particle energy is E, it decreases by an amount dE in travelling a further small distance dx. The rate of energy loss with distance $-dE/dx$, can be calculated. It is proportional to the number of electrons per unit volume of absorber.

Ionizing radiations are characterised by a quantity known as the linear energy transfer, LET, which is equal to $-dE/dx$. It is normally measured in kev/μm, 50 kev/μm is a high LET. The subscript ∞ is sometimes added, and LET_∞ signifies that secondary particles of all energies are included in the loss. When energy transfers due to collision are restricted to those less than some specified value Δ, so that secondary electrons of higher energy than Δ are counted as separate tracks, a subscript Δ is used LET_Δ. LET_{100} denotes that only secondary particles of up to 100 ev are included. LET is understood to mean LET_∞ unless otherwise

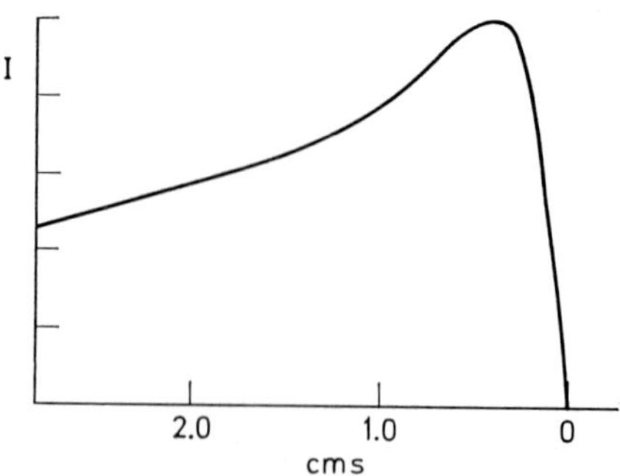

Fig. 10.50. Ionization of a single alpha particle as a function of its residual range in air. The ionization is proportional to the LET at any point. (After HOLLOWAY et al. (1958), by courtesy of authors and publishers)

specified. This quantity is equal to the stopping power of the medium. A detailed account of linear energy transfer is given in the ICRU Report 16 (1970). X-rays are an example of low LET radiation, alpha particles are high LET radiation. A plutonium alpha particle of energy 5.14 Mev and range 37 μm has an average LET over its range of 139 kev/μm, but the LET is not constant over the whole range. The energy loss per ion pair produced is nearly constant and is taken as equal to 35 ev. The number of ions produced at each point along the path is therefore to a first approximation proportional to the LET, or stopping power. Fig. 10.50 shows how the LET varies with the range for polonium 210 alpha particles. It is approximately constant for a large portion of the range, but passes through a maximum towards the end of the range, because the slowly moving alpha particle spends more time near a given atom, and can more easily remove electrons.

3. The Nature of Alpha Particle Ionization

The ionization produced by an alpha particle is confined to a region very close to the track of the particle. 90 percent of the ionization is contained in a cylinder of radius 0.01 μm whose axis lies along the track. Some delta rays spread out further and another 9 percent of the ionization lies between 0.20 and 0.01 μm from the axis. The density of ionization is therefore 4000 times higher in the central core of radius 0.01 μm than it is in the region out to 0.2 μm. It is customary to express absorbed doses for all ionizing radiations, neutrons, X-rays, alpha particles for example, in rads.

$$1 \text{ rad} = 100 \text{ ergs per gram.}$$

For soft tissues 1 g has a volume very close to 1 cm^3. The rad unit is used for x-rays where uniform doses over volumes of the order of 1 cm^3 can easily be obtained. For alpha particles, if the central core of radius 0.01 μm and length 57 μm (the range of the alpha particle from ^{239}Pu) is taken, the average absorbed dose is equivalent to 10^6 rads. For the region out to 2.5 μm which represents a typical radius of a cell nucleus the value is 100 rads. When a piece of tissue containing plutonium and many cells is considered and the number of disintegrations is multiplied by the alpha energy and divided by the tissue mass in the appropriate units a much lower "rad dose" is usually obtained, because there are regions between the alpha tracks which have virtually no absorbed dose. Alpha particle tracks are spatially very inhomogeneous and a dose in rads can be thought of as representing the probability of the passage of the alpha particle through a given small volume such as a cell nucleus. The actual dose to this volume is very high or zero according to whether the passage of an alpha particle has or has not occurred. Such a passage is called an event.

4. The Tissue Equivalent Sphere of Rossi and the Quantities Y and Z

Rossi (1967) has analysed the distribution of the energy of ionization in small volumes of the order of cell volumes and nuclear volumes. His approach is both theoretical and experimental and covers different types of radiation of different LET values.

Rossi defined two quantities Y and Z. Z is the energy deposited in a sphere divided by the mass of the sphere. It can be measured in rads. Z is the increment in local energy density due to a single event, and for small spherical volumes, the size of a cell nucleus, it is ΔZ that is important. $Z = \Sigma \Delta Z$ and for high LET radiations where several events can occur in such a volume, ΔZ can be very different from Z. This is easily understood for alpha radiation but it is also true for other radiations. Thus it is important to know the frequency distribution of high and low values of ΔZ for a given situation. Y is defined as the energy deposited divided by the sphere diameter. It has the same dimensions as LET. Rossi has calculated the relation between Y and ΔZ. If an event of magnitude Y kev/μm deposits an energy ΔZ in a sphere of diameter d μm then

$$\Delta Z = \frac{30 \cdot 6\, Y}{d^2}\, 10^2 \text{ ergs/g} = \frac{30 \cdot 6\, Y}{d^2} \text{ rads.}$$

Z cannot be measured directly in volumes of the size of a cell nucleus, but it has been shown (Segrè, 1951) that the energy loss is almost entirely due to interaction with electrons, and if the cell nucleus could be expanded to a diameter of about 10 cm, without altering the number of molecules contained in it, the energy loss of an alpha particle crossing the 10 cm sphere would be the same as the energy loss in crossing the original cell nucleus. Rossi and his collaborators made "tissue equivalent spheres" by filling a 10 cm sphere with a gaseous mixture of atoms and molecules similar to those in a cell nucleus, and fitted a proportional counter so that the loss of energy of the alpha particles could be measured. The pressure was varied from atmospheric pressure down to a pressure of a few millimetres of mercury. In this way, the mass of gas in the sphere is varied; a larger mass is equivalent to a larger cell volume. The spherical volumes considered are of radii 0.5 to 10 μm. They determined the relative proportions of high and low values of ΔZ and ΔY experimentally for a given Z and Y of the incident particles, and these have been compared with theoretical values. Allowing for the limitations of the theoretical calculations, the agreement is satisfactory. The energy deposited by a charged particle in a small sphere of tissue divided by the mass of this sphere can vary over a wide range (by a factor of 10^3) for both high and low LET. For low LET as for X-rays and for large doses, ΔZ is equal to the average absorbed dose over a macroscopic volume per unit mass. But for high LET and spherical volumes of the order of a cell nucleus ΔZ and ΔY can be very much larger than the average absorbed dose over a macroscopic volume and consequently while some elemental volumes are receiving a high dose many are receiving no dose at all. The importance of Rossi's experiments lies in the experimental determination of the variation in magnitude of energy deposition in volumes of the order of cell nuclei. The total absorbed dose averaged over a volume of the order of cubic centimetres may be produced by a few very large energy depositions in volumes of a few cubic microns, or a large number of much smaller energy depositions. The experiments determine the energy distribution for a given incident particle of known LET. ROSSI's work confirms the pattern of ionization of the alpha particle described in Sec. VI.B.3. The passage of a single alpha particle through the cell nucleus produces such intense ionization that cell death appears to be highly probable. At low average doses the probability of an alpha particle passing through a cell is lower, since there are fewer alpha particles in a given volume.

5. The Irradiation of Cells by Alpha Particles

The localised effect of the ionization of a heavy charged particle was confirmed by ZIRKLE and his co-workers, ZIRKLE and BLOOM (1953), BLOOM et al. (1955). They used a beam of protons of about 1 Mev in energy and only 7 μm diameter. With this narrow beam different parts of the cell could be irradiated. Cells from the hearts of newts were used, and were examined under the microscope for damaged and broken chromosomes after irradiation. A few dozen protons delivered to the "chromosomal region" produced many breaks, whereas thousands of protons delivered to the "edge of the region" produced only the occasional damaged chromosome.

The technique of cultivating certain cells from both human and animal tissues and using such cultures to study the effect of different forms of irradiation was established in the 1950's. Such cells used for tissue culture purposes in radiobiology have adapted to laboratory conditions and may be different from the cells of normal or malignant tissues from which they were originally derived. They can however be used for quantitative experiments by studying the growth of individual cells into clones of 50 or more cells by successive division. By counting the fraction of an experimental group of cells which produce clones, a quantitative measure of an effect, such as bombardment by alpha particles, can be made. The experiments determine the number of cells with the capacity to make five or more divisions. They cannot distinguish between cells which are killed and cells which survive but cannot divide (frequently called "reproductive death").

Experiments on human kidney cells in culture by BARENDSEN et al. (1960) showed that the surviving fraction of irradiated cells obeyed the simple exponential law

$$n = n_0 \exp(-SD)$$

where D is the dose of polonium alpha radiation, n is the number of clones grown from n_0 irradiated cells and S is a constant. This relation is obtained when a cell is damaged by a single alpha particle (ZIMMER, 1961), sometimes referred to as a "one-hit" process. BARENDSEN calculated that the sensitive area of cross-section of the cell, which is proportional to S in the above equation is about that of a cell nucleus. This means that the probability of an alpha particle passing through the nucleus without damaging the cell sufficiently to prevent it forming a clone is very low. This is what one would expect from ROSSI's experiments. The conclusion, is supported by the work of DEERING and RICE (1962) with alpha particles and lithium ions on HeLa cells and further experiments by BARENDSEN and WALTER (1964) and TODD (1968) show that for high LET alpha radiation there is no recovery from cell damage. Recovery can be detected by comparing the effects of a given dose of alpha radiation with the same total dose in two separate irradiations. When recovery is present two separate irradiations give larger survivals. At lower LET's, of about 25 kev/μm, there is some evidence of recovery for alpha particles, and the survival curve is not a simple exponential, but has the sigmoid shape characteristic of X-rays and low LET radiation. The theoretical interpretation is that in this case more than one "hit" is required to kill or incapacitate a cell.

BARENDSEN (1964) investigated the effect of different LET's for polonium alpha particles on the human kidney cell line. The effect has a maximum at about 150 kev/μm and an alpha passing through the *cell* has about a 10 percent chance of impairing its reproductive capacity. This is most likely to happen when the alpha particle passes through the nucleus. For P 388 murine leukaemia cells the maximum effect appears to occur at a lower LET than for the human kidney cells (BERRY, 1970). In general for cells in tissue culture alpha particle radiation is characterised as being a "one-hit" process with no repair at LET's above about 25 kev/μm and a very high probability that an alpha particle passing through the cell *nucleus* will either kill the cell or render it incapable of division.

It cannot be assumed however that the same criteria apply to cells in living tissue. Some of the problems that arise have recently been outlined by NEARY (1970). There is evidence that some types of cell in vivo can survive even when they contain sufficient plutonium to make alpha bombardment of the cell nucleus inevitable. Macrophages in spleen, lung and in the peritoneal cavity are among the less sensitive cell types now recognised (SANDERS, 1969, 1970; SANDERS and ADEE, 1969, 1970). SANDERS and SANDERS and ADEE have studied the uptake of PuO_2 particles by macrophages in the peritoneal cavity of rats. They found that such cells were able to continue ingesting latex beads though already laden with plutonium oxide. There was however no evidence that they were still capable of division and therefore of malignant transformation[4]. The cells containing plutonium were receiving dose rates of from 40 to 225 rads/hour, and some time later showed evidence of cellular disintegration and decreased phagocytic capacity but the phagocytic capability of the reticulo-endothelial system as a whole in the peritoneal cavity remained intact for as long as seven days, implying that a sufficient number of cells lived long enough to carry on until they were probably replaced by new cells.

In the bone marrow of rabbits given $Pu(NO_3)_4$ by intravenous or intramuscular injection VAUGHAN et al. (1967) observed many macrophages containing plutonium aggregates. The rate of emission of alpha particles from aggregates of about 9×10^6 plutonium atoms (cf. Table 10.41) under the experimental conditions described is rather less than one per day. No damaged cells or portions of cells were ever seen under the microscope. If one percent of the macrophages were damaged at any one time this should have been noticeable. It suggests that the average lifetime of a macrophage containing plutonium is at least a hundred times as long as the time taken for a damaged macrophage to be removed from the system. However at present the data are not available to enable any assessment to be made of the probability of marrow macrophages or endosteal cells being killed or of the latter undergoing malignant change from the passage of an alpha particle.

No reliable estimate can be made as to the relative probabilities of cell death to malignant change since the mechanism by which such a change comes about is not known. It is not known whether a cell can be made malignant by the passage of an alpha particle through the nucleus rather than the cytoplasm, but there is some evidence that in general two separate events are required before tumour production by ionizing radiation can take place, MOLE (1963, 1969), HULSE and

4 The macrophage is a fully differentiated cell and not a cell capable of proliferation. It would not be expected to undergo malignant transformation.

Mole (1969). It is reasonable to assume that this also applies to high LET radiation. Most of the experimental data in support of this hypothesis are for beta and gamma radiation, the event being the passage of a beta particle through the cell. Many of the cells receiving two alpha particles through the nucleus will be killed, but if the alphas are well separated in time the two hits may occur to two different cells, one of which is a descendent of the other, so that the damage is cumulative; or two events may occur, of which one is a hit, and the other a more indirect effect of the radiation. It is not even certain whether the sensitive region is confined to the nucleus or whether it may be in the cytoplasm. Malignant change may arise in the case where the delta rays on the fringe of the alpha track cause damage while the intense core of ionization misses the vital parts of the cell. Since alpha emitters such as plutonium are carcinogenic the probability of some such series of events must be finite.

6. The Experimental Approach

James and Kember (1970) and many other radiobiologists consider the cell nucleus to be the vital target and have described a method for the direct measurement of the frequency distribution of alpha particle incidence in targets of similar dimensions to cell nuclei close to the bone surface. They consider that it is the one-hit processes that are important for radiation damage. Using the standard technique of autoradiography on thin sections of rat bone and a microscope graticule which had circles of diameter corresponding to a cell nucleus, they counted the number of alpha tracks crossing a circle and computed the number of hits that a cell nucleus would receive, in vivo, adjacent to that particular section. For a uniform plutonium distribution within volumes small compared with the range of the alpha particle the incidence of one-hit events to cell nuclei can be calculated for a given amount of plutonium. James' measurements showed a higher incidence of one-hit events than one would expect from a uniform plutonium distribution in which the plutonium nuclei emit alphas in a random manner. For the lumbar vertebrae of rats given a dose of 5 μCi/kg of body weight, assuming no movement of cells, after 3 weeks nearly all the cells on the trabecular endosteal surface had received two or more hits, but on the neural wall of the endosteum some cells had received only one hit. This is interesting because there is some evidence that in rats with a plutonium body burden tumours develop near neural canal walls.

The alternative approach, used by Jee et al. (1961, 1971), Vaughan et al. (1967), is to count or measure densitometrically the tracks on an autoradiograph or on a foil with fission tracks, within a certain area, and by making assumptions about the spatial distribution of the plutonium to calculate the amount of plutonium present and the dose rate, or frequency of alpha particle incidence to particular regions of interest. The methods of calculation are described in the next section.

C. The Derivation of Formulae for the Calculation of Dose Rates

1. Introduction

Most of the formulae in the following section can be applied to either alpha tracks in autoradiographs or fission tracks in solid state detectors. For the radioactive decay of plutonium 239 the exposure time is always so short compared with the half-life that the number of plutonium atoms can be taken as constant.

Then in a time T, if there are N atoms and the decay constant is λ the number of alpha particles is

$$N_A = N\lambda T$$

For fission tracks, if the product of the fission cross section and the neutron flux is β and the irradiation time in the reactor T, from N nuclei of plutonium the number of fission tracks is

$$N_F = 2N\beta T.$$

Again the number of plutonium nuclei undergoing fission is small compared with N so that N can be taken as constant. The formulae involving track counts in the next sections can be applied to either alpha or fission tracks provided that the correct value of the range R is used and either N_A or N_F as appropriate. The word film is to be interpreted as photographic emulsion or polycarbonate foil, according to whether alpha tracks or fission tracks are being considered.

2. The Point Source

In Fig. 10.51 the region between AA and DD represents film and the tissue section containing plutonium is below it. If a point source (for example an aggregate of plutonium atoms such as gives rise to the star pattern in Fig. 10.41) is at 0, the chance of a particle reaching the film depends on its direction, that is, it depends on the angle Θ in the figure. Of all the particles emitted in all directions, the fraction which pass through the film must pass through the spherical

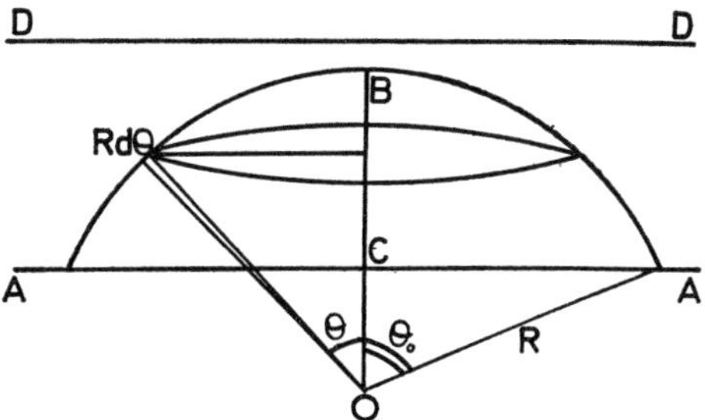

Fig. 10.51. Figure illustrating the calculation of the number of alpha particles reaching a film from a point source 0. (See text V.C.2)

cap ABA formed by the rotation of the figure about OCB. This cap is limited by AO, the range R of the particles. The fraction F reaching the film is the area of the cap ABA divided by the area of a sphere of radius R. The area of the elemental ring of radius $R \sin \Theta$ and width $R d\Theta$ is $2\pi R^2 \sin \Theta \, d\Theta$.

Therefore $F = \dfrac{2\pi R^2}{4\pi R^2} \displaystyle\int_0^{\theta_0} \sin \Theta d\Theta$

$$F = \tfrac{1}{2}(1 - \cos \Theta_0).$$

If $OC = Y$,

$$F = \tfrac{1}{2}(1 - Y/R) \qquad (1)$$

As Y increases F falls from 0.5 when point 0 is immediately below the film to 0 when $Y = R$. If sources such as 0 are scattered uniformly throughout the section, the average value for F is

$$F_{av} = \frac{1}{R}\int_0^R F\,dY = \frac{1}{2R}\int_0^R \left(1 - \frac{Y}{R}\right) dY = \frac{1}{4}. \qquad \text{(Spiers, 1953)}$$

If each source emits P tracks, on average $P/4$ tracks will reach the film from each one.

For a thin section, $Y=0$ to $Y=R$ if $d<R$,

$$F_{av}=\frac{1}{d}\int_0^d F\,d\,Y=\frac{1}{2}\left(1-\frac{d}{2R}\right). \tag{2}$$

A count of the number of tracks reaching the film from an aggregate enables the number of plutonium atoms in the aggregate to be calculated. For an aggregate distance Y below the film, the fraction of emitted alphas reaching the film is

$$\tfrac{1}{2}\,(1-Y/R)$$

from Eq. (1). Assuming that aggregates are randomly distributed, the average value of Y is $d/2$, where d is the thickness of the section. If an aggregate initially contains N_0 plutonium atoms, the decay constant is λ weeks^{-1}, and the exposure T weeks, $N_0\{1-\exp(-\lambda T)\}$ tracks are emitted in time T and

$$\tfrac{1}{2}\,N_0\{1-\exp(-\lambda T)\}\,(1-d/2\,R)$$

reach the film to form a star. For a four week exposure,

$$\lambda T=2.21\times 10^{-6} \quad \text{and} \quad \exp(-\lambda T)\approx 1-\lambda T.$$

Therefore the number of tracks in a star, S is

$$S=\tfrac{1}{2}\,N_0(1-d/2\,R)\,\lambda T \tag{3}$$

from which N_0 is obtained. A number of determinations of S must be made from different stars, and a histogram plotted as described in Sec. VI.C.6 to determine the correct value of S to use in Eq. (3).

3. The Uniform Volume Distribution

This is the same problem as the uniform distribution of sources which is given in the previous section. If the alpha particle sources such as aggregates of plutonium atoms are imagined to be spread out through a tissue section so as to produce an activity P per unit volume, then the number of tracks reaching an area A arise from a volume RA. The fraction of tracks reaching the film is 0.25, so that the track count is $PRA/4$ and the count per unit area is $S=PR/4$ from which P is obtained, if R is known. The dose rate D is the energy absorbed per unit mass of tissue per unit time. If the density is ϱ and the initial energy of the alpha particle is E, and T is the exposure time

$$D=PE/T\varrho=4SE/RT\varrho \tag{4}$$

$E=5.14$ Mev $=8.25\times 10^{-6}$ erg. $R=37$ µm in soft tissue.
$\varrho=1$ g/cm^3. If $S=10$ in an area of 200 µm^2 for 10 days exposure.
$D=0.22$ rad/day for a section for which $d>R$.
If $d<R$, say 7.5 µm,
$F_{av}=0.45$ from Eq. (3)
$S=PR\,(0.091)$,
$D=3.0$ rad/day.

4. The Uniform Plane Source Normal to the Film

A bone surface in an autoradiograph approximates to a surface deposit of plutonium on a plane surface which is infinite in length in both z directions in Fig. 10.52 and also in the y direction downward into the section. The x direction, OP, is at right angles to the yz plane. Bone sections cannot generally be cut very thin, and if the thickness of the section is greater than the particle range R, only the part of the section within a distance R of the film can contribute to the track count. Any section of thickness greater than R behaves as if its thickness were equal to R. It is assumed that the bone surface is vertical and in the yz plane in Fig. 10.52 and intersects the film which is in the xz plane at the line

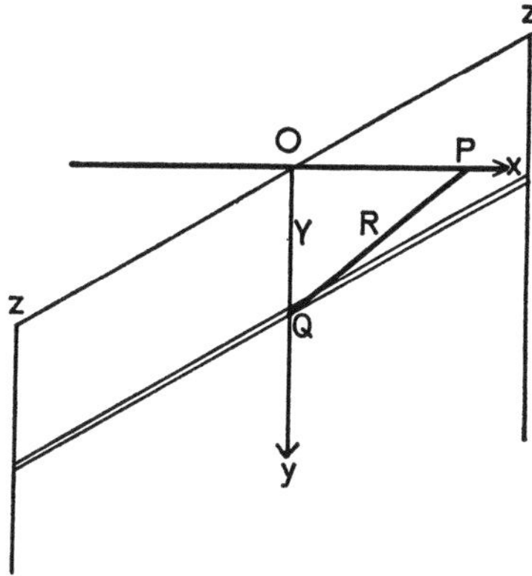

Fig. 10.52. Figure illustrating the calculation of the number of alpha particles reaching the film in the xz plane from a plane surface in the yz plane. The plane and film intersect along Oz, and the tracks reach out to a distance OR in the x direction, determined by the range R of the alpha particle. The plane extends to infinity in one y direction and both z directions

zOz. Tracks from the bone surface below the film reach the film on both sides of zOz out to a distance R along Ox for particles starting just below the film, but tracks from a point Q distance Y below the film can only reach a distance $(R^2 - Y^2)$ from zOz. For any element at Q of area $dY \cdot dz$, the fraction of particles reaching the film is $\frac{1}{2}(1 - Y/R)$ from Eq. (1) if the activity per unit area of the plane yz is B the total number of tracks reaching the film is

$$\int_0^R \frac{1}{2}\left(1 - \frac{Y}{R}\right) B\, dY\, dz = \frac{BR}{4}\, dz.$$

The total number of tracks per unit length of the plane yz in the z direction is $S = BR/4$.

The number S includes all the tracks per unit length of the plane out to a distance R on each side. If the ranges of the tracks on either side of the plane are R_1 and R_2 (usually the ranges in bone and bone marrow) then

$$S = \frac{B}{8}(R_1 + R_2). \tag{5}$$

If R_1 and R_2 are known, B can be found from the track count. The dose rate decreases rapidly as x increases from 0 to R because only particles close to the film can reach regions nearly R away from it.

To find how the dose rate in vivo varies with the distance x it must be remembered that then a cell in the yz plane receives a radiation dose from above as well as below. Fig. 10.53 shows the cell represented by an element of volume $dv = dA \cdot dx$ at a distance x from the yz plane which extends in both directions. Since the alpha range is R only alphas originating within the circle of radius $\sqrt{(R^2 - x^2)}$ can reach $dA \cdot dx$.

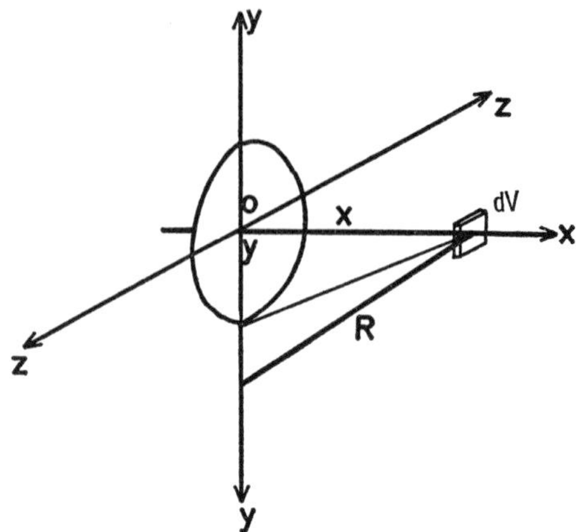

Fig. 10.53. Figure illustrating the calculation of the dose rate at a distance x from a plane distribution of plutonium in the yz plane. The plane extends to infinity in both y and both z directions

Consider an elementary ring of this circle of radius Y and width dY. For any point on this ring the fraction of alphas that pass through dA is $\frac{dA \cos\alpha}{4\pi r^2}$ where $r^2 = x^2 + y^2$. The path length of the alpha particle in the elemental volume $dA\, dx$ is $dx/\cos\alpha$ and the energy deposited by the alpha particle is

$$E\, dx/\cos\alpha.$$

The energy deposited by all the alphas from the ring of width dY is

$$dE = \frac{2\pi Y dY\ B\ dA \cos\alpha\ dx\ E}{4\pi r^2 \cos\alpha}.$$

For the whole disc, the energy deposited in the volume $dA\, dx$ is

$$\int_0^{\sqrt{(R^2-x^2)}} dE = \frac{BE\, dA\, dx}{2} \int_0^{\sqrt{(R^2-x^2)}} \frac{Y\, dY}{(x^2+y^2)} = \frac{BE\, dA\, dx}{2} \ln(R/x) \tag{6}$$

where x is a constant in the integration with respect to Y.

The dose rate in the volume $dA\, dx$ is the energy loss/mass per unit time, giving the dose rate

$$D = \frac{B\, dA\, dx\, E \ln(R/x)}{2T\, R\, \varrho\, dA\, dx} = \frac{B\, E \ln(R/x)}{2TR\varrho}. \tag{7}$$

This equation assumes that the LET, E/R is constant throughout the range of the alpha particle. B can be obtained from Eq. (5).

The change of D with x, dD/dx, is proportional to $1/x$, and becomes very large as x approaches zero. Near bone surfaces the region of importance is for values of x from 0 to 15 μm. The dose rates in Tables 10.35–10.38 in Sec. VII are calculated from Eq. (7) for $x = 5\ \mu m$, $R = 37\ \mu m$. ICRP consider that the average dose rate between 0 and 10 μm is an important quantity.

$$D_{av} = -\frac{1}{10} \int_0^{10} \frac{BE \ln(x/R)}{2TR\varrho} \cdot dx, \qquad D_{av} = \frac{BE}{2TR\varrho} \{1 + \ln(R/10)\}.$$

Taking $R = 37\ \mu m$

$$D_{av} = 2.31 \cdot \frac{BE}{2TR\varrho}.$$

The dose rate at 5 μm from the surface, which is given in the experimental results in Sec. VI is, from Eq. (7)

$$D_5 = 2.26 \cdot \frac{BE}{2TR\varrho}$$

so that D_{av} and D_5 are nearly equal.

$$D_{av} = 1.0\,2\,D_5. \tag{8}$$

The numerical relationship between the activity on the surface and the average dose rate over 10 μm at 10 μm intervals from the surface is shown in the following table for an activity B of 1 nanocurie (10^{-9} curie) per cm^2 after MAYS and TUELLER (1964) with an alpha range of 35.4 μm.

Distance (μm)	0–10	10–20	20–30	30–40
Dose rate (rad/day)	84.2	32.6	13.2	1.6

If an alpha particle range of 37 μm is used, the values of the dose rate only differ by a few percent except at the very end of the range. The stopping power and range of alpha particles in various media such as soft tissue has been calculated by WALSH (1970) and compared with the experimental and theoretical data of various workers. WALSH's conclusion is that the ranges of alpha particles in tissue are known to within 15%. His data lead to a value of 37.0 μm, the value used in this text, for the range of plutonium alphas in soft tissue, and an average stopping power, or LET, of 100 kev/μm.

5. The Approximations Involved in Calculating Dose Rates at Bone Surfaces

There are many approximations involved in calculating and applying these formulae to the dose from the alpha particle of plutonium. The problem of the spatial inhomogeneity of the alpha particle has already been discussed. In the derivation of Eq. (6) and (7) the assumption of an infinitely thin uniform plane distribution of plutonium is a crude approximation to the situation at the bone surface. The band of tracks should be $(R_1 + R_2)$ in width if the deposit is infinitely thin, but in practice it is from 6 to 14 μm wider than this (BLEANEY, 1969). One reason is that the mineral deposit does not end sharply at the bone surface, but there may be an osteoid border of varying width and mineral density, through which the plutonium is spread[5]. Also, when the marrow is not removed there is plutonium in the endosteal osteogenic cells along the bone surface. This layer of cells is often only one cell thick but can be two or three cells thick (cf. sec. IV A).

Fig. 10.54 shows a situation which approximates more to the distribution found in the shaft of the rabbit femur and discussed in Sec. IV A). In Fig. 10.54 the band of tracks on the film will extend from AA to BB, a distance R_1, the

5 Editors note. Also the plutonium surface may not always be perpendicular to the film.

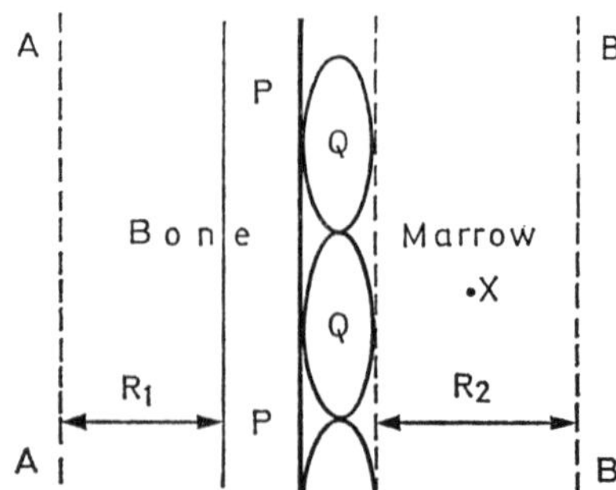

Fig. 10.54. Figure illustrating the position of the osteogenic cells QQ with respect to the mineral surface of bone and marrow. If plutonium spreads through the layer PP in bone and through the width of the cells QQ then tracks will be seen from AA to BB

range in bone beyond the plutonium deposit PP in bone, and R_2, the range in marrow beyond the plutonium in the osteogenic cells QQ. Half the plutonium represented by P in Eq. (7) is assumed to be in the layer PP 5 µm thick on the bone surface and half in the layer of cells QQ along the mineral surface. The dose rate at a point X, 10 µm, or about two cells width, from the mineral surface will differ from that given by Eq. (7). The contribution from the plutonium in the layer PP will be less, and the contribution from the plutonium in the layer QQ will be more than that calculated on the basis that the plutonium is at the mineral surface. The correct dose rate may be greater or less than Eq. (7), but as the exact distribution of the plutonium is variable along the surface, complicated calculations for particular distributions in bone, osteogenic cells and marrow are unlikely to be more accurate than the simple formula for a plane surface. The actual dose rate in the cells QQ in Fig. 10.54, which have plutonium either in the cell or on the surface, could be considerably higher than the dose rate at the point X. This is a particularly important consideration at low average dose rates.

6. The Poisson Distribution

Even if the plutonium is uniformly distributed in the tissue, when the number of tracks in a convenient area, 200 µm², is small enough to count under the microscope, the number in different 200 µm² areas will not be the same. This is because the emission of an alpha particle is a random event and each plutonium nucleus acts independently of the others. Over a long period of time the randomness averages out, and when there are many tracks the percentage variation in track count is less. The probability distribution for the number of tracks counted per unit area is a Poisson distribution, and the chance of counting k tracks per unit area is

$$P(k) = \frac{a^k e^{-a}}{k} \tag{9}$$

where a is an integer and is equal to the mean value of k. A typical Poisson distribution is shown in Fig. 10.55 fitted to an experimental histogram. For this distribution a is equal to the variance. If all the values of k are large, a is large and the curve is very spread out, with a low, broad maximum. For small values of k the curve is narrow with a high maximum value. The variance of the mean is a/N, so that the fractional standard deviation is

$$\frac{1}{a}\sqrt{\frac{a}{N}} = \frac{1}{\sqrt{aN}}$$

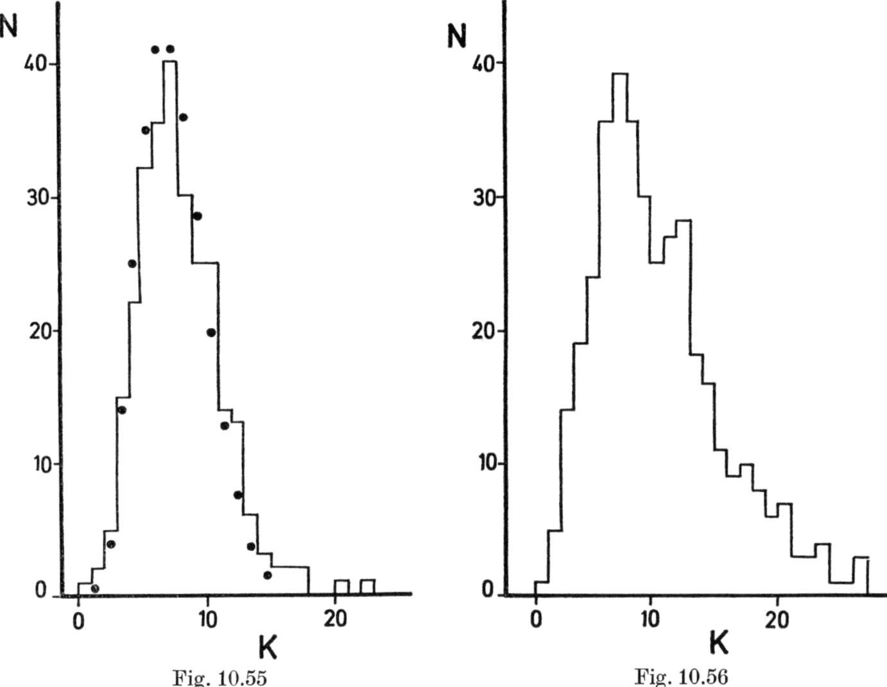

Fig. 10.55 Fig. 10.56

Fig. 10.55. An experimental histogram, the abscissae being the number of times a track count k occurs plotted against k for a bone surface. The circles • are the Poisson distribution obtained by multiplying the function $P(k)$ of Eq. (8) by the total number of count for $a = 7$. (BLEANEY and VAUGHAN, 1971)

Fig. 10.56. An experimental histogram as in Fig. 10.55 but for a bone surface with marrow *in situ*. The distribution cannot be fitted to a single Poisson distribution curve, showing that the plutonium distribution along the surface is not uniform. (BLEANEY and VAUGHAN, 1971)

and is smaller for larger values of N. In track counting N is limited by the amount of time available, but N should be as large as possible.

The k values and hence a cannot be very large in practice because otherwise errors in counting arise from overlapping tracks. $P(k)$ is a skew distribution, especially for low values of a. The circles in Fig. 10.55 are the values obtained from Eq. (9) by substituting the value of k at each point. putting $a = 7$, and multiplying by the total number of counts. This value of a is chosen as the best fit to the experimental histogram, and the close fit of the points to the histogram suggests that the plutonium distribution is uniform in this region. The value of a can also be found from the formula

$$a = \bar{k} = (a_1 + 2a_2 + \ldots r a_r + \ldots n a_n)/N \qquad (10)$$

where there are r counts of a_r tracks and N counts altogether. Fig. 10.56 is another experimental histogram, in which there is marrow along the bone surface, whereas in Fig. 10.55 the marrow has been removed. In Fig. 10.56 the histogram cannot be fitted to a single Poisson distribution curve, and this implies that the plutonium distribution varies along the surface. Fig. 10.44 is a histogram for the number of tracks from plutonium aggregates, and the fit to a Poisson distribution shows that the aggregates contain about the same number of plutonium atoms. The value of S in Eq. (4) and (5) must be found from Eq. (10) and N must be sufficiently large to reduce statistical errors to a satisfactory level.

7. The Relation Between Dose-Rate and "Hit-Frequency"

The unit of rads/day for dose rate is frequently used for radiation work. Since alpha particle dose is so inhomogeneous it is important to be able to relate average dose rate to the rate at which alpha particles hit a given cell nucleus, or the "hit-frequency". This is measured directly by the method of JAMES and KEMBER (1970) or can be estimated from the dose rate by making simple assumptions about the geometry. For example:

If a cell near a bone surface receives a dose rate of 10 rad/day, the number of alphas passing through unit area of the cell can be found from Eq. (6). If it is assumed for simplicity that the alpha particles cross the cell parallel to the x axis, the energy deposited by each one is $E\,dx$, where dx is the cell thickness. Taking dA as the cross-sectional area of the cell, the number of alphas passing through unit area is

$$\frac{BE\,dA\,dx}{2} \ln\left(\frac{R}{x}\right) \frac{1}{E\,dA\,dx} \text{ from Eq. (6).}$$

Eliminating B with Eq. (7) gives the number of alphas/day/area $=$ RD/E.

This is an upper estimate because of the approximation in assuming that all the alpha particles cross the cell in a direction parallel to dx. Putting $R = 37 \times 10^{-4}$ cm, $E = 8.23 \times 10^{-6}$ erg, and multiplying by the area of a 5 μm diameter nucleus which is 1.95×10^{-7} cm^2, the hit frequency is 0.08 day^{-1}, so that the nucleus is hit every 12.5 days, and the whole cell of roughly four times the cross-sectional area, every three days. This calculation does not include any alpha particles arising from plutonium in the marrow.

In bone marrow the hit frequency for a given dose rate will be less than at the bone surface by a factor which will depend on its location with respect to aggregates and fat spaces.

The formulae in this section are used to calculate the results shown in the tables in Sec. VII. It is useful to bear in mind that a dose rate near a bone surface of 10 rad/day means a hit to a nucleus about every 12.5 days, and the same average dose rate in marrow means a hit to a nucleus about every 12.5 days, but the estimate depends on the volume of the fat spaces. Applying this argument to man, for a 70 kg man with a body burden of 0.04 μCi, the maximum permissible burden for occupational exposure, the dose rate at the bone surface is equivalent to a hit to a nucleus of an endosteal cell every 60 years.

VII. Radiation Dose Measurements from Deposition of ^{239}Pu in the Skeleton

BETTY BLEANEY

A. Introduction

The previous section has indicated the great theoretical and technical difficulties that are involved in measuring the radiation dose from alpha particles. Even if a calculation can be made of the dose rate at one point in time i.e. when the animal is killed and the autoradiograph of the bone or marrow prepared the measurement will apply only to this point in time. In the case for instance when plutonium reaches the blood stream and therefore the skeleton from a wound the amount of plutonium irradiating the sensitive cells is always changing and it is difficult to obtain a reliable measurement of total dose.

Such measurements as are available from experimental studies will be discussed in the following pages.

B. Bone Surfaces

1. Intravenous Injections

Detailed studies of the distribution of plutonium on bone surfaces and estimates of dose rates to these surfaces have been made on rats (James and Kember, 1970; Jee and Arnold, 1962), rabbits (Bleaney, 1967, 1969; Bleaney and Vaughan, 1971) and dogs (Arnold and Jee, 1959; Twente, and Jee, 1961; Jee, Arnold et al., 1962; Jee, Stover et al., 1962; Stover and Jee, 1963; Jee, 1964). The amount of data is rather limited, because of the time-consuming nature of the measurements but a substantial start has been made. This section is an attempt to summarise the results and to compare the results for different animals. Exact comparison is seldon possible however, because of the difficulty of producing identical injection solutions of plutonium, and even if the repeatability of the injections were assured, the chemical state of the solution after injection depends on the actual number of μCi injected and the local pH, and this introduces unknown variables in comparisons between animals of different size[6].

a) Rats

James and Kember (1970) have measured the alpha particle incidence for cells near endosteal bone surface (Sec. VI.B.6). From these measurements they have calculated dose rates at bone surfaces, James (1970). For a target such as a cell the dose for the exposure time of the autoradiograph (80 hours), is found using the relation

$$D = \frac{\mathrm{LET}}{\varrho} \Phi$$

where Φ, the fluence, is the number of alpha particles incident on the area of the target, which has a circular cross-section of 5 μm diameter, and where LET and ϱ are as defined previously. The average dose in rads/day for all the targets counted, at the time of death is $\frac{24}{80} D$ with D in the appropriate units. James and Kember used 7 week old male rats weighing 100 g and injected 0.5 μCi of $Pu(NO_3)_4$, in 0.2 ml of 0.02 M nitric acid, in a tail vein.

The flux measurements were made at three distances from the mineral surface and the data are shown in Tables 10.34 and 10.35. The dose rates are in rads/day per μCi/kg of injected dose for all the tables in this section. In Tables 10.34 and 10.35 the dose rate increase slightly between one and four days after injection but then remains remarkably constant, and at 48 weeks there is more plutonium in the epiphysis of the femur than at one day after injection.

6 Editors note — While these statements are of course quite valid it should be pointed out that the amount of information available for plutonium exceeds that for any other bone seeking radionuclide with the possible exception of radium and it far exceeds in details relevant to local dose most non-radioactive substances encountered in general pharmacology and toxicology. Recognition of the problems is easier with radionuclides because of the ease of measurement of small quantites and the availability of techniques such as autoradiography. Application of these considerations of local dose to general problems in toxicology is clearly desirable.

Table 10.34. Dose rates in rads/day at various distances from the endosteal mineral surface in femurs of 7 week old rats given an intravenous injection of 5 μCi/kg of $Pu(NO_3)_4$. (Taken from JAMES, 1970)

Time after injection	5 μm	12.5 μm	20 μm	5/12.5[a]	5/20[a]
Epiphysis					
1 day	6.6	3.8	2.0	1.7	3.3
4 days	9.2	5.4	3.2	1.7	2.9
21 days	7.8	4.8	2.8	1.6	2.8
18 weeks	5.4	3.6	2.4	1.5	2.2
36 weeks	4.6	3.0	1.6	1.5	2.9
48 weeks	7.2	4.8	3.0	1.5	2.4
Diaphysis					
1 day	10.4	4.8	2.6	2.2	4.0
18 weeks	1.6	0.98	0.7	1.8	2.2

[a] The last two columns give the ratios of the dose rates at different distances from the surface.

Table 10.35. Dose rates in rads/day at various distances from the endosteal mineral surface of lumbar vertebrae in 7 week old rats given an intravenous injection of 5 μCi/kg of $Pu(NO_3)_4$. (Taken from JAMES, 1970)

Time after injection	5 μm	12.5 μm	20 μm	5/12.5[a]	5/20[a]
1 day	8.4	4.8	2.6	1.7	3.2
4 days	8.4	5.6	2.8	1.5	3.0
21 days	8.6	6.0	3.6	1.4	2.4
18 weeks	4.8	3.6	2.8	1.3	1.7

[a] The last two columns give the ratios of the dose rates at different distances from the surface.

The relative values of the dose rates at different distances from the bone surface are interesting. From equation 7 the ratios can be calculated for a uniform plane distribution of plutonium on the surface. For the ratios of the dose rates at distance 5 and 12.5 μm, and 5 and 20 μm,

$$D_5/D_{12\cdot 5} = 1.84 \quad \text{and} \quad D_5/D_{20} = 5.25.$$

JAMES' values range from 1.81 at one day to 1.3 at 18 weeks for the first ratio and 3.3 to 2.2 for the second except in the diaphysis of the femur where the one day values are higher. Most of the one day values agree well with the theoretical value from formula (7) and there is a systematic decrease with time after injection in the vertebrae and the diaphysis of the femur. For the epiphysis the dose rates remain approximately constant with time after injection. The decrease in the ratio of the dose rates with time is consistent with an increase in the proportion of plutonium in the endosteal cells and a decrease in the proportion on the mineral surface. The endosteal cells are about 2.5–4.4 μm thick in the adult human (Sec. VIII.A) and the layer is one to three cells wide, so that the range of from 5 to 20 μm from the mineral surface extends beyond the endosteal cells to the region of marrow cells, except where there are osteoclasts which can be 20 μm diameter. The flux measurements from which the dose rates are calculated show a random, single Poisson distribution at one day, but appreciable departures from this at later times after injection.

Table 10.36. Dose rates near endosteal surfaces of vertebrae of dogs

Injected activity (μCi/kg)	% of Pu per gram of bone (A)	Dose rate (rads/day)	Dose rate per μCi/kg of injected dose (B)	B/A
2.7	1.35	68	25	18
0.3	1.3	3.0	11	8.5
0.015	1.54	0.3	32	21

The dose rates in rads/day are for a distance of 5 μm from the endosteal mineral surface of trabecular bone in lumbar vertebrae of 18 month old dogs given an intravenous injection of plutonium citrate. The figures are for 28 days after injection and are taken from JEE (1964). The last column is the ratio of the dose rate in rads/day/μCi/kg of injected dose to the percentage of injected dose per gram of bone.

b) Dogs

Most of the experiments on beagle dogs in Utah were designed to study the effects of plutonium from injection as young adults aged about 18 months, until death from radiation damage or disease or scheduled sacrifice after disease. Such experiments cannot give direct information about the plutonium distribution at early times after injection. TWENTE and JEE (1961) have measured the plutonium deposits on the endosteal surfaces of thoracic vertebrae of dogs by both autoradiography and track-counting and micro-densitometry techniques. STOVER et al. (1959, 1962) describe the injection solution as "tetravalent plutonium in citric acid with a sodium buffer pH 3.5. Careful analysis indicates that not more than 2 per cent Pu(vi) was present and the solution was thought to be monomeric". JEE and TWENTE found that at least 75 per cent of the whole plutonium content of the vertebrate as measured by radiochemical methods was on the endosteal surfaces. The dose rate for measurements on lumbar vertebrae are given in Table 10.36. The figures give the average dose rates over the region counted at the time of death at 5 μm from the mineral surface and are calculated from data from JEE (1964). The dose rate is not proportional to the injected dose, and this fact is illustrated by the variation in the fourth column of figures which give the dose rate per μCi/kg of injected dose. The last column gives the ratio of dose rate per μCi/kg to percentage of injected dose per gram of bone, assuming that the skeletal weight is 10 percent of the body weight.

The ratio B/A in the last column of Table 10.36 is not constant because of the errors in the dose rate measurements and also radiochemical analyses have shown considerable variation in plutonium concentration from bone to bone and in individual bones, JEE, STOVER et al. (1962); it is listed for comparison with the same ratio in Table 10.38 and 10.40. The amount of plutonium in bone is sometimes converted into the dose rate in rads/day that would exist throughout the volume of the bone if the plutonium were uniformly distribution through it. It is called the "uniform dose rate" or "average bone dose rate". Such a quantity is purely fictitious, since the plutonium is almost entirely confined to bone surfaces, Haversion canals and vascular channels. It is defined here because it is quoted in the literature for comparison with elements such as radium which have a volume distribution in bone.

The actual dose rate at a given distance from the mineral surface varies for several reasons. Local regions of high dose rate can result from the uptake of plutonium in osteoclasts and macrophages near the surface, and on irregular surfaces of trabecular bone the contours may be of such a shape as to concentrate the radiation on to a particular cell. Also, (owing to the random nature of radioactive decay), a uniform deposit of plutonium gives a Poisson distribution of track counts (cf. Sec. VI.C.6) with a few very high and very low counts. The ratio of the highest to the lowest count is much higher than the ratio of the maximum to minimum densities of plutonium on the surface. The high counts are a small region of exceptionally high dose rate for that particular exposure time; they are a very small proportion of all the counts. The result of the inhomogeneity and the Poisson distribution is that particularly at very low dose levels there will be a few regions where some cells receive a dose well above average for the bone surface. JEE found a ratios of 8.6 between maximum and minimum dose rates in lumbar vertebrae, 28 days after injection. Unfortunately no figures are yet available for other times after injection, but the pattern of distribution in dogs appears to be very similar to that observed in mice, rats and rabbits.

c) *Rabbits*

Several series of rabbits have been given single injections of $Pu(NO_3)_4$ and killed at various times after injection. The first two series had intravenous injections of a solution that was 70 per cent ultrafilterable, mainly monomeric plutonium. Table 10.37 shows the data obtained for 6 week old weanling rabbits, from BLEANEY, TAYLOR and VAUGHAN (1971). The dose rates are for the endosteal surfaces of the metaphysis of the femur and the lumbar vertebra at 5 μm from the mineral surface at various times after injection. The rapid decrease in dose rate is to be expected in view of the amount of growth and remodelling in young animals. At 7 months after injection all the bone in the regions counted is bone laid down after injection. This bone is distinguished from bone laid down before injection by the low concentration of plutonium throughout the whole volume of new bone. The dose rate is probably largely due to endosteal cell plutonium. At one day after injection all the endosteal surface have a deposit of plutonium, and the ratio of maximum to minimum concentrations in these deposits is not more than 3 to 1. At later times after injection some regions of the surface have very little plutonium but about one fifth of the surface still has a significant dose rate.

The second series of rabbits was injected at 7 months when they were fully grown. Table 10.38 shows the dose rate measurements, and the decrease with time after injection is much less rapid than with growing rabbits, and is similar to the pattern in 7 week old rats. The last columns in Tables 10.36 and 10.38 give the ratio of the dose rate on vertebral endosteal surfaces to percentage uptake of plutonium in bone. This factor varies by a factor of nearly two but is usually

Table 10.37. Average dose rates in rads/day/μCi/kg of injected dose at 5 μm from the surface of trabecular bone in femurs and lumbar vertebrae of 6 week old rabbits given an intravenous injection of 1.25 μCi/kg of $Pu(NO_3)_4$

Time after injection	Dose rates in rads/day/μCi/kg of injected dose	
	Femurs	Lumbar vertebrae
1 day	20	13.5
8 days	10	9.3
112 days	6.8	2.6
7 months	2.0	2.0

Table 10.38. Average dose rates in rads/day/μCi/kg of injected dose at 5 μm from the surface of trabecular bone in femurs and lumbar vertebrae of 7 month old rabbits given an intravenous injection of $Pu(NO_3)_4$ of 1.1 μCi/kg, or for the two starred animals, 2.29 μCi/kg

Time after injection	% of Pu (per gram of ash)		Dose rate (rads/day)		Dose rate[a]
(days)	femur	vertebra	femur	vertebra	% of Pu
1	0.98	0.86	11	16	18
1	0.45	0.21	9	10	48
8	0.35	0.41	11	11	26
8	0.46	0.63	18	20	32
112	0.40	0.53	25	18	33
112	0.42	0.53	14	15	28
112*	0.30	0.45	21	15	37
112*	0.29	0.34	9	18	52
365	0.30	0.33	13	11	34
365	0.33	0.32	10	16	49

[a] The last column is the ratio of the dose rate to the percentage of injected dose per gram of bone ash for vertebrae (B/A).

between 20 and 30. Since the percentage of plutonium in bone ash can be easily measured, a rough estimate of the amount of plutonium or the dose rate on the surface can be obtained by multiplying it by the appropriate factor.

d) Comparison of Different Animals

Kember (1970) has compared the surface uptake of plutonium for different animals for a single intravenous injection of plutonium. The results are shown in Table 10.39. The uptake per μCi/kg injected appears to increase with the size of the animal and this agree with the findings of Bukhtoyarova (1961) and Rysina and Erokhin (1961) who reported that rabbits were more sensitive to plutonium then rats, and dogs more sensitive than rabbits per μCi/kg injected. Since for all these animals the alpha particle range is small compared with the dimensions of trabecular bone (unlike the case of strontium 90 where the mean range of the beta particles is from ^{90}strontium + ^{90}yttrium is about 5 mm) the problems of dosimetry are the same in each case, and the difference between species should be mainly due to physiological differences.

Table 10.39. Table showing the surface uptake per μCi/kg of injected dose for a single intravenous injection of plutonium for different animals (Taken from Kember, 1970)

Animal	Surface uptake ($nCi \times cm^{-2}$)	Method of measurement and source of data
Adult dog	0.50	Densitometry[a], 70 μm field
Weanling rabbit	0.16	Track counts[b]
Adult rabbit	0.25	Track counts[c]
Immature rat	0.14	Densitometry[d], 8.5 μm field

[a] Twente and Jee (1961) ^{239}Pu citrate. [b] Bleaney (1967) $^{239}Pu(NO_3)_4$. [c] Bleaney (1969) $^{239}Pu(NO_3)_4$. [d] James and Kember (1970) $^{239}Pu(NO_3)_4$.

2. Intramuscular Injections

a) Rabbits

The dose rates on endosteal surfaces of femurs and lumbar vertebrae of fully grown rabbits given an intramuscular injection in the upper forelimb are shown in Table 10.40. The pattern of uptake with increasing time after injection is different from that after intravenous injections. The dose rates increase to a maximum around seven to nine weeks after injection and then start to decrease slowly. The magnitude of the dose rates are less than those in Table 10.35 for

Table 10.40. Average dose rates in rads/day/μCi/kg of injected dose at 5 μm from the surface of trabecular bone in femurs and vertebrae of 7 month old rabbits given an intramuscular injection of $Pu(NO_3)_4$ of 2.61 μCi/kg

Time after injection	% of Pu (per gram of ash)		Dose rate		Dose rate[a]
(days)	femur	vertebra	femur	vertebra	% of Pu
8	0.076	0.113	2.5	3.8	33
28	0.070	0.089	3.8	2.5	28
49	0.23	0.34	10	7.5	22
112	0.14	0.22	6.2	3.8	17
280	0.18	0.36	6.2	6.2	17
365	0.17	0.14	5	3.8	27

[a] The last column is the ratio of the dose rate to the percentage of injected dose per gram of bone for the vertebrae (B/A).

the intravenous case, but not greatly so. One cannot assume that for plutonium introduced intramuscularly, most of it will remain at the injection site. The relative distribution of plutonium over the endosteal surface is similar to the case of intravenous injections, with a similar variation of dose rates over the surface for femurs and vertebrae at any particular time after injection. The ratio of dose rate to percentage uptake of plutonium in ashed bone is of the same order of magnitude as in Tables 10.36 and 10.38.

b) Bone Surfaces

All the data in Tables 10.36, 10.37, 10.38 and 10.40 have been calculated on the assumption that the plutonium is a uniform plane distribution, which is an infinitely thin layer on the mineral surface. The width of the band of plutonium can be measured from the width of the band of tracks, and varies from a few μm to about 14 μm for rabbit bones, Bleaney (1969a, b). It is known, Jee (1964) that osteoclasts contain plutonium and surface osteogenic cells have plutonium either in the cells or on the cell walls. This is due to the uptake of plutonium associated with these cells and the variable width of the osteoid border.

To find what proportion of surface plutonium is in the endosteal cells Bleaney and Vaughan (1971) measured the uptake on the endosteal surface of the shaft of the femur when the marrow was left in situ, and compared it with the other femur in which the marrow was extruded and measured separately (see Sec. 'V.A). This was done for two rabbits injected intravenously and two injected intramuscularly. The results have already been discussed in relation to the pattern of plutonium distribution in the skeleton (Sec. 'V.A), and are shown in Table 10.32. About 40 percent of the surface plutonium is associated with the endosteal cells. Also it was found that on the straight surface of the shaft of a fully grown rabbit femur the uptake of plutonium is very uniform. Fig. 10.55 is a histogram obtained from the endosteal surface from which the marrow had been removed for the second intramuscularly injected animal, and the counts lie close to a Poisson distribution. When a similar count is made on the femur in which the marrow was left in situ, the histogram in Fig. 10.56 was obtained, which is not a Poisson distribution. This suggests that the non-uniformity is due to the variable uptake in the osteogenic cells and the occasional osteoclast on the surface. These measurments were limited to four animals at a single time interval, 8 days, after injection. Observations of extruded marrows at later times after injection suggest that a significant proportion of the surface plutonium is in the endosteal cells, and JAMES' measurements on rats are consistent with an increase in this proportion with time after injection. His measurements are on trabecular bone, but trabecular bone is histologically similar to the shaft of the femur as regards the bone surface.

C. Marrow

1. Intravenous Injections

The distribution pattern for plutonium in bone marrow has been discussed in Sec. IV.B. The dosimetry is complicated by the fact that marrow is a mixture of cells and fat spaces. The plutonium is not found in the fat spaces. In the cellular regions, because of the very inhomogeneous distribution of the plutonium, it is difficult to estimate the total amount of plutonium present or to calculate a meaningful dose rate. In general the dose rates in marrow are much less than those at bone surfaces. When the plutonium distribution is fairly uniform with scattered single tracks, that is, a diffuse distribution such as is shown in Fig. 10.44 then one can count tracks over an area large compared with a fat space and use equation (4) to calculate the average dose rate to a marrow cell. When there are aggregates of plutonium the dose rate to neighbouring cells is comparatively large and cells 30 μm or more away are unaffected. If the aggregates are treated as a point source, the dose rate falls off with the square of the distance from the source. The cell containing the aggregate may be killed by the radiation and the plutonium will be engulfed by other cells, until they also die. One would expect therefore a more rapid turnover of cells, particularly macrophages in the marrow due to the presence of aggregates, and the consequent translocation of the plutonium tends to average out the non-uniformity in dose rate to the cells. The dose rate calculated on the assumption that there is a uniform volume distribution will represent the average dose rate to a cell over a period of time.

In young animals because of growth and remodelling, and especially because the marrow is very cellular, there is plutonium all through the marrow. The resorption at the end of the metaphyseal trabeculae due to growth gives rise to a band of aggregates right across the femur which remains in the marrow for a long time (Fig. 10.40). Typical dose rates in marrow for six week old rabbits given a single intravenous injection of 1.25 μCi/kg of $Pu(NO_3)_4$ are shown in Table 10.41. The first column is obtained by counting all the tracks in a given area. The second column is obtained by excluding tracks from aggregates which produce stars of three or more tracks. The dose rates, even for a neighbouring cell at 10 μm from a star centre are a factor of 10 less than those at the bone surface. At 7 months after injection, the size of the aggregates (given in the last column) is less, but nearly all the plutonium is now in aggregates of about 2×10^6 atoms, whereas initially there was a mixture of large aggregates and diffuse plutonium.

Table 10.41. Dose rates in rads/day/μCi/kg of injected dose, and intensity of deposits producing stars in marrow of femurs from 6 week old rabbits given a single intravenous injection of 1.25 μCi/kg of $Pu(NO_3)_4$

Time after injection	Dose rates			No. of stars in 10^6 cu μm	No. of Pu atoms in stars
	Total marrow, stars+diffuse	Diffuse only	10 μm from star centre		
1 day	4	~1.0	1.2	54	9×10^6
8 days	3	~1.0	1.2	40	8×10^6
7 weeks	1.1	~0.2	1.2	13	8×10^6
7 months	0.13	~0	0.32	35	2×10^6

There are no detailed measurements for fully grown rabbits given intravenous injections. In grown animals the marrow has a larger proportion of fat spaces and the amount of plutonium in the centre of the shaft of the long bones is much less than it is within 1000 μm of a bone surface. Near a surface the plutonium remains for a long time after injection and the pattern is similar to that in young animals except that as there is no growth large groups of aggregates stretching right across the section are not seen.

2. Intramuscular Injections

After intramuscular injections the pattern of distribution of plutonium is different. Large aggregates appear in significant numbers 112 days after injection. Initially there is a diffuse distribution and the total amount of plutonium in the marrow builds up to a maximum by 16 weeks after injection. The variation of the dose rate with time is similar to that on the bone surface. This is illustrated

Table 10.42. Diffuse dose rates in marrow in rads/day/μCi/kg of injected dose following a single intramuscular injection of $Pu(NO_3)_4$ to 7 month old rabbits

Time after injection (days)	First series		Second series	
	% of dose at injection site	dose rate (rads/day)	% of dose at injection site	dose rate (rads/day)
1			96	0.14–0.06
8	103.7	0.08–0.04		
28	105.2	0.12–0.03		
35			75	0.32–0.18
49	59.4	0.12–0.03		
56			52.4	0.48–0.14
63[a]	—	0.51–0.25		
112			47	0.70–0.35
112	61.9	1.0–0.53		
280	35.3	1.04–0.68		
288			29.3	0.20–0.10
365			34.8	0.12–0.10
365	67	0.46–0.12		
561			50.9	0.12–0.05

[a] Killed with lung infection.

in Table 10.42 which shows dose rates for intramuscular injections of 7 month old rabbits. After intramuscular injections the uptake in the marrow is dependent on the percentage of the dose retained at the site of injection. This tended to be higher for one series of rabbits than the other. The dose rates in Tables 10.41 and 10.42 refer to the region within 1000 μm of the endosteal surface. In the centre of the shaft of the femur the dose rate will be a factor of 10 or so lower.

The measurements in Tables 10.36, 10.38, 10.39 and 10.40 are for rabbits. Twente and Jee (1961) and Rosenthal and Lindenbaum (1967) have observed similar pattern of distribution in rats, mice and dogs, but the quantitative measurements now available are even more scanty than for bone surfaces. On the present evidence it seems probable that the importance of the uptake of plutonium in the marrow lies in the contribution which it makes to the dose to the endosteal cells at the bone surface, rather than in the dose to the cells in the marrow itself, in spite of the fact that a far greater number of cells are involved in the second case.

D. Conclusion

The relation between the dose rates in the tables and the frequency with which a given cell is hit by an alpha particle is discussed in Sec. VI.C.7. If the relation between hit-frequency and cell damage for a given cell were known, one could attempt to forecast incidence of malignant cells from dose rates, and total dose. In the absence of this information the direct experimental evidence obtained by injecting animals with relatively large doses of plutonium and observing tumour incidence must be used. This work is discussed in the next section. It can be said here however that the experimental evidence of the dose received by the osteogenic cells themselves from ^{239}Pu goes far to explain the high toxicity of this alpha emitter, compared for instance with ^{226}Ra, as demonstrated by Dougherty and Mays using death from tumour and total skeletal burden as indicators (Figs. 10.59 and 10.64).

On the other hand it is of considerable interest to note that with the dose rates calculated to have been received by the osteogenic cells in the young rabbits injected with ^{239}Pu described by Bleaney and Vaughan growth and remodelling appears to have been achieved. In ordinary histological preparations of both young and older bone no gross abnormality of the osteogenic or marrow cells was detected. Clearly further detailed work requires to be done on the effect of measured radiation dose from alpha particles on osteogenic cells.

VIII. Effects of ^{239}Pu Deposition in the Skeleton

Janet Vaughan

The most significant biological effects of plutonium as a bone-seeker is its ability to induce carcinogenic transformation in the cells at risk. It may also induce severe aplasia of marrow (Boyd, 1950; Suter and Boyd, 1950; Metcalf et al., 1950; Boyd and Fink, 1950; Clarke, 1962; Rosenthal et al., 1970) and dysplasia of bone (Metcalf et al., 1950; Jee, Stover et al., 1962; Taylor et al., 1962; Clarke, 1962; Jee et al., 1962). Such crippling degenerative changes have not been seen when the dose is less than the carcinogenic dose (International Commission on Radiological Protection, 1968).

A. Carcinogenesis

It is generally agreed that the cell at carcinogenic risk from radiation is the cell capable of proliferative activity (International Commission on Radiological Protection, 1968). In the skeleton such cells are found in 1) the osteogenic tissue; 2) the marrow which includes haemopoietic stem cells, blood vessels and connective tissue elements and 3) epithelial cells closely applied to bone (LOUTIT and VAUGHAN, 1971). In view of the fact that plutonium is actually deposited in or on the surface of the osteogenic tissue and in the marrow as well as in the mineral/matrix it is not surprising that all the available evidence suggests that, with the possible exception of ^{228}Th, it is the most carcinogenic of the bone-seeking radionuclides.

At this point something must be said about the character and distribution of the osteogenic cells and their possible distances from the mineral/matrix surface of bone. It is important in the first instance to differentiate between very young and mature bone. In the past, owing to the fact that it is technically easy to cut sections of relatively soft young bone and difficult to prepare good histological preparations of mature bone, too much emphasis has been put on the character of young osteogenic tissue. In the young skeleton where bone is being laid down in some parts and active remodelling is occurring in other parts the surfaces are largely covered with active osteoblasts and active osteoclasts, i.e. fully differentiated cells which do not divide. External to these cells are precursor cells, preosteoblasts and preosteoclasts, cells which are capable of proliferation (OWEN, 1970; BINGHAM et al., 1969). This means that in young bone the cell at carcinogenic risk is at least one cell removed from the mineral/matrix of bone where many radionuclides are deposited. In older bone where neither resorption nor apposition is active few osteoblasts and osteoclasts are found on the bone surface which is covered with a layer of flat spindle shaped cells of mesenchymal origin (VAUGHAN, 1970). These cells may overlap one another. They are probably capable of proliferation and differentiation given the appropriate stimulus, as for instance parathyroid hormone (BINGHAM et al., 1969). In the past they have often been called resting osteoblasts (PRITCHARD, 1956).

Using histological techniques (SISSONS and his colleagues (SISSONS et al., 1967) have measured the resting and active bone surface in the ilium and head of the femur and JOWSEY (JOWSEY et al., 1965) has done the same for the vertebrae using microradiographic methods. These workers agree that in the adult probably about 80–90 percent of bone surfaces are quiescent. The histological picture of such quiescent surfaces in adult human bone is shown in Figs. 10.57 and 10.58. Owing to the difference between soft tissues and hard mineralized bone matrix the osteogenic layer often separates from the endosteal bone surface remaining attached to the marrow. This occurs when the marrow is artificially extruded from the diaphysis of the long bones. SISSONS (1970) has made measurements of the cells and their distance from the surface and reaches the following conclusion "The thickness of the investing layer of cells in the various cases studied is not greater than 4.0–6.8 μm. Allowing an additional 25% for shrinkage during histological processing maximum values of 5.0–8.5 μ are obtained with mean values of 2.5–4.4 μm". VAUGHAN (1970b) found figures of the same order for sections of human vertebrae from accident cases loaned by DUNNILL (DUNNILL et al., 1967). She pointed out that in the case of both osteoblasts and osteoclasts which are differentiated cells the proliferating precursor lies rather further from the surface than the majority of cells and therefore gave a somewhat wider range but accepted a figure of 1–5 μm as representing the distance of the majority of cells at risk.

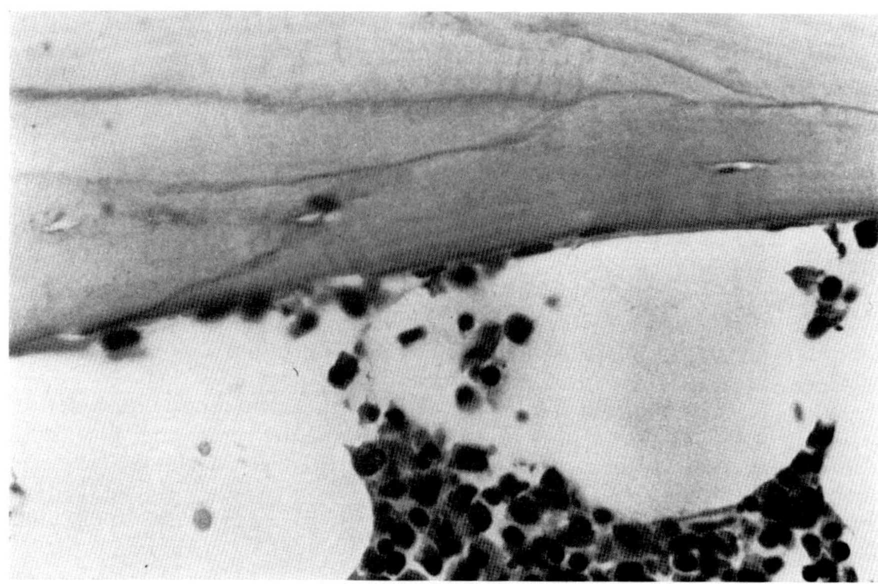

Fig. 10.57. 4 μ paraffin section of iliac crest. A thin layer of cells is present on the bone surface separating the bone from fatty and haematopoietic bone marrow. (SISSONS, 1970, by courtesy of author)

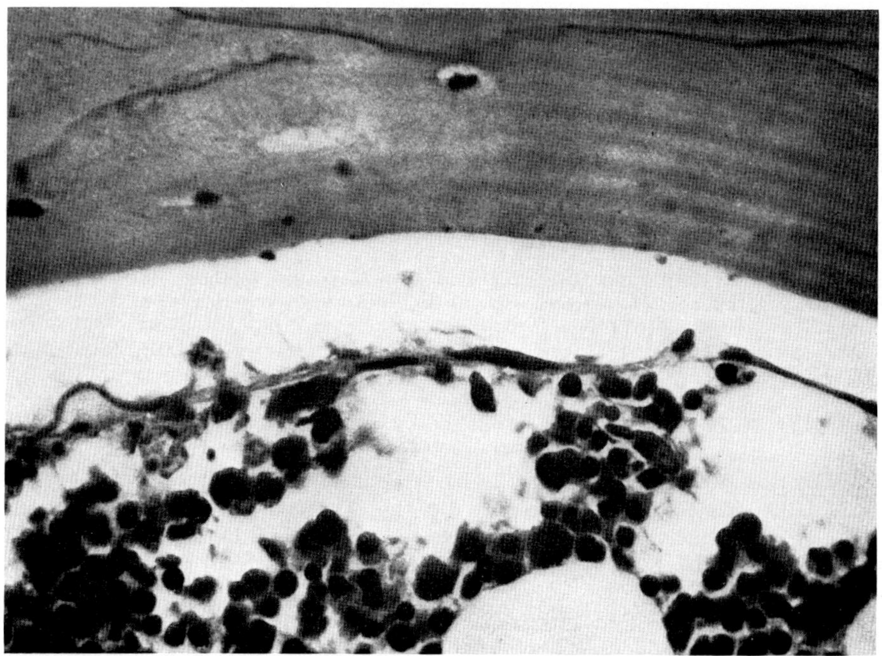

Fig. 10.58. 4 μ paraffin section of iliac crest. Note separation of bone marrow, and investing layer of flattened cells, from the bone surface. (SISSONS, 1970, by courtesy of author)

It is probable therefore that the vast majority of the osteogenic cells at risk in adult bone lie within 10 μm of the mineral surface (International Commission on Radiological Protection, 1968). This does not mean that other cells capable of osteogenic development are not present elsewhere in the marrow (Friedenstein, 1968; Owen, 1970).

1. Osteogenic Sarcoma

The term osteogenic sarcoma is here used to cover those malignant tumours which are generally agreed to arise from osteogenic tissue namely osteosarcoma, and chondrosarcoma. The present evidence suggests that the fibrosarcoma may arise from both osteogenic tissue and from marrow (Barnes and Khruschov, 1968; Barnes et al., 1971). The difficult giant cell tumour is not included (Sissons, 1966). Its origin is still controversial.

Recent experimental observations all confirm the early reports that with the exception of ^{228}Th, another surface seeker, ^{239}Pu is the most dangerous of the bone seeking radionuclides, with an extremely high incidence of osteogenic sarcoma at a relatively low body burden (Brues et al., 1947; Lisco et al., 1947; Finkel, 1953, 1956, 1959; Langham, 1959; Finkel and Biskis, 1962, 1968; Moskalev et al., 1969; Dougherty and Mays, 1969).

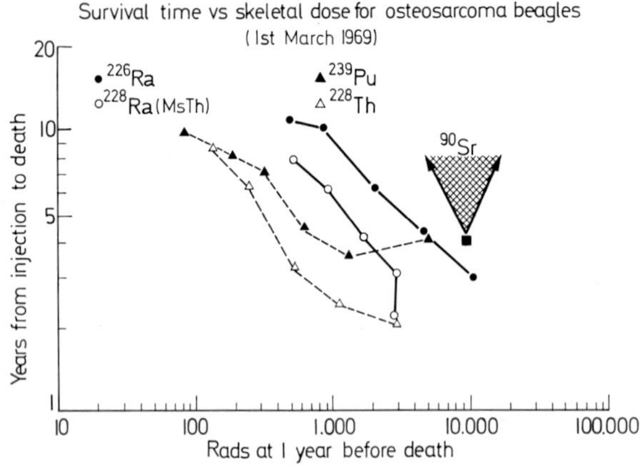

Fig. 10.59. Survival time for beagles developing osteosarcoma (March, 1969). Each point is the average for the osteosarcoma dogs at each injection level. In general the lower the dose, the longer the survival time. (Dougherty and Mays, 1969, by courtesy of authors and publishers)

In Fig. 10.59 are seen results from the Utah experiment in which young adult beagle dogs were injected with ^{228}Th, ^{239}Pu, ^{228}Ra and ^{226}Ra in citrate solution. Skeletal dose was calculated on the basis of skeletal burdens. Douoherty and Mays (1969) state that osteosarcoma was the chief cause of death. It is at once apparent that survival time for the same average skeletal dose is greatly reduced for both "surface seekers", ^{239}Pu and ^{228}Th, compared with the "volume seekers", ^{226}Ra and ^{90}Sr.

In 1947 when the first comparative result of the carcinogenic effect of some different bone-seeking radionuclides in rats, rabbits and mice was published (Brues et al., 1947; Lisco et al., 1947), the great capacity of plutonium to cause osteosarcoma particularly in spongy bone became apparent. These authors, in

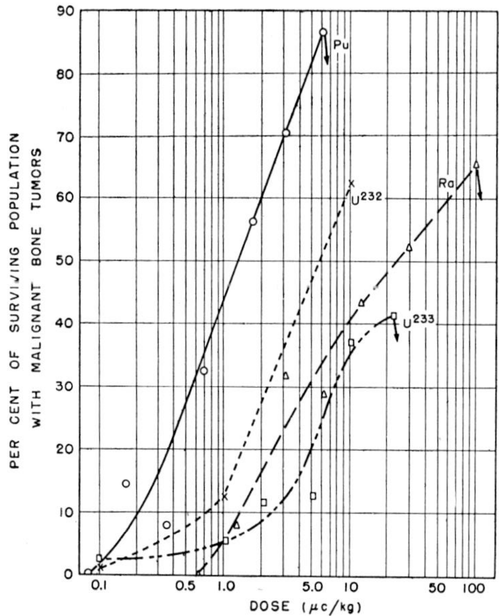

Fig. 10.60. Probability of developing a malignant bone tumour after injection of various isotope doses in female CFl mice if surviving the minimum latent period. (FINKEL, 1953, by courtesy of author and publishers)

another paper of the same date (BRUES et al., 1947), also commented on the fact that though much plutonium was deposited in the liver, liver tumours did not appear in these short lived rodents. [Later experiments however have indicated that liver tumours do indeed occur in beagles given a single injection of ^{239}Pu and may present a serious hazard. MAYS and his colleagues (MAYS et al., 1970) have speculated that the risk from liver tumours exceeds that from bone cancer if translocating ^{239}Pu redeposits about 50 percent in the liver and 50 percent in bone]. In 1953 further comparative studies were published by FINKEL (FINKEL, 1953) using CFl female mice[7]. ^{239}Pu nitrate, ^{210}Po chloride, ^{233}U nitrate and ^{226}Ra chloride were injected in isotonic salt solution at pH 2–4 into the tail vein. Again, plutonium was the most efficient producer of osteogenic sarcoma (see Fig. 10.60). In 1959, using ^{90}Sr and ^{45}Ca in addition to the four radionuclides mentioned, she commented on the fact that ^{239}Pu was the most potent skeletal carcinogen. In Fig. 10.61 is shown the percentage decrease in average survival time compared to controls as a function of injected dose (FINKEL, 1959).

a) Skeletal Site

As had originally been noted by LISCO and his colleagues (1947) FINKEL and BISKIS reported in 1962 that the majority of plutonium tumours in mice occurred in the spine. The skeletal location of these tumours are indicated in Table 10.43.

7 It must be noted that the CFl mice largely used in this and subsequent experiments by FINKEL et al. are known to have a skeletal dysplasia which may influence the carcinogenic response (LISCO, ROSENTHAL and VAUGHAN, 1971, personal communication). FINKEL herself notes that this strain has a higher incidence of spontaneous osteosarcoma than CBA mice (FINKEL and BISKIS, 1968), and that it also has a high spontaneous incidence of fibromas, osteomas and other bone lesions (FINKEL et al., 1969).

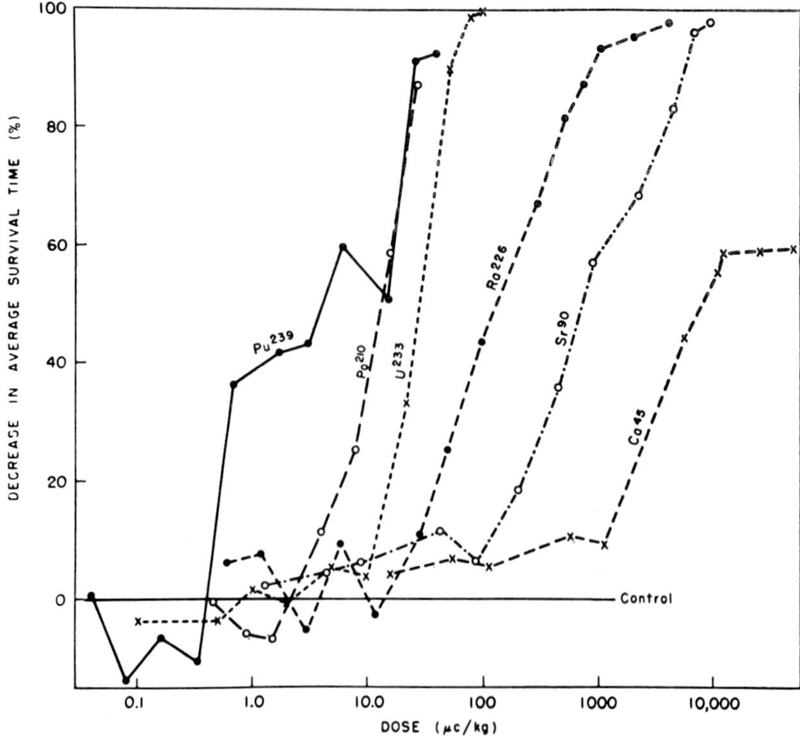

Fig. 10.61. Percentage decrease in average survival time of injected CFl mice compared to the controls as a function of injected dose. (FINKEL, 1959, by courtesy of author and publishers)

Table 10.43. Location of osteogenic sarcomas following injection of ^{239}Pu citrate to CF1 mice. (By courtesy of authors and publishers FINKEL and BISKIS, 1962)

Bone	No. tumours	Bone	No. tumours
Cervical vertebra	3	Humerus	4
Thoracic vertebra	29	Ulna	3
Lumbar vertebra	39	Metatarsal	1
Sacral vertebra	19	Ilium	6
Caudal vertebra	3	Scapula	1
Femur	11	Rib	5
Tibia	4	Head	6

Since she describes the mice used in this experiment as unhealthy the significance of the incidence of tumours in different tissues is difficult to assess. However, JEE and his colleagues published data on the Utah beagles in 1962 again indicating that at all injected dose levels except the very highest the majority of tumours occurred in spongy bone particularly that of the spine as shown in Fig. 10.62.

The probable reason for this not being true of the highest injected dose level in the Utah series is that this dose resulted in cell death and fibrosis of the osteogenic tissue rather than in carcinogenic change. It is known from dose measure-

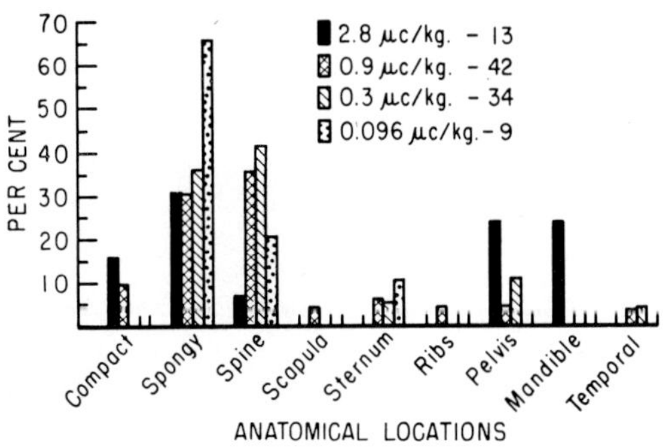

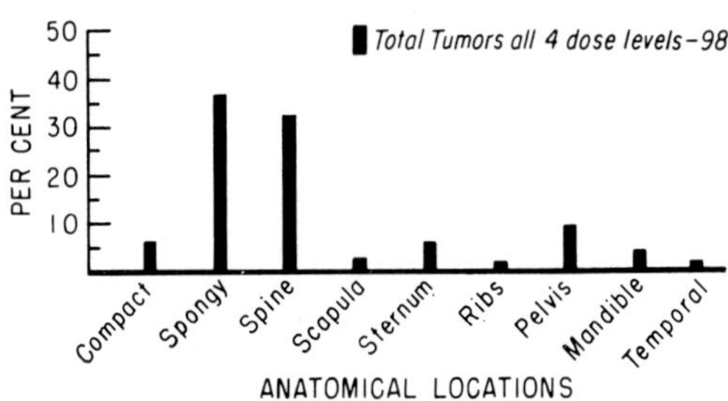

Fig. 10.62. Anatomic distribution of osteogenic sarcomas following intravenous injection of ^{239}Pu citrate in the dog. (Jee et al., 1962, by courtesy of authors and publishers)

ments at different sites in the long bones of rabbits given ^{90}Sr that in the case of an extremely high radiation dose the bone was dead but that malignant change occurred in adjacent sites where the radiation dose was lower (MacPherson et al., 1962). In Fig. 10.63 is shown a detailed autoradiograph of a trabeculus from a vertebrae of a 2.7 μCi/kg animal sacrificed 500 days post injection with a deposit of activity on the old endosteal surface buried by a layer of dense connective tissue (Arnold and Jee, 1962). Formation of such tissue probably follows the death of the osteogenic cells.

The only report that has been found of a higher incidence of plutonium induced tumours in long bones than in the spine comes from Bucktoyarova and Lemberg (1959). These workers used rats. The reason for the majority of tumours occurring in spongy bone as described by most observers is usually accepted to be the heavy concentration of plutonium on endosteal surfaces both in the endosteal cells themselves and in the bone matrix interface between cells and mineral tissue (Sec. IV.A). Jee and his colleagues (Jee et al., 1962) state that "although we observed a few missing osteocytes it is believed that this is not a significant cell type involved in tumour production". In man it is calculated that the endosteal surface in a vertebral body of approximately 40 cm^3 is of the order of 1000 cm^2 and the periosteal surface is 80 cm^2 (Dunnill et al., 1967; Sissons et al., 1967), so that the number of cells at risk on endosteal surfaces is far greater than on periosteal surfaces.

The importance of this endosteal deposition of plutonium in carcinogenesis is further evidenced by the experimental results of Taylor and his colleagues

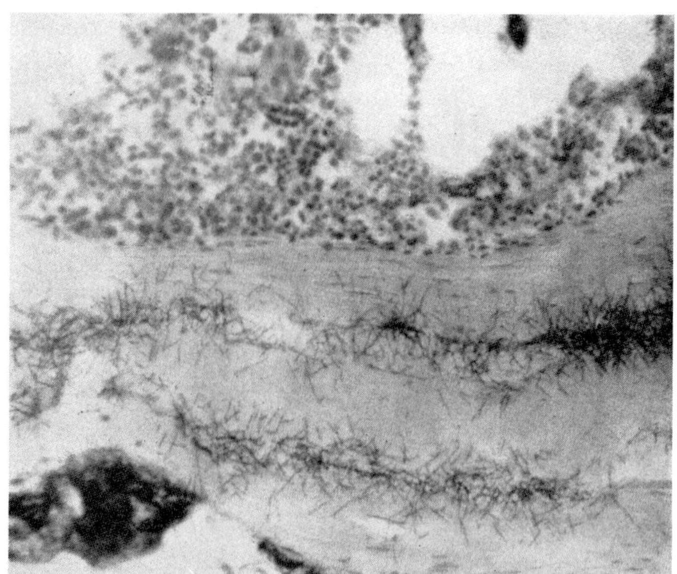

Fig. 10.63. Detailed autoradiograph of trabeculae from vertebra of beagle dog receiving 2.7 μCi/kg ^{239}Pu citrate, sacrificed 500 days post injection showing deposit of activity at old endosteal surface and now covered with dense connective tissue. (ARNOLD and JEE, 1962, by courtesy of authors and publishers)

and of ROSENTHAL and LINDENBAUM (1969) (TAYLOR et al., 1961; TAYLOR, 1962; BENSTED et al., 1965). TAYLOR has investigated the comparative metabolism and carcinogenicity of ^{239}Pu and ^{241}Am injected into rats. As already described (Sec. II.D) both ^{241}Am and tetravalent ^{239}Pu in 0.02 M HNO_3 or 0.4 per cent sodium citrate were deposited extensively in the skeleton in approximately equal quantities initially but the removal of ^{241}Am from the plasma was more rapid than that of ^{239}Pu. This difference was reflected in the rate of uptake of these nuclides into the proximal, distal and shaft regions of the femur. The half time of retention of these two nuclides was similar, of the order of 1500 days. Total skeletal retention was about 40 percent of the injected dose of ^{241}Am and 60 percent of the injected ^{239}Pu. Of the rats which survived for a year or more about eighty per cent developed bone tumours after the plutonium injection, whereas only twenty percent of the rats given americium developed tumours. The majority of the plutonium tumours were in the epiphysis. This difference in the carcinogenicity of the two radionuclides can be attributed to the fact that ^{239}Pu is particularly concentrated on endosteal surfaces while this is not true of ^{241}Am (Sec. IV). The difference in surface concentration, irrespective of the presence of ^{239}Pu in the osteogenic cells themselves, ensures that the sensitive proliferating cells will receive a higher radiation dose from ^{239}Pu than from ^{241}Am. Fractionation of a carcinogenic dose of ^{239}Pu does not appear to increase the number of osteogenic tumours as it does in the case of ^{32}P (BENSTED et al., 1961). However, BENSTED and his colleagues (BENSTED et al., 1965) conclude that further experimentation is necessary before drawing final conclusions about the effect of fractionation since the ^{239}Pu was given in a dose near the maximum level of tumour production.

When discussing both the metabolism of ^{239}Pu and its pattern of skeletal distribution the influence of the physico-chemical character of the plutonium solution was stressed (Sec. II and IV and Chap 9). The recent studies of ROSENTHAL and LINDENBAUM (1969) show that more monomeric than polymeric plutonium is deposited on endosteal surfaces and that monomeric plutonium is more carcinogenic than polymeric plutonium partly for the same reason and partly because it is more evenly distributed. Consequently, one might expect a higher probability of malignant change rather than cell death with this even distribution than with the large local doses produced by the aggregates of the polymeric solution. This will be likely to result in cell death. Sixty three percent of the mice given polymeric plutonium died with osteogenic sarcoma while 82 per cent of those given monomeric plutonium developed osteogenic sarcoma.

b) Microscopic Anatomy

JEE and his colleagues (JEE, STOVER et al., 1962) have described the microscopic anatomy of the osteogenic sarcoma occurring in the Utah beagles. The tumours varied over a wide range of sizes from 9×10 cm to microscopic lesions of 400 μm. Their histological appearance varied considerably in detail from lesion to lesion and from area to area within the same lesion with some showing relatively little and others considerable osteogenesis. Some showed largely osteoblastic, others chondroblastic or fibroblastic proliferation. Many tumours showed all the proliferative characteristics within one lesion. These authors emphasise that the tumours all seem to arise from osteogenic tissue on bone surfaces, and confirm what others have suggested, namely that gross tissue damage is not a necessary prelude to bone tumour production. MOSKALEV and his colleagues (MOSKALEV et al., 1969) consider that on the whole ^{239}Pu causes more osteolytic than osteoplastic tumours in rats though they report that BUCHTOYAROVA and LEMBERG (1969) found the reverse. They suggest that the difference may be due to the fact that they themselves used a citrate preparation while BUCHTOVAROVA used a nitrate. They also record chondrosarcoma and fibrosarcoma. This description of the plutonium osteogenic tumours agrees with the descriptions given for tumours arising in the osteogenic tissue from other radionuclides (NILSSON, 1962, 1970).

c) Relation of Bone Radiation Dose to Osteogenic Sarcoma

The question of radiation dose measurements to the sensitive cells in osteogenic tissue is discussed in detail in Sec. VII. There are some indirect measurements of radiation dose based on measurement by microdensitometric techniques of autoradiographs from bones containing ^{239}Pu at dosage levels at which osteosarcoma have occurred (TWENTE and JEE, 1961). TWENTE and JEE (1961) consider that this microdensitometric technique shows close agreement with an alpha track counting method. The dogs, whose bones were used for the study, were given 2.85 μCi ^{239}Pu/kg 92 days prior to sacrifice. It was calculated that the average localized dose rate to bone tissue and marrow cells on the endosteal surface of the vertebra was 40–60 rads per day. This figure is higher than that calculated by BLEANEY (1969) for the daily dose rate to the endosteal surface of the vertebra of rabbits given 2.3 μCi ^{239}Pu/kg and sacrified 112 days after injection. The maximum dose was then 24 rads per day. It is however not known whether the rabbits developed osteosarcoma at this injection dose level.

At present the measure of radiation dose used by MAYS and his colleagues (MAYS et al., 1970) in order to assess the hazards of osteogenic sarcoma is one calculated on the basis of average skeletal burden of plutonium at the time of death. Such a measure as average skeletal dose, though it is of value in making

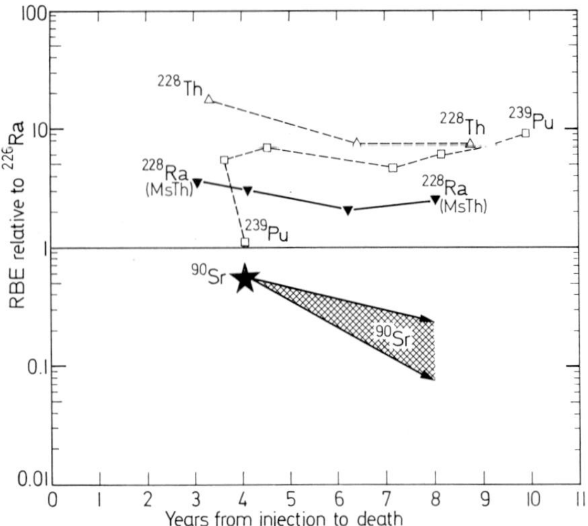

Fig. 10.64. R.B.E., for death with osteosarcomas in beagles (MARCH, 1969). The effectiveness of ^{239}Pu, ^{228}Th and ^{228}Ra (MsTh) is greater than that of ^{226}Ra whereas that of ^{90}Sr is less and decreases further at long survival times. (DOUGHERTY and MAYS, 1969, by courtesy of authors and publishers)

some assessment of the relative toxicity of different nuclides, does not give a measure of the radiation dose that is carcinogenic. In the case of plutonium the concentration of the nuclide is extremely uneven and possibly occurs in part actually in or on the cells at greatest risk while in the case of ^{226}Ra, for example, much of the radiation is distributed in the mineral/matrix far away from sensitive cells. DOUGHERTY and MAYS (1969) have shown that in dogs osteosarcoma have occurred 9.92 years after the injection of 0.0157 μCi/kg of ^{239}Pu citrate giving an average skeletal dose of only 78 rads one year before death. They suggest that 1 year is a reasonable period to allow for tumour growth from induction till death. It is perhaps of some interest that MOSKALEV and his colleagues (1969) who fed 0.5 μCi/kg of ^{239}Pu citrate daily to rats calculated that the average skeleton dose at death causing osteosarcoma was 57 rads. Both DOUGHERTY and MOSKALEV comment on the fact that as the dose is decreased so does the time for tumour production increase.

DOUGHERTY and MAYS (1969) have used their data on the survival time of osteosarcoma beagles given different bone seeking radionuclides as shown in Fig. 10.59 to calculate their R.B.E. i. e. relative biological effectiveness[8]. This is defined as the ratio of the dose from a *reference* radiation to that from a *tested* radiation when each produces an equal biological effect. The biological effect in that case was the average time to death with osteosarcoma, ^{226}Ra being chosen as the reference radiation because the permissible levels for bone-seekers are based on the observed toxicity of ^{226}Ra in humans. R.B.E. values, computed from doses scaled from Fig. 10.59 are shown in Fig. 10.64. At the longest time for which R.B.E. could be compared among all five radionuclides values relative to $^{226}\text{Ra} = 1$ were $^{239}\text{Pu} = 6$; $^{228}\text{Th} = 8$; $^{228}\text{Ra} = 2.5$ and $^{90}\text{Sr} = 0.09$–0.24. The authors further calculate predicted induction doses for man and compared them

8 Using the term in its broad sense i.e. not limiting the reference radiation to low LET — Editor.

with the calculated skeletal dose during a 50 year occupational exposure to a constant permissible burden. The latter was calculated on the basis of I.C.R.P. assumptions (1960), some of which may need revision, especially in the case of plutonium, since the skeleton probably contains less than 90 percent of the body burden of plutonium. The predicted radiation dose for induction of osteosarcoma was 200 rads and the dose from a permissible body burden during 50 years was 25 rads. It should however be kept in mind in assessing the significance of these calculations that they were based on mean skeletal dose rather than on dose to the sensitive osteogenic cells on endosteal surfaces. The most recent calculations of the distribution of ^{239}Pu between liver and bone following inhalation, ingestion or wounding, however, suggest that this distribution is more likely to be 50 percent to liver and 50 percent to skeleton (Mays et al., 1970 and Chap. 11) than the original I.C.R.P. assumption of 90 percent to skeleton. Mays has recently suggested that the RBE of ^{239}Pu relative to ^{226}Ra may be different in man, dog, and mouse depending on the degree of absorption of alpha particles emitted from bone structures of various thickness, the different rates of bone remodelling, the differences in dose for the same lifespan dose and other factors (Mays, 1971).

2. Cells at Carcinogenic Risk in Marrow

Bone marrow is an extremely complex tissue of mesenchymal origin giving rise to many different stem cells (Barnes and Loutit, 1967; Loutit, 1967; Friedenstein, 1968; Friedenstein et al., 1968; Owen, 1970) notably precursors of red cells, and cells of the granulocyte series, precursors of megakaryocytes, precursors of certain lymphocytes and possibly of fibroblasts (Barnes and Khruschov, 1968). It also contains blood vessels lined with endothelial cells, some of which at least are capable of proliferation and connective tissue cells. Malignant tumours of all these different cell types are recorded as occurring naturally in man, some rarely, sone more commonly (Jaffe, 1958; Willis, 1967). Some are known to occur in experimental animals with varying frequency following exposure to both external and internal radiation (Vaughan, 1973).

In the case of ^{90}Sr which does not concentrate in marrow but which decays to ^{90}Y, a long range beta emitter which may therefore irradiate marrow as well as osteogenic cells, malignant marrow dysplasias may be the predominant lesion under some circumstances (Vaughan, 1970b; Clarke et al., 1969; Goldman et al., 1969; Bustad et al., 1969) while tumours of mesenchymal origin, variously described as haemangioendothelioma, reticulum cell sarcomas, osteolytic sarcomas (Brues, 1949; Nilsson, 1962, 1970; Barnes et al., 1970) occur in a small number of animals as well as the more numerous classical bone forming osteosarcoma. Since ^{239}Pu deposits throughout the marrow various forms of malignancy arising in the marrow might be expected to occur.

a) Mesenchyme Tumours other than Leukaemia

A search through the published literature has given extremely few instances of tumours of mesenchymal origin other than osteogenic sarcoma similar to that found in the ^{90}Sr literature (Nilsson, 1962, 1970). Finkel and Biskis in 1962 listed two skeletal haemangioendotheliomas among 128 miscellaneous tumours in mice. Taylor and Bensted (1969) mention one fibrosarcoma of the jaw in a rat.

b) Leukaemia

The definition of the word leukaemia is difficult and means different things to different workers (Vaughan, 1970b). It is perhaps better to talk of a myeloproliferative disorder which may be defined as abnormal hyperplasia of marrow cells with subsequent metastases to other soft tissues particularly the spleen, liver and kidney. This is associated with severe anaemia of a leuco-erythroblastic type but not necessarily with a leucocytosis (VAUGHAN, 1970b). Such a condition might well be expected from the presence of ^{239}Pu in the marrow (VAUGHAN et al., 1967). Indeed it has been reported more frequently than with other alpha emitters such as radium and americium which are not found in marrow or concentrated particularly on endosteal surfaces.

In a small group of 26 rats which survived for 11 or more months following an intravenous injection of $^{239}Pu(NO_3)_4$ given in divided doses, BENSTED and his colleagues record 3 cases of myelogenous leukaemia at 46, 46, and 54 weeks after the last injection. There was only 1 case of leukaemia in 26 rats given a single dose of ^{241}Am. The strain used was not known to develop spontaneous leukaemia (BENSTED et al., 1965). Among the Utah beagles given ^{239}Pu to date, 52 dogs dying with osteosarcoma are reported and only one with leukaemia, at a dose of 0.0491 μCi/kg. One control dog also died with leukaemia. MOSKALEV and his colleagues however mention a rather higher relative incidence of tumours of the blood forming organs in rats compared with osteosarcoma following ^{239}Pu citrate (MOSKALEV et al., 1968). They emphasise that these neoplasms of the haemopoietic tissues are of great importance from the point of view of radiation hazards. They describe them as "leukoses of the reticulosis type; haemocytoblastoses with extensive leukaemic infiltrates in the bone marrow, liver, spleen, adrenals, kidneys, lymphatic nodes and lungs whereas pulmonary lymphosarcomas predominated in the controls".

It is difficult to interpret the significance of the reticuloendothelial tumours reported by FINKEL and BISKIS as occurring in mice following ^{239}Pu citrate injection since the CFl mice used are liable to such tumour development as they age without the intervention of radionuclides (LISCO et al., 1970, personal communication). In Table 10.44 is shown FINKEL's analysis of tumours of reticular tissue. It should be noted that 24.3 per cent of their control mice had some form

Table 10.44. Tumors of the reticular tissues following injection of ^{239}Pu citrate to CF1 mice. (By courtesy of authors and publishers FINKEL and BISKIS, 1962)

Amount injected	Percent with any reticular tumor	Tumor type					
		Lymphocytic	Granulocytic	Plasma cell	Reticulum cell sarcoma		Unclassified
					Type A	Type B	
(μc/kg)		(%)	(%)	(%)	(%)	(%)	(%)
40.6	0	0	0	0	0	0	0
26.5	0	0	0	0	0	0	0
15.6	3.8	3.8	0	0	0	0	0
6.3	4.5	4.5	0	0	0	0	0
3.1	4.4	4.4	0	0	0	0	0
1.7	8.5	5.1	1.7	0	1.7	0	0
0.69	12.4	6.8	1.4	1.4	1.4	0	1.4
0.34	23.4	6.7	5.6	8.9	1.1	1.1	0
0.16	31.7	12.5	4.5	12.5	1.1	0	1.1
0.08	28.6	8.6	1.9	13.3	1.0	3.8	0
0.04	21.7	7.5	4.2	7.5	2.5	0	0
Control	24.3	9.5	1.4	9.5	1.0	2.4	0.5
Total population	19.7	7.7	2.4	7.1	1.1	1.1	0.3

of reticular tumour while at only two dose levels was the incidence higher among the injected animals. Even then the differences were not striking. The high incidence in the control animals in FINKEL's series may have been due to the longer life span of the untreated animals.

3. Cells at Carcinogenic Risk in Epithelial Tissue

Long term follow up of men who have ingested ^{226}Ra has shown that a number of them may develop carcinomas of the epithelium overlying bone in the mastoid region and the sinuses of the skull (FINKEL et al., 1969; EVANS et al., 1969). This is attributed both to radiation from the underlying bone containing ^{226}Ra and to radon retained in the sinuses. No such tumours have been reported in patients given ^{224}Ra but they were injected only about 20 years ago. Since the latent period for sinus carcinoma is generally much longer than that for osteosarcoma this is not surprising. Further ^{224}Ra decays on the bone surface producing no radioactive noble gas (SPIERS and MAYS, 1970). ^{90}Sr has induced sinus carcinoma in rabbits (VAUGHAN and WILLIAMSON, 1969). ^{239}Pu might therefore be expected to have the same effect since much of the ^{239}Pu is retained on the mineral/matrix surface. It does not however produce radon and the range of the alpha particle is considerably shorter than that of ^{90}Sr–^{90}Y beta radiation. One sinus carcinoma is recorded in the Utah beagles given ^{239}Pu citrate. This occurred 2095 days after injection of 0.0996 μCi/kg. The skeletal dose was calculated to be 310 rads. In Fig. 10.65 is shown a photomicrograph of the wall of the frontal sinus of an adult beagle showing the proximity of the epithelial components of the mucosa to the underlying bone which brings the mucosa within a typical alpha range. The latent period in this instance was approximately equal to that

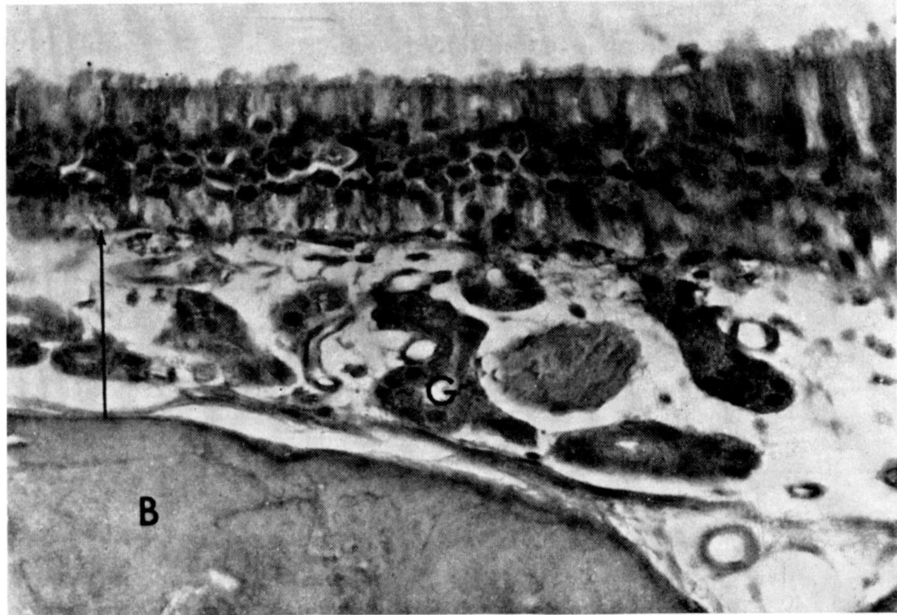

Fig. 10.65. Photomicrograph of the wall of the frontal sinus of an adult beagle showing the proximity of the epithelial components of the mucosa to the underlying bone (*B*), tubular glands (*G*). A typical α range. H and E. × 474. (TAYLOR et al., 1969, by courtesy of authors and publishers)

for the osteogenic sarcoma induction by comparable injected dose levels of ^{239}Pu but was significantly shorter than the period required for similar tumours to arise in the case of ^{226}Ra and ^{90}Sr (TAYLOR et al., 1969). There are too few such occurrences at present, however, to support this as a generalization. In mice FINKEL and BISKIS (1962) record one nasal epidermoid carcinoma.

B. Dysplasia

1. Bone

The effects of plutonium on the skeleton other than carcinogenesis have been studied by both radiological and histological techniques in many laboratories (HELLER, 1948; LANGHAM and CARTER, 1951; JEE, STOVER et al., 1962; CLARKE, 1962; TAYLOR et al., 1962; BUSTAD et al., 1962; FABRIKANT and SMITH, 1964).

All later workers have confirmed the early observations of HELLER (1948) who emphasised the preeminence of plutonium in interfering with bone growth and in causing death of osteocytes, though she concluded that bone reacted in a similar fashion to the all internally administered radionuclides. Differences observed depended principally on age, species, dose and mode of administration.

a) Radiological Changes

LANGHAM and CARTER (1951), who made an early comparative study of the effects of plutonium nitrate ($Pu(NO_3)_4$ and americium chloride ($AmCl_3$) injected intravenously into 35 days old rats, noted little difference in the skeletal changes brought about by the two radionuclides which they listed as: 1) generalized osteoporosis; 2) densities along epiphyseal growth lines; 3) bone necrosis; 4) marked thinning of the cortex of the long bones.

JEE and his colleagues (1962) found that relatively high doses of ^{239}Pu citrate given as a single intravenous injection to adult beagles induced a significant number of bone fractures. The incidence was highest in the ribs and the processes of the vertebrae and low in the limb bones. These fractures were unique in the minimal pain and inflammatory response associated with them and satisfactory healing occurred in many instances. Fracture distribution following a single intravenous injection are shown in Fig. 10.66. The radiograph of a rib cage showing multiple healed fractures is shown in Fig. 10.67. An intravenous dose in excess of 0.3 μCi/kg was required to induce a significant number of fractures. The highest incidence occurred between 1.0 and 3.0 μCi/kg doses.

In young miniature pigs early skeletal changes in males injected with 1.3 μCi $^{239}Pu(NO_3)_4$ were limited to a slight thickening of the cortical bone, outgrowths of cortical bone into the medullary canal, loss of the definite medullary canal and increased trabeculation in the spongiosa. The younger the animal the more severe the radiological lesions (BUSTAD et al., 1962). The effect on growth has been more carefully analysed by FABRIKANT and SMITH (1964). They made a comparative study of the effects on the skeleton of young growing rats given ^{239}Pu, ^{241}Am and ^{32}P intravenously. ^{239}Pu was in the form of $Pu(NO_3)_4$ at pH 5 and was given in either single doses of 2.95 μCi/kg or in fractionated doses to a total activity of 3.0 μCi. Plutonium was effective in reducing the rate of bone growth in a smaller dose than americium. One year following injection there was slight shortening of the humerus and tibia and up to 10 percent of the femur with little subsequent change. Failure to grow was not due to early fusion of the epiphyses since the plate remained intact, nor was it a transient lag in maturation since permanent shortening was present in the adult. Generalized osteosclerosis,

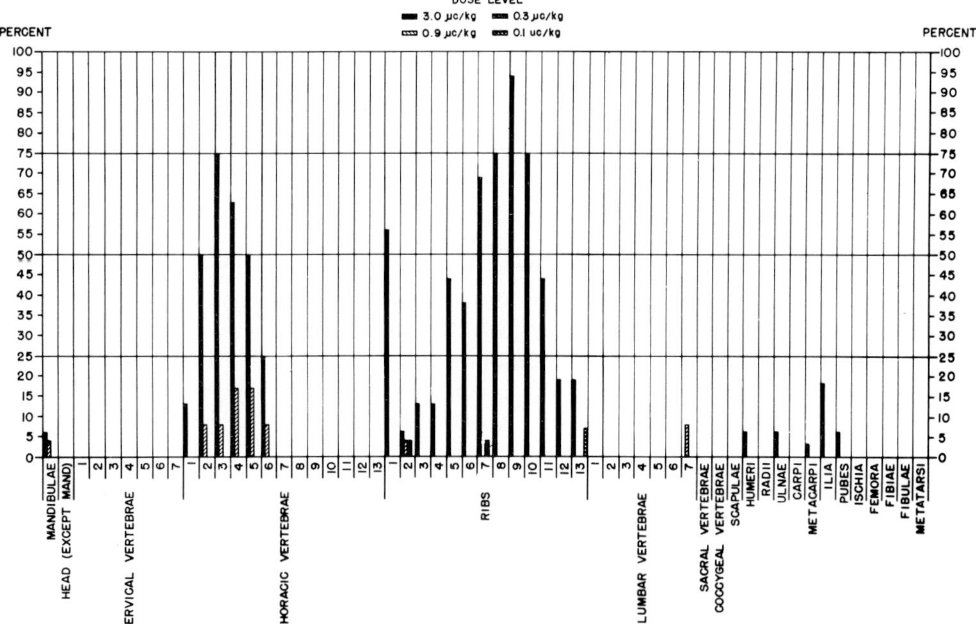

Fig. 10.66. Fracture distribution in beagles following a single intravenous injection of ^{239}Pu. (TAYLOR, et al., 1962, by courtesy of authors and publishers)

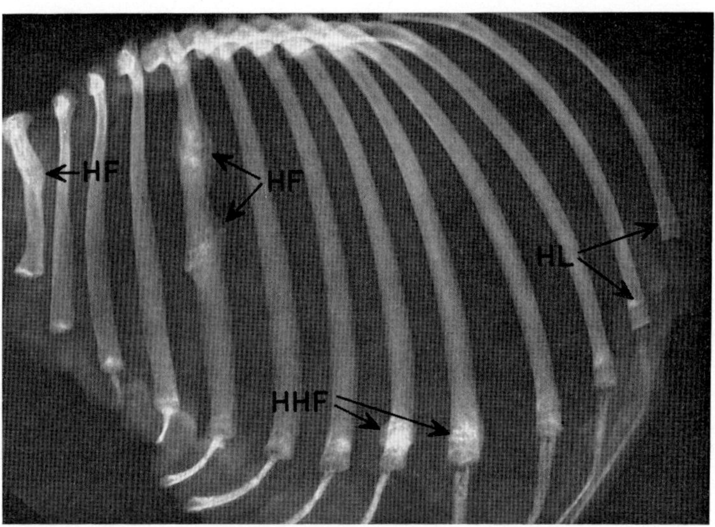

Fig. 10.67. Radiograph of rib cage of a dog taken 1192 days post injection, showing hot lines (*HL*), two healed "hot line" fractures (*HHF*) and three other healed fractures (*HF*). The dog received 2.69 μCi/kg ^{239}Pu via a single intravenous injection. (TAYLOR et al., 1962, by courtesy of authors and publishers)

frequently confined to the mid-shaft of the femur was characteristic of the animals receiving plutonium and was more pronounced with fractionation. Again, these changes were only seen in rats receiving a higher dose of americiun (5.8 μCi/kg).

Bilateral femoral narrowing due to cortical thinning, since the diameter of the medullary cavity was only minimally narrowed, often resulted in fracture. This was more common in the rats given americium and is compatible with the heavier deposition of americium around Haversian systems as seen in autoradiographs (Fig. 10.19). Tumours did not occur in these fracture sites.

b) *Histopathology*

The radiological changes in bone other than malignancies following exposure to plutonium are well explained by the detailed histology. When comparative studies have been made using ^{239}Pu, ^{226}Ra, ^{90}Sr, ^{32}P and ^{241}Am, it has been emphasised by all workers that irrespective of age at time of establishment of a body burden ^{239}Pu was the most damaging (CLARKE, 1962).

CLARKE (1962) studied the histopathology of ^{90}Sr, ^{226}Ra and ^{239}Pu in miniature pigs injected when 6 weeks, 6 months and 1 year old and killed 18 months later. With the exception of the pigs aged 6 months and 1 year given ^{90}Sr the average skeletal dose was calculated to be less than 5000 rads. All the animals injected at 6 weeks of age had thick, coarse and irregularly calcified hyaline trabeculae, disrupted columns of epiphyseal cartilage cells, large, coarse, poorly calcified primary trabeculae showing advanced necrosis and irregular disorganized lamellar Haversian systems. ^{90}Sr appeared to produce more epiphyseal damage while plutonium and radium caused more severe lamellar necrosis. Both plutonium and radium produced areas of complete Haversian system necrosis. CLARKE suggests that the more deleterious effect of plutonium could possibility be attributed to the extreme degree of fibrosis of the medullary blood vessel and resultant circulatory disruption. In animals six months of age at injection similarity of bone lesions between animals was restricted to increased calcification of the epiphyseal cartilage matrix, an increase in subperiosteal fibrous tissue and fibrous bone formation with disorganisation of lamellar bone.

With strontium bone necrosis was widespread; with radium necrotic changes were less severe and were associated with cellular degeneration while plutonium produced severe diffuse necrosis in secondary trabeculae along with extreme fibrosis of the blood vasculature. All the pigs injected at 1 year had cartilage remnants visible in their primary trabeculae, varying degrees of fibrosis and necrosis and widespread Haversian system necrosis. The plutonium animals showed numerous focal areas of osteon necrosis with an heavy reactive calcification of the Haversian canals. These findings in pigs are very similar to those recorded by JEE and his colleagues (JEE, STOVER et al., 1962) in adult beagles injected with ^{239}Pu citrate when the average accumulated dose to the skeleton varied from 326 ± 70 to 6470 ± 1590 rads. The damage was of course greater at the higher dose levels. Isolated empty lacunae in trabecular bone, together with canal plugs and loss of osteocytes in cortical bone were the only effects observed following a low dose of 0.096 μCi/kg. In Fig. 10.68 is shown a portion of cortical bone from a dog injected with 2.8 μCi/kg showing abnormal osteons, canal plugs and Haversian canals with patent vascular channels, while in Fig. 10.69 there is a large area of dead bone. All these bone changes are brought about by great changes in the vascular blood supply dependent on radiation damage from plutonium. The pattern of this vascular damage has been studied by JEE and his colleagues (JEE et al., 1969) and is illustrated in Fig. 10.70 which shows a drawing of spongy bone from a normal dog injected with Indian ink and a drawing of spongy bone from a beagle injected with 2.68 μCi of ^{239}Pu/kg and sacrificed after 777 days. The latter shows a marked reduction in vascularity, thickened trabec-

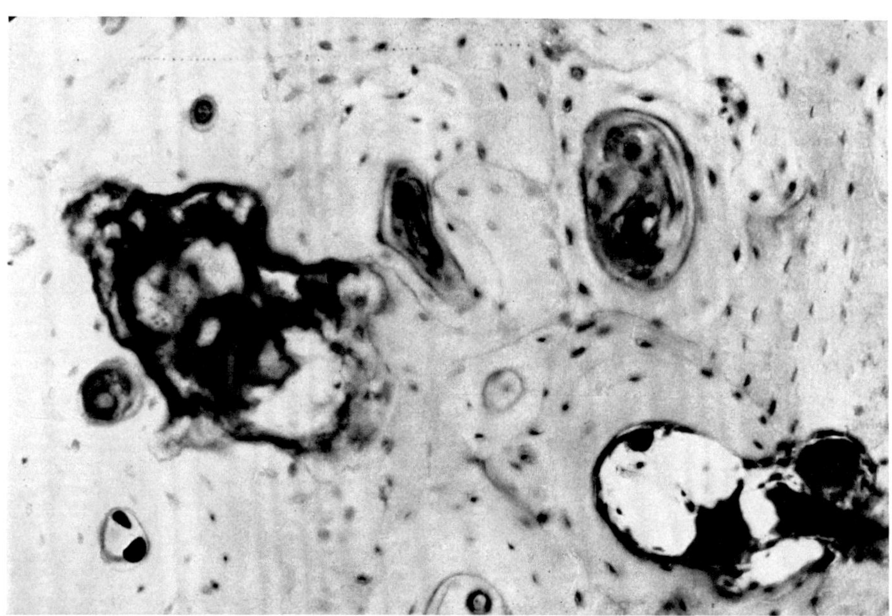

Fig. 10.68. A portion of the compacta from a dog injected with 2.8 μCi/kg ^{239}Pu citrate showing abnormal osteons, canal plugs and Haversian canals with patent vascular channels (Indian Ink). H and E. Colloidin section. (× 125). (Jee et al., 1962, by courtesy of authors and publishers)

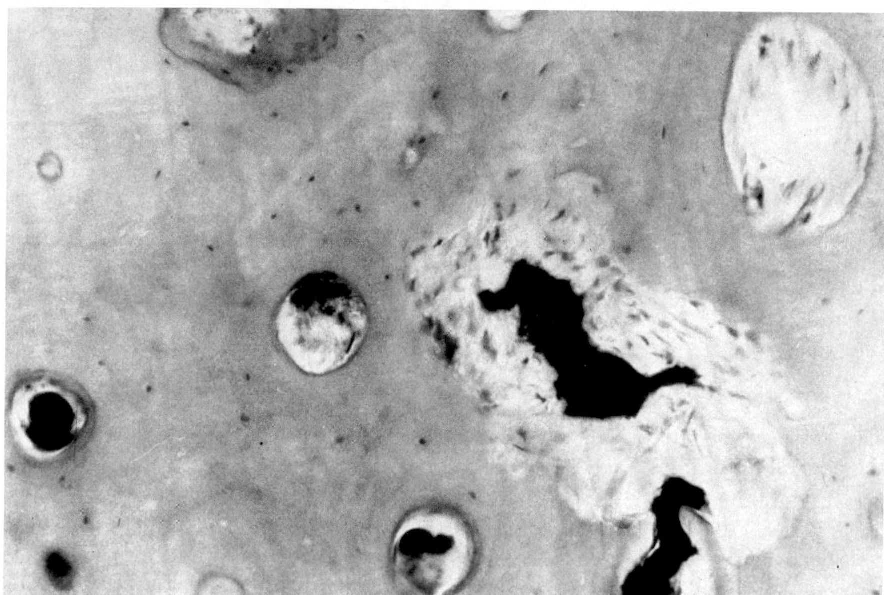

Fig. 10.69. Compacta of a dog injected with 2.8 μCi/kg ^{239}Pu showing a huge resorption cavity with fibrous connective tissue and pooled ink (blood vessels) viable and dead osteons with or without patent vascular channels (India ink in canals) and canal plugs. H and E celloidin section. (× 125). (Jee et al., 1962, by courtesy of authors and publishers)

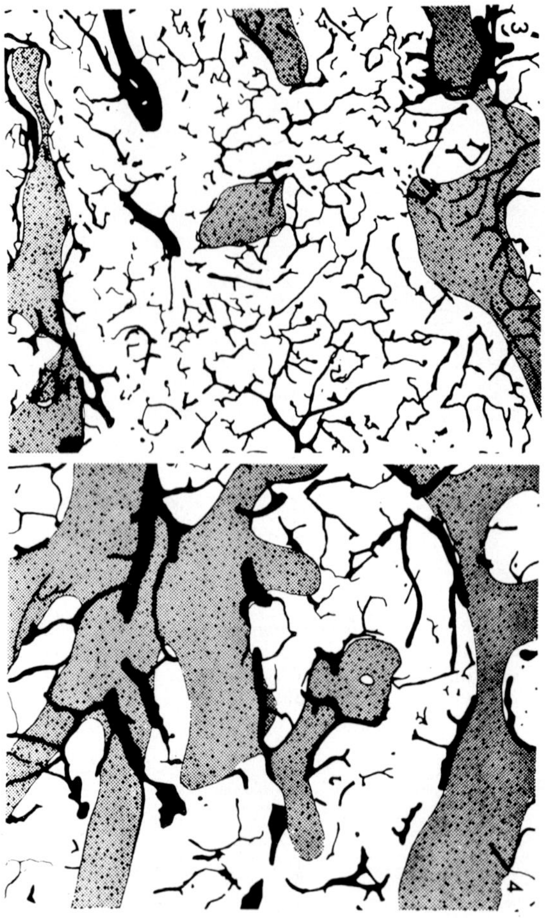

Fig. 10.70. Drawing of an area of spongy bone (metatarsal metaphysis) showing distribution of vessels: 3) a 3282 day old female beagle (central) showing the normal amount of vascularity. The bone is stippled. 4) Beagle injected with 2.68 μCi/kg of ^{239}Pu citrate and sacrificed after 777 days showing a marked reduction in vascularity, thickened trabeculae, higher density of large vessels and lower density of small vessels than control dogs. (JEE et al., 1969, by courtesy of authors and publishers)

ulae, higher density of large vessels and lower density of small vessels than in the control dog.

All radionuclides produce changes throughout both trabecular and cortical bone, which vary in degree depending on the radionuclide. JEE and his colleagues consider each has a characteristic pattern as seen in microradiographs, Fig. 10.71. High doses of plutonium are characterised by the heavy endosteal fibrosis due to extremely severe damage to endosteal cells, which are killed rather than becoming malignant. Except for ^{228}Th other radionuclides do not cause the same degree of endosteal injury since they are not particularly concentrated in endosteal cells.

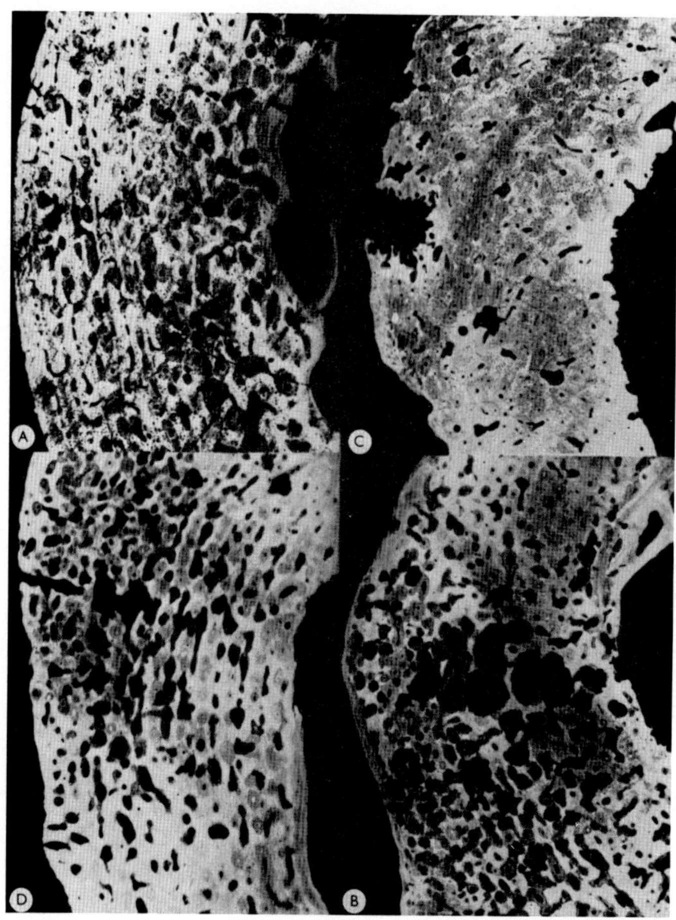

Fig. 10.71 a—d. Microradiograph of cross sections of the tibial diaphysis: a) compacta of control dog. Note that osteons are slightly less mineralized than interstitial lamellae. b) compacta of dog injected with 10 μCi/kg of ^{226}Ra. Note large resorption cavities, hypocalcified osteones and hypercalcified Haversian canal plugs (High calcification appears white on the photograph). c) compacta of dog injected intravenously with 0.9 μCi/kg of ^{228}Th. Note large area of periosteal resorption, rugged endosteal surface, few small cavities and bland mineralization. d) compacta of dog injected intravenously with 2.5 μCi/kg of plutonium. Note occasional large cavities, numerous hypocalcified osteones and a few hypercalcified plugs. (JEE and ARNOLD, 1960, by courtesy of authors and publishers)

2. Marrow

The earliest reports of the effects of ^{239}Pu indicated that given in large intravenous doses to rats it caused death from severe anaemia within a matter of weeks (BOYD, 1950; SUTER and BOYD, 1950; METCALF et al., 1950). Doses of the order of 300–3000 μg/kg were given. LANGHAM and CARTER (1951) noted that the haematological effects of plutonium nitrate were more severe than those of americium chloride. An injection of 0.063 μCi/g of plutonium induced a fatal anaemia while the rats given the same dose of americium survived with a low grade anaemia. With lower doses 0.032 μCi/g and 0.05 μCi/g both nuclides induced anaemia and persistent leucopaenia. The bone marrow was described as showing an increase in general cellularity. In 1954 DOUGHERTY reported that beagles given 0.27 μCi/kg of ^{239}Pu citrate showed a drop in white blood cell count within 30 days followed by a rapid return to a low normal count. On the other hand, a dose of 2.5 μCi/kg produced an acute fall in erythrocytes, leucocytes, heterophils and platelets with little tendency to return to normal. In 1960 DOUGHERTY and her colleagues (DOUGHERTY et al., 1960) noted that in the Utah experiment ^{239}Pu was the only radionuclide used that caused a mild anaemia. Later she compared the haematological effects of ^{226}Ra and ^{239}Pu (DOUGHERTY and ROSENBLATT, 1969) and stated that ^{239}Pu was approximately 4–8 times as effective as ^{226}Ra in depressing total white blood cells and segmented polymorphs in the peripheral blood but only 2–4 times as effective in the case of lymphocytes.

CLARKE (1962), who made a comparative histopathological study of the marrow in pigs given ^{226}Ra, ^{90}Sr and ^{239}Pu, found that in animals injected at one year marrow damage was not seen in the pigs receiving radium but was extensive with ^{90}Sr and ^{239}Pu in the primary trabecular zone. In general, plutonium appeared to have the most damaging effect in all age groups investigated. Marrow of the plutonium injected animals was less cellular and showed many bizarre vesiculated cells. Granulocytes and megakaryocytes were the most affected. Plutonium was given intravenously as citrate, 1.3 μCi/kg. It might therefore be expected to be largely monomeric and not be highly concentrated in the marrow. FINKEL and BISKIS (1962) report a fall in haemoglobin in mice given 15.6 μCi/kg ^{239}Pu citrate in acid solution; 6.1 μCi/kg had no effect. The effect on the leucocytes is not easy to interpret owing to the high normal incidence of reticular tumours (Sec. VIII. 2.b).

In a small number of rats given $^{239}Pu(NO_3)_4$ in a single injection, BENSTED et al. (1965) noted aplasia of the marrow in the epiphyses of the long bones but not in the metaphyseal and diaphyseal marrow. This abnormality ranged from diminished cellularity of the marrow to its replacement by loose and sometimes gelatinous connective tissue. In rats given a fractionated dose of ^{239}Pu the picture was approximately the same. In only one rat was peritrabecular fibrosis, such as that described by JEE, ARNOLD et al. (1962) noted. No such marrow changes following a comparable dose of americium were found (BENSTED et al., 1965).

It is clear from these rather scattered reports that plutonium has an effect on haemopoiesis in smaller doses than other bone-seeking radionuclides. However, as discussed (Sec. IV.B) the amount of plutonium in the marrow is largely dependent upon the form in which the plutonium reaches the blood stream i.e. whether it is largely monomeric or polymeric. In many of the recorded experiments ^{239}Pu citrate was used which is likely to be largely monomeric. Polymeric plutonium might be expected to produce much greater damage than the monomeric form which is more heavily concentrated on endosteal surfaces and indeed recent experiments reported by ROSENTHAL and her colleagues (LINDENBAUM et al., 1969; ROSENTHAL et al., 1970) confirm the great marrow toxicity of polymeric plutonium. They found that CFl/ANL mice died a protracted haemopoietic death following the injection of a polymeric plutonium of narrow range of particle size which deposited almost entirely in the reticuloendothelial organs. Six days following injection the liver, spleen and marrow contained 84 percent, 6 percent and ~2 percent of the injected amount respectively compared to only about 1 percent in calcified bone. After an injection of 7.8 μCi/kg all mice died of severe anaemia between 4 and 9 months. The mean terminal haemoglobin level was 1.9 g/100 ml compared to 17 g/100 ml in controls and the mean haematocrit was 5.4 percent compared to 46 percent in the controls. The marrow was generally aplastic and in some areas acellular though isolated areas of haemopoiesis were seen. In some later deaths myelopoiesis and erythropoiesis was seen in lymphatic and other tissues. Leukopaenia, splenic atrophy and hydropic centrolobular liver degeneration was present in all mice. With lower doses life span was increased and haemopoietic deaths were less common. Deaths from other causes including tumours of bone and liver were also seen at all dose levels.

In assessing the significance of these results it must be remembered that recent observations have indicated that mouse haemopoietic marrow is much more sensitive to gamma irradiation than rat haemopoietic marrow (Twentyman and Blackett, 1970). There is no reason to suppose that there may not be the same difference in sensitivity to alpha radiation. Further, the effect in larger animals and more particularly in man may again be different.

Until more precise information is forthcoming it is important to realize that osteogenic cells, all elements in the marrow capable of proliferation and epithelial tissues closely applied to bone have been shown to be at risk from plutonium deposited in the skeleton. Such deposition may result in both carcinogenesis and severe dysplasia. The possible presence of plutonium actually in or on the osteogenic cells at risk explains the extreme toxicity of this radionuclide and the high incidence of osteosarcoma at relatively low skeletal burdens.

IX. Removal of Internally Deposited Plutonium from the Skeleton

Janet Vaughan

In view of its toxicity it is not surprising that there has been extensive experimental work carried out in order to find agents able to prevent the absorption of plutonium into the body and to facilitate its removal, particularly from the skeleton.

A. Anion Exchange Resins

Gastrointestinal absorption is, as discussed, very slight but in workers exposed to continuous inhalation of very small amounts or a single large inhalation the amounts absorbed might become significant.

Russian workers (Belyaev, 1961) have employed anion exchange resins to prevent gastrointestinal absorption. Belyaev found that when anion exchange resins were administered to rats by stomach tube 5 to 10 minutes after stomach tube administration of ^{239}Pu, absorption decreased by as much as a factor of 10. This degree of effectiveness was only found with Pu(IV) citrate which is relatively well absorbed (Sec. II.C.2). Absorption of $Pu(NO_3)_4$ was only reduced by about a factor of 3.

Two rather different techniques using colloidal scavenging agents or chelating agents have been employed in an attempt to prevent skeletal deposition or remove plutonium from skeleton and soft tissues.

B. Colloidal Scavenging Agents

This technique depends, as postulated by Schubert and White (1950), on the fact that certain metals, such as zirconium, when at a physiological pH in the blood stream form small aggregates which are capable of adsorbing plutonium and related elements, and are small enough to be able to pass through the glomeruli

Table 10.45. Effect of zirconium and sodium citrate treatment on the urinary excretion of plutonium in rats[a]. (Taken from SCHUBERT, 1947). (Data expressed as average per cent of injected dose of plutonium excreted each day)

Days elapsed following Pu injection	Pu control	Treatment			
		Zirconium (100 mg on 2nd hr and 3rd day after Pu injection)	Zirconium (50 mg on 2nd and 24th hr after Pu injection)	Sodium citrate (4ml of 10% solution 2nd hr and 3rd day after Pu injection)	Sodium citrate (2 ml of 10% solution on 2nd and 24th hr after Pu injection)
0–1	0.75	8.2	5.1	3.2	3.1
2–3	0.027	0.032	0.060	0.037	0.041
4–5	0.045	0.022	0.060	0.016	0.023
7–10	0.021	0.064	0.083	0.020	0.017
12–14	0.016	0.064	0.053	0.026	0.022
Total Pu excreted in urine for 14-day period	1.1	9.6	6.5	3.7	4.0

[a] The rats (200-gram females) received 1.1 mg of Pu/kg.

into the urine, the final distribution and excretion pattern being that of zirconium. More recently it has been suggested that ion exchange or coprecipitation rather than adsorption on the colloid may be involved (ROSENTHAL and SCHUBERT, 1957). A similar mechanism probably explains the effectiveness of hexametaphosphate in reducing skeletal deposition by a factor of 3 which is discussed by SEMENOV and TREGUBENKO (1958). SCHUBERT first published experiments in 1947 that suggested that zirconium citrate was effective in removing plutonium particularly from skeleton and liver in both rats and dogs. Other metals, lanthanum and cerium were less effective. The results have an historic value and are shown in Table 10.45. It can be seen that only a small part of the heightened excretion is due to the citrate. Urinary excretion only was affected. The zirconium citrate was effective when given immediately after plutonium administration in reducing the plutonium content of both liver and skeleton by half but when given about 5 weeks after injection, though urinary excretion was increased roughly tenfold, the skeletal burden was unchanged. The liver content was, however, again reduced by half. The importance of prompt treatment with zirconium citrate was confirmed in the dog (JOFFE and TEMPLE, 1953).

In 1957 ROSENTHAL and SCHUBERT demonstrated again that even a small dose of zirconium, 50 mg/kg, given 30 minutes after intravenous ^{239}Pu citrate caused an extremely rapid and sustained fall in the level of ^{239}Pu in the blood. This was associated with an excretion of 25.1 percent of the injected plutonium in the urine compared with 1.4 percent in the controls in a 3 day period. The bulk of this excretion took place in less than 1 hour. This rapid removal of plutonium from the blood also accounts for the low skeletal content recorded in the earlier experiments. Further zirconium injections during one month had no effect on skeletal retention (HACKETT, 1953). The dose required to prevent bone deposition in rats (12 mg/kg) is much smaller than that required for maximum excretion (100 mg/kg or more) (SCHUBERT and WHITE, 1950). Doses which do not secure maximum excretion or use of zirconium citrate colloids of non optimal size may result in accumulation of plutonium in liver and soft tissued (FOREMAN et al., 1955; SCHUBERT, 1955; SCHUBERT and WHITE, 1950; HACKETT, 1953).

Limited studies in dogs suggest that in this larger animal zirconium citrate may be more effective in preventing both liver and skeletal deposition (THOMPSON, 1967) than it is in small animals.

Two human cases have been treated with zirconium at Hanford Laboratories in Richland, Washington. One received plutonium by inhalation and the other via a break in the skin. 600 mg of zirconium given in four repeated doses in the first case, and two doses in the second gave no therapeutic increase in excretion (THOMPSON, 1967). SANDERS (1961) also reports disappointing results in a woman with transdermal contamination with plutonyl nitrate in nitric acid who was treated 3 hours later with 1.2 g zirconium in citrate solution. Two cases treated with zirconium malate, 50 mg zirconium, repeated for a total of 15 doses over 100 days in a case of ingestion and five doses of 50 mg each over an 8 day period following a contaminated wound showed no effect on excretion but both cases developed a toxic labyrinthitis (THOMPSON, 1967). It must be admitted that the actual dose of zirconium was low. It has been suggested that the toxicity was due to heat sterilization of the injection fluid. However, other workers have found zirconium malate more toxic than zirconium citrate in the rat (KATZ et al., 1954). In view of the now proven use of chelating agents it is unlikely that zirconium will be used in future.

C. Chelating Agents

Sodium citrate was initially used in attempts to remove plutonium by complexing it in a soluble and therefore excretable form (PAINTER et al., 1946). Excretion was increased but not sufficiently to be of practical importance. Ethylenediamine tetraacetic acid was the first polyaminocarboxylic acid studied as a chelating agent for plutonium (FOREMAN and HAMILTON, 1961). It had proved worth while in the case of yttrium (VAUGHAN and TUTT, 1953). The structural formulae of some of the polyaminocarboxylic acids used to induce increased plutonium excretion are shown in Fig. 10.72. Ethylene-diamine-tetraacetic acid (EDTA or versine), was used first but has now been largely replaced by diethylene-triamine-pentaacetic acid (DTPA) or triethylene-tetraamine-hexaacetic acid (TTHA). A microbiologically produced chelating agent, desferrioxamine (DFOA), which is used with some success for the treatment of iron poisoning and iron storage diseases (BANNERMAN et al., 1962; SMITH, 1964–1965; TAYLOR, 1967) has also been used experimentally to remove ^{239}Pu. The formula for this agent is shown in Fig. 10.73 (KEBERLE, 1964). The free amino group accounts for the basic character of the compound which enables it to form salts with organic and inorganic acids as shown in Fig. 10.74. It removes iron from ferritin and haemosiderin and continues to do so until its maximum theoretical binding capacity is attained. Like other iron binding agents, for instance transferrin, it shows a strong affinity for plutonium and some for ^{228}Th but is much less effective with ^{241}Am and ^{227}Ac (TAYLOR, 1967). It also has much less affinity than DTPA for the alkaline earths and does not interact with manganese, zinc, cobalt and copper (KEBERLE, 1964).

1. Local Application to Skin and Wounds

The utility of chelating agents for decontaminating skin and wounds by local application has been questioned (THOMPSON, 1967). They are effective decontaminants but they are capable, at least under certain conditions, of increasing the absorption of ^{239}Pu from a wound (BALLOU, 1961; WILSON and SILKER, 1960).

POLYAMINOCARBOXYLIC ACIDS EMPLOYED IN THERAPEUTIC REMOVAL OF PLUTONIUM

Abbreviation	Name	Structural Formula
EDTA	Ethylenediaminetetraacetic acid	$(HOOCCH_2)_2NCH_2CH_2N(CH_2COOH)_2$
DTPA	Diethylenetriaminepentaacetic acid	$(HOOCCH_2)_2N{-}CH_2{-}CH_2{-}N(CH_2COOH){-}CH_2{-}CH_2{-}N(CH_2COOH)_2$
TTHA	Triethylenetetraminehexaacetic acid	$(HOOCCH_2)_2NCH_2CH_2{-}N(CH_2COOH){-}CH_2CH_2{-}N(CH_2COOH){-}CH_2CH_2{-}N(CH_2COOH)_2$
BAETA	2:2′—*bis*—[di(carboxymethyl)amino] diethyl ether	$(HOOCCH_2)_2N{-}CH_2CH_2{-}O{-}CH_2CH_2{-}N(CH_2COOH)_2$

Fig. 10.72. Polyaminocarboxylic acids employed in therapeutic removal of plutonium. (THOMPSON, 1967, by courtesy of author and publishers)

WILSON and SILKER (1960) describe the case of a man who cut his hand and a piece of highly polished plutonium oxide became imbedded in the wound. The wound was at once treated with a 10 percent solution of Na_4Ca EDTA. The urine passed within 3 hours of the accident had a very high ^{239}Pu count which the authors were fairly sure was not due to contamination of the specimen. Subsequent specimens contained much less ^{239}Pu. They attribute the high value to the possible formation of a plutonium chelate which was rapidly excreted on

1 Mol CH_3—COOH	Acetic acid
2 Mol HOOC—$(CH_2)_2$—COOH	Succinic acid
3 Mol NH_2—$(CH_2)_5$—NHOH	1-Amino-5-hydroxylamino-pentane

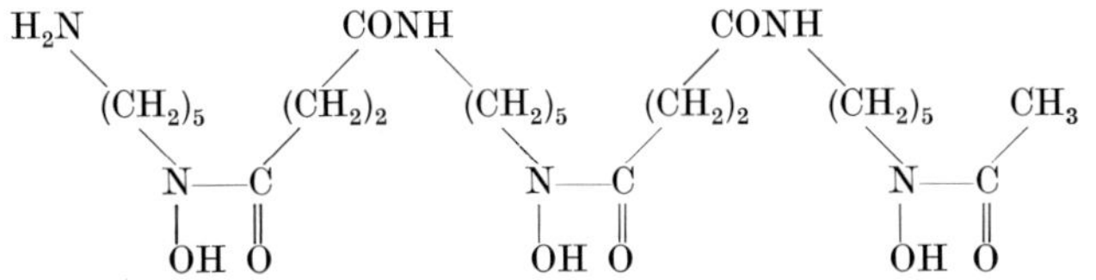

H2N
CONH
CONH
(CH2)5
(CH2)2
(CH2)5
(CH2)2
(CH2)5
CH3
N—C
N—C
N—C
OH O
OH O
OH O

Fig. 10.73. Desferrioxamine chemistry. (KEBERLE, 1964, by courtesy of author and publishers)

CO
H2N – (CH2)5
(CH2)2
N
C
O(-)
O
CH3 – C = O --- Fe3+ — O(-) – N – (CH2)5 · NH
O(-)
O
N
C
(CH2)5
(CH2)2
HN
CO

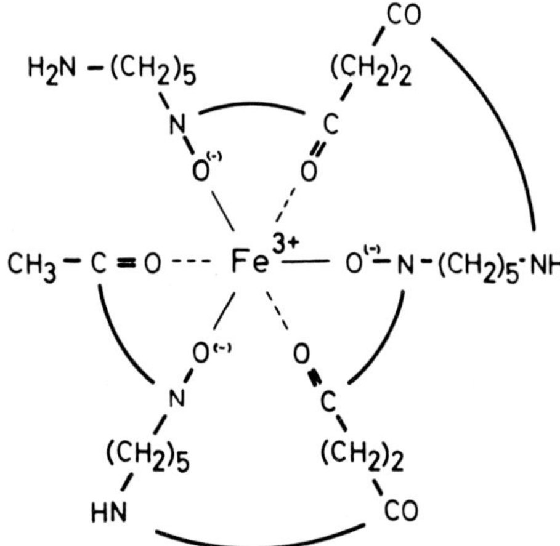

Fig. 10.74. Ferrioxamine B. (KEBERLE, 1964, by courtesy of author and publishers)

reaching the blood stream. The man was given further intravenous EDTA with insignificant effects on plutonium excretion. In vitro experiments which these authors report showed that metallic plutonium was more soluble in Na_4Ca EDTA than in distilled water.

Treatment of contaminated wounds with intravenous EDTA has indeed had varying results. FOREMAN and his colleagues (1958) (FOREMAN et al., 1958) report one case contaminated with ^{241}Am and two with ^{239}Pu. In the case of americium they calculated using LANGHAM's formula for ^{239}Pu excretion, the justification for which is questionable for americium, that as a result of treatment the body burden was reduced by more than nine tenths. Even if the actual figures are incorrect it is clear from the data given that EDTA did effectively increase urinary excretion of americium. The effect of treatment with Ca EDTA in the two cases of plutonium contamination was equivocal. SANDERS (1961) reports disappointing results in another case given both zirconium citrate and Ca EDTA. THOMPSON (1967) suggests, however, that the effect may have been complicated by enhanced absorption from the wound site.

The absorption of plutonium from an intramuscular injection site was considered not to be appreciably accelerated by intraperitoneal administration of EDTA in rats (FOREMAN et al., 1955). However, if DTPA was injected directly into an intramuscular deposit of $Pu(NO_3)_4$, pH 4–5, the retention of the intramuscular plutonium deposit was reduced by about one third. About 85 percent of the plutonium was absorbed and ultimately excreted (TAYLOR and SOWBY, 1962–1963). Since the experimental evidence already discussed (Sec. II.C.6) suggests there may be considerable translocation of ^{239}Pu from a wound site to both skeleton and liver, there would appear to be good reason for the immediate and continuous treatment of individuals with contaminated wounds with intravenous chelating agents. This should be accompanied by excision at the site of the deposit. Local treatment with chelating agents is probably of less value. The choice of chelating agents is discussed in the following sections.

2. Skeletal Deposition

a) EDTA

Prompt treatment with EDTA will reduce skeletal deposition of plutonium in the rat (HAMILTON and SCOTT, 1953; KATZ et al., 1954; FOREMAN et al., 1955; WEEKS et al., 1955; SEMENOV and TREGUBENKO, 1958; SMITH, 1958; FRIED et al., 1959; TAYLOR and SOWBY, 1962) and the dog (WAGER and TEMPLE, 1954; THOMPSON, 1956). The amount of plutonium deposited in the liver may be reduced by about a factor of 2 but the results are variable depending on the physical state of the injected plutonium (SCHUBERT et al., 1961). Repeated treatment is more effective than a single injection. It is not absolutely clear why EDTA should be more effective than zirconium citrate when given in repeated doses. The difference may well depend on the form in which the plutonium is present in the plasma at longer time intervals. Orally administered EDTA is relatively ineffective (HAMILTON and SCOTT, 1953).

FOREMAN and his colleagues in 1954 reported the use of EDTA in two human cases of plutonium contamination. One was a woman who cut her hand with broken glass from a flask of $Pu(NO_3)_4$ in 1 M HNO_3. The wound was debrided and she was given a continuous drip of 2.5 g Ca EDTA in 250 cm^3 saline twice daily. As seen in Fig. 10.75 there was a striking increase in excretion. In the second case where exposure had occurred 7 years previously from an inhalation accident no such spectacular results were obtained. There was a small increase in urinary excretion during the trial period. Given in large doses EDTA may give rise to renal symptoms in man (FOREMAN et al., 1958) but FOREMAN and his colleagues (FOREMAN et al., 1956) consider that 5 g a day given for not more than 5 days in succession and followed by 2 day rest periods is well within desirable safety limits. It has also been suggested that even with adequate rest periods treatment should not be continued indefinitely since it may lead to loss of other elements, notably zinc and manganese (NORWOOD et al., 1959; PERRY and SCHROEDER, 1957).

b) DTPA

In 1958 experiments in rats were reported by SMITH which gave encouraging data on the value of diethylene triamine pentaacetic acid (DTPA). This might be expected since it forms a chelate with plutonium which has a higher stability constant than that of EDTA with plutonium. SMITH gave 0.3 m mole intraperitoneally to rats together with 0.2 m mole calcium gluconate 1 hour after an intravenous injection of 1 μCi ^{239}Pu citrate. He noted that in animals killed 2 days later this had reduced skeletal retention from 61 percent of the administered dose in the controls to about 9 percent while EDTA had only reduced the skeletal retention to 44 percent. Rats that were given several treatments a week for

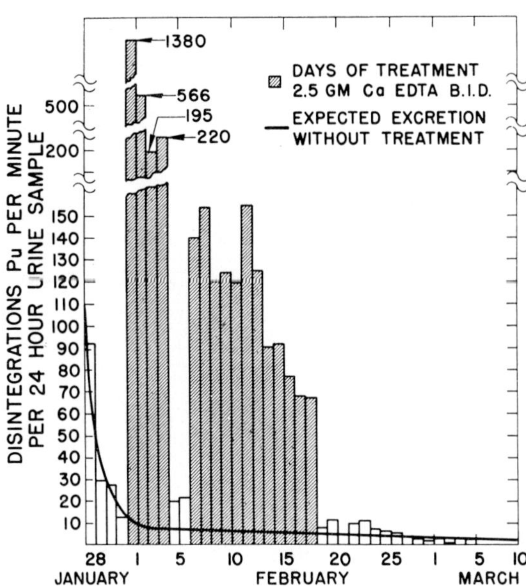

Fig. 10.75. Effect of Ca EDTA on urinary plutonium excretion in the human. (FOREMAN et al., 1958, by courtesy of authors and publishers)

4 weeks beginning 38 days after the plutonium injection lost about half the retained plutonium from the skeleton and there was also a reduction in soft tissue retention. These satisfactory results in rats were confirmed by FRIED et al. (1959) as shown in Table 10.46 and also in pigs whose metabolism, from the information available, is similar to that in the human (SMITH et al., 1961; CLARKE et al., 1962). The plutonium was administered intravenously to the pigs as the tetravalent citrate complex at pH 5. In one experiment animals were injected again intravenously with 40 ml of solution containing 9 g of the calcium disodium salt of DTPA at pH 7. In the second experiment animals injected with approximately 175 μCi of plutonium citrate were given 5 consecutive daily intravenous injection of $CaNa_3$ DTPA, 1 g the first day and 2 g on each of 4 successive days. The animals were all killed 6–7 days later. The effect of the prompt single dose of DTPA on excretion are shown in Table 10.47. Enhanced excretion is pre-eminently in the urine, the greatest effect being in the first 3 days. The effect on organ retention is shown in Table 10.48. As previously seen in rats, both liver and skeletal retention are favourably reduced. About 90 percent of the injected plutonium is excreted. The authors attribute the low recovery in the controls to sampling problems and analytical difficulties. They also found that DTPA given as long as 60 days after plutonium deposition appeared to have an appreciable effect in removing plutonium from the tissues. In the delayed effect excretion occurred largely in the faeces while prompt administration of DTPA increased urinary excretion.

It is however generally agreed that DTPA treatment is most effective when started as soon as possible after exposure and that a short course of repeated treatments is more effective than a single dose (TAYLOR and SOWBY, 1962; ROSENTHAL et al., 1969). It is important to note that TAYLOR and SOWBY (1962) found an optimum dose of 1.5 mM/kg body weight for rats. An increase

Table 10.46. Effect of chelating agents on tissue distribution and urinary excretion of ^{239}Pu in rats. (Taken from Fried et al, 1959)

Tissue	Percent injected dose per group[a]				Percent injected dose in treate percent injected dose in controls		
					Controls		
	Controls	EDTA	DTPA	BAETA	EDTA	DTPA	BAETA
Skeleton[b]	63.5	45.8	40.2	49.1	—	—	—
Liver	7.87	11.0	1.63	1.75	—	—	—
Spleen	0.36	0.42	0.15	0.25	—	—	—
Kidneys (both)	0.41	0.39	0.17	0.19	—	—	—
Urine collection (in days after Pu)							
0—1	1.2	1.2	1.3	1.2	1.0	1.1	1.0
1—4	0.40	0.46	0.45	0.47	1.1	1.1	1.2
4—5	0.045	0.045	0.041	0.045	1.0	0.91	1.0
5—6	0.094	0.088	0.079	0.073	0.94	0.84	0.78
6—7	0.035	0.93	5.2	5.7	27.0	148.0	164.0
7—8	0.061	0.36	1.6	1.5	5.9	26.6	24.1
8—9	0.031	0.14	0.59	0.62	4.5	19.0	20.0
9—11	0.042	0.098	0.48	0.57	2.3	11.4	13.6
11—12	0.062	0.22	0.57	0.54	3.5	9.2	8.7
12—13	0.062	0.56	2.9	3.0	9.0	46.2	47.6
13—14	0.020	0.11	0.40	0.45	5.5	20.0	22.5
14—15	0.044	0.25	0.60	0.65	5.7	13.6	14.8
15—18	0.093	0.37	0.67	0.75	4.0	7.2	8.1
18—19	0.024	0.046	0.14	0.15	1.9	5.6	6.3
19—20	0.018	0.031	0.086	0.11	1.7	4.8	6.1
20—21	0.033	0.12	0.10	0.16	3.6	3.0	4.8
21—22	0.016	0.40	1.5	1.8	25.0	95.6	113.
22—23	0.016	0.35	0.87	1.1	22.0	54.4	68.8
23—24	0.016	0.084	0.26	0.30	5.2	16.3	18.8
24—25	0.036	0.21	0.49	0.57	5.8	13.6	15.8
25—26	0.017	0.37	0.95	0.96	22.0	55.9	56.5
26—27	0.024	0.29	0.76	0.98	12.0	31.7	40.8
0—27	2.36	6.71	20.0	21.7			

[a] Values are expressed as the mean of 3 rats per group. [b] Total skeletal content was calculated from the content of the two femurs by multiplying by a factor of ten. Treatment administered 6th, 12th, 22nd, 25th, 26th day.

Table 10.47. Effect of single dose of DTPA on excretion of plutonium by pigs. (Taken from Smith et al., 1961)

Post injection (days)	Excretion (% dose/day)			
	Urine		Feces	
	Control[a]	DTPA treated[a]	Control	DTPA treated[a]
1	0.6	45.0	0.1	1.8
2	0.1	29.0	0.3	1.3
3	0.2	6.8	0.4	1.7
4	0.2	0.5	0.4	0.4
5	0.1	1.0	0.2	0.3
6	0.1	0.3	0.1	0.1
Total	1.3	83	1.5	5.6

[a] Average of results from 3 pigs.

Table 10.48. Effect of single prompt dose of DTPA on retention of plutonium by pigs. (Taken from Smith et al., 1961)

Tissue	Retention (% of dose/total tissue)	
	Control[a]	DTPA treated[a]
Liver	14.0	0.47
Kidneys	0.20	0.20
Spleen	0.24	0.013
Femur	2.5	0.29
Total skeleton	57.0[b]	5.9
Excreta:		
Urine	1.3	83.0
Feces	1.5	5.6
Total recovered	74.0	96.0

[a] Average of results from 3 pigs. [b] Data available from only 2 pigs.

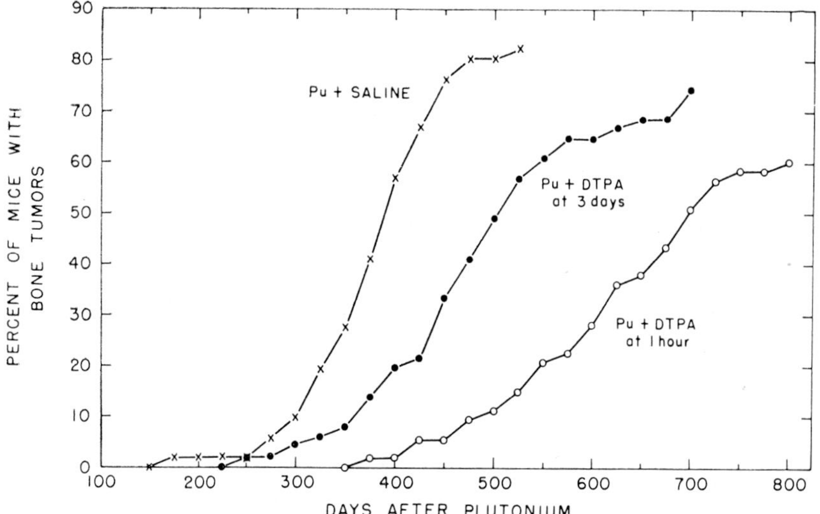

Fig. 10.76. Cumulative bone tumour incidence in mice expressed as percent of animals with bone tumours and plotted at 25 day intervals. Mice were given monomeric plutonium followed by DTPA for 12 days. (Rosenthal and Lindenbaum, 1967, by courtesy of authors and publishers)

above this did not increase excretion and this dose is well below the level at which acute toxic effects might be expected. Rosenthal et al. (1969) have found that in mice a dose of 10 μg/kg (equivalent to 1 g dose in man) given at 3 day intervals for 12 injections is adequate to remove one fifth of the skeletal plutonium and also appreciable amounts from the liver. This group has also published elegant experiments showing that mice treated with DTPA develop fewer osteosarcoma following plutonium injections than those receiving no treatment (Rosenthal et al., 1962; Rosenthal and Lindenbaum, 1967, 1969). These experiments are illustrated in Fig. 10.76 where the cumulative bone tumour in-

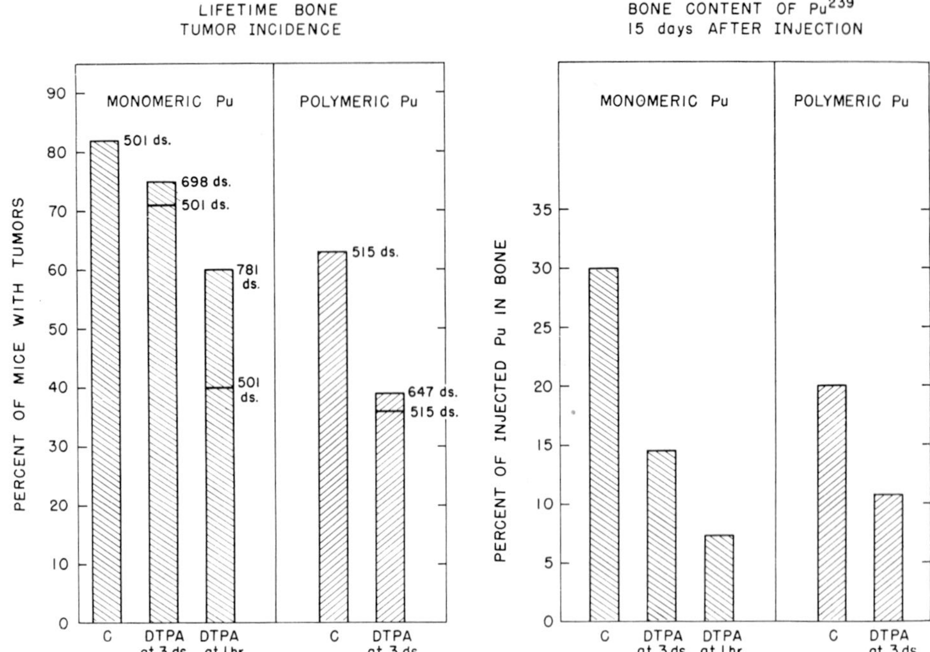

Fig. 10.77. Effects of twelve daily intraperitoneal injections of 500 mg of DTPA per kg initiated either 3 days or 1 hour after intravenous administration to mice receiving either 0.0643 μCi (~2.6 μCi/kg) of monomeric plutonium or 0.0587 μCi (~2.6 μCi/kg) of a polymeric form of plutonium. Control mice (*C*) received plutonium and twelve injections of saline solution. Bone tumour incidence in each DTPA-treated group is shown both at the time of death of the last plutonium control mouse (at 501 days for monomeric and at 515 days for polymeric plutonium) and at the time of death of the last treated mouse. These times are indicated at the top of each bar in the tumour incidence section of the figure. The amount of plutonium shown in the bone (right hand section) is that measured in two femurs × 10. (ROSENTHAL and LINDENBAUM, 1967, by courtesy of authors and publishers)

cidence is shown in controls and in mice given DTPA 1 hr and 3 days after an injection of plutonium. In Fig. 10.77 is shown the life time tumour incidence and the bone content of plutonium following injection of both monomeric and polymeric plutonium and the effect of DTPA treatment.

DTPA has proved useful in treating human cases of contamination if given early enough[9]. NORWOOD (1960, 1962a) reported some of the first cases. In one instance the average excretion of plutonium was increased by a factor of 55 in a man who had received an initial body deposition of about 0.4 μCi by inhalation 30 months previously. As treatment continued the effectiveness wore off until at the end of 50 weeks it was only 20 percent of that initially observed. Therapy was stopped for several months and then resumed. On resumption the daily output of plutonium was increased to approximately the initial value. The long period without therapy also enhanced the effect of DTPA, when started again, on faecal excretion which was increased by approximately a factor of 4. The overall picture is shown in Fig. 10.78. NORWOOD suggests that DTPA first removes

9 Further examples are presented in Chap. 11 which concerns primarily plutonium entering by the inhalation route. — Editor.

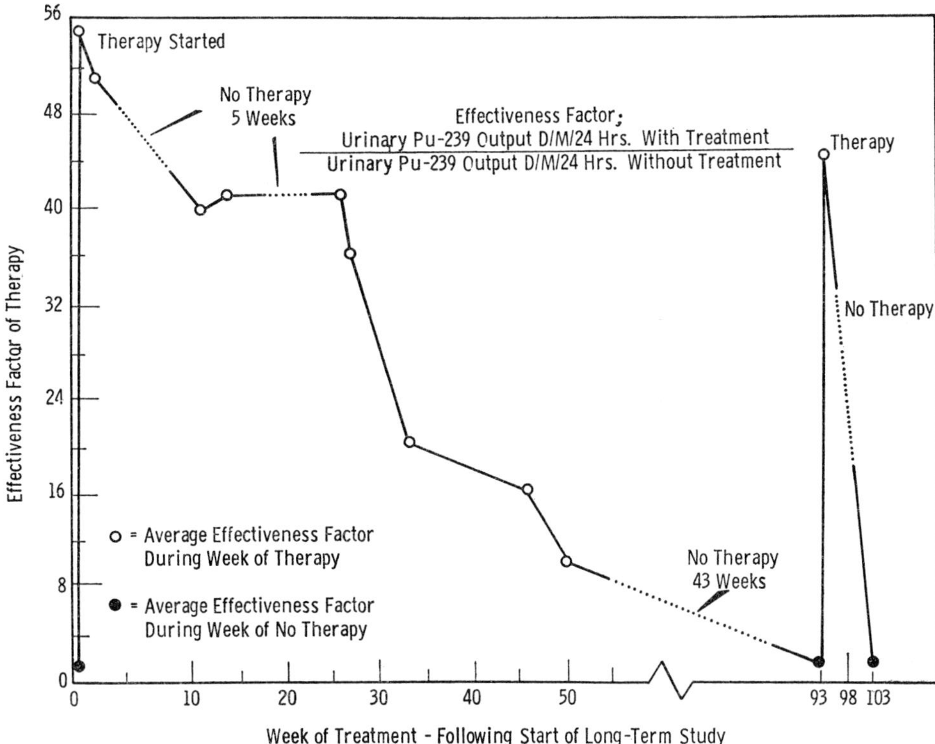

Fig. 10.78. Effectiveness of DTPA in human cases after 43 weeks without DTPA treatment which followed long period with treatment. (Norwood, 1962a, by courtesy of author and publishers)

some of the plutonium in the liver leaving the more refractory bone deposits and that after a period of no treatment some of the plutonium deposited elsewhere may be transferred to liver making it more amenable to the chelating agent.

He emphasised the importance of giving the smallest possible effective dose because of its potential toxic effect on the kidneys (Catsch, 1963–1964) and suggested a dose of 1 g DTPA given by slow drip 2 or 3 times a week for a 3 week period. Such a 3 week period of treatment should be alternated with a 3 week period without treatment (Norwood, 1962a, b). Observing this schedule of treatment there were no toxic effects (Norwood, 1960, 1962a, b). Norwood (1962b) noted no significant difference in results between daily doses of 0.9 and 1.6 g DTPA. Oral administration was only about 10 percent as effective in increasing urinary excretion as when given intravenously though it had some effect on faecal excretion (Norwood, 1962b).

On the other hand, Larson and his colleagues (Larson et al., 1968), reporting fully on a man previously mentioned by Wick (1964) who had a severely contaminated wound, found that DTPA given intravenously though it probably increased urinary excretion, had no effect on faecal excretion which they described as negligible in contrast to the finding of Lister and his colleagues (1963) (Sec. II.H). The effect on urinary excretion in this patient is illustrated in Fig. 10.79.

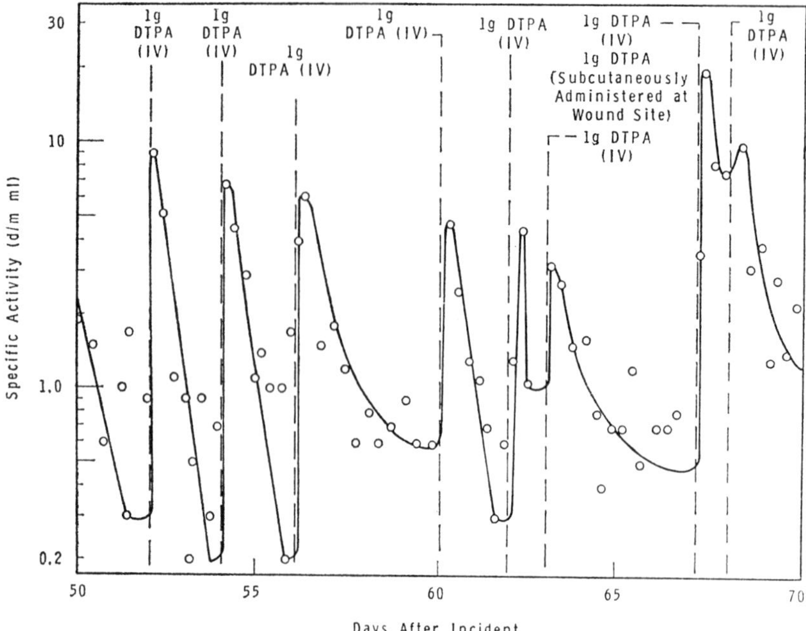

Fig. 10.79. Effect of DTPA on urinary excretion of ^{239}Pu in a man with a contaminated wound. (LARSON et al., 1968, by courtesy of authors and publishers)

They also give some data which suggest that while flushing the wound with DTPA solution may remove some plutonium in the flushing fluid it may also serve to increase absorption. LAGERQUIST and his colleagues record two cases (LAGERQUIST et al., 1965) both of which received prompt DTPA treatment within one hour with very different results. In case 1 an employee was sprayed with plutonium nitrate solution and plutonium entered the body via ingestion, inhalation and small acid burns on the skin. In case 2 the thumb was pierced by a wire contaminated with plutonium oxide. It was considered that removal was successful in case 1 but the effectiveness of treatment was inconclusive, in the authors' opinion, in the second case.

In Table 10.49 are shown the variable results obtained in treating three human cases who had received puncture wounds. Excision was undertaken in all cases and was either preceded or followed by intravenous DTPA (SCHOFIELD,

Table 10.49. A summary of the treatment of the three cases of wound contamination in man, showing the initial ^{239}Pu content of the wound, the surgical and DTPA treatment and the amount transferred to the rest of the body. (Taken from SCHOFIELD, 1969)

Case (1)	Initial wound content (μCi) (2)	Time to excision (min) (3)	Remaining after excision (nCi) (4)	Wound content after healing (nCi) (5)	Time to first DTPA treatment (min) (6)	Total DTPA given (g) (7)	Total Pu eliminated by DTPA (nCi) (8)	Estimated body content (nCi) (9)
1	1.0	690	90	90	690	0.55	90×10^{-3}	30
2	1.0	540	20	2	30	3.90	4.0	30
3	5.0	140	20	2	75	4.25	5.0	20

1969). The differences in effectiveness of DTPA in cases 2 and 3 could probably be explained by differences in the solubilities of the plutonium involved. Case 2 was primarily contaminated with plutonium nitrate in acid whereas case 3 was contaminated with plutonium metal which is very insoluble. Case 1 involved plutonium oxide. SCHOFIELD emphasises the importance of early excision of the wound. He considers that once the wound site has been excised, so removing the plutonium, excretion will be representative of the amount deposited elsewhere in the body. Excision is of course relatively easy with a small puncture wound but may be more difficult in other types of injury.

c) TTHA

TAYLOR and SOWBY (1962) made a comparative study of the effectiveness of DTPA and TTHA in rats and found a close similarity though on theoretical grounds TTHA might have been expected to be the more effective. The one possible advantage of TTHA is that some workers have found that it appears to be orally effective at a considerably lower dose level than DTPA (BALLOU, 1962a, b).

d) BAETA

2:2-bis (di[carboxymethyl]amino) diethyl ether is more effective in increasing plutonium excretion in rats than EDTA and probably about as effective as DTPA as illustrated in Figs. 10.80 and 10.81. A comparative study of the effect of these three chelating agents is shown in Table 10.44 (FRIED et al., 1959). Such comparative studies have not been made in man.

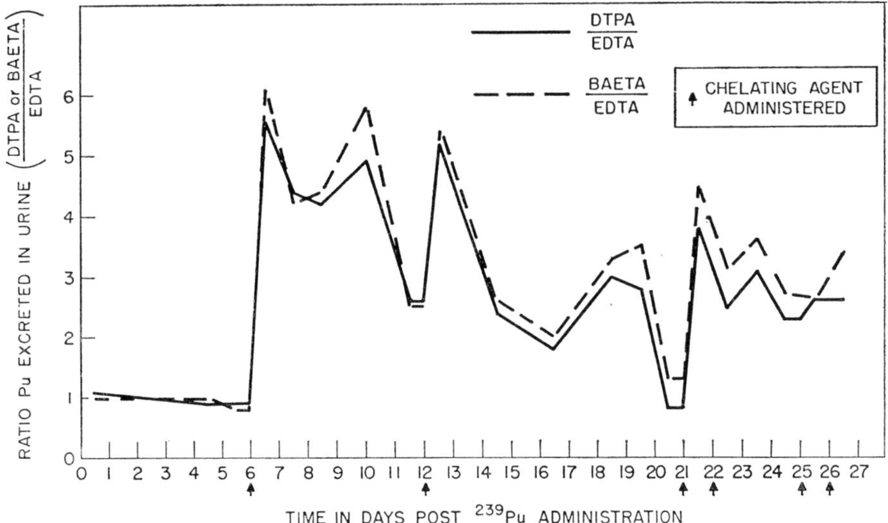

Fig. 10.80. The effectiveness of DTPA and BAETA relative to EDTA on the urinary excretion of ^{239}Pu in the rat. (FRIED et al., 1959, by courtesy of authors and publishers)

e) DFOA

DFOA has been given experimentally to both rats (SMITH, 1964; TAYLOR, 1967) and mice (ROSENTHAL and LINDENBAUM, 1964) burdened with plutonium. ROSENTHAL and LINDENBAUM administered the chelating agent 3 days after the plutonium and noted only a 9 percent increase in the urinary excretion of plu-

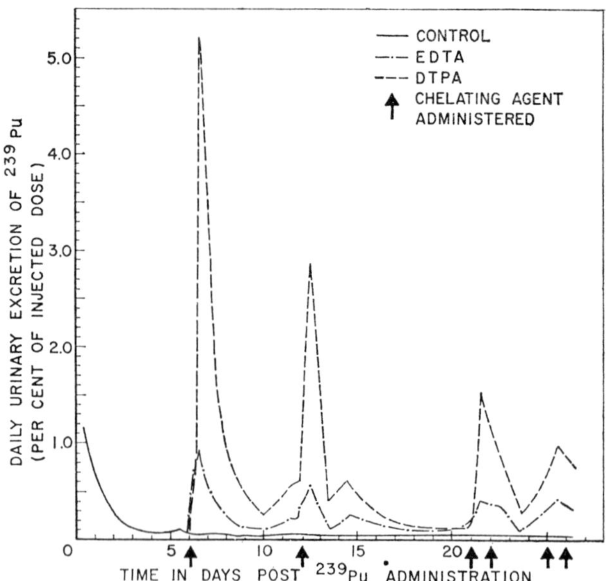

Fig. 10.81. Daily urinary excretion in percent of the injected dose of ^{239}Pu as affected by EDTA and DTPA in the rat. (FRIED et al., 1959, by courtesy of authors and publishers)

tonium and no reduction in the skeletal and liver burden. TAYLOR (1967) showed that in rats DFOA was extremely effective if administered intraperitoneally within an hour of exposure to the nuclide given either intravenously or intramuscularly as $Pu(NO_3)_4$ in 0.02 M HNO_3 but that if the chelating agent was given at longer time intervals it was far less effective. He also found that retention in the femur was greater in rats given DFOA than in rats given DTPA. However, both he and SMITH (1964) noted that if plutonium was given as the citrate rather than as the nitrate retention in the femur was reduced by the same amount by both chelating agents. TAYLOR (1967) suggests that DFOA probably acts by chelation of ^{239}Pu in the body fluids only and that it has a limited ability to remove the nuclide after its deposition in the bone and soft tissues. DFOA has two advantages over other chelating agents. Firstly, it can be given intramuscularly or intraperitoneally while DTPA must be given by slow drip. Secondly, it does not appear to be toxic to the kidney in high doses. Two other chelating antibiotics, vancomycin and cephalothin are without an effect on plutonium deposition (SMITH, 1966).

It is clear, from the few cases of human contamination that are reported in detail in the literature, that the effects of DTPA treatment are variable depending on the character of the contamination and possibly on the individual treated. Yet there is clear evidence of increased excretion under some conditions. It would appear that only too often in the past, treatment with EDTA, DTPA, TTPA has been withheld in the case of industrial accidents for fear that the chelating agent would have toxic results. The risks of later malignancy, if treatment is not started as soon as possible, would appear to be greater.

Further, the experimental evidence discussed would suggest that in view of its apparent specificity for plutonium and nontoxic qualities, DFOA might be given by intramuscular injection immediately after contamination followed at appropriate time intervals by DTPA.

f) Summary of Sec. IX

In summary, it may be said that it is doubtful whether any known agent will remove plutonium once it has been deposited in the skeleton. Chelating agents, particularly DTPA and TTPA, may prevent deposition in the skeleton. They must therefore be given immediately it is known that exposure may have occurred so that chelation results as plutonium reaches the blood stream. Repeated periods of treatment may prove useful in the case of wounds and perhaps of lung deposition when the plutonium is translocating continuously from the site of deposition to the skeleton.

X. Summary

Two plutonium isotopes ^{239}Pu and ^{238}Pu have practical applications. Since ^{239}Pu is so far the most useful its biological effects have been more extensively studied and are first discussed.

Analysis of the metabolic behaviour of ^{239}Pu in different species indicates that there are significant species differences. The behaviour of the radionuclide in adult dogs, pigs and sheep is probably not unlike its behaviour in man. If plutonium reaches the bloodstream in monomeric form it is initially largely deposited in the skeleton—if it reaches the bloodstream in polymeric form a greater proportion is initially deposited in the liver though it may subsequently translocate to the skeleton. Once it reaches the skeleton the retention of plutonium is long lasting. In man it is calculated that about 50 percent of the retained plutonium is present in the liver and about 50 percent in the skeleton at long time intervals after exposure. Plutonium may reach the bloodstream subsequent to ingestion, inhalation, external skin application and particularly from some form of wound. The mechanisms of translocation are still poorly understood. There is some experimental evidence that largely monomeric plutonium nitrate, given by intramuscular injection to simulate a wound, is initially translocated as a soluble complex, but later the deposition changes to a more colloidal pattern, suggesting that lymphatic drainage may then be involved.

The pattern of plutonium distribution in the skeleton is distinctive for this radionuclide. It is concentrated particularly on endosteal bone surfaces, where it may be found in or on the endosteal cells as well as on the mineral matrix surface. This distribution appears independent of the route of administration. The absolute amounts found on surfaces and in the bone marrow is dependent on the form in which the plutonium reaches the blood stream, the more polymeric forms appear in greater concentration in the marrow while the monomeric forms are found in greater concentration on the "bone surface".

There is good experimental evidence that binding to specific glycoproteins of bone as well as to mineral may account for the surface deposition on the mineral-matrix surface. The mechanisms of cellular fixation are still unknown in detail though it is thought that here again glycoproteins may play a part as well as cell lysosomes. Deposition in the marrow is largely associated with the haemosiderin of macrophages. The relationship of plutonium to iron pigments throughout the body raises interesting speculations. This pattern of plutonium distribution in both marrow and in the proliferating osteogenic cells themselves accounts for the great carcinogenic activity of this radionuclide compared with other α emitting isotopes which deposit in bone.

No actual measurements of radiation dose to sensitive tissues in bone in animals dying from osteosarcoma are available but many calculations have been made. Experimental evidence suggests that on the basis of estimates of total skeletal plutonium at the time of death from osteosarcoma in dogs the predicted osteosarcoma dose in man would be 200 rads to the whole skeleton and the dose resulting from a permissible skeletal burden during 50 years is 25 rads. Osteogenic sarcoma occurred in dogs injected with 0.0157 μCi/kg ^{239}Pu which leads to a calculated radiation dose to the whole skeleton of only 76 rads. In rats fed ^{239}Pu continuously 57 rads was the average skeletal dose calculated to have accumulated at the time of death with osteosarcoma. Owing to uneven distribution actual measurements of doses to the sensitive cells will of course be considerably higher as precise measurements in various isolated preparations has shown.

The most frequent tumour recorded in experimental animals is the osteoblastic sarcoma but leukaemia and other tumours arising from mesenchyme marrow elements and from epithelium closely in contact with bone have been described.

A number of chelating agents have been used in an attempt to remove plutonium from the skeleton. It is doubtful if this can be done but deposition may be prevented by achieving chelation in the blood stream. This is of importance when there is continuous translocation from some soft tissue deposit. It can best be attempted in the case of wounds by immediate debridement and immediate administration of either DTPA or TTHA intravenously. Treatment should not be given continuously because of possible toxic effects on the kidney but repeated periods of treatment lasting several days may increase urinary excretion and therefore reduce the body burden.

Certain differences have been observed in the metabolic behaviour of ^{238}Pu from that of ^{239}Pu. These can be ascribed generally to mass effects resulting from the large differences in the specific activities of the two isotopes, though the apparent greater solubility in vivo of ^{238}Pu is not easily explained in terms of mass effect. The evidence available suggests that considerable caution must be adopted in using biological data obtained with one plutonium isotope to predict the biological behaviour of another.

It would appear particularly important in the future to study with greater precision the mechanism of plutonium binding to both osteogenic cells and to the mineral/matrix elements of bone. Further, the forms in which plutonium translocates from soft tissues to the skeleton, merit further investigation. Comparative studies using plutonium and the actinides are likely to be of great interest in view of the differences already known between the binding, distribution and carcinogenicity of plutonium and americium.

We are grateful to the following for helpful discussion and criticism: Dr. ROGER BERRY, Dr. G. W. DOLPHIN, Dr. C. W. MAYS, Dr. ROBIN MOLE, Dr. RAY OLIVER and MARGARET WILLIAMSON.

References[10, 11]

ANDREWS, A. T. de B., HERRING, G. M.: Further studies on bone sialoprotein. Biochim. biophys. Acta (Amst.) **101**, 239–241 (1965).

10 It has not been possible to verify all the Russian references in the original. They are taken from other Russian bibliographies in the form quoted. The other references have been verified from original publications or circulated reports.

11 *Abbreviation Used:* AWRE, Atomic Weapons Research Establishment, Aldermaston, Berkshire, United Kingdom. BNWL, Battelle Northwest Laboratory, Richland, Washington 99352. UCRL, University of California Radiation Laboratory, Berkeley and Livermore, California. USAEC, United States Atomic Energy Commission (most titles still in print or information concerning them can be obtained from Microsurance, Inc. Oak Ridge Microreproduction Center, P.O. Box 3522, Oak Ridge, Tenn. 37830 or the National Technical Information Service, U.S. Department of Commerce, Springfield, Virginia, 22151).

ARNOLD, J. S.: Second Annual Conference of Plutonium, Radium and Mesothorium, Salt Lake City, Utah, T I D 7639, Division of Technical Information Extension USAEC (ed. STOVER, C. N.), p. 112–129 (1954).
ARNOLD, J. S., JEE, W. S. S.: Bone growth and osteoclastic activity as indicated by radioautographic distribution of plutonium. Amer. J. Anat. **101**, 367–394 (1957).
ARNOLD, J. S., JEE, W. S. S.: Autoradiography in the localization and radiation dosage of Ra^{226} and Pu^{239} in the bones of dogs. Lab. Invest. **8**, 194–204 (1959).
ARNOLD, J. S., JEE, W. S. S.: Pattern of long-term skeletal remodelling revealed by radioautographic distribution of ^{239}Pu in dogs. Hlth Phys. **8**, 705–707 (1962).
ATHERTON, D. R., MAYS, C. W., STOVER, B. J.: Radionuclide distribution in adult beagle bones. In: Research in Radiobiology Annual Report of Work in Progress on the Chronic Toxicity Program, USAEC Report COO-217, University of Utah, p. 118–125 (1958).
BAIR, W. J., NELSON, J. C., WILLARD, D. H.: Comparative metabolism of ^{237}Pu and ^{239}Pu in beagles. BNWL 1050, part 1, 3, 17–38 (1970).
BAIR, W. J., WILLARD, D. H., HERRING, J. P., GEORGE, L. A.: Retention, translocation and excretion of inhaled $Pu^{239}O_2$. Hlth Phys. **8**, 639–649 (1962).
BAIR, W. J., WILLARD, D. H., WEST, J. E.: Plutonium inhalation studies. In: Hanford Biology Research Annual Report for 1960, USAEC Report HW-69500, Hanford Atomic Products Operation, General Electric Co., Jan. 1961, p. 20.
BALLOU, J. E.: Effects of age and mode of ingestion on absorption of plutonium. Proc. Soc. exp. Biol. (N.Y.) **98**, 726–727 (1958).
BALLOU, J. E.: Wound decontamination with EDTA. In: Hanford Biology Research Annual Report for 1960, USAEC Report HW-69500, Hanford Atomic Products Operation, General Electric Co., Jan. 1961.
BALLOU, J. E.: Preliminary evaluation of several chelating agents for plutonium removal. Hlth Phys. **8**, 731–734 (1962a).
BALLOU, J. E.: Removal of deposited plutonium by triethylenetetraamine hexaacetic acid. Nature (Lond.) **193**, 1303–1304 (1962b).
BALLOU, J. E., OAKLEY, W. D.: Absorption and decontamination of plutonium in rats. In: Biology Research-Annual Report 1956, USAEC Report HW-47500, Hanford Atomic Products Operation, General Electric Co., Jan. 1957.
BALLOU, J. E., THOMPSON, R. C., CLARKE, W. J., PALOTOY, J. L.: Comparative toxicity of plutonium-238 and plutonium-239 in the rat. Hlth Phys. **13**, 1087–1092 (1967).
BANNERMAN, R. M., CALLENDER, S. T., WILLIAMS, D. L.: Effect of desferrioxamine and DTPA in iron overload. Brit. med. J. **1962 II**, 1573–1577.
BARENDSEN, G. W.: Impairment of proliferative capacity of human cells by alpha particles with differing energy transfer. Int. J. Radiat. Biol. **8**, 453–466 (1964).
BARENDSEN, G. W., BENSKER, T. W., VERGROESEN, A. J., BUDKE, L.: Effects of different ionizing radiations on human cells in tissue culture. Radiat. Res. **13**, 841–849 (1960).
BARENDSEN, G. W., WALTER, H. M. D.: Effects of different ionizing radiations on human cells in tissue culture. Radiat. Res. **21**, 314–329 (1964).
BARNES, D. W. H., CARR, T. E. F., EVANS, E. P., LOUTIT, J. F.: ^{90}Sr-induced osteosarcomas in radiation chimaeras. Int. J. Radiat. Biol. **18**, 531–537 (1970).
BARNES, D. W. H., KHRUSCHOV, N. G.: Fibroblasts in sterile inflammation: study in mouse radiation chimaeras, Nature (Lond.) **218**, 599–601 (1968).
BARNES, D. W. H., LOUTIT, J. F.: Haemopoietic stem cells in the peripheral blood. Lancet **1967 II**, 1138–1141.
BEACH, S. A., DOLPHIN, G. W.: Determination of plutonium body burdens from measurements of daily urine excretion. In: Assessment of radioactivity in man, p. 603–615. Vienna: International Atomic Energy Agency 1964.
BEACH, S. A., DOLPHIN, G. W., DUNCAN, F. P., DUNSTER, H. J.: A basis for routine urine sampling of workers exposed to plutonium 239. Hlth Phys. **12**, 1671–1682 (1966).
BELYAEV, YU. A.: Possible ways of influencing the elimination of plutonium from the animal organism. In: Biological effects of radiation and problems of radioactive isotope distribution. USAEC Report AEC-tr-5265, p. 180–187, translated from a publication of the State Publishing House of Literature in the Field of Atomic Science and Technology, Moscow 1961.
BELYAEV, YU. A.: In book "Plutonii 239", M. MEDGIZ (1962), p. 151, quoted by BULDAKOV et al. (1970), p. 10.
BELYAEV, YU. A., LEMBERG, V. K.: In book "Raspredelenie, biologicheskoe deistvie uskorenie vyvedeniys radioaktivnykh isotopov. M. MEDITSINA 1964, p. 343, quoted by BULDAKOV et al. 1970, p. 10.
BELYAEV, YU. A. et al.: p. 7, quoted BULDAKOV, L. A., et al. (1970), p. 13.
BENO, M.: A study of "haemosiderin" in the marrow of the femur of the normal young adult rabbit compared with that in rabbits 4 months after an intravenous injection of $^{239}Pu(NO_3)_4$. Brit. J. Haemat. **15**, 487–493 (1968).

Bensted, J. P. M., Blackett, N. M., Lamerton, L. F.: Histological and dosimetric consideration of bone tumour production with radioactive phosphorus. Brit. J. Radiol. **34**, 160–175 (1961).

Bensted, J. P. M., Taylor, D. M., Sowby, F. D.: The carcinogenic effects of americium 241 and plutonium 239 in the rat. Brit. J. Radiol. **38**, 920–925 (1965).

Berry, R. J., Survival of murine leukaemia cells in vivo after irradiation in vitro under aerobic and hypoxic conditions with monoenergetic accelerated charged particles. Radiat. Res. **44**, 237–247 (1970).

Bessis, M. C.: Hemosiderin and movements of iron within the body. Nouv. Rev. franç. Hémat. **2**, 153–158 (1962).

Bessis, M. C., Breton-Gorius, J.: Ferritin and ferruginous micelles in normal erythroblasts and hypochromic hypersideremic anemias. Blood **14**, 423–432 (1959).

Bessis, M. C., Breton-Gorius, J.: Iron metabolism in the bone marrow as seen by electron microscopy: A critical review. Blood **19**, 635–663 (1962).

Bingham, P. J., Brazell, I. A., Owen, M.: The effect of parathyroid extract on cellular activity and plasma calcium levels in vivo. J. Endocr. **45**, 387–400 (1969).

Bleaney, B.: Radiation dose-rates near bone surfaces in rabbits after an injection of plutonium. Phys. in Med. Biol. **12**, 145–160 (1967).

Bleaney, B.: The radiation dose-rates near bone surfaces in rabbits after intravenous and intramuscular injection of plutonium 239. Brit. J. Radiol. **42**, 51–56 (1969a).

Bleaney, B.: Plutonium deposition on bone surfaces and in bone marrow following intravenous and intramuscular injections. In: Delayed effects of bone-seeking radionuclides (ed. Mays, C. W., W. S. Jee, R. D. Lloyd, B. J. Stover, J. H. Dougherty and G. N. Taylor), p. 125–135. Utah: University of Utah Press 1969b.

Bleaney, B., Taylor, D., Vaughan, J.: Personal communication (1971).

Bleaney, B., Vaughan, J.: Distribution of ^{239}Pu in the bone marrow and on the endosteal surface of the femur of adult rabbits following injection of $^{239}Pu(NO_3)_4$. Brit. J. Radiol. **44**, 67–73 (1971).

Bloom, W. (Ed.): Histopathology of irradiation from external and internal sources. National Nuclear Energy Series. New York: McGraw-Hill 1948.

Bloom, W., Zirkle, R. E., Uretz, R. B.: Irradiation of parts of individual cells. Ann. N.Y. Acad. Sci. **59**, 503–513 (1955).

Bohr, N.: Velocity-range relation for fission fragments. Phys. Rev. **59**, 270–275 (1941).

Boocock, G., Popplewell, D. S.: Distribution of plutonium in serum proteins following intravenous injection into rats. Nature (Lond.) **208**, 282–283 (1965).

Boocock, G., Popplewell, D. S.: In vitro distribution of americium in human blood serum proteins. Nature (Lond.) **210**, 1283–1284 (1966).

Boocock, G., Popplewell, D. S.: Actinide interactions with blood serum proteins. AWRE Report No 0–15/67, p. 1–24 (1967).

Boyd, G. A.: Studies with plutonium. In: Biological studies with polonium, radium and plutonium (ed. R. M. Fink), National Nuclear Energy Series, p. 224–227. New York: McGraw Hill 1950.

Boyd, G. A., Fink, R. M.: Discussion of relative toxicities of polonium, plutonium and radium. In: Biological studies with polonium, radium and plutonium (ed. R. M. Fink), National Nuclear Energy Series, p. 247–257. New York: McGraw Hill 1950.

Brown, H., Moretti, E. S., Rosenthal, M. W., Lindenbaum, A.: Personal communication (1970).

Bruenger, F. W., Stover, B. J., Atherton, D. R.: Determination of plutonium in biological material by solvent extraction with primary amines. Analyt. Chem. **35**, 1671–1673 (1963).

Bruenger, F. W., Stover, B. J., Stevens, W., Atherton, D. R.: Exchange of ^{239}PuIV between transferrin and ferritin in vitro. Hlth Phys. **16**, 339–340 (1969).

Brues, A. M.: Biological hazards and toxicity of radioactive isotopes. J. clin. Invest. **28**, 1286–1296 (1949).

Brues, A. M., Lisco, H., Finkel, M. P.: Carcinogenic action of some substances which may be a problem in certain future industries. Cancer Res. **7**, 48 (1947).

Bukhtoyarova, Z. M.: The dynamics of the changes in the bone tissue of rabbits poisoned by plutonium 239. In: Biological effects of radiation and problems of radioactive isotope distribution, p. 153–161, edit. by Lebedinskii, A. V., and Moskalev, Y. I., AEC-tr-5265 (1961).

Bukhtoyarova, Z. M., Lemberg, V. K.: In: Vop. Onkol. **5**, 8, 140–148 (1959), quoted Moskalev et al. (1969).

Buldakov, L. A.: Radiobiologiya 8, 1, 65 (1968).

Buldakov, L. A., Lyubchanskii, E. R., Moskalev, Y. I., Nifatov, A. P., Atom Publications, Moscow 1969 LF-tr-41, UC-48, 1970. Problems of plutonium toxicology. (Translation and production prepared for U.S. Atomic Energy Commission.)

BULDAKOV, L. A., NIFATOV, A. P., TOLOCHKOVA, N. M., BUROV, N. I.: Absorption of ^{239}Pu through skin and from hypodermic cellular tissue of sucking-pigs. Radiobiologiya **7**, 591–601 (1967), quoted BULDAKOV et al. (1970), p. 39.

BUSTAD, L. K., CLARKE, W. J., GEORGE, L. A., HORSTMAN, V. G., MCCLELLAN, R. O., PERSING, R. L., SEIGNEUR, L. J., TERRY, J. L.: Preliminary observations on metabolism and toxicity of plutonium in miniature swine. Hlth Phys. 8, 615–620 (1962).

BUSTAD, L. K., GOLDMAN, M., ROSENBLATT, L. S., MCKELVIE, D. H., HERTZENDORF, I. I.: Haematopoietic changes in beagles fed ^{90}Sr. In: Delayed effects of bone-seeking radionuclides (ed. MAYS, C. W., W. S. S. JEE, R. D. LLOYD, B. J. STOVER, J. H. DOUGHERTY and G. N. TAYLOR), p. 279–290. Utah: University of Utah Press 1969.

CABLE, J. W., HORSTMAN, V. G., CLARKE, W. J., BUSTAD, L. K.: Effects of intradermal injections of plutonium in swine. Hlth Phys. 8, 629–634 (1962).

CARRITT, J., FRYXELL, R., KLEINSCHMIDT, J., KLEINSCHMIDT, R., LANGHAM, W., SAN PIETRO, A., SCHAFFER, R., SCHNAP, B.: The distribution and excretion of plutonium administered intravenously to the rat. J. biol. Chem. **171**, 273–283 (1947).

CATSCH, A.: Zur Toxikologie der Diäthylentriaminpentaessigsäure. Naunyn-Schmiedebergs Arch. exp. Path. Pharmak. **246**, 316–329 (1963–1964).

CHIPPERFIELD, A. R., TAYLOR, D. M.: Binding of plutonium to glycoproteins in vitro. Radiat. Res. **43**, 393–402 (1970a).

CHIPPERFIELD, A. R., TAYLOR, D. M.: The binding of americium and plutonium to bone glycoproteins. Europ. J. Biochem. **17**, 581–585 (1970b).

CLARKE, W. J.: Comparative histopathology of ^{239}Pu, ^{226}Ra and ^{90}Sr in pig bone. Hlth Phys. 8, 621–627 (1962).

CLARKE, W. J., HOWARD, E. B., HACKETT, P. L.: Strontium-90 induced neoplasia in swine. In: Delayed effects of bone-seeking radionuclides (ed. MAYS, C. W., W. S. S. JEE, R. D. LLOYD, B. J. STOVER, J. H. DOUGHERTY and G. N. TAYLOR), p. 263–277. Utah: University of Utah Press 1969.

CLARKE, W. J., PARK, J. F., PALOTOY, J. L., BAIR, W. J.: Plutonium inhalation studies. VII. Bronchiolo-alveolar carcinomas of the canine lung following plutonium particle inhalation. Hlth Phys. **12**, 609–613 (1966).

COPP, D. H., AXELROD, D. J., HAMILTON, J. G.: The deposition of radioactive metals in bone as a potential health hazard. Amer. J. Roentgenol. **58**, 10–16 (1947).

CROWLEY, J., LANZ, H., SCOTT, K., HAMILTON, J. G.: A comparison of the metabolism of plutonium in man and the rat. Argonne National Laboratory Report CH-3589 (1946).

DEERING, R. A., RICE, R.: Heavy ion irradiation of HeLa cells. Radiat. Res. **17**, 774–786 (1962).

DOBSON, E. L., GOFMAN, J. W., JONES, H. B., KELLY, L. S., WALKER, L. A.: Studies with colloids containing radioisotopes of yttrium, zirconium, columbium and lanthanum. J. Lab. clin. Med. **34**, 305–312 (1949).

DOLPHIN, G. W., JACKSON, S., LISTER, B. A. J.: Interpretation of bioassay data. In: Assessment of Radioactivity in man, p. 329–354. Vienna: International Atomic Energy Agency 1964.

DOUGHERTY, J. H.: Haematological response to internal irradiation in the beagle. In: Some aspects of internal irradiation (T. F. DOUGHERTY, ed.), Symposium, Heber, Utah 1961, p. 79–93. Oxford: Pergamon Press 1962.

DOUGHERTY, J. H., BECK, B., LESCHKE, U., NICOLAYEV, H.: Haematology report. In: Research in Radiobiology Annual Report of Work in Progress on the Chronic Toxicity Program, USAEC Report COO-220, University of Utah, p. 80–95 (1960).

DOUGHERTY, J. H., ROSENBLATT, L. S.: Leukocyte depression in beagles injected with ^{226}Ra or ^{239}Pu. In: Delayed effects of bone-seeking radionuclides (ed. MAYS, C. W., W. S. S. JEE, R. D. LLOYD, B. J. STOVER, J. H. DOUGHERTY and G. N. TAYLOR), p. 457–470. Utah: University of Utah Press 1969.

DOUGHERTY, T. F., MAYS, C. W.: Bone cancer induced by internally-deposited emitters in beagles. In: Radiation induced cancer. International Atomic Energy Agency, Vienna, p. 361–367 (1969).

DUNNILL, M. S., ANDERSON, J. A., WHITEHEAD, R.: Quantitative histological studies on age changes in bone. J. Path. Bact. **94**, 275–291 (1967).

DURBIN, P. W.: Plutonium in man: A twenty-five year review. UCRL 20850, (1971).

EVANS, R. D., KEANE, A. T., KOLENKOW, R. J., NEAL, W. R., SHANAHAN, M. M.: Radiogenic tumours in the radium and mesothorium cases studied at M.I.T. In: Delayed effects of bone-seeking radionuclides (ed. MAYS, C. W., W. S. S. JEE, R. D. LLOYD, B. J. STOVER, J. H. DOUGHERTY and G. N. TAYLOR), p. 157–191. Utah: University of Utah Press 1969.

FABRIKANT, J., SMITH, C. L. D.: Radiographic changes following the administration of bone seeking radionuclides. Brit. J. Radiol. **37**, 53–62 (1964).

Finch, C. A., Hegsted, M., Kinney, T. D., Thomas, E. D., Rath, C. E., Haskins, D., Finch, S., Fluharty, R. G.: Iron metabolism. The pathophysiology of iron storage. Blood **5**, 983–1008 (1950).
Finkel, A. J., Miller, C. E., Hasterlik, R. J.: Radium-induced malignant tumours in man. In: Delayed effects of bone-seeking radionuclides (ed. Mays, C. W., W. S. S. Jee, R. D. Lloyd, B. J. Stover, J. H. Dougherty and G. N. Taylor), p. 195–224. Utah: University of Utah Press 1969.
Finkel, M. P.: Relative biological effectiveness of radium and other alpha emitters in CFl female mice. Proc. Soc. exp. Biol. (N.Y.) **83**, 494–498 (1953).
Finkel, M. P.: Relative biological effectiveness of internal emitters. Radiology **67**, 665–672 (1956).
Finkel, M. P.: Late effects of internally deposited radioisotopes in laboratory animals. Radiat. Res., Suppl. **1**, 265–279 (1959).
Finkel, M. P., Biskis, B. O.: Toxicity of plutonium in mice. Hlth Phys. **8**, 565–579 (1962).
Finkel, M. P., Biskis, B. O.: Experimental induction of osteosarcoma. Progr. exp. Tumour Res. **10**, 72–111 (1968). (Homburger, F., ed. Basel: S. Karger.
Finkel, M. P., Biskis, B. O., Jinkins, P. B.: Toxicity of ^{226}Ra in mice. In: Radiation induced cancer. Vienna: International Atomic Energy Agency 1969.
Fleischer, R. L., Price, P. M., Walker, R. M.: Solid state physics: Applications to nuclear science and geophysics. Ann. Rev. nuclear Sci. **15**, 1–28 (1965).
Foreman, H.: Studies on the mechanism of plutonium uptake by bone. Hlth Phys. **8**, 713–716 (1962).
Foreman, H., Finnigan, C., Lushbaugh, C. C.: Nephrotoxic hazard from uncontrolled edathamil calcium-disodium therapy. J. Amer. med. Ass. **160**, 1042–1046 (1956).
Foreman, H., Fuqua, P. A., Norwood, W. D.: Experimental administration of ethylene diamine tetraacetic acid in plutonium poisoning. Los Alamos Scientific Laboratory Report AECU 2923 (1954).
Foreman, H., Fuqua, P. A. Norwood, W. D.: Experimental administration of ethylenediamine-tetra-acetic acid in plutonium poisoning. Arch. industr. Hyg. **10**, 226–231 (1954).
Foreman, H., Hamilton, J. G.: The use of chelating agents for accelerating excretion of radioelements. USAEC Report UCRL 1351, Radiation Lab., University of California, June (1961).
Foreman, H., Moss, W., Eustler, B. C.: Clinical experience with radioactive materials. Amer. J. Roentgenol. **79**, 1071–1079 (1958).
Foreman, H., Moss, W., Langham, W.: Plutonium accumulation from long term occupational exposure. Hlth Phys. **2**, 326–333 (1960).
Foreman, H., Post, J., Finnegan, C.: The effect of x irradiation on the absorption of plutonium in the gastrointestinal tract. Radiat. Res. **7**, 267–269 (1957).
Foreman, H., Trujillo, T. T., Johnson, O., Finnegan, C.: Ca EDTA and excretion of plutonium. Proc. Soc. exp. Biol. (N.Y.) **89**, 339–342 (1955).
Fried, J. F., Graul, E. H., Schubert, J., Westfall, W. M.: Superior chelating agents for the treatment of plutonium poisoning. Atompraxis **5**, 1–5 (1959).
Friedenstein, A. J.: Induction of bone tissue by transitional epithelium. Clin. Orthop. **59**, 21–37 (1968).
Friedenstein, A. J., Petrakova, K. V., Kurolesova, A. J., Frolova, G. P.: Heterotopic transplants of bone marrow. Analysis of precursor cells for osteogenic and haematopoietic tissues. Transplantation **6**, 230–247 (1968).
Gale, E., Torrance, J., Bothwell, T.: The quantitative estimation of total iron stores in human bone marrow. J. clin. Invest. **42**, 1076–1082 (1963).
Goldman, M., Della Rosa, R. J., McKelvie, D. H.: Metabolic, dosimetric and pathological consequences in the skeleton of beagles fed ^{90}Sr. In: Delayed effects of bone-seeking radionuclides (ed. Mays, C. W., W. S. S. Jee, R. D. Lloyd, B. J. Stover, J. H. Dougherty and G. N. Taylor, p. 61–77. Utah: University of Utah Press 1969.
Hackett, P. L.: Effect of repeated zirconium citrate injections on distribution of plutonium in the rat. Proc. Soc. exp. Biol. (N.Y.) **83**, 710–712 (1953).
Hamilton, E. I.: The registration of charged particles in solids. An alternative to autoradiography in the life sciences. Int. J. appl. Radiat. Isotopes **19**, 159–161 (1968).
Hamilton, J. G.: The metabolism of the fission products and the heaviest elements. Radiology **49**, 325–343 (1947).
Hamilton, J. G., Scott, K. G.: Effect of calcium salt of versene upon metabolism of plutonium in the rat. Proc. Soc. exp. Biol. (N.Y.) **83**, 301–305 (1953).
Hammond, S. E.: Plutonium excretion following acute accidental exposure. In: Proceedings of the Fourth Annual Meeting on Bioassay and Analytical Chemistry, USAEC Report, Washington 1023, University of Rochester (1959).

HEALY, J. W.: Estimation of plutonium lung burden by urine analysis. Amer. industr. Hyg. Ass. Quart. **18**, 261–266 (1957).
HELLER, M.: Bone. In: Histopathology of irradiation from external and internal sources (ed. W. BLOOM). National Nuclear Energy Series, p. 70–103. New York: McGraw Hill Book Co. 1948.
HERRING, G. M.: Studies on the protein bound chondroitin sulphate of bovine cortical bone. Biochem. J. **107**, 41–49 (1968).
HERRING, G. M.: Mucosubstances and ion binding in bone. Bibl. Nutr. Dieta (Basel) **13**, 147–154 (1969).
HERRING, G. M., VAUGHAN, J., WILLIAMSON, M.: Preliminary report on the site of localisation and possible binding agents for yttrium, americium and plutonium in cortical bone. Hlth Phys. **8**, 717–724 (1962).
HOLLOWAY, M. G., LIVINGSTONE, M. S.: Range and specific ionization of alpha particles. Physiol. Rev. **54**, 18–37 (1938).
HORSTMAN, V. G., PERSING, R. L., BUSTAD, L. K.: Effects of intradermal injection of plutonium in swine. In: Hanford Biology Research Annual Report for 1960, USAEC Report HW-69500, Hanford Atomic Products Operation, General Electric Co., Jan. 1961, p. 64–66.
HULSE, E. V., MOLE, R. H.: Skin tumour incidence in CBA mice given fractionated exposures to low energy beta particles. Br. J. Cancer **23**, 452–463 (1969).
International Commission on Radiological Protection Committee 2: Report on permissible doses for internal radiation. Hlth Phys. **3**, 1–380 (1960).
International Commission on Radiological Protection: A review of the radiosensitivity of the tissues in bone. ICRP Publication 11 (1968). Oxford: Pergamon Press.
International Commission on Radiological Protection: Metabolism of plutonium and related elements and their compounds. Report of a Task Group of ICRP in preparation (1972). Oxford: Pergamon Press.
International Commission on Radiation units and Measurements: Linear Energy Transfer. ICRU Report 16 (1970).
JAFFE, H. L.: Tumours and tumourous conditions of the bone and joints, p. 479–501 (1958). London: Henry Kimpton.
JAMES, A. C.: Ph. D. Thesis, London University (1970).
JAMES, A. C., KEMBER, N. F.: Alpha particle incidence in small targets. Phys. Med. Biol. **15**, 39–46 (1970).
JEE, W. S. S.: A critical survey of the analysis of microscopic distribution of some bone-seeking radionuclides and assessment of absorbed dose. In: Assessment of radioactivity in man, p. 369–393. Vienna: International Atomic Energy Agency 1964.
JEE, W. S. S.: Personal communication (1970).
JEE, W. S. S., ARNOLD, J. S.: Effect of internally deposited radioisotopes upon blood vessels of cortical bones. Proc. Soc. exp. Biol. (N.Y.) **105**, 351–356 (1960).
JEE, W. S. S., ARNOLD, J. S.: The failure of plutonium to deposit in the osteoid of rachitic rats. Hlth Phys. 8, 709–711 (1962).
JEE, W. S. S., ARNOLD, J. S., COCHRAN, T. H., TWENTE, J. A., MICAL, R. S.: Relationship of microdistribution of alpha particles to damage. In: Some aspects of internal irradiation (T. F. DOUGHERTY, ed.), Symposium, Heber, Utah 1961, p. 27–45. Oxford: Pergamon Press 1962.
JEE, W. S. S., BARTLEY, M. H., DOCKUM, N. L., YEE, J., KENNER, G. H.: Vascular changes in bones following bone-seeking radionuclides. In: Delayed effects of bone-seeking radionuclides (ed. MAYS, C. W., W. S. S. JEE, R. D. LLOYD, B. J. STOVER, J. H. DOUGHERTY and G. N. TAYLOR), p. 437–454. Utah: University of Utah Press 1969.
JEE, W. S. S., MILLER, L. G., DELL, R. B.: High resolution neutron induced auto-radiography of bone containing ^{239}Pu. Research in Radiobiology, Report COO 119, 240–249 (1971).
JEE, W. S. S., PARK, H. Z., BURGGRAAF, R.: Estimates of residence time of ^{239}Pu in trabecular bones of beagles. Research in Radiobiology Report COO-119, 240, p. 188–198 (1969).
JEE, W. S. S., STOVER, B. J., TAYLOR, G. N., CHRISTENSEN, W. R.: The skeletal toxicity of ^{239}Pu in adult beagles. Hlth Phys. 8, 599–607 (1962).
JOFFE, M. H., TEMPLE, L. A.: The effect of zirconium citrate on the distribution of intravenously administered plutonium in the dog. 1) Preliminary observations. In: Biology Research Annual Report 1952, USAEC Report HW-28636, Hanford Atomic Products Operation, General Electric Co., p. 92–97 (1953).
JOHNSON, D. R., BOYETT, R. H., BECKER, K.: Sensitive automatic counting of alpha particle tracks in polymers and its applications to dentistry. Hlth Phys. **18**, 424–426 (1970).
JOWSEY, J.: The use of the milling machine for preparing bone sections for microradiography and microautoradiography. J. scient. Instrum. **32**, 159–163 (1955).

JOWSEY, J., KELLY, P. J., RIGGS, B. L., BIANCO, A. J., Jr., SCHOLZ, D. A., GERSHON-COHEN, J.: Quantitative microradiographic studies of normal and osteoporotic bone. J. Bone Jt Surg. **47** A, 785–806 (1965).

JOWSEY, J., LAFFERTY, W., RABINOWITZ, J.: Analysis of distribution of ^{45}Ca in dog bone by a quantitative autoradiographic method. J. Bone Jt Surg. **47** A, 359–370 (1965).

KATZ, J.: Low level chronic plutonium absorption and deposition in the rat. In: Hanford Biology Research Annual Report for 1952, USAEC Report HW 25021, Hanford Atomic Products Operation, General Electric Co. (1953).

KATZ, J., KORNBERG, H. A.: Absorption and deposition in the skeleton and soft tissues of the rat of plutonium fed chronically as solutions of very low mass concentration. In: Hanford Biology Research Annual Report 1952, USAEC HW 28636, Hanford Atomic Products Operation, General Electric Co., Jan. (1953).

KATZ, J., KORNBERG, H. A., PARKER, H. M.: Absorption of plutonium fed chronically to rats. Amer. J. Roentgenol. **73**, 303–308 (1955).

KATZ, J., WEEKS, M. H.: Absorption of plutonium from the gastrointestinal tract of rats: Effect of plutonium concentration. In: Biology Research Annual Report 1953, USAEC HW 30437, Hanford Atomic Products Operation, General Electric Co., Jan., p. 106–108 (1954).

KATZ, J., WEEKS, M. H., OAKLEY, W. D.: Relative effectiveness of various agents for preventing the internal deposition of plutonium in the rat. In: Hanford Biology Research Annual Report for 1953, USAEC Report HW 30437, Hanford Atomic Products Operation, General Electric Co. (1954).

KATZ, J., WEEKS, M. H., THOMPSON, R. C.: Gastrointestinal absorption of plutonium. ii. Effect of plutonium concentration in solution fed. In: Hanford Biology Research Annual Report for 1953, USAEC Report HW 30220, Hanford Atomic Products Operation, General Electric Co. (1954).

KEBERLE, H.: The biochemistry of desferrioxamine and its relation to iron metabolism. Ann. N.Y. Acad. Sci. **119**, 758–768 (1964).

KEMBER, N. F.: Personal communication (1970).

KHODYREVA, M. A.: Penetration of ^{239}Pu through the skin. Med. Rad. **10**, Oct. 42–46 (1965) (from National Lending Library, Science and Technology RTS 5223).

KISIELESKI, W. E., WOODRUFF, L.: Studies of the distribution of plutonium in the rat. Argonne National Laboratory, Biology Division, Quarterly Report, Aug. to Nov., 1947, ANL 4108, p. 86–103 (1948). Also issued as a separate report, AECD 2009 F.

KORNBERG, H. A.: Proceedings of the Second Annual Conference on Plutonium, Radium and Mesothorium (ed. C. N. STOVER), Salt Lake City, p. 249–254 (1954).

LAGERQUIST, C. R., BOKOWSKI, D. L., HAMMOND, S. E., HYLTON, D. B.: Plutonium content of several internal organs following occupational exposure. In: Proceedings of the 13th Annual Bio-Assay and Analytical Chemistry Meeting held at Berkeley, California on October 12–13, 1967. Conf 671048, p. 103 (1968) (quoted MAYS et al., 1969).

LAGERQUIST, C. R., HAMMOND, S. E., PUTZIER, E. A., PILTINGSRUD, C. W.: Effectiveness of early DTPA treatments in two types of plutonium exposures in humans. Hlth Phys. **11**, 1177–1180 (1965).

LANGHAM, W. H.: The application of excretion analysis to the determination of body burden of radioactive isotopes. Brit. J. Radiol., Suppl. **7**, 95–113 (1957).

LANGHAM, W. H.: Physiology and toxicology of plutonium-239 and its industrial medical control. Hlth Phys. **2**, 172–185 (1959).

LANGHAM, W. H., BASSETT, S. H., HARRIS, P. S., CARTER, R. E.: Distribution and excretion of plutonium administered intravenously in man. Los Alamos Scientific Laboratory Report LA 1151 (1950).

LANGHAM, W. H., CARTER, R. E., BUSCH, E., JOHNSON, O., STRANG, V., TRUJILLO, I., VIER, M., WELLNITZ, J., WILLIAMS, A.: The relative physiological and toxicological properties of americium and plutonium. Los Alamos Scientific Laboratory Report LA-1309, Nov. 13, 1951.

LANGHAM, W. H., LAWRENCE, J. N. P., MCCLELLAND, J., HEMPELMANN, L. H.: The Los Alamos Scientific Laboratory experience with plutonium in man. Hlth Phys. 8, 753–760 (1962).

LARSON, H. V., NEWTON, C. E., BAUMGARTNER, W. V., HEID, K. R., CROOK, G. H.: The management of an extensive plutonium wound and the evaluation of the residual internal deposition of plutonium. Phys. in Med. Biol. **13**, 45–53 (1968).

LEDERER, C. M., HOLLANDER, J. M., PERLMAN, I.: Table of isotopes, 6th ed. New York: J. Wiley & Sons 1967.

LINDENBAUM, A., LUND, C. J.: Alpha counting by liquid scintillation spectrometry: Plutonium 239 in animal tissues. Radiat. Res. **37**, 131–140 (1969).

LINDENBAUM, A., LUND, C. J., SMOLER, M., ROSENTHAL, M, W,: Preparation, characterization and distribution in mouse tissues of graded polymeric and monomeric plutonium. Radiochemical and autoradiographic studies. In: Diagnosis and treatment of deposited radionuclides. Proceedings of Symposium, Richland 1967 (ed. H. A. KORNBERG and W. D. NORWOOD), Excerpta Medica Foundation, Monograph on Nuclear Medicine No 2, p. 56–64 (1968).

LINDENBAUM, A., ROSENTHAL, M. W., RUSSELL, J. J., MORETTI, E. S., SMYTH, M. A.: Metabolic and therapeutic studies of plutonium. V. Argonne National Laboratory Report ANL 7635, p. 186–190 (1969).

LISCO, H., FINKEL, M. P., BRUES, A. M.: Carcinogenic properties of radioactive fission products and of plutonium. Radiology **49**, 361–363 (1947).

LISCO, H., KISIELESKI, W. E.: The fate and pathological effects of plutonium metal implanted into rabbits and rats. Amer. J. Path. **29**, 305–321 (1953).

LISCO, H., ROSENTHAL, M., VAUGHAN, J.: Personal communication (1970).

LISTER, B. A. J., MORGAN, A., SHERWOOD, R. J.: Excretion of plutonium following accidental skin contamination. Hlth Phys. **9**, 803–815 (1963).

LOUTIT, J. F.: Grafts of haemopoietic tissue: the nature of haemopoietic stem cells. J. clin. Path. **20**, Suppl. 2, 535–539 (1967).

LOUTIT, J. F., VAUGHAN, J. M.: The radiosensitive tissues in bone. Brit. J. Radiol. **44**, 815 (1971).

MACPHERSON, S., OWEN, M., VAUGHAN, J.: The relation of radiation dose to radiation damage in the tibia of weanling rabbits injected with ^{90}Sr. Brit. J. Radiol. **35**, 221–234 (1962).

MAGNO, P. J., KAUFFMAN, P. E., GROULX, P. R.: Radiol. Hlth Data Rep. **10**, 47 (quoted MAYS et al., 1969).

MARKLEY, J. F., ROSENTHAL, M. W., LINDENBAUM, A.: Distribution and removal of monomeric and polymeric plutonium in rats and mice. Int. J. Radiat. Biol. **8**, 271–278 (1964).

MARSHALL, J. H.: The retention of radionuclides in bone. In: Delayed effects of bone-seeking radionuclides (ed. MAYS, C. W., W. S. S. JEE, R. D. LLOYD, B. J. STOVER, J. H. DOUGHERTY and G. N. TAYLOR), p. 7–27. Utah: University of Utah Press 1969.

MAYS, C. W.: Personal communication (1971).

MAYS, C. W., TAYLOR, G. N., JEE, W. S. S., DOUGHERTY, T. F.: Speculated risk to bone and liver from ^{239}Pu. Hlth Phys. **19**, 601–610 (1970).

MAYS, C. W., TUELLER, A. B.: Determination of localised alpha dose in soft tissue near radioactive bone. Research in Radiobiology, COO-119-229, p. 199–205 (1964).

METCALF, R. G., CASARETT, G., BOYD, G. A.: Pathology studies on rats injected with polonium, plutonium and radium. In: Biological studies with polonium, radium and plutonium (ed. R. M. FINK), National Nuclear Energy Series, p. 257–282. New York: McGraw Hill 1950.

MIDDLESWORTH, L. VAN: Study of plutonium metabolism in bone. USAEC Report MDDC-1022, University of California, 1947.

MOLE, R. H.: Cancer production by chronic exposure to penetrating gamma irradiation. National Cancer Inst. Monogr. **14**, 271–289 (1964).

MOLE, R. H.: Bone tumour production in mice by strontium 90: Further experimental support for a two-event hypothesis. Brit. J. Cancer **17**, 524–531 (1963).

MORITZ, A. R., HENRIQUES, F. C., JR.: Studies of thermal injury. ii. The relative importance of time and surface temperature in the causation of cutaneous burns. Amer. J. Path. **23**, 695–720 (1947).

MORROW, P. E., GIBB, F. R., DAVIES, H., MITOLA, J., WOOD, D., WRAIGHT, N., CAMPBELL, H. S.: The retention and fate of inhaled plutonium dioxide in dogs. Hlth Phys. **13**, 113–133 (1967).

MOSKALEV, Y. I., BULDAKOV, L. A., KOSHURNIKOVA, N. A., NIFATOV, A. P., RESKETOV, G. N.: Combined effect of ^{90}Sr, ^{144}Ce and ^{239}Pu on the rat organism. (Report 1) 441–452; (Report 2) 453–462. In: Distribution and biological effects of radioactive isotopes by MOSKALEV, Y. I., Atomizdat, Moscow. United States Atomic Energy Commission, Division of Technical Information 1968.

MOSKALEV, Y. I., STRELTSOVA, V. N., BULDAKOV, L. A.: Late effects of radionuclide damage. In: Delayed effects of bone-seeking radionuclides (ed. MAYS, C. W., W. S. S. JEE, R. D. LLOYD, B. J. STOVER, J. H. DOUGHERTY and G. N. TAYLOR), p. 589–609. Utah: University of Utah Press 1969.

NEARY, G. J.: Track structure in relation to radiobiology. In: Charged particle tracks in solids and liquids. Proc. Second L. H. GRAY Conference, Cambridge 1969 (ed. G. E. ADAMS, D. K. BEWLEY, J. W. BOAG) (London Inst. of Physics and the Physical Society), 1970.

NILSSON, A.: ^{90}Sr induced bone and bone marrow changes. Uppsala 1962.

NILSSON, A.: Pathologic effects of different doses of radiostrontium in mice. Dose effect relationship in ^{90}Sr induced bone tumours. Acta radiol. (Stockh.) **9**, 155–176 (1970).

NORWOOD, W. D.: DTPA effectiveness in removing internally deposited plutonium from humans. J. occup. Med. **2**, 371–376 (1960).
NORWOOD, W. D.: Long term administration of DTPA for plutonium elimination. A follow up study in one patient. J. occup. Med. **4**, 130–132 (1962a).
NORWOOD, W. D.: Early diagnosis and treatment of individuals who have excessive depositions of radioisotopes. J. occup. Med. **4**, 373–382 (1962b).
NORWOOD, W. D.: Therapeutic removal of plutonium in humans. Hlth Phys. **8**, 747–750 (1962c).
NORWOOD, W. D., FUQUA, P. A., WILSON, R. H., HEALY, J. W.: Case studies. Treatment of plutonium inhalation. In: Medical sciences. Progress in nuclear energy series VII, vol. 2 (J. C. BUGHER, J. COURSAGET and J. F. LOUTIT, eds.), p. 105–113. New York: Pergamon Press 1959.
OAKLEY, W. D., THOMPSON, R. C.: Further studies on percutaneous absorption and decontamination of plutonium in rats. In: Biology Research-Annual Report 1955, USAEC Report HW-41500, Hanford Atomic Products Operation, General Electric Co., Feb. (1956).
OWEN, M.: The origin of bone cells. Int. Rev. Cytol. **28**, 213–238 (1970).
PAINTER, E., RUSSELL, E., PROSSER, C. L., SWIFT, M. N., KISIELESKI, W., SACHER, G.: Clinical physiology of dogs injected with plutonium. USAEC Report AECD-2042, University of Chicago, June 1946.
PARK, J. F., BAIR, W. J., BUSCH, R. H.: Progress in beagle dog studies with transuranium elements at Battelle North West. Hlth Phys. **22**, 803–810 (1972).
PARK, J. F., BAIR, W. J., HOWARD, E. B., CLARKE, W. J.: Chronic effects of inhaled $^{239}PuO_2$ in beagles. BNWL 1050, Part 1, p. 3.3–3.5 (1970a).
PARK, J. F., HOWARD, E. B., BAIR, W. J.: Acute toxicity of inhaled $^{238}PuO_2$ in beagles. BNWL 1050, Part 1, p. 3.6–3.13 (1970b).
PEARSE, A. G. E.: Histochemistry theoretical and applied. London: J. & A. Churchill Ltd. 1960.
PERRY, H. M., JR., SCHROEDER, H. A.: Lesions resembling vitamin B complex deficiency and urinary loss of zinc produced by ethylene diamine tetraacetic. Amer. J. Med. **22**, 168–172 (1957).
PRITCHARD, J. J.: The osteoblast. In: The Biochemistry and physiology of bone (ed. BOURNE, G. H.), p. 179–212. New York: Academic Press 1956.
Radiological Health Data Report **10**, **47** (quoted MAYS, et al., 1969).
RIGGS, B. L., JR., BASSINGTHWAITE, J. B., JOWSEY, J., PEQUEGNAT, E. P.: Autoradiographic method for quantitation, deposition and distribution of radiocalcium in bone. J. Lab. clin. Med. **75**, 520–528 (1970).
ROGERS, A. W.: Techniques of autoradiography. Amsterdam: Elsevier Publishing Co. 1967.
ROSENTHAL, M. W., LINDENBAUM, A.: Effect of desferrioxamine-B-methane sulphonate (DFOM) on removal of plutonium in vitro and in vivo. Proc. Soc. exp. Biol. (N.Y.) **117**, 749–750 (1964).
ROSENTHAL, M. W., LINDENBAUM, A.: Influence of DTPA therapy on long-term effects of retained monomeric plutonium: comparison with polymeric plutonium. Radiat. Res. **31**, 506–521 (1967).
ROSENTHAL, M. W., LINDENBAUM, A.: Osteosarcomas as related to tissue distribution of monomeric and polymeric plutonium in mice. In: Delayed effects of bone-seeking radionuclides (ed. MAYS, C. W., W. S. S. JEE, R. D. LLOYD, B. J. STOVER, J. H. DOUGHERTY and G. N. TAYLOR), p. 371–384. Utah: University of Utah Press 1969.
ROSENTHAL, M. W., LINDENBAUM, A., FRITZ, T. E., REHFELD, C. E.: Haematopoietic deaths after deposition of plutonium in bone marrow. Proc. IVth Int. Cong. Radiat. Res., in the press (1970).
ROSENTHAL, M. W., MARKLEY, J. F., LINDENBAUM, A., SCHUBERT, J.: Influence of DTPA therapy on long-term effects of retained plutonium. Hlth Phys. **8**, 741–745 (1962).
ROSENTHAL, M. W., MARSHALL, J. H., LINDENBAUM, A.: Autoradiographic and radiochemical studies of the effect of colloidal state of intravenously injected plutonium on its distribution in bone and marrow. In: Diagnosis and treatment of deposited radionuclides. Proceedings of Symposium, Richland 1967 (ed. H. A. KORNBERG and W. D. NORWOOD), Excerpta Medica Foundation, Monograph on Nuclear Medicine 2, p. 73–80 (1968).
ROSENTHAL, M. W., RUSSELL, J. J., MORETTI, E. S., LINDENBAUM, A.: Effective dose of DTPA, spaced at 3 day intervals in removal of skeletal plutonium. Hlth Phys. **16**, 806–807 (1969).
ROSENTHAL, M. W., SCHUBERT, J.: Kinetics of body distribution of plutonium as influenced by zirconium. Radiat. Res. **6**, 349–354 (1957).
ROSSI, H. H.: Energy distribution in the absorption of radiation. In: Advanc. biol. med. Phys. **2**, 27–85 (1967).
ROTONDI, E.: Energy loss of alpha particles in tissue. Radiat. Res. **33**, 1–9 (1968).

RUSSELL, E. R., NICKSON, J. J.: The distribution and excretion of plutonium in two human subjects. University of Chicago, Metallurgical Laboratory Report, Argonne National Laboratory Report CH 3607, p. 1–18 (1946).

RYSINA, T. N., EROKHIN, R. A.: The distribution and elimination of plutonium in dogs in the long term periods after introduction. AEC-tr-5265, p. 117–126 (1961). Translation from a publication of the State Publishing House of Literature in the Field of Atomic Science and Technology Moscow. Biological effects of radioactive isotope distribution (ed. A. V. LEBEDINSKI and MOSKALEV, Y. I.).

RYSINA, T. N., YEROKHIN, R. A.: Distribution and excretion of plutonium in dogs at long intervals following administration, p. 8–18. In: Plutonium-239. Its distribution, biological effect and accelerated elimination. FTD-TT 63-559 AD 430440. Gosudarstvennoye Izdatel'stvo Meditsinky Literatury Medgiz-1962-Moskva. U. S. Air Force Command Foreign Technology Division. Wright-Patterson Air Force Base, Ohio.

SANDERS, C. L.: The biological behaviour of $^{239}PuO_2$ particles: role of the peritoneal mononuclear phagocyte. Radiat. Res. **38**, 125–139 (1969).

SANDERS, C. L.: Maintenance of phagocytic function following $^{239}PuO_2$ particle administration. Hlth Phys. **18**, 82–85 (1970).

SANDERS, C. L., ADEE, R. R.: The ultrastructure of mononuclear phagocytes following intraperitoneal administration of $^{239}PuO_2$ particles. J. reticuloendothel. Soc. **6**, 1–23 (1969).

SANDERS, C. L., ADEE, R. R.: Ultrastructural localization of inhaled $^{239}PuO_2$ particles in alveolar epithelium and macrophages. Hlth Phys. **18**, 293–295 (1970).

SANDERS, S. M., JR.: Plutonium excretion. Arch. environmental Hlth **2**, 474–483 (1961).

SCHOFIELD, G. B.: Comparison in the medical management of three cases of plutonium contaminated wounds. In: Handling radiation accidents, p. 163–172. Vienna: International Atomic Energy Agency 1969.

SCHUBERT, J.: Treatment of plutonium poisoning by metal displacement. Science **105**, 389–390 (1947).

SCHUBERT, J.: Removal of radioelements from the mammalian body. A. Rev. nuclear Sci. **5**, 369–412 (1955).

SCHUBERT, J., FINKEL, M. P., WHITE, M. R., HIRSCH, G. M.: Plutonium and yttrium content of the blood, liver and skeleton of the rat at different times after intravenous administration. J. biol. Chem. **182**, 635–642 (1950).

SCHUBERT, J., FRIED, J. F., ROSENTHAL, M. W., LINDENBAUM, A.: Tissue distribution of monomeric and polymeric plutonium as modified by a chelating agent. Radiat. Res. **15**, 220–226 (1961).

SCHUBERT, J., WHITE, M. R.: The effect of different dose levels of zirconium on the excretion and distribution of plutonium and yttrium. J. biol. Chem. **184**, 191–196 (1950).

SCOTT, K. G., AXELROD, D. J., FISHER, H., CROWLEY, J. F., HAMILTON, J. G.: The metabolism of plutonium in rats following intramuscular injection. J. biol. Chem. **176**, 283–293 (1948).

SEGRÈ, E.: Experimental nuclear physics, vol. 1 (ed. E. SEGRÈ). New York: John Wiley & Sons, Inc. 1952.

SEMENOV, D. I., TREGUBENKO, I. P.: The action of chelating compounds on tissue storage and excretion from the living organism of radioyttrium, radiocerium and plutonium. Biochemistry USSR **23**, 55, quoted THOMPSON, R. C. (1967) (1958).

SIKOV, M. R., MAHLUM, D. D.: Plutonium in the developing animal. Hlth Phys. **82**, 707–712 (1972).

SISSONS, H. A.: Tumours of bone. In: Systematic pathology (eds. A. PAYLING WRIGHT and W. ST. CLAIR SYMMONS), p. 1396–1427. London: Longmans 1966.

SISSONS, H. A.: Dimensions of cells covering bone surfaces. Medical Research Council Committee on Protection against Ionizing Radiation. PIRC/PL/70/4 (1970).

SISSONS, H. A., HOLLEY, K. J., HEIGHWAY, J.: Normal bone structure in relation to osteomalacia. In: L'Osteomalacie, Tours 1965 (ed. HIOCO, D. J.), p. 19–37. Paris: Masson & Cie. 1967.

SMITH, R. S.: Chelating agents in the diagnosis and treatment of iron overload in thalassemia. Ann. N.Y. Acad. Sci. **119**, 776–788 (1964–1965).

SMITH, V. H.: Removal of internally deposited plutonium. Nature (Lond.) **181**, 1792 (1958).

SMITH, V. H.: Prevention of plutonium deposition by desferrioxamine-B. Nature (Lond.) **204**, 899–900 (1964).

SMITH, V. H.: Removal of internally deposited radionuclides. BNWL 280, p. 81–285 (1966).

SMITH, V. H., BALLOU, J. E., CLARKE, W. J., THOMPSON, R. C.: Effectiveness of DTPA in removing plutonium from the pig. Proc. Soc. exp. Biol. (N.Y.) **107**, 120–125 (1961).

SPIERS, F. W.: Alpha ray dosage in bone containing radium. Brit. J. Radiol. **26**, 296–301 (1953).

STEVENS, W., BRUENGER, E. W., STOVER, B. J.: In vivo studies on the interactions of PuIV with blood constituents. Radiat. Res. **33**, 490–500 (1968).

STOVER, B. J., ATHERTON, D. R., BRUENGER, F. W., BUSTER, D. S.: Further studies of the metabolism of ^{239}Pu in adult beagles. Hlth Phys. **8**, 589–597 (1962).

STOVER, B. J., ATHERTON, D. R., KELLER, N.: Metabolism of Pu^{239} in adult beagle dogs. Radiat. Res. **10**, 130–147 (1959).

STOVER, B. J., JEE, W. S. S.: Some effects of alpha irradiation on the composition and structure of bone. Hlth Phys. **9**, 267–275 (1963).

STUART, B. O.: Comparative distribution of ^{238}Pu and ^{239}Pu in rats following inhalation of the oxide. BNWL 1050, part 1, p. 3.19–3.20 (1970).

SUTER, G. M., BOYD, G. A.: Haematology studies of rats injected with radium and plutonium. In: Biological studies with polonium, radium and plutonium (ed. R. M. FINK), National Nuclear Energy Series, p. 282–290. New York: McGraw Hill 1950.

TAYLOR, D. M.: Some aspects of the comparative metabolism of plutonium and americium in rats. Hlth Phys. 8, 673–677 (1962).

TAYLOR, D. M.: The effects of desferrioxamine on the retention of actinide elements in the rat. Hlth Phys. **13**, 135–140 (1967).

TAYLOR, D. M.: The metabolism of plutonium in adult rabbits. Brit. J. Radiol. **42**, 44–50 (1969).

TAYLOR, D. M.: Personal communication (1970).

TAYLOR, D. M., BENSTED, J. P. M.: Long term biological damage from plutonium 239 and americium 241 in rats. In: Delayed effects of bone-seeking radionuclides (ed. MAYS, C. W., W. S. S. JEE, R. D. LLOYD, B. J. STOVER, J. H. DOUGHERTY and G. N. TAYLOR), p. 357–370. Utah: University of Utah Press 1969.

TAYLOR, D. M., CHIPPERFIELD, A. R.: The comparative binding of plutonium and americium to bone proteins. Medical Research Council Committee Protection against Ionizing Radiations PIRC/PL/70/3 (1970a).

TAYLOR, D. M., CHIPPERFIELD, A. R.: The mode of fixation of plutonium 239 and americium 241 in bone: a possible explanation of their different carcinogenicity, p. 215–217, ed. A. M. JELLIFFE and B. STRICKLAND. Symposium Ossium 1968. London: E. & S. Livingstone 1970b.

TAYLOR, D. M., CHIPPERFIELD, A. R.: The binding of transplutonium elements to proteins of bone in Proceedings of the Seminar on Radiation Protection Problems relating to Transuranic elements. EUR 4612 d-f-e, p. 187–204. Luxembourg 1971.

TAYLOR, D. M., SOWBY, F. D.: The removal of americium and plutonium from the rat by chelating agents. Phys. in Med. Biol. **7**, 83–90 (1962–1963).

TAYLOR, D. M., SOWBY, F. D., KEMBER, N. F.: The metabolism of americium and plutonium in the rat. Phys. in Med. Biol. **6**, 73–86 (1961).

TAYLOR, G. N., CHRISTENSEN, W. R., JEE, W. S. S., REHFELD, C. E., FISHER, W.: Anatomical distribution of fractures in beagles injected with ^{239}Pu. Hlth Phys. 8, 609–613 (1962).

TAYLOR, G. N., DOUGHERTY, T. F., SHABESTARI, L., DOUGHERTY, J. H.: Soft tissue tumors in internally irradiated beagles. In: Delayed effects of bone-seeking radionuclides (ed. MAYS, C. W., W. S. S. JEE, R. D. LLOYD, B. J. STOVER, J. H. DOUGHERTY and G. N. TAYLOR), p. 323–335. Utah: University of Utah Press 1969.

THOMPSON, R. C.: Research at Hanford on therapy for internally deposited plutonium. In: Therapy of Radioelement Poisoning, USAEC Report ANL 5584, Argonne National Laboratory, August 1956.

THOMPSON, R. C.: Biological Factors. In: Plutonium handbook, vol. 2 (ed. O. J. WICK), p. 785–829. New York: Gordon & Breach 1967.

TODD, P.: Fractionated heavy ion irradiation of cultured human cells. Radiat. Res. **34**, 378–389 (1968).

TODD, R., LOGAN, R.: Plutonium-237 labelling. A new technique for plutonium metabolic studies in animals. Int. J. appl. Radiat. Isotopes **17**, 253–255 (1966).

TSEVELEVA, I. A.: The plutonium content in the protein fraction of rat tubular bone. Biochemistry **25**, 636–639 (1960).

TURNER, G. A.: Studies on the transport of plutonium and other actinide elements in the blood. Ph. D. Thesis, University of London (1967).

TURNER, G. A., TAYLOR, D. M.: The transport of plutonium, americium and curium in the blood of rats. Phys. in Med. Biol. **13**, 535–546 (1968).

TURNER, R. C., MAYNEORD, W. V., RADLEY, J. M.: The alpha ray activity of human tissues. Brit. J. Radiol. **31**, 397–406 (1958).

TWENTE, J. A., JEE, W. S. S.: The determination of localized concentration of ^{239}Pu in bone. Hlth Phys. **5**, 142–148 (1961).

TWENTYMAN, P. R., BLACKETT, N. M.: Red cell production in the continuously irradiated mouse. Brit. J. Radiol. **43**, 898–902 (1970).

VAUGHAN, J. M.: The physiology of bone, p. 51–60. Oxford: Clarendon Press 1970a.

VAUGHAN, J. M.: Radiation and myeloproliferative disorders in man. In: Symposium, Myeloproliferative Disorders of Animals and Man. Atomic Energy Agency (1970b).

VAUGHAN, J. M.: Note on character of cells on trabecular bone surfaces in adult human vertebrae. Medical Research Council Committee on Protection against Ionizing Radiations. PIRC/PL **40**, 1 (1970c).

VAUGHAN, J. M.: The effects of irradiation on the skeleton:: Clarendon Press 1973.

VAUGHAN, J. M., BLEANEY, B., WILLIAMSON, M.: The uptake of plutonium in bone marrow — a possible leukaemic risk. Brit. J. Haemat. **13**, 492–502 (1967).

VAUGHAN, J. M., TUTT, M. L.: Use of ethylene diamine tetraacetic acid (versene) for removing fission products from the skeleton. Lancet **1953** II, 856–859.

VAUGHAN, J. M., WILLIAMSON, M.: ^{90}Sr in the rabbit. The relative risks of osteosarcoma and squamous cell carcinoma. In: Delayed effects of bone-seeking radionuclides (ed. MAYS, C. W., W. S. S. JEE, R. D. LLOYD, B. J. STOVER, J. H. DOUGHERTY and G. N. TAYLOR), p. 337–353. Utah: University of Utah Press 1969.

WAGER, R. W., TEMPLE, L. A.: Comparison of Ca EDTA with zirconium citrate in promoting excretion of plutonium from the dog. In: Biology Research Annual Report 1953, USAEC Report HW 30437, Hanford Atomic Products Operation, General Electric Co., p. 114–121 (1954).

WALSH, P. J.: Stopping power and range of alpha particles. Hlth Phys. **19**, 312–316 (1970).

WEEKS, M. H., KATZ, J., OAKLEY, W. D., BALLOU, J. E., GEORGE, L. A., BUSTAD, L. K., THOMPSON, R. C., KORNBERG, H. A.: Further studies on the gastrointestinal absorption of plutonium. Radiat. Res. **4**, 339–347 (1956).

WEEKS, M. H., OAKLEY, W. D.: Absorption of plutonium through the living skin of the rat. Rates of penetration and methods of skin contamination. In: Biology Research Annual Report 1952, USAEC Report HW 28636, Hanford Atomic Products Operation, General Electric Co., p. 85–91 (1953).

WEEKS, M. H., OAKLEY, W. D.: Absorption of plutonium through the living skin of the rat. In: Biology Research Annual Report 1953, USAEC Report HW 30437, Hanford Atomic Products Operation, General Electric Co., p. 102–108 (1954a).

WEEKS, M. H., OAKLEY, W. D.: Influence of plutonium concentration on effectiveness of therapeutic agents. In: Biology Research Annual Report 1953, USAEC Report HW 30437, Hanford Atomic Products Operation, General Electric Co., p. 109–113 (1954b).

WEEKS, M. H., OAKLEY, W. D.: Percutaneous absorption of plutonium solutions in rats. In: Biology Research Annual Report 1954, USAEC Report HW 35917, Hanford Atomic Products Operation, General Electric Co., p. 56–63 (1955).

WEEKS, M. H., OAKLEY, W. D., KATZ, J., THOMPSON, R. C.: Absorption of plutonium through skin of the rat: Rates of penetration and methods for skin contamination. In: Biology Research Annual Report 1952, USAEC Report HW 30232, Hanford Atomic Products Operation, General Electric Co., p. 1–20 (1953).

WEEKS, M. H., OAKLEY, W. D., THOMPSON, R. C.: Influence of plutonium concentration on effectiveness of therapeutic agents. Radiat. Res. **2**, 237–239 (1955).

WHITE, M. R., SCHUBERT, J.: The action of salts of zirconium and other metals on plutonium and yttrium distribution and excretion. J. Pharmacol. exp. Ther. **104**, 317–324 (1952).

WICK, O. J.: Report of Accidental Plutonium Deposition. USAEC Report HW 83457, parts 1 and 2, August 7 (1964).

WILLIAMS, P. A., PEACOCKE, A. R.: The physical properties of a glycoprotein from bovine cortical bone. Biochim. biophys. Acta (Amst.) **101**, 327–335 (1965).

WILLIAMS, P. A., PEACOCKE, A. R.: The binding of calcium and yttrium ions to a glycoprotein from bovine cortical bone. Biochem. J. **105**, 1177–1185 (1967).

WILLIAMSON, M.: The sites of deposition of certain radioactive isotopes which concentrate in bone. Thesis, Oxford (1963).

WILLIAMSON, M., VAUGHAN, J.: A preliminary report on the sites of deposition of Y, Am and Pu in cortical bone and in the region of the epiphyseal cartilage plate. In: Bone and Tooth Symposium (ed. H. J. J. BLACKWOOD, p. 71–83. Oxford: Pergamon Press 1964.

WILLIAMSON, M., VAUGHAN, J.: Histochemistry of the mucosaccharides in the epiphyseal plate of young rabbits. Nature (Lond.) **215**, 711–714 (1967).

WILLIS, R. A.: Pathology of tumours, 4th ed., p. 658–659, 682–685. London: Butterworths 1967.

WILSON, R. H.: Decontamination of human skin after immersion in a plutonium solution. USAEC Report HW-42030, Hanford Atomic Products Operation, General Electric Co. (1956).

WILSON, R. H.: A review of significant Hanford plutonium deposition cases. Proc. 4th Annual Meeting on Bioassay and Analytical Chemistry, No 3 and 4, 1958, USAEC Report WASH-1023, University of Rochester, p. 71–83 (1959).

Wilson, R. H., Silker, W. B.: Plutonium contaminated injury case study and associated use of Na_4EDTA as a decontaminating agent. USAEC Report HW 66309, Hanford Atomic Products Operation, General Electric Co. (1960).
Zimmer, K. G.: Studies on quantitative radiation biology. Edinburgh: Oliver and Boyd (1961).
Zirkle, R. E., Bloom, W.: Irradiation of parts of individual cells. Science **117**, 487–493 (1953).
Zndanov, D. Z.: Obschaya anatomiya i fisiologiya limfaticheskoi sistemy. M. L., Medgiz (1952).

Note added in Proof

Since this section on the distribution, excretion and effects of plutonium as a bone seeker was delivered to the editors in 1971 there have been a large number of further publications covering this subject. Only those that alter or greatly amplify the conclusions of the original manuscript are included briefly in this postscript.

II. Metabolism of ^{239}Pu with Special Reference to the Skeleton

Lagerquist and his colleagues have recently reviewed a series of autopsy results on humans occupationally exposed to plutonium (Lagerquist, Hammond, Bokowski and Hylton, 1972). They comment on the fact that distribution of the radionuclide was variable, depending on the mode of entry, the chemical form of the plutonium and the length of time since exposure. There was no typical distribution but on average the tracheal-bronchial lymph nodes were more contamined than the skeleton.

A. Species Differences

Rosenthal and her colleagues (Rosenthal, Lindenbaum, Russell, Moretti and Chladek, 1972) have made a comparison of the metabolism of monomeric and polymeric plutonium in the Dutch belted rabbit and in CFl ANL mice. Mice showed a higher percent of the injected monomeric plutonium in the femurs and a lower percent in the liver than the rabbits. How far this is due to the known bone dyscrasia resulting in a high degree of trabeculation in this strain of mouse (Lisco, Rosenthal and Vaughan, 1970) is not yet clear. The mice also showed a higher uptake of polymeric plutonium in the marrow and a higher uptake of both forms in the spleen and lungs than did the rabbits.

Lloyd and Marshall (1972) have drawn attention to the likelihood of extremely significant differences existing in the toxicity of ^{239}Pu in man and dog because of the great differences in bone structure in the two species. The surface/volume ratio of bone mineral in trabecular bone of the dog is higher by a factor of 2 than the corresponding values in man. The burial rate of surface deposits of ^{239}Pu in a 1.5 year old dog is at least an order of magnitude greater than in an adult man. This burial rate in dogs appears to be critically dependent on age and decreases rapidly after the age of 1.5 years, the age at which most of the toxicity studies have been performed. The percentage of the marrow volume irradiated by alpha-emitting bone surface seekers is higher in the dog—about 25 percent compared with about 10 percent in man.

B. Age Differences

The importance of age in affecting both absorption from the gut and retention in different organs continues to receive experimental confirmation. Sikov and Mahlum (1972b) have shown that in the newborn rat there was an equal fraction of the administered dose of monomeric and polymeric plutonium in the femur at one day after injection. In the foetus exposed to ^{239}Pu at 19 days gestation the plutonium concentration in the foetal femur exceeded that in the femur of the dam, approximating the concentration in the liver of the dam. Extremely important differences in distribution and retention in relation to the size of administered dose and age are discussed in detail in Sec II.G and Sec. IV.A (STOVER, ATHERTON and BUSTER, 1972; JEE, 1972a, b).

C. Routes of Entry with Special Reference to Skeletal Uptake

BALLOU and his colleagues (BALLOU, PARK and MORROW, 1972) have recently made a comparison of the metabolic equivalence of ingested, injected and inhaled ^{239}Pu citrate in beagle dogs which is of some importance since the results all come from the same laboratory. They conclude that it is probably justifiable to use plutonium citrate given intravenously as a model system to describe the metabolism of plutonium absorbed by less direct routes and that plutonium nitrate should also be considered "a soluble" compound in this connotation. Inhalation exposure of course led to a greater relative retention in the lung. The tendency of plutonium to concentrate in lymph nodes emphasises the role of the lymphatic system in normal plutonium clearance, however administered, and the potential for plutonium to concentrate in relatively small volumes. They consider, however, that effects on bone, liver and lung are, at present, more likely to present a radiation hazard. Approximately 0.08% of the oral dose was retained after 3 days, 50% in the skeleton and 25% in the liver. After intravenous administration 32–45% was found in the liver and 29% in the skeleton during the 100 day post injection period. Following inhalation after 100 days, 36% of the initial deposit was retained; 33% of this in the lung, 44% in the skeleton and 17% in the liver.

D. Plasma Transport

DURBIN and her colleagues (DURBIN, HOROVITZ and CLOSE, 1972) have recently presented a kinetic model for plutonium deposition kinetics in the rat. They consider that probably the plutonium transferrin complex is most likely broken down in the red marrow on the surface of the reticulocytes by the same mechanism as breaks down the iron transferrin complex (KATZ, 1970). The Pu atoms so released may a) combine again with other transferrin molecules and recirculate; b) can reenter the circulation as unbound plutonium or c) they can diffuse to the nearest bone surface and be bound there. STOVER and her colleages (STEVENS, ATHERTON, STOVER and BRUENGER, 1972) have shown that the strictly monomeric plutonium transferrin complex is more slowly removed from the plasma than Pu(IV) in citrate buffer. Polymeric plutonium initially leaves the blood at a rapid rate. No protein binding of polymeric plutonium appears to occur. These data suggest that different factors are involved in the removal of monomeric plutonium and plutonium transferrin complex from the circulation than are involved in removal of polymeric plutonium.

F. Excretion

A digital computer programme for the estimation of the body content of plutonium from urine data has recently been published (BEACH, 1973). It gives data on one man over a three year period which illustrates the very great day to day variation in excretion that may occur over a long time period with no obvious reason, and therefore the importance, if a measure of whole body retention, is wanted of not placing reliance on single observations.

G. Retention

STOVER and her colleagues have recently reviewed their data on the retention of ^{239}Pu IV in beagle dogs and have included important results obtained on dogs given a low dose (0.016 μc/kg) which was not previously available (STOVER, ATHERTON and BUSTER, 1972). They now conclude that in dogs receiving a low dose, the rate of decrease of retention in the skeleton is greater during the first two years than at later times, due to the greater rate of remodelling that occurs in the young dog, with loss of ^{239}Pu. In dogs given a high dose the remodelling is disturbed by radiation damage and plutonium is not therefore lost or buried so readily. In consequence the early fall in skeletal retention is not seen. Determination of approximate skeletal retention in one dog 4375 days after injection confirmed the two phases of decrease in retention for the entire skeleton at low dose levels. These observations suggest that the skeletal toxicity of a given dose of ^{239}Pu in an older dog is likely to be greater than the same dose given to a young dog owing to differences in the patterns of remodelling and therefore of exposure of the sensitive osteogenic cells. A gross effect of dose level on total excretion was not seen indicating that the effect of dose on skeletal retention is the opposite to the effect on hepatic retention.

III. Other Plutonium Isotopes

The translocation of $^{238}Pu(NO_3)_4$ and $^{239}Pu(NO_3)_4$, at pH 1 to 1.5, have been studied by MORIN, NENOT and LAFUMA (1972) following intramuscular injection or inhalation of 56 ng ^{238}Pu and 16 μg ^{239}Pu into rats. The rate of translocation of ^{238}Pu from the injection site was more rapid than that of ^{239}Pu, but the fractions of the *absorbed* nuclide deposited in bone and other tissues were similar for the two nuclides. The urinary and faecal excretion of ^{238}Pu was greater than that of ^{239}Pu by factors of 3.5 and 2.4 respectively at 90 days. Following inhalation the translocation of ^{238}Pu from lung was slower than that of ^{239}Pu but the fraction of the absorbed ^{238}Pu in bone was about three times greater than that of ^{239}Pu at 45 days. The slower rate of translocation of ^{238}Pu observed in these experiments after inhalation of a nitrate aerosol at pH is in contrast to the observations of STUART (1970) who found more rapid translocation of ^{238}Pu than ^{239}Pu after inhalation of PuO_2 particles and probably reflects different patterns of cellular uptake in lung. MORIN et al. (1972) observed that the uptake of ^{239}Pu by lung macrophages was twice that of ^{238}Pu with the reverse ratio in the "surfactant" fraction.

IV.

STOVER and her colleagues (ATHERTON, STOVER, JEE, STEVENS and BRUENGER, 1972) present evidence to suggest that skeletal retention of polymeric plutonium in beagles is not associated with bone *per se* but with the amount of red marrow a bone contains. This is confirmed by TAYLOR, JEE, DELL, WILLIAMS and SHABESTARI (1972). LINDENBAUM reports that in both mice and beagles the marrow content of polymeric plutonium (in long bones) at 6 days after injection was approximately twice that of a monomeric preparation.

IV. A. Pattern of ^{239}Pu Distribution in the Skeleton

JEE (1972) has recently reported the skeletal distribution of ^{239}Pu in a dog that inhaled PuO_2 and died 2565 days later with a body burden of 1.5 μCi, 5% of which was in the skeleton. Most of the plutonium of the femur and lumbar vertebral body was on bone surfaces, though a little was buried by a thin layer of new bone. He concludes that much of the translocated plutonium reached the bone when it was remodelling at an extremely low rate. STOVER and her colleagues (STOVER, ATHERTON and BUSTER, 1972) point out that in dogs given ^{239}Pu in a low dose (0.016 μCi/kg) at 1.5 years of age the initial deposition of plutonium in bone, which is predominantly on the surfaces, decreases rapidly with a relative increase in buried plutonium since at this age the rate of remodelling is high but later remodelling decreases rapidly and after three years is extremely slow. This means that the toxicity of a given injected low dose of plutonium given to 3–5 years old dogs is likely to be considerably higher than the same dose given to dogs injected at 14–18 months old. The danger of equating data on bone distribution obtained in dogs with that in man, due to differences in the pattern of bone trabeculation and bone remodelling, cannot be over emphasised (LLOYD and MARSHALL, 1972).

VI. B. 6. The Experimental Approach

Recently JAMES (1972) has reported measurements on eight week old rats given a single intravenous injection of soluble $Pu(NO_3)_4$ at a dose level of 4.5 μCi/kg. He measured the dose rate as in the paper of JAMES and KEMBER (1970). The results confirm the pattern of variation of local dose rate with time, due to remodelling, by a factor of 2 either way in many cases. Using the results of BENSTED, TAYLOR and SOWBY (1965) he estimated that a cumulative α dose of about 2000 rads (delivered to primitive cells at trabecular surfaces over a period of 36 weeks) may be associated with a 10% probability of developing a tumour in an individual femoral epiphysis.

VIII. A. 1. c)

STOVER and her colleagues (STOVER, ATHERTON, BRUENGER and STEVENS, 1972) have recently made calculations of skeletal dose rates in the beagle as a function of dose levels using the average skeletal burden of plutonium to calculate radiation dose. Hitherto they have assumed wet skeletal weight to be 10 percent of the weight of the dog at injection but for these recent calculations they have determined, by better technique, that skeletal weight is 7.5 ± 0.9 percent of the body weight at injection. This new figure may well alter many calculations about radiation dose from the Utah laboratory. Except at a high dose (2.8 μCi/kg) level the average dose to the liver exceeds that to the skeleton after 1300 days.

VIII. B. 2

A complete analysis of the haematological effect of a single intravenous injections of ^{239}Pu to beagles (0.016–2.8 μCi/kg) has recently been reported by DOUGHERTY (DOUGHERTY, 1972). The changes observed were dose dependent and were most marked in the granular leucocytes which were maximally depressed at 2–3 weeks. A sustained lymphopaenia occurred at the two highest dose levels. A transient early thrombocytopenia and anaemia were also found in the higher dose dogs. A moderate anaemia reappeared in the terminal stage. There was myaloid metaplasia in liver and spleens at autopsy.

References

ATHERTON, D. R., STOVER, B. J., JEE, W. S. S., STEVENS, W., BRUENGER, F. W.: Skeletal retention and distribution of polymeric and monomeric ^{239}Pu in beagles. In: Research in Radiobiology Annual Report of Work in Progress on the Chronic Toxicity Program, USAEC Report COO-119-246, p. 126–136. University of Utah 1972.

BALLOU, J. E., PARK, J. F., MORROW, W. G.: On the metabolic equivalence of ingested, injected and inhaled ^{239}Pu citrate. Hlth Phys. **22**, 857–862 (1972).

BEACH, S. A.: Sebeach, a digital computer program for the estimation of body content of plutonium from urine data. Hlth Phys. **24**, 9–16 (1973).

DOUGHERTY, J. H.: The haematologic changes induced by ^{239}Pu in beagles. In: Radiobiology of plutonium (eds. STOVER, B. J., and JEE, W. S. S.), p. 75–86. Salt Lake City: The J. W. Press, University of Utah 1972.

DURBIN, P. W., HOROVITZ, M. V., CLOSE, E. R.: Plutonium deposition kinetics in the rat. Hlth Phys. **22**, 731–741 (1972).

JAMES, A. C.: Dose to osteogenic cells from plutonium 239 deposited in rat bone. Radiat. Res. **51**, 654–673 (1972).

JEE, W. S. S.: Distribution and toxicity of ^{239}Pu in bone. Hlth Phys. **22**, 583–595 (1972).

KATZ, J. H.: Transferrin and its function in the regulation of iron metabolism. In: Regulation of haematopoiesis (ed. A. S. GORDON), vol. 1, p. 539. New York: Appleton-Century-Crofts 1970.

LAGERQUIST, C. R., HAMMOND, S. E., BOKOWSKI, D. B., HYLTON, D. B.: Distribution of plutonium and americium in occupationally exposed humans as found from autopsy samples. Hlth Phys. **23**, 418 (1972).

LINDENBAUM, A., BAXTER, D. W., ROSENTHAL, M. W.: Comparison of blood clearance and 6 day tissue distribution of monomeric and polymeric ^{239}Pu in the beagle. Hlth Phys. **51**, 540 (1972).

LLOYD, E., MARSHALL, J. H.: Toxicity of ^{239}Pu relative to ^{226}Ra in man and dog. In: Radiobiology of plutonium (eds. STOVER, B. J., and JEE, W. S. S.), p. 377–383. Salt Lake City: The J. W. Press, University of Utah 1972.

MORIN, M., NENOT, J. C., LAFUMA, J.: Metabolic and therapeutic study following administration to rats of ^{238}Pu nitrate, a comparison with ^{239}Pu. Hlth Phys. **23**, 475–480 (1972).

ROSENTHAL, M. W., LINDENBAUM, A., RUSSELL, J. J., MORETTI, E., CHLADEK: Metabolism of monomeric and polymeric plutonium in the rabbit; comparison with the mouse. Hlth Phys. **23**, 231–238 (1972).

SIKOV, M. R., MAHLUM, D. D.: Age-dependence of ^{239}plutonium metabolism and effect in the rat. In: Radiobiology of plutonium (ed. STOVER, B. J., and JEE, W. S. S.), p. 261–272. Salt Lake City: The J. W. Press, University of Utah 1972b.

STEVENS, W., ATHERTON, D. R., STOVER, B. J., BRUENGER, F. W.: The effect of the physical chemical state of ^{239}Pu on its early retention in plasma and selected soft tissues of beagles. In: Research in Radiobiology Annual Report of Work in Progress in the Chronic Toxicity Program, USAEC Report COO-119-246, p. 137–147. University of Utah 1972.

STOVER, B. J., ATHERTON, D. R., BRUENGER, F. W., STEVENS, W.: Comparison of skeletal and hepatic dose rates for ^{239}Pu in the beagle as a function of dose level. In: Research in Radiobiology Annual Report of Work in Progress in the Chronic Toxicity Program, USAEC Report COO-119-246, p. 167–192. University of Utah 1972.

STOVER, B. J., ATHERTON, D. R., BUSTER, D. S.: Retention of ^{239}Pu (IV) in the beagle. In: Radiobiology of plutonium (eds. STOVER, B. J., and JEE, W. S. S.), p. 149–169. Salt Lake City: The J. W. Press, University of Utah 1972.

TAYLOR, G. N., JEE, W. S. S., DELL, R., WILLIAMS, J., SHABESTARI, L.: Anatomic distribution of monomeric and polymeric ^{239}Pu. In: Research in Radiobiology Annual Report of Work in Progress on the Chronic Toxicity Program, USAEC Report COO-119-246, p. 106–125. University of Utah 1972.

Chapter 11

Plutonium in Soft Tissues with Emphasis on the Respiratory Tract[1]

W. J. BAIR, J. E. BALLOU, J. F. PARK, and C. L. SANDERS

With 43 Figures

I. Introduction

The previous chapter discussed plutonium in bone and in those soft tissues such as liver that would accumulate plutonium following deposition in wounds. Plutonium in the respiratory tract was also mentioned. However, because experience has shown the respiratory tract to be the most common route for entry of plutonium into the body, this chapter will expand upon the disposition and biological effects of inhaled plutonium.

Plutonium compounds occurring as airborne contaminants are either very insoluble in body fluids, e.g., PuO_2, or are relatively soluble and transportable in the body, e.g., plutonium nitrate. Relatively soluble plutonium compounds translocate from the respiratory tract to bone and liver, etc., though generally not at a rapid rate. Insoluble plutonium translocates from the respiratory tract very slowly—over many years, and thus, imparts most of the emitted radiation energy to lungs and associated tissues such as thoracic lymph nodes. Because insoluble plutonium is inhaled as particles and remains in particle form in the body for a relatively long time, the energy emitted by the plutonium is absorbed in discrete volumes of tissues surrounding the particles. Because of the short range of alpha particles, this volume of tissue is relatively small. Thus, the distribution of absorbed alpha energy throughout tissue-containing insoluble plutonium particles is very heterogenous. This is an example of what has been referred to as the "particle problem". Although both "soluble" and "insoluble" forms of plutonium will be discussed, the particulate characteristics of insoluble plutonium constitute radiobiological problems different from those discussed in previous chapters and will be emphasized herein.

To visualize the problem of plutonium particles deposited in tissue, it is helpful to consider the range of alpha doses emitted by particles of different size and of different isotope composition, Table 11.1. A 0.1 μm $^{238}PuO_2$ particle is equivalent to a 0.64 μm $^{239}PuO_2$ particle in terms of alpha dose delivered to a 100 μm sphere of tissue. A 1 μm $^{238}PuO_2$ particle is a very significant radiation source. Consider, too, that such particles are of a size that they can be readily contained in macrophages or other cells of the respiratory tract.

An electron micrograph prepared from a sample of a plutonium dioxide aerosol, Fig. 11.1, illustrates the "typical" particle shapes and size distribution.

---

1 This work was performed under the United States Atomic Energy Commission Contract AT(45-1)-1830 with Battelle, Pacific Northwest Laboratories, Richland, Washington.

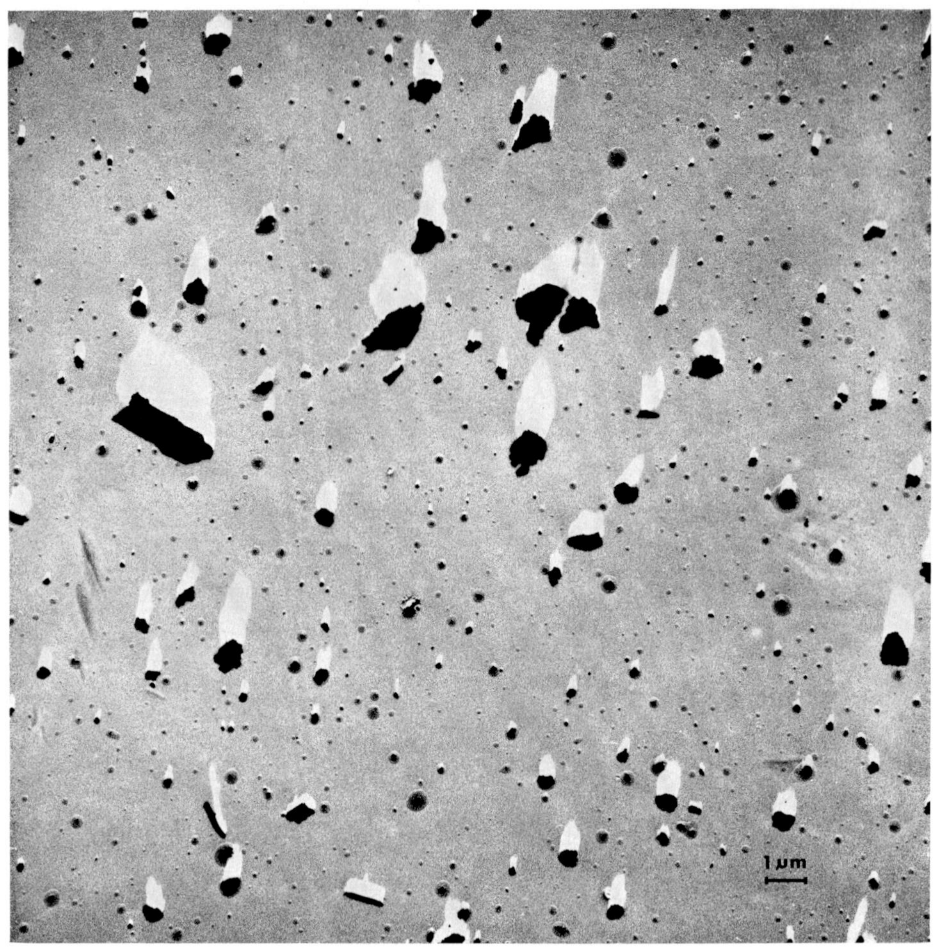

Fig. 11.1. Electronmicrograph of $^{239}PuO_2$ particles

Table 11.1. Comparative radiological properties of ^{238}Pu and ^{239}Pu

PuO_2 particle size (µm)	Mass of Pu (g)	^{238}Pu µCi	^{238}Pu daily dose to 100 µm sphere of tissue[a] (rads)	^{239}Pu µCi	^{239}Pu daily dose to 100 µm sphere of tissue[a] (rads)
0.01	4.8×10^{-18}	5.63×10^{-11}	0.03	2.6×10^{-13}	0.00013
0.1	4.8×10^{-15}	5.63×10^{-8}	30	2.6×10^{-10}	0.13
0.25	7.5×10^{-14}	8.82×10^{-7}	470	4.07×10^{-9}	2
0.5	6.02×10^{-13}	7.04×10^{-6}	3770	3.25×10^{-8}	16
1.0	4.8×10^{-12}	5.63×10^{-5}	30000	2.6×10^{-7}	130

[a] Assuming all alpha energy is absorbed with the 100 µm sphere.

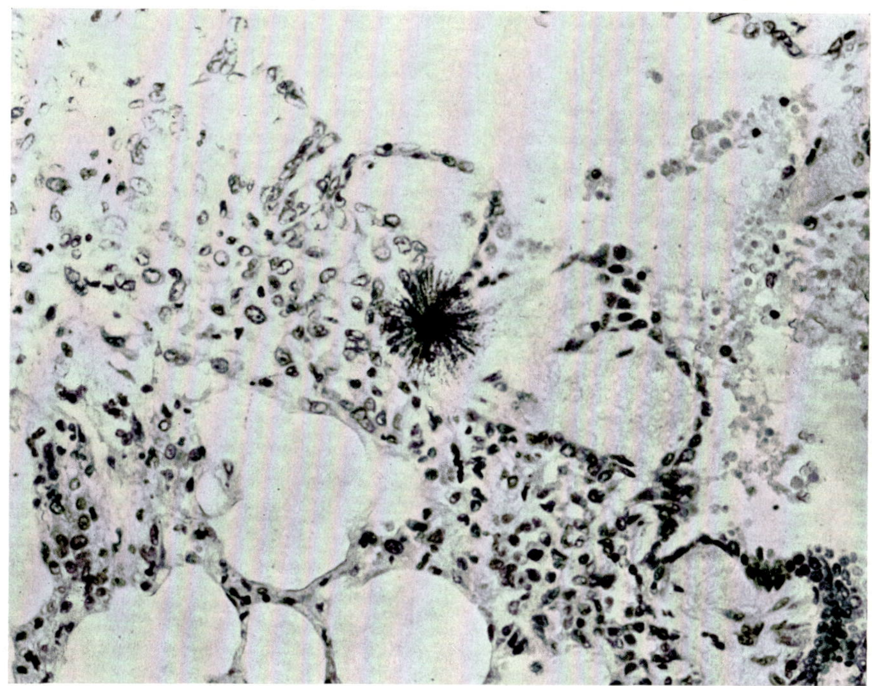

Fig. 11.2. Autoradiograph of section from dog lung after inhalation of $^{239}PuO_2$

Crystalline plutonium dioxide is cubic, but aerosols of plutonium particles seldom contain cubic particles. Some particles appear to be parts of fractured cubes, however. In any case the particle shape and, to a degree, size and density will vary depending upon how the particles were formed; and plutonium particles can be formed under many physical and chemical conditions ranging from oxidation of metal in air to explosive situations involving high temperature and pressures.

Fig. 11.2 is an autoradiograph of plutonium dioxide particles. The preparation of particles in lung was exposed 14 days to a photographic emulsion. The number of alpha tracks seen on the autoradiograph is related to the quantity of plutonium present in the particle. Under the conditions of this autoradiograph, each alpha track represents about 9×10^{-10} μg $^{239}PuO_2$, or a particle about 0.054 μm in diameter.[2]

The autoradiograph also provides an impression of the volume around the plutonium particle in which the alpha energy is absorbed. The range of the 5.5 MeV alpha particles from ^{239}Pu is about 50 μm in tissue. The tracks in the emulsion are about 27 μm long. Thus, it is readily seen that plutonium particles are point sources of alpha radioactivity irradiating volumes of tissue approximately 100 μm in diameter. However, since the range of alpha particles in air is about 3.8 cm (STP), and because the lung is composed of many air-filled structures, cells at distances up to about 300 μm from plutonium particles can be irradiated.

The foregoing is intended to illustrate some of the physical-chemical complexities that confound an understanding of plutonium particles in the lung. The following will introduce biological complexities.

2 Ed. Note. Techniques for such calculations appear in Chap. 10, p. 435.

II. Disposition of Inhaled Plutonium

A. Clearance of Plutonium from the Lung

The clearance of inhaled plutonium from the lung and the body has been characterized in animal experiments and to a lesser extent in man following accidental inhalation exposures. Because of technical difficulties associated with detecting plutonium in the living human body, and because of the infrequency of human exposure, most of the available data on plutonium has been obtained in animal experiments. An example of a clearance curve for inhaled plutonium is given in Fig. 11.3. Authors have experimented with several techniques to construct these curves, ranging from using total daily excretion data and content of plutonium at sacrifice to various methods of *in vivo* counting (BAIR and WILLARD, 1963; MORROW et al., 1967). The fraction of plutonium deposited in the body is usually plotted against time in days after the inhalation exposure. Within the first week to 10 days after exposure, a fraction of the deposited plutonium is cleared from the respiratory tract and excreted. In the case of PuO_2, all plutonium cleared from the respiratory tract is excreted in feces except for a very small fraction that may be solubilized, absorbed into the circulating blood, and if not deposited in another tissue, cleared by the kidneys. The magnitude of the fraction cleared from the body within the first week or 10 days depends upon the fraction of readily soluble material present and also upon the distribution of the deposited plutonium within the respiratory tract. Plutonium particles deposited on ciliated epithelium are trapped in mucus and propelled from the respiratory tract. Particles deposited on the nonciliated epithelium below the terminal bronchioles and in the alveoli are not readily transported from the lung and are available for interaction with cellular and noncellular constituents of the respiratory tract. On the basis of an exponential model as seen in Fig. 11.3, the early clearance of particles, primarily those deposited on ciliated epithelium, occurs with a half-time of about 12 to 36 hours (MORROW et al., 1967). Plutonium deposited below the ciliated epithelium of the terminal bronchioles is cleared from the body very slowly. This fraction, which is not readily available for early clearance by ciliary processes, can be estimated by extrapolating the long-term whole-body clearance curve to the ordinate at zero time. The slope of this curve provides an estimate of the rate of clearance. Over the first several months after exposure to plutonium, the clearance of alveolar-deposited plutonium from the body usually can be described by an exponential equation of the form shown in Fig. 11.3. A retention half-time can be derived from this equation. Over several years' period, the whole-body clearance rate is observed to change slowly and is best described mathematically by a power function equation (STUART et al., 1968).

The kinetics of clearance of plutonium from the lung are difficult to determine because the only accurate measurement of the lung burden is that obtained at death when the lung tissue can be dissected and analyzed. During life, the measurements of plutonium lung burdens are complicated not only by the difficulty in quantitatively assessing the plutonium from the low energy X-ray emissions but by the translocation of plutonium to pulmonary and other thoracic lymph nodes. When the lung burden is obtained at death, it is generally assumed that clearance from the lung was exponential and a retention half-time is calculated. This is indicated in Fig. 11.3.

The retention half-time values for the whole body and for lung, just described, obtained from animal experiments are compared for several plutonium compounds in Table 11.2. Plutonium deposited in the lung alveoli is retained in the body

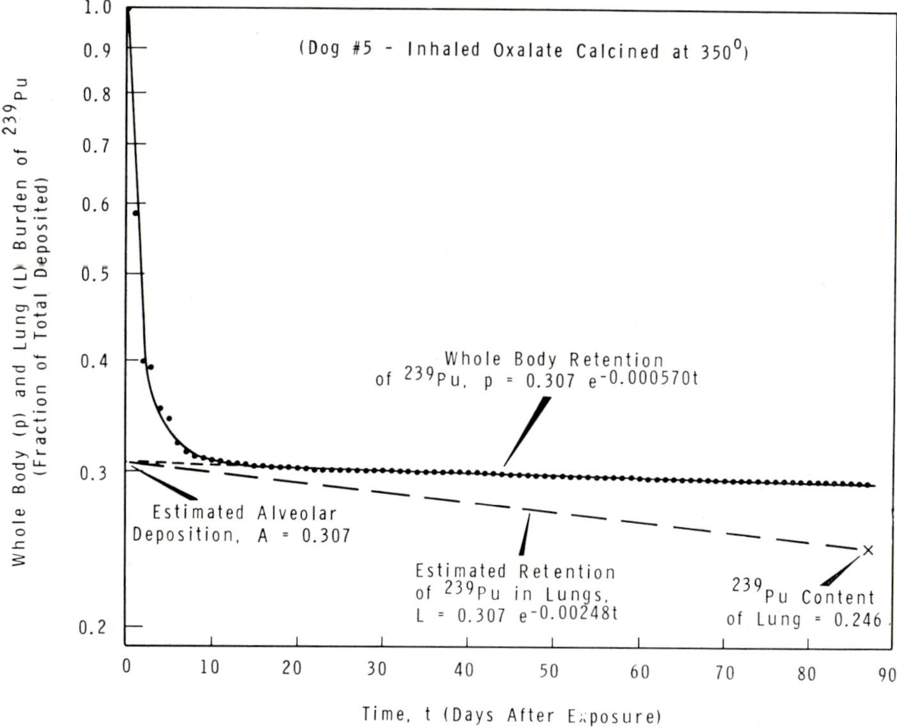

Fig. 11.3. Whole body and lung retention of inhaled $^{239}PuO_2$ in a beagle dog. (Bair and Park, 1968)

for a long time; half-time values range from about 200 to over 5000 days. The retention of plutonium in the lung is highly dependent upon the chemical compound inhaled and, in the case of PuO_2, on the particle size and process by which the oxide was formed. For transportable forms of plutonium, the pulmonary retention half-times range from 30 to about 200 days. Lowest values were reported for plutonium nitrate and citrate aerosols. The organic plutonium compounds studied in the USSR had pulmonary retention half-times of about 150 to 200 days. In these studies, while experimental conditions varied considerably, the clearance of plutonium from lungs of rodents seemed to be more rapid than in dogs. However, the greatest discrepancies among these data generally can be related to the duration of the observations. Longest half-times were reported when the animals were observed for the longest periods. This was true for both soluble and insoluble compounds.

In a long-term study of inhaled PuO_2 in dogs, the best estimate of the pulmonary retention half-time is about 1000 days. In shorter studies of 3 months' to a year's duration, the estimates for the same PuO_2 (oxalate calcined at 350°) were 350 to 390 days for dogs and 460 days for mice. An oxide prepared at 1000° and another prepared by heating the metal to 450° C showed retention half-times of 800 days in comparable short-term experiments. Calcining at high temperatures reduces the percentage of oxalate present and probably decreases the surface area and thus its solubility rate. The median particle size of the 450° metal-oxide was relatively large and may have contributed to its longer retention.

Table 11.2. Retention of alveolar deposited plutonium. (Values are means or ranges when more than one individual were observed)

Compound	Particle size (μm)[a] CMD	Particle size (μm)[a] MMD	Animal species (number)	Duration of study (days)	Retention half time (days) Whole body	Retention half time (days) lung	Reference
$Pu(NO_3)_4$	(0.1–10)		Rats	32		30	Tregubenko (1966)
	0.12	0.6	Dogs (11)	100–300	600	120	Park et al. (1968)
		0.9[b]	Dogs (14)	100	500–600	250	Ballou and Park (1972)
	(0.1–10)		Rats	256		212	Lyubchanskii (1967)
		0.3–0.5	Rats (6)	80		26–37	Suzuki (1971)
Pu citrate	(0.1–10)		Rats	64–256	—	137–173	Lyubchanskii (1964, 1967)
		1.45[b]	Dogs (14)	100	900	200	Ballou and Park (1972)
Sodium plutonyl-triacetate	(0.1–10)		Rats	256	—	209	Lyubchanskii (1964, 1967)
Pu(III) chloride	(0.1–10)		Rats	256	—	169	Lyubchanskii (1964, 1967)
Ammonium plutonium-penta-carbonate in 10% carbonate	(0.1–10)		Rats	256	—	167	Lyubchanskii (1964, 1967)
Ammonium plutonium-(IV)-pentacarbonate in H_2O	(0.1–10)		Rats	230	—	122	Lyubchanskii (1964, 1967)
PuF_4	~0.2		Dogs (6)[d]	90	240	180	Dilley (1970)
$^{239}PuO_2$							
Fume			Rat	—	—	180	Abrams (1946)
350° oxalate			Mice (160)	500	—	180–460	Bair et al. (1961)
		1.1–4.9[c]	Dogs (16)	125–468	—	~400	Morrow et al. (1967)
450° oxalate (dust)	0.60	4.3	Dogs (8)	65–105		>350	West and Bair (1964)
450° oxalate	0.60	4.3	Dogs (4)	270	1600	>1000	Bair and McClanahan (1961)
1000° oxalate[d]	0.51	2.8	Dogs (3)	90	1200	780	Bair and Park (1968)

350° oxalate	0.45	2.8	Dogs (3)	90	1350	370	Bair and Park (1968)
450° metal	0.50	4.8	Dogs (3)	90	2700	840	Bair and Park (1968)
123° metal	0.46	1.3	Dogs (3)	90	1600	700	Bair and Park (1968)
900° oxalate	0.05	0.12	Dogs (3)	150	700	470	Bair (1970)
350° oxalate	0.3–0.5	3.0	Dogs (∼100)	∼12 yr	—	1000	Park et al. (1972)
$^{239}PuO_2$							
Field studies [e]		10–20	Dogs (84)	0–456	—	174	Wilson and Terry (1968)
Field studies		10–20	Sheep (132)	14–930	—	399	Wilson and Terry (1968)
Field studies		10–20	Burros (84)	7–456	—	155	Wilson and Terry (1968)
$^{238}PuO_2$							
700° oxalate	0.05	0.1	Dogs (12)	27–185	900	300	Park et al. (1970)
		4.4–4.9 [c]	Dogs (3)	65	—	380	Morrow et al. (1967)
Crushed microspheres	0.18	0.64	Dogs (6)	50–106	2200	600	Willard and Park (1970)
Crushed microspheres	0.18	0.64	Dogs	2 yr	5600	1100	Willard and Park (1970)
PuO_2—ZrO_2	0.12	0.26	Dogs (3)	90	400	150	Willard and Park (1970)
PuO_2—ThO_2	0.13	0.34	Dogs (3)	90	3600	310	Willard and Park (1970)

[a] Count Median Diameter (CMD) and Mass Median Diameter (MMD) of the aerosol. [b] Aerodynamic median activity diameter of the aerosol. [c] Mass Median Aerodynamic Diameter estimated from Mass Median Diameter and density of plutonium compound. [d] Three of six dogs treated with DTPA, without affecting retention. [e] These animals were exposed to a cloud from the high explosive detonation of a plutonium weapon. The average density of the particles was 4.9 g/cm³ and the mean plutonium content of each particle was estimated to be slightly more than 10%. Other constituents include metal oxides from the device, uranium dioxide, and a minor component from the desert environment. The authors suggest particles less than 1 μm had a higher density and were composed of uranium and plutonium oxide. (Editor's note—see also Chap. 15).

The above pertains to plutonium compounds prepared under reasonably well controlled laboratory conditions. Such pure compounds may not be the form in which plutonium is encountered by man in accident situations. One potential accident is the high explosive detonation of plutonium such as in a nuclear weapon. Inhalation of plutonium released from high explosive detonations was studied by exposing dogs, sheep, and burros to the plutonium contaminated cloud (Stewart and Wilson, 1968; Wilson and Terry, 1968). The plutonium-containing particles had an average density of 4.9 g/cm³ compared with about 11.5 g/cm³ for plutonium dioxide. Thus, the particles contained other material from the explosion—probably nitrates, soil, and metallic parts of the weapon including some uranium. Sheep showed an early clearance of more than 90% of the deposited plutonium with a half-time of 3.3 days, and burros cleared over 40% with a half-time of 4 days. This early clearance was not seen in dogs. The retention half times for the long-term components are given in Table 11.2. The values for dogs and burros were comparable, 174 and 155 days, respectively. The value for sheep was 399 days, but sheep were studied for 930 days compared to 456 days for dogs and burros. Again, this difference between species may be due to the difference in the duration of the observation period because rate of clearance of particles from the lung appears to decrease with time after exposure. Because the inhaled particles contained only about 10% plutonium, the retention half-times may be more characteristic of the major constituents of the particles than the plutonium.[3] In fact, the values are not too different from those reported for uranium oxides (Morrow et al., 1964; Bair et al., 1970).

There are no human data quantitatively suitable for estimating pulmonary retention half-times for plutonium compounds. In the few documented accident cases, either the inhaled material was not identified, the aerosol was a mixture of compounds, or the individual was exposed on more than one occasion.

In summary, experimental animal data suggest that the pulmonary retention half-time is about 150 days for plutonium chloride and ammonium plutonium-

Table 11.3. Disposition of inhaled soluble plutonium compounds.

	Pu citrate							Pu nitrate	
	Rats				Dogs			Rats	
	Lyubchanskii (1964)				Ballou (1972)			Tregubenko (1966)	
Days after exposure	1	32	128	256	1	30	100	1	32
Tissue									
Total body									
Lung	13	5	3	1.5	75	37	29	34	9.7
Liver	1	1	0.3	0.3	1	5	16	0.3	0.6
Skeleton	5	10	8	8	6	38	38	1.5	5.7
Lymph nodes[d]	—	—	—	—	0.05	0.20	0.4	—	—
Turbinates	1[b]	1	—	—	0.6	0.4	0.8	—	—
Kidney	—	—	—	—	0.3	0.5	0.4	0.1	0.2
Spleen	—	—	—	—	0.1	0.4	0.2	0.08	0.2
All other tissues	—	—	—	—	—	—	—	—	—
Urine	—	—	—	—	0.6	2.0	2.0		
Feces	—	—	—	—	99	156	208		

[a] Dose was computed as the total amount deposited less the amount excreted during the first 6 days (G.I. tract burden at 1 day) in Ballou's study, as percent of initial lung

3 Ed. Note. Further discussion of this and other field experiments appears in Chap. 15.

pentacarbonates, 200 days for plutonium fluoride and citrate, 250 to 300 days for plutonium nitrate, and up to 1000 days for PuO_2.

B. Translocation to Other Tissues

Plutonium may be removed from the lung via bronchial and tracheal mucociliary processes, introduced into the gastrointestinal tract, and excreted. This accounts for only part of the plutonium cleared from the respiratory tract. Another fraction is accumulated by the lymphatics and deposited in lymph nodes while another fraction, probably dissolved by tissue fluids, may be transported by blood and lymph to other tissues of the body such as bone and liver. Some of this plutonium is excreted in urine, and some may be excreted via the bile duct and the intestine. The distribution of plutonium in the body is essentially the same for all plutonium compounds, but may differ quantitatively depending upon the chemical and physical state of the plutonium deposited in the body and the route of entry. These phenomena are discussed in detail in Chaps. 9 and 10.

Data for a number of the more soluble plutonium compounds are shown in Table 11.3. These compounds were administered as aerosols to either rats or dogs. The data are expressed as percentages of the estimated amount of plutonium initially deposited in the alveolar region of the lung with the exception of those reported by SUZUKI. Depending upon the method selected by the authors, the amount of plutonium deposited in the alveoli may be a value extrapolated from excretion data as described earlier or it may be taken as the body burden at 5 to 7 days after exposure when the plutonium initially deposited in the upper respiratory passages and the gastrointestinal tract has been eliminated from the body.

Plutonium citrate and plutonium nitrate are gradually translocated from lung to bone and to a somewhat lesser extent, liver. Smaller quantities are deposited in soft tissues such as lymph nodes, kidney, and spleen. The translocation from

Percent of administered dose[a]

Pu nitrate								Pu nitrate[c]			PuF_4
Dogs								Rats			Dogs
BAIR et al. (1962)	PARK (1968)				BALLOU (1971)			SUZUKI (1971)			DILLEY (1970)
30	75–103	109–138	172–236	303	1	30	100	1	32	80	90
91	88	86	80	80							76
61	51	46	35	32	88	32	41	95	85	66	70
12	9.5	8.6	16	12	0.32	9	10	0.1	2	2	0.2
18	22	26	24	29	2	43	28	4	11	21	0.55
0.5	2.6	1.5	1.5	1.6	0.06	0.4	0.6	0.5	0.2	0.2	4.8
—	—	—	—	—	0.85	0.85	0.43	2[b]	2	8	—
—	—	—	—	—	0.1	0.5	0.2	0.02	0.2	0.2	—
—	—	—	—	—	0.016	0.25	0.14	0.01	0.1	0.2	—
—	—	—	—	—	—	—	—	—	—	—	1.8
0.8	1	1.6	2.1	2.2	0.9	1.7	2.3				3.6
8.6	11	13	18	18	47	92	81				20

burden in Russian studies, and as percent of alveolar-deposited plutonium in BAIR, PARK and DILLEY's studies. [b] Head.
[c] Percent of body burden at death. [d] T.B. NODES.

Table 11.4. Concentration of ^{239}Pu in tissues of dogs after a single inhalation exposure to plutonium citrate, nitrate, or fluoride (nCi/gram)

Time after exposure, days	Pu citrate[a] BALLOU (1972)			Pu nitrate BAIR et al.[b] (1962)	Pu nitrate PARK et al. (1968)					PuF_4 DILLEY[c] (1970)
	30	62	100	30	75	96	103	236	303	90
Tissue										
Tracheobronchial lymph nodes	0.78	1.7	3.7	61	2000	3900	420	160	160	26
Lung	10	14	19	75	420	390	140	36	51	3
Liver	0.31	2.1	2.8	7.4	20	31	6	5	4	0.006
Bone	0.96	1.8	2.0	2	13	16	5	2	1	0.004
Turbinates	2.6	1.7	4.9	—	—	—	—	—	—	—
Terminal body burden μCi	1.4	2.8	3.3		47	5	65	9	8	

[a] 3–12 μCi deposited.
[b] About 10 μCi body burden (mean of 3 dogs).
[c] Mean of two dogs having similar body burdens, 0.2 and 0.3 μCi.

lung to bone and liver was more rapid for plutonium citrate than plutonium nitrate. Both compounds were translocated more rapidly and apparently excreted more rapidly in rats than in dogs. The data of SUZUKI are given as percent of body burden and cannot be compared directly with the other data shown in Table 11.3.

The difference between plutonium nitrate and citrate is relatively small considering that unlike plutonium nitrate, the citrate-complexed plutonium is less likely to form aggregates or radiocolloids. After about 100 days, the percentages translocated to liver were similar for both plutonium citrate and nitrate in dogs. However, about 10% more citrate than nitrate was translocated to skeleton.

Although the data are very limited, plutonium fluoride appears to be less readily translocated from the respiratory tract than citrate and nitrate, with the exception of the accumulation in the tracheobronchial lymph nodes.

The concentrations of plutonium (as nCi/gm) in several tissues of dogs after inhalation of plutonium nitrate, plutonium citrate, and plutonium fluoride are compared in Table 11.4. The body burdens varied so that the data are only comparable on a relative basis. Lung had the highest mean concentration after inhalation of plutonium citrate, several times that of liver and skeleton, and the turbinates which had the next highest concentration. After inhalation of plutonium nitrate, the mean concentrations in the tracheobronchial lymph nodes exceeded those in the lungs after one month and reached levels three to ten times the lung level during the first year after exposure. Unfortunately, data are not available at longer times after exposure when concentrations in bone might be expected to increase.

Inhalation of plutonium fluoride resulted in mean concentrations of plutonium in the tracheobronchial lymph nodes almost ten times that in the lung after 90 days. The concentrations in liver and bone were lower by several orders of magnitude, but, again, data are not available to indicate whether this persists for long periods after exposure.

Autoradiographs of tissue sections from dogs exposed to plutonium nitrate aerosols indicate why retention in lung is relatively long and may also suggest

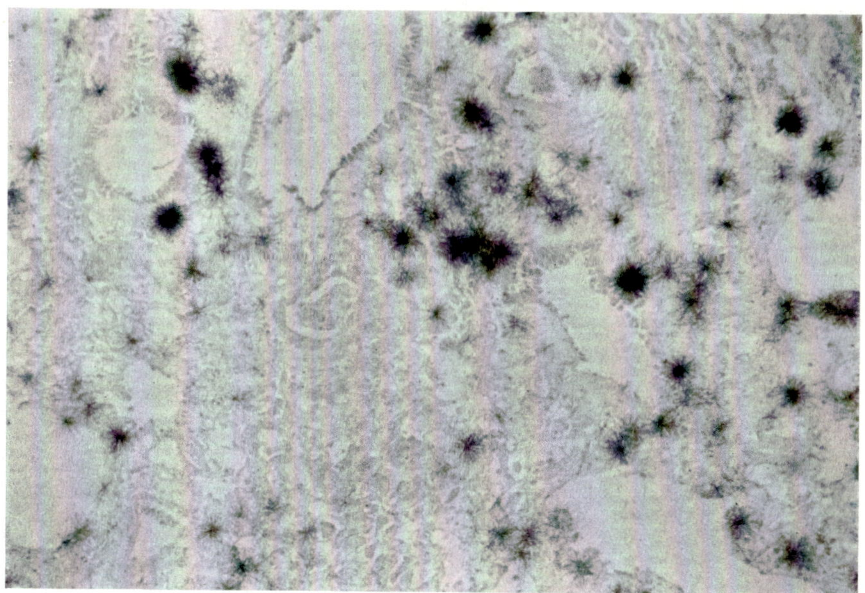

Fig. 11.4. Autoradiograph of lung section from a dog 138 days after inhalation of $^{239}Pu(NO_3)_4$. (BAIR, 1970)

something about the mechanism of translocation. Fig. 11.4 is an autoradiograph of a lung section from a dog after inhalation of plutonium nitrate. The presence of large alpha "stars" indicates the plutonium either existed as large aggregates in the inhaled aerosol or was accumulated by cellular action. There are also single tracks indicative of small particles or plutonium in monomeric form. Fig. 11.5 is an autoradiograph of a tracheobronchial lymph node showing the plutonium nitrate concentrated in a portion of the node. The concentration of plutonium in this portion of the node is many times the mean value that would be obtained by digesting and analyzing the total node. After inhalation of plutonium, only single alpha tracks have been seen in bone tissue, Fig. 11.6. This is in contrast to the microdistribution of intravenously injected plutonium. Particles or aggregates have been seen in liver, Fig. 11.7, after inhalation of plutonium nitrate.

An autoradiograph of lung and tracheobronchial lymph node sections from a dog 90 days after inhalation of plutonium fluoride, Fig. 11.8 and 11.9, indicates that the plutonium was present as particles or aggregates. Only a few single tracks can be seen. This shows the near absence of ionic plutonium.

A few other plutonium compounds have been studied in the USSR. These data are taken from LYUBCHANSKII (1967), Table 11.5. All of these compounds were studied in rats. Plutonium citrate, sodium plutonyltriacetate, ammonium plutonium pentacarbonate, and plutonium chloride all appear to be more readily translocated from lung to other tissues, bone and liver, than plutonium nitrate. These rat data for plutonium citrate and plutonium nitrate show less total retention than has been observed in dogs, Table 11.3.

Plutonium dioxide is the most extensively studied plutonium compound in respect to the inhalation route of entry. All of the many experiments with mice, rats, and dogs have led to the same general conclusions. Inhaled plutonium dioxide is retained in the periphery of the lung where part of it is gradually

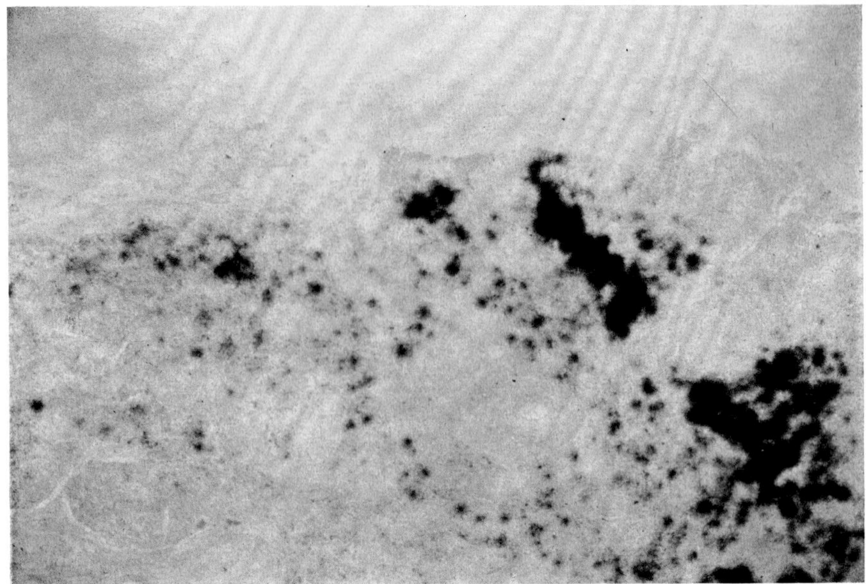

Fig. 11.5. Autoradiograph of tracheobronchial lymph node section from a dog 303 days after inhalation of $^{239}Pu(NO_3)_4$. (Bair, 1970)

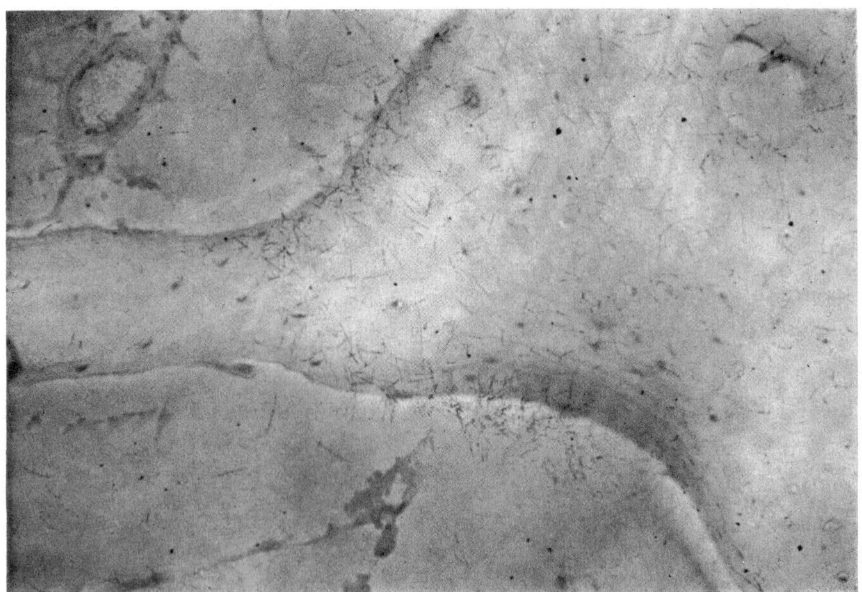

Fig. 11.6. Autoradiograph of bone section from dog 75 days after inhalation of $^{239}Pu(NO_3)_4$. (Bair, 1970)

accumulated by the lymphatic vessels and transported to the regional nodes of the tracheobronchial region and to the relatively more distant nodes in the mediastinum. The residence time for plutonium in lymph nodes appears to be

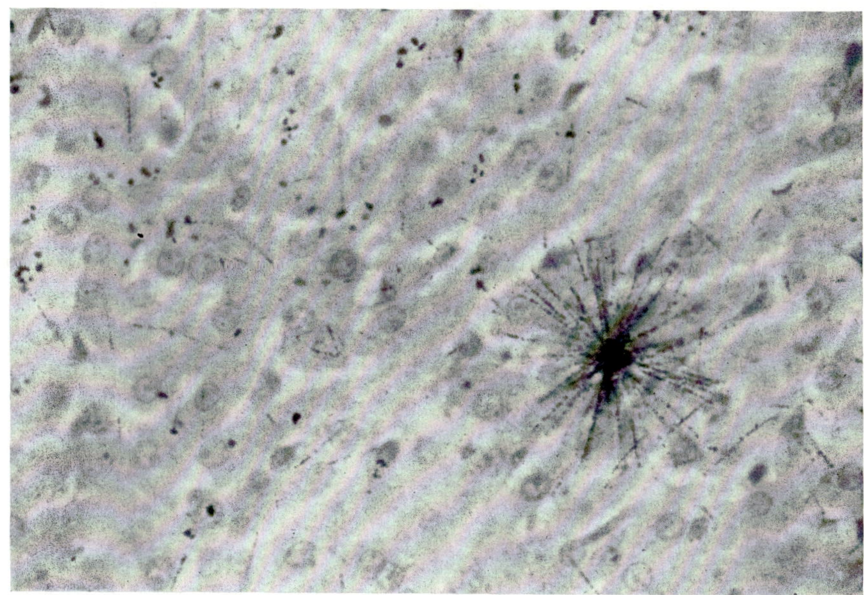

Fig. 11.7. Autoradiograph of liver section from dog 117 days after inhalation of $^{239}Pu(NO_3)_4$. (BAIR, 1970)

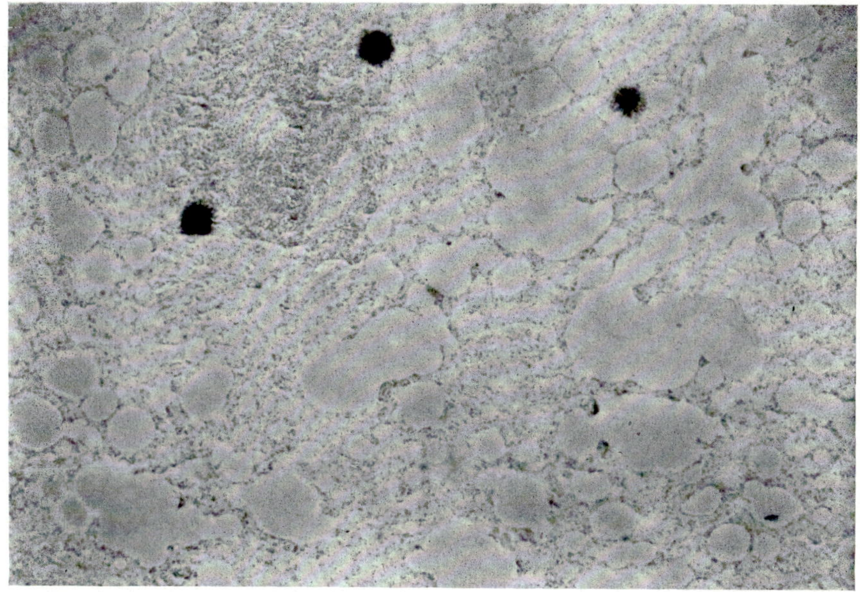

Fig. 11.8. Autoradiograph of lung section from dog 90 days after inhalation of $^{239}PuF_4$. (BAIR, 1970)

very long; there is no direct evidence for clearance of plutonium particles from thoracic lymph nodes. Indirect evidence includes the occasional observation of a particle in liver or spleen and the gradual accumulation of plutonium particles

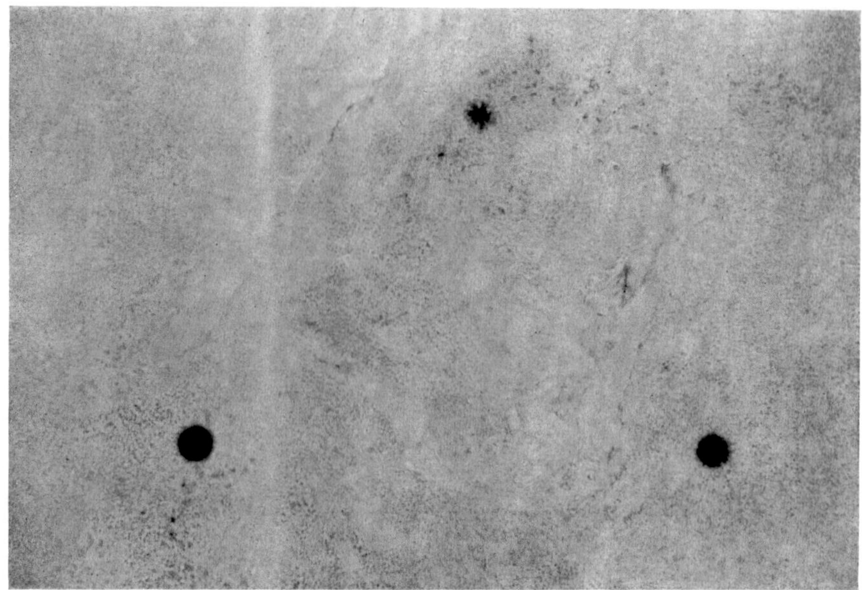

Fig. 11.9. Autoradiograph of tracheobronchial lymph node section from a dog 90 days after inhalation of $^{239}PuF_4$. (BAIR, 1970)

in hepatic lymph nodes. The route by which these particles were transported from lung might have included the tracheobronchial lymph nodes. The very slow buildup of plutonium in bone and some of the soft tissues may be due to slow solubilization of plutonium in lung and, probably to a lesser extent, in the lymph nodes.

The results of an early mouse experiment are shown in Table 11.6. Over a period of about 1 year, the plutonium in lungs and associated lymph nodes (thorax) decreased from 96% to about 20% of the amount estimated to have been deposited in the alveoli. The skeleton contained about 5% (the 5-day value is high because the nasal turbinates were analyzed with the bone) and the liver about 2%. The other tissues contained only trace quantities of plutonium.

Table 11.5. Comparison of plutonium content of rat lungs after inhalation of several plutonium compounds (LYUBCHANSKII, 1967). (Percent of initial lung burden)

Pu compound	Valence	pH	1 day		256 days	
			Lung	SIO[a]	Lung	SIO[a]
Plutonium citrate (2% citrate)	IV	6.5	47	20	4	22
Sodium plutonyltriacetate	VI	6.5	70	26	4	26
Ammonium plutonium pentacarbonate 10% $(NH_4)_2CO_2$	IV	8	117	11	4	20
Ammonium plutonium pentacarbonate in H_2O	IV	7.4	85	3	6	11
Plutonium chloride	III	2	88	14	7	24
Plutonium nitrate	IV	2	74	3	9	10

[a] Sum of Internal organs.

Table 11.6. Translocation of inhaled $^{239}PuO_2$ in mice[a] (BAIR et al., 1961). (Percent of estimated alveolar deposited plutonium[b])

Tissue	Time after exposure (days)					
	5	21	63	133	357	490
Whole body	100	80	31	42	29	20
Thorax	96	71	29	24	19	7
Bone	2.4	1.6	1.3	3.8	4.9	8.0
Liver	0.7	0.89	0.21	0.4	2.0	0.36
Spleen	0.005	0.1	0.01	0.08	0.7	0.2
Kidney	0.08	3.6	0.24	0.058	0.13	0.006
Ovary	0.04	0.008	0.02	0.02	0.006	0.02
Cervical lymph nodes and salivary glands	0.08	0.3	0.007	0.05	0.4	0.07

[a] Values are means of 15–20 mice. [b] 5-day body burden less amount in G.I. tract.

More extensive data have been obtained in dog experiments (PARK et al., 1972; WEST et al., 1964; BAIR et al., 1970). Nearly 100 dogs were given single exposures of about 30 minutes duration to $^{239}PuO_2$ aerosols with a CMD of 0.3 to 0.5 μm and 3 μm MMD. The $^{239}PuO_2$ was prepared by calcining plutonium oxalate at about 350° C. The deposited dose varied from less than a microcurie to about 50 microcuries. At the highest doses dogs died early of the radiation effects of the plutonium. Others died of radiation injury at longer times after exposure and others were sacrificed before pathology developed. Data obtained on the disposition of plutonium within the body of the dogs are plotted in Fig. 11.10. The plutonium contents of tissues of dogs are expressed as percent of the amount initially deposited in the lower respiratory tract or alveoli. After the first hundred days, plutonium appeared to be eliminated from lung with a half-time of about 1000 days. After 9 to 10 years, about 10% of the alveolar-deposited plutonium remained in the lungs. A large fraction of plutonium eliminated from lung was accumulated in the thoracic lymph nodes. After nearly 10 years, these lymph nodes contained about 50% of the total plutonium initially deposited. Other lymph nodes in the body also accumulated plutonium—in particular, abdominal lymph nodes located near liver accumulated about 5%. About 15% of the deposited plutonium was translocated to liver and about 4 to 5% to skeleton. About 85% of the plutonium initially deposited was retained in the dogs 9 to 10 years after exposure.

The plutonium content of thoracic lymph nodes in these dogs did not appear to reach an equilibrium state. The rate of accumulation of plutonium in the lymph nodes even after 9 to 10 years appeared to exceed the rate of elimination of plutonium from the lymph nodes, if, indeed, elimination occurred at all. Fig. 11.11 shows a logarithmic relationship between the plutonium content of thoracic lymph nodes and time. The radiobiological significance of the accumulation of plutonium in thoracic lymph nodes is not as yet fully understood.

Data from four of the dogs which survived 85 to 110 months after exposure illustrate the relative mean concentrations of plutonium in several tissues, Table 11.7. The highest concentration was in the thoracic lymph nodes and the next highest in the abdominal lymph nodes. The mean concentrations in lung and liver were similar, about 1/2000 of that in the thoracic lymph nodes. The concentrations of plutonium in skeleton and spleen were about one-tenth that of lung and liver. The mean radiation doses to these several tissues would be relatively the same as the concentrations of plutonium. Thus, the lymph

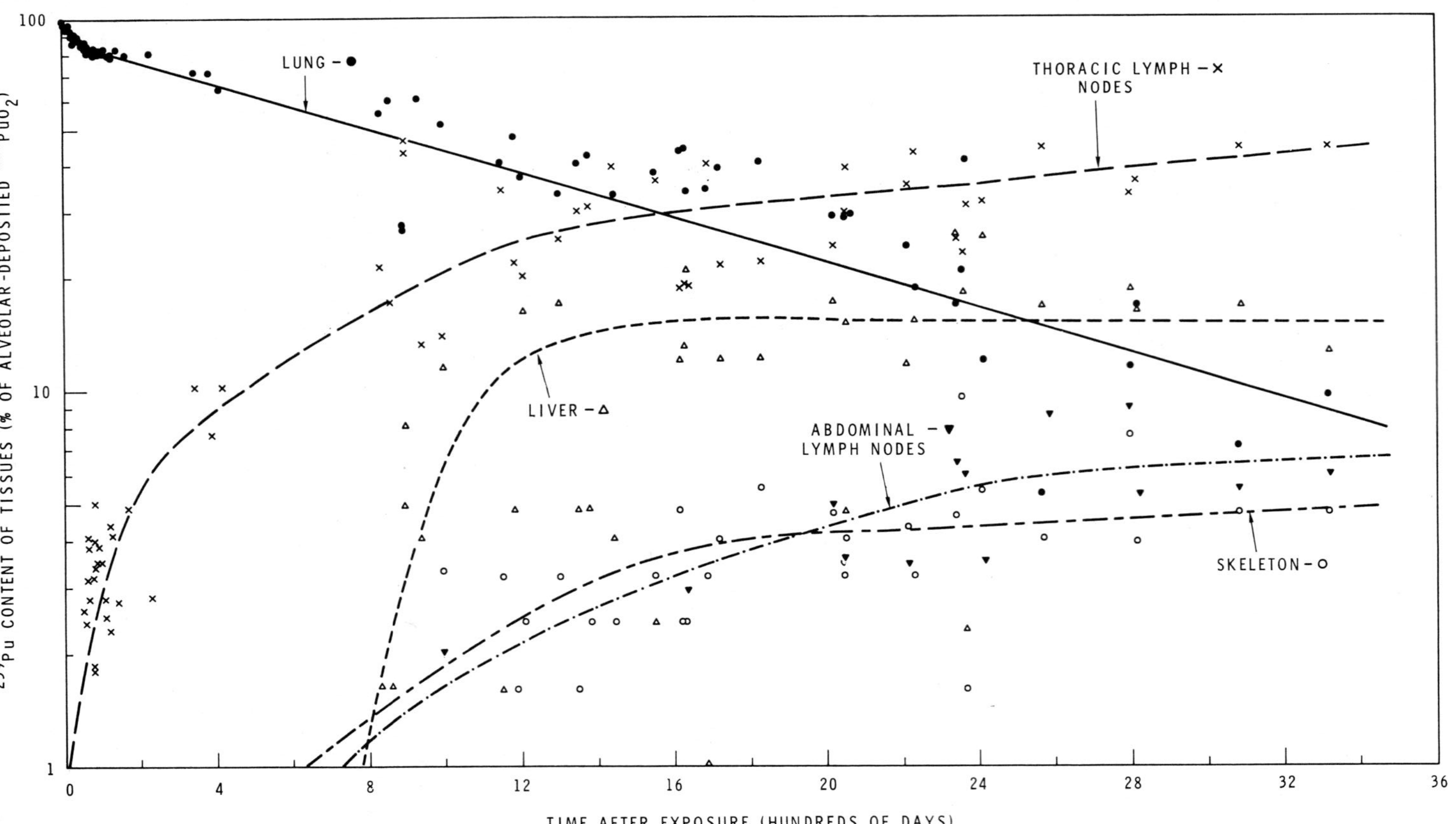

Fig. 11.10. Retention and translocation of alveolar-deposited $^{239}PuO_2$ in dogs (MMD = ~3 μm). (PARK et al., 1972)

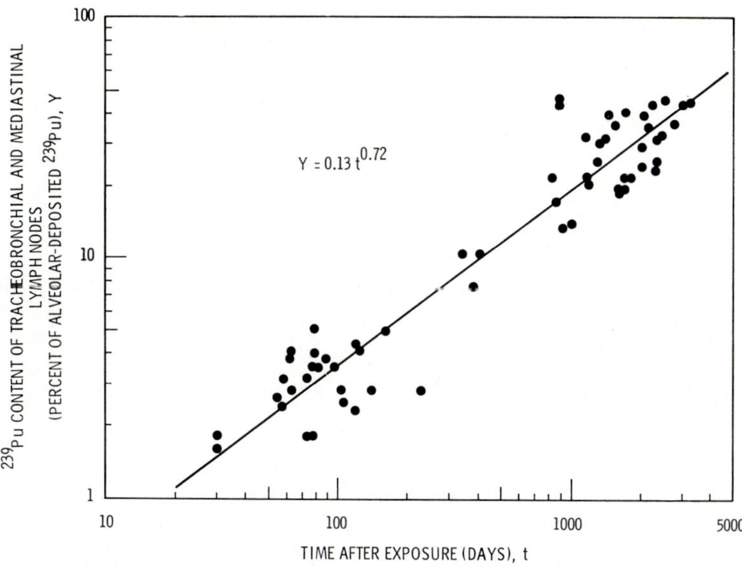

Fig. 11.11. Accumulation of ^{239}Pu in tracheobronchial and mediastinal lymph nodes in dogs after inhalation of PuO_2. (BAIR, 1970)

nodes received a much higher radiation exposure than other tissues in the body. However, the mean concentrations of plutonium in the tissues may not be as important as the localized doses. Autoradiographs show plutonium particles concentrated within areas of lung; the peripheral lymphatics in Fig. 11.12, alveolar septal cells in Fig. 11.13, and in alveolar epithelial cells in Fig. 11.14. The radiation dose to these structures could be very high depending upon the degree of mobility of the particles in the lung. Plutonium is also concentrated in lymph nodes; within the medullary sinus in Fig. 11.15 and around, but generally not in, germinal centers as in Fig. 11.16. In individual animals some thoracic and abdominal lymph nodes may contain little if any plutonium while other nodes may contain high concentrations. Plutonium particles were also identified in liver but not in bone. Both liver and bone tend to concentrate ionic plutonium to form "hot spots" of radiation dose (see Chaps. 9 and 10). Thus, the microscopic distribution of radiation dose varied over a very wide range in all tissues that contained an appreciable fraction of the body content.

Table 11.7. Plutonium in tissues of dogs after inhalation of $^{239}PuO_2$. (PARK et al., 1970)

Dog	Survival time (months after exposure)	Terminal body burden (μCi)	Pu concentration (μCi per gram wet tissue)					
			Lungs	Liver	Skeleton	Spleen	Thoracic lymph nodes	Abdominal lymph nodes
273	85	1.4	1.1 (7)	1.1 (21)	0.11 (5)	0.3 (0.6)	3700 (56)	160 (9)
278	93	0.8	1.2 (14)	0.5 (23)	0.11 (10)	0.4 (1.4)	1200 (41)	210 (10)
254	94	0.6	1.5 (21)	0.4 (21)	0.03 (5)	0.2 (0.7)	2600 (45)	92 (6)
109	110	0.4	0.7 (13)	0.3 (16)	0.04 (6)	0.1 (0.6)	420 (56)	90 (7)

The percentage of the body burden of plutonium in each tissue is given in parentheses.

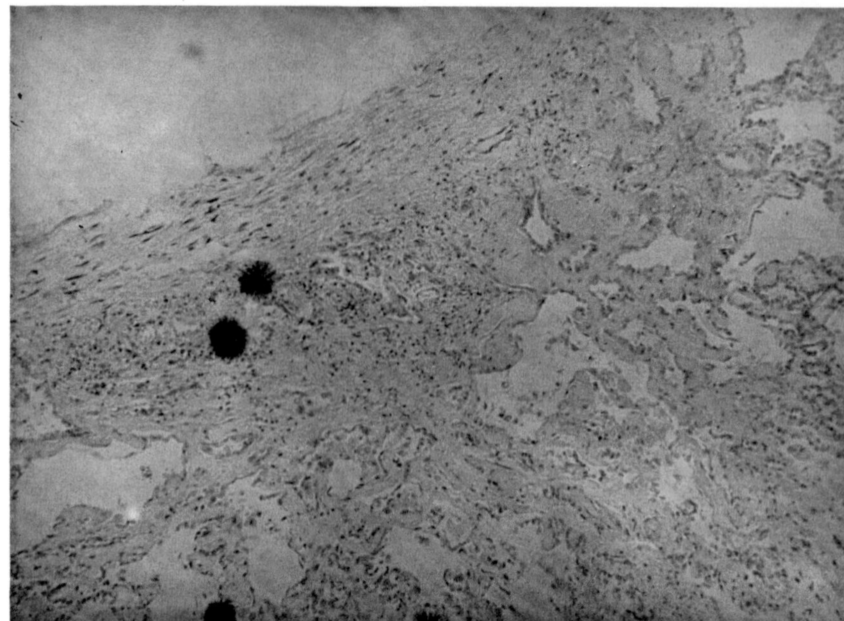

Fig. 11.12. Autoradiograph showing localization of plutonium in the area of subpleural lymphatics in a dog 933 days after inhalation of $^{239}PuO_2$ (88 ×). (PARK, unpublished)

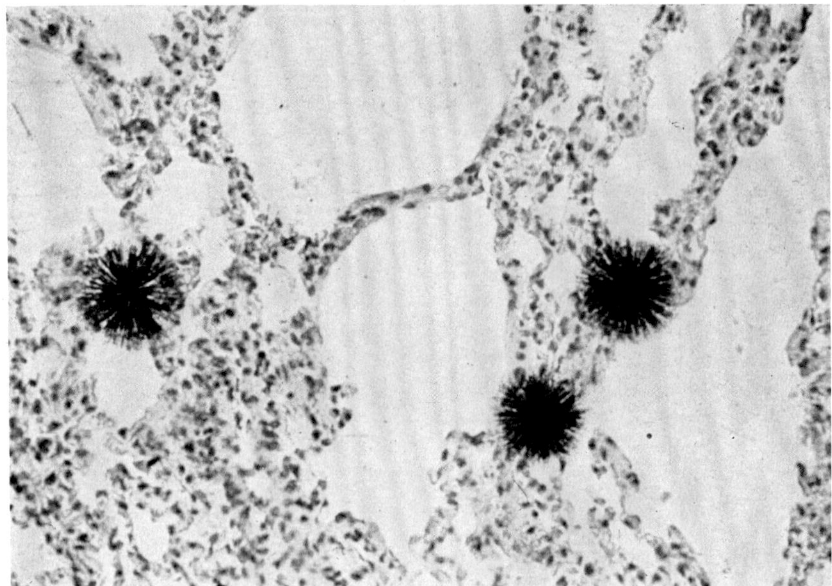

Fig. 11.13. $^{239}PuO_2$ particles in alveolar septa 7 days after exposure (autoradiograph, hematoxylin and eosin) (224 ×). (CLARKE and BAIR, 1964)

While the data from the long-term dog study provide a basis for development of a model for inhaled PuO_2, results from other experiments indicate that deviations from this model can be expected depending upon physical properties of

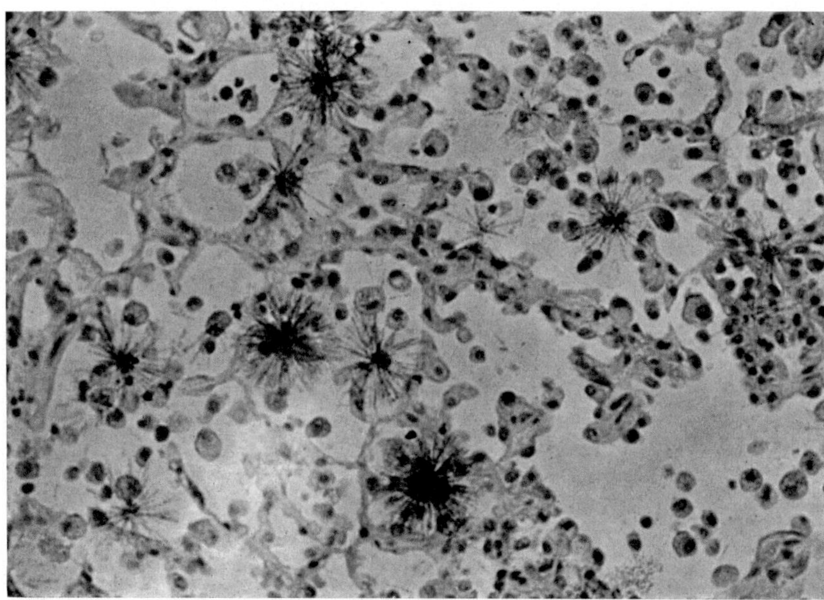

Fig. 11.14. PuO_2 particles in alveolar macrophages, 79 days postexposure (autoradiograph, hematoxylin and eosin) (224 ×). (CLARKE and BAIR, 1964)

the aerosol inhaled. For example, plutonium can undergo oxidation under a multitude of conditions of temperature, pressure, and environment. Four plutonium dioxides prepared under four different conditions were studied in dogs after a single inhalation exposure (BAIR and PARK, 1968). Three months after exposure greater than 90% of alveolar-deposited plutonium was retained in the lungs of all dogs except those that inhaled the oxalate calcined at 350° C, Table 11.8. In these dogs only 84% was retained in the lungs, the difference being due to a greater fraction accumulating in the bronchial and mediastinal lymph nodes. This plutonium dioxide was prepared similarly to that used in the long-term experiment described above. While this comparative study was done on relatively few dogs, the results are suggestive that the conditions under which plutonium is oxidized may affect the disposition of the inhaled PuO_2.

The particle size of the inhaled aerosol is also important. Data from dog experiments are plotted in Fig. 11.17. Thirty-day values for lung retention, amounts in the thoracic lymph nodes and other tissues and total excreted are given for five $^{239}PuO_2$ aerosols having different particle size distributions. The data are expressed as percent of alveolar-deposited plutonium; amounts deposited in upper respiratory tract are excluded. For all aerosols, the majority of the plutonium cleared from the lungs was excreted in the feces. Assuming that there was no secretion of systemic plutonium into the gastrointestinal tract, the amount excreted in the feces provides an estimate of the plutonium cleared from the deep respiratory tract by mucus-ciliary processes. This route was most effective for the inhaled aerosol with a MMD of 0.23 μm and least effective for aerosols with MMD's of 2 to 4 μm. These data indicate that particle size may influence the accumulation of plutonium in lymph nodes.

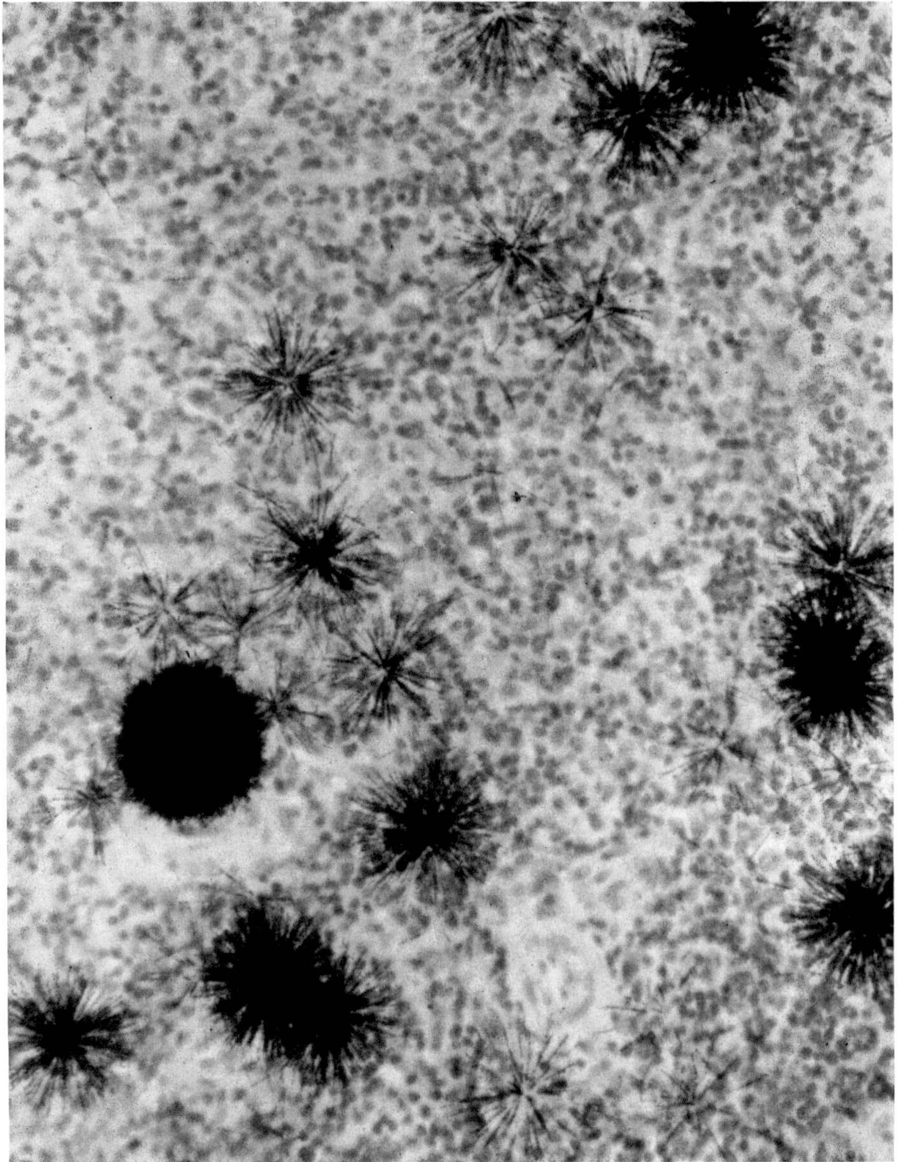

Fig. 11.15. Autoradiograph of tracheobronchial lymph node section from dog 30 days after inhalation of $^{239}PuO_2$. (Bair et al., 1964)

Translocation to tissues outside the respiratory tract and associated lymph nodes was also a function of particle size of the inhaled aerosol. A greater percentage of plutonium was translocated in dogs that inhaled the aerosol composed of the smallest particles. Since the translocation of plutonium from the respiratory tract probably involves transport by blood, it is reasonable to conclude that solubilization of PuO_2 particles in tissue fluids is certainly related to the size as well as to other surface properties of the particles.

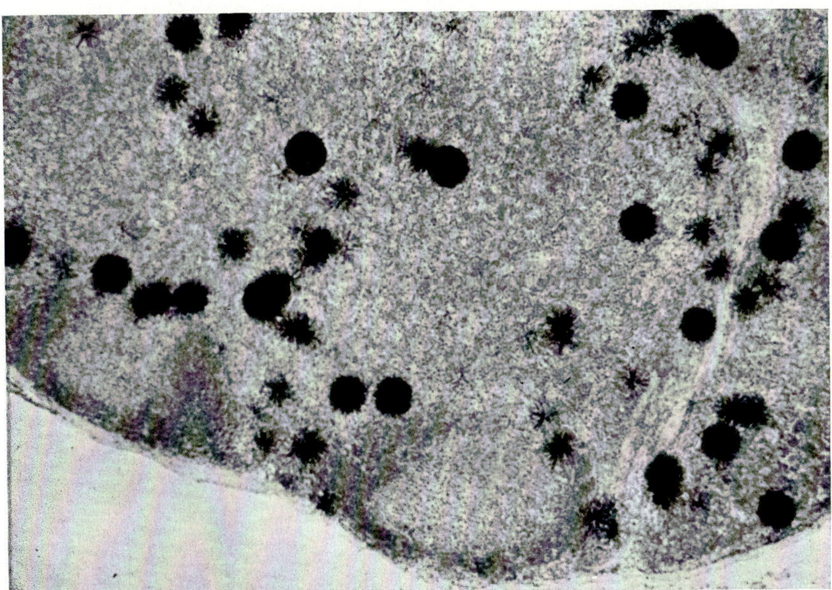

Fig. 11.16. PuO_2 particle deposition in the lymphatic sinuses and nodule peripheries of a bronchial lymph node, 14 days postexposure (autoradiograph, hematoxylin and eosin) (88 ×). (CLARKE and BAIR, 1964)

Data from a study of animals exposed to plutonium in a cloud from a high explosive detonation of a nuclear weapon are summarized in Table 11.9. The translocation of plutonium from the respiratory tract was qualitatively similar to that observed in experiments with relatively pure PuO_2; however, there were quantitative differences. For example, in all three animal species, the lymph node level was considerably less than reported for pure PuO_2 at comparable times after exposure and the kidney level was higher. Only data for one femur were reported. However, assuming the concentration of plutonium in the total skeleton was the same as in the femur, the total skeleton burden is estimated to be high relative to the other tissues. This suggests that some of the plutonium

Table 11.8. Disposition of four different plutonium dioxides in dogs (3 months after exposure). (BAIR and PARK, 1968)

PuO_2 inhaled	% of alveolar-deposited plutonium[a]					
	Excreted		Total body burden	Lung	Bronchial and mediastinal lymph nodes	All other tissues
	Urine	Feces				
I. Oxalate calcined at 1000° (2.8 μm MMD)	0.015	5.4	94.6	92	2.4	0.16
II. Oxalate calcined at 350° (2.8 μm MMD)	0.043	4.4	95.6	84.4	10.7	0.50
III. Metal oxidized at 450° (4.8 μm MMD)	0.039	2.4	97.6	92.5	4.7	0.48
IV. Metal oxidized at 123° (1.3 μm MMD)	0.013	5.1	94.9	91.3	3.3	0.36

[a] Values are means of three dogs.

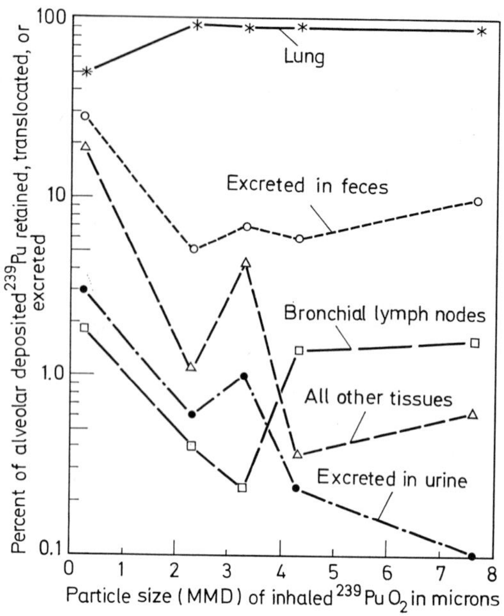

Fig. 11.17. Effect of particle size on retention, translocation and excretion of inhaled $^{239}PuO_2$ 30 days after exposure. (Each value is mean of three dogs.) (Bair et al., 1964)

contained in the inhaled particles might not have been an oxide, but more soluble compounds such as nitrates. These data, too, indicate that the physical and chemical form of inhaled plutonium is important to the long-term disposition in the body.

All of the above pertains to ^{239}Pu. A number of experiments with $^{238}PuO_2$ suggest that it behaves differently in the body than $^{239}PuO_2$; in particular, ^{238}Pu seems to be translocated more readily from the lung to other tissues in the body. An example of this is given in Table 11.10 showing the comparative distribution of ^{238}Pu and ^{239}Pu in rats (Stuart et al., 1968) exposed to PuO_2 aerosols. The particle size characteristics of the two aerosols were quite similar. The rats exposed to $^{238}PuO_2$ showed a relatively high rate of translocation to skeleton,

Table 11.9. Plutonium tissue burdens of animals exposed to cloud from high explosive detonation of a plutonium weapon. (Wilson and Terry, 1968)

	Expressed as percent of respirable (< 10 μm) aerosol											
Days after exposure	Dogs				Sheep				Burros			
	0	7	195	456	0	7	195	456	0	7	195	456
Tissue												
Hilar lymph node	0.60	1.0	1.7	0.80	0.05	0.01	0.11	0.06	0.04	0.0	0.08	0.03
Lung	22	24	7.9	3.5	9.4	1.9	0.42	0.68	18	9.3	2.3	2.5
Kidney	0.10	0.80	1.2	0.44	0.05	0.01	0.06	0.06	0.08	0.10	0.14	0.03
Liver	0.57	0.78	3.5	1.4	0.14	0.11	0.08	0.12	4.0	4.3	2.7	1.1
Femur	0.15	0.45	2.0	1.0	0.04	0.23	0.29	0.10	0.15	0.35	2.9	0.04

Table 11.10. Comparative distribution of ^{238}Pu and ^{239}Pu in rat tissues after inhalation of PuO_2. (STUART et al., 1968)

Days after exposure	Percent of terminal body burden										
	$^{239}PuO_2$[a]				$^{238}PuO_2$[a]						
	13	28	68	113	20	48	78	127	320	465	481
Tissue											
Thoracic lymph nodes	0.3	1.4	1.8	—	0.4	0.6	0.8	3.4	3.3	0.1	3.3
Lung	97	96	97	99	82	84	77	78	64	40	68
Spleen	0	0.02	0.04	0	0.13	0.16	0.2	0.2	0.7	0.6	0.5
Kidneys	0	0	0	0	0.5	0.4	0.4	1.1	0.5	3.3	0.5
Liver	0.02	0.4	0.3	0.3	2.4	1.3	1.9	1.7	1.6	3.6	3.6
Skeleton	0.01	0.1	0.1	0.16	11	12	23	15	30	49	23

[a] Characteristics of PuO_2. Oxalate calcined in air at 350° C. Aerosol CMD = 0.1 (± 0.02) for both ^{238}Pu and ^{239}Pu. Standard geometric deviation was 1.5–1.9 for ^{238}Pu and 1.7–2.0 for ^{239}Pu.

Table 11.11. Comparative distribution of ^{238}Pu and ^{239}Pu in dog tissues after inhalation of PuO_2

	Percent of terminal body burden[a]					
	$^{239}PuO_2$ (BAIR, 1970)			$^{238}PuO_2$ (PARK et al., 1970)		
Days after exposure	150	150	150	94	125	180
Tissue						
Thoracic lymoh nodes	2.4	4.1	8.0	3	9	4
Lung	96	94	91	91	80	77
Spleen	0.007	0.005	0.02	0.3	—	—
Liver	0.7	0.6	0.3	1	3	5
Skeleton	0.4	0.5	0.3	4	7	13

Characteristics of PuO_2:
^{238}Pu — Oxalate calcined in air at 700° C. CMD = 0.04 μm, MMD = ~0.1 μm.
^{239}Pu — Oxalate calcined in air at 900° C. CMD = 0.05 μM, MMD = ~0.12 μm.
[a] Values are for individual dogs.

11% of the body burden at 20 days and over 20% at times of a year or more after exposure. The translocation to spleen and kidney was also greater than in rats exposed to $^{239}PuO_2$. Accumulation of plutonium in thoracic lymph nodes was about the same for the two isotopes.

Similar observations were made in dogs (PARK et al., 1969; BAIR, 1970), Table 11.11. Data from two experiments are compared. The particle size characteristics in the experiments chosen for comparison were similar for the $^{238}PuO_2$ and $^{239}PuO_2$ aerosols. Again the dogs exposed to $^{238}PuO_2$ showed a higher rate of translocation to skeleton, up to 13% of the body burden after 6 months, compared with less than 1% for the ^{239}Pu dogs. Liver and other tissues also reflected this difference between the two plutonium isotopes. This relatively high rate of translocation from lungs to skeleton and other tissues is a general finding for $^{238}PuO_2$ prepared under a variety of conditions, Table 11.12, and was even more apparent two years after exposure.

Table 11.12. Distribution and retention of plutonium in dogs after inhalation of $^{238}PuO_2$ in various forms. (WILLARD and PARK, 1970)

Aerosol	Time post-exposure (months)	Retention half-time (days)		Terminal body burden (μCi)	Plutonium distribution (Percent of terminal body burden)			
		Whole-body	Lungs		Lungs	Thoracic lymph nodes	Skeleton	Liver
$^{238}PuO_2$–ZrO_2[a]	3	400	150	0.1–0.2	72	11	9	3
$^{238}PuO_2$–ThO_2[a]	3	3600	310	3.3–6.5	83	13	3	1
$^{238}PuO_2$–C.M.[b]	2–3	2200	600	33–162	93	5	1	1
$^{238}PuO_2$–C.M.	22	5600	1100	3.1	72	7	12	7
$^{238}PuO_2$–M[c]	22	—	—	0.8	28	46	15	9
$^{238}PuO_2$–700° C[d]	1–6	900	300	25–261	90	4	4	2
$^{238}PuO_2$–350° C[e]	23	—	—	3.0	4	4	64	23

[a] Mean of 3 dogs. [b] Mean of 6 dogs exposed to crushed microspheres (C.M.).
[c] Inhaled intact 50 μm microspheres, only fragments retained.
[d] Mean of 12 dogs exposed to $^{238}PuO_2$ calcined at 700° C. [e] Calcined at 350° C.

Thus, experimental animal data indicate that $^{238}PuO_2$ is relatively readily transported from the respiratory tract to other tissues in the body—that it is more transportable than $^{239}PuO_2$ even when both compounds were prepared under similar conditions, exhibit similar aerosol characteristics, and are indistinguishable by all tests including X-ray diffraction, but do differ in specific activity. The specific activity is 17.4 Ci/g for ^{238}Pu and 6.17×10^{-2} Ci/g for ^{239}Pu. It has been proposed that the relatively high specific activity of ^{238}Pu causes spallation from the surface of the particles—that $^{238}PuO_2$ particles are relatively unstable disintegrating to smaller particles of increased solubility in tissue fluids. Also, because of the energy in each plutonium particle, the microenvironment around the particle is at a relatively high temperature. This could increase the rate of solubility of the $^{238}PuO_2$ particles. This is consistent with the observed accumulation of ^{238}Pu in bone after inhalation of $^{238}PuO_2$.[4]

C. Excretion of Plutonium

Because of severe limitations of *in vivo* counting technology for plutonium, extensive efforts have been made to correlate the rate of plutonium excretion with body burden. This is dealt with in detail in Chap. 13. Extensive excretion data have been collected in animal experiments to guide those concerned with interpreting human bioassay data and to provide a better understanding of the disposition of various inhaled plutonium compounds.

As mentioned previously, the excretion of plutonium indicates the relative roles of the several mechanisms active in clearing plutonium from the body. Plutonium excreted in feces after an inhalation exposure can be assumed to be largely composed of plutonium cleared from the lung via the mucus-ciliary process. However, care must be exercised in interpreting excretion data. For example, data have been obtained to indicate that, depending upon the com-

4 Plutonium-238 and plutonium-239 also behave differently when given by other routes of administration, Chap. 9 and 10. Editor.

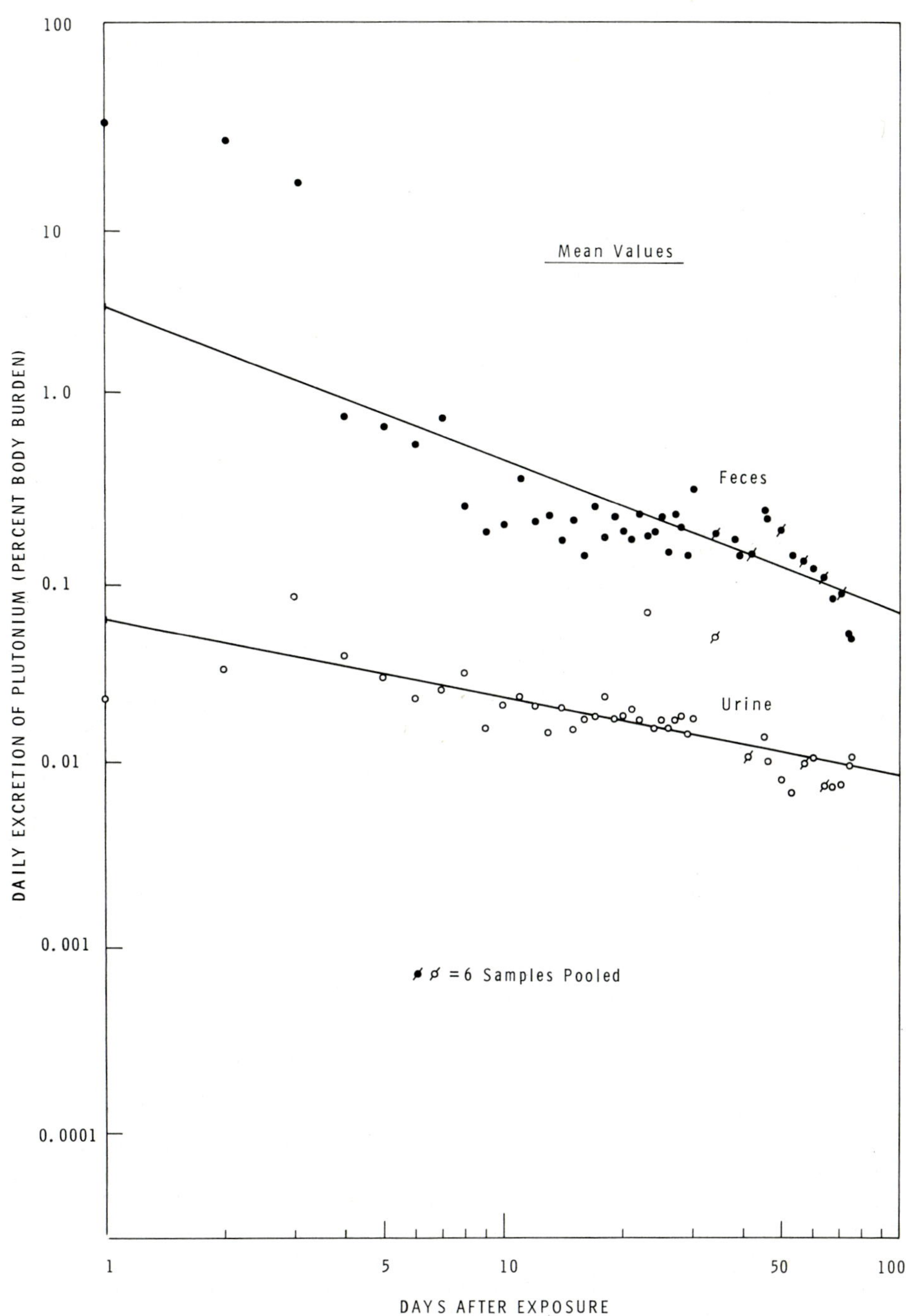

Fig. 11.18. Excretion of plutonium nitrate in dogs. (Stuart et al., 1968)

pound inhaled, plutonium can be secreted into the gastrointestinal tract by the bile (Matsusaka and Bair, 1968). This could in some circumstances comprise an appreciable fraction of the total plutonium excreted in feces.

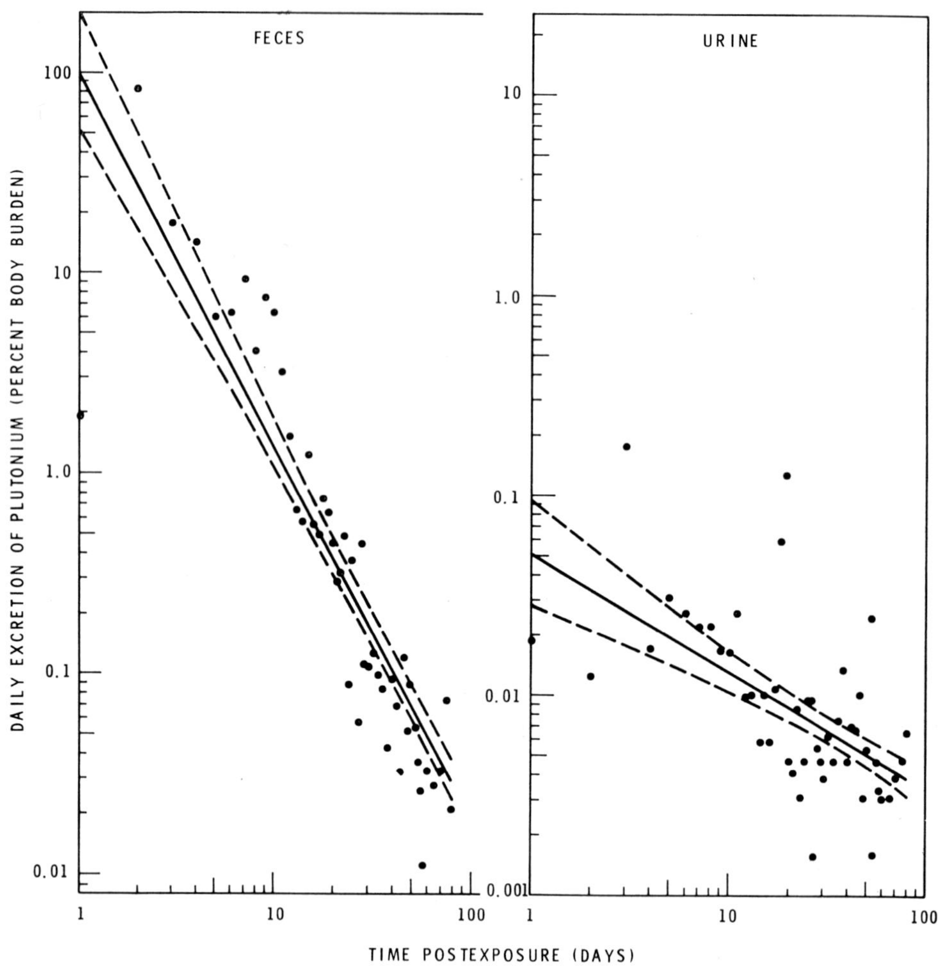

Fig. 11.19. Daily excretion of ^{239}Pu after inhalation of $^{239}PuF_4$ (dog 429). (BAIR, 1970)

As far as is known, plutonium appearing in urine is derived from solubilized plutonium entering the blood from the respiratory tract or other tissues in which plutonium is deposited.

The daily excretion of plutonium following inhalation of plutonium nitrate, plutonium fluoride, and plutonium dioxide is illustrated in the next three figures taken from reports of dog experiments. The amount of plutonium excreted each day is expressed as percentage of the total body burden at the beginning of that day. An example of the daily excretion of plutonium following inhalation of $^{239}Pu(NO_3)_4$ is shown in Fig. 11.18. The rate of fecal excretion is an order of magnitude higher than urinary excretion, but decreases more rapidly. Although the data are unavailable, it would appear that after several months, the rate of excretion in urine would exceed that in feces. This appears to happen much earlier in the case of PuF_4 inhalation, Fig. 11.19. Within 50 days after exposure, the daily excretion of plutonium in urine equaled that in feces. The rates of urinary excretion of plutonium are similar for both $Pu(NO_3)_4$ and PuF_4 in the

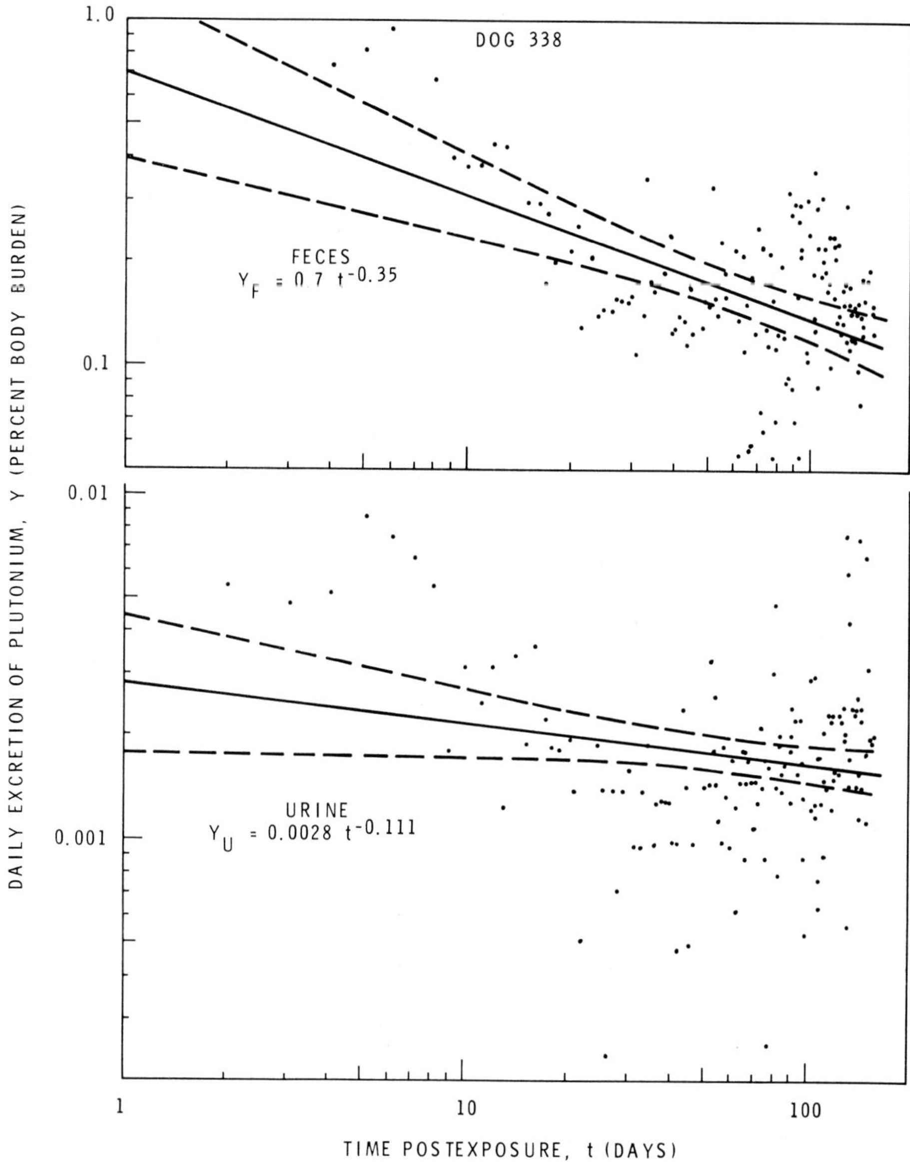

Fig. 11.20. Daily excretion of ^{239}Pu after inhalation of $^{239}PuO_2$ (CMD = 0.05 μm, MMD = 0.12 μm). (Bair, 1970)

examples shown. On the other hand the excretion of plutonium in urine after inhalation of PuO_2 is almost an order of magnitude less than for $Pu(NO_3)_4$ and PuF_4, and almost three orders of magnitude less than excretion of plutonium in feces, Fig. 11.20. In the long-term dog experiment with $^{239}PuO_2$ referred to previously, urinary and fecal excretion rates are about the same after several years, with a tendency for urinary levels to be slightly higher than fecal elimination rates.

Table 11.13. Constants for Pu excretion equations[a] (0 to ~90 days). (BAIR, 1970)

Plutonium compound		Particle size		Urine		Feces	
		CMD	MMD	A	b	A	b
$^{239}PuO_2$	1000° oxalate	0.51	2.8	0.02	−1.3	4	−1.1
	350° oxalate	0.45	2.8	0.03	−1.1	16	−1.5
	450° metal	0.50	4.8	0.1	−1.5	16	−1.7
	123° metal	0.46	1.3	0.03	−1.4	18	−1.5
	900° oxalate	0.05	0.12	0.01	−0.5	1	−0.7
$^{238}PuO_2$	750° oxalate	0.05	0.1	0.01	−0.3	3	−1.2
$^{239}PuF_4$		0.2		0.2	−0.6	80	−1.8
$^{239}Pu(NO_3)_4$		0.12	0.6	0.06	−0.4	3	−0.8

[a] $\gamma = At^b$, γ = Percent of body burden excreted on day, t.

In Table 11.13, constants for excretion equations are given for several plutonium compounds. These were all calculated over the period 0 to 90 days after exposure. Power function equations were fitted to the data. High intercept values and low slope values are characteristic of relatively high levels of excretion. Conversely, low intercept and high slope values characterize low levels of excretion. Values for urinary excretion of plutonium compounds can be separated into two groups. The low excretion rate group is comprised of four oxides having a particle size >1 to about 5 μm, MMD. The high rate group includes $^{239}PuO_2$ and $^{238}PuO_2$ with a particle size of about 0.1 μm, MMD, PuF_4 and $Pu(NO_3)_4$. For all forms of plutonium except PuF_4 and the 450° C PuO_2, the intercept values were remarkably similar. Thus, the major difference was the rate at which the daily excretion rate decreased. Thus, urinary excretion of plutonium appears to be quite dependent upon the chemical form and particle size of the inhaled material.

The high fecal excretion of plutonium observed after inhalation of PuF_4, Fig. 11.19, is reflected in the high intercept value in Table 11.13. The 900° C oxalate comprised of a distribution of small particles and $Pu(NO_3)_4$ showed very similar levels of excretion in feces.

The data available from animal experiments on excretion of plutonium after an inhalation exposure clearly indicate an important influence of chemical and physical properties of the inhaled aerosol. These results have significant implications on the interpretation of human bioassay data. Measurement of a single parameter such as rate of excretion of plutonium in urine does not seem to be an adequate basis on which to estimate plutonium lung burdens, especially in those cases where little is known about the material to which an individual was exposed.

III. Interaction of Plutonium Particles with Cells

The disposition of plutonium particles in the body and the resulting biological effects depend upon the interaction of the particles with the cellular elements with which they come in contact. The alveoli of the lung contain reticuloendothelial cells derived from circulating monocytes which originate in the bone marrow (PINKETT et al., 1966). These reticuloendothelial cells consist of mononuclear cells and histiocytes within the septal walls and alveolar macrophages in the air spaces. Macrophages are capable of phagocytizing plutonium particles deposited in the alveoli within a few hours after exposure to

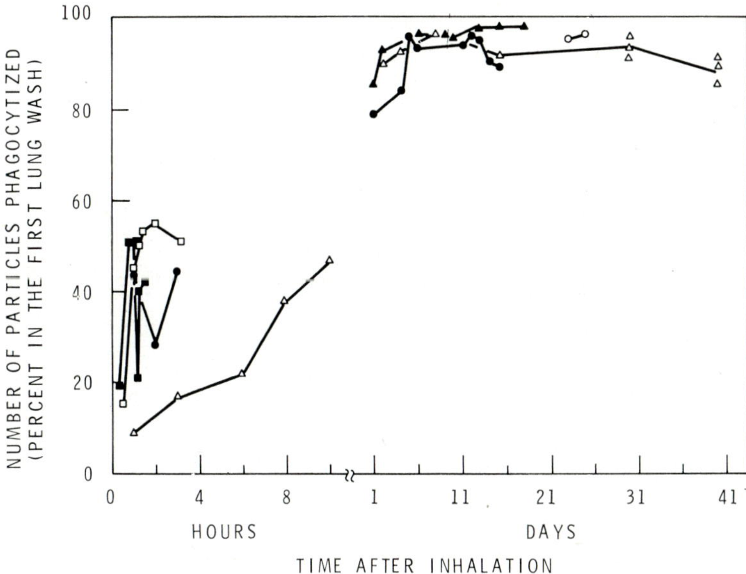

Fig. 11.21. Phagocytosis of inhaled $^{239}PuO_2$ particles in rat lung. (Results derived from lung washes. Each symbol represents a different ^{239}Pu lung burden ranging from 0.01 to 0.3 μCi with the exception of those represented by the open triangles which were 1 to 4 μCi.) (SANDERS, 1969)

a plutonium aerosol, Fig. 11.21 (SANDERS, 1969). Phagocytized plutonium particles are rapidly localized in the phagolysosomes of reticuloendothelial cells. This has been observed in alveolar macrophages (SANDERS and ADEE, 1970), in peritoneal macrophages (SANDERS and ADEE, 1969), and in liver Kupffer cells (RAHMAN and LINDENBAUM, 1964). Phagocytosis of plutonium particles by macrophages is essentially completed within 4 hours after deposition (SANDERS, 1969). The fraction of alveolar macrophages engulfing PuO_2 particles was proportional to the amount of plutonium deposited in the alveoli. Nearly all PuO_2 in saline washes from rat lungs is found in macrophages. It has been estimated that over 50% of alveolar deposited PuO_2 particles is phagocytized by alveolar macrophages within the first few weeks after exposure (SANDERS, 1972). While the mechanism is not clearly known, the alveolar macrophage is capable of transporting plutonium particles from the alveoli to the ciliated epithelium of the bronchioles. These cells containing plutonium particles can then be removed from the lung in the mucus blanket propelled up the respiratory passages by ciliary action. This constitutes an important mechanism for removal of plutonium and other particles from the lung.

Plutonium particles not phagocytized by alveolar macrophages and removed from the lung are found in Type I alveolar epithelial cells, in peribronchiolar vascular areas, and in pulmonary lymphatics. Type I alveolar epithelial cells phagocytize plutonium particles rapidly, within a few hours, after deposition (SANDERS and ADEE, 1970). Fig. 11.22 shows plutonium particles within rat Type I alveolar epithelial cells 60 minutes, 90 minutes, and 1 week after inhalation of plutonium. The fate of particles phagocytized by Type I epithelial cells is not known. These cells appear to be relatively radioresistant. Fine structural studies of lung exposed to radiation doses up to 13000 rads from ^{239}Pu showed no evidence

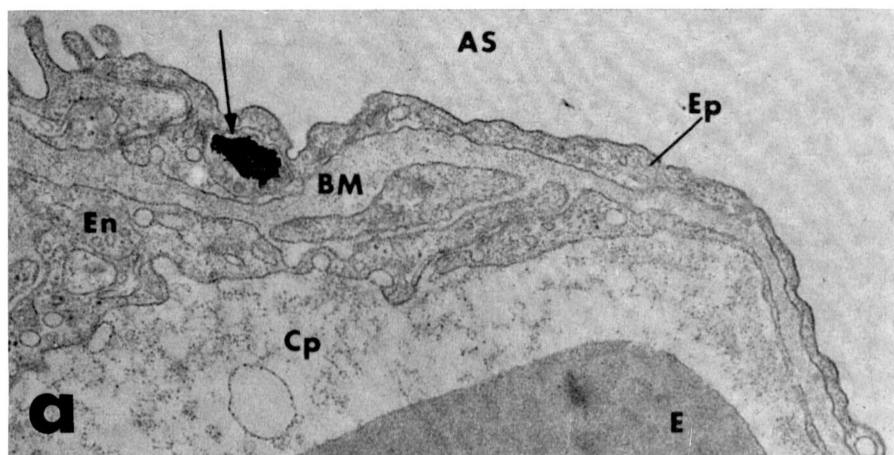

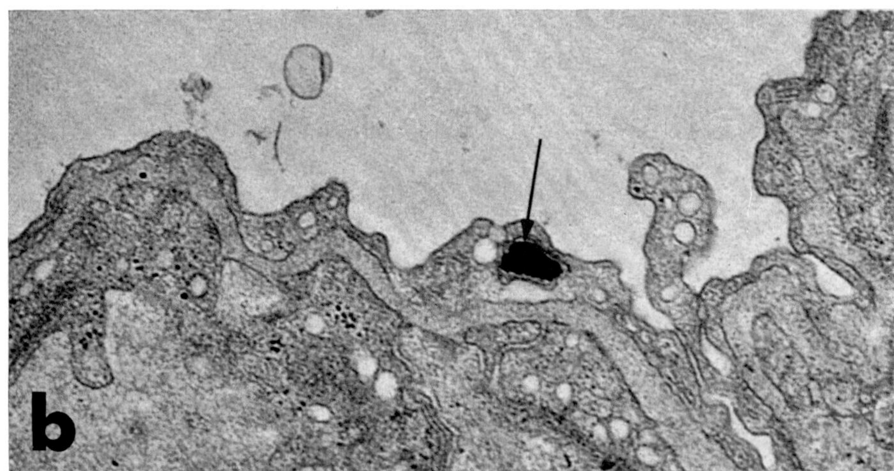

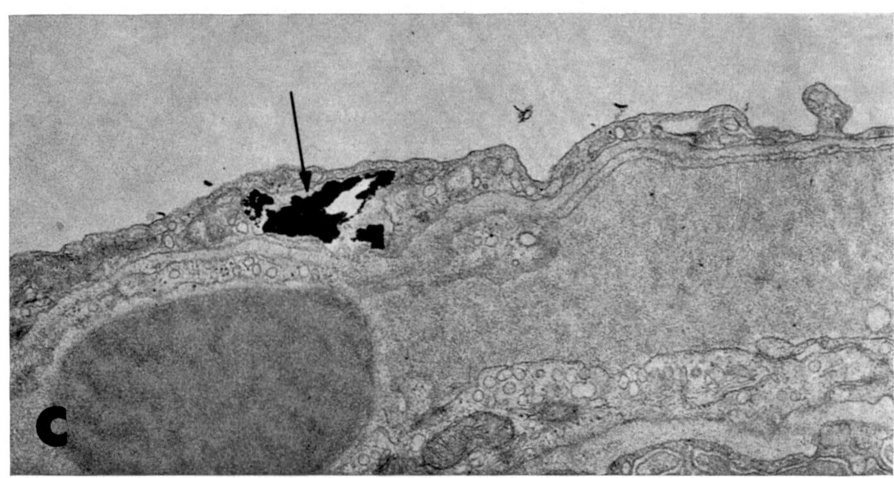

Fig. 11.22 a–c

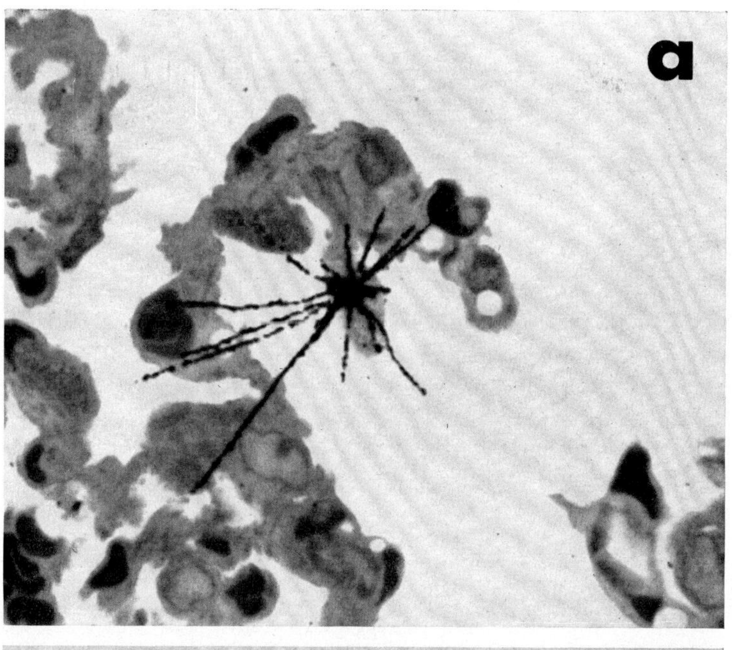

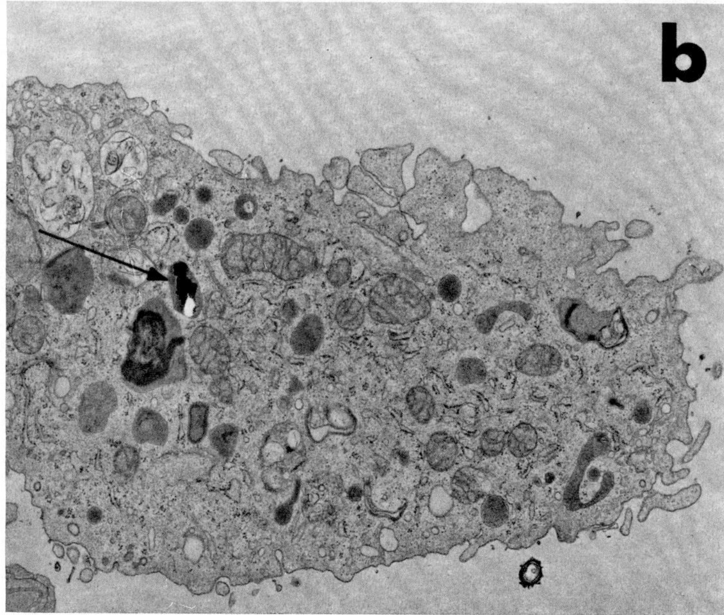

Fig. 11.23. a Autoradiograph showing the presence of plutonium within an alveolar macrophage; epon section, methylene blue and azure II stains (1300 ×). b Electron micrograph taken from a thin section immediately adjacent to the thick section in (a). The plutonium particle (arrow) is shown localized within a granular inclusion. (10000 ×). (SANDERS, 1970)

Fig. 11.22. a Plutonium particle localized within type I alveolar epithelium (arrows) 60 min after exposure. *AS* alveolar sac, *Ep* epithelial cell, *BM* basement membrane, *En* endothelial cell, *Cp* capillary space, *E* erythrocyte. (20000 ×). b Plutonium particle localized within type I alveolar epithelium (arrow) 90 min after exposure. (30000 ×). c Plutonium particles localized within type I alveolar epithelium (arrow) 1 week after exposure. (15000 ×). (SANDERS, 1970)

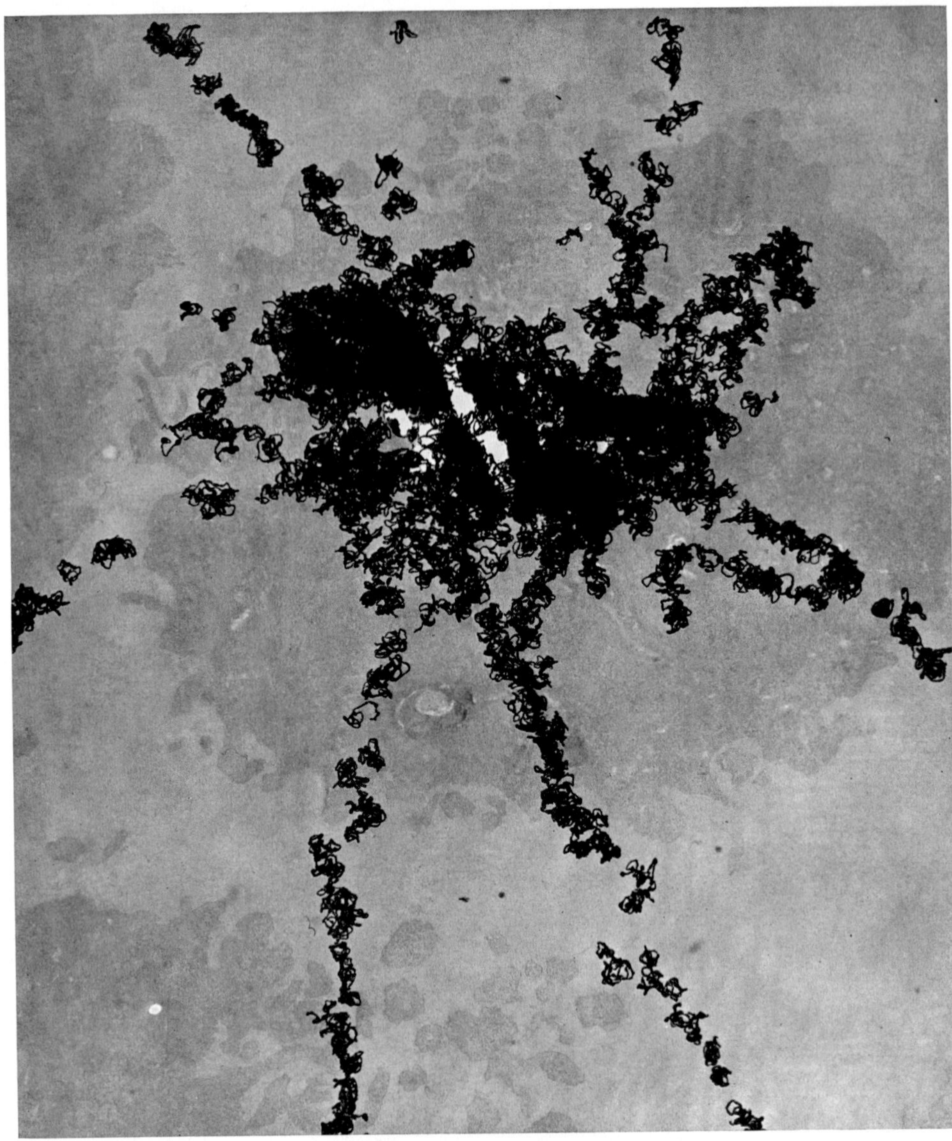

Fig. 11.24. Electron micrograms of an autoradiogram of a rat macrophage containing ^{239}Pu particles. (The rat was sacrificed 2 hours after intraperitoneal administration of $^{239}PuO_2$.) (Adee et al., 1968)

of Type I cell detachment or removal (Sanders et al., 1971). If particles contained in Type I cells are not readily removed from the lung by physiological process, it could account for the long retention time of PuO_2, Table 11.2.

Type II alveolar epithelial cells, or the so-called "granular pneumonocytes" do not phagocytize plutonium particles or other particles such as India ink, latex beads, or metal oxides (Sanders et al., 1970, 1971; Esterly and Faulkner, 1970) and, thus, are not directly involved in clearance of particles from the lung.

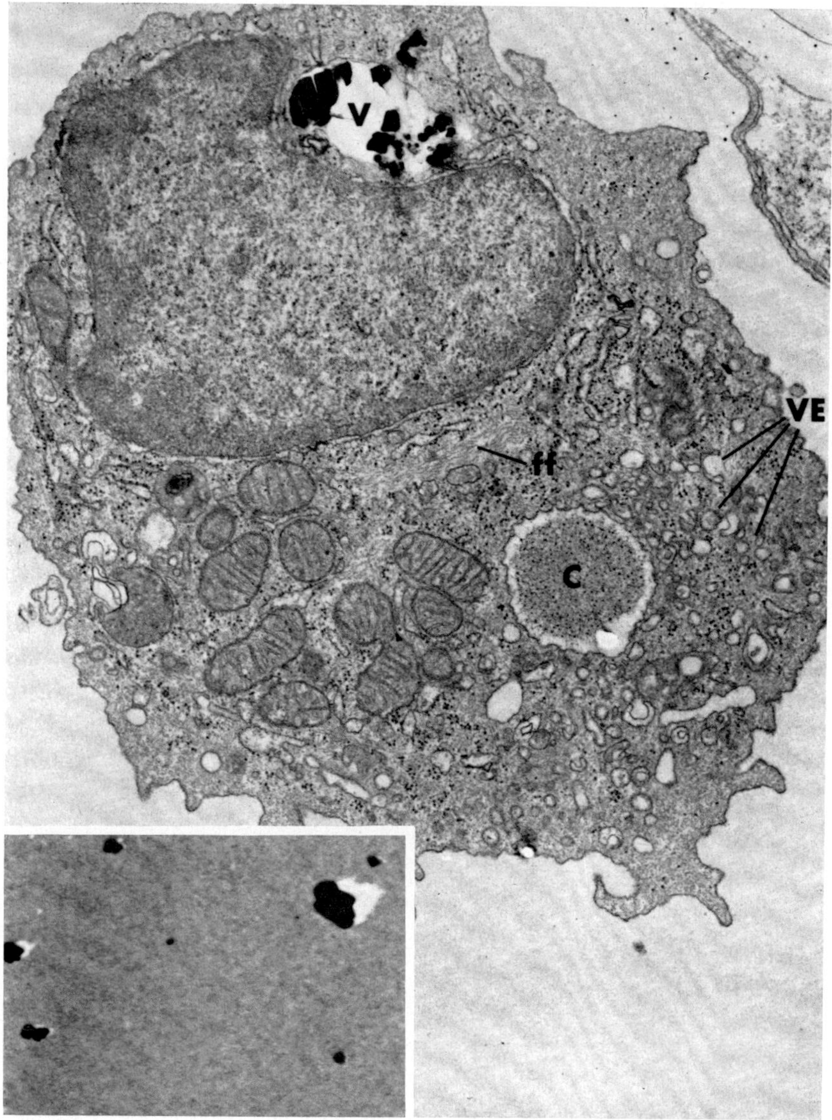

Fig. 11.25. Engulfed $^{239}PuO_2$ particles in a rat mononuclear phagocyte 24 hours after intraperitoneal injection. (The insert shows plutonium particles in the injected suspension.) (34000 ×.) (SANDERS and ADEE, 1969)

Yet, in rats, inhaled plutonium caused proliferation, desquamation, and clearance of Type II epithelial cells out to the ciliated bronchioles (SANDERS et al., 1971).

The intracellular localization of plutonium particles within pulmonary macrophages has been demonstrated by autoradiography of smears and lung sections (SANDERS, 1969). Fig. 11.23a is an autoradiographic (light level) demonstration of the phagocytosis of plutonium particles by lung macrophages. Fig. 11.23b is an electron micrograph of an adjacent section showing a plutonium particle within a granular inclusion.

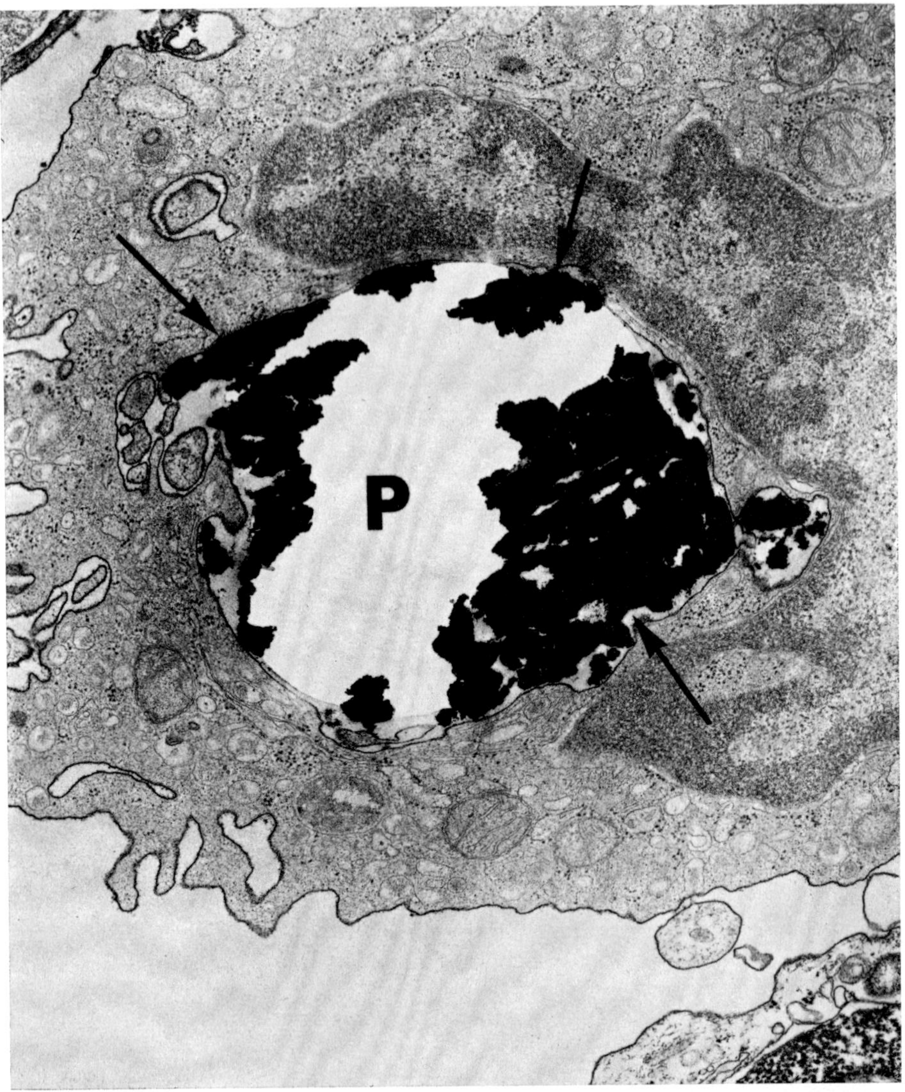

Fig. 11.26. Peritoneal mononuclear phagocyte at 1 hour after administration of 19 μCi of $^{239}PuO_2$ particles. A large mass of very electron-dense plutonium particles (arrows) is concentrated within a large, membrane-bound phagocytic vacuole (*P*). The clear area in the center of the vacuole is an artifact. (23000 ×). (SANDERS, 1969)

Plutonium particles were located and identified within macrophages at the subcellular level by the technique of electronmicroscopic autoradiography, Fig. 11.24. The paths of alpha particles emitted from the plutonium within the cell formed a "star" of tracks of reduced silver halide grains.

Fig. 11.25 is an example of several plutonium particles aggregated within a phagosome near the cell surface. Insert is an electronmicrogram of particles from a saline suspension (1.4 μCi/ml) deposited directly on an electron microscope grid.

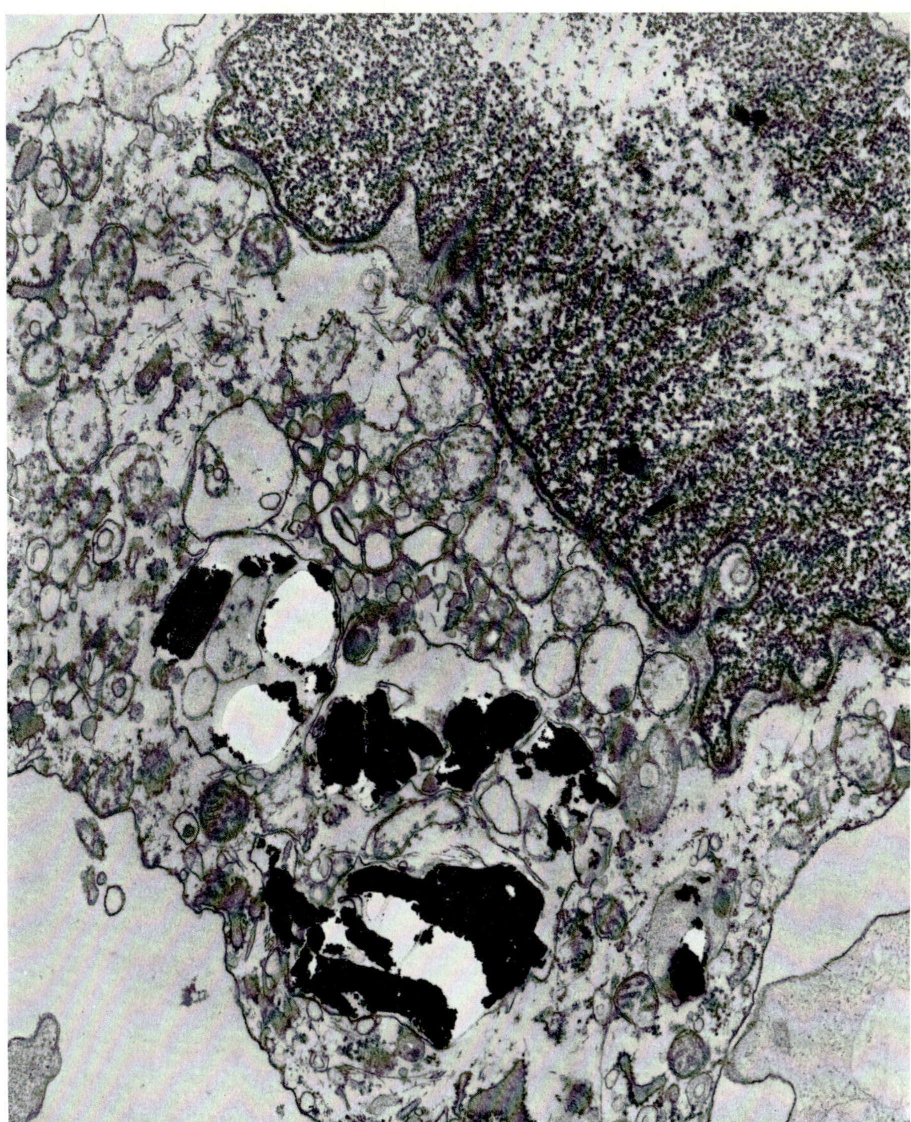

Fig. 11.27. Peritoneal mononuclear phagocyte at 1 hour after administration of 19 μCi of $^{239}PuO_2$ particles. Well-defined individual plutonium particles are present within several phagocytic vacuoles. This phagocyte displays lethal alteration changes leading to the release of engulfed plutonium particles. (23000 ×). (SANDERS, 1969)

The marked accumulation of plutonium particles within phagosomes of macrophages 1 hour after intraperitoneal injection of 19 μCi is shown in Fig. 11.26.

Lethal alterations to macrophages may occur as early as 1 hour after phagocytosis of large amounts of plutonium, Fig. 11.27. However, most cells do not show such early effects of the large radiation dose delivered by a large mass of plutonium located within the cell. Calculated radiation dose rates delivered to cells containing plutonium particles of several sizes are given in Table 11.14.

Table 11.14. Radiation dose to phagocytes containing $^{239}PuO_2$ particles[a]

Particle diameter (μm)	Alpha radiation dose (rad/hour)
0.05	0.04
0.10	0.3
0.25	5
0.50	40
1.0	320

[a] Assuming a point source ^{239}Pu particle emitting alphas with a LET of 100 keV/μm located in the center of a 12 μm diameter mononuclear phagocyte; thus, 12% of the emitted alpha energy is absorbed by the phagocyte

A 1 μm $^{239}PuO_2$ particle will deliver over 300 rads per hour to the cell in which it is contained. Significant alterations of macrophage structure occurred within 7 days in rats given intraperitoneal injections of 1.4 μCi $^{239}PuO_2$. These included a marked increase in cell size, increased surface activity, and marked phagolysosomal activity due to engulfment of cellular debris.

Among the alterations in macrophages attributable to plutonium by different authors are a decreased incorporation of ^3H-thymidine into macrophages (Masse, 1970), a reduced phagocytosis of latex test particles by $^{239}PuO_2$-laden macrophages (Sanders, 1970), an alveolar-depressed clearance of iron oxide dust following a prior exposure to ^{239}Pu aerosol (Nenot, 1971), and a net decrease in the numbers of saline extractable alveolar macrophages following inhalation of ^{239}Pu (Sanders, 1971; Masse, 1971).

Table 11.15. Biological effects of inhaled plutonium

Biological effect	Mean survival time of dose group (days)			Number of animals in dose group			Effect incidence (%)		
	Controls	Exposed		Controls	Exposed		Controls	Exposed	
		+	−		+	−		+	−
Ammonium ^{239}Plutonium pentacarbonate									
LSS	575±9	488±14	563±11	244	83	121	——	——	——
PN	575±9	571±21	——	244	48	——	0.4	4.2	——
PS	575±9	571±21	——	244	48	——	38.2	91.8	——
OS	575±9	571±16	571±21	244	102	48	0	1.0	0
PL	Inconsistent changes								
RI	——	——	——	——	——	——	——	——	——
Plutonium citrate									
LSS	575±9	476±16	576±13	244	100	132	——	——	
PN	575±9	616±13	——	244	152	——	0.4	4.4	——
PS	575±9	616±13	——	244	152	——	38.2	60.9	——
OS	575±9	616±13	——	244	152	——	0	1.3	——
PL	——	——	——	——	——	——	——	——	——
RI	——	——	——	——	——	——	——	——	——

[a] Buldakov et al., 1969. [b] Buldakov et al., 1970. [c] Koshurnikova et al., 1968. [d] Antonchenko et al., 1969.

+ Positive effect dose group, +D Smallest dose at which the specified biological effect *WAS* observed.

Many of the fine structural alterations observed in the lung following inhalation of ^{239}Pu (SANDERS et al., 1971) are simiar to those described following external irradiation of the lung. General features were characterized by the accumulation of a membranous-proteinaceous exudate in the air spaces, disruption of the capillary endothelium, and accumulation of collagen, elastin, fibroblasts, plasma cells, and mast cells in greatly thickened septal walls. The response of the type II epithelium is particularly interesting since these cells may serve as precursors for epithelial neoplasms induced by inhaled plutonium (CLARKE et al., 1966; HOWARD, 1970).

IV. Biological Effects of Inhaled Plutonium

The most needed information about the toxicity of a potential environmental contaminant such as plutonium is its low-level, long-term effect on human life. Such information has to be derived from animal experimentation. To do this, several laboratories have studied the biological effects of very high doses of inhaled plutonium as well as those effects occurring at moderately low doses. The long-term experiments required to provide the information at low doses are costly, difficult to perform, and even more difficult to interpret. The latter occurs because in long-term studies there is a possibility of other influencing factors such as common disease processes and extraneous environmental forces. Hopefully, the results can be extrapolated satisfactorily to the very low doses which man is most likely to encounter. The following summarizes knowledge of the biological effects of inhaled plutonium gained from numerous experiments with rodents and dogs. This information is compiled in Tables 11.15–11.19. The tables

citrate and ammonium plutonium pentacarbonate in rats

Dose												References
Whole body (μCi/kg)		Lung (μCi/g)		Total radiation dose (rads)				Dose rate (rads/day)				
				Lungs		Skeleton		Lungs		Skeleton		
+D	−D	+D	−D	+D	−D	+D	−D	+D	−D	+D	−D	
0.9	0.3	0.154	0.045	1080	327	247	76	2.2	0.58	0.51	0.14	a,b,c,d
0.024	?	0.004	?	27	?	7	——	0.05	?	0.01	——	c,d
0.024	?	0.004	?	27	?	7	——	0.05	?	0.01	——	c,d
0.04	0.024	0.007	0.004	58	42	14	7	0.10	0.05	0.025	0.01	c,d
												a
1.8	0.5	0.3	0.09	1100	<1100	——	——	11	<11	——	——	a
0.9	0.24	0.15	0.04	880	240	265	76	2.0	0.4	0.56	0.13	a,b,c,d
0.05	?	0.008	?	43	?	15	——	0.07	?	0.024	——	c,d
0.05	?	0.008	?	43	?	15	——	0.07	?	0.024	——	c,d
0.05	?	0.008	?	43	——	15	?	——	——	0.024	?	c,d
1.8	0.3	0.3	0.06	——	——	130	<130	——	——	——	——	a
1.8	0.5	0.3	0.09	1100	<1100	——	——	11	<11	——	——	a

— Negative effect dose group, −D Largest dose at which the specified biological effect *WAS NOT* observed.
? Largest dose that does not cause the specified biological effect is unknown.
—— Information not available
LSS = Life-span Shortening, PN = Pulmonary Neoplasia, PS = Pneumosclerosis, OS = Osteosarcoma, PL = Panleukopenia, RI = Respiratory insufficiency.

Table 11.16. Biological effects

Biological effect	Mean survival time of dose group (days): Controls	Exposed +	Exposed −	Number of animals in dose group: Controls	Exposed +	Exposed −	Effect incidence (%): Controls	Exposed +	Exposed −
^{239}Pu nitrate (intratracheal injection) in rats									
LSS	586±20	540±15	618±13	42[a]	87	93	——	——	——
PN	586±20	584±16	612±17	42[a]	96	93	2.4	4.2	2.2
PS	586±20	618±13	584±16	42[a]	93	96	64.3	90.3	78.3
OS	586±20	618±13	584±16	42[a]	93	96	0	1.1	0
STT	672± 8	633±15	——	248	89	——	58.0	74.0	——
ET	672± 8	633±15	——	248	89	——	58.0	74.0	——
0.3 ml 0.01 N HNO_3 (intratracheal injection) in rats									
LSS	672±8	586±20	——	248	42[a]	——	——	——	——
PN	672±8	586±20	——	248	42[a]	——	0	2.4	——
PS	672±8	586±20	——	248	42[a]	——	23.4	64.3	——
OS	672±8	——	586±20	248	——	42[a]	0	0	0
$^{239}Pu(NO_3)_4$ in dogs									
LSS	>300	175–300	——	6	4	——	——	100	——
PN	>300	——	75–300	6	——	4	0	0	——
PF (PS)	>300	175–300	——	6	4	——	0	100	——
OS	>300	——	75–300	6	——	4	0	0	——
PL	>300	175–300	——	6	4	——	0	100	——
RI	>300	175–300	——	6	4	——	0	100	——
LYMDS	>300	175–300	——	6	4	——	0	100	——
BAHM	>300	175–300	75–300	6	11	——	0	100	——

[a] Data from the 0.01 N HNO_3 injected animals used as control values for plutonium nitrate study. [b] Buldakov et al., 1969. [c] Buldakov et al., 1970. [d] Erokhin et al., 1969. [e] Koshurnikova et al., 1968. [f] Park et al., 1968.

\+ Positive effect dose group, +D Smallest dose at which the specified biological effect *WAS* observed.

— Negative effect dose group, —D Largest dose at which the specified biological effect *WAS NOT* observed.

Table 11.17. Biological effects

Biological effect	Mean survival time of dose group (days): Controls	Exposed +	Exposed −	Number of animals in dose group: Controls	Exposed +	Exposed −	Effect incidence (%): Controls	Exposed +	Exposed −
LSS	693–738[a]	——	661–732[a]	144	——	138	——	——	——
PN[b]									
LSS[c]	335–500	——	323–500	86	——	11	——	——	——
PN[c]	335–500	400	500	86	17	21	0	12	0
PF[c]	335–500	400	——	86	21	——	——	high	——
BAHM[c]	335–500	400	——	86	21	——	——	high	——
TS	335–500	500	——	86	21	——	0	5	——

[a] Average life-span. [b] None reported but not all animals were examined. [c] Intratracheal injection. [d] Bair et al., 1962. [e] Temple et al., 1960.

\+ Positive effect dose group, +D Smallest dose at which the specified biological effect *WAS* observed.

of inhaled plutonium nitrate

Dose												References
Whole body (μCi/kg)		Lung (μCi/g)		Total radiation dose (rads)				Dose rate (rads/day)				
				Lungs		Skeleton		Lungs		Skeleton		
+D	−D	+D	−D	+D	−D	+D	−D	+D	−D	+D	−D	
0.6	0.3	0.10	0.05	580	283	106	56	1.1	0.46	0.2	0.09	b, c, d, e
0.2	0.06	0.03	0.01	182	59	35	11	0.3	0.1	0.06	0.02	b, c, d, e
0.3	0.2	0.05	0.03	283	182	56	35	0.46	0.31	0.09	0.06[b]	b, c, d, e
0.3	0.2	0.05	0.03	283	182	56	35	0.46	0.31	0.09	0.06	b, c, d, e
0.003	?	0.0004	?	3	?	0.5	?	0.0004	?	0.0001	?	b
0.003	?	0.0004	?	3	?	0.5	?	0.0004	?	0.0001	?	b
0	0	0	0	0	0	0	0	0	0	0	0	b, c, d, e
0	0	0	0	0	0	0	0	0	0	0	0	b, c, d, e
0	0	0	0	0	0	0	0	0	0	0	0	b, c, d, e
0	0	0	0	0	0	0	0	0	0	0	0	b, c, d, e
1.0	?	0.1	?	——	?	——	?	——	?	——	?	f
?	?	?	?	?	?	——	——	?	?	——	——	f
1.0	?	0.1	?	——	?	——	——	——	?	——	——	f
?	?	?	?	——	——	?	?	——	——	?	?	f
1.0	?	0.1	?	——	?	——	?	——	?	——	?	f
1.0	?	0.1	?	——	?	——	?	——	?	——	?	f
1.0	?	0.1	?	——	?	——	?	——	?	——	?	f
1.0	?	0.1	?	?	?	——	——	?	?	——	——	f

? Largest dose that does not cause the specified biological effect is unknown.
—— Information not available.

LSS = Life-span shortening, PN = Pulmonary neoplasia, PS = Pneumosclerosis, OS = Osteosarcoma, PL = Panleukopenia, RI = Respiratory insufficiency, STT = Soft tissue tumors (other than pulmonary), ET = Endocrine tumors, PF = Pulmonary fibrosis, LYMDS = Lymphadenosclerosis (thoracic lymph nodes), BAHM = Bronchiolar, alveolar, hyperplasia and metaplasia.

of inhaled $^{239}PuO_2$ in mice

Dose												References
Whole body (μCi/kg)		Lung (μCi/g)		Total radiation dose (rads)				Dose rate (rads/day)				
				Lungs		Skeleton		Lungs		Skeleton		
+D	−D	+D	−D	+D	−D	+D	−D	+D	−D	+D	−D	
?	0.03	?	0.005	?	20	?	——	?	0.04	?	——	d
												d
?	6.4	?	1.0	?	——	?	——	?	——	?	——	e
2.4	0.12	0.4	0.02	2300	115	——	——	6	0.23	——	——	e
0.12	?	0.02	——	115	?	——	——	0.23	?	——	——	e
0.12	?	0.02	——	115	?	——	——	0.23	?	——	——	e
0.12	?	0.02	——	115	?	——	——	0.23	?	——	——	e

— Negative effect dose group, −D Largest dose at which the specified biological effect *WAS NOT* observed.
? Largest dose that does not cause the specified biological effect is unknown.
–– Information not available.

LSS = Life-span shortening; PN = Pulmonary neoplasia; PF = Pulmonary fibrosis; BAHM = Bronchiolar, alveolar, hyperplasia and metaplasia; TS = Thoracic sarcomas.

Table 11.18. Biological effects

Biological effect	Mean survival time of dose group (days)			Number of animals in dose group			Effect incidence (%)		
	Controls	Exposed		Controls	Exposed		Controls	Exposed	
		+	−		+	−		+	−
LSS	——	200–600	——	——	——	——	——	——	——
PN	——	200–600	——	——	——	——	——	50	——
BAHM	——	200–600	——	——	——	——	——	——	——
LYMDS	——	200–600	——	——	——	——	——	——	——
LSS	293[d]	156	——	24	12	——	——	——	——
PN	——	——	156	24	——	12	0	0	0
PF	293[d]	156	——	24	12	——	——	——	——
BAHM	293[d]	156	——	24	12	——	——	——	——
PE	293[d]	156	——	24	12	——	——	——	——
LEU	77[d]	77[d]	77[d]	6	6	6	——	——	——
LYM	77[d]	77[d]	77[d]	6	6	6	——	——	——
NEUT	77[d]	77[d]	77[d]	6	6	6	——	——	——
LYMPDS	293[d]	156	——	24	12	——	——	——	——
PN[c]	400[d]	27–400[d]	——	19	16	——	0	32	——
PF	400[d]	27–400[d]	——	19	16	——	——	100	——
PM	400[d]	27–400[d]	——	19	16	——	——	63	——

[a] Focal dose to 100 milligrams lung tissue. [b] Dose calculated as average dose to the lung. [c] Transthoracic injection. [d] Euthanized. [e] LISCO, 1959. [f] STUART et al., 1968. [g] SANDERS and PARK, 1970.

+ Positive effect dose group, +D Smallest dose at which the specified biological effect *WAS* observed.

— Negative effect dose group, —D Largest dose at which the specified biological effect *WAS NOT* observed.

give the smallest doses associated with specific biological effects reported and the largest doses at which specific effects have not been observed.

A. Modes of Death

As in the case of most toxic materials, the criterion of greatest interest is the shortening of life-span. Depending upon the dose, inhaled plutonium will shorten the life-span. The mode of death varies from the acute to the long-term or chronic.

1. Acute Toxicity

Two modes of acute toxicity and death have been observed. In the first, deposition of large amounts of plutonium in the lungs can cause death within a week. This acute respiratory death is characterized by severe inflammatory reaction, edema, hemorrhage, and necrosis destroying the functional tissue of the lung. The animal essentially drowns in its own fluids.

The second mode of acute death occurs at somewhat lower doses and at times ranging from about 1 month to several months after exposure. This mode of death differs from the first in that the development of extensive fibrosis is a contributing factor to the loss of functional tissue in the lung. Death is a result of respiratory insufficiency and is preceded by rapidly increasing respiratory rates and by high arterial blood CO_2 and low O_2.

of inhaled $^{239}PuO_2$ in rats

Dose														References
Whole body (μCi/kg)		Lung (μCi/g)		Total radiation dose (rads)				Dose rate (rads/day)						
				Lungs		Skeleton		Lungs		Skeleton				
+D	−D	+D	−D	+D	−D	+D	−D	+D	−D	+D	−D			
~1.0	?	~0.2	?	——	?	——	?	——	?	——	?			e
~1.0	?	~0.2	?		?	——	?	——	?	——	?			e
~1.0	?	~0.2	?	——	?	——	?	——	?	——	?			e
~1.0	?	~0.2	?	——	?	——	?	——	?	——	?			e
1.6	?	~0.4	?	~10000	?	——	?	~64	?	——	?			f
?	1.6	?	0.4	?	~10000	?	—	——	~64	——	——			f
1.6	?	~0.4	?	~10000	?	——	?	~64	?	——	?			f
1.6	?	~0.4	?	~10000	?	——	?	~64	?	——	?			f
1.6	?	~0.4	?	~10000	?	——	?	~64	?	——	?			f
12.0	1.6	3.0	0.4	~20000	~10000	——	—	~130	~64	——	——			f
12.0	1.6	3.0	0.4	~20000	~10000	——	—	~130	~64	——	——			f
12.0	1.6	3.0	0.4	~20000	~10000	——	—	~130	~64	——	——			f
1.6	?	~0.4	?	~10000	?	——	?	~64	?	——	?			f
3.0	?	0.7	?	~200000[a] ~20000[b]	?	——	?	~400[a] ~40[b]	?	——	?			g
3.0	?	0.7	?	~200000[a] ~20000[b]	?	——	?	~400[a] ~40[b]	?	——	?			g
3.0	?	0.7	?	~200000[a] ~20000[b]	?	——	?	~400[a] ~40[b]	?	——	?			g

? Largest dose that does not cause the specified biological effect is unknown.
-- Information not available.

LSS = Life-span shortening; PN = Pulmonary neoplasia; BAHM = Bronchiolar, alveolar, hyperplasia and metaplasia; LYMDS = Lymphadenosclerosis (Thoracic lymph nodes); PF = Pulmonary fibrosis; PE = Pulmonary edema; LEU = Leucopenia; LYM = Lymphopenia; NEUT = Neutrophilia; PM = Pulmonary metaplasia.

2. Subacute Toxicity

The third mode of death, subacute, is also a respiratory death but occurs at lower plutonium doses. The syndrome is similar to that of the second mode of acute toxicity except that fibrosis develops more gradually; and in dogs, death may occur from 1 year to 5 years after exposure. Death is preceded for about 2 months by a rapidly increasing respiratory rate. In dog studies, transient increased respiratory rates were observed at intervals over a 1- or 2-year period.

3. Carcinogenic Death

Neoplasia has been shown to be a consequence of inhaling plutonium. Pulmonary carcinogenesis occurred in dogs that died a respiratory death 3 to 5 years after exposure (Park, 1971) and was the cause of death in dogs which survived 6 to 11 years. Pulmonary carcinogenesis has also been reported in rodents after inhalation of soluble and insoluble forms of plutonium (Temple et al., 1960; Bair, 1960; Lisco, 1959; Koshurnikova et al., 1968; Antonchenko et al., 1969; and Buldakov et al., 1970). Most of these animals also showed pulmonary fibrosis and other pulmonary lesions. Inhaled soluble plutonium caused bone tumors as well.

Table 11.19. Biological effects

Biological effect	Mean survival time of dose group (days)			Number of animals in dose group			Effect incidence (%)		
	Controls	Exposed		Controls	Exposed		Controls	Exposed	
		+	—		+	—		+	—
LSS	>3300	850–3300	——	24	35	——	——	——	——
PN	>3300	1100–3300	——	24	35	——	——	69	——
PF	>3300	850–3300	——	24	35	——	0	100	——
BAHM	>3300	850–3300	——	24	35	——	0	100	——
RI	>3300	850–3300	——	24	35	——	0	100	——
LYM	>3300	850–3300	——	24	35	——	0	100	——
LEU	>3300	850–3300	——	24	35	——	0	100	——
LS	>3300	850–3300	——	24	35	——	0	100	——
TS	>3300	1600–2800	——	24	35	——	0	10	——
LYM	——	335[a]	——	——	12	——	0	75	0
LEU	——	335[a]	——	——	12	——	0	66	0
LYMPHA	——	306[a]	320[a]	——	7	11	0	100	0
BAHM	——	320[a]	320[a]	——	9	10	0	100	0
PF	——	320[a]	320[a]	——	9	10	0	100	0

[a] Euthanized. [b] Park et al., 1972. [c] Howard, 1970. [d] Park et al., 1970. [e] Bair et al., 1966. [f] Yuile et al., 1970.

\+ Positive effect dose group, +D Smallest dose at which the specified biological effect *WAS* observed.

— Negative effect dose group, —D Largest dose at which the specified biological effect *WAS NOT* observed.

B. Dose-Mortality Relationships

Sufficient data have been obtained at several laboratories to begin establishment of the quantitative relationship between the amount of plutonium deposited in the lung and mortality or life shortening.

Soluble forms of plutonium, plutonium citrate, and ammonium plutonium pentacarbonate have been studied in about 1500 rats (Buldakov et al., 1969), Fig. 11.28. In comparison with 224 control rats, no effects of inhaled plutonium on life shortening were observed at doses less than about 0.6 nCi/gram body weight which gave a calculated dose to the lung of 1 rad/day, or a total of about 575 rads. Deposition of more than about 4 nCi/gram body weight caused death within 2 to 3 months at total lung doses of 3000 to 6000 rads.

Plutonium nitrate and sodium plutonyltriacetate given by intratracheal injection to about 800 rats were less effective at high doses in causing life shortening than plutonium citrate given by inhalation (Buldakov et al., 1969), Fig. 11.29. Although this difference may be due to somewhat greater translocation of plutonium citrate than the other compounds, it may also be due to the fact that inhaled plutonium is more uniformly distributed throughout the lungs than intratracheally injected plutonium which tends to localize in the lung. In the latter case, the radiation injury to the lung would be localized in contrast to the case where the plutonium is uniformly distributed throughout the lung irradiating a much larger fraction of the functional tissue.

In dogs, deposition of 6 to 74 μCi of inhaled $^{239}Pu(NO_3)_4$ caused mortality from 75 to 303 days after exposure (Park et al., 1968). The lung retained 40 to 70% of the body burden of plutonium. The cause of death in all dogs was cardio-

of inhaled $^{239}PuO_2$ in beagle dogs

Dose														References
Whole body (μCi/kg)		Lung (μCi/g)		Total radiation dose (rads)				Dose rate (rads/day)						
				Lungs		Skeleton		Lungs		Skeleton				
+D	−D	+D	−D	+D	−D	+D	−D	+D	−D	+D	−D			
0.02 0.4	?	0.003–0.04	?	2000–12000	?	——	?	0.5–8.0	?	——	?			b,c,d
0.02–0.4	?	0.003–0.04	?	2000–12000	?	——	?	0.5–8.0	?	——	?			b,c,d
0.02–0.4	?	0.003–0.04	?	2000–12000	?	——	?	0.5–8.0	?	——	?			b,c,d
0.02–0.4	?	0.003–0.04	?	2000–12000	?	——	?	0.5–8.0	?	——	?			b,c,d
0.10–0.4	?	0.013–0.04	?	4000–12000	?	——	?	2.0–8.0	?	——	?			e
0.02–0.4	?	0.003–0.04	?	2000–12000	?	——	?	0.5–8.0	?	——	?			b
0.02–0.4	?	0.003–0.04	?	2000–12000	?	——	?	0.5–8.0	?	——	?			b
0.02–0.4	?	0.003–0.04	?	2000–12000	?	——	?	0.5–8.0	?	——	?			c
~0.1	?	~0.013	?	~4000	?	——	?	~2	?	——	?			c,d
0.01–1.4	?	0.002–0.2	?	100–14000	?	——	?	0.5–40	?	——	?			f
0.01–1.4	?	0.002–0.2	?	100–14000	?	——	?	0.5–40	?	——	?			f
0.5–1.4	0.3	0.07–0.2	0.03	2000–14000	2000	——	——	16–40	7	——	——			f
0.2–1.4	0.1	0.03–0.2	0.02	1650–14000	1000	——	——	5–40	3	——	——			f
0.2–1.4	0.1	0.03–0.2	0.02	1650–14000	1000	——	——	5–40	3	——	——			f

? Largest dose that does not cause the specified biological effect is unknown.
—— Information not available.

LSS = Life-span shortening; PN = Pulmonary neoplasia; PF = Pulmonary fibrosis; BAHM = Bronchiolar, alveolar, hyperplasia and metaplasia; RI = Respiratory insufficiency; LYM = Lymphopenia; LEU = Leukopenia; LS = Lymphosclerosis (thoracic lymph nodes); TS = Thoracic sarcomas; LYMPHA = Lymphadenopathy (thoracic lymph nodes).

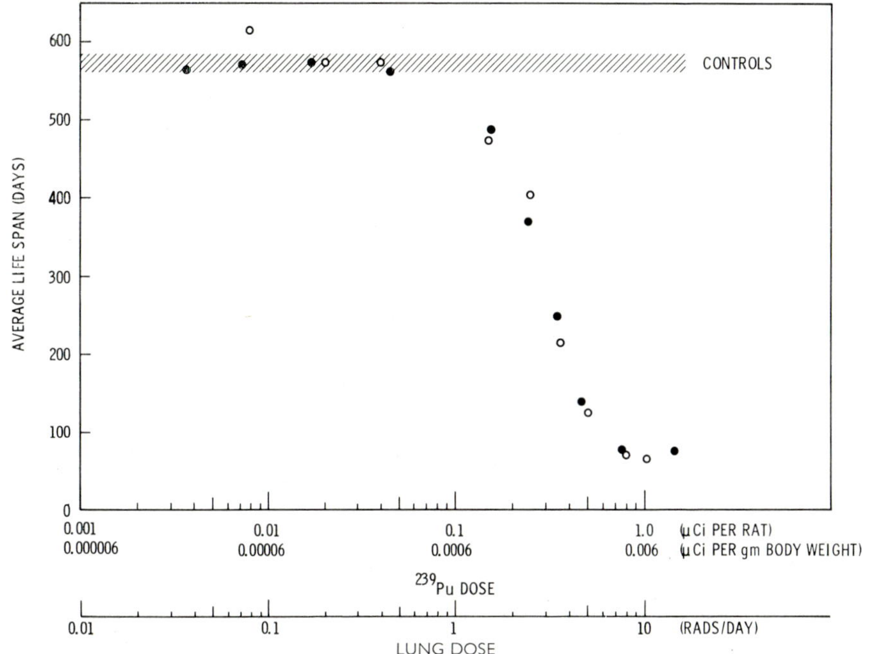

Fig. 11.28. Survival time of rats after inhalation of plutonium citrate or ammonium plutonium pentacarbonate. ○ Plutonium Citrate; ● Ammonium Plutonium Pentacarbonate. (BULDAKOV et al., 1970)

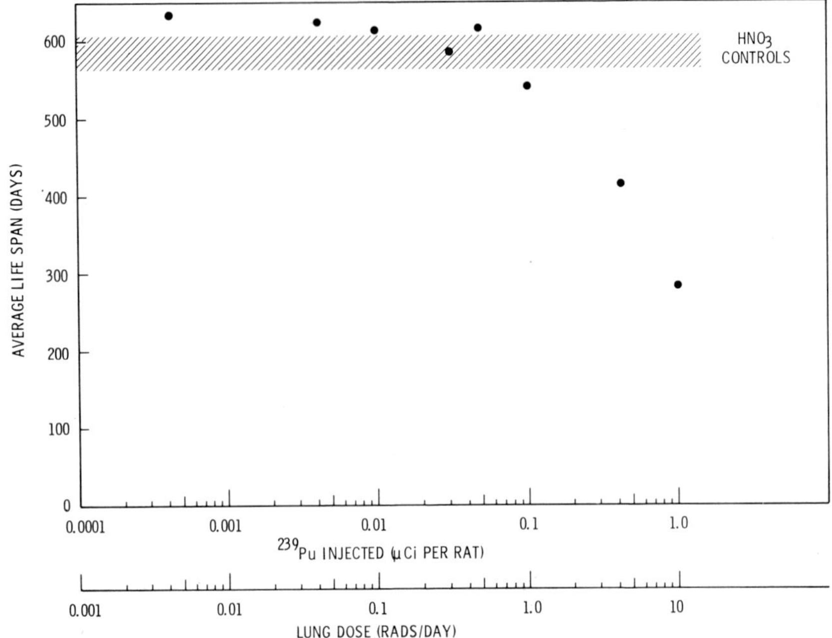

Fig. 11.29. Survival time of rats after a single intratracheal injection of $Pu(NO_3)_4$. (BULDAKOV et al., 1970)

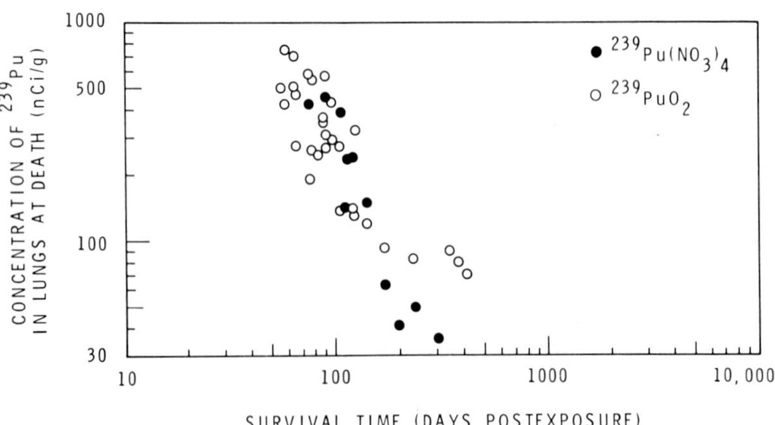

Fig. 11.30. Survival time of beagles after inhalation of $^{239}Pu(NO_3)_4$ and $^{239}PuO_2$. (BAIR, 1970)

pulmonary insufficiency. Mortality in these dogs is compared with $^{239}PuO_2$ data in Fig. 11.30. At the doses compared, $^{239}Pu(NO_3)_4$ and $^{239}PuO_2$ were similarly effective in causing death. Fig. 11.31 compares the survival times of dogs that inhaled $^{239}Pu(NO_3)_4$ with those which were given plutonium citrate intravenously at the University of Utah. Comparing initial doses of 3 to 4 μCi/kg, the dogs that inhaled $^{239}Pu(NO_3)_4$ died after 100 days while those given plutonium citrate intravenously survived 1000 to 2000 days, more than 10 times longer than those

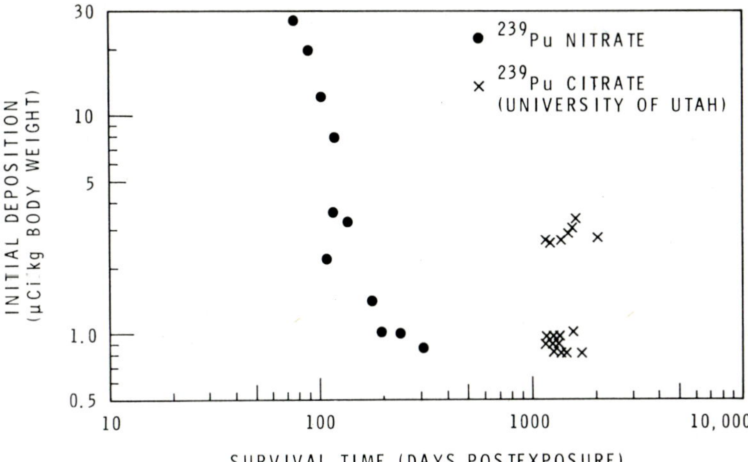

Fig. 11.31. Survival time of beagles after inhalation of ^{239}Pu nitrate and injection of ^{239}Pu citrate. (BAIR, 1970)

inhaling $Pu(NO_3)_4$. Thus, on a μCi/kg basis, inhaled $Pu(NO_3)_4$ is more toxic than intravenously injected plutonium citrate.

Of possible insoluble forms of plutonium, the oxide has received the greatest interest. A total of 748 female BAF mice, varying in age from 122 to 369 days, were exposed to $^{239}PuO_2$ aerosols, depositing 0.038 to 0.72 nCi in their lungs (BAIR et al., 1962). Comparison of mean life-splan with 144 control mice showed no significant shortening of life-span due to the inhaled $^{239}PuO_2$ and no lung tumors were seen. The highest level of exposure in this experiment gave an average accumulated dose over a 500-day period of about 20 rads assuming uniform distribution of the plutonium throughout the lung. The maximum levels used in this experiment were equivalent to about 200 times the maximum permissible lung burden for man.

In a study of beagle dogs given single 10- to 30-minute exposures to $^{239}PuO_2$, it was observed that deposition of more than 0.1 μCi/g lung caused death within about a year due to respiratory insufficiency resulting from severe irradiation injury of the respiratory tissues (BAIR and WILLARD, 1961). The survival times of 65 dogs are plotted as a function of the estimated amount of plutonium initially deposited in the alveolar lungs of the dogs expressed as nCi/g of bloodless lung (PARK et al., 1972), Fig. 11.32. The curve, fitted to all the data, by least squares analyses, can be extrapolated to 15 years postexposure (about 16 years of age, the maximum expected life-span of these dogs), suggesting that deposition of more than 5 nCi/gram of lung might be expected to cause premature death due to pulmonary neoplasia and/or pulmonary fibrosis-induced respiratory insufficiency. Thirty-six of the dogs died between 55 and 1600 days postexposure due to plutonium-induced pulmonary edema, fibrosis, and bronchiolar and alveolar epithelial hyperplasia and metaplasia, which resulted in severe respiratory insufficiency characterized by progressive hypercapnia and hypoxemia (PARK et al., 1964). Of the 29 dogs that survived more than 1000 days after exposure, 24 had pulmonary neoplasia in addition to the fibrotic and metaplastic lesions. Twenty of the 21 dogs that survived more than 1600 days postexposure had pulmonary neoplasia. None of the control dogs had pulmonary neoplasia. The incidence of

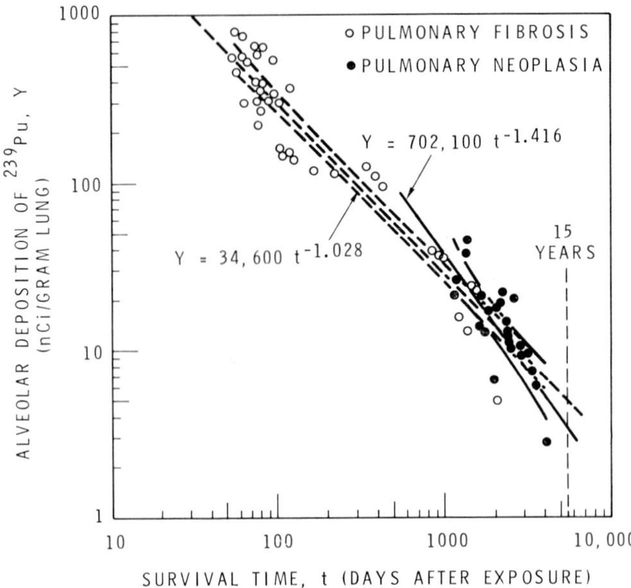

Fig. 11.32. Relationship between the quantity of $^{239}PuO_2$ deposited and survival time of dogs. (Park et al., 1972)

primary pulmonary neoplasia in dogs has been reported to be 0.2 to 0.6% (Nielsen, 1970). Approximately 39% (24/62) of the dogs exposed to $^{239}PuO_2$ showed pulmonary neoplasia; however, 82% (24/29) of the exposed dogs that survived at least 1000 days had pulmonary neoplasia, and nearly 100% of those that survived more than 1600 days had pulmonary neoplasia.

Since mortality was due to two causes, pulmonary fibrosis and neoplasia, a curve was also fitted by least squares analysis to the data for just the dogs that showed pulmonary neoplasia, Fig. 11.32. The slopes of the curves for all dogs and for just the tumor dogs were different; the intercepts at 15 years post-exposure were about 5 and 3 nCi/g, respectively. However, considering the spread of the data points and the uncertainty in extrapolating the curves to 15 years, a conservative estimate is that deposition of more than 1 nCi/g might be expected to cause premature death due to pulmonary neoplasia, or neoplasia and pulmonary fibrosis-induced respiratory insufficiency.

The estimated initial alveolar deposition in the dogs with plutonium-induced pulmonary tumors was 0.2 to 3.3 μCi or 3 to 45 nCi/g bloodless lung. This amount of plutonium is 100 to 1500 times the estimated maximum permissible lung burden (the quantity at equilibrium resulting in a mean dose of 0.3 rem/week) for man, which is 0.016 μCi or about 0.03 nCi/g lung, assuming that the bloodless lung of man weighs 500 g. The conservative estimate of 1 nCi/g initial deposition ($\simeq$ 70 nCi in the total lung) causing premature death in the dog is more than 30 times the concentration equivalent to the maximum permissible lung burden for man. This assumes that the induction of pulmonary tumors at low doses is related to plutonium concentration and not to total plutonium. If tumor induction is related to the total amount of plutonium, total number of particles, or total cells at risk, then the size of the human lung compared to the dog lung may not be important; and 70 nCi in the dog lung, causing premature death due to

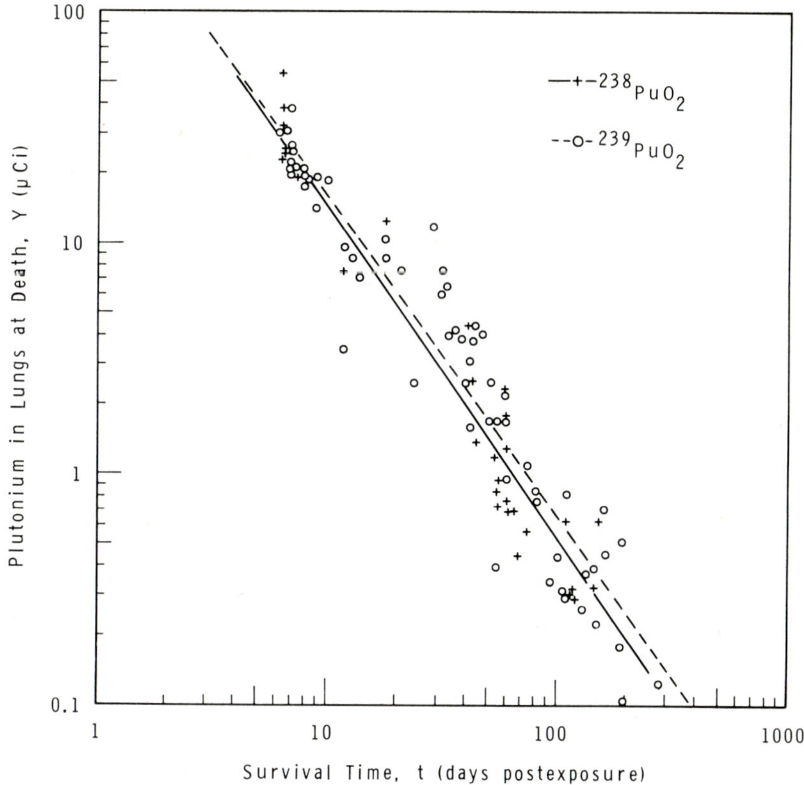

Fig. 11.33. Comparative effects of inhaled ^{239}Pu and ^{238}Pu on survival time of rats. (Stuart et al., 1968)

pulmonary cancer, may be only about 5 times the amount equivalent to the maximum permissible lung burden for man.

Considerable interest has been expressed in the relative hazard of ^{239}Pu and ^{238}Pu. Since ^{239}Pu has a specific activity of 6.17×10^{-2} Ci/g and ^{238}Pu, 17.4 Ci/g, radiological injury might be expected to be 282 times more severe for ^{238}Pu when the two isotopes are compared on an equal weight basis. Identical radioactivity levels of the two plutonium isotopes are comprised of 282 times more ^{239}Pu than ^{238}Pu, which could cause different distributions of particles and alpha-irradiation patterns in the lung and possibly different toxic responses to the two isotopes. Fig. 11.33 shows the relationship between the amount of plutonium in the lung at death and the time of death for the two isotopes in a comparative study of acute toxicity (Stuart et al., 1968) in rats. There was no significant difference between the two isotopes in causing acute mortality. Similar results were obtained in dogs (Park et al., 1969). Longer term studies have not been reported. It is possible that over a longer period of time and at low dose levels the difference in volume of lung tissue irradiated and the difference in dose per particle could cause one or the other of the two isotopes to be more hazardous. Also, as discussed previously, $^{238}PuO_2$ appears to be translocated from lung to liver and bone at a greater rate than $^{239}PuO_2$. This could affect the long-term toxicity.

C. Clinical Changes

A variety of clinical measurements have been made on animals with plutonium lung burdens. Some of the clinical signs are specific for plutonium localized in the lung and will be emphasized.

1. Hematology

The first indication of a pathologic change in animals that have inhaled plutonium dioxide is lymphopenia (BAIR, 1960; BAIR and WILLARD, 1962; WEST and BAIR, 1964). Fig. 11.34 shows the mean leukocyte values of 14 dogs with body burdens of 0.2 to 1.0 μCi (PARK et al., 1972). The mean lymphocyte count of the plutonium-exposed dogs was less than the control dogs, 6 months after exposure, and continued to be 30 to 50% of the control-dog values for the subsequent period. The mean total leukocyte count of the plutonium-exposed dogs was frequently less than that of the control dogs, primarily due to the decrease in lymphocytes. Leukocyte counts were frequently elevated prior to death when

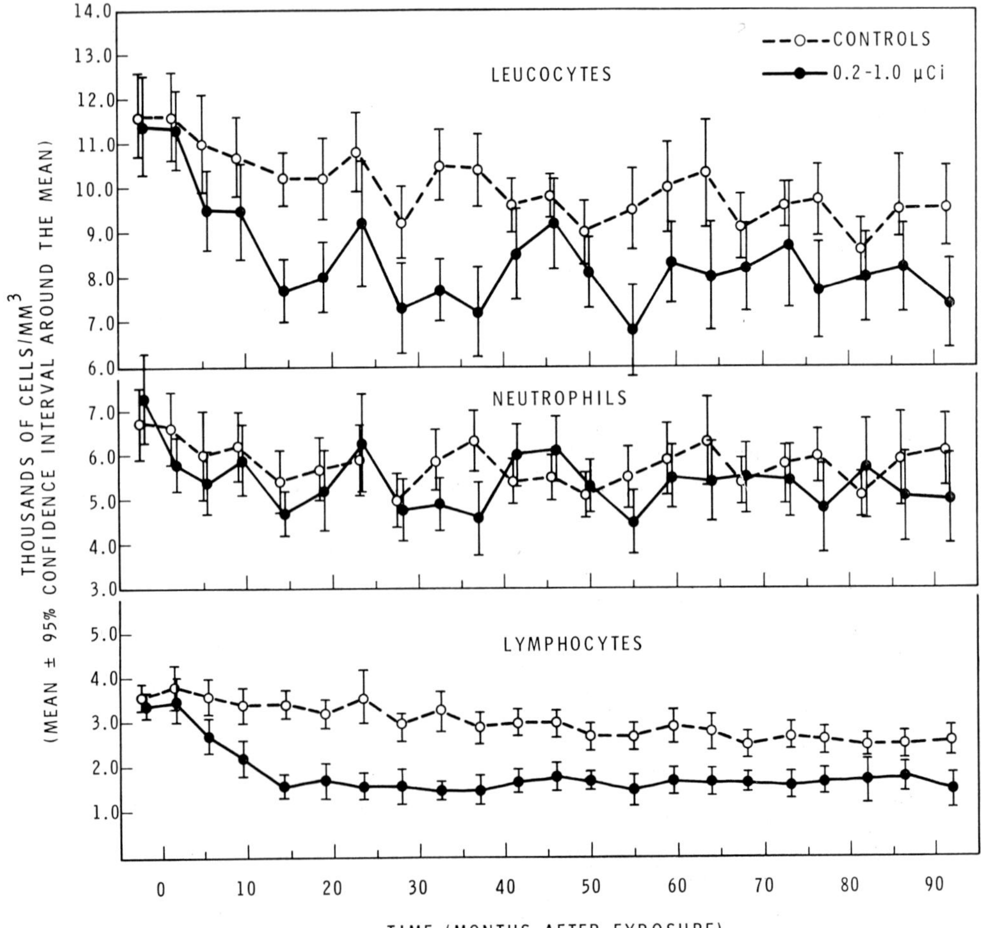

Fig. 11.34. Leukocyte values of beagles after inhalation of $^{239}PuO_2$. (PARK et al., 1972)

the lungs were severely damaged due to pulmonary neoplasia. All exposed dogs showed persistent lymphopenia and leukopenia throughout their life-span except two dogs sacrificed 900–1700 days after deposition of less than 0.003 and 0.0002 μ Ci/g of lung, respectively (Bair et al., 1966). No other significant changes in the hemograms of these dogs were observed. The critical tissue related to lymphopenia is unknown. Yuile et al. (1970) suggest plutonium in the lung and lymph node irradiating circulating cells rather than depression of hematopoiesis and/or lymphopoiesis is responsible for the lymphopenia. In this study lymphopenia and leukopenia were observed at initial deposition greater than 0.002 μCi $^{239}PuO_2$/g lungs of dogs euthanized up to 100 days postexposure after accumulating a total integrated dose of 100 rads in lungs and lymph nodes. Severe lymphopenia and leukopenia were consistently observed in dogs with initial depositions higher than 0.16 μCi/g of lung, accumulating total doses of 12400 rads to lung and 15000 rads to lymph nodes. However, lymphopenia was diagnosed early after exposure when only a fraction of these doses had accumulated. The ability of rats and dogs to form antibodies against sheep red blood cells was impaired following inhalation of $^{239}PuO_2$ (Dilley, 1968 and 1970). There is a possible relationship between

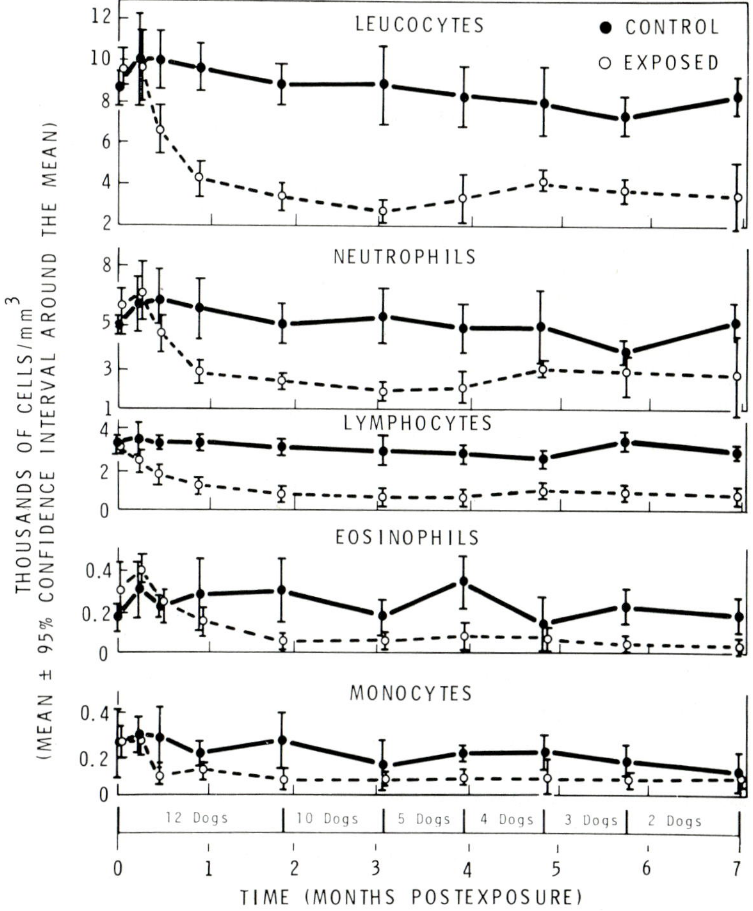

Fig. 11.35. Leukocyte counts in beagle dogs after inhalation of $^{239}Pu(NO_3)_4$. (Park et al., 1968)

plutonium-induced lymphopenia, lymph node pathology, decreased immunological capability, and the pathogenesis of plutonium-induced pulmonary neoplasia.

Inhalation of soluble forms of plutonium results in other hematologic changes. Both lymphocyte and neutrophil levels decreased after inhalation of plutonium citrate (BULDAKOV et al., 1969), plutonium nitrate (PARK et al., 1968; BALLOU, 1972), and ammonium plutonium pentacarbonate (BULDAKOV et al., 1969). The panleukopenia seen in these studies is due to the translocation of plutonium to skeleton and other hematopoietic tissues. Insufficient data are available to establish a quantitative relationship between these hematologic changes and amount of plutonium inhaled; however, the qualitative changes are well documented. An example in Fig. 11.35 gives leukocyte values for 12 dogs with lung burdens of 6 to 74 μCi $^{239}Pu(NO_3)_4$ (PARK et al., 1968). In these dogs, panleukopenia appeared the second week after exposure. Leukocyte counts dropped to 25% of control values. Neutrophils and lymphocytes showed the most significant depression, but all leukocytic cells were depressed. Erythrocyte, hemoglobin, and hematocrit values were normal.

In contrast to inhaled $^{239}PuO_2$, $^{238}PuO_2$ has been found to cause a decrease in circulating neutrophils as well as lymphocytes (PARK et al., 1970). This occurs when ^{238}Pu is translocated to skeleton and other hematopoietic tissues as illustrated previously.

2. Respiratory Physiology

Changes in respiratory function have been observed in rodents and dogs responding to extensive radiation damage of lung tissue containing ^{238}Pu and ^{239}Pu. The time of onset of respiratory dysfunction depends upon the quantity of plutonium in the lung. In animals with acute lethal doses, respiratory rates increase very rapidly, to nearly 200 per minute in dogs from a normal rate of 20–30. Increased respiratory rates occur more gradually and are often transient in lower dose level animals.

As respiratory pathology develops, changes in arterial blood gases also occur. These changes become more pronounced just before death. For example, blood pO_2 was depressed before death and blood pCO_2 increased in dogs in studies of $^{239}PuO_2$ (PARK and BAIR, 1966), $^{239}Pu(NO_3)_4$ (PARK et al., 1968), and $^{238}PuO_2$ (PARK et al., 1969).

D. Pathology and Carcinogenesis

Pathologic changes in animals after inhalation of insoluble plutonium (e.g., $^{239}PuO_2$) are almost entirely confined to lungs and thoracic lymph nodes. Translocation of small quantities might cause localized tissue damage in liver, abdominal lymph nodes, and skeleton; but such changes are rarely observed. On the other hand, inhaled soluble plutonium [e.g., $Pu(NO_3)_4$, plutonium citrate, etc. and $^{238}PuO_2$ when it translocates from lung to other tissues] may cause significant lesions in not only lung and thoracic lymph nodes, but also in bone, liver, and and possibly in spleen and kidney.

Gross pathologic changes in animals that die due to fibrosis-induced respiratory insufficiency are severe. The lungs are edematous, enlarged, firm, and usually hemorrhagic. Thoracic lymph nodes generally show evidence of radiation damage—enlarged and hemorrhagic. At long times after exposure, they may be small and indurated. Enlargement of the right heart also occurs if the cardiopulmonary insufficiency has persisted for several months.

Histopathology in animals exposed to plutonium aerosols has been described in detail (CLARKE et al., 1964; CLARKE and BAIR, 1964; CLARKE et al., 1966; HOWARD, 1970; and WEST and BAIR, 1964) and will not be repeated here. Instead, attention will be directed to the carcinogenic response of tissues to irradiation by alpha particles emitted by plutonium isotopes.

In rats exposed to ammonium plutonium pentacarbonate or plutonium citrate pulmonary neoplasia occurred in 4% of the rats in which 0.004 to 0.008 μCi/g lung was deposited compared to 0.39% in controls, Table 11.15. The 4% incidence followed a pattern of decreasing incidence which was as much as 20 to 30% at 0.04 to 0.08 μCi/g of lung. The tumors were malignant, squamous cell carcinomas and adenocarcinomas. There was also an increased incidence of hemangiosarcomas in the lungs. An increased incidence of pneumosclerosis also occurred at the lowest dose levels studied. Pathology in the thoracic lymph nodes was not mentioned in these studies. Several of the plutonium-treated rats had lymphosarcoma, but the incidence was not higher than the controls.

The incidence of osteosarcoma was about 1% with deposition of 0.007 to 0.008 μCi/g of lung with both materials while rats with deposition of 0.004 μCi/g of lung of ammonium plutonium pentacarbonate showed no osteosarcomas. A 1% incidence may not seem significant; however, in 500 control rats there were no bone tumors. Deposition of larger quantities, 0.15 to 0.25 μCi ^{239}Pu/g of lung, resulted in 18% osteosarcomas.

Intratracheal injection of plutonium nitrate, pH = 2, in the rat caused an increased incidence of pulmonary neoplasia, pneumosclerosis and osteosarcoma at levels of plutonium in lung about ten times more than was required for similar tumor incidence after inhalation of plutonium citrate, Table 11.16. No significant increases in tumors were seen in the lung with injection of 0.01 μCi/g of lung or in the skeleton with 0.031 μCi/g of lung. This study was complicated by an increased incidence of pulmonary neoplasia after injection of 0.01 N HNO_3 solution. A number of other soft tissue tumors, especially endocrine gland tumors, were also observed to increase at levels of 0.003 to 0.03 μCi/g lung.

Beagle dogs died due to respiratory insufficiency caused by severe pulmonary fibrosis, bronchiolar and alveolar hyperplasia, and pulmonary edema 300 days after deposition of about 0.1 μCi $^{239}Pu(NO_3)_4$/g lung (PARK et al., 1968), Table 11.16. The dogs showed no neoplasia. The tracheobronchial lymph nodes of these dogs showed sclerotic changes associated with plutonium concentrations higher than in any other tissue at the time of death. No changes were reported in lymph nodes of rats intratracheally injected with plutonium nitrate (BULDAKOV, 1970).

In the case of soluble plutonium, increased incidences of pulmonary and skeletal neoplasia were observed at lower doses than the dose that caused a decrease of mean life-span. Although the average plutonium rat lived as long as the average control rat, some of the plutonium-exposed rats that lived as long or longer than the mean life-span of control rats showed plutonium-induced neoplasia.

It is apparent that both lung and skeleton should be considered among the critical tissues after inhalation of soluble plutonium. This is contrary to a common belief that skeleton is the critical tissue for inhaled soluble plutonium. The lowest level reported to produce neoplasia in the lung of rats was 0.004 μCi/g of lung and in the skeleton of rats 0.007 μCi/g of lung. This was equivalent to a total radiation dose (assuming uniform distribution of dose) of 42 rads (0.08 rads/day) in the lung and 4 rads (0.006 rads/day) in the skeleton. No tumors were observed with doses of 1.8 rads in the skeleton. However, there was an increased lung tumor incidence at all levels of inhaled soluble plutonium studied.

Biological effects of PuO_2 in rodents are summarized in Tables 11.17 and 11.18. It was mentioned previously that no life-span shortening or neoplasia was seen in mice having lung burdens 0.005 μCi $^{239}PuO_2$/g lung (BAIR et al., 1962). In another study (TEMPLE et al., 1960), after intratracheal injection of 0.16 μCi $^{239}PuO_2$ (1 μCi/g lung) (mean diameter 0.5 μm) in 41 mice, one bronchiolar carcinoma occurred. Of 17 mice given 0.06 μCi (0.4 μCi/g lung), two developed squamous cell carcinomas and of 21 given 0.003 μCi (0.02 μCi/g lung), one developed a fibrosarcoma. In another experiment of 73 mice given 0.1 μCi $^{239}PuO_2$ by inhalation, one developed a bronchiolar carcinoma (BAIR, 1960). The mice showed pulmonary fibrosis with bronchiolar and alveolar hyperplasia and metaplasia. These studies showed that carcinoma could be produced in the mouse lung with plutonium.

Pulmonary neoplasia occurred in 50 percent of rats that survived 200 to 600 days after inhalation of $^{239}PuO_2$ smoke (0.2 μm mean diameter), depositing about 0.2 μCi/g of lung (LISCO, 1959). These rats showed pulmonary fibrosis and bronchiolar and alveolar metaplasia. The tumors were epidermoid carcinomas, adenocarcinomas, and hemangioendotheliomas and were associated with localized areas of fibrosis mainly in the periphery of the lung. The thoracic lymph nodes contained plutonium in the central sinuses but not in the lymphatic tissue. The plutonium was frequently associated with scars in the lymph nodes. Also, the plutonium was frequently within large phagocytes containing pigment in the scars within the lymph node and lung. No primary malignant tumors of lymph nodes were seen in the mediastinum.

Rats initially depositing about 0.4 μCi $^{239}PuO_2$ (CMD 0.1 μm) per gram of lung died approximately 274 days postexposure (STUART et al., 1968). No neoplasia was observed, but all other lesions were similar to those described previously for rats. Lung dose was about 10000 rads. The rats also showed lymphopenia, leukopenia, and neutrophilia.

The effects of localized deposits of plutonium in lungs were studied in rats given 0.7 μCi $^{239}PuO_2$ through the thorax wall. The plutonium was localized primarily in 50 to 100 milligrams of tissue in subpleural areas (SANDERS and PARK, 1970). Five of 16 rats developed tumors by the 400th day postexposure. The tumors were two squamous cell carcinomas, three adenocarcinomas, and one endothelioma. One rat had both adenocarcinoma and squamous cell carcinoma associated with the plutonium-induced fibroplasia. The estimated dose to the 100 milligrams of tissue was about 200000 rads. If this were distributed throughout the lung, the mean dose would have been 20000 rads and the rats would likely have died due to pulmonary fibrosis within 100 days and before tumors developed. This illustrates induction of pulmonary neoplasia by localized high doses of plutonium in the lung.

The long-term biological effects of inhaled "insoluble" plutonium in beagle dogs and the incidence of pulmonary neoplasia were mentioned previously (PARK et al., 1972), Table 11.19. The survival times of these dogs are shown in Fig. 11.32. All of the dogs had adenocarcinomas of the lung, Fig. 11.36 and 11.37. Metastasis occurred to thoracic lymph nodes, Fig. 11.38, and to many systemic organs. Two dogs had peripheral squamous cell carcinomas and three dogs had epidermoid carcinomas (HOWARD, 1970). Two dogs had lymphangiosarcomas and one dog a capillary hemangioma in the lung.

The tumors appeared to originate in the peripheral region of the lung, Fig. 11.39, associated with plutonium particles, Fig. 11.40. The thoracic radiograph shows a well demarcated peripherally located tumor in the dorsal part of the right diaphragmatic lobe 50 months postexposure. This particular tumor was first

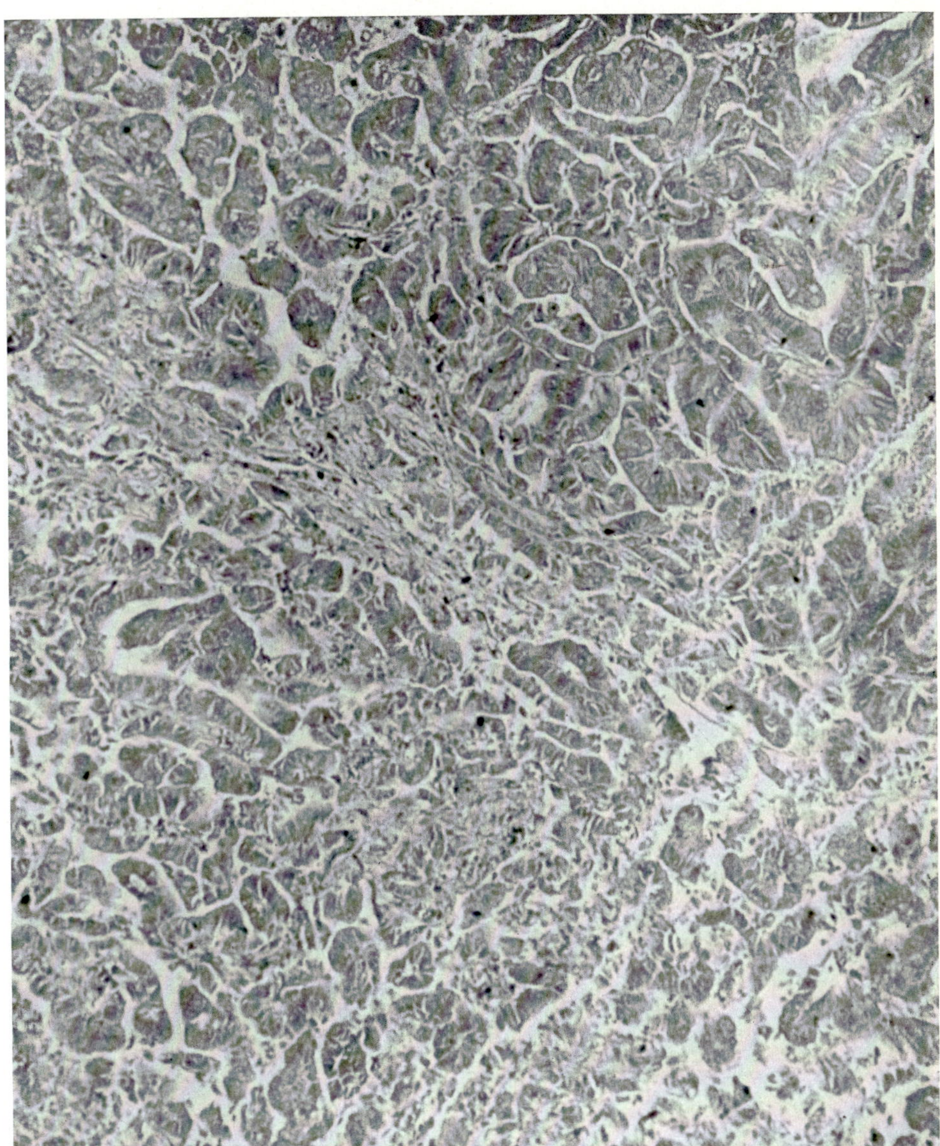

Fig. 11.36. Bronchiolar-alveolar carcinoma with papillary character in a dog 45 months after inhalation of $^{239}PuO_2$ (hematoxylin and eosin). (82 ×, reduced to $^7/_9$). (CLARKE et al., 1969)

identified 37 months postexposure. By 60 months postexposure, the tumor occupied nearly all of the right diaphragmatic lobe.

The tracheobronchial and mediastinal lymph nodes showed severe sclerotic lesions associated with the plutonium, Fig. 11.41 to 11.43, as did a few hepatic nodes which were examined (PARK et al., 1970).

Three dogs had lesions that appeared to be derived from the pleura and mediastinum. These generally have been classified as thoracic sarcomas. The cells were pleomorphic and tumors have been variously classified as lymph

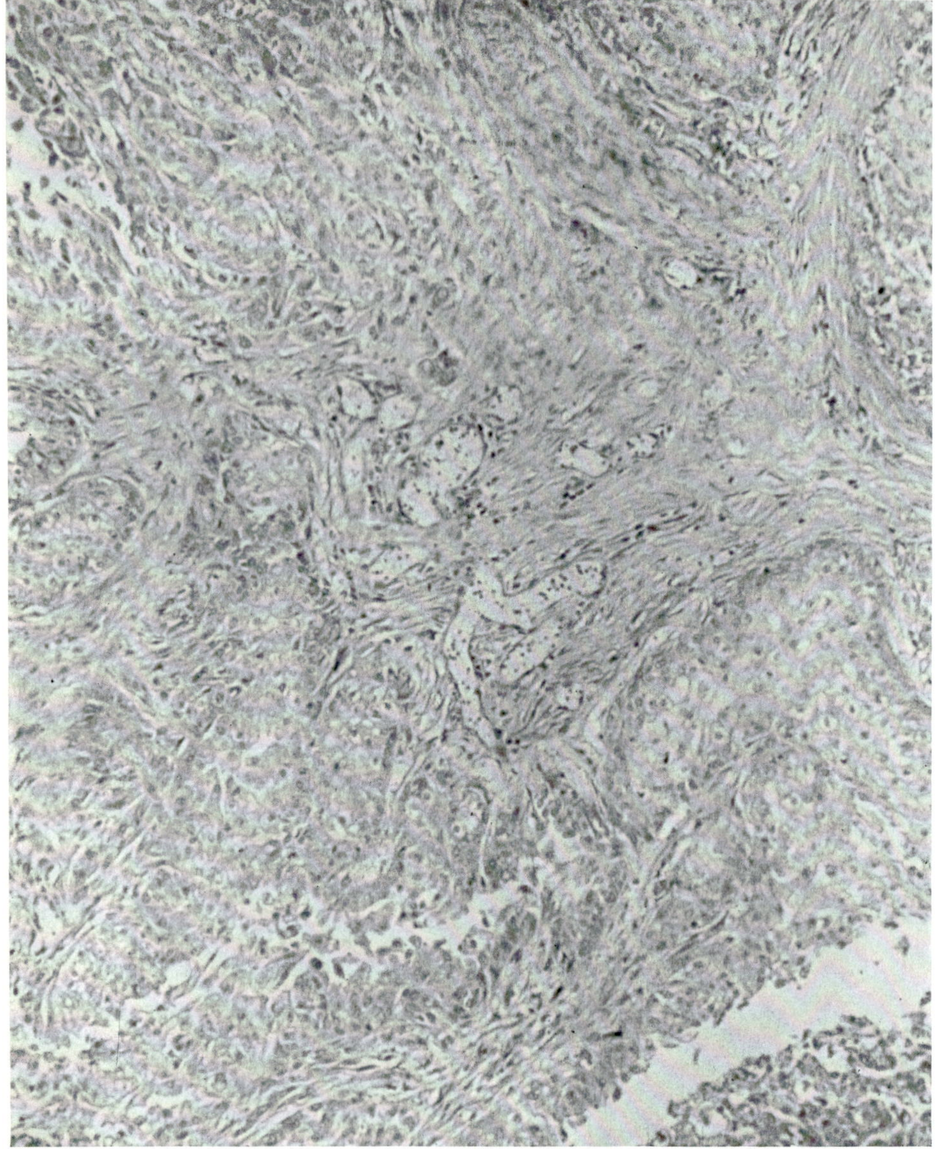

Fig. 11.37. Bronchiolar-alveolar carcinoma with glandular character in a dog 45 months after inhalation of $^{239}PuO_2$ (hematoxylin and eosin). (82 ×, reduced to $^7/_9$). (Clarke et al., 1965)

angiosarcoma, capillary hemangioma, endothelioma or mesothelioma (Howard, 1970). The results suggest that tumors were caused by the plutonium concentrated in the mediastinal lymph nodes and lymphatics. These dogs also had adenocarcinomas of the lung.

In another study with beagle dogs, the degree of lung pathology appeared to be related to total dose while lymph node pathology correlated best with dose rate (Yuile et al., 1970). No pulmonary or lymph node tumors were observed.

Fig. 11.38. Metastases of primary lung tumor to left bronchial lymph node in a dog 45 months after inhalation of $^{239}PuO_2$ (hematoxylin and eosin). (82 ×, reduced to $^7/_9$). (CLARKE et al., 1965)

The highest dose not causing lymph node pathology 400 days postexposure was deposition of 0.028 μCi/g lung which resulted in a total mean lymph node dose of 2000 rad (4.5 rad/day). Lymph node pathology occurred after deposition of 0.04 μCi/g lung which resulted in a total mean dose of 3000 rads (7.0 rad/day) to the lymph nodes. Deposition of 0.016 μCi/g lung giving a total mean dose of 1000 rads to the lung caused no pathology in the lung; however, 0.03 μCi/g lung (2000 rads) caused pathology. However, in another study (BAIR et al., 1966),

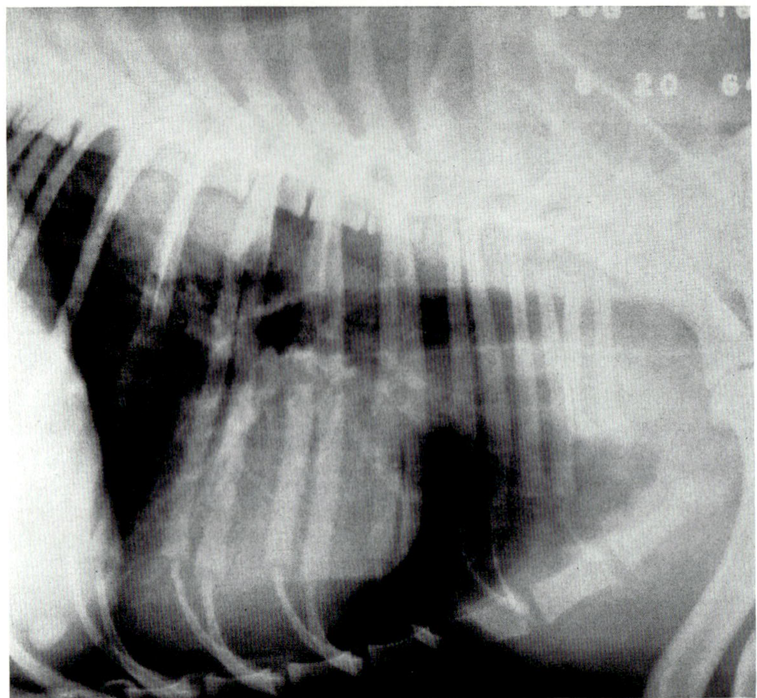

Fig. 11.39. Lateral radiograph of dog 50 months after inhalation of $^{239}PuO_2$. Tumor in right diaphragmatic lobe (lung burden ~0.25 μCi). (PARK et al., 1966)

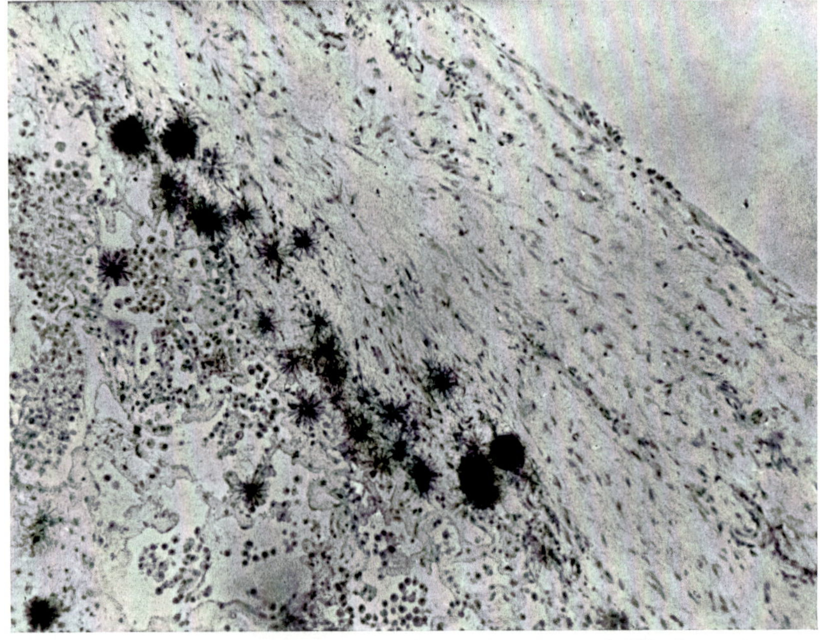

Fig. 11.40. Pleural fibrosis associated with a heavy deposition of $^{239}PuO_2$ in the region of the subpleural lymphatics, 384 days postexposure (autoradiograph, hematoxylin and eosin). (88 ×). (CLARKE and BAIR, 1964)

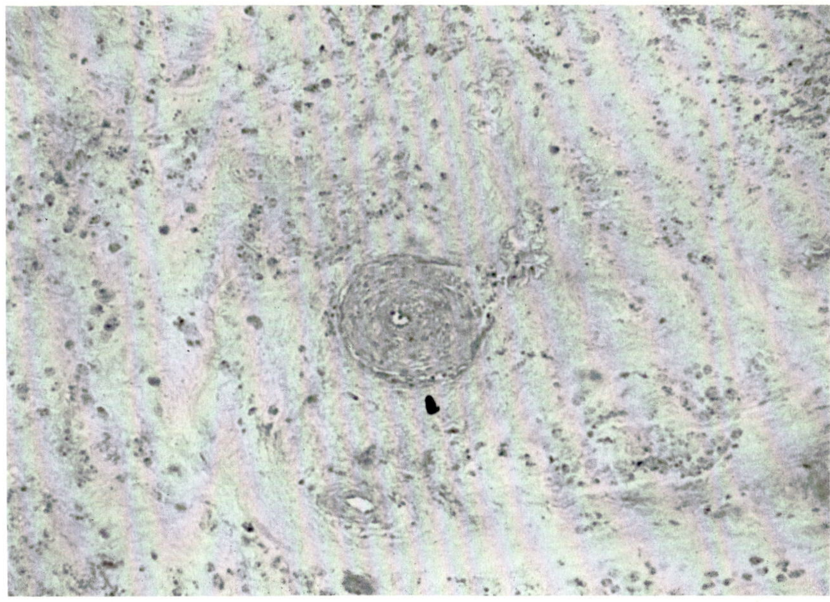

Fig. 11.41. Complete replacement of lymphatic elements by fibrous connective tissue in bronchial lymph gland of a dog 855 days after inhalation of $^{239}PuO_2$ (hematoxylin and eosin). (88 ×). (CLARKE and BAIR, 1964)

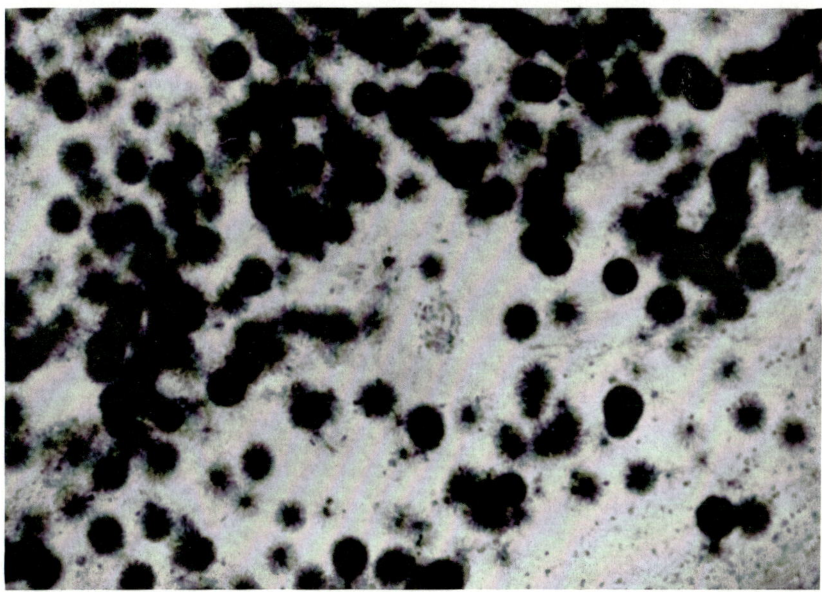

Fig. 11.42. $^{239}PuO_2$ deposition and fibrosis in a bronchial lymph node of dog 855 days post-exposure (autoradiograph, hematoxylin and eosin). (88 ×). (CLARKE and BAIR, 1964)

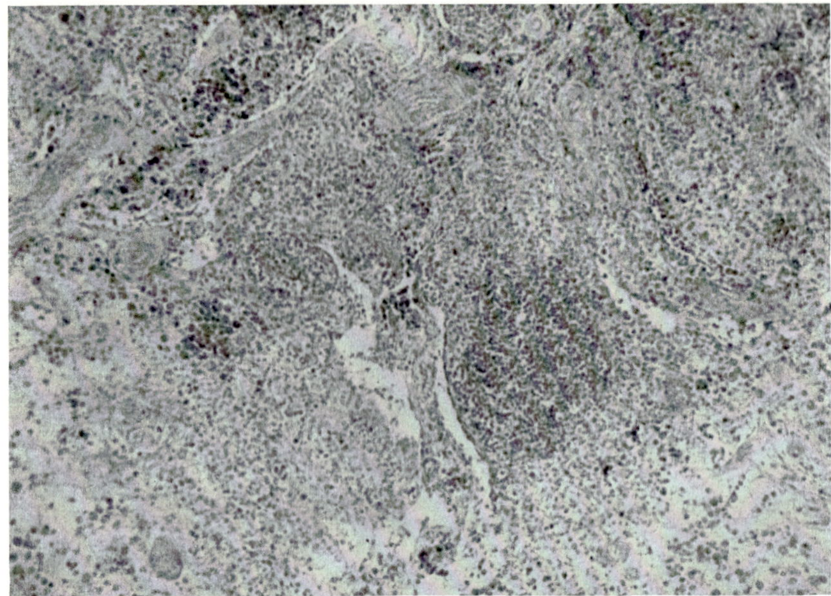

Fig. 11.43. Focal edema and necrosis surrounding a small island of residual lymphatic tissue in bronchial lymph node in a dog 124 days after inhalation of $^{239}PuO_2$ (hematoxylin and eosin). (88 ×). (CLARKE and BAIR, 1964)

no significant lesions were observed in the lungs of a dog sacrificed 1700 days postexposure with alveolar deposition of 0.0002 μCi/g of lung but this dog had fibrotic lesions in the thoracic lymph nodes. The lowest level of exposure in which lesions were observed in the lung was a dog sacrificed 4068 days after deposition of 0.003 μCi/g lung (PARK et al., 1972).

The acute toxicity and biological effects of inhaled $^{238}PuO_2$ (CMD 0.05 to 0.1) in rats (STUART et al., 1968) to 300 days postexposure and dogs (PARK et al., 1969 and 1970) to 185 days postexposure were similar to those caused by $^{239}PuO_2$ at comparable radiation dose levels. However, because ^{238}Pu translocates from the lung faster than ^{239}Pu, the initial deposition level required to produce comparable long-term effects may differ. No tumors were observed in the rats. Three dogs showed carcinoma *in situ*. Several dogs with single 50 to 300 μm ^{238}Pu microspheres in the lungs were alive after 5 years and showed no clinical effects (WILLARD, 1970). At postexposure times of up to 24 months after inhalation of $^{238}PuO_2$ (CMD <0.1), dogs continued to show biological effects similar to the dogs exposed to inhaled $^{239}PuO_2$ (WILLARD, 1970). Osteogenic sarcomas are beginning to appear at times of about 5 years postexposure, however.

E. Summary

The direct effects of plutonium dioxide in the lungs of dogs include pulmonary fibrosis and pulmonary neoplasia. Irradiation of blood within the lung may contribute to leukopenia and lymphopenia. Irradiation of blood directly as it passes through the lung or depression of lymphocytes or perhaps the irradiation of other constituents of blood may decrease the immunological capabilities of the animal and in turn make it more susceptible to neoplasia.

The direct effects of plutonium dioxide in the lymph nodes of dogs is fibrosis, scarring, and loss of lymphatic nodules and viable lymphocytes. The plutonium in the lymph nodes and lymphatics may also contribute to the induction of thoracic sarcomas related to the concentration of plutonium near the endothelial and mesothelial tissues. Two dogs (HOWARD, 1970) that inhaled PuO_2 and several rats that inhaled "soluble" plutonium developed malignant lymphoma, but the incidence is probably not greater than in controls. The plutonium was not located in the lymphopoietic tissue. Lymphoma was reported in the regional lymph nodes of one dog after subcutaneous injection in the foot of air oxidized plutonium (JOHNSON and WATTERS, 1970; WATTERS and LEBEL, 1972) and in the hepatic lymph nodes of a pig after injection of plutonium nitrate in the skin (MCCLANAHAN et al., 1968). Plutonium in the lymph nodes may also contribute to the lymphopenia and leukopenia.

The lowest exposure level at which these effects occurred following the inhalation of $^{239}PuO_2$ in dogs was after deposition of about 0.003 μCi/g (blood free) lung with accumulated radiation dose to the lung of about 2000 rads. The highest level that will not produce some of these effects in dogs is not known but is probably less than 0.001 μCi/g of lung initially deposited in the deep lung (alveoli).

The biological effects of inhaled soluble plutonium in the lungs of rats include pulmonary neoplasia and fibrosis similar to that described for dogs after inhalation of $^{239}PuO_2$. Panleukopenia was observed including a decrease in neutrophils which was not observed after inhaled $^{239}PuO_2$. The thoracic lymph nodes of the rats were either not severely damaged or not observed in detail. No thoracic sarcomas were reported in rats. The most outstanding difference between the biological effects of inhaled $^{239}PuO_2$ and plutonium nitrate is the high incidence of osteosarcomas after inhalation of plutonium nitrate and the complete absence of osteosarcomas after inhalation of $^{239}PuO_2$.

The level at which pulmonary neoplasia and osteosarcoma were observed in rats (0.004 to 0.008 μCi/g lung) was similar to the level at which pulmonary neoplasia was observed in the dog (0.003 μCi/g lung); however, the radiation dose to the lungs of the rats was 40 to 2500 rads while the dose to the dogs was 2000 to 12000 rads. This is in part due to the relatively long retention of $^{239}PuO_2$ in the lung of the dog that may live 12 years compared to the rat with a life-span of 2 to 3 years. Also, the lowest dose that will cause a pulmonary tumor in a dog has not been determined. However, the distribution of dose in the lung related to many small particles of soluble plutonium may differ from the dose distribution due to small discrete $^{239}PuO_2$ particles. The estimated dose responsible for induction of neoplasia relative to the volume of lung tissue at risk was studied by SANDERS (1970) in rats given transthoracic injections of small $^{239}PuO_2$ particles and by RICHMOND (1970) with large single (100 to 200 μm) $^{238}PuO_2$ microspheres. Accumulated doses to rat lung tissue 75 μm from the $^{238}PuO_2$ microspheres were about 106 rads 211 days after injection into pulmonary vasculature. The only significant effects in rats sacrificed at 600 days postexposure were collagenous microlesions around the microspheres.

This summarizes the biological effects of inhaled plutonium in experimental animals. Throughout the world a number of inhalation exposure accidents have occurred. There are no documented effects of inhaled plutonium in man.

V. Therapeutic Removal of Plutonium from the Lung

Attempts to remove plutonium from the lung have taken four general approaches based on (1) treatment to stimulate the normal physiological processes of lung clearance; (2) treatment with chelating agents to form stable, readily excreted complexes of plutonium; (3) direct physical removal of plutonium from the lung as by pulmonary lavage; and (4) combinations of the first three approaches.

Physiologically active agents which at first hand might be expected to be the most effective at promoting plutonium clearance and excretion have been discouragingly ineffective. TOMBROPOULOS (1964) concluded after reviewing the evidence at that time, that treatment to stimulate normal lung clearance mechanisms was largely unsuccessful in decreasing the lung burden of PuO_2. Among the agents tested were bronchodilators, mucolytic agents, surfactants, phagocytic stimulants, and expectorants (Table 11.20). The reader is referred to a more recent review of this work by BAIR and SMITH (1969) for a detailed discussion of the general problem of radionuclide removal. Recent studies by DILLEY and MCDONALD (1970) claim some success in removing inhaled PuO_2 from rats by treatment with promazine HCl, Halotestin, Estradiol, Phenergan, Progesterone, Diuril, Diamox, and Miltown. Of these agents, the latter three reduced the lung burden appreciably but had the undesirable side effect of increasing the deposition in other potentially radiosensitive tissues. Subsequent treatment with a chelating agent diethylenetriaminepentaacetate (Na_3Ca DTPA) effectively reduced the lung burden as much as 25 to 50% and decreased the systemic burden 45 to 90%, illustrating the rationale and efficacy of combined treatment therapy. According to the authors, the initial drug treatment appeared to make PuO_2 more amenable to chelation by DTPA. Since FOREMAN (1960) has shown DTPA does not normally enter the cell, it is conceivable that the mechanism of drug action was to alter cellular permeability to the extent that DTPA could gain access to intracellular PuO_2 or that the soluble fraction of PuO_2 could more readily penetrate to the extracellular spaces. MARKLEY (1963), for example, was able to demonstrate increased removal of apparently intracellular deposits of polymeric plutonium by combined treatment with DTPA and the pentaethyl ester of DTPA. Esterification of the chelate presumably increased its lipid solubility and therefore the potential for intracellular penetration of the DTPA ester.

Chelating agents of the polyamino-polycarboxylic acid type, have proved the most successful at increasing the urinary excretion of plutonium in man following industrial exposure to plutonium compounds (NORWOOD, 1962). Of these agents, the calcium trisodium salt of diethylenetriaminepentaacetic acid (Na_3Ca DTPA) has been of most practical value. DTPA has the advantage of low toxicity while it forms soluble, stable complexes with the mobile fraction of plutonium. These complexes are subsequently voided with the urine without further metabolism. Plutonium compounds which resist solubilization with DTPA are expected to be unresponsive to DTPA therapy.

The usual chelation treatment regimen has employed repeated doses of 1g of the drug administered by intravenous drip on alternate days for short treatment periods of a few days duration. No harmful side effects have been observed when adhering to this treatment schedule even when periodic chelate treatment has continued for years. Use of the drug, however, is contraindicated where renal function may be impaired (NORWOOD, 1960). Chelation treatment is considered to be most effective when administered within hours after plutonium incorporation since the radionuclide may become tissue bound or otherwise become in-

Table 11.20. Tests for removal of inhaled $^{239}PuO_2$ (TOMBROPOULOS, 1964)

Experiment	Treatment	Duration of experiment (days)	Retained in lungs[a] (μCi)	Controls (%)	Retained in carcass[a,b] (nCi)	Controls (%)
$^{239}PuO_2$[c]	Controls	29	1.5 ±0.34	100	12 ± 3.6	—
	25% DPTA[e] + 25% Pluronics F68[f]		1.8 ±0.28	120	13 ± 6.6	110
	$NaHSO_3$ + KI (intraperitoneally)		1.9 ±0.41	120	75 ±61.5	620
	$NaHSO_3$ + KI (in drinking water)		2.1 ±0.19	140	20 ± 3.04	170
	Aerosol TR[g]		1.9 ±0.32	120	22 ± 6.6	180
$^{239}PuO_2$[c]	Controls	21	2.8 ±0.57	100	29 ±11.58	—
	20% Pluronics F68 + 25% DTPA + hyaluronidase		3.6 ±0.54	130	36 ±15.05	130
	Tween-80[h] + hyaluronidase		3.0 ±0.50	110	26 ±17.23	90
	Globulin + 20% Pluronics F68 + 25% DTPA + hyaluronidase		3.2 ±0.47	120	38 ±17.38	130
	Epinephrine 1st day Globulin + 25% DTPA + 20% Pluronics F68 + hyaluronidase the second day and thereafter		3.5 ±0.60	130	270 ±320	950
$^{239}PuO_2$[c,d]	Control (air through water)	7	1.3 ±0.36	100		
	Air + CO_2 through water		1.8 ±0.43	140		
	Alevaire[i]		1.4 ±0.31	110		
	Isuprel[j] 1:100		1.2 ±0.31	90		
	Isuprel 1:100 + Alevaire		1.2 ±0.36	91		
	Isuprel 1:100 + Alevaire + 400 mg NH_4Cl/100 ml of drinking water		1.1 ±0.15	85		
	Isuprel 1:100 + NH_4Cl (400 mg/100 ml of drinking water)		0.97 ±0.38	74		
	Isuprel 1:100 + NH_4Cl (400 mg/100 ml of drinking water)		0.97 ±0.38	74		
	Alevaire + NH_4Cl (400 mg/100 ml of water)		1.1 ±0.34	87		
$^{239}PuO_2$[d]	Controls	7	0.9 ±0.094	100		
	Isuprel 1:100		0.9 ±0.094	100		
$^{239}PuO_2$[d]	Controls	7	0.12 ±0.015	100		
	Dornavac		0.12 ±0.015	100		
$^{239}PuO_2$[d]	Controls	7	2.8 ±1.37	100		
	Sodium polystyrene sulfonate		3.6 ±1.40	130		

All treatments were started immediately after exposure.

[a] Mean + 95 percent intervals around the mean. [b] Other than lungs and skin. [c] Treatments were given daily for 7 days, then on alternate days for duration of experiments. [d] Treatments were given twice daily for 5 days. [e] DTPA-Calciumtrisodium salt of diethylenetriaminepentaacetic acid. [f] Pluronics F68 — Polypropylenoglycolethylene oxide polymer. [g] Aerosol TR — Sodium bis(tridecyl)-sulfosuccinate. [h] Tween-80 — Polyoxyethylene (20) sorbitan monooleate. [i] Alevaire-p-isooctylpolyoxyethylenephenol polymer. [j] Isuprel-α-(isopropylaminomethyl) protocatechuyl alcohol.

accessible to chelation. Animal studies have demonstrated a marked decrease in effectiveness when the drug was administered by injection or inhalation more than 4 hours after plutonium incorporation (SMITH et al., 1961; LYUBCHANSKII, 1966). Administration of DTPA by inhalation appears to have certain advantages following industrial accidents since the usual exposure to plutonium involves the respiratory tract. Furthermore, according to PODA (1971), inhalation therapy not only expedites the early administration of the drug but also eliminates the inconvenience of the intravenous drip procedure. While experimental data are few, treatment of rats with negatively charged 1% NaCl aerosols for several hours daily caused a 8–34% reduction of ^{239}Pu lung burden compared with control values (WEHNER, 1971).

The most recent development in removal therapy for inhaled plutonium is pulmonary lavage. Although this procedure has been successfully employed in the past as a supportive therapy in severe respiratory disease, it has only recently been attempted as therapy following industrial exposure to plutonium (MCCLELLAN, 1971). The use of this technique as an industrial hygiene procedure was preceded by studies in dogs having lung burdens of fission products by PFLEGER et al. (1969) and MUGGENBURG et al. (1970) to demonstrate the efficacy and low potential hazard in removing inhaled radioactive particles. Among the procedures investigated were both unilateral and bilateral lavage comparing the effectiveness of washing with physiological saline or DTPA solutions. DTPA lavage of both lungs proved to be the most effective treatment investigated. In his cytology studies, SANDERS (1969, 1970) used pulmonary lavage in rats to remove macrophages containing plutonium from lung. MCDONALD et al. (1972) estimated that about one-half of the inhaled PuO_2 lung burden in dogs could be removed by repeated pulmonary lavage. Their study employed a series of 19 weekly unilateral lavages with physiological saline. The plutonium concentration in the washed and untreated lung were compared at necropsy. A similar, although less protracted, series of washes administered to baboons by the French workers also showed promising results in PuO_2 removal (KUNZLE-LUTZ et al., 1970). Although the effect of unilateral lavage on total lung clearance is not yet known and the procedure in general requires more exhaustive investigation, it is clearly one of the more promising approaches to effective removal of insoluble inhaled plutonium. Whether this seemingly heroic treatment would meet the acceptance of the medical profession charged with routine industrial hygiene in the plutonium industries is questionable at this time. It is expected, however, that pulmonary lavage will be seriously considered along with the more conservative approach of drug therapy in future decorporation therapy for accidental plutonium inhalation exposures.

References

ABRAMS, R., SEIBERT, H. C., FORKER, L., GREENBERG, D., LISCO, H., JACOBSON, L. O., SIMMONS, E. L.: Acute toxicity of intubated plutonium. Metallurgical Laboratory, Univ. of Chicago, Biology Division, CH-3875 (1946).

ABRAMS, R., SEIBERT, H. C., POTTS, A. M., FORKER, L. L., GREENBERG, D., POSTEL, S., LOHR, W.: Metabolism and distribution of inhaled plutonium in rats. Document MDDC-677 (1947) (also CH-3655, 1947).

ADEE, R. R., SANDERS, C. L., BERLIN, J. D.: Subcellular localization and indentification of alpha emitters by electron microscopic autoradiography. Health Physics **15**, 461–463 (1968).

ANTONCHENKO, G. P., KOSHURNIKOVA, N. A., LYUBCHANSKII, E. R.: Morphological changes in the lungs of rat following inhalation of large doses of soluble plutonium-239 compounds. Radiobiologiya **9** (1) 75–80 (1969) [also Radiobiology, USSR, Engl. tr. **9** (1), 97–104 (1969)].

BAIR, W. J.: Radioisotope toxicity: from pulmonary absorption, p. 431–448. In: R. S. CALDECOTT and L. A. SNYDER (eds.), Radioisotopes in the biosphere, publ. by the Center for Continuation Study, University of Minnesota, Minneapolis (1960).

BAIR, W. J.: Plutonium inhalation studies. (A series of lectures given in Japan in 1969 at the invitation of the Japanese Atomic Energy Commission.) BNWL-1221, Battelle-Northwest, Richland, Washington (1970).

BAIR, W. J.: Toxicology of inhaled plutonium — experimental animal studies. Proceedings of the Seminar on Radiation Protection Problems Relating to Transuranium Elements, Karlsruhe, W. Germany, EUR 4612 d-f-e (1971).

BAIR, W. J., MCCLANAHAN, B. J.: Plutonium inhalation studies. II. Excretion and translocation of inhaled $Pu^{239}O_2$ dust. Arch. environm. Hlth **2**, 648–655 (1961).

BAIR, W. J., PARK, J. F.: Comparative disposition of four types of plutonium dioxides inhaled by dogs, p. 181–197. In: Proceedings of the First Internat. Congr. of Radiation Protection. Oxford: Pergamon Press 1968.

BAIR, W. J., PARK, J. F., CLARKE, W. J.: Long-term study of inhaled plutonium in dogs. Technical Report No AFWL-TR-65-214. Air Force Weapons Laboratory, Kirtland Air Force Base, New Mexico (1966).

BAIR, W. J., SMITH, V. H.: Radionuclide contamination and removal, p. 157–223. In: A. M. FRANCIS DUHAMEL (ed.), Health physics, vol. 2, part 1, Progress in nuclear energy series XII. Oxford: Pergamon Press 1969.

BAIR, W. J., STUART, B. O., PARK, J. F., CLARKE, W. J.: Factors affecting retention, translocation and excretion of radioactive particles, p. 253–274. In: Radiological health and safety in mining and milling of nuclear materials, vol. 1. Vienna, Austria: IAEA 1964.

BAIR, W. J., WATSON, C. R., WILLARD, D. H., POWERS, G. J., OLSON, R. J.: Exposure of beagles to Oak Ridge U_3O_8 ash. Pacific Northwest Laboratory Annual Report for 1968 to the USAEC Division of Biology and Medicine, BNWL-1050, vol. 1, part 1, p. 3.46–3.49 (1970).

BAIR, W. J., WIGGINS, A. D., TEMPLE, L. A.: The effect of inhaled $Pu^{239}O_2$ on the life span of mice. Hlth Phys. 8, 659–663 (1962).

BAIR, W. J., WILLARD, D. H.: Plutonium inhalation studies. IV. Mortality in dogs after inhalation of $^{239}PuO_2$. Radiat. Res. **16**, 811–821 (1962).

BAIR, W. J., WILLARD, D. H.: Plutonium inhalation studies. III. Effect of particle size and total dose on deposition, retention, and translocation. Hlth Phys. **9**, 253–266 (1963).

BAIR, W. J., WILLARD, D. H., HERRING, J. P., GEORGE, L. A. II: Retention, translocation and excretion of inhaled $Pu^{239}O_2$. Hlth Phys. 8, 639–649 (1962).

BAIR, W. J., WILLARD, D. H., MARKS, S., HACKETT, P. L.: Preliminary observations on the pharmacodynamics and biological effects of inhaled plutonium oxide in dogs. Radiat. Res. **12**, 419 (1960) (Abstr.).

BAIR, W. J., WILLARD, D. H., TEMPLE, L. A.: Plutonium inhalation studies. I. The retention and translocation of inhaled $Pu^{239}O_2$ particles in mice. Hlth Phys. **7**, 54–60 (1961).

BALLOU, J. E., PARK, J. F.: Disposition of ingested, injected and inhaled ^{239}Pu citrate and ^{239}Pu nitrate. Pacific Northwest Laboratory Annual Report for 1971 to the USAEC Division of Biology and Medicine, BNWL-1650, vol. 1, part 1, p. 146–152 (1972).

BALLOU, J. E., PARK, J. F., MORROW, W. G.: On the metabolic equivalence of ingested, injected and inhaled ^{239}Pu citrate. Hlth Phys. **22**, 857–862 (1972).

BULDAKOV, L. A., LYUBCHANSKII, E. R.: Experimental basis for maximum allowable load (MAL) of Pu-239 in the human organism, and maximum allowable concentration (MAC) of Pu-239 in air at work locations. Original source unknown, ANL-TRANS-864 (1970).

BULDAKOV, L. A., LYUBCHANSKII, E. R., MOSKALEV, Y. I., NIFATOV, A. P.: Problems of plutonium toxicity. Moscow: Atomic Publication 1969 (Translated as USAEC Document LF-TR-41, p. 149–178, 1970).

CLARKE, W. J., BAIR, W. J.: Plutonium inhalation studies. VI. Pathologic effects of inhaled plutonium particles in dogs. Hlth Phys. **10**, 391–398 (1964).

CLARKE, W. J., PARK, J. F., BAIR, W. J.: Plutonium particle-induced neoplasia of the canine lung. II. Histopathology and conclusions, p. 345–355. In: Lucio Severi (ed.), Lung tumours in animals. Perugia, Italy: University of Perugia 1966.

CLARKE, W. J., PARK, J. F., PALOTAY, J. L., BAIR, W. J.: Bronchiolo-alveolar tumors of the canine lung following inhalation of plutonium particles. Amer. Rev. resp. Dis. **90**, 963–967 (1964).

CLARKE, W. J., PARK, J. F., PALOTAY, J. L., BAIR, W. J.: Plutonium inhalation studies. VII. Bronchiolo-alveolar carcinomas of the canine lung following plutonium particles inhalation. Hlth Phys. **12**, 609–613 (1966).

DILLEY, J. V.: Heterophile antibody formation in rats and beagles after $^{239}PuO_2$ inhalation. Pacific Northwest Laboratory Annual Report for 1967 to the USAEC Division of Biology and Medicine, BNWL-714, vol. 1, p. 3.17–3.20 (1968).

DILLEY, J. V.: Effect of DTPA on inhaled $^{239}PuF_4$ in beagles. Pacific Northwest Laboratory Annual Report for 1968 to the USAEC Division of Biology and Medicine, BNWL-1050, vol. 1, part 1, p. 5.14–5.17 (1970).

Dilley, J. V.: Immunocompetence of beagles exposed to $^{238}PuO_2$ and $^{239}PuO_2$. Pacific Northwest Laboratory Annual Report for 1968 to the USAEC Division of Biology and Medicine. BNWL-1050, vol. 1, part 1, p. 3.14–3.16 (1970).

Dilley, J. V., McDonald, K. E.: Therapeutic removal of inhaled plutonium. Pacific Northwest Laboratory Annual Report for 1969 to the USAEC Division of Biology and Medicine, BNWL-1306, vol. 1, part 1, p. 51–52 (1970).

Durbin, Patricia W.: Plutonium in man: A twenty-five year review. UCRL-20850, UC-41 Health and Safety, TID-4500 (58th ed.) June 1971.

Erokhin, R. A., Koshurnikova, N. A., Lemberg, V. K., Nifatov, A. P., Puzyrev, A. A.: Extrapulmonal tumors in rat following intratracheal introduction of soluble plutonium compounds. Gig. Tr. prof. Zabol. **10**, 33 (1969).

Erokhin, R. A., Koshurnikova, N. A., Lemberg, V. K., Nifatov, A. P., Puzyrev, A. A.: Lung tumors in rat following intratracheal introduction of soluble plutonium-239 compounds. Gig. Tr. prof. Zabol. **5**, 61 (1969).

Esterly, J. R., Faulkner, C. S. II: The granular pneumonocyte. Absence of phagocytic activity. Amer. Rev. resp. Dis. **101**, 869–876 (1970).

Foreman, H.: The pharmacology of some useful chelating agents, p. 82–94. In: M. J. Seven and L. A. Johnson (eds.), Metal binding in medicine. Philadelphia: J. P. Lippincott 1970.

Howard, E. B.: The morphology of experimental lung tumors in beagle dogs, p. 147–160. In: P. Nettesheim, M. G. Hanna, Jr., and J. W. Deatherage, Jr. (eds.), Morphology of experimental respiratory carcinogenesis (AEC Symposium Series No 21). Springfield, Virginia: NTIS 1970.

Johnson, L. J., Watters, R. L., Lagerquist, C. R., Hammond, S. E.: Relative distribution of plutonium and americium following experimental PuO_2 implants. Hlth Phys. **19**, 743–749 (1970).

Koshurnikova, N. A., Lemberg. V. K., Lyubchanskii, E. R.: Long-term consequences of inhaling soluble compounds of plutonium-239. Papers from the symposium: Long-term consequences of the effect of ionizing radiation on the organism. Moscow: Institute Biofiziki MZ SSR 1968.

Koshurnikova, N. A., Lemberg, V. K., Nifatov, A. P., Puzyrev, A. A.: Pneumosclerosis in rat following intratracheal introduction of soluble Pu^{239} compounds. Gig. Tr. prof. Zabol. 11, 27–32 (1968).

Kunzle-Lutz, M., Metivier, H., Nolibe, D.: Effects of lung washing on the retention of inhaled particles of plutonium oxide. Arch. Mal. prof. **31**, 513–516 (1970).

Kunzle-Lutz, M., Nolibe, D., Petaby, G., Rannaud, J. C.: Retention and lung clearance of plutonium oxide. Structure of alveolar macrophages. Arch. Mal. prof. **31**, 185–196 (1970).

Lisco, H.: Autoradiographic and histopathologic studies in radiation carcinogenesis of the lung. Lab. Invest. 8, 162–170 (1959).

Lyubchanskii, E. R.: The behaviour of plutonium-239 citrate injected intraperitoneally once or repeatedly into rats, p. 117–123. In: The distribution, biological action, and accelerated excretion of radioactive isotopes. Moscow: Meditsina 1964.

Lyubchanskii, E. R.: Use of the Na_3Ca salt of DTPA (Pentacine) for removing Pu^{239} from the rat organism in inhalation affection. Med. Radiol. **10** (1), 45–49 (1965).

Lyubchanskii, E. R.: The behavior of Pu-239 in rats after a single inhalation of some of its chemical compound. Radiobiology, USSR (Engl. tr.) **7**, No 4, 84–96 (1967) [Radiobiologiya **7**, No 4, 541–547 (1967)].

Markley, J. F.: Removal of polymeric plutonium from mice by combined therapy with the calcium chelate and penta-ethyl ester of DTPA. Int. J. Radiat. Biol. **7**, 405–407 (1963).

Masse, R.: Etude cytologique comparee de l'influence du plutonium et de la silice inhales sur le comportement du macrophage alveolaire, p. 247–259. In: C. N. Davies (ed.), International Symposium on Inhaled Particles, vol. III, Sept. 1970. Oxford: Pergamon Press 1972.

Matsusaka, N., Bair, W. J.: Biliary excretion of plutonium after pulmonary deposition in rats. Pacific Northwest Laboratory Annual Report for 1967 to the USAEC Division of Biology and Medicine, BNWL-714, vol. 1, p. 3.20–3.22 (1968).

McClanahan, B. J., Howard, E. B., Ragan, H. A., Beamer, J. L.: Late effects of intradermally administered plutonium in swine. Pacific Northwest Laboratory Annual Report for 1967 to the USAEC Division of Biology and Medicine, BNWL-714, vol. 1, p. 1.12–1.14 (1968).

McClellan, R. O.: Progress in studies with transuranic elements at the Lovelace Foundation. Hlth Phys. (in press).

McDonald, K. E., Park, J. F., Olson, R. J., Busch, R. H., Sanders, C. L.: Removal of inhaled $^{239}PuO_2$ from beagle dogs by pulmonary lavage. Pacific Northwest Laboratory

Annual Report for 1971 to the USAEC Division of Biology and Medicine, BNWL-1650, vol. 1, part. 1, p. 157–163 (1972).

Morrow, P. E., Gibb, F. R., Davies, H., Mitola, J., Wood, D., Wraight, N., Campbell, H. S.: The retention and fate of inhaled plutonium dioxide in dogs. Hlth Phys. **13**, 113–133 (1967).

Morrow, P. E., Gibb, F. R., Johnson, L.: Clearance of insoluble dust from the lower respiratory tract. Hlth Phys. **10**, 543–555 (1964).

Muggenburg, B. A., Pfleger, R. C., McClellan, R. O.: Bilateral bronchopulmonary lavage with DTPA solution for removing inhaled $^{144}CeCl_3$. Annual Report of Fission Product Inhalation Program, Lovelace Foundation for Medical Education and Research, LF-43, p. 286–294 (1970).

Nenot, J. C.: Etude de l'influence de l'irradiation sur l'epuration pulmonaire p. 239–246. In: C. N. Davies (ed.), International Symposium on Inhaled Particles, vol. III, Sept. 1970. London: Pergamon Press 1972.

Nielsen, S. W.: Pulmonary neoplasia in domestic animals, p. 123–145. In: P. Nettesheim, M. G. Hanna, Jr., and J. W. Deatherage, Jr. (eds.), Morphology of experimental respiratory carcinogenesis (AEC Symposium Series No 21). Springfield, Virginia: NTIS 1970.

Norwood, W. D.: DTPA-effectiveness in removing internally deposited plutonium from humans. J. occup. Med. **2**, 371–376 (1960).

Norwood, W. D.: Long-term administration of DTPA for plutonium elimination. A followup study in one patient. J. occup. Med. **4**, 130–132 (1962).

Norwood, W. D.: Plutonium (^{239}Pu) toxicity diagnosis and therapy. J. occup. Med. **14**, 37–44 (1972).

Park, J. F., Bair, W. J., Busch, R. H.: Progress in beagle dog studies with transuranium elements at Battelle-Northwest. Hlth Phys. **22**, 803–810 (1972).

Park, J. F., Bair, W. J., Howard, E. B.: Acute toxicity of inhaled plutonium-239 nitrate in beagles. Pacific Northwest Laboratory Annual Report for 1967 to the USAEC Division of Biology and Medicine, BNWL-714, vol. 1, p. 3.22–3.26 (1968).

Park, J. F., Clarke, W. J.: Chronic toxicity of inhaled plutonium in dogs. Hanford Biology Research Annual Report for 1962, HW-76000, p. 118–125 (1963).

Park, J. F., Clarke, W. J., Bair, W. J.: Plutonium particle-induced neoplasia of the canine lung. I. Clinical and gross pathology, p. 331–344. In: Lung tumours in animals. Perugia, Italy: University of Perugia 1966.

Park, J. F., Howard, E. B., Bair, W. J.: Acute toxicity of inhaled $^{238}PuO_2$ in beagle dogs. Technical Report No AFWL-TR-69-75, Air Force Weapons Laboratory, Kirtland Air Force Base, New Mexico (1969).

Park, J. F., Howard, E. B., Bair, W. J.: Chronic effects of inhaled $^{239}PuO_2$ in beagles. Pacific Northwest Laboratory Annual Report for 1969 to the USAEC Division of Biology and Medicine, BNWL-1306, vol. 1, part 1, p. 44–47 (1970).

Park, J. F., Howard, E. B., Bair, W. J.: Acute toxicity of inhaled $^{238}PuO_2$ in beagles. Pacific Northwest Laboratory Annual Report for 1968 to the USAEC Division of Biology and Medicine, BNWL-1050, vol. 1, part 1, p. 3.6–3.13 (1970).

Park, J. F., Howard, E. B., Stuart, B. O., Wehner, A. P., Dilley, J. V.: Cocarcinogenic studies in pulmonary carcinogenesis, p. 417–436. In: P. Nettesheim, M. G. Hanna, Jr., and J. W. Deatherage, Jr. (eds.), Morphology of experimental respiratory carcinogenesis (AEC Symposium Series No 21). Springfield, Virginia: NTIS 1970.

Pfleger, R. C., Wilson, A. J., McClellan, R. O.: Pulmonary lavage as a therapeutic measure for removing inhaled "insoluble" materials from the lung. Hlth Phys. **16**, 758–763 (1969).

Pinket, M. O., Cowdrey, C. R., Nowell, P. C.: Mixed hematopoietic and pulmonary origin of "alveolar macrophages" as demonstrated by chromosome markers. Amer. J. Path. **48**, 859–867 (1966).

Poda, G. A.: Personal Communication. (1971).

Rahman, Y. E., Lindenbaum, A.: Lysosome particles and subcellular distributions of polymeric tetravalent plutonium-239. Radiat. Res. **21**, 575–583 (1964).

Richmond, C. R., Langham, J., Stone, R. S.: Biological response to small discrete highly radioactive sources. II. Morphogenesis of microlesions in rat lungs from intravenously injected $^{238}PuO_2$ microspheres. Hlth Phys. **18**, 401–408 (1970).

Sanders, C. L.: The distribution of inhaled plutonium-239 dioxide particles within pulmonary macrophages. Arch. environm. Hlth **18**, 905–912 (1969).

Sanders, C. L.: The biological behavior of $^{239}PuO_2$ particles: role of the peritoneal mononuclear phagocyte. Radiat. Res. **38**, 125–139 (1969).

Sanders, C. L.: Maintenance of phagocytic function following $^{239}PuO_2$ particle administration. Hlth Phys. **18**, 82–85 (1970).

SANDERS, C. L.: Deposition patterns and the toxicity of transuranium elements in lung. Hlth Phys. **22**, 607–615 (1972).

SANDERS, C. L., ADEE, R. R.: The ultrastructure of mononuclear phagocytes following intraperitoneal administration of $^{239}PuO_2$ particles. J. reticuloendoth. Soc. **6**, 1–23 (1969).

SANDERS, C. L., ADEE, R. R.: Ultrastructural localization of inhaled $^{239}PuO_2$ particles in alveolar epithelium and macrophages. Hlth Phys. **18**, 293–295 (1970).

SANDERS, C. L., ADEE, R. R , JACKSON, THELMA A.: Fine structure of alveolar areas in the lung following inhalation of $^{239}PuO_2$ particles. Arch. environm. Hlth **22**, 525–533 (1971).

SANDERS, C. L., JACKSON, THELMA A., ADEE, R. R., POWERS, G. J., WEHNER, A. P.: Distribution of inhaled metal oxide particles in pulmonary alveoli. Arch. intern. Med. **127**, 1085–1089 (1971).

SANDERS, C. L., PARK, J. F.: Pulmonary distribution of alpha dose from $^{239}PuO_2$ and induction of neoplasia in rats and dogs, p. 489–498. In: C. N. DAVIES (ed.), International Symposium on Inhaled Particles, vol. III, Sept. 1970. Oxford: Pergamon Press 1972.

SCOTT, K. G., AXELBROD, D. J., CROWLEY, J., HAMILTON, J. G.: Deposition and fate of plutonium, uranium, and their fission products inhaled as aerosols by rats and man. Arch. Path. **48**, 31–54 (1949).

SMITH, V. H.: Therapeutic removal of internally deposited transuranium elements. Hlth Phys. **22**, 765–778 (1972).

SMITH, V. H., BALLOU, J. E., CLARKE, W. J., THOMPSON, R. C.: Effectiveness of DTPA in removing plutonium from the pig. Proc. Soc. exp. Biol. (N.Y.) **107**, 120–123 (1961).

STEWART, K., WILSON, R. H.: Final evaluation of the biological measurements on operation Roller Coaster (Joint US/UK Field Experiments) UKAEA Atomic Weapons Research Establishment AWRE Report No 0-76/67 (1968).

STUART, B. O., BAIR, W. J., CLARKE, W. J., HOWARD, E. B.: Acute toxicity of inhaled plutonium oxide-238 and -239 in rats. Technical Report No AFWL-TR-68-49, Air Force Weapons Laboratory, Kirtland Air Force Base, New Mexico (1968).

STUART, B. O., BAIR, W. J., PARK, J. F.: Interpretation of excretion data from beagle dogs after ^{239}Pu inhalation, p. 243–255. In: H. A. KORNBERG and W. D. NORWOOD (eds.), Diagnosis and treatment of deposited radionuclides Amsterdam: Excerpta Medica Foundation 1968.

SUZUKI, M., OKABAY ASHI, H., WATANABE, S., HONGO, S., OHNO, S.: Studies on the inhalation of submicron plutonium nitrate aerosols in wistar adult rats. Final Report of Research Project of Plutonium Hazards, April 1966–March 1970, Chiba, Japan. NIRS-Pu-8, p. 66–75 (1971).

TEMPLE, L. A., MARKS, S., BAIR, W. J.: Tumours in mice after pulmonary deposition of radioactive particles. Int. J. Radiat. Biol. **2**, 143–156 (1960).

TOMBROPOULOS, E. G.: Review of therapeutic procedures for removal of inhaled radioactive materials. Hlth Phys. **10**, 1251–1257 (1964).

TREGUBENKO, I. P.: Inhalation of plutonium by rats. Tr. Inst. Biol., Akad. Nauk. SSSR, Ural. Filial, No 46, 89–97 (1966).

WATTERS, R. L., LEBEL, J. L.: Progress in the beagle studies at Colorado State University. Hlth Phys. **22**, 811–814 (1972).

WEHNER, A. P.: Negatively charged aerosols, effect on pulmonary clearance of inhaled $^{239}PuO_2$ in rats. Chest **60**, 468–471 (1971).

WEST, J. E., BAIR, W. J.: Plutonium inhalation studies. V. Radiation syndrome in beagles after inhalation of plutonium dioxide. Radiat. Res. **22**, 489–506 (1964).

WILLARD, D. H., PARK, J. F.: Long-term effects of inhaled $^{238}PuO_2$ in beagles. Pacific Northwest Laboratory Annual Report for 1969 to the USAEC Division of Biology and Medicine, BNWL-1306, vol. 1, part 1, p. 66–67 (1970).

WILSON, R. H., TERRY, J. L.: Plutonium uptake by animals exposed to a non-nuclear detonation of a plutonium-bearing weapon stimulant. Operation Roller Coaster Project Officers Report—Project 4.1 POR-2512 (WT-2512) (1968).

WILSON, R. H., THOMAS, R. G., STANNARD, J. N.: Biological and medical studies associated with a field release of plutonium. Operation Plumbbob, WT-1511, University of Rochester, Rochester, N.Y. (1961).

YUILE, C. L., GIBB, F. R., MORROW, P. E.: Dose-related local and systemic effects of inhaled plutonium-238 and plutonium-239 dioxide in dogs. Radiat. Res. **44**, 821–834 (1970).

Chapter 12

Maximum Permissible Body Burdens and Concentrations of Plutonium: Biological Basis and History of Development[1]

W. H. LANGHAM* and J. W. HEALY

I. Introduction

Any attempt to write an account of the early events and decisions leading to the current protection limits for plutonium will be controversial at best. Many of the early actions and decisions and the reasoning on which they were based were documented only as memoranda among the principal personnel of the various projects, institutions and organizations involved, as monthly progress reports and in the minutes of the various project, group and section council meetings. All exchanges of ideas, information and actions, of necessity, were carried out under strict security regulations.

The present account will be subject to even more controversy because most of the principal participants in these early happenings are still very much alive and each will have his own opinions, views and notes regarding what went on. After 25 years or so, it would be surprising indeed if all who participated would remember the events in exactly the same way and assign to them the same order of importance. Perhaps there is a lesson to be learned by those now faced with making similar decisions that cannot wait for accumulation of unequivocal data. Documentation of the reasoning involved (even if, or particularly if, based on inadequate data) will enable the next generation to more fully understand the inherent uncertainties and basis of the decisions at the time and to proceed compassionately with appropriate modifications and refinements without regarding previous work either as dogma passed down from the most high or as inadequate, irrelevant and immaterial and, therefore, to be completely ignored and rediscovered.

Despite the controversy that may arise, we feel that an attempt to document briefly the early biological basis and history of the development of current protection standards for plutonium is worthwhile, since studies with plutonium and similar work with strontium have constituted the basis for present radiation limits for radionuclides (except radium) that deposit in bone. No attempt will be made to compile a complete review of all excellent research that has been directed toward better understanding of the physiology, toxicology and industrial medical control of plutonium. Previous and following chapters are devoted to such in-depth coverage. Rather, it is our intent to cover in approximate chronologic fashion the evolution of protection standards for plutonium, emphasizing those biological observations that had the most impact on the derivation of

[1] Work supported by U. S. Atomic Energy Commission under Contract at the Los Alamos Scientific Laboratory of the University of California, Los Alamos, New Mexico 87544.

* Deceased.

current values. This emphasis seems appropriate as biological considerations and relevant data become somewhat obscure and depersonalized by the formal equations currently used to derive maximum permissible body burdens and concentrations in air and water. In taking the above approach, we extend our profound apologies both to those whom we may annoy by misrepresentation of their views or those we many overlook.

II. Radiation Protection Criteria Prior to 1943

Experience with the biological effects of radiation began within months after the discovery of X-rays and radium. RÖNTGEN discovered X-rays in 1895, and a case of roentgen-dermatitis was reported less than a year later. Several more cases were reported by 1900. PIERRE and Madam CURIE isolated radium in 1898, and in 1900 the occurrence of chronic radium-dermatitis was documented. The first case of occupational roentgen-cancer as well as the first fatality from this disease was reported from Germany in 1902.[2]

After World War I, a number of committees were organized to recommend protective measures against radiation exposure. These committees were the forerunners of the International Commission on Radiological Protection (ICRP) and the U. S. National Committee on Radiation Protection (NCRP). (Now "Council"—Editor.) The ICRP had its beginning in the Committee on X-ray and Radium Protection authorized by the Second International Congress of Radiology in 1928 (Int. Congr. Radiol. 1928) and the NCRP in the organization in 1929 of a U. S. Advisory Committee on X-ray and Radium Protection (TAYLOR, 1958a). The function of the latter committee was to provide input from the U. S. to the International Committee. It became the NCRP in 1946.

The early history and major recommendations of the NCRP and ICRP have been reviewed by L. S. TAYLOR (1958a and b). Although these groups provided guidance on radiation protection measures, they did not recommend a tolerance dose in their first publications. In 1934 and again in 1937 the International Committee published a recommendation of 0.2 r per day or 1 r per week as the tolerance dose. The U. S. Advisory Committee on X-ray and Radium Protection recommended a tolerance value for X-rays of 0.2 r per day in 1931 (NCRP, 1931), which was subsequently changed to 0.1 r per day in 1936 (NCRP, 1936) and extended to include both X- and gamma-rays in 1938 (NCRP, 1938). The roentgen (r) had been adopted as the international unit of X-ray exposure at the Stockholm Congress of Radiology in 1928. The definition of the roentgen was modified later at the Chicago Congress of Radiology in 1937 to make it include gamma-rays.

During the decade of the 1920's when the various committees were concerned with establishing protection criteria for X-rays and gamma-rays of radium, another human experience attesting to the dangers of radiation and radioactive materials came to light. In 1922, 1923, and 1924, nine girls died who had been employed for several years as luminous watch dial painters in a New Jersey factory (MARTLAND, 1929). No investigations were made as to the cause of death in any of these cases. In September 1924 BLUM (1924), a New York dentist, made the first report suggesting an occupational poisoning associated with the same plant from which the early deaths occurred. He reported before the American Dental Association a case of unusual and intractable osteomyelitis of the mandible and, knowing nothing of the nature of the luminous paint, thought from the unusual clinical behavior that some sort of occupational exposure existed. In May and September 1925, HOFFMAN (1925a and b) reported on his investigations of the New Jersey plant and offered the opinion that an unusual occupa-

2 These data and incidences quoted from HUEPER (1942).

tional poisoning existed and thought that mesothorium in the luminous dial paint might be mainly responsible. In the August issue of the Journal of Industrial Hygiene, CASTLE et al. (1925) reported on "Necrosis of the Jaw in Workers Employed in Applying Luminous Paint Containing Radium". Their report was a result of an investigation of conditions in the New Jersey plant conducted in March 1924 at the request of plant executives, but the findings had not been revealed immediately at company request.

On October 8, 1925, HARRISON S. MARTLAND presented a paper at the New York Pathological Society meeting on "Some Unrecognized Dangers in the Use and Handling of Radioactive Substances", and the paper was published in full in the Journal of the American Medical Association (MARTLAND, 1925) in the same issue as Hoffman's paper on "Radium (Mesothorium) Necrosis". For the next several years MARTLAND and others vigorously pursued the radium poisoning problem and the New Jersey cases in a highly scientific and thorough manner, and within four years MARTLAND and HUMPHRIES (1929) reported the first cases of radium-induced osteogenic sarcoma. By 1932 it was apparent also that industrial uses of radium were not the only potential source of radium poisoning. Radium was being administered by the medical profession and peddled by the nostrum vendors in the form of radium tonics (AMA Bureau of Investigation, 1932; GETTLER and NORRIS, 1933), the most famous being "Radithor." The status of the radium poisoning problem as of 1933 was reviewed by EVANS (1933).

For the next several years efforts were directed toward studies of radium uptake, distribution and excretion and toward development of vastly more sensitive methods of measurement, including quantitation of radium content and elimination rate of living persons (EVANS, 1935, 1937; AUB et al., 1938). Only by knowing the body burdens in exposed individuals with and without overt signs and symptoms of radium poisoning could an approach be made toward establishing a harmful level. In 1934 R. D. EVANS became Assistant Professor of Physics at the Massachusetts Institute of Technology, which resulted in a collaborative effort by EVANS, AUB, and MARTLAND to carefully measure the body burdens of fixed radium in known exposure cases. EVANS has continued the follow-up and measurement of known cases of radium exposure and currently has observations on over 650 cases (1971). By 1941 the observations were that signs of poisoning and even death may occur with body burdens as low as 1 μg of radium fixed in the bone for several years. Seven cases with more than 0.02 but less than 0.5 μg after 7 to 25 years were completely symptom-free. These observations were not referenced until 1943 (EVANS, 1943); however, they were used in 1941 as the basis for the first recommendation of 0.1 μg as the tolerance dose for radium (National Committee on Radiation Protection, 1941)—*a value that is still accepted*. It was recommended also that the radon concentration in the atmosphere of workrooms should not exceed 10^{-11} Ci per liter. Report No. 5 (NBS-27) was prepared, in response to a request from organizations concerned with the dial painting industry, by a special committee[3] appointed by LYMAN BRIGGS, Director of the National Bureau of Standards.

3 The committee membership was Dr. L. F. CURTISS, Chairman, representing the National Bureau of Standards, Washington, D.C.; Dr. R. D. EVANS, Physicist, Massachusetts Institute of Technology, Cambridge, Mass.; Dr. G. FAILLA, Physicist, Memorial Hospital, New York City; Dr. FREDERICK B. FLINN, Director of Industrial Hygiene, Columbia University, New York City; Dr. HARRISON S. MARTLAND, Chief Medical Examiner of Essex County, Newark, New Jersey; Dr. J. E. PAUL, representing the U. S. Radium Corporation, New York City; Dr. J. S. ROGERS, representing the Division of Labor Standards, Department of Labor, Washington, D.C.; Captain CHARLES S. STEPHENSON (MC), representing the Bureau of Medicine and Surgery, U. S. Department of the Navy, Washington, D.C.; and Dr. G. T. TAYLOR, representing the Radium Chemical Company, New York City.

The quantitative studies of early radium exposure cases during the 1930's by EVANS, AUB and MARTLAND are the most important factor in establishment of present exposure guides for plutonium as well as all other bone-seeking radionuclides. Historically, it is interesting to note that the recommendation of 0.1 μg as the tolerance dose for radium (which provided the basis for presently accepted protection standards for plutonium) occurred only two months after the discovery of plutonium and only 18 months prior to the first demonstration of a sustained nuclear reaction.

At the beginning of 1943 (when the Manhattan District's Plutonium Project was just getting underway), only three tolerance values for occupational exposure to radiation and radioactive materials had been established: 0.1 r per day for external X- and gamma-radiation, 0.1 μg of radium as the maximum allowable body deposition and 1×10^{-11} Ci of radon per liter in the air of workrooms. The unit of external radiation exposure was the roentgen and applied only to X- and gamma-rays.

III. Plutonium Occupational Protection Criteria (1943—1946)

During 1943 the Plutonium Project's Health Division under Dr. R. S. STONE had little time to worry about the potential hazards of plutonium *per se*. The external radiation hazards associated with neutrons and gamma-rays from the large-scale chain-reacting piles being planned and the radiation hazards that would be imposed by the large quantities of fission products in the uranium slugs from which the plutonium had to be separated were staggering and had to receive first priority. The hazards imposed by fission products were both external and internal, and little or nothing was known about body uptake, deposition, distribution and excretion of most of them. Dr. G. T. SEABORG (1970) aptly summed up the status of the plutonium hazard problem during 1943 by pointing out that "up to the fall of 1943 the cyclotron-produced plutonium in existence amounted to only 2 mg, a quantity distributed throughout the program over a period of one and a half years—and a quantity so precious we couldn't afford to ingest any of it". The situation changed fast, however; with the start-up of the Clinton (Site X, Clinton, Tennessee) reactor (November 4, 1943), milligram amounts of plutonium became available by the first of the year. Production of plutonium grew to gram amounts by March 1, 1944, and with the start-up of the Hanford piles[4] (Site W, Hanford, Washington) to kilogram amounts by mid-1945.

Concern over the hazards of plutonium processing developed even more suddenly than did production of plutonium itself; Dr. SEABORG, Head of the Metallurgical Laboratory's Chemistry Division, wrote the following letter (1944 a) to Dr. R. S. STONE, Medical Director:

R. S. STONE
G. T. SEABORG

January 5, 1944
(Declassified, July 18, 1969)

Physiological Hazards of Working with Plutonium

It has occurred to me that the physiological hazards of working with plutonium and its compounds may be very great. Due to its alpha radiation and long life it may be that the permanent location in the body of even very small amounts, say one milligram or less, may be very harmful. The ingestion of such extraordinarily small amounts as some few tens of micrograms might be unpleasant, if it locates itself in a permanent position. In the handling of the relatively large amounts soon to begin here and at Site Y, there are many conceivable methods by which amounts of this order might be taken in unless the greatest care is exercised.

---

4 Operation of the first Hanford pile began in September 1944.

In addition to helping to set up safety measures in handling so as to prevent the occurrence of such accidents, I would like to suggest that a program to trace the course of plutonium in the body be initiated as soon as possible. In my opinion such a program should have the very highest priority.

G. T. S. (signed)

GTS : EES
c/c to S. K. ALLISON
A. H. COMPTON
T. R. HOGNESS
Central Reading File

This letter was the first sounding of the alarm and the first time anyone dared to make a guess as to how much plutonium might be physiologically harmful. In a letter to one of the authors granting permission to use his letter of January 5, 1944, Dr. SEABORG (1971) added the following comment: "In retrospect, it is quite clear that plutonium proved to be physiologically harmful in even smaller amounts than I suspected when I issued my early warning, at which time I was considered by some to be an alarmist. However, I take some satisfaction in the fact that the danger was recognized before very many of our chemists began to handle plutonium in quantity so that our laboratory rebuilding program and changes in working philosophy took place in time". Dr. SEABORG's satisfaction is due largely to the rapidity with which he and the others responsible took action once the hazard of plutonium was recognized.

Ten days later SEABORG (1944b) wrote to STONE again saying that he was even more concerned about the hazard of plutonium, having read the article on radium toxicity by R. D. EVANS in the September issue of the Journal of Industrial Hygiene and Toxicology (1943), and felt that 10 mg of plutonium for metabolic studies should be made available to Dr. J. G. HAMILTON, Director of the Radiation Laboratory of the University of California, as soon as possible. On January 19, 1944, Dr. STONE reported the plutonium toxicity problem before a Project Council meeting (Clinton Laboratories, 1944), pointing out that 1 to 2 μg of fixed radium in the body had been fatal and that plutonium may be less toxic by a factor of about 50. On this basis, 5 μg (0.3 μCi) of stored plutonium was proposed as the tolerance level. This appears to be the first documented report of a direct comparison between radium and plutonium toxicity, and undoubtedly it was made on the basis of equivalent alpha energy deposition based on the ^{226}Ra decay scheme at the time[5], assuming about 50 percent exhalation of radon by man (EVANS, 1937). By the end of January (25 days after Dr. SEABORG's original letter), plans and actions to modify plutonium laboratories and working

5 Radium-226 decay scheme.

$$\mathrm{Ra} \xrightarrow[1600\ \mathrm{yr}]{\alpha\ (4.8\ \mathrm{Mev})} \mathrm{Rn} \xrightarrow[3.8\ \mathrm{d}]{\alpha\ (5.5\ \mathrm{Mev})} \mathrm{RaA} \xrightarrow[3.0\ \mathrm{m}]{\alpha\ (6.0\ \mathrm{Mev})} \mathrm{RaB} \xrightarrow[27\ \mathrm{m}]{\beta}$$

$$\longrightarrow \mathrm{RaC}\ (19.7\ \mathrm{m}) \begin{cases} \xrightarrow{\alpha\ \ 0.04\ \%} \mathrm{RaC''} \xrightarrow[1.3\ \mathrm{m}]{\beta} \\ \xrightarrow[19.8\ \mathrm{m}]{\beta\gamma\ \ 99.96\ \%} \mathrm{RaC'} \xrightarrow[10^{-6}\ \mathrm{s}]{\alpha\ (7.7\ \mathrm{Mev})} \end{cases} \longrightarrow \mathrm{RaD} \xrightarrow[22\ \mathrm{yr}]{\beta}$$

$$\longrightarrow \mathrm{RaE} \xrightarrow[5\ \mathrm{d}]{\beta} \mathrm{RaF} \xrightarrow[140\ \mathrm{d}]{\alpha\ (5.2\ \mathrm{Mev})} \mathrm{RaG}\ \text{(stable lead)}$$

Plutonium tolerance level: Calculated on basis 0.1 μg radium-226 and Pu alpha energy = 5.15 Mev.

$$(\mathrm{TL})_{\mathrm{Pu}} = 0.1\ \mu\mathrm{g} \times \frac{24500}{1600}\left[\frac{4.8}{5.15} + \frac{0.5\,(5.5)}{5.15} + \frac{0.5\,(6.0)}{5.15} + \frac{0.5\,(7.7)}{5.15}\right] \cong 5.0\ \mu\mathrm{g}$$

practices were already underway at both the Chicago and Clinton laboratories. By January 29, 11 mg of plutonium had been allocated (for shipment on February 1) to Dr. HAMILTON at Berkeley for animal uptake, distribution and excretion studies (PETERSON, 1944). The material was delivered on February 10 (HAMILTON, 1944), and the Metallurgical Laboratory's progress report for February (ALLISON 1944a) contained the following brief announcement:

"*Product Studies*—Oral absorption of all valence states is less than 0.05%; lung retention high; absorbed material predominantly in the skeleton; excretion very small in urine and feces."

This curt announcement coming only a few days after receipt of the plutonium reflects the dedication of Dr. HAMILTON and his group and was the first contribution of plutonium animal experimental data to the establishment of current protection criteria. The importance of these first animal experiments to the biological basis for present plutonium protection criteria can hardly be overemphasized. They delineated the battlefield: plutonium, like radium, concentrated in bone and would be expected to produce similar types of bone lesions, including osteogenic sarcoma, its initial rate of elimination may be much slower than radium, its fixation in the lungs may be much higher and its absorption in all valence states may be much less. Subsequent experiments by HAMILTON's group and others confirmed these experiments and provided more quantitative information.

What appears to be the first value for a dangerous level of plutonium in air was proposed in a letter dated March 1 (1944b), from Dr. S. K. ALLISON (Director of the Chicago Metallurgical Laboratory) to Dr. J. R. OPPENHEIMER (Director of Site Y), in which he pointed out that the plutonium problem would exist at Site Y (Los Alamos, New Mexico), and Site X (Clinton Laboratories, Clinton, Tennessee) and Site W (Hanford, Washington). ALLISON proposed that it would be dangerous to work 48 hours per week for two years in an atmosphere that contained 2×10^{-15} g/cm^3 (1.2×10^{-10} μCi/cm^3). Evidently he made the calculation himself based on 5 μg of plutonium deposited in the lungs as being dangerous. This assumption must have come from an analogy with the 1 μg harmful level of fixed radium in the skeleton of dial painters. If about 50 μg of plutonium in the skeleton was equivalent in energy deposition to 1 μg of radium and the lungs weighed about one-tenth as much as the skeleton, then 5 μg of plutonium might be a dangerous level if deposited in the lungs. He assumed a breathing rate of 10 liters per minute and evidently 100 percent deposition in the lung with no elimination and translocation. On the basis of these assumptions, the dangerous air concentration for 48-hour occupancy for two years would be:

$$\frac{5 \times 10^{-6}\ \text{g}}{10^4\ \text{cm}^3/\text{min} \times 2.8 \times 10^5\ \text{min}/2\ \text{yr}} = \sim 2 \times 10^{-15}\ \text{g}/\text{cm}^3.$$

ALLISON also proposed some data on an average (standard) man that might be of interest in the plutonium hazard evaluation problem: "In a male of average weight 160 pounds, the weight of the skeleton, where radium is deposited, is 21.6 pounds. The weight of the lungs in this hypothetical average man is 4.34 pounds when they are full of fluid, and 1 pound when the normal fluids are removed. We have taken a factor of 10 as roughly representing the ratio of skeletal weight to lung weight." This tentative air tolerance value was reported at the Project Council Information Meeting (Health) held on March 7, 1944 (NICKSON, 1944). It is of more than casual historical interest that a heated discussion of the effect of particle size on lung deposition and retention ensued in which it was intimated that the majority of particles of less than 1 μm would stay in the lung. At the same meeting, Dr. STONE suggested the use of "rep" for radiation other than X- and gamma-rays (α, β, n, etc.) and "rem" for roentgen equivalent mouse or man. These concepts had been transmitted to him by H. M. PARKER, then at the Clinton Laboratories.

In the Clinton Laboratories report for the month ending April 29, 1944 (Metallurgical Laboratory, 1944a), H. M. PARKER proposed a tentative plutonium air tolerance of 5×10^{-10} µg/cm^3 (3×10^{-11} µCi/cm^3) for a 1-year exposure based on a comparison with 0.1 r per day tolerance exposure for X- and gamma-rays. He assumed 50 percent retention of dust from the inhaled air. For heavy-particle radiation he assumed the tolerance dose to be 0.01 rep per day (twice as damaging as neutrons, 10 times as damaging as X-rays). Conceptually, this was probably the first use of an RBE (relative biological effectiveness) of 10 for alpha particles, although he did not call it RBE. From these assumptions and the assumption of uniform distribution over the entire lung surface ($\sim 8 \times 10^5$ cm^2), he calculated that the lung tissue would receive 0.01 rep per day when the lung burden reached 0.64 µg (0.04 µCi). Assuming no elimination mechanism from the lungs, a person could breathe air containing 4.3×10^{-10} µg/cm^3 for one year before building up to a lung exposure of 0.01 rep per day. A direct quote of the conclusions is interesting: "Thus, a provisional level of 5×10^{-10} µg of Pu/cm^3 in air should be a safe temporary limit. In the meantime, improved knowledge of metabolism of the product and of dust particle retention in the lungs should lead to a better value. The tentative figure above is extremely conservative in all respects except for the assumption of uniform distribution through the lung". The value of 5×10^{-10} µg/cm^3 remained the air concentration guide for the remainder of the Plutonium Project period. J. E. ROSE (1945) reexamined the air tolerance level in the fall of 1945. Working from first principles and preliminary animal data as a basis for more realistic lung deposition and clearance rates, he concluded that there was no compelling justification for changing the air tolerance from PARKER's original value.

With increased availability of plutonium from the Clinton pile, a number of important animal toxicity experiments were started. R. D. FINKLE and E. PAINTER started injecting toxic amounts of plutonium into rats, mice, rabbits and dogs. The first injections at Chicago were on May 22, 1944, and were the beginning of attempts to compare the metabolism and toxicity of plutonium with radium and, through animal experimentation, to provide better guidance for the protection of those working with plutonium. A parallel effort to compare the acute and subacute toxicity of ^{226}Ra, ^{210}Po and ^{239}Pu was started shortly thereafter by FINK et al. (1950) at the Manhattan Project Laboratory (later Atomic Energy Project) of the University of Rochester School of Medicine and Dentistry, Rochester, New York. These early studies and subsequent observations of low-dose, long-term effects (primarily at Chicago) involved many people[6] and eventually provided the empirical comparison of chronic toxicity of plutonium and radium that is the basis for the currently accepted 0.04 µCi maximum permissible plutonium body burden for man.

By the end of June 1944, studies of deposition, retention, absorption and elimination of plutonium administered to rats via inhalation and tracheal intubation had been or were being initiated both in K. S. Coles' Biological Research

6 Among those making contributions to these early studies were R. ABRAMS, W. BLOOM, G. BOYD, A. M. BRUES, G. CASSARETT, K. S. COLE, R. FINK, M. FINKEL, R. FINKLE, D. GARDNER, L. O. JACOBSON, W. KISIELESKI, B. LAWRENCE, H. LISCO, R. METCALF, R. MURRY, W. P. NORRIS, E. PAINTER, C. L. PROSSER, E. RUSSELL, G. SACHER, J. SCHUBERT, H. SILBERTSTEIN, E. L. SIMMONS, R. E. SNYDER, M. N. SWIFT, R. ZIRKLE, and others.

Section (R. ABRAMS et al., 1946) at the Metallurgical Laboratory and in J. G. Hamilton's group (K. G. SCOTT et al., 1945) at the Radiation Laboratory of the University of California. These studies were to provide much of the biological basis for current maximum permissible plutonium air concentrations through their contribution to current models of kinetics of inhaled particulates. These and the early toxicity studies mentioned above were reported from time to time in progress reports and summarized (to date) in a report of a conference on plutonium held in Chicago on May 14–15, 1945 (NICKSON, 1945); however, they did not result in specific reports and analysis of data until the 1946–1950 period.

By the first of March 1945, urine assays were being applied both at Chicago and Los Alamos in attempts to estimate exposure of personnel to plutonium. Body burden was estimated on the assumption that the excretion rate in man was the same as that for the rat and rabbit and at about 20 days after exposure reached a steady state at 0.01 percent of the body burden per 24 hours. In April, tracer studies in subjects with short life expectancy were initiated through both the Chicago and Los Alamos Laboratories (and a little later at the Radiation Laboratory at Berkeley) to established the human urinary excretion rate (NICKSON, 1945). The first urinary assay results on personnel at Los Alamos suggested that some of the workers in the plutonium recovery group already might have approached or exceeded a body burden of 1 μg. Late in March a meeting was held at Los Alamos with Dr. HYMER FRIEDELL representing Dr. STAFFORD WARREN (Medical Director of the Manhattan District) to discuss these early results. Other participants were Drs. L. H. HEMPELMANN (Leader of the Los Alamos Health Group), J. W. KENNEDY (Leader of the Los Alamos Chemistry Division), and W. H. LANGHAM. Based largely on apprehension over the autoradiographic studies of J. G. HAMILTON's group at Berkeley (showing plutonium distribution in bone was much more non-uniform than was radium), the decision was made to introduce a safety factor of 5 and to lower the maximum allowable plutonium body burden from 5 μg to 1 μg (0.06 μCi). S. T. CANTRIL and H. M. PARKER (then at Hanford) chose to introduce a safety factor of 10 and introduced a provisional body burden of 0.5 μg (0.03 μCi) for the Hanford Operations. In a memorandum report to CANTRIL, PARKER (1945) proposed a tolerance concentration for plutonium in plant and village drinking water. The tolerance concentration in village drinking water ($\sim 10^{-5}$ μg/cm^3, 6×10^{-7} μCi/cm^3) was based on a maximum allowable body burden of 0.5 μg, a 60-year effective exposure time, a plutonium absorption rate of 0.05 percent and a water intake of 5 liters per day. The plant water tolerance (5×10^{-5} μg/cm^3 or 3×10^{-6} μCi/cm^3) was based on the same allowable body burden, a working exposure time of 30 years and a daily water intake of 2 liters.

In summary, during the period 1943 through July 1946 (when the AEC was established) health protection standards for the Manhattan District's plutonium operations evolved from no standards at all to tentatively accepted values for the maximum allowable body burden and tolerance concentrations in air and water. These values were as given in Table 12.1. The maximum allowable body burden was based on comparison of energy deposition from plutonium and 0.1 μCi of radium with a safety factor of 5 or 10 as a precaution against the difference in distribution between radium and plutonium in bone. The comparison axiomatically designated bone as the critical organ. The air tolerance concentration was based on a comparison of dose to the lung from uniformly-deposited plutonium with X- or gamma-ray exposure in which it was assumed that 0.01 rep per day of alpha radiation reached in 1 to 2 years was equivalent in effect to 0.1 r per day of X- or gamma-rays. The water tolerance concentration was based

Table 12.1. Plutonium occupational protection criteria derived during the Manhattan District's plutonium project (1943–1946)

Maximum allowable body burden	
Manhattan District	1.0 μg (0.06 μCi)
Hanford operations	0.5 μg (0.03 μCi)
Air tolerance concentration	
1–2 year exposure	5×10^{-10} μg/cm³ (3×10^{-11} μCi/cm³).
Water tolerance concentration	
Community	1×10^{-5} μg/cm³ (7×10^{-7} μCi/cm³)
Plant	5×10^{-5} μg/cm³ (3×10^{-6} μCi/cm³)

on rate of intake that would result in the maximum allowable body burden in the appropriate time frame. It is interesting to compare the above values with those in current use (see Sec. V).

IV. Plutonium Occupational Protection Criteria (1946—1950)

After the war had ended and about the time of enactment of the Atomic Energy Act (transferring responsibilities from the Manhattan District to the U. S. Atomic Energy Commission) the general urgency relaxed, but studies relevant to the hazardous properties of plutonium continued. HAMILTON's group at Berkeley continued metabolic and inhalation studies, and the Biology Division at Chicago (then a part of the Argonne National Laboratory) continued and expanded studies comparing the chronic or delayed toxicity of ^{89}Sr, ^{226}Ra and ^{239}Pu under the direction of Dr. AUSTIN BRUES.

In 1946 the toxicity, inhalation and metabolic studies initiated earlier began to bear fruit in the form of classified summary and interpretive reports (SCOTT et al., 1945, 1949; ABRAMS et al., 1946; PAINTER et al., 1946; FINKLE et al., 1946; BOYD et al., 1946a and b; FINK, 1950). At about the same time relaxation of security and classification restrictions began, and plutonium could be mentioned outside hallowed halls. Perhaps the credit for the first open disclosure of plutonium as a potential industrial hazard goes to Drs. AUSTIN BRUES, HERMANN LISCO and MIRIAM FINKEL. This disclosure was in a manuscript entitled "Carcinogenic Action of Some Substances which may be a Problem in Certain Future Industries", declassified on July 31, 1946 (BRUES et al., 1946). The paper was presented at a meeting of the Radiological Society of North America (December 1–6, 1946) and published in a very condensed form in a Special Plutonium Project issue of Radiology in September 1947 (LISCO et al., 1947). A direct quote from the concluding remarks of the uncondensed version is interesting indeed: "It is noteworthy, however, that two general principles have been derived from animal experiments which, if true, greatly facilitate extrapolation to longer times: (1) a linear relation between dose, time after latency and tumor expectancy and (2) a logarithmic, or a least gradual, change in latent time with dose. It is of interest that this scheme postulates a true tolerance dose where the latent time exceeds the life span, but in man this would be singularly low (e.g., log dose = −100)". The above quote is referred to affectionately by some as BRUES' Law.

In December 1946 the U. S. Committee on X-Ray and Radium Protection reorganized, extended its scope to respond to the rapid expansion in the radiation field and renamed itself the National Committee on Radiation Protection (TAYLOR, 1958a). During the next three or four years much of the material collected during

Table 12.2. Plutonium occupational protection criteria resulting from the Chalk River permissible doses conference

US and UK preliminary versions (November 1949)	
Maximum permissible amount in body	0.1 μg (0.006 μCi)
Maximum permissible air concentration soluble and insoluble compounds	5×10^{-12} μg/cm³ (3×10^{-13} μCi/cm³)
Maximum permissible water concentration	4×10^{-6} μg/cm³ (3×10^{-7} μCi/cm³)
Canadian final report (May 1950)	
Maximum permissible amount in body	0.5 μg (0.03 μCi)
Maximum permissible air concentration (24-hour day)	2.5×10^{-11} μg/cm³ (1.5×10^{-12} μCi/cm³)
Maximum permissible drinking water concentration	2×10^{-5} μg/cm³ (1.2×10^{-6} μCi/cm³)

and immediately after the Manhattan Project was issued, with some modification, in unclassified form giving rise to the MDDC and AECD series of documents and appeared in national journals and various volumes of the National Nuclear Energy Series. This multiplicity of reporting of the same material with minor modifications increased immeasurably the problem of reconstruction of the biological bases and history of development of radiation protection criteria. Perhaps the best bibliographical compilation with respect to internal radiation (including plutonium) is that given in the 1959 report of ICRP Committee II on Permissible Dose for Internal Radiation (ICRP, 1960). After the USAEC assumed responsibility from the Manhattan District for the plutonium operations, there were few or no adjustments in plutonium exposure criteria with the self-imposed Hanford levels being approximately a factor of 2 less than elsewhere.

On September 29 and 30, 1949, the first Tripartite (USA, UK, Canada) Permissible Doses Conference[7] was held at CHALK RIVER, Ontario. At this meeting Dr. BRUES reported that comparative toxicity studies in mice and rats at the Argonne National Laboratory suggested a toxicity ratio between equal microcurie amounts of plutonium and radium of approximately 15 to 1, a very large factor of difference from the earlier anticipated ratio based on energy release (see p. 573). On this basis, the conference adopted a plutonium body burden of 0.1 μg (0.006 μCi) and calculated the corresponding maximum levels for air and water. The US and UK reports of the conference were issued promptly and gave the occupational maximum permissible levels for plutonium given at the top of Table 12.2. Immediately a wave of technical correspondence and telephoning ensued, initiated by Dr. W. LANGHAM and involving Drs. SHIELDS WARREN, AUSTIN BRUES, R. D. EVANS, and K. Z. MORGAN. It was Dr. LANGHAM's contention that (1) the Chalk River values were too restrictive and should not be adopted as the official policy for USAEC operations, (2) comparative chronic toxicity studies of plutonium and radium in dogs should be initiated (at Los Alamos) immediately and (3) chronic plutonium *inhalation* studies in dogs should be initiated (at Los Alamos) also.

This wave of correspondence culminated in a meeting in Washington, D.C., on January 24, 1950, called by Dr. WARREN (Director of the AEC's Division of

7 Participants were: from the United States, Dr. SHIELDS WARREN (Chairman) and Drs. A. M. BRUES, G. FAILLA, J. G. HAMILTON, L. HEMPELMANN, F. DE HOFFMAN, W. H. LANGHAM, K. Z. MORGAN, H. M. PARKER, L. S. TAYLOR and B. S. WOLF and Colonel C. A. NELSON; from Canada, Drs. E. A. BRAATEN, H. CARMICHAEL, A. J. CIPRIANI, G. H. GUEST, G. C. LAURENCE, W. B. LEWIS, G. E. MCMURTRIE and E. RENTON; from the United Kingdom, Professor J. S. MITCHELL, Drs. E. F. EDSON and G. J. NEARY and Mr. A. C. CHAMBERLAIN.

Biology and Medicine) to consider whether the Chalk River values would be adopted. A list of those attending this meeting does not seem to be available, but the principals were Drs. BRUES, LANGHAM and WARREN. Dr. LANGHAM gave a 2-hour presentation of all the reasons why the proposed Chalk River values should not be adopted. Dr. BRUES gave a 5-minute presentation in which he pointed out that the 15 to 1 toxicity ratio of plutonium to radium (on a μCi basis) indicated by the Argonne rodent studies was based on injected dose; however, plutonium retention in the rodent was 75 percent compared to 25 percent for radium, and radon retention in the rodent was 15 to 20 percent compared to about 50 percent in man. Assuming the minimal damaging dose of fixed radium in man was 1 μCi and taking the above retention parameters into consideration, the comparable fixed minimal damaging dose of plutonium would be about 6 μg. That is:

$$1\,\mu\text{Ci Ra} = \frac{1}{15} \times \frac{0.75}{0.25}\left[\frac{4.8 + 0.5\,(5.5 + 6.0 + 7.7)}{4.8 + 0.15\,(5.5 + 6.0 + 7.7)}\right] = \sim 0.4\,\mu\text{Ci}\ (6\,\mu\text{g})\ \text{Pu},$$

where the numbers in brackets represent the ratio of energies imparted to man and rodent from radium and its retained decay products (see Footnote 5 for radium decay scheme—Ed.). Dr. WARREN took 5 minutes to summarize the discussion and announced that the AEC operations could proceed on the basis of a maximum permissible plutonium body burden of 0.5 μg (equivalent in effect to about 0.1 μg radium) and maximum air and water concentrations derived accordingly. The meeting lasted 2 hours and 10 minutes and would have lasted only 10 minutes had Dr. BRUES spoken first.

In May 1950 the Canadian version of the Chalk River Permissible Doses Conference was issued (MCMURTRIE, 1950). This report was considered to be the final as it was modified in accordance with suggestions received from the various delegates, including Dr. BRUES' modification of the 15 to 1 toxicity ratio of plutonium to radium. The report gave the maximum permissible levels for plutonium shown at the bottom of Table 12.2.

At a meeting at Buckland House (near Harwell, England) called by Sir JOHN COCKCROFT (August 3–5, 1950) and attended by R. D. EVANS, L. MARINELLI, S. WARREN and J. LOUTIT among others[8], BRUES' derivation of the 0.04 μCi was agreed to. In the Argonne National Laboratory Quarterly Report for February, March and April 1951, BRUES (1951) summarized as follows the animal data on which the 0.04 μCi maximum permissible plutonium body burden for man was based: "The toxicity ratio between radium and plutonium has been evaluated from the data of a large number of experiments. The best available ratios, in terms of injected μc per kg radium to plutonium, are:

(1) for acute toxicity to small animals, 15
(2) for chronic survival, 10
(3) for formation of bone tumors in rats and mice, 15
(4) for formation of bone tumors in rabbits, 8
(5) for bone fractures in rats and rabbits, about 10.

"Making appropriate allowance for relative retention of the two elements in rodents and for the greater retention of radon in man, a maximum permissible retained dose of 0.04 μc plutonium in man is the best value available from present biological information. This value depends ultimately on the corresponding permissible dose of radium, presently established as 0.1 μc."

On the basis of the relative biological effects of plutonium and radium as observed in animal experiments, the 1950 Recommendations of the ICRP (see

8 A complete list of attendees is not available.

ICRP-ICRU, 1951) call attention to the use by the US, UK and Canada of 0.04 μCi as the maximum permissible amount of ^{239}Pu fixed in the body.

In 1950 experiments to compare chronic or delayed toxicity of plutonium and radium in dogs were initiated at the University of Utah School of Medicine under the direction of Dr. J. Z. BOWERS and the guidance of Drs. AUSTIN BRUES, W. CLAUS, R. D. EVANS, and W. LANGHAM. Radium-228 ($MsTh_1$) was added to the protocol because of its role in exposure of the early radium dial painting cases. Chronic plutonium inhalation experiments in dogs were being initiated also at the University of Rochester School of Medicine and Dentistry and later at Hanford by Dr. W. J. BAIR and colleagues. These efforts marked the end of the era of deriving maximum permissible values for internally-deposited radionuclides through establishing simple empirical relationships and the beginning of the NCRP-ICRP efforts to generalize the methods of deriving such values through calculation of dose to the critical organ and relating this to effect, taking into consideration relevant biological data.

V. Plutonium Protection Criteria (1950—1971)

The value for the maximum permissible plutonium body burden of 0.04 μCi, as derived by biological comparison following the Chalk River Permissible Doses Conference, has not been altered in subsequent years, although attempts to fit the number into the overall framework of dose calculations for internal emitters have tended to becloud somewhat the original concepts described above. The ICRP-NCRP have made changes in the application of the body burden value to derivation of maximum permissible concentrations in air and water and consideration of the dose to the lung. The remainder of this chapter will deal primarily with these changes.

A. The Standard Man

An important development in derivation of maximum permissible quantities was agreement, among those involved, on values of organ weights and other parameters of the "standard man". This agreement provided uniform values for air and water that could be applied to exposure control in general with adequate accuracy. The first formal steps toward agreement were taken at the Chalk River Permissible Doses Conference in 1949 (*op cit*), where earlier work by CIPRIANI and LISCO served as the basis of discussion. Although some detailed changes were made in later years, the values adopted at this meeting have held up well.

Table 12.3. Parameters of the standard man of importance to standards for plutonium

Mass of the bone[a]	7000 g
Mass of the lung[a]	1000 g
Breathing rate	
total day[a]	20 m³
8 hours at work [after 1960[b]]	10 m³
Water intake	
total day [until 1953[a]]	2.5 l
total day [after 1953[c]]	2.2 l
8 hours at work [after 1960[b]]	1.1 l
Time of occupational exposure [until 1960[a]]	70 yr.
Time of occupational exposure [after 1960[b, c]]	50 yr.

[a] Chalk River Conference, 1949; reported by MCMURTRIE (1950). [b] ICRP Committee II Report of 1959 (ICRP, 1959). [c] NCRP Report No. 11 (NBS Handbook 52, 1953).

Table 12.3 gives the values which are of primary importance to the derivation of maximum permissible values for plutonium, with indication of the changes which took place over the period under discussion.

B. Changes in Values and Concepts

A summary of the important metabolic constants and resulting MPC's for plutonium as derived from the reports of individual conferences and reports of the NCRP and ICRP is given in Table 12.4. Of necessity, such a summary table must omit some of the concepts, and these will be discussed below. The nomenclature used in Table 12.4 is primarily that introduced by the Subcommittee on Permissible Internal Dose of the NCRP in their 1953 report and later adopted by the corresponding subcommittee of the ICRP. These terms are:

T_b — the biological half life of the material in the organ of concern. The elimination constant ($0.693/T_b$) is additive with the decay constant of the radioactive isotope to give an overall effective elimination rate. In this table it refers to bone for "soluble" forms and lung for "insoluble" forms.

f_1 — the fraction of the material ingested which passes from the gastrointestinal tract to the blood.

f_2 — the fraction of the material present in the whole body which is in the organ of concern. This can change with time depending upon the radioactive half-life of the isotope and the relative rates of uptake and elimination of the isotope.

Table 12.4. Summary of MPC's and Maximum Permissible Body Burdens for Plutonium-239 Showing Changes in Parameters and Derived Figures Over the Period 1949–1960.

	Chalk River (1949)	ICRP (1950)	NCRP No. 11 (NBS-52) (1953)	Harriman Conference (1953)	ICRP (1955)	ICRP[a] (1959)
"Soluble" Forms						
MPBB[b]	0.5 μg: C (0.03 μCi)	0.04 μCi	0.04 μCi	0.04 μCi	0.04 μCi	0.04 μCi
T_b (bone)	6930 d[d]	6930 d[d]	4.3×10^4 d	—	4.3×10^4 d	7.3×10^4 d
f_1	—	—	1.4×10^{-4} 7.0×10^{-5}	—	1×10^{-4}	3×10^{-5}
f_2	—	—	0.75	—	0.75	0.9
f_2'	—	—	0.7	—	0.7	0.8
f_w	1×10^{-3}	1×10^{-3}	1×10^{-4}	1×10^{-3}	1×10^{-4}	2.4×10^{-5}
MPC_w [e]	1.2×10^{-6}	1.5×10^{-6}	1.5×10^{-6}	—	6×10^{-6}	5×10^{-5}
f_a	0.1	0.1	0.18	0.25[f]	0.18	0.2
MPC_a [e]	1.5×10^{-12}	2×10^{-12}	2×10^{-12}	—	2×10^{-12}	6×10^{-13}
"Insoluble" Forms						
MPLB[g]	—	—	0.008 μCi	—	0.02 μCi	—
T_b (lung)	—	139 d[h]	360 d	—	360 d	1 yr
f_2	—	—	1.0	—	1.0	—
f_a	—	0.25	0.12	0.10	0.12	0.12
MPC_a [e]	—	7.5×10^{-12} [i]	2×10^{-12}	—	2×10^{-12}	10^{-11}

[a] Continuous occupational exposure MPC's given. [b] MPBB = Maximum Permissible Body Burden (with bone as the critical organ). [c] Actual comparison gave 0.6 μg which, for purpose of this recommendation, was rounded off to 0.5 μg. [d] Given as a mean life of 10^4 days. [e] In units of μCi/cm³. [f] Retention of soluble material in alveoli. If multiplied by f_1 used by the NCRP (1953), f_a becomes 0.18. [g] MPLB = Maximum Permissible Lung Burden. [h] Given as a mean life of 200 days. [i] Not used. Value for insoluble was recommended to be the same as for soluble (see text).

For this reason, f_2 is normally taken at a long time when the equilibrium state is reached.

f_2' — the fraction of the material in the blood which passes to the organ of concern.

f_w — the fraction of the material which reaches the organ of concern following ingestion of the isotope. Since this will equal the product of the quantity passing from the gastrointestinal tract to the blood and the fraction passing from the blood to the organ of concern, $f_w = f_1 f_2'$

f_a — the fraction of the material inhaled which reaches the organ of concern. This will be the sum of the material absorbed directly from the lung and that which is swallowed and absorbed from the gastrointestinal tract.

Discussions at the Chalk River Permissible Doses Conference (*op cit.*) have been summarized earlier with emphasis on derivation of the maximum permissible body burden. The value for the MPC in air was based on a soluble compound with 10 percent retention in the lung, and the material retained with a mean life of 10^4 days once it had passed into the body and been deposited in bone. Thus, the MPC was:

$$\mathrm{MPC}_a = 0.5\,\mu\mathrm{g} \times \frac{1}{10^4} \times \frac{1}{0.1} \times \frac{1}{2 \times 10^7} = 2.5 \times 10^{-11}\,\mu\mathrm{g/cm^3}\ ^{9}$$
$$= 1.5 \times 10^{-12}\,\mu\mathrm{Ci/cm^3}.$$

Similarly, the MPC for water was derived by assuming 0.1 percent absorption in low concentrations and ingestion of 2.5 liters of water per day:

$$\mathrm{MPC}_w = 0.5 \times \frac{1}{10^4} \times \frac{1}{10^{-3}} \times \frac{1}{2.5 \times 10^3} = 2 \times 10^{-5}\,\mu\mathrm{g/cm^3}\ ^{10}$$
$$= 1.2 \times 10^{-6}\,\mu\mathrm{Ci/cm^3}.$$

These calculations were performed to obtain the MPC which would yield the maximum permissible body burden under equilibrium conditions since, with the elimination constants chosen, the body would be more than 90 percent of the way to equilibrium after the 70-year life-span chosen as the basis of calculation.

No MPC was derived at this meeting based on lung dose, although there was considerable discussion of the problem of lung retention and dose to the lung. In connection with present concern over the single particle problem, it is of interest to note the following quotation from the minutes of the meeting: "Dr. HAMILTON pointed out that the cells in the immediate neighborhood of a dust particle containing 1 or 2% of plutonium would be subjected to a dose of about 400 r/day. The general opinion which emerged from the discussion was that the carcinogenic effect per unit volume is probably considerably less for irradiation of small masses of tissue than for large". Again a prophetic foreshadowing of present concerns is given in the statement: "A brief discussion of the proportion of insoluble particulate material transported from the lung to the lymph nodes merely served to indicate that this factor is rather dependent on the nature and size of the particles". Thus, on a subject which the standards-setting bodies have been accused of ignoring, we find serious discussion in 1949 with, however, the conclusion that more information is needed—the same situation in which we find ourselves today!

An important outcome of the Chalk River Permissible Doses Conference was the formulation of a lung model which, while crude, served as the basis for further

9 The fractions in this equation represent in order, (a) mean life t (i.e. $1.44 \times T_b$), (b) f_a and (c) ml of air breathed per day.

10 The fractions in this equation represent in order, (a) *mean life*, (b) f_w and (c) ml of water ingested per day.

development. This lung model was given as: "...if specific data were lacking, the convention be adopted that 50% of any aerosol reaches the alveoli of the lungs. If the particles are soluble they are considered to be totally absorbed; if insoluble then the 50% amount is to be regarded as retained for 24 hours, after which only half of it, i.e., 25% of the inhaled amount, is retained *in situ* indefinitely; further, that the particles be assumed spherical." The wording of this model indicates that the originators felt a strong need to give guidance but also felt some unease in providing a definitive statement. Unfortunately, the preface to the statement relating to use when specific data are unavailable seems to have been overlooked or deemphasized in many later considerations of lung dose, although the statement still survives.

The first presentation of these concepts for general use was in the 1950 ICRP recommendations published as ICRU Report No. 6 (NBS Handbook 47) issued June 29, 1951 (op cit.). The values for internal emitters were prefaced by the statement of the Commission that they did not consider that there was sufficient information at the time to make firm recommendations; however, they did indicate that they were bringing to the notice of users values which were commonly used in the US, Canada and Great Britain. The statement on plutonium is of considerable interest and is quoted below:

> On the basis of the relative biological effects of plutonium and radium, as observed in animal experiments, it is accepted that
>
> a) *The maximum permissible amount of ^{239}Pu fixed in the body is 0.04 microcurie.*
>
> For soluble compounds of plutonium in the atmosphere, it is estimated that 10 percent of the inhaled material is absorbed, with a mean life of 10^4 days. The maximum permissible concentration in air is, therefore, 2×10^{-12} microcurie/ml
>
> For insoluble compounds, it is estimated that the mean life in the lung is 200 days. If the irradiation of the lungs by alpha rays were limited to the biological equivalent of 0.3 r/week, the corresponding concentration of the plutonium in air would be 7.5×10^{-12} microcurie/ml. In view of the possibility of the transference of some of the insoluble material from the lungs to the skeleton, it is suggested that
>
> b) *The maximum permissible concentration of ^{239}Pu in air is 2×10^{-12} microcurie/ml, for soluble and insoluble compounds.*
>
> c) *For ^{239}Pu in liquid media, assuming that 0.1 percent of the ingested amount is retained in the skeleton with a mean life of 10^4 days, the maximum permissible concentrations is 1.5×10^{-6} microcurie/ml.*

The body burdens and MPC's given above are essentially those from the Canadian Chalk River report with the addition of a calculation for "insoluble" plutonium based on dose to the lung, which was not accepted for implementation at that time.

In 1953 the NCRP published the report of their Subcommittee on Internal Emitters as National Committee on Radiation Protection Report No. 11 (NBS Handbook 52) (NCRP, 1953). Although their MPC values remained the same as were derived following the Chalk River Conference and as given by the ICRP, there were some changes in the listed metabolic constants for plutonium as shown in Table 12.4. Chief among these were an increase in half-life for "soluble" forms and a more detailed lung model which is very close to that adopted for and used many years for ICRP calculations of MPC's. This latter is given as: "In dealing with the inhalation of radioisotopes, unless information specific to the radioisotope is available, it is assumed in the case of *soluble* compounds that 25 percent is retained in the lower respiratory tract. From this tract it goes to the blood stream, and a part of this goes to the critical organ within a few days. Fifty percent is held up in the upper respiratory tract and swallowed, so a fraction of that swallowed also reaches the critical organ. In the case of *insoluble* compounds, it is assumed that 12 percent is retained in the lower respiratory tract, which is usually taken as the critical organ when considering the inhalation of insoluble compounds. The rest is eliminated by exhalation and swallowing." Use of this model resulted in changes in the values of f_a for both the "soluble" and "insoluble" materials.

Use of these modified metabolic factors would have led to some changes in the MPC's. Water and air values for soluble plutonium can be calculated from these factors, assuming that exposure is long enough for equilibrium to be reached, as follows:

$$MPC_a = 0.04 \times \frac{0.693}{4.3 \times 10^4} \times \frac{1}{0.18} \times \frac{1}{2 \times 10^7} = 2 \times 10^{-13}\ \mu Ci/cm^3$$ [11]

$$MPC_w = 0.04 \times \frac{0.693}{4.3 \times 10^4} \times \frac{1}{10^{-4}} \times \frac{1}{2200} = 3 \times 10^{-6}\ \mu Ci/cm^3$$ [11].

However, with the increased retention time for plutonium in the body, the assumption of equilibrium is no longer valid since only about 34 percent of the equilibrium value would be achieved in 70 years. Therefore, the MPC_w could be increased to $9 \times 10^{-6}\ \mu Ci/cm^3$ and the MPC_a could be increased over the calculated values to $5 \times 10^{-13}\ \mu Ci/cm^3$. For "insoluble" plutonium, use of these metabolic factors and a lung burden of 0.008 μCi would give:

$$MPC_a = 0.008 \times \frac{0.693}{360} \times \frac{1}{0.12} \times \frac{1}{2 \times 10^7} = 6 \times 10^{-12}\ \mu Ci/cm^3.$$

In this case, the equilibrium assumption is valid because of the assumed half life in the lung of 360 days.

It is of interest that the actual MPC's recommended in this document were unchanged from the recommendation of the Chalk River Permissible Doses Conference, in spite of the change in factors utilized to describe the behaviour of Pu in the body. Also, use of the same MPC for soluble and insoluble plutonium indicates that the concern for possible transfer from lung to the body still existed.

The Harriman Tripartite Conference in March 1953 (Tripartite Conference, 1953), about the time of issuance of the NCRP report, gave scant attention to the problem of plutonium *per se* beyond affirming the value of 0.04 μCi derived after the Chalk River Conference. However, an important change was made at this meeting in the RBE factor for alpha particles *viz:* a lowering of the value from 20 to 10 as based on the possibility of carcinogenesis. There also ensued a lengthy discussion on the apparent discrepancy between the bone limit, as based on the comparison with radium, and as based on calculations from the external dose limit assuming a uniform distribution of isotope and an amount limited to give a dose of 0.3 rem per week as was used for other organs. This resulted in a statement concerning the calculation of such doses: "Radium is assumed, for purposes of calculation, to be uniformly distributed. Other alpha emitting bone seekers are assumed to be non-uniformly distributed by a factor of 5". This statement can be interpreted in several ways. First, it could mean that the size of the critical organ should be considered as 1/5 as great as that chosen for radium. In other words, since the calcified portion of bone is taken as 7000 g, the statement could be interpreted to mean that the

11 Editors note: These calculations illustrate the change in MPC values resulting from the slightly different metabolic factors and use the format of previous calculations in this chapter. They do not show the actual approach as in the quoted NBS Handbook. See for example equation G 5 (in NCRP No. 11, NBS, 1953). Also, the values would refer to continuous occupational exposure as in the earlier examples.

critical organ for these calculations should be 1400 g. It can also be interpreted that the alpha radiation is 5 times as damaging for the other alpha emitters as for radium and that the effect considered should be 5 times as large but dispersed in the same size of critical organ. In view of the origin of the factor of 5 in animal experiments with plutonium and its explanation on the basis of non-uniform distribution in bone, the former interpretation would seem to be more reasonable, but later trends seemed to favor the latter interpretation.

Revised recommendations of the ICRP were published in 1955 (ICRP, 1955). Values of the metabolic constants were largely as given in the NCRP report of 1953 (op cit.). The MPC in water for soluble plutonium was revised as a result of the use of these metabolic values, and the maximum permissible lung burden was changed from 0.008 μCi in the NCRP report to 0.02 μCi, to reflect the acceptance of the RBE of 10 rather than 20. In spite of these changes, however, the MPC in air for both soluble and insoluble plutonium remained at a single value as was originally recommended at the Chalk River Permissible Doses Conference (2×10^{-12} μCi/cm³, rounded off from the original value at Chalk River of 1.5×10^{-12} μCi/cm³).

It is of interest that the introduction to the report of the Subcommittee on Internal Emitters states: "In the case of all bone-seeking radioisotopes (with the exception of Ra, ^{32}P and radioisotopes that emit only X- or γ radiation) a *factor of safety* of 5 is applied to the calculations to take into account the uneven distribution of the radioactive material within the bone, ..." (emphasis added). Thus, the factor for the difference in effectiveness of Pu versus Ra defined by biological experimentation described earlier was described as a "factor of safety". This report also introduced the combination of energy, the RBE, and the distribution factor into one term written as $\Sigma E(RBE)N$ so that this value could be used in place of the energy, with the resulting dose coming out directly in "rems". It may be noted that at this time the ICRU had adopted the rad as an official unit but had not as yet adopted the rem (ICRU, 1954). Therefore, protection organizations such as the ICRP and NCRP were using it as a unit of convenience rather than an officially defined and accepted unit. The definition adopted by the ICRP was that of their Subcommittee on External Dose (ICRP, 1955) and was given as "the rem is the absorbed dose of any ionizing radiation which has the same biological effectiveness as one rad of X- radiation, ...".[12]

An additional feature of this report was the derivation of MPC values as based on the dose to the gastrointestinal tract. Models for the mass of material in the gastrointestinal tract and time of transit through each section were added to the standard man, and the dose was calculated based upon the dose rate at the surface of a semi-infinite mass of material. The concentration was equal to the quantity of the contents of the portion of the gastrointestinal tract of interest which dilutes the amount of radioisotope taken in per day. At this time, the calculations were made on the basis of the full energy of the alpha particle, and the recommended MPC_w based on this dose was 3×10^{-6} μCi/cm³ or one-half of the value based on

12 It is noted that this definition implies only the factor of RBE due to differences in specific ionization of the radiations and does not overtly include other factors such as non-homogeneous distribution. Inclusion of this factor in the internal dose energy term had the effect of producing an *ad hoc* definition of the rem for use in estimating the effective dose from bone-seekers. In other words, it seemed to be this step which produced the definition of the non-uniform distribution factor as effectively multiplying the effectiveness of the energy absorbed rather than limiting the size of the critical organ as discussed above. In the cited report, however, it is still clear that the values are based directly on a comparison with radium and that the factor of 5 is still primarily a non-uniform distribution factor, although no description of the derivation of the factor for biological equivalence is given.

uptake into the body. An MPC_a based on exposure of the gastrointestinal tract from material transferred from the upper respiratory tract and lung was calculated also. This value was 5×10^{-10} μCi/cm^3, considerably larger than the value which had been accepted for some years based on transfer from lung to body. In addition, supplementary maximum permissible body burdens based on the gastrointestinal tract were calculated, assuming all of the material to be located there by multiplying the 0.04 μCi based on bone as critical organ by the ratio of the MPC based on the gastrointestinal tract to that based on bone. Values obtained were 0.02 μCi for ingestion and 10 μCi for inhalation. The pertinence and usefulness of these values were not indicated but occasions when they might become the controlling figure can be visualized.

In 1956 the ICRU (1957) accepted the concept of the rem and defined a quantity known as the "RBE dose" which was defined as: "RBE dose is equal numerically to the product of the dose in rads and an agreed conventional value of the RBE with respect to a particular form of radiation effect. The standard of comparison is X- or gamma radiation having a LET in water of 3 kev/μ delivered at a rate of about 10 rad/min." The unit of the RBE dose was taken as the rem with the note that it had the same inherent looseness as the RBE and, in addition, assumed conventional and not necessarily measured values of RBE.

The current values recommended by both the ICRP and NCRP were derived in 1959 and published by both ICRP and NCRP in 1959 (ICRP, 1959; NCRP Report 22 NBS Handbook 69, 1959). As would be expected from the composition of the subcommittees, there is a striking similarity in the numbers, although there are a few differences in philosophy. In discussing the basic standards, the NCRP report indicates for internal emitters: "For bone-seekers the maximum permissible limit is based on the distribution of the deposit, the RBE, and a comparison of the energy release in the bone with the energy release delivered by a maximum permissible body burden of 0.1 μg ^{226}Ra plus daughters". The rem is defined as given by the ICRU or equal to rads times RBE with no additional factor for other mechanisms which could result in differences in damage.

In the ICRP report on internal radiation (1959), in the section discussing the basic standards, reference is made to the 1958 recommendations of the main commission (ICRP, 1958) which provided a basic limitation on the dose accumulated in the gonads, the blood-forming organs, and the lenses of the eye, at any age over 18 years, of 5(N-18) rems where N is the age of the individual in years. The report[13] then interpreted this limitation as: "The effective RBE dose delivered to the bone from internal or external radiation during any 13 week period averaged over the entire skeleton shall not exceed the average RBE dose to correspond to the skeleton due to a body burden of 0.1 μc of ^{226}Ra. This is considered to a dose rate of 0.56 rem/week in the case of ^{226}Ra (derived from a dose rate of 0.06 rad per week, an RBE of 10 and $n=1$). In computing the effective RBE dose to the skeleton, all absorbed energy shall be weighted by a relative damage factor, n. The relative damage factor is taken as one for all energy absorbed from external radiation and for all internal emitters when the element taken into the body is an isotope of radium. If the isotope taken into the body is not an isotope of radium, the relative damage factor, n, is taken as 1 for all energy absorbed from X- or γ-radiation and as 5 for all other energy components". Later when discussing the basis for their calculations, they state[14]: "The first method[15] is

13 Page 3 of ICRP (1959).
14 Page 12 of ICRP (1959).
15 Refers to method of comparison to radium in contrast to calculation of dose.

the result of a calculation designed to determine (i) the amount (μc) deposited in the bone that will deliver the same effective RBE dose as delivered by 0.1 μc of ^{226}Ra and its daughter products and (ii) the amount (μc) deposited in bone that will result in damage comparable to that observed from known deposits of ^{226}Ra in the bone". Thus, it would appear that this subcommittee interpreted the limitation based on blood-forming organs to apply to the skeleton even though the marrow, which is instrumental in blood formation in the bone, is not included in the mass of the organ and the real limitation from human experience with radium would seem to be production of bone cancer rather than effects on the blood-forming organs. It appears that use of the *ad hoc* definition of the rem was continued in this context even though the ICRU at this time had officially recognized the unit as applying only to LET effects. This further implies that the Committee considered the increased effectiveness of these bone-seekers to be due to an enhancement of the effectiveness of the alpha particle rather than a decrease in size of the organ affected, even though they did clearly spell out that the effect was probably due to non-homogeneity in the bone. In fact, the equation used for calculating the body burden simply used the ratios of the quantities and energies, although the factor for 5 was included in the energy term.

It may be noted that the ICRU did later revise their concepts of RBE dose and rem (ICRU, 1968) to correspond to the definition implied by Subcommittee II of the ICRP in 1959. Here they defined the dose equivalent as: "...the product of absorbed dose, (D), Quality Factor, (QF), absorbed dose distribution factor, (DF), and other necessary modifying factors". In addition, several further changes were made in this report in the metabolic constants for soluble plutonium, as is indicated in Table 12.4 (last column). Of particular interest is the increase in fraction of plutonium in bone of that in the total body to 0.9 as a result of the observed distribution of material absorbed from the gastrointestinal tract. This change is of some interest, since it is now known that deposition of plutonium in various organs varies depending upon route of administration and also upon the compound administered (Chap. 9). For example, the dog experiments at Utah using intravenous administration of plutonium as the citrate (Mays et al., 1970) and the plutonium oxide inhalation experiments with dogs at Hanford (Park et al., 1968) both show sizable depositions in the liver, as do the few human autopsy cases available (Shipman et al., 1961). The result of these changes for metabolic parameters, plus the decision for the first time to use these revised values in reassessing the maximum permissible concentrations, resulted in a lowering of the MPC in air for soluble plutonium by about a factor of 3 and an increase in the MPC in water value by a factor of about 10.

One of the more interesting changes resulted from the decision to list separate MPC's in air for soluble and insoluble isotopes. These were calculated on the basis of the lung model described earlier and reproduced in Table 12.5. Of particular interest is the footnote which states that the $12^1/_2\%$ retained in the lung for a long period is "...taken up into body fluids". The definition of body fluids and ultimate fate of the material seem to be unclear. If one assumes that this uptake is by solubilization and eventual passage to the bone, then breathing air at the recommended level for continuous exposure of 10^{-11} $\mu Ci/cm^3$ for 50 years would amount to a total intake in this retained fraction of $10^{-11} \times 365 \times 20 \times 10^6 \times 0.125 \times 50 = 0.46$ μCi[16]. While this calculation ignores the slow elimination of

16 The factors in this calculation are the MPC for insoluble plutonium, the number of days per year, the quantity of air breathed per day by the standard man expressed in ml, the fraction of the inhaled plutonium retained for a long period in the lung, and the 50 year time of exposure.

Table 12.5. "Particulates in the Respiratory Tract of the Standard Man. Retention of particulate matter in the lungs depends on many factors such as size, shape and density of the particles, the chemical form and whether or not the person is a mouth breather; however, when specific data are lacking, it is assumed the distribution is as shown below

Distribution	Readily soluble compounds (%)	Other compounds (%)
Exhaled	25	25
Deposited in upper respiratory passages and subsequently swallowed	50	50
Deposited in the lungs (lower respiratory passages)	25 (this is taken up into the body)	25[a]

[a] Of this, half is eliminated from the lungs and swallowed in the first 24 hours, making a total of $62^1/_2$% swallowed. The remaining $12^1/_2$% is retained in the lungs with a half life of 120 days, it being assumed that this portion is taken up into body fluids."
Taken from ICRP Pub. 2 (1959), Table 10, p. 153.

such material, it can be seen that this assumption would lead to an accumulation of considerably more than a maximum permissible body burden. While the paths of elimination of material in the lung are variable and not always well-established in individual instances, it appears from both animal studies and autopsy results (PARK et al., 1968; SHIPMAN et al., 1961) that the major route of elimination is via the lymphatic system with deposition and long-term retention in the lymph nodes. However, this may vary with different particle sizes of materials and different degrees of "solubility", a term which is difficult to define and which seems to relate more to the metabolic behavior of the material in the body than to the more familiar chemical concept of solubility[17].

The MPC's calculated on dose to the gastrointestinal tract following either ingestion or inhalation were increased over those in the earlier ICRP report because of experiments on mice and rats in which large quantities of the oxide or nitrate were administered orally with little or no indication of damage or histological change (SULLIVAN and THOMPSON, 1957). As a result of these findings, it was concluded that the alpha particle did not penetrate to the sensitive cells of the intestinal lining, and thus only 1 percent of the energy of alpha emitters was used in calculating the dose to the gastrointestinal tract. Parenthetically, it might be noted that this factor was generalized to all alpha emitters, although the work was done only with plutonium. It is not at all certain that other alpha emitters, which are more soluble and less prone to complex, may not penetrate the barrier and produce some damage to the intestinal wall.

In addition to these changes, the 1959 ICRP report also provided MPC's based on exposure to a number of other organs, including the total body where the doses were assessed by assuming uniform distribution in the full 70-kg mass of the standard man. Also, MPC's based on exposure 40 hours per week for 50 weeks per year were suggested for use in occupational situations rather than continuous exposure basis utilized earlier.

17 Editors note: This problem has been examined in detail by an ICRP Task Force (ICRP, 1966) and a revised lung model with more detailed breakdown of parameters is all but officially adopted.

VI. Current Situation (1971)

The maximum permissible body burden and the maximum permissible concentrations of plutonium derived by the ICRP-NCRP in 1959 are still the currently recommended values. Both organizations have continued deliberations on the hazards of internal radiation exposure, and revised publications of their deliberations are in preparation. It is expected that such revisions will show no compelling reasons for drastic changes in the 1959 recommendations—they have served their purpose well despite the few inconsistencies in rationale pointed out in the preceding pages. The MPC's have been incorporated into the federal regulations of the AEC not only as limitations on air and water concentrations for occupational exposure but also, when reduced by appropriate factors, for exposure of population groups as limitations on effluents and environmental concentrations. In spite of sporadic criticism of the values and intensive work on plutonium over the past years, there have been no steps taken prior to the preparation of this chapter to make any significant changes in the maximum permissible body burden. In fact, the data from the long-term dog experiments at the University of Utah now indicate little necessity for change in the non-uniform distribution factor (a "relative hazard factor") derived several decades ago by informed judgement based on rodent experiments despite the great difference in life span of the two species.

These safety levels derived from apparently meager information have served as the basis for a protection program leading to the remarkable safety record (despite the unsupported apprehensions of a few) of the extensive plutonium handling operations of the AEC and, more recently, industry in general. There has not been a single known case of damage among those who have worked with plutonium over the past three decades even though many (perhaps 50 to 100) are known to have accumulated approximately "one maximum permissible body burden (0.04 μCi)" and, in a few cases, considerably more. This is a record in sharp contrast to that of the early radium industry, or a number of other industries where only the appearance of injury led to institution of proper control. Constant vigilance must be maintained, however, if the projected role of plutonium in the world's future power economy is to become reality without unacceptable risk. Because of the sensitivity of detection of plutonium in air, water and other segments of the environment and the magnitude of discrimination factors along the food chain from soil to man (absolute minimum total discrimination of the order of 10^8), it is almost inconceivable that environmental contamination would be allowed to approach harmful levels from ingestion[18]. Chronic inhalation of material discharged directly to the atmosphere or resuspended from accumulated deposition and the long-term effects of such inhaled material on the lungs, lymph nodes and liver are the pressing problems for the immediate future. The oncogenic risk of long-term retention of discrete plutonium particles in these tissues cannot be assessed unequivocally at the present time, although the problem was recognized at the Chalk River Permissible Doses Conference in 1949 and has been considered by the NCRP and ICRP in subsequent years. The future safe handling and use of plutonium, the element frequently and somewhat erroneously called the "most hazardous material known to mankind", are both feasible and necessary and will be accomplished as long as vigilance is maintained and the depth of experience, knowledge and compassion for mankind exhibited by those pioneers is extended and applied by those responsible for future risk assessment and exposure control.

18 See also Chap. 8.

References

Abrams, R., Seibert, H. C., Forker, L., Greenberg, D., Lisco, H., Jacobsen, L. O., Simmons, E. L.: Metabolism and distribution of inhaled plutonium in rats. Report CH-3655 or MDDC-677 (Oct. 1946).

Allison, S. K. (Director): Metallurgical Laboratory Progress Report for Feb. 1944. Report MUC-NH-807 (1944a).

Allison, S. K.: Memorandum to R. J. Oppenheimer. Report MUC-SKA-516 (March 1, 1944b).

American Medical Association, Bureau of Investigation: Radium as a patent medicine. J. Amer. med. Ass. **98**, 1397–1399 (1932).

Aub, J. C., Evans, R. D., Gallagher, D. M., Tibbetts, D. M.: Effects of treatment on radium and calcium metabolism in the human body. Ann. intern. Med. **11**, 1443–1463 (1938).

Blum, T.: Osteomyelitis of the mandible and maxilla. Read before the American Dental Association (Sept. 1924).

Boyd, G. A., Silberstein, H. E., Fink, R. M., Frenkel, A., Minto, W. L., Metcalf, R. G., Casarett, G., Suter, C. M.: Comparative intravenous lethal dose pilot studies for polonium, radium and plutonium in rats. Report M-1878 (June 1, 1946a); also R. F. Fink et al. (1950).

Boyd, G. A., Williams, A., Minto, W. L., Tiedeman, D. V., Fink, R. M., Casarett, G., Metcalf, R. G.: Comparative lethal dosage study of radium, plutonium and polonium in rats with clinical, pathological and hematological observations. Report M-1902 (June 22, 1946b); also R. F. Fink et al. (1950).

Brues, A. M.: Comparative chronic toxicities of radium and plutonium. Argonne National Laboratory Quarterly Report (Feb., March and April 1951), ANL-4625, p. 106–124.

Brues, A. M., Lisco, H., Finkel, M.: Carcinogenic action of some substances which may be a problem in certain future industries. Report MDDC-145 (July 1946).

Castle, W. B., Drinker, K. R., Drinker, C. K.: Necrosis of the jaw in workers employed in applying a luminous paint containing radium. J. industr. Hyg. **7**, 371 (Aug. 1925).

Clinton Laboratories: Project Council policy meeting. Report CS-1262 (Jan. 19, 1944).

Evans, R. D.: Radium poisoning, a review of present knowledge. Amer. J. publ. Hlth **23**, 1017–1023 (1933).

Evans, R. D.: Apparatus for the determination of minute quantities of radium, radon and thoron in solids, liquids and gases. Rev. Sci. Instrum. **6**, 99–112 (1935).

Evans, R. D.: The quantitative determination of the radium content of living persons. Amer. J. Roentgenol. **37**, 368–378 (1937).

Evans, R. D.: Protection of radium dial workers and radiologists from injury by radium. J. industr. Hyg. **25**, 253–269 (1943).

Evans, R. D.: Personal communication to W. Langham (1971).

Fink, R. F., ed.: Biological studies with polonium, radium and plutonium. National Nuclear Energy Series IV-3. New York: McGraw-Hill Book Co. 1950.

Finkle, R. D.: The toxicity and metabolism of plutonium in laboratory animals. Report CH-3783 (Aug. 1946).

Gettler, A. O., Norris, C.: Poisoning from drinking radium water. J. Amer. med. Ass. **100**, 400–402 (1933).

Hamilton, J. G.: Metabolism of product. Report CN-2383 (Nov. 20, 1944).

Hoffman, F. L.: Radium (mesothorium) necrosis. Read before the Section on Preventive and Industrial Medicine and Public Health, Industrial Hygiene Association (May 29, 1925a).

Hoffman, F. L.: Radium (mesothorium) necrosis. J. Amer. med. Ass. **85**, 961–965 (1925b).

Hueper, W. C.: Occupational tumors and allied diseases, p. 224–281. Baltimore, Md.: Chas. H. Thomas 1942.

International Congress of Radiology: Second Congress. Brit. J. Radiol. **1**, 349–365 (1928).

International Commission on Radiological Units: Recommendations of the International Commission on Radiological Units. Amer. J. Roentgenol. **71**, 139–142 (1954).

International Commission on Radiological Units: Report of the International Commission on Radiological Units and Measurements (ICRU) 1956. ICRU Report No 8 (National Bureau of Standards Handbook 62, April 10, 1957).

International Commission on Radiological Protection and International Commission on Radiological Units: Recommendations of the International Commission on Radiation Protection and the International Commission on Radiological Units 1950. ICRU Report No 6 (National Bureau of Standards Handbook 47: June 29, 1951).

International Commission on Radiological Protection: Recommendations of the International Commission on Radiological Protection (revised Dec. 1, 1954). Brit. J. Radiol. Suppl., **6**, 1–92 (1955).

International Commission on Radiological Protection: Recommendations of the International Commission on Radiological Protection (adopted Sept. 9, 1958). ICRP Publication 1. London: Pergamon Press 1959.

International Commission on Radiological Protection: Report of ICRP Committee II on Permissible Dose for Internal Radiation (1959). ICRP Publication No 2. London: Pergamon Press (1959; also Hlth Phys. **3**, 1–380 (1960).

International Commission on Radiological Protection, Task Group on Lung Dynamics: Deposition and retention models for internal dosimetry of the human respiratory tract. Hlth Phys. **12**, 173–208 (1966).

International Commission on Radiation Units and Measurements: Radiation quantities and units. ICRU Report 11, Washington, D. C. (Sept. 1, 1968).

Lisco, H., Finkel, M., Brues, A.: Carcinogenic properties of radioactive fission products and plutonium. Radiology **49** (**3**), 361–363 (1947).

Martland, H. S.: Occupational poisoning in manufacture of luminous watch dials. J. Amer. med. Ass. **92**, 466 (1929).

Martland, H. S., Conlon, P., Knef, J. P.: Some unrecognized dangers in the use and handling of radioactive substances. J. Amer. med. Ass. **85**, 1769–1776 (1925).

Martland, H. S., Humphries, R. E.: Osteogenic sarcoma in dial painters using luminous paint. Arch. Path. **7**, 406–417 (1929).

Mays, C. W., Taylor, G. N., Jee, W. S. S., Dougherty, T. F.: Speculated risk to bone and liver from ^{239}Pu. Hlth Phys. **19**, 601–610 (1970).

McMurtrie, G. E. (Secretary): Permissible Doses Conference Held at Chalk River, Ontario (Sept. 1949). Report RM-10 (May 1950).

Metallurgical Laboratory: Project Report for Month Ending April 29, 1944a. Report CH-1433.

Metallurgical Laboratory: Monthly Health Report on Problems Relating to Product for Period Ending June 30, 1944b. Report CN-1910.

National Committee on Radiation Protection and Measurements: X-Ray Protection. NCRP Report No 1 (National Bureau of Standards Handbook 15: 1931).

National Committee on Radiation Protection and Measurements: X-Ray Protection. NCRP Report No 3 (National Bureau of Standards Handbook 20: 1936).

National Committee on Radiation Protection and Measurements: Radium Protection. NCRP Report No 4 (National Bureau of Standards Handbook 23: 1938).

National Committee on Radiation Protection and Measurements: Safe handling of radioactive luminous compound. NCRP Report No 5. (National Bureau of Standards Handbook 27: 1941).

National Committee on Radiation Protection and Measurements: Maximum permissible amounts of radioisotopes in the human body and maximum permissible concentrations in air and water. NCRP Report No 11 (National Bureau of Standards Handbook 52, Washington: March 20, 1953).

National Committee on Radiation Protection and Measurements: Maximum permissible body burdens and maximum permissible concentrations of radionuclides in air and water for occupational exposure. NCRP Report No 22. (National Bureau of Standards Handbook 69: June 5, 1959).

Nickson, J. J. (Recording Secretary): Project Information Meeting (Health) (March 7, 1944).

Nickson, J. J.: Report of a Conference on Plutonium — May 14 and 15, 1945. Report CH-3167 (July 23, 1945).

Painter, E., Russell, E., Prosser, C. L., Swift, M. N., Kisielski, W., Sacher, G.: Clinical physiology of dogs injected with plutonium. Report CH-3858 (June 1946).

Park, J. F., Bair, W. J., Clarke, W. J., Howard, E. B.: Chronic effects of inhaled plutonium dioxide in dogs. Battelle-Northwest Laboratory Report BNWL-714 (1968), p. 3.3–3.4.

Parker, H. M.: Tolerance concentration of product in drinking water. Memorandum to S. T. Cantril, AECD-2906 (August 4, 1945).

Peterson, A. V.: X-49 Deliveries. Memorandum to A. H. Compton, War Department NDH-5028 (Jan. 29, 1944).

Rose, J. E.: Product tolerance. Memorandum to R. S. Stone, MUC-JER-128 (Sept. 1945).

Scott, K. G., Axelbrod, D. J., Crowley, J., Hamilton, J. G.: Studies of the inhalation of fissionable materials. Report CH-3590 (Dec. 31, 1945); Deposition and fate of plutonium, uranium and their fission products inhaled as aerosols by rats and man. Arch. Path. **48**, 31–54 (1949).

Seaborg, G. T.: Physiological hazards of working with plutonium. Memorandum to R. S. Stone, MUC-GTS-384 (Jan. 5, 1944a).

Seaborg, G. T.: Health hazards in working with plutonium. Memorandum to R. S. Stone, MUS-GTS-399 Jan. 15, 1944b).

Seaborg, G. T.: Plutonium revisited. AEC Release S-34-70 (Oct. 8, 1970).

SEABORG, G. T.: Personal communication to W. H. LANGHAM (May 12, 1971).

SHIPMAN, T. L., LUSHBAUGH, C. C., PETERSEN, D. F., LANGHAM, W. H., HARRIS, P. S., LAWRENCE, J. N. P.: Acute radiation death resulting from an accidental nuclear critical excursion. J. occup. Med., Suppl. **3** (3), 146–192 (March 1961).

SULLIVAN, M. F., THOMPSON, R. C.: Absence of lethal radiation effects following massive oral administration of plutonium. Nature (Lond.) **180**, 651–652 (1957).

TAYLOR, L. S.: Brief History of the National Committee on Radiation Protection and Measurements (NCRP) Covering the Period 1929–1946. Hlth Phys. **1**, 3–10 (1958a).

TAYLOR, L. S.: History of the International Commission on Radiological Protection (ICRP). Hlth Phys. **1**, 97–104 (1958b).

Tripartite Conference on Permissible Doses: Report of a Meeting Held at Arden House, Harriman, New York (March 30–31 and April 1, 1953).

Addendum to Chapter 12

Added in Proof with Permission of J. HEALY

In January 1973 the results of a twenty-seven year study of selected Los Alamos plutonium workers was released as a Los Alamos Scientific Laboratory Informal Report. Since these results have a direct bearing on the validity of the standards for maximum permissible body burden and concentrations in air and water described in this chapter, the abstract of the Los Alamos Report is reproduced below. These findings should be considered also as a supplement to Sec. V, VIII, and Appendix A of Chap. 14 (Editor).

Abstract

Twenty-five male subjects who worked with plutonium during World War II under extraordinarily crude working conditions have been followed medically for a period of 27 years. Within the past year, 21 of these men have been examined at the Los Alamos Scientific Laboratory, and 3 more will be studied in 1973. In addition to physical examinations and laboratory studies (complete blood count, blood chemistry profile, and urinalysis), roentgenograms were taken of the chest, pelvis, knee, and teeth. The chromosomes of lymphocytes cultured from the peripheral blood and cells exfoliated from the pulmonary tract were also studied. Urine specimens assayed for plutonium gave a calculated current body burden (excluding the lungs) ranging from 0.005 to 0.42 μCi, and low-energy radiation emitted by internally deposited transuranic elements in the chest disclosed lung burdens probably of less than approximately 0.01 μCi. To date, none of the medical findings in the group can be attributed definitely to internally deposited plutonium. The bronchial cells of several of the subjects showed moderate to marked metaplastic change, but the significance of these changes is not clear. Diseases and physical changes characteristic of a male population entering its sixth decade were observed. Because of the small body burdens on the order of the maximum permissible level in these men so heavily exposed to plutonium compounds, we conclude that the body has protective mechanisms which are effective in discriminating against these materials following some types of occupational exposures. This is presumably explained by the insolubility of many of its compounds. Plutonium is more toxic than radium if deposited in certain body tissues, especially bone; however, from the practical point of view, plutonium seems to be less hazardous to handle.

Abstract of Los Alamos Scientific Laboratory Informal Report, LA-5148-MS entitled "A Twenty-Seven Year Study of Selected Los Alamos Plutonium Workers" by L. H. HEMPELMANN, C. R. RICHMOND, and G. L. VOETZ.

Chapter 13

Bioassay of Plutonium

Anne de G. Low-Beer

With 2 Figures

I. Introduction

Guidelines have been promulgated by the *International Commission for Radiological Protection* and by national commissions in many countries, for the protection of workers in establishments where atomic energy is employed. These guidelines require personnel monitoring for each individual who has a reasonable possibility of acquiring a dose that exceeds a small fraction of the maximum permissible doses established by these regulatory bodies. Such monitoring includes the use of film badges, thermoluminescent dosimeters, monitoring of working areas, and radiochemical analysis of excreta and other biological materials together with *in vivo* counting. The two latter procedures constitute a bioassay program.

Permissible dose recommendations defined by NCRP (1971) are related to "(1) exposure at low dose rates or from sporadic small exposures which may occur in a random pattern with some semblance of daily, weekly, or monthly repetition; (2) accidental exposures which may result in high level acute exposure, and which may occur as rare exceptions in a general area of good control." It is the function of a bioassay program to evaluate the effect on personnel of both types of risk.

The foregoing principles apply to the hazard of internal contamination by an radionuclide, and therefore hold true for the transplutonium elements as well as for plutonium. The general considerations that underlie a bioassay program for any of the transuranium elements are set forth in this chapter and are not repeated in Chap. 20. These include protocols for routine monitoring and for accidental exposures, methods of preparation of samples for radiochemical analysis, and instrumentation for detection of alpha radioactivity and L X-rays. Interpretation of bioassay results relies on models that have been developed for evaluation of plutonium body burdens and is, therefore, included in this chapter. Finally, radiochemical procedures designed to isolate plutonium when it is the only nuclide of interest will be found in this chapter, while those in which plutonium is one of several actinide isotopes that may be present in a given sample, are included in Chap 20. Methods of gamma spectrometry which are applicable to many of the transplutonium elements, but not to plutonium, are also discussed in Chap. 20. Thus, these two chapters should be regarded as complementary parts of a single subject. An arbitrary division of material has been adopted in the hope of avoiding redundancy. Nevertheless, some regrettable but inevitable overlapping and duplication will be noted.

II. Scope and Frequency of Sampling for Routine Monitoring

The vast majority of routine bioassay determinations are performed on urine. Reasons for the selection of urine rather than feces are discussed in the section on Interpretation of Bioassay Results. Fecal samples, nose swipes, and blood are frequently analyzed in cases of accidental acute exposure.

Persons included in a routine program should include those who work directly with radioactive materials, and those whose proximity to such work involves a possibility of exposure. Among the latter are monitors, maintenance personnel, custodians, and animal caretakers.

Frequency of routine sampling should depend on the degree of risk inherent in the occupation. The shortest interval reported by any laboratory is three months for persons at high risk, the longest is one or more years for those whose chance of exposure is minimal. HARLEY (1964) has suggested that experience should dictate the scope of a routine bioassay program. Where satisfactory environmental control has been demonstrated over a long period, and determinations are consistently negative, persons shown to be at very low risk may be dropped from the routine program.

III. Collection and Initial Handling of Samples

Comparability of results requires accurate knowledge of the time covered by the collection. For detection of very low levels of radioactivity, this should be 24 hours for a urine sample. Adequacy of the collection period may be determined by having the subject record the starting and ending times, by measuring the volume (assuming excretion of approximately 1.5 liters per 24 hours), or by performing creatinine determinations on all samples. Quality control is essential to insure consistently accurate results. This is accomplished by making reagent blanks for all determinations and by use of frequent spikes or tracers to check on chemical recovery.

Sample collections should be made away from the laboratory to avoid contamination of the sample or container by nonmetabolized radionuclides. Accordingly, many laboratories require two overnight collections, and subjects are requested to remove work clothing and to bathe before starting the collection.

Because of the phenomenon of surface adsorption at neutral or alkaline pH, collections should be made in plastic rather than glass containers. If use of glass cannot be avoided, urine should be acidified to approximately 0.1 N H^+ with either nitric or hydrochloric acid at the time of collection. Samples that are stored for any length of time should be acidified even when plastic containers are used.

Urine samples to be analyzed radiochemically are transferred to beakers of suitable size, and the urine volume is measured. The sample may be digested by one of two methods: (All operations must be carried out in hood).

Wet Ashing. Enough concentrated nitric acid is added to change the color of the sample to dark brown. One or two ml of octyl alcohol are added to prevent foaming. The beaker is covered with a speedy vap and set on a hot plate to evaporate. When the sample is dry, it should be reheated to dryness with 10 ml increments of concentrated nitric acid until a pure white ash remains. The residue is dissolved by addition of 10 ml concentrated nitric acid and 35 ml distilled water, heated gently on a hot plate. The contents of the beaker is transferred quantitatively with repeated small rinses of 2 N nitric acid to a 90 ml pyrex centrifuge tube. The sample is then ready for the bismuth phosphate, lanthanum fluoride precipitation procedure.

Alkaline Phosphate Precipitation (LASL, 1958). The sample is made 0.1 N in H^+ by addition of a suitable amount of concentrated hydrochloric acid. The beaker is set on a hot plate and equipped with a mechanical or magnetic stirrer. A thermometer is placed in the sample and the beaker is covered with a watch glass. The sample is heated gently with continuous stirring to 85° C. Six ml of 6 M phosphoric acid are added, and by dropwise addition of concentrated ammonium hydroxide, the pH is brought to 8 (determined by Universal Hydrion paper). The temperature should not be allowed to exceed 90° C. The heat is reduced to bring the temperature of the sample to approximately 65° C and stirring is continued for several hours. The stirrer and thermometer are removed and the covered beaker is placed overnight in an oven at 65° C. The following morning the supernatant is carefully aspirated taking care not to disturb the precipitate. The remaining suspension is transferred to one or two 90 ml pyrex centrifuge tubes and centrifuged for five minutes at 2000 rpm. The supernatant is discarded and the precipitate is washed with distilled water and centrifuged for five minutes at 2000 rpm. The supernatant is discarded and the precipitate is dissolved in a few ml of concentrated nitric acid and transferred quantitatively to the original beaker, using several small rinses of 2 N nitric acid. The beaker is set on the hot plate, covered with a speedy vap and evaporated to dryness. Evaporation with small increments of concentrated nitric acid is continued until a pure white ash is obtained. Two or three such evaporations usually suffice.

In this procedure, monovalent cations remain in solution and are eliminated when the supernatant is aspirated. All other cations are precipitated. The divalent cations that are normally present in urine serve as carriers. With very dilute urines, however, it may be necessary to add a small amount of calcium or magnesium chloride to bring about precipitation. Alkaline phosphate precipitation is a quicker method than wet ashing, and the elimination of monovalent cations is an advantage in some radiochemical procedures, such as ion exchange chromatography. LOW-BEER and PARKER (1964) have shown that recovery of actinide elements by this method is as good or better than by wet ashing, except in urines from subjects who have received DTPA (or other chelating agent). At pH 8, the metal chelate complex is tightly bound, so that 90% of the transuranium elements are lost in the supernatant. HORM (1971) has shown that this caveat applies also to extraction from untreated urine. In the presence of DTPA, HDEHP extracts only free americium from raw urine. Only wet ashing destroys all organic material, and so destroys the complex and prevents its reformation if the pH is elevated.

Fecal samples, tissues, blood, and nose swipes are dried in an oven at approximately 70° C for 24 to 48 hours and are then heated in a muffle furnace at 500° C for 12 to 16 hours. The resulting ash is then digested with concentrated nitric acid, as in the wet ashing method for urine.

The presence of relatively large amounts of iron in such samples interferes significantly with recovery of actinides by established methods. GOLCHERT and SEDLET (1972) have developed a modification of the anion exchange method (Sec. V.B.1) for the analysis of environmental samples. The method has proved equally effective in removing iron from samples of feces, blood, and tissues. The essential features of the method will be found in Sec. V.B.2.

IV. Radiochemistry of Plutonium

The chemistry of plutonium and industrial methods for its separation, conversion and recovery have been discussed comprehensively by CLEVELAND (1970). A review of the chemical and physical properties of plutonium will be found in

Chap. 9 of this volume. Those aspects of plutonium chemistry that are of particular interest to the analytical radiochemist have been summarized by COLEMAN (1965) in a publication of the Committee on Nuclear Science of the National Academy of Sciences, National Research Council (1965). From the point of view of the bioassayist, the most important feature of plutonium chemistry is the fact that in aqueous solution ions may be maintained selectively in either the +III or +IV valence states. In the +III state, plutonium will undergo all the reactions common to the transplutonium elements, while in the naturally predominating +IV state, the greater stability of plutonium complexes can be exploited in ion exchange and solvent extraction procedures. The tendency of +IV ions to polymerize as a result of hydrolysis should be mentioned although it need not be a problem in bioassay, since polymerization is favored by high concentration of plutonium, by low acidity, and by increased temperature. The avoidance of polymerization in analytical work is important because the chemical properties of polymers are different from those of ions and would, therefore, give midleading results.

V. Procedures for Isolation of Plutonium

The literature is replete with methods for determination of plutonium in biological and environmental samples. Most of these methods are variations of basic procedures which utilize the precipitation, ion exchange, and complex forming properties of plutonium. In the following sections, detailed procedures are presented to illustrate the application of chemical principles to the bioassay of plutonium. All of these methods were designed originally for the isolation of plutonium only, although several of them are applicable to the transplutonium elements as well. Thus, no distinction is made between methods that employ the +III or +IV valence states of plutonium.

A. Precipitation Methods

1. Coprecipitation with $BiPO_4$ and LaF_3 (SCHUBERT et al., 1951). This was the earliest precipitation method for recovery of plutonium from biological material. It is now used widely for all the actinide elements except uranium and radium and is presented in Chap. 20.

2. Coprecipitation with Cupferride (BROOKS, R.O.R., 1965).

a) Dissolve the ashed residue of a 24 hour urine sample in ~30 ml of 4 N HCl and transfer to a 250 ml beaker. Use alternate washings of 4 N HCl and water to complete transfer to a final volume of 100 ml and acidity of about 2 N. Stir until solution is complete.

b) Add 5 ml of $FeCl_3$ solution (145 mg $FeCl_3$/liter) and 10 ml of 5% hydroxylamine hydrochloride followed by a few drops of cresol-red indicator.

c) Adjust the pH to 1 with 2 N NH_4OH added dropwise with constant stirring. Avoid formation of phosphate precipitation as this would reduce recovery. Allow the titrated solution to stand for one hour to insure reduction of plutonium to III.

d) Transfer the solution to a 500 ml separatory funnel and add 2 ml of 5% cupferron solution (make fresh weekly). Shake vigorously and allow to stand for 45 minutes to allow complete formation of iron cupferride.

e) Add 30 ml chloroform and shake vigorously. When the chloroform layer has settled, drain it through a 7 cm Whatman 41 filter paper into a 100 ml separatory funnel. Wash the chloroform by shaking with 20 ml of distilled water and run the chloroform into a 250 ml round bottom flask through filter paper as above. Return the water to the original aqueous solution and add 5 ml of $FeCl_3$ solution and 2 ml of 5% cupferron as a scavenge. Shake and let stand for 45 minutes.

f) Extract the cupferrides with three 15 ml increments of chloroform, or continue until the chloroform remains colorless. Filter each such chloroform fraction into the 100 ml separatory funnel. Wash the filter paper with chloroform using a pipette, to remove cupferrides adhering to the paper. Add washings to the separatory funnel.

g) Add 20 ml distilled water to the chloroform extracts. Shake and allow to settle. Filter the chloroform layer into the 250 ml round bottom flask and wash the filter paper with chloroform.

h) Evaporate the excess chloroform in a fume cupboard using a heat mantle. Remove the final drops by blowing in air.

i) Add 3 ml of concentrated H_2SO_4 and 1 ml of concentrated HNO_3 to the residue and heat. Evaporate final traces of H_2SO_4 with a stream of air. If the residue is not white, add increments of acid and take to dryness again.

j) Dissolve the residue in 3 ml of concentrated HCl and transfer by means of a pipette to a platinum planchet. Dry by infra-red lamp. Use two further aliquots of HCl to insure quantitative transfer.

k) Take up the dried residue in a few drops of distilled water and add enough 2 N NH_4OH to insure complete precipitation of ferric hydroxide. Spread the precipitate evenly over the planchet using a glass rod with a fine tip.

l) Dry the planchet and flame over a bunsen burner to form the red oxide Fe_2O_3.

Note. This method which is one of the earliest procedures for the determination of plutonium (it was developed originally in 1956) was first used in large scale separations. It is still in use as an analytical procedure in some laboratories, but it is given here because of its historical interest. Many of the determinations made in older experiments employed this method—a fact that should be borne in mind when results of such early work are being considered.

B. Ion Exchange

The prototype method given here owes much of its development to the work of many earlier investigators. In its present form, the method is widely used as it has the advantages of a high degree of sensitivity (0.05 dpm per 24 hour urine sample), reproducibility, and simplicity. It is specific for plutonium as compared with transplutonium elements, as it depends on adsorption of the anionic plutonium nitrate complex. Such a complex is formed only by plutonium in oxidation states greater than III. Recovery is said to be $84 \pm 14\%$.

1. Anion Exchange (Campbell and Moss, 1965)

a) Prepare a chromatographic column of following dimensions: a tube 3 inches long by $^5/_{16}$ inch inner diameter, topped with a reservoir $2^5/_8$ inches long by $1^3/_{32}$ inches inner diameter. Pack the tube to a height of 2.5–3.0 inches with Dowex AG 1X-2, 50–80 mesh anion exchange resin. Introduce the resin in a slurry of distilled water. Convert the resin to the nitrate form with two 5 ml washes of 8 N HNO_3.

b) Dissolve the ashed residue of the alkaline phosphate digestion procedure in 80 ml of 8 N HNO_3. Add the solution to the column and allow it do drain completely. Wash the container with three 10 ml portions of 8 N HNO_3, adding each one in turn to the column and allowing it to drain thoroughly before adding the next increment.

c) Wash the walls of the reservoir with 5 ml of 8 N HNO_3 and allow it to drain.

d) Add 3 ml of concentrated HCl to the top of the column, taking care not to disturb the resin. Allow this to drain completely and discard effluent and all washings.

e) Add 2 ml of 0.5 N HCl to the top of the column and collect all but the first few drops in a 15 ml centrifuge tube.

f) Elute the column with two 5 ml portions of 0.5 N HCl and collect eluate in the centrifuge tube from step "e".

g) Add a few crystals of hydroxylamine hydrochloride[1] to the top of the column and follow with 2 ml of hyrdiodic hydrochloric acid solution (1 ml hydriodic acid to 9 ml concentrated HCl, prepared fresh daily). Allow the column to drain completely and collect the effluent in the centrifuge tube.

h) Evaporate the contents of the centrifuge tube after adding 1 ml of concentrated HNO_3. This will oxidize and expel residual iodine.

i) The sample may then be slurried and plated for direct counting, or prepared for electrodeposition (vide infra).

2. Procedure for Removal of Iron (GOLCHERT and SEDLET, 1972)

a) The activity is coprecipitated on $Ca_3(PO_4)_2$ which carries iron as well.

b) The precipitate is dissolved in $Al(NO_3)_3$ HNO_3 (8 N).

c) The dissolved material is placed on the column described in procedure 1 of this section and the remaining steps of that procedure are carried out. Iron is removed in the effluent and washes.

C. Solvent Extraction

Some of the earliest separations of plutonium were performed by extraction into organic solvents containing suitable ligands. Tributyl phosphate was a commonly used complexing agent. In more recent times, thenoyl trifluoroacetone (TTA) has been more widely used because of the greater stability of the resulting complex. Solvent extraction has the advantages of speed and simplicity but its sensitivity is significantly less than that of ion exchange chromatography. The use of solvent extraction for isolation of plutonium from urine or other biological material usually requires an initial precipitation step. This may be the $BiPO_4$, LaF_3 coprecipitations described in Chap. 20, or the fluoride precipitation may be done with suitable carrier directly from the ashed residue of the sample. The latter method is used in the procedure given below.

1. Isolation of Plutonium by TTA Extraction[2] (SCHWENDIMAN and HEALY, 1958).

a) Cerium fluoride precipitation—carried out with ashed residue of a 24 hour urine sample, dissolved in 65 ml 2 N HNO_3 and transferred to a 100 ml centrifuge tube.

1. Add 0.5 gram hydroxylamine hydrochloride to the solution. Stir until dissolved. Use *teflon* rods for all stirring operations.

2. Add 1 ml cerium nitrate carrier solution (approximately 20 mg Ce^{+3} per ml). Transfer the solution to a *new* 100 ml *lusteroid* centrifuge tube with a small amount of distilled water. (Use the pyrex tubes as holders for the teflon rods during the following centrifuging operations).

Note. Carry out the following fluoride precipitations in a hood.

1 Added to prevent immediate oxidation of hydriodic acid.

2 The method given here is a modification of the original procedure used at Argonne National Laboratory.

3. Add 10 ml 1:1 hydrofluoric acid. Use a transparent *bakelite* graduated cylinder for measuring the hydrofluoric acid. Stir the solution for 1 minute; rinse off the teflon rod into the tube with distilled water; and let the solution sit for 5 minutes.

4. Centrifuge for 5 minutes at approximately 2000 rpm. Discard the supernatant, being careful not to disturb the precipitate.

5. Slurry the precipitate in 1 to 2 ml concentrated nitric acid. Add 10 ml distilled water and stir vigorously.

6. Add 2 N nitric acid in 10 ml portions until there is approximately 65–70 ml total volume, stirring well after each addition.

7. Add 10 ml 1:1 hydrofluoric acid and repeat the above precipitation and centrifuging procedures.

b) Solvent Extraction.

1. After pouring off the supernatant from the second fluoride precipitation, slurry the precipitate with 1 to 2 ml aluminium nitrate solution (740 grams/liter solution, including 20 ml concentrated nitric acid). Then add aluminum nitrate solution in 10 ml portions to a total volume of approximately 40 ml, with vigorous stirring after each addition. Continue stirring until all of the precipitate is in solution.

Note. The solution may sit overnight at this point, if necessary.

2. Add 100 milligrams sodium nitrite. Stir until dissolved. Set timer for 15 minutes. Transfer the solution to a 125 ml Squibb separatory funnel with 10 ml distilled water. (The separatory funnel is mounted in the wrist-action shaker. Clamp the funnel in such a way that the stopcock of the funnel is immediately below the lower jaw of the clamp. If mounted in this way, the funnel will not become loose during the shaking operations). Save the lusteroid tube, temporarily, and use it as a waste receiver during the extraction steps.

3. *Fifteen minutes after the addition of the sodium nitrite*, add 10 ml of 0.25 M TTA-benzene solution. Insert the vented teflon stopper. Shake well for 30 minutes.

4. Let stand for 5 minutes. After separation of the layers, remove the lower (aqueous) layer, collecting it in the lusteroid tube. This layer may then be discarded.

5. Add 20 ml distilled water to the funnel and shake for 10 minutes. Let stand for 5 minutes. Remove and discard the lower (aqueous) layer.

6. Add 10 ml distilled water to the funnel and shake for 10 minutes. Let stand for 5 minutes. Remove and discard the lower (aqueous) layer. The lusteroid tube should now be discarded.

7. Add 10 ml 8 N hydrochloric acid to the funnel and shake for 20 minutes. Let stand for 5 minutes. Collect the lower (hydrochloric acid solution) layer in a 30 ml beaker.

8. Add 5 ml 8 N hydrochloric acid to the funnel and shake for 5 minutes. Let stand for 5 minutes. Collect the lower (hydrochloric acid solution) layer in the same 30 ml beaker.

Note. The solution may sit overnight at this point, if necessary.

9. Discard the benzene solution remaining in the funnel by pouring it into the *organic* liquid waste container.

10. Evaporate the total HCl strip to incipient dryness. The residue may then be slurried, plated on a planchet for direct counting, or prepared for electrodeposition.

D. Electrodeposition of Plutonium

The apparatus of SCHWENDIMAN and HEALY (1958) is used for electrodeposition of all actinide elements. It is described briefly in Chap. 20 together with methods of electrodeposition of trivalent cations including reduced plutonium. For plutonium in higher states of oxidation, typically +IV, the following procedure is necessary.

1. Evaporate the final product of any of the chemical procedures to incipient dryness.

2. Rinse down the sides of the beaker with a small amount of distilled water. Add 2–3 drops of phenophthalein indicator solution. Neutralize the solution by addition of 12 *N* potassium hydroxide solution with a capillary-tipped medicine dropper. The neutralization is complete when the solution becomes slightly brown or *very faint pink*.

Note. Do not pass the end-point of this neutralization, as plutonium IV will precipitate from alkaline solution. If the end-point is inadvertently passed, as indicated by the deep pink color of the solution, quickly add 1:19 hydrochloric acid to the solution until the pink color just disappears.

3. Add 2 ml 5% sodium hypochlorite solution.

4. Add 5 ml 2 *N* potassium hydroxide solution.

Note. The sample may be left overnight at this point, if necessary. If this is to be the case, bring the solution to a boil, cover with a watch glass, and after cooling, cover with parafilm. Do not carry out the following evaporation until the next day, when the sample is to be left overnight.

5. Evaporate the solution to half volume, and then cool to room temperature.

6. After evaporation and cooling of the sample, transfer the solution to the electrodeposition cell with distilled water, to give a total volume of 10 ml (to the red line marked on the cell wall).

7. Electroplate under these conditions—12 volts and 300 milliamperes—for 5 hours.

8. Remove cell(s) from the apparatus without turning off the power supply. Immediately decant the solution and rinse the cell with several portions of distilled water.

9. Dismantle the cell, rinse the electroplated disc with distilled water and dry in air.

10. The sample may then be counted in a proportional counter, an alpha pulse height analyzer, or radioautographed for nuclear track counting.

E. Nuclear Track Counting

The technique of nuclear track counting offers an alternative to electronic detection of radioactivity. At a time when few other methods had attained detection limits lower than 0.2 dpm per 24 hour urine sample, nuclear track counting had a sensitivity of 0.05 dpm. The method and the instruments that support it were developed by SCHWENDIMAN and HEALY (1958). The principal features of the system are:

1. Thin layers of nuclear track emulsion (Kodak NTA) spread on 1×3 inch microslides.

2. Elimination of background counts in the emulsion by prior exposure to H_2O_2 vapor under dessication at room temperature.

3. A "camera" designed to maintain electroplated discs in contact with the emulsion during exposure.

4. Exposure of emulsion to samples for 168 hours at 5° C.

5. Developing, fixing, and washing of slides without exposure to light.

6. Storage of exposed slides at 5° C.

7. Microscopic counting by a scanning system so that tracks within a total area of 4 mm^2 are counted.

8. Conversion of the number of tracks per unit area to dpm using the factors of exposure time, film efficiency, and exposure area.

9. Routine checking of film efficiency with a standard source.

VI. Detection of Radioactivity

An exhaustive discussion of instruments used to evaluate internal contamination by actinide elements is beyond the scope of this article. References cited in following sections provide the means of pursuing this subject in greater depth.

A. Alpha Counting

The principles and methodology of alpha counting have been reviewed in a comprehensive article by JAFFEY (1954).

1. Total Alpha Counting

Instruments used for total alpha counting are predominantly either pulse counters—usually of the proportional gas flow type—or scintillation counters. The former operate by collecting secondary ionizations of a gas which is contained between two electrodes at high potential. The electric field sweeps apart secondary ions formed in the gas by alpha particles, and the resulting pulse is measured. The initial charge from an alpha particle can, in this way, be multiplied by factors from 10 to 10^5. These counters are operated with methane or an argon methane mixture.

Scintillation counters depend on the conversion of secondary electron energy into light by means of a solid or liquid scintillator, and the conversion of light flashes into electrical pulses by a photo-multiplier tube. Because these conversions are proportional to energy, scintillation counters have spectrometric properties. They are useful, however, for total alpha counting, as they have the advantage of very low background. A method for routine alpha counting of bioassay samples using a zinc sulfide phosphor has been described by HALLDEN and HARLEY (1960), and is used in many laboratories.

To insure low detection limits and reproducibility of alpha counting, the following requirements must be satisfied. Counting instruments should approach 50% efficiency (2π geometry); background must be low and stable; planchets must be of platinum or stainless steel and must be thoroughly clean and free of scratches in which a pile-up of the sample can occur; the mass of the sample mus be as small as possible, and it must be spread evenly over the surface of the planchet in order to minimize self-absorption; planchets are dried under an infrared lamp, and are flamed to cherry red; counts should be made for at least 120 minutes to insure a small standard deviation. Proportional gas flow counters require frequent checks of counting efficiency, and they should be calibrated at regular intervals to insure that voltage and gain settings are optimal.

Total alpha activity of a sample is calculated by:

$$\frac{(\text{net cpm of sample} \pm \sigma) \times 100}{\text{C.E.} \times \text{C.R.} \times \text{S.A.} \times \text{aliquot}} = (\text{dpm} \pm \sigma)/\text{whole sample}$$

C.E. = counting efficiency; C.R. = chemical recovery; S.A. = self absorption (determined by incorporating a spike in the slurry or residue of the final product of the chemical procedure carried out with only the reagents).

The count of the reagent blank constitutes the background.

2. Alpha Spectrometry

Isotopes of the actinide elements emit alpha particles with energies in the range of about 4–8 Mev. There are fifteen isotopes of plutonium with energies from 4.89 to 6.58 Mev. Of these, only ^{238}Pu and ^{239}Pu, with energies of 5.50 and 5.16 Mev, respectively, are ordinarily encountered in bioassay programs. The transplutonium isotopes of greatest significance in bioassay have energies from 5.4 Mev (^{241}Am) to 6.12 Mev (^{252}Cf). Instruments described by Jaffey (1954) for the identification of these nuclides by alpha energy measurements have been largely superseded by the development in recent years of semiconductor detectors. The use of these instruments for alpha spectrometry has been discussed by Hofker et al. (1969). The essential features are a semiconductor silicon diode with a reverse bias across it, an amplifier, and a multichannel pulse-height analyzer. Ionization occurs when a nuclear particle penetrates the depletion layer of the detector, caused by the reverse bias. Pairs of charge carriers are freed and, under the influence of an electrical field, they travel towards two electrodes thereby producing a pluse. The number of charge carriers so freed is a function of the energy of the incident particle, and the pulse height is, therefore, proportional to this energy. Because of the very short range of alpha particles, surface barrier detectors give best results by causing the depletion layer to form at the surface of the detector. A surface barrier consists of a thin coating of gold which has been evaporated on the surface of the silicon diode. These counters must be operated under vacuum to avoid energy losses due to collision of alpha particles with gas molecules. The superiority of these detectors is due to their low background and their high resolution.

The preparation of samples for alpha spectrometry is as important as the counting instrument itself. To insure high resolution, samples must be virtually weightless, and must be free of all interfering material. They must also be as nearly point sources as possible. These objectives are attained by electeodeposition (see Sec. V).

B. Gamma Spectrometry

Because of the exceedingly short range of alpha particles (Walsh, 1970) *in vivo* evaluation of alpha contamination depends on detection of gamma energy. In the case of plutonium, this is not feasible by the traditional method of whole body counting which requires the presence of abundant gamma rays of energies above approximately 0.1 Mev (see Chap. 20). Until the development of instruments capable of detecting low energy-photons (vide infra), the bioassayist was wholly dependent on radiochemical analysis of excreta followed by alpha counting, for evaluation of internal contamination by plutonium. An exception to this has been reported by Heid et al. (1971) who have utilized the 59.5 Kev gamma ray of ^{241}Am, the daughter product of ^{241}Pu, in whole body counting. Quantitation of internally deposited plutonium by this method requires knowledge of the Pu/Am ratio of the ingested material. It also assumes that the metabolic behavior of the two nuclides is identical. Since the validity of this assumption is questionable (Chap. 18), this method can only approximate the body burden of plutonium, and it may underestimate it significantly.

C. Low Energy Photon Spectrometry

Development of instruments capable of detecting the L X-rays that are characteristic of all heavy elements has, in the words of Hollander and Perlman (1966), created a "revolution in nuclear radiation counting". L X rays of plutonium and the transplutonium elements have energies in the range 14 to 26 kev. Their values have been tabulated by Fine and Hendee (1970). It is obvious that instruments capable of detecting the L X-rays from internally deposited ^{239}Pu would circumvent the limitations of conventional whole body counting with respect to this nuclide. In practice attenuation of L X-ray counts due to absorption in bone and soft tissue poses serious problems. Plutonium deposited in bone is not detectable by this technique. Deposition in lungs or other soft tissues can be determined quantitatively, but only after correlation of attenuation with chest thickness and body size. This question has been discussed by Newton et al. (1972) and Morsy et al. (1972). The first instruments developed for this purpose were proportional counters consisting of two or more large area detectors and employing a mixture of argon-methane gas. They have been described by Fessler et al. (1961), Lanisart and Morucci (1964), Ehret et al. (1964) and Tomitani and Tanaka (1970) and Sharma et al. (1972). In order to eliminate background counts, these counters are operated in anti-coincidence mode, and may also be designed to eliminate all counts except those from the nuclide of interest. Their chief disadvantage is their rather low sensitivity. A detection limit of 12 nCi in 30 minutes counting time is given by Ehret et al. (1964) and this value closely approaches the 16 nCi maximum permissible lung burden for ^{239}Pu given by Heid et al. (1971).

L X-rays can be measured also by scintillation counters consisting of NaI crystals of thicknesses of a few millimeters. Absorption of low energy photons is almost complete within 0 25 mm of the surface of the crystal, and background is significantly reduced due to the thinness of the crystal. Such counters are being used increasingly for the measurement of lung burdens of ^{239}Pu. Typically the counter consists of two or more detectors operated in anti-coincidence mode. Swinth and Griffin (1970) have described an array of 52 one millimeter thick NaI crystals coupled to low-noise phototubes. While this system is promising, the authors have noted the need for further improvement to attain greater sensitivity. The recent development of Phoswich detectors appears to offer greater potential for *in vivo* counting of low energy photons. These detectors consist of two different scintillators, typically a thin NaI crystal mounted on a thicker crystal of C_sI. Because of the difference in scintillation decay time of the two crystals, the detectors discriminate between X-ray/gamma, beta/gamma and neutron/gamma signals. Evaluations of their clinical performance have not yet appeared in the literature.

While thin NaI crystals provide a fairly sensitive means of counting L X-rays, their resolving power leaves much to be desired. The most spectacular advances in low energy photon spectrometry have come about through the development of semiconductors. These are solid-state devices which employ semiconductor diodes of silicon or germanium drifted with lithium. They have been described by Bowman et al. (1966) who report counting efficiencies of 100%, and resolution

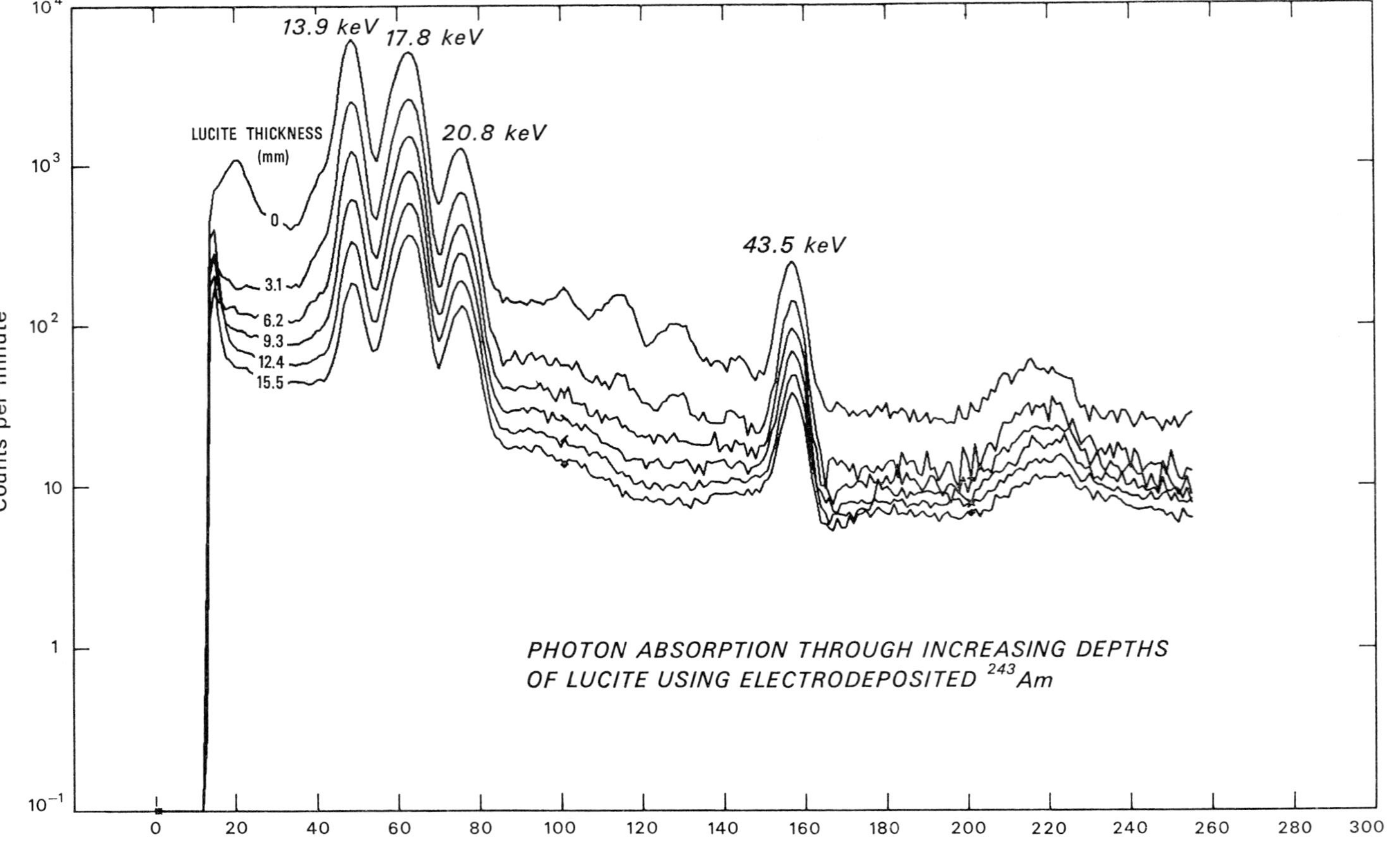

Fig. 13.1. Computer-drawn plots of spectra from a ^{243}Am source counted on a Si(Li) semiconductor with various depths of tissue-equivalent liquid between the source and the detector. Counts per minute are plotted semi-logarithmically against channel number covering an energy range from 1 to 129 keV. The difference in absorption for the different X-ray peaks is clearly illustrated. (PARKER et al., 1968)

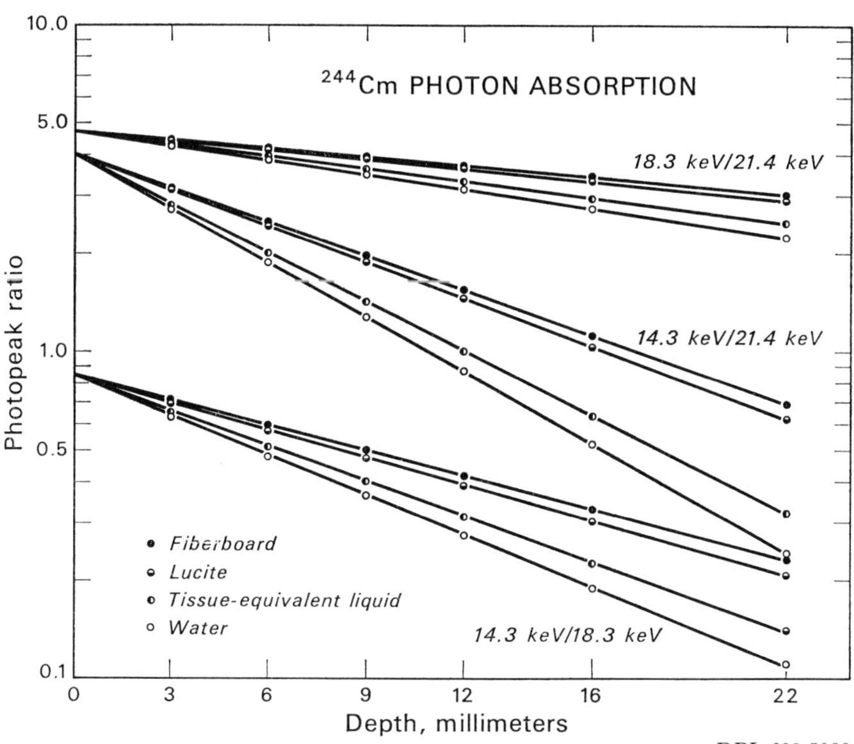

Fig. 13.2. Curves of X- and γ-ray peak ratios of ^{244}Cm plotted semilogarithmically against depth of absorber for fiberboard, Lucite, water, and tissue equivalent liquid. Ratios between various peak regions were taken from data like those shown in Fig. 13.1. (PARKER et al., 1968)

of such a high order that photons which differ in energy by as little as 1.1 keV can be distinguished with germanium detectors. The spectrometry attainable with germanium detectors has also been reported by DREXLER and PERZL (1967). Silicon detectors have best resolution for the low energies of the L X-rays region, while germanium has equally good resolution over the entire range of gamma energies. Both detectors must be maintained at cryogenic temperature to insure stability and resolution. Fig. 13.1 illustrates the resolution of a point source of ^{243}Am by a Si(Li) detector, and also shows the attenuating effect on peak height of various thickness of an absorbing material. Fig. 13.2 demonstrates the effect on peak ratios of ^{244}Cm of several different absorbing materials of the same thickness. PARKER et al. (1968) have shown that by using such calibration methods an unknown point source can be quantitated, and localized with respect to depth in an absorbing material. The implications of this capability for wound counting in cases of human exposure are obvious.

ANDERSON et al. (1970a, b) have described the design of germanium systems for wound counting, lung counting, and whole body counting and have compared their performance to that of the sodium iodide crystals of conventional scintillation counters. The authors report a detection limit of 0.73 nCi for plutonium deposited in a wound at a depth no greater than 6 mm in a 10 minute period of counting for the germanium detector. Although the detection limit is higher by a factor of 7 than that attained with a thin NaI crystal, it is considered adequate for routine wound monitoring. The lower sensitivity of semiconductors, as compared with NaI crystals of appropriate thickness, may be partially offset by using an array of several germanium detectors. Nevertheless, high detection limits and long periods of counting limit the usefulness of semiconductors in lung or whole body counting. Their greatest value in such applications is due to their high resolution. Activity detected by a NaI scintillator appears spectrometrically as a broad peak, the components of which are indistinguishable. The semiconductor, on the other hand, is capable of resolving broad energy bands into discrete peaks separated by very small differences in energy, and thus affords a means of identifying the radioisotopes that are present. Semiconductors are, therefore, instruments of great potential for the comprehensive investigation of internally deposited radionuclides.

Developing methods of direct measurement and identification of radionuclides in biological materials and in the living organizm hold great promise for the future of bioassay programs. It seems probable that much of the time consuming and laborious effort now expended on the preparation of samples for alpha counting may be eliminated as modern and sophisticated counting instruments become generally available. When *in vivo* counting is feasible, the difficult problem of estimating the body burden from the observed excretion rate is also eliminated.

VII. Interpretation of Bioassay Data

Quantitation of human internal contamination by plutonium and other actinide elements is a complex problem when evaluation depends on alpha determinations of excreta and other biological materials. Many investigations have elucidated patterns of excretion and retention in a number of animal species. Extrapolation of animal data to man is, however, fraught with uncertainties. Moreover, in the uncontrolled circumstances of human exposure, many variables must be taken into account.

These have been discussed by POCHIN (1964), BEACH and DOLPHIN (1964), and by many others. They include the nature of the actinide-containing compound, the route of intake, the metabolic fate of the material, and the variability of excretion rates among individuals and in the same individual during different time periods. LANGHAM et al. (1950) were able to study the excretion of a group of human subjects who had received doses of citrated plutonium by intravenous injection. They proposed a power function of the form

$$Y = a\,t^{-c}$$

to describe long-term excretion of plutonium, and to account for differences in rate of excretion as plutonium is distributed to various body compartments. In this equation, Y is the excretion rate in fraction of injected dose excreted per day, t is the time after exposure in days, and a and c are constants. When the equation was fitted to the excretion curves of the human subjects, a and c were found to have the following values for excretion rates over a period of five years:

$$Y u = 0.002\,t^{-0.74}$$

$$Y u + f = 0.0079\,t^{-0.94},$$

where $Y u$ and $Y u + f$ are the fractions of injected doses of plutonium excreted per day in the urine and urine plus feces, respectively, and t is the time in days after injection. Using these values, LANGHAM (1957, 1964) developed a series of equations from which he was able to determine the original body burden, fractional retention at any given time, coefficients of elimination in urine and feces, and the ratio of urinary to fecal excretion at any given time. BEACH and DOLPHIN (1964) have suggested the inclusion of an exponential term in LANGHAM'S basic equation to express the early and rapid urinary excretion of presumably unmetabolized plutonium that occurs when citrated plutonium is administered intravenously. The equation proposed by BEACH and DOLPHIN is:

$$Y u(t) = 0.410 \exp(-0.67 t) + 0.16\, t^{-0.68}$$

The second component of the equation is a power function, and represents the excretion of metabolized plutonium. DURBIN (1971) has reexamined the Langham equations and has concluded that involvement of plutonium with the iron transport system modifies the intermediate steps in deposition, turnover, and excretion, without impairing the validity of the major aspects of Langham's original analysis.

BEACH and DOLPHIN (op. cit.) have pointed out that intravenous injection of citrated plutonium is not representative of the typical industrial exposure where internal contamination occurs most frequently by inhalation or by puncture wound. In either case, the rate of movement of plutonium from thesite of initial deposition governs the rate at which systemic contamination will occur. BEACH and DOLPHIN have suggested the following equation to account for this process:

$$E u(t) = \int_0^t q\, 1\, (\tau) Y u(t-\tau) d\tau,$$

where $E u(t)$ is the daily urinary excretion of plutonium, q is the amount initially retained at the site of entry, and the amount released to the blood at any time is $q l(t)$. $Y u(t)$ is the power function component of the basic equation. It is obvious that urinary excretion before equilibrium is established will underestimate the body burden. Accordingly, evaluation based on urinary excretion is unreliable earlier than about 100 days after exposure.

In the typical puncture wound, plutonium is in colloidal form due to complexing with the anions of strong acids. Its rate of diffusion from the wound is, therefore, markedly slower than that of plutonium in ionic form, as in the citrate complex.

LAFUMA et al. (1972) have proposed a model which postulates that plutonium in colloidal form diffuses slowly from the site of entry and that a fraction of the diffused activity is excreted in the urine without prior deposition in bone or soft tissue—a probability which significantly alters the interpretation of early bioassay results.

DOLPHIN (1972) has called attention to the relatively new technique of analysis of chromosome aberrations as an additional tool for evaluation of internal contamination.

The distribution of inhaled plutonium in the compartments of the respiratory tract and its release from the pulmonary region of the lungs are determined by particle size and solubility of the compound. The *Task Group on Lung Dynamics* (1966) has developed a model to describe the kinetics of inhalation exposures. According to this model, about 75% of the inhaled plutonium is rapidly expelled from the nasopharyngeal and tracheobronchial regions by ciliary action, and is excreted in the feces during the first four to seven days after exposure. The 25% of the inhaled material that reaches the pulmonary region diffuses into the adjacent lymph nodes at a rate that is proportional to the solubility of the particles. Typically, these are oxides of plutonium, and their extreme insolubility accounts for the half-life of plutonium in the lung.

The importance of determining fecal excretion immediately following an inhalation exposure has been emphasized by DOLPHIN and JACKSON (1964) and by EAKINS and MORGAN (1964). A high fecal to urinary ratio at this time establishes the existence of a lung burden, and provided that all fecal specimens for the first few days are analyzed, the total burden can be approximated by use of the model. It is evident, however, that systemic contamination is more reliably estimated from urinary excretion rates, due in part to the relation of urine collections to definite time periods.

It will be seen from the foregoing that internal contamination by plutonium can be estimated with reasonable accuracy when the time of exposure is known. In the case of chronic or sporadic exposure at unknown times, the problem is more difficult. This is the kind of exposure encountered in routine programs of surveillance. A number of investigators have developed codes based on the LANGHAM equations that are designed to apprehend any increase in activity found in routine monitoring of urine samples. The codes depend on computerization of analytical data, and each analytical result represents an interval defined by the last previous sampling. A significant increase in activity is presumed to indicate an additional intake during the interval. Such codes have been reported by LAWRENCE (1962), SNYDER (1962, 1964), BEACH et al. (1966) and by BEACH (1973). BEACH et al. (1966) have suggested establishment of reference and investigation levels for routine monitoring. Excretion values at or below the reference level may be regarded as insignificant, while those approaching or exceeding the investigation level require medical judgment regarding treatment or possible change in work assignment. They point out that frequent monitoring is advisable in situations that have a high potential for exposure, since a number of small exposures may produce a cumulative excretion rate that exceeds the investigation level. In order to overcome the problem of daily fluctuation of excretion rate, these authors advocate collection of several samples on successive days rather than a single 24 hour sample, and analysis of an aliquot of the pooled samples.

SNYDER (1972) has developed a computer code which adjusts the parameters a and c of the Langham equation to fit the observed excretion curve of an exposed individual. By this means misinterpretation due to fluctuations within an overall trend are minimized. The author points out that the method is most useful when a large number of samples are analyzed.

In view of the many uncertainties that beset the determination of the relation of excretion to body burden, the thoroughness and accuracy of bioassay procedures assume great importance. The necessity for frequent monitoring of individuals at high risk has been mentioned, and has been discussed by HOLLIDAY et al. (1970). In all monitoring, whether routine or in cases of accidental acute exposure, it is advisable to follow an established protocol. Only in this way can the physician be assured of receiving comparable and consistent information on which to base his decisions. In accidental exposure cases, investigation should include whole body, wound, and/or lung counting, determination of all urinary and fecal excretion for several days following the accident, analysis of nose swipes and blood samples taken immediately after exposure. A prototype procedure encompassing all of these measures is practiced at the Battelle Northwest Laboratory, and has been reported by HEID et al. (1971). The methods and procedures followed at HARWELL have been reviewed by JACKSON and TAYLOR (1964).

Summary

The principles on which bioassay programs are based have been reviewed and related to recommendations of the International Commission for Radiological Protection.

Methods of collection and preparation of biological samples for radiochemical analysis have been presented.

Methods for the isolation and identification of plutonium in biological materials are given in detail.

Detection of radioactivity is discussed in terms of alpha counting and low-energy photon spectrometry. The limitations of conventional whole-body counting for detection of plutonium are elucidated. Interpretation of bioassay results is discussed, and equations relating excretion rates to body burden are presented.

References

Anderson, B. V., Bramson, P. E., Unruh, C. M.: Comparison of germanium and sodium iodide in *in vivo* measurement systems. IAEA-SM 143/35 (1970).

Assessment of radioactive contamination in man. Vienna: IAEA 1972.

Beach, S. A.: Sebeach, a digital computer program for the estimation of body content of plutonium from urine data. Hlth Phys. **24**, 9–16 (1973).

Beach, S. A., Dolphin, G. W.: Plutonium body burdens. In: Assessment of radioactivity in man, vol. II, p. 603–615. Vienna: IAEA 1964.

Beach, S. A., Dolphin, G. W., Duncan, K. P., Dunster, H. J.: A basis for routine urine sampling of workers exposed to plutonium-239. Hlth Phys. **12**, 1671–1682 (1966).

Bowman, H. R., Hyde, E. K., Thompson, S. G., Jared, R. C.: Application of high resolution semiconductor detectors in X-ray emission spectrography. Science **151**, 362–368 (1966).

Brooks, R. O. R.: Determination of plutonium in urine. In: The radiochemistry of plutonium. National Academy of Sciences, National Research Council, NAS-NS3058, p. 150–151 (1965).

Campbell, E. E., Moss, W. D.: Determination of plutonium in urine by anion exchange. Hlth Phys. **11**, 737–742 (1965).

Cleveland, J. M.: The chemistry of plutonium. New York: Gordon & Breach, Science Pubishers 1970.

Coleman, G. H.: The radiochemistry of plutonium. National Academy of Sciences, National Research Council, NAS-NS3058, p. 4–103 (1965).

Dolphin, G. W.: Some problems in interpretation of bioassay data. In: Assessment of radioactive contamination in man, p. 425–435. Vienna: IAEA 1972.

Dolphin, G. W., Jackson, S.: Interpretation of bioassay data. In: Assessment of radioactivity in man, vol. I, p. 329–354. Vienna: IAEA 1964.

Drexler, G., Perzl, F.: Spectrometry of low-energy γ and x-rays with Ge (Li) detectors. Med. Instr. and Meth. **48**, 332–334 (1967).

Durbin, P. W.: Plutonium in man: a twenty-five year review. University of California Lawrence Berkeley Laboratory Report No. UCRL-20850, 1971. 141 p.

Eakins, J. D., Morgan, A.: Fecal analysis in bioassay. In: Assessment of radioactivity in man, vol. I, p. 231–244. Vienna: IAEA 1964.

Ehret, R., Kiefer, H., Maushart, R., Möhrle, G.: Performance of an arrangement of several large-area proportional counters for the assessment of ^{239}Pu lung burdens. In: Assessment of radioactivity in man, vol. I, p. 141–149. Vienna: IAEA 1964.

Fessler, H., Kiefer, H., Maushart, R.: Zur Messung von Quantenstrahlung im Energiebereich von 3–30 Kev mit großflächigen Proportionalzählrohren. Atompraxis **7**, 401 (1961).

Golchert, N., Sedlet, J.: Detection of plutonium in environmental samples. To be published (1972).

Halden, N., Harley, J. R.: An improved alpha-counting technique. Analyt. Chem. **32**, 1861–1863 (1960).

HARLEY, J. H.: Sampling and analysis for assessment of body burdens. In: Assessment of radioactivity in man, vol. 1, p. 155–167. Vienna: IAEA 1964.

HEID, K. R., JECH, J. J., BRAMSON, P. E.: Routine and emergency evaluations of internal exposure at Hanford. BNWL-SA-3840 (1971).

HOFKER, W. K., NIEUHUIS, K., POST, J. C.: α-particle spectrometry with semi-conductor detectors. Philips Tech. Rev. **30**, 13–22 (1969).

HOLLANDER, J. M., PERLMAN, I.: The semiconductor revolution in nuclear radiation counting. Science **154**, 84–93 (1966).

HOLLIDAY, B., DOLPHIN, G. W., DUNSTER, H. J.: Radiological protection of workers exposed to airborne plutonium particulate. Hlth Phys. **18**, 529–540 (1970).

HORM, I.: Techniques for separating americium from urine after DTPA therapy. Hlth Phys. **21**, 41–46 (1971).

International Commission for Radiological Protection. Reports Nos. 6, 7, 8, 9, 10, 12.

JACKSON, S., TAYLOR, N. A.: A survey of the methods used in the United Kingdom Atomic Energy Authority for the determination of radionuclides in urine. In: Assessment of radioactivity in man, vol. I, p. 169–194. Vienna: IAIA 1964.

JAFFEY, A. H.: Radiochemical assay by alpha and fission measurements. The Actinide Elements, National Nuclear Energy Series, Div. IV, Vol. 14A, SEABORG and KATZ, eds., p. 596–732. New York: McGraw Hill 1954.

LAFUMA, J., NENOT, J. C., MARIN, M.: Problemes poses par l'utilisation des donnees d'excretion urinaire pour l'evaluation de la charge corporelle. In: Assessment of radioactive contamination in man, p. 235–245. Vienna: IAEA 1972.

LANGHAM, W. H.: The application of excretion analysis to the determination of body burden of radioisotopes. Brit. J. Radiol. Suppl. **7**, 95–113 (1957).

LANGHAM, W. H.: Physiological properties of plutonium and assessment of body burden in man. In: Assessment of radioactivity in man, vol. II, p. 565–581. Vienna: IAEA 1964.

LANGHAM, W. H., BASSETT, S. H., HARRIS, P. S., CARTER, R. E.: Distribution and excretion of Pu administered intravenously to man. LA1151 (1950).

LANISART, A., MORUCCI, J.-P.: Nouveau compteur proportionnel destine a la detection *in vivo* de traces de plutonium dans les poumons. In: Assessment of radioactivity in man, vol. I, p. 131–140. Vienna: IAEA 1964.

LAWRENCE, J. N. P.: PUQFUA, An IBM 704 code for computing plutonium body burdens. Hlth Phys. **8**, 61–66 (1962).

Los Alamos Scientific Laboratory, Analytical Procedures of the Industrial Hygiene Group. LA 1858, 2nd Ed. (1958).

LOW-BEER, A. DE G., PARKER, H. G.: Decreased recovery of gross alpha determinations by alkaline phosphate precipitation in the presence of DTPA. Hlth Phys. **11**, 61–62 (1965).

MORSY, S. M., EL-ASSOLY, F. M., ALOUSH, A. A.: Direct methods for the assessment of plutonium-239 and uranium-235 body burdens. In: Assessment of radioactive contamination in man, p. 115–127. Vienna: IAEA 1972.

National Council on Radiation Protection and Measurements Reports Nos. 33 (1968) and 39 (1971), Washington, D.C.

National Research Council, National Academy of Sciences. The Radiochemistry of Plutonium. NAS–NR 3058 (1965).

NEWTON, D., FRY, F. A., TAYLOR, B. T., EAGLE, M. C.: Factors affecting the assessment of plutonium-239 in vivo by external counting methods. In: Assessment of radioactive contamination in man, p. 83–96. Vienna: IAEA 1972.

PARKER, H. G., WRIGHT, S. R., LOW-BEER, A. DE G.: Actinide Element Studies with a Si(Li) "wound counter". Semiannual Report Biology and Medicine, Donner Laboratory, Lawrence Berkeley Laboratory, UCRL-18793, 128–134 (1968).

POCHIN, E. E.: Formulation of relationships between the radiation exposure of tissues and the excretion rate of nuclides. In: Assessment of radioactivity in man, vol. I, p. 3–14. Vienna: IAEA 1964.

SCHWENDIMAN, L. C., HEALY, J. W.: Nuclear track technique for low level Pu in urine. Nucleonics **16**, 78–82 (1958).

SHARMA, R. C., NILSSON, I., LINDGREN, L.: A twin large-area proportional flow counter for the assay of plutonium in human lungs. Aktiebolaget Atomenergi Studsvik, Nyköping, Sweden (1972) AE-463, 35 p.

SNYDER, W. S.: A method of interpreting excretion data which allows for statistical fluctuation of the data. In: Assessment of radioactive contamination in man, p. 485–494. Vienna: IAEA 1972.

SNYDER, W. S.: Major sources of error in interpreting urinalysis data to estimate the body burden of ^{239}Pu: a preliminary study. Hlth Phys. **8**, 767–772 (1962).

SNYDER, W. S.: On the estimation of a systemic body burden of plutonium. In: Assessment of radioactivity in man, vol. II, p. 583–588. Vienna: IAEA 1964.

SWINTH, K. L., GRIFFIN, B. I.: A developmental scintillation counter for detection of plutonium *in vivo*. Hlth Phys. **19**, 543–550 (1970).

Task Group on Lung Dynamics, Deposition and retention models for internal dosimetry of the human respiratory tract. Hlth Phys. **12**, 173–207 (1966).

TOMITANI, T., TANAKA, E.: Large area proportional counter for assessment of plutonium lung burden. Hlth Phys. **18**, 195–206 (1970).

WALSH, P. J.: Stopping power and range of alpha particles. Hlth Phys. **19**, 312–316 (1970).

Chapter 14

Plutonium. Industrial Hygiene, Health Physics and Related Aspects

H. M. Parker

With 14 Figures

I. Introduction

Due to its unique history as a man-made element, plutonium passed very quickly from the stage of laboratory experimentation in trace amounts to manufacture as a military product in substantial quantities. If man is entitled to any satisfaction from a previous grievous experience in industrial hygiene, it would be that the tragedy of the earlier experience of the radium dial painters paved the way for the adequate protection of plutonium workers.

The broad principles of protection from the predominantly alpha-particle emitting plutonium consisted simply in isolating the material from contact with the human body by performing all operations in closed glove boxes with a secondary defense of protective clothing. From the start it was realized that the environment needed protection by the best possible filtration of exhaust air from the glove boxes.

These defenses were intended to eliminate:

1. Inhalation. Three separate hazards were recognized from the start:

Direct irradiation of the lung tissue, an effect forewarned by the experience in uranium and radium mining.

Transfer to the circulating blood and eventual residence in liver, kidney, or bone, with a quite early determination that bone deposition was one presumptive limiting factor.

Transfer from nasal and bronchial passages to the digestive tract, with secondary absorption to the bloodstream.

2. Ingestion. The prior radium experience had signalled the need for prevention of ingestion. While gross errors similar to brush-tipping would be unthinkable in this case, considerable attention was given to potential transfer to the mouth from food-handling or smoking with contaminated fingers.

3. Contamination of Intact Skin. The hazards contemplated were:

Relatively permanent deposition in a deep skin layer.

Transmission through skin to access with circulating blood.

With plutonium contamination on the *outer* surface of skin, the thickness of the horny layer is sufficient to absorb all the alpha particles. A third case, absorption into the skin, but not through it, was added later (Newberry, 1964; Dunster, 1967).

4. Contamination of Wounds. This was an obvious concern in two aspects:

Direct access of soluble plutonium compounds to the blood stream.

Implantation of a small piece, for example, of plutonium metal with subsequent chronic leaching to the blood stream.

In retrospect, protection against these factors has been generally successful. There has been no identifiable case of plutonium transfer to the bone in sufficient quantity to demonstrably injure an individual—although this falls short of assurance that *no* late-effects case will develop. There has been a continuing history of the need to remove portions of tissue, predominantly from the hands, following wounds unresponsive to decontamination. Experience has shown that direct ingestion can be dismissed as a practical hazard (less than one case per 10000 man-years of plutonium work). Conversely, inhalation continues as a major source of difficulty, involving one case for 40 man-years at some level of contamination, and one case per 450 man-years at 5% or more of the currently accepted occupational deposition limit in a typical major facility (HEID and JECH, 1969).

The strong attention to prevention of internal deposition by no means completes the catalog of needed defenses. Chemical manipulation of the actinide element, plutonium, frequently involves a fluoride stage. Powerful alpha-emitters react with many of the light elements to generate neutrons, and the yield from plutonium fluoride is particularly significant. The yield from plutonium oxide, which is 30 times less, must also be considered. Here, the learning experience from radium failed because concern for the alpha-neutron reaction for it was essentially confined to the deliberate mixture of radium and beryllium as a neutron source. Had the conventional compound for early radium use happened to be the fluoride, there would have been a long-unknown neutron component in its radiation effects.

Neutron generation by spontaneous fission of plutonium must also be considered. In addition, plutonium, although an alpha-emitter, does emit X-rays and gamma rays. With plutonium in industrial quantities, protection from these penetrating radiations is required. The need for these steps was correctly inferred from prior radium experience.

One unique factor with fissionable materials such as plutonium is the ability to create a critical mass, with major releases of penetrating radiation, and the creation of a host of residual radioactive nuclides. Steps to preserve subcriticality are an essential part of the health-physics protocol.

All that has been said so far inherently equates plutonium with ^{239}Pu, the particular nuclide that is predominantly produced in the initial chain reaction of natural uranium in which the ^{235}U is the fissionable component, and ^{238}U the fertile component, producing ^{239}Pu through the intermediate step of neptunium. Especially in the last decade, long term irradiation in nuclear reactors has generated substantial quantities of the higher plutonium isotopes ^{240}Pu and ^{241}Pu[1], a battery of other transuranium elements, americium, curium, californium, berkelium, and einsteinium, as well as a particularly radioactive form of plutonium, ^{238}Pu. The latter also may be generated separately for use as a heat source, now widely used in space applications, and in coming use for implanted artificial heart engines.

It is no longer possible uniquely to define the radiation hazards of plutonium. One must be cognizant of the particular complex of plutonium and other transuranium nuclides involved in any laboratory or factory process. As long as the radionuclides are successfully isolated from the body, the main changes are in recognition of the various neutron, X-ray, and gamma ray components and, of

1 ^{241}Pu is primarily a beta-gamma emitter; there are some cases in which the controlling hazard of "plutonium" ceases to be due to the alpha-particle irradiation (see Appendix A).

course, in criticality control. Yet it is clear that current practices fall short of perfection of this isolation. When these complex mixtures are deposited, their proper measurement and comparison with acceptable standards, and application of corrective treatment procedures present new difficulties to radiation hygienists and physicists. Thus, in 1972, the challenges to health physicists in the plutonium field are greater than ever before. Moreover, they are not limited to the industrial hygiene aspects of inplant protection. Environmental releases, which have currently been reasonably well controlled, with the possible exception of spectacular incidents with military weapons and aborted space devices, can be expected to increase as the total usage of plutonium escalates, unless improved control of releases to the atmosphere and long term release of wastes to ground or water can be achieved. Contemplation of a projected 6 tons of plutonium per year going to waste storage by the year 2000 is a clear signal for the highest possible quality assurance of plutonium hazard control in the future.

II. Protection against Intake of Plutonium

A. Sealed Enclosures

Plutonium work, except at trivial levels, should be conducted in sealed enclosures such as glove boxes. By trivial levels, one means such things as routine bioassay samples which can be safely processed in conventional laboratory hoods. Hood operation with 1 mg Pu is undesirable or even intolerable. A switch at not more than 0.1 mg seems more appropriate because of the inherently greater risk of escape from a hood.

Conventional sealed enclosures will normally be of stainless steel with viewing windows of transparent plastic or glass. These windows may be permanently sealed or may better be gasketted to permit safer replacement. Materials passed into or out of the enclosure go through some form of airlock so that the interior space never communicates directly with the outside. In Europe, particularly, well-engineered proprietary ports are used. In the United States, it is common practice to use plastic bags with provision to seal an exiting contaminated part in an interior plastic bag. Similarly, in good practice, the manipulating gloves are attached to the box by flanges which themselves constitute a double barrier, so that a defective glove may be removed without exposure of the glove box interior to the room. The atmosphere of the box may be an inert gas, whenever the plutonium is to be handled in a pyrophoric state. Extreme precautions are needed to minimize explosions or fires in glove boxes as these are the readiest means of breaching enclosure integrity either to the operating room or to the environment. Generally, a glove box is continuously exhausted so that a breach of a glove or other part will provide an in-flow of not less than 150 linear feet of air per minute (Unruh, 1970). Fig. 14.1 is a schematic drawing of a typical glove type hood illustrating the arrangement of machined glove ports, the box filtration system, the double airlock and its separate air control system. Fig. 14.2 shows an actual process line used for development of plutonium oxide reactor fuel. Interesting features are the absence of clutter which almost invariably appears in standard chemical hoods. This reduces the risk of casual contamination and most importantly reduces the fire risk from spontaneous combustion, a frequent cause of which is getting both oily substances and acids on the same wiping rag or paper. In this hood can be seen a narrow necked detachable metal bottle containing a carefully size-graded special sand. This provides an excellent method of quenching a small fire in a process hood. These boxes are of circa 1960 vintage, and detailed

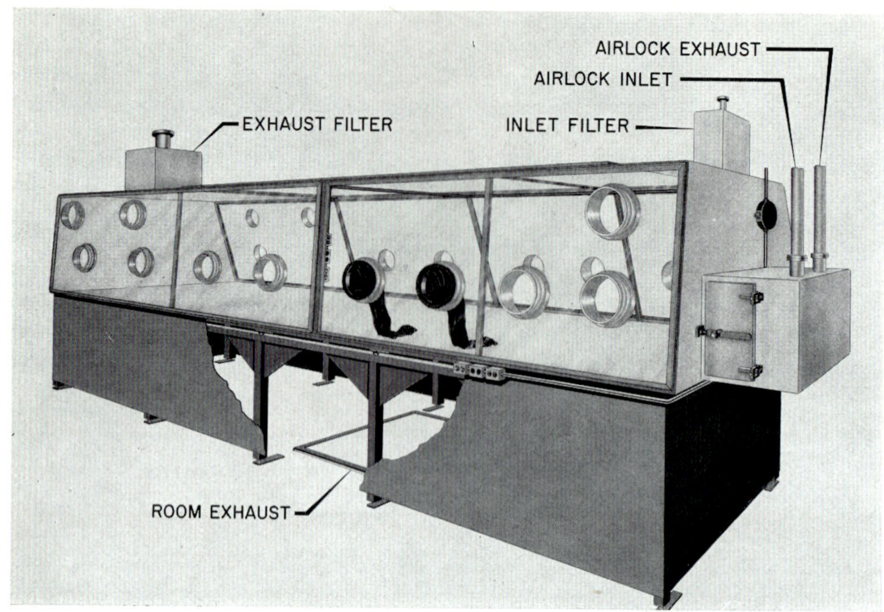

Fig. 14.1. Conventional design of process hood. Drawing of a typical glove type process hood in operations requiring protection against alpha particle contamination only. The smoothly machined circular ports permit leak-tight attachment of the long rubber gloves. The box interior will be maintained slightly below ambient air pressure by exhaust through the exhaust filter. The inlet filter prevents blowback of particulate contamination. All materials enter or leave the box through the air lock with its own air supply

improvements are being continuously made in newer installations. Some interesting examples are:

a) *Application of Human Engineering to the Spacing and Height of Glove Ports.* Fig. 14.1 and 14.2 suggest that comfortable working for long hours at these glove ports is more appropriate to the physique of a gorilla than of a man. Rather belatedly, considerable attention has been given to optimizing the placement—height, spacing, and even angling of the ports—especially where there is one designated task at an individual glove station. In addition to operator comfort, these principles contribute to safety by reducing the risk of accidents potentially leading to wounds or breakage of gloves.

b) *Replacement Absolute Filters.* The older filters were fabricated as rectangular blocks. Newer ones are cylindrical. They are gasketted into cylindrical guides so that a replacement filter ejects the old one into a plastic bag for "bagging-out" with about the same ease as replacing an injector-type razor blade.

c) *Filter Penetration Testing.* Modern boxes have installed facilities for performing the D.O.P.[2] penetration test in place. Such care eliminates gross contamination of the main exhaust ducts, which is a particular hazard if the plutonium used is in a pyrophoric form, and in all cases is a major problem if maintenance work is required on the ducts. Additionally, extra care with box filtration adds one more barrier against environmental contamination should there be trouble with the main exhaust system.

2 Di-octyl phthalate.

Fig. 14.2. Process line for plutonium fuels. An actual box used in the peparation of mixed uranium oxide-plutonium oxide fuel rods with Zircaloy cladding. The taller central section contains metal working equipment, such as a swage, for reducing the rods to a prescribed diameter. Note that the operator's gloves and any loose gloves balloon inward due to pressure differential. Other gloves are loosely tied off in pairs to prevent this, while some unused ports may be blanked off

Fig. 14.3 illustrates a permissible method of performing a once-only operation with a substantial quantity of plutonium, in this particular case the pressing and encapsulation of a ^{238}Pu source. A lucite glove box is prepared with complete sealing, conventional glove ports on the face, a bagging-out port on the left side, filtered exhaust on the top, and containing all the tools and equipment needed for the work. It is then placed in a standard laboratory hood (which would be a disaster area if used itself for this work) to utilize the needed services, such as argon for inert atmosphere. The plutonium material is bagged-in, the operations performed, and the finished piece bagged-out. The conventional hood services are crimped off and detached, maintaining the isolation. The whole lucite box (in the now disreputable condition shown in Fig. 14.3b) is enclosed in a steel box, welded tight, and dispatched to an approved storage area.

A well designed permanent box or process line has two fire-resistant so-called absolute filters in series. The design standard for first class filters calls for an efficiency of not less than 99.95% at rated flow for standard test conditions. These tests are made in situ by measuring the penetration of D.O.P. particles of prescribed diameter (e.g., 0.3 μm). The in situ test is essential because there is probably more risk of defect in the gasketting than in the fiter per se. The

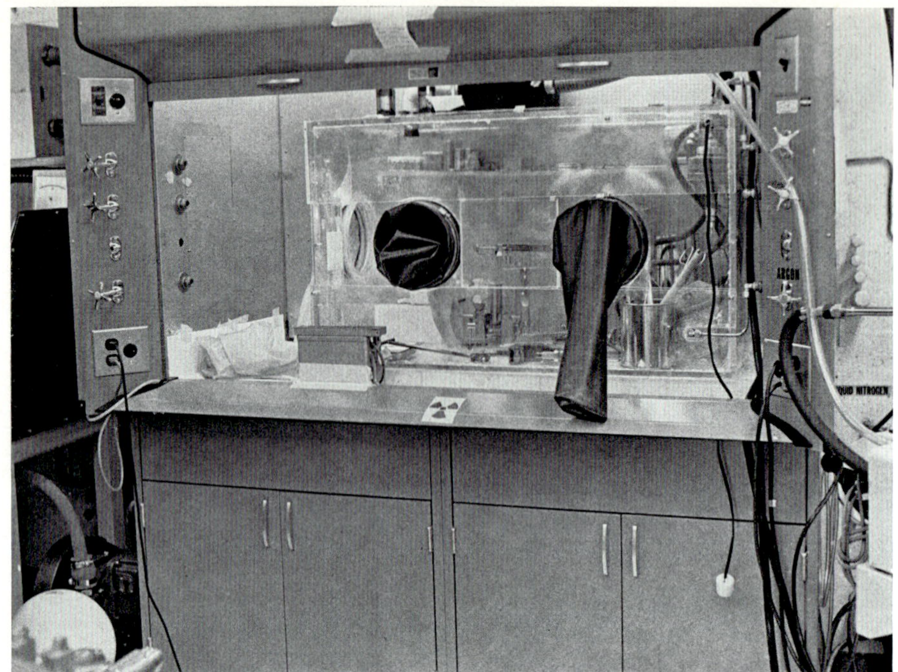

a

b

Fig. 14.3a and b. Special purpose lucite glove box. A carefully prepared sealed lucite glove box is used for special single operations not provided for in permanently installed glove type hoods. a Shows the box coupled to services in a conventional laboratory hood, and containing all the equipment for the special task. b The condition of the hood at the conclusion of the task, prior to preparation for integral removal as contaminated waste. (Courtesy: Donald W. Douglas Laboratories of the McDonnell-Douglas Corp.)

Fig. 14.4. Neutron-shielded process line. A plutonium process line with neutron shielding wall. The large viewing ports are plate glass windows with the interior space filled with water to moderate the fast neutrons from the plutonium sources. (Courtesy: Atlantic Richfield Hanford Company, Richland)

two filters in series are not necessarily claimed to achieve a compounded release as good as 0.05% of the 0.05% that penetrates the first filter. The aerosol or dust that impinges on the first filter has a spectrum of particle sizes. Those that successfully penetrate the first filter will have a new spectrum less easily removed by the second filter. Rather, the intent is to have a backup filter against the failure of the first one or its gasket mounting to the exhaust duct framework. In all major operations, an air monitor is installed after the second filter to provide a signal should both filters fail. A continuous air sample is drawn from between the two to provide warning of first filter defect. It is these practices which have led to the relatively good experience in environmental releases.

Fig. 14.4 shows a modern automated or semi-automated plutonium process line where the composition and quantity of plutonium require neutron protection for the operators. This is achieved by the use of water walls maintained in place by plate glass sheets. The walls shown are 12 inches thick. Water is effective by slowing down the fast neutrons. In some cases, soluble boron compounds are added to absorb the slow neutrons with minimal production of penetrating radiation.

It will be recognized that a 12-inch wall on a glove box would have the unwieldy effect of shortening the arms by 12 inches. In the example illustrated, the shielding wall is separated from the process line which has glove ports only for maintenance and special adjustments at times when the plutonium load is low, which, together with short exposure time in the maintenance zone, adequately limits the neutron exposure.

The complete filtration, monitoring, and air exhaust system of a plutonium plant constitutes a major integral portion of a process facility. Even in a small

Fig. 14.5. Air exhaust and filtration system of a small licensed plutonium facility. Shown during installation is a bank of eleven "absolute filters" each $2' \times 2' \times 1'$, with a total air flow of 8800 cfm. Each aperture is sealed by a plastic bag held by the gasket-lined covers with four corner screw-downs. Not installed at this stage were small bleed-off lines for continuous proportional sampling of the air contamination. (Courtesy: Donald W. Douglas Laboratories of the McDonnell-Douglas Corp.)

licensed facility the apparatus is substantial, as shown in Fig. 14.5. The dimensions can be inferred by considering that each of the eleven TV screen-shaped ports for the absolute filters measures about $30'' \times 20''$.

The almost universal practice of manipulating through rubber gloves developed in the early days when small quantities of ^{239}Pu were involved. Today, there are two sound reasons why it would be preferable to use mechanical manipulators:

1. Large masses of plutonium, especially those rich in the higher isotopes, lead to undesirable or directly damaging contact irradiation levels.

2. A substantial fraction of operator exposures occurs either through puncture wounds with the hands in the gloves or through room contamination following defects in the gloves (Fig. 14.6).

a

b

Fig. 14.6a and b. The two main causes of personnel contamination in a glove line. a A small plutonium-contaminated wire has penetrated the thick glove of the box, the thin surgical glove on the operator's hand, and the operator's finger. Essentially no contaminated air escapes, and the pin prick wound may be so small as to escape detection visually or by search for surface contamination on the finger. Yet the imbedded amount may require surgical removal. b A typical cause of room air contamination is a small glove rupture as in this glove held here for demonstration by an operator using other glove ports

The appeal of direct manual dexterity is such that a change to manipulators may be difficult to promote. Yet the experience to date, and the expected proliferation of plutonium operations related to advanced reactor fuel preparation and other tasks calls for its serious consideration.

In processing factories, design engineers prefer to develop their automatic or semi-automatic systems in long lines. Health physics pressure should be to have each work station a separate enclosure connected sequentially through air locks incorporating fire doors. While the continuous system is cheaper and simpler to operate, the penalty in the event of a fire transmitted along a line is too high a price to pay, both for internal cleanup and for potential environmental release.

B. Control in Other Working Spaces

When it is conceded that sealed enclosures are imperfect, it becomes imperative to provide secondary contamination protection throughout the operating buildings. Conventional aspects of this are:

1. Protective Clothing for the Operator

This includes rubber gloves, coveralls, work shoes or shoe covers, and ready access to full face air masks in the event of detected air contamination.

2. Continuous Air Monitoring in the Process Rooms

The customary method of measuring air concentration is to pull air preferably through a thin filter (such as a molecular pore filter) at a known rate for a known time. The alpha activity of the material on the filter face is then measured after the decay of the radon and thoron daughter products which represent a naturally occurring interference. Where a prompt warning is required there are two options:

a) Alarm response at a relatively high concentration level, sufficient to stand out above the interfering radon and thoron daughters.

b) Continuous compensation for the daughter products by reading off a beta-alpha coincidence signal from the daughters, or by a variety of other methods of compensation.

One British method uses the β-α coincidence correction, while another uses the specific signal in a CsI crystal from the 5.1 MeV alpha particle of plutonium, compensated for the degraded alpha particles above 5.1 MeV that are deduced from the total activity of alpha particles above 6 MeV originating in the radium and thorium series (IREDALE and HINDER, 1962). A French system distinguishes dusts by granulometric differences with selective filters in series (BILLARD et al., 1961).

Solid state detection is becoming increasingly effective and popular (LINDEKEN and LAKIN, 1967). A sound approach is to combine a sensitive recorder, even if its accuracy is somewhat variable, as the alarm meter, with a reliable non-alarm meter to provide a permanent record of air contamination. Electrostatic deposition of the room aerosols on aluminum plates is one method of achieving the second objective (WINKLER et al., 1969).

When contamination occurs in a process room, it is likely to be highly uneven. Opinion is sharply divided on the relative merits of having an adequate number of detection devices coupled to a central exhaust system (POMAROLA et al., 1966), where better electronics and system maintenance can be achieved, at the expense of possibly measuring at the wrong points versus the alternative of having a simple portable system on each operator with the receptor on the forehead or other point as close to the nose and mouth as possible. The data on the relatively high frequency of inhalation events would suggest that the controversy is best resolved by employing both methods. This would indeed have considerable value in contamination case follow-up if the actual intake could be determined by the monitor on the person. This is not so because of individual variations in breathing pattern. The specific respiratory cycle, the average volume intake, the nature of the inhaled aerosol or dust, and the proportion of nose and mouth breathing all affect the deposition.

A desirable feature practised in most of the principal installations is to measure the characteristics of the radioactive particles in process rooms that experience plutonium contamination. This provides an insight into the percentage of any accidental air contamination that might present a respirable hazard, by reference to standard industrial hygiene surveys of intake and retention of particles of various aerodynamic diameters. Such data are especially useful if the spectrum of sizes is shown to be fairly consistent. In the event of an accidental inhalation case, the size distribution of particles on a nose smear can perhaps be compared with the known ambient size spectrum to improve the prompt estimate of amount inhaled. The most dangerous atmosphere is one in which pure plutonium particles (often as oxide) concentrate in the size range of optimum (or more correctly "maximum" because it is the reverse of optimum for safety) respirability. Paradoxically, the safest atmosphere is an industrially dirty one in which much of the plutonium may attach to larger inert particles which do not penetrate to the lung. One is not aware of conscious attempts to promote such an atmosphere, the effort being better applied to minimizing the primary contamination.

The literature is rich in measurements of room air contamination. It is not referenced here because the recommended course for those needing it is to refer to practices in their part of the world.

3. Change Rooms

Incoming operators change into the complete protective apparel. On leaving, they discard the work clothing according to a step-off pad ritual designed to minimize accidental cross-contamination. The hands and feet are monitored for plutonium contamination; the worker may be detained for on-the-spot decontamination should the readings on these surfaces exceed prescribed levels.

4. Zoned Air Change

A process building is compartmented into zones of relative cleanliness, with reduced air pressure in the direction of higher contamination potential. The conventional plan is:

Zone 1. Offices, lunchrooms, and the "clean" side of the change rooms.

Zone 2. Corridors between change rooms and laboratory or process rooms.

A double barrier air lock is provided between Zones 1 and 2. Conceptually, the transition between Zone 1 and Zone 2 is at the point of removal of the operators' protective clothing. In practice it is at the air lock, with the clothing removal ritual used to prevent significant contamination of Zone 1. Zone 2 is normally essentially uncontaminated.

Zone 3. Laboratory and process rooms.

Zone 4. The interior of process lines and glove boxes.

In smaller facilities, there is little or no differentiation between Zone 2 and 3. The essentials of good control are to maintain the integrity of the Zone 4 to Zone 3 separation so that Zone 3 is clean except for rare and promptly recognizable accidents, and to maintain high standards in the change room procedure so that the complete cleanliness of Zone 1 is assured. A current weakness of some designs is that it is necessary to provide emergency exits from Zone 2 or even Zone 3 to the outside in the event immediate building evacuation is needed. Although these exits are normally sealed with breakable tape, it is superior practice to have each one built as a double barrier air lock.

5. Maintenance Access

Experience throughout the nuclear industry has shown that the greatest risk to operating personnel frequently arises during maintenance operations. At the same time, the risk of spreading contamination to others tends to be maximized. This is certainly true in a plutonium processing operation. Maintenance personnel should have complete protective covering, including a mask either with supplied air or with self-generating air supply. In principle, the maintenance worker is *inside* the process line or glove box; his personal protective layer should provide isolation from plutonium as nearly equivalent as possible to that of a closed glove box. In some countries, the process equipment is built in the form of a rectangle with the interior forming a sort of Zone 4 A accessible only to maintenance workers clothed in the equivalent of a "frogman" suit. This type of isolation of prescribed maintenance zones seems to be a highly effective system.

6. Personnel Monitoring Stations

A major station is maintained at the change room interface. In addition, good practice calls for installed alpha-monitors at the door of each process room, used to check the hands and feet *every time* a man leaves a Zone 3. Individual monitors at the actual work station have been found useful. Periodic checks by the operator lead to prompt recognition of minor contamination usually long before it achieves a level sufficient to trigger the air monitoring devices.

Historically, the alpha sensitive surface monitors were troublesome because they required a very thin and hence fragile window over either an ion chamber or proportional counter. Development of devices such as scintillation counters has provided dependable instrumentation at reasonable cost. Here, the thin lightproof film (aluminized Mylar, for example) can be applied directly to the flat solid scintillator face. It is standard practice at each fixed station monitor to provide high- and low-activity alpha particle sources to test the proper functioning of the equipment before each test.

7. Material Transfers

All plutonium bearing material removed from process lines of glove boxes is removed through the "bagging" or equivalent devices enclosed in two separately sealed containers. The double wrapping ensures that at least the outer surface is contamination-free. Special precautions are sometimes needed to prevent damage to the package by self-heating of the sources or by radiolytic decomposition. This is especially the case if separated highly radioactive transuranium elements are involved.

Storage of removed products in special vaults is required, with provision for detection of leaks and for decontamination if necessary.

8. Waste Handling

The valuable plutonium materials for nuclear fuels, military parts, space or biomedical heat engines, or whatever, naturally get careful attention and control. It is a disconcerting fact that the small percentage that degenerates into some form of waste requires equal or even greater control with respect to contamination.

Liquid wastes are routed to prescribed storage tanks. Solid wastes such as metal chips or powder residues are collected in sealed containers and either stored or prepared for recovery. Nuisance wastes, while lower in activity, probably cause more trouble. They include large volumes of paper wipes and the like used for cleanup inside glove boxes. These are either stored in sealed containers in "permanent" repositories or further treated. Spontaneous combustion of some types of waste is a threat that must be overcome.

A principal problem is the projection of these storage demands into the future, with a proliferating nuclear industry. With material of half-life about 24000 years for ^{239}Pu, and necessary storage time of hundreds of thousands of years, "permanence" of a surface repository creates a legitimate question in the public mind. In the United States at the present time, serious consideration is being given to storage in deep underground structures such as salt mines. European practice is in some respects ahead in this field. Compression of the solid wastes to reduce volume is a first step. A desirable first step is the separation of wastes into two categories—combustible and noncombustible. It is the latter that are then profitably compressed in a baling press. A volume reduction by a factor of 2 to 3 is conventional unless very powerful presses are used; a volume reduction by a factor of 5 would be considered good. More recently, successful incineration of the combustible plutonium wastes has been demonstrated. It is anticipated that this approach will become widespread, as it already is for conventional fission product wastes. French practice is probably the most advanced in this area. For example, at the Marcoule plant of the C.E.A. the main production incinerator fires 900 kg of waste per hour at 900° C, with an afterburner at 1200° C. The off-gases are cleaned in three stages, and the incinerated product cooled and compacted in standard steel drums. The drum is filled up with loose gravel, sealed and taken to storage. By contrast, the incinerator for recovery of plutonium from combustible solids copes with only 1 kg per hour although it is nearly half the size of the main incinerator, reflecting the extra care needed in the plutonium case. Considerable attention has been given to the risk of fire in the loading stages, which are done in a nitrogen atmosphere. Most of the plutonium waste arrives packaged in polyvinyl chloride bags or sealed sheets for contamination control. Burning of such material releases chlorides which can adversely affect metal parts. The Marcoule installation is designed to cope with this hazard. Criticality is a serious risk in such an incinerator. It is controlled by completely emptying the system after each 300 g Pu have been treated. Accurate knowledge of the waste content of each incoming package is hard to obtain. Here it is monitored by its weak photonic radiations between 370 keV and 440 keV, with results from standard size packages good to about $\pm 30\%$. Volume reduction to 3% of the original is achieved, making chemical recovery practical. As an example of the care needed in such work, the cooling water around the final parts of the equipment is loaded with 20 g boron per liter as salts to suppress the criticality enhancement of the water blanket.

C. Control of Off-Standard Conditions

A natural order of graded response to the degree of contamination develops. These details need not be considered here. The essential feature of control is that any demonstrable level of plutonium contamination requires investigation of its origin and appropriate clean-up action.

D. Major Accidental Releases

The required action is evacuation of the operators by the best means available to them to minimize spread of contamination to other work areas or the environment. Damage control is performed by fully protected personnel (as for maintenance). The often long and laborious rehabilitation is similarly attacked.

E. Incidental Releases

Whenever plutonium contamination of the process room is detected, work is interrupted, and the affected area defined by surface monitoring equipment. For all except the most trivial cases, the operators use air masks at this stage[3]. Cleanup is begun at the periphery, with simple scrubbing with soap and water often being effective. Walls and floors of the process rooms are usually finished in special paints or other surfaces to facilitate this. Stainless steel and other parts may be treated with stronger reagents. However, the cleanup is notably different from "mopping the kitchen floor". Great care is needed to prevent further spreading of contamination. As each area is cleaned, it is covered with paper to minimize its recontamination. Also, the mopping liquid and tools must be collected and sent to plutonium waste storage.

The customary detection limit for contamination is about 500 alpha-disintegrations per minute over a 100 cm^2 area. Tested surfaces must be dry, for a water layer would itself be thick enough to absorb the alpha particles. Speedier identification of contamination can be done by meters sensitive to the emitted X-rays or gamma-rays, but the sensitivity is usually not competitive with the alpha-monitor. It is also clear that safe decontamination of a porous surface is virtually impossible, because the activity would be shielded from the alpha-monitor. It is a salutory reflection that a work room measuring $4\,m \times 4\,m \times 3\,m$, if uniformly contaminated, could be left with 8 μCi Pu if all surfaces were left contaminated at levels just below the detection limit. Should the material subsequently rub off and become airborne, one-five-hundredth of this amount would constitute a maximum permissible lung burden. For such reasons, it is common practice to paint the decontaminated surfaces with a sealing paint. All such surfaces subsequently require some degree of "perpetual care". When the painted surface is outside a process room (i.e., where there is normally no contamination), the sealant should be a bright yellow paint or an equivalent warning signal. British work on the physics of resuspension of contamination, and on the related hazards, has been notably thorough. Contributions of H. J. Dunster, I. S. Jones and S. F. Pond, K. Stewart, and others, with further references to other work is conveniently grouped in a Gatlinburg symposium report (Fish, 1967).

III. Contamination of Personnel

Comprehensive reviews of the management of plutonium contaminated personnel are available from many sources.

One systematic approach to decision levels on surface contamination includes the importance of resuspension and such matters as contamination of home and

3 Installations often have established limits of air contamination up to which a full face fitting mask is deemed adequate, whereas at higher levels, a mask with supplied air (piped in or chemically generated) is mandatory. The switch point is usually 10 or 20 MPC_a.

environment from residual contamination of workers (HEALY, 1971). Each proposed limit has a symbol identifying the mechanism leading to the lowest value according to the code

A — skin absorption,
H — inhalation from clothing or skin,
R — resuspension,

with other codes for gamma emitters. For the alpha-emitting plutonium isotopes, the signal is H for contamination of skin or clothing, and often R for transfers to home. Interestingly, the signals for ^{241}Am and all the higher transuranium elements switch to A for skin and transfers, remaining H for worker clothing. To some extent, this is based on the known eagerness of plutonium to form complexes, compared with americium and higher transplutonium elements. HEALY (private communication) reports that the switch is perhaps an artifact due to current paucity of knowledge about the behavior of the higher elements. His decision levels are meant to be a basis for organized discussion of realistic action levels.

Here, the highlights of systems will be described. This account will relate primarily to the Hanford system for convenience, with no implication that other systems are not equally useful. In particular, the British writings on the topic are universally highly regarded. As an example, a report of specific considerations of protection against airborne particulates by HOLLIDAY et al. (1969, 1970) is a model of conciseness and clarity. Indeed, there is no major group that has engaged in plutonium handling that has not made some contribution worthy of adoption by the others, and of eventual incorporation into an international protocol. Since the best results in this phase of industrial hygiene often hinge on apparently minor details, it is recommended that newcomers to the field both review the literature in depth and personally inspect operating methods at a plutonium site.

A. Skin Contamination

The extent of the contaminated area is defined and surrounding areas protected from contamination spread. Loose material responds to removal with gauze and mild detergents. More obstinate cases are then scrubbed with soap and water, including quite abrasive soaps if necessary. Recalcitrant cases are treated chemically. One scheme uses a 10% solution of the chelating agent Na_4 EDTA (ethylene diamine tetraacetic acid). This is contraindicated if there is a break in the skin in the affected part (WILSON and SILKER, 1960). The customary last resort is the use of saturated potassium permanganate solution followed by a 4% solution of sodium bisulfite. In effect, this chemically removes the outer layer of skin. An alternative method is to apply a sticky tape or a plastic spray-on film, which upon removal brings some contamination (and possibly superficial skin debris) with it. This dry method has the merit of reducing the volume of contaminated wastes.

The simpler steps are commonly accepted as non-medical treatments. Persistent cases with abrasion of substantial skin areas are preferably regarded as medical cases (JECH et al., 1969a). National customs vary widely in this respect; health physicists who go beyond the accepted boundaries acquire dangerous liability.

B. Eye, Nose, or Mouth Contamination

Irrigation with water or a normal saline solution is the usual limit of decontamination.

C. Wounds

For this purpose, in addition to major wounds, cuts, punctures and severe abrasion are included in the category. The conventional surface counter is no longer effective for detection of the source. Instead, a wound counter consisting of a thin slice of sodium iodide crystal with suitable electronics as a scintillation counter is used (ROESCH and BAUM, 1958). The chosen slice thickness, usually 1 mm, is an optimizing compromise between absorbing the low energy L X-rays of ^{239}Pu (13.6, 17.2, and 20.2 keV) to give a strong signal, and rejecting as much background signal from external natural radiation as possible by minimizing the crystal mass. The minimum detection amount for ^{239}Pu in a 10-minute counting period is about 0.1 nCi. For more complex mixtures of plutonium it is more realistic to claim a limit of 0.1 nCi to 0.6 nCi depending on isotope mix and depth of deposition. Precise location of the imbedded contamination can be achieved (except as to depth) by using a thin metal diaphragm as a collimator; the signal from the soft X-ray occurs only when the hole is over the plutonium. Reasonably fast and accurate localization contributes materially to the physician's confidence when he is called on to make excisions close to a part that would damage finger dexterity, for example.

There are two reasons for gaining the highest possible removal from a wound. There may be a possibility of radiation damage at the site, with possible malignancy. Alternatively, residual material may be transmitted to the bone, probably the more cogent of the two reasons. There is a reluctance to leave at the wound site an appreciable fraction of a potentially transmissible maximum permissible body burden.

For related reasons, there is a continuing effort to improve the sensitivity of detection of contamination in wounds and their speedy location. In one method employed by WRIGHT LANGHAM et al., at Los Alamos, a dual scintillation counter circuit is employed. A thin sodium iodide crystal to read the L X-ray signal is backed by a thick cesium iodide crystal which accurately reads the ambient radiation background signal and subtracts the appropriate background from the thin crystal signal.

In one development, a lithium-drifted silicon detector is used to measure the depth of a contamination source (SWINTH, 1968, PALMER et al., 1968). Depth discrimination is possible because a silicon or germanium detector can resolve the three L X-ray lines at 13.6, 17.2, and 20.2 keV. The relative intensity of the lines varies with source depth. For a point source, discrimination can be achieved to within ± 0.2 mm. The technique is ideal where an identifiable plutonium piece has been imbedded. For routine use it suffers two disadvantages:

a) Discrimination is gained at the expense of sensitivity.

b) Puncture wounds typically have contamination along the whole track.

In another application at Hanford, the hand is examined by 26 thin slice NaI crystals above and below the hand. The sensitivity of this version is between 0.6 nCi and 1.2 nCi plutonium depending on isotope mix, average depth, and dispersion of the deposition. In a series of 200 routine examinations, no previously undiscovered depositions were found, whereas all known depositions, run as a blind test, were correctly identified (ANDERSEN, 1972).

D. Lung Contamination

Modern technology leads to more and more plutonium work in powder form, frequently as an insoluble oxide. This increases the inherent risk of inhalation and of the deposition of insoluble particles in the lung. The biological evidence of the long retention time of such particles, the experimental production of neoplasms of the animal lung, the notable concentration of the particles in lymph nodes, and the difficulty of removing known deposits, all lead to some concern about the reliability of present maximum permissible lung burdens. There is no health physics area in which full attention to the concept of lowest practical exposure is more relevant.

This concern is matched by continuing improvements in the early detection of airborne contamination, improvements in means of measuring lung burdens more accurately, and improvements in the very limited steps to effect removal of lung burdens.

E. Detection of Airborne Contamination

The use of continuously operating air samples has been referred to in Sec. II.B.2. Within the last decade, the sensitivity of such devices has increased by about one order of magnitude. Unless there is a breakthrough in technology, further gains will be increasingly difficult. Good recognition at the present time involves one accepted maximum permissible concentration (MPC_a) in one hour, when there are no interferences from conflicting radionuclides other than the natural radon and thoron daughters. Since the occupational limits are predicated on a 40-hour week, there is some safety margin here. Yet this is offset by the risk of appreciably higher concentration at the breathing zone. As the detection limit is driven downwards, reliable measurement is reduced because of the random nature of the particle size distribution. Circumstances are reported in which it takes 2000 m^3 of air to get a representative sample accurate to 0.1 MPC_a (SHERWOOD and STEVENS, 1965). Man himself, breathing 40 to 50 m^3 in a workweek, is not a representative sampler at that concentration level. The sample activity may be almost totally contained on one single active particle, and it may be such a particle that the operator inhales.

1. Detection of Contamination by Nasal Smears

Whenever a high air concentration is observed, or if there is any possibility of leakage from a ruptured glove box, nasal smear samples should be taken. With commonly employed materials such as plutonium oxide powder, there is believed to be a reasonable correlation between the nasal smear and the lung deposition. In 30 oxide inhalation cases, the following relationships, each good within a factor of ± 5 were found (HEID and JECH, 1969):

Burden per 1 nCi nasal smear	
Initial pulmonary	1 nCi
Pulmonary at 7 days	0.6 nCi
Initial lung	3 nCi
Ultimate systemic	0.1 nCi

With other aerosols, there is no assurance that such relationships apply. In fact, with very fine particles there seems to be a finite possibility of lung deposition without a nasal smear signal.

2. Detection of Lung Burdens by External Means

The whole body counter is a versatile tool for the detection of various radionuclide burdens in the body. It is not at its best when applied to plutonium.

The characteristic X-rays of ^{239}Pu, used as signal generator in the wound and hand counters, are ineffective here because of their poor penetration of the chest walls (Waite et al., 1970). Reliance is placed on measuring the 60 keV radiation of ^{241}Am, present as a daughter product of ^{241}Pu. Conventional counters for this purpose employ thin NaI crystals of large cross-section (Lidén and McCall, 1962). A double crystal system using a thin CsI crystal for the soft radiation and a thicker NaI in anticoincidence improves the background by 60% permitting measurement without a steel shielding room (Laurer and Eisenbud, 1969). Several other systems are in the development stage; for example, at Los Alamos. An alternative detector, the Maushart proportional counter, has enjoyed considerable popularity in European practice (Kiefer and Maushart, 1962). The Hanford crystal counter employs four crystals approximately 9.5 mm thick and 127 mm diameter. It provides a detection capability of 0.15 nCi to 0.6 nCi ^{241}Am, depending on chest wall thickness. In the absence of more specific data, it is provisionally assumed that the plutonium burden is 15 times this. However, the true burden is a function of both the plutonium isotopic mix and the age of the sample since separation and in-growth of the daughter ^{241}Am. In a plant or laboratory using a constant source of plutonium, it is desirable to reserve a sample for actual measurement of these relative amounts at any subsequent time. Failing that, the activity in the nose smear may be enough to permit analysis. By coupling these data with phantom measurements simulating the build of the contaminated individual, amounts comparable with the maximum permissible lung burden or better are measurable. This at least provides a reasonable signal for attempted therapy of substantial depositions. One would like to have a system that reveals chronic buildup of minor depositions at the level of, say, 5% of the MPLB. This is impossible with present sensitivities. In any case, over time, the chronic burden would probably involve an aggregate of many plutonium mixes, with various unknown ages of americium growth. In addition, there is the possibility of biological separation of americium from plutonium in the lung.

3. Detection in Feces

An inhalation event will normally be accompanied by deposition in the upper respiratory passages, part of which by ciliary action will be transferred to the digestive tract. Most of this material will appear in the feces, especially for relatively insoluble material such as the oxide. Even for soluble materials, the fecal elimination will be initially some six times higher in feces than urine. After an inhalation incident, fecal samples are usually analyzed; if only one is done, the second-day sample is expected to give the peak value (Lister, 1966, 1968). A representative schedule for feces sampling at Hanford is as follows (Heid and Jech, 1968; Heid et al., 1968):

Schedule 1. Obtain five daily fecal samples within the first seven days post intake.

If any of the following circumstances arise:

A. Nasal smears exceed 500 d/m.
B. Nasal smears $>100<500$ d/m and exposure duration >5 min.
C. Nasal smears $>5<100$ d/m and exposure duration >50 min.
D. Exposure to fumes from a fire.
E. Air sample results exceed 2×10^{-10} μCi/cm^3 for an 8-hour period and exposure duration >1 hour.

Schedule 2. Obtain one fecal sample on the second day post intake.

If any of the following circumstances arise:

A. Nasal smears are positive but do not meet criteria in Schedule 1.
B. Any other person in the same incident meets the criteria in Schedule 1.
C. Air sample results exceed 2×10^{-11} μCi/cm³ for an 8-hour period.
D. Widespread skin contamination in a dry form or facial contamination >1000 d/m.
E. Clothing contamination in a dry form >5000 d/m.
F. Possible plutonium inhalation is suspected for other reasons.

In Schedule 1, a correspondence, again within a factor of ± 5, is claimed between 5 times the nasal smear activity and the integrated fecal excretion in the first five days post intake, but some non-conforming cases are known. Nevertheless, the combination of nasal smear, lung count, and fecal examination does seem to provide a reasonable basis for early identification of substantial lung burdens that would prompt removal attempts.

4. Detection in Urine[4]

Internal contamination by plutonium at the levels of interest cannot be detected by the weak photonic irradiation emitted, in most cases. Thus, the initial health physics effort when plutonium was first used concentrated on means of estimating body deposition from radiochemical analysis of available excreta. Parallel work with animals at that time was exclusively with soluble plutonium compounds, and it became clear that the activity of urine provided a reasonable basis for detection of body burdens. The program was placed on a sound quantitative basis following the work of LANGHAM (1956), who introduced known amounts of plutonium into a number of selected people with terminal illnesses, judged not likely to materially affect the normal excretion rate. LANGHAM also measured fecal elimination rate. Although the fecal signal is stronger in the first few days after intake, the urine signal is thereafter higher, and in any case tends to be more consistent. Combining this with the greater acceptability of urine sampling by both worker and analytical technicians, urine analysis became the universal method of choice. With the later growing use of plutonium in oxide form and the prevalence of intake by the inhalation route, the relative significance of fecal sampling has perhaps steadily increased. Yet urine analysis remains as the major tool for routine detection and attempted quantification of all "systemic burdens" of plutonium; in fact, insoluble plutonium deposits in lung and lymph nodes are arbitrarily excluded from the terminology of systemic burden, and are sought either by direct measurements of emitted photonic radiation or deduced from pertubations of the estimated systemic burden over time. Regularly scheduled urine analyses have successfully pointed to increments of systemic burden from transfer from both lung and contaminated wound sites (see Appendix A). These observations indirectly support the interpretability of urine samples for standard cases not obfuscated in part by such long-term accretions from specific compartments.

Routine analysis for plutonium in urine has thus become standard practice in all major plutonium operations. Several methods are available of more or less comparable effectiveness. The particular method used at Hanford is briefly as follows:

Home collection: Voidings one-half hour before retiring and one-half hour after rising on two successive days closely approximate a total 24-hour sample, and minimize inconvenience to the workers.

4 The systematic ordering of this chapter would have called for notes on urine analysis after 3 or 4 of the previous sections. Instead, they have been grouped together here.

Analytical method: Chemical extraction with TTA (thenoyl trifluoracetone) followed by electrodeposition on a stainless steel plate is used. The plutonium activity is measured by counting alpha particle tracks on nuclear track emulsion (SCHWENDIMAN et al., 1951).

Sensitivity: The sample counting detection limit is 0.05 dis/min. For soluble plutonium in a single intake the sample activity can be related to the initial intake by the now classical LANGHAM equation (LANGHAM, 1956).

$$\frac{\text{Daily excretion rate}}{\text{Initial intake}} = 0.002\, t^{-0.74}$$

where t is the time in days after intake. Thus the sample sensitivity converts to a sliding scale of inferred body burden. For ^{239}Pu only, the maximum permissible body burden (MPBB) is taken as 0.04 μCi. Detection levels become:

Time post single intake	Fraction of MPBB
30 days	0.005
90 days	0.01
1 year	0.03
5 years	0.08

For known accident cases, a single voiding, say, 2 hours after the event provides an adequate preliminary sample analyzed by less efficient rush procedures to signal the need for removal therapy for gross intakes.

In the more general case of chronic minor intakes over a working career it is impossible to convert the bio-assay reading to known values of MPBB, because the excretion rate becomes the sum of a series of power functions of time, where each time is unknown. However, an upper limit can be set by assuming that each incremental intake occurred the day after the previous sample was taken. On this basis, routine schedules for frequency of bio-assay of personnel, depending on the anticipated severity of their chronic intakes, can be set. A frequency of sampling scaled to detect 1% of the MPBB is recommended (WILSON, 1967). This equates to a monthly or quarterly schedule in the regular plutonium plant staff. Some installations plan schedules responding only to 5% MPBB or more. An annual sample for occasional workers gives 3% MPBB under the upper limit assumption. Those with access to a plutonium facility but never engaged in plutonium work should prudently be sampled at 5 year intervals. The pessimistic limit for that group is 8% MPBB and the more probable limit (based on median time exposure) is 6% MPBB. Naturally, these long-interval samples are expected to show no contamination. The complete overview of bioassay results from all interval classes from one month to five years does provide one of the better indices of the effectiveness of the defenses against plutonium contamination (WILSON, 1967).

The British group has challenged the rationale of a comprehensive program of this type, "provided that the environmental control and monitoring are adequate" (BEACH et al., 1966). Depending on the precise interpretation given to the word "adequate" one can say that this begs the question; obviously a no-defect containment system would lead to zero deposits. More generally, their arguments merit consideration because the only time at which treatment has worthwhile effect is shortly after the intake. This group recommends five base-line samples before commencing routine plutonium work and a series of five samples at 6-month intervals. This indeed tends to eliminate panic reaction if a single result on a regular sampling program throws a high reading due to cross-contamination. Clearly the total sampling programs under the two regimes are about the same size. Adherents of the monthly or bimonthly mode expect that trends indicating the beginning of off-standard conditions that escape the monitoring network will be suggested earlier in their system. Demonstration of this in a given system would require statistical analysis of many sample results. The British attitude, while not explicitly based on the assumption that exposures are nowadays predominantly due to inhalation of insoluble oxides, seems to lean strongly that way. For such cases, in which the urine signal is very slowly accumulated and probably not well formulated in models, their points are very well taken. BEACH and DOLPHIN (1964) have given revised formulae for fractional excretion of soluble plutonium compounds incorporating a more rapid component for the first few days. These are:

$$\text{Urine:}\quad \frac{\text{Daily excretion}}{\text{Initial intake}} = 0.0041\ \mathrm{e}^{-0.67\,\mathrm{T}} + 0.0016\ \mathrm{T}^{-0.68}$$

$$\text{Feces:}\quad \frac{\text{Daily excretion}}{\text{Initial intake}} = 0.004\ \mathrm{e}^{-0.46\,\mathrm{T}} + 0.004\ \mathrm{T}^{-1}$$

where T is the elapsed time in days. These equations have been referenced by the ICRP Report of Committee IV (1968), although it is not clear whether they wholly endorse them as a substitute for the LANGHAM equations. BEACH and DOLPHIN (1964) also presented an expression for urinary excretion which includes a factor for slow mobilization of plutonium initially deposited in the lung, as follows:

Urinary excretion rate $= Q(b-1)\ 0.0016 \int_1^T \tau^{-b}(\tau - T)^{0.68}\, d\tau$ where Q (in μCi) is the amount of inhaled plutonium eventually transferred across lung epithelium. The power function τ^{-b} is perhaps an improvement on the Healy method of the next paragraph. b was arbitrarily taken as 1.177, corresponding with 50% transfer from the lung in 25 days, 75% in $3^1/_2$ years, and 84% in 50 years.

When the intake, either acute, accidental or chronic, involves relatively insoluble material, the Langham equations break down. HEALY (1957) has presented a mathematical compartment model for the case of lung deposition with consideration of the subsequent time variation of activity in urine, feces, and blood, all of which can be measured to get a more or less empirical fit to the data. The Healy model is based on an assumed constant elimination rate from the supply pool. More recently, HEID et al. (1970a and b), have applied the ICRP lung model, including some recent tentative modifications (ICRP, 1971) to incorporate multiple mobilization half-lives up to as long as 10000 days. Inferred lung burden is still probably not highly reliable, but it is the best that can be done at levels below the sensitivity of the lung counter. The present state of knowledge does not indicate a need for treatment of these cases. Rather, they comprise an information bank for eventual determination of the true deposition at autopsy, and for a concurrent search for possible malignant changes or precursor changes through the mechanism of the U.S. Transuranium Registry (see Sec. V, p. 635).

The uncertainties of measurement of lung body burden are well recognized by those active in the field. The perturbation of bioassay results when DTPA has been administered is well known. A comparison has been made between the interpretations of three major U. S. locations—Los Alamos, Oak Ridge, and Hanford—based on six well-documented Hanford cases (BRAMSON et al., 1972).

Los Alamos uses a computer code (Lawrence, 1962) and Oak Ridge a more complex code accounting for statistical variations in following the Langham equations (Snyder et al., 1967); Hanford still uses manual calculation. In general, the Los Alamos and Hanford interpretations are in close agreement. The authors wisely point out that this is no proof that they are superior to the Oak Ridge results. Some of the points may not be resolved until enough cases come to autopsy through the Transuranium Registry. These intercomparisons, in any case, relate only to data manipulation. More realistic intercomparisons are being organized to test lung counter signals. These will include the circulation of loaded chest phantoms, and ultimately the circulation of contaminated individuals, if possible. With no reflection on the three laboratories cooperating above, there are locations where the reports of sensitivity and detection limits seem to be a product of local enthusiasm. Considerable effort can be wasted at other locations by prematurely installing a new system that will not match the originator's claim. Cross comparison is desirable not just for lung counters, but for all systems used in plutonium hazard evaluation, where the needed results press the limits of resolution.

IV. Treatment of Plutonium Contamination Cases

The field of health physics stops at the boundary at which information on the nature and extent of body depositions is transmitted to the physician to permit an informed basis for his actions. Current approaches will be superficially included here. A comprehensive summary of the state of knowledge to 1967, although naturally weighted heavily by U.S. contributions, can be found in the Proceedings of a Symposium on Diagnosis and Treatment of Deposited Radionuclides held at Richland, Wash., the Hanford atomic site (Kornberg and Norwood, 1968)[5]. There is no agreed protocol. In fact, it is proper to leave an area for the physician's judgment and it is socially desirable in some borderline cases for the patient himself to participate in a decision which balances some present impairment and discomfort against a very remote risk of future radiation injury.

As a guide, for wound cases one finds a readiness to use DTPA locally to remove contamination. Excision is usually not employed if the residual amount is assuredly less than 1 nCi Pu. Conversely, if excision is started it may be continued until there is no measurable residue (0.1 nCi to 0.6 nCi).

In the few cases in which a substantial fraction of a permissible systemic burden of soluble plutonium is known to be in the body from any mode of entry, there is no hesitation to use DTPA to enhance removal via the urine. National practice has to be responsive to the respective drug codes. In the United States, the Federal Drug Administration regards DTPA as an experimental drug requiring special blood tests prior to administration. This poses problems because the key feature of effective chelating therapy is to begin treatment at the earliest opportunity, ideally within about 30 minutes after plutonium intake. It is this factor, above all others, that makes it desirable to have a battery of special tests for fast recognition of the general magnitude of an accidental intake. Currently, such steps as urine analysis are fast if completed in two hours, unless the signal is strong enough for detection of plutonium or americium by the photonic radiations. Similarly, for lung counting, it is essential first to be sure that there is no external contamination on the patient, which might completely vitiate the lung counter

---

5 See also Chap. 8, Sec. III. F, and Chap. 10, Sec. IX of this volume. (Editor.)

reading, and then to move him from the accident location to a central testing laboratory. This accounts in part for the strong interest in nasal smears that can be immediately read.

For lung deposits, especially with insoluble materials, there is more controversy. The extreme range covers nebulized DTPA for any known deposit to total rejection of the value of the method (SCHULTE and WHIPPLE, 1968). Other methods discussed, but not standardized, are mild lung irritants such as 10% saline solution to stimulate coughing, which is probably more useful to give a confirmatory positive sputum sample than to remove plutonium, and the use of lavage.

Lavage is a procedure not to be undertaken lightly. The expected mortality is comparable with that in anesthesia ($\sim$1 in 4000 cases). The incubation or latent period for lung cancer may be of the order of 10 to 40 years. Lung lavage for a person over age 50 may well be a poor tradeoff of competing risks. For a person age 20 to 30 after one week for thorough investigation of the deposit, lung lavage for 50 times the MPLB would appear to be a good risk. Others might make this determination at 10 $\times$ MPLB.

V. The U.S. Transuranium Registry

During the summer of 1968 the Atomic Energy Commission authorized the establishment, under the directorship of the Hanford Environmental Health Foundation, Richland, Washington, of the U.S. Transuranium Registry to serve as a national focal point for the accumulation, evaluation, interpretation and dissemination of data on the uptake, retention, distribution and biological effects of plutonium and other transuranium elements in occupationally exposed workers. The purpose is to identify hazards to the worker that may exist from internal depositions of plutonium and other transuranium elements, which exhibit long latent periods between the deposition within the body and the possible demonstration of adverse effects. The number of workers potentially exposed has been estimated at 20000 in facilities currently cooperating with the Registry. With the expected expansion of the nuclear industry, this number will continue to increase.

The tissues and organs from some recent autopsy cases of exposed plutonium workers indicate the health physics estimates of internal systemic depositions to be conservative, ranging from 1 to 13 times the depositions extrapolated from the autopsy samples when the amount in the lung is not included (NORWOOD et al., 1972). The periodic estimates of employee systemic depositions generally made by health physics personnel from urinalyses data, do not reflect the insoluble plutonium deposited in the lungs. Recent cases analyzed by the Registry show that these lung depositions can be significant and have been as much as ten times greater than the estimated systemic burden.

The distribution of plutonium found at autopsy is greatly different from the original assumptions made by the National Council on Radiation Protection and Measurements (NCRP, 1953, 1959) and the International Commission on Radiological Protection (ICRP, 1960, 1964, 1968). The autopsy data indicate the percentage of the total skeletal burden to range from 0 to 77 with a mean of 32, while the liver ranged from 0 to 80 with a mean of 20. This compares with the assumption used in determining the maximum permissible body burden that 80% or more goes to the skeleton and 7% to the liver. In studies of 12 of the 14 cases that have come to autopsy so far, the ratio of *concentration* in skeleton to that in liver ranged from 0.05 to 5.2 with a mean value of 1.0. The Registry is now concerned that its bone samples are too small (usually ~100 g), and that extrapolation to the total skeletal burden in 10000 g is unrealistic. This is particularly troublesome because the ratio of plutonium concentrations in various representative bones is far more variable than anticipated. For example, in five cases, the concentration ratios rib/sternum were 30, 2.6, 1.25, 0.49, and 0.19. Rib/vertebrae ratios in 3 cases were 0.74, 0.30, and 0.25.

Existing models have assumed that the contribution to total body burden from muscle, fat, skin, and other organs not included in the standard "search list" of lung, lymph nodes, skeleton, liver, spleen, and kidney is negligible. By scaling up from the lowest measured concentration in such tissues in one case, the Registry came to a value of 27%, considerably higher than either liver or lungs in the specific case. This triggered a search for deposits in muscle in two available cases, which scaled up to 2.5% and 4.0% body burden, respectively. The Registry prudently notes the anomaly of scaling from less than 50 g to the total muscle taken as 28000 g in standard man. Equally prudently, it is preparing to undertake major sampling of muscle (as well as bone) in future. Ideally, of course, total body sampling of a substantial number of higher exposure cases is needed. This runs into two social problems. First, the families of those enlightened workers who have willed their tissues to the Registry are making an irreplaceable contribution to science. Yet the American ethic has historically been sensitive to major destruction of the willed body, and the Registry must properly respect this sensitivity. What is needed most, and unfortunately does not currently seem probable, is a major increase in the analytical sensitivity of nondestructive testing for the photonic radiations of plutonium. Second, the detractors of the nuclear energy establishment could easily imply that certain organizations do not wish to cooperate with the Registry, because it may be shown that their practices are less than perfect, or more despicably, they may be accused of wanting to avoid evidence supporting compensation. It can be stated categorically that none of the cases to date shows any evidence of causation between the plutonium burdens and death. It would be absurd, however, to expect any valid epidemiological conclusions from 14 retrospective studies. Unfortunately for those seeking the truth, the nuclear detractors cannot be said to have confined themselves to *valid* epidemiological conclusions in other contexts.

It seems abundantly clear that the complete story of plutonium in man cannot be written without the enlightened cooperation of families of plutonium workers, management of plutonium operations, and dedicated physicians and scientists. Given this support, the results to date clearly indicate that the Registry could eventually become a key factor in establishing rational protocols for the management of plutonium contaminated individuals, comparable with the understanding of radium cases developed over more than three decades by R. D. Evans (1933, 1943, 1972).

VI. Protection against Criticality

The need to protect against an accidental nuclear chain reaction—criticality—is unique to operation with the fissionable elements, including plutonium. The mystique of this form of protection has been vested in a coterie of nuclear physicists and engineers; the detailed rites are foreign to the industrial hygienist or health physicist. It is well that this should be so, because a half-informed practitioner is an invitation to disaster.

In basic principle, though, the methods of control are simple, and two in number.

A. Safe Geometry

Whenever plutonium operations can be conducted in specific vessels and pipes, it is possible to keep their dimensions below the size that could allow criticality including all feasible accidental alterations to the vessel's geometry. This is the preferred modus operandi whenever applicable.

B. Administrative Controls

The control principle is to permit not more than 45% of the critical mass appropriate to the particular configurations and materials used to be present at one time. Accidental double-batching then preserves a small safety margin. For example, a plutonium-water system has a minimum critical mass of about 500 g, from which one conventional batch limit of 220 g is derived.

The art of criticality protection is to derive from basic experimentation and theory the correct critical mass for all proposed plutonium isotope mixtures and chemical and physical forms to be used. Similar precautions extend to transportation and storage of plutonium-bearing materials.

An excellent summary of such work appears in Chap. 27 of the AEC Plutonium Handbook, together with 68 references (Clayton and Reardon, 1967). The approaches in many countries are well described, although at an earlier date, in a European Nuclear Energy Agency report of a Karlsruhe Symposium (ENEA, 1961). When the claim is made that the principal hazard from plutonium is intake in the body, and especially inhalation, the hazard from radiation generated by plutonium criticality is being overlooked. This is the one circumstance that has already led to at least one fatality (Shipman et al., 1961).

VII. Protection from External Radiations

A. Nature of the Radiations

As indicated in the introduction, with the increasing interest in and use of isotopes other than ^{239}Pu, protection from plutonium radiations has progressed rapidly from simple protection against alpha radiation to a rather complex situation involving shielding against X-rays, gamma rays, and neutrons at levels otherwise potentially damaging or even lethal to the operators. As a graphic example of the change, let it be assumed that a mythical country with a dynasty of King Pluto's created a coinage from plutonium metal, equal in size to a U.S. silver dollar. A coin collector holds one commemorating Pluto the 238th in his left hand, and one commemorating Pluto the 239th in his right hand. The coins are physically indistinguishable except for the markings, ^{238}Pu and ^{239}Pu. The right hand receives 1 rem/hour from X-rays and gamma rays, with an insignificant 3×10^{-5} rem/hour from neutrons. Meanwhile the left hand receives about 330 rem/hour from X-rays and gamma rays, plus 3 rem/hour from neutrons. Once the feasible doses from plutonium surfaces are recognized, there are no particularly unique features to providing effective shielding, testing the emitted radiation with survey meters, and supplementing the protection with personnel meters. The relevant factors appear in Chap. 25 of the AEC Plutonium Handbook (Larson, 1967). However, they have been modified by newer experiments and calculations (Faust et al., 1971), the major change being a reduction from 960 rads/hour to 330 rads/hour for the surface dose of pure ^{238}Pu. In addition, the changes in dose with changing composition of the plutonium mix have been codified, so that designers can incorporate appropriate shielding into proposed processing facilities for handling mixed UO_2-PuO_2 nuclear fuels and the like.

The conventional starting fuel for present-day commercial nuclear reactors is uranium oxide enriched in ^{235}U. As the ^{235}U is consumed by fission, some of the ^{238}U, the main constituent of natural uranium, is converted to ^{239}Np by neutron absorption, and this decays with a half-life of 2.35 days to ^{239}Pu. This plutonium in situ continues to be bombarded with neutrons. Some of it undergoes fission but some captures another neutron to yield ^{240}Pu, which in turn is partially boosted to ^{241}Pu and ^{242}Pu. Meanwhile, through other secondary chains and radioactive decay, lesser amounts of ^{236}Pu and ^{238}Pu are created. The plutonium isotopic content of the fuel thus changes continuously with fuel exposure, and is conditioned also by the details of the fuel disposition, choice of neutron moderator (whether graphite, light water, or heavy water) and other factors. In the context of the reactor industry, fuel exposure, which is a measure of the integrated neutron bombardment per unit mass of uranium, is stated indirectly in the units of megawatt-days of power per metric ton of uranium (MWd/MT or MWd/MTU, a confusing nomenclature because the two M's have different connotations).

Table 14.1 shows representative isotopic compositions that would arise after three degrees of exposure, which have arbitrarily been taken as:

Brief exposure	about 140 MWd/MT,
Moderate exposure	about 13000 MWd/MT,
Heavy exposure	about 33000 MWd/MT.

Table 14.1. Plutonium compositions in reactor fuel

Pu isotope	Weight percent of the isotope		
	Brief reactor exposure	Moderate reactor exposure	Heavy reactor exposure
236	—	9.10^{-7}	6.10^{-6}
238	—	0.3	2.0
239	~99.44	79.0	55.5
240	~ 0.5	14.8	25.3
241	—	4.9	12.1
242	—	1.0	5.1
Total	~100	~100	~100

Upon reprocessing the spent fuels, the plutoniums as a family are separated from the residual uranium and the americium and higher transplutonium elements present. If the plutonium were then prepared as a metal block the surface dose-rate will change with time due to the ingrowth of ^{237}U and ^{241}Am, the two radioactive daughters of ^{241}Pu, in the following pattern.

These figures apply to the metallic form. The reconstitution for further fuel use is now generally done in the form of plutonium oxide, for which the dose-rates are approximately one-half (usually specified as 1/1.93), due to the lower density of the oxide form.

Some of the protection considerations, in brief form, are:

Most of the plutonium isotopes and the immediate daughter products are alpha-emitters. Among the beta-emitters, such as ^{241}Pu, none has a vigorous gamma emission comparable with those that provide the strong gamma radiation in the radium decay series, used in clinical radiation therapy.

The sources of X-rays and gamma rays to be considered are:

a) Gamma rays coincident with alpha decay: For the plutonium isotopes, these are weak, except for ^{238}Pu. They are the principal radiations of ^{241}Am and ^{237}U. It is the 60 keV emission of ^{241}Am that facilitates its recognition in the lung counter.

b) X-rays from internal conversion: These are the principal photonic radiations of ^{238}Pu, ^{239}Pu, ^{240}Pu, and ^{242}Pu. The conversion X-rays give 98% of the very high surface dose-rate of ^{238}Pu, and 92% of the modest surface dose-rate of ^{239}Pu. They have the merit of falling in the energy range of 17 keV (^{239}Pu spans 13.6 to 20.2 keV, which is practically equivalent). Thus they are easily shielded. The percent transmission is essentially zero through a typical glove box wall, and 30% through a typical viewing port. An additional lead glass sheet or even ordinary glass completes the shielding.

Similarly, their transmission to the critical organs, except possibly the testes, is insignificant. In any case, it is inexcusable not to have this radiation confined to the glove box, except at the glove ports, which are themselves shielded when in use by the hands and forearms at the expense of irradiation of those parts.

c) Gamma rays from spontaneous fission: These are of relatively low intensity in all cases.

d) Gamma rays from fission products generated by spontaneous fission:

While the energies of these radiations are high, the total yield from most conventional plutonium mixes is quite low, being largely controlled by the insignificant contribution from ^{239}Pu. The high energy implies that thick shields are required for its elimination. This would be considered only for such special operations as prolonged manipulation of a fabricated ^{238}Pu piece, for which the dose-rate at the unshielded body might be of the order of several millirads per hour.

e) Gamma rays secondary to neutrons: Conceptually, both the neutrons from spontaneous fission and those generated by the alpha-neutron reaction in light elements, as in PuF, will give rise to gamma rays as they are absorbed in intervening materials, and will also induce activation products, some of which will continue to emit gamma radiation. These are not normally classed as practical hazards. If the levels were appreciably higher, the gradually increasing induced radioactivity of all glove boxes, shipping containers, etc., would have to be taken into account.

The activity of plutonium samples containing ^{241}Pu is not constant because of the in-growth of the daughter products ^{241}Am and ^{237}U, both of which make notable contributions to the surface dose rate, when, as is the usual case, there is little ^{238}Pu present.

The ^{237}U activity rises rapidly in the first few days, and then levels off after two months to a steady plateau, being governed by the relationship $(1-e^{-0.102t})$, where t is the post-purification time in days. The growth of the ^{241}Am is virtually linear with time at least for several years. As measured by the X-ray and gamma-ray emissions, ^{241}Am overtakes the ^{237}U after 3 months (Note: the 4 months cross-over shown in many curves still repeated in the literature is based on superseded 1958 data), and rapidly becomes the leading component, e.g., 40 times the ^{237}U component after one year. Operators must be constantly aware of this changing pattern whenever radiations from these daughters are not fully shielded. Some illustrative numbers have been included above (Table 14.2).

Another variable hazard is presented by spreading a plutonium source. Since most of the photonic radiation is in the low energy range, most of it is self-shielded in a compact source. The effect at a distance then depends on the solid angle

Table 14.2. Contact photonic dose-rate (rads/hour)

Time after purification-days	Brief reactor exposure	Moderate reactor exposure	Heavy reactor exposure
0	1	2.2	10.3
100	1	2.6	11.2
1000	1	5.9	19.1

subtended by the source at a point on the body. When it is spread, the effective solid angle increases. Poor housekeeping in glove boxes can lead to a thin layer being spread over the box surfaces, materially increasing the dose to unshielded part of the body.

The sources of neutron radiation to be considered are:

a) Neutrons from Spontaneous Fission

The dose rate per unit flux density is reported as 0.115 mrem/hr. ROESCH has derived a formula for the neutron dose rate at the surface of a plutonium sphere from which the coin examples were calculated (ROESCH, 1958). ^{238}Pu, ^{240}Pu, and ^{242}Pu are the isotopes of concern, in relative contributions of 3.3:1:1.7. In practical terms of reprocessing irradiated fuel elements, the neutron dose-rate of plutonium surfaces rises from virtually zero for short-term irradiation to 700 mrem/hr after reactor exposure to 10000 MW days per ton.

b) Neutrons from α-n Reactions

With plutonium fluoride in a safe batch quantity, the neutron dose-rate close to the source is of the order of 5000 mrem/hr. More realistically at a glove box face it is about 20 mrem/hr, intolerable without shielding. Thus glove boxes for fluoride extractions and even for all PuO_2 operations need neutron shielding in the form of hydrogenous materials. Such materials as water, pressed wood, or plastics are suitable. Special formulations incorporating boron are convenient. The hydrogenous materials slow the fast neutrons (initially 1 to 10 MeV) which are then rapidly absorbed by the boron.

The practical absorption half-value thicknesses of these materials are in the range of 5 to 10 cm. Thus a useful water wall tends to be a cumbersome attachment to a glove box. Nevertheless, water may be the material of choice because of the fire propagation propensities of some of the other materials. It also serves conveniently for the viewing ports. The possibility of leakage of such systems and a resultant drastic change in criticality parameters has to be considered at all times.

B. Survey Instrumentation

For the photonic radiations conventional gamma-ray survey meters are adequate. However, if at any step the 17 keV X-radiation is not contained, special meters responsive at this low energy are needed. The neutron dose is generally obtained by moderation in specially sized hydrogenous cylinders with measurement of the moderated neutrons in a BF_3 counter or thermal neutron sensitive scintillation counter. The moderator size is empirically adjusted to cause a response that can be interpreted as a dose-equivalent rate in rems/hr. The original form of these meters utilized two coaxial cylinders of paraffin, the inner one functioning essentially as a neutron flux meter, and the combination as a dosimeter (DE PANGHER, 1959). A successful commercial instrument achieves the same result more conveniently by judicious placement of partial boron shielding with the moderator (TRACERLAB, 1968).

C. Personnel Monitoring

Records of personnel exposure throughout the nuclear industry have been customarily obtained from standard film badges worn on the outside of the protective clothing. Suitably chosen filters over the film flatten the response curve so that there is approximately equal sensitivity to gamma rays over the range 50 keV to 1.25 MeV (UNRUH et al., 1963). Routine calibration can thereafter be done with one calibration source, optionally radium, ^{137}Cs or ^{60}Co. Most film badges were not intended to measure such soft X-rays as the 17 keV internal conversion groups, but they suffice if special filter systems and calibration are made with a suitable fluorescent X-ray source (LARSON et al., 1954). Separation of the dose for the soft component needs to be continued because only about 35% of its total contribution is properly assigned as a whole body dose.

The neutron component has normally been measured by nuclear track film, a tedious procedure when relatively low doses are to be evaluated.

In recent years, measurement of both photonic and neutronic components has been improved by thermoluminescent dosimetry (TLD) (ATTIX, 1967; KORBA and HAY, 1970).

A representative receptor as used at Hanford includes five LiF chips and special filters to provide dosimetry for both photons and neutrons (KOCHER et al., 1971). Two of the five LiF chips are used for the photon dosimetry giving a measure of the penetrating component and the non-penetrating component (17 keV) or soft component. The neutron dosimetry is provided by three chips including chips of ^{6}LiF and ^{7}LiF. These systems work on albedo neutrons from the body. Their response is dependent on the neutron energy spectrum, hence it is necessary to calibrate such systems with neutron spectra closely duplicating field exposure conditions. With such calibration, this type of TLD system is probably the best available for personnel neutron dosimetry, although the current performance capabilities of these systems do not fully meet the desired levels of performance. Some groups use dose-rate monitoring instruments and time-keeping to calculate a neutron dose for entry in the personnel exposure records in lieu of using personnel dosimeters. AEC Workshop Reports review the current status of personnel neutron dosimeter recent development (UNRUH, 1969, 1971).

Plutonium fluoride neutron sources are commonly used for dosimeter calibration when a plutonium fluoride component may be encountered in the plutonium work. Plutonium beryllium sources may be used when only plutonium metal or plutonium oxides are present.

The ratio of gamma to neutron personnel dose-equivalents in plutonium facilities may vary widely depending on the plutonium material form, dispersal and local shielding. Gamma to neutron dose-equivalent ratios varying from 1:1 to 1:5 are not uncommon. Neutron dosimetry is clearly an important part of a modern radiation protection program in plutonium operations.

Measurement of the dose to hands and forearm is particularly important with close work with "hot" sources. Film in wrist badges and in small finger rings has long been used, but the finger rings particularly have a poor performance record. Substitution of TLD offers improvement (ENDRES, 1968). The TLD finger ring dosimeters are preferred over the film system because they are more resistant to environmental damage such as might arise from light leaks, heat and humidity. The lower energy dependence properties of the TLD material also offer the opportunity for better dose interpretations. Generally, though, the records of finger doses lack precision; it is impossible to have the finger rings at the maximum dose-rate position, and the neutron component is usually ignored. Presumably, some sort of neutron dose estimate could have been made with nuclear track films. The obvious application of combined ^{6}LiF and ^{7}LiF chips encounters a problem. It is effective for the whole body dose because it actually measures slow neutrons moderated in the body. There is not enough moderating mass in the hands to achieve reliable results.

Operator protection is often maintained by calculating or measuring both photonic and neutronic doses in various geometries and then establishing daily time limits for specific operations. Custom has allowed hand and finger doses to come much closer to permissible limits than is the case for whole body exposure. Special limits for hands and forearms are also more generous. Recent recommendations of the NCRP suggest some degree of interest in reducing such doses (NCRP, 1971). While there is legitimate room for a range of opinion, the intuitive feeling that if some late effects occur on the hands they will be correctable, and by no means comparable with possible whole body effects such as leukemia from external radiation or bone tumor from deposition of plutonium is not the whole story. The conservative view is that impairment of function of the hands is troublesome and cosmetically displeasing. In addition, surgery or electrocoagulation used to treat finger lesions is exceptionally painful because nature designed fingers to be sensitive. Demonstrably deleterious effects on the hands are not being found to date. The extreme escalation of surface dose in modern plutonium sources and its wide variability set the stage for future problems if handling practices are not correspondingly improved. Once the need for, and feasibility of, remote manipulation are accepted, there could be a major reduction in hand exposures to levels more responsive to the recommended concept of lowest practicable level of exposure in all operations.

D. Protection against Criticality

Meters, preferably neutron meters but sometimes gamma meters, set to give an audible alarm, provided with redundant circuitry to minimize failures, and tested on a regular schedule are mandatory. A fast pair of legs to respond to an alarm is the other necessary equipment.

Health physics personnel responding to a criticality incident should have a Geiger counter or other equally sensitive meter that can be held against the stomach of the evacuated personnel. A strong signal from the ^{24}Na activity, induced by slow neutrons moderated in the body exposed to a fast neutron flux, fortunately arises at levels well below those at which serious radiation injury will result (WILSON, 1962). Such a quick-sort is invaluable to direct medical attention to critical cases. Almost more importantly, negative results promptly relieve the anxiety of the unexposed personnel. The knowledge that criticality accidents have caused painful deaths is foremost in the mind of one who sees the "Blue-flash" of a criticality incident until he can be confidently reassured that his exposure was clearly sub-lethal.

Re-entry to an evacuated area is made under full protective clothing and air masks, and carrying high-reading dose-rate survey meters as well as those of conventional range. It is absolutely essential that neutron meters be used, because the most insidious situation is to have a criticality event that continues as a nuclear reactor without recognition.

VIII. Summary of Plutonium Internal Deposition Experience

USAEC experience from 1957–1966 reported cases in excess of 25% of the permissible body burden (Ross, 1968). There were 136 cases, rising to 203 through 1970 (personal communication). Twenty-nine of these exceeded the limit (37 through 1970). About 65% represented inhalation, and 25% wounds, the remainder being uncertain. Chelating agents were used in about 25% of the cases; wound excision covered about 18%. Ross recognizes the uncertainties in the measurements. The experience with plutonium is not conspicuously good, even measured against full utilization of accepted maximum permissible burdens. It is less satisfactory when measured against admittedly arbitrary standards of "lowest practicable level" achieved today in many other areas of radiation protection. On the other hand, as measured by the conventional standards of industrial hygiene, the performance to date has been outstanding. With the exception of criticality accidents, injury has been limited to removal of small amounts of tissue around wound sites, and the trauma of occasional lung lavage. There has been no significant deleterious effect of chelating agents, such as DTPA, used to promote removal of plutonium from the body. Above all, there seems to be no identifiable case of malignancy of bone, lung, or other critical organs and tissues. None of this is cause for complacency in view of the anticipated long latent period for tumor induction, and the lack of proof that these carcinogenic effects show a real or practical dose threshold.

Hanford experience[6] for 1946 to 1967 was as follows (Jech et al., 1969b; Heid and Jech, 1969):

Injury cases potentially contaminated	230	(310)
Cases containing measurable plutonium	136	(162)
Cases surgically decontaminated	86	(100)
Deposition above 5% MPBB	15	(16)
Deposition above 50% MPBB	5	(5)
Plutonium inhalation cases	1140	(1501)
Chelating agent used	12	(13)
Deposition above 5% MPBB	98	(101)
Deposition above 50% MPBB	12	(14)

Figures in parentheses represent a private communication updating through 1971. They demonstrate an encouraging stabilization of the cases above 5% MPBB in the face of considerably improved measurement capability since 1967.

Of the 136 cases reported by Ross, 132 could be identified as to year of origin. Sixty-three of these, or about 48%, had occurred in the last 3 years of his reporting decade. He was concerned that this might indicate an accelerating trend. Actually, 67 cases have been added in the remaining four years of record through 1970. Thus the USAEC data continue to show a fairly high incidence rate, but not an explosively expanding one at this time. An incidence rate based

6 For easier comparison with the total AEC experience which is cut off at 25 percent MPBB the Hanford cases with depositions equal to or greater than this amount are—injury cases, 9 through 1967 and 10 through 1971; inhalation cases, 29 and 31 respectively.

on the number of injuries per kilogram of plutonium used, or alternatively, per man-hour of plutonium operations, would inherently seem to have more interest, but it cannot be derived from published data. In any case, such an index might be expected to be a function of the form of plutonium used, with an expectation of higher incidence rate for inhalation with plutonium in oxide form as for nuclear fuels, and higher incidence rate for wounds with plutonium in metal form as for space heaters and other applications.

If one uses the percentage of cases exceeding the maximum permissible annual dose commitment as a crude index of severity, it will be seen that the severity rate has fallen from 21% in the prime reporting decade to about 12% in the added report period. Possibly the number of cases is too low and their occurrence too sporadic to justify a rigorous conclusion that the severity rate has materially fallen.

AEC Licensees'[7] experience to 1966 shows one plutonium case and one americium case above 25% MPBB, but the workload must have been relatively light in this field (ROEDER, 1968).

Autopsy determinations of lung burden among Hanford plutonium workers, non-radiation workers, and residents found results compatible with air concentrations typical of plutonium facilities for the first group, and accountable from world wide fallout for the other two groups (NEWTON et al., 1968).

A recent updating of the above material extends the cases to 1971, with a total of 356 autopsy cases (NELSON et al., 1971). Some observations on these results are included in Appendix A, which also describes the learning experience from a number of specific plutonium contamination cases, some of which have been followed for almost 20 years. This experience is mainly in confirming or refining metabolic pathways in man. The absence of deleterious effects to date precludes analysis of dose-effect relationships in man.

IX. Environmental Protection for Plutonium[8]

As indicated in Chap. 15, environmental considerations for plutonium have not had the worldwide attention given to such nuclides as ^{90}Sr and ^{131}I. Many factors may have contributed to this—the origin of plutonium production as a highly classified operation, long continuation of certain security measures that conditioned scientists to minimize publication (whereas the contrary policy might have won public confidence), the fact that plutonium in the environment is re-concentrated to a much lesser degree than such well-studied nuclides as ^{32}P, ^{90}Sr, and ^{131}I, and the fact that committed atomic explosions have provided a long term artificial background of plutonium. Current worldwide attention to environmental problems has stimulated research plans to fill in gaps for the transuranium elements as a class. Meanwhile, competent reviews of environmental studies, even though they omit significant masses of classified data that would now be releasable, show a modest to reasonable breadth beginning with 1948 (PRICE, 1971, 1972). The referenced 1971 report addresses itself specifically to the transuranium elements in one section. In addition, it has the merit of providing abstracts of other papers not specifically analyzed in the text. Other

7 In the United States the U.S. Atomic Energy Commission controls the use of reactor-produced nuclides through a regulatory and licensing system. An "AEC licensee", as used here, is an individual or organization permitted to use specified nuclides under these controls.
8 This section takes up in some detail those aspects of environmental protection most pertinent to the nuclear energy industry. More general aspects are considered in the following Chapter (15). (Editor.)

useful source material includes reports of three AEC Radioecology Symposia (1st, SCHULTZ and KLEMENTS, 1963; 2nd, NELSON, EVANS et al., 1969; 3rd, Oak Ridge—in press).

The worldwide surface contamination peaks between latitudes 40° N and 50° N at about 1.5–2 nCi $^{239}Pu/m^2$ (HARLEY, 1971, and earlier reports). Separation of plutonium from soil samples at these low levels is tedious and subject to variable yield. Good chemistry in one series gave a recovery from soil of 67.9% with a standard deviation of 32.2%, decidedly inferior to air filter yields (79.9% $\pm$15.5%) (DE BORTOLI, 1967). It is now standard practice to spike soil samples with trace ^{236}Pu so that the yield can be experimentally found, making its variability less important. One concern with this technique, which applies to a similar application to urine analysis, is that there is no absolute assurance that all chemical and physical forms of plutonium will be extracted in the same ratio as the tracer ^{236}Pu. With the vigorous chemistry resorted to in some forms of soil analysis this may not be a major factor. Various U.S. methods of analysis for plutonium in soil were reviewed at a recent AEC Environmental Plutonium Symposium at Los Alamos. (Proceedings of Environmental Plutonium Symposium, 1971.) Wide variability of sample recovery by laboratories recognized to be highly skilled in such work was noted. In addition to the chemical problem, the problems of standardizing the method of taking a soil sample, and particularly the inherent randomness when the total contaminant may be present as a few specific particles were noted.

Representative air contamination in mid-latitudes, circa 1965–1966, was 0.1 pCi/1000 m^3. Measured lung burdens were ~0.45 pCi/kg (MAGNO et al., 1967), ~0.5 pCi/kg (NEWTON et al., 1968) and ~0.15 pCi/kg (SMORODINTSEVA et al., translation 1969). These are compatible with the air concentrations, except that the Russian air concentration seemed low by an order of magnitude. They probably fairly represent lung contamination throughout the northern mid-latitudes at that period of time. (See Appendix A, Part 1.)

Environmental sampling and some degree of ecological study have been applied to the U.S. Nevada test site, and the atomic bomb incidents at Palomares and Thule. As an example, the Danish studies at Thule represented a sound ecological approach, which revealed no notable surprises, found no conditions of major concern, and correctly reserved judgment on some topics to be pursued later (AARKROG, 1971).

A major plutonium fire at Rocky Flats in 1969 led to concerned outside scientists measuring plutonium in soil samples, and questioning the adequacy of the protective and monitoring systems. There was eventual agreement on the soil sampling competency. It was determined, however, that the particular samples arose from spread from a temporarily substandard storage of contaminated oils and solvents in drums. It is no surprise to health physicists to find that the most frequent cause of environmental contamination is neither the operation of the obviously most dangerous facility, nor the controlled storage of the obviously most dangerous major wastes. Rather, as in this case, it comes from an intermediate class of wastes, temporarily held in seemingly adequate control, but with failure to recognize and correct a deteriorating storage condition in time. Examples from other sites would include such aberrations as a contaminated waste ditch, safely operated for years and then abandoned without adequate immobilization of radioactive sediments and restriction of access by wildlife. PRICE (1971) has pointed out the large number of animals which are attracted to food, shelter, and water around a pond in the advanced stages of plant community succession, with increased potential for contamination spread. Despite the popular trend to improve the "quality of life" by making industrial facilities attractive, the use of surface disposal ponds for dilute radioactive wastes is one case in which ecological common sense demands that the facility be established and maintained in a simple, early succession stage, or even sterilized.

Review of the Rocky Flats case led to publication of the following data applying to the total experience to date (HAMMOND, 1971):

Controlled releases	airborne effluents	42 mCi
	liquid effluents	90 mCi
Accidental releases	1957 fire	~0.06 Ci
	1969 fire	~0.0002 Ci
Waste-contaminated soil	transference	~2.6 Ci

The almost incredibly low releases from the Rocky Flats fires stimulated a review of all reported release incidents and development of theory to assist accident analysis (HUNT, 1971a, b). The crux of the matter is that the greatest mass of plutonium is attached to large particles that do not escape from the vicinity of the accident. Such large particles would in any case not contribute to an inhalation hazard. However, one cannot dismiss the possibility that they would subsequently disintegrate by weathering to a transportable respirable size.

Soil samples from the environs of the Hanford plant show evidence of local contamination in the vicinity of the plutonium working facilities, dubious results at two miles from the stack, which is still 10 miles within the controlled boundary, and none off-site (CORLEY et al., 1971). When the surface activity is plotted against distance from the stack, there is randomness in the data with no clear relationship between activity and distance. By contrast, had there been steady minor stack releases averaged over years of operation, the reliable meteorologcal data at this site would predict a fall of over two orders of magnitude over the range covered, perhaps somewhat modified by redistribution. Because of the random errors in sampling, probably due to real variations in point to point settling of discrete particles, it is difficult to discern that distance at which plant discharges demonstrably add to existing fallout. The question was ingeniously resolved by measuring the ratio of plutonium to ^{137}Cs in the samples. This ratio is distinctly higher for the local particles whereas the fallout particles are relatively rich in ^{137}Cs and other nuclides. The results clearly show that only the immediate inner area has a local infestation. In other words, living within two miles of this major plant throughout its operating career would not have added significantly to the body burden of plutonium (however, see Appendix A, Part 1). As a point of perspective, the highest plutonium alpha activity close to the Hanford plutonium facility was 7.5% of the natural alpha activity of the uranium and thorium series in typical soil.

Wind transference of activity was seen to be the major cause of difficulty at Rocky Flats. Hanford is by no means deficient in wind. The possibility of resuspension was foreseen over 20 years ago, and studies were made in preparation for an untoward event (HEALY and FUQUAY, 1958, 1959). The work is a fascinating exercise in atmospheric physics, but the parameters have never yielded fully to quantification. Despite the one experience at Rocky Flats, the transference is usually not severe, because, as it were, the fine radioactive particles hide behind larger inert particle neighbors on the ground surface or adhere to them. They are essentially in a quiescent boundary layer. Thus the problem is principally one of movement when a whole layer is picked up in a dust storm. There is some evidence suggestive of increased air concentrations during dust storms, but the extreme randomness of sampling for particulate matter under such conditions contraindicates a sound statistical analysis unless many samples are taken. To relate the increased air contamination to augmented inhalation risk requires knowledge of the particle size of the actual contaminated particles. Elementary theory, as given above suggests that when the surface layer is resuspended, much of the activity may be attached to larger non-respirable particles. A major restudy of resuspension problems is currently underway (HEALY, unpublished). With vegetative cover, the problem of resuspension is trivial.

The demonstrably low releases of plutonium to the atmosphere point clearly to the health physics control of the problem at its source. The solution is as follows:

Installation and maintenance of the highest grade filtration on all exhaust ducts.

Selection of fire resistant filters and frames in duplicate. A fire presents probably the greatest risk of breaching the exhaust.

Continuous sampling between the two filters so that failure or reduced efficiency of the first is promptly detected.

Alarm type sampling on the exit side of the second filter.

Air-locking of all entrances and exits including the emergency exits.

It is less easy to list the elements of a sound waste management program. The high level wastes must have first class retention with capability to protect against a tank leakage. The third line of defense results in material being deposited in the ground. Retention on sediments is remarkably good so that movement into ground water is minimal. Nevertheless, uncontained deposits can never be dismissed because of the long half-lives involved. The general renaissance of interest in man's obligation not to prejudice the environmental future has included a most vigorous consideration of the ultimate issues for transuranium wastes. Calculations of waste composition 30 million years from now must surely be man's most forward look to date (BELL and DILLON, 1971). It is of passing interest to note that reprocessing wastes (not plutonium plant wastes) have the transuranium elements, plutonium and americium, as the key offenders between ages 300 and 30000 years. Curiously, the learning experience between plutonium and radium that started plutonium work on the right foot, comes full circle. From about 50000 to 500000 years the key offender in nuclear wastes will be none other than ^{226}Ra. Even more surprisingly, the most-unwanted title thereafter passes until the 30 millionth year to none of the heavy element decay chains but to a radioiodine, ^{129}I. Meanwhile, improved engineered storage can and will be developed for a more modest time span—of the order of 100 years.

At the present time, the most cogent environmental risk in the plutonium industry (excluding nuclear weapons incidents) probably involves substandard handling and transportation of the secondary wastes, the discarded equipment, the contaminated paper, and so on. At this writing, this aspect is receiving renewed attention now that the transuranium elements, as industrial products and wastes, are here to stay in more than one sense. As an example of the care taken in disposal of solid wastes, one can cite the example of a licensed laboratory close to the Hanford Works, which is using ^{238}Pu in the development of heat sources for space applications, artificial hearts, and the like. Casual waste from the glove box operations is collected in small cans, which are separately sealed, and then sealed in larger cans. A number of these cans are then placed in a stainless steel box, which is welded hermetically tight and tested by helium leakage. This box is then transported to a major burial site within the Hanford reservation with virtually no risk of escape of contamination en route.

The questions of safe packaging, intra-plant shipment, off-site shipment of completed pieces such as PuO_2 fuels, and off-site shipment of wastes to central repositories have not been covered in this review. Considerable talent has gone into engineered safety to protect packages from damage under severe impact conditions and when subjected to severe fires. The International Atomic Energy Agency has been productive in establishing uniform regulations for safe transport of all radioactive materials. The experience to date has been good (D. E. Patterson and V. P. De Fatta, 1962). Yet, with a markedly expanding frequency of shipment, the assurance of absolute perfection is absent.

All the major nations involved with plutonium wastes are actively strengthening their plans for ultimate storage. In the United States their commitment to a disused salt mine, previously considered only for disposal of the highly radioactive mixed fission product wastes, is under active discussion. In West Germany, plans are well advanced for retention of virtually all that nation's nuclear wastes in the Aase Salt Mine, near Braunschweig. Other nations less favored by underground formations that offer effective isolation from the biosphere are seeking the best approaches open to them.

Lastly, as a matter of practicality, use of nuclear materials at all will involve some residual releases to the environment, occasionally from accidents but steadily from a diverse battery of individually trivial sources. Every contaminated plutonium worker is a source of environmental pollution, and he cannot be kept in engineered storage. More pertinently, ground disposal of so-called decontaminated equipment, which is obviously still contaminated in lesser degree, will have to continue. Criteria governing this will probably hinge on intuitive arguments that increments of one or a few percent addition to worldwide plutonium fallout are inconsequential, or more generally that similar increments to the natural radioactivity (radium etc.) of soils, plants, and animals are acceptable. As in all radiation protection problems in which there may be no threshold of absolute safety, the acceptable solutions are not wholly contained in science, but include social, economic, and political vectors.

The issues of plutonium contamination of the environment are far from resolved. The current experience does not suggest substantial present risks to man or any other life form; neither is it comprehensive enough to dismiss the risks as insignificant. Importantly, there is a trend to improve present practices, and in some cases to recover and relocate some past releases. More importantly, there is a responsiveness to the expected exponential increase of plutonium availability, coupled with higher sensitivity to environmental impacts. With these signals, it is unlikely that plutonium misuse, short of nuclear war, will ever rank with man's major mistakes.

Acknowledgments. The author is greatly indebted to C. M. Unruh and K. R. Heid, Battelle-Northwest Laboratory, for collaboration in selecting source material and for manuscript review. Former colleagues in this Laboratory, B. V. Andersen, L. A. Carter, J. P. Corley, J. J. Fuquay, H. V. Larson, C. E. Newton, Jr., J. M. Nielsen, and J. M. Selby, were most helpful in providing material, as were also Dr. P. A. Fuqua and Dr. J. A. Norcross, Hanford Environmental Health Foundation; Mr. M. L. Smith, Donald W. Douglas Laboratories; and Mr. W. C. Schmidt, Atlantic Richfield Company, all of Richland, Wash.

Work performed under AEC Contract AT-45-1-1830.

Appendix A

Some Learning Experiences from Autopsy Data and Selected Personnel Contamination Cases

Part 1. Autopsy Data

The data from 356 autopsy cases should provide a wealth of information on the outcome of casual incidental intake and substantial accidental intake of plutonium (NELSON et al., 1971). However, as the authors point out, there are several defects in the data which have gradually been reduced by learning experience.

Had NELSON et al., been measuring the concentration of plutonium in moon rocks, they would doubtless have had access to the most sensitive analytical methods feasible. As it was, they used good, but not outstanding methods, which steadily improved. The minimum sample activity which has a 95% probability of being distinguishable from zero was about 1 pCi prior to 1954, and about 0.02 pCi thereafter. The early data were difficult to interpret in an overall sense because so many readings were at or below the minimum detectable level. Usually, in any sampling program, positive results not far above minimum detectable level have wide error limits. These limits are not quoted in the report. Neither is there a record of analytical yield which might have been obtained at greater expense by salting the samples with trace amounts of ^{236}Pu. Differentiation of fallout plutonium from industrial plutonium could have been made by isotopic analysis. At an approximate cost of $ 400 per sample that, too, was beyond the resources of this program. A great improvement was made circa 1963 by radically increasing the size of the autopsy samples.

The cases were divided into four categories as follows:

Hanford Plutonium Workers	71
Others having had Hanford work experience	89
Local residents of more than 3 years	146
Individuals not residing locally	50

The second and third categories may be polluted by accidental inclusions of a former atomic energy worker, since the cross-checking of autopsy cases against former employment is not complete.

As a result of these problems, together with the risk of cross-contamination in the laboratory, the total mass of data is relatively uninformative, beyond the fact that the plutonium workers clearly have higher burdens on the average. Only two of the workers were known to have positive burdens from routine bioassay samples. Thus there is at least presumptive evidence of generalized deposition below conventional detection methods during employment.

With such data, a so-called inferential average concentration for each class can be obtained as follows:

If p_i is the ith person with a positive result P_i fCi/g and n_i is the ith person with no detectable burden at the detection limit N_i fCi/g, the inferential average concentration lies between:

$$\frac{\Sigma P_i}{\Sigma p_i + \Sigma n_i} \text{ f Ci/g and } \frac{\Sigma P_i + \Sigma N_i}{\Sigma p_i + \Sigma n_i} \text{ f Ci/g.}$$

For the whole data, these averages are shown in Table 14.3.

Table 14.3. So-called inferential average concentrations (f Ci/g)

Class		Organ or tissue		
		lung	liver	bone
I	Pu workers	4.4–5.4	2.3–2.9	3.5–7.7
II	Other work experience	0.4–1.5	0.5–1.3	1.7–6.9
III	Local residents	0.2–1.0	0.2–1.1	0.6–4.5
IV	Non-residents	0.1–1.0	0.2–1.3	0.2–4.2

The data on activity of tracheo-bronchial lymph nodes has been disregarded because of the small sample size. However, it is clear that some of these concentrations were the highest of any recorded. This may further complicate the lung data where node separation was not performed completely.

Such inferential averages are not informative except for the plutonium workers. As a first step in clarifying the data, the average sample weight, rounded to grams, is shown in Table 14.4, for Class I and Class III. Class II is quite similar to Class I as is Class IV to Class III.

Table 14.4. Average sample weights (g) by 5-year intervals

Interval	Lung		Liver		Bone	
	Class I	Class III	Class I	Class III	Class I	Class III
1951–1955	34 (6)	37 (6)	30 (5)	33 (6)	7 (1)	9 (2)
1956–1960	41	41	33	25	7	5
1961–1965	84	78	57	95	10	11
1966–1970	437	367	199	215	146	66

The numbers in parentheses are hypothetical weights made more comparable to subsequent data by accounting for the lower analytical sensitivity prior to 1954.

It is now clear that more interpretable data should arise by discarding the first decade. This notably increases the percentage of positive samples as in Table 14.5.

Table 14.5. Percentage of positive samples by class and decade

Class	Decade	Lung	Liver	Bone
I	1951–1960	26	48	4
	1961–1970	71	55	12
II	1951–1960	30	23	7
	1961–1970	50	47	10
III	1951–1960	17	20	2
	1961–1970	46	38	11
IV	1951–1960	20	0	7
	1961–1970	31	38	4

The recalculated inferential averages appear in Table 14.6.

Table 14.6. Inferential average concentrations (f Ci/g) 1961–1970

Class	Organ or tissue		
	Lung	Liver	Bone
I	5.8–5.9 (1.8–1.9)	2.2–2.3 (1.2–1.3)	5.4–7.3 (0.2–2.2)
II	0.4–0.5 (0.4–0.5)	0.4–0.5 (0.3–0.5)	3.6–6.7 (0.4–3.5)
III	0.3–0.4 (0.2–0.3)	0.2–0.4 (0.2–0.4)	1.0–3.6 (0.4–3.0)
IV	0.1–0.3 (0.1–0.3)	0.2–0.4 (0.1–0.4)	1–3.6 (0.4–3.7)

In some cases, the range of the inferential average is sufficiently closed up to persuade one that the average is meaningful. Unfortunately, this is not necessarily the case. Without resorting to complex statistical analysis, one can see that the average in a large enough class should not be materially affected by eliminating the extremes, i.e., the highest and lowest concentrations in the class. The numbers in parentheses in Table 14.6 are so derived. Many of the entries are so weighted by a single case that the apparent averages change significantly. The date for bone are seen to be relatively uninstructive. Even the apparently stable value for lungs of plutonium workers is grossly distorted by a single case.

The ratio of concentrations in lung, liver, and bone for a given class should be more or less constant if the plutonium is being acquired through the same mechanism. This might be expected to be inhalation, supplemented by some minor intake through cuts and abrasions for the plutonium workers. The data are not incompatible with this assumption.

Some tentative conclusions can be drawn from the lung data. There is other evidence that atmospheric contamination for local residents is insignificantly greater than that due to plutonium fallout. Thus Classes III and IV can be added to give a total of 103 cases showing an average lung concentration of (0.27 ± 0.07) fCi/g (or 0.27 pCi/kg). This is probably the best available figure for lung burdens in mid-latitude over the last decade.

Only 42% of the 103 cases showed positive results; treatment of the below detection limit data is plausible, but arbitrary at best. Therefore, it is more reasonable to quote an average concentration of (0.3 ± 0.1) fCi/g. Presumably, the lung deposition occurs as discrete particles possibly with a wide range of activity. A small sample of lung may show quite non-representative activity. The actual observations cover two decades of activity. If the non-worker lung cases are reduced to those with a sample size of about 100 g or more, 47 cases are left (34 female and 13 male), and about equally divided above and below age 55. In this limited class, the women had three times the activity of the men, and the older group twice the activity of the younger. It is difficult to assign statistical significance to these variations, but they should be watched with interest as the sub-classes expand.

If 0.3 fCi/g is a reasonable number for the unexposed population, it would suggest that Hanford front-line plutonium workers have an additional lung burden of about 1.6 fCi/g on the average, from 40 cases, and that other Hanford workers have a real but small increment of about one-half of the fallout background. The total lung burden equates to 1.6 pCi for the typical plutonium worker, orders of magnitude below that detectable by external means. Only the accumulation of massive autopsy data can prove or deny this tentative conclusion.

The biological consequences of these deposits is uncertain, but at most it must be very small relative to other available insults. In terms of dose, the apparent population lung dose equates to 0.3 mrem per year, which is trivial compared with natural variations in background radiation. Yet if the "hot-spot" theory of particulate contamination has merit, it cannot be stated categorically that no detriment will arise by combination of this insult with some other cancer-promoting condition.

Part 2. Selected Deposition Cases

General Interpretation

Hanford methods of interpretation of deposition data have been given by HEID et al. (1970a, b) and JECH et al. (1971, 1972). The prompt actions which permit rational medical treatment have been briefly described in Sec. III. In this appendix, the interest is in what can be learned about plutonium behavior by long term follow-up. Only rarely do the late observations contribute to the welfare of a particular contaminated worker (see Cases 1 and 6). Plutonium has so often (and wrongly) been described as the most toxic of the radioactive elements that it is natural that an affected worker should become concerned about his prognosis when long-continued tests are made. NORWOOD and FUQUA (1969) refer to a case in which therapy was discontinued due to the emotional make-up of an individual. The relentless search for additional scientific data, especially when therapy is not involved, is an improper assignment of priorities.

Fig. 14.7—14.11 are taken from the Hanford papers and internal protocols with the consent of the authors and of the Battelle-Northwest Laboratories.

Fig. 14.7 illustrates the fit of urinary excretion rate to the Langham equation over a 60-day period, for a known soluble deposition which can be calculated. An *essential* parameter is knowledge of the date of intake. Interpretation of a positive urinary excretion in a routine bio-assay program is inevitably uncertain when the specific cause cannot be identified. Since the excretion rate drops by a factor of 8 in the first month, there are two sound reasons for preferring a monthly sampling interval over the 6-month interval advocated by others:

a) There is a higher probability of obtaining a reliable demonstrably positive sample.

b) The ultimate estimate of intake date cannot be wrong by more than one month.

Fig. 14.8 shows the fit of excretion data to the Healy model (1957) of mobilization of plutonium from initially insoluble plutonium. For example, the derived value of Q_0 could be a measure of a lung burden of plutonium oxide (minus that initially cleared by ciliary action).

Fig. 14.9 shows a 7-year follow-up of urine activity originating from a wound source of plutonium nitrate, with the wound cleaned to eliminate a continuing source. After some 200 days, the scatter of the individual low activity samples is perhaps more impressive than their fit to the curve. The visual impression can be improved by grouping the data in sets of 3 or even 5 if the interval between samples is small compared with the time post intake. Computer analysis of such data is surprisingly consistent. It can be set to give the best position of a straight

ANALYTICAL DATA

d/m Pu per Sample	0.48	0.32	0.19	0.12	0.10	0.060	0.065	0.040	0.035
Days Post Intake	2	4	7	13	20	30	40	50	60

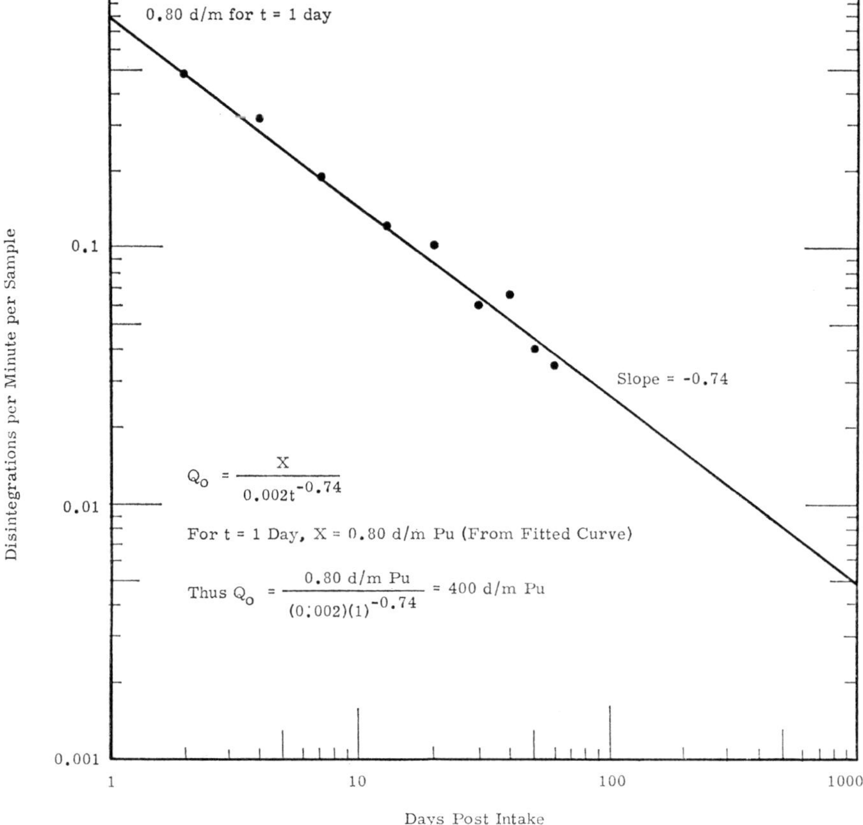

Fig. 14.7. Calculation of initial systemic burden, Q_0, from application of the Langham model to urinary excretion of initially soluble plutonium with a known intake date

line known to have the power slope $t^{-0.74}$ or to determine the best fit to any power slope, which might give values between about -0.72 and -0.78 for the power exponent. In the subject case, the best fit gave -0.76, which changes the intercept at day 1 from 120 dis/min to 130 dis/min.

Fig. 14.10 demonstrates the problems of interpretation when DTPA is used to chelate the systemic burden. The enhancement ratio of urinary excretion was 100 on treatment days. In other individuals it is variously between 30 and 100. A value of 50 is conventionally used to estimate the systemic burden during treatment. Jech et al., (1971, 1972), point out that the estimate of deposition is likely to be too high for some time after treatment. Here it was calculated as 18 n Ci for the period 25 to 120 days post treatment (130 to 220 days post intake). It reverted to 10 nCi in the period 200–9120 days post intake, confirmed later by a continuing Langham fit over 8 years. A cluster of points

ANALYTICAL DATA

d/mPu per Sample	0.27	0.28	0.29	0.34	0.35	0.35	0.37	0.40	0.37	0.38	0.42	0.38	0.40	0.36	0.33	0.33	0.30	0.27	0.21	0.18	0.15	0.13
Days Post Intake	20	30	40	50	70	90	100	150	190	250	300	350	480	600	800	1000	1200	1500	1800	2000	2600	3300

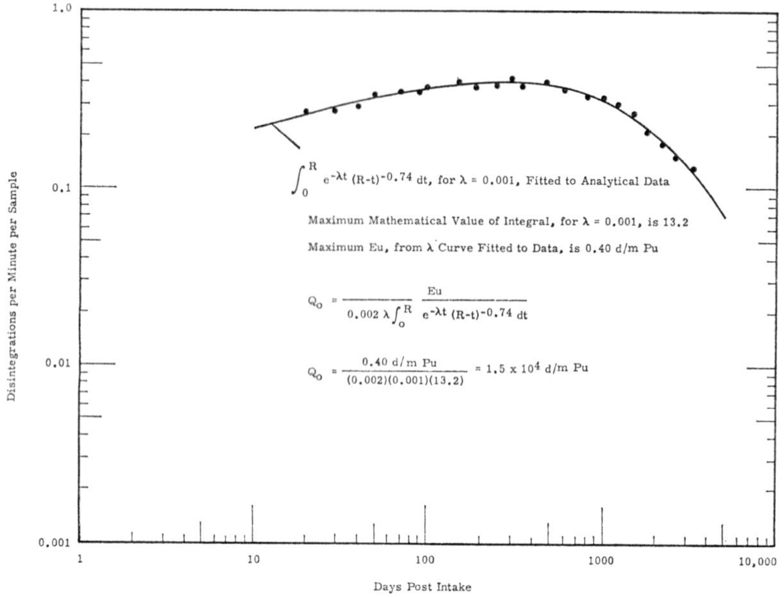

1.0
0.1
0.01
0.001
Disintegrations per Minute per Sample
1
10
100
1000
10,000
Days Post Intake
$\int_0^R e^{-\lambda t}(R-t)^{-0.74}\,dt$, for $\lambda = 0.001$, Fitted to Analytical Data
Maximum Mathematical Value of Integral, for $\lambda = 0.001$, is 13.2
Maximum Eu, from λ Curve Fitted to Data, is 0.40 d/m Pu
$Q_o = \dfrac{Eu}{0.002\,\lambda \int_0^R e^{-\lambda t}(R-t)^{-0.74}\,dt}$
$Q_o = \dfrac{0.40 \text{ d/m Pu}}{(0.002)(0.001)(13.2)} = 1.5 \times 10^4$ d/m Pu

Fig. 14.8. Application of the Healy model to calculation of systemic burden slowly acquired from initial deposition of nominally insoluble plutonium in a specific compartment such as the lung

100
10
1
0.1
D/M PER SAMPLE
1
10
100
1000
DAYS POST INTAKE
LANGHAM MODEL
$E_u = 0.002\,Q_0 t^{-0.74}$

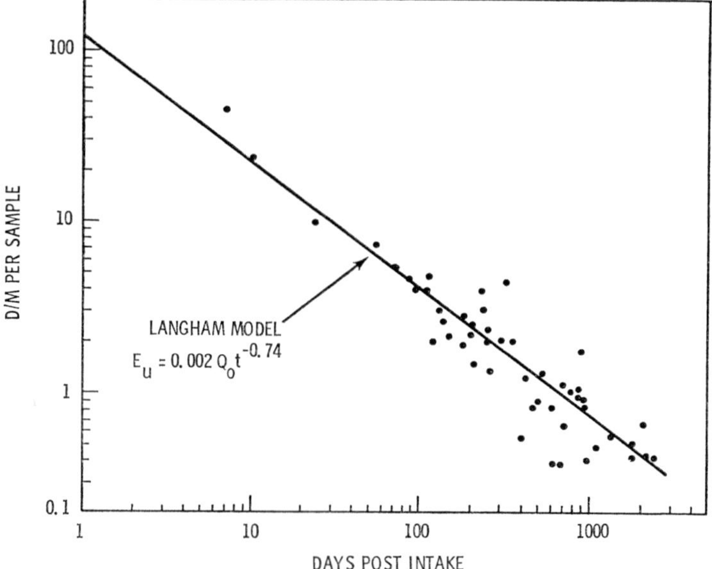

Fig. 14.9. Fit of urinary excretion data to the Langham model following plutonium nitrate intake from a wound

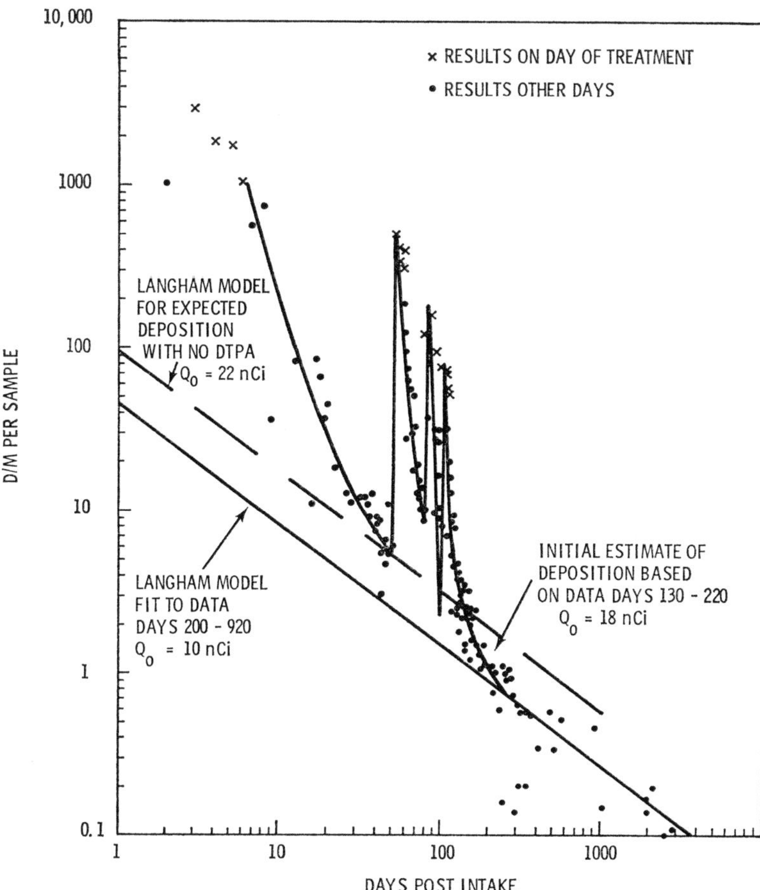

Fig. 14.10. Distortion of urinary excretion resulting from a plutonium nitrate intake, treated sporadically with DTPA over the first $3^1/_2$ months. At the end of 7 months, the estimated systemic burden is nearly twice the finally accepted value

is seen well below the fitted line at about 120 days. A review of all Hanford data shows this to be the usual occurrence (HEID, personal communication). If so, the temporary suppression merits study as a possible DTPA effect, and for practical analysis, these data points should be rejected. The LANGHAM line for 22 n Ci burden refers to the calculated deposition had DTPA not been used. Thus DTPA prevented 60% of the potential burden, and this is shown to be typical.

Fig. 14.11 shows the fit of urinary excretion data arising from a plutonium oxide inhalation case to a combined Langham model for transportable plutonium and a Healy model for the non-transportable fraction slowly mobilized from the lung. Oxide inhalation cases treated with DTPA present serious difficulties in modeling, due in part to loss of interpretability during the treatment period, and uncertainty with respect to time of return to an undisturbed excretion curve (JECH et al., 1971, 1972) (see also Case 7).

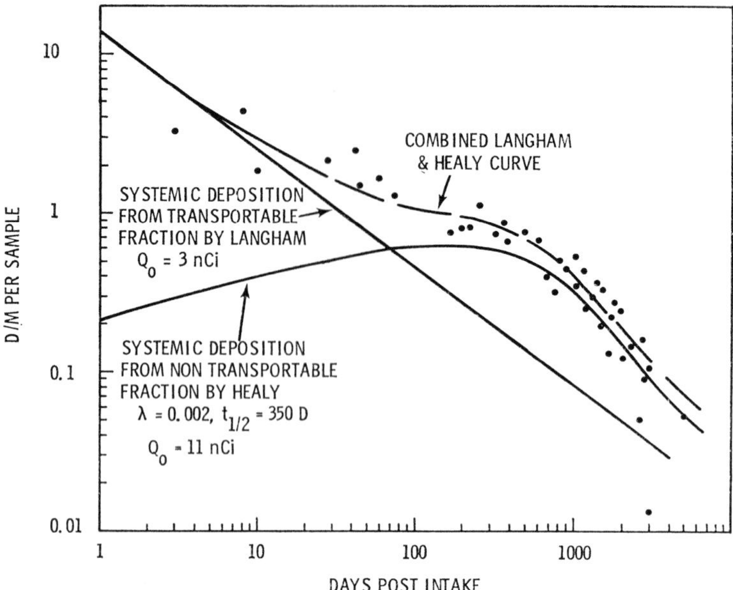

Fig. 14.11. Application of combined Langham and Healy equations to the transportable and non-transportable fractions of the ultimate systemic burden, following inhalation of plutonium oxide particles. With the low activity of the samples, the curve-fitting over the period 50—2000 days post intake is reasonably good

Refinement of Body Burden

The conventional maximum permissible body burden of plutonium (0.04 μCi) was originally based on ^{239}Pu only. With more mature plutonium, ^{240}Pu is the first to grow in. Both are customarily measured together (e.g., in urine) by their alpha-emissions, and they have the same permissible limits. With even more mature plutonium, all the isotopes from ^{238}Pu to ^{242}Pu are present. For practical purposes, the alpha count is proportional to the body burden for the 4 alpha-emitters. ^{241}Pu has to be accounted for by other means. Hanford refers to a normal plutonium, which by weight is 93.5% ^{239}Pu, 6.0% ^{240}Pu, and 0.5% ^{241}Pu. The measured alpha MPBB of this composition is 81% due to ^{239}Pu and 19% due to ^{240}Pu. In addition, the beta radiation of ^{241}Pu contributes 35% of a MPBB. It will be clear that mixtures much richer in ^{241}Pu will need correction for the unmeasured ^{241}Pu. For the heavy exposure plutonium of Table 14.1 (p. 638) at the MPBB, ^{238}Pu contributes 79% of the alpha activity, while ^{241}Pu constitutes an unmeasured contribution of $14 \times$ MPBB. The picture is further complicated in the body by the growth of the alpha-emitting ^{241}Am. The biological effect of the beta and gamma radiations from ^{241}Pu may be sharply different from that of the alpha-emitting plutoniums.

Specific Cases of Interest

Case 1. Late Detection of Minor Wound

An operator received a puncture wound in 1952 while using a steel brush on plutonium in a glove box. Hand contamination of 0.003 μCi plutonium removed easily from the external skin surface. No bioassay sampling was requested at the time and by chance one week later the employee transferred to another assignment which did not routinely require bioassay. The first indication of intake was obtained three years later when the employee again changed work and a routine bioassay examination gave positive results.

Under questioning the employee stated that the area around the wound had itched for several months and that approximately six months after the incident he noticed two small black spots which showed up near the spot where the skin had been punctured. When these spots disappeared, the itching stopped. An examination using a plutonium wound counter in 1956 ($3^1/_2$ years after the injury was received) indicated 0.02 μCi remaining at the wound site. Excision of tissue reduced the contamination at the wound site to 0.0001 μCi.

Based on an evaluation utilizing urine data, the employee had an internal deposition estimated at 0.048 μCi of plutonium initially, and now about 0.04 μCi. Excretion data 20 years after the event still follow a Langham curve.

The incident emphasizes the need to investigate even minor puncture wounds. Had this wound deposition been detected promptly, practically all the systemic burden could have been avoided by minor surgery.

Case 2. Plutonium Metal Fragment in Wound

A technician was compressing a plutonium specimen inside a glove box when an explosion occurred. Fragments of the plutonium specimen were ejected from the jaws of the press, some of which inflicted a puncture wound in the arm of the technician. Efforts to decontaminate skin surfaces were effective except at the wound site where 0.05 μCi activity remained. A wound counter measuring the 17 keV radiation of plutonium showed initial activity of 36 μCi, with 8.5, 2.6, 2.8 and 0.1 in four stages of excision. These are uncorrected for tissue absorption. The third excision removed a plutonium metal flake (Fig. 14.12) which contained 7100 μCi (LARSON et al., 1968). Activity in excised tissue, other than the flake totaled 3.2 μCi. Faced with readings which went down and then up again, the authors concluded that the counter malfunctioned due to the extremely high activity of the metal flake.

As an exercise in health physics, it can be shown that the results could well have been real. The removed piece has a surface area of 16.8 mm^2; from the known activity of 7100 μCi, a mass of 112 mg and hence a thickness of 0.3 mm can be derived. The initial depth of the flake was about 19 mm. The 17 keV radiation escapes only from a range thickness of the metal. Therefore, the appar-

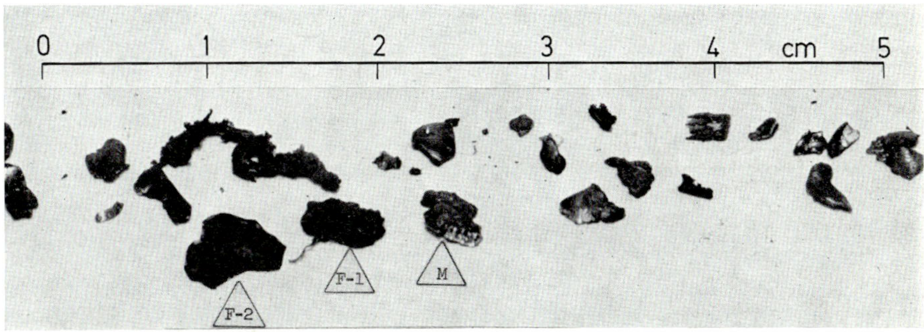

Fig. 14.12. Plutonium metal fragments, ejected in an explosion. The piece F-1 was protruding from the injured arm. F-2 is the deeply embedded flake removed by surgery

ent activity from the face would be ~190 μCi, and from the edge only ~19 μCi. The shielding factor for tissue depth is about 0.15. Thus the wound counter would see between 30 μCi and 3 μCi, depending on the orientation of the flake. That orientation could have changed with the probing at excision. The initial counts and counts after excision are compatible with this explanation. The interest in the exercise is the practical application of plutonium metal dosimetry.

The patient was given chelation therapy for 50 weeks, during which the systemic burden was estimated to be about 0.16 μCi with a DTPA enhancement ratio of 30. Experience with other cases suggests that this would overestimate the burden. The current burden, 8 years later, is estimated to be about 0.1 μCi. In addition, 0.06 μCi remained at the wound site. Measurements are continuing after 8 years.

Case 3. Plutonium Oxide Inhalation from Improperly Adjusted Mask

An employee was changing the primary filter from a glove box when a spread of plutonium occurred. The filter area had previously been completely enclosed in a plastic greenhouse or tent arrangement. The employee was exposed to the contaminated atmosphere inside the greenhouse for an estimated 15 minutes. Upon exit from the greenhouse, a personal survey indicated extensive skin and nasal contamination up to 0.02 μCi plutonium, ascribed to an ill-fitting mask. A lung examination detected 0.003 μCi of ^{241}Am on the day of the incident, 0.001 μCi on the following day, and 0.0005 μCi a week later. Sputum samples collected immediately following the incident contained 3×10^{-5} μCi Pu/ml. Fecal samples collected during the first 5 days post intake contained about 0.2 μCi plutonium. The employee was administered DTPA on eight separate occasions. From current evaluations, the employee incurred an initial deep lung burden of 0.018 μCi of plutonium (based on an initial burden of 0.03 μCi clearing to 0.018 μCi in 7 days) and a projected ultimate systemic burden of 0.016 μCi or 40% of the MPBB, with bone as reference.

This was a 1968 case reflecting, in the primary records, increasing sophistication in using auxiliary methods to refine the analysis. A nasal smear removed 20 nCi, enough to permit an accurate isotopic analysis which read:

	Atom percent	Percent of total alpha	Wt percent Pu only
^{238}Pu	0.0261	5.6	0.026
^{239}Pu	92.2	72.4	92.2
^{240}Pu	6.88	19.4	6.91
^{241}Pu	0.782	—	0.79
^{242}Pu	0.078	<0.1	~0.079
^{241}Am	0.0521	2.6	—

The initial ratio of $\frac{\text{Pu alpha}}{\text{Am alpha}} = 37.5$.

The particle size in the smear had activity mean aerodynamic diameter (AMAD) of 7.5 μm. Its solubility in normal saline was less than 0.9% in 2 hours. Smears from the mask had AMAD of 4.5 μm, the size used to fit lung model data. The true AMAD of the original aerosol could have been smaller.

Feces samples in the first five days yielded 165 n Ci. Adding the nasal smear as another mode of early clearance, conventional lung models for 4.5 μm particles gave

Initial pulmonary burden	26 nCi,
Pulmonary burden at 7 days	16 nCi,
Total systemic deposition	3.5 nCi.

Delayed feces samples, too few for good evaluation, gave a tentative initial pulmonary burden of 15–18 nCi.

Lung counting of the ^{241}Am radiation, using equations derived by NELSON (1968) from the ICRP lung model (MORROW, 1966) gave these results:

	nCi ^{241}Am	nCi Pu (alpha-emitter)
Initial pulmonary burden	0.75	30
Pulmonary burden at 7 days	0.45	18
Lung burden at 2.7 yrs	0.6	8

Fig. 14.13 shows the measured lung burden of ^{241}Am and the inferred plutonium burden. After the rapid fall in 7 days, the americium count slowly rises, due to growth of new americium offset by removal of americium and the parent ^{241}Pu. The inferred plutonium burden in the lung fell slowly. Such calculations can only be done when the original isotopic composition is known, and the continually changing Pu/Am ratio can be established. Determination of the systemic burden by urine analysis continues to show 10.2 nCi after 3 years. The model projects an ultimate burden of about 16 nCi. However, there is a continuing discrepancy between the present systemic burden deduced from the lung model (4 nCi) and the straightforward excretion data.

The case is particularly instructive in showing the general reliability of crude initial guides, at least to the extent of signalling rational treatment. The comparison as given by HEID et al. (1970a, b) is:

	Evaluation technique	Systemic deposition nCi Pu	Initial pulmonary deposition nCi Pu
Prompt	nasal smear guide	2.1	21
	initial in vivo guide		
	for assumed Puα/Amα = 15	1.3	13
	for true Puα/Amα = 40	3.3	33
	initial urine — Langham model (corrected for DTPA influence)	1.3–6	—
Interim	fecal — lung model		
	early ^{241}Am and Puα/Amα data	3.2	24
	final Pu data	3.3	25
	in vivo — lung model early data	4	33
	urine — Langham early data	2–10	—
Formal	in vivo — lung model	4	33
	urine — Langham	10	—
	urine — Healy	7.5	—

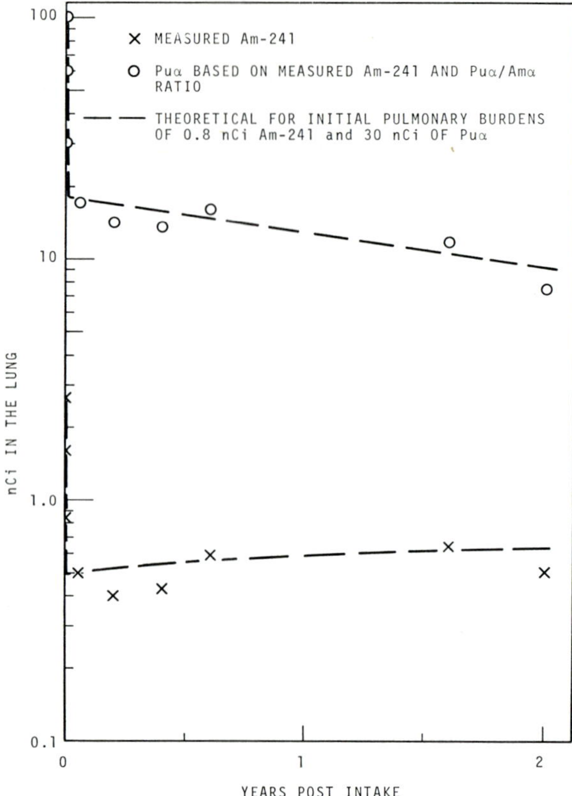

Fig. 14.13. Measured ^{241}Am lung burden and calculated Puα lung burden after plutonium oxide inhalation (Case No 3). The precipitous drop in the first seven days is virtually a vertical line on this scale. The main interest is the good fit of the derived falling plutonium burden to a theoretical model in the presence of an increasing burden of ^{241}Am

Case 4. Lung Burden Unrelated to a Specific Accident and below Detection Limit of Routine Bioassay

From retrospective study of the employee's work schedule it is believed that the inhalation occurred while he was performing work which involved welding on special plutonium fuel elements. Routine bioassay sampling did not detect any plutonium in urine samples collected during or immediately following this work. A routine lung examination some 8 years later detected 0.004 μCi of ^{241}Am and approximately 0.008 μCi of ^{239}Pu in the thorax region. The latter measurement is somewhat questionable since it is close to the minimum detectable for the equipment used. A Pu/Am ratio of 2 is in reasonable agreement with the ratio currently being observed in his fecal samples. Based on evaluation of these data, this employee incurred an initial intake to the deep lung of 0.1 μCi and received a systemic burden estimated at 0.001 μCi of plutonium or 2.5% of the MPBB, with bone as reference. The transference from lung to systemic burden is not compatible with existing models. The lung burden is close to existing detection limits. The employee was sent to two other major laboratories to check the lung burden, with these results:

	nCi ^{241}Am	nCi Pu
Hanford	4.5 ± 0.6	8 (dubious)
Lab A	5.7 ± 0.9	<4
Lab B	4.1 ± 0.5	<13

This is a fair measure of the present state-of-the art.

Case 5. Repetitive Depositions of Plutonium Compounds

The immediate incident involved an employee who was bagging material out of a glove box using a plastic bag which ruptured in the process. The employee remained in the contaminated atmosphere for an estimated 15 minutes after the rupture occurred; however, he was wearing a mask during most of this period. At the completion of work, personal surveys indicated nasal smears of 0.015 μCi plutonium and a lung examination indicated 0.004 μCi of ^{241}Am. A week later this had decreased to 0.0004 μCi. Utilizing a Pu/Am ratio of 22 and particle size of 7 microns as determined for a sample of the material, one can calculate an initial deep lung burden of 0.009 μCi of plutonium. Evaluation of the urine excreta data indicates the intake resulted in a systemic burden of 0.013 μCi plutonium or 33% of the MPBB with bone as reference. In addition, it was determined that there was a systemic burden of 0.105 μCi ^{241}Pu, which is 12% of the relevant MPBB. In total, then, the case had a 50% MPBB. Nine DTPA treatments were administered.

Two items are of interest in this case:

a) Fecal excretion provided a more definitive signal than did urinary excretion, because of high activity in the early samples.

b) The individual was involved in six separate incidents with plutonium in different solubility forms over a span of 7 years. Fortunately, all except the one reported above gave very low depositions. Otherwise, the interpretation of an additional burden would have been troublesome. Carrying this to the limit of more or less continuous chronic exposure at very low levels (as implied in Part I of Appendix A), it would be difficult to detect a systemic burden, and virtually impossible to quantify it.

The case also raises a fundamental social problem. The experience of this individual was so at variance with the norm over 50000 man-years of plutonium work that he can properly be classed as incident-prone. Should such a man be removed from hazardous work for his own protection or the protection of fellow workers? In a large institution in which he could be found useful employment in non-radiation work, one would tend to answer in the affirmative. If displacement of an older worker with no other special training would lead to probable unemployment [9], the balancing of benefits versus risks is a difficult task. If the risk were confined to the individual, he could properly be made a party to discussing his options. Should the risks to fellow employees be judged to be significant, management would be compelled to weight that factor heavily.

Case 6. Undetected Wound in Experimental Laboratory

An employee had worked with several special enrichments of plutonium during November of 1966. No unusual conditions were reported during this work

9 Although one knows of no social studies bearing on the point, it appears that the incident-prone worker is not necessarily a careless irresponsible type, but is rather the willing worker always pressing to do more than his fair share. Unemployment for him is a devastating morale breaker.

nor was the employee involved in any known radiation incident. However, a result of 2×10^{-6} μCi plutonium was reported for a routine urine sample collected on January 1 of the following year. At first it was assumed that the intake had occurred through inhalation, the usual mode of intake when first detected by routine bioassay sampling. However, the urinary excretion pattern observed during the next few weeks did not follow the expected pattern. Consequently, a search was made for contaminated wounds on 2/21/67, using a plutonium wound counter. A wound was found in the employee's hand which contained 0.01 μCi of plutonium. This was reduced through surgical excision of tissue at the wound site to 0.0004 μCi. Based on isotopic analysis of the material removed from the wound, 11/9/66 was estimated as the most probable date of intake, since the employee had been handling material of this composition at that time. Sixteen DTPA treatments were administered. Evaluation of the systemic burden, utilizing urinary excretion data, was 0.02 μCi of plutonium, or 50% of the MPBB, with bone as reference. Had this employee not been working in a Critical Mass Laboratory where a complete, accurate inventory of plutonium compositions with their times of use is kept, it would have been difficult to assign a deposition date to support the evaluations. The 8.9 nCi removed from the wound was a sufficient source to measure the 4.6% abundance of ^{240}Pu. This compares with the precisely known $4.62 \pm 0.3\%$ ^{240}Pu content of the material used only for a few days.

The material was a dry nitrate so that the mobilization data are representative of transfer from a soluble deposit in fleshy tissue. As a result of this and a few other nearly-missed cases, the practice is now encouraged of supplementing bioassay samples with routine photonic radiation counts of hands and lungs. On the whole, there is always a good chance of suspecting inhalation from signals of air contamination. The small puncture wounds can clearly occur with so little surface contamination that it is not detectable by the conventional alpha-counting hand counters.

Case 7. Puncture Wound with DTPA Treatment

An employee was removing stainless steel piping from a solvent extraction hood when he received a puncture wound in his right hand. Personal surveys upon exit from the work area detected external skin contamination of 0.005 μCi of plutonium. A blood sample collected at that time indicated 0.0002 μCi plutonium/ml. An examination of the wound site using a sodium iodide plutonium wound counter detected 0.12 μCi plutonium which was reduced by surgical excision to 0.01 μCi. The wound was flushed with DTPA before closure and a total of 47 DTPA treatments were administered, Based on evaluation utilizing urine excreta data the employee received a systemic burden estimated at 0.015 μCi plutonium as alpha emitters or 40% of the MPBB, with bone as reference. In addition, a systemic burden of 1.2 μCi ^{241}Pu is derived, or 130% of the formal MPBB. This is a relatively recent case illustrating growing attention to the usually ignored ^{241}Pu burden. The excretion pattern is shown in Fig. 14.14, which has some data points averaged to reduce the clutter. As pointed out by JECH et al. (1971, 1972) the DTPA enhancement ratio was 60–80 although the apparent ratio between days of treatment and other days was only about 5–10. Thus the DTPA was decidedly more effective than would have been judged at the time. This case maintained elevated excretion rates for 80 days post-treatment, and then shows the marked suppression below the eventual Langham fit. There was a typical elimination of 60% of the otherwise expected systemic burden.

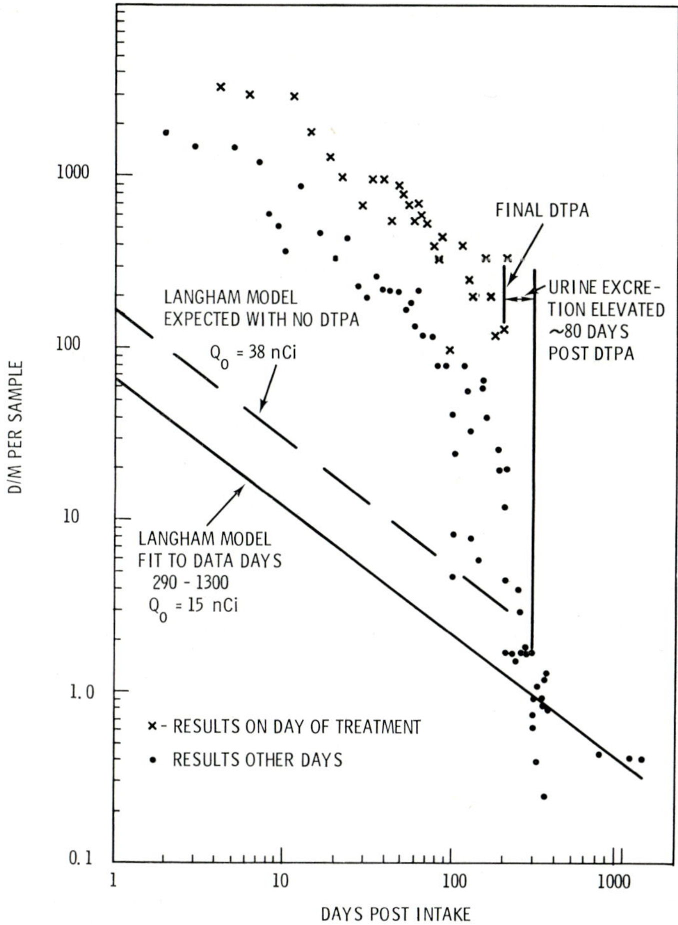

Fig. 14.14. Prolonged elevation of urinary excretion of a soluble salt (with no supplementation from the initial wound site), following intensive DTPA therapy. Some of the data points have been grouped together for clarity

Case 8. Long Term Reliability of Langham Excretion Formula

A simple case of deposition of soluble plutonium occurred in 1952. It was studied with frequent bioassay samples from January 1, 1953, to July 26, 1961. The best fit to a Langham curve of exponent —0.74 showed an estimated initial systemic deposition of 26 nCi. In early 1959, DTPA was administered over 12 consecutive days during which the enhancement ratio averaged 40. This demonstrates the effectiveness of late chelation in terms of enhancement ratio. As a practical therapy for a 7-year deposition, it is questionable because the elimination rate is still trivial. In the subject case, the enhanced elimination was 0.65 nCi of the then existing deposition of 23.5 nCi (a 10% reduction from the original burden through 6% elimination by urine and 48% by feces according to Langham equations), or less than 0.3% removal by the late chelation.

The principal interest of the case is that it has now been followed for 20 years, and the power function of $t^{-0.74}$ is still unchanged. This does not prove that the calculated deposition is correct because this involves also the numerical constant 0.002 in the Langham equation. There is a preliminary suggestion that deposits are lower than calculated but this can only be ascertained by total body analysis of well-documented cases at autopsy. It is remarkable enough that Langham was able to derive so powerful a tool from his limited observations.

In another Hanford case of a lung burden of soluble plutonium, the elimination rate was successfully reconciled to a combined Langham and Healy model over the first 1000 days. Now after 4500 days with the lung burden essentially removed the total systemic burden of only 26 nCi is faithfully following the $t^{-0.74}$ function.

References[10]

Aarkrog, A.: Radioecological investigations of plutonium in an arctic marine environment. Hlth Phys. **20**, No 1, 31–48 (1971).

Andersen, B. V.: Battelle Northwest Laboratories—private communication (1972).

Attix, F. H. (ed.): Luminescence dosimetry. Int. Conf. on Luminescence Dosimetry, USAEC, 1-514 CONF-650637 (1967).

Beach, S. A., Dolphin, G. W.: Determination of plutonium body burdens from measurements of daily urine excretion. In: Assessment of radioactivity in man, vol. II, p. 603–615. Vienna: I.A.E.A. 1964.

Beach, S. A., Dolphin, G. W., Duncan, K. P., Dunster, H. J.: A basis for routine urine sampling of workers exposed to plutonium-239. Hlth Phys. **12**, 1671–1682 (1966).

Bell, M. J., Dillon, R. S.: The long-term hazard of radioactive wastes produced by the enriched uranium, Pu–^{238}U, and ^{233}U–Th fuel cycles. USAEC-ORNL-TM-3548 (1971).

Billard, F., Miribel, J., Pradel, J.: Controle de la contamination atmospherique par "Double Mandarin." French CEA. CEA-2085 (1961).

Bortoli, M. C. de: Radiochemical determination of plutonium in soil and other environmental samples. Analyt. Chem. **39**, 375–377 (1967).

Bramson, P. E., Jech, J. J., Andersen, B. V.: Comparison of manual and computer evaluation of systemic plutonium burden. Hlth Phys., **22**, 911–922 (1972).

Clayton, E. D., Reardon, W. A.: Nuclear safety and criticality of plutonium, 875–919 in AEC Plutonium Handbook. Ed. O. J. Wick. New York: Gordon & Breach Science Publ. 1967.

Corley, J. P., Robertson, D. M., Brauer, F. P.: Plutonium in surface soil in the Hanford Plant environs. Proceedings of Environmental Plutonium Symposium, LA-4756, 85–88 (1971).

Dunster, H. J.: The concept of derived working limits for surface contamination. Surface contamination (B. R. Fish, ed.), p. 139–147. Oxford: Pergamon Press 1967.

Endres, G. W. R.: Characteristics of teflon-LiF thermoluminescent dosimeters. BNWL-741 (1968).

ENEA: Criticality control in chemical and metallurgical plants. Karlsruhe Symposium, p. 1–622 (1961).

Evans, R. D.: Radium poisoning: a review of present knowledge. Amer. J. publ. Hlth **23**, 1017–1023 (1933).

Evans, R. D.: Protection of radium dial workers and radiologists from injury by radium. J. industr. Hyg. **25**, 253–269 (1943).

Evans, R. D.: Radiogenic effects in man. In: Radiobiology of plutonium (Stover, B. J., and Jee, W. S. S., eds.), p. 431–468. Salt Lake City: The J.W. Press, Univ. of Utah 1972.

Faust, L. G., Brackenbush, L. W., Smith, R. C., Nichols, L. L., Brite, D. W.: Radiation dose rates from UO_2-PuO_2 thermal reactor fuels. BNWL-SA-3661 (1971).

Fish, B. R. (ed.): Surface contamination. Proceedings of a symposium held at Gatlinburg, Tennessee. Oxford: Pergamon Press 1967.

Hammond, S. E.: Industrial-type operations as a source of environmental plutonium. Proceedings of Environmental Plutonium Symposium, LA-4756, 25–35 (1971).

Harley, J. H.: Worldwide plutonium fallout from weapons tests. Proceedings of Environmental Plutonium Symposium, LA-4756, 13–19 (1971).

Healy, J. W.: Estimation of plutonium lung burden by urine analysis. Amer. industr. Hyg. Ass. Quart. **18**, 261–266 (1957).

Healy, J. W.: Surface contamination: decision levels. USAEC-LA-4558-MS (1971).

10 See note in References for Chap. 8 regarding availability of governmental and other "in house" documents (Editor).

HEALY, J. W., FUQUAY, J. J.: Wind pickup of radioactive particles from the ground. Proceedings of the 2nd Internat. Conference on Peaceful Uses of Atomic Energy **18**, Paper P/391, 291–295 (1958).

HEALY, J. W., FUQUAY, J. J.: Wind pickup of radioactive particles from the ground. Progress in nuclear energy series XII, vol. 1 — Health physics. Oxford: Pergamon Press 1959.

HEID, K. R., HENLE, R. C., SELBY, J. M.: Prompt mitigatory action after accidental exposure to radionuclides; in the monograph No 2 on nuclear medicine and biology "Diagnosis and treatment of deposited radionuclides" (eds. KORNBERG, H. A., and NORWOOD, W. D.), p. 593–599. Amsterdam: Excerpta Medica Foundation 1968.

HEID, K. R., JECH, J. J.: Assessing the probable severity of plutonium inhalation cases, p. 1–17. BNWL-SA-1595 (1968).

HEID, K. R., JECH, J. J.: Prompt handling of cases involving accidental exposure to plutonium. BNWL-SA-2062 (1969).

HEID, K. R., JECH, J. J., ANDERSEN, B. V.: Interpretation of data on internal contamination. BNWL-SA-3486 (1970a). Also in the Proceedings of the Seminar on Radiation Protection Problems Relating to Transuranium Elements, Karlsruhe (W. Germany), Sept. 21–25, 1970. EUR 4612 d-f-e (1970b).

HEID, K. R., JECH, J. J., ANDERSEN, B. V.: Interpretation of human urinary excretion of plutonium for cases treated with DTPA. BNWL-SA-4071 (1971) and in Hlth Phys. **22**, 787–792 (1972).

HOLLIDAY, B., DOLPHIN, G. W., DUNSTER, H. J.: Radiological protection of workers exposed to airborne plutonium particulate, p. 1–14, AHSB(RP)R 96 U.K.A.E. (1969). Also read as Paper No 1 of the 2nd Congr. of I.R.P.A. Brighton, England (1970).

HUNT, D. C.: Restricted release of plutonium—Part 1—Observational data. Nuclear Safety **12**, 85–89 (1971a).

HUNT, D. C.: Restricted release of plutonium—Part 2—Theory. Nuclear Safety **12**, 203–216 (1971b).

I.C.R.P. Publication 2: Report of Committee II on permissible dose for internal radiation. Oxford: Pargamon Press 1960.

I.C.R.P. Publication 6: Recommendations of the Internat. Commission on Radiological Protection (as amended 1959 and revised 1962). Oxford: Pergamon Press 1964.

I.C.R.P. Publication 10: Report of Committee IV on evaluation of radiation dose to body tissues from internal contamination due to occupational exposure. Oxford: Pergamon Press 1968.

ICRP Publication 10A: The assessment of internal contamination resulting from recurrent or prolonged uptakes. Oxford: Pergamon Press 1971.

IREDALE, P., HINDER, G.: The detection of airborne plutonium hazards. AERE-R-3783 (1962).

JECH, J. J., ANDERSEN, B. V., HEID, K. R.: Interpretation of human urinary excretion of plutonium for cases treated with DTPA. BNWL-SA-4071 (1971) and in Hlth Phys. **22**, 787–792 (1972).

JECH, J. J., HEID, K. R., CROOK, G. H.: Interaction between health physicists and industrial physicians immediately following radiation incidents. Read at Western Industrial Health Conference, San Francisco. BNWL-SA-2663 (1969a).

JECH, J. J., HEID, K. R., LARSON, H. V.: Prompt assessments and mitigatory action after accidental intake of plutonium. IAEA Symposium on Handling of Radiation Accidents, p. 77–93, IAEA (1969b).

KIEFER, H., MAUSHART, R.: Determination of Plutonium-239 body burden using gamma spectrometry with proportional counters. Proceedings of the symposium on whole-body counting, p. 289–293. Vienna: IAEA 1962.

KOCHER, L. F., ENDRES, G. W. R., NICHOLS, L. L., SHIPLER, D. B., HAVERFIELD, A. J.: The Hanford thermoluminescent multipurpose dosimeter. BNWL-SA-3955 (1971).

KORBA, A., HAY, J. E.: A thermoluminescent personnel neutron dosimeter. Hlth Phys. **18**, 581–584 (1970).

KORNBERG, H. A., NORWOOD, W. D. (eds.): Monographs on nuclear medicine and biology, No 2—Diagnosis and treatment of deposited radionuclides. Amsterdam: Excerpta Medica Foundation 1968.

LANGHAM, W. H.: Determination of internally deposited radioactive isotopes from excretion analyses. Amer. industr. Hyg. Ass. Quart. **17**, 305–318 (1956).

LARSON, H. V.: Factors in controlling personnel exposure to radiations from external sources, p. 845–857—AEC Plutonium Handbook, ed. O. J. WICK. New York: Gordon and Breach Science Publ. 1967.

LARSON, H. V., MYERS, I. T., ROESCH, W. C.: A wide beam fluorescent X-ray soucre. HW-31781 (1954) declassified (1956).

Larson, H. V., Newton, C. E., Jr., Baumgartner, W. V., Heid, K. R., Crook, G. H.: The management of an extensive plutonium wound and the evaluation of the residual internal deposition of plutonium. Phys. in Med. Biol. **13**, 45–53 (1968).

Laurer, G. R., Eisenbud, M.: In vivo measurement of nuclides emitting soft penetrating radiations. New York University Medical Center, AD690243 (1969).

Lawrence, J. N. P.: Hlth Phys. 8, 61 (1962).

Lidén, K. V. H., McCall, R. C.: Low-energy photon detectors for whole-body counting. Proceedings of symposium on whole-body counting, p. 145–166. Vienna: IAEA 1962.

Lindeken, C. L., Lakin, R. W.: Improvements in the solidstate plutonium-alpha air monitor. UCRL-50228 (1967).

Lister, B. A. J.: Early assessment of the seriousness of lung contamination by insoluble alpha and low energy beta emitting materials after inhalation exposure. UKAEA Report AERE-R-5292 (1966). Also in Proceedings of the First Internat. Congr. of Radiation Protection, Rome 1966, vol. 2, p. 1191–1198. Oxford: Pergamon Press 1968.

Magno, P. J., Kauffman, P. E., Schleien, B.: Plutonium in environmental and biological media. Hlth Phys. **13**, 1325–1330 (1967).

Morrow, P. E. (Chairman): ICRP Task Group on lung dynamics: Deposition and retention models for internal dosimetry of the human repiratory tract. Hlth Phys. **12**, 173–207 (1966).

N.C.R.P. Report 11: Maximum permisible amounts of radioisotopes in the human body and maximum permissible concentrations in air and water (N.B.S. Handbook 52) (1953).

N.C.R.P. Report 22: Maximum permissible body burdens and maximum permissible concentrations of radionuclides in air and in water for occupational exposure (N.B.S. Handbook 69) (1959).

N.C.R.P. Report 39: Basic radiation protection criteria. National Council on Radiation Protection and Mesurements (1971).

Nelson, D. J., Evans, F. C., (eds.): Symposium on radioecology. Proc. 2nd Nat. Symp. Ann Arbor, Michigan. CONF-670503 (1969).

Nelson, I. C.: Theoretical excretion of plutonium in urine based on the new ICRP lung model. In: Monographs on nuclear medicine and biology, No 2—Diagnosis and treatment of deposited radionuclides, p. 266–278. Amsterdam: Excerpta medica Foundation 1968.

Nelson, I. C., Heid, K. R., Fuqua, P. A., Mahoney, T. D.: Plutonium in autopsy tissue samples. BNWL-SA-4077 (1971).

Newberry, G. R.: Measurement and assessment of skin doses from skin contamination. In: Radiation and skin, AHSB (RP), R **39**, 44–67 (1964).

Newton, C. E., Jr., Larson, H. V., Heid, K. R., Nelson, I. C., Fuqua, P. A., Norwood, W. D., Marks, S., Mahoney, T. D.: Tissue analysis for plutonium at autopsy. In: Kornberg, H. A., W. D. Norwood (eds.) Monographs on nuclear biology and medicine series, No. 2 p. 460–470. Amsterdam: Excerpta Medica Foundation 1968.

Norwood, W. D., Fuqua, P. A.: Medical care for accidental deposition of plutonium (^{239}Pu) within the body. IAEA-SM-119/25 in Proceedings of a symposium—Handling of radiation accidents, p. 147–162. Vienna: IAEA 1969.

Norwood, W. D., Norcross, J. A., Newton, C. E., Jr., Hylton, D. B., La Gerquist, C.: Preliminary autopsy findings in U.S. Transuranium Registry Cases. Read. May, 1972 and to be published in the Proceedings of the 12th Annual Hanford Biology Symposium (Radionuclide Carcinogenesis) (1972).

Palmer, H. E., Wogman, N. A., Cooper, J. S.: The determination of the depth and amount of ^{238}Pu in wounds with Si(Li) detectors. Proc. of the symposium on Diagnosis and Treatment of Deposited Radionuclides, Richland, Washington, p. 164–170 (May 1967). Amsterdam: Excerpta Medica Foundation 1968.

Pangher, J. de: Double moderator neutron dosimeter. Nucl. Instr. and Methods **5**, 61–64 (1959).

Patterson, D. E., De Fatta, V. P.: A summary of incidents involving USAEC shipments of radioactive material 1957–1961. U.S.A.E.C. TID-16764 (1962) (and a series of updating supplements such as TID-16764 Suppl. 1, not all by the same two authors).

Pomarola, J., Risselin, A., Feliers, P.: Assessment of individual risk during atmospheric contamination by plutonium. CEA Fontenay-aux-Roses, NP-18254 (1966).

Price, K. R.: A critical review of biological accumulation, discrimination and uptake of radionuclides important to waste management practices. (1943–1971), p. 1–67. BNWL-B-148 (1971).

Price, K. R.: A review of transuranic elements in soils, plants, and animals. (This is the transuranium portion of the above in-house document.) To be published in Journal of Environmental Quality (1972).

Proceedings of environmental plutonium symposium, Los Alamos Scientific Laboratory. USAEC LA-4756, 19 papers, 1–119 (1971).

ROEDER, J. R.: A statistical summary of USAEC licensees' internal exposure experience. (1957–1966), p. 435–450. In: KORNBERG, H. A., and W. D. NORWOOD, Monographs on nuclear medicine and biology, No 2. Amsterdam: Excerpta Medica Foundation 1968.

ROESCH, W. C.: Surface dose from plutonium. Proceedings of the Second United Nations Conference on the Peaceful Uses of Atomic Energy, Geneva **23**, 339–345 (1958). Geneva: United Nations 1958.

ROESCH, W. C., BAUM, J. W.: Detection of plutonium in wounds. In: Proceedings of the Second United Nations International Conference on the Peaceful Uses of Atomic Energy, Geneva **23**, 142–143 (1958). Geneva: United Nations 1958.

ROSS, D. M.: A statistical summary of U.S. AEC contractors' internal exposure experience (1957–1966), p. 427–434. In: KORNBERG, H. A., and W. D. NORWOOD (eds.), Monographs on nuclear medicine and biology, No 2. Amsterdam: Excerpta Medica Foundation 1968.

SCHULTE, H. F., WHIPPLE, H. O.: Chelating agents in plutonium deposition—a minority view, p. 587–592. In: KORNBERG, H. A., and W. D. NORWOOD, Monographs on nuclear medicine and Biology, No 2 (1968).

SCHULTZ, V., KLEMENT, A. W., JR. (eds.): Radioecology. Proc. 1st Nat. Symposium on Radioecology, Colorado State Univ. New York: Reinhold Publishing Corp., and Am. Inst. Biol. Sci. Wash. D.C. 1963.

SCHWENDIMAN, L. C., HEALY, J. W., REID, D. L.: The application of nuclear track emulsions to the analysis of urine for very low level plutonium. USAEC Report HW-22680 (1951).

SHERWOOD, R. J., STEVENS, D. C.: Some observations on the nature and particle size of airborne plutonium in the Radiochemical Laboratories, Harwell. Ann. occup. Hyg. **8**, 93–108 (1965). Also as AERE-R-4672 (1963, declassified 1964).

SHIPMAN, T. L., LUSBAUGH, C. C., PETERSEN, D. F., LANGHAM, W. H., HARRIS, P. S.: Acute radiation death resulting from an accidental nuclear critical excursion. Special Suppl. J. occup. Med. **3**, 147–192 (1961).

SMORODINTSEVA, G. I.: Study of uptake of airborne Pu-239 by the human organism. U.N. Scientific Committee Document A/AC.82/G/L.1301, HASL Translation (Nov. 1969).

SNYDER, W. S., FORD, M. R., WARNER, G.: Proceedings of the 13th Annual Bioassay and Analytical Chemistry Meeting (Oct. 1967).

SWINTH, K. L.: Wound counting with solid state detectors. Pacific Northwest Laboratory Annual Report for 1968 to the USAEC Division of Biology and Medicine. Part 3, Instrumentation, p. 1–6, BNWL-1051 (1968).

Tracerlab (now Trapelo West) Technical Bulletin S-6 (1968). (Ascribes design basis to ANDERSON, I. O., and J. BRAUN: A neutron REM counter AE-132 and further work by UCLRC. Evaluation of a neutron REM dosimeter. Hazards Control Quarterly Report No 18. UCLRL-12167.)

UNRUH, C. M.: AEC workshop on personnel neutron dosimetry. BNWL-1340 (1969).

UNRUH, C. M.: Transuranium nuclides—a manual of good practice. BNWL-SA-3075, 99 pps. (1970). (A revised draft in press will be available from C. M. UNRUH, Battelle Northwest Laboratories, Richland, Wash. USA.)

UNRUH, C. M.: Second AEC workshop on personnel neutron dosimetry. BNWL-1616 (1971).

UNRUH, C. M., KOCHER, L. F., BRAMSON P. E.: The new Hanford film badge dosimeter. HW-76944 (1963).

WAITE, D. A., ANDERSEN, B. V., BRAMSON, P. E.: A projection chest phantom for in vivo dosimetry. A.I.H.A. Journ. **31**, 322–326 (1970).

WILSON, R. H.: A method for immediate detection of high level neutron exposure by measurement of sodium-24 in humans. HW-73891 Rev. (1962).

WILSON, R. H.: Controlling and evaluating plutonium deposition in humans, p. 831–844 AEC Plutonium Handbook, ed. O. J. WICK. New York: Gordon and Breach Science Publ. 1967.

WILSON, R. H., SILKER, W. B.: Plutonium contaminated injury case study and associated use of Na_4 EDTA as a decontaminating agent. USAEC Report HW-66309 (1960).

WINKLER, R., HOETZL, H., SANSONI, B.: Rapid identification and activity determination of long-lived alpha emitters in air. CONF-690540, 184–196 (1969).

Note: See addendum to chapter 12, page 592 (Editor).

Chapter 15

Plutonium in the Environment[1]

J. N. STANNARD[2]

I. Introduction

During the years of World War II and the development by the Manhattan District of the U.S. Corps of Engineers, the possibility of loss of radionuclides to the environment was constantly in mind. Each major installation developed an environmental surveillance capability and made regular measurements of air and water, and sometimes soil or other media both on and to a lesser extent, off-site. In the volume summarizing the Industrial Medicine of the Plutonium Project, CANTRIL and PARKER (1951) had the following to say about the Columbia River:

> There may be some concern that radioactive water released from the plant to the Columbia River will so contaminate the river that its water will be dangerous to man, fish, or fowl. It is true that the water which passes through the units to act as a coolant does become radioactive. But it is also true that the greatest part of this radioactivity is lost before the waste water is ever put back into the river. A continuous record is maintained of all water which flows from the plants to the river, and at no time has the combined waste water been in excess of that which would cause an overtolerance radiation exposure to any living thing immersed in it. The voluminous further dilution that occurs in the river brings the quantity down to one which cannot be measured below the plants. But in spite of this, the river water is periodically analyzed, with a negative result. As for the addition of heat to the river by the dumping of large quantities of coolant waters into it—there is heat added, but the dilution factor of the river itself is so vast that it requires the most delicate thermometers to register this small increase in temperature. This too is dissipated before the river flows through the limits of the reservation[3].

The "activity" referred to was, of course, primarily activation products and secondarily fission products arising from three sources—fission of natural uranium in the cooling water, fission of so-called "tramp" uranium left as trace contamination on the outer surfaces of fuel pieces, and very sporadically by escape of material from failed fuel elements. Only a very small fraction of the activity, if any, was due to plutonium isotopes[4]. At this same time a formal study was undertaken of the effects, primarily of "pile effluent" on fish fingerlings, embryos and eggs of various species at the Hanford Works, and on a more general scale in collaboration with the Fisheries Laboratory at the University of Washington. The results were gradually published and issued as declassified reports in the later 1940's and early 1950's (e.g. GAMERTSFELDER, 1947; OLSON, 1948; HERDE,

---

1 Work supported in part by Contract No. AT (11-1) 3490 between the U.S. Atomic Energy Commission and the University of Rochester and has been assigned Report No. UR-3490-117.

2 Secs. III A and B of this chapter were prepared in collaboration with Professor ROBERT H. WILSON, University of Rochester and he reviewed most of Sec. III and part of Sec. II. His collaboration and counsel are gratefully acknowledged.

3 CANTRIL and PARKER, *op cit.*, p. 482–483.

4 A current survey of the character and amount of radioactivity in the Hanford environs can be found in the paper by CORLEY et al. (1971).

1948; Coopey, 1948; Herde and Cline, 1949) and documentation is now both copious and beginning to have the perspective of time (Watson et al., 1969; Jennings and Osterberg, 1969). In fact from these and similar studies at Oak Ridge and later at Brookhaven a formal subdivision of radiobiology known as "Radioecology" developed, and is currently vigorous and important. While much of the work continues to appear through in-house project reports there have been several full-scale symposia (Schultz and Klement, 1963; Iaea, 1966; Aberg and Hungate, 1967; Nelson and Evans, 1969) and open literature publications are now numerous.

The pharmacologist-toxicologist of the future should expect to be involved in these aspects of radiotoxicology or, at least, have a general familiarity with their techniques and philosophy. In fact much of the controversy surrounding nuclear power station activity involves possible environmental effects (Iaea, 1970). For these reasons a short survey[5] of our knowledge vis-a-vis plutonium in the environment is considered appropriate to this volume. A recent symposium on "Environmental Plutonium" (LA-4756, 1971) can be consulted for further detail, particularly regarding methodology, as well as summaries by Romney and Davis (1971) and by Noshkin (1971).

II. Modes of Release of Plutonium to the Biosphere

Plutonium may be introduced into the biosphere in at least four principal ways: ablative destruction of plutonium-fueled space power-source during reentry, minor or major release from a nuclear industry facility (fuel reprocessing or fabricating plant or nuclear reactor), dispersal by detonation of a nuclear explosive device with or without a contribution to the energy release from the nuclear components, or a transportation accident involving plutonium sources, fuel elements, weapon components, etc. From the first of these ways dissemination may be hemisphere- or even world-wide, so that even though tens of kilocuries may be dispersed the contribution to unit deposition will be low. Dispersal from a nuclear facility may range from trace amounts to a few kilograms[6], depending on source strength and mode of release, the former exemplified by the minute but continuous loss from plutonium fabrication operations, the latter by the unlikely cataclysmic failure of a plutonium-fueled reactor. The dispersal range from a nuclear facility would be limited (a few kilometers unless the release were associated with a high-order explosion), but air and ground concentrations near the scene could be very high. Release from an above-ground nuclear explosion is with sufficient yield comparable to that of a space power-source except that the amount of plutonium would ordinarily be very much less. The debris (both fission products and plutonium) is elevated to high altitudes and becomes distributed over much of the earth's surface. The detonation of only the high explosive components of a nuclear weapon leads to complete disintegration and dispersal of the contained plutonium, much of it in the respirable size range. The amount released depends on the number and design of the weapons involved, the extent of dispersal depends on the weather regime at the site. In any event, it will be on the order of tens of kilometers down-wind, and maximum airborne

5 Complete coverage is not attempted since this would be outside of the purview of this volume but the symposium and review will provide a reasonable start along with the coverage in Chap. 14 of this volume (Ed).

6 The reactor fuel, mixed uranium-plutonium, is in a form very unlikely to vaporize even under the conditions of such a failure.

concentrations at ground level will be found at one to a few kilometers from the point of release.

Other transportation accidents with plutonium must be considered possible as plutonium inventories build up and plutonium as a fuel increases in attractiveness. However, even though large amounts of plutonium could be involved, the hazard is of a lower order than nuclear weapons accidents. Being a highly reactive metal, elemental plutonium is almost invariably encapsulated for purely physical reasons. In the dioxide form it is highly stable. Alloyed, its reactivity is much decreased. Thus, its availability for dispersal in a transportation accident in which no explosion takes place is limited. Furthermore the energy release, while it may be very significant to those intimately involved is ordinarily very low, so that dispersal will be quite limited—tens to a few thousands of meters. Additionally, such evidence as is available indicates that only a tiny fraction would enter the biosphere, the bulk being of a size or form unavailable to organisms.

Most plutonium production, handling, and fabrication facilities have until recently been under the direct control of the Atomic Energy Commission and other governmental agencies. These facilities have been located in relatively remote areas with access denied to the general public. With the increasing number of commerical reactors fueled by plutonium or uranium which proceeds to grow plutonium and the possibility of locating reprocessing facilities less far removed from the reactors than has generally been the case heretofore both reactors and plants may move closer to centers of population. (The location of the reactors will be the more important of the two since, fortunately, the number of reprocessing plants needed is much less than the number of reactors; thus prime sites can be more easily selected for them.) In locating such facilities environmental and meteorological factors are considered in depth. But the philosophy is primarily to build such plants to provide maximum containment although this cannot be absolute. As stated by MERKER (1967) "Historical design and operational experience show this concept to be practicable"[7]. To date we have essentially no data on large releases of plutonium from a nuclear reactor or other facility since there have been none of note (see Sec. IV for discussion of the Rocky Flats events). But there have been local contamination incidents from other sources and formal experiments to study the behavior of plutonium in the environment. These will be discussed first and more detail on general environmental considerations postponed to Sec. IV.

III. Local Contamination with Plutonium

The possibilities for local contamination from an accidental release of plutonium have led to a variety of field trials to evaluate the consequences. Also there have been at least two weapons accidents: the crash of a bomber carrying nuclear weapons (Thule, Greenland), and a refueling accident over Palomares, Spain, where the nuclear weapons were dropped and plutonium from some was released on impact. Weapon design is such that a nuclear contribution to the energy of the explosion is virtually impossible, but the high explosive components may detonate dispersing the plutonium over a restricted but significant area. Major field experiments have been conducted around these possibilities and will be discussed prior to further consideration of the actual incidents.

7 Op cit. p. 687., also Chap. 14. Sometimes the term "complete containment" is used but this implies "zero release" which is of course an overstatement since small planned releases are part of every such operation.

A. Test Group '57

The first formal environmental contamination experiments with plutonium were carried out under AEC auspices at the Nevada Test Site. In 1956 a test (Project '56) was carried out to study the behavior of released plutonium in air, soil, and the desert environment. There was no biological work in this experiment. In 1957, as part of "Operation Plumbbob" a fullscale biomedical program was undertaken by a group from the University of Rochester (Wilson et al., 1961) along with more elaborate studies of aerosols, cloud physics, decontamination and area monitoring studies by a group from Sandia Corp (Cowan, 1960) and the United States Air Force. This group was known as "Test Group '57" ("T.G. '57").

The primary purpose of the biomedical program was the direct determination of the uptake and retention of plutonium by a variety of animals as a function of time of exposure and the relation of the amounts found to air and ground levels. There were two major phases and groups of exposed animals—those exposed during cloud passage and those exposed during a long-term post-event period. Rats[8] and dogs were used in the former; dogs, sheep and burros for the latter. An array of air samplers and a program of soil sampling were set up so that exposures to resuspended plutonium could be correlated ultimately with tissue contents.

For the cloud passage phase the animals were placed from 500 to 2000 feet down-wind from Ground Zero. For the resuspension phase nearly 100 animals in three major groups were placed a few days after detonation in locations where the prevailing winds would be likely to bring contamination. The levels of ground contamination at the locations of the three groups of animals were of the order of 2.6, 40 and 560 micrograms per square meter of desert floor soil. A Casella Mark I cascade impactor was located near each site to give a crude estimate of the average plutonium aerosol to which the animals were exposed. These animals remained in place for periods ranging from 4 to 160 days. The plan was to measure the accumulation of plutonium in the lung and other tissues as a function of time and various aerosol parameters.

The results of this operation were somewhat surprising. It had been expected that the continued exposure phase would result in larger lung and body burdens than the brief exposure to the cloud at time of release. Actually the dogs exposed to the cloud passage at the time of detonation showed generally higher burdens than the animals exposed to resuspended plutonium for periods as long as 160 days even though the former were not at the range of maximum airborne concentrations at ground level. Also there was no appreciable increase in lung burden with time and very little difference in burden in animals placed at the several locations despite the large factor of difference in surface contamination in the long-term exposure phase.

Partial explanation for these results may reside in the very low concentrations found in the post-exposure animals and in the difficulties of radiochemical analysis at such low concentrations. But the air sampling data provided a clue that it was not all a matter of measureability. Despite a 200 fold difference in ground concentrations the air concentrations at the different locations varied by only a factor of about seven even under extremely windy conditions. This reflects the cumulative nature of resuspension and its lack of first order dependence on surface levels in the immediate environs. The gross lung activities were

8 The rats contributed so few data that they were not reported upon.

too low and variable to show real differences with this relatively small spread of air concentrations. The lack of build-up of lung burden with time probably reflects two factors—firstly the air concentration fell off fairly rapidly with time ($T^1/_2$ *ca.* 35 days); secondly lung clearance times seemed to be more rapid than anticipated from laboratory exposure data.

None of the animals in this test, either cloud passage or long-term exposure, accumulated sufficient plutonium to prognose a biological effect. Thus the experiment, despite its magnitude, must be considered as primarily a metabolic study[9]. In general results pointed to lower amounts in the body than were predicted after such a long term of exposure. This may reflect partly the fact that much of the resuspended material in the desert environment may have been attached to non-respirable particle sizes, and partly the fact that resuspension from the surface was somewhat less than expected.

B. Operation Roller Coaster

Partly because of the unexpectedly larger lung burdens found after cloud passage in the TG '57 work and partly because of the need for further studies to clarify discrepancies between animal and sampler results a new and rather different project was developed. This was a joint United States—United Kingdom venture and was carried out in a remote area of north-central Nevada. There were four test firings and the emphasis was placed on inhalation of plutonium from cloud passage (Wilson and Terry, 1965, 1968)[10]. Much more emphasis was placed on close proximity of animal and air sampler locations and the correlation of measured aerosol and animal tissue burdens. Also larger animals (sheep and burros) played a much more central role than in TG '57.

A total of over 300 animals was exposed to the plutonium-containing clouds (non-nuclear detonation) of two of the trials. The animals were sacrificed serially at times ranging from one hour to two and one half years after the single exposures. There was no chronic exposure phase. Analytical procedures and aerosol measurements were much more elaborate and the results turned out to be much more consistent than in the earlier work. Portions of this rather considerable effort can be reviewed in unclassified reports (Wilson and Terry, 1965, 1968; Stewart et al., 1965) but the final conclusions are not yet publicly available.

Nevertheless, even without the final papers it is clear that this undertaking has provided reliable and consistent data of considerable interest. There was consistent agreement between the aerosol parameters and the measured lung burdens providing suitable adjustments were made for the respirable fraction in each aerosol.

The initial lung concentrations for the three species were very similar despite wide differences in lung size, weight, breathing rate, etc. The ratios were 1.0: 0.900: 0.917 for dogs: sheep: burro respectively (Wilson and Terry, 1968). There was, however a marked difference in clearance pattern among the three species with each species forming a statistically consistent set. The lung clearance pattern for the dog closely resembled that found in laboratory studies with inhaled PuO_2, thus negating the often repeated statement that field and laboratory

9 Much non-biological information was obtained which has been reported elsewhere (e.g. Dick and Baker, 1967).
10 R. H. Wilson, Univ. of Rochester was the Project Officer. A large number of U.S. and U.K. scientists were actively involved and a large support team was assembled through the DOD, Defense Atomic-Support Agency, Atomic Weapons Research Establishment and, of course, the AEC.

results never or practically never agree. The clearance halftime for dog was found to be 180 days; clearance in sheep and burro corresponded more closely to power functions, with exponents of -0.416 and -0.242, respectively, than to the exponential clearance observed in dogs (WILSON, 1972). For such a function there is of course no comparable clearance halftime. The exponents indicate, however that clearance in sheep was rapid and extensive in comparison to the other two species, while the integrated clearance in burros over the course of a few years is reasonably comparable to that of dogs. Of the three species, burro compares most favorably with estimates for man (MORROW, 1960); sheep are poor subjects for modelling human exposure (WILSON, 1972; WILSON and TERRY, 1968). Thus for most purposes the accumulated data from the field experiments show reasonable correlation with the large body of information gained in laboratory experiments, particularly those employing the dog.

Translocation to other organs was measured (WILSON and TERRY, 1968) but was never sufficient to cast doubt on the choice of lung as critical organ for these exposure situations. Translocation to lymph nodes, considered in detail in Chap. 11, was definite but not as marked as seen in chronic exposures in either field (WILSON et al., 1961); or laboratory (STANNARD, 1958; HODGE and THOMAS, 1958; BAIR et al., this volume and 1962).

Because of the similarity of anatomical structure of the lung of the burro to that of man and the similarity of deposition and clearance patterns between burro and laboratory data for man it was suggested by WILSON and WILSON and TERRY (op cit.) that the figures for the burro be used as a guide for man with respect to inhalation sequelae, although the burro may not be a particularly appropriate model for man in other respects.

An interesting feature of the Roller Coaster operation was the introduction of one test shot in a typical earth-covered high explosive magazine. There was relatively much more inert dust in the cloud at this event. The lung clearance rates found in the limited number of animals exposed to this cloud were significantly higher than in the other events where less inert dust was present. The difference approached a factor of three (WILSON and TERRY, 1968). This could be attributed to the more rapid clearance of mineral particles to which plutonium was attached as a tracer as compared to PuO_2 itself or mixtures with a smaller mineral phase component.

C. Guide Lines for Local Contamination with Plutonium

The data from the field tests combined with controlled laboratory experiments give a reasonable start on assaying the hazard from accidental area contamination with plutonium in moderate to high concentrations. It was determined early that some arbitrary lines needed to be drawn to guide action in case of a contaminating incident[11]. Based in part on results of the Test Group '57 experiment and in part on a series of assumptions (LANGHAM, 1972) an informal convention was developed of using 10, 100, and 1000 μg $^{239}Pu/m^2$ surface contamination as operating contour lines. It was planned that access would be prohibited inside the 1000 μg/m^2 line and full decontamination procedures instituted, including removal of surface soil. Restrictions on occupancy and operations would apply

11 Primarily an incident like those for which the field tests are most relevant. Obviously an incident under very different conditions might require rather different guidelines for action. Also political, sociological and economic factors may modify action determined on purely technical grounds.

between 1000 and 100 μg/m² lines, and only surveillance with advisory precautions outside of the 10 μg/m² contour.

In actuality these somewhat arbitrary numbers seem to have been subject to several variations both in the interpretation of the basic contributory data (LANGHAM, 1971, 1972; JORDAN, 1971) and their application to any given situation. No agreed national or international standards exist and none of the guides have appeared with full documentation outside of the DOD literature. Also it may be argued that surface contamination expressed as units per square meter may not be an appropriate unit. HEALY (1972) for example points out that the depth to which the plutonium extends may be important in case of wind velocities sufficient to resuspend more than the "surface" layer. However, the present unit is convenient to measure in a field situation.

A report on this subject from outside the DOD has been prepared by KATHREN (1968). He reviews the basic factors entering the decisions and gives a summary of maximum permissible alpha surface contamination levels, primarily for places of work, in several countries. The figures he quotes from U.S. DOD sources are higher than most of these. He then presents a set of interim maximum permissible figures for environmental PuO_2 which range from 40 μCi/m² for remote or controlled limited—access areas to 0.04 μCi/m² for urban, suburban, or recreation areas.

It is very doubtful if life-time occupancy would even be considered for civilians under present conditions for any area with significant plutonium surface contamination (e.g. of the order of 10 μg ^{239}Pu/m² or more) although temporary occupancy by the military for urgent reasons could occur. The reasons for this gap between theory and practice are numerous. The amount of resuspension by wind of surface-deposited plutonium is very variable and its quantitative role subject to large differences in opinion among the working groups. Also in peacetime operations the application of the philosophy of reducing all radiation exposures to the *lowest practicable level*, always recommended by the standard setting bodies, would be applied fully to any local contamination incident in the public domain.

The importance of transfer processes and data on the movement of plutonium in the environment will be presented in Sec. IV. Meanwhile a short review of what was actually done in the two major contamination incidents will illustrate the application of the control principles outlined above.

D. Actual Incidents of Local Contamination

While there have been releases from and continue to be possibilities of local contamination with plutonium from nuclear facilities these are in no way comparable to the two major reported incidents involving nuclear weapons ("Broken Arrow") already cited *viz*: those at Palomares, Spain and Thule, Greenland. In each of the cases contamination contours were established as outlined above, although the choice of figures must have been based as much on what had to be done as on a scientific hazard analysis. At Palomares crops were removed from any area reading above 5 μg/m² (0.32 μCi/m², the lower limit of detection of the portable instruments available at the time) at the surface. Removal of vegetation and plowing to a depth of at least 10 inches was carried out in all accessible areas between this contour and that with readings of 500 μg/m² (32 μCi/m²) to reduce the probability of resuspension[12]. Areas with readings

12 It was originally proposed to plow only between 500 μg/m² and 50 μg/m² but with equipment on hand it was decided to go to the lower contour (JORDAN, 1971).

above 500 μg/m² (about 5.4 acres) were stripped of vegetation and topsoil and all contaminated material placed in containers for shipment to and burial in the United States (Langham, 1968). The total area plowed was approximately 550 acres (Langham, 1972).

In the incident at Thule, Greenland surface contamination contours were also established to aid in the operation, carried out under extremely severe conditions of the Arctic night. However in this instance *all* contaminated material that it was possible to move was removed, including some ice and snowpack, placed in huge containers adapted for the purpose and shipped back to the continental United States after the ice melted in the following summer (Langham, op cit.; Jordan, op cit.). One reason for the difference in the Thule operation was probably the fact that the contamination rested on ice and could become more widely disseminated on entering the water of the bay and the open sea than if confined to a terresterial environment (see subsequent discussion of the binding of Pu to soil).

Further details regarding the Thule operation were reported in the journal "USAF Nuclear Safety" (1970) with papers by each of the principal groups taking part including the Danish Group's General Scientific and Health Physics activities.

Although there was some personnel monitoring in these incidents, especially of the crews involved in decontamination work, the reports carry little reference to data from estimation of actual or potential human body burdens or data from animal sampling although it is understood that such programs were recommended and attempted[13].

Langham (1972) has reported on a revisit to both Thule and Palomares and summarizes the work since the incidents. At Thule the follow-up studies carried out by Danish investigators showed an elevated level of plutonium in bottom sediments of North Star Bay as well as in crustacea and bivalves which live on the bottom out to approximately 20 kilometers from the point of initial contamination. Higher animals (seals, walrus, birds) showed levels not significantly different from the worldwide fallout figure (see Sec. IV). This applied also to the ingesta of these higher forms even though their diets include some of the lower organisms characteristic of the bay bottom. No measurements were reported on man.

In the follow-up of the Spanish incident one hundred of the potentially exposed residents of Palomares were counted for plutonium in the lung counter facility at Madrid. No positive counts were found in any of these 100 individuals at the lower limit of detection of the counter, about 40 nCi. This was true also when some individuals were returned later after modification of the equipment had lowered the limit of detection of the counter to about 16 nCi. Urine samples (24 hour collections) were likewise negative although this does not indicate that

13 In this connection the recommendations of Wilson and Terry (1968) are of interest. They present curves for the decrease in lung concentration with time from the Roller Coaster test data and suggest that using such curves the lung *concentrations* in animals collected a few days or weeks after an accident might be used to estimate air concentrations of respirable aerosol where the animal was located at the time of cloud passage. There are obviously many sources for error in such an extrapolation but it is frequently likely to provide the only means to estimation of human exposure from such an accident. Wilson and Terry further suggest that since the initial lung concentrations of the three species (dog, sheep, burro) correspond so closely, it is reasonable to assume, in the absence of better information, that initial lung concentrations in other animals (e.g., goats, cows, horses) will also correspond and that for the first few days, at least, an extrapolation to initial lung concentrations and thence to human exposures can be made from sampling indigenous large animals.

there was *no* contamination. Also 16 nCi is well above the limit normally applied to non-occupational situations (HEALY, 1972).

An air sampling program in the village indicated measureable amounts of plutonium in occasional samples taken under conditions of high wind velocity. The mean Pu air values in the village were 0.38×10^{-15} μCi/cm^3 in 1966 and 0.09×10^{-15} μCi/cm^3 in 1967 (LANGHAM, 1972). (The maximum permissible air concentration for ordinary occupational exposure is 2×10^{-12} μCi/cm^3 and for continuous occupational exposure 6×10^{-13} μCi/cm^3.) Measureable amounts of this airborne material were found to have deposited on vegetation in and around the contaminated site almost entirely on the surfaces of leaves and stems. None of the amounts approached MPC levels but they were not difficult to measure[14].

Of great interest in the report of LANGHAM's return visit is the fact that the *primary* difference seen between the present town of Palomares and the one he saw when he arrived soon after the accident is the ugly scar on the landscape where all topsoil had been removed. Ecological recovery in these arid climates is slow indeed.

E. Conclusions Regarding Local Contamination Experiments and Incidents

The field tests and incidents described in this section have in general confirmed the prediction that plutonium released to the environment behaves approximately as expected from laboratory experiments. Inhalation is the most significant route of entry. About 15 percent of the inhaled plutonium is deposited in the lower respiratory tract and this leaves very slowly ($T^1/_2$ 1 year, up to perhaps 500 days or longer). Good correlation between biological uptake and aerosol parameters can be expected if, as in all such cases, the sampling devices really "see" the same aerosol as the exposed organism.

The approximate areas involved in each of the actual incidents or in the tests was from a few to several hundred acres. In isolated locations such as most of these were such areas are comfortably small and can be regarded as "local". But if such an incident occurred in a location of high population density the areas involved would be far from comfortably small. Thus the need for some ready plan and criteria for action is evident. LANGHAM (1968) summarizes some of these and makes some interim suggestions. Others are considered in Sec. III C above. Of great importance to the audience to which this chapter is directed is the realization that experience has been gained and trained personnel are immediately available and in readiness for use if needed. But further work, particularly covering the movement of plutonium in the environment is needed and a better public information system must be extant to prevent a local contamination incident from inducing unwarranted fears and stimulating overaction while at the same time there is enough concern to allow application of the necessary measures for isolation and decontamination. This point is spoken to in more detail by JORDAN (1971).

Nevertheless there are some important and perplexing scientific problems still unsolved. Chief among these is the "hot particle" problem which appears over and over again in consideration of radioisotope inhalation toxicology and is

14 Some of this follow-up work has been presented by the Spanish workers at meetings but is not generally available by journal publication. These are for example, IRANZO (1968), IRANZO and RAMOS (1969), IRANZO and SALVATERE (1970). Copies, particularly of the second publication and of a Los Alamos report by MOSS (1966) may be available in some libraries. Also ODLAND et al. (1968) have reported on exposures of participants in the cleanup.

discussed in some detail elsewhere in this volume (Chaps. 11, 12). This refers to the evaluation of biological effects and hazards from small but highly radioactive particulates which may be inhaled. The dose rate around such sources may be locally quite high. Will this simply kill the surrounding cells or become a focus for future neoplasia? A definite solution to this dilemma eludes us still although current data suggest the uniform source is worse than the highly discrete one. As long as such problems remain unsolved ultra-conservative approaches to contamination events can be expected. Definitive research to resolve the biological aspects of this "hot particle" problem continues (Brues, 1970) but somehow does not seem to get done on the scale needed to answer the questions.

IV. Ecological and General Environmental Consideration of Plutonium Contamination

A. Dimensions

We turn now to a consideration of more general environmental contamination and of the possible consequences to the environment from long-standing or long-developing plutonium contamination. Langham (1968) estimates that various nuclear energy activites, including those discussed above, have contributed about 500 kilocuries of plutonium (mostly ^{239}Pu) to the world environment. The largest single contribution has come from atmospheric testing of nuclear weapons, about 300 kilocuries (Harley, 1971).

In the period of major concern for radioisotopes in the environment i.e., during and after the period of atmospheric testing the release of fission products took precedence. In terms of activity released, solubility and mobility in the biosphere, and thus potential hazard to man the fission products are of far greater importance initially. However with its very long physical and biological half-life and high relative toxicity plutonium cannot be ignored as a general environmental contaminant. This is not to say that the average levels from fall-out to date are or are approaching hazardous ones; they are far below hazardous levels to even the most sensitive organism or stage of development. But they are easily measurable and must be considered both at the time of release and distribution, and cumulatively.

Hardy and Krey (1971) give total ^{239}Pu in soil at various sampling sites where no contamination is known to have occurred except that from generalized atmospheric distribution. ^{239}Pu averaged 2.3 mCi per km$^2 \pm 13$ percent for New York area sites. Accumulated deposits were similar in mid-west sampling locations but lower than these (about 1.0 mCi/km^2) in Florida and southern Texas, and less than 1.0 in Los Angeles. These are consistent with the distribution of fission products in fall-out both with respect to latitudinal variations and precipitation patterns. The amounts look large when integrated over an area as large as a square kilometer but they amount to only a few d/min above background in a 1 gram soil sample and, as indicated in Chap. 14, this is a small fraction of the naturally occurring alpha-particle activity of soil.

Plutonium-238 has entered the environment through the reentry and burn-up in the upper atmosphere of a SNAP-9A device (see Chap. 8) over the Indian Ocean in 1964. Harley (1971) describes the behavior of this material and the fact that the ^{238}Pu/^{239}Pu ratio can be used to study the behavior of such plutonium in the atmosphere. About 17 kCi of ^{238}Pu were contained in the original satellite. This slowly reached the ground, much of it in the southern hemisphere over the next several years with a peak in early 1966 in the southern hemisphere, somewhat later in the northern hemisphere.

B. Movement into the Biosphere and Man

1. Terrestrial Environment

The low solubility of plutonium in water and biological fluids and its tendency to remain fixed in soil or other media permits prediction of large factors of discrimination against its movement into living organisms and to man. LANGHAM and BECKER (1969) and LANGHAM (1971) have summarized data bearing on this point utilizing earlier published papers and declassified reports. Plutonium uptake by plants from soil or other growth media is of the order of 10^{-3} to 10^{-4} of that in the media on a dry weight basis. Transfer from plant to man or animal involves a gut absorption factor of less than 10^{-4} in the adult. Even in the very young infant absorption from the G.I. tract is only about 0.01. Only after inhalation are the factors of absorption and retention large enough to support prediction of significant body burdens from any ordinary contamination source[15].

The contrast in movement into the biosphere between plutonium and other common radionuclides is marked. This is seen in the voluminous literature on the movement of fission products in the environment a small fraction of which includes the transuranic elements. For example, JACOBSON and OVERSTREET (1948) found that the activity transported from clay suspensions to leaves of barley ranged from 1.60 percent of dose for ^{89}Sr to 0.01–0.00045 percent of dose for Pu in various valence states. Few other nuclides tested showed as low a translocation factor as plutonium although few were as high as Sr. Amounts initially fixed in the clay or taken up by the root were both much larger and more alike for the several nuclides. In studies of the content of fall-out radionuclides in hay KLOKE and LUDWIEG (1962) found factors of between 23 and 100 between ^{90}Sr and ^{239}Pu contents over the period 1954 to 1960. Some of this of course, is due to differences in the amounts of the elements released. REDISKE et al. (1955) measured the concentration factors (concn. isotope in leaves/concn. in soil) in barley plants for a large number of fission products and found the largest factor for ^{90}Sr. In the one test involving plutonium the concentration factor (Ephrata loamy sand to plant) was 1.7 for ^{90}Sr, 1.05 for ^{131}I, 0.1 for ^{137}Cs, and only 0.0009 for ^{239}Pu.

On this basis it can be concluded that plutonium is very much less likely to pass through terrestrial food chains to man than a majority of the fission products or many other radionuclides potentially present in the environment. This large factor of contrast is, however, not so large that concern for plutonium in the terrestrial environment can be written off entirely.

ROMNEY et al. (1970) have conducted long-term cropping experiments under glasshouse conditions. The results, while showing relatively little transfer from soil to plants, showed a small but consistent increase in accumulation of Pu in

15 The importance of the inhalation route was emphasized in earlier Sections of this Chapter. In terms of accumulated amounts from fall-out to date the average amount in or on soil and likely to be inhaled as dust is small enough to be little cause for concern. But again (see p. 677/678) the "hot spot" problem enters. SCHIAGER (1964), for example, measured alpha-active fallout particles over a $2^1/_2$ year period following atmospheric weapons tests in 1961 and found the size range of suspended particles to be one to four microns, well within the range expected to be inhaled and deposited in upper respiratory passages and the lung. He estimated that the average amount of plutonium retained in the lungs of exposed individuals as 2–6 pCi. This is orders of magnitude below the maximum allowable lung burden but there arises aqain the concern that highly active particles could be hazardous even though the average concentration over an entire organ, like lung, is very low. This question is still unresolved as discussed earlier in this Chapter. The experimental data in animals were summarized in Chap. 11. Eventually the U.S. Transuranium Registry should help establish the effectiveness in man from actual analyses of lung tissue at autopsy.

Table 15.1. ^{239}Pu in successive crops of ladino clover grown on contaminated soil under glasshouse conditions. (ROMNEY et al., reproduced in part from Health Physics, **19**, 487–491, 1970 by permission of the Health Physics Society and the author)

Cropping period	Crop yield g/dry weight	^{239}Pu in dry plant material	
		d/min/crop	d/min/g
1958	66	205	3.1
1959	110	737	6.7
1960	194	1377	7.1
1961	310	3565	11.5
1962	316	7142	22.6

The growth was in a 1:1 mixture of Yolo soil and NTS Area 11 soil. The residual ^{239}Pu activity was 1.62×10^5 d/min/g (1.18 μg/g soil).

plant tissue during a 5-year period of successive croppings. The data, taken from their publication are reproduced in Table 15.1.

There was an apparent increase from about 3 d/min/g to about 23 d/min/g over a 5 year cropping period. The authors suggest that some of this increase may reflect continuing development of the root system until the crop was fully established. Another factor might be the role of natural chelating agents in the soil making the plutonium more available to the plants, and the authors produced experimental evidence that some chelating agents can enhance plutonium uptake by plants. However, on balance it should be observed that the uptakes are still very low even after the five years of cropping and major alterations of availability are not presaged.

Another way to measure ecological and environmental behavior of plutonium is by the collection of flora and fauna over extended periods. Examination of Nevada Test Site environment began very early after the events described earlier (1956, 1958, 1966) and some of the data are summarized by ROMNEY, MORK and LARSON (1970) and ROMNEY and DAVIS (1971). The kangaroo rat and jackrabbit, the ubiquitous residents of these areas, were collected in a trapping grid close to sampling stations. Their tissues showed continued presence of plutonium, more in the jackrabbit than in the kangaroo rat although far from enough to produce biological change, as pointed out by PAGLIA (1968). Easily measurable amounts were present in bone, G.I. tract, and lung. There was more in bone than in lung and it was only in bone that there was a reasonably consistent tendency for accumulation with time to occur. The amounts were higher in animals from areas of higher residual contamination. In the other tissues reported from animals collected from the same location at different times (8 to 10 years apart) there was no consistent evidence of accumulation (ROMNEY et al. op. cit., Table 4). The total number of samples is, however, too small to establish this point beyond doubt. Gross anatomic lesions were "minimal" and non-specific (ROMNEY and DAVIS, 1971) and no hematologic changes were found.

The amounts in G.I. tract of the animals collected some time after the contaminating event were much greater than in lung. This suggests that most of the ingress of plutonium in these later periods of exposure may be via contaminated foliage or fallout particles swallowed when eating or ingested while preening fur after dust baths or digging. Thus for long-term exposure in these environments inhalation may be less important than expected in the shorter-term exposure situation. However since man would ingest much less of the soil and dust than these small animals this difference with time may not be so large as the animal data suggest.

The amounts present in plants *in situ* have not been reported in much detail (in contrast to experimental uptake studies). The information available is consistent with the notion that any plutonium found is primarily on the surfaces of leaves, needles, or stems (OLAFSON et al., 1957) and that it tends to remain thereon for rather long periods after deposition. LANGHAM (1972) makes a similar statement regarding the follow-up studies at Palomares.

2. Aquatic Environment

Information on this subject has, until very recently, been widely scattered and intermingled with data on fission products. We are indebted to NOSHKIN (1972) for a timely summary of the movement of plutonium in the aquasphere and into aquatic life forms. The primary source of plutonium for such studies was fall-out and the average concentrations are therefore very low. But NOSHKIN also presents some data for nuclear facilities (Nuclear Fuel Services, West Valley, New York and Windscale, England).

While concentrations of plutonium are consistently below those of ^{90}Sr and ^{137}Cs the factor of difference does not appear to be as large as those found in the terrestrial environment. Also quite significant concentration factors (ratio of amounts in organisms to amounts in surrounding water) are reported for some marine invertebrates and algae. These range from a mean value of 24 for the scallop adductor muscle (sample from Cape Cod area) to 21000 for the Sargassum weed (North Atlantic), the latter an unexpected and unexplained high affinity. (NOSHKIN points out that since the Sargassum weed is transported in quantity out of the Sargasso Sea it could be an important mode for transport of plutonium from one water mass to another.)

Amounts of plutonium in marine vertebrates showed behavior similar to that in terrestrial vertebrates with primary concentrations in bone, liver, and gut. Concentration factors were lower than the maximum seen in invertebrates and algae. In bone they ranged from 21 in the blue shark to 570 in the blue fish; in liver from 14 in bass to 64 in tautog; in muscle from 1 in flounder to 13 in the maho shark. The highest concentration factor reported was 1060 for gut in the tautog.

Such concentration factors must be viewed in relation to their base line, usually less than 1 fCi/l. Thus actual concentrations are less than 1 pCi/kg wet weight in the marine invertebrates and algae and somewhat less in the vertebrates collected in areas not locally contaminated. These are biologically unimportant. But as the base line rises so does the amount in organisms, as data from specimens collected near Thule, Bikini, Guam, etc. attest. Thus if the amount of plutonium in the aquasphere should rise markedly we would expect the food chains in the aquatic environment to present a greater potential hazard to man than the terrestrial food chains described in Sec. IV B 1 above. Fortunately the large gut discrimination factor in man would still hold, as would the fact that the most usually consumed parts of marine food sources have the lowest concentrations[16].

Some of the most illustrative figures for marine forms are *summarized in Tables 15.2* and 15.3, adapted from Noshkin's paper.

16 An interesting side-light on this feature is quoted by BOWEN et al. (1971). They refer to WARD's (1966) study of plutonium uptake by the lobster. As in other marine invertebrates the highest concentration of ^{239}Pu is in the calcified shell (about 90 percent). All fine except for the devotees of lobster bisque—a soup thickened by a paste prepared from lobster shell! As in all such cases the eating habits of the population are frequently important.

Table 15.2. Plutonium-239 concentrations and concentration factors in marine invertebrates and algae. [Selected from data summarized by Noshkin (1972) with permission of author, publishers and symposium editors. Original sources cited by Noshkin][a]

Organism	Location	Tissue	Mean values		Date
			pCi/kg wet	Concn. factor[b]	
Blue mussel	Cape Cod	body	0.26	300	1970
	Cape Cod	shell	0.42	490	1970
Soft shell clam	Cape Cod	body	0.37	440	1970
Oyster	Cape Cod	body	0.11	130	1970
Scallop	Cape Cod	adductor	0.02	24	1970
	Cape Cod	body	0.49	520	1970
	Cape Cod	shell	0.59	600	1970
Crustacea	Dannish inner waters	body	1.5	1765	1968
Zooplankton (mixed)	Atlantic	—	2.03	2300	1961
Sargassum (6 samples)	Atlantic	—	20.7	21000	1965–1970
Blue mussel	Windscale area	body	2450	—	1964
Bivalves (mixed)	Thule (zone I)	body	8000	—	1968
Bivalves	Thule (zone II)	body	40	—	1968
Zooplankton	Thule	body	4.8	—	1968
Porphyra	Windscale area	body	1470	—	1969

[a] Modified from Health Physics **22**, 537–550 (1972) by permission of the Health Physics Society. [b] Concentration factor = activity per unit weight of organism/activity per unit weight of sea water.

Table 15.3. Plutonium-239 concentrations and concentration factors in marine vertebrates. [Selected for data summarized by Noshkin (1972) with permission of author, publisher and symposium editors; original sources cited by Noshkin].[a]

Organism	Location	Tissue	Mean values		Date
			pCi/kg wet	Concn. factor[b]	
Blue shark	Cape Hatteras	bone[c]	0.02–0.03	21	1970
		muscle	0.001–0.005	4	1970
Blue fin tuna	Cape Cod	bone[c]	0.1	90	1970
		muscle	0.003	2	1970
Blue fish	Cape Cod	bone	0.6	570	1969
		muscle	0.005	5	1969
		liver	0.03	26	1969
Flounder	Cape Cod	bone	0.05	50	1970
		muscle	0.002	1	1970
		liver	0.06	57	1970
		gut	1.2	160	1970
Striped bass	Cape Cod	bone	0.2	160	1970
		muscle	0.005	4	1970
		liver	0.02	14	1970
		gut	0.04	36	1970

[a] Modified from Health Physics **22**, 537–550 (1972) by permission of the Health Physics Society. [b] Activity per unit weight of organism/activity per unit weight of sea water. [c] Backbone.

3. Concluding Comments on Sec. IV. B

How much plutonium has actually entered the biosphere? Information is very limited except for specific sites such as those described above and the world-wide concentrations. HARLEY (1971) quotes figures obtained by MAGNO et al. (1967) for total diet, human lung and bone in 1965–1966 at the maximum of deposits from atmospheric testing. Dietary intake was estimated to average 7×10^{-3} pCi/day. The amounts in lung samples were 0.45 pCi/kg average, and in bone 0.04 to 0.12 pCi/kg. HARLEY (op cit.) (summarized by PARKER in Chap. 14 of this volume) quotes data developed by workers in the USSR (SMORODINTSEVA et al., 1969) giving lung concentrations of the order of 0.15 pCi/kg with much higher concentrations in pulmonary lymph nodes (not unexpected, STANNARD, 1958). These are all minute concentrations and far below those associated in present work at least, with biological change. They are consistent with the statements made above regarding the large discrimination factor against plutonium entering terrestrial food chains in quantity. In a sense they are a testimony to the tremendous sensitivity of techniques for measuring radioactivity and radionuclides. Yet they do show that measurable amounts of man-made element[17], plutonium, have entered the biosphere and man. The highest concentration factors appear in the aquatic environment.

C. Movement of Plutonium in the Environment

A rather considerable effort has been and is being expended to determine the general characteristics of plutonium transport in the environment. In determining world-wide movements the same techniques are applied as for fission product transport and fall-out. Frequently ratios of ^{239}Pu to ^{90}Sr or to ^{137}Cs provide a useful index (HARLEY, 1971; CORLEY et al., 1971) particularly for determining any unique behavior of plutonium, in either terrestrial or aquatic environments.

More work has been done and more importance has been attached to the terrestrial movement of plutonium in relatively localized areas. Again the Nevada Test Site has provided and will provide the best controlled and most thoroughly studied area. In fact quite intensive further study is presaged by the organization by the AEC of a Nevada Applied Ecology group. This group will consider also the requirements of the National Environmental Policy Act of 1969 (Public Law 91–190) and related statutes. An important feature of this is a "Plutonium Environmental Studies Program" which was described in general terms by ROMNEY and DAVIS (1971). The current studies are much more sophisticated than earlier ones and much more extensive by virture of new techniques for low level (helicopter) aerial surveys using the FIDLER probe (for americium) developed by Lawrence Radiation Laboratory and used, among others, by Reynolds Electrical and Engineering Co. staff. Also the generally less laborious methods now available for plutonium analyses make such work more practical to carry out than heretofore. (Some brief remarks on current methodology are contained in Chap. 14.)

Details of the results are quite beyond the purview of this book but a few generalizations may be of interest. Firstly movement of plutonium is largest, as might be expected, at or soon after the time of its release. Various fall-out models for this phase have been constructed and tested. Once deposited on ground, foliage, structures, etc. it does not move about very readily although it can be

17 Recently the occurrence of a very long-lived isotope of plutonium, ^{244}Pu, has been reported in a Precambrian age carbonate from Southern California (HOFFMAN and LAWRENCE, 1971) and it would be expected that ambient neutron fluxes would produce small quantities of ^{239}Pu from natural uranium in soils but the amounts are so small as not to negate the significance of the statement made.

removed by suitable decontamination procedures. In a desert environment with high winds, movement of the surface contamination clearly occurs but it usually falls off rapidly with distance from the source. In one set of surveys for example (Mork, 1970) residual plutonium in one area which was at 2000 μg/m² at ground zero fell to 4 μg/m² at a distance downwind of 6 miles. In another area the figures were from 780 μg/m² at ground zero to less than 1 μg/m² in soil at four miles. Yet wind transference is not to be considered unimportant. Witness the fact that the major movement of plutonium from the Rocky Flats plants is ascribed to wind transference (see Chap. 14)[18].

Plutonium does not appear to enter terrestrial surface or near-surface waters to any appreciable extent from blown dust or fixed contamination since it remains generally well-fixed to soil components.

These facts indicate that plutonium found at distances of many miles from suspected sources (e.g. Olafson et al., 1957, and current PHS "off-site" surveys, Bliss and Dunn (1971) may have come from fall-out at the time of initial release, deposition from world-wide fall-out, or slow transfer of small quantities from a moderate to high level source by wind and general erosion processes.

On an even more local scale it is interesting to note that in the desert environment there is a marked tendency for the surface contaminants to accumulate around clumps of vegetation. Since these are miniature ecosystems of a sort there is undoubtedly some concentration of nuclides in the biota from this process. Its magnitude is difficult to quantify.

Few actual measurements exist for movement of plutonium in other types of terrestrial environment. There is little in the experimental work which would predict transfer factors higher than those seen in the desert studies. Indeed it may well be that, as predicted, the desert environment maximizes the chances for transfer by causes other than the activities of man.

It might be expected that plutonium fires would be a formidable source of area contamination. In general this does not seem to have been true. Obviously most of these events have occurred within facilities where containment is the watchword in design and operation (Mishima, 1964). But even events where containment has obviously been breached or the events which occurred in the open do not seem to have produced a finely dispersed widely disseminatable aerosol (Hammond, 1971 and Chap. 14).

In a recent detailed summary of the Rocky Flats area in Colorado Krey and Hardy (1970) and Hammond (op cit.) found the bulk of plutonium-related activity was downwind of the prevailing wind vector, *not* in the direction it would have been carried by the winds on the day of the major fire in May 1969. Leaking barrels of plutonium-laden cutting oil stored near a corner of the site are considered the most likely source of the contaminant with a possible small contribution from other sources. Rocky Flats plutonium may be distinguished from weapons test fall-out plutonium as far as 30 to 40 miles east of the site but confirmation of this depends on further analyses including correlation with ^{90}Sr analyses[19]. The 3 mCi/km² contour is currently the lowest one at which this

18 Much effort has been expended to determine an entity called the "resuspension factor" in various environments and under various conditions. The factor is defined as the ratio of air concentration (μg/m³) above a contaminated surface to the ambient surface contamination level (μg/m²). This factor is not used in this chapter since it has not proven to be either very useful or easily measured. However for the record let it be entered that the values found tend to be low, e.g. 10^{-5} to 10^{-7} in the environment of the Nevada Test Site with an extreme range from as high as 10^{-2} to as low as 10^{-11} (Langham, 1971).

19 As detailed in Chap. 14 the distinction between locally originating material and that from worldwide fallout can be based also on the plutonium to ^{137}Cs ratio.

distinction can be made with reasonable confidence. This extends about 8 miles to the east and southeast of the plant and much closer in other directions. Most of this is thought to have been transferred at times of high wind velocities (strong gusty westerly winds exceed 70–80 mph several days each year). In this study movement down into the soil seemed to be somewhat greater than at NTS—*viz*: as far as 13 cm with as much as 60 percent of the plutonium below 5 cm in some cases. However, the depth distributions showed wide variations due in part to differences in the depth of soil atop underlying rock and the character of the soil.

Again the amounts measured while easily detectable are, except close upon "ground zero", far below the maximum allowable amounts for population exposure by current or anticipated criteria.

Movement of plutonium in aquatic environments has been summarized by Noshkin (1972). As expected from the general behavior of elements in its class it is relatively insoluble and goes relatively quickly to the bottom sediments. There is thus a geochemical separation from other constituents of fallout or nuclear facility effluents. Cores taken from the bottom of Buzzards Bay in Massachusetts confirm this view as do samples at various depths in the open ocean and various estuaries. Thus plutonium tends to be rather rapidly depleted from surface waters unless held by a surface organism. The same holds for fresh water as the studies of Magno et al. (1970) at the NFS reprocessing facility demonstrate. Thus, as found, the concentrations in organisms are higher in those feeding on surfaces or sediments.

However, the summarized data (Noshkin, 1972) indicate greater mobility of plutonium in the marine environment than anticipated and it is more involved with life processes. Thus the potential for hazard in the event of any conspicuous rise in worldwide levels or local contamination events must be viewed conservatively until more information is obtained from the aquatic environment[20].

V. Summary[20]

In conclusion, let it be stated that the potential for environmental contamination with plutonium was realized very early in the development ofnuclear energy. Major installations developed area surveillance procedures and have assiduously applied very conservative criteria over many years with a guiding philosophy of maximum containment. With plutonium becoming a more common article of commerce the opportunities for local or small area contamination rise. To anticipate these several overt tests have been conducted in isolated locations involving animal exposures as well as a battery of physical measurements. These have shown less build-up of plutonium during living in a contaminated area than was predicted; on the other hand, more than expected deposition occurred in lungs and other tissues from the passage of the initial cloud. Criteria based in part on these field tests and in part on laboratory data can be and have been applied to actual events such as those at Palomares, Spain and Thule, Greenland.

Much of the plutonium currently residing in soil, water, and the biota of the world arose from the atmospheric testing of nuclear weapons. A small fraction is plutonium-238 from the burn-up of a SNAP device on reentry of a space vehicle. The balance is from several sources including minute but long-term releases from nuclear energy facilities.

Movement of plutonium into the biosphere through terrestrial food chains, water supply, etc. is of a low order because of its low solubility and poor absorp-

20 Attention directed toward environmental contamination with plutonium should not obscure the fact that many sources will contain also americium, which appears to transfer much more readily in the environment. See Chap. 20.

tion from the gastrointestinal tract. Entry *via* inhalation results in more retention and more absorption than *via* other routes. However the amount of plutonium which is airborne falls off with time to such an extent that long exposures may result in greater burdens in the G.I. tract than in the lung. Only bone shows a clear tendency to accumulate plutonium over years of exposure of small animals in the field. In test animals even the 160 day test period was too short to demonstrate this effect if it occurs at all in larger species.

In aquatic environments there is evidence of greater mobility of plutonium and greater concentration in some forms than anticipated from its behavior in terrestrial environments. This aspect requires more study.

Future Portents. Movement of plutonium in the environment and its potential as a hazard to man and the environment is currently under intensive study as part of current activities growing out of the National Environmental Policy Act of 1969. Newer methods allow much more rapid surveys and analytical problems are much less severe than even a few years ago. While nothing in recent results negates the general conclusions that the amounts now extant are far below those clearly hazardous to the present generation there are still large gaps in our fund of information. While hardly to be expected from plutonium we cannot eliminate genetic change as a possible long-term result of contamination of the biosphere with other radionuclides. Experimental work in this area is almost non-existent, in contrast to the extensive studies with external radiation. Also, as pointed out above certain lower aquatic forms of life are known to concentrate plutonium by fairly significant factors. With an element of very long half-life and demonstrated high effectiveness in producing biological damage we can only proceed with efforts to mount additional ecological studies for both our own satisfaction as scientists and our ability to reassure the public that some important factor, route of transfer or effect has not been overlooked.

References

Aberg, B., Hungate, F. P.: Radioecological concentration processes. Proc. Internat. Symposium held in Stockholm, 25–29 April, 1966. New York: Pergamon Press 1967.

Bair, W. J., Willard, D. H., Herring, J. P., George, L. A., II: Retention, translocation and excretion of inhaled $^{239}Pu\ O_2$. Hlth Phys. 8, 639–649 (1962).

Bliss, W., Dunn, L.: Measurement of plutonium in soil around the Nevada Test Site: p. 89–92 in: Los Alamos Scientific Laboratory Reference LA-4756 (1971).

Bowen, V. T., Olsen, J. S., Osterberg, C. L., Pavera, J.: Ecological interactions of marine radioactivity. In: Radioactivity in the marine environment. National Academy of Sciences, 1971.

Brues, A. M.: Mechanisms of carcinogenesis. Argonne National Laboratory, Report ANL-1770, Annual Report of Division of Biological and Medical Research, Dec. 1970.

Cantril, S. T., Parker, H. M.: Status of health and protection at the Hanford Engineer Works: Paper 9 in: Industrial medicine of the plutonium project (R. S. Stone, ed.), p. 476–484. Natural Nuclear Energy Science. New York: McGraw Hill 1951.

Coopey, R. W.: The accumulation of radioactivity as shown by a limnological study of the Columbia River in the vicinity of the Hanford Works, Preliminary Report. USAEC document HW-11662 (1948).

Corley, J. P., Robertson, D. M., Brauer, F. P.: Plutonium in surface soil in the Hanford Plant environs. In: Proc. of Environmental Plutonium Symposium, held at LASL, Aug. 4–5, 1971, p. 85–88, LA-4756 (1971).

Cowan, M.: Plutonium contamination from one-point detonation… WT-1510, Sandia Corporation, Albuquerque, New Mexico, Aug. (1960).

Dick, J. L., Baker, T. P., Jr.: Monitoring and decontamination techniques for plutonium fallout on large-area surfaces. Operation Plumbbob, Defense Atomic Support Agency, Sandia Base, Albuquerque, New Mexico, Report WT-1512, March (1967).

Environmental Plutonium Symposium, Proceedings of: See Los Alamos Scientific Laboratory.

Gamertsfelder, C. C.: Fish laboratory report for first half of 1947. AEC report, HW-8080 (1947).

HAMMOND, S. E.: Industrial-type operations as a source of environmental plutonium, p. 25–35, in: Los Alamos Scientific Laboratory Report, LA-4756 (1971).
HARDY, E. P., KREY, P. W.: Determining the accumulated deposit of radionuclides by soil sampling and analysis, p. 37–42, in: Los Alamos Scientific Laboratory Report, LA-4756 (1971).
HARLEY, J. H.: Worldwide plutonium fallout for weapons tests, In: Environmental plutonium symposium. Los Alamos, New Mexico, Aug. (1971), Los Alamos Scientific Laboratory Report, LA-4756 (1971).
HEALY, J. W.: Personal communication to the author, June, 1972.
HERDE, K. E.: A one-year study of radioactivity in the Columbia River fish. USAEC report HW-11344 (1948).
HERDE, K. E., CLINE, J. F.: Radioactivity in Pekin ducks on the Columbia River. USAEC Report, HW-12079 (1949).
HODGE, H. C., THOMAS, R. G.: Health hazards from the inhalation of insoluble uranium and thorium dioxide. Proc. of Second Intl. Conf. on Peaceful Uses of Atomic Energy, United Nations, Geneva, **23**, 302–305 (1958).
HOFFMAN, D. et al.: Isotopes—detection of plutonium 244 in nature. Nature (Lond.) **234**, p. 132–134, Nov. (1971).
International Atomic Energy Agency: Disposal of radioactive wastes into seas, oceans and surface waters. Proceedings of a Symposium. Vienna: IAEA 1966.
International Atomic Energy Agency: Environmental aspects of nuclear power stations. Proc. of a Symposium held in New York 10–14 Aug. (1970). Vienna 1971.
IRANZO, E.: First results for the program of action following the Palomares accident. Symposium on radiological protection of the public in a nuclear disaster (IRPA). Interlaken, Switzerland (June 1968).
IRANZO, E., RAMOS, E.: Measures to determine the risk to which a population can be subjected as a result of a nuclear accident generating radioactive aerosols. In: Environmental contamination by radioactive materials. Vienna: IAEA 1969.
IRANZO, E., SALVADOR, S.: Inhalation risks in the people living near a contaminated area. Second Internat. Congr. of the Internat. Radiation Protection Association. Brighton, England (May 1970).
JACOBSON, L., OVERSTREET, R.: The uptake by plants of plutonium and some products of nuclear fission adsorbed as soil colloids. Soil Sci. **65**, 129–134 (1948).
JENNINGS, C. D., OSTERBERG, C.: Sediment radioactivity in the columbia river estuary. In: Symposium on radioecology (D. J. NELSON and F. C. EVANS, ed). USAEC document, Conf.-670503 (March 1969).
JORDAN, H. A.: Distribution of plutonium from accidents and field experiments, p. 21–24, in: Los Alamos Scientific Laboratory Report LA-4756 (1971).
KATHREN, R. L.: Towards interim acceptable surface contamination levels for environmental PuO_2. Battelle Northwest Laboratories, Report BNWL-SA-1510 (April 1968).
KLOKE, A., LUDWIEG, F.: Strontium-90, cesium-137 und plutonium-239 in heuproben var 1953 bis 1961. Naturwissenschaften **49**, 65–67 (1962).
KREY, P. W., HARDY, E. P.: Plutonium in soil around the Rocky Flats Plant. U.S.A.E.C. Health and Safety Laboratory, N.Y., Report HASL-235, Aug. (1970).
LANGHAM, W. H.: The problem of large-area plutonium contamination. Seminar paper 002, Bureau of Radiological Health, Public Health Service, U.S. Dept. of HEW (1968). (Copies may be obtained from Clearing House for Scientific and Technical Information, Springfield, Virginia, 22151.)
LANGHAM, W. H.: With editorial assistance of JAMES F. BECKER: Biological considerations of nonnuclear incidents involving nuclear warheads. Lawrence Radiation Laboratory Report UCRL-50639. April (1969).
LANGHAM, W. H.: Plutonium distribution as a problem in environmental science, p. 3–11 in: Los Alamos Scientific Laboratory Report LA-4756 (1971).
LANGHAM, W. H.: Personal communication to J. N. STANNARD, Jan. (1972).
LANGHAM, W.H.: Los Alamos Scientific Laboratory, Report LA-4756, Environmental Plutonium Symposium, Aug. 1971, issued Dec. 1971. Available from National Technical Information Service, U.S. Dept. Commerce, 5285 Port Royal Road, Springfield, Virginia 22151.
MAGNO, P. J., KAUFFMAN, P. E., SCHLEIN, B.: Plutonium in environmental and biological media. Hlth Phys. **13**, 1325–1330 (1967).
MAGNO, P. J., REAVEY, T., APENDIANAKIS, J.: Liquid waste effluents from a nuclear fuel processing plant. Proc. 5th Ann. Health Physics Soc. Midyear Topical Symposium, p. 208–220 (1970).
MERKER, L. G.: Design of equipment and facilities. Chap. 26, in: Plutonium Handbook — A guide to the technology (O. J. WICK, editor), vol. 2, p. 859–873. New York: Gordon and Breach 1967.

Mishima, J.: Review of research on plutonium releases during overheating of fires. Hanford Atomic Products Operation, Richland, Washington, Report-HW-83668, Aug. (1964).

Mork, H. M.: Redistribution of plutonium in the environs of the Nevada Test Site. UCLA Laboratory of Nuclear Medicine and Radiation Biology, Unnumbered report dated Aug. (1970).

Morrow, P. E.: Some physical and physiological factors controlling the fate of inhaled substances I. Hlth Phys. **2**, 366–378 (1960).

Moss, W. D.: Report on bioassay laboratory in Madrid, Spain. Los Alamos Scientific Laboratory, Report H5MR66-1 (Sept. 1966) (Cit. by W. Langham, 1972).

Nelson, D. J., Evans, F. C.: Symposium on Radioecology, USAEC document, Conf-670503 (1969) available from Clearing House for Federal Scientific and Technical Information, U.S. Department of Commerce, Springfield, Virginia 22151, U.S.A.

Noshkin, V. E.: Ecological aspects of plutonium dissemination in aquatic environments: What has plutonium to tell us about other transuranics? Proc. of Eleventh Hanford Biology Symposium, Sept. 1971. Hlth Phys. **22**, 537–550 (1972).

Odland, L. T., Farr, R. L., Blackborn, K. E., Clay, A. J.: Industrial medical experience associated with the Palomares nuclear accident. J. occup. Med. **10**, 356–362 (1968).

Olafson, J. H., Mishita, H., Larson, K. H.: The distribution of plutonium in the soils of central and northern New Mexico as a result of the atomic bomb test of July 16, 1945. Report UCLA-406. Univ. of California at Los Angeles, Atomic Energy Project, Sept. (1957).

Olson, P. H.: Some effects of pile area effluent water on young silver salmon. USAEC Report HW-8944 (1948).

Paglia, D. E.: Hematopathologic surveys of kangaroo rats (Dipodomys microps) populating plutonium contaminated regions of the Nevada Test Site. Hlth Phys. **15**, 493–498 (1968).

Rediske, J. H., Cline, J. F., Selders, A. A.: The absorption of fission products by plants. Hanford Atomic Products Operation, Report-HW-36734 (May 1955).

Romney, E. M., Davis, J.: Ecological aspects of plutonium dissemination in terrestrial environments: What have we learned from Nevada Test Site data? Paper presented at Eleventh Hanford Biology Symposium, Richland, Washington, Sept. 1971. Hlth Phys. **22**, 551–557 (1972).

Romney, E. M., Mork, H. M., Larson, K. H.: Persistence of plutonium in soil, plants, and small animals. Hlth Phys. **19**, 487–491 (1970).

Schiager, K. J.: Alpha-active fallout particles; physical characteristics related to pulmonary exposure. Ph.D. Thesis, University of Michigan (1964).

Schultz, V., Klement, A. W., Jr.: Radioecology. Proc. 1st National Symposium on Radioecology, Colorado State University, Fort Collins, Colorado, Sept. 10–15 (1961). Reinhold Publishing Corporation and the American Institute of Biological Sciences, U.S.A. (1963).

Smorodintseva, G. I., et al.: Study of uptake of airborne Pu-239 by the human organism. U.N. Scientific Committee Document A/AC.82/G/L.1301, HASL Translation, Nov. 1969.

Stannard, J. N.: An evaluation of inhalation hazards in the nuclear energy industry. Proc. Second Intl. Conf. on Peaceful Uses of Atomic Energy, United Nations, Geneva **23**, 306–312 (1958).

Stewart, K., Thomas, D. M. C., Terry, J. L., Wilson, R. H.: A preliminary evaluation of the biological measurements on Operation Roller Coaster (Joint US/UK Experiments). UK Atomic Energy Authority. Report AWREO-29/65, AWRE, Aldermasten, Berks. July (1965).

USAF Nuclear Safety: Vol. 65 (Part 2), special edition—Project Crested Ice. Jan. Feb. Mar. (1970), Directorate of Nuclear Safety (AFINST), Kirkland AFB, New Mexico, 87117.

Ward, E. E.: Uptake of plutonium by the lobster, *Humarus vulgaris*. Nature (Lond.) **209**, 625–626 (1966).

Watson, D. O., Cushing, C. E., Coutant, C. C., Templeton, W. L.: Effect of Hanford reactor shutdown on Columbia River biota. In: Symposium on Radioecology (D. J. Nelson and F. C. Evans, ed.). USAEC Document, Conf-670503 (March 1969).

Wilson, R. H.: Personal communication concerning reanalysis of the lung clearance data (1972).

Wilson, R. H., Terry, J. L.: Plutonium uptake by animals exposed to a non-nuclear detonation of a plutonium-bearing weapon simulant. Part I. Field Operations, University of Rochester Atomic Energy Project, Report No. UR-665, June (1965).

Wilson, R. H., Terry, J. L.: Plutonium uptake by animals exposed to a non-nuclear detonation of a plutonium-bearing weapon simulant. Report POR-2512 (WT-2512). Operation Roller Coaster, Defense Atomic Support Agency, Sept. (1968)[21].

Wilson, R. H., Thomas, R. G., Stannard, J. N.: Biological and medical studies associated with field release of plutonium. Operation Plumbbob, WT-1511, University of Rochester, Rochester, N.Y. (1961).

21 Inquiries should be addressed to Director, Defense Atomic Support Agency, Washington, D.C. 20305, U.S.A.

The Transplutonic Elements

Preface

Among the isotopes of the transplutonium actinides are found some of the most toxic elements known; the hazards are entirely ascribable to the radioactivity. Comparing the alpha specific activities (i.e., in this sense—the toxicities) of various isotopes carries a certain academic interest, i.e., the "most toxic" nuclide can be selected, but there may be also practical value in pointing out the relative toxicities at least in orders of magnitude among the transplutonium nuclides finding increasingly wide use. The nuclides with the most intense alpha activities have short half-lives, sometimes vanishingly short, and therefore are usually not quantitatively important. Amazingly minute amounts of some nuclides produce one curie of alpha activity: the smallest calculated mass is 32.4 nanograms of ^{244}Cf; 1.9 micrograms of ^{238}Cm, and 2.22 micrograms of ^{232}Pu are also fantastically small quantities. By comparison, the mass to give one curie of alpha activity of the technologically important nuclides is one or two orders of magnitude greater: 1.4 mg of ^{252}Cf, 308.2 mg of ^{241}Am, 57.45 mg of ^{238}Pu and 16.28 g of ^{239}Pu[1]. To keep a sense of proportion, these figures may be set beside the maximum permissible body burdens of ^{226}Ra plus daughters —0.1 ug, or of ^{239}Pu —0.64 ug (0.04 μCi). To paraphrase Dr. SEABORG's 1944 prophecy, the hazards of working with plutonium and the transplutonic elements are very great.

For these elements, rigorous safe handling practices must be specified. Both industrial safety and biomedical uses obviously require a quantitative understanding of the biological effects and the metabolic properties. In this section of the Handbook, most of the available information has been collected.

A brief history of the discovery of elements 95 to 105 serves as the introduction. The apparently conflicting evidence and the controversy that attended some of the early work has been resolved authoritatively, dispassionately, and with admirable clarity.

The next chapter describes the physical and chemical properties of these elements and their compounds giving considerable detail on americium, curium, berkelium, and californium, (little is known about the other elements) and stressing the properties of biological importance. The severe handicaps to experimental work imposed by the short-lives, the intense radioactivity, and the sometimes vanishingly small samples have been overcome through the use of elegant ultramicro techniques.

Fortunately for the purposes of this Handbook, the Eleventh Hanford Biology Symposium held at Richland, Washington, 27–29 September 1971 brought together more than 60 papers on The Biological Implications of the Transuranium Elements. It was appropriate that attention centered again on the topic, The Biology of the Transuranic Elements, selected for the First Hanford Biology Symposium a decade earlier. More than half of the 1971 papers specifically named "elements other than plutonium" and the published data from this meeting more than doubled the total information on a) the metabolic fate of the transplutonium elements, and b) their biological effects on living systems. As a con-

1 The radioactivity and the radiation toxicology of plutonium is discussed in detail in the Plutonium Section, e.g., in Chaps. 8 and 12.

sequence, Chap. 18 contains not only selections of the new data but also recalculations of new data incorporating them with older data into coherent presentations, together with hitherto unpublished research from the Lawrence Berkeley Laboratory. A critical summary of unprecedented scope has thereby been drawn from the totality of information.

The current standards for permissible exposures to americium, curium, berkelium, and californium are next discussed centering on the selection of maximum permissible concentrations in air and in water and of maximum permissible body burdens.

The effectiveness of the procedures safeguarding the health of those exposed to these elements rests squarely on the sensitivity and reliability of their bioassay. Specific analytical methods are given in detail in Chap. 20. This chapter, however, is not intended to stand alone, but should be read in conjunction with Chap. 13 which gives something of the rationale for these methods patterned as they are on the basic work with plutonium.

Finally, the current medical applications are reviewed. For example, americium-241 and californium-252, despite their extraordinary toxicity, serve as useful sources of neutrons or of monochromatic gamma rays.

The identification of the first man-made element, hailed a quarter century ago as a triumph of modern science, led promptly to the preparation of the actinides through lawrencium (element 103) and recently to rutherfordium (element 104) and hahnium (element 105). Nor will scientists concede that the limit has been reached; new superheavy elements, perhaps in the range of atomic numbers 110 to 126, may possess sufficient stability to be characterized or used. The production of gram to kilogram amounts, e.g. of americium or curium, set the stage for what are described as revolutionary changes in technology which permitted the use of their extraordinary properties as sources of electric power and in various biomedical applications. Benefits already demonstrated foretell an era of growing accomplishments for the good of mankind. Benefits in this era will be meaningless, however, unless atomic weapons are controlled so that the awful threats of disaster and destruction are forever removed.

Chapter 16

A History of the Transplutonic Elements[1]

A. Ghiorso

With 11 Figures

In the last twenty seven years it has been my good fortune to be associated with the discovery of all of the elements beyond plutonium. This time span has witnessed the development of experimental and theoretical techniques which encompass a very wide range—from the first almost intuitive gropings to find new elements to the modern sophisticated approaches where on occasion even single atoms can be identified with certainty. Now we are on the threshold of possibly opening up a vast new storehouse of knowledge if the predictions of a "magic island" of stability are borne out by experiment and this subject will be covered lightly at the conclusion of this narrative.

I. Americium and Curium (Elements 95 and 96)

Early in 1944 Glenn T. Seaborg had pushed to near completion the fantastic development of the chemical processes to be used in the Hanford plant that was to produce plutonium for weapons use. It was a natural step for him to undertake the task of trying to move further up the periodic table. For this purpose he assigned two of his young scientists, Ralph A. James and Leon O. Morgan, to the arduous chemical tasks that were to become necessary and asked me to concern myself with developing and using various detection devices needed in the experiments. The location of the discovery of the first two elements to be found, element 96, curium, and element 95, americium, was in the Metallurgical Laboratory of the University of Chicago and took place in the wartime years of 1944 and 1945.

First attempts were utter failures. Small amounts of ^{239}Pu in the milligram range were bombarded by neutrons and deuterons but no new alpha particle emitters were to be found, partly because of the relatively insensitive techniques that we employed at that time and partly because the wrong chemical properties were assumed for element 95. In July, 1944, we tried a different approach and bombarded ^{239}Pu with 32 MeV helium ions accelerated by the Berkeley 60″ cyclotron. The irradiated target was flown to Chicago and was chemically processed in accordance with the newly-developed actinide concept. Seaborg had come to the conclusion that the elements beyond plutonium should have a characteristic stable (III) oxidation state and thus be readily separable from the plutonium target material by using a suitable chemical procedure.

On July 14, 1944 a new longer range alpha particle activity was found in the expected chemical fraction. The identification, though crude, was definite. A range curve was measured in a 2π air-filled pulse ionization chamber by placing weighed

1 Work done under the auspices of the U.S. Atomic Energy Commission.

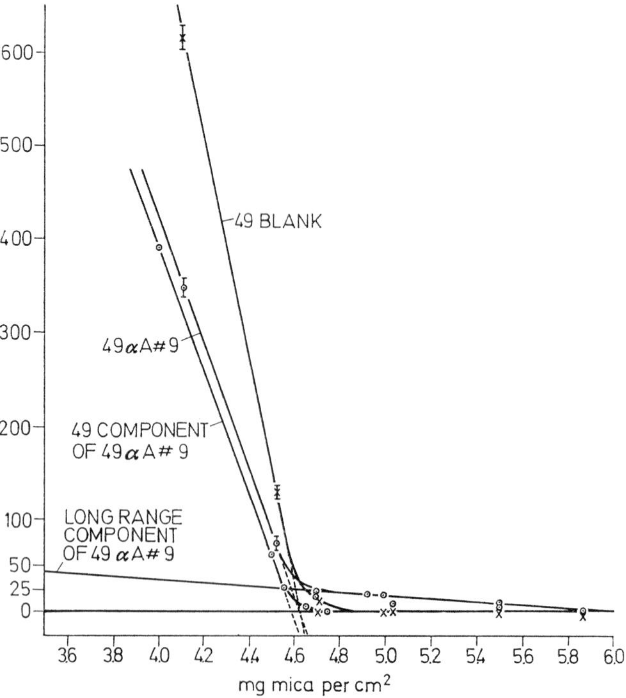

Fig. 16.1. Original mica absorption data showing presence of $^{242}96$ alpha particles in helium-irradiated ^{239}Pu

mica sheets over the sample and energetic alpha particles with a range of 5.86 mg/cm² were found. At the time we could not know whether the activity was due to element 95 or 96 since the reactions that were possible could produce isotopes of both elements.

$$\begin{array}{ll}
(1) & {}^{239}_{94}\mathrm{Pu} + {}^{4}_{2}\mathrm{He} \longrightarrow {}^{242}95 + {}^{1}_{1}p \\
(2) & {}^{239}_{94}\mathrm{Pu} + {}^{4}_{2}\mathrm{He} \longrightarrow {}^{241}95 + {}^{1}_{1}p + {}^{1}_{0}n \\
(3) & {}^{239}_{94}\mathrm{Pu} + {}^{4}_{2}\mathrm{He} \longrightarrow {}^{242}96 + {}^{1}_{0}n \\
(4) & {}^{239}_{94}\mathrm{Pu} + {}^{4}_{2}\mathrm{He} \longrightarrow {}^{241}96 + 2\,{}^{1}_{0}n
\end{array}$$

We later found that reaction (3) is the one that was producing the new activity. The range of 6.1-MeV ^{242}Cm corresponds closely to the value measured. Later work showed that this isotope of element 96 could also be produced in neutron reactors by the series of reactions:

$$\begin{array}{ll}
(1) & {}^{239}\mathrm{Pu} + n \longrightarrow {}^{240}\mathrm{Pu} + \gamma \\
(2) & {}^{240}\mathrm{Pu} + n \longrightarrow {}^{241}\mathrm{Pu} + \gamma \\
(3) & {}^{241}\mathrm{Pu} \longrightarrow {}^{241}95 + \beta^- \\
(4) & {}^{241}95 + n \longrightarrow {}^{242}95 + \gamma \\
(5) & {}^{242}95 \longrightarrow {}^{242}96 + \beta^-
\end{array}$$

During the rest of 1944 more vigorous bombardments of plutonium were made with deuterons. The targets were separated into fractions according to the newly learned chemistry and finally another alpha activity showed up. These new alpha

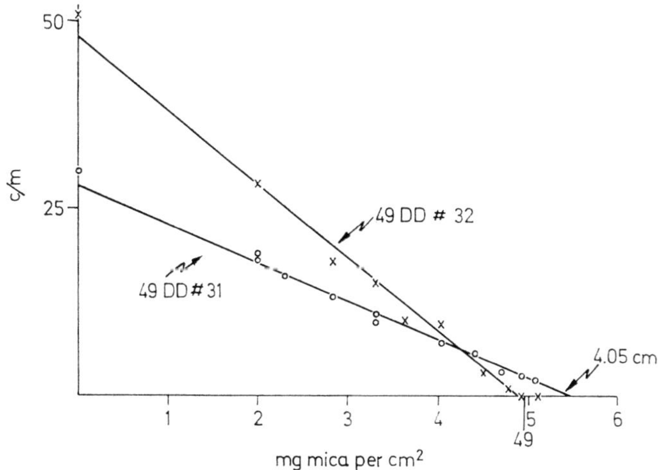

Fig. 16.2. Original mica absorption data showing presence of $^{241}95$ alpha particles in deuteron-irradiated plutonium

particles had a range only about 0.5 mg/cm² longer than that of the ^{239}Pu alphas but the activity definitely followed the actinide chemistry. The counting rate of the new activity was very small by the standards of the time, perhaps a hundred alpha counts per minute. In retrospect we believe that we were observing the 5.5-MeV alpha particles from $^{241}95$ produced in part from the beta particle decay of the ^{241}Pu which was already present in the target material as a minor constituent. The characterization of this new activity was made in October, 1944 (see Fig. 16.2).

By January of the following year we had established the fact that both of these new activities were produced in neutron-bombarded plutonium. The picture portrayed above by the series of reactions thus gradually began to evolve. Subsequent experiments proved that this was indeed the correct interpretation when the isotope ^{238}Pu was identified as the alpha decay daughter of $^{242}96$ and when $^{242}96$ was produced by bombarding $^{241}95$ with neutrons. A further proof was furnished by the Los Alamos mass spectrometer group when they discovered that the mass 241 fraction in a sample of plutonium could be fractionated. This came about because americium metal was more volatile than plutonium metal, thus enabling a separation of $^{241}95$ from its parent, ^{241}Pu.

The publication of this work (Seaborg et al., 1949) came in April, 1946, and at that time it was suggested that element 95 be named americium after the Americas in analogy to its rare earth homolog, europium. Element 96 was named curium after Pierre and Marie Curie, in analogy to its homolog, gadolinium, named after J. Gadolin.

Isotopes of americium are now known from mass 237 to 246. The two most important isotopes are ^{241}Am and ^{243}Am with alpha half-lives of about 500 to 8000 years respectively, and are available in kilogram quantities.

Isotopes of curium are presently known from mass 238 to 250. The most important isotopes are represented by masses 244 to 248, ranging in alpha half-life from 20 to 10^7 years and are available in subkilogram amounts.

II. Berkelium and Californium (Elements 97 and 98)

By the early months of 1946 Seaborg and many of his group had returned to the University of California's Radiation Laboratory and had begun to establish a powerful nuclear chemistry division. Much of the period until 1949 was spent in consolidating the chemical and nuclear knowledge of the elements through atomic number 96. We learned how to produce much greater quantities of the two new elements and new detection methods were studied to give greater sensitivity in finding new atoms. The development of gridded ionization chambers and multi-channel pulse analyzers now made it possible to characterize alpha particles which were emitted at rates as low as 1 per hour.

By 1949 we had accumulated a few milligrams of ^{241}Am from reactor bombardment of plutonium and begun attempts at creating element 97. In December of that year Stanley G. Thompson, Seaborg, and I tried the following reaction (S. G. Thompson et al., 1950a):

$$^{241}_{95}\mathrm{Am} + ^{4}_{2}\mathrm{He} \longrightarrow {}^{243}97 + 2\,^{1}_{0}n$$

using 35 MeV helium ions from the 60″ cyclotron.

The isotope was expected to have a short half-life so that a relatively fast chemical procedure was needed to enable identification of its alpha particle radiation. For this purpose we used a cation-exchange column for the first time in the heavy element regions. It had been shown that ion exchange chromatography was eminently suitable for separating the rare-earth elements—elements whose chemical properties were so similar that it was difficult to use ordinary aqueous chemical methods. Seaborg's actinide hypothesis suggested that this method should be equally applicable to the new rare-earthlike series of (III) elements. In addition, the technique lent itself nicely to the making of thin samples for alpha particle measurement, an essential requirement in their energy analysis.

Following a bombardment and a chemical separation that took a few hours, we had what we believed was a fraction which should contain transcurium elements and examined it in our alpha pulse analyzer. We were very excited to discover that we had produced several tens of counts per minute of an activity with a totally new alpha spectrum. Three distinct peaks in the range of 6.5 to 6.8 MeV were observed and found to decay with a half-life of 4.5 hours. We subsequently found that the principal mode of decay was by orbital electron-capture and that we were only detecting about 0.1% of the decaying atoms of element 97. This meant that even though we only had about a thousandth as much ^{242}Cm to bombard, we might be able find to element 98 if we could produce an isotope which decayed principally by alpha decay.

We did not waste much time in trying this new experiment. In February, 1950, Thompson, Kenneth Street, Jr., Seaborg, and I bombarded a few micrograms of curium which had been produced for us in the high-flux Canadian reactor. The reaction again used 35-MeV helium ions from the 60″ cyclotron.

$$^{242}_{96}\mathrm{Cm} + ^{4}_{2}\mathrm{He} \longrightarrow {}^{245}_{98}\mathrm{Cf} + ^{1}_{0}n.$$

The very first experiment was successful. An alpha-particle activity with an energy of 7.1 MeV and a 44-min half-life was found to elute early from the ion exchange column in the predicted position for element 98. Several thousand counts were clearly observed and a very few experiments were sufficient to prove that element 98 had been discovered (Thompson et al., 1950b).

Element 97 was named berkelium and element 98 was christened californium to honor the city and university where the discoveries were made.

Isotopes of berkelium have been identified from mass 243 to 250. ^{249}Bk is a very low-energy beta emitter with a half-life of about 300 days, decaying to the 500 year alpha emitter, ^{249}Cf. It is produced now in subgram amounts in the United States reactor actinide production program. More suitable long-lived alpha-emitting isotopes with half-lives in the thousands of years are ^{247}Bk and ^{248}Bk but these can only be produced by accelerated positive ions. It is hoped that a national program can be started to produce these isotopes in milligram quantities by using the surplus beam from one of the 200-MeV proton injectors which are a part of the high-energy accelerators at Brookhaven and Batavia. Bombardment of gram quantities of ^{248}Cm should be capable of making the transmutations in a period of a year or two.

Isotopes of californium are now known from mass 242 to 255. Probably the most important of these is ^{252}Cf. It is a 2.6-year alpha emitter but decays by spontaneous fission 3% of the time. It is the neutrons from this decay which make it so important as a possible therapeutic agent in needles implanted in patients. This nuclide was discovered by the Berkeley group in the "Mike" debris in connection with the discovery of elements 99 and 100 which is described in the next section.

III. Einsteinium and Fermium (Elements 99 and 100)

On November, 1952, the first test thermonuclear explosion, code-named "Mike", was set off by the U.S. in the Pacific Ocean. This awesome event was the source of multi-neutron captures in the ^{238}U material which was included in the device and the new isotope ^{244}Pu was soon found in the mass spectrometer at the Argonne National Laboratory. This knowledge became known in our laboratory about a month after the explosion and we determined to see if even higher masses had been formed which would then beta-decay to elements of higher atomic number. Within a couple of days after receiving one of the filter papers which had been flown through the radioactive cloud we had indeed found a new alpha-emitter. This activity had an alpha energy of 6.6 MeV and a half-life of about 20 days and turned out to be ^{253}99. Again ion-exchange chromatography was used to identify a new element. It turned out to have a 17-day beta-emitting precursor ^{253}Cf and this was probably formed in the explosion in the following manner:

$$^{238}\mathrm{U}(n,\gamma)^{239}\mathrm{U}(n,\gamma)^{240}\mathrm{U}(n,\gamma)^{241}\mathrm{U}(n,\gamma)^{242}\mathrm{U}\;\ldots\ldots\ldots\ldots\;^{251}\mathrm{U}(n,\gamma)^{252}\mathrm{U}(n,\gamma)^{253}\mathrm{U}$$

$$^{253}\mathrm{U}\xrightarrow{\beta^-}{}^{253}\mathrm{Np}\xrightarrow{\beta^-}{}^{253}\mathrm{Pu}\xrightarrow{\beta^-}{}^{253}\mathrm{Am}\xrightarrow{\beta^-}{}^{253}\mathrm{Cm}\xrightarrow{\beta^-}{}^{253}\mathrm{Bk}\xrightarrow{\beta^-}{}^{253}\mathrm{Cf}$$

The neutron captures would have taken place in a fraction of a microsecond before the explosive separation of the material in the device and the subsequent beta decays from ^{253}U to ^{253}Cf would have taken place in seconds as the atoms were dispersed afterwards.

A few months later we were able to identify element 100 by working up large quantities of coral containing nuclear debris which had fallen out onto a nearby island. The isotope found was the 1-day 7.1-MeV ^{255}100 which had been kept alive by a longer-lived isotope of element 99, the 40-day ^{255}99. This isotope had been formed as the result of the addition of seventeen neutrons to the ^{238}U target atoms. Fig. 16.3 is an elution curve for one of the experiments in which element 100 was discovered.

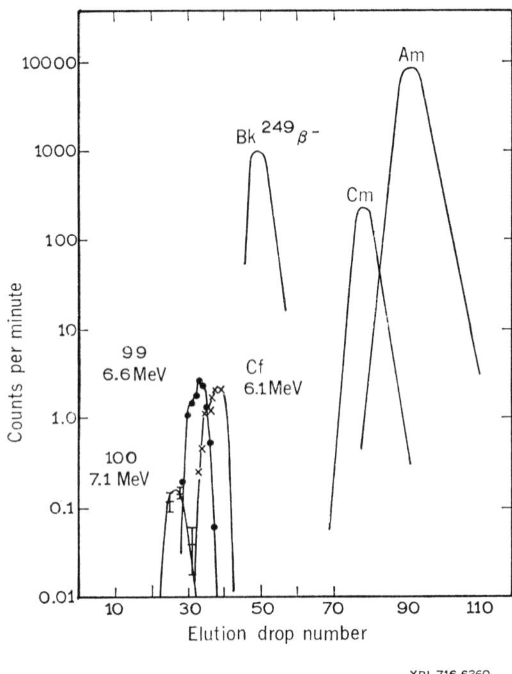

Fig. 16.3. Elution curve for element 100. From second chemical identification (March 1, 1953). Except for ^{249}Bk the detected activities are due to alpha particles. Resin used was Dowex-50 at 87° C with citrate elution

The work on these elements was finally published (Ghiorso et al., 1955a) by a large team of scientists from Berkeley, Argonne and Los Alamos who had participated in various ways in the exciting work. They suggested the names einsteinium and fermium for elements 99 and the symbols Es and Fm were eventually adopted.

One of the very important discoveries that emerged from this work was the disclosure of a nuclear subshell at 152 neutrons (Ghiorso et al., 1954). This first became apparent when we found that the alpha energy of ^{252}Cf was almost a hundred kilovolts higher than that of ^{250}Cf. If the normal trend had continued its energy would have been lower by several hundred kilovolts. This is a very sensitive indication by even-even isotopes and eventually this shell structure was found to extend all the way to element 105. How much further it will persist is still a matter of speculation.

Isotopes of einsteinium have been identified from mass 243 to 256. The 20-day isotope with mass 253 is the one normally encountered at the moment but within a few years it is expected that the 6.44-MeV 276-day ^{254}Es will be available in multi-microgram quantities. Isotopes of fermium are known from mass 244 to 258. The longest-lived isotope is 6.53-MeV 80-day ^{257}Fm, but unfortunately it will probably never be possible to manufacture it in more than submicrogram quantities by means of neutron reactors. It is possible to make it in milligram quantities in an underground nuclear explosion by the rapid-neutron-capture method but the costs of the explosion and the chemical recovery of the material from the debris are very high.

IV. Mendelevium, Md (Element 101)

The discovery of mendelevium was probably the most dramatic in the entire sequence of the evolution of the transuranium elements. It marked the first time in which a new element was produced and identified one atom at a time.

Within a year after the discovery of einsteinium and fermium in the "Mike" explosion we found that we could produce the same nuclides by means of high flux neutron reactors. By 1955 we had transmuted a substantial percentage of our initial plutonium starting material into the heaviest isotopes of curium so that the equilibrium amount of ^{253}Es that was available for experiments amounted to about 10^9 atoms. It occurred to us in a very optimistic moment that it might be possible to form element 101 by the reaction

$$^{253}_{99}\mathrm{Es} + ^{4}_{2}\mathrm{He} \longrightarrow ^{256}101 + ^{1}_{0}n.$$

The obvious problem that we had to overcome was that the target material had only a 20-day half-life. In order to perform many experiments quickly we needed a nondestructive technique of using the target material over and over again. We decided to rely on the recoil energy that was imparted to a transmuted atom in the collision of a high speed projectile with a target atom, a technique which had never been used before. Since the recoil energy was very small in the reaction that we had selected, many new problems had to be overcome before we were successful, but finally we were able to devise a procedure which enabled us to make many bombardments before the target became useless.

Again we used the old reliable 60″ cyclotron, this time with 40-MeV helium ions. The recoiling product atoms coming from the very thin einsteinium target were caught in a gold catcher foil which was then dissolved and subjected to the usual cation-exchange chromatographic separation. At first we concentrated on experiments which were aimed at finding an alpha emitter since we expected that to be the most likely mode of decay. We sped up our procedures fast enough to be able to "see" a 5-minute activity but without a sign of high energy alpha particles and were beginning to despair of finding anything. Finally, for we had nothing to lose, one of the irradiations was prolonged in the hope that maybe something longer-lived in the range of an hour or two might show up. We had been concentrating in the range of 5–10 minutes. Still no new alpha counts showed up in the alpha pulse analyzer but after a very long interval a single event occurred that showed up on a paper chart recorder. The recorder had been set up to measure the height of very large ionization pulses that would occur in the ion chamber whenever a spontaneous fission took place in the sample. Somewhat later, after another long bombardment, another such "big kick" was detected.

At first we were puzzled by these events but then we suddenly realized that they must be the key to the whole problem of why we could not see any new alpha activities. On the basis of these two spontaneous fission counts we hypothesized that ^{256}101 had a half-life in the hour range and decayed by electron-capture to ^{256}Fm. This daughter isotope would then decay by spontaneous fission with a half-life also in the hour range. These were farreaching conclusions to draw on the basis of just two events. They turned out to be quite correct. The proof of our conjecture was provided by a classic experiment performed on February 18, 1955 in which we performed three successive experiments and combined the products in the chemical separation step. The chromatographic elution gave a fraction in the predicted position for element 101 which exhibited five spontaneous fissions with a half-life of roughly three hours while the element 100 fraction which

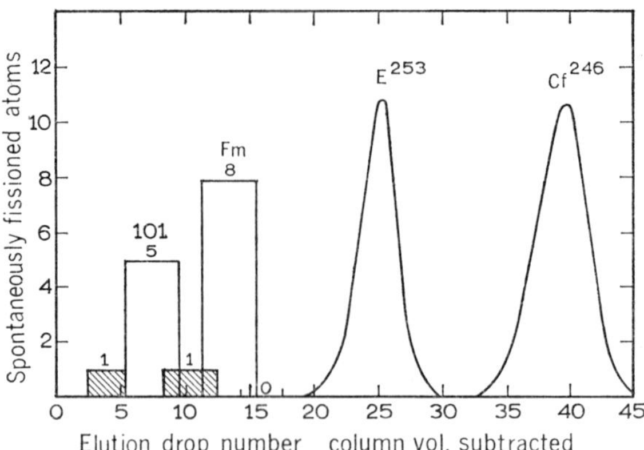

Fig. 16.4. Elution data showing the chemical identification of element 101 by means of its daughter, the 2.7-hr SF-emitting ^{256}Fm. The ^{253}Es and ^{246}Cf activities are due to alpha particles

eluted somewhat later than element 101 showed eight spontaneous fission events with about the same half-life. Fig. 16.4 shows the original elution data.

Further experiments conducted before our target had decayed too much yielded further corroboration and by the time that we published our work (Ghiorso et al., 1955b) we had successfully separated a total of seventeen atoms of element 101 with an average of about one per experiment. In this paper the team of Bernard G. Harvey, Greggory R. Choppin, Thompson, Seaborg and I suggested that the new element should be called mendelevium after the great Russian chemist, Dimitri Mendeleev, who had made such great contributions to chemistry. His periodic law accurately predicted the chemical properties of known and unknown elements and was the basis for the discovery of many of them.

Isotopes of mendelevium are known from mass 248 to 258. ^{258}Md has an 80-day alpha half-life and is made by bombardment of ^{255}Es with helium ions.

V. Nobelium, No (Element 102)

The First of the "Controversial" Elements

Temporarily we had come to the end of the road as far as making new elements was concerned. The heaviest target material that we had was element 99 so that we had to have a projectile with at least three protons to make element 102. We had been experimenting with the acceleration of heavier particles such as carbon ions in the 60″ cyclotron but the beams were very small and did not have a well-defined energy. We had been able to make very small amounts of isotopes of Cf, Es, and Fm so that it did appear feasible to go further up the atomic scale, but we needed a better accelerator for the purpose. Following a suggestion by Luis Alvarez a study was set up in collaboration with Yale University with the object of designing twin heavy ion linear accelerators (HILAC) to be built at the two research centers. These machines turned out to be the first in the world to take advantage of magnetic strong-focussing. Their characteristics were set by the requirements of the experiments envisioned at that time and by the availability of funds. The maximum ion energy was limited to 10 MeV per nucleon. The maximum Z that could be accelerated was around argon. The maximum average beam current was a few microamperes. Construction of the Berkeley HILAC was completed early in 1957 and after a long de-bugging stage experiments were envisioned looking toward the production of element 102. Before these could be carried out we were surprised by an announcement from Sweden in the summer of 1957.

A group of workers from the Argonne National Laboratory, the Atomic Energy Establishment at Harwell in England, and the Nobel Institute for Physics in Stockholm claimed (FIELDS et al., 1957) the discovery of an isotope of element 102 as a result of research done at the Nobel Institute. The group reported that in irradiations of ^{244}Cm with ^{13}C ions accelerated in the Nobel Institute's cyclotron they had found an 8.5-MeV alpha emitter with a half-life of about 10 minutes. They claimed that it had been identified as being an isotope of element 102 by the chromatographic ion exchange method and proposed that the new element be given the name nobelium to honor the Nobel Institute. The name was accepted within a matter of months by the Commission on Atomic weights of the International Union of Pure and Applied Chemistry which had become the international arbiter of chemical nomenclature.

This acceptance of the claim to discovery of element 102 turned out to be premature. All attempts to duplicate the Stockholm experiment failed to corroborate their findings. At Berkeley we devoted many months of time with the new HILAC using research techniques orders of magnitude more sensitive than those used in Sweden and yet failed to find the reported activity (GHIORSO et al., 1958a). Later we were to find out that experiments performed in Dubna in the USSR at about the same time also did not verify the discovery but these could not rule it out because of the use of a different target and projectile combination. Although we used the same kind of target and projectile as used in Stockholm without success the authors of the "discovery" paper refused to withdraw their claim feeling that there was some possibility, even though small, that their work was correct. Even this remote possibility was completely discounted in 1968 when we found that element 102 was found only in the (II) state in aqueous solution. This meant that it would not appear in the expected elution position in a cation exchange column and thus their chemical proof as well as their physical proof for element 102 vanished.

After our unsuccessful attempts to find the Stockholm alpha particles we finally began to look in various ways for element 102 alpha activities with other characteristics. We felt that the half lives of the isotopes that we could make at that time would be in the seconds region. Since the technology was not yet available for the efficient direct detection of the short-lived alpha activities themselves, we resorted to a new method. In a paper (GHIORSO et al., loc. cit.) published in the summer of 1958 by TORBJORN SIKKELAND, JOHN R. WALTON, SEABORG, and the author a novel "double recoil" technique was described which permitted the identification by both physical and chemical means of an isotope of element 102. A mixture of curium isotopes was bombarded with 68-MeV ^{12}C ions provided by the new Berkeley HILAC. The following reactions were assumed to take place:

$$^{246}\text{Cm} + {}^{12}_{6}\text{C} \longrightarrow {}^{254}102 + 4\,{}^{1}_{0}n$$

$$^{254}102 \xrightarrow[\text{secs}]{\alpha} {}^{250}_{100}\text{Fm} \xrightarrow[\text{30 min}]{\alpha}$$

As shown in Fig. 16.5, a thin curium target with 95% ^{244}Cm and 4.5% ^{246}Cm was set up in such a manner that the recoiling element-102 atoms produced in the bombardment were stopped in helium gas and then attracted electrically as

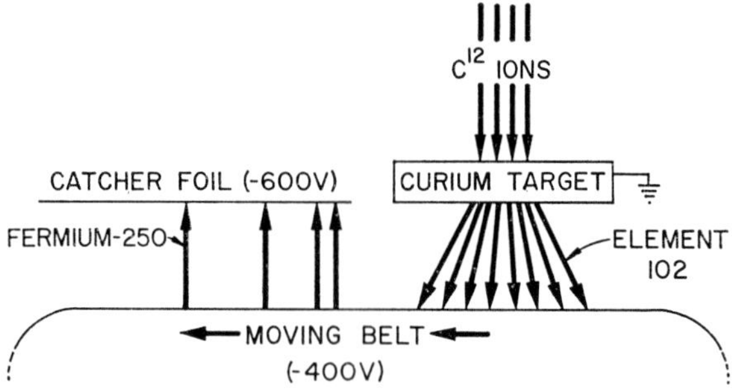

Fig. 16.5. Double-recoil experiment used in the detection of ^{250}Fm as the alpha-decay daughter of ^{254}102

positive ions to a moving metallic belt mounted nearby. In a second recoil action approximately half of the 30-min ^{250}Fm atoms from the alpha decay of ^{254}102 were captured by a negatively-charged foil adjacent to the belt. The discovery of element 102 was based on showing that the doubly-recoiled atoms had the atomic number of fermium. For this crucial test it was necessary to use all of the transferred atoms in each experiment to make the chemical identification and thus no half-life information was possible.

The half-life of the element 102 parent was deduced by a completely physical method. From the speed of the belt and the variation of the ^{250}Fm activity along the catcher foil, we obtained information that was consistent with a half-life of approximately 3 seconds for the 102 parent of this alpha emitter. About a year later in experiments in which element-102 atoms were attracted electrically to a stationary metallized tape which was then periodically moved into counting position within a gridded alpha chamber we found an alpha activity of 8.3 MeV and about 3 seconds half-life. We also observed spontaneous fissions with the same half-life in 30% abundance relative to the alpha emissions. Since the half-life was the same as that observed in the earlier experiments we quite naturally assumed that these characteristics corresponded to ^{254}102 which we had established by the chemical milking of ^{250}Fm.

Before this work could be brought to a satisfactory conclusion it was abruptly halted in 1959 by a disastrous radiation incident. A helium-cooled window which was interposed in the beam line between the HILAC and the curium target was ruptured by an accidental momentary overpressure. Unfortunately, the foil on the target side failed and an explosively propelled jet of helium impinged upon the fragile ^{244}Cm target. The target contained about 10^{12} alpha disintegrations per minute and this was carried into the experimental room in micron-sized particles. Unfortunately, at that time there was no secondary containment in the room and the activity was carried out into the high-bay area. The very dangerous radioactive dust was soon deposited uniformly throughout the structure. Fortunately, I detected the spill almost immediately and gave the alarm to evacuate everyone in the building. Miraculously no one ingested a significant amount of curium and it did teach us an extremely important lesson. It was estimated that I inhaled possibly 10^5 α dis/min and I was the closest to the explosion. By an heroic effort of Health Chemistry personnel the building was ready for occupancy again in about three weeks but for many years curium continued to be found in small quantities in obscure places in the building.

The incident not only disposed of all of our precious curium that was available at the time but understandably caused us to be quite gun-shy in the use of highly active alpha-emitting targets. We might have pursued the element-102 work by using plutonium targets and oxygen ions but we knew that the smaller reaction cross sections would make the problem even more difficult so we decided to push on to try to find an isotope of element 103.

In 1961 during work which led to the successful identification of element 103 we reported finding another isotope of element 102. This activity, produced by the HILAC bombardment of a mixture of californium isotopes by boron projectiles, decayed by the emission of 8.2-MeV alpha particles with a half-life of approximately 15 secs. We assigned this activity to $^{255}102$ on the basis of comparative yields of this and other product activities and the variation of its yield with energy of the bombarding particles. This mass assignment was not definite since the excitation function was very broad, both because it was the product of a (^{11}B, pxn) reaction and because the target was not monoisotopic.

Let us now return to 1957 to look at the parallel line of research being undertaken by large scientific teams in the USSR. The director of this work was GIORGI N. FLEROV, a pioneering nuclear physicist who had discovered spontaneous fission in ^{238}U, following OTTO HAHN's famous report on the discovery of neutron-induced fission. FLEROV and his men did not have curium available to them at this time for the production of element 102. Even if they had it is doubtful that they would have used it since they had been similarly "burned" by a radioactive incident with the 500-year ^{241}Am. They chose instead to bombard the relatively inactive isotopes of plutonium with ^{16}O and ^{18}O ions. At the Kurchatov Institute in Moscow the cyclotron had been converted to accelerate heavy ions for the purpose of making new transuranium elements and they were hard on the heels of the Berkeley group.

Their first interesting result (FLEROV, 1958) was obtained in the autumn of 1957 in bombardments of ^{239}Pu with ^{16}O ions in which they found an alpha activity with an energy of 8.8 ± 0.5 MeV having a half-life in the range of 2 to 40 seconds. The Soviet scientists bombarded targets containing 100–200 micrograms/cm^2 of ^{241}Pu with intense currents of ^{16}O ions. The product atoms were caught in vacuum in an aluminium catcher foil that was cyclically shuttled a distance of 200 cm in 2 seconds and placed next to a nuclear emulsion (see Fig. 16.6). Optical scanning of the emulsion and a measure of track lengths gave them a rough capability for detection of alpha particles (Fig. 16.6).

Much or all of the 8.8-MeV activity turns out to be caused by an isomer of polonium, the 25-sec 8.87-MeV ^{211m}Po, produced by heavy ion interaction with a tiny lead impurity in the target material. This same activity plagues all experiments in this field even today since it simulates some isotopes of 103, 104, 105 and probably 106 and 107. It is likely that some of their alpha counts were indeed due to 2.3-sec 8.43-MeV $^{252}102$ but the very crude resolution afforded by their use of nuclear emulsions for detectors makes it impossible to discriminate against the background activities (another is 8.43-MeV ^{214}Fr kept alive by 3-sec ^{214}Ra produced from lead impurity). In Fig. 16.7 is shown their gross alpha-particle spectrum obtained from ten 3-hour bombardments. Although it certainly is possible that element 102 was first produced in these experiments since historically they preceded the Berkeley efforts, I do not believe that they can be considered as a discovery of the element since they do not persuade any knowledgeable person that such an identification was made.

Soviet experimenters soon moved to the new Dubna Laboratory, an international research center set up cooperatively by the Socialist countries. There they began using a very large 3.1 meter cyclotron which had been constructed specifically for heavy ion acceleration. They started a new series of experiments

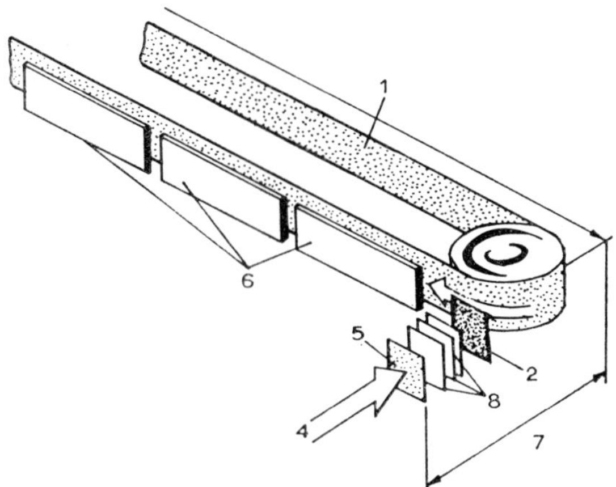

Fig. 16.6. Experimental setup of the first Dubna attempts to identify an alpha particle emitting isotope of element 102. The beam (*4*) of ^{16}O ions traverses a collimator plate (*5*) and energy-degrading foils (*8*) to knock the transmutation recoils out of a ^{241}Pu target (*2*) onto a metal ribbon (*1*) which then carries the atoms in front of nuclear emulsions (*6*). The emulsions serve as alpha particle track detectors; the range of each track gives a rough measure of the alpha-particle energy and the position gives an indication of the half life

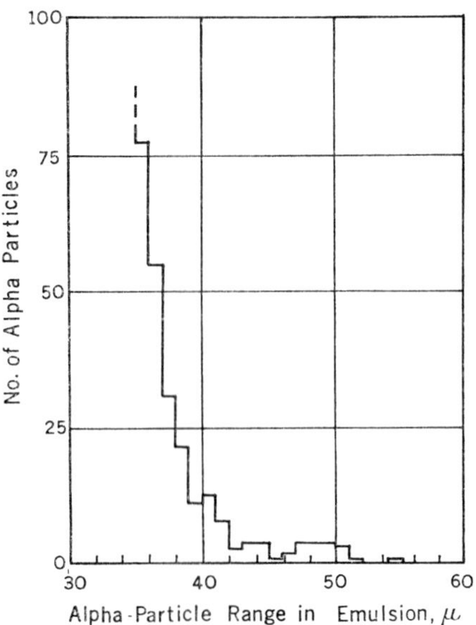

Fig. 16.7. Alpha-particle range distribution as detected by nuclear emulsions in the early U.S.S.R. experiments aimed at producing element 102 by the bombardment of ^{239}Pu with ^{16}O ions. The broad peak at ~50 microns is mostly due to 25-sec ^{211m}Po

which culminated in 1963 with the successful identification of a new element-102 isotope (DONETS et al., 1964; DRUIN et al., 1964). In bombardments of ^{238}U with ^{22}Ne ions they produced $^{256}102$. The method was based on the Berkeley double-recoil technique with the chemical identification of the alpha emitting daughter, ^{252}Fm. The half-life of $^{256}102$ was deduced to be about 8 seconds from the relative amounts of daughter activity transferred by alpha recoil to the catcher foil. (A better value of 3.2 seconds was later determined by our group.)

In December, 1965, Flerov and his group identified (Donets et al., 1966) $^{254}102$ by the double-recoil method and arrived at the surprising conclusion that its half-life was 50 ± 10 secs, not 3 seconds as indicated in our 1958 work, and this result was soon verified by another method of production. At about this time they also began to use a new method, excellent for alpha spectroscopy, which had been pioneered at Berkeley by MacFarlane and the author a few years before.

The method was the so-called "gas-jet" technique to transport recoil atoms, which were stopped in a carrier gas such as helium, to a place of collection where they could be examined by sensitive semi-conductor alpha particle detectors. In the Soviet version of this method the jet of helium gas impinged upon one arm of a collector shaped like a Maltese cross. After a suitable bombardment, the cross was periodically rotated 90° to put the collected heavy atoms in front of a detector while another arm was simultaneously placed to collect a new batch of jet-borne atoms from the target chamber. With a two-parameter analyzer the decay and energy of the various alpha activities could be followed by noting the time of arrival and pulse size of each event after each rotation of the cross.

With the aid of this modern technique and the employment of a second heavy-ion accelerator, a new 1.5 meter sector-focussed heavy-ion cyclotron at Dubna, a Soviet group (Zager et al., 1966) identified the alpha particles from $^{254}102$ and found that they had an energy of 8.10 MeV and a half-life of 30 to 40 seconds, again much longer than the old Berkeley value. The mass assignment was proved when they detected the 30-min ^{250}Fm daughter nuclei which recoiled into the detector face after alpha decay of $^{254}102$. They looked for spontaneous fissions and concluded that this isotope did not exhibit a prominent branch decay by this mode as indicated by the 1959 Berkeley work. Later another group identified $^{252}102$ as emitting 8.41 MeV alpha particles with a half-life of 4.5 ± 1.5 secs. Spontaneous fissions which were observed in this case with about the same half-life were attributed to a lower-Z element. The isotope $^{255}102$ was found to have an 8.08 MeV energy and a two minute half-life and thus this nuclide also did not correspond to our assignment. Two other activities were found so that by October, 1966, at an International Heavy Ion Conference held at Dubna Flerov's groups were able to announce the seemingly complete identification of isotopes of element 102 from mass 252 to 256.

Responding to this challenge by the Soviet scientists, whose results seemed to be at variance with all of the earlier Berkeley research on element 102, we set out in November 1966 to repeat the old experiments with the objective of clearing up the discrepancies. This time we had the advantages of greatly improved HILAC beams and superior detection capabilities. In a relatively short time we were able to produce and identify all of the isotopes of element 102 from 251 to 257 (Ghiorso et al., 1967a) and a few years later we identified masses 258 and 259. For most of these experiments we used a gas-jet system to transport the transmutation products to the periphery of a wheel. The wheel acted as a conveyor to successively place the collected atoms in front of a series of four semiconductor alpha particle detectors for energy analysis. The relative numbers of atoms observed at each detecting station gave a measurement of the half-life of each species detected.

To span the range of the element-102 isotopes, we bombarded at various energies monoisotopic targets of ^{244}Cm and ^{248}Cm with concentrated beams of ^{12}C and ^{13}C ions. Compound nucleus reactions with these target-ion combinations had the distinct advantage, as compared with the combinations used by the Dubna groups, of much larger cross sections to produce the same nuclide. This enabled us to make substantially more accurate measurements of alpha half-lives and energies and more definitive observation of spontaneous fission branching decay.

The first observation that we made was that $^{254}102$ did not indeed have a 3-second half-life as claimed in our 1958, 1959 papers; a value of 55 ± 5 sec was found in agreement with Dubna. On the other hand its formation cross section was about a microbarn, consistent with the value found in 1958 in our milking experiments when we chemically isolated its daughter, ^{250}Fm. Clearly there was some error in the milking experiments which implied a 3-second half-life by the physical identifications of "^{250}Fm".

The next step in unraveling the puzzle was finding that $^{252}102$ had a half-life of 2.3 ± 0.3 secs and an alpha energy of 8.41 ± 0.02 MeV (the Dubna values are 4.5 ± 1.5 sec and 8.41 ± 0.03 MeV). In addition, a branching ratio for spontaneous fission of about 30% was found and this important fact was in excellent agreement with the work of 1959 when we used a grid chamber as our detector. The old 8.3-MeV energy measurement can be corrected to 8.4 MeV by using the modern value for the alpha energy of ^{248}Fm as an internal standard and the old 3-sec half-life is within the standard deviation errors in the two measurements.

It is impossible to be certain as to the reason why our original experiments gave the *wrong half-life* for $^{254}102$ but it is likely that the confusion was caused by $^{252}102$. Since there was 20 times as much ^{244}Cm in the old target as ^{246}Cm, it is possible that the 25-min 7.22-MeV ^{244}Cf granddaughter could have been mistaken for 30-min 7.43-MeV ^{250}Fm.

$$^{252}102 \xrightarrow[\alpha]{2{\cdot}3\text{ sec}} {}^{248}\mathrm{Fm} \xrightarrow[\alpha]{36\text{ sec}} {}^{244}\mathrm{Cf}.$$

In those early experiments we were plagued by relatively poor energy resolution and drift problems in the multi-chamber system used to analyze the small amounts of activity available at that time. In the chemical milking experiments we were not bothered by this problem since ^{244}Cf was separated from the ^{250}Fm by the ion-exchange column.

The isotope, $^{257}102$, was identified in later experiments and found to have prominent groups at 8.22, and 8.32 MeV with a half-life of 23 ± 2 secs. These data correspond reasonably well with the 15-sec 8.2-MeV alpha activity observed in 1961 in connection with the experiments on element 103.

We believe that it is fair to draw the following chronological conclusions regarding the discovery of element 102:

a) The 1957 work at the Nobel Institute was completely in error.

b) The 1957 work at Dubna was so vague and inconclusive as to be virtually meaningless although, of course, it is possible that the radiations from an isotope of element 102 were included as a small part of their range spectrum.

c) In 1958 at Berkeley the isotope, ^{250}Fm, was *chemically* shown to exist in a recoil-milked fraction thus proving the atomic number of the $^{254}102$ parent.

d) In 1959 at Berkeley the isotope, $^{252}102$, was discovered as shown by its alpha energy, half-life, and spontaneous fission branch decay.

e) In 1961 at Berkeley the isotope, $^{257}102$, was discovered as shown by its alpha energy and half-life.

f) In 1963 at Dubna for the first time in *their* laboratory the isotope, $^{256}102$, was positively identified by the chemical milking of the recoil daughter, ^{252}Fm.

At the time this definitive work was completed ten years had passed since the original Stockholm experiments and the name nobelium had come into common usage. As a consequence of this situation, we decided not to contest the name. In a review paper (Ghiorso et al., 1967 b) on the element 102 research we suggested that it be retained, but the symbol No for the element will always be for us a sharp reminder of its somewhat painful history.

Quite recently a relatively long-lived isotope of nobelium has been produced. This is ^{259}No, an alpha emitter with a half-life of about an hour, which should make possible more extensive studies of the chemistry of element 102. The important discovery of the divalent characteristics of nobelium in aqueous solutions was made using the 3-min ^{255}No isotope.

VI. Lawrencium (Lr) (Element 103)

In 1961 Sikkeland, Almon E. Larsh, Robert M. Latimer, and the author reported the discovery (Ghiorso et al., 1961) of element 103. In this case it was not possible to identify chemically a known daughter isotope and thus prove the atomic number of the new activity, so instead we had to rely on purely physical arguments. The recoiling atoms from the bombardment of a mixture of californium isotopes by either ^{11}B or ^{10}B ions accelerated by the HILAC were electrically attracted to a metallized mylar tape which was pulled successively in front of four semi-conductor alpha detectors. The art of making and using solid state detectors was in its infancy so there were many problems to be overcome. The great value of these sensitive detectors was proved in these experiments since the amount of activity that we could make was very small. Only about 100 total alpha events were observed in all of the many experiments so that there were many possibilities for error. We reported an 8.6-MeV alpha activity with a half-life of about 8 seconds. By cross bombardment and substitution experiments we came to the conclusion that the new activity was due to element 103 and that its mass number was probably 257 although 258 and possibly even 259 were also possibilities. We suggested the name lawrencium for the new element in honor of E. O. Lawrence, the pioneering inventor of the cyclotron.

In 1965 the Dubna groups under Flerov again challenged our claim to the discovery of a new element. They found an alpha emitter, 45-sec ^{256}Lr, which they identified chemically through its granddaughter, ^{252}Fm, using the same double-recoil system that had been successful in the identification of ^{254}No and ^{256}No (Donets et al., 1965). During the next several years they looked for ^{257}Lr and finally concluded (Flerov et al., 1967) that it had the same alpha energy and half-life as ^{256}Lr. Since they did not find our 8.6 MeV alpha activity, they assumed that it must be due to some error on our part.

The Soviet experiments were not too surprising to us and we deduced from them that the mass number of our lawrencium isotope was most likely 258. Faith in our work was not misplaced for by 1968 we had proven quite conclusively that this was indeed the correct assignment (Ghiorso et al., 1969) and that its half-life was closer to 4 seconds than to 8 (Eskola et al., 1971). Since it is the atomic number that is essential in the determination of a new element, we take the position that a wrong assignment of mass number is of no consequence—in fact, is almost to be expected in the first exploratory measurements. In the new experiments, contrary to the Dubna work, we found that ^{257}Lr did *not* have the same characteristics as ^{256}Lr, but instead had a half-life of 0.6 sec and an energy of 8.87 MeV (Eskola et al., loc. cit.). This proved to be a crucial factor in the identification of isotopes of element 105.

In 1970 our group (Silva et al., 1970a), in a tour-de-force of chemistry, succeeded in showing that lawrencium reverted to the III state in aqueous solution as expected by the actinide hypothesis. This extremely difficult experiment employed the 35-sec ^{256}Lr isotope so that speed was vitally necessary. A solvent extraction method was used to distinguish between 2+, 3+, and 4+ oxidation states. On the average one atom was detected in each manually-performed experiment and several hundred experiments were eventually conducted.

By 1971 our group was able to greatly extend the list of known isotopes of lawrencium from mass 253 to mass 260. ^{260}Lr is an alpha emitter with a half-life of 3 minutes and thus the longest-lived isotope known.

VII. Element 104—Rutherfordium (Rf) or Kurchatovium (Ku)?

There is certainly no question that G. N. Flerov and his Dubna colleagues made the first claim (Flerov et al., 1964) to the discovery of element 104. In 1964 the isotope ^{242}Pu was bombarded with ^{22}Ne in the 3.1-meter heavy ion cyclotron with the objective of making a spontaneously-fissioning isotope of element 104. They found a new 0.3-second activity which they concluded was probably due to ^{260}104 but the number of fission events observed was so small that it was extremely difficult to be certain as to the identity of the atomic number involved. At best, even with a lot of activity, it is difficult to define the atomic number of a spontaneous-fission emitter since in the heavy element region the energy and mass distributions of the fragments do not differ greatly from one element to another.

Starting in 1966, efforts were made at Dubna to support their claim by chemical experiments designed to show that the spontaneous-fission activity came from a non-actinide element. It was expected that element 103 would be the last of the 3^+ actinide elements; element 104 should thus have chemical properties similar to hafnium and zirconium and form a relatively more volatile chloride compared to the actinides. I. Zvara et al. (1966) at Dubna set up a very complicated system in which the nuclear reaction products from the cyclotron bombardment were chlorinated by reaction with $NbCl_5$ and $ZrCl_4$ vapors at 200° to 300° C. A small fraction of the total spontaneous-fission activity was able to pass through the hot filter and it was deduced that only the more volatile 104-tetrachloride would be able to do so in accordance with predictions. The apparent half-life of the activity which survived the chemical treatment seemed to agree with the 0.3-sec value obtained in the purely physical experiments but the number of detected events was small and the variation very large.

Since the chemical experiments seemed to confirm the physical experiments, the Dubna groups decided that element 104 had indeed been discovered and suggested that it be given the name kurchatovium (Ku) in honor of Igor Kurchatov, a pioneering Soviet nuclear physicist. The international nomenclature committee has not accepted this name since the Dubna work was soon questioned by the Berkeley group. Suddenly, for a change, we were the challengers rather than the challenged.

At first we were inclined to accept the Soviet work since it *sounded* convincing and we were deeply immersed in the problems associated with the nobelium isotopes. Our doubts were aroused, however, when we discovered that ^{258}No had a spontaneous-fission half-life of only a millisecond. Over a period of time we had formulated an empirical spontaneous-fission systematics (Ghiorso et al., 1958b) which indicated that there was a maximum half-life for this mode of decay for

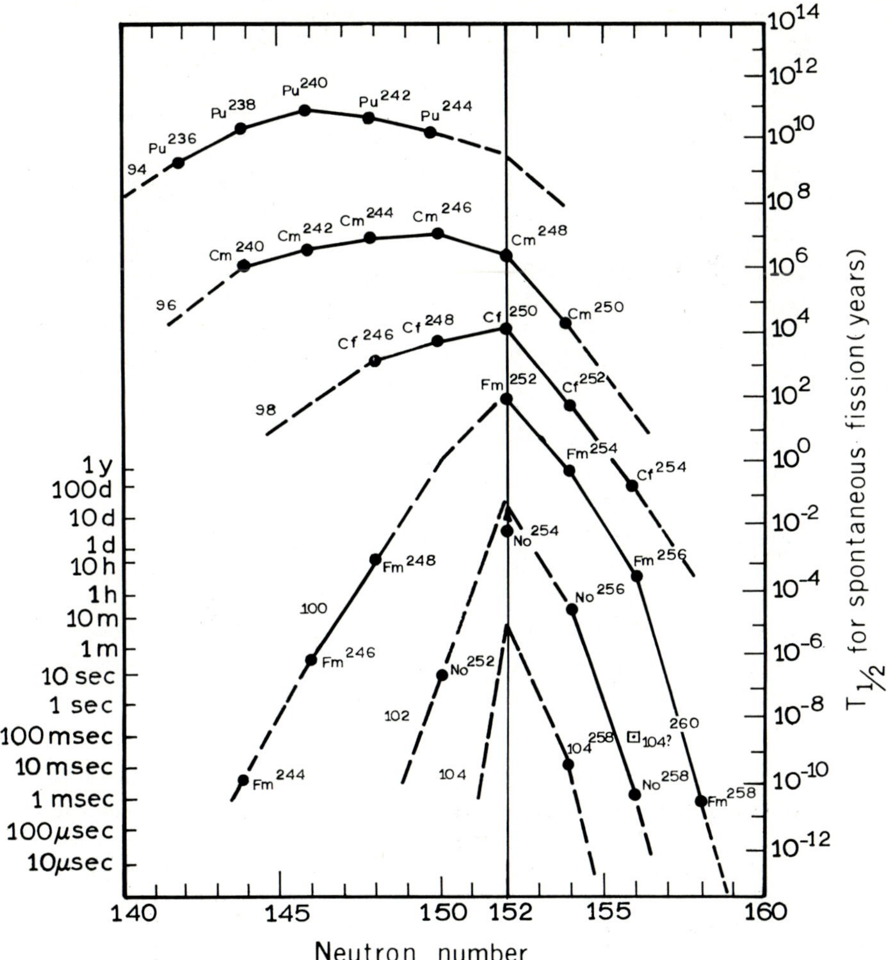

Fig. 16.8. Empirical correlation between spontaneous fission half-life and neutron number for even-even nuclides. Notice the sudden decreases beyond the 152-neutrons subshell

those nuclides with 152 neutrons and that as the atomic number was increased the half-lives became drastically shortened (see Fig. 16.8). Following this view we would expect that $^{260}104$ with the same number of neutrons (156) as ^{258}No should have an even shorter half-life, perhaps in the microsecond region[2]. This argument would not rule out the possibility that the Dubna activity was due to $^{261}104$ or $^{259}104$ since nuclides with an odd number of neutrons tend to be highly hindered as to spontaneous-fission decay.

We set to work to see if we could produce the 0.3-sec activity by bombarding einsteinium, californium, and curium isotopes with ions from ^{11}B to ^{18}O. In experiments which should have been successful we did not see any such short-lived spontaneous fissions but our efforts were hindered by a considerable background from ^{256}Fm, the 2.7-hr SF emitter. After expending much effort in this direction

2 Considerable strength to this argument was added in 1970 by the discovery that 158-neutron ^{258}Fm had an SF half-life of only 300 microseconds (Hulet et al., 1971).

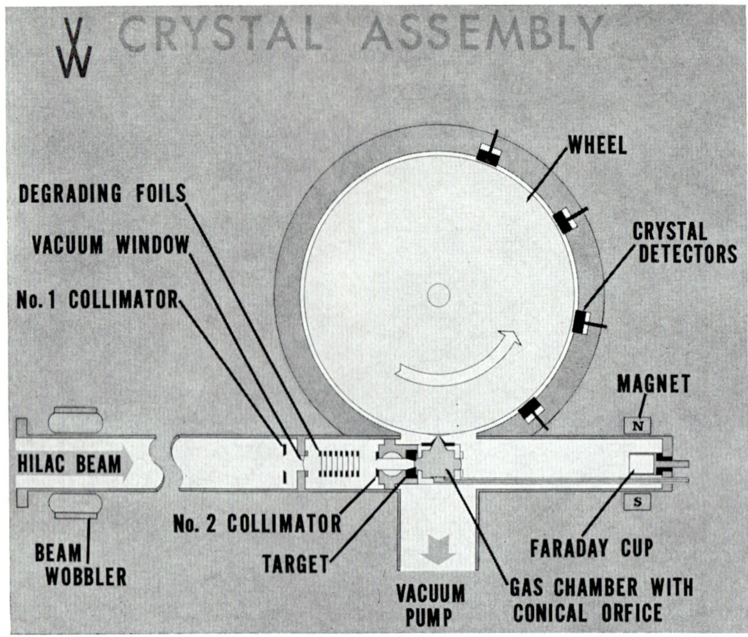

Fig. 16.9. Early 4-station system used for the detection of the 4.5-sec alpha-emitting isotope, $^{257}104$, as well as other heavy nuclides

without success we decided to reserve judgment and continue in our efforts to find an alpha emitter which could be assigned without ambiguity to element 104.

An atom that disintegrates by undergoing spontaneous fission provides little in the way of a "fingerprint" since one fission differs little from another and only the half-life can be used to distinguish them. On the other hand an atom that decays by the emission of an alpha particle with a distinctive energy and half-life to form a daughter atom which in turn decays with a different alpha energy and half-life can be uniquely identified. With this technique as our ultimate goal we proceeded with our search for element 104 and succeeded in developing over a period of time a marvelously sensitive apparatus that was vital to the discovery of the alpha-emitting isotopes of elements 104 and 105.

Early in 1968 enough ^{249}Cf became available to allow us to bombard it effectively with carbon ions and hope to see short-lived alpha activities. After some initial experiments which disclosed that there was 3–5 sec alpha emission in the region of 8.8 MeV, we began the development of a system which used a vertically-mounted large wheel as a conveyor for the jet-deposited new atoms (see Fig. 16.9). This new system was mechanically and electrically superior to our old apparatus used for the investigation of the nobelium isotopes. In addition we equipped it with auxiliary detectors which would allow us to identify the element-102 daughter products simultaneously with the identification of the element-104 mother atoms. A small computer was installed to make it possible to handle the wealth of information that was to ensue.

These mother-daughter measurements were accomplished in the following way. Just as in the double-recoil experiments which were successful in the case of element 102, we took advantage of the separation of daughter from mother afforded by the recoil energy after alpha decay. By periodically moving the detecting crystals which were "looking at" the alpha particles emitted by the mother atoms on the conveyor wheel to positions off the wheel, we were able to measure the activities from those atoms which had been embedded in the crystals by alpha recoil. To increase the likelihood of detection of these daughter atoms the shuttled-positions of the crystals were next to a similar set of detectors to establish almost a 4π geometry. Ultimately we employed seven detecting stations around the wheel. To avoid the loss of detection of element-104 atoms while measuring the recoil daughters, we used a duplicate system so that we could measure the mother and daughter simultaneously. This meant the use of four detectors at each station, each with its own preamplifier-amplifier system. The 28-detector system, though highly complex, performs its functions with great precision and stability. The reproducibility of a given experimental measurement is excellent and it is possible to run the system for days at a time with little more than minimal attention. As the work on element 104 progressed, the system, which started with only four crystals, was gradually improved until, with the conclusion of the element-105 work, we had built it up to use twenty-eight detectors.

With the advent of the new equipment we were able to identify (GHIORSO et al., 1969) unambiguously the isotope, $^{257}104$, in ^{12}C bombardments of ^{249}Cf (see Fig. 16.10). It was found to have a complex alpha spectrum with lines ranging from 8.70 to 9.00 MeV and a half-life of 4.5 ± 1.0 sec. Its known daughter, ^{253}No, was found to be transferred by alpha recoil as expected. The same target when bombarded by ^{13}C ions produced another isotope of element 104, this one with mass 259 having two prominent groups at 8.77 and 8.86 MeV and a somewhat shorter half-life of 3 seconds. Again the recoil daughter was identified, this time the isotope, ^{255}No. By using a different method we discovered an 11-msec SF activity which we tentatively assigned to $^{258}104$. This assignment agreed well with our predictions for this nuclide and further cast doubt on the origin of the 0.3-sec SF's. The members of the research team publishing these results, MATTI J. NURMIA, JAMES A. HARRIS, KARI ESKOLA, PIRKKO ESKOLA, and the author suggested that element 104 be named rutherfordium with the symbol Rf in honor of Lord ERNEST RUTHERFORD.

About a year later in bombardments of ^{248}Cm with ^{18}O ions, a much longer-lived isotope of element 104 was found, $^{261}104$, with a half-life of 65 ± 10 seconds and an alpha energy of 8.3 MeV. Once more the recoil daughter, this time ^{257}No, was shown to be transferred into the detecting crystals as they viewed the wheel. These findings seemed to reinforce our empirical criticisms of the assignment of the 0.3-sec SF activity to element 104 since now its mass could not be 259 or 261.

After the publication of our research, the Dubna groups endeavored to repeat their work on element 104. Now they found that the half-life of the SF activity was about 0.1 sec and that these recoil products tended to be ejected in a cone with a sharply-peaked forward distribution. The angular distribution experiment would rule out the possibility that the 0.1-sec activity could be due to an isomer of a nuclide near the target nucleus in atomic or mass number but could not rule out one in the region of nobelium or lawrencium. The smaller value for the half-life also meant that the old chemical experiments could not have worked since there was a minimum time of about a second necessary for the atoms to transit the hot filter. Very recently they have measured a 4.5 ± 1.5-sec component in their $^{242}Pu + {}^{22}Ne$ bombardments which they attribute to $^{259}104$ and claim that this must be the SF activity which they observed in their chemical experiments. It is certainly possible that this is a correct interpretation since our old experiments were not set up to rule out such a possibility. It would be very surprising,

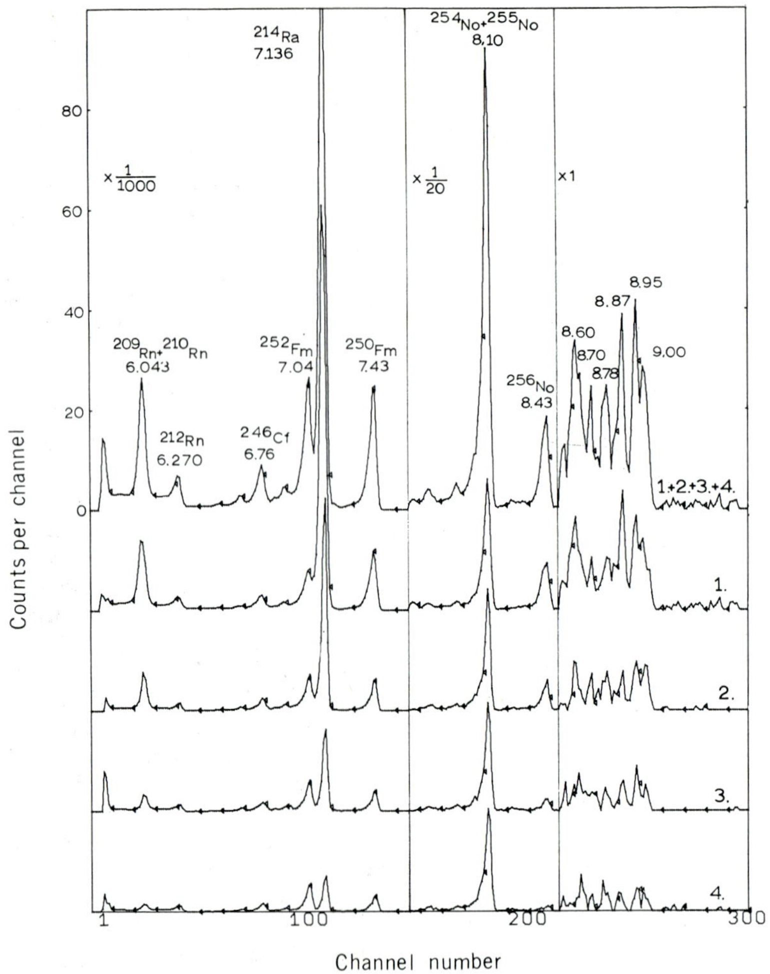

Fig. 16.10. Alpha spectrum of ^{257}104. The lines at 8.70, 8.78, 8.95, and 9.00 MeV are due to this isotope. That at 8.87 MeV is due to 0.6-sec ^{257}Lr and 25-sec ^{211m}Po and that at 8.60 MeV is due to 4-sec ^{258}Lr. The spectra shown are from each detecting station around the wheel with their sum plotted topmost

however, if this turns out to be the case since ^{261}104 certainly does not have a prominent SF-branching ratio. One possibility is that the new SF activity is actually due to ^{252}No produced by the reaction ^{238}Pu(^{22}Ne,α4n) since ^{238}Pu is present in their ^{242}Pu target to the extent of about 1%.

In 1970 the first aqueous chemistry (Silva et al., 1970b) was performed with rutherfordium. Although the half-life of the isotope used, the 65-sec ^{261}Rf, was longer than that available for the lawrencium chemical experiment, the number of events was an order of magnitude less. As a consequence of this difficulty we were not able to make the tests as complete as one would like. After several hundred individual experiments aimed at the separation of element 104 from the actinide elements by means of a cation-exchange column, we had a total of some seventeen events which showed that rutherfordium had an aqueous chemistry similar to that of hafnium. Further chemical tests are intended when we have successfully completed an automatic system which can repeat experiments and over again. It is just too difficult to do this sort of operation manually.

VIII. Hahnium (Ha) — Element 105

Again the Soviet groups were the first to make a claim (Flerov et al., 1968) on the discovery of element 105 and again we were to challenge the validity of their experiments. The system used at Dubna consisted of catching the jet-borne atoms on the flange of a wheel which was viewed by an annular semiconductor detector through which the helium jet passed. Whenever an interesting event ocurred the wheel would be stopped for an interval long enough to allow detection of an alpha particle emitted by the element-103 daughter atom. In bombardments of ^{243}Am with ^{22}Ne projectiles they found a miniscule number of "correlated" alpha events which they attributed to ^{260}105 and ^{261}105. As indicated above they looked for delayed coincidences of alphas in the range 8.8 to 10.3 MeV succeeded by those in the range 8.35 to 8.6 MeV. They tentatively concluded that there was a positive effect beyond statistical variation at 9.4 and 9.7 MeV. The extremely low rate involved, approximately one per day, made it very difficult to refine these data; they reported that the half-life for the first activity was between 0.1 and 3 seconds and that of the second was probably somewhat greater than 0.01 seconds.

The Dubna conclusions depended completely on the assumption that the 8.35 to 8.6 MeV window in the element-103 daughter alpha spectrum included the emissions for ^{256}Lr and ^{257}Lr and that their half-lives were both about 35 seconds. Now we know that ^{257}Lr has an energy of about 8.9 MeV and a half-life of only 0.6 seconds and would not have been detected in the above experiment, thus ruling out the detection of ^{261}105. The identification of ^{260}105 in this work is invalid since, as will be shown shortly, its alpha energy is several hundred kilovolts less than 9.4 MeV.

Returning to the Berkeley work now I will review the sequences of our element-105 research. In November 1968 after we had found our first traces of rutherfordium we tried bombarding our ^{249}Cf target with ^{15}N ions in the hope that the production cross section might be high enough for us to also see a trace of element 105. In a couple of long runs we observed 8.9-MeV alpha particles with a short half-life and, in ten times smaller abundance, a peak at 9.1 MeV. We were discouraged from pursuing the matter to a conclusion when we discovered that the 8.9-MeV peak could also be produced by ^{13}C bombardment and thus could not possibly be due to element 105. Quite understandably we threw out the 9.1-MeV peak also since it was only marginally detected. We assumed that the 9.1-MeV activity was due to some artifact and temporarily abandoned the search for element 105 in order to find out more about the isotopes of lawrencium and rutherfordium. We had not yet eliminated all of the "bugs" from our new vertical-wheel system and thus had not yet built up a high degree of confidence in its ability to ferret out rare activities. By early 1970 we had discovered that the 8.9-MeV peak was due to 0.6-sec ^{257}Lr and with a vastly improved detection system once more we set out to find element 105.

On March 4, 1970 the experiment was begun again and within a few hours we knew that we had succeeded (Ghiorso et al., 1970) once more for we were seeing alpha counts at 9.1-MeV decaying with a half-life of a second or two. We realized somewhat sheepishly that this was the same peak that had been observed more than a year earlier. In rather short order we found alpha groups at 9.06, 9.10, and 9.14 MeV with a half-life of 1.6 ± 0.3 secs and were able to

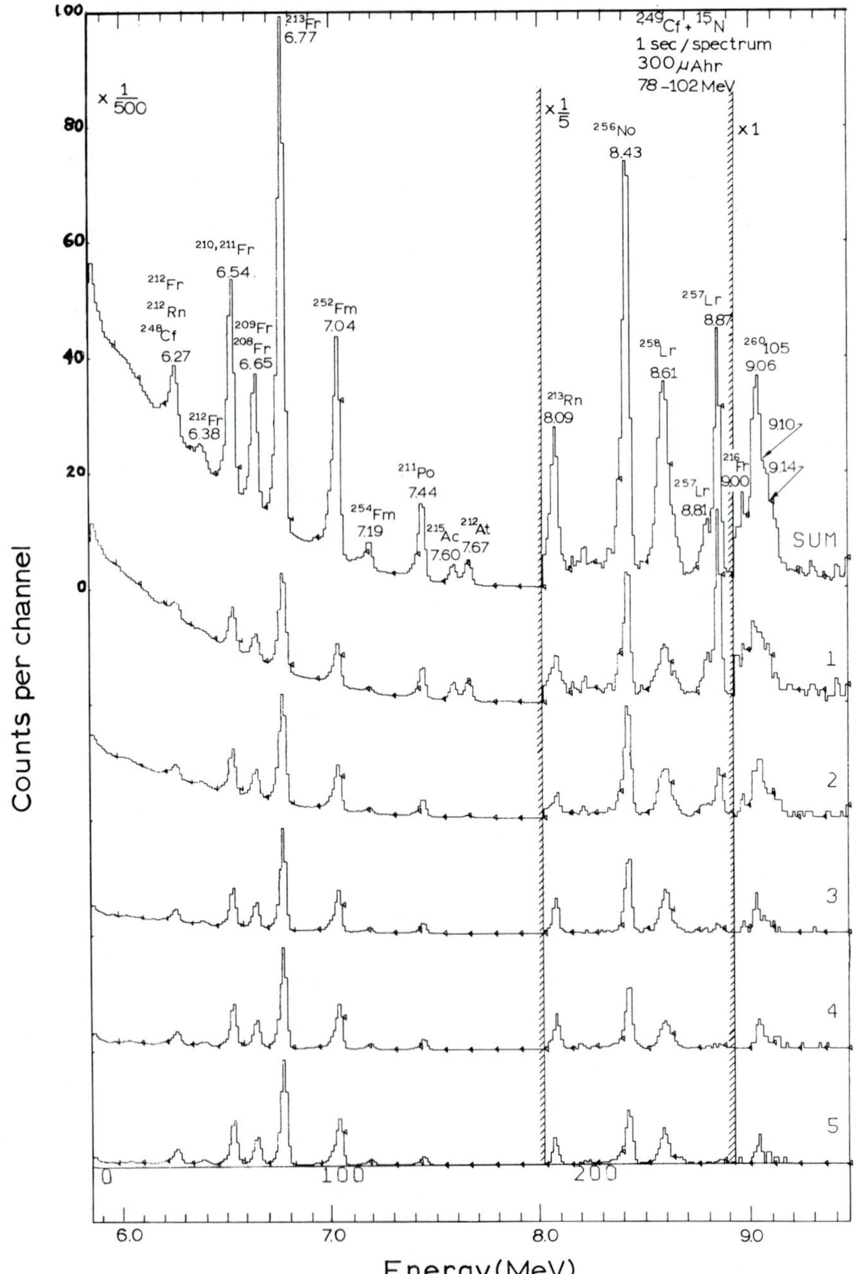

Fig. 16.11. Alpha spectrum of $^{260}105$. The lines at 9.06, 9.10, and 9.14 MeV are due to this isotope. The five-station system was used in these experiments and the individual spectra are shown with their sum plotted topmost

establish beyond a reasonable doubt the atomic number of the new nuclide (see Fig. 16.11). The mother-daughter recoil-milking experiment in this case showed that the 35-sec ^{256}Lr was transferred to the mother crystals and the amount thus observed at each detecting station decreased with the same half-life as the 9.1-MeV alpha activity. In addition to this rather good proof, we tried a time-correlation technique which promises to be important for the detection of even heavier elements.

The seven detecting stations were arranged at 45° intervals around the wheel so that only eight collection spots were produced. This was done in order to measure a time correlation between the emission of a mother alpha particle and that of its daughter. Thus, when an atom of element 105 emits an alpha particle which is detected by one of the mother crystals, it follows that the newborn element-103 daughter atom must be kicked into the surface of the wheel. If the spot on the wheel where this emission takes place is tracked by the computer, a watch can be made for the subsequent emission from that spot of an alpha particle from the daughter. If the amount of background activity is not too large this method can be very successful, so periodically new spots on the wheel are used automatically by advancing the wheel a few steps to reduce the buildup of longer-lived interfering activities. The technique when applied to the $^{260}105$–^{256}Lr case worked out beautifully and gave a further confirmation of our discovery. Since we felt that we had identified element 105 in a completely unambiguous fashion, the discovery team, which was the same as for element 104, suggested the name hahnium with the symbol Ha to honor the late OTTO HAHN, the discoverer of fission.

At about this same time the Dubna group reported finding a 2-sec SF emitter which they thought could be due to element 105. Gas chromatography experiments of the same type used in their element-104 work were cited to show that the 2-sec activity behaved like Ta but their evidence was not conclusive. Our work shows that it cannot be due to ^{260}Ha since we find less than a 25% branching ratio for fission of that isotope. There is some possibility, however, that the SF activity is due to a branch decay of ^{261}Ha. In later experiments we were able to identify this nuclide as an 8.93–MeV 1.7 ± 0.8-sec alpha emitter. The daughter, ^{257}Lr, was found both by recoil-milking and time-correlation experiments. We take the position that even if the SF activity is due to a branch decay of ^{261}Ha this does not imply discovery of the element by the Dubna group. It seems to us that the discoverer is the group that first proves to its scientific peers that it has indeed done so.

Very recently we have produced a third isotope of hahnium, ^{262}Ha. This was found to have an 8.45-MeV alpha energy and a half-life of about 42 seconds Again its daughter, ^{258}Lr, was identified by recoil-milking and time-correlation techniques. No chemical experiments have yet been performed by the Berkeley group, but we hope to be able to do so with our automatic chemistry system when the SuperHILAC resumes operation late in 1971.

IX. Future Heavy Elements

We believe that it will be possible to produce and identify alpha-emitting isotopes of elements 106 and 107. Our first choices for element 106 are to look for $^{261}106$ and $^{263}106$ by bombarding ^{249}Cf with ^{16}O and ^{18}O ions. The energies of these two isotopes should be in the range of ~ 9 MeV and the half-lives are expected to be roughly a few seconds. The daughters and grand-daughters are now known so that it should be possible in principle to identify a mother-daughter-grand-daughter sequence of alpha emission and thus be entirely certain of the results. The main stumbling block is that of their production cross sections. If they do not undergo significant branch decay by spontaneous fission, they should be observable at a rate of one or two alpha counts per day.

For element 107 we anticipate that the best choice may either be ^{19}F on ^{249}Cf to produce $^{264}107$ or ^{16}O on ^{253}Es to produce $^{266}107$. Here the cross sections will be even smaller and the work will be very difficult indeed. It is conceivable that the best approach may be to use two or more targets in tandem through which the beam passes simultaneously so that the data rate can be increased significantly.

In February 1971 the Dubna group reported finding a ca. 2-sec SF activity by bombarding ^{243}Am with ^{26}Mg ions. Conceivably the activity could be due to element 107 but the experiment suffers from the same uncertainties found in the element-104 and element-105 work.

X. Superheavy Elements

Since 1964 a great deal of the theoretical work has been published in which there seems to be a consensus that there could well be an island-of-stability in a region of nuclides near neutron number 184 and proton numbers 114 and possibly 124 or 126. Although the predictions of half-lives for nuclides on this island vary enormously, it seems that there is an excellent chance that elements in this region can live long enough for detection. The problem is—how does one make such superheavy nuclei?

At the present time we believe that there is a possibility of reaching the island by bombardments of the heavy elements with superheavy ions. There have been a few unsuccessful attempts to produce suitable reactions using Ar, Zn, and Kr ions at Berkeley, Orsay, and Dubna, but the tests were not of the type that could be regarded as conclusive—new accelerator capability is vitally necessary. To this end the Berkeley HILAC was shut down in February 1971 for an extensive modification which will make it possible to accelerate any atom up to energies as high as 8.5 MeV/nucleon. The new machine, which we designate by the name Super-HILAC, will be operating by the end of 1971 and we should soon know whether the "magic" island exists.

References

Donets, E. D., Schegolev, V. A., Ermakov, V. A.: Synthesis of the isotope of element 103 (lawrencium) with mass number 256. Atomnaya Energ. **19**, 109–113 (1965).

Donets, E. D., Schegolev, V. A., Ermakov, V. A.: Synthesis of the element 102 of mass number 256. Atomnaya Energ. **16**, 195–207 (1964).

Druin, V. A., Skobelev, V. A., Fefilov, B. V., Flerov, G. N.: Spontaneous fission of element 102 isotope with mass number 256, JINR Preprint P-1580 2 p. (1964), English.

Eskola, K., Eskola, P., Nurmia, M., Ghiorso, A.: Studies of lawrencium isotopes with mass numbers 255 through 260. University of California Lawrence Radiation Laboratory Publication UCRL 20442 (1971) (to be published in Physical Review).

Fields, P. R., Friedman, A. M., Milsted, J., Atterling, H., Forsling, W., Holm, L. W., Astrom, B.: Production of the new element 102. Physic. Rev. **107**, 1460–1462 (1957).

Flerov, G. N.: Heavy ion research. Proceedings of the 2nd Internat. Conference on Peaceful Uses of Atomic Energy, **14**, United Nations, New York, 151–157 (1958).

Flerov, G. N., Druin, V. A., Demin, A. G., Lobanov, Yu. V., Skobelev, N. K., Akap'ev, G. N., Fefilov, B. V., Kolesov, I. V., Gavrilov, K. A., Kharitonov, Yu. P., Chelnokov, L. P.: Experiments on search for alpha-radioactive isotopes of element 105. Joint Inst. Nucl. Res. (Dubna) Preprint P7-3808 (15 pp., 1968).

Flerov, G. N., Korotkin, Yu. S., Mikheev, V. L., Miller, M. B., Polikanov, S. M., Schegolev, V. A.: Synthesis and physical identification of the isotope of element 103 with mass number 260. Nucl. Phys. A **106**, 476–480 (1967).

FLEROV, G. N., OGANESYAN, YU. TS., LOBANOV, YU. V., KUZNETSOV, V. I., DRUIN V. A., PERELYGIN, V. P., GAVRILOV, K. A., TRETIAKOVA, S. P., PLOTKO, V. M.: On the nuclear properties of the isotopes $^{256}103$ and $^{257}103$. Soviet J. At. Energy **17**, 1046–1048 (1964) and Phys. Letters **13**, 73 (1964).

GHIORSO, A., HARVEY, B. G., CHOPPIN, G. R., THOMPSON, S. G., SEABORG, G. T.: New element mendelevium, atomic number 101. Physic. Rev. **98**, 1518–1519 (1955b).

GHIORSO, A., NURMIA, M., ESKOLA, K., HARRIS, J., ESKOLA, P.: New element hahnium, atomic number 105. Phys. Rev. Letters **24**, 1498–1503 (1970).

GHIORSO, A., NURMIA, M., HARRIS, J., ESKOLA, K., ESKOLA, P.: Positive identification of two alpha-particle-emitting isotopes of element 104. Phys. Rev. Letters **22**, 1317–1324 (1969).

GHIORSO, A., SIKKELAND, T.: Heavy ion reactions with heavy elements. Reprint from 2nd UN Geneva Conference, 158–164 (1958b).

GHIORSO, A., SIKKELAND, T.: The search for element 102. Physics Today **20**, 25–32 (1967b).

GHIORSO, A., SIKKELAND, T., LARSH, A. E., LATIMER, R. M.: New element, lawrencium, atomic number 103. Phys. Rev. Letters **6**, 473–475 (1961).

GHIORSO, A., SIKKELAND, T., NURMIA, M. J.: Isotopes of element 102 with mass 251 to 258. Phys. Rev. Letters **18**, 401–404 (1967a).

GHIORSO, A., SIKKELAND, T., WALTON, J. R., SEABORG, G. T.: Attempts to confirm the existence of the 10-minute isotope of 102. Phys. Rev. Letters **1**, 18–21 (1958a).

GHIORSO, A., THOMPSON, S. G., HIGGINS, G. H., HARVEY, B. G., SEABORG, G. T.: Evidence for subshell at $N = 152$. Physic. Rev. **95**, 293–295 (1954).

GHIORSO, A., THOMPSON, S. G., HIGGINS, G. H., SEABORG, G. T., STUDIER, M. H., FIELDS, P. R., FRIED, S. M., DIAMOND, H., MECH, J. F., PYLE, G. L., HUIZENGA, J. R., HIRSCH, A., MANNING, W. M., BROWNE, C. I., SMITH, H. L., SPENCE, R. W.: New elements einsteinium and fermium, atomic numbers 99 and 100. Physic. Rev. **99**, 1048–1049 (1955a).

HULET, E. K., WILD, J. F., LOUGHEED, R. W., EVANS, J. E., QUALHEIM, B. J.: Spontaneous-fission half-life of ^{258}Fm and nuclear instability. Physic. Rev. Letters **26**, 523–526 (1971).

SEABORG, G. T., JAMES, R. A., MORGAN, L. O.: National Nuclear Energy Series, Plutonium Project Record, vol. 14 B, The transuranium elements: Research Papers (New York: McGraw-Hill Book Company, Inc. 1949), Paper No 22.1. The new element americium (atomic number 95); SEABORG, G. T., JAMES R. A., and MORGAN L. O., Paper No 22.1. The new element curium (atomic number 96).

SILVA, R., HARRIS, J., NURMIA, M., ESKOLA, K., GHIORSO, A.: Chemical separation of rutherfordium. Inorg. Nucl. Chem. Letters **6**, 871–877 (1970b).

SILVA, R., SIKKELAND, T., NURMIA, M., GHIORSO, A.: Tracer chemical studies of lawrencium. Inorg. Nucl. Chem. Letters **6**, 733–739 (1970a).

THOMPSON, S. G., GHIORSO, A., SEABORG, G. T.: Element 97. Physic. Rev. **77**, 838–839 (1950a).

THOMPSON, S. G., GHIORSO, A., SEABORG, G. T.: The new element berkelium (atomic number 97). Physic. Rev. **80**, 781–789 (1950a).

THOMPSON, S. G., STREET, K. ,JR., GHIORSO, A., SEABORG, G. T.: Element 98. Physic. Rev. **78**, 298–299 (1950b).

THOMPSON, S. G., STREET, K., JR., GHIORSO, A., SEABORG, G. T.: The new element californium (atomic number 98). Physic. Rev. **80**, 790–796 (1950b).

ZAGER, B. A., MILLER, M. B., MIKHEEV, V. L., POLIKANOV, S. M., SUKHOV, A. M., FLEROV, G. N., CHELNOKOV, L. P.: The properties of the isotope 102^{254}. Atomnaya Energiya **20**, 230–232 (1966).

ZVARA, I., CHUBURKOV, YU. T., TSALEIKA, R., ZVAROVA, T. S., SHALAEVSKII, M. R., SHILOV, B. V.: Chemical properties of element 104. Atomnaya Energiya **21**, 83–84 (1966).

Chapter 17

Chemical and Physical Properties of the Transplutonium Elements

David M. Taylor

With 2 Figures

I. Introduction

In contrast to plutonium, which is now one of the best understood elements in the Periodic Table, our knowledge of the elements beyond plutonium—the transplutonic elements—is not yet very extensive. The chemistry of americium, and to a lesser extent curium, has been studied in some detail but at present knowledge of the chemical properties of elements 97 to 103 is rather scanty. This unbalanced knowledge of the chemistry of the transplutonium elements is not surprising in view of the fact that the production of the higher elements is difficult and expensive. Further, most of the available isotopes are short lived and consequently of low mass. The combination of short half life, intense radioactivity and very small mass poses many problems in the study of chemical properties. The elegant ultramicro and tracer methods which have been devised to permit chemical investigations to be made on as little as a few tens of atoms of material has made the study of the chemistry of the transplutonic elements one of the most fascinating areas of modern inorganic chemistry.

In this chapter the chemistry of americium, curium, berkelium and californium will be reviewed with special reference to those properties which may be important in determining their biological behaviour. For the sake of completeness what is known concerning the chemistry of einsteinium, fermium, mendelevium, nobelium and lawrencium will be discussed, although at the present time insufficient quantities of these elements have been prepared to pose any biological hazard.

The general aspects of the chemistry of the transplutonium elements have been discussed in the now classic, but outdated, book "The Chemistry of the Actinide Elements" by Katz and Seaborg (1957) and, more recently, by Pascal (1970). Certain aspects of the chemistry of these elements have also been discussed in reviews by Penneman and Keenan (1960); Higgins (1960); Cunningham (1964), Bagnall (1964), and Seaborg (1968).

A. The Position of the Transplutonium Elements in the Periodic Table

The existence of an actinide series, created by the successive filling of the 5 f electron shell, starting with element 89, actinium and ending with element 103,

Table 17.1. The actinide series

Atomic number	Element	Ionic radius M^{3+} (nm)	Electronic configuration	Oxidation states
89	Actinium	0.1153	(Rn) 6d $7s^2$	3^b
90	Thorium	0.1080	(Rn) $6d^2$ $7s^2$	3^a 4^b
91	Protactinium	0.1050	(Rn) $5f^2$ 6d $7s^2$ or $5f^3$ 6d $7s^2$	3^a 4 5^b
92	Uranium	0.1040	(Rn) $5f^3$ 6d $7s^2$	3 4 5 6^b
93	Neptunium	0.1029	(Rn) $5f^4$ 6d $7s^2$ or $5f^5$ $7s^2$	3 4 5^b 6
94	Plutonium	0.1014	(Rn) $5f^6$ $7s^2$	(2) 3 4^b 5 6 7
95	Americium	0.1006	(Rn) $5f^7$ $7s^2$	2^a 3^b 4^a 5 6
96	Curium	0.0996	(Rn) $5f^7$ 6d $7s^2$	3^b 4^a
97	Berkelium	0.0982	(Rn) $5f^8$ 6d $7s^2$ or $5f^9$ $7s^2$	3^b 4^a
98	Californium	0.0976	(Rn) $5f^{10}$ $7s^2$	3^b
99	Einsteinium	0.097	(Rn) $5f^{11}$ $7s^2$	3^b
100	Fermium	?	(Rn) $5f^{12}$ $7s^2$	3^b
101	Mendelevium	?	(Rn) $5f^{13}$ $7s^2$	2 3^b
102	Nobelium	?	(Rn) $5f^{14}$ $7s^2$	2^b 3
103	Lawrencium	?	(Rn) $5f^{14}$ 6d $7s^2$	3^b

[a] Indicates oxidation state unstable in solution. [b] Indicates most stable oxidation state.

was suggested by Seaborg (1954). The known or predicted chemical properties of elements numbers 95 to 103 are consistent with their being assigned to fill the nine positions beyond plutonium in the actinide series, see Table 17.1. From their predominantly trivalent oxidation states and general chemical properties the nine highest members of the actinide series exhibit closer analogies to the corresponding members of the lanthanide series than is shown by plutonium.

B. Electronic Structure

The predicted electronic structures for neutral gaseous atoms of the transplutonic elements in the ground state are illustrated in Table 17.1. Physical measurements of magnetic properties and other parameters have confirmed these predictions for americium, curium, berkelium and californium (Cunningham, 1964).

C. Isotopes of the Transplutonic Elements

The isotopes of americium, curium, berkelium and californium are listed in Tables 17.2–17.5 which also shows the mass of element per curie of alpha activity for each alpha emitting isotope, and the principal production reactions. The isotopes of elements 99 to 103, einsteinium, fermium, mendelevium, nobelium and lawrencium, together with typical production reactions are listed in Table 17.6.

It will be seen that many of the nuclides listed in these tables have short half lives and consequently their specific activities are very high. This is especially apparent with the isotopes of the last four transplutonic elements (Table 17.6).

Table 17.2. Americium isotopes. (Lederer, Hollander and Perlman, 1969)

Isotope	Half-life	Mass per curie of α-radiation	Principal radiations[a]	Principal production reactions
^{237}Am	1.3 hr	9.83 μg	α 6.02	^{239}Pu(P 3n) ^{237}Am, ^{239}(d 4n) ^{237}Am
^{238}Am	1.9 hr	—	γ Pu X-rays 0.36, 0.58, 0.98, 1.35	^{239}Pu (P 2n) ^{238}Am, ^{239}Pu (d 3n) ^{238}Am, ^{237}Np (α 3n) ^{238}Am
^{239}Am	12.1 hr	18.44 mg	α 5.78 (0.005); EC (99+ %); γ Pu X-rays 0.21, 0.23, 0.28; e^- 0.02–0.04, 0.048, 0.088, 0.106, 0.156	^{239}Pu (pn) ^{239}Am, ^{239}Cu (d2n) ^{237}Am ^{237}Np (α2n) ^{239}ÅAm
^{240}Am	51.0 hr	—	EC, γ Pu X-rays 0.90, 1.00; e^- 0.022, 0.038, 0.079, 0.094	^{239}Pu (dn) ^{240}Am, ^{239}Pu (αp2n) ^{240}Am, ^{237}Np (αn) ^{240}Am
^{241}Am	458 yr S.F. 2×10^{14}yr	308.4 mg	α 5.49 (85 %), 5.44 (13 %); γ NpL X-rays 0.069, 0.101, 0.208, 0.335, 0.37, 0.663, 0.722; e^- 0.022, 0.038, 0.054	daughter ^{241}Pu
^{242}Am	16.01 hr	—	β 0.67 max (84 %); EC (16 %); γ Pu X-rays; Cm L X-rays; e^- 0.021, 0.037	^{241}Am (nγ) ^{242}Am or ^{238}U or ^{239}Pu by multiple n capture
^{243m}Am	152 yr	21.4 mg	α 5.21 (0.48 %); IT 99+ %; γ Am L X-rays; Np X-rays 0.049; 0.087, 0.110, 0.163	^{241}Am (nγ) ^{242m}Am
^{243}Am	7.8×10^3 yr	5.40 g	α 5.28 (87 %), 5.23 (11 %); γ NpL X-rays 0.044, 0.075; e^- 0.011, 0.023, 0.052, 0.069	^{238}U or ^{239}Pu by multiple n capture
^{244}Am	10.1 hr	—	β^- 0.39 max; e^- 0.030, 0.037, 0.077, 0.094; γ Cm X-rays 0.10, 0.15, 0.75, 0.90	^{243}Am (nγ) ^{244}Am
^{244m}Am	26 m	—	β^- 1.50 max (99+ %); e^- 0.020, 0.037; γ Cm L X-rays	^{243}Cnγ) ^{244m}Am
^{245}Am	2.07 hr	—	β^- 0.91 max; e^- 0.125; v Cm X-rays 0.253	daughter ^{245}Pu
^{246}Am	25.0 m	—	β^- 2.10 max (7 %), 1.60 max; v Cm X-rays 0.80, 1.07	daughter ^{246}Pu

S.F. = Spontaneous fission.

Table 17.3. Curium isotopes. (Lederer, Hollander and Perlman, 1969)

Isotope	Half-life	Mass per curie of α-radiation	Principal radiations[a]	Principal production reactions
^{238}Cm	2.5 hr	1.90 μg	α 6.51 (10%); EC (90%)	$^{239}Pu(\alpha\ 5n)\ ^{238}Am$; $^{238}Pu(\alpha\ 4n)^{238}Cm$
^{230}Cm	2.9 hr	—	EC, γ Am X-rays 0.188	$^{239}Pu(\alpha\ 4n)\ ^{239}Cm$
^{240}Cm	26.8 d	49.23 μg	α 6.29 (72%), 6.25 (28%)	$^{239}Pu(\alpha\ 3n)^{240}Cm$
^{241}Cm	35 d	6.46 mg	α 5.94 (1%); EC (99%); Am X-rays 0.47, 0.60; e^- 0.123, 0.350	$^{239}Pu(\alpha\ 2n)\ ^{241}Cm$
^{242}Cm	162.5 d	301 μg	α 6.12 (74%), 6.07 (26%); γ Pu LX-rays, 0.044, 0.10, 0.16, 0.58, 0.89; e^- 0.022, 0.039	daughter ^{242m}Am multiple n capture from ^{238}U, ^{239}Pu
^{243}Cm	32 yr	21.7 mg	α 6.06, 5.99, 5.79, 5.74; γ Pu LX-rays, 0.21, 0.23, 0.28; e^- 0.02–0.04, 0.048, 0.088, 0.106, 0.156	multiple n capture from ^{238}U or ^{239}Pu
^{244}Cm	17.6 yr	12.0 mg	α 5.81, 5.31; γ Cm LX-rays 0.043, 0.10, 0.15, 0.26, 0.59, 0.82; e^- 0.022, 0.038	multiple n capture from ^{238}U or ^{239}Pu
^{245}Cm	9.3×10^3 yr	6.37 g	α 5.36, 5.31; γ Pu X-rays 0.13, 0.17	multiple n capture from ^{238}U, ^{239}Pu, ^{244}Cm
^{246}Cm	5.5×10^3 yr	3.78 g	α 5.39, 5.34; γ Pu LX-rays	multiple n capture from ^{238}U, ^{239}Pu, etc.
^{247}Cm	1.6×10^7 yr	11 kg		multiple n capture from ^{238}U, ^{239}Pu, etc.
^{248}Cm	4.7×10^5 yr	365.9 g	α 5.08, 5.04; γ Pu LX-rays; SF	daughter ^{252}Cf multiple n capture from ^{238}U, etc.
^{249}Cm	64 m	—	β 0.9 max	$^{248}Cm(n\gamma)\ ^{249}Cm$ multiple n capture from ^{238}U, etc.
^{250}Cm	SE $T^1/_2$ 1.7×10^4 yr	—	S. F.	multiple n capture from ^{238}U

S.F. = Spontaneous fission resulting in the emission of neutrons, γ-rays, electrons and radiations from daughter nuclides or fission fragments.

Table 17.4. Berkelium isotopes. (Lederer, Hollander and Perlman, 1969)

Isotope	Half-life	Mass per curie of α-activity	Principal radiations [a]	Principal production reactions
^{243}Bk	4.6 hr	2.38 μg	α (0.15%) 6.76, 6.72, 6.57, 6.54, 6.21; E.C. (99+%); γ Cm X-rays 0.75, 0.84, 0.95	^{241}Am (α 2n) ^{243}Bk, ^{242}Cm (αn) ^{243}Bk
^{244}Bk	4.6 hr	5.71 mg	α (0.006%) 6.67, 6.62; E.C. (99+%); Cm X-rays 0.14, 0.19, 0.22, 0.33, 0.49, 0.89, 0.92, 1.16	^{243}Am (α 3n) ^{244}Bk
^{245}Bk	4.98 d	8.49 mg	α (0.11) 6.36, 6.32, 6.15, 6.12, 5.89; E.C. (99+%); γ Cm X-rays 0.25, 0.39; e^- 0.125	^{243}Am (α 2n) ^{245}Bk, ^{242}Cm (αp) ^{245}Bk
^{246}Bk	1.8 d	—	E.C.; γ Cm X-rays 0.80, 1.07	^{244}Cm (αpn) ^{246}Bk, ^{243}Am (αn) ^{246}Bk
^{247}Bk	1.4×10^3 yr	966 mg	α 5.68, 5.52; γ Am X-rays 0.089, 0.27	daughter ^{247}Cf; ^{244}Cm (αp) ^{247}Bk
^{248}Bk	16 hr	—	E.C. 30%; γ Cm X-rays; β^- 0.65 max (70%)	^{247}Bk (nν) ^{248}Bk, ^{245}Cm (αp) ^{248}Cm ^{246}Cm (αpn) ^{248}Bk
^{248}Bk	9 yr	—		^{246}Cm (αpn) ^{248}Bk
^{249}Bk	314 d S.F. 6×10^8 yr	2.72 g	α (0.0022%) 5.42; β^- (99+%) 0.125 max; γ 0.32	multiple n-capture from ^{238}U ^{239}Pu, etc.
^{250}Bk	193.3 m	—	β^- 1.76 max, 0.73 max; e^- 0.019, 0.036; γ Cf LX-rays 0.99, 1.03	^{249}Bk (nγ) ^{259}Bk daughter ^{254}Es

S.F. = Spontaneous fission.

Table 17.5. Californium isotopes. (LEDERER, HOLLANDER and PERLMAN, 1969)

Isotope	Half-life	Mass per curie of α-activity	Principal radiations[a]	Principal production reactions
^{244}Cf	25 m	32.4 ng	α 7.18	$^{244}Cm(\alpha 4n)\,^{244}Cf$; $^{238}U(C^{12}6n)\,^{244}Cf$
^{245}Cf	44 m	191.0 ng	α (30%) 7.12; E.C. (70%)	$^{244}Cm(\alpha 3n)\,^{245}Cf$; $^{238}U(C^{12}5n)\,^{245}Cf$
^{246}Cf	35.7 hr	2.80 µg	α 6.76, 6.72; γ Cm LX-rays	$^{244}Cm(\alpha 2n)\,^{246}Cf$; $^{238}U(C^{12}4n)\,^{245}Cf$
^{247}Cf	2.5 hr	—	E.C.; γ Bk X-rays 0.295, 0.417, 0.460; e^- 0.164	^{244} Cm $(\alpha n)\,^{247}Cf$; $^{238}U(N^{14}p4n)\,^{247}Cf$
^{248}Cf	350 d S.F. 21.5×10^4 yr	664 µg	α 6.27, 6.22; γ Cm LX-rays	$^{238}U(N^{14}p3n)^{248}Cf$; daugther ^{248}Bk
^{249}Cf	360 yr S.F. 1.5×10^9 yr	251.5 mg	α 5.81; γ Cm X-rays 0.33, 0.39	daughter ^{249}Bk, multiple n capture from ^{238}U, ^{239}Pu, etc.
^{250}Cf	13.2 yr S.F. 1.7×10^4 yr	9.2 mg	α 6.03, 5.99; e^- 0.023, 0.038; γ Cm LX-rays	multiple n capture from ^{238}U, etc. daughter ^{250}Bk ^{254}Fm
^{251}Cf	800 yr	561 mg	α 5.85, 5.67; γ Cm X-rays 0.18	multiple n capture from ^{238}U
^{252}Cf	2.646 yr S.F. 85 yr	1.9 mg	α (96.9%) 6.12, 6.08; e^- 0.022, 0.038; γ Cm LX-rays; S.F. (3.1%)	multiple n capture from ^{238}U, ^{239}Pu, etc.
^{253}Cf	17.6 d	11 mg	α (0.31) 5.98 β^- (99+%) 0.27 max	multiple n capture from ^{238}U, etc.
^{254}Cf	60.5 d	58.8 mg	α (0.2) 5.84 S.F. (99+%)	multiple n capture from ^{238}U, etc.

S.F. = Spontaneous fission resulting in the emission of neutrons, γ rays, electrons and radiations from daughter nuclides and fission fragments.

Table 17.6. Isotopes of elements 99 to 103. (Lederer, Hollander and Perlman, 1963)

Isotope	Half-life	Principal radiations[a]	Production reaction
Einsteinium			
^{245}Es	1.3 m	α (17%) 7.70[a], E.C. (83%)	^{235}U (^{14}N 4n) ^{245}Es
^{246}Es	7.5 m	α (10%) 7.33, E.C. (90%)	^{238}U (^{14}N 6n) ^{246}Es
^{247}Es	5.0 m	α (7%) 7.33, E.C. (93%)	^{238}U (^{14}N 5n) ^{247}Es
^{248}Es	25 m	α (0.3%) 6.88, E.C. (99+%)	^{249}Cf (α 3n) ^{248}Es
^{249}Es	2 hr	α (0.13%) 6.77, E.C. (99+%)	^{249}Bk (α 4n) ^{240}Es
^{250}Es	8 hr	E.C.	^{249}Bk (α 3n) ^{250}Es
^{251}Es	1.5 d	α (0.53) 6.49, E.C. (99+%)	^{249}Bk (α 2n) ^{251}Es
^{252}Es	140 d	α 6.64, 6.58	^{252}Cf (α 2n) ^{252}Es
^{253}Es	20.4 d S.F. 6.4×10^5 yr	α 6.64	daughter ^{253}Cf
^{254}Es	276 d S.F. 7×10^5 yr	α 6.44	multiple n capture ^{238}U, etc.
^{254m}Es	39.3 hr	β^- (99+%) 1–13 max	multiple n capture ^{238}U, etc.
^{255}Es	38.3 d	α (8.5%) 6.31, β^- 91.5%	multiple n capture ^{238}U, etc.
^{256}Es	short		^{255}Es (nγ) ^{256}Es
Fermium			
^{248}Fm	0.6 m	α	^{240}Pu (C^{12} 4n) ^{248}Fm
^{249}Fm	2.5 m	α 7.9	^{238}U (O^{15} 5n) ^{249}Fm
^{250}Fm	30 m	α 7.44	^{249}Cf (α 3n) ^{250}Fm
^{251}Fm	7 hr	α (1%) 6.89, E.C. 99%	^{249}Cf (α 2n) ^{251}Fm
^{252}Fm	22.7 hr	α 7.05	^{249}Cf (α n) ^{252}Fm
^{253}Fm	3 d	α (11%) 6.96, 6.91; E.C. (89%)	^{252}Cf (α 3n) ^{253}Fm
^{254}Fm	3.24 h S.F. 246 d	α (99%) 7.20, 7.16 S.F. (0.055%)	daughter ^{254m}Es
^{255}Fm	20.1 h S.F. 10^4 y	α 7.03	daughter ^{255}Es
^{256}Fm	2.7 hr	α (3%) 6.86, S.F. (97%)	^{255}Es (nγ) ^{256}Es (β) ^{256}Fm
^{257}Fm	80 d S.F. 100 yr	α 6.53	multiple n uptake from ^{242}Pu
^{258}Fm	11 d	S.F.	multiple n capture from ^{244}Cm
Mendelevium			
^{255}Md	0.6 hr	α (10%) 7.34, E.C. (90%)	^{253}Es (α 2n) ^{255}Md
^{256}Md	1.5 hr	α (3%) 7.18, E.C. (97%)	^{11}β, ^{12}C ^{13}C on ^{252}Cf
^{257}Md	3 hr	α (8%) 7.25, 7.08; E.C. (92%)	^{11}β, ^{12}C ^{13}C on ^{252}Cf
Nobelium			
^{251}No	0.8 s	α 8.68, 8.58	^{244}Cm (^{12}C 5n) ^{251}No
^{252}No	2.1 s	α (70%) 8.41 S.F. (30%)	^{239}Pu (^{18}O 5n) ^{252}No
^{253}No	95 s	α 8.02	^{242}Pu (^{16}O 5n) ^{253}No
^{254}No	55 s	α 8.10	^{246}Cm (^{12}C 4n) ^{254}No
^{255}No	180 s	α 8.11	^{239}U (^{22}Ne 5n) ^{255}No
^{256}No	2,7 s	α 8.43	^{246}Cm (^{13}C 3n) ^{256}No
^{257}No	20 s	α 8.27, 8.23	^{248}Cm (^{13}C 4n) ^{257}No
Lawrencium			
^{256}Lw	45 s	α	^{243}Am (^{18}O 5n) ^{256}Lw
^{158}Lw	8 s	α 8.6	^{10}B, ^{11}B on $^{250-252}$Cf
^{259}Lw			

S.F. = Spontaneous fission.

[a] Energies in MeV.

II. Transplutonium Elements and their Compounds

Americium, curium and californium are now being produced on kilogram, gram and milligram scales respectively in the United States and in some other countries. Americium-241 arises as the daughter of ^{241}Pu and ^{242}Cm is produced by neutron irradiation of ^{241}Am. The longer lived ^{244}Cm is produced by neutron irradiation of plutonium using a step wise irradiation and recycling process which leads to ^{242}Pu, ^{243}Am and ^{244}Cm. Irradiation of these nuclides in high neutron flux reactors leads to ^{252}Cf by multiple neutron capture (FERGUSON and BIGELOW, 1969). Irradiation of ^{252}Cf with neutrons has been used to produce microgram quantities of ^{253}Es.

In contrast to the above nuclides, fermium, mendelevium, nobelium and lawrencium have been prepared on a very small scale mainly in the United States and the U.S.S.R. The scale of production has ranged from a few atoms of lawrencium to about 10^9 atoms of fermium.

A. The Metallic States

1. Americium

Americium metal is a silvery white, malleable and ductile metal which melts at $994 \pm 7°C$ (MCWHAN et al., 1962).

The metal oxidises slowly in air and dissolves readily in dilute hydrochloric acid.

2. Curium

Curium metal is a silvery, hard, brittle metal with a melting point of $1350 \pm 60°C$ (EUBANKS and THOMPSON, 1969).

Metallic curium oxidises very rapidly in the presence of traces of oxygen, the rate of oxidation being much more rapid than that of plutonium or americium. This more rapid oxidation may be due to self-heating by radioactive decay, which is sufficient to make pure curium metal glow red hot.

3. Berkelium

Small quantities of berkelium metal have been prepared and used for magnetic susceptibility studies (FUJITA, 1969), but no other data have been reported.

4. Higher Elements

None of the other transplutonic elements have yet been prepared in the metallic state. Attempts to prepare californium metal were reported to be unsuccessful due to the high vapour pressure of elemental californium (FUJITA, 1969).

B. Oxidation States

1. Americium

It has now been claimed that americium exhibits all oxidation states from (II) to (VII), but of these six states the trivalent state is the most common.

Divalent americium has been demonstrated, by spectroscopic and paramagnetic studies, only after stabilisation of the Am^{2+} in a crystal of calcium fluoride (EDELSTEIN et al., 1966). The existence of heptavalent americium, Am(VII), has been inferred by ZAITSEVA (1969) from studies of the disproportionation of Am(VI). The Am(VII) ion is said to be stable in fairly concentrated solutions in 0.3 M $KClO_4$.

2. Curium

The trivalent state is the most stable oxidation state of curium, but Cm(IV) is present in some compounds. The existence of Cm(VI) has been claimed by AKATSU (1965), but other workers have been unable to confirm this claim (BAGNALL and EASEY, 1966; HOLCOMB, 1967; HAISSINSKY et al., 1968) and it seems unlikely that any oxidation state higher than Cm(IV) can exist.

3. Higher Transplutonic Elements

For the elements berkelium, californium, fermium and einsteinium, the trivalent state appears to be the stable oxidation state although BkO_2 representing the tetravalent state Bk(IV) has been prepared (PETERSON and CUNNINGHAM, 1967). Recently the formation of Es(II) has been demonstrated after stabilisation in a CaF_2 crystal (EDELSTEIN et al., 1970). Tracer studies have suggested that there may be a moderately stable divalent state of californium but attempts to reduce $CfCl_3$ to $CfCl_2$ have been unsuccessful (MILLIGAN, 1970). In contrast, mendelevium has been shown to exist as both Md(II) and Md(III) in aqueous solution, (HULET et al., 1967) and No(II) has been shown to be more stable than No(III) in aqueous solutions (MALY et al., 1968). Recent studies on the last member of the actinide series suggest that Lr(III) is the most stable oxidation state of this element.

This pattern of stability of the oxidation states of the higher transplutonic elements is consistent with increased 5 f electron binding as the 5 f shell approaches completion, thus No(II) would be expected to be the most stable state for nobelium in aqueous solution. However, lawrencium should contain one 6 d electron which would be bound with a similar energy to the two 7 s electrons, but less than that of the 5 f electrons, thus Lr(III) would be expected to be the most stable state in aqueous solution.

C. Compounds

Up to the present time relatively few compounds of the transplutonic elements have been prepared in pure form and characterised. However, compounds representative of the most important types of inorganic compound have been prepared or detected for a number of the elements.

1. Oxides and Hydroxides

Trivalent Oxides. Sesquioxides, M_2O_3, of americium, curium, berkelium and californium have been prepared and studied crystallographically (KATZ and SEABORG, 1957; HAUG, 1967; PETERSON and CUNNINGHAM, 1967; GREEN, 1965). The curium sesquioxide, Cm_2O_3, appears to be remarkably stable towards self-irradiation (HAUG, 1967).

Tetravalent Oxides. The dioxides, MO_2, of americium, curium and berkelium have been reported (KATZ and SEABORG, 1957; PETERSON and CUNNINGHAM, 1967) and these are important representatives of the very small group of compounds which are known containing transplutonic elements in the tetravalent state.

Americium dioxide, AmO_2, is a black solid, stable up to about 1000°C. The reddish brown Am_2O_3 results from the heating of AmO_2 in hydrogen at 600°C. The white, or faint tan coloured, Cm_2O_3 is prepared in a similar way by heating CmO_2 in hydrogen at 700–1055°C (HAUG, 1967).

The americium and curium oxides are insoluble in water, but soluble in acids.

Americium is known to form two hydroxides, pink $Am(OH)_3$ and the black $Am(OH)_4$ (PENNEMAN et al., 1961); $Cm(OH)_3$ and $Bk(OH)_3$ are also known. The hydroxides are insoluble in water but soluble in hydrochloric, perchloric and other acids. The solubility products of $Am(OH)_3$ and $Cm(OH)_3$ have been calculated to be 1.9×10^{-21} and 2×10^{-20} respectively (LATIMER, 1952).

2. Hydrides

Americium metal disintegrates in hydrogen at 50° C to yield a black powder with the mean composition $AmH_{2.7}$ (WESTRUM and EYRING, 1951). A lower hydride AmH_2 has been reported and it seems likely that two hydrides of probable composition AmH_2 and Am_4H_{15} exist (KATZ and SEABORG, 1957).

Crystallographic studies of a curium hydride of unknown composition, but presumably CmH_{2+x}, have been described by BANSAL and DAMIEN (1970).

3. Halogen Compounds

Trihalides. Trihalide compounds of most of the transplutonic elements have been prepared, at least in solution, and some oxyhalides such as BkOCl (PETERSON, 1967) and EsOCl, EsOF, CfOBr and CfOI (FUJITA et al., 1969; FUJITA, 1969) have been studied.

Trifluorides of americium, curium, berkelium and californium have been prepared and characterised; these are all sparingly soluble. In contrast to the trifluorides, the trichlorides, tribromides and triiodides are soluble in water.

Tetrahalides. The tetrafluorides of americium, curium and berkelium have been prepared and their crystal structures determined (KEENAN and ASPREY, 1969). These compounds have been prepared by treatment of the dioxides or trifluorides with elemental fluorine at high temperatures (ASPREY et al., 1957; KEENAN and ASPREY, 1969). The addition of water to AmF_4 or CmF_4 results in the oxidation of water and reduction to Am(III) and Cm(III) respectively (PENNEMAN and KEENAN, 1960). A crystalline fluoride complex of Am(IV), Rb_2AmF_6, has been prepared by KRUSE and ASPREY (1962).

Higher halides. The higher halides AmF_5 and AmF_6 do not occur, but double fluoride salts of Am(V), $RbAmO_2F_2$ and K_2AmO_2F, are known (PENNEMAN and KEENAN, 1960), and an americyl fluoride, AmO_2F_2 has been reported (PENNEMAN, 1968). Complex chlorides of Am(V), probably $Cs_8(AmO_2)_3Cl$, and Am(VI), $Cs_2AmO_2Cl_4$, have been described by BAGNALL et al. (1967).

Other halogen compounds. An insoluble americium iodate is known (PENNEMAN and KEENAN, 1960).

4. Other Compounds

Americium sesquisulphide, Am_2S_3, has been prepared by heating AmO_2 with carbon disulphide and hydrogen sulphide at about 1500° C (ZACHARIASEN, 1954), and americium monosulphide, AmS, has been identified by DAMIEN (1971). A berkelium sesquisulphide has been described by COHEN et al. (1968).

Insoluble oxalates and phosphates of americium and curium are known as well as three double sulphates of americium and potassium, for example $K_8Am_2(SO_4)_7$ (PENNEMAN and KEENAN, 1960). Double carbonate salts of Am(V), for example $RbAmO_2CO_3$ and $NH_4AmO_2CO_3$, have been prepared and characterised (NIGON et al., 1954).

III. The Solution Chemistry of the Transplutonic Elements

A. Absorption Spectra

Like the corresponding members of the lanthanide series, the actinide elements show characteristic absorption spectra in solution.

The major oxidation states of americium exhibit such major differences in their spectra that all four oxidation states may be differentiated from each other and determined quantitatively by spectrophotometry. The information on the absorption spectra of Am(III), Am(V) and Am(VI) has been reviewed and discussed by several authors (KATZ and SEABORG, 1957; PENNEMAN and ASPREY, 1955; YAKOVLEV and KOSYAKOV, 1955). Trivalent americium is characterised by a number of absorption bands with a sharp peak, consisting of four very sharp lines, at 503 nm. The Am(V) spectrum shows two sharp peaks at 513 and 715 nm while Am(VI) is distinguished by a minor peak at 666 nm and a sharp peak at 995 nm. The absorption spectrum of Am(IV) has been measured, after stabilizing the ion in 13 M ammonium fluoride, by ASPREY and PENNEMAN (1962) who found a characteristic sharp peak at 456 nm. The absorption spectrum of Am(VII) has been reported by ZAITSEVA (1969) to show characteristic peaks at 520 and 780 nm.

The spectrum of Cm(III) is confined to the ultraviolet and near ultraviolet end of the spectrum. Three major peaks at 375, 381, and 396 nm may be used for quantitative determinations (CARNALL et al., 1958).

The absorption spectra of Cf(III) and Es(III) have been studied by most elegant micro methods utilising a few nanolitres of solution or a few nanograms of compound adsorbed onto a single bead of ion-exchange resin (GREEN and CUNNINGHAM, 1966). For Cf(III) the major peak was found to lie at 473 ± 2 nm (GREEN and CUNNINGHAM, 1966) and for Es(III) the main peak was located at 495 nm (FUJITA et al., 1969).

B. Hydrolytic Reactions

The general features of hydrolysis and complex ion formation have been discussed in Chap. 9 and will not be repeated here.

The smaller ionic potential, resulting from the lower ionic charge and larger ionic radius, indicates that the trivalent transplutonic ions will hydrolyse much less readily than Pu(IV) ions. However, as with Pu(III), hydrolytic reactions and also complex formation are still important in relation to the solution chemistry of the transplutonic elements.

The hydrolysis of Am(III) nitrate at a concentration of 10^{-9} M americium in nitric acid has been studied by STARIK and GINZBERG (1960) in relation to the adsorption of colloidal species on quartz surfaces. They concluded that Am(III) was present in ionic form from pH 1 to pH 4.5, but above pH 4.5 hydrolysis occurs leading to the formation of colloidal species at pH values greater than 7.

Limited studies on the hydrolysis of Cm(III) have been carried out by TAYLOR (1970) who studied the ultrafilterability of solutions of ^{244}Cm at various pH values. A solution containing 5×10^{-10} M ^{244}Cm in 0.02 M nitric acid was prepared by dilution of a stock solution, and aliquots of this solution were adjusted to pH values ranging from 2 to 12, measured with a glass electrode, by the careful addition of NaOH to the rapidly stirred solution. After standing overnight the solutions were filtered through a 25 nm pore diameter "Millipore" filter. The ultrafilterability decreased from 100 percent at pH 2 to 68 percent at pH 4.5 and to 6 percent at pH 5.4; at pH 10 to 12 the filterability was 10 to 12 percent. These results suggest that at this concentration the formation of large polymeric species occurs mainly above pH 4.5, a conclusion which is similar to the observations of STARIK and GINZBERG (1960) on americium.

The first hydrolysis constants of Am(III), Cm(III), Bk(III) and Cf(III) have been studied by DESIRE et al. (1969) who found that the acidity of the cations increases slightly with increasing atomic number; this increase in acidity is predicted by the actinide theory (SEABORG, 1954). The hydrolytic behaviour of the higher transplutonic elements has not yet been described. Since most of the higher transplutonic nuclides are of short half life and consequently very high specific activity, most work will be carried out with extremely low concentrations of the element. At such concentrations the hydrolytic behaviour may be considerably modified by the presence of even tracer amounts of complexing anions in the solutions.

C. Complex Formation

The chemistry of the complexes formed by transplutonium elements with a variety of ligands has attracted considerable attention due to the importance of complexes such as lactate, tartrate, citrate and α-hydroxyisobutyrate in the separation of the various elements by ion exchange or solvent extraction.

1. Complexes with Inorganic Anions

Americium and curium have been found to form complexes with chloride ions in strong hydrochloric acid, Cm(III) being complexed to a lesser extent than Am(III) (WARD and WELCH, 1956). Soluble carbonate complexes of Am(III) and Cm(III) are known (KATZ and SEABORG, 1957) and complex formation between Am(III) and NO_3^- and SO_4^{2-} in acid solutions is suggested by spectral changes (HALL and HERNIMAN, 1954).

2. Acetate Complexes

The acetate complexes of Am(III), Cm(III), Bk(III) and Cf(III) have been studied by CHOPPIN and SCHNEIDER (1970) by solvent extraction methods. The stability constants of the mono-acetato complexes were determined at 25°C in 2.0 M $NaClO_4$, and the following values for K_1 were found:

$$\mathrm{Am} = 93 \pm 14; \qquad \mathrm{Cm} = 107 \pm 9;$$
$$\mathrm{Bk} = 116 \pm 6 \quad \text{and} \quad \mathrm{Cf} = 128 \pm 14.$$

3. Lactate Complexes

The composition and stability constants of the lactate complexes of Am(III), Cm(III), Cf(III) and Fm(III) were studied by ERMAKOV and STARY (1965). Using a solvent extraction method the stability constants for the tri-lactato complexes were found to be log $\beta_3 = 5.71$ for Am; 5.76 for Cm; 6.09 for Cf and 6,36 for Fm. An ion exchange method yielded values of log $\beta_3 = 5.78$ for Cm, 6.08 for Cf and 6.27 for Fm.

4. Aminopolycarboxylic Acids

Complex formation between various aminopolycarboxylic acids and transplutonium elements has been studied by a number of workers.

The complexing of americium by eight substances ranging from aminodiacetic acid to tetraethylenepentamineheptaacetic acid has been studied by DELL SITE

Table 17.7. Stability constants of transplutonium complexes with aminopolyacetic acids

	DCTA[a] log β	EDTA[b] log β	DTPA[c] log β
Am (III)	18.79	18.16	22.92
Cm (III)	18.81	18.45	22.98
Bk (III)	19.16	18.88	22.79
Cf (III)	19.42	19.09	22.57
Es (III)	19.44	—	22.62
Fm (III)	19.56	—	22.70

[a] BAYBARZ (1966). [b] FUGER (1958, 1961). [c] BAYBARZ (1965).

and BAYBARZ (1969). These authors used spectrophotometric methods in a detailed study of the composition and stability in relation to pH for some twenty two complexes of Am(III). Diethylenetriaminepentaacetic acid (DTPA) was found to form a 1:1 metal ligand complex, but ethylenediaminetetraacetic acid (EDTA) was found to form three complexes corresponding to AmL, Am_2L_3 and AmL_2. Three complexes were also formed with triethylenetetraminehexaacetic acid (TTHA) corresponding to Am_2L, AmL and Am_2L_3.

The stability constants for the complexes of Am(III), Cm(III), Bk(III), Cf(III), Es(III), and Fm(III) with EDTA, DTPA and 1:2 diaminocyclohexane-tetraacetic acid (DCTA) are listed in Table 17.7. The stability constants for DTPA and DCTA, as well as those for the acetate and lactate complexes discussed above, show an increase in stability with increasing atomic number. This increase in stability with increasing atomic number is also observed in the lanthanide series (KRUMHOLZ, 1964). However, with DTPA the stability constants remain virtually constant from Am(III) to Fm(III); a similar phenomenon is also seen with the DTPA complexes of the corresponding lanthanides.

5. Other Complexes

In the separation of transplutonic elements by ion exchange α-hydroxy-isobutyrate has been shown to be a very effective complexing agent (CHOPPIN et al., 1956).

Thenoyltrifluoroacetone and tri-n-butyl phosphate have found important applications in the isolation of americium and curium (KATZ and SEABORG, 1957).

D. Oxidation States in Solution

In contrast to plutonium the oxidation states of the transplutonic elements in solution present a simple picture. For all the elements except nobelium, the trivalent state is the most stable oxidation state in solution. Nobelium is unique among the actinide elements in that the divalent state No(II) is the most stable in solution (MALY et al., 1968). The Md(II) ion is stable in acid solution but Md(III) is the predominant oxidation state in solution (HULET et al., 1967).

Although No(II) and Md(II) are the only divalent states to be observed in solutions of transplutonic elements, values for the standard oxidation potentials of the M^{2+}/M^{3+} couple have been measured or calculated for the elements from americium to nobelium; these are illustrated in Table 17.8 (NUGENT et al., 1969).

Apart from americium, none of the transplutonic elements has been shown to exhibit higher oxidation states than 3+ in solution.

Table 17.8. Standard oxidation potentials for M^{2+}/M^{3+} couples

Z	Element	Oxidation potential (V)	
		measured	calculated
95	Americium	—	2.6
96	Curium	—	5.0
97	Berkelium	—	3.4
98	Californium	1.9	2.0
99	Einsteinium	1.6	1.6
100	Fermium	—	1.3
101	Mendelium	0.15 ± 0.05	—
102	Nobelium	-1.45 ± 0.05	—
103	Lawrencium	—	—

1. Americium

The formal potentials for the oxidation and reduction of americium ions in acid or basic solutions are illustrated in Fig. 17.1. The values for acid solution are taken from the data of PENNEMAN and ASPREY (1956) and CUNNINGHAM (1964).

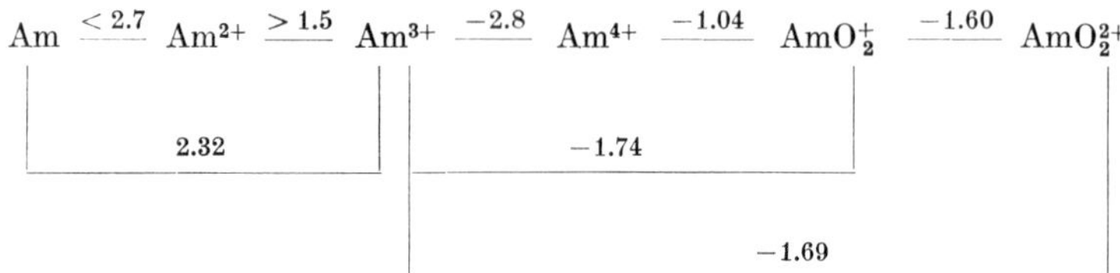

a) Acid solution

$$Am \xrightarrow{2.71} Am(OH)_3 \xrightarrow{0.4} Am(OH)_4 \xrightarrow{-0.7} AmO_2(OH) \xrightarrow{-1.1} AmO_2(OH)_2$$

b) Alkaline solution

Fig. 17.1. Formal potentials for americium couples (volts)

The data in basic solutions are taken from the work of LATIMER (1952). The formal potentials of the Am^{3+}/Am^{4+} and AmO_2^+/AmO_2^{2+} couples have been measured directly in high concentrations of phosphoric acid a medium in which all four ionic species can exist (YANIR et al., 1970). These authors found values of 1.75v for the Am^{3+}/Am^{4+} couple and 1.43v for the AmO_2^+/AmO_2^{2+} couples.

The Am(II) ion is unknown in solution and has only been demonstrated after stabilisation in a crystal of calcium fluoride (EDELSTEIN et al., 1966).

Americium(III) is the common oxidation state of americium in solution and strong oxidising conditions are required to oxidise Am(III) to higher states. The chemical properties of the Am(III) ion are similar to those of the trivalent rare earths, it is precipitated from solution by hydroxide, fluoride, oxalate and phosphate.

Solutions of tetravalent americium have been prepared by dissolving freshly prepared $Am(OH)_4$ in acid but the Am^{4+} ion is highly unstable towards disproportionation. PENNEMAN et al. (1961) found that in nitric or perchloric acid solutions disproportionation occurs by the following reaction

$$2\,Am^{4+} + 2\,H_2O \rightleftharpoons AmO_2^+ + Am^{3+} + 4\,H^+;$$

in sulphuric acid solution at least half of the AmO_2^+ produced by the simple disproportionation of Am^{4+} is consumed by a consecutive reaction

$$Am^{4+} + AmO_2^+ \rightleftharpoons Am^{3+} + AmO_2^{2+}.$$

The Am^{4+} ion may be stabilised by complexing in strong solutions of ammonium fluoride (ASPREY and PENNEMAN, 1961). In 15 M NH_4F solutions Am(IV) was found to be stable towards disproportionation and reduction by water, but slow reduction to Am(III) occurred as a result of alpha radiation from ^{241}Am (ASPREY and PENNEMAN, 1962).

Pentavalent americium exists in solution as the AmO_2^+ ion which is unstable and disproportionates to Am(III) and Am(VI). This disproportionation may be envisaged as occurring by the two reactions

$$2\,AmO_2 + 4\,H^+ \rightleftharpoons AmO_2^{2+} + Am^{4+} + H_2O$$

$$AmO_2^+ + Am^{4+} \rightleftharpoons AmO_2^{2+} + Am^{3+}$$

the overall reaction being described by the net reaction

$$3\,AmO_2^+ 4\,H^+ \rightleftharpoons 2\,AmO_2^{2+} + Am^{3+} + 2\,H_2O$$

The kinetics of this reaction have been studied by COLEMAN (1963) using the long lived isotope ^{243}Am in order to reduce the effects of radiolysis. These experiments confirm the strong dependence of the disproportionation rate on hydrogen ion concentration. COLEMAN (1963) found that at room temperature the rate of disproportionation of Pu(V) increased 450 times as the perchloric acid concentration was raised from 3 M to 8 M. In 2 N nitric, hydrochloric and sulphuric acids the rate of disproportionation was found to be 4, 4.6, and 24 times greater than the rate in 2 N perchloric acid.

The $^{241}AmO_2^+$ ion is reduced to Am^{3+} by the products of the radiolysis of water by ^{241}Am alpha particles. The rate of reduction is influenced by impurities in the solution, and by the addition to the solution of hydrochloric acid or $Zr(ClO_4)_4$, which forms a complex with H_2O_2.

The americyl ion, AmO_2^{2+}, is a strong oxidising agent in aqueous solution and is rapidly reduced to Am^{3+} by reducing agents such as hydrogen peroxide, hydrazine, chloride and iodide ions.

ZAITSEVA (1969) has reported that the stability of Am(VI) in solution was greatly reduced by lowering the activity of the solution. This was assumed to result from disproportionation of Am(VI) to Am(V) and Am(VII), the latter being rapidly reduced. The heptavalent Am(VII) was said to be stable for several minutes in 0.3 M $KClO_4$. The following reaction scheme

$$2\,AmO_2 + 3\,H_2O \rightleftharpoons 2\,AmO_2^+ + H_3AmO_5 + 3\,H^+$$

was proposed for the disproportionation in weakly acid or neutral solutions.

2. Curium

Curium(III) is the only stable oxidation state in solution which is known. Attempts to prepare Cm(IV) in aqueous solution by the oxidation of Cm(III) with the strongest oxidising agents have been unsuccessful (KATZ and SEABORG, 1957; HOLCOMB, 1967).

IV. Interactions of Transplutonium Elements with Proteins and other Substances of Biological Importance

It was shown in Chap. 9 that hydrolysis and complex formation play important roles in determining the biological behaviour of plutonium. These reactions play equally important roles in governing the metabolic behaviour of americium and curium, and they may be expected to be similarly important in relation to the higher transplutonic elements.

Like plutonium, americium and curium entering the vascular compartment in soluble form rapidly become associated with the plasma proteins (TURNER and TAYLOR, 1968). However, the rate of removal of Am(III) and Cm(III) from the blood is much more rapid than that of plutonium and the proportion of the total plasma Am(III) and Cm(III) which becomes proteinbound within 1 minute of injection into rats is considerably less than that of Pu(IV), although by 1 hour all three metals are almost entirely protein-bound (TURNER and TAYLOR, 1968).

Following intramuscular contamination, Am(III) and Cm(III) are deposited in poorly soluble form at the entry site from which they are slowly translocated to distant parts of the body. The rate of translocation of Am(III) and Cm(III) following intramuscular injection is more rapid than that of both Pu(IV) and Pu(III) (SCOTT et al., 1948a; SCOTT et al., 1949; SCOTT et al., 1948b).

Apart from limited studies with Cf(III), Bk(III) and Es(III) (PARKER et al., 1962, 1972, ATHERTON and LLOYD, 1972, HUNGATE et al., 1972) our knowledge of the biological behaviour of the transplutonium elements releates mainly to Am(III) and Cm (III). The behaviour of the higher elements up to lawrencium may be expected to be generally similar to americium and curium, probably showing the progressively greater deposition in the skeleton and smaller liver uptake with increasing atomic number which has been predicted by DURBIN (1960) by analogy with the lanthanides. An exception may be found with nobelium which is predominantly divalent in solution. Because of its smaller ionic potential No(II) would be expected to show less tendency to hydrolyse than the M(III) ions; this may be reflected in its biological behaviour leading, for example, to a more rapid translocation of No(II) from the site of intramuscular contamination than that which has been observed with Am(III) (SCOTT et al., 1948a).

A. Interactions with Plasma Proteins

The distribution of Am(III) and Cm(III) amongst the serum proteins has been studied by gel filtration and by electrophoretic methods (BOOCOCK and POPPLEWELL, 1966; POPPLEWELL and BOOCOCK, 1968; BRUENGER et al., 1969; TURNER and TAYLOR, 1968). These studies, carried out on sera labelled *in vivo* or *in vitro*, demonstrated that both Am(III) and Cm(III) were protein-bound but failed to indicate the association of either metal with a specific plasma protein. The electrophoretic studies of TURNER and TAYLOR (1968) revealed a more or less uniform distribution of Am(III) and Cm(III) amongst the α-, β- and γ-globulins of sera collected at time intervals ranging from 1 minute to 1 day after intravenous injection of the metals in essentially soluble form. The majority of the metal in the serum was associated with the globulins but 6 to 21 percent of the Am(III) and 5 to 7 percent of the Cm(III) was found in the albumin band.

In studies with canine and human blood BRUENGER et al. (1969) demonstrated binding of Am(III) to transferrin, albumin and a third unidentified protein which could be an α-globulin. Pre-saturation of the sera with iron reduced the binding of Am(III) in the transferrin region. The binding of Am(III) and Cm(III) to human transferrin *in vitro* was studied by CHIPPERFIELD and TAYLOR (1971) using a gel filtration technique to separate protein-bound and nonprotein-bound metal. These experiments failed to demonstrate any binding of Cm(III) to transferrin and only very weak binding of Am(III). A similar weak binding of Am(III) to bovine γ-globulin was also observed (Table 17.9).

On the basis of the information available at present it must be concluded that Am(III) and Cm(III) form only very weak complexes with serum proteins, so

Table 17.9. Binding of plutonium, americium and curium to proteins *in vitro* at pH 7.4. (Binding is expressed at the percentage of protein bound metal recovered from Sephadex gel columns.) (From CHIPPERFIELD and TAYLOR, 1970, 1972)

Protein	Percent of protein bound metal		
	Pu(IV)	Am(III)	Cm(III)
Bone sialoprotein	54.7 ± 9.3	10.4 ± 2.9	12.2 ± 1.1
Chondroitin sulphateprotein	49.2 ± 3.0	14.9 ± 6.6	10.0 ± 1.6
CPC-soluble glycoprotein	36.6 ± 5.1	4.6 ± 2.9	8.7 ± 1.2
Glycoprotein I	30.0 ± 12.4	2.8 ± 0.9	37.5 ± 12.6
Glycoprotein II	50.3 ± 9.1	5.1 ± 2.0	8.2 ± 1.3
Soluble collagen	23.3 ± 4.3	0.6 ± 0.2	0.9 ± 0.3
Human transferrin	18.9 ± 6.9	0.2 ± 0.0	0.0
Bovine γ globulins	13.8 ± 5.1	0.7	—
Polyglutamic acid	68.7 ± 0.5	8.0 ± 3.0	27.2 ± 5.5
Chondroitin sulphate	13.2 ± 5.7	0.3 ± 0.4	0.3 ± 0.4

weak in fact that breakdown of the complex and redistribution of the metal may occur during even relatively mild separation procedures such as gel filtration and electrophoresis. The most important binding protein for these metals in the plasma cannot therefore be identified; however, in view of its role in the transport of a number of other metals transferrin might be expected to play a major role in the transport of americium, curium and probably other transplutonium elements.

B. Binding to Ferritin

The association of Am(III) has been demonstrated in the cytosol of canine liver by STOVER et al. (1970) using gel permeation chromatography, and these authors consider that ferritin is the major americium binding species in the dog and that the binding of Am(III) to ferritin is irreversible.

Studies of the subcellular distribution of Am(III) and Cm(III) in rat liver have been reported by POPPLEWELL et al. (1971) who found that both elements became associated with ferritin in the cytosol within 1 hour of their intravenous injection in soluble form. However at longer times after injection the majority of the Am(III) and Cm(III) had migrated from the cytosol into the formed elements of the cells, principally into lysosomal structures.

The rapid association of both Am(III) and Cm(III) with ferritin in the cytosol is very different from the observations with Pu(IV) in which association with ferritin was not observed until about 3 days after injection (BOOCOCK et al., 1970). POPPLEWELL et al. (1971) have suggested that this rapid association of Am(III) and Cm(III) with ferritin may be due to adsorption of hydrolysed americium, or curium, species onto the hydrated iron oxide-phosphate complex of the iron micelles of ferritin, as well as covalent or electrostatic bonding to the protein moiety.

C. Binding to Bone Proteins

The binding of Am(III) and Cm(III) to soluble collagen and five glycoprotein fractions isolated from bovine cortical bone has been studied by CHIPPERFIELD and TAYLOR (1970, 1972; TAYLOR and CHIPPERFIELD, 1971) using a gel filtration procedure. The relative binding of Am(III) and Cm(III) by bone proteins, expressed as the percentage of protein-bound metal recovered from the Sephadex columns, is compared in Table 17.9 which also lists the comparable data for

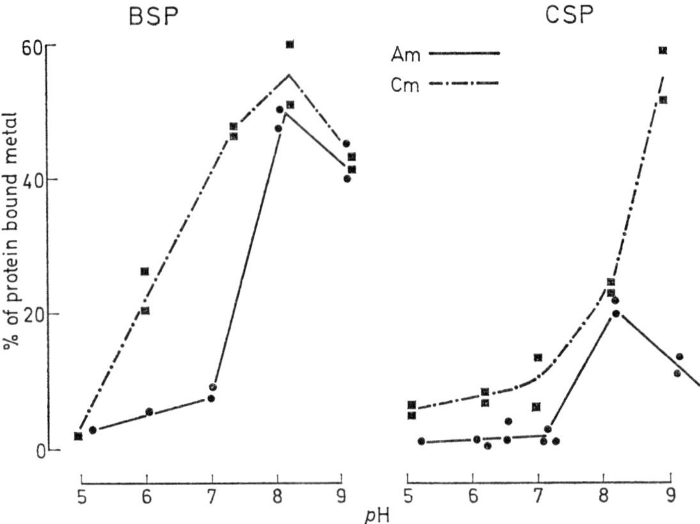

Fig. 17.2. The effect of pH on the binding of Am(III) and Cm(III) to bone sialoprotein (BSP) and the bone chondroitin sulphate protein-complexes (BSP) as determined by gel filtration on Sephadex G.50

Pu(IV). The nature of the various glycoprotein fractions studied has been discussed in Chap. 9. The data in this table show that, with the exception of the binding of Cm(III) to glycoprotein I, the binding of Am(III) and Cm(III) to the various proteins is of a similar magnitude. With all the proteins studied the binding of Am(III) and Cm(III) was less than that of Pu(IV).

Studies of the effect of pH on the binding of Am(III) and Cm(III) to bone sialoprotein (BSP) show that maximum binding occurs at a pH of about 8, Fig. 17.2. Similar studies with the bone chondroitin sulphate-protein complexes (CSP) show a similar pH of maximum binding for Am(III) but the binding of Cm(III) shows no maximum up to pH 9 (Fig. 17.2). These results are again in contrast to those obtained with Pu(IV) which showed maximum binding to both BSP and CSP at pH 6. This suggests that carboxyl groups are probably not important in the binding of Am(III) or Cm(III) to these glycoproteins, although Am(III) and Cm(III) can bind to the carboxyl groups of poly-L-glutamic acid (CHIPPERFIELD and TAYLOR, 1970, 1972). Detailed studies of the binding of Am(III) to poly-L-glutamic acid, protein-free chondroitin sulphate, dextran and amino acids, and of the states of americium and BSP at high pH do not permit a definitive identification of the mechanism of binding of Am(III) by BSP (CHIPPERFIELD and TAYLOR, 1970). Similar, but less detailed, studies have been made with Cm(III). On the basis of the evidence obtained so far it seems likely that the Am(III)-BSP and Cm(III)-BSP species are ill-defined complexes between glycoprotein and hydrolysis products of Am(III) and Cm(III) (CHIPPERFIELD and TAYLOR, 1970, 1972; TAYLOR and CHIPPERFIELD, 1971).

The significance of binding to bone glycoproteins in relation to the fixation of Am(III) and Cm(III) in living bone is not yet understood, but both BSP and CSP can markedly reduce the adsorption of both metals onto bone mineral *in vitro* (TAYLOR and CHIPPERFIELD, 1971).

D. Binding to Other Substances

Two other Am(III) and Cm(III) binding proteins have been observed in the cytosol of rat liver (POPPLEWELL et al., 1971) but these have not been identified. The association of Am(III) with lipofuscin and an unidentified substance of low molecular weight in the cytosol of canine liver has been reported by STOVER et al. (1970).

The association of Am(III) with connective tissue in the thyroid gland of the dog has been reported by Stevens et al. (1969) who have suggested that the metal may be bound by the anionic groups of mucopolysaccharides. Weak binding of both Am(III) and Cm(III) by chondroitin sulphate has been reported by CHIPPERFIELD and TAYLOR (1970, 1972).

The lactate complexes of Am(III), Cm(III), Cf(III) and Fm(III) have been discussed in Sec. III, C.3 and citrate complexes are known although they do not appear to have been studied in detail.

Attempts to demonstrate amino acid complexes of Am(III) by gel filtration were unsuccessful and it must be concluded that their stability would be very low (CHIPPERFIELD and TAYLOR, 1970).

The elements from Am(III) to Fm(III) form complexes of high stability with DTPA (Sec. III, C.4) and this substance has been shown to be effective in enhancing the rate of elimination of both Am(III) and Cm(III) from experimental animals (TAYLOR, 1967).

As in the case of plutonium, further information on the interactions of the transplutonium elements with components of cells and tissues is needed in order to increase our understanding of the biological behaviour of these elements.

References

AKATSU, J.: Possible curium(VI) in aqueous solution. Radiochim. Acta **4**, 58–60 (1965).

ASPREY, L. B., ELLINGER, F. H., FRIED, S., ZACHARIASEN, W. H.: Evidence for quadrivalent curium. II. Curium tetrafluoride. J. Amer. chem. Soc. **79**, 5825 (1957).

ASPREY, L. B., PENNEMAN, R. A.: First observation of aqueous tetravalent americium. J. Amer. chem. Soc. **83**, 2200 (1961).

ATHERTON, D. R., LLOYD, R. D.: The distribution and retention of ^{249}Cf in beagle soft tissue. Hlth. Phys. **22**, 675–678 (1972).

BAGNALL, K. W.: The transuranium elements. Sci. Progr. **52**, 66–83 (1964).

BAGNALL, K. W., EASEY, J. F.: The existence of Curium (VI). Radiochim. Acta **5**, 180 (1966).

BAGNALL, K. W., LAIDLER, J. B., STEWART, M. A. A.: Americyl(V) and Americyl(VI) chloro-complexes. Chem. Comm. **1**, 24–25 (1967).

BANSAL, B. M., DAMIEN, D.: Curium hydride. Inorg. Nucl. Chem. Lett. **6**, 603–606 (1970).

BAYBARZ, R. D.: Dissociation constants of the transplutonium element chelates with diethylenetriaminepentaacetic acid (DTPA) and the application of DTPA chelates to solvent extraction separations of transplutonium elements from the lanthanide elements. J. Inorg. Nucl. Chem. **27**, 1831–1839 (1965).

BAYBARZ, R. D.: Dissociation constants of the transplutonium element chelates of 1:2 diaminocyclohexanetetraacetic acid. J. Inorg. Nucl. Chem. **28**, 1055–1061 (1966).

BOOCOCK, G., DANPURE, C. J., POPPLEWELL, D. S., TAYLOR, D. M.: The subcellular distribution of plutonium in rat. Radiat. Res. **42**, 381–396 (1970).

BOOCOCK, G., POPPLEWELL, D. S.: In vitro distribution of americium in human blood serum proteins. Nature (Lond.) **210**, 1283–1284 (1966).

BRUENGER, F. W., STEVENS, W., STOVER, B. J.: Americium-241 in the blood: *In vivo* and *in vitro* observations. Radiat. Res. **37**, 349–360 (1969).

CARNALL, W. T., FIELDS, P. R., STEWART, D. C., KEENAN, T. K.: The absorption spectrum of aqueous curium (III). J. Inorg. Nucl. Chem. **6**, 213–216 (1958).

CHIPPERFIELD, A. R., TAYLOR, D. M.: The binding of americium and plutonium to bone glycoproteins. Europ. J. Biochem. **17**, 581–585 (1970).

CHIPPERFIELD, A. R., TAYLOR, D. M.: The binding of Thorium(IV), Plutonium(IV), Americium(III) and Curium(III) to the constituents of bovine cortical bone *in vitro*. Radiat. Res. **51**, 15–30 (1972).

CHOPPIN, G. R., HARVEY, B. G., THOMPSON, S. G.: A new eluant for the separation of the actinide elements. J. Inorg. Nucl. Chem. 2, 66–68 (1956).
CHOPPIN, G. R., SCHNEIDER, J. K.: The acetate complexing by trivalent actinide ions. J. Inorg. Nucl. Chem. 32, 3283–3288 (1970).
COHEN, D., FRIED, S., SIEGEL, S., TANI, B.: Preparation and crystal structure of some berkelium compounds. Inorg. Nucl. Chem. Lett. 4, 257–260 (1968).
COLEMAN, J. S.: The kinetics of the disproportionation of americium (V). Inorg. Chem. 2, 53–57 (1963).
CUNNINGHAM, B. B.: Chemistry of the actinide elements. Ann. Rev. Nucl. Sci. 14, 323–346 (1964).
DAMIEN, D.: α-form of americium sesquisulphide and americium monosulphide. Inorg. Nucl. Chem. Lett. 7, 291–297 (1971).
DELL SITE, A., BAYBARZ, R. D.: Spectrophotometric study of the complexing of Am^{3+} with aminopolyacetic acids. J. Inorg. Nucl. Chem. 31, 2201–2233 (1969).
DESIRE, B., HUSSONNOIS, M., GUILLAUMONT, R.: Détermination de la premier constant d'hydrolyse de l'americium, du curium, du berkelium et du californium. C. R. Acad. Sci. (Paris), Ser. C 269, 448–451 (1969).
DURBIN, P. W.: Metabolic characteristics within a chemical family. Hlth Phys. 2, 225–238 (1960).
EDELSTEIN, N., CONWAY, J. G., FUJITA, D., KOLBE, W., MCLAUGHLIN, R.: Formation and characterization of divalent einsteinium in a CaF_2 crystal. J. chem. Phys. 52, 6425–6426 (1970).
EDELSTEIN, N., EASLEY, W., MCLAUGHLIN, R.: Formation and characterization of divalent americium in CaF_2 crystals. J. chem. Phys. 44, 3130–3131 (1966).
ERMAKOV, V. A., STARY, I.: Izuchenie kompleksoobrasovaniya Am, Cm, Cf I, Fm S molochnoi kislotoi. (Joint Int. Nucl. Res. Dubna USSR, Lab. Nucl. Reactions) Dep. mn. 9 p. (1965).
EUBANKS, I. D., THOMPSON, M. C.: Preparation of curium metal. Inorg. Nucl. Chem. Lett. 5, 185–191 (1969).
FERGUSON, D. E., BIGELOW, J. E.: Production of californium-252 and other transplutonium isotopes in the United States. Actinide Rev. 1 (3), 213–221 (1969).
FUGER, J.: Ion exchange behaviour and dissociation constants of americium, curium and californium complexes with ethylene diaminetetraacetic acid. J. Inorg. Nucl. Chem. 5, 332–338 (1958).
FUGER, J.: Actinide and lanthanide ion exchange separation studies. II. Separations by aminopolyacetic acids. J. Inorg. Nucl. Chem. 18, 263–269 (1961).
FUJITA, D. K.: Some magnetic, spectroscopic and crystallographic properties of berkelium, californium and einsteinium. UCRL-19507 (Nov. 1969).
FUJITA, D. K., CUNNINGHAM, B. B., PARSONS, T. C.: Crystal structures and lattice parameters of einsteinium trichloride and einsteinium oxychloride. Inorg. Nucl. Chem. Lett. 5, 307–313 (1969).
FUJITA, D. K., CUNNINGHAM, B. B., PARSONS, T. C., PETERSON, J. R.: The solution absorption spectrum of Es^{3+}. Inorg. Nucl. Chem. Lett. 5, 245–250 (1969).
GREEN, J. L.: The absorption spectrum of Cf^{3+} and crystallography of californium sesquioxide and californium trichloride. UCRL-16516 (1965).
GREEN, J. L., CUNNINGHAM, B. B.: Absorption Spectrum of Cf^{3+}. Inorg. Nucl. Chem. Lett. 2, 365–371 (1966).
HAISSINSKY, M., DRAN, J. C., KLEIN, R., PLUCHET, E.: Oxydation anodique du curium. J. Chim. phys. 65, 2119–2125 (1968).
HALL, G. R., HERNIMAN, P. D.: The separation and purification of americium-241 and the absorption spectra of tervalent and quinquevalent americium solutions. J. chem. Soc. 2214–2221 (1954).
HAUG, H. O.: Curium sesquioxide Cm_2O_3. J. Inorg. Nucl. Chem. 29, 2753–2758 (1967).
HIGGINS, G. H.: The radiochemistry of the transcurium elements. NAS-NS 3031 National Academy of Sciences, National Research Council, Washington (1960).
HOLCOMB, H. P.: A test for oxidation of actinides in concentrated CsF solutions. J. Inorg. Nucl. Chem. 29, 2885–2888 (1967).
HULET, E. K., LOUGHEED, R. W., BRADY, J. D., STONE, R. E., COOPS, M. S.: Mendelevium: Divalency and other chemical properties. Science 158, 486–488 (1967).
HUNGATE, F. P., BALLOU, J. E., MAHLUM, D. D., KASHIMA, M., SMITH, V. H., SANDERS, C. L., BAXTER, D. W., SIKOV, M. R., THOMPSON, R. C.: Preliminary data on ^{253}Es and ^{249}Bk metabolism in rats. Hlth. Phys. 22, 653–656 (1972).
KATZ, J. J., SEABORG, G. T.: The chemistry of the actinide elements. London: Methuen 1957.
KEENAN, T. K., ASPREY, L. B.: Lattice constants of actinide tetrafluorides including berkelium. Inorg. Chem. 8, 235–238 (1969).

KRUMHOLZ, P.: In: Progress in the science and technology of the rare Earths (ed. L. Eyring), p. 110–138. Oxford: Pergamon Press 1964.

KRUSE, F. H., ASPREY, L. B.: A crystalline fluoride complex of tetravalent americium. Inorg. Chem. **1**, 137–139 (1962).

LATIMER, W. M.: Oxidation potentials, 2nd ed. New York: Prentice Hall 1952.

LEDERER, C. M., HOLLANDER, J. M., PERLMAN, I.: Table of Isotopes, 6th ed. New York: J. Wiley È Sons 1967.

McWHAN, D. B., CUNNINGHAM, B. B., WALLMAN, J. C.: Crystal structure, thermal expansion and melting point of americium metal. J. Inorg. Nucl. Chem. **24**, 1025–1038 (1962).

MALY, J., SIKKELAND, T., SILVA, R., GHIORSO, A.: Nobelium: Tracer chemistry of the divalent and trivalent ions. Science **160**, 1114–1115 (1968).

MILLIGAN, W. O.: Crystal structure and morphology of hydrous oxides and hydroxides in the lanthanide and actinide series. ORO 3955-1 (Feb. 1970).

NIGON, J. P., PENNEMAN, R. A., STARITZKY, E., KEENAN, T. K., ASPREY, L. B.: Alkali carbonates of Np (V), Pu (V) and Am (V). J. phys. Chem. **58**, 403–404 (1954).

NUGENT, L. J., BAYBARZ, R. D., BURNETT, J. L.: Electron transfer spectra and the II–III oxidation potentials of some lanthanide and actinide halides in solution. J. phys. Chem. **73**, 1177–1178 (1969).

PARKER, H. G., LOW-BEER, A. DE G., ISAAC, E. L.: Comparison of retention and organ distribution of Am^{241} and Cf^{252} in mice: The effect in vivo of DTPA chelation. Hlth Phys. **8**, 679–684 (1962).

PARKER, H. G., WRIGHT, S. R., LOW-BEER, A. DE G., YAEGER, D. J.: The metabolism of ^{253}Es in mice. Hlth. Phys. **22**, 647–651 (1972).

PASCAL, P. (ed.): Nouveau traité de chimie minerale, vol. XV Transuraniens, 5th ed. Paris: Masson Cie. 1970.

PENNEMAN, R. A.: In: The encyclopedia of the chemical elements (ed. C. A. Hampel), p. 18–21. New York: Reinhold 1968.

PENNEMAN, R. A., ASPREY, L. B.: A review of americium and curium chemistry. Proc. Intern. Conf. Peaceful Uses Atomic Energy, Geneva 1955, **7**, 353–362 (1956).

PENNEMAN, R. A., COLEMAN, J. S., KEENAN, T. K.: Alkaline oxidation of americium: Preparation and reactions of Am (IV) hydroxide. J. Inorg. Nucl. Chem. **17**, 138–145 (1961).

PENNEMAN, R. A., KEENAN, T. K.: The radiochemistry of americium and curium. NAS-NS 3006 — National Academy of Sciences, National Research Council, Washington (1960).

PETERSON, J. R.: The solution absorption spectrum of Bk^{3+} and the crystallography of berkelium dioxide, sesquioxide, trichloride, oxychloride and trifluoride. UCRL-17875 (Oct. 1967).

PETERSON, J. R., CUNNINGHAM, B. B.: Crystal structures and lattice parameters of the compounds of berkelium. 1. Berkelium dioxide and cubic sesquioxide. Inorg. Nucl. Chem. Lett. **3**, 327–336 (1967).

POPPLEWELL, D. S., BOOCOCK, G.: Distribution of some actinides in blood serum proteins. In: Diagnosis and treatment of deposited radionuclides (W. D. NORWOOD and H. A. KORNBERG, eds.). Amsterdam: Excerpta med. Found. 1968.

POPPLEWELL, D. S., BOOCOCK, G., TAYLOR, D. M., DANPURE, C. J.: The subcellular distribution of americium and curium in rat liver. In: Radiation protection problems relating to transuranium elements, p. 205–222. Luxembourg: Commission of the European Communities, Centre for Information and Documentation 1971.

SCOTT, K. G., AXELROD, D. J., FISHER, H., CROWLEY, J. F., HAMILTON, J. G.: The metabolism of plutonium in rats following intramuscular injection. J. biol. Chem. **176**, 283–293 (1948a).

SCOTT, K. G., AXELROD, D. J., HAMILTON, J. G.: The metabolism of curium in the rat. J. biol. Chem. **177**, 325–335 (1959).

SCOTT, K. G., COPP, D. H., AXELROD, D. J., HAMILTON, J. G.: The metabolism of americium in the rat. J. biol. Chem. **175**, 691–703 (1948b).

SEABORG, G. T.: Correlation of properties as actinide transition series. Chap. 17, in: The actinide elements IV.14.A. (G. T. SEABORG and J. J. KATZ, N.N.E.S.), p. 733–768. New York: McGraw Hill 1954.

SEABORG, G. T.: Transuranium elements. In: The encyclopedia of the chemical elements (C. A. HAMPEL, ed), p. 739–763. New York: Reinhold 1968.

STARIK, I. YE., GINZBERG, F. L.: State of microquantities of radioelements in solution — XIV Investigation of the state of americium in aqueous solutions. Radiochemistry (USSR) **1**, 215–218 (1960).

STEVENS, W., STOVER, B. J., BRUENGER, F. W., TAYLOR, G. N.: Some observations on the deposition of americium-241 in the thyroid gland of the beagle. Radiat. Res. **39**, 201–206 (1969).

STOVER, B. J., BRUENGER, F. W., STEVENS, W.: Association of americium with ferritin in canine liver. Radiat. Res. **43**, 173–186 (1970).

TAYLOR, D. M.: The effects of desferrioxamine on the retention of actinide elements in the rat. Hlth Phys. **13**, 135–140 (1967).

TAYLOR, D. M.: Unpublished observations (1970).

TAYLOR, D. M., CHIPPERFIELD, A. R.: The binding of transplutonium elements to proteins of bone. In: Radiation protection problems relating to transuranium elements, p. 187–204. Luxembourg: Commission of the European Community, Centre for Information and Documentation 1971.

TURNER, G. A., TAYLOR, D. M.: The transport of plutonium, americium and curium in the blood of rats. Phys. in Med. Biol. **13**, 535–546 (1968).

WARD, M., WELCH, G. A.: The chloride complexes of trivalent plutonium, americium and curium. J. Inorg. Nucl. Chem. **2**, 395–402 (1956).

WESTRUM, E. F., JR., EYRING, L.: The preparation and some properties of americium metal. J. Amer. chem. Soc. **73**, 3396 (1951).

YAKOVLEV, G. N., KOSYAKOV, V. N.: Spectrophotometric studies of the behaviour of americium ions in solution. Proc. Intern. Conf. Peaceful Uses Atomic Energy, Geneva 1955, **7**, 363–368 (1956).

YANIR, E., GIVON, M., MARCUS, Y.: Direct determination of the formal potential of the Am (V)—Am (III) and Am (VI)—Am (V) couples in phosphoric acid. J. Inorg. Nucl. Chem. Lett. **6**, 415–419 (1970).

ZACHARIASEN, W. H.: Crystal chemistry of the 5 f elements. Chap. 18, The Actinide Elements. Natl. Nuclear Energy Series Div. IV 14A, p. 769–796 (1954).

ZAITSEVA, V. P.: O disproportsirovanii Am (VI) I poluchenii semivalentnogo ameritsiya. Dokl. Akad. Nauk SSSR **188**, 826–829 (1969).

Chapter 18

Metabolism and Biological Effects of the Transplutonium Elements[1]

PATRICIA W. DURBIN

With 40 Figures

I. Introduction

The transplutonium elements (SEABORG et al., 1949a, b)—formed from plutonium by sequential neutron captures and beta-particle decays—are among the most hazardous by-products of nuclear power generation (E. D. ARNOLD, 1970). In his preliminary reports of the first biological studies with americium and curium, HAMILTON (1947, 1948) pointed out the potential dangers of the actinide elements; "They share the radioactive and metabolic characteristics of radium—alpha-particle emission, long physical half-lives, and selective deposition in and prolonged retention in the skeleton." When these experiments were begun in 1946 (SCOTT et al., 1948a, 1949), only a few milligrams of americium and curium existed. They were scientific curiosities, and the time when they might be serious biological problems seemed very remote.

A. Americium and Curium in Power Reactor Wastes

Scarcely 10 years later (1957) pilot-model nuclear power plants were generating electricity. Anticipating the problems that would arise in the management of reactor wastes, and based on the projections of nuclear power development at that time, E. D. ARNOLD (1957) calculated that the U.S. inventory of americium and curium in the year A.D. 2000 would be about 300 kg[2].

Large, water-cooled, uranium-fueled, power reactors are now being designed with thermal power ratings up to 3120 MW (PIGFORD, 1971); and reactors fueled with mixtures of uranium and plutonium, the so-called fast-breeder reactors, are being designed or are under construction in the United States and the Soviet Union (GILLETTE, 1971). SEABORG (1968) estimates that production of separated americium and curium isotopes in the United States alone will be about 300 kg/yr by A.D. 1980—more than 20 times E. D. ARNOLD's prediction made 11 years before. If nuclear power development proceeds according to present plans, total production of americium and curium isotopes in the next 25 years will be measurable in metric tons, and hundreds of megacuries of these elements will have to be processed for recycling or storage.

1 Prepared for the U.S. Atomic Energy Commission under Contract No. W-7405-eng-48.

2 E. D. ARNOLD's (1957) estimate assumed construction in the United States alone of 1000 nuclear reactors fueled with natural uranium (slightly enriched with ^{235}U), each operating at a thermal power level of 700 MW.

B. Production of Berkelium, Californium and Einsteinium

As long as accelerators were the sole source of the transplutonium elements of atomic number 97 and greater (see Chap. 16 of this volume), the small amounts in existence were not of general biological concern. However, the world's supply of these heavier elements has also risen in recent years, and will continue to soar, as high-flux isotope reactors (HFIR) become fully operative[3] (KELLER, JR., 1967; SEABORG, 1968). At the Oak Ridge National Laboratory, the HFIR and its adjacent chemical processing facility alone, have the capacity to make and purify annually, grams of californium, hundreds of milligrams of berkelium, and milligrams of einsteinium. Elements 97 and greater are hard to contain and handle. Weight for weight, they are many times more dangerous—both as external and internal radiation sources—than plutonium, americium, or curium (DENHAM, 1969). They have higher specific activities, higher rates of spontaneous fission, and a large number of the disintegrations are accompanied by ejection of energetic photons (see Table 17.1).

C. Practical Application of Transplutonium Isotopes

Several of the transplutonium isotopes possess special radiation and radioactive decay characteristics that make them especially suitable as selfcontained thermoelectric power units (KALLFELZ et al., 1970), neutron sources (BUSHONG et al., 1970; OLIVER, 1971), and long-lived sources of monoenergetic photons (CAMERON, 1970). See also SEABORG (1968) and Chap. 21 of this volume. Production in the near future of large amounts of transplutonium isotopes and the eventual widespread use of some of them in radiation and power sources create the potential for accidental contamination even outside of controlled areas.

II. Chemical and Biological Similarities of the Actinides and Lanthanides

Before the transuranium elements were synthesized, chemistry textbooks showed a simple filling of the 6*d* shell of electrons in the elements of atomic number 89 through 92. The very stable (III) state exhibited by elements 95 (americium) and 96 (curium), along with other chemical evidence, convinced SEABORG (1944, 1945, 1946) of the existence of an actinide series (with filling of the 5*f* subshell) analogous in valence electron configuration to the lanthanide series (in which 14 electrons are added to the 4*f* subshell).

A. Similarities Due to a Common (III) Valence State

The *f* electrons do not participate in bond formation. Thus, the lanthanide elements and the actinide elements beginning with americium exhibit a principle (III) state[4]. The common (III) state confers a general uniformity upon the biological behavior of these two groups of elements. The general features of the metabolism of yttrium, the lightweight lanthanides, and americium and curium, namely, low absorption with deposition principally in the liver and skeleton, were established in tracer studies with rats (HAMILTON, 1947, 1948; SCOTT et al.,

3 In addition to the HFIR, at least three other reactors were producing or were planning to produce transplutonium elements as of 1966: plutonium production reactors at Savannah River, U.S.; the BR-2 reactor at Mol, Belgium; and the EL3 reactor at Saclay, France (KELLER, JR., 1967).

4 MALY et al. (1968) found that the most stable state of element 102, nobelium, is (II).

1948a, 1949). HAMILTON (1948) was immediately impressed "by the fact that actinium, americium, and curium behave in an almost indistinguishable manner from the lanthanide rare earths".

B. Differences Related to Atomic Size

Filling of the *f* shells is accompanied by a slow contraction of atomic dimensions (GOLDSCHMIDT et al., 1925; ZACHARIASEN, 1954; TEMPLETON and DAUBEN, 1954; PETERSON, 1967) which leads, as the atomic number increases, to increased hydrolysis of aqueous ions, poorer solubility of compounds, and greater stability of complex ions (MARTELL and CALVIN, 1952; EYRING, 1964; SILLEN and MARTELL 1964; see also Chap. 17 of this volume).

Although the individual distribution patterns of all the lanthanides in the rat are generally similar, they are not identical (DURBIN et al., 1956). There is a regular shift in the initial distribution pattern, which appears to be related to ionic size (DURBIN, 1962). Liver deposition is high and skeletal deposition low for the lightweight lanthanides of larger ionic radius (cerium through gadolinium); whereas, skeletal deposition is high and liver deposition is low for the heavier lanthanides of smaller ionic size (terbium through lutetium). The initial partitions of americium and curium between liver and skeleton are closer to those of the lanthanides of comparable ionic radius than to the distributions of the lanthanides with the same number of *f* electrons (DURBIN, 1962).

C. Biology of the Actinides Related to Both Chemistry and Radiation

The movement of the actinide elements in biological systems, that is, their absorption into, distribution within, and elimination from organisms, is determined by their chemistry; but their biological effects are caused by radiation damage rather than chemical interactions. All of the transplutonium isotopes are radioactive, and with rare exceptions their physical half-lives are shorter than that of ^{239}Pu (24,360 yr). Thus, they all have very high specific activities[5], and the mass required to produce demonstrable biological effects is usually too small to be weighed. For this reason units of radioactivity, microcuries (μCi), and concentrations of radioactivity, microcuries per gram of tissues (μCi/g), will be used throughout this chapter.

Nearly all of the isotopes of the transplutonium elements decay with the emission of alpha-particles or by spontaneous fission (see Table 17.2). The radiation doses to tissue are largely from these densely ionizing particles. The high levels of radiation damage inflicted by even small amounts of these elements make it difficult to study their metabolism uncomplicated by radiation effects.

III. Interactions of Multicharged Cations with Life Forms Other than Mammals

The bulk of information about the biology of the transplutonium elements deals with their behavior in a few laboratory mammals. For the time being, their transfer from the environment into food chains and their behavior in living creatures other than mammals must be inferred from what is known of the biology of chemically similar elements, specifically yttrium, the lanthanides, and plutonium.

5 In this chapter "specific activity" is defined as radioactive intensity per unit weight of element (Ci/g), and is the inverse of the mass per curie term shown in Table 17.2.

A. Data Sources on Plutonium and Lanthanide Biology

There is a substantial body of data on the biology of the lanthanides, both stable and radioactive; and since its discovery in the 1940's, ^{239}Pu has been under constant investigation, because of its importance as a nuclear energy source. (See Chap. 10 and 11 of this volume.)

STEIDLE (1935) summarized the early biological and pharmacological studies of the lanthanide elements. More recent investigations have been reviewed by KYKER (1962). These elements have no known function in living systems, and although they are abundant in nature, they appear to be excluded from most organisms. Indeed, yttrium and the lanthanides are present in organisms in such small amounts, that until the recent development of very sensitive analytical chemical methods (GOFMAN et al., 1961; KAMEDA, 1962) their presence went largely undetected.

Unfortunately, the classical pharmacological and toxicological studies of the lanthanides reviewed by STEIDLE (1935) and KYKER (1962) are not especially suitable for predicting the biological behavior of the actinides. The mean lethal dose at 30 days ($LD_{50/30}$) of several stable lanthanides in mammalia (COCHRAN et al., 1950) and amphibia (STEIDLE, 1935) is about 100 to 300 mg/kg (simple salts injected intraperitoneally). Solutions of 0.0015 M $CeCl_3$ are lethal to protozoa (STEIDLE, 1935). These toxic effects appear to be due to at least two mechanisms: (a) changes in pH resulting from hydrolysis of lanthanide ions, and (b) reactions with proteins, such as chelation, conjugation, and coagulation.

On the other hand, as was pointed out above, the toxicity of actinide isotopes with high specific activity is due to radiation. The mass of actinide associated with an acutely lethal radiation dose is exceedingly small. The $LD_{50/30}$ in the rat of ^{241}Am, the longest-lived readily-available actinide isotope, is 110 μCi/kg—only 32 μg/kg (ZALIKIN et al., 1968). Thus the mass of ^{241}Am associated with a lethal radiation dose is 0.0003 to 0.0009 of the mass of the $LD_{50/30}$ of the stable lanthanides. The radiation dose rate in a solution of 0.0015 M ^{241}Am is 240 rad/min—sufficient to deliver a lethal dose to small, suspended organisms in a few minutes. Therefore, the following sections on the uptake and distribution of lanthanide radioisotopes by various organisms were considered more appropriate to the actinide case.

Many radioactive isotopes of yttrium and of the lightweight lanthanides are created in nuclear fission, and they account for about one-half of the fission-product radioactivity (Plutonium Project, 1946). Thus, reactor wastes and nuclear test debris contain large quantities of high-specific-activity isotopes of several trivalent elements, which are easily detectable because of their radioactivity. International concern about the consequences of radioactive fallout from nuclear weapons tests engendered the many studies of the occurrence of fission products in a variety of organisms and environments. Much of this informations has been collected in reviews and conference proceedings (U.S. Congress, 1957, 1959; United Nations, 1958; SCHULTZ and KLEMENT, 1961; HUNGATE, 1965; NELSON and EVANS, 1967; National Research Council, 1971).

B. Microorganisms

Many microorganisms assimilate trivalent elements during growth. RICHARDS and TROUTMAN (1940) found that yeasts accumulated lanthanum. G. T. JOHNSON and KYKER (1960) studied the uptake of ^{144}Ce by pure cultures of yeasts, molds, and bacteria. Significant ^{144}Ce uptake was found in all cultures in which adequate growth was obtained, and those of two bacteria, one yeast, and five molds each accumulated more than 85% of the added ^{144}Ce during 4 to 10 days of growth.

C. Algae

Many algae concentrate multicharged cations, often achieving concentrations 2000 to 4000 times that in the suspending medium (RICE, 1961). Radioisotopes of lanthanum and cerium have been detected in freshwater algae collected from the Columbia River near discharge points of the Hanford reactors (DAVIS et al., 1958); and radioisotopes of yttrium, cerium, and plutonium from fallout have been found in marine algae collected in the Gulf of Mexico (SIMEK et al., 1967) and near the Rongelap, Eniwetok, and Bikini atolls (HELD, 1961; WELANDER, 1967).

At the pH of natural waters, multicharged cations tend to hydrolyze, to aggregate, and to be adsorbed onto particulates. Fallout radioactivities have been shown to be largely associated with particulates (GREENSDALE and BALLOU, 1954). Most of the ^{90}Y and ^{144}Ce associated with algae, appear to be inside cells (RICE, 1961; SIMEK et al., 1967), and ZLOBIN and MOKANER (1970) suggest that assimilation of plutonium (and presumably chemically similar elements also) by algae involves active transport of colloidal particles. RICE (1961) demonstrated that several algae accumulated ^{90}Y more efficiently in the colloidal form than in the ionic.

D. Seaweeds and Lichens

Fallout ^{144}Ce has been identified as the principal gamma-emitting radionuclide not only in seaweeds harvested from such widely separated locations as the Black Sea (GLAZUNOV et al., 1963) and the Gulf of Mexico (SHELBY, 1963) but also in lichens harvested from granite outcrops in the high rainfall region of the Georgia, U.S., piedmont (PLUMMER, 1967). Only small amounts of the ^{144}Ce could be washed from the lichens, leading PLUMMER (1967) to conclude that most of the ^{144}Ce had been assimilated into the cells.

E. Plants

Uptake by plants of ^{241}Am, ^{242}Cm, ^{239}Pu, and radioisotopes of several lanthanides has been studied under a variety of growth conditions. Multicharged cations are excluded from plants (MENZEL, 1965). When soluble ^{241}Am was uniformly mixed with well-fertilized soil, uptake in soybean stems and leaves was 0.003% (WALLACE, 1972). Bean plants grown in loam soil took up 0.07% of ^{242}Cm (W.A. THOMAS and JACOBS, 1969).

Accumulation by a species varies with the element in the order: yttrium > cerium, americium, curium > plutonium. The quantities taken up by bean plants grown in nutrient solutions or clay suspensions were: ^{91}Y, 0.10% and ^{144}Ce, 0.03% (JACOBSON and OVERSTREET, 1948); ^{242}Cm, 0.03% (W. A. THOMAS and JACOBS, 1969); ^{239}Pu (IV), 0.002% and ^{239}Pu (III), 0.00045% (JACOBSON and OVERSTREET, 1948).

Accumulation in plants varies not only with species but also in the different parts of the plant; the uptake of ^{144}Ce is in the order: roots > leaves > stems > fruit > seeds (JACOBSON and OVERSTREET, 1948).

Increased availability of ^{239}Pu for assimilation by plants with continually developing root structures has been demonstrated. ROMNEY et al. (1970) found progressively larger amounts of ^{239}Pu in successive croppings of Ladino clover or alfalfa grown in soil contaminated with fallout particles.

WALLACE (1972) demonstrated that some of the ^{241}Am deposited in shoots and leaves of bush beans was translocated to new growth after plants grown in labeled soil were transplanted into uncontaminated soil.

Plants grown in highly contaminated soils (0.2 μCi/g soil of either ^{90}Y or ^{144}Ce) assimilated sufficient amounts to produce direct radiation injury of roots and indirect injury of tops, presumably because of impairment of water and nutrient transport (JACOBSON and OVERSTREET, 1948).

Plant uptake of ^{239}Pu and ^{241}Am can be changed by treatment of the soil. Assimilation of ^{239}Pu by oat plants was slightly increased by addition of nitrate and phosphorus fertilizers (CUMMINGS and BANKERT, 1971). Addition to the soil of the chelating agent, diethylenetriaminepentaacetic acid (DTPA), increased ^{241}Am uptake by soybeans to 25 times control values (WALLACE, 1972), and increased ^{239}Pu uptake by alfalfa to four times control values (ROMNEY et al., 1970).

F. Aquatic Invertebrates

As predicted by RICE (1961), lanthanide radioisotopes and ^{239}Pu have been detected most frequently in filter-feeding and bottom-feeding marine invertebrates. Six years after the last nuclear tests at Eniwetok and Bikini atolls, the combined concentrations of ^{144}Ce and ^{155}Eu in marine invertebrates were 6.2 to 6.7 pCi/g dry weight—about 20% of the concentrations in algae and phytoplankton, their chief sources of food (WELANDER, 1967). Four years after the fallout accident at Rongelap atoll, bottom-feeding holothurians and marine snails still contained detectable amounts of lanthanide radionuclides and ^{239}Pu (HELD, 1961). Several pelecypod species were collected in 1963–1964 from the lower Trent and Neuse rivers of eastern North Carolina (WOLFE and SCHELSKE, 1967). Samples were taken over a 30-mile range with salinities varying from 0.1 to 15%. In all species sampled from all locations, ^{144}Ce contributed more than half of the detectable gamma-ray activity.

Lanthanide radioisotopes are also assimilated into the tissues of aquatic invertebrates directly from water. GONG et al. (1956) incubated a species of clam in labeled sea water. Lanthanide radioisotopes continued to accumulate in soft tissues over a 6-day period attaining concentrations as much as 280 times that of the water. KHMELEVA (1962) demonstrated tissue uptake of ^{144}Ce and ^{91}Y

Table 18.1. Uptake of ^{144}Ce and ^{91}Y from sea water by a crab and a mollusc (percent of body isotope content). (Data of ZESENKO, 1966)

Carcinus meanes			Mytilus galloprovincialis		
Tissue	^{144}Ce	^{91}Y	Tissue	^{144}Ce	^{91}Y
Muscle	2.6	0.6	Muscle	0.1	0.3
Liver	0.1	0.3	Viscera	4.7	9.8
Stomach	0.3	0.02	Foot	0.03	0.03
Heart	0.004	0.001	Mantle	1.1	0.2
Gonads	0.005	0.3	Gills	0.6	3.0
Gills	12.0	24.2			
Total soft tissues	12.8	25.7	Total soft tissues	7.4	12.5
Chitin	87.3	74.1	Byssus[a]	17.7	39.1
			Shell	75.5	47.9

[a] A tuft of filaments by which a bivalve mollusc attaches itself to a surface.

from water by an amphipod (*Gammarus lacustris O. Sars.*). Uptake was depressed by addition of the chelating agent, ethylenediaminetetraacetic acid (EDTA), to the water.

Zesenko (1966) measured the uptake of ^{144}Ce and ^{91}Y from sea water by a species of crab (*Carcinus meanes*) and a species of mollusc (*Mytilus galloprovincialis*). The tissue distribution data are summarized in Table 18.1. The large amounts of these lanthanides found in the gill, shell, and byssus of the mollusc are more likely the result of surface phenomena than metabolic activity.

G. Fish

Stable lanthanide elements have been detected in fish. Using neutron activation analysis, Kameda (1962) detected lanthanum in concentrations of 0.1 to 10 parts per million (ppm) in bones, scales, gills, and fins and in concentrations of 0.001 to 0.01 ppm in organs and muscle. Other lanthanides found (praseodymium, samarium, europium, dysprosium, and holmium) were located most frequently in bony structures.

Lanthanide isotopes have not been detected in fish taken from the Columbia river below the Hanford reactor outfall (Davis et al., 1958), nor were they detectable in fish caught at Eniwetok or Bikini atolls 4 to 6 years after cessation of nuclear testing (Welander, 1967), nor in fish caught at Rongelap atoll 4 years after the fallout accident (Held, 1961). However, ^{239}Pu, which has a much longer radioactive half-life than any lanthanide fission product, was present in all fish taken at Bikini and Eniwetok atolls at levels of 1 to 7 pCi/g dry weight (Welander, 1967).

Cerium-144 has been measured in tissues of the nonmigratory thornback ray (*Raia clavata*) taken from the Irish Sea near the Windscale reactor waste disposal area (Mauchline and Taylor, 1964). These rays live their entire lives in the presence of a nearly constant concentration of fission products. They feed chiefly on small crustaceans, and, as noted in the preceeding section, these bottom-feeding, filter-feeding invertebrates can accumulate significant amounts of lanthanide isotopes from both diet and water. The ^{144}Ce concentration of the gastrointestinal contents in the rays was 11 pCi/g wet weight. Tissues concentrations ranged from 0.04 pCi/g in flesh to 0.14 pCi/g in liver. The concentration in skin was high, 2.2 pCi/g, but some of this may have been related to surface phenomena. If the average tissue concentration is taken to be 0.07 pCi/g and the gastrointestinal tract and contents constitute 5% of the body weight, gastrointestinal absorption can be estimated to be about 3% of ingested ^{144}Ce.

Prosser et al. (1946) studied the uptake of fission products by nonfeeding goldfish. Lanthanide isotopes were taken up by gill and oral membranes, transported by the blood, and accumulated in viscera and in new or regenerating bone. Boroughs et al. (1956) administered ^{91}Y orally to a small marine fish (*Tilapia mossambica*); about 2% of the dose was absorbed by the gastrointestinal tract. Two weeks later the viscera contained 43% of the absorbed dose; and muscle, skeleton, and gills and integument contained 29, 16, and 12% of the dose, respectively. Based on the 2-week observation period, the biological half-life of ^{91}Y in the whole body was about 30 days.

Ivanov (1966) studied the effect of several radioisotopes on the development of fish eggs. Isotopes of cerium, yttrium, and iron accumulated on the egg surfaces. Chromosome aberrations and developmental anomalies in the embryo were observed and were related to both the radiation dose and the youth of the embryos at the start of radiation.

H. Insects

Radioisotopes of cerium and lanthanum were present in insect larvae collected from the Columbia River downstream from the Hanford reactors (DAVIS et al., 1958), and VOLKOVA (1963) reported that dragonfly larvae "like other aquatic insect larvae" accumulated ^{144}Ce. However, neither report makes clear whether the isotopes were assimilated into larval tissues or were present in the digestive tracts.

The toxicity of ^{239}Pu has been demonstrated in adult wasps (*Habrobracon juglandis*). Virgin adult female wasps were fed ^{239}Pu in a single meal of a citric acid-sugar solution (ERDMAN, 1962). An oral dose of 0.11 to 0.14 μCi/wasp was lethal to all test hosts in 8 days. Lower doses induced temporary sterility (0.04 μCi/wasp) and reduced fecundity (0.02 μCi/wasp). Chromosome damage was observed in germ cells, and some ^{239}Pu was found in eggs at all three dose levels. These results are reported to be comparable to the results with X-irradiation or ingestion of beta-particle emitters.

I. Birds

DAVIS et al. (1958) were not able to detect lanthanide radioisotopes in ducks taken from the environs of the Columbia river near the Hanford reactors. Neither lanthanide radioisotopes not ^{239}Pu were found in sea birds caught at Eniwetok or Bikini atolls (WELANDER, 1967). However, both ^{144}Ce and ^{239}Pu were present in significant amounts on bird skin and feathers, presumably because these birds "dust bathe" in the contaminated top soil. Cerium-144 was found in the tissues of ptarmigan (*Lagopus lagopus*) and squaw ducks (*Clangula hyemalis*) caught in Alaska in 1959, in amounts equal to those of ^{90}Sr (DAVIS et al., 1961).

MRAZ et al. (1964) conducted a laboratory investigation of the gastrointestinal absorption and tissue distribution of ^{91}Y, ^{140}La, ^{144}Ce, ^{142}Pr, ^{147}Nd, and ^{147}Pm in mature White Leghorn pullets; these isotopes were given as chlorides either orally or by injection into a superficial pectoral muscle. Eggs were collected and the hens were killed 4 days later. Excreta were not analyzed. The isotope concentrations (percent administered dose per gram wet weight) reported for intramuscularly injected birds and the organ and tissue weights of White Leghorn hens published by MITCHELL et al. (1931) were used to calculate the isotope content of whole tissues, and the fraction of the body content in each tissue as shown in Table 18.2. Neglecting excretion, the absorption of each nuclide following intramuscular injection was calculated to be as follows: ^{142}Pr, 7%; ^{144}Ce, 10%; ^{147}Nd, ^{147}Pm, ^{91}Y, each 30%; and ^{140}La, 64%. More than 40% of the body content of all the lanthanide isotopes was calculated to be in the skeleton (including marrow), and 3.5 to 22% was estimated to be in the liver.

In the studies in which the isotope was administered orally, the concentrations were reported only for blood, liver, muscle, tibia shaft, and egg components. The concentrations in the remainder of the tissues, which contributed more than one-third of the body weight, were assumed to be the same as those in blood. Tissue isotope contents were calculated as above, and the range of lanthanide radioisotopes absorbed from the G.I. tract of the hen was estimated to be as follows: ^{144}Ce, ^{147}Pm, and ^{91}Y, 0.07 to 0.1%; and ^{140}La, ^{142}Pr, and ^{147}Nd, 0.3 to 0.6%. Their distribution in the hen's tissues was unusual; a surprisingly large percentage of the body content (56 to 70%) was estimated to be present in skeletal muscle. The skeleton contained from 20 to 37% of the body content, while the liver contained no more than 5% of the body content—in both cases less than when these same radioisotopes were administered intramuscularly.

Table 18.2. Distribution of lanthanide fission products in tissues and eggs of laying hens[a] 4 days after intramuscular injection. Calculated from isotope concentrations reported by MRAZ et al. (1964) and organ weights of Leghorn chickens reported by MITCHELL et al. (1931)

Tissue	Percent administered dose						Percent of body content					
	^{140}La	^{144}Ce	^{142}Pr	^{147}Nd	^{147}Pm	^{91}Y	^{140}La	^{144}Ce	^{142}Pr	^{147}Nd	^{147}Pm	^{91}Y
Blood	1.6	0.07	0.22	1.1	1.0	0.87	3.5	0.97	4.4	5.2	4.2	3.7
Liver	10.2	0.26	0.71	1.6	1.7	3.7	22.3	3.4	14.2	7.4	6.7	16.0
Kidneys	0.71	0.08	0.09	0.21	0.35	0.26	1.6	1.1	1.8	0.98	1.4	1.11
Spleen	0.12	0.06	0.04	0.14	0.08	0.08	0.26	0.83	0.80	0.65	0.32	0.34
Heart	0.08	0.18	0.01	0.26	0.06	0.05	0.18	2.5	0.20	1.2	0.24	0.20
Lungs	0.30	0.06	0.03	0.33	0.10	0.11	0.66	0.83	0.60	1.5	0.40	0.47
Brain[b]	0.02	<0.01	<0.01	0.01	0.01	0.01	0.04	<0.14	<0.20	0.05	0.04	0.04
Gizzard	1.0	0.16	0.06	0.29	0.19	0.22	2.2	2.2	1.2	1.4	0.76	0.94
Muscle[c]	0.81	0.81	0.32	0.97	0.64	0.73	1.7	10.7	6.4	4.5	2.5	3.1
Ovaries[aa]	0.48	0.04	0.04	0.15	0.17	0.13	1.0	0.55	0.80	0.70	0.68	0.55
Repr. tract[aa,ab]	0.84	0.14	0.21	0.39	0.26	0.25	1.8	1.9	4.2	1.8	1.0	1.1
Bone[ac]	22.2	3.9	2.1	12.7	18.0	14.8	48.5	51.3	42.0	59.0	72.0	63.2
Residue[ba]	7.2	1.8	1.2	3.4	2.3	2.2	15.8	23.7	24.0	15.8	9.2	9.4
Body	45.7	7.6	5.0	21.5	25.0	23.4						
Egg components (total of 4 eggs)												
Shell	0.01	<0.01	0.03	<0.01	<0.01	0.07						
Albumin	<0.01	<0.01	0.02	<0.01	0.03	<0.01						
Yolk	18.1	2.7	1.8	6.1	5.0	6.1						

[a] Body weight 1800 g.
[b] Weight not given by MITCHELL et al. (1931); assumed to be 10 g.
[c] Concentration used was the average for pectoral and gastrocnemius muscles.
[aa] MITCHELL et al. (1931) reported the weight of the entire reproductive tract as 44 g; ovary is assumed to be 5% or 2.2 g, and the rest of the tract, 42 g.
[ab] Concentration used was the average for infundibulum, magnum, isthmus, and uterus which were assayed separately.
[ac] Concentration of isotopes was assayed only in tibia shaft, which weighs 5.5 g. ^{241}Am in tibia shaft of the rat is 0.01 of ^{241}Am in whole skeleton (DURBIN et al., 1972c). Skeleton isotope = (conc. %/g) × 5.5 g × 100.
[ba] MITCHELL et al. (1931) reported the weight of skin, pancreas, and intestine as 330 g. Concentration used was the average of blood, gizzard, muscle, and reproductive tract (excluding ovaries).

Table 18.3. Summary of experimental conditions for mammalian studies with

Animal, isotope and chemical form	Dose adm. (μCi/kg)	Routes of administration[a]	Animal data: strain[b]	sex	age	weight
						Americium
Rat						
$^{241}AmCl_3$, pH = 5	3.6	i.m., s.t.	W.	♀	—[aa]	275 g
$^{241}AmCl_3$, pH = 5	30	i.v.	S.D.	♂	47 day	150 g
$^{241}AmCl_3$, pH = 5	32–189	i.v.	S.D.	♂	35 day	95 g
^{241}Am citrate	69	i.m.	—	♀	—	225 g
^{241}Am citrate	0.5–7	i.v.	M.A.	♂	49 day	105 g
^{241}Am citrate	1.0–1.9	i.v.	M.A.	♂	49 day	105 g
^{241}Am citrate, 5%	0.4	i.v.	alb.	—	—	~250 g
^{241}Am citrate, 0.4%	1–3	i.v., s.t.	M.A.	♂	49 day	105 g
$^{241}Am(NO_3)_3$, pH = 1.7	—	i.v., s.t.	M.A.	♂	49 day	105 g
$^{241}AmCl_3$	5–10	i.v., s.t.	alb.	♀	~150 day	200 g
$^{241}AmCl_3$	0.25–200	i.v.	alb.	♀	~150 day	200 g
^{241}Am citrate, 0.6%	0.37	i.m.	C.R.	♀	110 day	275 g
$^{241}Am(NO_3)_3$, pH = 3.8	5.9	i.p.	W.	—	—	170 g
$^{241}AmCl_3$, pH = 4	7	i.p., i.v.	—	♂	—	170 g
^{241}Am citrate	40	i.p.	S.D.	—	10 day	~25 g
^{241}Am citrate	variable	i.v.	W.	♂, ♀	adult	—
$^{241}AmCl_3$	2.5–25	i.v.	W.	♂, ♀	—	180 g
$^{241}Am(NO_3)_3$, pH = 2	~1 μCi/rat	i.m.	S.D.	♂	~60 day	300 g
^{241}Am citrate, pH = 7	—	i.v., i.p.	S.D.	♀	—	—
^{241}Am citrate, 1%	1.0	i.v.	H.	♀	—	190 g
Chinese Hamster						
^{241}Am citrate, pH = 6	1–100	i.p.	—	—	adult	—
Mouse						
^{241}Am citrate, 3%	0.7	i.m.	Swiss	♂	adult	30–35 g
$^{241}Am(NO_3)_3$	50	i.v.	CBA	♂, ♀	adult	20–35 g
^{243}Am citrate	5	i.m.	Swiss	♀	30–35 day	21–25 g
Dog						
^{241}Am citrate, pH = 3.5	0.00179–4.49	i.v.	beagle	♂, ♀	~15 mo	11 kg
$^{241}AmCl_3$	up to 60	i.v.	—	—	—	—

isotopes of the transplutonium elements: Parenteral and oral administration

Objectives and length of study	References[c]
Studies	
Metabolism, 256 days	SCOTT et al. (1948a), HAMILTON (1947, 1948)
Metabolism, 48 days Toxicity, 315 days	LANGHAM and CARTER (1951)
Bone autoradiographs, 1 day	ASLING et al. (1952)
Metabolism, 18 months Toxicity, life span	D. M. TAYLOR et al. (1961), FABRIKANT and SMITH (1964), BENSTED et al. (1965), BENSTED and TAYLOR (1969), D. M. TAYLOR and BENSTED (1967)
Removal by tetracycline, DTPA[ab], desferrioxamine, 21 days	D. M. TAYLOR and SOWBY (1962), SOWBY and TAYLOR, (1960), D. M. TAYLOR (1967), D. M. TAYLOR et al. (1971)
Removal by DTPA[ab], 30 days	FOREMAN (1962b)
Metabolism, 4 days Metabolism, 4 days and bone uptake, 1 day	D. M. TAYLOR (1962)
Metabolism, 64 days Toxicity, life span	ZALIKIN et al. (1968, 1969)
Metabolism, life span	DURBIN et al. (1967, 1972c)
Metabolism, 9 months	TSEVELEVA and YEROKHIN (1969)
Removal by DTPA[ab] and other chelating agents, 180 days	BELYAEV (1969)
Autoradiographs, 4 days	HAMMARSTRÖM and NILSSON (1970b)
Reproductive transfer, to F_1 and F_2 generations	OVCHARENKO (1970, 1972), MOSKALEV et al. (1969)
Toxicity, life span	RUDNITSKAYA and MOSKALEV (1970), MOSKALEV and RUDNITSKAYA (1970)
Metabolism and removal by DTPA[ab] 4 months	LAFUMA et al. (1970), NÉNOT et al. (1971b, 1972)
Metabolism in pregnancy and lactation, 43 days	HOLLINS and DURAKOVIC (1972)
Removal by DTPA[ab], 13 days	SEIDEL and VOLF (1972)
Metabolism and toxicity, 362 days	MCKAY et al. (1969, 1972)
Metabolism and removal by DTPA[ab], 14 days	PARKER et al. (1962)
Autoradiographs, 60 days	HAMMARSTRÖM and NILSSON (1970a)
Metabolism, 14 days	WRIGHT (1972)
Metabolism and toxicity, life span	LLOYD et al. (1967a, b, 1968, 1969, 1970), STEVENS et al. (1969a, b), G. N. TAYLOR et al. (1969a, b), ATHERTON et al. (1968, 1969a), J. H. DOUGHERTY (1970)
Toxicity, $>$ 30 days	ZALIKIN et al. (1969)

Table 18.3

Animal, isotope and chemical form	Dose adm. (μCi/kg)	Routes of administration[a]	Animal data			
			strain[b]	sex	age	weight
						Americium
Primate						
Cynomolgus monkey (*Macaca fascicularis*)						
^{241}Am citrate, 3%	0.44–0.87	i.m., i.v.		♀	$>$6.5 yr	2.5–4.5 kg
Baboon (*Papio anubis*)						
^{241}Am citrate, pH = 3.5	0.2	i.v.	—	♀	7–9 yr	—
Swine						
$^{241}Am(NO_3)_3$	—	cut.	—	—	56 day	—
Sheep						
^{241}Am citrate, pH = 4	150 μCi/ewe	i.v.	Suffolk	♀	adult	—
						Curium
Rat						
$^{242}CmCl_3$, pH = 5	4.4	i.m., s.t.	alb.	♀	$>$120 day	225 g
$^{242}CmCl_3$, pH = 3–5	5.0	i.m., i.v.	alb.	♂, ♀	50–120 days	185–250 g
$^{242}CmCl_3$, pH = 4	1.8	i.m., i.v.	S.D,	♀	$>$110 day	250 g
^{242}Cm citrate, 1.5%	1.8	i.m., i.v.	S.D.	♀	$>$110 day	250 g
$^{242}Cm(NO_3)_3$, pH = 1.5	8.3–83	i.m., i.v.	W.	♂	~60 day	~300 g
^{242}Cm citrate, 1%	1.0	i.v.	H.	♀	—	190 g
$^{244}CmCl_3$, pH = 3	5–400	i.v., s.t.	alb.	♂	—	190 g
Sheep						
^{244}Cm citrate, pH = 4	140 μCi/ewe	i.v.	Suffolk	♀	adult	—
						Berkelium
Rat						
$^{249}BkCl_3$, pH = 2	6.25	i.v., s.t.	W	♂	3 mo	400 g
Dog						
^{249}Bk citrate	—	i.v.	beagle	♂, ♀	15 mo	9–13.5 kg
						Californium
Rat						
$^{249-252}Cf$ citrate, 0.6%	0.33	i.m.	C.R.	♀	136 day	275 g
^{252}Cf citrate, pH = 6	7.7	i.p.	—	♂	—	130 g
Mouse						
^{252}Cf citrate, 3%	0.7	i.m.	Swiss	♂	adult	30–35 g
^{252}Cf citrate	4.0	i.m.	Swiss	♀	35–42 day	25 g
Chinese Hamster						
^{252}Cf citrate, 1.2%	0.056–45	i.p.	—	—	80–120 day	—

(continued)

Objectives and length of study	References[c]
Studies	
Metabolism, 973 days	Durbin et al. (1961, 1972c)
Metabolism, 3 months	Rosen et al. (1972)
Skin absorption, 168 hr	Osanov et al. (1971)
Milk secretion, 10 days	McClellan et al. (1962b)
Studies	
Metabolism, 256 days	Scott et al. (1949), Hamilton (1947, 1948)
Removal by EDTA[ab], 40 days	K. G. Scott (unpublished)
Metabolism, 1 day	Williams et al. (1961)
Metabolism, 1 day	
Metabolism and removal by DTPA[ab], 4 months	Nénot et al. (1970, 1972), Lafuma et al. (1970)
Removal by DTPA[ab], 13 days	Seidel and Volf (1972)
Metabolism and toxicity, 360 days	Semenov (1971a, b)
Milk secretion, 10 days	McClellan et al. (1962b)
Studies	
Metabolism, 21 days	Hungate et al. (1972)
Autoradiographs	G. N. Taylor et al. (1972)
Studies	
Metabolism, 91 days	Durbin (1962), Durbin et al. (1972d)
Metabolism, 64 days	Mewhinney et al. (1971, 1972)
Metabolism and removal by DTPA[ab], 14 days	Parker et al. (1962)
Metabolism, 122 days	S. R. Wright (unpublished)
Metabolism and toxicity	Mewhinney et al. (1972), Brooks et al. (1972)

Table 18.3

Animal, isotope and chemical form	Dose adm. (μCi/kg)	Routes of administration[a]	Animal data			
			strain[b]	sex	age	weight
Dog						
^{249}Cf citrate, pH = 3.5	0.28, 2.8	i.v.	beagle	♂, ♀	> 15 mo	9–13.5 kg
^{252}Cf citrate, pH = 3.5	0.29	i.v.	beagle	♂, ♀	> 15 mo	9–13.5 kg
						Einsteinium
Rat						
$^{253}EsCl_3$, pH = 2	6.25	i.v., i.p.	W	♂, ♀	3 mo	400 g
$^{253}Es(NO_3)_3$, pH = 2	3.3	i.v., i.m., s.c.	W	♀	—	285 g
Mouse						
^{253}Es citrate, 4.6%	1.3	i.m.	Swiss	♀	30–35 day	21–25 g

[a] Intravenous, i.v.; intraperitoneal, i.p.; intramuscular, i.m.; subcutaneous, s.c.; oral intubation, s.t.; skin application, cut.
[b] Wistar, W.; Sprague-Dawley, S.D.; Marshall-August, M.A.; albino unstated strain, alb.; Charles River Sprague-Dawley, C.R.; Heiligenberg, H.

IV. The Metabolism and Distribution of the Transplutonium Elements in Mammals: Absorption

The research objectives and experimental conditions of the animal experiments with the transplutonium elements are summarized in Tables 18.3 and 18.4. The results of some experiments performed in this Laboratory are reported in this Chapter for the first time. The experimental conditions are included in Table 18.3. These experiments have been included in the bibliography as well: studies of the metabolism and microscopic distribution of ^{241}Am in rats and cynomolgus monkeys (Durbin et al., 1972c); studies of the distribution of $^{249-252}Cf$ in the rat (Durbin et al., 1972d); studies of the weights and microscopic anatomy of the tissues and bones of the female rat from birth to senescence (Durbin et al., 1972e).

A. Absorption from the Gastrointestinal Tract

Absorption following oral administration of the trivalent transplutonium elements has been studied only in rats. Single doses of slightly acid solutions of chlorides were administered by gavage. Scott et al. (1948a, 1949) reported the subsequent absorption of $^{241}AmCl_3$ to be $<0.01\%$ and of $^{242}CmCl_3$, $<0.05\%$. Recalculation of their original data suggests that those values are low, and that absorption from oral administration of either $^{241}AmCl_3$ or $^{242}CmCl_3$ was probably between 0.05 and 0.1%. So, absorption of these elements by the rat after oral inoculation has been restudied: Based on analyses of selected tissues, Zalikin et al. (1968) estimated absorption of $^{241}AmCl_3$ to be 0.03%, and Semenov (1971b) estimated absorption of $^{244}CmCl_3$ to be 0.05%. After oral administration of $^{249}BkCl_3$ and $^{253}EsCl_3$ to rats, Hungate et al. (1972) found 0.009% of the ^{249}Bk and 0.036% of the ^{253}Es in bone and tissues. Correcting for urinary and endogenous fecal excretion, total absorption by rats, subsequent to oral administration of these elements, was about 0.01% of ^{249}Bk and 0.06% of ^{253}Es.

(continued)

Objectives and length of study	References [c]
Metabolism and toxicity Metabolism and toxicity	LLOYD et al. (1972b, e), ATHERTON et al. (1972), G. N. TAYLOR et al. (1972), STEVENS and BRUENGER (1972)
Studies	
Metabolism, reproductive transfer, skin absorption, 21 days	HUNGATE et al. (1972)
Metabolism and removal DTPA[ab], 24 days	V. H. SMITH (1972a, b)
Metabolism and removal by DTPA[ab], 14 days	PARKER et al. (1972)

[c] Reference cited first contains original or most detailed description of study; others provide ongoing or ancillary information about the same or similar groups of animals.
[aa] Dash indicates information not available. This convention is followed in all tables.
[ab] DTPA, diethylenetriaminepentaacetic acid; EDTA, ethylenediaminetetraacetic acid.

Studies have been made of the gastrointestinal absorption of ^{144}Ce(III) and ^{239}Pu(IV) in very young animals, and of the influence on absorption of the presence of citrate ion and multiple vs. single feedings. No comparable studies have been made with the transplutonium elements, so the effects of these factors on absorption of the trivalent actinides must be inferred from experience with chemically similar elements.

Absorption following a single oral dose of $^{144}CeCl_3$ in rats was originally reported to be 0.02% (SCOTT et al., 1947). GRIGORYAN (1966) fed to rats 120 consecutive daily doses, each 0.8 μCi of $^{144}CeCl_3$, and from analysis of tissues calculated the average overall absorption to be 0.1 to 0.15%. Administration of an easily measurable amount of ^{144}Ce and the long feeding period insured (a) accurate tissues analyses, (b) good mixing of the isotope with the gastrointestinal contents, and (c) compensation for the daily fluctuations of gastrointestinal tract function.

The absorption by 1.5-year-old sheep following oral dosage of $^{144}CeCl_3$ remained constant—0.03% of the daily dose throughout periods of daily feeding varying from 18 to 1140 days (BULDAKOV and BUROV, 1967). These same authors fed single doses of $^{144}CeCl_3$ to lambs about 25 and 180 days old and obtained absorption values of 0.15% and 0.07%, respectively.

The greater absorption of multicharged cations by the intestinal tract of the very young animal has also been demonstrated in rats and dogs. SIKOV and MAHLUM (1972a) found that 1-day-old rats absorbed nearly four times more orally administered ^{239}Pu(IV) citrate than did 20- to 35-day-old rats, and 3-day-old pups absorbed more than four times as much ^{239}Pu(IV) citrate as did adult dogs (BULDAKOV et al., 1969).

The absorption of oral doses of ^{239}Pu(IV) is enhanced if it is administered with an excess of sodium citrate. See the review by BULDAKOV et al. (1969). For example, the rat's body content of ^{239}Pu(IV), after various feeding regimens, ranged from 0.002 to 0.012% if ^{239}Pu(IV) was given in a nitrate solution of pH 1 to 2, but was 0.04 to 0.2% after feeding ^{239}Pu(IV) in a sodium citrate solution of pH 4 to 6.5.

Table 18.4. Summary of experimental conditions for mammalian studies with isotopes of the

Animals, Isotope	Chemical form	Particle size	Method of administration	Strain[a]	Sex	Age
						Americium
Rat						
^{241}Am	citrate, 5%	solution	intubation	alb.	♀	—
^{241}Am	chloride	solution	intubation	alb.	♀	~150 day
^{241}Am	nitrate, pH = 3.8	solution	intubation	W.	—	—
^{241}Am	nitrate, pH = 0.6	mist	inhalation	S.D.	♂	~60 day
Dog						
^{241}Am	oxide	0.82–0.99 μm AMAD	inhalation	beagle	♂–♀	446 day
^{241}Am	nitrate, pH = 1.5–2.0	—	inhalation	mongrel	♂–♀	—
						Curium
Rat						
^{242}Cm	chloride, pH = 0.6	mist	inhalation	S.D.	♂	~60 day
^{242}Cm	nitrate, pH = 0.6	mist	inhalation	S.D.	♂	~60 day
Dog						
^{244}Cm	chloride, CsCl carrier	1.6 μm AMAD	inhalation	beagle	♂–♀	11–47 mo
^{244}Cm	oxide ($CmO_{1.73}$)	0.5 μm AMAD	inhalation	beagle	♂–♀	11–47 mo
						Einsteinium
Rat						
^{253}Es	chloride, pH = 2	solution	intubation	W.	♂	3 mo
^{253}Es	hydroxide, pH = 12	suspension	intubation	W.	♂	3 mo

[a] Albino unstated strain, alb.; Wistar, W.; Sprague-Dawley, S.D.

The best available estimates of intestinal absorption of multicharged cations by the rat are as follows: ^{144}Ce(III), 0.1 to 0.15%; the trivalent actinides, 0.01 to 0.06%; and Pu(IV), 0.003 to 0.012%. Intestinal absorption of actinide (III) chlorides is about five times as great as that of $^{239}Pu(NO_3)_4$ and one-tenth to one-half as much as the absorption of $^{144}CeCl_3$.

Based on experience with ^{144}Ce(III) and ^{239}Pu(IV) it could be anticipated that (a) the fraction of actinide (III) absorbed during a period of continuous ingestion would not be too different from the fraction absorbed from a single feeding; (b) absorption of an oral dose of actinide (III) elements by newborn or very young animals would be about four times as great as that absorbed by adults; and (c) intestinal absorption of actinide (III) elements would probably be increased 10- to 20-fold if they were fed in a sodium citrate solution.

B. Absorption from an Intramuscular Injection

The intramuscular injection route was used in the earliest studies of the biology of the lanthanides and actinides because these easily hydrolyzable elements could be introduced into the body in a way that avoided formation of colloidal aggregates in the circulation (Scott et al., 1947, 1948a, b, 1949). Intramuscular absorption *per se* was not studied purposefully, but the information is available as an experimental by-product. The circulation of tissue fluids in the leg muscles of the rat is not the same as that in the fleshy parts of the human hand—the

transplutonium elements: Inhalation and intratracheal administration

Wt	Initial body burden	References[b]
Studies		
~250 g	0.1 μCi	Foreman (1962b)
200 g	0.6 μCi	Zalikin et al. (1968)
170 g	0.3 μCi	Tseveleva and Yerokhin (1969)
~300 g	4–6 μCi	Nénot et al. (1971a–c, 1972), Lafuma et al. (1970)
8.4–10.6 kg	38–51 μCi	McClellan and Rupprecht (1968), R. G. Thomas (1970), R. G. Thomas et al. (1969, 1972), Barnes (1970)
—	≤1.5 μCi/kg	Buldakov et al. (1972), Nifatov et al. (1972)
Studies		
~300 g	4–6 μCi	Nénot (1971a, 1972), Lafuma et al. (1970)
~300 g	—	
5.7–12.8 kg	1.4–3.4 μCi	McClellan et al. (1972), Gallegos et al. (1970)
5.7–12.8 kg	0.8–1.8 μCi	
Studies		
400 g	0.21–47.2 μCi	Ballou and Morrow (1972), Hungate et al. (1972)
400 g	2.5 μCi	

[b] Reference cited first contains original or most detailed description of study; others provide ongoing or ancillary information about the same or similar groups of animals.

most common site of accidental lacerations and puncture wounds—however, the patterns of absorption of these elements, after intramuscular injection in animals, represents nearly all the information upon which to base judgments of the consequence of contaminated wounds.

The data available on the intramuscular absorption of simple salts of the trivalent actinides administered intramuscularly to rats are collected in Table 18.5. The intramuscular absorption of multicharged cations is complex and is related to the administered mass, the chemical form of the element, and the pH and volume of solution injected. In the case of long-lived alpha-particle emitters, there is the added complication of local tissue damage if a significant amount remains at the injection site for a long enough time.

Absorption of intramuscularly administered lanthanide elements in the rat is summarized in Fig. 18.1.[6] The dependence on the administered mass is apparent. When the amount injected was less than 0.01 μg, about 50% was absorbed in the first day or so; another 40% was absorbed with a half-time of about 25 days; and the remaining 10% left the injection site with a half-time of 100 to 200 days.

6 The curves in Fig. 18.1 were drawn from a composite of many observations made in the course of metabolic studies of radioisotopes of yttrium and the 14 lanthanide elements administered as chlorides. These experiments were performed at the Crocker Laboratory, University of California, in the years 1943 to 1957 under the general direction of J. G. Hamilton, K. G. Scott, and P. W. Durbin. Some of the information was presented in Laboratory progress reports, but much is unpublished.

Table 18.5. Absorption of various salts of trivalent actinides following intramuscular injection in rats

Administered nuclide and injection conditions					Absorption at days after injection (percent of administered dose)											Reference
Compound	Activity (μCi)	Mass (μg)	Volume (ml)	pH	1	4	7	10	16	20	30	64	90	120	256	
$^{241}AmCl_3$	1.0	0.3	1.0	5	58	84	—	—	92	—	—	93	—	—	98	SCOTT et al. (1948a)
$^{241}Am(NO_3)_3$	0.01–5.0	0.03–1.5	0.1	0.5–2.5	10	—	46	55	—	—	57	—	84	85	—	LAFUMA et al. (1970), NÉNOT et al. (1971a, 1972)
$^{241}Am(SO_4)_3$	0.1	0.03	0.1	0.5–2.5	24	—	—	—	—	—	—	—	92	—	—	NÉNOT et al. (1971a)
$^{242}CmCl_3$	1.0	0.0003	1.0	5.0	90	92	—	—	92	—	—	98	—	—	95	SCOTT et al. (1949)
$^{242}Cm(NO_3)_3$	5.0	0.0015	—	1.5	18	—	50	—	—	83	—	89	—	—	—	NÉNOT et al. (1972)
$^{242}Cm(NO_3)_3$ + $La(NO_3)_3$	0.1–1.6	≤200 (of La)	0.1	0.5–2.5	6.5	—	—	—	—	30	—	55	—	—	—	LAFUMA et al. (1970), NÉNOT et al. (1970, 1972)
$^{253}Es(NO_3)_3$	0.95	<0.001	0.2	2.0	—	—	—	—	—	95	—	—	—	—	—	V. H. SMITH (1972b)
$^{239}PuCl_3$	1.0	15	1.0	2.5	—	33	—	—	42	—	—	60	—	—	76	SCOTT et al. (1948b)

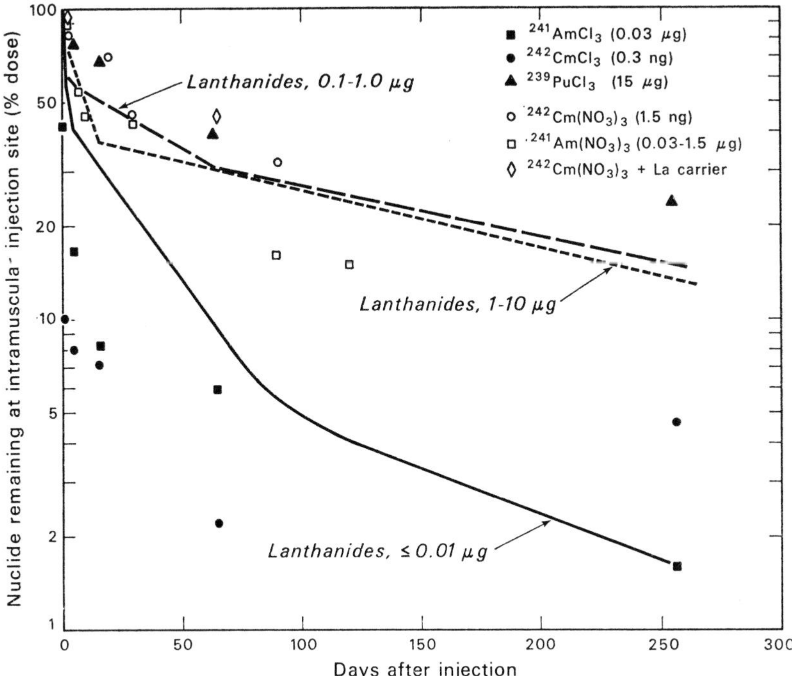

Fig. 18.1. Absorption of chlorides or nitrates of Pu(III), Am(III), and Cm(III) intramuscularly injected into rats (shown as data points), and the mean absorption of varying masses of intramuscularly injected chlorides of several lanthanides (shown as curves). Data are presented as percent of administered dose remaining at the injection site; Absorption (%) = (100%—Injection site). References: $^{241}AmCl_3$, SCOTT et al. (1948a); $^{242}CmCl_3$, SCOTT et al. (1949); $^{239}PuCl_3$, SCOTT et al. (1948b); $^{241}Am(NO_3)_3$, LAFUMA et al. (1970) and NÉNOT et al. (1971b); $^{242}Cm(NO_3)_3$, NÉNOT et al. (1972); $^{242}Cm(NO_3)_3$ + La carrier, LAFUMA et al. (1970) and NÉNOT et al. (1971b, 1972)

As the amount of the injected mass was increased, the fraction absorbed in the first few days declined, and the amounts associated with the slower absorption rates increased. If 100 μg or more were injected, only 5 to 10% was absorbed during the first few days, and very little was absorbed thereafter.

The mechanisms by which multicharged cations are absorbed from parenteral injection sites are not known exactly, but it is reasonable to suggest that those fractions rapidly absorbed (half-times: hours to a few days) represent passage of soluble isotope into the lymph. The slower migrating fractions are probably subject to a combination of mechanisms: the consolidation of hydrolysis products and their subsequent sequestration in tissue phagocytes. Absorption of ^{239}Pu from contaminated wounds is characterized by a component with a half-time of the order of years, which appears to be the result of the encapsulation of the unabsorbed material by a fibrous tissue reaction (LANGHAM et al., 1962; HAMMOND and PUTZIER, 1964).

The absorption of $^{241}AmCl_3$, $^{242}CmCl_3$, or $^{239}PuCl_3$, administered intramuscularly (the solid points in Fig. 18.1), was comparable to absorption of lanthanides of the same specific activity. The $^{241}AmCl_3$ and $^{242}CmCl_3$ were each absorbed more

Table 18.6. The effect of DTPA on absorption of intramuscularly injected $^{241}Am(NO_3)_3$[a] and $^{242}Cm(NO_3)_3$[b]

Isotope	Absorption at time after isotope injection (Percent of administered dose)							Reference
	minutes			days				
	0	1	30	1	10	20	90	
$^{241}Am(NO_3)_3$[a]	—	—	—	—	65.1	—	95.7	Lafuma et al. (1970), Nénot et al. (1971 a)
$^{242}Cm(NO_3)_3 + La(NO_3)_3$	100	92	61.6	57	—	74	—	Lafuma et al. (1970), Nénot et al. (1970)

[a] Compare with experiment $^{241}Am(NO_3)_3$ in Table 18.5; 50 mM/kg of $CaNa_3$ DTPA injected into leg opposite ^{241}Am-injected leg beginning 24 hr after the ^{241}Am and 2 times each week thereafter. Rats killed 24 hr after the last DTPA injection.
[b] Compare with $^{242}Cm(NO_3)_3 + La(NO_3)_3$ in Table 18.5; 50 mM/kg of $CaNa_3$ DTPA injected into the ^{242}Cm injection site; 0-, 1-, and 30-min and 1-day groups were each given one DTPA injection at those times after the ^{242}Cm injection and were killed 24 hr later. The 20-day group received the first DTPA injection 24 hr after the ^{242}Cm, and two DTPA injection each week thereafter.

easily than the $^{241}Am(NO_3)_3$ or $^{242}Cm(NO_3)_3$[7], and variable amounts up to 200 μg of stable $La(NO_3)_3$ carrier[8] depressed the uptake of ^{242}Cm still further.

Absorption of intramuscularly injected lanthanides (Durbin et al., 1956) and actinides is greatly accelerated if they are administered as citrate complexes. The amounts of trivalent actinides absorbed in the first 4 days after intramuscular injection of solutions containing 0.4 to 5% of sodium citrate were: in rats, 97% of ^{241}Am (Durbin et al., 1967), 97.3% of ^{242}Cm (Williams et al., 1961), and 97.9% of $^{249-252}$Cf (Durbin et al., 1972d), in mice, 98% to 99% of ^{241}Am and ^{252}Cf (Parker et al., 1962) and 96.5% of ^{253}Es (Parker et al., 1972); and in monkeys, 98.4% of ^{241}Am (Durbin et al., 1962).

Citrate complexes of the transplutonium elements are unstable in body fluids, and their behavior in the body is identical to that observed after intramuscular or oral administration of simple salts. The behavior of the stable complex with DTPA is different. Within 24 hours after intramuscular injection of $^{242}CmDTPA^{-2}$, the ^{242}Cm was completely absorbed and was completely excreted in the urine (Nénot, 1971 a).

Lafuma et al. (1970) and Nénot et al. (1970, 1971 a) investigated the influence of DTPA on the absorption of intramuscularly injected $^{241}Am(NO_3)_3$ or $^{242}Cm(NO_3)_3 + La(NO_3)_3$; their results are summarized in Table 18.6. Introduction of DTPA into the same muscle one minute after the injection of ^{242}Cm + La promoted nearly complete absorption and excretion of ^{242}Cm. If the interval between injections was either 30 minutes or as long as 24 hours, the ^{242}Cm absorption was 10 times as great as the control values (Table 18.5). About 40% of the administered dose remained at the injection site—an indication of the speed with which trivalent cations are hydrolyzed in tissue fluids and are rendered less available for complexing. Repeated injections of DTPA increased the ^{242}Cm absorption to

7 It is possible that the apparently better absorption of the chlorides, compared to the nitrates, was caused by introduction of the chlorides in a larger volume, 1 ml as compared to 0.1 ml of the nitrate solutions.

8 The experiments described by Lafuma et al. (1970) and Nénot et al. (1970) dealt with many of the same animals. The ^{242}Cm solutions used in those experiments contained some stable lanthanum (Nénot et al., 1972), and the mass of lanthanum given to each rat varied with the injected volume.

Table 18.7. Calculation of initial rate of absorption of ^{241}Am from a puncture wound

	^{241}Am activity at time following accident (dis/min)	
	3 days	134 days
Removed surgically (on day 3)	21×10^5	(21×10^5)
Hand	4.5×10^5	2.0×10^5
Body	2.3×10^5 (calc.)	0.8×10^5 (meas.)
Urine	$< 1 \times 10^5$	4.0×10^5
Total in wound at $t = 0$	27.8×10^5	27.8×10^5

Fraction absorbed in 2.5 hr before surgery $\dfrac{2.3 \times 10^5}{27.8 \times 10^5} = 0.083$

Calculated absorption rate, k $k = \dfrac{0.083 \times 24 \text{ hr/day}}{2.5 \text{ hr}} = 0.8/\text{day}$

Calculated absorption half-time, $T_{1/2}$ $T_{1/2} = \dfrac{0.693}{0.8/\text{day}} = 0.87$ day

74% of the dose in 20 days compared with 30% absorption by controls. The absorption of ^{241}Am increased by only an additional 10% of the injected dose (Nénot et al., 1971a) when repeated systemic DTPA treatments were given intramuscularly in the opposite leg or by inhalation as an aerosol (see Sec. XI.C.7.).

Absorption of ^{241}Am from a puncture wound in an accident case was followed for 2 years by measurement of excreta and external monitoring of the contaminated hand (R. L. Dobson, 1955; R. L. Dobson, H. G. Parker, M. C. Wales, and M. Chupp, unpublished). A chemist was performing an experiment with a concentrated source of ^{241}Am in a closed gloved box. He struck the end of a thin glass pipette which pierced his glove, inflicting a small puncture wound between the thumb and forefinger. The pipette had not dipped into the ^{241}Am solution but had been used to bubble NH_3 gas over the solution and had become contaminated by splash. Therapy with EDTA (0.12 to 0.25 mM/kg/day of $CaNa_2$ EDTA) was started 1.5 hr after the accident. The bulk of the ^{241}Am (20×10^5 dis/min) was removed surgically 2.5 hr after the accident. Two days later an additional 1×10^5 dis/min were removed surgically leaving 4.5×10^5 dis/min in the hand. The EDTA treatment was continued for 60 days.

The measurements of the ^{241}Am in the wounded hand are shown in Fig. 18.2. The ^{241}Am content of the body was measured 134 days after the accident and was estimated to be 0.8×10^5 dis/min (Rowland et al., 1956). At that time the hand still contained 2×10^5 dis/min. During the first 134 days approximately 4×10^5 dis/min were excreted in the urine.

The initial rate of ^{241}Am absorption was estimated from these data to be 80%/day (as shown in Table 18.7), and is of the same order of magnitude as the initial rate of ^{241}Am absorption from an intramuscular injection in the rat. The parameters of the two-component exponential curve drawn to the external measurements of the hand are given in Fig. 18.2. The half-time of the first component was 13 days. Absorption at this rate is assumed to have begun 2.5 days after the injection, when the wounded hand was first measured. Once scar tissue formed, the absorption of ^{241}Am slowed markedly. The half-time of the long-lived component was 1870 days. At the time the body content was remeasured, 810 days after the accident, what had previously been judged to be a diffuse deposit

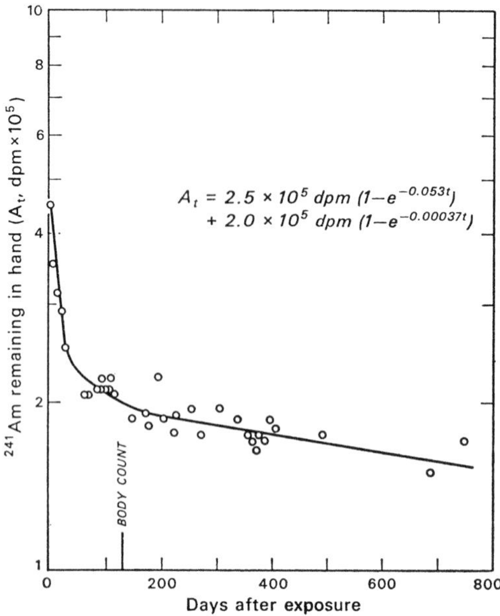

Fig. 18.2. Absorption of ^{241}Am (in an alkaline solution) from a puncture wound in the hand. [Data of R. L. DOBSON (1955) and R. L. DOBSON, H. G. PARKER, M. C. WALES and M. CHUPP (unpublished)]

of ^{241}Am in the hand had consolidated into a well-circumscribed knot of fibrous tissue (MILLER, 1957).

It seems unlikely that the EDTA treatments had influenced ^{241}Am absorption; EDTA was being given at a relatively low level, 0.12 mM/kg/day. A daily dose of EDTA 10 times as large, and given for 15 days, did not change the absorption of intramuscularly injected ^{242}Cm in rats (K. G. SCOTT, M. H. WILLIAMS, and H. I. JOHNSON, unpublished). A much larger quantity of a more efficient complexing agent, 50 mM/kg/day of DTPA, increased intramuscular absorption of ^{241}Am by only an additional 10% (NÉNOT et al., 1971a).

C. Absorption Through the Intact Skin

The only trivalent actinide for which cutaneous absorption has been studied is ^{253}Es (HUNGATE et al., 1972). The $^{253}Es(NO_3)_3$ in 0.01 N HNO_3 was applied to washed rat skin from which the hair had been previously clipped. The solution was removed after 30 min. Tissue analyses 7 days later indicated an absorption of 0.03%/cm^2 of the applied ^{253}Es; 0.049%/cm^2 if a correction was applied for urinary excretion.

Some clues to the behavior of the less soluble, lighter weight actinides can be obtained from experiments with ^{144}Ce(III). GRIGORYAN et al. (1966) applied a slightly acid solution of $^{144}CeCl_3$ to rat skin that had been shaved 4 days before. At the end of a 3-hr exposure period, the ^{144}Ce was washed off. Absorption was 0.07 to 0.15%, calculated from tissue analyses of animals killed 1 or 7 days after the ^{144}Ce application. The amount of ^{144}Ce absorbed was increased 2 to 3 times if the rat skin had just previously been exposed to 60 μR-min of ultraviolet light. Apparently the effect of the UV treatment was to increase capillary circulation.

D. Absorption of Inhaled Transplutonium Elements

Internal burdens of the transplutonium elements have been acquired most frequently as a result of accidental inhalation of particles or droplets (FOREMAN et al., 1958; CALDWELL, 1966; BRODSKY et al., 1966, 1969; MAY, 1968; JEANMAIRE and BALLADA, 1970; SANDERS, 1971; ROSEN et al., 1971). A general model of lung dynamics has been developed to predict for man the lung deposition and retention and the systemic absorption of radioactive aerosols of varying sizes and solubility properties (ICRP, Task Group on Lung Dynamics, 1966). The quantities of inhaled airborne particles selectively deposited in the several regions of the respiratory tract depend largely upon the size and density of the particles. The quantities eliminated from the lungs, either by absorption into the body or by various clearance mechanisms, depend on the chemical properties of the particles, chiefly their solubility (HATCH and GROSS, 1964).

1. Deposition of Particles

The toxicity of inhaled radionuclides is a function of the radiation doses delivered to the regions of the respiratory tract and to the organs in which the absorbed radionuclide is deposited. The radiation doses, in turn, are functions of the amounts of radionuclide initially deposited in each tissue and the time courses of their release.

The initial body burden (IBB) of an inhaled radionuclide is

$$\mathrm{IBB} = (\mathrm{N}-\mathrm{P}) + (\mathrm{T}-\mathrm{B}) + (\mathrm{P}) \qquad \text{Eq. (1)}$$

where all terms are in units of radioactivity or are expressed as fractions of the amount inhaled. The ICRP Task Group (1966) divided the respiratory tract into three regions: naso-pharynx (N—P), tracheobronchial tree (T—B), and pulmonary (P). The schematic model of the particle deposition sites and clearance processes in man, developed by the ICRP Task Group (1966), is shown in Fig. 18.3; for unit density particles with AMAD[9] of 1 μm, the IBB $= 0.59$, N—P $= 0.20$, T—B $= 0.09$, and P $= 0.30$ of the total amount inhaled. The IBB can be obtained experimentally from either (a) the relationship between air concentration (μCi/liter) and the total volume breathed, (b) the immediate in vivo measurement of the whole-body radioactivity, or (c) the sum of the radioactivity in the body and in all the excreta accumulated to the time of the body measurement.

2. "Solubility" of Compounds

The ICRP Task Group (1966) also considered the influence of the chemical properties of inhaled materials. Simple chemical compounds were classified according to available information on their solubility in water, dilute alkali, or in protein solutions:

Class D is comparable to the designation "soluble" of the lung model in current use (ICRP, 1959). These compounds exhibit clearance and absorption half-times of the order of 1 day or less. No compounds of the lanthanides or actinides are included in Class D; however, Class D would include lanthanide and actinide complexes such as those with citrate and the aminopolycarboxylic acids, EDTA and DTPA.

Class W compounds approximate the presently used classification of "insoluble" (ICRP, 1959) and are expected to exhibit clearance half-times ranging from a few weeks to a few months. Lanthanide and actinide halides, nitrates,

9 AMAD, activity mean aerodynamic diameter.

carbonates, phosphates, and presumably also sulfates and sulfides are considered Class W compounds.

The biological half-times (T_i's) shown in Fig. 18.3 are those for Class W compounds.

Class Y contains those compounds that are nearly insoluble in biological fluids. They are tenaciously retained in the lung, and are poorly absorbed into the body. They manifest clearance half-times of the order of months to years. Lanthanide and actinide hydroxides, oxides, fluorides, and carbides comprise the bulk of simple chemical compounds of Class Y.

3. Clearance of Particles from the Lung

The cilia of the nasopharynx and tracheobronchial tree (to the level of the terminal bronchioles) continuously move a layer of mucus toward the esophagus. The early phase of rapid clearance of particles from the upper respiratory tract is usually considered to be the result of the muco-ciliary transport process. Most of the particulate material initially deposited in the N—P and T—B regions is cleared to the G. I. tract within an hour; 50% of soluble aerosols (Class D), 90% of poorly soluble aerosols (Class W), and 99% of insoluble aerosols (Class Y) so deposited are cleared with biological half-times of 4 to 10 min (ICRP, 1966).

4. Evaluation of Initial Lung Deposition in Dogs

The immediate purpose of inhalation studies of radioactive material is to ascertain the relationships between local radiation doses and observed toxicity, and the ultimate goal is prediction of the chronic toxicity of a given level of inhaled radioactivity. Therefore, the material initially deposited in the upper repiratory tract, and which is cleared rapidly, is usually not considered to contribute significantly to the total radiation dose to the lungs or other organs. There is an exception in the case of short-lived radionuclides that deliver most of their radiation dose to the upper respiratory tract structures and to the gastrointestinal tract during the early clearance phase.

It is also customary to eliminate from long-term consideration that fraction of the initial pulmonary deposit that is cleared by migratory phagocytes and subsequently entrained in the muco-ciliary transport process during the first few days after exposure (pathway f in Fig. 18.3, $t_f = 1$ day). Long-term results are then discussed in terms of the initial lung burden (ILB) which is defined as

$$\mathrm{ILB} = (\mathrm{P} - \mathrm{f}) = (\mathrm{P} - 0.4\,\mathrm{P}) = 0.6\,\mathrm{P} \qquad \text{Eq. (2)}$$

where P is the initial pulmonary deposit, f is the fraction of that deposit cleared with a 1-day half-time. All terms are expressed in radioactivity units or as percent of the amount inhaled.

In their studies with dogs, R. G. THOMAS et al. (1972) and MCCLELLAN et al. (1972) define ILB as in Eq. (2) except that t_f is taken to be 7 or 8 days after exposure. These workers determine ILB in two ways: (a) from an immediate whole-body measurement less the first 7 days' feces, or (b) from the sum of the radioactivity of all the tissues and excreta at sacrifice less the first 7 days' feces.

5. Deposition of Inhaled Actinides in the Dog Lung

The ICRP Task Group report (ICRP, 1966) was completed in mid-1965. At that time there was only one brief report of whole-body rentention of ^{241}Am citrate administered intratracheally to rats (FOREMAN, 1962b). In some very new studies beagles have been exposed by inhalation to precisely-known quantities of well-characterized particles of $^{241}AmO_2$ (R. G. THOMAS et al., 1972), $^{244}CmCl_3$, and $^{244}CmO_{1.73}$ (McCLELLAN et al., 1972). Some of the experimental details and some of the results of these studies are shown in Tables 18.4 and 18.8 respectively.

The measured values of the IBB and ILB for the various particle sizes are compared in Table 18.8 to those predicted by the ICRP model for man (ICRP, 1966). The agreement is excellent for the 1.6-μm-AMAD particles of $^{244}CmCl_3$ and the 0.5-μm-AMAD particles of $^{244}CmO_{1.73}$, but nearly twice as much of the ^{241}Am radioactivity from the 0.9-μm-AMAD particles of $^{241}AmO_2$ was deposited in the dog lung as was predicted by the human deposition model.

6. Body Absorption of Actinides Deposited in the Dog Lung

It was necessary to recalculate the data as originally reported by R. G. THOMAS et al. (1972) and McCLELLAN et al. (1972) in order to estimate the amount of the actinide compounds deposited in dog lung that was subsequently absorbed into the body.

R. G. THOMAS et al. (1972) expressed the important results of their inhalation studies of $^{241}AmO_2$ as μCi of ^{241}Am; these included the total amount inhaled, IBB, ILB, and total urinary and fecal excretion. Tissue contents of ^{241}Am were presented as a percent of the sacrifice body burden (SacBB). The percent of the ILB was calculated from the relationship

$$\mathrm{ILB}(\%) = [\mathrm{SacBB}(\%)]\ [\mathrm{SacBB}(\mu\mathrm{Ci})/\mathrm{ILB}(\mu\mathrm{Ci})]. \qquad \text{Eq. (3)}$$

McCLELLAN et al. (1972) expressed the major results of their inhalation studies with $^{244}CmCl_3$ and $^{244}CmO_{1.73}$ as μCi of ^{244}Cm. The results included the total amount inhaled, IBB, and ILB. Urinary excretion was presented as the percent ILB per day, and tissue-^{244}Cm content was reported as percent ILB per gram wet tissue. The tissue results were converted to percent ILB in whole tissues and organs using the beagle tissue weights given by CUDDIHY et al. (1970) and HOBBS et al. (1970) of that same Laboratory. The recalculated data agreed with preliminary results reported by GALLEGOS et al. (1970).

The lung-clearance model (ICRP, 1966) designates actinide halides and nitrates as Class W compounds, and predicts that Class W compounds will be eliminated from the lung with an average half-time of 90 days by a combination of absorption into the body (pathways e and h, Fig. 18.3) and clearance to the gastrointestinal tract (pathway g). The model predicts that the ultimate disposition of the ILB of Class W compounds will be as follows: 33% of the ILB (6% of the inhaled dose) will be absorbed into the body while 67% of the ILB (12% of the inhaled dose) will be cleared by phagocytic and ciliary action to the gastrointestinal tract. Actinide oxides are designated Class Y compounds, for which the predicted average lung elimination half-time is 360 days—only one-fourth as fast for Class W compounds—although the ultimate distribution of the ILB between absorption into the body and gastrointestinal tract clearance is presumed to be the same as for Class W compounds.

Elimination of $^{244}CmCl_3$ (a Class W compound) from the dog lung (Fig. 18.4) is the complex resultant of several processes, but most of these processes seem to involve solubilization and absorption into the body rather than clearance via the gastrointestinal tract. The curve of elimination from the lung for $^{244}CmCl_3$ can be described approximately by 4 exponential terms: (1) 20% of ILB, $T_{1/2} =$ 4 hr; (2) 60% of ILB, $T_{1/2} = 2.5$ days; (3) 12% of ILB, $T_{1/2} = 21$ days; (4) 9% of ILB, $T_{1/2} \geq 200$ days. The rates at which ^{244}Cm was simultaneously being

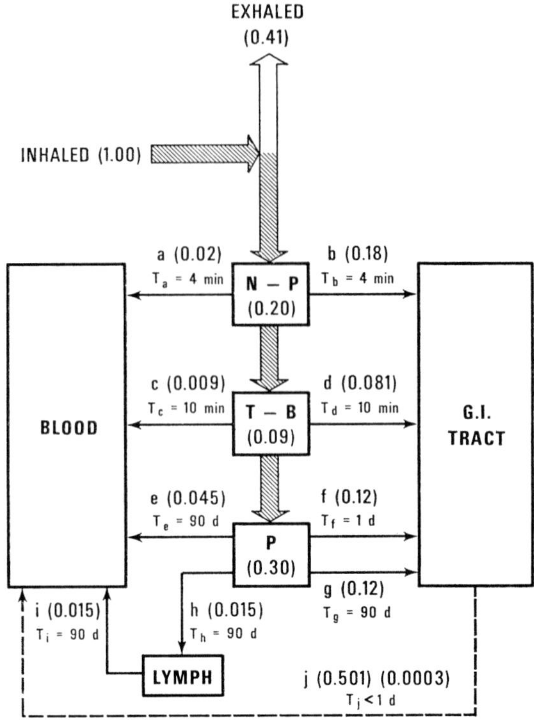

Fig. 18.3. Kinetic model of the deposition and clearance of inhaled particles from the respiratory tract. Deposition values were read from Fig. 11 of the ICRP Task Group report (ICRP, 1966) for an aerosol of AMAD = 1 μm and a tidal volume of 1450 ml. Transfer coefficients (a–j) and transfer half-times (T_a—T_j) were taken from Table 4 of the above report for poorly soluble compounds (Class W)

deposited in skeleton and liver and eliminated in urine (half-times of 4 to 9 hr and 2 to 3 days) corresponded roughly to rates of the two rapid phases of elimination from the lung.

Elimination of $^{244}CmO_{1.73}$ (a Class Y compound) from the lung was only a little slower than that of $^{244}CmCl_3$ and also could be approximated by 4 exponential terms: (1) 14% of ILB, $T_{1/2} = 1$ hr; (2) 31% of ILB, $T_{1/2} = 0.9$ days; (3) 37% of ILB, $T_{1/2} = 5.4$ days; (4) 18% of ILB, $T_{1/2} \geq 80$ days. Uptake in skeleton and liver and excretion in urine were all characterized by half-times of 0.6 day and 3 to 4 days—again close to the early rates of elimination from lung.

Absorption from the lung of the moderately soluble $^{244}CmCl_3$ was almost the same as the water-insoluble compounds, $^{241}AmO_2$ and $^{244}CmO_{1.73}$, at the time when the lung burden had been reduced to 10% of the ILB (LB_{10})[10]. As shown in Table 18.9 the absorption of all three actinide compounds (85% of the ILB at

10 Most studies of inhalation or intratracheal instillation do not extend over a long enough interval to achieve complete lung clearance, but most such studies, especially of relatively soluble compounds, are carried out long enough to achieve 90% lung clearance. For purposes of comparing the results of studies of varying duration in which varying degrees of lung clearance were achieved, an artificial common time has been adopted, designated LB_{10}, the time after exposure at which the lung burden has been reduced to 10% of the ILB.

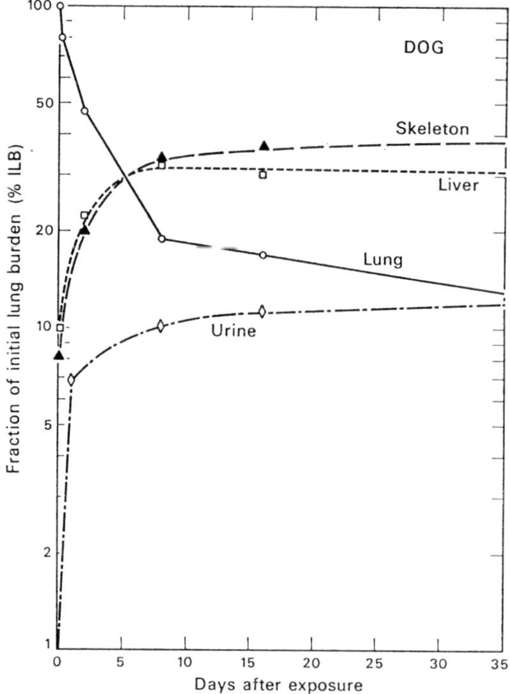

Fig. 18.4. Distribution of $^{244}CmCl_3$ inhaled by dogs (AMAD = 1.6 μm, administered with a CsCl vector). Terminal slopes were determined using later data points (not shown). (Data of McClellan et al., 1972)

LB_{10}) was much greater than expected [33% of the ILB is predicted by the lung model (ICRP, 1966)]. By definition, 10% of the ILB remanied in the lungs, and less than 5 to 8% of ILB was probably eliminated from the lungs by a clearance mechanism coupled to gastrointestinal tract excretion (the lung model predicts 67% of ILB elimination by such clearance mechanisms). The brief report by Buldakov et al. (1972) of their studies of the inhalation of $^{241}Am(NO_3)_3$ aerosols by mongrel dogs did not provide enough details to be included in Table 18.9. However, they reported that elimination of ^{241}Am from the lungs was 96% complete by 414 days after exposure at which time skeleton and liver combined contained 82% "of the initial deposit".

7. Inhalation and Intratracheal Instillation Studies in Rats

Nénot et al. (1971a–c) exposed rats by inhalation to a mist containing $^{241}Am(NO_3)_3$. The mean diameter of the particles was stated to be less than 5 μm, but the diameters were not measured. The deposition parameters were calculated for this aerosol (shown in Table 18.8) using the ICRP lung model (ICRP, 1966) and an average particle diameter of 2.5-μm AMAD, but the biological behavior was more like that of a particle of 1.6-μm AMAD.

Nénot et al. (1971a–c) defined ILB as

$$\text{ILB} = \text{IBB} - \sum_{t=0}^{t=3} \text{Fe} \qquad \text{Eq. (4)}$$

Table 18.8. Deposition of actinide aerosols in the respiratory tract of rat and beagle dog compared to the deposition predicted by the report of the Task Group on Lung Dynamics (ICRP, 1966)

Compound	Animal	Particle size AMAD (μm)	IBB[a] / Inhaled (%)	ILB[b] / Inhaled (%)	ILB / IBB (%)	Reference
$^{241}Am(NO_3)_3$	rat	<5[c]	—	—	37	NÉNOT et al. (1971 a–c)
ICRP, predicted[aa]		2.5	82	19	24	
$^{241}AmO_2$	dog	0.9	75	45	60	R. G. THOMAS et al. (1972)
ICRP, predicted		1.0	42	26	62	
$^{244}CmCl_3$	dog	1.6	64	33	52	MCCLELLAN et al. (1972)
ICRP, predicted		1.5	62	25	41	
$^{244}CmO_{1.73}$	dog	0.5	35	27	77	MCCLELLAN et al. (1972)
ICRP, predicted		0.5	38	28	73	

[a] IBB, initial body burden; the total amount in the animal at $t=0$.

[b] ILB, initial lung burden; the total amount deposited in the lungs, less (by convention) the fraction cleared promptly via pathway f (ICRP, 1966).

[c] Particles stated to be less than 5 μm in diameter.

[aa] All predicted values are means of the top and bottom of the range of the appropriate curve of depositon vs. particle diameter shown in Fig. 11 of the ICRP report (1966).

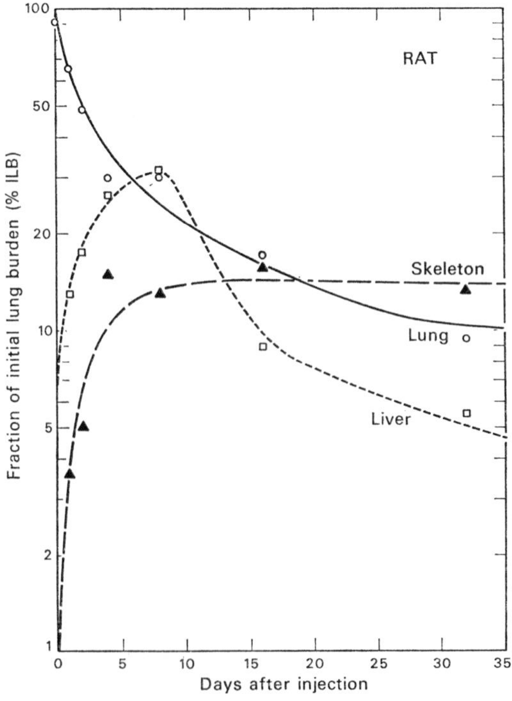

Fig. 18.5. Distribution of $^{241}AmCl_3$ instilled in the lungs of rats. Initial lung burden = initial body burden corrected for prompt regurgitation of 23% of the administered dose. Terminal slopes were determined using later data points (not shown). (Data of ZALIKIN et al., 1968)

where Fe is fecal excretion, and t is days after exposure. The IBB was determined by whole-body counting immediately after cessation of exposure. Tissue content and excretion of the inhaled $^{241}Am(NO_3)_3$ aerosol were reported as percent of the administered dose (NÉNOT et al., 1971c), which is assumed to be equal to the IBB. Fecal excretion data were available only for the first day postexposure. For the present calculations, the ILB was assumed to be equal to the IBB less the ^{241}Am excreted in the feces during the first day (day-1 feces also included the ^{241}Am in the contents of the gastrointestinal tract which would have been discharged on day 2). The data calculated in this way compared favorably to other results expressed by these authors as a percent of the ILB (NÉNOT et al., 1971a, b).

Experiments in which solutions or suspensions are instilled directly into the lungs are useful in the determination of the solubilities of various compounds in the lung. Lung retention and tissue uptake of chlorides and nitrates of ^{241}Am (ZALIKIN et al., 1968; TSEVELEVA and YEROKHIN, 1969) and ^{253}Es (HUNGATE et al., 1972) have been studied after instillation of slightly acid solutions in the lungs of rats. The data from these investigations and from the inhalation studies mentioned above have been recalculated[11] in terms of the percent ILB. Some of the results of a study in which $^{241}AmCl_3$ was administered to rats by intratracheal instillation (ZALIKIN et al., 1968) appear in Fig. 18.5.

8. Absorption of Actinides from the Lungs of Rats after Inhalation or Instillation

Assessment of absorption of actinide compounds from rat lung is complicated. Of the americium or curium isotopes absorbed by the rat, a large percentage (45 to 65%) is deposited in the liver and is almost quantitatively excreted in the feces within 30 days (SCOTT et al., 1948a, 1949; DURBIN et al., 1967). A large fraction of $^{253}EsCl_3$, about 40% of that in the blood, is promptly excreted by kidney (HUNGATE et al., 1972). If only the tissue content is measured, and urinary and fecal excretion of absorbed nuclide are neglected, total absorption will be seriously underestimated. Urinary excretion of $^{253}EsCl_3$ can be estimated, if one assumes that the rat kidney excretes the same fraction of $^{253}EsCl_3$ after it is absorbed, whether from the lung or from parenteral injection,

$$\text{Ab}\,(\%) = \sum \text{tissues}\,(\%) + 0.4\ \text{Ab}. \qquad \text{Eq. (5)}$$

A more complicated method was required to describe the simultaneous deposition in, and elimination of actinide nuclide from the liver, and to evaluate the fraction of the total quantity of nuclide that was absorbed from the lung and that had subsequently passed through the rat liver. A description of the mathematics and of the calculations are given in Appendix B.

Table 18.9 shows the time required to clear 90% of the ILB (LB_{10}); the fractions of ILB in the tissues and excreta, and the total absorption into the body at $t = LB_{10}$; and an estimate of the fraction of ILB cleared by phagocytosis coupled to gastrointestinal tract excretion. The mean LB_{10} was 30 days in rats for ^{241}Am instilled as a solution or inhaled as droplets of nitrates or chlorides. Reduction of the lung burden in 30 days to 10% of its initial value corresponds to an effective or average elimination half-time (T_{eff}) of 10 days. For elimination of $^{253}EsCl_3$ $T_{eff} = 6$ days. The elimination of transplutonium elements inhaled by dogs as dry particles was somewhat slower; $T_{eff} = 15$ days for $^{244}CmCl_3$ and $T_{eff} = 20$ days for $^{244}CmO_{1.73}$. In every experiment, elimination of trivalent actinides

11 The details of these calculations are given in Appendix A.

Table 18.9. Fate of the initial lung burden (ILB) of ^{241}Am, ^{244}Cm, and ^{253}Es compounds of the time required to eliminate 90% of ILB (LB_{10}) and the fractions of ILB leaving the

Compound	Animal	Mode of exposure	LB_{10} (days)	Absorbed (% ILB)[a]		
				skeleton	liver	kidney and spleen
^{241}Am citrate	rat	instilled[c]	—	—	—	—
$^{241}AmCl_3$	rat	instilled[c]	30	13	5.5	0.49
$^{241}Am(NO_3)_3$	rat	instilled[c]	<30	31	10.5	0.83
$^{253}EsCl_3$	rat	instilled[c]	20	30	2.0	1.4
$^{241}Am(NO_3)_3$	rat	inhaled	40	22	3.3	—
$^{241}AmO_2$	beagle	inhaled	<127	36	31	0.3
$^{244}CmCl_3$	beagle	inhaled	50	38	31	1.5
$^{244}CmO_{1.73}$	beagle	inhaled	64	34	36	1.3

[a] The absorbed fraction is the sum of the radioactivity in the tissues (except lung), the urine, and, in the case of the rat studies, the fecal excretion of absorbed nuclide (Fe_{end}). See Appendix B.

[b] Slow pathway (g) only. The prompt clearance pathway (f) is, by convention, not included in ILB in inhalation experiments.

from the animal lung was more efficient than was predicted by the lung model for Class W or Class Y compounds in man (ICRP, 1966). In addition, the lung clearance model, based largely on experience with compounds of plutonium, predicts that two-thirds of the ILB of these relatively insoluble materials will be eliminated by clearance processes and one-third by absorption. The experiments do not agree with the model. Absorption into the body accounted for 75 to 85% of the americium compounds and 90% of the $^{253}EsCl_3$ initially present in the lung. Less than 15% (5 to 15%) could be presumed to have been cleared and excreted in the feces.

9. ^{241}Am in the Human Lung

In reporting accidental inhalations of ^{241}Am compounds, several authors have commented on the disappearance of ^{241}Am from the lung and its subsequent redistribution in the body (MAY, 1968; RUNDO et al., 1970; ROSEN et al., 1971). In all cases, either the contamination levels were very low or chelation therapy was started as soon as possible, so these results are difficult to evaluate. It appears, however, that ^{241}Am compounds inhaled as dry dusts left the human lung more rapidly than did dusts of plutonium compounds, but perhaps not as rapidly as was measured in the animal studies discussed above. In vivo measurements of small amounts of ^{241}Am were imprecise, but it seems clear that a large fraction of the ^{241}Am leaving the human lung was absorbed into the body.

The ^{241}Am contaminant of modern plutonium[12] has been used to estimate, by in vivo measurement, ^{239}Pu lung burdens in dogs (MORROW et al., 1967; BAIR, 1970) and in man (LAGERQUIST et al., 1967). The body absorption and lung clearance results shown in Table 18.9 were obtained in experiments using total masses of transplutonium compounds of 10^{-6} to 10^{-9} g. If ^{241}Am is present as a minor contaminant of an insoluble matrix, the behavior of ^{241}Am in the lung seems to be the same as that of the matrix. Neither BAIR (1970) nor MORROW

12 ^{241}Am is the daughter of 13-year ^{241}Pu which is produced from ^{239}Pu by successive neutron captures. The amount of ^{241}Am present in a ^{239}Pu sample depends on the neutron flux of the reactor, the length of bombardment, and the elapsed time since the isotopes were separated.

administered to rats or dogs by intratracheal instillation or by inhalation expressed in terms lung by absorption (pathways e and h, Fig. 18.3) and by clearance (pathway g) ICRP (1966)

			Retained or cleared (% ILB)		Reference
urine	Fe_{end}	total	lung	cleared[b]	
—	—	>76	—	—	FOREMAN (1962b)
4[aa]	54	77	9.4	<13	ZALIKIN et al. (1968)
8.8[ab]	42	86	5.6	<2.0	TSEVELEVA and YEROKHIN (1969)
35[ab]	4.4	88[ab]	11	0	HUNGATE et al. (1972)
12	39	76	10	13.4	NÉNOT et al. (1971a–c)
17	—	>85	7.4	<7.8	R. G. THOMAS et al. (1972)
13	—	>84	10	<5.6	MCCLELLAN et al. (1972)
13	—	>84	10	<5.4	MCCLELLAN et al. (1972)

[c] In the instillation studies ILB is assumed to equal IBB. In two of the three studies cited here (ZALIKIN et al., 1968; TSEVELEVA and YEROKHIN, 1969) it was necessary to correct the data for the 25 to 35% of the administered dose that leaked out of the lung immediately after administration. See Appendix A.

[aa] Estimated from parenteral administration of same compound.

[ab] Absorbed total includes a soft tissue component of about 16%.

et al. (1967) were able to detect dissociation of ^{241}Am from inhaled $^{239}PuO_2$ or $^{239}PuF_4$, but absorption of ^{241}Am in excess of ^{239}Pu was detected if the inhaled matrix was the more soluble $^{239}Pu(NO_3)_4$.

V. Circulatory Transport

Many multicharged cations are transported in mammalian plasma in association with proteins. About one-half of serum calcium is associated with protein, mainly albumin (MCLEAN and HASTINGS, 1935). Almost all of plasma iron is bound to the β_1-globulin, transferrin (KATZ, 1970). More than 90% of plasma copper is bound to the α_2-globulin, ceruloplasmin (ADELSTEIN and VALLEE, 1962). Plasma zinc is associated with several proteins in varying amounts and with variable stabilities, but most of the plasma zinc is rather firmly bound to globulins (VALLEE, 1962).

A. Circulatory Clearance

In vivo and in vitro studies with rat, dog, horse, and human plasma have demonstrated conclusively that in normal or iron-deficient plasma, Pu(IV) is bound firmly and almost quantitatively to transferrin, and that Pu(IV) and Fe(III) occupy the same binding sites (POPPLEWELL and BOOCOCK, 1967; STOVER et al., 1968; TURNER and TAYLOR, 1968a, b; STEVENS et al., 1968a). The slow disappearance from the blood of intravenously injected, soluble Pu(IV) (PAINTER et al., 1946; LANGHAM et al., 1951; STOVER et al., 1959) and its equally slow deposition in tissues (SCHUBERT et al., 1950; D. M. TAYLOR, 1962) are consequences of the stability of the Pu-transferrin complex in the circulation.

In contrast, clearance of the trivalent actinides from the blood is prompt, and tissue uptake is likewise prompt. As shown in Fig. 18.6, intravenously injected ^{241}Am or ^{244}Cm is 90% cleared from the blood in 0.25 to 1 hr and 99% cleared in 2.5 to 10 hr in rats, beagles, lactating Suffolk ewes, and cynomolgus monkeys. Circulatory clearance of ^{249}Cf and ^{252}Cf in beagles proceeds somewhat more slowly (Fig. 18.7).

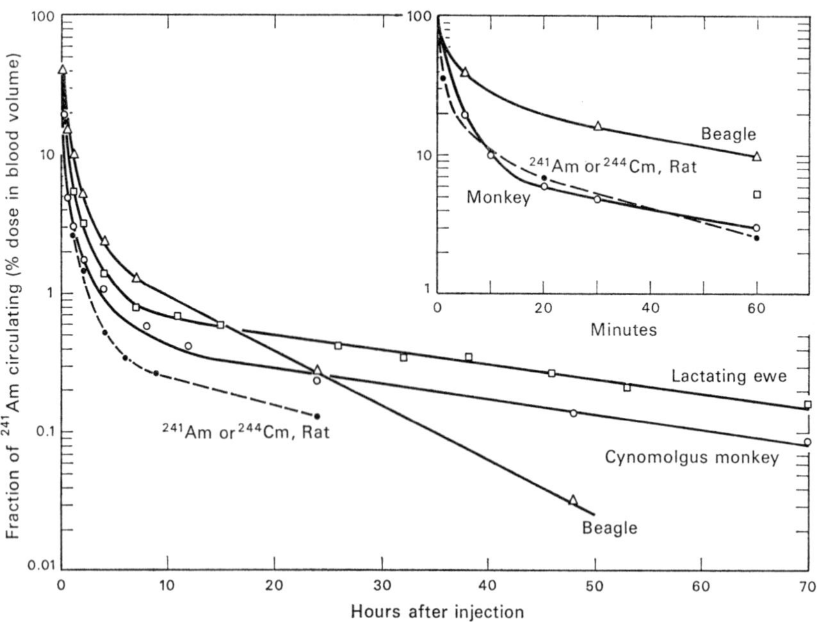

Fig. 18.6. Disappearance of ^{241}Am from the circulation after intravenous injection in beagles, BRUENGER et al. (1969a) nad STEVENS and BRUENGER (1972); lactating ewes, McCLELLAN et al. (1962a, b); young rats, TURNER and TAYLOR (1968b); and cynomolgus monkeys, DURBIN et al. (1972c)

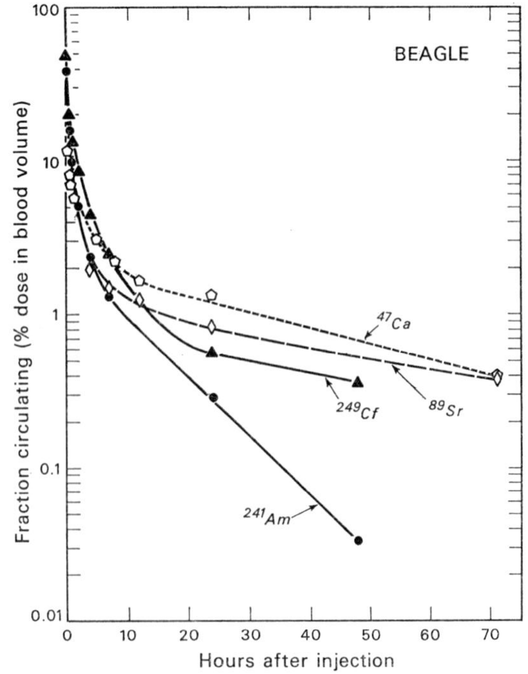

Fig. 18.7. Clearance from the blood of the actinides, ^{241}Am and ^{249}Cf, as compared to the alkaline earths, ^{47}Ca and ^{89}Sr, after intravenous injection in beagles. ^{241}Am and ^{249}Cf from BRUENGER et al. (1969a) and STEVENS and BRUENGER et al. (1972); ^{47}Ca from COLLINS et al. (1962); ^{89}Sr from SWIFT and PROSSER (1947)

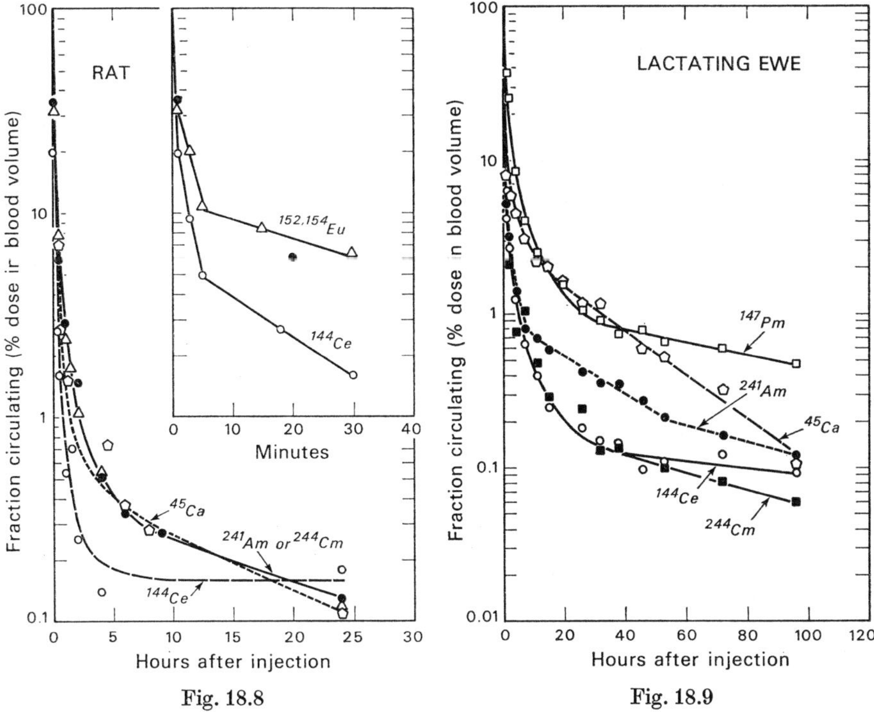

Fig. 18.8 Fig. 18.9

Fig. 18.8. Comparison of the blood clearance of the actinides, ^{241}Am or ^{244}Cm, the lanthanides, ^{144}Ce and ^{152,154}Eu, and the alkaline earth, ^{45}Ca, after intravenous injection in the rat. ^{241}Am and ^{244}Cm from TURNER and TAYLOR (1968b); ^{144}Ce and ^{152,154}Eu from DURBIN et al. (1955); ^{45}Ca from COPP et al. (1951)

Fig. 18.9. Comparison of the blood clearance of the actinides, ^{241}Am and ^{244}Cm, the lanthanides, ^{144}Ce and ^{147}Pm, and the alkaline earth, ^{45}Ca, after intravenous injection in the lactating Suffolk ewe. (Data from MCCLELLAN et al., 1962a, b, 1963)

The time course of the circulatory clearance of the trivalent actinides is similar to that of radioisotopes of calcium, strontium and several lanthanides in the beagle (Fig. 18.7), the rat (Fig. 18.8), and the lactating ewe (Fig. 18.9)[13]. Although the circulatory clearance rates of the trivalent actinides are faster than that of Pu(IV), which exists in plasma as a stable transferrin complex, they are much slower than the clearance rate of colloidal material (E. L. DOBSON et al., 1949; TURNER and TAYLOR, 1968b). After intravenous injection of slightly acid solutions of chlorides or nitrates of americium or curium in rats, dogs, or monkeys,

13 Data regarding blood clearance were reported from the following papers as percent of injected dose in plasma volume or blood volume, and values read from tables or figures could be used without further manipulation: ^{241}Am and ^{244}Cm in rat (D. M. TAYLOR, 1962; TURNER and TAYLOR, 1968b); ^{144}Ce, ^{152,154}Eu, and ^{160}Tb in rat (DURBIN et al., 1955); ^{45}Ca in rat (COPP et al., 1951); ^{241}Am and ^{249}Cf in the beagle (STEVENS and BRUENGER, 1972). In beagles, data for ^{89}Sr (SWIFT and PROSSER, 1947) and ^{47}Ca (COLLINS et al., 1962) were reported as percent dose per milliliter of blood or of plasma, respectively. Values for volumes of total plasma or total blood of the beagle given by STOVER et al. (1959) and J. H. DOUGHERTY and SEYMOUR (1957) were used to convert these data to percent of nuclide circulating. Data for all of the nuclides in the blood of sheep (MCCLELLAN et al., 1962a, b, 1963) were converted from percent dose per ml of plasma, or of plasma × body weight using a plasma volume of 3.75% of the body weight for the sheep (V. G. HORSTMANN and L. K. BUSTAD, unpublished) and a body weight of 56 kg for the adult ewe (GRIGORYAN, 1966).

neither the gross nor the microscopic deposition patterns resembled those of colloidal material (D. M. Taylor, 1962; Williams et al., 1961; Zalikin et al., 1968; Belyaev, 1969; Lloyd et al., 1970; Rudnitskaya and Moskalev, 1970; K. G. Scott, J. G. Hamilton, and M. H. Williams, unpublished). Usually the injections solutions of these high-specific-activity nuclides are so dilute (less than 1 μg/ml) and circulatory mixing is so rapid that intravascular aggregation of primary hydrolysis products is not likely to occur.

B. Protein Binding

Prompt blood clearance without intravascular formation of colloidal aggregates suggests that the trivalent actinides exist in plasma as complexes or loose associations with either inorganic plasma constituents or plasma proteins or both. The interaction of these elements with serum proteins has been studied using the techniques developed to elucidate the protein binding of Pu(IV).

Gross separation of a protein-bound nuclide from low molecular weight, largely inorganic compounds (gel filtration with Sephadex G 25 or G 50) indicates that at least 60% of americium, curium, or californium is associated with protein after incubation in vitro with human plasma (Popplewell and Boocock, 1967; Bruenger et al., 1969a; Stevens and Bruenger, 1972). Interpretation of the results for americium and curium is complicated, because they react strongly with the gel.

Separation of the plasma proteins by size and weight (gel filtration with Sephadex G 200 or G 100-G 200 combined) indicates that americium, curium, and californium elute from the gel primarily with the protein peak that contains albumin, transferrin and α-globulins (molecular weight about 70000). A smaller amount eluted with the protein peak containing lipoproteins and macroglobulins (molecular weight more than 200000) (Popplewell and Boocock, 1967; Bruenger et al., 1969a, 1971; Stevens and Bruenger, 1972).

Resolution of the albumin-transferrin fraction (anion-exchange chromatography with Sephadex DEAE A-50 cellulose) showed that the radioactivity peaks of americium, curium, and californium fall either between, and slightly overlapping, the small transferrin and the larger albumin peaks (Popplewell and Boocock, 1967), or are associated to some extent with both albumin and transferrin peaks (Bruenger et al., 1969a; Stevens and Bruenger, 1972). Association of americium with transferrin was more pronounced if competition with Fe(III) was reduced by the use of iron-deficient plasma (Bruenger et al., 1969a). The behavior of californium was more like that of Pu(IV); that is, more of the ^{249}Cf radioactivity was more closely associated with the transferrin peak. But unlike Pu(IV) the ^{249}Cf and transferrin peaks did not coincide (Stevens and Bruenger, 1972).

Two explanation have been offered: First, Popplewell and Boocock (1967) suggested that the actinides(III) form complexes with transferrin, but that the unstable complexes do not survive the separation processes intact. Second, Bruenger et al., (1969a) and Stevens and Bruenger (1972) suggested that the complexing protein may not be transferrin, but an as-yet-unidentified plasma constituent of about the same size as transferrin and with a slightly greater effective charge.

The status of americium and curium has been studied in the plasma of living animals (Turner and Taylor, 1968b; Bruenger et al., 1969a). An estimate of the extent of protein binding of intravenously injected ^{241}Am and ^{244}Cm in rats and ^{241}Am in dogs is shown in Table 18.10. One minute after injection, protein

Table 18.10. Protein-bound fraction of intravenously injected $^{241}Am(III)$, $^{244}Cm(III)$, and $^{239}Pu(IV)$ in the blood of rats, and $^{241}Am(III)$ in the blood of dogs. (Data of TURNER and TAYLOR, 1968b, and BRUENGER et al., 1969a)

Dose distribution (%) at time following injection	^{241}Am citrate		$^{241}Am(NO_3)_3$ (rat)	$^{244}Cm(NO_3)_3$ (rat)	^{239}Pu citrate (rat)
	(rat)	(dog)			
1 minute					
% dose in plasma	14.9	65	45.4	45.2	92.4
% dose bound	5.6	36.4	5.9	26.0	67.0
% dose unbound	9.3	28.6	39.5	19.2	25.4
10 minutes					
% dose in plasma	—	28.0	—	—	—
% dose bound	—	17.9	—	—	—
% dose unbound	—	10.7	—	—	—
1 hour					
% dose in plasma	2.2	9.8	1.4	4.3	68.9
% dose bound	2.1	(9.3)[a]	(1.3)[a]	4.3	58.4
% dose unbound	0.1	(0.52)[a]	(0.1)[a]	0	10.5
Initial rate of clearance (1 min to 1 hour)					
Bound (%/min)	1.6	2.3	2.5	3.0	0.22
Unbound (%/min)	7.5	6.7	6.8	>10.0	1.5

[a] Estimated from ^{241}Am citrate in rat.

binding of the actinides(III) ranged from 13 to 58% of the circulating nuclide compared with 72.5% of Pu(IV) protein-bound. After 1 hr, 85 to 95% of americium or curium still circulating was, like Pu(IV), associated with proteins. As expected, the protein-bound fraction of the actinides(III) and Pu(IV) left the blood more slowly than the fraction which was not associated with protein. During the first hour after injection, protein-bound and unbound ^{241}Am left the circulation at rates of 1.6 and 6.7%/min, respectively; protein-bound and unbound ^{244}Cm left the circulation at rates of 3 and 10%/min, respectively. Clearance of both protein-bound and unbound Pu(IV) was much slower—0.22 and 1.5%/min, respectively.

VI. Deposition Kinetics

Tissue deposition kinetics of the transplutonium elements have not been studied in detail, but the general features of an uptake model can be obtained by combining the available data with what is known about the accumulation of the lanthanides and Pu(IV).

A. Kinetic Experiments with ^{241}Am

D. M. TAYLOR (1962) demonstrated that the uptake pattern of intravenously injected $^{241}Am(NO_3)_3$ in the ends and shaft of the femur of 7-week-old male rats was more like that of Ca(II) than that of Pu(IV). BELYAEV (1969) injected $^{241}AmCl_3$ intravenously in 160 to 180 g male rats[14] and measured the simultaneous disappearance from blood and uptake in liver and skeleton.

14 Growth curves of two albino strains and one hooded strain indicate that the rats used by BELYAEV (1969) were 6.5- to 7.5-weeks old.

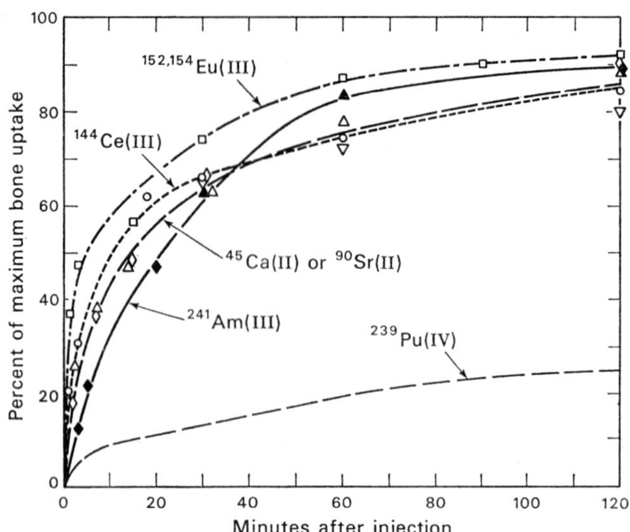

Fig. 18.10. Comparison of the deposition in the bone of rats of intravenously injected $^{241}Am(NO_3)_3$, solid triangle ▲, D. M. TAYLOR (1962); $^{241}AmCl_3$, solid diamond ◆, BELYAEV (1969); ^{144}Ce citrate, open circle ○, and $^{152,154}Eu$ citrate, open rectangle □, DURBIN et al. (1955); ^{45}Ca citrate, open diamond ◊, L. PALMER and E. PALMER (unpublished); $^{45}Ca(NO_3)_2$, inverted open triangle ▽, D. M. TAYLOR (1962); ^{90}Sr citrate, open triangle △, G. D. BARR and P. W. DURBIN (unpublished); and $^{239}Pu(IV)$ citrate, SCHUBERT et al. (1950) recalculated by DURBIN et al. (1972a)

The Am(III) bone-uptake data from these two experiments are shown in Fig. 18.10 along with comparable data for Ca(II), Sr(II), Ce(III), Eu(III), and Pu(IV). In each case the data have been recalculated as a percent of the maximum bone uptake, in order to permit direct comparisons of results originally reported in several different ways. As expected, because experimental animals of the same sex and age were used, the results of the two ^{241}Am studies were superimposable, and the data have been combined.

B. Recalculation of ^{241}Am Kinetic Data

In addition to information on uptake of Am(III) in bone and liver, the BELYAEV (1969) study provides an opportunity to estimate the movement of Am(III) in extracellular fluid (ECF). Values for ^{241}Am in blood, liver, and skeleton at times ranging from 3 min to 16 hr after intravenous injection, were read from the published curves. Liver and skeleton values were corrected for the ^{241}Am in their contained plasma as described by DURBIN et al. (1972a) assuming that skeleton and liver constitute 9 and 3% of the body weight, respectively. The corrected data were replotted linearly (not shown) to interpolate additional values intermediate between the widely spaced data points.

Assuming that injection and radioanalytical techniques were accurate, the ^{241}Am content of soft tissues plus extracellular fluid (St + ECF) was estimated as

$$(\mathrm{St} + \mathrm{ECF})\ (\%) = 100 - (\mathrm{Li}_t + \mathrm{Sk}_t + \mathrm{Bl}_t + \mathrm{Ex}_t), \qquad \text{Eq. (6)}$$

where Li_t, Sk_t, Bl_t, and Ex_t are the percent of injected dose in liver, skeleton, blood, and accumulated excreta, respectively, at time t after injection. In this experiment early excretion was not measured, so (ST + ECF) also includes the small amount of ^{241}Am excreted.

Accumulation of ^{241}Am in liver and skeleton was calculated according to the method described by COMAR (1955) in which A_t and A_{eq} represent tissue accumulation at time t and maximum tissue uptake, respectively:

$$\text{Accumulation} = A_t/A_{eq}. \qquad \text{Eq. (7)}$$

C. Accumulation and Clearance Equations

In BELYAEV'S (1969) experiment, clearance of ^{241}Am from blood and (ST + ECF) and its accumulation in liver and skeleton can be described by the semilogarithmic curves in Fig. 18.11 and by the following equations (t is min):

$$\text{Blood}_t\,(\%) = 62\,e^{-0.33t} + 32\,e^{-0.024t} + 5\,e^{-0.0035t}; \qquad \text{Eq. (8)}$$

Uptake in soft tissues and extracellular fluid, $t = 0$ to $t = 5$ min

$$(\text{ST} + \text{ECF})\,(\%) = 43\,(1 - e^{-0.46t}); \qquad \text{Eq. (9)}$$

Clearance of soft tissues and extracellular fluid, $t > 5$ min

$$(\text{ST} + \text{ECF})\,(\%) = 23\,e^{-0.06t} + 14\,e^{-0.003t} + 9; \qquad \text{Eq. (10)}$$

$$\text{Liver}_t\,(\%) = 20\,(1 - e^{-0.14t}) + 19\,(1 - e^{-0.019t}) + 22\,(1 - e^{-0.0033t}); \qquad \text{Eq. (11)}$$

$$\text{Skeleton}_t\,(\%) = 8\,(1 - e^{-0.058t}) + 16\,(1 - e^{-0.0033t}) + 6\,(1 - e^{-0.0034t}). \qquad \text{Eq. (12)}$$

The equations describing accumulation of Am(III) in liver and skeleton, Eq. (11) and (12) respectively, have rate terms in common with each other and also in common with the blood- and (ST + ECF)-clearance equations, Eqs. (8) and (10), respectively. The common rate terms in the Am(III) liver and bone accumulation equations indicate that Am(III) is being deposited in both tissues

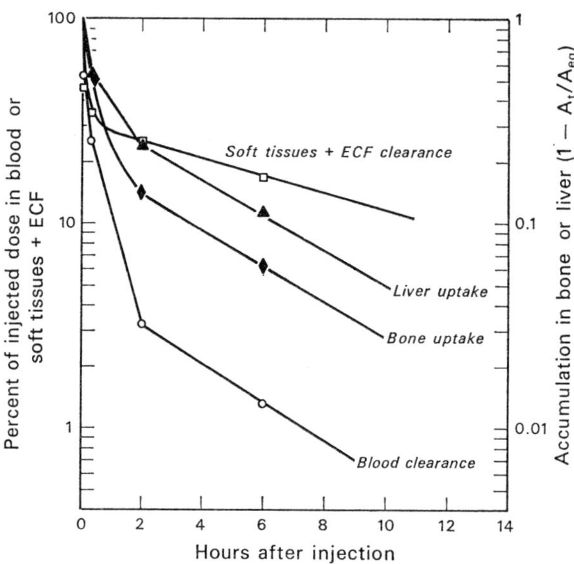

Fig. 18.11. Clearance of ^{241}Am from the blood or soft tissues plus extracellular fluid (ECF) and accumulation of ^{241}Am in liver and bone. Young male rats injected intravenously with $^{241}AmCl_3$; data of BELYAEV (1969) recalculated as described in text

from a common pool. The common rate terms in the transport equations, blood and (ST + ECF), and the accumulation equations, liver and skeleton, suggest that the rate-limiting reactions are in the transport system rather than in the liver and skeleton. The rate of accumulation of the transplutonium elements in the major organs of deposition seems to depend on (a) the rates of transfer of protein-bound and unbound nuclide between blood and ECF, and (b) the rates of protein binding and release in blood and ECF.

D. Kinetic Model for Uptake of Multicharged Cations

A kinetic model embodying the above features has been proposed to describe the transport and deposition of Pu(IV) in the rat (DURBIN et al., 1972a). The

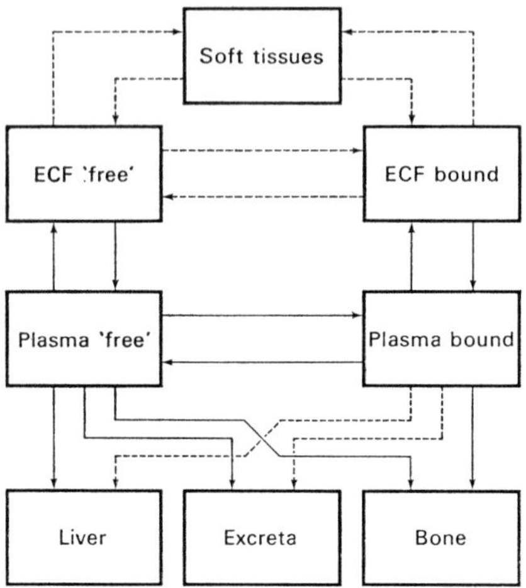

Fig. 18.12. Kinetic model of initial transport and deposition of trivalent elements

model, shown in Fig. 18.12, consists of four interconnected transport compartments, protein-bound and unbound, both in blood and in ECF; and four effectively nonreturning accumulation compartments, liver, skeleton, soft tissue, and excretion.

1. Movement in Extracellular Fluid

As shown in Table 18.11, within the first 5 min after intravenous injection of Ca(II), Ce(III), Eu(III), Am(III), and Pu(IV), nearly one-half of the injected nuclide leaves the circulation but cannot be accounted for by excretion or by uptake in liver or bone. Balance studies with ^{45}Ca (L. PALMER and E. PALMER, unpublished), ^{90}Sr (G. D. BARR and P. W. DURBIN, unpublished), and several lanthanides (DURBIN et al., 1955) demonstrated that the missing nuclide was present in soft tissues, though only transiently. Only 20 to 25% of the nuclide that was present in soft tissue at 5 min after injection remained for 24 hr, indicating that the nuclides were associated with ECF rather than with cellular components.

The amount of nuclide leaking into the ECF depends on the stability of protein binding. The more stable the nuclide-protein moiety, the more rapidly it forms—the more will remain in the circulation, and the less will pass into ECF

Table 18.11. Distribution of ^{45}Ca, ^{144}Ce, $^{152,154}Eu$, ^{241}Am, and ^{239}Pu 5 minutes after intravenous injection in the rat. (Percent of injected dose in tissue)

Nuclide (form administered)	$^{45}Ca(II)$ [a] chloride	$^{144}Ce(III)$ [b] citrate	$^{152,154}Eu(III)$ [b] citrate	$^{241}Am(III)$ [c] chloride	$^{239}Pu(IV)$ [aa] citrate
Blood volume	9	5	11	37	52
Skeleton [ab]	26	13	17	6	4
Liver [ab]	0	29	16	14.5	3
Excreted	0	0.8	0.7	43 (Excreted + Extracellular fluid and soft tissue)	0.1
Extracellular fluid and soft tissue [ac]	65	52	55		40

[a] ^{45}Ca, 60-day-old, female, Long-Evans (Thomson, 1953)
[b] ^{144}Ce and $^{152,154}Eu$, 3- to 4-month-old, female, Sprague-Dawley (Durbin et al., 1955).
[c] ^{241}Am, 170-g male, Wistar (Belyaev, 1969). Skeleton and liver weights assumed to be 9 and 3% of the body weight, respectively.
[aa] ^{239}Pu, 250-g female, Sprague-Dawley (Schubert et al., 1950; recalculated by Durbin et al., 1972a).
[ab] Nuclide content of skeleton and liver corrected for nuclide in contained plasma as described by Durbin et al. (1972a).
[ac] Calculated by difference from 100%.

or be immediately deposited in liver and bone. Pu(IV)-transferrin is an example of a stable, rapidly-forming, nuclide-protein complex. Leakage of Pu(IV) into the ECF has been calculated to range from 20 to 40% of the injected dose depending upon whether $Pu(NO_3)_4$ or Pu(IV) citrate, respectively, was injected (Turner and Taylor, 1968b; Durbin et al., 1972a). At the other extreme are Ca(II) and Ce(III), which are only loosely and transiently associated with serum proteins: 50 to 60% of the injected dose of these nuclides passes into the ECF in the first 5 min after intravenous injection.

The Pu-transferrin complex is so stable that the degree of protein binding was the same after injection of $Pu(NO_3)_4$ or Pu(IV) citrate (Turner and Taylor, 1968b). In the case of the trivalent actinides, the serum protein complex (formed most likely with transferrin) or complexes are much less stable. In the rat there was less protein binding of ^{241}Am injected intravenously as a citrate complex, than of either ^{241}Am or ^{244}Cm injected as nitrates.

Clearance of Am(III) from the (ST + ECF) compartment is represented by at least two exponential terms and one constant term (Eq. 10). The rapidly disappearing component has a rate constant of 6%/min. The rate constant of the same term in the (ST + ECF)-clearance equations of Ca(II), Ce(III), and Eu(III) ranged from 2.5 to 3.2%/min (see references cited in Table 18.11). The slower component of Eq. (10) has a rate constant of 0.3%/min compared to a range of 0.12 to 0.22%/min for the other elements listed above.

Although the protein in the ECF is quite dilute, it includes about 50% of the human transferrin (Katz, 1970); thus, protein binding is possible in the ECF as well as in the blood. The components of the (ST + ECF)-clearance equation may represent the separate return to the circulation of unbound and protein-bound Am(III), respectively. Multiple rates of ECF clearance may also be the result of the many different rates of ECF and lymph flow in the various tissues (Winchell et al., 1964). The real situation is probably a combination of free and protein-bound Am(III) returning from a variety of ECF sites at a variety of rates.

Tseveleva and Yerokhin (1969) administered $^{241}Am(NO_3)_3$ intraperitoneally—effectively placing all of the ^{241}Am in the ECF-unbound compartment at

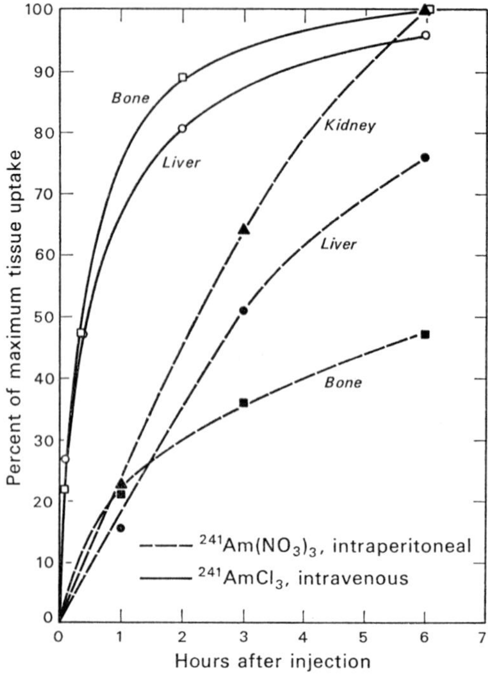

Fig. 18.13. Accumulation of ^{241}Am in the skeleton, liver, and kidneys of rats after intravenous injection of $^{241}AmCl_3$ (Belyaev, 1969), or intraperitoneal injection of $^{241}Am(NO_3)_3$ (Tseveleva and Yerokhin, 1969). Data are expressed as percent of the maximum uptake in each tissue

$t=0$. The liver uptake data (shown also in Fig. 18.13) were recalculated[15] as described above, yielding an accumulation equation (t is in min),

$$\text{Liver}_t\,(\%) = 16\,(1-e^{-0.0088\,t}) + 49\,(1-e^{-0.0022\,t}). \qquad \text{Eq. (13)}$$

The slower of the two rate constants in Eq. (13) (0.22%/min), obtained in an experiment in which all of the Am(III) was initially in the ECF, is very nearly the same as the slowest rate constant of the clearance equation of the (ST + ECF) compartment (0.3%/min, Eq. 10) determined in an experiment in which $^{241}AmCl_3$ was injected intravenously, placing 100% of Am(III) in the unbound compartment of plasma at $t=0$ (Belyaev, 1969). Thus, there is independent support for the view that this term is associated with return of the nuclide from the ECF to the circulation.

2. Deposition of Am (III) in Bone and Liver

The bone uptake of Am(III) (shown in Fig. 18.10) occupies a position closer to the rapid accumulation of the lanthanides and alkaline earths than to the slowly accumulating Pu(IV). The difference between bone uptake of Am(III) and the other (II) and (III) elements shown was greatest during the first few minutes after injection. Five min after injection Am(III) uptake in bone was only 19% of maximum uptake compared with 38 to 50% of maximum uptake calculated

15 The bone data were not amenable to recalculation, because A_{eq} could not be determined with sufficient precision.

Table 18.12. Distribution of isotopes of the transplutonium elements[a] in the adult rat after intravenous or intramuscular injection

Tissue	Percent of injected or absorbed dose[b] at days after injection										
	^{241}Am[b]			^{242}Cm[b]			^{249}Bk	$^{249-252}Cf$[b]		^{253}Es	
	4 days	15 days	64 days	4 days	16 days	64 days	21 days	4 days	91 days	7 days	24 days[b]
Skeleton[c]	33.8	34.0	32.7	30.7	30.2	27.2	38.0	61.6	62.3	68.8	56.9
Incisors (four)	1.4	1.4	0.1	—	—	—	—	2.7	0.2	—	—
Liver	41.8	10.5	0.6	35.8	20.0	2.3	4.0	13.9	1.2	7.8	1.7
Kidneys	1.2	0.7	0.6	1.5	1.1	0.8	1.3	1.3	0.5	—[aa]	—[aa]
Soft tissue balance	1.5	1.0	0.7	6.8	5.1	2.5	11.5	2.2	1.8	2.0	9.4
G.I. tract and contents	2.1	1.7	0.2	5.7	1.6	0.2	—[aa]	1.7	0.4	—[aa]	—[aa]
Urine	10.0	10.2	11.5	1.2	3.6	67.0 (Urine and Feces)	12.6	7.3	10.7	10.9	16.4
Feces	7.5	36.2	53.8	18.2	38.4		12.6	9.5	22.7	10.5	15.6
Recovered	96.7	94.1	103.6	107.3	96.6	97.6	80.0	99.3	99.3	—	—
Unabsorbed	1.9	1.9	1.2	7.5	7.0	2.4	—	2.1	0.5	—	4.7
Femur, tibia-fibula (one each)[ab]	3.1	2.6	2.4	3.1	3.0	2.6	—	5.2	5.1	5.4	4.2

[a] ^{241}Am: DURBIN et al. (1967, 1972c); ^{242}Cm: SCOTT et al. (1949); ^{249}Bk: HUNGATE et al. (1972); $^{249-252}Cf$: DURBIN et al. (1972d); ^{253}Es: V. H. SMITH (1972b).
[b] Results of intramuscular studies are expressed as percent absorbed dose and are corrected to 100% recovery.
[c] Includes incisor teeth unless incisors are shown as a separate sample.
[aa] Included in soft tissue balance.
[ab] Included in total skeleton.

Table 18.13. Distribution of isotopes of the transplutonium elements[a] in the immature rat after intravenous or intraperitoneal injection

Tissue	Percent of injected dose at days after injection										
	^{241}Am (high dose)			^{241}Am (low dose)			^{242}Cm[b]		^{252}Cf		
	4 days	16 days	48 days	4 days	21 days	62 days	15 days	50 days	4 days	16 days	64 days
Skeleton[c]	30.8	34.7	45.5	47.8[aa]	41.6[aa]	39.4[aa]	34.3	34.5	56.3[ab]	55.6[ab]	53.4[ab]
Liver	38.3	12.6	11.9	21.9	12.8	12.7	18	5.5	4.3	2.5	0.6
Kidneys	2.7	1.6	2.5	1.0	0.5	0.5	1.0	0.9	1.6	0.6	0.6
Soft tissue balance	5.9	6.4	6.2	—	—	—	4.0	3.5	5.0	4.3	4.3
G.I. tract and contents	2.7	0.9	1.1	—	—	—	1.5	0.6	1.4	0.9	0.9
Urine	5.6	8.0	9.9	11.8	30.7	44.5	7.9	11.3	—	—	—
Feces	10.1	27.6	20.6	—	—	—	33.2	43.6	—	—	—
Recovered	96.5	91.7	97.6	82.2	87.3	94.1	88.0	100.5	—	—	—
Femur, tibia-fibula (one each)	—	—	—	4.1	3.5	3.3	3.0	3.5	5.5	5.4	5.2

[a] ^{241}Am (high dose): Langham and Carter (1951); ^{241}Am (low dose): D. M. Taylor et al. (1961); ^{242}Cm: K. G. Scott, M. H. Williams, and H. I. Johnson (unpublished); ^{252}Cf: Mewhinney et al. (1972).
[b] Results corrected to 100% recovery.
[c] Includes incisor teeth.
[aa] Estimated from the average of the following relationships between isotope content of whole skeleton and of leg bones: ^{241}Am, adult skel (%)/fem(%) = 23.1 and skel(%)/fem-tib(%) = 11.8; ^{242}Cm, 45 days old, skel(%)/fem-tib(%) = 10.8; ^{45}Ca, 43 days old, skel(%)/fem(%) = 22.2
[ab] Estimated by author, skel(%)/fem(%) = 18.7.

for the lanthanides and alkaline earths. After the first 30 min, except for Pu(IV), all of the bone uptake curves were quite close together. The lower, very early, bone uptake of Am(III) is probably the results of greater, early protein binding compared with the other (II) and (III) elements studied.

Liver accumulation of Am(III) apparently is also influenced by the somewhat more stable, early, protein binding of this element compared with the lanthanides and alkaline earths. Five min after injection, the Am(III) in the liver was 24% of the maximum uptake (see Fig. 18.13), and Eu(III) and Ce(III) in liver were 42 and 50%, respectively, of the maximum uptake (DURBIN et al., 1955).

VII. Distribution and Retention of Transplutonium Elements in the Skeleton

In all of the mammals studied thus far, the transplutonium elements are deposited primarily in liver and skeleton with lesser amounts initially present in kidney and spleen. The early distributions, 4 to 60 days after administration of soluble noncolloidal solutions of the transplutonium elements, are given in Tables 18.12 through 18.17: ^{241}Am, ^{242}Cm, ^{249}Bk, $^{249-252}$Cf, and ^{253}Es in adult rats—Table 18.12; ^{241}Am, ^{242}Cm, and ^{252}Cf in immature rats—Table 18.13; ^{243}Am, ^{252}Cf, and ^{253}Es in adult mice—Table 18.14; ^{241}Am and ^{252}Cf in adult Chinese hamsters—Table 18.15; ^{241}Am and ^{249}Cf or ^{252}Cf in adult beagles—Table 18.16; and ^{241}Am in adult cynomolgus monkeys and adult baboons—Table 18.17. All of the nuclide distribution data in Tables 18.12 through 18.17 are presented as either percent of injected dose (intravenous or intraperitoneal administration) or the equivalent, percent of absorbed dose (intramuscular administration). In the text the nuclide content of specific organs is given as percent of dose, and normalization to 100% absorption and 100% recovery is implied.

Table 18.14. Distribution of isotopes of the transplutonium elements[a] in the adult mouse after intramuscular injection

Tissue	Percent of absorbed dose at days after injection					
	^{243}Am		^{252}Cf		^{253}Es	
	4 days	14 days	4 days	14 days	4 days	14 days
Skeleton and teeth	~44[b]	~39[b]	~43[b]	~45[b]	46.5	39.7
Liver	29	19	7.5	1.5	3.3	1.8
Kidneys	1.4	0.3	—	—	—	0.2
Residual soft tissues[c]	~3[b]	~3[b]	~6[b]	~5[b]	3.2	3.9
Urine	21	22	5.8	6	30	26.7
Feces	6.6	16	18	21	1.4	2.0
Unabsorbed	0.5	0.8	~2.5[aa]	~2.5[aa]	3.5	2.7
Recovered	105.4	98.9	82.8	81	87.9	77.0
Hind leg bones	1.6	1.8	~2.5[aa]	~2.5[aa]	2.3	1.8
Carcass	46.9	41.8	49.0	50.0	49.7	43.6

[a] ^{243}Am: WRIGHT (1972); ^{252}Cf: PARKER et al. (1962); ^{253}Es: PARKER et al. (1972).
[b] Skeleton and residual soft tissues were assayed as a unit. Their separate ^{243}Am and ^{252}Cf contents were estimated assuming Sk_{est} (%) = (Carc-ST), where soft tissue (including G.I. tract) was assumed to be the mean of the measured soft tissue values in rats and hamsters; 3.3 and 2.7% of ^{243}Am, and 5.7 and 4.7% of ^{252}Cf at 4 and 14 days, respectively.
[c] Other soft tissues including G.I. tract and contents.
[aa] Data available for injected leg only, and one-half each assigned to the hind-leg bones and the unabsorbed fraction.

Table 18.15. Distribution of isotopes of the transplutonium elements[a] in the adult Chinese hamster after intraperitoneal injection

Tissue	Percent of injected dose at days after injection					
	^{241}Am			^{252}Cf		
	4 days	16 days	64 days	4 days[b]	16 days	64 days
Skeleton[c]	19	22	26	28	34	29
Liver	55	53	52	27	18	18
Kidneys	6	1	<1	7.7	1.5	0.9
Soft tissue balance	3	3	3	6[aa]	5[aa]	8[aa]
G.I. tract and contents	—	—	—	1.5	2.3	1.3
Whole body[ab]	83	79	81	70	61	57
Femur (one)	—	—	—	1.7	2.3	1.9

[a] ^{241}Am: McKAY et al. (1969, 1972); ^{252}Cf: MEWHINNEY et al. (1972).
[b] Values shown for 4 days are means of the published values for 2 and 8 days.
[c] Includes incisor teeth.
[aa] Soft tissue balance = (whole body)—(measured tissues).
[ab] Measured by whole-body counting.

Table 18.16. Distribution of isotopes of ^{241}Am and $^{249-252}$Cf in the young adult beagle after intravenous injection

Tissue	Percent of injected dose at days after injection			
	^{241}Am[a]		^{249}Cf or ^{252}Cf[b]	
	1 to 22 days	1510 to 1917 days	7 to 13 days	21 to 36 days
Skeleton	31.0	32.7	46.6	54.7
Liver	48.7	46.4	17.4	17.4
Kidneys	0.6	0.6	1.1	0.8
Spleen	0.03	0.06	0.2	0.1
Soft tissue balance	3.7	1.8	10.1	5.4
G.I. tract and contents	0.6	0.1	1.6	0.6
Urine	9.7	9.7[c]	15.9	16.4[c]
Feces	1.4	1.4[c]	7.3	7.5[c]
Recovered	95.7	93.5	100.2	102.9

[a] ^{241}Am: LLOYD et al. (1970, 1971, 1972c, d); 1 to 22 days, 4 dogs (T15W5.5, T32W5.5, T33W5.5, T16W5) 2.8 to 4.5 μCi/kg; 1510 to 1917 days, 3 dogs (F6W3, F7W2, M5W3) 0.095 to 0.31 μCi/kg.
[b] ^{249}Cf or ^{252}Cf: ATHERTON and LLOYD (1972); LLOYD et al. (1972b, e): 7 to 13 days, 2 dogs (T2G5, T2F5) 2.8 μCi/kg; 21 to 36 days, 2 dogs (T3G5, T1F5) 2.8 μCi/kg.
[c] Excretion collections to 21 days only.

A. Gross Deposition

1. Americium

The initial deposition of ^{241}Am or ^{243}Am in the adult mammalian skeleton ranges from a low of 19% of the dose in Chinese hamsters (McKAY et al., 1972) to a high of 58% in young rats (D. M. TAYLOR et al., 1961, recalculated). In 4 of the 6 species studied, initial skeletal deposition of ^{241}Am was about 30% of the dose. The low initial deposition of ^{241}Am in the Chinese hamster skeleton has not yet been clarified. Eleven metabolic studies of rats parenterally injected with

Table 18.17. Distribution of ^{241}Am in adult female primates[a] after intramuscular or intravenous injection

Tissue	Percent of injected or absorbed dose at days after injection				
	Cynomolgus monkey[b]			Baboon	
	8 days	21 to 42 days	81 to 112 days	32 days	86 days
Skeleton[c]	27.2	35.5	42.2	44.2[aa]	36.8[aa]
Liver	54.4	44.7	29.0	13.6[aa]	22.2[aa]
Kidneys	0.2	0.06	0.2	0.1	0.08
Soft tissue balance	6.6	4.0	3.6	11.0	9.0
G.I. tract and contents	1.3	0.9	0.4	0.9	1.6
Urine	8.3	11.0	11.6	9.6	13.8
Feces	1.0	3.0	12.6	6.5	14.4
Recovered	100.2	101.2	100.5	86.4	98.3
Unabsorbed	1.6	0.6	0.4	—	—
Femur, tibia, and fibula (one each)	1.8	2.3	2.8	2.5	2.6
Humerus (one)	0.9	1.2	1.2	1.6	1.0

[a] Cynomolgus monkey: Durbin et al. (1972c); baboon: Rosen et al. (1972).
[b] Results of intramuscular studies are expressed as percent absorbed dose corrected to 100% recovery.
[c] Includes teeth.
[aa] Apparent decrease in ^{241}Am contents in skeleton and its increase in liver are the results of greater initial skeleton uptake and less initial liver uptake by the animal sampled at 32 days. Partial-body counting and liver biopsies demonstrated constant skeleton retention and declining liver retention of ^{241}Am by both animals over the 86-day period of observation.

^{241}Am include measurement of the ^{241}Am content of the skeleton. Within a single experiment the standard deviations (S.D.) ranged from 9 to 18% of measured bone values. In the largest group studied, 20 young adult female rats that were killed 1 to 16 days after intramuscular injection of ^{241}Am citrate, deposition in the skeleton and teeth was $35.7 \pm 4.1\%$ (Durbin et al., 1967, 1972c). The S.D. was 11% of the measured mean for skeleton, and the range of ^{241}Am in the skeletons of individual rats was 28 to 45% of the dose.

Variability within a species is therefore, as great as or greater than the differences among species. In addition to individual variations among animals, other causes of the variability of results from the same species include the age of the animals and experimental and analytical techniques.

Three of the 11 ^{241}Am investigations in rats shown in Table 18.18 were balance studies; that is, all excreta were collected and all tissues were analyzed. Experimental and analytical techniques were refined until approximately 100% of the injected nuclide was recovered (Scott et al., 1948a; Langham and Carter, 1951; Durbin et al., 1967, 1972c). Reports of three other ^{241}Am studies with rats include enough information to allow material balances to be calculated (D. M. Taylor et al., 1961; Zalikin et al., 1968; Belyaev, 1969). The results of Nénot et al. (1971b) are not included with the balance studies; although excreta were collected, experimental recoveries were not reported, some tissues were not analyzed, and details of the calculation of ^{241}Am in skeleton were not included.

In five studies with rats, four of which can be classified as balance studies, ^{241}Am was measured in the whole skeleton or in selected bones and the residual carcass (Scott et al., 1948a; Langham and Carter, 1951; D. M. Taylor et al., 1961; Durbin et al., 1967; Hollins and Durakovic, 1972). In the other six

Table 18.18. Summary of initial bone and liver deposition of ^{241}Am in rats

Experiment	Chemical form and mode of administration[a]	Sex and age (days)	Days after injection	Percent injected or absorbed dose ± S.D.	
				bones and teeth	liver and feces and G.I. contents
Scott et al. (1948a)	Chlor., i.m.	♀, > 120	1[b]	23.2[c]	62.3
Langham and Carter (1951)	Chlor., i.v.	♂, 47	4[b]	31.9[c]	53.0
D. M. Taylor et al. (1961)	Citr., i.v.	♂, 49	4[b]	58.1 ± 5.2[c]	32.8 ± 9.1
Durbin et al. (1967)	Citr., i.m.	♀, 110	1[b]	35.7 ± 4.1[c]	49.4 ± 4.6
Zalikin et al. (1968)	Chlor., i.v.	♀, 150	1[b]	20.3	68.6
Belyaev (1969)	Chlor., i.v.	♂, ~49[aa]	1	30.4	61[ab]
Belyaev (1969)	Chlor., i.p.	♂, ~49[aa]	1[b]	22.4	68
Tseveleva and Yerokhin (1969)	Nitr., i.p.	—, —	1	7.1	64.8[ab]
Nénot et al. (1971b)	Nitr., i.v.	♂, ~60[aa]	1[b]	10	71
Nénot et al. (1971b)	Nitr., i.m.	♂, ~60[aa]	10[b]	30.6	39.8
Hollins and Durakovic (1972)	Citr., i.v.	♀, 140	7	19.4 ± 3.4[c]	—

[a] Chlor., chloride; nitr., nitrate; citr., citrate; i.m., intramuscular; i.v., intravenous; i.p., intraperitoneal.
[b] Whole body and excreta measured, data corrected for absorption and 100% recovery (see text).
[c] Whole skeleton or all bones measured. In all other studies, skeleton was apparently calculated from measurement of femur or other specific bones, but the details of those calculations were not reported.
[aa] Age estimated from Sprague-Dawley rat growth curve.
[ab] Liver only; excreta and G.I. contents not reported.

studies, skeletal ^{241}Am was apparently estimated as the product of a predetermined or assumed proportionality factor and the ^{241}Am content of one bone or a group of bones (Zalikin et al., 1968; Belyaev, 1969; Tseveleva and Yerokhin, 1969; Nénot et al., 1971b)[16]. Unfortunately, none of these reports include the details of any calculations. For the purposes of this review, ^{241}Am measured in whole skeleton has been considered more reliable.

There are no systematic studies of the variations of ^{241}Am metabolism as a function of injection route, chemical form, or experimental animal age. The 11 studies of the rat (Table 18.18) can be grouped roughly by animal age as follows: adult (Scott et al., 1948a; Durbin et al., 1967; Zalikin et al., 1968; Hollins and Durakovic, 1972), immature (Langham and Carter, 1951; D. M. Taylor et al., 1961; Belyaev, 1969), and age not reported—probably young or intermediate between young and adult (Tseveleva and Yerokhin, 1969; Nénot et al., 1971b).

In the most complete studies, between 80 and 90% of parenterally administered ^{241}Am is initially partitioned between skeleton and liver (Table 18.18).

16 Several experiments in this Laboratory yielded ^{241}Am measurements in whole skeleton and leg bones (K. G. Scott, unpublished; Durbin et al., 1972c). In adult female rats 1 to 180 days after intramuscular injection of $^{241}AmCl_3$ or ^{241}Am citrate, the ^{241}Am in whole skeleton was 23.2 times one femur, 26.4 times one tibia, and 12.4 times femur-tibia-fibula. In young male rats (D. M. Taylor et al., 1961), ^{241}Am in whole skeleton was 21.3 times one femur, 26.2 times one tibia-fibula, and 12.1 times femur-tibia-fibula.

For purposes of this discussion ^{241}Am in the gastrointestinal contents and passed feces was assumed to have originated in the liver, and was added to the measured liver value to estimate maximum tissue uptake. Within the age groupings above, it appears that initial skeletal deposition of ^{241}Am was greater (and deposition in liver smaller) (a) after administration of the ^{241}Am citrate complex, compared with nitrate or chloride; (b) when ^{241}Am was injected intramuscularly, compared with intravenous administration; (c) in rats with immature skeletons as compared with adults; (d) in those studies in which analytical techniques yielded close to 100% material recoveries, and (e) in those studies in which whole skeletal ^{241}Am was measured rather than estimated.

2. Curium

Skeletal deposition of ^{242}Cm in the rat is nearly the same as that of ^{241}Am. In both young and adult rats, 30.7 and 34.3%, respectively, of intramuscularly injected ^{242}Cm was initially deposited in the skeletons (Scott et al., 1949; K. G. Scott, M. H. Williams, and H. I. Johnson, unpublished). Distribution of curium in a larger species (dogs) has been studied only after inhalation (Table 18.9). Fifty to 64 days after beagles were exposed to an aerosol of $^{244}CmCl_3$ or $^{244}CmO_{1.73}$, their skeletons contained 38.5 and 33.8%, respectively, of the initial lung burden (McClellan et al., 1972); about 85% of the ILB had been absorbed, and skeletal burdens represented slightly more than 40% of the absorbed dose. In comparable inhalation studies with $^{241}AmO_2$, the skeleton of one beagle killed 127 days after exposure contained 36.2% of the ILB, about 42% of the absorbed dose (R. G. Thomas et al., 1972).

Nine studies of ^{242}Cm in rats are compiled in Table 18.19. All but two (Nénot et al., 1970) were complete balance studies in which ^{242}Cm was measured in the whole skeleton[17]. Like ^{241}Am in rats, 80 to 90% of parenterally administered

Table 18.19. Summary of deposition of ^{242}Cm in bone and liver of rats

Experiment	Chemical form and mode of administration[a]	Sex and age (days)	Days after injection	Percent injected or absorbed dose ± S.D.	
				bones and teeth	liver and feces and G.I. contents
Scott et al. (1949)	Chlor., i.m.	♀, ~110	1[b]	27.7[c]	60.9
K. G. Scott and M. H.	Chlor., i.v.	♂, >110	4[b]	25.5[c]	61.6
Williams (unpublished)	Chlor., i.m.	♂, ~60	15[b]	34.3[c]	52.8
Williams et al. (1961)	Chlor., i.v.	♀, ~110	1[b]	22.2 ± 2.0[c]	59.6 ± 1.2
Williams et al. (1961)	Chlor., i.m.	♀, ≥110	1[b]	28.5 ± 1.2[c]	50.0 ± 0.8
Williams et al. (1961)	Citr., i.v.	♀, ≥110	1[b]	24.3 ± 2.0[c]	58.5 ± 3.5
Williams et al. (1961)	Citr., i.m.	♀, ≥110	1[b]	27.8 ± 1.3[c]	53.5 ± 0.6
Nénot et al. (1970)	Nitr., i.v.	♂, ~60[aa]	1	20	68
Nénot et al. (1970)	Nitr., i.m.	♂, ~60[aa]	1	21	48

[a] Chlor., chloride; nitr., nitrate; citr., citrate; i.v., intravenous; i.m., intramuscular.
[b] Whole body and excreta measured; data corrected for absorption and 100% recovery. See text.
[c] Whole skeleton measured.
[aa] Age estimated from Sprague-Dawley rat growth curve.

17 In this Laboratory ^{242}Cm in the whole adult rat skeleton was 11.3 times the ^{242}Cm in one femur-tibia-fibula.

^{242}Cm is initially divided between liver and skeleton (Table 18.19). The initial partition of ^{242}Cm between these two tissues seems to be similarly influenced—but to a lesser degree—by the same factors that influence the initial distribution of ^{241}Am in this species (see Sec. VII.A.1, above). In adult female rats early skeletal deposition of ^{242}Cm was the same after intramuscular injection of either $^{242}CmCl_3$ or ^{242}Cm citrate, but was slightly reduced if either compound were given intravenously. Early skeletal deposition seems to be greater in the younger rats, but the effect is not certain, because the age of the young rats is not known exactly.

3. Berkelium

There is only one study of the metabolic behavior of berkelium. Hungate et al. (1972) reported that at 4 hr and 21 days after intravenous injection of $^{249}BkCl_3$ in rats, the skeleton contained 30 and 38% of the injected dose, respectively. The amount contained in the total skeleton was calculated using 20 times the measured value for one femur. The femur contents of ^{241}Am and $^{249-252}$Cf in the rat were 0.043 and 0.045 of the total skeletal nuclide, respectively. Using the average femur content of these two closely related elements, the calculated values for ^{249}Bk in the rat skeleton are 34.4 and 43.5% at 4 hr and 21 days after injection, respectively. The recalculated values for ^{249}Bk in skeleton were used to prepare Table 18.12.

Although excreta were collected and analyzed, as were most tissues, the calculated recovery of the injected ^{249}Bk 21 days after injection was only 84.5%. A low recovery implies that the stated values of all tissues including skeleton, are low, unless all of the losses were incurred in collection of excreta. The soft tissues (excluding liver) still contained 11.5% of the dose 21 days after injection, and undoubtedly contained a larger fraction of the dose immediately after injection. During the same period, the skeleton probably accumulated additional ^{249}Bk released from soft tissues, thus the amount of ^{249}Bk in the skeleton 21 days after injection may overestimate the initial deposition in bone.

4. Californium

Biological studies have been conducted with ^{249}Cf, ^{252}Cf, or a mixture of californium isotopes designated herein as $^{249-252}$Cf (the ^{250}Cf and ^{252}Cf isotopes contribute more than 75% of the weight of the original mixture). Even though isotopic designations have been retained for clarity, this usage does not imply any differences in the metabolic behavior of the different isotopes in the absence of radiation damage.

Initial skeletal deposition of californium (see Tables 18.12 through 18.17) ranged from 28% of ^{252}Cf in Chinese hamsters (Mewhinney et al., 1972) to 64.3% of $^{249-252}$Cf in adult rats (Durbin et al., 1972d). In all species examined to date, initial skeletal deposition of californium has been greater than that of americium, curium, or berkelium. In the three most thoroughly studied species, the early bone depositions of both californium and americium were in the order: rat > beagle > Chinese hamster. It is not known at the present time whether the larger amounts of californium and americium deposited in the skeletons of the rats (compared with other species) are caused by less efficient renal excretion or greater skeletal affinity or both.

In all of the studies to date, californium has been administered as a citrate complex, so there is no information on whether the nature of the compound influences distribution. The greater solubility of the hydrolysis products of californium compared with those of with americium and curium, suggest that the distribution pattern of californium will not be significantly changed by its introduction as a potentially hydrolyzable salt.

The data in Tables 18.12 and 18.13 indicate a somewhat lower initial bone deposition of ^{252}Cf in growing rats (56.3%) than of $^{249-252}$Cf in adults (64.3%)—the reverse of expectation. However, the amount of $^{249-252}$Cf in the hind-leg bones was 5.5% in the growing rats and 5.2% in the adults 4 days after injection. Some of the difference in total skeletal content may have resulted from the proportionality factor (18.7 × one femur) used by MEWHINNEY et al. (1972) to estimate total skeletal ^{252}Cf. In the adult rats $^{249-252}$Cf was measured in all of the bones, and the entire skeleton contained 21.8 times the $^{249-252}$Cf in one femur, 28.1 times that in one tibia, and 12.3 times that in the femur-tibia-fibula (DURBIN et al., 1972d).

5. Einsteinium

Some preliminary studies have been made of ^{253}Es metabolism in rats and mice. In both species the initial skeletal deposition of ^{253}Es was slightly greater than or equal to that of $^{249-252}$Cf (Tables 18.12 and 18.14). Seven days after intravenous injection of $^{253}Es(NO_3)_3$ the rat skeleton contained 68.8% of the dose (V. H. SMITH, 1972b) compared to 64.3% of $^{249-252}$Cf (DURBIN et al., 1972d). Four days after intramuscular injection of ^{253}Es citrate, skeletal deposition in the mouse was 46.5% compared to 43% of ^{252}Cf (PARKER et al., 1962, 1972).

When $^{253}EsCl_3$ was administered to rats, initial urinary excretion of ^{253}Es was higher and skeletal deposition lower than when $^{253}Es(NO_3)_3$ was injected. The ^{253}Es in the skeleton and in accumulated urine and feces was 31.3%, 41,3%, and 4.4%, respectively, 21 days after intravenous injection of $^{253}EsCl_3$ (HUNGATE et al., 1972). [Compare these results with the data shown in Table 18.12 for intravenously injected $^{253}Es(NO_3)_3$.] If the raw data from the $^{253}EsCl_3$ study are normalized to take account of the 4-fold greater urinary excretion of $^{253}EsCl_3$, the initial skeletal deposition becomes 50% of the normalized dose—closer to, but still less than the initial skeletal uptake of ^{253}Es administered as the nitrate.

The ^{253}Es content of several separate bones of the rat was measured 24 days after injection of $^{253}Es(NO_3)_3$, and the total ^{253}Es in the bony tissues was found to be 23.9 times the ^{253}Es content of one femur.

6. Trends in Skeletal Deposition of the Actinides

As noted in Sec. II.B, a study of the metabolism in rats of citrate complexes of the lanthanides (DURBIN et al., 1956) revealed several trends that could be related to the decrease in ionic radius that accompanies the filling of the $4f$-shell of electrons—greater initial skeletal uptake, greater early renal excretion, and reduced initial liver uptake. The metabolic behavior of ^{241}Am and ^{242}Cm resembled that of lanthanides of comparable ionic size (DURBIN, 1962). Two more actinides, ^{249}Bk and ^{253}Es, have recently been investigated, and more reliable information is now available for ^{241}Am, ^{242}Cm, and $^{249-252}$Cf. It seems appropriate to reexamine the dependence of the metabolic behavior of the trivalent actinides on their ionic radii.

Table 18.20 summarizes the initial partitions of the lanthanide and trivalent actinide elements among skeleton (including incisors), liver (L_{max}), and residual soft tissues (S.T.) of the rat. The data have been recalculated to reduce the variability caused by (a) differences in material recovery, (b) less than complete absorption in intramuscular studies, and (c) greater renal excretion of the heavier members of both series. In 16 of 19 studies, tissues distribution data were available

Table 18.20. Partition of trivalent lanthanide and actinide isotopes among bone (including teeth), liver (L_{max})[a], and residual soft tissues (S.T.) 4 days after injection into rats[b]

No. of f-electrons	Lanthanide series (4f)					Actinide series (5f)				
	symbol	ion[c] radius (Å)	nuclide content (% normalized dose)			symbol	ion[c] radius (Å)	nuclide content (% normalized dose)		
			bone	L_{max}[a]	S.T.			bone	L_{max}[a]	S.T.
1	Ce	1.03	30	63	8	Th	(1.08)	—	—	—
2	Pr	1.01	28	60	11	Pa	(1.05)	—	—	—
3	Nd	1.00	40	48	12	U	1.03	—	—	—
4	Pm	0.98	40	53	7	Np	1.01	—	—	—
5	Sm	0.96	38	55	7	Pu	1.00	—	—	—
6	Eu	0.95	43	43	14	Am	0.98	39	57	4
7	Gd	0.94	57	28	16	Cm	0.98	28	60	13
8	Tb	0.92	72	16	12	Bk	0.95	60	23	16
9	Dy	0.91	79	12	9	Cf	0.94	69	27	4
10	Ho	0.89	71	20	10	Es	0.93	77	20	2
11	Er	0.88	78	12	10	Fm	—	—	—	—
12	Tm	0.87	82	9	9	Md	—	—	—	—
13	Yb	0.86	72	12	16	No	—	—	—	—
14	Lu	0.85	80	12	8	Lr	—	—	—	—

[a] L_{max} defined by Eq. 23 (Sec. VIII.A.1.).
[b] Data have been corrected for material recovery and absorption (intramuscular studies), and have been normalized to take account of the variable urinary excretion using Eq. 14. Studies were 4 days except ^{242}Cm, ^{249}Bk, and ^{235}Es which were 1-, 21-, and 7-day studies, respectively. Lanthanide studies are those of Durbin et al. (1956); actinide studies, except ^{242}Cm, are those shown in Table 18.12; ^{242}Cm data of Williams et al. (1961).
[c] Ion radius calculated for the cubic sesquioxide; lanthanide series, Templeton and Dauben (1954); Pu(III), Am(III), Cm(III), Zachariasen (1954); Bk, Cf, Peterson (1967); Es, Fujita (1969) as calculated by B. B. Cunningham (private communication).

4 days after injection, and results have been expressed as percent of the body burden at 4 days (BB)[18].

The behavior of lanthanides and actinides with the same ionic radii, in chemical reactions that depend on the ionic radius, is similar (Cotton and Wilkinson 1966). On the basis of their chemistry and on the earlier observation of the behavior of ^{241}Am and ^{242}Cm, it was expected that the initial distributions of the heavier actinides would also resemble those of the lanthanides of the same ionic radii but with fewer f-electrons. It can be seen from Table 18.20 that the initial distribution pattern of ^{241}Am was, as expected, nearly the same as that of ^{147}Pm (same ionic radius, 0.98 Å and two less f-electrons). There is no ready explanation, except for the great stability of the $5f^7$ configuration of ^{242}Cm, to account for its unusual metabolic behavior which most closely resembled that of cerium and praseodymium (with larger ionic radii and 5 to 6 fewer f-electrons). The distributions of the three heavier actinides, ^{249}Bk, $^{249-252}$Cf, and ^{253}Es, most closely resembled lanthanides of slightly smaller ionic radii and one to two fewer f-electrons.

The trivalent lanthanides and actinides form stable complexes with many organic acids (Chap. 17, this volume). The formation constants of the complexes

18 $$BB = Sk + L_{max} + S.T. = 100 - \sum U \qquad \text{Eq. (14)}$$

where BB is the body burden 4 days after injection corrected for material recovery and absorption; Sk and S.T. are the skeletal and residual soft tissue nuclide contents at day 4; $\sum U$ is the amount of nuclide excreted in urine during the first 4 days after injection, and L_{max} is the maximum liver uptake calculated from Eq. 23 (Sec. VIII.A.1.).

of the trivalent elements and the aminopolycarboxylic acids, EDTA, DCTA, and DTPA, have been measured (lanthanides, MACKEY et al., 1962; MOELLER and THOMPSON, 1962; MOELLER and HSEU, 1962: actinides, FUGER, 1958, 1961; BAYBARZ, 1965, 1966). The stabilities of the EDTA-actinide chelates are much greater, and those of the DCTA-actinide and DTPA-actinide chelates are somewhat greater than the chelates of the lanthanides of the same ionic radii. It appears that the initial uptake of the heavier actinides in bone fits the pattern of the stabilities of the actinide-organic acid chelates.

In the 10 years since it was suggested that the distributions in mammals of the lanthanides and actinides were dependent upon their ionic size, much information has accumulated on the mechanisms of transport of these elements in blood and of their binding in bone and liver.

The reactions between trivalent cations and bone surfaces probably do not control the overall distribution of these elements, although they may influence it. Bone surfaces accumulate many cations present in the circulation that can neither be utilized nor excreted. The mechanisms of accumulation in bone may include such reactions with bone mineral as adsorption, ion exchange, and ionic bond formation (NEUMAN and NEUMAN, 1958) and (or) the formation of complexes with proteins or other organic bone constituents (D. M. TAYLOR, 1972; Chap. 17, this volume).

The uptake of lanthanides and actinides by bone mineral appears to be independent of the ionic radius. D. M. TAYLOR et al. (1971) have shown that the in vitro uptake on powdered bone ash of ^{241}Am(III) (ionic radius 0.98 Å) and of ^{239}Pu(IV) (ionic radius 0.90 Å) were nearly the same, 97% $\pm$ 1.6 and 98.3% $\pm$ 0.7, respectively. In vitro experiments by FOREMAN (1962a) suggested that Pu(IV) accumulates on powdered bone or bone ash by adsorption, a relatively nonspecific reaction. On the other hand, reactions with organic bone constituents appear to be dependent on ionic radius. The complexes of the smaller Pu(IV) ion and any of the organic bone constituents tested thus far were much more stable (as determined by gel filtration) than the complexes with Am(III) or Cm(III) (D. M. TAYLOR, 1972).

The mechanisms by which the liver accumulates noncolloidal trivalent cations are not yet understood, but because they involve membrane transport, they are likely to be specific and dependent upon ionic size. It has been shown in the case of iron (FAWWAZ et al., 1966) and suggested in the case of Pu(IV) (DURBIN et al., 1972a) that liver accumulates only nuclide not bound to plasma protein. If that interpretation is correct, liver uptake would depend on the degree of plasma protein binding, which is in turn dependent upon ionic size. Further, the smaller the ionic radius of a lanthanide (and presumably of an actinide also) the greater is the tendency to hydrolyze; however, the hydrolysis products of the heavier lanthanides are soluble[19]. The liver does not accumulate most divalent cations, and the primary hydrolysis products of trivalent cations are of the form, $M(OH)^{+2}$. Thus, hydrolysis would also tend to reduce liver uptake of the smaller, heavier members of both series in a regular way that was also dependent upon ionic radius.

Binding to plasma proteins, which has been shown to be dependent upon ionic size (TURNER and TAYLOR, 1968b; BRUENGER et al., 1971; STEVENS and BRUENGER, 1972), is obviously important and may well be the determining factor, in the partition of the multivalent cations among bone, liver, soft tissues, and renal excretion. Both a high degree of plasma protein binding and formation of

19 The greater solubility of primary hydrolysis products of the heavier lanthanides may account for their greater early renal excretion.

soluble, less highly-charged hydrolysis products would inhibit liver uptake and lead to greater deposition in bone as well as greater renal excretion.

B. Intraskeletal Distribution

The ^{241}Am, $^{249-252}$Cf, or ^{253}Es contents of selected bones of rats, the ^{241}Am or ^{249}Cf contents of separate bones of beagles, and the ^{241}Am contents of individual bones of cynomolgus monkeys and baboons are shown in Table 18.21. Data are presented as percent of the total skeletal nuclide. Also included is the percent of total skeletal ash in the various bones of rats, beagles, and cynomolgus monkeys.

Except for the larger ^{241}Am content of the skull and mandible of the monkey, the intraskeletal distributions of ^{241}Am were nearly the same in the two primate species. The intraskeletal distributions of ^{241}Am, $^{249-252}$Cf, and ^{253}Es were similar in the rats; ^{241}Am and ^{249}Cf distributions were similar in the beagles. In all four species the fraction of skeletal nuclide was about the same as the skeletal ash content of scapulae, clavicles, cervical vertebrae, ribs, humeri, femora[20], pelvis, and patellae. The skeletal nuclide was greater than the ash fraction of the thoracic, lumbar, and caudal vertebrae and sternum. Skeletal nuclide content was less than the ash fraction of skull, mandible, distal long bones[20], and paw bones.

C. Microscopic Distribution

The local deposition of ^{241}Am in trabecular, endosteal, and periosteal surfaces and around cortical vascular channels in the long bones of mature and growing rats was demonstrated autoradiographically almost as soon as the nuclide was available for biological study (HAMILTON, 1947, 1948; SCOTT et al., 1948a; LANGHAM and CARTER, 1951; ASLING et al., 1952). The detailed autoradiographic studies of J. S. ARNOLD and JEE (1957) using ^{239}Pu and of HERRING et al. (1962) using ^{241}Am show conclusively that the ultimate site of deposition of these cations is the mineralized bone surface immediately beneath the cellular layer of trabeculae, endosteum, periosteum, or cortical vascular channels.

The early observations of ^{241}Am in rat bone have been confirmed and extended by others using improved bone sectioning techniques and special nuclear emulsions (D. M. TAYLOR et al., 1961; DURBIN et al., 1967; NÉNOT et al., 1972). The microscopic distribution of ^{241}Am has more recently been studied in the bones of other species: dog (HERRING et al., 1962; LLOYD et al., 1972a), mouse (HAMMARSTRÖM and NILSSON, 1970a, b), and cynomolgus monkey (DURBIN et al., 1972c). Curium-242 and $^{249-252}$Cf have been studied only in the bones of rats. The microscopic distribution of these two actinides in the long bones and vertebrae of rats was indistinguishable from that of ^{241}Am (HAMILTON, 1947, 1948; SCOTT et al., 1949; NÉNOT et al., 1972; MEWHINNEY et al., 1971; DURBIN et al., 1972d).

1. Initial Distribution of ^{241}Am and $^{249-252}$Cf in Rat Long Bones

Typical autoradiographs are shown of the initial deposition of ^{241}Am in the femur (Fig. 18.14a) and of $^{249-252}$Cf[21] in the tibia of the young adult rat (Fig. 18.15a). The animals were killed 4 days after the intramuscular injection of the nuclides as citrate complexes; absorption was nearly complete. As HAMILTON (1947, 1948) and SCOTT et al. (1948a, 1949) described for comparable preparations with ^{241}Am or ^{242}Cm, the nuclides shown in Figs. 18.14a and 18.15a were deposited most

20 The nuclide content of the rat tibiae was the same as the ash fraction.
21 The ^{252}Cf isotope had largely decayed by the time the autoradiographs were prepared.

Table 18.21. Intraskeletal distribution of ^{241}Am, $^{249-252}Cf$, ^{253}Es, and bone ash in several animal species expressed as percent total skeletal nuclide or ash content

	Adult rat				Beagle dog			Cynomolgus monkey		Baboon
	^{241}Am [a]	$^{249-252}Cf$ [a]	^{253}Es [b]	Ash [a]	^{241}Am [c]	^{249}Cf [c]	Ash [c]	^{241}Am [a]	Ash [a]	^{241}Am [aa]
Skull [ab]	—	—	—	14.7	7.77	8.64	17.7	15.7	15.8	{ 15.2 (skull and mandibles)
Mandibles [ab]	2.79	2.52	—	4.72	2.60	3.23	9.0	5.19	5.90	
Cerv. vert.					7.91	6.41	7.6			
Thor. vert.	{ 23.8 (cerv.–lumb. vert.)	23.3	24.1	20.3	18.1	13.4	6.1	{ 30.6 (cerv.–caud. vert.)	21.4	{ 29.2 (cerv.–lumb. vert.)
Lumb. vert.					19.0	13.3	6.0			
Caud. vert.	—	—	—	11.9	1.81	1.04	0.9			—
Scapulae	—	—	—	{ 2.65 (scapulae and clavicles)	3.5	4.97	3.9	3.55	3.02	4.34
Clavicles	—	—	—		—	—	—	0.59	0.57	0.43
Ribs	3.98	3.71	4.9	5.42	8.52	12.0	8.4	5.92	4.53	7.68
Sternum	—	—	3.0	2.64	5.95	4.26	0.3	1.47	0.89	1.18
Humeri		—	—	—	6.19	7.62	6.2	6.46	7.17	6.42
Radii	{ 9.64 (humeri–ulnae)	—	—	8.63	1.11	1.29	2.7			1.89
Ulnae		—	—	—	1.34	1.49	2.9	{ 15.7 (radii–tib.-fib.)	{ 23.7 (radii–tib.-fib.)	2.16
Femora	8.48	8.95	8.6	8.48	5.81	7.36	6.6			8.46
Tib.-Fib.	7.61	7.35	—	6.53	3.34	3.50	6.4			4.89
Pawbones	—	—	4.6	7.25	2.02	4.56	10.0	6.16	8.60	5.97
Pelvis	5.75	—	6.7	6.77	4.88	7.00	5.3	8.43	7.68	10.7
Patellae	—	—	—	—	—	—	—	0.22	0.27	—
Teeth	3.91	4.20	—	—	—	—	—	1.32	—	—
Skeletal nuclide (%)	35.6	64.3	56.9	—	31.0	48.6	—	31.9	—	40.5
Ash wt. skeleton(g)	—	—	—	8.9	—	—	327	—	153	—

[a] ^{241}Am measured in 20 adult female rats 1 to 15 days after injection. $^{249-252}Cf$ measured in 3 adult female rats 4 days after injection. Bone ash measured in 50 or more female rats killed between the ages of 100 and 300 days; total skeletal ash measured in 50 female rats 100 to 130 days old. ^{241}Am measured in 8 adult female cynomolgus monkeys killed 1 to 63 days after injection. Bone ash measured in 19 adult female cynomolgus monkeys (DURBIN et al., 1972c–e).

[b] ^{253}Es measured in 8 adult female rats 24 days after injection (V. H. SMITH, 1972b).

[c] ^{241}Am measured in 6 beagles 1 to 21 days after injection (LLOYD et al., 1971, 1972a). Skeletal ash measured in 11 adult beagles of both sexes in the same colony (B.W. = 8.47 to 13.1 kg) (ATHERTON et al., 1969b). ^{249}Cf measured in two beagles killed 7 and 21 days after injection (LLOYD et al., 1972b).

[aa] ^{241}Am measured in two adult female baboons killed 32 and 86 days after injection (ROSEN et al., 1972).

[ab] In the studies with dogs and baboons, skull and mandibles include teeth.

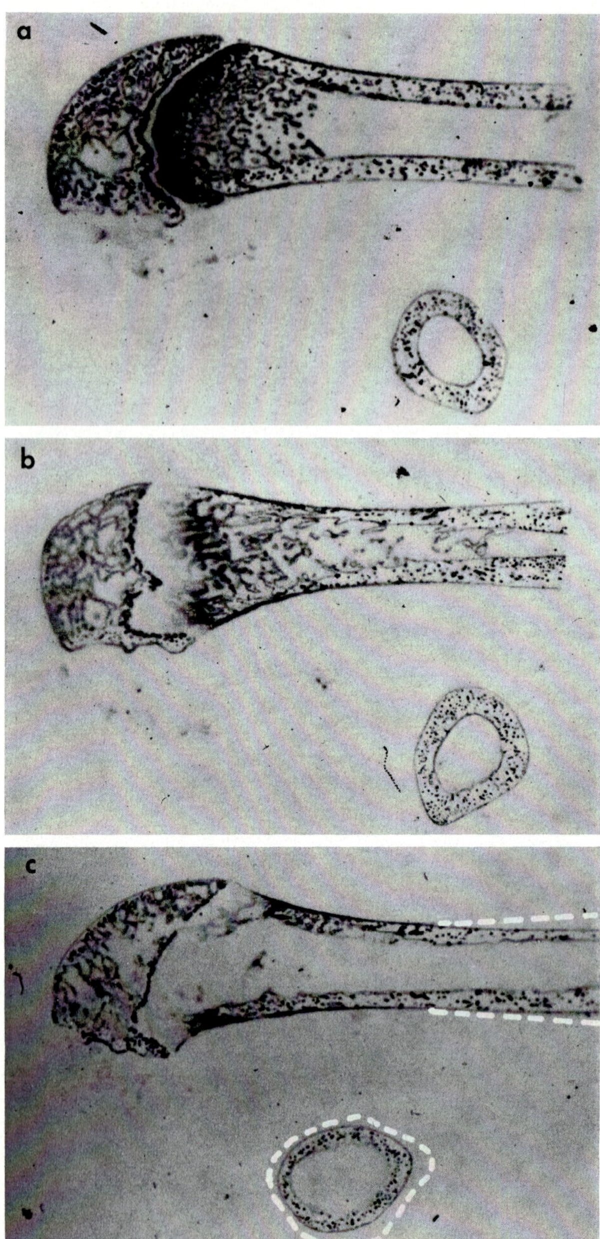

Fig. 18.14a—c. Gross autoradiographs of bones from female rats injected intramuscularly with 0.37 μCi/kg of ^{241}Am citrate at 110 days of age. Bones and autoradiographs were prepared as described by DURBIN et al. (1967) using Kodak NTA plates. ×5. a A central longitudinal section and mid-diaphyseal cross section of the femur of a rat killed 4 days after injection. Plate exposed 1556 days. b The femur of a rat killed 95 days after injection. Plate exposed 1556 days. c The femur of a rat killed 739 days after injection. The perimeter of the bone at death is shown as a broken white line. Plate exposed 1620 days

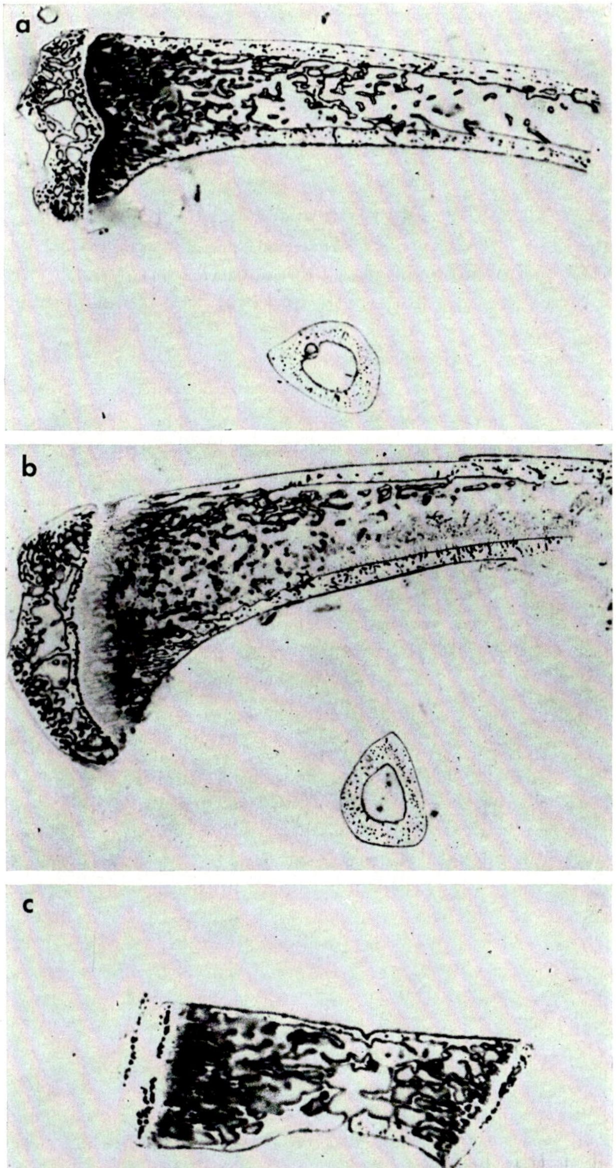

Fig. 18.15 a—c. Gross autoradiographs of bones from female rats injected intramuscularly with 0.33 μCi/kg of $^{249-252}Cf$ citrate at 136 days of age. See Fig. 18.14 for method of preparation of plates, which were exposed 2050 days. a A central longitudinal section and mid-diaphyseal cross section of the tibia of a rat killed 4 days after injection. × 5. b The tibia of a rat killed 91 days after injection. × 5. c A central longitudinal section of a vertebra (L1) of a rat killed 91 days after injection. × 7.5

heavily about the trabeculae of the primary spongiosa and on epiphyseal and metaphyseal trabeculae. Less intense deposits were found on the endosteal and periosteal surfaces of the shaft and in spots throughout the cortex corresponding to the locations of blood vessels. The local bone depositions of ^{241}Am and ^{242}Cm were described by HAMILTON (1948) as indistinguishable from the depositions of the trivalent nuclides, ^{144}Ce, ^{147}Pm, and ^{227}Ac, but different from the depositions of the tetravalent nuclides, ^{95}Zr, ^{228}Th, and ^{239}Pu, in that the +4 nuclides were not deposited in significant amounts around cortical blood vessels.

The autoradiographs in Figs. 18.14a and 18.15a, prepared by more refined techniques, permit the demonstration of additional initially heavy concentrations of ^{241}Am or $^{249-252}$Cf in the long bones of female rats 3.5 to 4.5 months old—on the surfaces of newly forming trabeculae beneath the articular cartilage and on the bony plate just above the proliferating cartilage of the epiphyseal plate. As in the early gross preparations, the bone spicules of the primary spongiosa are too small and too crowded to be resolved.

Using the method of alpha-track counting, D. M. TAYLOR et al. (1961) estimated local radiation dose rates at various locations in the femora of 7-week-old male rats. The average dose rates 1 to 8 days after intravenous injection of 3.3 µCi/kg of ^{241}Am citrate were as follows: in the epiphysis, ossification centers below the articular cartilage and above the epiphyseal cartilage, 102 to 108 rad/day; in the metaphysis, above and below the active band in the primary spongiosa, 30 to 65 rad/day; in the active band, 136 rad/day; beneath the periosteum, 101 rad/day; and in the diaphysis, endosteum, 20 rad/day; periosteum, 35 rad/day; and perivascularly, 36 rad/day.

2. Initial Deposition of ^{241}Am in Monkey Bone

The distribution of ^{241}Am in several bones of cynomolgus monkeys is shown grossly in Figs. 18.16 and 18.17. In the autoradiographs prepared from the bones of adult females (more than 6 years of age[22]) killed 8 days after intravenous injection of 0.56 µCi/kg of ^{241}Am citrate, bone surfaces—trabecular, endosteal, periosteal, and perivascular—were labeled almost uniformly. HERRING et al. (1962) showed that the microscopic distribution of ^{241}Am on the bone surfaces of growing dogs varied with the growth activity of those surfaces. In decreasing order ^{241}Am deposition was: resorbing surfaces $\geq$ inactive (resting) surfaces $>$ active (growing) surfaces. LLOYD et al. (1972a) measured alpha-track densities of endosteal and periosteal surfaces of bones of young adult beagles killed 21 days after injection of 2.8 µCi/kg of ^{241}Am. The intensity of labeling of epiphyseal trabeculae in the distal femur was 1.5 times that of periosteal surfaces, and in the vertebral body the trabecular label was twice as intense as that of the periosteal surfaces. At age 6 years or more one would not expect a significant fraction of the total bone surface of the monkeys either to be growing or resorbing at any given time. Although the overall impression of the ^{241}Am distribution in the adult monkey bones is that of uniform labeling, closer inspection revealed gradations in the intensity of the label, as predicted by the studies of HERRING et al. (1962) and LLOYD et al. (1972a).

Autoradiographs were also prepared from bone sections taken from an adolescent monkey (2.5 years of age) killed 8 days after intravenous injection of 0.56 µCi/kg of ^{241}Am citrate (see Figs. 18.16c, d and 18.17a, c). Distribution of ^{241}Am

22 The ages of the monkeys were estimated from the dentition-age data of HURME and VAN WAGENEN (1961) and skeletal roentgenograms and the bone-age data of VAN WAGENEN and ASLING (1958).

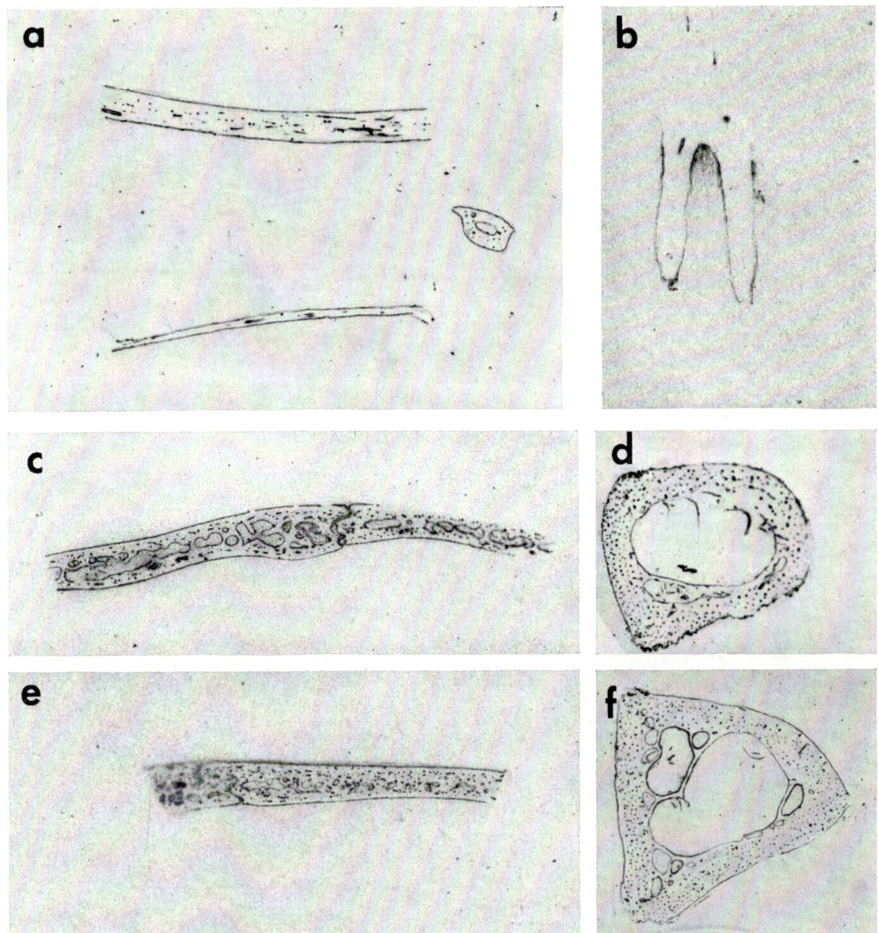

Fig. 18.16a—f. Gross autoradiographs of ^{241}Am in bones of female cynomolgus monkeys. See Fig. 18.14 for preparation of plates, which were exposed 1400 days. a Longitudinal and cross sections from the central one-third of the 5th rib of adult no. 42 (more than 6 years old, killed 8 days after an intravenous injection of 0.56 μCi/kg of ^{241}Am citrate). ×2.7. b A central section of a premolar of adult no. 8 (more than 6 years old, killed 8 days after an intramuscular injection of 0.56 μCi/kg of ^{241}Am citrate). ×2. c A cross section of the calvarium of adolescent no. 41 (2.5 years old, killed 8 days after an intravenous injection of 0.56 μCi/kg of ^{241}Am citrate). ×2.3. d A cross section of the mid-diaphysis of the humerus of adolescent no. 41 described in Fig. 18.16c. ×2.7. e A cross section of the calvarium of adult no. 42 described in Fig. 18.16a. ×2,3. f A cross section of the mid-diaphysis of the humerus of adult no. 8 described in Fig. 18.16b. ×2.7

in the adolescent skeleton was less uniform than in the adult skeleton and resembled the distribution of ^{241}Am in bones of young adult beagles (LLOYD et al., 1972a). Especially heavy concentrations of ^{241}Am were present on ragged (presumably resorbing) surfaces of the humerus midshaft (Fig. 18.16e), below the articular cartilage and in the primary spongiosa of the proximal humerus (Fig. 18.17a), and on apparently resorbing surfaces of the spinous process of the lumbar vertebra (Fig. 18.17c). Some lightly labeled surfaces on the trabeculae in the proximal humerus and vertebral body were also apparent.

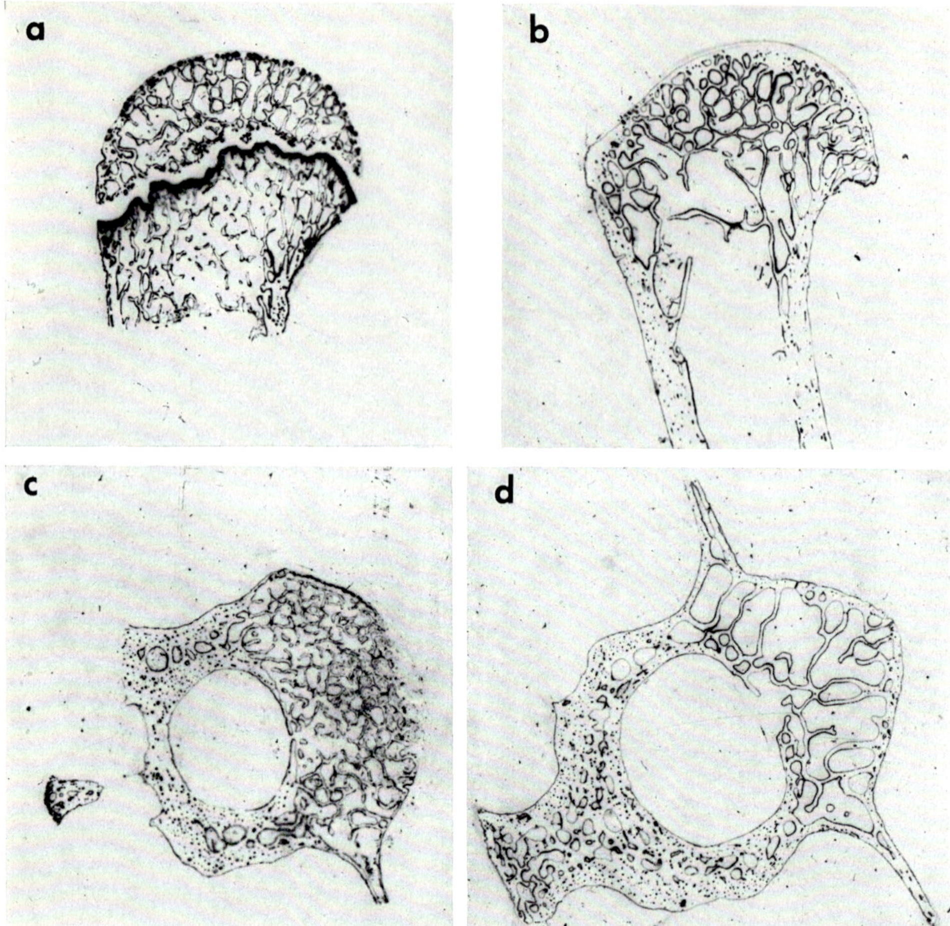

Fig. 18.17a—d. Gross autoradiographs of ^{241}Am in bones of female cynomolgus monkeys. See Fig. 18.14 for method of preparation of plates, which were exposed 1400 days. ×3. a A central longitudinal section of the proximal humerus of adolescent no. 41 described in Fig. 18.16c. b A central longitudinal section of the proximal humerus of adult no. 8 described in Fig. 18.16b. c A central cross section of vertebra (L1) of adolescent no. 41 described in Fig. 18.16c. d A central cross section of vertebra (L1) of adult no. 42 described in Fig. 18.16a

Using thin sections of undecalcified rat long bone permanently affixed to nuclear track plates, LANGHAM and CARTER (1951) showed that the intense band of radioactivity just below the epiphyseal cartilage (see Figs. 18.14a, 18.15a, 18.17a) was the result of ^{241}Am deposited on the many tiny newly calcified spicules of the primary spongisa. Interdigitated with capillary tufts, these fine spicules of bone constitute a large, well-vascularized, readily available mineralized surface. A high power view of the calcifying cartilage of the proximal humeral epiphysis of an adolescent monkey is shown in Fig. 18.18a.

JEE and ARNOLD (1960) demonstrated the accumulation of ^{228}Th and ^{239}Pu on the dentinal surfaces of the pulp cavity and the cemental surfaces of the roots of the teeth of young adult dogs. HAMMARSTRÖM and NILSSON (1970a, b) used

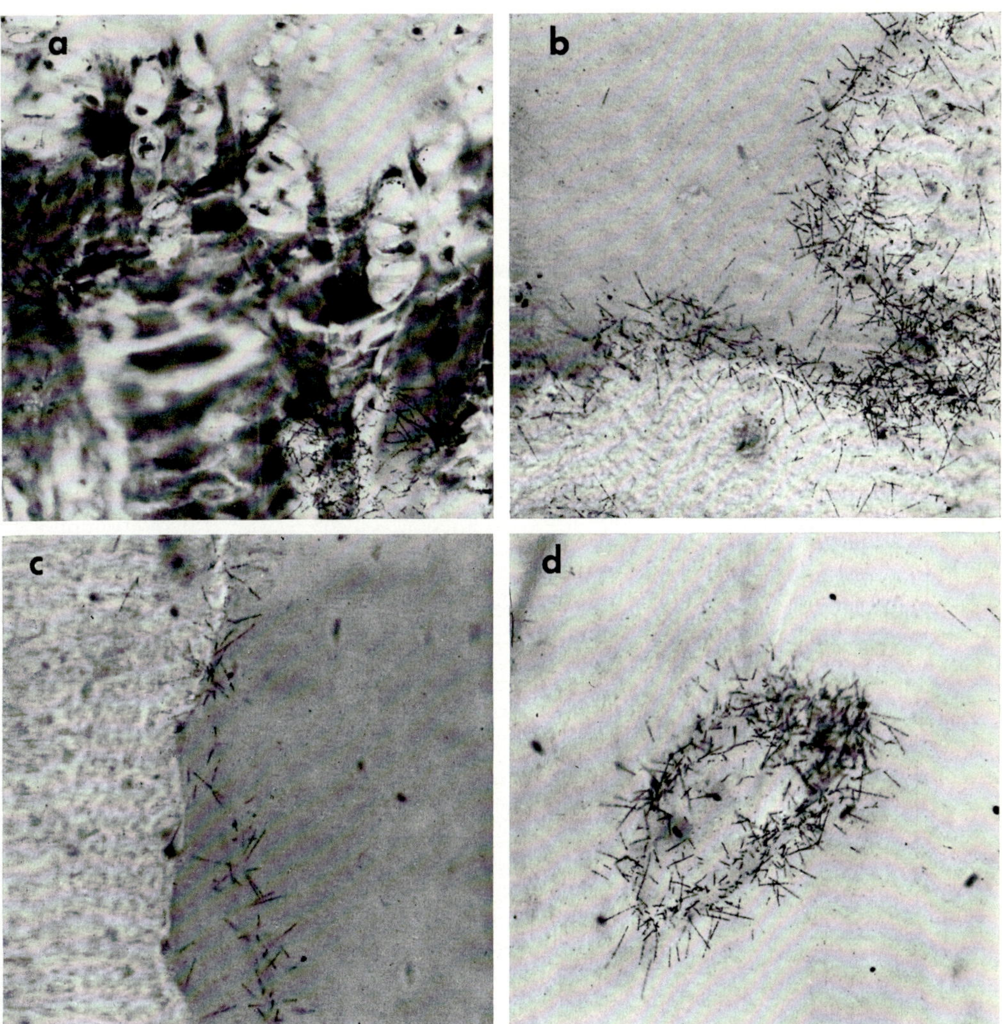

Fig. 18.18a—d. Microautoradiographs of ^{241}Am in undecalcified sections of bones of female cynomolgus monkeys. Bone sections and autoradiographs were prepared as described by J. S. ARNOLD and JEE (1954) and J. S. ARNOLD (1954) using Kodak NTA liquid emulsion, which was exposed 62 days. ×267. a The growth cartilage of the proximal humerus of adolescent no. 41 described in Fig. 18.16c. Americium-241 is associated with the spicules of recently calcified cartilage. b A trabeculus in the vertebral body (L1) of adult no. 45 (more than 6 years old, killed 48 weeks after an intramuscular injection of 0.44 μCi/kg of ^{241}Am citrate). Americium-241 is shown on a surface unaltered during the 48-week post-injection interval. c The metaphyseal cortex of the proximal humerus of adult no. 45 described in Fig. 18.18b. New bone has been laid down during the 48-week postinjection interval. d The cortex of vertebra (L1) of adult no. 45 described in Fig. 18.18b. Labeled Haversian canal surface has been filled with new bone during the 48-week postinjection interval

whole-body sections of adult mice and of suckling rats to demonstrate the presence of ^{241}Am in the pulp of the incisors and the developing molars (as well as on the dentinal surfaces) during the first few hours after injection. Presumably the presence of ^{241}Am in the dental pulp is temporary as is the presence of ^{241}Am in bone marrow (Hammarström and Nilsson, 1970a). Deposition of ^{241}Am on the cemental surface of the premolar of an adult monkey is shown in Fig. 18.16b. The section was not central, and two small dark spots at the level of the gum line are all that can be seen of the pulp cavity.

3. The Effect of Bone Remodeling on Distribution

As a whole, the mammalian skeleton is not static. During the life span the skeleton undergoes a great deal of structural change (Frost, 1963). During growth, remodeling is extensive and rapid and is accompanied by a net increase in bone mass. In adult life, remodeling proceeds slowly and involves only a fraction of the skeleton at any one time; there is little net change in skeletal mass. In some bones of some mammals during the last one-half to one-third of the life-span, resorptive activity exceeds bone replacement leading to a net loss of skeletal mass.

Most of the information concerning the effect of skeletal remodeling on the distribution of the transplutonium elements in bone has been obtained from the study of ^{241}Am in rats. J. S. Arnold and Jee (1957) described the effects of growth and remodeling on the distribution pattern of ^{239}Pu in the bones of rats. In general, the long-term distribution and redistribution of the transplutonium elements is much like that of ^{239}Pu (D. M. Taylor et al., 1961; Durbin et al., 1967). Surfaces labeled on the day of injection may be unchanged (Fig. 18.18b); they may be partially or completely resorbed or they may be buried by new bone laid down on trabeculae (Figs. 18.14b and 18.15b), beneath endosteum or periosteum (Figs. 18.15c and 18.18a), or on the surfaces of vascular channels (Fig. 18.18d).

Skeletal maturation in the rat is different from that of most mammals. The epiphyses of the long bones comprising the knee and elbow joints—those bones that have been studied most—do not undergo true union (Simpson et al., 1950). Although longitudinal growth ultimately ceases, and the epiphyseal cartilage becomes thin and inactive, degeneration of chondrocytes and formation of new primary spongiosa proceed slowly in both sexes well into middle life (6 to 12 months beyond sexual maturity). In spite of the unusual persistence of un-united epiphyses in the rat, it is a useful animal in which to study bone growth and remodeling: It is a well-characterized species; its life span is only about 3 years—compressing growth, adulthood, and senility into a short span of time.

The most obvious effect of bone remodeling on ^{241}Am distribution in growing rats is the prompt resorption of the active band of ^{241}Am deposited in the primary spongiosa on day 1. In their studies of ^{241}Am distribution, D. M. Taylor et al. (1961) examined the femora of male rats injected at 7 weeks of age with 3.3 μCi/kg of ^{241}Am citrate and killed at intervals up to 21 days later[23]. Twenty-one days after injection, the primary spongiosa elaborated on day 1 had been resorbed except for that portion that had been incorporated into the shaft. The ^{241}Am deposited beneath the articular cartilage had been covered by new bone. Some of the endosteal deposit in the upper metaphysis had been resorbed, and the periosteal deposit in the diaphysis had been buried by new bone.

Alpha-track counting yielded the following local dose rates 21 days after injection: in the epiphysis, the ossification center beneath the articular cartilage, 43 rad/day—about one-half the dose rate delivered to that structure 1 to 8 days after injection; in the bony plate above the epiphyseal cartilage, 157 rad/day—

23 The skeletons of male and female rats grow at about the same rate during the early weeks of life. Between the ages 49 and 70 days the femora of female Charles River rats gained 18% in length, 29% in wet weight, and 113% in ash weight.

50% greater than the initial dose rate; in the metaphysis the radioactivity was gone from the primary spongiosa and this structure was no longer being irradiated; beneath the periosteum of the upper metaphysis, in the remnant of the day-1 spongiosa, 144 rad/day—about 40% more than the early dose rate; in the diaphysis—in the endosteum, periosteum, and perivascularly—21, 24, and 38 rad/day, respectively—about the same dose rates in these structures as were measured 1 to 8 days after the injection.

In more mature female rats (injected with 0.37 μCi/kg of ^{241}Am citrate at 110 days of age) longitudinal growth was proceeding slowly, and 95 days after injection some of the day-1 spongiosa was still present (Fig. 18.14b). Remnants of the day-1 spongiosa persisted through 180 days, but these were completely removed by 300 days after injection (DURBIN et al., 1967). The autoradiographs shown in Figs. 18.15a, b were prepared from bones of female rats injected with 0.33 μCi/kg of $^{249-252}$Cf citrate at 136 days of age. During the 91 days after injection there was little growth in length; most of the day-1 spongiosa was still present. In the studies of MEWHINNEY et al. (1971) in which 7.7 μCi/kg of ^{252}Cf were injected into 37-day-old male rats, longitudinal growth continued, but some of the day-1 spongiosa was still present—associated with a bony plate in the upper diaphysis suggesting that resorption had been disturbed by the high radiation dose (HELLER, 1948; DURBIN et al., 1958; FABRIKANT and SMITH, 1964).

A significant fraction of ^{241}Am (DURBIN et al., 1967) and ^{239}Pu (J.S. ARNOLD and JEE, 1957) released from bone destroyed by remodeling recirculates and is redeposited on bone surfaces; some of these surfaces were present at the time of injection, some were formed later. If recirculated ^{241}Am is deposited on pre-injection surfaces, their autoradiographic image will be augmented. If recirculated ^{241}Am is deposited on surfaces formed after the injection, their autoradiographic image will be lighter and less distinct than the surfaces labeled on day 1. In Figs. 18.15b, c, prepared from bones of rats killed 91 days after injection of $^{249-252}$Cf, californium can be seen deposited on recently formed primary spongiosa, on the endosteal surface of the dorsal metaphysis of the tibia, and on some new vertebral trabeculae. If the new surfaces are in a region undergoing continuous accumulation of new bone, the bone laid down after injection will be diffusely labeled throughout (COPP et al., 1947; J. S. ARNOLD and JEE, 1957; JEE, 1972).

The long-term effects of slow continuous remodeling and redeposition on the distribution of ^{241}Am are seen in the autoradiographs in Fig. 18.15c prepared from a rat killed 739 days after injection of ^{241}Am citrate at 110 days of age. Radiochemically, the femur of this rat contained two-thirds of the ^{241}Am radioactivity present in the femur of a rat killed at 4 days. There are fragmentary remnants of the day-1 spongiosa in the dark bands in the upper metaphyseal cortex and in isolated spots in the marrow cavity of the diaphysis. Outlines of the epiphyseal trabeculae are light and woolly, indicating deposition on trabeculae formed after the ^{241}Am injection. The outline of the entire cortex is thin, and the endosteal deposit is light, suggesting that the original endosteal deposit had been resorbed, and that the ^{241}Am now present was redeposited ^{241}Am that had been laid down on previously unlabeled surfaces. As the outlines of the inner and outer margins of the bone from which Fig. 18.15c was made show, most of the original endosteal deposit in the diaphysis had been resorbed, and the entire circumference of the day-1 periosteal deposit had been covered by a thick layer of new bone. In the epiphysis, the trabecular outlines are pale and indistinct, indicating deposition on structures formed after the injection. The ^{241}Am deposit which was laid down just beneath the articular cartilage had been removed. The only identifiable remnant of the bone as it existed the day of injection was the bony plate above the epiphyseal plate, and even here the marrow had penetrated in several places in an approach to epiphyseal union.

D. Retention of ^{241}Am and ^{242}Cm

1. Rat

The original studies with ^{241}Am and ^{242}Cm in rats indicated that, like ^{239}Pu and the lanthanides, the fraction of these transplutonium nuclides deposited in the skeleton would very likely remain there for a long time. HAMILTON (1947, 1948) and SCOTT et al. (1948a, 1949) could not detect any loss of ^{241}Am or ^{242}Cm from the rat skeleton over a period of 256 days after injection. However, both nuclides had been administered intramuscularly as chlorides, and 4 to 6% of the injected dose of each nuclide was absorbed between the 16th and 256th days. A small loss from bone could have been masked by replenishment from the injection site. More recent long-term studies in rats demonstrate that ^{241}Am is lost very slowly from the skeleton (DURBIN et al., 1967; 1972c).

D. M. TAYLOR et al. (1961) reported a biological half-time of 1600 days for ^{241}Am in the rat femur when ^{241}Am citrate was injected intravenously in 7-week-old male rats at a level of 3.3 μCi/kg. Femurs were obtained from 37 rats that died or were killed from 21 to 539 days after injection. DURBIN et al. (1969) analyzed ^{241}Am in several bones, the incisors, and the entire skeletons of 150 young adult female rats killed from 1 to 779 days after a single intramuscular injection of 0.33 μCi/kg of ^{241}Am citrate. The complexed ^{241}Am was absorbed rapidly (97% in 24 hr), so the results were not perturbed by a persistent reservoir. The standard deviations of the mean values of the skeletons and the sampled bones were of the order of 10% at 1, 4, and 8 days after injection. In some cases the S.D. exceeded the amount of ^{241}Am apparently lost during the entire 779-day observation period. The scatter of the data was reduced by eliminating from calculations the data from 39 rats for which the measured or estimated material recovery deviated more than one S.D. from the mean recovery for all 150 rats.

The smoothed data for retention of ^{241}Am in the whole skeleton (incisors excluded) and for 20 × femur are shown in Fig. 18.19, which also includes a curve prepared from the data of D. M. TAYLOR et al. (1961) for retention of ^{241}Am in the femora of young male rats. Retention curves[24] were fitted by the method of least squares as follows (t is days after injection):

Femora, $F_t\,(\%) = 0.44\,e^{-0.013t} + 2.92\,e^{-0.0002t}$ Eq. (15)

Tibiae, $T_t\,(\%) = 0.30\,e^{-0.015t} + 2.45\,e^{-0.00026t}$ Eq. (16)

Vertebrae[25], $V_t\,(\%) = 0.32\,e^{-0.017t} + 1.48\,e^{-0.00027t}$ Eq. (17)

Ribs[25], $R_t\,(\%) = 0.10\,e^{-0.013t} + 0.32\,e^{+0.00015t}$ Eq. (18)

Mandibles, $M_t\,(\%) = 0.98\,e^{+0.00009t}$ Eq. (19)

Skeleton, $Sk_t\,(\%) = 2.0\,e^{-0.017t} + 33\,e^{-0.00015t}$ Eq. (20)

There was a small but measurable early loss of ^{241}Am from long bones, vertebrae, ribs, and the whole edentate skeleton—half-times ranged from 40 to 50 days. However, the major losses occurred more slowly: half-times were 2500, 2700, and 3500 days in lumbar vertebrae, tibiae, and femora, respectively. The slopes

24 The parameters of the retention equations, Eq. 15 through 20, are from DURBIN et al. (1967).

25 Samples consisted of 3 lumbar vertebrae (L4, 5, 6) and 5 large ribs.

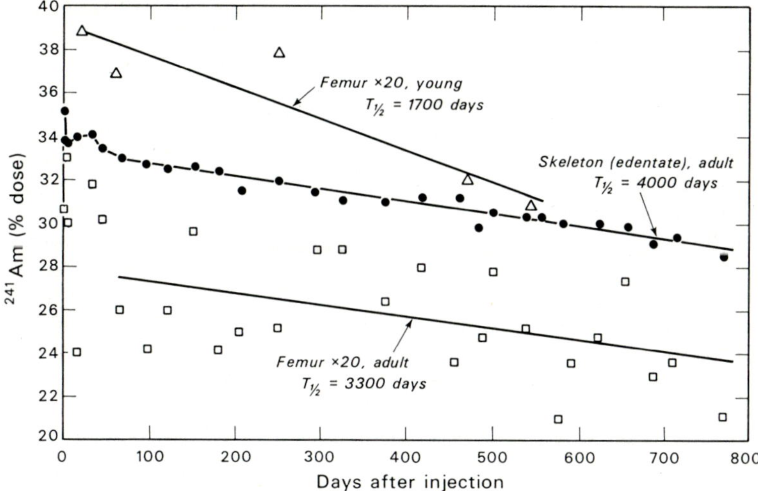

Fig. 18.19. Retention of ^{241}Am in the femora and whole edentate skeleton of adult female rats, DURBIN et al. (1967, 1972c) and in the femora of young male rats, D. M. TAYLOR et al. (1961)

of the ^{241}Am retention curves for ribs and mandibles were positive. Retention of ^{241}Am in the whole skeleton is the summation of the retentions in the separate bones, and the half-time of retention of ^{241}Am in the entire skeleton was 4600 days.

There is a two-fold difference between the biological half-times of ^{241}Am measured in the femora in the two experiments. DURBIN et al. (1967, 1972e) used rats in which the growth of the long bones was 90% complete. The ^{241}Am dose was low; there were no detectable hematological changes, and, except for some transient morphological changes in liver and kidney, no detectable consequences of radiation in the tissues and bones. On the other hand, D. M. TAYLOR et al. (1961) used rapidly growing rats. They were injected at a level of 3.3 μCi/kg of ^{241}Am, a dose which was shown subsequently by the same investigators to disturb the growth of the long bones—with osteoporosis in the ends and osteosclerosis in the shafts—and to induce bone tumors in 20% of the rats within 525 days of the injection (FABRIKANT and SMITH, 1964; BENSTED et al., 1965). Either greater resorption during growth, or bone loss consequent to radiation damage, could lead to more rapid elimination of ^{241}Am than was observed in the more mature rats given the smaller amount of ^{241}Am.

J. S. ARNOLD and JEE (1957, 1959) proposed that ^{239}Pu does not participate in exchange reactions, but was only lost from bone surfaces when they were physically destroyed by resorption. DURBIN et al. (1967) showed that ^{241}Am was lost from long bones and vertebrae of rats (bones that lose ash weight after middle life), but was not lost from the mandibles (bones that achieve constant ash weight late in life), and continued to accumulate in the ribs (bones that gain ash weight during the entire life span). These findings were interpreted as evidence that ^{241}Am, and presumably the other transplutonium elements, are lost from bone only in the course of remodeling.

The autoradiographs in Figs. 18.14a–c and 18.15a, b show that during the life span of the female rat the long bones undergo extensive remodeling, which appears to involve most of the initially labeled trabecular surfaces. Yet during the 779 days of the experiment, that portion of the initial ^{241}Am deposit that had been eliminated was only 26% in the femora and 35% in the lumbar vertebrae (Durbin et al., 1967).

There is a persistent but declining level of ^{241}Am in all of the soft tissues and in the continually growing incisors of the rat long after the initial deposits of ^{241}Am have been cleared from these sites. The autoradiographs in Figs. 18.14b, c and 18.15b also show that both ^{241}Am and $^{249-252}Cf$ continue to be deposited in bone. The source of the persistent low-level supply of these nuclides can only be the skeleton itself. Using the material balance of ^{241}Am in the skeleton, tissues, and excreta, Durbin et al. (1967) calculated that at least 40% of ^{241}Am released from bone was redeposited in bone and that the measured biological half-time in bone, $T_{1/2\,obs}$, was related to the physiological half-time of the labeled surfaces, $T_{1/2\,phys}$, by the eqaution

$$T_{1/2\,obs} = T_{1/2\,phys}/(1-fr), \qquad \text{Eq. (21)}$$

where fr is the fraction of circulating nuclide that is redeposited. In general, redeposition has the least effect on the skeletal retention of the alkaline earths. The presence of the large excess of stable calcium dilutes the radionuclide, promotes excretion, and reduces reutilization. The loose protein binding also facilitates excretion. Redeposition has the greatest influence on retention of Pu(IV). The strong protein binding promotes reutilization and impedes excretion. The trivalent lanthanides and actinides lie somewhere between these extremes.

2. Chinese Hamster

McKay et al. (1972) calculated the skeletal content of ^{241}Am from analyses of bones taken 225 and 360 days after injection of 1 μCi/kg of ^{241}Am citrate into Chinese hamsters. The half-time of ^{241}Am in the Chinese hamster skeleton, based on these two measurements, is at least 1450 days.

3. Monkey

In an ongoing long-term study of ^{241}Am metabolism in cynomolgus monkeys, 19 adult females have been injected with 0.44 to 0.87 μCi/kg of ^{241}Am citrate (Durbin et al., 1972c). Eighteen complete skeletons have been analyzed (taken from monkeys killed 1 to 973 days after injection). The ^{241}Am content of the skeleton of one living animal (no. 75) has been estimated from continuous excretion measurements (Fig. 18.20) and from the liver and soft tissue ^{241}Am-retention curves (Fig. 18.21). The skeleton of monkey no. 75 was estimated to contain 35% of the dose 1400 days after injection. Determination of the retention of ^{241}Am in the skeleton of this species is complicated by (a) individual variation in initial skeletal uptake and (b) by the continuous skeletal accumulation of ^{241}Am during the 300-day interval of soft-tissue and liver clearance (Fig. 18.20).

After 100 days the skeletal retention curve appears to consist of 2 components. The slope of the long-term component, defined by only 4 animals killed between 700 and 1400 days, is not different from zero. If it is assumed that the liver and soft tissues continue to lose ^{241}Am at the rates shown in Fig. 18.21, the sum of the liver and soft-tissue ^{241}Am can be calculated and then subtracted from the whole-body retention shown in Fig. 18.20. Using only the data from monkey no. 75, the half-time of skeletal retention of ^{241}Am in the monkey is estimated to be 12600 days (34.5 years). If a straight line with this slope is fitted to the whole-body retention curve in Fig. 18.20, the slope of the first component can

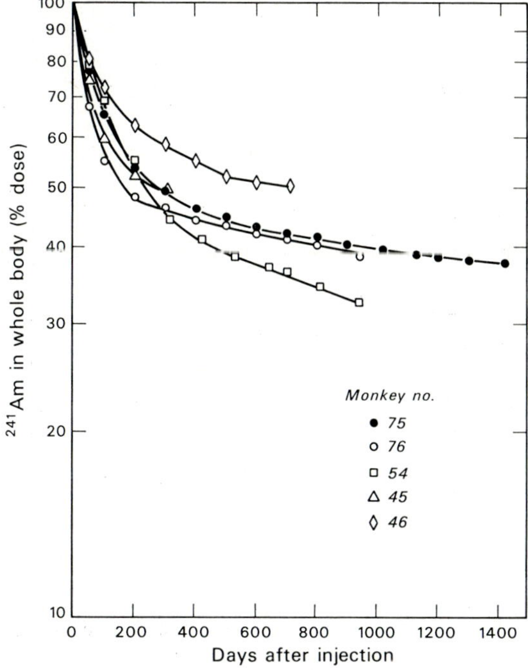

Fig. 18.20. Retention of ^{241}Am in the whole body of adult monkeys after intravenous or intramuscular injection of 0.44 to 0.87 μCi/kg of ^{241}Am citrate. Retention was calculated by subtraction of summed excretion from 100%. (Data of DURBIN et al., 1972c)

be obtained graphically. The retention equation for ^{241}Am in the monkey skeleton is estimated to be (t is days after injection)

$$Sk_t(\%) = 11\,e^{-1.6\times 10^{-3}t} + 36\,e^{-5.5\times 10^{-5}t}. \qquad \text{Eq. (22)}$$

4. Dog

Long-term experiments with ^{241}Am in beagles have been underway for several years (LLOYD et al., 1967a, b). As of March 1972, 101 beagles had each been given a single intravenous injection of ^{241}Am citrate at dose levels ranging from 0.002 to 4.5 μCi/kg (LLOYD et al., 1970, 1971, 1972c; T. F. DOUGHERTY, 1972). Unfortunately, retention of ^{241}Am in the dog skeleton cannot now be estimated from analysis of the bones of individual dogs. At the 2.8 μCi/kg level, radiation damage in the liver was so severe that by 100 days after injection there was massive destruction of liver tissues followed by release of a large fraction of the ^{241}Am initially deposited in that organ. Loss of ^{241}Am from the liver was delayed, but became significant 1 year after injection of 0.9 μCi/kg and 2 years after injection of 0.3 μCi/kg (LLOYD et al., 1970, 1972a, c, d). The ^{241}Am released from the liver was partly excreted and partly recirculated to other soft tissues and to the skeleton. The skeletons of the dogs in the 0.9 and 2.8 μCi/kg groups contained 5 and 19% more of the injected dose 850 and 480 days, respectively, after injection than they were estimated by in vivo measurement to have contained on the day after injection. Thus far most of the analyzed skeletons are those of high-dose dogs in which ^{241}Am was apparently accumulating in bone faster than it was being released.

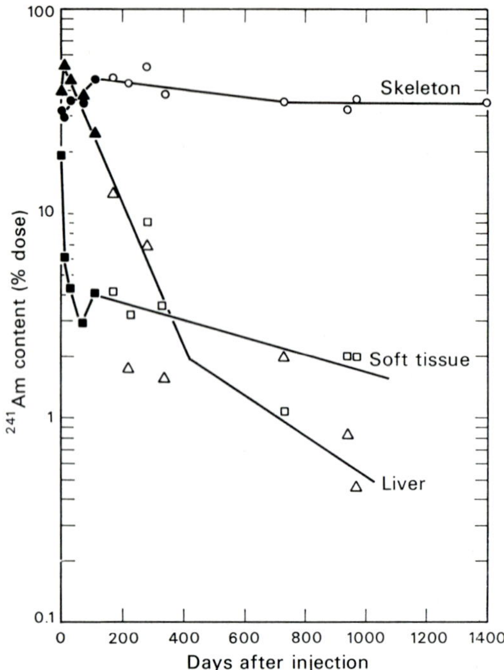

Fig. 18.21. Retention of ^{241}Am in the skeleton, liver, and residual soft tissues of adult cynomolgus monkeys after intravenous or intramuscular injection of 0.44 to 0.87 μCi/kg of ^{241}Am citrate. Solid symbols represent pairs of monkeys, and open symbols, single animals. (Data of Durbin et al., 1972c)

Whole-body and partial-body in vivo measurement techniques were developed to study separately the retention of ^{241}Am in liver and nonliver tissues. These measurements have been reported through 850 days after injection (Lloyd et al., 1970). Americium-241 in nonliver tissues was found primarily, but not exclusively, in bone: 28.8%/33.8% = 0.85 of the nonliver burden was in bone 1 to 22 days after injection. It can be assumed that during the first 200 days after injection the decline in nonliver ^{241}Am in the lower-dose level dogs (two dogs each at 0.02, 0.1 and 0.3 μCi/kg) was largely loss from the soft tissues. The decline in nonliver ^{241}Am during the interval 200 to 850 days could be assumed to represent loss from both bone and soft tissues, but mostly from bone. Release of ^{241}Am from the livers of the 0.3 μCi/kg dogs began at about 2 years. At that time the best estimate of ^{241}Am retention in the dog skeleton was the slope of that portion of the nonliver retention curve lying between 200 and 600 days; the half-time was about 18 years.

E. Retention of $^{249-252}$Cf and ^{253}Es

Skeletal retention of the elements heavier than curium has not been measured directly in any species—the experiments with rats, in which bone content of $^{249-252}$Cf was measured, were too short to characterize long-term retention (Mewhinney et al., 1971, 1972; Durbin et al., 1972d). In rats and mice—animals that promptly eliminate multicharged cations from liver and soft tissues—whole-

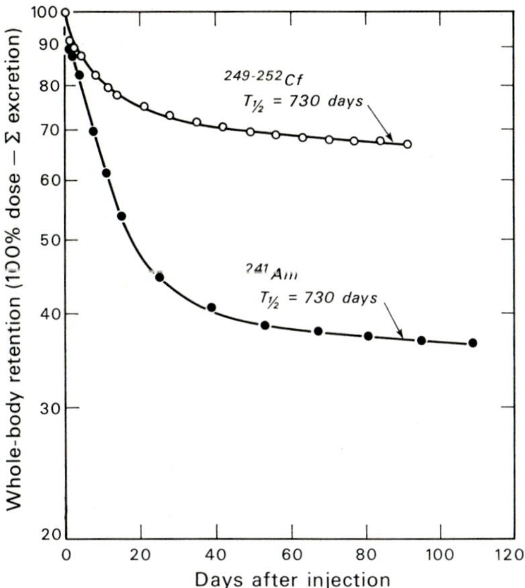

Fig. 18.22. Retention of ^{241}Am or $^{249-252}$Cf in the whole bodies of adult female rats after intramuscular injection of 0.37 μCi/kg of ^{241}Am citrate or 0.33 μCi/kg of $^{249-252}$Cf citrate. Retention was calculated by subtraction of summed excreta from 100%. (Data of DURBIN et al., 1967, 1972c, d)

body retention of $^{249-252}$Cf was nearly the same as skeletal retention at times longer than 30 days after injection. S. R. WRIGHT (unpublished) measured the whole-body retention of ^{252}Cf in adult mice up to 120 days after injection of 4 μCi/kg of ^{252}Cf citrate and estimated the half-time to be 900 days. MEWHINNEY et al. (1971, 1972) measured whole-body retention of ^{252}Cf up to 64 days after injection of 7.7 μCi/kg of ^{252}Cf citrate into 35-day-old rats and estimated the half-time to be 1000 days. Later, MEWHINNEY and HARRIS (1972, and unpublished) measured whole-body retention of ^{252}Cf up to 130 days after injection of 5 μCi/kg of ^{252}Cf citrate into adults of several species. Whole-body half-times of ^{252}Cf in rats and mice were 490 and 410 days, respectively.

Whole-body retention measurements of ^{252}Cf in Chinese hamsters, Syrian hamsters, and beagles included the large contribution of the ^{252}Cf retained in the liver as well as that in bone; measured half-times for these animals were 285, 580, and 630 days, respectively. Whole-body and partial-body measurements up to 400 days after injection indicated that the retention half-time of ^{249}Cf or ^{252}Cf in nonliver tissues (mostly bone) of the beagle is at least 11 years (LLOYD et al., 1972e).

Whole-body retention can also be calculated from the difference between 100% and the summed excreta as is shown for $^{249-252}$Cf and ^{241}Am in Fig. 18.22 (DURBIN et al., 1972c, d). Beginning 50 days after injection, when most of the $^{249-252}$Cf has been cleared from the liver and soft tissues of the rat and, like ^{241}Am, the $^{249-252}$Cf deposited in the incisors during the first few days has been lost by functional attrition (DURBIN et al., 1967), the slopes of the ^{241}Am and $^{249-252}$Cf whole-body retention curves were the same ($T_{1/2}$ = 730 days). From these findings californium appears to be retained in the rodent skeleton at least as long as is ^{241}Am.

Retentions of ^{249}Bk and ^{253}Es have not been studied beyond 24 days after injection. For ^{241}Am, ^{242}Cm, and $^{249-252}$Cf the half-times of whole-body retention in rats (equivalent to skeletal retention in rats at times longer than 30 days after injection) have been found to be similar as far as available measurements permit comparison. One may infer that the half-time of retention of ^{249}Bk in the rat skeleton would also be similar, inasmuch as ^{249}Bk occupies the position between ^{242}Cm and $^{249-252}$Cf. The physical half-life of ^{253}Es, the major einsteinium isotope, is only 20.5 days, and except for rapid loss from short-lived bony structures, this is also its effective half-time in bone.

VIII. Deposition and Retention of Transplutonium Elements in Liver

One of the distinguishing features of the metabolism of the lighter weight members of the lanthanide and actinide series is their large initial uptake by the mammalian liver (HAMILTON, 1947, 1948). In controlled experiments in rats it has been shown that, like their lanthanide homologues, the amounts of the individual trivalent actinides taken up by the liver were related directly to the ionic radius and inversely to the atomic number (DURBIN, 1962).

A. Gross Deposition

1. Americium

Four to 22 days after parenteral administration, the liver uptake of ^{241}Am ranged from 29% of the injected dose in mice to 54% in cynomolgus monkeys and Chinese hamsters (Tables 18.12 through 18.17). The time at which liver ^{241}Am reached a maximum value varied among the species; in rats and mice this was 6 to 24 hours after injection of ^{241}Am citrate (Fig. 18.13). Between days 1 and 4 after injection, about 10% of the injected dose was eliminated from the livers of these rodents and appeared in the feces, so that the maximum liver content, L_{max}, of ^{241}Am as calculated from Eq. (23) was about 40% of the dose for the mice and 50% for the rats.

$$L_{max}(\%) = L_t + (\text{G.I. contents})_t + \sum Fe_t; \qquad \text{Eq. (23)}$$

where L_t and $(\text{G.I. contents})_t$ are the amounts of ^{241}Am in liver and G.I. contents, respectively, at time t; and $\sum Fe_t$ is the cumulative fecal excretion to time t. The initial deposition of ^{241}Am in the Chinese hamster liver appeared to be complete by 24 hr and changed little in the succeeding 64 days. The soft tissues (apart from liver and kidneys) of the dogs and monkeys contained appreciable amounts of ^{241}Am for several days after injection. Americium-241 continued to accumulate in the livers of these animals apparently as the result of the return to the circulation of the ^{241}Am cleared from soft tissues. The ^{241}Am in dog liver increased from 48 to 54% between days 7 and 22 (ATHERTON et al., 1968; LLOYD et al., 1970; ATHERTON and LLOYD, 1972). Between days 1 and 8 the ^{241}Am content of the monkey livers rose from 40 to 54% (DURBIN et al., 1972c).

Within-group variability of the initial liver uptake of ^{241}Am was not large (Table 18.18). In an experiment in which a large group of female rats of the same age and body weight were each given a single injection of 0.37 μCi/kg of ^{241}Am citrate, the S.D. was 10 to 15% of initial liver uptake. The mean liver ^{241}Am of 5 rats killed 1 day after injection was 50.1% $\pm$7.4% (DURBIN et al., 1967); the liver ^{241}Am (L_{max}) of 30 rats killed 1 to 46 days after injection was calculated from Eq. (23) to be 49.4% $\pm$4.6% (DURBIN et al., 1972c).

In the rat experiments, initial liver uptake of ^{241}Am also varied to some extent with the experimental conditions (Table 18.18). Those factors that influenced initial ^{241}Am deposition in skeleton and liver—the age of the rats, the chemical form of the ^{241}Am, and the route of administration—were discussed in Sec. VII.A.1. Inasmuch as the sum of the initial deposits of ^{241}Am in the rat liver and skeleton was nearly constant, those conditions that favored skeletal uptake of ^{241}Am—active bone growth, administration of ^{241}Am as the citrate complex, and administration by intramuscular injection, by inhalation, or by mouth—tended at the same time to divert ^{241}Am from the liver. Conversely, those experimental conditions that favored liver uptake of ^{241}Am absence of active bone growth, and administration of readily hydrolyzable salts by intravenous or intraperitoneal injection—tended at the same time to suppress ^{241}Am deposition in bone. In addition to the above factors, ZALIKIN et al. (1969) reported that initial deposition of ^{241}Am in liver was greater (by 11% of the injected dose) and skeletal deposition less by nearly the same amount when a large quantity of $^{241}AmCl_3$, 200 μCi/kg, was administered intravenously.

2. Curium

To date, the metabolism of parenterally administered curium has been studied only in rats (Table 18.12 and 18.19). In 7 experiments with rats L_{max} calculated from Eq. (23) was 56.7% ± 4.2%—somewhat lower than L_{max} calculated for ^{241}Am (63.6% ± 7.2% also in 7 studies with adult rats). Curium-244 metabolism has been investigated in beagles exposed to aerosols of $^{244}CmCl_3$ or $^{244}CmO_{1.73}$ (McCLELLAN et al., 1972). The ^{244}Cm was rapidly absorbed into the body from these aerosols. Eight days after exposure, 80% of the ILB of $^{244}CmCl_3$ and 70% of the IBL of $^{244}CmO_{1.73}$ had been absorbed. The livers of 10 dogs with an average ILB of 0.26 μCi/kg of $^{244}CmCl_3$ contained 28.8% ± 4.1% of the ILB (31% of the absorbed dose) from 8 to 256 days after exposure. The livers of 6 dogs with an ILB of 0.14 μCi/kg of $^{244}CmO_{1.73}$ contained 31.5% ± 4.1% of the ILB (34% of the absorbed dose) during the interval 8 to 64 days after exposure. The deposition of ^{241}Am in the livers of 2 dogs with an average ILB of 3 μCi/kg of ^{241}Am was 34% of the ILB (38% of the absorbed dose) 127 and 256 days after exposure to an $^{241}AmO_2$ aerosol (R. G. THOMAS et al., 1972).

The within-group variability of the initial ^{242}Cm liver uptake in separate rat experiments was less than was found for ^{241}Am; S.D.'s ranged from 1 to 6% of observed mean values. As was noted above in Sec. VII.A.2, the sum of the initial depositions of ^{242}Cm in liver and skeleton was nearly constant. Examination of Table 18.19 suggests that the factors that influenced the initial partition of ^{241}Am between liver and bone also influenced, but to a lesser degree, the initial partition of ^{242}Cm between liver and bone. The fraction of ^{242}Cm originally deposited in the livers of rats was only slightly influenced by the growth status of the animals, the chemical form of the ^{242}Cm, or the route of administration.

3. Berkelium

HUNGATE et al. (1972) reported that the accumulation of ^{249}Bk in the liver of the rat was 23% of the dose 4 hr after intravenous injection of $^{249}BkCl_3$. By day 21 the liver contained only 4% of the injected ^{249}Bk (Table 18.12).

A systematic study of the metabolism of the lanthanide elements in the rat demonstrated that a significant reduction in liver uptake occurred at gadolinium ($4f^7$; ionic radius, 0.94 Å) midway through the 14-element series (DURBIN et al.,

1956; DURBIN, 1962). The $5f$ shell of the actinide series is half filled at curium, but the ionic radius of curium is 0.98 Å (ZACHARIASEN, 1954). The sharp downward shift in liver deposition occurs in the actinide series at berkelium ($5f^8$; ionic radius, 0.954 Å, PETERSON, 1967). (See also Sec. VII.A.6.)

4. Californium

Distribution of $^{249-252}Cf$ in the mouse, rat, Chinese hamster, and beagle is shown in Tables 18.12 through 18.16. The liver content of $^{249-252}Cf$ 4 to 21 days after parenteral administration of the citrate complex varied from 7.5% of the injected dose in mice to 27% of the dose in Chinese hamsters. L_{max} of $^{249-252}Cf$ calculated from Eq. (23) was 25% of injected $^{249-252}Cf$ in rats and 26% of ^{252}Cf in mice—very close to the 4-day liver content of ^{252}Cf in Chinese hamsters and ^{249}Cf in beagles.

The L_{max} of $^{249-252}Cf$ in the rat was about the same as the 4-hr liver uptake, which constituted 23% of the injected ^{249}Bk, as reported by HUNGATE et al. (1972). In all species studied, $^{249-252}Cf$ uptake in liver was significantly less than the liver uptakes of ^{241}Am or $^{242-244}Cm$.

There are not enough experiments with one species to examine the possible influence of experimental conditions on the initial deposition of $^{249-252}Cf$. It was noted above, in the discussions of the factors influencing the deposition patterns of newly injected ^{241}Am and ^{242}Cm, that the distribution of ^{242}Cm seemed to be less dependent on experimental conditions than was that of ^{241}Am. If that trend is real and continues, one would anticipate that the gross initial distribution of $^{249-252}Cf$ in a single species would be even less affected by the growth status of the animals or the chemical form or route of administration of the $^{249-252}Cf$.

5. Einsteinium

The initial liver uptake of ^{253}Es is low in both mouse and rat. Four days after intramuscular injection of ^{253}Es citrate, the mouse liver contained 3.3% of the injected dose; 1.4% was found in passed feces, and 12% was not recovered (PARKER et al., 1972; Table 18.14). If the unrecovered ^{253}Es had been in feces, L_{max} of ^{253}Es in the mouse would lie between 5 and 17%.

Four hours after intravenous injection, rat liver contained 18% of the ^{253}Es administered as $^{253}EsCl_3$ (HUNGATE et al., 1972) and 33.5% of the ^{253}Es administered as $^{253}Es(NO_3)_3$ (V. H. SMITH, 1972b). Neither report included measurements of ^{253}Es in plasma or in accumulated urine during the first 4 hr, so it was not possible to account for the unusually high urine output of $^{253}EsCl_3$ which tended to divert ^{253}Es from all tissues [during the first 24 hr after injection 12% of $^{253}Es(NO_3)_3$ and 36% of $^{253}EsCl_3$ were excreted in the urine, see Sec. XI.A.1.]. The L_{max} for ^{253}Es was calculated from Eq. (23) and the data of V. H. SMITH (1972b) for rats killed 1, 7, and 24 days after injection of $^{253}Es(NO_3)_3$ and was 26, 23.5, and 22.2%, respectively. The average of the three estimates of L_{max} for ^{253}Es was 23.4%, very close to the 25% calculated above for $^{249-252}Cf$.

B. Microscopic Distribution

The microscopic localization of ^{241}Am or $^{249-252}Cf$ in the livers of several mammals has been studied autoradiographically. The distribution of these elements within hepatic cells has been examined by fractionation of homogenates of canine liver.

1. Initial Distribution of ^{241}Am, ^{249}Bk, and $^{249-252}$Cf

Within hours after administration in soluble form ^{241}Am, ^{249}Bk, and $^{249-252}$Cf are distributed almost uniformly throughout liver tissue, and are associated with hepatic cells (Fig. 18.23a). The experimental animals in which this was observed include the mouse, ^{241}Am (HAMMARSTRÖM and NILSSON, 1970a); rat, ^{241}Am (ZALIKIN et al., 1968; RUDNITSKAYA and MOSKALEV, 1970; DURBIN et al., 1972c), $^{249-252}$Cf (DURBIN et al., 1972d); beagle, ^{241}Am (G. N. TAYLOR et al., 1969a), ^{249}Bk, and ^{249}Cf (G. N. TAYLOR et al., 1972); and monkey, ^{241}Am (DURBIN et al.,

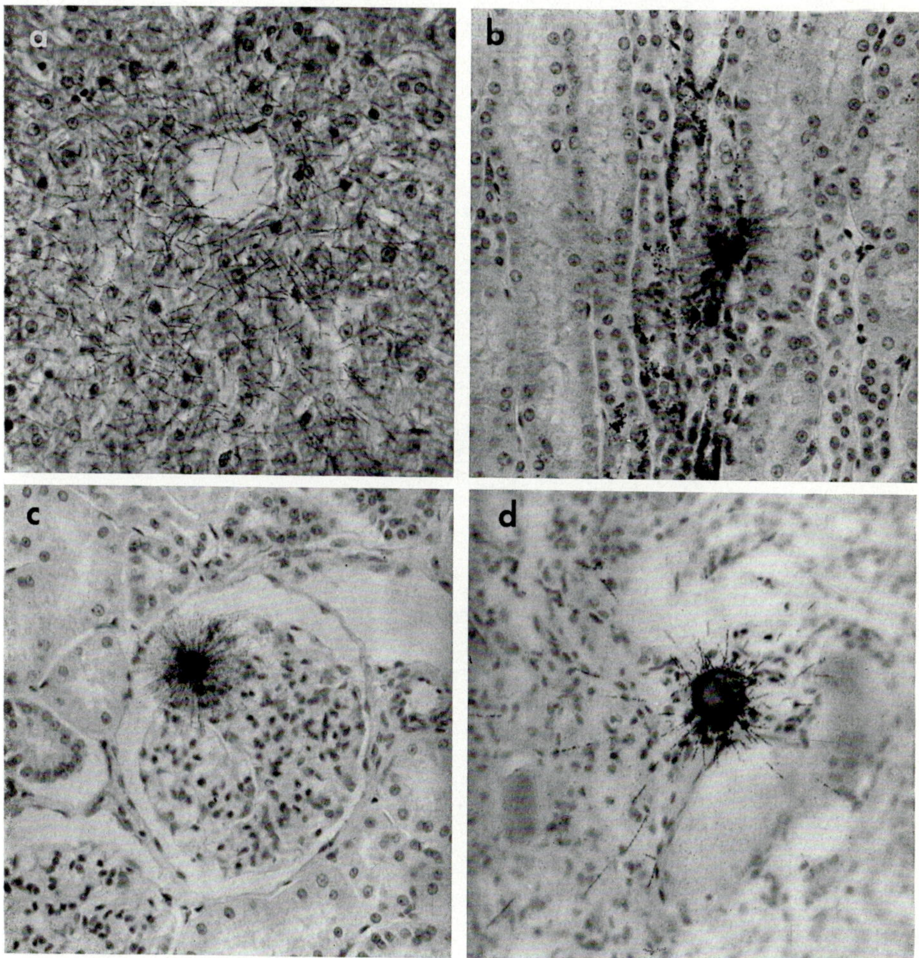

Fig. 18.23a—d. Microautoradiographs of ^{241}Am in liver or kidneys of monkeys or rats. Autoradiographs were stained with hematoxylin-eosin and prepared as described by BOYD (1955) using Kodak NTA plates. ×267. a In the liver of adult monkey no. 8, described in Fig. 18.16b, the ^{241}Am is more concentrated near the central veins 8 days after injection. Plate exposed 540 days. b In the kidney of adult monkey no. 45, described in Fig. 18.18b, single tracks and track clusters are associated with brown pigment in renal tubules and intertubular connective tissue 336 days after injection. Plate exposed 887 days. c In the kidney of adult monkey no. 55 (more than 6 years old, killed 21 days after intramuscular injection of 0.71 μCi/kg of ^{241}Am citrate), aggregates of alpha tracks can be seen in the glomerular tuft. Plate exposed 545 days. d In the kidney of a rat (described in Fig. 18.14) killed 712 days post-injection, ^{241}Am alpha tracks emanate from the surface of a renal stone. Plate exposed 274 days

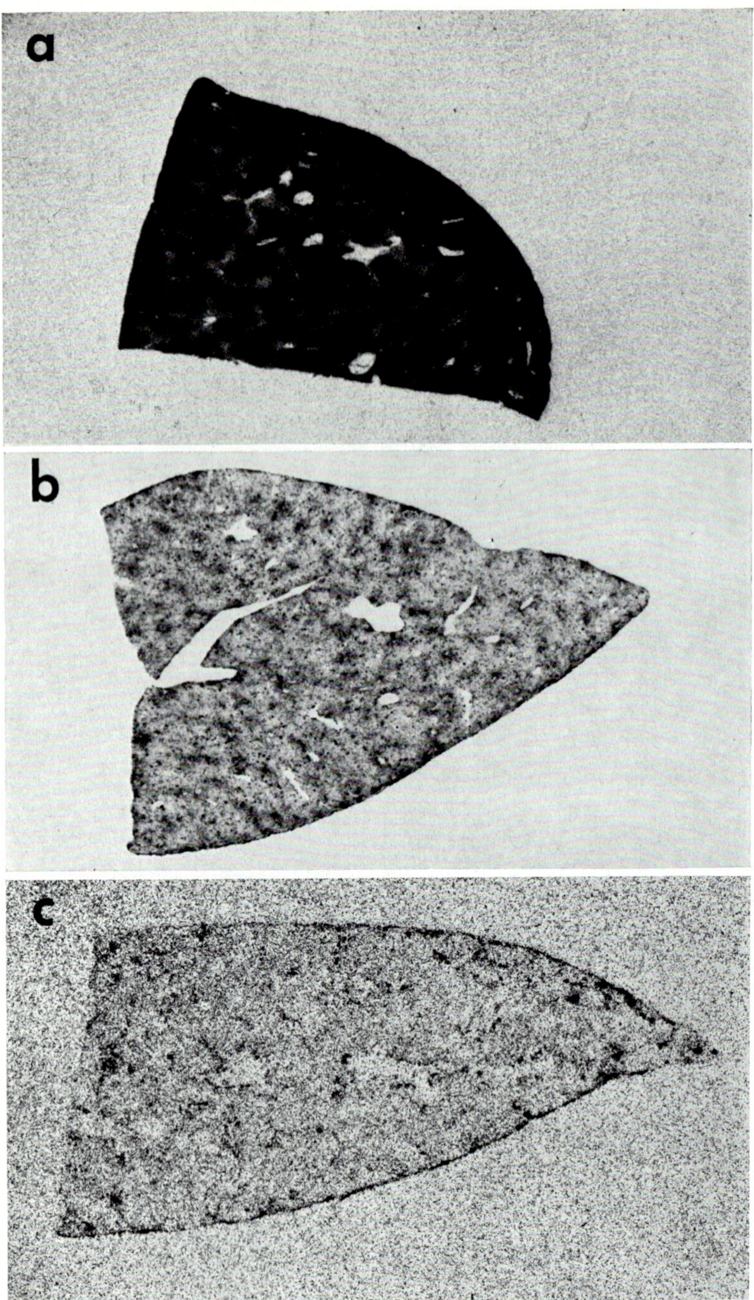

Fig. 18.24a—c. Americium-241 in the liver of adult monkeys. Gross autoradiographs were prepared as described by BOYD (1955). × 5. Note: The gross autoradiographs shown in Figs. 18.24, 18.25, 18.28, and 18.29 were prepared over a 12-year interval. During that time Kodak medical X-ray film was changed and Dupont film became available. For visual comparisons Kodak (red-base) sensitivity equals 2.5, Kodak (green-base) sensitivity equals 5, and Dupont Cronar (low graininess) sensitivity equals 1. a Adult no. 7 (killed 1 day after intramuscular injection of 0.67 μCi/kg of ^{241}Am citrate). Kodak single-coated (green-base) medical X-ray film was exposed 369 days. b Adult no. 73 (killed 168 days after intravenous injection of 0.69 μCi/kg of ^{241}Am citrate). Dupont Cronar single-coated medical X-ray film was exposed 766 days. c Adult no. 54 (killed 938 days after intramuscular injection of 0.69 μCi/kg of ^{241}Am citrate). Kodak single-coated (red base) X-ray film was exposed 1633 days

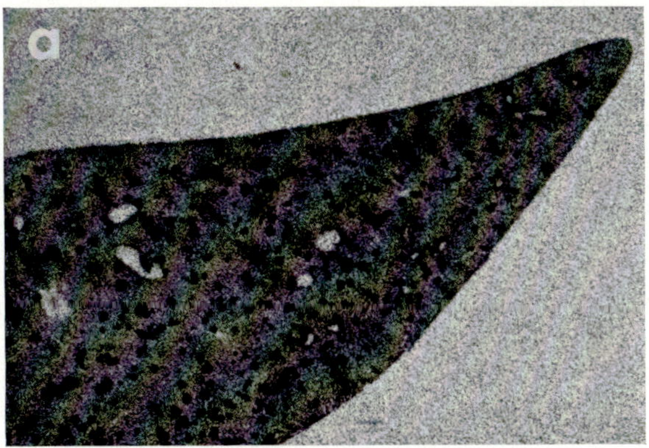

Fig. 18.25a—c. Gross autoradiographs of ^{241}Am in the livers of rats described in Fig. 18.14. Gross autoradiographs were prepared as described in Fig. 18.24. ×8. a At 1 day after injection random foci of high concentration are superimposed on a generally uniform distribution. Kodak single-coated (green-base) medical X-ray film was exposed 383 days. b At 15 days after injection intense foci are no longer present, and slightly higher concentrations near central veins persist. Kodak single-coated (red-base) medical X-ray film was exposed 1939 days. c At 32 days after injection radioactivity is barely detectable above film background. Kodak single-coated (red-base) medical X-ray film was exposed 1939 days

1972c). The initial distribution of the transplutonium elements in mammalian liver appeared to be independent of the dose, within the limits of the doses that have been studied.

Cell fractionation showed that within the first few days after their administration as citrate complexes, ^{241}Am and ^{249}Cf in the liver of the dog are primarily bound to soluble ferritin, the iron-storage protein of the liver, but within a few weeks they had been shifted to insoluble ferritin incorporated into or on the surface of cellular organelles (STOVER et al., 1970, 1972; STEVENS et al., 1971; BRUENGER et al., 1972).

There are some early minor variations in the microscopic distribution of ^{241}Am in the livers of rats and monkeys, and these seem to be related to local fluctuations of the hepatic circulation. The overexposed autoradiographs of monkey liver shown in Figs. 18.23a and 18.24a (monkeys killed 1 and 8 days after intramuscular injection of ^{241}Am citrate) reveal concentrations of ^{241}Am near central veins about twice the concentration in portal areas.

The earliest autoradiographs of rat liver (Fig. 18.25a): rat killed 1 day after intramuscular injection of ^{241}Am citrate, show variable numbers of intense foci of alpha tracks associated with clusters of 10 to 20 liver cells. In these foci the ^{241}Am concentration was 5 to 20 times the general hepatic background. It was not possible in these preparations to distinguish among ^{241}Am tracks emanating from hepatic cells, sinusoids, or bile ducts; however, the foci of greater ^{241}Am concentration did not appear to involve uptake in Kupfer cells.

Four days after intramuscular injection of $^{249-252}$Cf citrate the concentration of $^{249-252}$Cf in rat liver was somewhat greater around central veins, but there were no foci of high concentration (DURBIN et al., 1972d).

In the liver of the beagle (as in all other beagle tissues examined) the connective tissue of the small arterioles was an important secondary site of deposition of ^{241}Am, ^{249}Bk, and ^{249}Cf (G. N. TAYLOR et al., 1969a, 1972).

2. Sequential Changes in Microscopic Distribution

After initial deposition is complete, a number of changes occur in the microscopic distribution of ^{241}Am and $^{249-252}$Cf in the livers of rats, dogs, and monkeys. These changes, which vary with both the species and the injected dose, are the result of (a) excretion of the element from the liver into the gastrointestinal tract and (b) redistribution within the liver and (or) elimination into the circulation following radiation death of hepatic cells.

a) Rat. Americium-241, $^{242-244}$Cm, ^{249}Bk, and $^{249-252}$Cf were promptly and quantitatively excreted from the liver into the feces of rats and mice if the injected dose was less than 5 μCi/kg (PARKER et al., 1962; ZALIKIN et al., 1968; HUNGATE et al., 1972; LINDENBAUM and ROSENTHAL, 1972; and Tables 18.12 through 18.14). If the injected dose was large, excretion of these nuclides from the livers of rats and mice was delayed, and a fraction of the liver nuclide content was released into the circulation as shown by continued accumulation in the skeleton (TSEVELEVA and YEROKHIN, 1969; and Tables 18.13 and 18.14).

The changes in the microscopic distribution of ^{241}Am in the rat liver can be summarized from the low-dose experiment: 0.37 μCi/kg of ^{241}Am citrate (DURBIN et al., 1967, 1972c). Decline of the ^{241}Am radioactivity in liver sections was visually detectable as soon as 4 days after injection, and excretion of ^{241}Am from the rat liver was nearly complete by the 64th day. Thereafter, microautoradiographs showed a faint overall dusting of single alpha tracks and occasional reticuloendothelial cells from which several alpha tracks emerged. As early as

15 days after injection, and even at this low dose level, radiation damage (followed by hepatic cell death) could be demonstrated in the foci of the greatest ^{241}Am concentration (Fig. 18.25a). The total radiation doses in the low- and high-intensity regions were estimated to be 9 and 90 rads, respectively. By the 15th day all of the ^{241}Am that had been initially associated with the high-intensity foci had been excreted (Fig. 18.25b, c), and all of the damaged cells were gone. Evidence of cell destruction persisted—brown pigment, presumably hemosiderin (not observed heretofore) was present in many reticuloendothelial cells, and some of these pigment-laden cells also emitted multiple alpha tracks. The only evidence of repaired cellular damage (compared to age controls) was a slight elevation of the number of polyploid and binucleated hepatic cells.

The initial distribution of $^{249-252}$Cf in rat liver was more uniform than was that of ^{241}Am. Excretion of $^{249-252}$Cf from rat liver was prompt and complete, and 91 days after injection only occasional single alpha tracks could be seen. There was no evidence of residual or repaired cell damage.

In the rat, although high doses of ^{241}Am were lost from the liver both by excretion into bile and by recirculation, the net rate of loss of ^{241}Am was slower than in low-dose studies (LANGHAM and CARTER, 1951; D. M. TAYLOR et al., 1961; TSEVELEVA and YEROKHIN, 1969). RUDNITSKAYA and MOSKALEV (1970) calculated that 5 μCi/kg of ^{241}Am delivered a total radiation dose of about 300 rads to the rat liver. Thus, at dose levels greater than 5 μCi/kg of ^{241}Am, radiation damage in the liver was severe, and removal of cell debris containing ^{241}Am was inefficient. As long as 200 days after injection of large doses of ^{241}Am, RUDNITSKAYA and MOSKALEV (1970) found that rat liver contained significant numbers of clusters of alpha tracks, presumably emanating from reticuloendothelial cells. Deaths that occurred later than 200 days after injection in the high-level groups were attributed to liver insufficiency.

b) Monkey. The livers of monkeys and baboons lost ^{241}Am, but more slowly than did the livers of rats and mice given comparable doses (ROSEN et al., 1972; DURBIN et al., 1972c; and Table 18.17). The dose range used in primate studies, 0.20 to 0.87 μCi/kg, was estimated to deliver total radiation doses ranging from 95 to 415 rads, respectively, to the liver. At the higher doses, loss of ^{241}Am from monkey liver was a combination of excretion into feces and elimination into the circulation.

The general distribution of ^{241}Am throughout the monkey liver, with slightly greater early concentration near central veins (Fig. 18.24a) persisted almost unchanged for 42 to 70 days. After that time the decline in ^{241}Am concentration could be seen in the autoradiographs (Fig. 18.24b). Concentrations of alpha tracks emanating from clusters of hepatic cells were found in some but not all portal areas 42 to 70 days after injection—coincident with the period of maximum fecal excretion. These clusters of ^{241}Am-bearing cells had disappeared by 168 days.

Between 42 and 70 days, aggregations of alpha tracks emanating from reticuloendothelial cells (which also contained granules of brown pigment) could be found beneath the capsule, in portal connective tissue, and occasionally near central veins. The number of these multiply-labeled pigment-laden cells rose to a maximum between 112 and 168 days (Fig. 18.24b) and declined thereafter. After 336 days the general hepatic background was low, and the number of aggregations of tracks associated with pigmented reticuloendothelial cells was small. The aggregates of ^{241}Am were most often located beneath the capsule or near central veins (Fig. 18.24c).

Radiation damage was either mild or, to a large extent, repairable. Although the livers of the 3 monkeys killed between 336 and 938 days might be considered slightly atrophic, the changes were so minimal that they were not distinguishable from normal aging.

c) Dog. The dog liver excretes little, if any, ^{241}Am (LLOYD et al., 1970). At low doses (less than 0.1 μCi/kg) ^{241}Am is retained; at high doses it appeared to be redistributed within the liver coincident with its release into the circulation following the radiation death of hepatic cells (G. N. TAYLOR et al., 1967, 1969a). Late uptake of ^{241}Am in the spleens and lymph nodes of the high-dose ^{241}Am dogs suggested that at least part of the redistribution of a high dose of ^{241}Am within the dog liver—a general shift from hepatic sinusoids to portal connective tissue—takes place via recirculation of cell debris. Both redistribution of ^{241}Am within the dog liver and recirculation of ^{241}Am to bone and other tissues were more rapid and more dramatic, the larger the injected dose (LLOYD et al., 1970).

The time-dependent changes of the microscopic distribution of ^{241}Am in the canine liver have been studied only in high-dose animals (2.8 μCi/kg of ^{241}Am). Biopsy specimens of liver taken 150 days after injection showed a generally uniform hepatic distribution similar to that seen in specimens taken from 1 to 22 days after injection. At 150 days occasional aggregations of alpha tracks could be found emerging from hemosiderin-laden reticuloendothelial cells in hepatic sinusoids. By 250 days, 27% of the original liver burden (13% of the injected dose) had been lost, and more than one-half of the remaining ^{241}Am was associated with reticuloendothelial cells in hepatic sinusoids. By the 401st day 54% of the original liver burden (26% of the injected dose) had been eliminated, and most of the remaining ^{241}Am was associated with aggregations (some quite large) of alpha tracks emanating from pigmented reticuloendothelial cells. These cells (containing both hemosiderin and ^{241}Am) were located for the most part interstitially in portal areas, frequently near small interlobular veins (G. N. TAYLOR et al., 1967, 1969a). At that time, the amount of radiation absorbed was large: 5500 to 7600 rads; the degree of hepatic cell damage (and presumably of reticuloendothelial cells also) was great; there was no regeneration. The livers of the 400-day dogs were only one-half normal weight. There was also evidence of damage in connective tissue and vasculature. Liver failure was the primary cause of death (G. N. TAYLOR et al., 1969a; LLOYD et al., 1970). Liver function tests indicated significant liver injury 1 to 2 years after 0.9 μCi/kg of ^{241}Am, and as of March 1972, liver insufficiency was the cause or a major contributor to the cause of deaths of 2 of the 12 beagles in that dose level group (LLOYD et al., 1972a). Cellular degeneration, fatty changes, and fibrosis were noted in the livers of two beagles killed 512 and 1022 days after receiving an average ILB of 2.8 μCi/kg from exposure to an aerosol containing $^{241}AmO_2$ (R. G. THOMAS et al., 1972).

C. Retention

Low doses of all the actinide elements are excreted almost quantitatively from the livers of rats and mice (Tables 18.12–18.14). The livers of the cynomolgus monkey and baboon excrete the major fraction of a low-dose ^{241}Am burden. At doses greater than 0.2 μCi/kg some ^{241}Am is released from damaged liver cells and is recirculated (ROSEN et al., 1972; DURBIN et al., 1972c). The livers of beagles and Chinese and Syrian hamsters did not excrete significant amounts of any actinide element studied (STOVER et al., 1959, 1971; LLOYD et al., 1970, 1972e; McKAY et al., 1972; MEWHINNEY et al., 1972; MEWHINNEY and HARRIS, 1972; and Tables 18.15 and 18.16). Some of the actinide burden in the dog liver was released into the circulation subsequent to the death of labeled hepatic parenchymal and reticuloendothelial cells.

1. Rat and Mouse

Rapid clearance of ^{241}Am (Fig. 18.26) and ^{242}Cm from the rat liver and the quantitative recovery of the cleared nuclide in feces were demonstrated in the first animal experiments with these elements (SCOTT et al., 1948a, 1949). The rapid decline in whole-body retention of ^{241}Am and $^{249-252}$Cf between days 1 and 45 (Fig. 18.22) represents largely excretion from the liver.

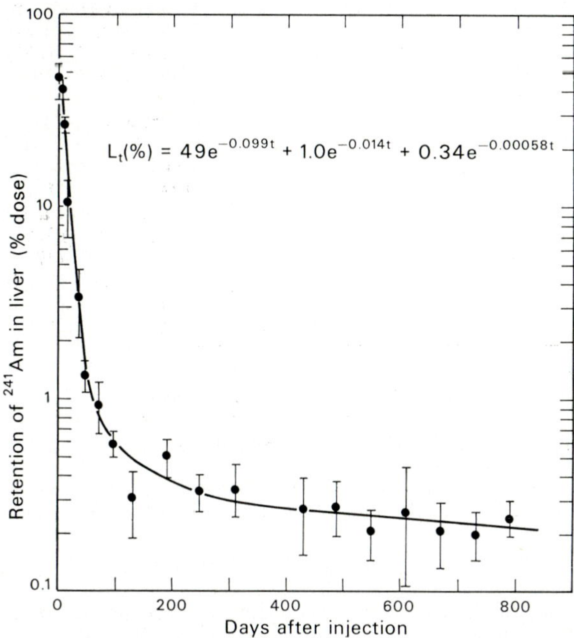

Fig. 18.26. Retention of ^{241}Am in the liver of the rat. Data of DURBIN et al. (1967, 1972c). Each point represents the mean for five or more rats; bars represent the S.D.

The ability of rat and mouse liver to clear the trivalent actinides was not measurably impaired if the nuclide was given in soluble form and if the injected dose was less than about 5 μCi/kg. The amount of radiation absorbed was apparently low enough so that local cellular damage could be repaired or compensated. Larger doses induced more damage, and biliary secretion—presumed to be the process by which monomeric cations are excreted from the liver—became slower and less efficient, prolonging liver retention (LANGHAM and CARTER, 1951). Polymeric or hydrolyzed actinides are lost from the rat and mouse liver at least in part by some other mechanism which proceeds more slowly than biliary excretion of unaggregated forms (TSEVELEVA and YEROKHIN, 1969; LINDENBAUM and ROSENTHAL, 1972).

The half-times of retention of soluble actinides in rat liver calculated from low-dose experiments for ^{241}Am, ^{242}Cm, $^{249-252}$Cf, and ^{253}Es are compared with retention half-times of three light-weight lanthanides, ^{144}Ce, ^{147}Pm, and ^{152,154}Eu in Table 18.22. The liver retention half-times were calculated either directly from measurements of dissected livers or indirectly from cumulative fecal excretion[26].

26 Calculation of liver retention from cumulative fecal excretion assumes that (a) between days 1 and 45 (after correction for attrition of incisors) radioactivity in the feces represents biliary excretion into the intestine, and (b) no radioactivity is reabsorbed from the gastrointestinal tract into the body.

Table 18.22. Clearance of ^{241}Am, ^{242}Cm, $^{249-252}Cf$ and four lanthanides from the liver of the rat

Isotope	Liver clearance half-time (days)			Reference
	injection route[a]	liver	fecal excretion	
$^{241}AmCl_3$	i.m.	7.5	7.3	SCOTT et al. (1948a)[b]
$^{241}AmCl_3$	i.v.	16[c]	—	LANGHAM and CARTER (1951)
^{241}Am citrate	i.v.	—	10.8[c]	D. M. TAYLOR et al. (1961)
^{241}Am citrate	i.m.	7	7	DURBIN et al. (1972c)
$^{241}AmCl_3$	i.v.	8	—	ZALIKIN et al. (1968)
$^{241}Am(NO_3)_3$	i.p.	10[c]	—	TSEVELEVA and YEROKHIN (1969)
$^{242}CmCl_3$	i.m.	8.5	9.0	SCOTT et al. (1949)[b]
$^{242}CmCl_3$	i.v.	—	9.2	K. G. SCOTT, M. H. WILLIAMS and H. I. JOHNSON, unpublished[b]
$^{249-252}Cf$ citrate	i.m.	—	8.2	DURBIN et al. (1972d)
^{252}Cf citrate	i.p.	8	—	MEWHINNEY et al. (1972)
$^{253}Es(NO_3)_3$	i.v.	5.6	—	V. H. SMITH (1972b)
^{144}Ce citrate	i.m.	—	8.3	DURBIN et al. (1956)[b]
^{147}Pm citrate	i.m.	—	7.0	DURBIN et al. (1956)[b]
$^{152,154}Eu$ citrate	i.m.	—	6.5	DURBIN et al. (1956)[b]
^{160}Tb citrate	i.m.	5.6	—	DURBIN et al. (1956)[b]

Average liver half-time:
^{241}Am, 7.5 $\pm$0.5 days ^{242}Cm, 9 $\pm$ 0.3 days $^{249-252}Cf$, 8.1 days

[a] i.m., intramuscular; i.v., intravenous; i.p., intraperitoneal.
[b] Calculated from the original data.
[c] Not included in average because injected dose exceeded 5 μCi/kg (LANGHAM and CARTER, 1951; D. M. TAYLOR et al., 1961), or because ^{241}Am was administered in a readily hydrolyzable form.

Whenever possible, both methods of calculation were used. On the average, the hepatic clearance half-time of soluble forms of three of the lighter and larger actinides (ionic radius $\geqq$ 0.94 Å) was 8.2 days for the rat. This compares favorably to the clearance half-times of ^{144}Ce and ^{147}Pm (8.3 and 7.0 days, respectively) in the rat and to 9 days for ^{241}Am citrate in the mouse (LINDENBAUM and ROSENTHAL, 1972). Liver clearances of ^{253}Es and the heavier, smaller lanthanides, $^{152,154}Eu$ and ^{160}Tb (ionic radii $\leqq$ 0.95 Å), were faster, ranging from 5.6 to 6.5 days.

2. Monkey

Early clearance of ^{241}Am from the monkey liver (Fig. 18.21) is complex. At the dose levels studied, 0.44 to 0.87 μCi/kg, ^{241}Am appeared to be lost from the monkey liver by both biliary excretion and by recirculation following liver cell damage. Hepatic clearance of ^{241}Am in rat and mouse began immediately; maximum daily rates of liver loss and fecal excretion both occurred between days 1 and 4 postinjection. In the monkey, on the other hand, there was a delay of 4 to 8 weeks after injection (the delay time for 10 of 12 monkeys was 4.5 weeks) during which time only 5% of the injected dose of ^{241}Am was lost from the liver and only 3% of the dose had been excreted in feces. After the quiescent period, nearly all of the initial ^{241}Am liver burden was lost with a half-time of 70 days (calculated both from measurements of dissected livers and from cumulative fecal excretion of several individual monkeys).

It is not yet known how much of the variation in the ^{241}Am content of the monkey skeleton (Fig. 18.21) was the result of the initial partition of ^{241}Am between skeleton and liver and how much was the result of late uptake of ^{241}Am

recirculated from the liver. If it is assumed that (a) the initial deposition of ^{241}Am in the monkey liver, L_{max}, is 52.5% of the injected dose, and (b) $\sum Fe_t$, the cumulative fecal excretion, represents biliary loss from liver, then the fraction of L_{max} which was lost by recirculation can be estimated:

$$\text{Amount recirculated} = L_{max} - (L_t + \sum Fe_t). \qquad \text{Eq. (24)}$$

Fecal excretion of ^{241}Am by 11 monkeys killed 63 to 973 days after injection accounted for 43 to 100% of the average L_{max} (40.6 to 52.5% of the injected dose). The average fecal excretion was 0.77 L_{max}. Loss of ^{241}Am from the monkey liver in the dose range 0.44 to 0.87 μCi/kg, as a consequence of recirculation of ^{241}Am from damaged cells, was about 23% of L_{max}.

In two baboons that received 0.2 μCi/kg of ^{241}Am citrate (one-fourth to one-half the dose received by the monkeys) the clearance half-time of ^{241}Am from the liver was determined to be 152 days, based on periodic biopsies and in vivo measurements up to 86 days after injection (Rosen et al., 1972). There was no indication in these animals that ^{241}Am originally deposited in the liver was released into the circulation.

3. Dog

Biliary excretion of ^{241}Am from dog liver has not been demonstrated, but it may contribute to the small fraction of a low dose of ^{241}Am that is lost slowly. The sequential changes in the microscopic distribution of ^{241}Am in the livers of high-dose dogs (G. N. Taylor et al., 1969a) and the time course of the uptake in the skeletons and soft tissues of these same dogs (Lloyd et al., 1970) indicate that nearly all of the observed loss from liver is the result of release of ^{241}Am from damaged cells.

The amount of nuclide lost from the dog liver, and the rate of that loss, are both dependent on the radiation dose in liver tissue, that is, proportional to the administered dose. At an injection level of 2.8 μCi/kg of ^{241}Am, the dog liver began to lose ^{241}Am about 50 days after injection. The rate of loss of ^{241}Am continued to accelerate until the dogs died 401 and 448 days after injection. At that time the liver contained only 27% of its initial ^{241}Am deposit. The average half-time of liver retention was estimated to be less than 240 days. After injection of 0.9 μCi/kg of ^{241}Am, very little ^{241}Am was lost from the liver during the first 200 days; but subsequently ^{241}Am was lost from the liver with a half-time of about 750 days.

The net loss of ^{241}Am from the livers of dogs given still lower doses (0.02 to 0.3 μCi/kg) was small, but appeared to proceed at a constant rate during the 850-day observation period. The liver retention half-time was about 3500 days, which is equivalent to a rate of loss of 0.02% of the liver burden per day.

Retention of ^{249}Cf or ^{252}Cf in beagle liver has been measured by a combination of whole-body and partial-body counting (Lloyd et al., 1972b, e). The average ^{249}Cf content of the livers of 7 beagles was 27% 1 to 2 days after injection (3 beagles each injected with 2.8 μCi/kg of ^{249}Cf citrate and 4 beagles injected with 0.3 μCi/kg). By the 14th day the average ^{249}Cf content of the livers of 6 of these beagles (one was killed before the 14th day) had declined to 17.8%—a loss of 9.2%. Between days 2 and 14, 4.3% was excreted in the feces providing suggestive evidence for early biliary excretion of ^{249}Cf by the beagle. The fact that all of the ^{249}Cf lost from the liver could not be accounted for by fecal excretion suggests further that some of the initial ^{249}Cf liver deposit was loosely bound and could be released into the circulation. Radiation damage of liver cells would not seem to be involved in such rapid release even at the high, 2.8 μCi/kg dose level. The declining concentrations of ^{249}Cf in contained blood and ECF were probably minor contributors to the reduction of total ^{249}Cf in liver, because by the end of day-1 the total blood volume contained only 1% of the injected dose.

Measurements of 4 beagles each injected with 0.3 μCi/kg of ^{249}Cf and 4 beagles injected with 0.3 μCi/kg of ^{252}Cf have continued for 400 days. The liver retentions of the two californium isotopes were identical. The half-time of retention of ^{249}Cf or ^{252}Cf in the beagle liver was estimated to be 3.3 years, which is only about one-third of the retention half-time of 9.6 years determined for low doses of ^{241}Am in this species.

4. Chinese Hamster

Unlike the rat and mouse, the liver of the Chinese hamster retains both ^{241}Am and ^{252}Cf, and the liver of the Syrian hamster retains ^{252}Cf (McKAY et al., 1972; MEWHINNEY et al., 1972; MEWHINNEY and HARRIS, 1972). At a dose level of 1 μCi/kg of ^{241}Am, measurement of Chinese hamster livers dissected 225 and 360 days after injection indicated a retention half-time of about 3400 days (McKAY et al., 1972).

Retention of ^{252}Cf in the Chinese and Syrian hamster livers has been measured only at a dose level of 5 μCi/kg—five times the dose used in the ^{241}Am studies. Measurements of livers from Chinese hamsters killed 1 to 64 days after injection indicated a retention half-time greater than 125 days (MEWHINNEY et al., 1972). During the 64-day postinjection interval, 8% of the dose (30% of the initial liver burden of ^{252}Cf) was lost from the liver, while the ^{252}Cf content of the carcass (mainly skeleton) increased by 5% of the injected dose. Simultaneous loss from liver and uptake in the skeleton suggested that the mechanism of liver loss of ^{252}Cf in the Chinese hamster was release from radiation-damaged cells.

D. Long-Term Content of ^{241}Am in Rat and Monkey Liver

The initial clearance of low doses of ^{241}Am from the livers of rats, mice, and monkeys is rapid and efficient, but it is never complete. As can be seen from Figs. 18.21 and 18.25 a small amount of ^{241}Am is present in rat liver as long as 2 years after injection and in the monkey liver for as long as 3 years.

Sequential autoradiographs of rat livers (see Sec. VIII.B.2.) indicated that significant numbers of reticuloendothelial cells, which contained aggregations of ^{241}Am tracks associated with hemosiderin, were still present 70 days after injection but were rarely seen after 300 days. The 50-day component of the ^{241}Am retention curve probably represents, at least in part, the elimination of these ^{241}Am-labeled reticuloendothelial cells.

According to the general model of long-term metabolism of ^{241}Am proposed by DURBIN et al. (1967) and shown in Fig. 18.27 the longest-term components of the liver retention curve (half-times 500 days and 1200 days in monkey and rat livers, respectively) represent feeedback to the circulation of ^{241}Am released from the skeleton in the course of normal bone remodeling.

Only two components could be extracted from the retention curve of ^{241}Am in the monkey liver (half-times of 70 and 500 days): The short-term component is presumed to represent a combination of several clearance mechanisms—biliary excretion, release to the circulation from damaged cells, and loss from labeled reticuloendothelial cells—which appear to proceed at rates that cannot be differentiated. The longer-term component is assumed to represent largely feedback from other soft tissues and bone.

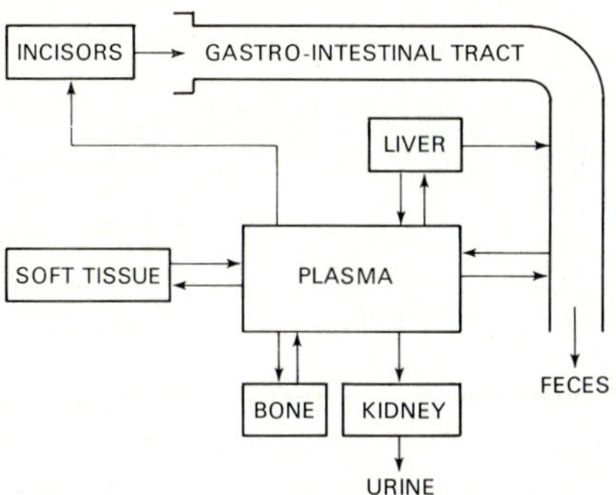

Fig. 18.27. Kinetic model of the metabolism of trivalent actinide elements in mammals modified from DURBIN et al. (1967). The incisor teeth compartment is present only in rodents

IX. Deposition and Retention of Transplutonium Elements in Kidney

A. Initial Deposition

The kidney is an important early deposition site of the transplutonium elements in mammals. In most species studied, more than one-half of the 60-day urinary excretion of these elements is passed in the first 24 hr when the plasma nuclide concentration is high. The kidneys of animals killed 1 day after injection still contain significant amounts of the nuclides. During the first few days after injection, the concentrations of transplutonium elements in the kidneys of all of the mammals studied (except the Chinese hamster) are nearly the same as the concentrations in bone, but only about one-fifth those in the liver.

One day after injection of chlorides or citrates of ^{241}Am, ^{242}Cm, $^{249-252}$Cf, or ^{253}Es the kidneys of rats and mice contained 1.5 to 4.5% of the injected dose (SCOTT et al., 1948a, 1949; ZALIKIN et al., 1968; BELYAEV, 1969; DURBIN et al., 1967, 1972c; WILLIAMS et al., 1961; PARKER et al., 1962, 1972; V. H. SMITH, 1972b). The kidneys of dogs and monkeys contained 0.5 to 1.5% of ^{241}Am or ^{249}Cf 1 day after injection (ATHERTON et al., 1968; ATHERTON and LLOYD, 1972; DURBIN et al., 1972c). The kidneys of Chinese hamsters contained 11 to 12% of the dose 1 day after intraperitoneal injection of ^{241}Am or ^{252}Cf citrate (MCKAY et al., 1972; MEWHINNEY et al., 1972). During the first week after injection the concentrations of ^{241}Am or ^{252}Cf in Chinese hamster kidney were nearly the same as those in the liver.

B. Microscopic Distribution

1. Rat

By autoradiographic analysis, most of the initial deposit of a low dose of soluble ^{241}Am is found as single tracks in the cortex in association with the convoluted tubules (Fig. 18.28a). The rat kidney does not retain ^{241}Am. At postinjection times greater than 16 days, long autoradiographic exposures are required to demonstrate the presence of ^{241}Am (Fig. 18.28b).

Some point sources of multiple alpha tracks emanating from reticuloendothelial cells could be seen in the connective tissue of the cortex of most kidney sections from 8 to 32 days after injection. The number of these labeled cells gradually declined, but they did not disappear. At postinjection times greater

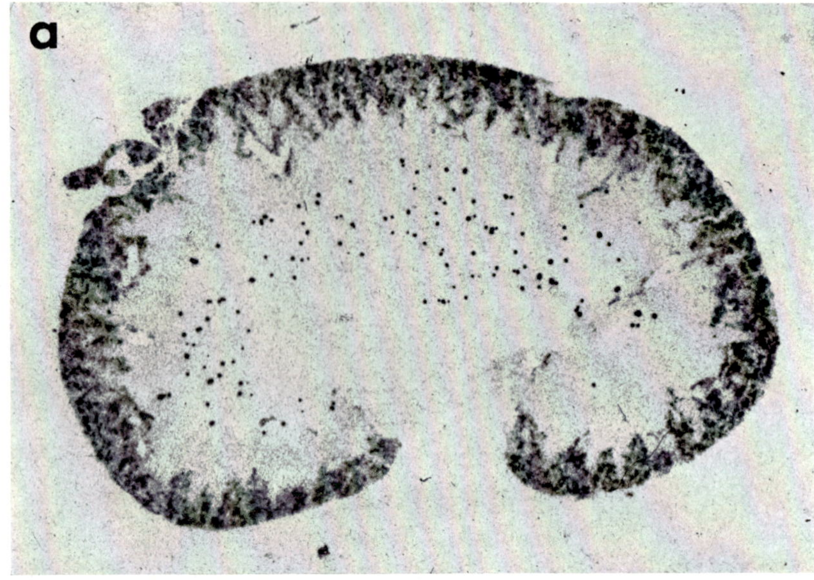

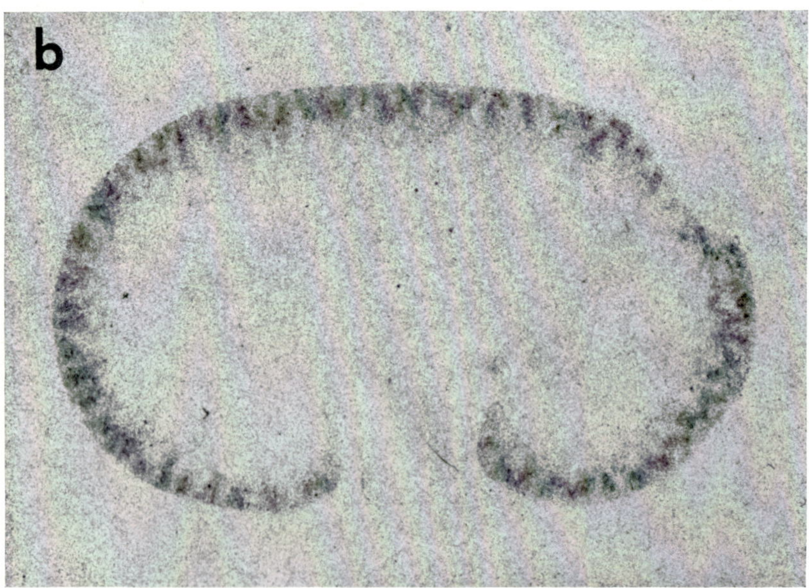

Fig. 18.28a and b. Americium-241 in the kidneys of rats described in Fig. 18.14. Gross autoradiographs were prepared as described in Fig. 18.24. Kodak single-coated (red-base) medical X-ray film was exposed for 1957 days. $\times$ 5. a Four days after injection the ^{241}Am is localized in the cortex and on renal stones in the medulla. b At 46 days after injection, cortical localization is much reduced, and labeling of renal stones persists

than 1 year most kidney sections still contained a few small foci of alpha tracks associated with brown pigment granules in reticuloendothelial cells.

At all times after injection, most rat kidney sections contained variable numbers of concretions, which stained light blue to deep purple with hematoxylin-eosin. These were usually located in the outer zone of the medulla or near the cortico-medullary junction (Fig. 18.28a). At short postinjection times, when the rats were young and relatively free from chronic renal disease, these concretions (presumable renal stones) were small and their ^{241}Am label was intense. With the passage of time, the small intensely labeled stones were either dissolved or passed, because at postinjection times longer than 1 year, renal stones were only lightly labeled on their surfaces (Fig. 18.23d), even though they tended to be larger. The stones found in the kidneys of the older rats apparently were formed long after the injection and acquired their label at a time when the ^{241}Am concentration in the blood was low.

At the dose level used in this study (0.37 μCi/kg of ^{241}Am) early kidney damage could not be clearly demonstrated. Later in the study, any early damage that might have been present was repaired or obscured by the gradual onset of chronic pyelonephritis, which is endemic in the Charles River rat (Durbin et al., 1972e).

In the high-dose studies of Rudnitskaya and Moskalev (1970) ^{241}Am was still detectable in the renal cortex, associated with both glomeruli and tubules, as long as 200 days after injection. After 30 days, clusters of ^{241}Am tracks could also be found in the medulla. The injection levels, 2.5 to 25 μCi/kg of ^{241}Am, were calculated to deliver average radiation doses to the kidney of 17 to 170 rads, respectively. Early tissue damage included destruction of glomerular capillary tufts and focal destruction of tubules. The tissue damage was accompanied by accumulations of large clusters of alpha tracks in the intertubular connective tissue. Cysts in the corticomedullary zone and sclerotic foci were common findings in the kidneys of the 200-day survivors.

Rat kidney containing $^{249-252}$Cf was available only from two groups of rats killed 4 and 91 days after intramuscular injection of $^{249-252}$Cf citrate (Durbin et al., 1972d). The microscopic distribution of $^{249-252}$Cf in rat kidney was indistinguishable from that of ^{241}Am at the same postinjection intervals.

2. Monkey

In the monkey kidney, ^{241}Am was initially deposited for the most part in the tubules, although not all segments of the tubules appeared to be involved. Neither the proximal convoluted tubules (identifiable by their brush borders) nor the collecting tubules in the medulla contained significant numbers of alpha tracks. The ^{241}Am was most prominent in the straight tubules, presumably the limbs of the loops of Henle, in the medullary rays (Fig. 18.29a).

Some tubules in most, but not all, of the monkey kidneys contained granules of brown pigment. Shortly after injection, the ^{241}Am associated with the renal tubules seemed unrelated to the presence of pigment. However, after 112 days, pigment-containing cells of the renal tubules and of the tubular interstitial tissue were the dominant sites of ^{241}Am deposition (Fig. 18.23b).

Although the number of alpha tracks associated with the renal tubules and the interstitial tissue declined with time (Fig. 18.29b), such sites still contained ^{241}Am 938 days after injection (demonstrable by very long autoradiographic exposures).

More ^{241}Am was associated with the glomeruli of monkey kidney than was found in the rat kidney. During the first few days after injection, a light dusting of single tracks could be detected over most glomeruli. From 8 to 42 days some glomeruli contained point sources of multiple alpha tracks, "small stars" (Fig. 18.23c). Smaller stars of tracks could occasionally be found in the capsular membrane. The number of glomerular stars declined, but they did not disappear.

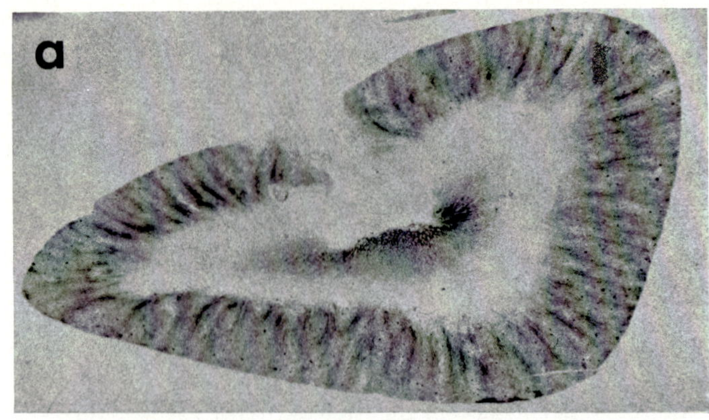

Fig. 18.29a and b. Americium-241 in the kidneys of monkeys. The gross autoradiographs were prepared as described in Fig. 18.24; Kodak single-coated (red-base) medical X-ray film was exposed for 805 days. ×3. a Adolescent no. 41, described in Fig. 18.16c, killed 8 days after injection. The heaviest concentrations of ^{241}Am are in the medullary rays and the papillae. b Adult monkey no. 81 (more than 6 years old, killed 81 days after intramuscular injection of 0.56 μCi/kg of ^{241}Am citrate). The cortical deposit is greatly reduced. Concentration in papillae is still high, and some of this deposit is associated with large numbers of small intertubular concretions

By 112 days single tracks over the glomeruli were more concentrated than the tracks over adjacent areas of tubules. Midsagittal sections of whole kidneys taken from monkeys killed 1 to 2 years after injection usually contained 1 to 5 small stars.

Shortly after injection, the greatest concentration of ^{241}Am in the monkey kidney was in the intertubular connective tissue in the renal papillae (Fig. 18.29a). As in other kidney structures, the amount of ^{241}Am at this site decreased (Fig. 18.29b) but was still detectable 938 days after injection.

The kidneys of some monkeys contained concretions which were usually located in the connective tissue between the distal collecting tubules in the renal papillae. When renal stones were present, they were heavily labeled (Fig. 18.29b).

As G. N. TAYLOR et al. (1969a, 1972) had noted in the dog kidney, ^{241}Am was selectively localized in the media of arterioles in the monkey kidneys. Compared with the ^{241}Am concentrations in other kidney structures, the vascular deposit was not especially prominent at first. The impression was that of a persistent vascular labeling, even though there was some decline with time. Thus, at long postinjection times, ^{241}Am was more concentrated in the walls of arterioles than elsewhere in the kidneys (except for special sites such as pigment-containing tubules and renal stones).

The monkeys in these studies were of variable age (6 or more years old at the time of injection); and postinjection times ranged from 1 day to 3 years. The animals were caught in the wild, and their health histories were unknown. Histologically, their kidneys ranged from apparently good health to combinations of mild interstitial fibrosis, occasional glomerular fibrosis, occasional tubular plugging, and small numbers of renal stones—changes that could be related to age or disease history as well as to radiation damage.

3. Dog

Some of the features of the early deposition of ^{241}Am, ^{249}Bk, and ^{249}Cf in the kidney of the dog resembled those described above for the monkey (G.N. Taylor et al., 1969a, 1972; Lloyd et al., 1970). Single alpha tracks were prominent in the straight portions of renal tubules, especially in medullary rays, and large aggregations of alpha tracks were associated with concretions and connective tissue in the renal papillae. The concentrations of these nuclides in the media of arterioles was greater than was found in the monkey kidney. The most important difference between the distributions of these nuclides in the kidneys of dogs and monkeys was the concentrated, and apparently permanent deposition in the basement membranes of the tuft and capsule of the glomeruli in the dog kidney. The level of ^{241}Am deposition in the dog glomeruli was sustained (or possibly even augmented) during 401 to 448 days after injection of 2.8 μCi/kg of ^{241}Am citrate, and led to average radiation doses in kidney tissue of 800 to 850 rads. At death, the 2 dogs in this dose group exhibited renal hypertension; other dogs in lower dose groups showed varying levels of proteinuria. One dog injected at a level of 0.9 μCi/kg of ^{241}Am died of renal failure 1533 days after injection (Lloyd et al., 1972a). Microscopically, there were great reductions in the size of the glomerular tufts and in the number of glomeruli. Local necrosis was observed in the cells adjacent to the "hot spots" of ^{241}Am on renal stones in the papillae, but the contribution of these radiation-induced lesions to reduced kidney function has not been evaluated, nor has it been possible to evaluate the contribution of irradiation of arteriole walls to the overall impairment of kidney function. Generalized glomerulosclerosis with fibrous replacement was found in the kidneys of 2 beagles killed 512 and 1022 days after exposure to average ILB of 2.8 μCi/kg from an aerosol containing $^{241}AmO_2$ (R. G. Thomas et al., 1972).

C. Retention

1. Rat

As shown in Fig. 18.30, 3 exponential terms (half-times 8, 25, and 1870 days) were needed to describe ^{241}Am retention in rats kidneys[27]. The autoradiographs

27 Variable numbers of stones in the kidneys of rats of all ages perturbed the kidney retention data. In all age groups kidneys with large numbers of stones contained more ^{241}Am than kidneys with fewer stones. In some age groups, mean ^{241}Am in the kidney was less than the S. D. Gross X-ray autoradiographs of kidney sections of the 150 rats in the low-dose study (Durbin et al., 1967, 1972c) were examined for the presence of radioactive "hot spots" in the medulla, which were judged to be images of ^{241}Am-labeled kidney stones. Rats were excluded from calculation of the mean kidney ^{241}Am if their autoradiographs showed more than twice the average number of "hot spots" determined for the age group.

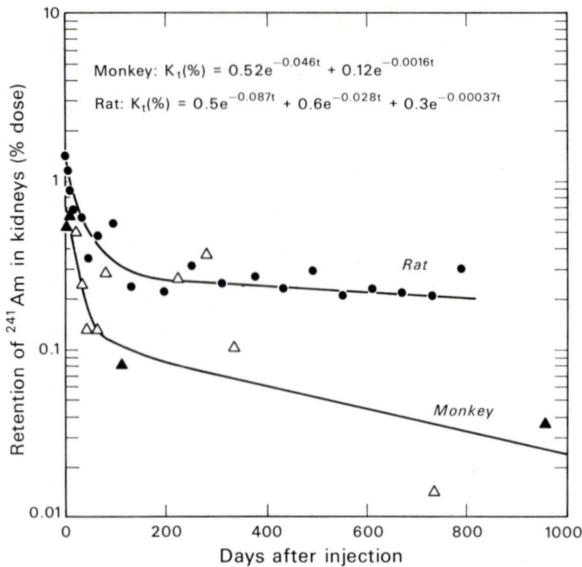

Fig. 18.30. Retention of ^{241}Am in kidneys of rat and monkey. Female rats, 110-days old, were injected intramuscularly with 0.37 μCi/kg of ^{241}Am citrate: solid circles (●) represent mean kidney retention of groups of 5 or more rats. Adult female cynomolgus monkeys were injected intravenously or intramuscularly with ^{241}Am citrate, 0.44 to 0.87 μCi/kg: solid triangles (▲) represent pairs of monkeys, open triangles (△) represent individual monkeys. (Data of DURBIN et al., 1972c)

(Fig. 18.28a, b) demonstrate the rapid disappearance of ^{241}Am initially associated with tubules in the renal cortex. Clearance of ^{241}Am from renal tubules is likely to be the major mechanism underlying the 8-day component of the ^{241}Am kidney retention curve.

During the interval 8 to 32 days postinjection, some of the ^{241}Am present on day 1 was cleared and some was apparently shifted from renal tubules to connective tissue and to pigment-laden reticuloendothelial cells. Clearance of ^{241}Am from connective tissue and the turnover of labeled reticuloendothelial cells probably contribute to the 25-day component of ^{241}Am retention.

According to the general metabolic model proposed by DURBIN et al. (1967) for the long-term behavior of multicharged cations, the 1870-day component of the kidney retention curve should represent feedback to the circulation of ^{241}Am released in the course of skeletal remodeling (see Fig. 18.27).

The available information on the retention in the rat kidney of ^{241}Am at various injection levels is collected in Fig. 18.31. Kidney retention curves determined in low-dose (DURBIN et al., 1972c) and moderate-dose (ZALIKIN et al., 1968) studies agreed reasonably well. The retention curve of ^{241}Am in the kidneys of the rats in the early experiments of SCOTT et al. (1948a) was parallel to those obtained in the later investigations, but at all postinjection times the amount of ^{241}Am in the kidneys was almost twice that found in the recent studies. The heavy kidney deposition in the early study does not seem to be an artifact of experimental technique (intramuscular absorption was 90% complete by 4 days), nor does it seem to be related to the chemical form administered ($^{241}AmCl_3$ in a recent study gave the same results as ^{241}Am citrate). The heavy kidney deposi-

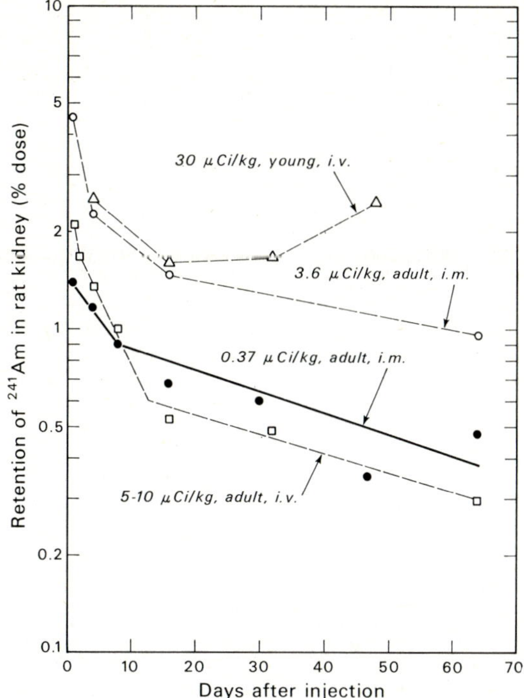

Fig. 18.31. The influence of experimental conditions on the retention of ^{241}Am by the rat kidney: 0.37 μCi/kg ^{241}Am citrate, adults, DURBIN et al. (1967, 1972c); 5 to 10 μCi/kg $^{241}AmCl_3$, adults, ZALIKIN et al. (1968); 3.6 μCi/kg $^{241}AmCl_3$, adults, SCOTT et al. (1948a); 30 μCi/kg $^{241}AmCl_3$, growing rats, LANGHAM and CARTER (1951)

tion in the early ^{241}Am and ^{242}Cm studies (SCOTT et al., 1948a, 1949) may have been caused by the presence of early pyelonephritis and renal stones[28].

The kidneys of the young rats in the high-dose studies of LANGHAM and CARTER (1951) initially lost ^{241}Am at about the same rate as the older rats that received the lower ^{241}Am doses. However, after the 16th day, the kidneys (like the bone and some other soft tissues) were accumulating the ^{241}Am released into the circulation from the damaged liver.

Retention of ^{242}Cm, ^{252}Cf, and ^{253}Es in the rat kidney is shown in Fig. 18.32. The retention of ^{242}Cm determined by SCOTT et al. (1949) was like that of ^{241}Am measured in a similar experiment by the same workers (SCOTT et al., 1948a).

Kidney retention of ^{252}Cf (administered intraperitoneally as a citrate, MEWHINNEY et al., 1972) by young male rats was very much like that of ^{241}Am (administered as a citrate intramuscularly) in adult females.

Intratracheally administered $^{253}EsCl_3$ (data corrected for the amount remaining in the lung), behaved somewhat differently in the rat kidney (HUNGATE et al., 1972; BALLOU and MORROW, 1972). The amount of ^{253}Es initially deposited in the kidney was 2 to 4 times greater than that of the other three actinides, and

28 Modern methods of controlling endemic diseases in rodent colonies—caesarean-derivation, sterilization of food, bedding, and cages, and rigorous hygiene in colony rooms—were not introduced until the late 1950's. In the absence of such measures, renal infections were probably common even in young rats in most colonies.

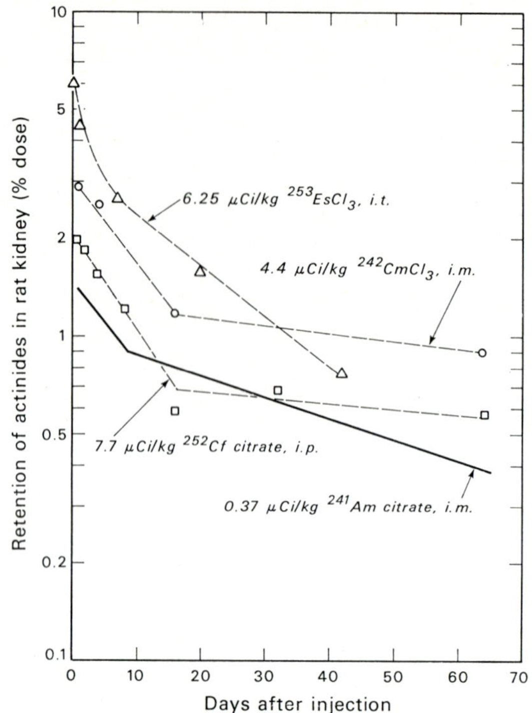

Fig. 18.32. Retention of trivalent actinide elements in rat kidney: ^{241}Am citrate, DURBIN et al. (1967, 1972c); $^{242}CmCl_3$, SCOTT et al. (1949); ^{252}Cf citrate, MEWHINNEY et al. (1972); $^{253}EsCl_3$, HUNGATE et al. (1972), and BALLOU and MORROW (1972)

the initial rate of clearance of ^{253}Es from the kidney was 3 times faster than that of ^{241}Am or ^{252}Cf (^{253}Es half-time, 1.5 days; ^{241}Am and ^{252}Cf half-time, 4 to 5 days).

2. Monkey

As shown in Fig. 18.30, the retention in the monkey kidney of a low dose of ^{241}Am can be adequately described by two exponential terms with half-times of 15 and 430 days. During the first few days after injection, ^{241}Am concentrations in monkey kidney were associated, in ascending order, with (a) media of arterioles, (b) loops of Henle, (c) connective tissue in the renal papillae, and (d) renal stones. Retention of ^{241}Am was greater and more prolonged in the kidneys that contained large numbers of renal stones[29]. All of the above listed soft-tissue deposits of ^{241}Am diminished, although not necessarily at the same rate, and all apparently contributed to the average initial renal clearance half-time of 15 days.

The data were not sufficient and the amount of ^{241}Am involved was too small to distinguish a separate component for the clearance (between 30 and 112 days) of ^{241}Am from the "stars" associated with the glomeruli and pigment-containing tubules.

29 The kidneys of the 3 monkeys killed at 81, 224, and 280 days contained many intensely labeled stones. The amount of ^{241}Am in these kidneys was unduly high, and the data from these animals were not included in the fitting of the slower component of the kidney retention curve. Histologic sections of the kidneys of the 4 other long-term monkeys did not contain any stones.

DURBIN et al. (1972b) reported a half-time of 470 days for the longest-observed components of the renal and gastrointestinal excretion curves of these monkeys (measured between 500 and 1400 days postinjection). The longest half-times observed for ^{241}Am in monkey liver and residual soft tissues were 440 days and 880 days, repsectively. At the low dose level used in these studies it seems reasonable to assume that the long-term component of the kidney retention curve represented an equilibrium with ^{241}Am that was being fed back into the circulation from other soft tissues and from bone.

3. Dog

The retention of ^{241}Am in the dog kidney has been studied only at high dose levels and is complicated by radiation injury sustained both from the initial ^{241}Am deposit in kidney and from the later accumulation of ^{241}Am released from the radiation-damaged liver. Late accumulation of ^{241}Am in the kidney was also observed in rats given a high dose of 30 μCi/kg (LANGHAM and CARTER, 1951).

Shortly after injection, ^{241}Am appeared to leave the dog kidney. Reported values for the ^{241}Am in the kidneys of individual dogs were as follows: 1 day, 1.3%; 7 days, 0.475%; 8 days, 0.453%; and 22 days, 0.362% (ATHERTON et al., 1968; ATHERTON and LLOYD, 1972). The kidneys of other dogs given 2.8 μCi/kg of ^{241}Am contained 1.28% at 401 days and 1.13% at 448 days—nearly equal to the 1-day ^{241}Am content. Between 50 and 400 days after injection of 2.8 μCi/kg of ^{241}Am, 30% of the dose was released from the liver, of which two-thirds (20%) was redistributed to soft tissues and bone, and one-third (10%) was presumed to have been excreted.

By 401 to 448 days, the kidneys of these dogs had sustained substantial irreparable radiation damage (LLOYD et al., 1969a, 1970). The radiation dose from the initial ^{241}Am deposit only, is calculated to be 25 rads at 30 days and, if no other processes intervene, 55 rads at infinite time[30]. It would appear that more than 90% of the damaging radiation dose (estimated above to be 800 to 880 rads) was delivered by the ^{241}Am taken up later in the course of its release from the liver. The late delivery of the radiation dose in the kidneys would help to account for the slow development of the renal lesions (R. G. THOMAS et al., 1972).

4. Mouse

The vasculature of the kidney was clearly outlined in whole-body autoradiographs of mice killed 5 min after intravenous injection of $^{241}Am(NO_3)_3$ (HAMMARSTRÖM and NILSSON, 1970a). The kidney exhibited a moderate amount of radioactivity throughout the 60-day study, and small foci of high concentration were seen in the cortex in some animals. There was no special accumulation of ^{241}Am in the renal pelvis, but the radioactivity over the ureter and urethra persisted up to 60 days. Measurements of dissected tissues indicate that a large fraction of the 1-day deposit of ^{241}Am in the mouse kidney was cleared with a half-time of about 4.5 days (WRIGHT, 1972).

5. Chinese Hamster

The retention patterns of ^{241}Am and ^{252}Cf were the same in the Chinese hamster kidney (McKAY et al., 1972; MEWHINNEY et al., 1972). Ten to 11% of the injected dose of both nuclides was cleared from the hamster kidneys with a half-time of about 5 days. The remaining 1% of each nuclide was not significantly diminished

30 Radiation dose was calculated using the following retention function,

$$Dog\ K_t(\%) = 0.92e^{-0.27t} + 0.53e^{-0.017t}.$$

during the 64-day experiments. Radiation damage in the hamster kidneys has not been investigated, but the calculated average dose in kidney of Chinese hamsters given 5 μCi/kg of either ^{241}Am or ^{252}Cf is calculated to be about 180 rads in 30 days. A radiation dose of this magnitude was shown to produce substantial tissue damage in the rat kidney (RUDNITSKAYA and MOSKALEV, 1970).

X. Deposition and Retention of Transplutonium Elements in Soft Tissues Other than Liver and Kidney

The description of the behavior of the transplutonium nuclides in soft tissues other than liver and kidney is far from complete. Table 18.23 summarizes the nuclide concentrations in soft tissues sampled a few days after injection of ^{241}Am or ^{249}Cf in the beagle (ATHERTON et al., 1968; LLOYD et al., 1970, 1972d, e; ATHERTON and LLOYD, 1972) and of ^{241}Am in the baboon (ROSEN et al., 1972). Soft tissues, especially the small organs and glands, have been analyzed only

Table 18.23. Concentrations of ^{241}Am and ^{249}Cf in selected soft tissues of the beagle and the baboon[a], at days after injection (%/kg)

Tissue	^{241}Am in beagle[b]			^{249}Cf in beagle[c]		^{241}Am in baboon[aa]
	2.8 to 4.5 μCi/kg 1 to 21 days	2.8 μCi/kg 401 to 448 days	0.09 to 0.3 μCi/kg 1510 to 1917 days	2.8 μCi/kg 7 to 13 days	2.8 μCi/kg 21 to 36 days	0.2 μCi/kg 32 days
Thyroid	55	110	37	46	22	3.4
Dura mater	8.5	—	5.2	20	14	—
Lymph nodes	8.5	17	1.3	12	10	6.5
Trachea	7.1	29	—[ab]	—[ab]	—	17
Aorta	4.8	5.2	2.0	5.0	3.5	15
Lung	4.6	4.9	3.0	12	7.6	18
Larynx	3.3	6.4	2.1	—	—	6.1
Salivary glands	3.2	—	1.2	<1.8	5.7	2.0[ac]
Adrenals	3.2	3.4	1.0	5.9	4.7	3.2
Pituitary	2.6	14.0	1.2	3.2	8.0	—
Prostate	1.9	5.1	0.7	4.2	5.4	—
Eyes	1.5	1.4	0.9	1.0	<1.4	—
Heart	0.8	1.7	2.3	1.9	1.2	4.0
Urinary bladder	0.8	—	0.7	—	—	—
Testes	0.8	1.0	0.4	1.4	0.8	—
Pelt	0.8	1.0	0.3	2.9	1.3	1.9
Pancreas	0.5	0.6	2.1	<2.5	1.4	3.1
Muscle	0.3	0.6	0.2	0.95	0.5	0.9
Fat	0.09	—	—	—	—	0.2
Thymus	<0.4	—	<1.3	<1.5	2.5	—
Brain	<0.02	0.1	0.3	<0.15	0.2	0.06
Ovaries	—	—	1.0	4.9	3.4	4.8
Uterus	—	—	—	<0.6	—	1.9
Mammary glands	—	—	—	—	—	3.9[ac]

[a] Tissues and organs shown in Tables 18.16 and 18.17 are not included. Nuclide concentrations in those tissues may be obtained from the nuclide data in Tables 18.16 and 18.17 and the average weights in Appendix C.
[b] 1 to 21 days—dogs T15W5.5, T32W5.5, T16W5, T33W5.5; 401 to 448 days—dogs T1W5, T2W5; 1510 to 1917 days—dogs F6W3, F7W2, M5W3 (LLOYD et al., 1970, 1972d).
[c] 7 to 13 days—dogs T2G5, T2F5; 21 to 36 days—dogs T3G5, T1F5 (LLOYD et al., 1972d).
[aa] Baboon No. B-416 (ROSEN et al., 1972).
[ab] Trachea and lung combined.
[ac] Baboon No. B-412 killed 86 days after injection.

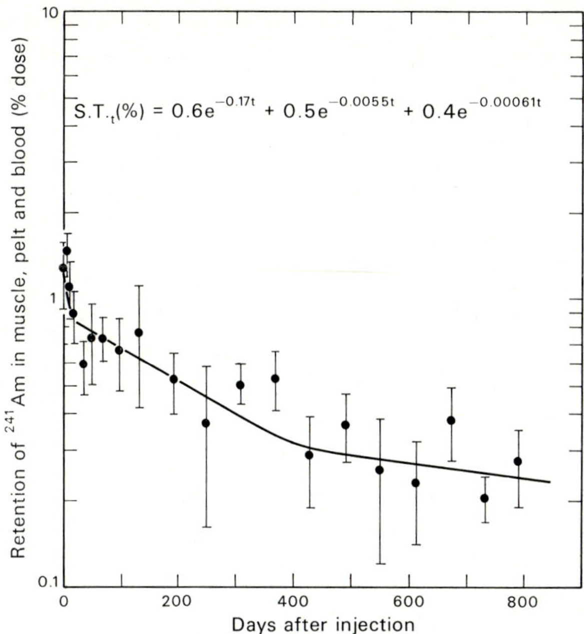

Fig. 18.33. Retention of ^{241}Am in the blood, muscle, and pelt (residual soft tissues) of the rat. (Data of DURBIN et al., 1972c)

occasionally in rodent studies. However, in rats, the spleen, lung, muscle, and pelt have been analyzed in several studies of ^{241}Am, ^{242}Cm, ^{249}Bk, $^{249-252}$Cf, and ^{253}Es. These data are collected in Table 18.24 [31].

The total amount of trivalent actinide in the soft tissues other than liver and kidney 1 to 2 days after injection ranged from 1.5 to 20% depending on the element, the species, and to some extent on the experimental conditions. The initial soft tissue burden was usually, but not always, greater in studies in which (a) the injected compound was easily hydrolyzed, (b) the route of injection was intravenous or intraperitoneal rather than intramuscular or intratracheal, (c) the injected dose was of the order of 25 μCi/kg or more, (d) the experimental subject was a large animal rather than a rodent. In similar experiments with dogs, rats, mice, and Chinese hamsters, the initial deposition of ^{241}Am in soft tissues has been found to be less than that of $^{249-252}$Cf.

From Tables 18.12 through 18.17 it is clear that the presence of trivalent actinides is temporary in most soft tissues. The bulk of the initial soft tissue content of ^{241}Am in rats (Fig. 18.33) and in monkeys (Fig. 18.21) is either excreted or redistributed to more permanent sites within a few days after administration. Thereafter, the nuclide content of the soft tissue is low and proportional to feedback from bone.

A. Spleen

Americium-241, ^{242}Cm, ^{249}Bk, $^{249-252}$Cf, and ^{253}Es have been measured in the spleen in a variety of rat studies (Table 18.24). In similar experiments the amounts

31 Nuclide concentrations in the tissues of rats, mice, and Chinese hamsters may be obtained from the tissue nuclide contents in Tables 18.12 through 18.15, and 18.24, and the average tissue weights in Appendix C.

Table 18.24. Deposition and retention of trivalent

Isotope and compound	Injn. route[a]	Spleen				Lungs			
		days	%	days	%	days	%	days	%
$^{241}AmCl_3$	i.m.	4	0.10	256	0.06	1	0.23	256	0.08
$^{241}AmCl_3$	i.v.	4	0.26	48	0.45	4	0.31	48	0.50
^{241}Am citrate	i.v.	4	0.38	62	0.24	4	0.22	62	0.19
$^{241}AmCl_3$	i.v.	1	0.22	64	0.30	1	0.13	64	0.10
$^{241}Am(NO_3)_3$	i.p.	3	0.91	270	0.22	1	1.13	270	0.17
$^{241}AmCl_3$	i.p.	3	0.84	180	0.18	1	0.26	30	0.07
^{241}Am citrate	i.m.	1	0.04	293	0.08	—	—	—	—
$^{242}CmCl_3$	i.m.	1	0.09	256	0.10	1	0.29	256	0.09
^{242}Cm citrate	i.v.	1	0.08			—	—	—	—
$^{249}BkCl_3$	i.v.			21	0.42	—	—	—	—
^{252}Cf citrate	i.p.	1	0.29	64	0.10	—	—	—	—
^{252}Cf citrate	i.m.	4	0.10	91	0.21	—	—	—	—
$^{253}Es(NO_3)_3$	i.v.	7	0.42	24	0.14	7	0.24	24	0.20
$^{253}EsCl_3$	i.v.			21	0.62	—	—	—	—
^{253}Es citrate (mouse)	i.m.	4	0.03	14	0.04	—	—	—	—

[a] i.m., intramuscular; i.v., intravenous; i.p., intraperitoneal.

of the various nuclides in rat spleen were the same. If the administered dose was small, and the injected compound was not readily hydrolyzed, the amount of ^{241}Am or $^{249-252}$Cf in spleens of several species (Tables 18.16, 18.17, 18.24) was 0.1 to 0.2% at 1 to 4 days postinjection. If the nuclide had been given intravenously or intraperitoneally in easily hydrolyzable form, the initial deposition in rat spleen was as much as 0.9%. In low-dose studies with rats, monkeys, and beagles, retention of trivalent actinides in spleen was characterized by either no change with time or a slow loss. In the high-dose rat study (Langham and Carter, 1951) the spleen lost ^{241}Am during the first 32 days, but accumulated ^{241}Am thereafter. In the dogs given 2.8 μCi/kg (Lloyd et al., 1970) ^{241}Am was lost from the spleen during the first 21 days, but by 401 to 448 days after injection, the spleen contained three times as much ^{241}Am as on day 1.

Americium-241 was initially distributed as a uniform dusting of single alpha tracks thoughout the red pulp of the spleens of mice (Hammarström and Nilsson, 1970a) and rats (Rudnitskaya and Moskalev, 1970; Durbin et al., 1972d). One day after injection the ^{241}Am in monkey spleen was associated for the most part with the connective tissue of the capsule and trabeculae and with the media of the small arterioles. By 8 days this selective localization had largely disappeared. Subsequently, ^{241}Am distribution in the monkey spleen appeared to be nearly uniform over both red pulp and connective tissues, but very little was present in white pulp (Durbin et al., 1972c).

In the high-dose rat studies of Rudnitskaya and Moskalev (1970) by 100 days after injection and in the beagle studies of G. N. Taylor et al. (1969a) by 401 to 448 days after injection, aggregates of ^{241}Am were found in splenic red pulp associated with hemosiderin-laden macrophages. There were fibrous changes in both red and white pulp in the rat spleens 200 days after injection. Pathological descriptions of the dog spleens have not been published; however, 401 to 448 days after injection of 2.8 μCi/kg of ^{241}Am the spleens were only one-fourth normal size (Lloyd et al., 1970). Foci of extramedullary hematopoiesis were noted in the spleens of the 4 beagles killed 127 to 1022 days after inhalation of an ILB of 2.8 μCi/kg of $^{241}AmO_2$ (R. G. Thomas et al., 1972).

actinide elements in selected soft tissues of the rat

Muscle				Pelt				Reference
days	%	days	%	days	%	days	%	
1	1.5	256	1.2	1	3.3	256	0.22	SCOTT et al. (1948a)
—	—	—	—	4	1.6	48	0.92	LANGHAM and CARTER (1951)
—	—	—	—	—	—	—	—	D. M. TAYLOR et al. (1961)
2	1.3	64	2.6	2	0.77	32	1.12	ZALIKIN et al. (1968)
—	—	—	—	—	—	—	—	TSEVELEVA and YEROKHIN (1969)
1	3.0	14	0.80	—	—	—	—	BELYAYEV (1969)
—	—	—	—	—	—	—	—	DURBIN et al. (1972c)
1	2.2	256	0.47	1	1.9	256	0.09	SCOTT et al. (1949)
—	—	—	—	—	—	—	—	WILLIAMS et al. (1961)
—	—	—	—	—	—	—	—	HUNGATE et al. (1972)
—	—	—	—	—	—	—	—	MEWHINNEY et al. (1972)
—	—	—	—	—	—	—	—	DURBIN et al. (1972d)
7	1.5	24	1.2	—	—	—	—	V. H. SMITH (1972b)
—	—	—	—	—	—	—	—	HUNGATE et al. (1972)
—	—	—	—	—	—	—	—	PARKER et al. (1972)

B. Lymphatic Tissue

Initial concentrations of ^{241}Am in lymph nodes of primates were 6.5 and 10%/kg in baboon and monkey, respectively, nearly the same as the ^{241}Am concentrations in spleen (DURBIN et al., 1972c; and Table 18.23). The initial ^{241}Am concentration in the lymph nodes of dogs was 8.5%/kg, about 6 times that of spleen. The ^{241}Am or ^{253}Es concentrations in rat lymph nodes 1 to 7 days after injection were 77 and 100%/kg, respectively, much greater than in the larger animals (DURBIN et al., 1972c; V. H. SMITH, 1972b).

Concentrations of ^{241}Am in the lymph nodes of the primates did not change for as long as 81 to 86 days after injection. In rat, ^{241}Am concentrations in the lymph nodes declined to 19%/kg by 256 days postinjection. At dose levels of 0.09 to 0.3 μCi/kg, the average ^{241}Am concentration in the lymph nodes of the beagles decreased; by 1510 to 1917 days after injection it was one-fourth of the initial value. In the beagles injected at the 2.8 μCi/kg level, the ^{241}Am concentration in the lymph nodes declined during the first 22 days. Thereafter, either ^{241}Am was accumulated in the lymph nodes or the lymph nodes were reduced in size, because by 401 to 448 days after injection the ^{241}Am concentration was twice as great as it had been 1 day after injection (LLOYD et al., 1970, 1972d).

RUDNITSKAYA and MOSKALEV (1970) reported an initially diffuse distribution of ^{241}Am in rat lymph nodes, but by 100 days after injection there was a tendency for most of the ^{241}Am present to be located in sinuses. Destruction of germinal centers was seen 200 days after injection of large doses of ^{241}Am in rats. At the late postinjection interval ^{241}Am in the lymph nodes of dogs was principally aggregated in macrophages in association with hemosiderin, though aggregates were fewer here in number than in the spleen (G. N. TAYLOR et al., 1969a).

C. Lung

With a single exception (TSEVELEVA and YEROKHIN, 1969), the initial lung deposition of ^{241}Am, ^{242}Cm, or ^{253}Es in rats ranged from 0.13 to 0.31% (Table 18.24). The average lung concentration was 100%/kg, about one-half that in rat spleen and five times that in skeletal muscle. The early concentration of ^{241}Am in dog lung was not as great as the concentration in kidney or trachea, but was higher than that in most tissues: for example, about 15 times the concentration in skeletal muscle. The ^{241}Am concentration in baboon and monkey lung was about 20%/kg 1 to 32 days after injection: nearly the same as in trachea and 20 times that of muscle.

One day after intravenous injection of $^{241}AmCl_3$, rat lung contained single alpha tracks associated with alveolar epithelium and perivascular and peribronchiolar tissue (RUDNITSKAYA and MOSKALEV, 1970). Shortly after injection ^{241}Am could be found associated with the tracheal cartilages of mice (HAMMARSTRÖM and NILSSON, 1970a); 401 to 448 days after injection, autoradiographs of dog lung showed ^{241}Am localized in cartilaginous structures and occasionally associated with macrophages in alveolar epithelium (G. N. TAYLOR et al., 1969a). In the dog lung most of the ^{249}Cf present 7 to 21 days after injection was in or adjacent to the perichondrium of cartilaginous plates and to a lesser degree in the lamina propria of bronchioles. An occasional spot of intense radioactivity was found in the submucosa of bronchioles (G. N. TAYLOR et al., 1972).

D. Muscle and Pelt

Skin and skeletal muscle constitute more than one half of the body weight of the mammals that have been used in transplutonium studies. The actinide concentrations in these tissues were not great at 1 to 7 days after injection, ranging from 0.3%/kg in dog muscle to 16%/kg in rat muscle, and from 0.8%/kg in dog pelt to 63%/kg in rat pelt. Although the concentrations in these two tissues are low, their bulk is large, and most of the transplutonium nuclide content of the residual soft tissues is contributed by the pelt and skeletal muscle. Therefore, retention of the transplutonium elements in muscle and pelt dominates the retention in residual soft tissue which is shown for rat and monkey in Figs. 18.21 and 18.33.

HAMMARSTRÖM and NILSSON'S (1970a) whole-body autoradiographs of ^{241}Am in mice showed a significant but very transient uptake of ^{241}Am in the follicles of the vibrissa (whiskers): ^{241}Am was not demonstrable by this technique 24 hr postinjection. Seven to 21 days after injection of ^{249}Cf, a high concentration was found in the glassy membrane of primary hair follicles of the dog pelt. A low-level diffuse distribution of tracks was seen over the rest of the tissue (G. N. TAYLOR et al., 1972).

E. Heart

The transplutonium elements were more concentrated in heart than in skeletal muscle in all of the species in which the radioactivity of the heart was measured separately. One to 7 days after injection, ^{241}Am, ^{242}Cm, or ^{253}Es concentrations in rat heart were 120%/kg, nearly seven times as much as in skeletal muscle (SCOTT et al., 1948a, 1949; LANGHAM and CARTER, 1951; V. H. SMITH, 1972b). The amounts of these nuclides in rat heart declined so that 256 days after injection the concentrations were 18%/kg, only twice that of muscle.

The concentration of ^{241}Am in dog heart 1 to 21 days after injection (0.8%/kg) was much lower than in rat heart, but it was greater than that of dog skeletal muscle (LLOYD et al., 1970, 1972d). In the beagles given 2.8 μCi/kg, concentrations of ^{241}Am in heart were the same 401 to 448 days after injection as they were after only 1 to 22 days. However, the apparently prolonged retention of a high dose of ^{241}Am in heart, and presumably in several other tissues also, was more likely the net result of an early decline in ^{241}Am content followed by a late period of accumulation.

Americium-241 concentrations in baboon heart were higher than those in dog heart: 4 and 2%/kg at 32 and 86 days after injection, respectively (ROSEN et al., 1972; and Table 18.23). During that interval the concentrations of ^{241}Am in both baboon heart and muscle declined with a half-time of about 50 days.

Autoradiographs of rat heart 3 days after injection of ^{241}Am showed single alpha tracks "projecting onto cardiac muscle fibers" RUDNITSKAYA and MOSKALEV, 1970). One to 3 weeks after injection, ^{249}Cf was selectively localized in the connective tissue of the coronary artery as well as in that of the small arterioles, and some large aggregations of ^{249}Cf were found in the circumflex branch of the coronary artery (G. N. TAYLOR et al., 1972). Californium-249 was also concentrated in the connective tissue of the endocardium of the mitral valve. The transplutonium elements deposited in these locations in connective tissue of the heart and its blood vessels appear to remain there for some time. The pattern of distribution of ^{241}Am in dog heart 401 to 448 days after injection was almost identical to that of ^{249}Cf at 7 to 22 days after injection. Superimposed on a low-level diffuse background of alpha tracks, ^{241}Am was selectively localized in the media of arterioles and the coronary artery and in the subendocardium of the mitral valve (G. N. TAYLOR, 1969a).

R. G. THOMAS et al. (1972) observed fibrous intimal thickening of many medium-size to large branches of the coronary arteries in the heart of the dog that died 1022 days after inhalation of an ILB of 2.7 μCi/kg of $^{241}AmO_2$. The hearts of the 200-day survivors of the high-dose study of $^{241}AmCl_3$ in rats showed sclerotic changes in both muscle and blood vessels (RUDNITSKAYA and MOSKALEV, 1970).

F. Aorta and Other Arteries

The media of all small arterioles, principally those with one or two muscle layers, was the most common deposition site of ^{249}Cf in the soft tissues of the dog 7 to 21 days after injection (G. N. TAYLOR et al., 1972). Deposition of ^{249}Cf was remarkably high in the media of at least two large arteries, the aorta and coronary artery. In dog liver, kidney, spleen, heart, lung, and prostate (other tissues were not reported), a high ^{241}Am deposition in the small arterioles was demonstrable autoradiographically 401 to 448 days after injection (G. N. TAYLOR, 1969a). The concentrations of ^{241}Am in the dog aorta, determined radiochemically, were higher than those of most other soft tissues at all postinjection times and all dose levels (LLOYD et al., 1972d).

The aorta and carotid artery were major sites of ^{241}Am concentration in the baboon: 15 and 12%/kg, respectively, 32 days after injection (ROSEN et al., 1972; and Table 18.23). These concentrations were about the same as those of lungs and trachea, and as much as one-half the concentration in bone. The ^{241}Am concentrations in these arterial structures were maintained at least until 86 days after injection. Autoradiographs indicate that small arterioles in monkey liver, kidney, and spleen were significant deposition sites of ^{241}Am 1 to 112 days after injection (DURBIN et al., 1972c).

G. Thyroid Gland

The concentrations of ^{241}Am, ^{244}Cm, and ^{249}Cf in the thyroid gland of the beagle (shown in Table 18.25) exceeded those in any other tissues except liver and bone (LLOYD et al., 1970, 1972d; ATHERTON and LLOYD, 1972; MCCLELLAN

Table 18.25. Concentration of transplutonium nuclides in the thyroid gland of the beagle

Days	Concentration at days after injection (% dose/kg)				Concentration at days after inhalation (% ILB/kg)		
	^{241}Am citrate[a]			^{249}Cf citrate[b]	$^{241}AmO_2$[c]	$^{244}CmCl_3$[aa]	$^{244}CmO_{1.73}$[aa]
	2.8 to 4.5 μCi/kg	0.91 μCi/kg	0.09 to 0.3 μCi/kg	2.8 μCi/kg	2.7 μCi/kg	0.11 μCi/kg	0.10 μCi/kg
1–2	115	—	—	—	—	130; 420[ab]	110; 130
7–8	42; 47	—	—	58	—	42; 70	80
13–16	46; 56	—	—	34	—	41; 42	48; 49
21–22	20	—	—	16	—	—	—
36	—	—	—	28	—	—	—
64	—	—	—	—	—	52; 77	—
127–256	—	—	—	—	10; 37	12; 15	—
401–448	80; 134	—	—	—	—	—	—
512–633	—	144	—	—	28		—
1022–1132	—	276	—	—	35	—	—
1323–1510	—	89; 157	69	—	—	—	—
1779–1917	—	67	10; 31	—	—	—	—

[a] Data of LLOYD et al. (1972d).
[b] Data of LLOYD et al. (1972e).
[c] Data of R. G. THOMAS et al. (1972)
[aa] Data of MCCLELLAN et al. (1972).
[ab] When two values are shown they are individual measurements from pairs of dogs.

et al., 1972; R. G. THOMAS et al., 1972). The initial thyroid deposit was not large, usually less than 0.1% of the dose. However, the thyroid gland of the beagle weights only about 1 g, and the net result was that 1 to 2 days after injection the concentrations of these nuclides in dog thyroid were of the order of 100%/kg.

Autoradiographic studies show that most of the ^{241}Am or ^{249}Cf in the dog thyroid was located in the connective tissue surrounding the follicles (G. N. TAYLOR et al., 1969b, 1972; LLOYD et al., 1970). The connective tissue in which the nuclides were bound was argyrophilic and gave a positive periodic acid-Schiff (PAS) reaction. Differential centrifugation of tissue homogenates confirmed the stability of the binding of ^{241}Am to the connective tissue of the dog thyroid gland (STEVENS et al., 1968b, 1969b; STOVER et al., 1972).

It appears from Table 18.25 that the transplutonium nuclide content of the dog thyroid declines during at least the first 8 months after injection. The high concentrations of ^{241}Am at 401 and 448 days in thyroids of dogs receiving 2.8 μCi/kg and at 633 and 1533 days in those receiving 0.91 μCi/kg may be partly the result of radiation-induced loss of thyroid tissue (thyroid glands were one-third to one-half normal weight LLOYD et al., 1970, 1972a), and partly the result of accumulation of late-circulating ^{241}Am released from the liver.

Autoradiographs of the 401- and 448-day dog thyroids showed some foci of intense ^{241}Am concentration that were not seen in the thyroids of the ^{241}Am- or ^{249}Cf-injected dogs killed 1 to 22 days after injection. Most of these aggregates of ^{241}Am appeared to coincide with foci of greater amounts of argyrophilic PAS-positive connective tissue. The small size of the thyroid glands of these two dogs (injection level, 2.8 μCi/kg of ^{241}Am) suggest loss of glandular tissue with subsequent condensation of the remaining connective tissue framework. However, as of 448 days after injection, no definite thyroid lesions had been observed, nor had these beagles shown any clinical symptoms of reduced thyroid function (G. N. TAYLOR et al., 1969b). The thyroid glands of two beagles killed 512 and

1022 days after exposure to an $^{241}AmO_2$ aerosol (average ILB, 2.7 μCi/kg) were atrophic; there was substantial fibrous replacement of the center of the gland, and only a few hyperplastic follicles persisted at the periphery (R. G. THOMAS et al., 1972). In a dog that died 1533 days after injection of 0.9 μCi/kg of ^{241}Am, the epithelial component of the thyroid gland was markedly reduced and no well-defined follicles remained (LLOYD et al., 1972a).

ZALIKIN et al. (1968) detected significant amounts of ^{241}Am in the thyroid glands of rats, but the tissue concentrations were uncertain. SEIDEL and VOLF (1972) confirmed that observation and demonstrated the presence of high concentrations of both ^{241}Am and ^{242}Cm in the rat thyroid. At 6 and 13 days after intravenous injection, the thyroid contained 0.021 ± 0.002 and 0.028 ± 0.006% respectively, of the dose of ^{241}Am citrate, and 0.024 ± 0.004 and 0.022 ± 0.002% respectively, of the dose of ^{242}Cm citrate. The average weight of the rats was 190 g, and the thyroid weight estimated from Appendix C was 17 mg. Thus, the average concentration of ^{241}Am or ^{242}Cm, 6 to 13 days after injection was of the order of 1400%/kg, nearly 10 times that found in beagle thyroids shortly after injection of either ^{241}Am or ^{249}Cf.

Autoradiographs of the thyroids of rats given 12.5 μCi/kg of $^{241}AmCl_3$ and killed 3 days later were reported to show alpha tracks "projected onto the colloid", but whether these emanated from endothelial cells, connective tissue, or blood vessels is not clear. At this high dose level some epithelial cell hyperplasia and edema of the connective tissue stroma were noted 200 days after injection (RUDNITSKAYA and MOSKALEV, 1970).

At a much lower dose of ^{241}Am (0.037 μCi/kg) no effects of radiation could be detected in the thyroid glands of rats through 790 days after injection. The weights of the glands were the same as those of age controls, and the microscopic structures of the thyroids from injected rats and controls were indistinguishable (DURBIN et al., 1972c, e).

Whole-body autoradiographs of ^{241}Am in mice revealed only low concentrations in thyroid at all postinjection times (HAMMARSTRÖM and NILSSON, 1970a). McKAY et al. (1972) were not able to detect ^{241}Am in the thyroid glands of Chinese hamsters killed 32 days after injection.

ROSEN et al. (1972) reported only small concentrations of ^{241}Am in baboon thyroid, similar to those in other endocrine glands. The thyroid glands of monkeys were not analyzed for ^{241}Am content. However, the thyroid glands of the monkeys killed 2 to 3 years after injection of 0.44 to 0.87 μCi/kg of ^{241}Am were of normal weight and showed no microscopic signs of reduced follicular size or increased amounts of fibrous of fatty tissue (DURBIN et al., 1972c).

H. Adrenal Gland

The concentrations of ^{241}Am and ^{249}Cf in the adrenal glands of baboons and beagles were small, 3 to 6%/kg, 8 to 32 days after injection (Table 18.23) declining to 1%/kg by 1917 days after injection.

In the adrenal glands of the rat, the concentrations of ^{241}Am, ^{242}Cm, or ^{253}Es measured 6 to 8 days after injection have been reported to range from 90 to 180%/kg (ZALIKIN et al., 1968; V. H. SMITH, 1972b; SEIDEL and VOLF, 1972). Whole-body autoradiographs of adult mice showed concentrations of ^{241}Am in the adrenal cortices equivalent to those detected in liver and spleen (HAMMARSTRÖM and NILSSON, 1970a). Somewhat contradictory results were obtained by

RUDNITSKAYA and MOSKALEV (1970) who detected more alpha tracks in the medulla than in the cortex of the rat adrenal gland 200 days after injection of 2.5 μCi/kg of ^{241}Am.

I. Testes and Prostate Gland

The initial concentrations of ^{241}Am or ^{249}Cf in the testes of the beagles were low, about the same as those in pelt and heart and only two to three times the concentration in skeletal muscle (LLOYD et al., 1970, 1972d, e; ATHERTON and LLOYD, 1972; and Table 18.23). In the prostate gland the concentrations were two to four times those in testes. The concentrations of ^{241}Am in beagle testes and prostate declined to about one-half their initial values by 1917 days after injection.

The testes of 47-day old rats given 30 μCi/kg of $^{241}AmCl_3$ contained 0.08 and 0.04% of the dose 4 and 16 days after injection, respectively (LANGHAM and CARTER, 1951). Based on the weight of the testes of the adult rat (Appendix C) these amounts of ^{241}Am correspond to concentrations of 28 and 14%/kg, respectively—appreciably greater than those measured in the beagle.

Whole-body autoradiographs revealed a low, but increasing concentration of ^{241}Am in the interstitial tissue of mouse testes (HAMMARSTRÖM and NILSSON, 1970a). Americium-241 appeared to be present in parenchymal as well as stromal elements of rat testes (RUDNITSKAYA and MOSKALEV, 1970). In the dog prostate gland the ^{241}Am was associated with glandular epithelium and to a lesser degree with the media of small arterioles (G. N. TAYLOR et al., 1969a).

Testicular atrophy was observed 9 months after injection of 20 μCi/kg of ^{241}Am in adult rats (OVCHARENKO, 1972). The pituitaries of some male rats showed castration changes 200 days after injection of a high dose of ^{241}Am (RUDNITSKAYA and MOSKALEV, 1970). HAMMARSTRÖM and NILSSON (1970a) also observed testicular atrophy in long-term studies of ^{241}Am in rodents.

J. Ovary, Uterus, and Mammary Gland

In the baboons and beagles, ^{241}Am and ^{249}Cf were initially present in ovaries at concentrations of 3 to 5%/kg, five to ten times those in skeletal muscle (Table 18.23). Concentrations of ^{241}Am, ^{242}Cm, or ^{253}Es in rat ovaries, 6 to 8 days after injection, have been reported to range from 60 to 230%/kg—two to ten times those in skeletal muscle (ZALIKIN et al., 1968; V. H. SMITH, 1972b; SEIDEL and VOLF, 1972).

In whole-body autoradiographs of pregnant mice, HAMMARSTRÖM and NILSSON (1970a) noted high ^{241}Am concentrations in some ovarian follicles, a moderate concentration in ovarian interstitial tissue, and low concentrations in corpora lutea. In the nonpregnant dog ovary, 7 days after injection, ^{249}Cf was detected extracellularly in the deeply-staining zona pellucida of some of the more mature primary oocytes (Graafian follicles) (G. N. TAYLOR et al., 1972).

Histological changes in rat ovaries 200 days after injection of high doses of ^{241}Am included the presence of some follicles without oocytes and some focal stromal leuteinization (RUDNITSKAYA and MOSKALEV, 1970). Similar changes also occur in the normally aging female rat (DURBIN et al., 1972e).

Americium-241 has been measured in the uterus of the nonpregnant baboon, and the concentration was about the same as in the pelt (ROSEN et al., 1972; and Table 18.23). Whole-body autoradiographs of pregnant mice (19th day of gestation) showed detectable concentrations of ^{241}Am in fetal membranes (but not in placentae) and significant concentrations in mammary tissue (HAMMARSTRÖM and NILSSON, 1970a). Mammary glands of non-pregnant, nonlactating baboons concentrated ^{241}Am to about the same degree as adrenal gland and pancreas and about four times more than skeletal muscle (Table 18.24).

MCCLELLAN et al. (1962b) studied the passage of intravenously injected ^{241}Am or ^{244}Cm into the milk of the lactating ewe. Concentrations of these nuclides in milk were 2.3 to 3.2 times their concentrations in plasma. In the same experiment the ratio of the concentration of ^{45}Ca in milk to its concentration in plasma was 35. A milk : plasma concentration ratio greater than one suggests that there was a tendency towards active secretion of the trivalent actinides by the mammary gland of the sheep.

XI. Excretion of the Transplutonium Elements

Early rates of excretion of ^{241}Am, ^{242}Cm, ^{249}Bk, $^{249-252}Cf$, and ^{253}Es in the urine and feces of the rat are shown in Figs. 18.34 and 18.35, respectively. Long-term excretion rates of ^{241}Am, ^{242}Cm, and $^{249-252}Cf$ appear in Fig. 18.36 for urine and Fig. 18.37 for feces. The original data used to prepare Figs. 18.34 through 18.37 are collected in Table 18.26 as accumulated excretion in urine and feces to facilitate independent reanalysis by the interested reader.

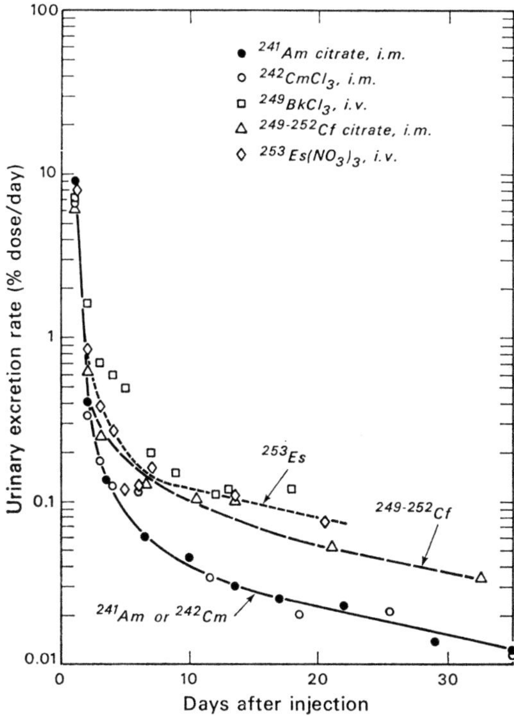

Fig. 18.34. Early excretion rates of trivalent actinide elements in the urine of the adult rat: ^{241}Am and $^{249-252}Cf$, DURBIN et al. (1967, 1972c, d); ^{242}Cm, SCOTT et al. (1949); ^{249}Bk, HUNGATE et al. (1972); ^{253}Es, V. H. SMITH (1972b). See also Table 18.26

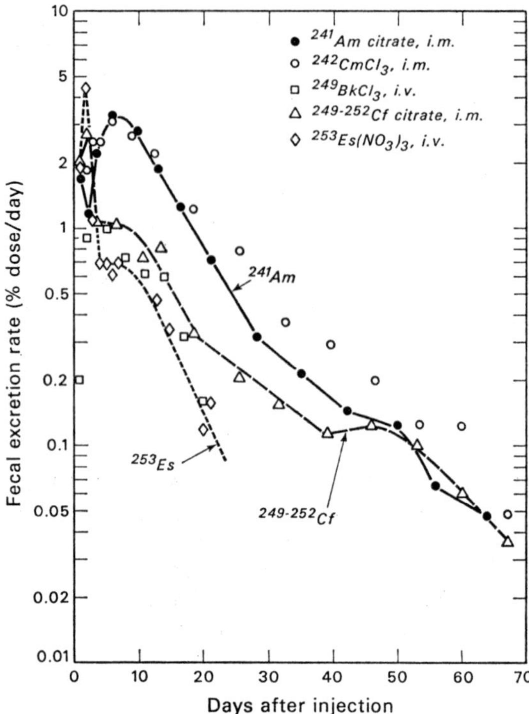

Fig. 18.35

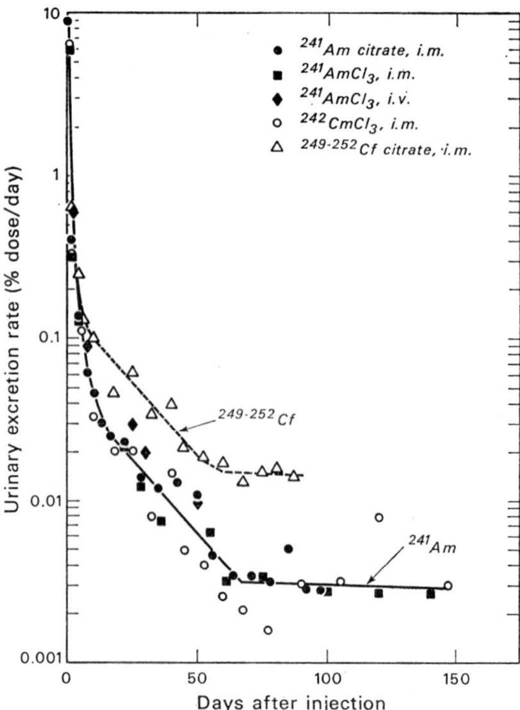

Fig. 18.36

Table 18.26. Excretion of trivalent actinide elements by the adult rat

Day after injection (t)	Cumulative excretion[a] (% dose)											
	$^{241}AmCl_3$[b]		^{241}Am cit.[c]		$^{242}CmCl_3$[aa]		$^{249}BkCl_3$[ab]		$^{249-252}$Cf cit.[ac]		$^{253}Es(NO_3)_3$[ba]	
	Σ Ur[bb]	Σ Fe[bc]	Σ Ur	Σ Fe	Σ Ur	Σ Fe	Σ Ur	Σ Fe	Σ Ur	Σ Fe	Σ Ur	Σ Fe
1	6.1	1.7	9.0	1.7	6.3	0.8	7.0	0.2	6.2	2.1	12.1	2.1
2	6.4	3.3	9.4	2.8	6.6	2.6	8.6	1.1	6.8	4.7	13.0	6.6
4	6.7	8.5	9.7	7.2	6.9	7.6	9.9	3.1	7.3	6.9	13.6	8.4
7	6.8	14.1	10.0	17.2	7.3	17.0	11.0	6.0	7.8	11.0	14.0	14.0
14	6.9	31.8	10.2	34.3	7.5	24.9	11.9	10.4	8.4	15.6	14.8	14.0
21	7.1	41.5	10.4	42.1	7.7	33.7	12.7	12.4	8.8	17.8	15.3	15.6
28	7.2	47.3	10.6	46.0	7.8	42.2			9.2	19.3		
35	7.3	50.4	10.7	47.9	7.9	47.8			9.4	20.4		
49	7.5	54.2	10.8	50.2	8.0	52.4			9.8	22.0		
63	7.6	56.6	10.9	50.8	8.1	53.8			10.1	23.2		
77	7.7	57.8	11.0	51.5	8.1	55.8			10.3	24.7		
91	7.8	59.3	11.0	51.8	8.1	56.6			10.5	25.1		
105	7.8	60.6	11.1	52.1	8.2	57.6						
133	8.0	62.0		52.4[ca]	8.4	58.5						
161	8.2	63.4		52.8	8.5	59.4						
189	8.4	65.4		53.4	8.7	60.3						
217	8.6	67.4		53.8	8.9	61.1						
245	8.8	69.4		54.0	9.1	61.6						
301				54.3								
357				54.6								
413				55.0								
469				55.4								
525				55.8								
581				56.3								
637				56.6								
693				57.1								
749				57.4								

[a] These data are readily manipulated as follows: to calculate whole-body retention, $\text{Retention}_t = 100\% - \Sigma(\text{Ur}+\text{Fe})_{0-t}$; to calculate excretion in an interval, $\text{Ur}(\%)_{t_2-t_1} = \Sigma\,\text{Ur}_{t_2} - \Sigma\,\text{Ur}_{t_1}$; to calculate excretion rate, $\text{Ur}(\%/\text{day})_{t_1-t_2} = \frac{\text{UR}(\%)_{t_2-t_1}}{t_2 - t_1}$.

[b] $^{241}AmCl_3$, 3.6 μCi/kg, intramuscular (Scott et al., 1948a); original data recalculated.

[c] ^{241}Am citrate, 0.37 μCi/kg, intramuscular (Durbin et al., 1967, 1972c).

[aa] $^{242}CmCl_3$, 4.4 μCi/kg, intramuscular (Scott et al., 1949; K. G. Scott et al., unpublished); original data recalculated.

[ab] $^{249}BkCl_3$, 6.25 μCi/kg, intravenous (Hungate et al., 1972).

[ac] $^{249-252}$Cf citrate, 0.33 μCi/kg, intramuscular (Durbin et al., 1972d).

[ba] $^{253}Es(NO_3)_3$, 3.3 μCi/kg, intravenous (V. H. Smith, 1972b)

[bb] Ur = urine.

[bc] Fe = feces.

[ca] Urine and feces combined after 106 days.

Fig. 18.35. Early rates of excretion of trivalent actinide elements in the feces of the adult rat. (Data sources are the same as given in Fig. 18.34)

Fig. 18.36. Long-term rates of excretion of trivalent actinide elements in the urine of the adult rat: ^{241}Am citrate, Durbin et al. (1967, 1972c); $^{241}AmCl_3$, intramuscular injection, Scott et al. (1948a); $^{241}AmCl_3$, intravenous injection, Zalikin et al. (1968); $^{242}CmCl_3$, Scott et al. (1949); $^{249-252}$Cf citrate, Durbin et al. (1972d). See also Table 18.26

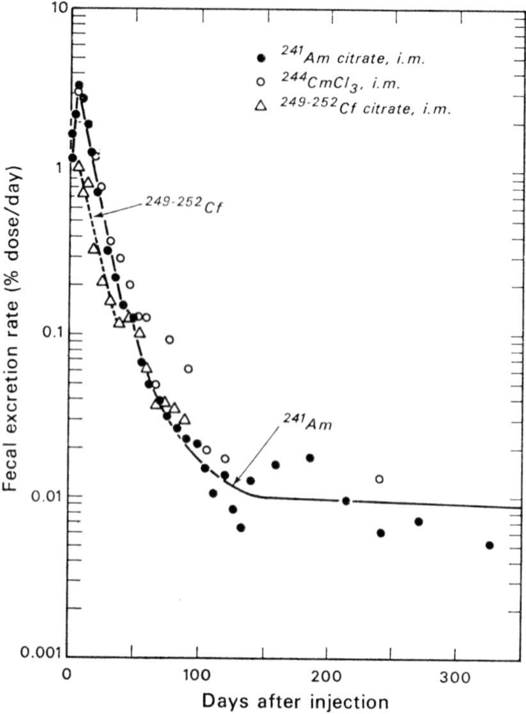

Fig. 18.37

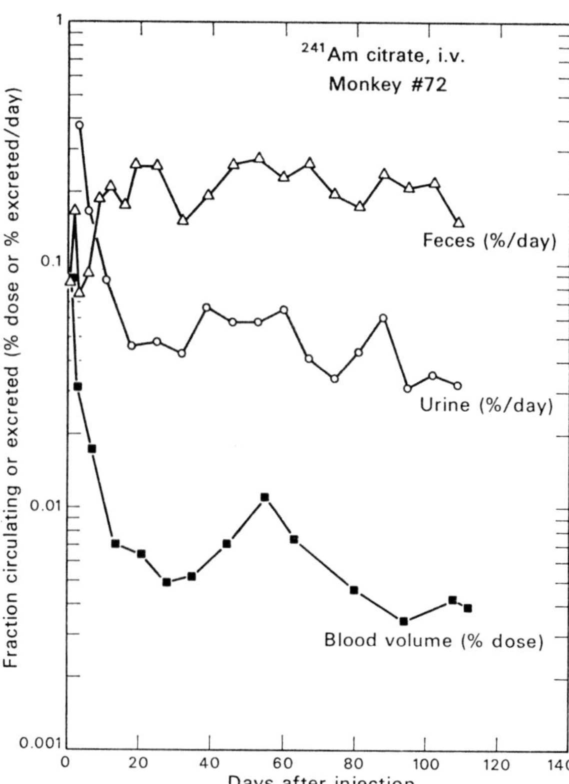

Fig. 18.38

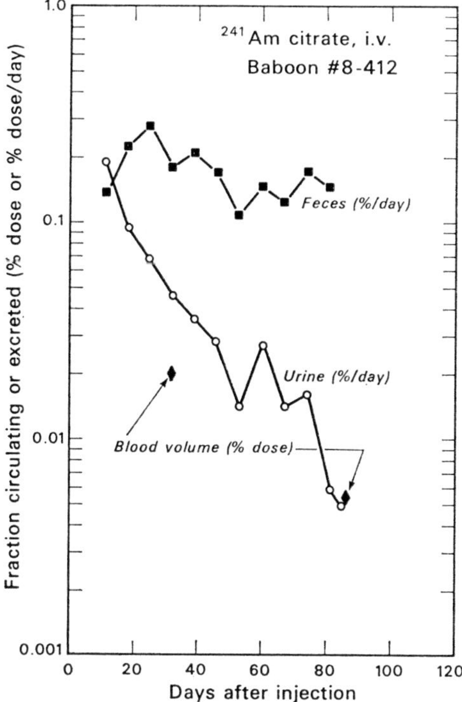

Fig. 18.39. Early excretion rates (urine and feces) of ^{241}Am in baboon. Data of ROSEN et al (1972). See also Tables 18.27 and 18.28

The early rates of excretion of ^{241}Am in the urine and feces and the concentration of ^{241}Am in the plasma of an individual cynomolgus monkey and a baboon are shown in Figs. 18.38 and 18.39, respectively. Long-term rates of excretion of ^{241}Am by an individual monkey are given in Fig. 18.40. The original data (compiled as the average accumulated excretion of several animals) are presented in Tables 18.27 and 18.28.

The excretion of ^{241}Am or ^{249}Cf by the beagle during the first 3 weeks after injection is compared to ^{241}Am excretion by the monkey and baboon in Table 18.27.

Excretion of ^{241}Am, ^{252}Cf, and ^{253}Es by the mouse was reported only as accumulated totals 4 and 14 days after injection; these data appear in Table 18.14. Published reports of the metabolism of ^{241}Am or ^{252}Cf in the Chinese hamster did not include measurements of excreta.

Fig. 18.37. Long-term rates of excretion of trivalent actinide elements in the feces of adult rats. (Data sources are the same as given in Fig. 18.36)

Fig. 18.38. Early excretion rates (urine and feces) and plasma content of ^{241}Am in cynomolgus monkey no. 72. Data of DURBIN et al. (1972b, c). See also Tables 18.27 and 18.28

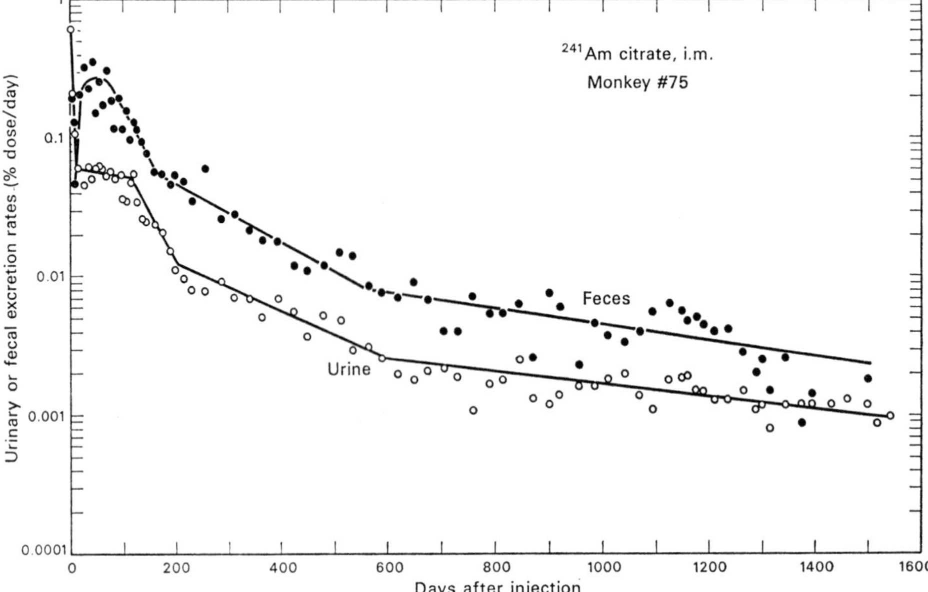

Fig. 18.40. Long-term rates of excretion of ^{241}Am in urine and feces of cynomolgus monkey no. 75. Data of DURBIN et al. (1972b, c). See also Table 18.27 and 18.28

Table 18.27. Excretion by monkey, baboon, and dog of ^{241}Am and ^{249}Cf during the first 3 weeks after injection

Days after injection	Excretion in interval (% dose)							
	^{241}Am, monkey[a]		^{241}Am, baboon[b]		^{241}Am, beagle dog[c]		^{249}Cf, beagle dog[aa]	
	urine	feces	urine	feces	urine	feces	urine	feces
1	6.01	0.10			8.03	0.22	11.36	3.22
2	0.87	0.11	8.70 (days 1–7)	0.71 (days 1–7)	0.34	0.51	1.23	1.42
3	0.41	0.12			0.34	0.14	1.50	1.15
4–7	1.10	0.42			0.51	0.33	1.14	1.07
8–14	0.70	0.96	1.25	1.50	0.33	0.14	0.76	0.41
15–21	0.51	1.10	0.51	1.45	0.17	0.06	0.48	0.22
Total	9.60	2.81	10.46	3.66	9.72	1.45	16.47	7.49

[a] Mean of 14 to 20 cynomolgus monkeys (DURBIN et al., 1972c).
[b] Data from 2 baboons (ROSEN *et al.*, 1972).
[c] Mean of 4 dogs (LLOYD et al., 1970).
[aa] Mean of 11 dogs (LLOYD et al., 1972b).

A. Renal Excretion

Transplutonium nuclides in the urine are derived entirely from the plasma. Urine is essentially an ultrafiltrate of plasma, and the amount of nuclide excreted in urine is proportional to the fraction of circulating nuclide that can be filtered by the kidney. The mechanisms by which the kidney excretes multivalent cations

Table 18.28. Cumulative excretion of ^{241}Am by the cynomolgus monkey and the baboon

Days after injection	Cynomolgus monkey[a]			Baboon[b]		
	no. of animals	% dose		no. of animals	% dose	
		urine	feces		urine	feces
28	16	10.1	4.3	2	10.9	5.6
56	12	11.6	10.1	1	11.7	10.3
84	10	12.8	14.7	1	12.2	14.5
112	9	13.6	18.0			
140	7	14.2	20.2			
196	6	15.5	24.2			
252	6	16.3	26.7			
308	5	16.7	28.0			
364	5	17.0	29.1			
420	4	17.3	29.8			
476	4	17.7	30.8			
532	4	18.0	31.8			
588	4	18.3	32.5			
644	4	18.5	33.0			
742	4	18.8	33.8			
854	2	19.0	34.8			
966	2	19.2	35.5			
1022	1	19.3	35.7			
1078	1	19.4	35.9			
1190	1	19.5	36.6			
1302	1	19.7	36.9			
1414	1	19.8	37.1			
1526	1	19.9	37.2			
1638	1	20.0	37.4			

[a] Data of DURBIN et al. (1972c)
[b] Data of ROSEN et al. (1972).

have not yet been investigated, and the discussion that follows—based on calculated renal clearances, autoradiographic studies of localization in kidney, and on renal physiology—is largely the opinion of the reviewer.

1. Renal Clearance

The excretory function of the mammalian kidney is accomplished by a combination of glomerular filtration and secretion by the proximal convoluted tubules. Proteins with molecular weights greater than 68000 and polysaccharides with molecular weights greater than 55000 are not filtered by the normally functioning kidney. Many compounds and electrolytes that are filtered from the plasma through the glomerulus are not actively secreted by the tubules. However, there are no materials known to be excreted by the tubules that are not also filtered through the glomerulus of the mammalian kidney.

The kidney also functions as a homeostatic regulator, and many essential materials are reabsorbed from the glomerular filtrate also by the proximal tubules. Some passive reabsorption of water and electrolytes takes place in the loops of Henle and the distal convoluted tubules (H. W. SMITH, 1951; BERLINER, 1961).

Renal excretion can be described quantitatively in terms of the renal clearance (C_x), the volume of plasma from which substance (x) is removed per unit time, as follows:

$$C_x(\text{ml/min}) = \frac{U_x(\%/\text{ml}) \times V_x(\text{ml/min})}{P_x(\%/\text{ml})}, \qquad \text{Eq. (25)}$$

where V_x is the rate of urine flow, and U_x and P_x are the concentrations in urine and plasma, respectively. In this discussion concentrations are expressed as percent dose per ml. If the measured quantity is total daily urine output (%/day), and the average plasma concentration during the 24-hr interval is ($\overline{P_x}$), Eq. (25) becomes,

$$C_x(\text{ml/min}) = \frac{\overline{U_x}(\%/\text{day})}{1440\,(\text{min/day}) \times \overline{P_x}(\%/\text{ml})}, \qquad \text{Eq. (26)}$$

and V_x and U_x need not be evaluated.

The carbohydrate inulin, and in some species the normal metabolite, creatinine, are filtered by the glomerulus but are not secreted by the tubules. The rate of inulin (or creatinine) clearance is equal to the glomerular filtration rate (GFR). The GFR is given for several species in Appendix D. For materials that are actively secreted by the tubules, the renal clearance is

$$C_x(\%/\text{ml}) = \text{GFR}\,(\text{ml/min}) + \frac{T_x(\%/\text{min})}{P_x(\%/\text{ml})} \qquad \text{Eq. (27)}$$

where T_x is the rate of tubular secretion. Clearance of materials secreted by the renal tubules exceeds the GFR. Tubular reabsorption can also be calculated from Eq. (27) except that T_x is negative in sign, and C_x is less than the GFR. Determination of the true renal clearance of materials that circulate bound to proteins, Ca(II) for example, must take account of the extent to which they can be filtered through the glomerulus.

Renal clearances [calculated using Eq. (26)] for ^{241}Am in the rat, beagle, and monkey; for ^{249}Cf in the beagle; and for ^{239}Pu in the rat, beagle, and man are shown in Table 18.29; the average plasma concentrations and daily urinary excretion rates are shown in Table 18.29, and the appropriate values of GFR and total plasma volume ($\sum V_p$) are shown in Appendix D.

Renal clearance of total circulating ^{241}Am and ^{249}Cf ranged from 4.3 plasma volumes per day in the beagle to 7.3 plasma volumes per day in the rat. Clearances of ^{239}Pu were only one-fiftieth as great, ranging from 0.05 to 0.16 plasma volumes per day in human beings and rats, respectively. These clearances of total circulating nuclide did not include any correction for the large fractions of these nuclides that circulate bound to plasma proteins (see Sec. V.B.). Depending on the method of separation, measurements of plasma protein binding of multivalent cations lie between 85 and 100% (LEWIN, 1954; DURBIN et al., 1955; STEVENS et al., 1968a; BRUENGER et al., 1969a, 1971; TURNER and TAYLOR, 1968b). Gel filtration studies indicate that the fraction of nuclide not associated with plasma protein (or not remaining associated during separation) could be recovered with a mixture of low-molecular-weight compounds which are presumed to be filterable. Reliable values of protein binding are not yet available for the various nuclides in different species; the Law of Mass Action predicts a higher degree of protein binding in larger species with larger plasma volumes and a larger total number of potential binding sites[32].

Such low clearances (<10%) of total circulating nuclide make it seem quite unlikely that tubular secretion is involved in the renal excretion of multivalent cations. However, low renal clearance can be the net result of filtration combined with efficient reabsorption as well as the result of inhibited filtration. Mathemat-

32 The small plasma volumes of the mouse and Chinese hamster have limited total potential binding capacities, which may partially explain the high excretion of ^{241}Am and ^{252}Cf by these animals in the urine passed during the first 24 hr after injection (PARKER et al., 1962, 1972; WRIGHT, 1972; MEWHINNEY and HARRIS, 1972).

Table 18.29. Plasma concentrations (P_x), urinary excretion rates (U_x), renal clearances (C_x), and calculated filterable fraction in plasma of $^{241}Am(III)$, $^{249}Cf(III)$, and $^{239}Pu(IV)$ in several mammals

Nuclide	Species	Days after injection	P_x (% dose/ml)	U_x (% dose/min)	C_x (ml/min)	C_x (Plasma volumes per day)	Filterable fraction (% of plasma content)	References
$^{241}Am(III)$	Rat[a]	1	1.4×10^{-1}	3.4×10^{-3}	2.4×10^{-2}			D. M. TAYLOR et al. (1961)
		2	2.3×10^{-2}	4×10^{-4}	1.7×10^{-2}			TURNER and TAYLOR (1968b)
		3–4	8×10^{-3}	2.7×10^{-4}	3.4×10^{-2}			
		Mean			2.5×10^{-2}	7.3	3.8	
	Beagle	1	4.6×10^{-3}	5.5×10^{-3}	1.2			STEVENS and BRUENGER (1972)
		2–4	1.1×10^{-4}	1.6×10^{-4}	1.4			LLOYD et al. (1970)
		Mean			1.3	4.3	3.2	
	Monkey	1	5.8×10^{-3}	3.4×10^{-3}	5.9×10^{-1}			DURBIN et al. (1972c)
		2	1.3×10^{-3}	6.7×10^{-4}	5.2×10^{-1}			
		3–4	5.4×10^{-4}	2.7×10^{-4}	5×10^{-1}			
		14–110[b]	5.8×10^{-5}	3.3×10^{-5}	5.7×10^{-1}			
		Mean			5.4×10^{-1}	6.5	4.5	
$^{249}Cf(III)$	Beagle	1	5.2×10^{-3}	7.8×10^{-3}	1.5			STEVENS and BRUENGER (1972)
		2–4	5.4×10^{-4}	6.4×10^{-4}	1.2			LLOYD et al. (1972b)
		Mean			1.35	4.3	3.4	
$^{239}Pu(IV)$	Rat[a]	2–3	5×10^{-1}	1.6×10^{-4}	4.4×10^{-4}			D. M. TAYLOR et al. (1961)
		7–14	7.1×10^{-2}	5×10^{-5}	6.9×10^{-4}			TURNER and TAYLOR (1968b)
		Mean			5.7×10^{-4}	1.6×10^{-1}	8×10^{-2}	
	Beagle	3	9×10^{-3}	1.2×10^{-4}	1.4×10^{-2}			STOVER et al. (1959)
		60–180	1.6×10^{-4}	5.2×10^{-6}	3.4×10^{-2}			
		Mean			2.4×10^{-2}	8.5×10^{-2}	6×10^{-2}	
	Man	1–6	—	—	3.8×10^{-2}			LANGHAM et al. (1950); DURBIN (1972)
		19–24	—	—	1.7×10^{-1}			
		Mean			1.0×10^{-1}	5×10^{-2}	8×10^{-2}	

[a] Body weight of 6- to 8-week-old male rats, 110 g.
[b] Body weight of monkey No. 72, 2.5 kg.

ical demonstration of $T_x < 0$ in Eq. (27) is not sufficient proof of the existence of a renal tubular reabsorption mechanism for a given substance, because the value calculated for T_x is extremely sensitive to the value chosen for the filterability of the substance through the glomerulus. Independent evidence, preferrably demonstration of the substance present at a much higher concentration in glomerular filtrate than in passed urine, is required to establish whether a substance is actively reabsorbed.

Nearly all of the protein-bound ^{239}Pu and probably a significant fraction of protein-bound ^{241}Am and ^{249}Cf are associated with transferrin, the iron-transport protein. One ought then to examine what is known about the renal excretion of iron. Only small amounts of iron are normally excreted in the urine (0.1 to 0.2 mg/day by adult human beings); renal clearance of circulating iron is about 3% (MOORE and DUBACH, 1962; BOTHWELL and FINCH, 1962). Body iron is sustained by regulation of absorption and efficient binding to plasma protein. There is no evidence of the existence of a renal reabsorption pathway for iron. In fact, iron injected in excess of the transferrin binding capacity is filtered into the urine.

Renal reabsorption systems are quite specific. Except for iron, multivalent cations are foreign to the mammalian kidney. In order to be reabsorbed, they would have to participate in an existing transport system. Such a transport system apparently does not exist for iron, the essential mineral most similar chemically to the lanthanide and actinide elements. The possibility that multivalent cations could penetrate the renal reabsorption system of Ca(II) seems unlikely.

On the other hand, there is evidence that the transplutonium elements are *not* actively reabsorbed. The renal clearance of the actinide elements is greatest during the first few hours after injection when protein binding is not as complete as it becomes later on (Table 18.10). The renal clearance of the actinide elements by the rat with its small plasma volume and limited binding capacity is greater than the renal clearances of these elements by larger animals with their more efficient binding potential. Both secretion and reabsorption occur in the proximal convoluted tubules. If either mechanism were involved in the renal excretion of ^{241}Am, one would expect to find some radioactivity localized in that particular structure. However, autoradiographs of dog and monkey kidney[33] showed ^{241}Am or ^{249}Cf to be most concentrated in the medullary rays where they seem to be associated with the straight portions of the loops of Henle (G. N. TAYLOR et al., 1969a, 1972; DURBIN et al., 1972c; Figs. 18.23a, 18.29a, b). The concentrations of ^{241}Am or ^{249}Cf in the cells of the Henle loops and of neighboring interstitial tissue may be a consequence of the drastic changes that occur in the composition of the urine as it passes through the tubules. First, after passage through the proximal tubules the water content of the glomerular filtrate has been reduced by 80% and most of the major anionic complexing species—including bicarbonate, amino acids, and citrate—have been almost completely reabsorbed. Thus, there may be local enhancement of hydrolysis of multivalent cations leading to adsorption on cell surfaces. The tendency towards hydrolysis may be counteracted to some extent by the lower pH of urine as it passes into the descending loop. Second, there is a tendency for all remaining electrolytes to diffuse back into the circulation from the more concentrated fluid in the tubules, thus some passive reabsorption is possible.

---

33 Autoradiographs of rat kidney showed ^{241}Am associated to some extent with convoluted tubules in the renal cortex, but the preparations were not good enough to distinguish between proximal and distal tubules nor among tracks emanating from tubule cells, tubular lumens, or interstitial tissue (DURBIN et al., 1972c).

Without evidence that multivalent cations are secreted or reabsorbed by the renal tubules, one may infer that their presence in urine is due to filtration alone, and that the amount filtered is an approximation of their in vivo binding to large molecules. The filtered fraction of circulating nuclide was estimated from the ratio of renal clearance and the GFR as shown in Table 18.28. The filtered fractions of circulating ^{241}Am or ^{249}Cf were 3.2 to 4.5% in rat, beagle, and monkey[34]. Renal filtration of ^{239}Pu was much lower—0.06 to 0.08% in rat, beagle, and man—as might be expected from the greater stability of the ^{239}Pu(IV)-transferrin complex.

2. Rates of Renal Excretion

The general features of renal excretion of multicharged cations can be inferred from the kinetic models of deposition (Fig. 18.12) and long-term metabolism (Fig. 18.27). As discussed above, the multivalent nuclide in urine has been derived exclusively from the blood. Thus, the rate of renal excretion depends on the net rate at which nuclide is entering and leaving blood. During deposition (immediately after an injection or an accidental exposure) the rate of change of the blood level is the resultant of all the rates at which the nuclide is being cycled through ECF and being taken up by bone and liver. Later, renal excretion rates will reflect the rates at which the nuclide is returning to the circulation from soft tissue and short-lived bone structures. Still later, after the nuclide has been cleared from soft tissues (see the discussion of long-term retention in Sections VIII, IX, and X), the renal excretion rate will depend on the rate of release from bone.

All of the features of the long-term model in Fig. 18.27 have not been tested in one species; it was constructed from observations in different animals. The experimental findings pertinent to the renal excretion aspects of the model include the following: (a) The initial distributions of ^{241}Am or ^{242}Cm in the rat were nearly the same, and the urine curves of these two elements were superimposable (Figs. 18.34 and 18.36); (b) In the rat, at times longer than 20 days after injection, the urine curves of ^{241}Am or $^{249-252}$Cf had the same slope suggesting a common source of plasma nuclide (Figs. 18.34 and 18.36); (c) In the monkeys, from 20 to 500 days after injection of ^{241}Am, during which time some of the initial liver burden was being released into the circulation, the curves of liver retention and renal excretion had the same slope, indicating that the liver was the source of the urinary nuclide (Figs. 18.21 and 18.40); (d) At long times after injection, when nearly all of the ^{241}Am remaining in the body of the rat was in the skeleton, the slopes of the retention curves in liver, kidney, residual soft tissues, and combined urine and feces were all much alike. The skeleton was the only available source of nuclide entering the central plasma compartment from which the peripheral compartments were being replenished.

---

34 One to 4 days after injection, the transplutonium nuclide content of the kidneys of rats was 13 to 42% of the nuclide in urine and kidney combined; in beagles and monkeys, kidney represented 6.7 to 8.1% of the combined total. After the first few hours little nuclide would have been present in the kidney's contained blood. Soluble (complexed) nuclide in formed urine would probably have been removed during fixation and processing of kidney sections, and would not have registered on autoradiographs. It appears from the autoradiographs that nuclide in the kidney at these early postinjection times is largely associated with straight tubules, interstitial tissue, and structures in interstitial tissue. If there is either adsorption of filtered nuclide on tubule cell surfaces or passive diffusion, some of the nuclide combined with these kidney structures was derived from glomerular filtrate. Thus, some of the nuclide in kidney may have been filtered but not passed in urine leading to underestimation of the renal clearance.

3. The Quantity of Transplutonium Nuclide Excreted in Urine

The amount of transplutonium nuclide excreted in urine depends on how much of the circulating nuclide can be filtered by the kidney (as discussed above). After deposition is complete, the amount of nuclide in the circulation depends, in turn, on the amounts released from the returning compartments. Mathematical analysis of models of tracer behavior assumes that the rate of change of the amount of tracer in a depleting compartment (dq_2/dt) is a first-order reaction and proportional to the amount present in the compartment (q_2),

$$dq_2/dt = -q_2 k_{12} \qquad \text{Eq. (28)}$$

where k_{12} is the rate of loss of tracer from the peripheral compartment (2) to the plasma compartment (1). Thus, the amount of nuclide returned to the plasma depends, in its turn, on the initial partition of the nuclide between compartments which can return nuclide to the circulation (bone and soft tissues other than liver) and those compartments which are partly or wholly nonreturning (the incisors of rodents and the liver of several species).

After injection of transplutonium nuclides at doses low enough to avoid significant radiation damage in the liver, the initial liver deposit in rat and mouse (and apparently also in monkey and baboon) was eliminated directly into the gastrointestinal tract and did not contribute to renal excretion. The initial transplutonium deposit in the liver of the beagle and Chinese hamster was almost completely retained (see Sec. VIII.C.). The rate of release of ^{241}Am from the beagle liver approached that of release from bone (LLOYD et al., 1970, 1972c). It is not known whether any portion of the ^{241}Am or ^{252}Cf deposited in the liver of the Chinese hamster was either recirculated or excreted (A. L. BROOKS, private communication). If the injected dose is large enough to cause liver damage, the liver becomes a partly-returning compartment. Some of the initial nuclide deposit is released into the circulation, and during that time elevated urine levels have been observed in studies with monkeys (DURBIN et al., 1972b, c; Fig. 18.38) and rats (LANGHAM and CARTER, 1951).

The initial partition of ^{241}Am or ^{242}Cm in most species was 30 to 35% of the dose in hard tissues and 50 to 60% in liver; that of $^{249-252}$Cf or ^{253}Es was the reverse; 60 to 65% of the dose in hard tissues and 25 to 30% in liver (Tables 18.12 through 18.19). In rats the initial deposit of $^{249-252}$Cf or ^{253}Es in soft tissues was nearly twice as great as that of ^{241}Am or ^{242}Cm leading to higher early nuclide levels in urine. Compared with the lighter weight actinides, the greater amount of $^{249-252}$Cf in the rat skeleton, resulted in higher long-term levels in the urine[35] (Figs. 18.34 and 18.36). In the beagle, the larger initial deposits of ^{249}Cf (compared with ^{241}Am) in soft tissues other than liver (and probably also in short-lived bone structures) are presumed to be the sources of the greater amounts of ^{249}Cf excreted in the urine (Table 18.27).

B. Gastrointestinal Excretion

The long-term metabolic model shown in Fig. 18.27 includes several routes by which transplutonium elements might enter the gastrointestinal tract to be

35 During the first 50 days after injection in rats, the level of $^{249-252}$Cf in urine was two to three times that of ^{241}Am, in accord with the deposition of $^{249-252}$Cf in bone and soft tissues other than liver, which was two times higher. However, at postinjection times longer than 50 days, the urine level of $^{249-252}$Cf was at least five times that of ^{241}Am. The possibilities that $^{249-252}$Cf was released more readily from bone, or was redeposited in bone to a lesser degree than ^{241}Am or ^{239}Pu, or that $^{249-252}$Cf was recirculated in a more readily filterable form are important for purposes of establishing radiation protection criteria and deserve investigation.

excreted in feces. The following mechanisms appear possible in all mammals: (a) excretion from liver into bile, (b) elimination in digestive secretions other than bile, and (c) functional shedding of intestinal epithelium containing nuclide complexed with transferrin. In rodents there is an additional source of fecal nuclide—the swallowing of labeled mineral in the course of functional attrition of the constantly growing incisors.

1. Biliary Excretion

Early in the study of the biology of the transplutonium elements, SCOTT et al. (1948a) observed that the initial deposit of ^{241}Am in the liver of the rat could be recovered entirely in the feces, and they suggested that the ^{241}Am excreted by the gastrointestinal tract had been eliminated from the liver in the bile. The quantitative excretion of the initial liver burden in feces has been observed for many other multivalent cations administered in soluble form to rats and mice (DURBIN et al., 1956, 1960; SCHUBERT et al., 1961; WRIGHT, 1972). The view that the multivalent nuclide present in feces had been derived by biliary excretion from the liver, was reinforced by the measurement of common slopes in the curves of liver retention and fecal excretion of ^{241}Am, ^{242}Cm, and $^{249-252}$Cf during the first 30 days after injection in rats, and of ^{241}Am during the first 200 days after injection in cynomolgus monkeys (Sec. VIII.C, Table 18.22, and Figs. 18.21, 18.26, 18.34, and 18.40).

Bile appears to be by far the most important route of entry of multicharged cations into the gastrointestinal tracts of rats, mice, and monkeys. The possible role of biliary secretion in the excretion of ^{241}Am or ^{249}Cf from the liver of the beagle was discussed in Sec. VIII.C.

Americium-241 in the monkey: At autopsy bile was collected (1 to 4 ml from the gall bladder) from two monkeys killed 8 days after injection of ^{241}Am citrate, and from two monkeys killed 21 and 42 days after injection. At 8 days, when fecal excretion was low (average for 16 monkeys, 0.09%/day), the ^{241}Am concentration in bile was 0.0035 to 0.008%/ml—the same as, or up to twice as great as, the concentration in skeletal muscle. Between days 21 and 42, when fecal excretion had risen to 0.17%/day (average for 14 monkeys), the ^{241}Am concentration in bile was 0.013 to 0.017%/ml—four to five times the concentration in muscle (DURBIN et al., 1972c). If the amount of gall-bladder bile produced daily by the monkey is similar to the amount estimated to be elaborated by man[36] (HAWK et al., 1947), fecal excretion of ^{241}Am by the monkey can be accounted for almost completely by biliary secretion.

Plutonium-239 in the rat: BALLOU and HESS (1972) administered ^{239}Pu(IV) citrate intravenously to rats, perfused the intestine, and collected bile by cannulation of the bile duct. During the first 5 hr after injection, about one-half of the ^{239}Pu recovered from the instestinal contents was derived from bile.

Cerium-144 in the rat: Rats were injected with ^{144}Ce citrate and 1 hr later their bile ducts were ligated. At autopsy, 3 days later, the ^{144}Ce content of the liver was lower while the ^{144}Ce in the skeleton, soft tissues, and accumulated urine was greater than in sham operated controls, suggesting that bile was being

36 Hepatic bile flow in normal adult human beings is estimated to be between 500 and 1000 ml/day. Gall bladder bile flow is less, because 12 to 25% of the water is reabsorbed: The flow is estimated to be 450 to 900 ml/day (6.4 to 12.8 ml/kg per day for a 70-kg man). If all of the ^{241}Am in feces is assumed to have been derived from gall bladder bile, bile flow in the cynomolgus monkey would be 6 ml/kg per day, not too different from the lower range of the human estimate.

regurgitated into the circulation. In another experiment bile was collected continuously through an indwelling catheter in the bile duct for 24 or 48 hr beginning on the 4th day after injection of ^{144}Ce. Bile volumes averaged 14 ml/day (75 ml/kg per day). The average concentration of ^{144}Ce in the bile of 10 rats was $0.022 \pm 0.008\%$/ml (M. H. WILLIAMS, N. JEUNG, and P. W. DURBIN, unpublished).

Iron in man: Secretion in bile accounts for about 40% of normal human fecal excretion of iron (0.25 mg of a total of 0.6 mg of fecal iron daily) according to GREEN et al. (1968). That group of investigators injected ^{59}Fe into normal persons and noted that biliary secretion of iron rose to a maximum on the 3rd day after injection and then declined to a negligible level by the 5th day. They suggested that the iron remaining in the liver had reacted to form insoluble forms of ferritin.

The pattern of fecal excretion of ^{241}Am, and to a lesser degree of $^{249-252}Cf$, by the beagle is similar to that of iron excretion in man, and the very stable association of ^{241}Am or $^{249-252}Cf$ with ferritin in the canine liver has been amply demonstrated (BRUENGER et al., 1969b, 1972).

2. Secretion into Digestive Fluids Other than Bile

In studies with rodents, clearance of the original liver burden in bile to the feces accounts for all or nearly all the fecal excretion up to 30 days after injection, even in the most carefully controlled balance studies. However, the possibility should not be overlooked that a portion of these nuclides might enter the gastrointestinal tract in digestive secretions other than bile, especially during the first few hours after injection while significant amounts are still circulating.

Calcium is a normal constituent of the gastrointestinal secretions of rats and is present in them at a level proportional to that of the ionized calcium in plasma (GRAN, 1960). The digestive secretions of normal human adults, which include bile, contribute to the intestinal contents 130 to 227 mg/day of endogenous calcium, of which 130 ± 47 mg/day is excreted (HEANEY and SKILLMAN, 1964).

About 40% of normal human fecal iron (0.25 mg/day) is believed to have been derived from digestive secretions other than bile (POLLYCOVE, 1966; MOORE and DUBACH, 1962).

The above observations suggest that a portion of multivalent cation in the circulation (not bound to protein) might be able to enter the gastrointestinal lumen by diffusion along with other electrolytes. Whether such a mechanism is normally operable in intact animals is not known; however, there is some evidence suggesting gastrointestinal secretion of ^{239}Pu or ^{144}Ce in rodents with obstructed bile ducts or biliary fistulas.

BALLOU and HESS (1972) found that nearly one-half of ^{239}Pu entering the perfused intestine during the first 5 hr after injection was not of biliary origin. Fluids secreted into the duodenum and jejunum accounted for nearly 40% of the ^{239}Pu in the gastrointestinal contents.

In a study of ^{144}Ce excretion, rats were injected intravenously with ^{144}Ce citrate; their bile ducts were ligated immediately thereafter. During the first hour after the ^{144}Ce injection (and obstruction of the bile duct) 0.4 and 0.6% of the dose, respectively, was secreted into the contents of the stomach and small intestine. The total ^{144}Ce recovered in the gastrointestinal contents was slightly greater than in sham operated controls (M. H. WILLIAMS, N. JEUNG, and P. W. DURBIN, unpublished).

3. Shedding of Intestinal Epithelium

The adult human intestine desquamates between 50 and 80 g of epithelial cells each day. Transferrin has been detected in intestinal epithelium, and CROSBY (1970) suggests that the degree of iron saturation of the transferrin in the intestinal epithelium is a major factor in the regulation of iron absorption. GREEN et al. (1968) have calculated that 0.1 mg/day of iron (17% of total fecal iron) is lost in feces with these dead cells.

The metabolic studies with lanthanide and actinide elements in which the radioactivity of the completely empty gastrointestinal tract has been measured show that the intestine always contains some nuclide (DURBIN et al., 1956; ZALIKIN et al., 1968; LLOYD et al., 1972d, e). Nuclide concentrations are greatest immediately after injection, and decline at about the same rates as do the concentrations in muscle and pelt.

If some of the nuclide in intestinal tissue is present as a transferrin complex inside epithelial cells, that amount will be eliminated in feces as the cells are lost by normal attrition. The level of nuclide-transferrin complex in intestine will be proportional to the nuclide level in blood. Therefore, except for the first few days after injection when significant amounts of these nuclides are circulating, loss with shed intestinal cells would make a minor contribution to total fecal excretion.

4. Functional Attrition of Rodent Incisors

The incisors of rodents grow, calcify, and erupt throughout the life of the animal[37]. They are continuously worn away at their incisal edges by gnawing, and complete replacement takes 40 to 50 days (SCHOUR and MASSLER, 1949). The ground mineral is evidently swallowed and can be completely recovered in feces (as discussed below).

Whole-body autoradiographs have demonstrated the uptake of ^{239}Pu and ^{241}Am in the incisors of adults and in the developing molars of very young rats and mice (ULLBERG et al., 1962; HAMMARSTRÖM and NILSSON, 1970a, b). DURBIN et al. (1967, 1972c, d) have shown that 4 days after intramuscular injection of the citrates of ^{241}Am or $^{249-252}$Cf, the incisors of adult rats contained 1.4 and 2.7% of the dose, respectively. The uptake of these two nuclides in incisors was 4.7 and 4.4%, respectively, of their total uptake in the edentate skeleton (Table 18.12).

The incisors continued to take up ^{241}Am. During the first week after injection, no ^{241}Am was lost from the incisors, but the level of circulating ^{241}Am (from clearance of soft tissues and turnover of very short-lived bone structures) was high enough so that the maximum incisor content of ^{241}Am (1.6%) was not reached until the eighth day.

About 45 days are required for an externally visible label, a metabolic or dye marker, to progress from the base of the tooth to complete erosion. It can be seen from the anatomy of the rat incisors and from the autoradiographs of HAMMARSTRÖM and NILSSON (1971a, b) that although nuclide deposition on day 1 was heaviest at the base of the incisors just anterior to the odontogenic epithelium, there was some deposition over all of the surface of the pulp cavity which extends in an elongated cone of decreasing diameter from the base nearly to the tooth tip. Therefore, as expected, some nuclide loss is observed sooner than 45 days. To further complicate matters, the incisors continue to accumulate nuclide from the blood so that between 32 and 46 days, even as the bulk of the day-1 deposit

37 Rats, mice, and other rodents that eat a mixed diet have molars with limited growth and distinct roots. Those rodents that are completely herbivorous (rabbits and guinea pigs) have molars that grow from permanent pulp.

at the incisor tips was being lost, the remainder of the incisors contained an amount of nuclide proportional to the integral of the plasma concentration from days 8 to 46 (BAUER et al., 1961).

In metabolic studies with rats, if a careful collection is made not only of the feces, but also of the dust, hair, and food scraps from the cage bottoms, and the daily fecal excretion rate is plotted on a large scale (Fig. 18.35), a distinct "hump" can be found between days 40 and 60 after injection of any bone-seeking nuclide. The hump in Fig. 18.35 for $^{249-252}$Cf is more prominent than that for ^{241}Am not only because of the larger amount of $^{249-252}$Cf deposited in the incisors, but also because the amount of $^{249-252}$Cf deposited in and subsequently eliminated from the liver is smaller. Graphic integration of the area between the hump in the fecal curve and a smooth curve connecting the portions of the fecal curves before and after it results in a calculated amount of nuclide excreted in feces equal to the amount of nuclide taken up by the incisors between the time of injection and the acquisition of the maximum deposit about 8 days later.

Because the incisors continue to grow, a fraction of circulating nuclide will be deposited in them each day—to be lost some 45 days later. As indicated in the metabolic model, eventually the incisors should be accumulating and losing nuclide at the same rate at which it is being returned to circulation from long-lived bone structures. DURBIN et al. (1967) calculated by graphic integration of the incisor vs. time curve that between days 120 and 790 after injection of ^{241}Am in rats 0.37% (7.5% of total fecal nuclide passed in that interval) had been taken up and then lost by the incisors.

5. Species Differences in Fecal Excretion Patterns

Rats and Mice. The livers of rats and mice eliminated their initial burdens of actinide elements in about 30 days. By the 60th day after injection the nuclide accumulated in the incisors in the first few days has been worn away and excreted (Fig. 18.35). This constitutes a delay of about 30 days between nuclide uptake in liver and its elimination in feces, and an approximate 45-day delay between nuclide uptake in and loss from the incisors. So, at any time, the amount of nuclide in the liver is proportional to the concentration in blood during the preceding 30 days, and the nuclide content of the incisors represents the average blood concentration in the 45 days preceding sampling. These delays between nuclide uptake and loss from the liver and the incisors obscure relationships between current blood concentrations and fecal excretion of these nuclides by rats as long as the blood concentration is changing rapidly (for about the first 120 days after injection).

After 5 to 6 months the skeleton is almost the only source of circulating nuclide, and the rates of loss of nuclide through both the liver and incisors approach an equilibrium with the slow rate of nuclide release from bone. By then the nuclide concentration in blood is changing so slowly that the 30- to 45-day lag between liver and incisors uptake and the loss of nuclide to feces becomes relatively unimportant. The rate of excretion of nuclide in feces ultimately approaches the rate of change of the concentration in blood, and the amount of nuclide excreted daily by the gastrointestinal tract becomes almost proportional to the concentration in blood at the time of sampling of feces.

Presumably the liver continues to accumulate about 50% of circulating ^{241}Am or ^{242}Cm and about 25% of $^{249-252}$Cf or ^{253}Es, and incisors continue to acquire 2 to 4% of circulating nuclide; so that for rats and mice the gastrointestinal tract is always the most important excretory organ. At times long after injection, the daily output of multivalent cations in feces is 2 to 5 times that excreted in urine.

Monkey. The initial deposit of ^{241}Am in the monkey liver was lost in about 200 days, and that in the baboon liver would be expected to have been reduced to 6.25% of its initial value in about 600 days (4 times the measured clearance half-time of 150 days). In the monkey the amounts of nuclide excreted daily in urine and feces were changing at the same rate at times longer than 200 days after injection. The gastrointestinal tract was the major organ of elimination of ^{241}Am by the monkey presumably because close to one-half of circulating ^{241}Am continued to be deposited in, and subsequently lost from, the liver. The daily loss of ^{241}Am in feces was 2 to 3 times that excreted in urine after 600 days (Fig. 18.40).

Dog. At injected doses low enough to avoid significant radiation damage (less than 0.1 μCi/kg of ^{241}Am) most of the initial liver burden of ^{241}Am or $^{249-252}$Cf left the liver very slowly. Excretion of these nuclides by the beagle has been reported only for the first 3 weeks after injection. It seems likely that during the life-time of the nuclide reservoir in the liver, fecal excretion would be unrelated to the concentration in blood at the time of fecal sampling.

C. Acceleration of the Excretion of Actinide Elements

It was noticed in the first metabolic studies of the fission products and radioactive heavy elements that the behavior of many of these newly-made radionuclides was similar to that of ^{226}Ra (Stone, 1944; Copp et al., 1947). These early workers were aware of the toxicity of ^{226}Ra, and the problem of removal of long-lived radionuclides from bone was given prompt attention (Copp et al., 1946; Painter et al., 1946; Schubert, 1947).

The intravenous injection of zirconium citrate soon after nuclide administration was found to reduce the bone uptake of ^{239}Pu or of lanthanide nuclides (Schubert, 1949; Kawin et al., 1950; see also the reviews of Rosenthal, 1955, 1959).

Prompt administration of the aminopolycarboxylic acid, EDTA, was found to be effective in preventing accumulation of ^{239}Pu or the radioactive lanthanides in bone, and prolonged treatment promoted elimination of these multivalent cations from soft tissues (Foreman, 1950; Foreman and Hamilton, 1951; Hamilton and Scott, 1953; Foreman et al., 1955; Foreman and Finnegan, 1957). The results of animal studies were sufficiently encouraging to lead to the use of EDTA in the treatment of several persons accidentally exposed to ^{239}Pu or ^{241}Am (Foreman et al., 1954, 1958; R. L. Dobson, 1955). However, it was soon learned that EDTA produced nephrosis at the dosage levels and over the prolonged treatment schedules required for significant enhancement of radionuclide excretion (R. L. Dobson, 1955; Foreman, 1959).

At a conference on radioelement removal, Kroll (1955) reported briefly that a newly synthesized aminopolycarboxylic acid, diethylenetriaminepentaacetic acid (DTPA), chelated multicharged cations, and that the stability constants of the DTPA chelates were about 100 times greater than those of the corresponding EDTA chelates. Shortly thereafter, intravenously injected DTPA chelates with ^{90}Y and ^{140}La were found to be quantitatively excreted in the urine of human subjects (Kroll et al., 1957), and V. H. Smith (1958a, b) reported that in rats DTPA was superior to EDTA in preventing initial bone deposition of ^{239}Pu and in promoting excretion of deposited ^{239}Pu. Since that time the effectiveness of DTPA in accelerating excretion of ^{239}Pu and the trivalent transplutonium elements has been amply demonstrated in animals (V. H. Smith, 1959, 1972a; Sowby and Taylor, 1960; Schubert et al., 1961; see also the references listed in Tables 18.30 and 18.31). DTPA is the treatment of choice in reducing the burden of ^{239}Pu or ^{241}Am of contaminated persons (Norwood, 1960; Lagerquist et al., 1967; Brodsky et al., 1966, 1969; Fasiska et al., 1971.

1. Some Pharmacological Characteristics of EDTA and DTPA

Although this review cannot explain all of the phenomena associated with in vivo chelation with EDTA and DTPA, some of their general properties have been summarized in order to provide a better understanding of the mechanisms of action.

Both EDTA and DTPA form stable complexes with calcium, consequently their rapid introduction into the blood causes a precipitous and dangerous fall in serum calcium. They are commonly administered as the mixed salts, $CaNa_2$ EDTA or $CaNa_3$DTPA. The British group administer H_5DTPA simultaneously with an equimolar quantity of calcium gluconate (Sowby and Taylor, 1960; D. M. Taylor and Sowby, 1962; D. M. Taylor, 1967; D. M. Taylor et al., 1971; James and Taylor, 1971).

Foreman (1959) studied the distribution and excretion of ^{14}C-labeled EDTA and DTPA in rats and their disappearance from the circulation, and the renal excretion of ^{14}C-EDTA in man. Absorption of these compounds from the gastrointestinal tract was low—only 2 to 4% was absorbed in 24 hr. In rats, for either compound the rates of clearance from the blood to urine were not different after intravenous, intraperitoneal, or intramuscular injection. Both compounds were excreted rapidly by the kidneys—61 to 85% of EDTA or DTPA in 90 min and 96 to 99% of EDTA in 24 hr; there was some residual DTPA (8 to 13%) in the rats after 24 hr. Renal clearance calculations indicated that both compounds were excreted by the renal tubules as well as being filtered through the glomeruli.

Half-times of EDTA and DTPA clearance from the blood and of renal excretion ranged from 33 to 44 min in rats. Clearance of EDTA from the blood of human subjects was slower; the half-time was 66 to 69 min. In the human subjects, 60% of intravenously administered EDTA and 30% of intramuscularly administered EDTA was excreted in the urine within 90 min after injection. Excretion of EDTA was nearly complete in 12 to 24 hr.

Immediately after injection in the rats, EDTA or DTPA left the plasma compartment and was diluted in body fluid volumes equal to 26.5 and 30% of the body weight, respectively. The most recent estimate of the volume of ECF in the rat is 27.7% of the body weight (bromide space measured by R. N. Pierson, Jr., and D. C. Price, unpublished). The amounts of either compound present in tissues were low at all postinjection times, and erythrocytes were found to be impermeable to them. These observations led Foreman (1959) to conclude that EDTA and DTPA were able to pass through the capillaries and to equilibrate with ECF, but were not able to penetrate cells. It follows, then, that neither chelating agent can act on metals inside cells such as the ^{239}Pu, ^{241}Am, or ^{252}Cf associated with ferritin in the hepatic cells of the beagle (Stover et al., 1970, 1972; Bruenger et al., 1972) or colloidal ^{239}Pu confined within liver phagocytes (Lindenbaum and Rosenthal, 1972).

Large and frequent doses of EDTA in rats have been found to produce degeneration of the proximal renal tubules: After 16 daily doses of 0.67 mM/kg of EDTA the tubular lesions were reversible, but were apparently irreversible after 16 daily doses of 8 mM/kg (Foreman, 1959). Similar lesions were apparently also induced by DTPA, but the dosages and severity of the lesions were not reported. However, there were no reports or grossly visible lesions in later studies in which as many as 50 daily doses of 0.46 mM/kg of DTPA were administered to rats (FOREMAN, 1962)b, or after a series of intermittent doses of 3 mM/kg of DTPA, totaling 27 mM/kg (D. M. TAYLOR and SOWBY, 1962). Amounts of DTPA ranging from 3 to 7 g/week have been tolerated by human beings under treatment for extended periods for removal of ^{239}Pu or ^{241}Am (LAGERQUIST et al., 1967; FASISKA et al., 1971).

FOREMAN (1959) also studied the in vivo stability in rats of EDTA and DTPA chelates with selected metals. The chelates with yttrium, cobalt, chromium, zinc, and lead were stable in the body and were excreted almost completely in the urine within 48 hr. The EDTA chelate with Fe(III) was not stable; 84% of the Fe(III)-EDTA chelate dissociated and was fixed in the body. The Fe(III)-DTPA chelate has been reported to be stable to dissociation in vivo (M. RUBIN and J.V. PRINCIOTTO, unpublished, quoted by BRICK and RATH, 1963). Chelation with DTPA is regarded as a fairly efficient method of accelerating iron excretion in various iron-storage diseases, but is not the treatment of choice (BRICK and RATH, 1963). The ability of DTPA to compete with the normal plasma iron-carrier, transferrin, for ferric ion suggests a mechanism by which DTPA, even though excluded from cells, might be able to promote elimination of metals from intracellular deposits in liver and spleen. By chelating some iron in serum, DTPA may act indirectly to mobilize iron from ferritin and hemosiderin in storage depots and promote its subsequent release in soluble form into body fluids. It is reasonable to assume that other metals, such as ^{239}Pu or ^{241}Am, when combined with ferritin, would be released into body fluids simultaneously with the iron and would then be amenable to chelation.

2. General Description of Animal Studies of DTPA Effectiveness

For the most part, rats and mice have been used to assess the effectiveness of EDTA and DTPA in reducing bone burdens of the trivalent actinides. Studies with EDTA have been abandoned since the demonstration by V. H. SMITH (1958a, b) and D. M. TAYLOR and SOWBY (1962) of the much greater effectiveness of DTPA in reducing bone burdens of ^{239}Pu and ^{241}Am. The rodent studies of DTPA treatment for removal of transplutonium elements, summarized in Tables 18.30 and 18.31, can be divided into two general categories, which for the purposes of this review have been designated prompt nuclide deposition studies and persistent reservoir studies.

(a) Prompt nuclide deposition studies. Nuclides usually have been administered as citrate complexes, which accomplished prompt and complete distribution. The three exceptional experiments in which the nuclide was administered in a form other than a citrate complex were: (1) $^{241}AmCl_3$ injected intraperitoneally (BELYAEV, 1969); (2) $^{253}Es(NO_3)_3$ injected intramuscularly (V. H. SMITH, 1972a); and (3) $^{241}Am(NO_3)_3$ given by inhalation[38] (NÉNOT et al., 1971c). In all of these experiments DTPA was given by intraperitoneal or intramuscular injection. In some studies DTPA injections were started as soon as 1 hr after the nuclide injection to assess the ability of DTPA to prevent nuclide deposition in bone. In other experiments DTPA injections were delayed as long as 25 days to examine the ability of DTPA to remove trivalent nuclides already deposited in the skeleton. Treatment intervals varied from daily to weekly, and the amounts of DTPA

38 This inhalation study was included in Table 18.30 because by the time DTPA was begun, 20 days after exposure, nearly all of the ^{241}Am had been absorbed.

Table 18.30. Design of experiments using DTPA to promote excretion of promptly deposited trivalent actinides in rodents, and amount of nuclide remaining in bone after treatment

Isotope and route of administration[a]	Length of study (days)	DTPA treatment: treatment schedule (time after nuclide injn.)	number of treatments	DTPA per treatment[b] (mM/kg)	total DTPA given (mM/kg)	Amount of nuclide in bone (% of control)	Reference
Studies using rats							
^{241}Am citrate, i.v.	21	1 hr, 7, 14 d	3	3.0	9.0	17.5	SOWBY and TAYLOR (1960b)
	21	6 hr, 7, 14 d	3	3.0	9.0	25	
	21	7, 14 d	2	3.0	6.0	50	
^{241}Am citrate, i.v.	30	2 hr, 1–29 d	30	0.46	13.8	17[c]	FOREMAN (1962)
	30	2 m, 1–29 d	30	0.028, s.t.	0.84	33.8[c]	
	30	7–29 d	23	0.46	10.6	36[c]	
	30	14–29 d	16	0.46	7.4	52[c]	
^{241}Am citrate, i.t.	30	2 hr, 1–29 d	30	0.46	13.8	21[c]	
	50	2 hr, 1–49 d	50	0.46	23.0	14.3[c]	
^{241}Am citrate, i.v.	21	7–11, 14–17 d	9	3.0	27.0	50.5	D. M. TAYLOR and SOWBY (1962)
$^{241}AmCl_3$, i.p.	3	1 d	1	1.5	1.5	52	BELYAEV (1969)
	16	1–13 d	13	0.41	5.7	22	
	40	25–37 d	13	0.41	5.7	63	
^{241}Am citrate, i.v.	13	6, 8, 11 d	3	1.0	3.0	77	SEIDEL and VOLF (1972)
^{242}Cm citrate, i.v.	13	6, 8, 11 d	3	1.0	3.0	67	
$^{253}Es(NO_3)_3$, i.m.	21	1 h, 1–3, 5, 7, 9, 12, 15, 17 d	10	0.33	3.3	42.3	V. H. SMITH (1972a)
$^{241}Am(NO_3)_3$, aer.	40	20 d, 2/week	6	0.12, i.m.	0.72	82	NÉNOT et al. (1971c)
	90		20	0.12, i.m.	2.4	42	
	124		30	0.12, i.m.	3.6	26	
Studies using mice							
^{241}Am citrate, i.m.	4	2 h, 1–3 d	4	0.11	0.44	34	PARKER et al. (1962)
	14	2 hr, 1–4, 7–11 d	10	0.11	1.10	34	
^{252}Cf citrate, i.m.	4	2 hr, 1–3 d	4	0.11	0.44	46	
	14	2 hr, 1–4, 7–11 d	10	0.11	1.10	38.5	
^{253}Es citrate, i.m.	1	2 hr	1	0.11	0.11	76.5	PARKER et al. (1972)
	4	2 hr, 1–3 d	4	0.11	0.44	68.5	
	14	2 hr, 1–4, 7–11 d	10	0.11	1.10	50	

[a] i.v., intravenous; i.t., intratracheal; i.m., intramuscular; i.p., intraperitoneal; s.t., stomach tube; aer., aerosol.
[b] DTPA administered by intraperitoneal injection unless otherwise specified.
[c] Based on whole-body measurement at 30 days postinjection; at that time liver and soft tissues have been cleared.

Table 18.31. Design of experiments using DTPA to mobilize ^{241}Am or ^{242}Cm from a persistent reservoir in muscle or lung of rats and to prevent their deposition in bone

Isotope and route of administration[a]	Length of study (days)	DTPA treatment					Amount of absorbed nuclide in bone (% of control)	Reference
		treatment schedule (time after nuclide adm.)	number of treat- ments	DTPA per treatment (mM/kg)	total DTPA given (mM/kg)	route of administration of DTPA[b]		
$^{242}Cm(NO_3)_3$, i.m.[c]	1	0.5 hr	1	0.12	0.12	i.m., inj.leg	3.6	Nénot et al. (1970)
	20	1, 5, 10, 15 d	4	0.12	0.48		12	
$^{242}Cm(NO_3)_3$, i.m.	42	21 d, 2/week	6	0.12	0.72	i.m., opp.leg	45	Nénot et al. (1972)
	81		16	0.12	1.92		50	
	111		24	0.12	2.88		35	
$^{241}Am(NO_3)_3$, i.m.	10	1, 7 d	2	0.12	0.24	i.m. opp.leg	21	Nénot et al. (1971b)
	90	1 d, 1/week	13	0.12	1.56		24	
	10	1, 7 d	2	0.0012	0.0024	aerosol	84	
$^{241}Am(SO_4)_3$, i.m.	90	1 d, 1/week	13	0.12	1.56	i.m., opp.leg	13	
$^{241}Am(NO_3)_3$, aer.	90	1 d, 1/week	13	0.12	1.56	i.m.	2.8	
	90		13	0.0012	0.016	aerosol	5.8	
$^{241}Am(NO_3)_3$, aer.	60	1 d, 1/week	8	0.12	0.96	i.m.	9	Nénot et al. (1971a)
	60		8	0.0012	0.01	aerosol	18	
$^{253}EsCl_3$, i.t.	7	1 hr	1	0.2	0.2	i.m.	7	Ballou and Morrow (1972, and unpublished)

[a] i.m., intramuscular; aer., aerosol.; i.t., intratracheal instillation.
[b] Intramuscular injections of DTPA were given either in the same leg into which the nuclide was injected (inj.leg) or in the opposite leg (opp.leg).
[c] In the first studies with ^{242}Cm, the $^{242}Cm(NO_3)_3$ solutions contained variable amounts of stable lanthanum carrier. (See Tables 18.5 and 18.6, Section IV.B.).

given in a single treatment ranged from 0.11 mM/kg in the mouse studies (PARKER et al., 1962, 1972) to 3 mM/kg in the experiments with growing rats (SOWBY and TAYLOR, 1960; D. M. TAYLOR and SOWBY, 1962).

(b) Persistent reservoir studies. In the French experiments (summarized in Table 18.31), ^{241}Am or ^{242}Cm nitrates or sulfates were administered intramuscularly or as aerosols. The aim of the studies was to find the optimal treatment of persons accidentally contaminated with transplutonium elements by puncture wound or by inhalation. In either case—contaminated muscle or lung—a reservoir was established that continuously fed radioactivity into the circulation over periods lasting from a few days to several months. DTPA treatment was given either by intramuscular injection (0.12 mM/kg per injection) or as an aerosol. The investigators estimated that the amount of DTPA aerosol deposited in the lung was about 1% of the intramuscular DTPA dose. DTPA treatments were started 1 or 21 days after the nuclide administration and were repeated bi-weekly or weekly for intervals of 10 to 120 days.

3. Chelation of Actinides in Soft Tissues

The amounts of actinide nuclides initially deposited in the soft tissues of rodents are small, and the soft tissues are cleared quickly (see Sec. X and Fig. 18.33). In the absence of a chelating agent, some of the nuclide released from soft tissue is redeposited in bone. As discussed in Sec. VIII, the trivalent actinides, when administered at low doses to rodents, are cleared promptly from the liver without therapeutic intervention.

DTPA has been found to hasten the clearance of actinide nuclides from soft tissues. A short course of DTPA injections reduced the soft tissues contents of ^{241}Am, ^{252}Cf, or ^{253}Es in the mouse, and the ^{241}Am content of rat tissues to 10% of control levels, and accelerated liver clearance (D. M. TAYLOR and SOWBY, 1962; PARKER et al., 1962, 1972). Less frequent DTPA treatments or smaller amounts of DTPA per treatment were less effective in promoting loss of ^{241}Am or ^{242}Cm from the soft tissues and livers of rats (SEIDEL and VOLF, 1972).

4. Chelation of Circulating Nuclide

Shortly after DTPA became available, it was recognized that if treatment were initiated within a few minutes to a few hours after parenteral injection of soluble ^{239}Pu(IV), DTPA chelated and promoted efficient excretion of the large amount of ^{239}Pu still circulating in the body fluids, and effectively prevented its deposition in liver and bone (V. H. SMITH, 1958a, b). V. H. SMITH (1959) also showed that a minimum dose of 0.2 mM/kg of DTPA given 1 hr after intravenous injection of ^{239}Pu citrate into rats could limit the ^{239}Pu uptake in bone to the amount taken up before the DTPA injection (20% of the ^{239}Pu content of bones of control rats).

The ability of DTPA to prevent skeletal deposition of the trivalent actinides is more restricted, because like the lanthanide and alkaline earth elements, they are deposited very quickly after injection. It is apparent from Figs. 18.6 through 18.9 and from Table 18.10 that 1 hr after injection of soluble ^{241}Am or ^{242}Cm only 3 to 10% of the dose was still circulating and 28% was in soft tissues and ECF. It can be seen from Figs. 18.10, 18.11, and 18.13 that 1 hr after injection the deposition of ^{241}Am in bone was 76% complete; deposition in the liver was 68% complete.

5. Chelation of Transplutonium Elements in Marrow or Loosely Associated with Bone

In a short-term single-dose study, BELYAEV (1969) gave DTPA to rats 1 day after administration of ^{241}Am; the rats were killed 2 days later. Even though deposition of ^{241}Am was complete at the time of the DTPA injection, the DTPA treatment reduced the ^{241}Am content of bone to 50% of the control value.

Several studies involve a short series of intermittent DTPA treatments begun 1 to 24 hr after administration to rats of ^{241}Am (BELYAEV, 1969; SOWBY and TAYLOR, 1962), or ^{253}Es (V. H. SMITH, 1972a), or after administration to mice of ^{241}Am, ^{252}Cf, or ^{253}Es (PARKER et al., 1962, 1972). In all cases the amount of actinide remaining in the skeleton 14 to 21 days after the nuclide injection was less than the amount expected. If, at the time of the first DTPA injection, all of the nuclide in bone were firmly fixed, the only expected loss from bone would have been the quantity of nuclide released from bone that was resorbed during the DTPA treatment (see paragraph 6 below) plus any nuclide in tissue fluids or soft tissues that was prevented from depositing in bone. The dose-response study of V. H. SMITH (1959) showed that if the amount of DTPA (given 1 hr after a ^{239}Pu injection) were increased above 0.2 mM/kg, the bone content was reduced to less than the amount of ^{239}Pu deposited in 1 hr (20% of control). A single dose of 4 mM/kg of DTPA reduced the ^{239}Pu in bone to 10% of the control value. Apparently about-one-half of the ^{239}Pu in bone 1 hr after injection could be removed, if the DTPA dose were large.

These results suggest that for 24 to 48 hr after nuclide injection, up to one-half of trivalent actinide considered to be deposited in bone (based on analysis of whole long bones including marrow) is either in the marrow or is only loosely bound to the bone surfaces and is still amenable to chelation by DTPA. Some of the 1-hr deposit of ^{239}Pu in bone appears also to be amenable to chelation, if the amount of DTPA is large enough.

6. Chelation of Actinides Deposited in Bone

At times longer than 1 to 2 days after nuclide injection, any nuclide that is still in the marrow has been deposited in bone, and nuclide initially only loosely bound at bone surfaces has become more firmly fixed. Growing surfaces are covered by new bone. In rapidly growing rats some of the nuclide deposited on metaphyseal trabeculae has already been resorbed and redistributed (D. M. TAYLOR et al., 1961).

The effect of delayed DTPA treatment on removal of ^{241}Am or ^{242}Cm from the bones of young adult rats, estimated to be 60 to 90 days old at the time of nuclide injection, has been the subject of several investigations (FOREMAN, 1962b; BELYAEV, 1969; SEIDEL and VOLF, 1972; NÉNOT et al., 1971c; Table 18.30). Delayed DTPA treatment of ^{241}Am deposited in the bones of 7-week-old rats has been studied by SOWBY and TAYLOR (1960) and D. M. TAYLOR and SOWBY (1962).

JAMES and TAYLOR (1971) used quantitative autoradiography to investigate the influence of DTPA on the microscopic distribution of ^{239}Pu in the ends of the long bones of growing rats. Five treatments of DTPA (1.5 mM/kg each) were given during a 14-day period beginning 7 days after injection of a solution of 85% ultrafilterable $^{239}Pu(NO_3)_4$ to 7-week-old rats. They concluded that the observed reduction in the ^{239}Pu bone burden was due to chelation of ^{239}Pu released into tissue fluids during osteoclastic resorption of labeled bone surfaces in the normal course of growth remodeling.

In young rats, resorption of trabeculae elaborated by centers of endochondral ossification in ribs, vertebrae, and the ends of long bones appears to proceed fast enough and to involve a large enough fraction of the total bone nuclide to account for the amount of nuclide lost from bone during the first 10 to 30 days of a course of DTPA treatments begun 7 to 14 days after nuclide injection. Using quantitative autoradiography, D. M. Taylor et al. (1961) showed that ^{241}Am deposited on the metaphyseal trabeculae in the ends of the femora of 7-week-old male rats was almost completely resorbed within 21 days after injection—equivalent to a constant removal rate of 5%/day. The rate of resorption of labeled metaphyseal trabeculae of older rats appears to be slower. The autoradiographs shown by Durbin et al. (1967) can be interpreted to represent about 80% resorption in 42 days leading to a calculated resorption rate of about 2%/day (see Figs. 18.14 and 18.15).

These estimates of bone surface resorption agree with the rate of resorption of calcium in adult rat bone. Bauer et al. (1961) calculated from ^{45}Ca kinetic data that calcium in the tibia of the 90-day-old rat was being resorbed at a rate of 0.13 mg/hr, which is equivalent to resorption of 3%/day of the mass of the tibia[39].

Assuming that 3 and 4%, respectively, of the labeled surfaces of the bones of adult and growing rats is resorbed daily, the calculated amounts of nuclide remaining in bone agree with the amounts measured by Foreman (1962b), Seidel and Volf (1972) Sowby and Taylor (1960), and D. M. Taylor and Sowby (1962), in which DTPA treatments were begun 6 to 14 days after nuclide injection and lasted for 7 to 23 days. In the studies of Belyaev (1969) and Nénot et al. (1971c), in which DTPA treatment was delayed for 20 days or longer after nuclide administration, the amounts of nuclide lost from bone were less than expected at a constant resorption rate of 3%/day. Apparently, by the time the DTPA treatments were started in the latter experiments, a large fraction of the initially labeled trabecular surfaces had already been resorbed, and the nuclide had been redistributed to other surfaces with lower probabilities of resorption during the short course of treatments. The studies of Nénot et al. (1971c) indicated however, that if DTPA treatments were continued long enough (in this case 104 days), enough bone surface, even though with low nuclide concentration, would eventually be resorbed to reduce the nuclide burden in the whole skeleton to about 25% of the control level.

The average rate of resorption of surfaces in the adult human skeleton has been estimated to be about 5%/year or 0.014%/day (Durbin, 1972). Thus, as was pointed out by James and Taylor (1971), DTPA therapy can be expected to have only a marginal effect in removing deposited nuclides from the human skeleton. Consequently, prevention of the systemic distribution of nuclides in contaminated wounds or in the lungs appears to be the only realistic therapeutic approach.

7. Diversion of Nuclide from a Persistent Reservoir

The studies of Lafuma et al. (1970) and Nénot et al. (1970, 1971a–c, 1972), in which DTPA was administered intermittently while ^{241}Am or ^{242}Cm was being absorbed from a reservoir in muscle or lung, are practical demonstrations of the usefulness of DTPA in chelating nuclides as they enter the circulation; the

39 The ash weight of the tibia of the 90-day-old female Sprague-Dawley rat is 260 mg (Durbin et al., 1972e), and bone ash is about 40% calcium by weight. Therefore,

$$\text{Resorption (\%/day)} = \frac{0.13 \text{ mg Ca/hr} \times 24 \text{ hr/day} \times 100}{260 \text{ mg ash} \times 0.4 \text{ mg Ca/mg ash}} = 3\%/\text{day}.$$

nuclides are diverted to urinary excretion, and are effectively prevented from being deposited in bone and liver. DTPA (0.12 mM/kg in each treatment) administered intramuscularly to rats every fifth day or twice weekly starting 1 day after nuclide administration and continued for 2 to 4 months, prevented deposition of 75 to 90% of the administered nuclide.

BALLOU and MORROW (1972, and unpublished) determined that DTPA effectively removes ^{253}Es ($^{253}EsCl_3$ instilled intratracheally in rats) from the lung and prevents deposition of ^{253}Es in bone. The DTPA was given as a single injection (0.2 mM/kg) 1 hour after the nuclide administration. The rats were killed 1 week after the instillation at which time the lungs contained 4% of the instilled ^{253}Es (18% of untreated controls) and the skeleton contained 3% of the ^{253}Es (7% of untreated controls based on absorbed dose).

Not only did DTPA efficiently prevent the nuclides from depositing in bone and tissues, but the radioactivity was cleared more rapidly from both the lung and muscle reservoirs[40] reducing the potential radiation dose in those tissues.

It is of interest to note that although the amount of DTPA administered as an aerosol was too small to have a significant effect on the behavior of intramuscularly injected ^{241}Am, it was very effective in chelating ^{241}Am deposited in the lungs. In fact, repeated inhalation of small amounts of DTPA (about 0.0012 mM/kg in each treatment) was almost as effective in clearing ^{241}Am from the lungs and diverting it from bone as a similar series of intramuscular injections of 0.12 mM/kg of DTPA (NÉNOT et al., 1971a).

8. DTPA Treatment of Persons Exposed to ^{241}Am

In general, human experience agrees with the above experimental findings. In the case of a wound contaminated with ^{241}Am (described in Sec. IV.B and Table 18.7) it was calculated that nearly two-thirds of the ^{241}Am in the wound site was chelated and excreted in urine during a 60-day course of EDTA injections (R. L. DOBSON, 1955). It appears from Fig. 18.2 that mobilization of ^{241}Am from the wound site declined at about the time EDTA treatments were stopped.

BRODSKY et al. (1969) and FASISKA et al. (1971) have reported the prolonged treatment with DTPA of a man who had inhaled ^{241}Am oxides over a period of several years before his contamination was discovered. Weekly injections of 1 g DTPA for 15 months removed about one-half of the body burden which, measured by whole-body counting, was 2 μCi at the start of the DTPA treatments. A combination of partial-body and whole-body measurements indicated that the DTPA treatments had removed about 1 μCi of ^{241}Am, mostly from the lungs and liver, leaving about 1 μCi in the skeleton. Weekly or bi-weekly DTPA treatments of 0.5 g each were continued for another year, and an additional 0.15 μCi was removed.

Some preliminary results have been reported of the metabolism in human subjects of an isotope of the artificially produced lanthanide, promethium, ^{143}Pm (PALMER et al., 1968a, b). Since the initial tissue distribution of ^{241}Am in rats most closely resembled that of the lanthanide of the same ionic radius, ^{147}Pm (see Sec. V.A.6), the behavior of ^{143}Pm in human beings is expected to resemble that of ^{241}Am. Well-characterized experiments with ^{143}Pm should provide insight into the problems of dealing with persons contaminated with ^{241}Am or isotopes of the other trivalent actinides. The stability constant of the ^{241}Am-DTPA chelate

---

40 The intramuscular injection site could be cleared even more rapidly, if the DTPA were injected into the same location as the nuclide. See Sec. IV.B.

is greater than that of the ^{143}Pm-DTPA chelate (see the references in Sec. VII.A.6, and Chap. 17 of this volume). Chelates of ^{241}Am in bone, blood, and tissue constituents may also be more stable than those of ^{143}Pm, so there may be some small differences in the in vivo chelation behavior of these two, otherwise similar, elements.

In the study of Palmer et al. (1968a, b) single or multiple injections of 1 g each of DTPA (equivalent to 0.014 mM/kg in a 70 kg man) were given at various times after intravenous injection of 0.1 μCi of $^{143}PmCl_3$. Retention of ^{143}Pm was measured by whole-body counting 6 days after its injection. A single DTPA injection given 5 min prior to the ^{143}Pm completely prevented deposition; given 30 min afterwards, the body content was reduced to 14% of the control, while at 5 or 24 hr the reduction was 58 or 85%, respectively. When given 80 days afterwards, only 5% of the body content of ^{143}Pm was eliminated in the ensuing week. Three DTPA injections 1, 5, and 24 hr after the ^{143}Pm injection reduced the body content to 24% of control, and four DTPA injections, 1, 24, 48, and 72 hr after the ^{143}Pm removed all but 12% of the control body content.

As one would expect from the 100-fold difference in the amount of DTPA administered, the single injection of 0.014 mM/kg at 24 hr was not as effective in removing ^{143}Pm from man as was a single dose of 1.5 mM/kg of DTPA in removing ^{241}Am from rats; 58% of the control body content of the ^{143}Pm remained, compared with 38% of the control body content of the ^{241}Am in the rats (Belyaev, 1969). Four injections of DTPA starting at 1 hr were more effective in reducing the body content of ^{143}Pm in man (88% removed) than was a comparable course of DTPA treatments in removing ^{241}Am from rats, which had been reduced to 18% after 7 daily injections of 0.46 mM/kg of DTPA starting 2 hr after the ^{241}Am administration (Foreman, 1962b).

9. Suitability of ^{241}Am as a Tracer for ^{239}Pu

In the plutonium processing industry, wounds of the hands and fingers are frequently contaminated with metallic plutonium or $^{239}PuO_2$. Assessment of the level of contamination and of the efficacy of treatment is difficult and often inaccurate, and requires special sophisticated detection equipment because low-energy L X-rays are the only photons emitted by an ordinary mixture of plutonium isotopes. Americium-241, the radioactive daughter of ^{241}Pu and a contaminant of all modern plutonium (Sec. IV.D.9), emits a 60 keV γ-ray which is accurately and easily measured in vivo by currently available photon detecting equipment. The ^{241}Am γ-ray has been used to monitor suspected contaminations of plutonium workers (Hammond and Putzier, 1964; Lagerquist et al., 1967).

The question of whether ^{241}Am is a good tracer for ^{239}Pu in the body was examined in a series of experiments with beagles (L. J. Johnson, 1969, which contains all of the original data; L. J. Johnson et al., 1970a, b, 1972; Watters and Lebel, 1972). A colloidal suspension of $^{239}PuO_2$[41] (mean count diameter, 7 μm) was implanted subcutaneously above the left metacarpus of each of 40 beagles. Twenty of these beagles also each received DTPA by intravenous injection (0.05 mM/kg per treatment) at 2 hr, on days 1 through 6, days 9, 12, and 14, and then on three consecutive days each month for up to 1 year after the $^{239}PuO_2$ injection. Groups of dogs were killed at intervals from 2 weeks to 1 year postinjection.

41 In the discussion of these experiments the designation ^{239}Pu represents the mixture of plutonium isotopes with a composition by weight of 0.035% ^{238}Pu, 86.63% ^{239}Pu, 11.69% ^{240}Pu, 1.51% ^{241}Pu, and 0.135% ^{242}Pu. All of the isotopes except ^{241}Pu, which is a beta emitter, contribute to the alpha radioactivity of the mixture.

The migration and/or solubilization of the implanted particles appeared to be characteristic of the $^{239}PuO_2$. External measurement of the photons of energy greater than 10 keV (those reaching the detector were chiefly the 60 keV γ-rays of ^{241}Am) demonstrated that a fraction of the implanted nuclides migrated to the superficial cervical lymph nodes (the major nodes draining the implant area). During the first 2 weeks, about 3% of the implanted nuclides migrated to the lymph nodes, and by the end of one year, the cervical lymph nodes contained an average of 17% of the implanted dose. Later, separate measurements of the alpha radioactivity of the two nuclides in the dissected nodes showed that 100 days after implantation they contained about 8% of each nuclide. In the DTPA-treated beagles the externally measurable radioactivity of the lymph nodes was less at all times after injection than in the untreated controls, but the differences were not always significant.

On the average, 0.65% of the implanted ^{241}Am and 0.27% of the ^{239}Pu migrated in soluble form out of the implant site and/or the lymph nodes and were distributed through the body. The average amounts implanted were 0.55 μCi of ^{241}Am and 5.8 μCi of ^{239}Pu. At autopsy 100 days after implantation, the lymph nodes contained 0.044 μCi of ^{241}Am and 0.464 μCi of ^{239}Pu, and the measured tissues contained 0.0036 μCi of ^{241}Am and 0.016 μCi of ^{239}Pu.

Any changes occurring in the nuclide content of the tissues between 2 weeks and 1 year after the $^{239}PuO_2$ implantation were obscured by the great variability of the data and by the net flow of nuclide into and out of the tissues. The experimenters concluded that there were no well-defined trends in the tissue distributions with time after injection. The data were pooled, without regard for the length of time between injection and sacrifice, to form two large groups—untreated and DTPA-treated.

The measured body content of ^{241}Am, expressed as percent of body content 100 days after implantation (excluding lymph nodes and neglecting unmeasured soft tissues), was distributed 60% in liver, 34% in bone[42], 2% in lung, 1.7% in kidneys, 0.8% in spleen, 0.43% in thyroid (equivalent to 430%/kg), and 0.12% in the adrenals. The distribution of ^{239}Pu was 44% in liver, 51% in skeleton, 2% each in spleen and lung, 0.5% each in kidneys and adrenal glands, and 0.03% in the thyroid gland. These tissue distributions are about what one would expect from the studies of beagles intravenously injected with ^{241}Am or ^{239}Pu (Stover et al., 1959, 1967, 1971; Lloyd et al., 1970, 1972a, c). In the untreated beagles, a greater fraction of the implanted ^{241}Am than ^{239}Pu was absorbed in the body and of the absorbed ^{241}Am, a greater fraction was deposited in the liver, kidneys, thyroid gland and other soft tissues; while a smaller fraction of the absorbed ^{241}Am was deposited in the skeleton.

The average ^{241}Am contents of the livers, bones, kidneys, and thyroid glands of the DTPA-treated dogs were 36, 52, 55, and 61%, respectively, of those tissues in untreated controls. The DTPA treatment did not change, or actually seemed to increase the ^{241}Am content of spleens, lungs, and adrenal, which were 167, 92, and 200%, respectively, of the amounts in the control tissues. Although the DTPA treatments significantly increased the amounts of ^{239}Pu excreted in urine

---

42 The total ^{241}Am or ^{239}Pu in the skeleton was calculated using the proportionality factor of 20 times the nuclide content of one femur (L. J. Johnson et al., 1972). According to the data of Lloyd et al. (1972a) based on analyses of complete skeletons of beagles injected intravenously with either ^{241}Am or ^{239}Pu, one femur contains 3 or 4% of total skeletal ^{241}Am or ^{239}Pu, respectively, and the appropriate proportionality factors are 33.3 or 25, respectively. The tabulated values of ^{241}Am and ^{239}Pu in the skeleton were recalculated for this review using these proportionality factors.

(as well as increasing urinary excretion of ^{241}Am), the amounts of ^{239}Pu in the livers and bones of the DTPA-treated beagles were almost the same as those of the controls.

The concentration of ^{241}Am in the blood of untreated beagles was highest during the first few weeks after the ^{241}Am-labeled PuO_2 was implanted, and the DTPA treatments were given almost daily during the first 2 weeks after the implantation, so it is reasonable to assume that the major action of the DTPA was to prevent deposition of some of the ^{241}Am as it was being absorbed.

The experimenters concluded, that, except for early external monitoring of the wound site itself and possibly of the major lymph nodes draining the site, ^{241}Am was not a suitable tracer for the metabolism of absorbed ^{239}Pu, and that the differential behavior of the two elements was exaggerated still further by the administration of DTPA.

XII. Toxicity of the Transplutonium Elements

A. Lethal Radiation Injury

The mean lethal dose (LD_{50}), which is the injected amount required to kill one-half of the animals at some specified time, has been determined in adult rats at times from 2 weeks to 1 year after injection of ^{239}Pu, ^{241}Am, or ^{244}Cm; in growing rats after injection of ^{239}Pu or ^{241}Am; and in Chinese hamsters and dogs after injection of ^{241}Am. The LD_{50}'s expressed as μCi/kg injected or inhaled are shown in Table 18.32 (rodents) and in Table 18.33 (dogs) as are the toxicities of the β-emitting bone-seekers, ^{90}Sr-^{90}Y and ^{144}Ce-^{144}Pr.

Table 18.32. Acute and subacute toxicity of ^{241}Am and ^{244}Cm in the adult and young rat[a] and the Chinese hamster compared with that of ^{239}Pu, ^{144}Ce, and ^{90}Sr

Days after injection	LD_{50} (μCi/kg) ≤17	30	60	120–150	250	300–350	References
Adult rat[a]							
^{241}Am	—	110	60	43	25	—	ZALIKIN et al. (1968, 1969)
^{244}Cm	—	110	60	38	18	4.4	SEMENOV (1971a)
^{239}Pu	125	90	60	30	—	—	FINKLE et al. (1946)
^{144}Ce	—	1830	—	872	826	452	MOSKALEV et al. (1966)
^{99}Sr	—	1185	—	979	629	480	MOSKALEV et al. (1966)
Young rat[a]							
^{241}Am	189	95	64	49	47	34	LANGHAM and CARTER (1951)
^{239}Pu	—	71	63	59	52	30	LANGHAM and CARTER (1951)
Chinese hamster							
^{241}Am	—	100	—	10	—	5	MCKAY et al. (1972)
^{144}Ce	5000	—	2500	1250	—	—	BROOKS et al. (1969)
^{90}Sr	3000	—	20000	—	1000	—	BROOKS and MCCLELLAN (1969)

[a] Adult rat, more than 60 days old or more than 130 g body weight; young rat, 30 to 60 days old, body weight approximately 100 g.
[b] Isotopes were administered intravenously or intraperitoneally as chloride, nitrate, or citrate.

Table 18.33. Acute and chronic toxicity of ^{241}Am in dogs compared with the toxicity of ^{239}Pu, ^{144}Ce, and ^{90}Sr. Toxicity is expressed as 50% lethal dose or 100% lethal dose at varying times after exposure

Isotope, chemical form and method of administration[a]	Dose (μCi/g)	50% lethal	100% lethal	References
		Days after injection		
$^{241}AmCl_3$, i.v.	10–60	—	21	Zalikin et al. (1969)
^{241}Am citrate, i.v.	2.8	—	450	T. F. Dougherty et al. (1972)
	0.9	~1800	—	
$^{241}Am(NO_3)_3$, inhal.	1.5	—	~180	Buldakov et al. (1972), Nifatov et al. (1972)
$^{241}AmO_2$, inhal.	4.7[b]	~300	—	R. G. Thomas (1970), R. G. Thomas et al. (1972)
^{239}Pu(VI)citrate, i.v.	22	—	16	Painter et al. (1946)
	13	—	90	
^{239}Pu(IV)citrate, i.v.	2.9	1325	—	T. F. Dougherty et al. (1972)
	0.9	1725	—	
^{239}Pu(IV) nitrate, polymeric, inhal.	1.5	—	~180	Buldakov et al. (1972), Nifatov et al. (1972)
^{144}Ce citrate, i.v.	~110	30	—	Fritz et al. (1966)
	50	1200	—	
$^{144}CeCl_3$, inhal.	370[b]	30	—	McClellan et al. (1970)
	120[b]	~1600	—	
^{90}Sr citrate, i.v.	100	1225	—	T. F. Dougherty et al. (1972)
$^{90}SrCl_2$, inhal.	270[b]	30	—	McClellan et al. (1970)
	120[b]	150	—	

[a] i.v., intravenous; inhal., inhalation.
[b] Initial body burden.

1. Acute Toxicity

The acutely toxic doses at 30 days ($LD_{50/30}$) of the three long-lived actinides, ^{239}Pu, ^{241}Am, and ^{244}Cm, were about the same for rodents, 100 μCi/kg, on the average. Pilot experiments with $^{253}EsCl_3$ indicated that in rats injected as adults, 21-day-old weanlings, or on the day of birth, 150 μCi/kg was 50% lethal in 90 to 100 days (Mahlum and Sikov, 1972 and unpublished). If one takes into account the high initial urinary excretion of $^{253}EsCl_3$ (about 40% is excreted in the urine in the first week after injection, Hungate et al., 1972) and the 20.5-day physical half-life of ^{253}Es, the total amount of α-particle energy available per μCi of ^{253}Es at 30 and 60 days after injection is only about 40 and 25%, respectively, of that available from the long-lived actinides. Therefore, the $LD_{50/30}$ of ^{253}Es is probably more than 250 μCi/kg.

At post-injection times of 60 through 350 days, the LD_{50}'s of ^{239}Pu, ^{241}Am and ^{253}Es were greater in the young rats than in the adults, suggesting a greater resistance of the growing animals to the lethal effects of these nuclides. The LD_{50} of ^{241}Am, determined 120 to 180 days after injection, was less in the Chinese hamsters than in the adult rats. Multivalent cations were eliminated rapidly from the rat liver, but were retained in the liver of the hamster where additional injury very likely contributed to the greater lethality of ^{241}Am in the hamsters.

The acute LD_{50} of ^{241}Am or ^{239}Pu is not well defined in the dog; injection of 10 to 60 μCi/kg of either nuclide was 100% lethal in less than 30 days (Table 18.33). The $LD_{50/30}$ of ^{241}Am or ^{239}Pu in dogs seems to be about one-half that for adult rats.

Deaths occurring earlier than 60 days after injection of a large dose of an actinide were associated primarily with bone marrow injury. Delayed deaths (those occurring 120 to 180 days after injection of somewhat lower doses) also appeared to be associated with bone marrow depression. The radiation damage that accumulated in the livers of Chinese hamsters and dogs, as well as in the lungs of those animals exposed by inhalation or intratracheal instillation, probably also contributed to the delayed deaths. Late deaths, 1 year or more after exposure, appeared to involve radiation injury in several organs—bone marrow, liver, kidneys, and lungs (after inhalation exposure)—or the consequences of radiation injury—myelogenous leukemia and bone tumors.

2. Toxicity Relative to β-emitting Bone Seekers

The distribution of ^{144}Ce-^{144}Pr is sufficiently similar to that of the actinides, ^{241}Am, ^{244}Cm, and ^{239}Pu, to provide a basis for comparing the toxicities of α-emitting and β-emitting nuclides that are deposited primarily on bone surfaces. The average $LD_{50/30}$ of the α-emitters in the rodents was about 100 μCi/kg, and in the dog the median acute lethal dose was about 35 μCi/kg. The $LD_{50/30}$ of ^{144}Ce-^{144}Pr for Chinese hamsters, rats, and dogs, was 3750, 1830, and 110 μCi/kg respectively; on the basis of injected doses, the relative effectiveness[43] for acute lethality was, respectively, 38, 18, and 3.

If acute lethality is due primarily to radiation damage of the bone marrow, a better estimate of relative effectiveness would be one based on the amount of radiation absorbed in bone. The path length of α-particles is less than 100 μm in soft tissue or bone, and all of the energy of α-emitters deposited in bone is absorbed in bone. However, some of the energy of the long-range β-emitters such as ^{90}Sr-^{90}Y and ^{144}Ce-^{144}Pr is dissipated outside of bone. PARMLEY et al. (1961) measured the energy absorption in the skeleton of ^{90}Sr-^{90}Y as a function of skeletal size. They found that 32, 48, and 79% of the β-particle energy of deposited ^{90}Sr-^{90}Y was absorbed in the skeletons of mice (nearly the same size as Chinese hamsters), rats, and dogs, respectively. The average energy of the α-particles emitted by ^{241}Am, ^{244}Cm, and ^{239}Pu was 5.5 MeV. The average energy of the β-particles ($\bar{E}$) emitted by the ^{144}Ce-^{144}Pr pair was 1.1 MeV and that of the ^{90}Sr-^{90}Y pair was 1.0 MeV. The average energy absorption in the skeletons of Chinese hamsters, rats, and dogs, from the β-particles of ^{144}Ce-^{144}Pr was 0.41, 0.53, and 0.87 MeV, respectively. The relative effectiveness for direct cell killing (acute lethality) based on energy absorbed in bone[44] was 2.8 for Chinese hamsters[45], 1.8 for rats, and 0.5 for dogs. The relative effectiveness for bone marrow damage of the α-emitting bone-surface seekers appeared to be inversely related to body size, suggesting that in the animals with larger bones some marrow cells escape irradiation by the short-range α-particles. Marrow that was apparently intact

43 For the purposes of this discussion relative toxicity or relative effectiveness in producing some biological endpoint is defined as the inverse of the ratio of the injected doses of the α-emitter and the β-emitter.

44 For example, the relative effectiveness of the α-emitters based on approximate energy absorbed in bone of rats was $[(100 \times 5.5)/(1830 \times 0.53)]^{-1} = 1.8$.

45 MCKAY et al. (1972) calculated more precisely using Chinese hamster tissue distribution data,that the relative effectiveness was 3.2 for acute lethality of ^{241}Am compared with ^{144}Ce–^{144}Pr or ^{90}Sr–^{90}Y.

was found in the shafts of the long bones of mice and rats injected with lethal doses of ^{239}Pu (Bloom, 1948; Langham and Carter, 1951), ^{241}Am (Langham and Carter, 1951), and ^{223}Ra (Durbin et al., 1958).

3. Delayed Lethality

In the adult rats, the relative effectiveness for lethality of the α-emitters was 1.8, at 120 to 180 days after injection, the same as it was for acute lethality. This concurs with the view that delayed deaths were also associated with bone marrow injury. In the Chinese hamsters 120 to 180 days postinjection, the relative effectiveness of ^{241}Am was greater, (7.3) than it was for acute lethality, suggesting the introduction of another mode of radiation injury for which the more densely ionizing α-particles were more efficient: most likely cell death in the liver. At times longer than 1 year the α-emitters were eight times more efficient than the β-emitters in inducing deaths among adult rats. Except for myelogenous leukemia and bone tumors, the causes of the late deaths among rats were not specified; however, it seems clear that for these endpoints the α-emitters were more effective than they were for direct killing of bone marrow cells.

Some late deaths among the dogs were attributable to accumulated damage in the liver and kidneys. If it is assumed for simplicity that one-half of the injected dose of either the α-emitting or the β-emitting nuclides was deposited in soft tissues where absorption of the β-particle energy was complete, and one-half was deposited in bone where 79% of the β-particle energy was absorbed, the average relative effectiveness of the α-emitters was about 9 for late deaths in the dog.

B. The Formed Elements of the Blood

The hemorrhagic syndrome (pancytopenia and massive internal bleeding caused by bone marrow destruction, Lorenz and Congdon, 1954) was the apparent cause of death of rats, Chinese hamsters, or dogs given large doses of transplutonium nuclides. Bone marrow aplasia and severe internal hemorrhage were observed in rats and Chinese hamsters dying within 30 days after injections of 50 µCi/kg or more of ^{241}Am or ^{244}Cm, and in dogs dying within 3 weeks after injection of 10 to 60 µCi/kg of ^{241}Am (Langham and Carter, 1951; Zalikin et al., 1968, 1969; Semenov, 1971a; McKay et al., 1972).

With sublethal injected doses, bone marrow was still being irradiated, but at a lower dose rate, by nuclides deposited on the inner surfaces of bone. Depletion of the marrow, especially the erythropoietic elements, was a major contributor to lethality in mice continously exposed to external radiation at dose rates greater than 10 R/day (Sacher, 1963). Marrow depletion was likely to be the most important factor in the later deaths of animals bearing moderate burdens of actinide elements in their bones.

The depression of erythrocytes, polymorphonuclear leukocytes (PMN), and thrombocytes (platelets) was the result of radiation damage of stem cells in the marrow. The cause of the reduction in the number of circulating lymphocytes is less clear. The concentrations of transplutonium elements in lymphatic tissues were low, and the depression of lymphocytes was, therefore, presumed to be due primarily to irradiation of mature cells circulating through organs with high nuclide deposits—bone, liver, and lung (in the case of inhalation exposure).

The hematological effects of ^{241}Am were generally similar to those observed with equal doses of ^{239}Pu in growing rats (Langham and Carter, 1951) or in young adult beagles (J. H. Dougherty, 1972a). In beagles, in the long-term low-dose studies, both ^{241}Am and ^{239}Pu were 2 to 6 times more effective than ^{226}Ra in reducing the numbers of circulating leukocytes (J. H. Dougherty, 1972a, b)[46]. Preliminary observations in beagles injected with ^{249}Cf or ^{252}Cf suggest that for long-term hematological effects the toxicity of the californium isotopes may equal or exceed that of ^{239}Pu (J. H. Dougherty, 1971).

1. Leukocytes

a) Acutely lethal doses in rodents. Injected doses of ^{241}Am or ^{244}Cm greater than 100 μCi/kg elicited maximum hematological effects in rodents. Within 7 to 10 days postinjection, the numbers of circulating PMN's and lymphocytes fell to 10 to 20% of preinjection levels and recovered to only 25 to 40% of control levels at the time of death (Langham and Carter, 1951; Zalikin et al., 1968); Semenov, 1971a). Immature PMN's and immature and unusual forms of lymphocytes considered to be characteristic of radiation damage (binucleated, enlarged, or vacuolated cells) were present in the peripheral blood 2 to 3 weeks after injection (Langham and Carter, 1951; Zalikin et al., 1968).

b) Acutely lethal doses in dogs. Dosages of 10 to 60 μCi/kg of ^{241}Am were acutely lethal in dogs. The numbers of PMN's and lymphocytes declined immediately after injection, and at death 10 to 20 days later, they were only 20% of preinjection values (Zalikin et al., 1969).

c) Intermediate doses in rats. Dosages in the range of 10 to 60 μCi/kg of ^{241}Am or ^{244}Cm were 50% lethal to rats in 60 to 300 days. Within 10 to 14 days the numbers of PMN's and lymphocytes fell to between 20 to 30% and 40 to 60%, respectively, of control levels. Recovery was variable. In 6-month survivors, the PMN's and lymphocytes had recovered to 50 and 75% of preinjection levels, respectively. The numbers of PMN's and lymphocytes were within the lower limits of normal in rats that survived for 1 year (Langham and Carter, 1951; Zalikin et al., 1968, 1969; Semenov, 1971a).

d) Moderate to low doses in beagles and rats. Doses of ^{241}Am or ^{249}Cf ranging from 2 to 10 μCi/kg severely depressed the leucocyte levels of beagles, and 2 to 10 μCi/kg of ^{241}Am or ^{244}Cm reduced these levels by about 50% in adult rats.

In beagles, by 30 days after injection of 2.8 μCi/kg of ^{241}Am (100% lethal in 15 months) the PMN's had fallen to about 15% of the preinjection value. There was only partial recovery; after 3 months the PMN's were 35 to 50% of control values, and they remained at that low level for the next 12 months. The same dose of ^{249}Cf depressed the PMN's to less than 10% of the control value within 2 to 3 weeks after injection. By 8 weeks after injection the PMN's had recovered to 30% of their initial value.

The lymphocytes of the beagles were less affected. Thirty days after injection of 2.8 μCi/kg of ^{241}Am, the lymphocytes had declined to 50% of the control level, and remained well below normal until the deaths of the dogs at 13 and 15 months. After injection of 2.8 μCi/kg of ^{249}Cf, the level of lymphocytes decreased slowly, and by 8 weeks was only 30% of the control level (J. H. Dougherty, 1970, 1971).

Doses of 5 to 10 μCi/kg of ^{241}Am or ^{244}Cm (lethal to 50% of the rats within 1 year) depressed the PMN's to 40 to 50% of preinjection levels within 10 to

46 J. H. Dougherty (1972a, b) calculated relative hematological effectiveness from the ratio of skeletal dose rates from various radionuclides and ^{226}Ra at a time when the various leukocytes had been depressed to 50% of their preinjection levels.

14 days. Recovery was variable: In the ^{241}Am study, PMN's returned to within normal limits by 6 months after injection. Lymphocytes declined gradually so that 1 year after injection they were only 50% of the control value (ZALIKIN et al., 1968, 1969). In the ^{244}Cm study, the PMN's were still only 60 to 70% of normal 1 year after injection. Lymphocytes declined to 70% of the control value by the 10th day, and there was only a slight recovery, to 80% of normal by 1 year (SEMENOV, 1971a).

e) Low doses in dogs and rats. Doses of ^{241}Am or ^{244}Cm less than 2.5 μCi/kg had little consistent effect on the leukocytes levels of rats up to 18 months after injection (ZALIKIN et al., 1969; SEMENOV, 1971a; DURBIN et al., 1972c).

In the beagles the effects on the leukocytes of 0.9 μCi/kg of ^{241}Am were about the same as those of the larger dose of 2.8 μCi/kg. There were no early changes in numbers of circulating lymphocytes or PMN's after injection of 0.1 or 0.3 μCi/kg of ^{241}Am; however, by 18 months after injection, both types of leukocytes had declined to 75 to 80% of preinjection levels (J. H. DOUGHERTY, 1970). The hematological effects of 0.28 μCi/kg of either ^{249}Cf or ^{252}Cf have been reported up to 8 weeks after injection. The average energy per decay of ^{252}Cf is nearly twice that of ^{249}Cf[47]. The decline in PMN's and lymphocytes occurred sooner after injection of ^{252}Cf, but the effects on the white cell counts were not significantly different over the short observation period. The numbers of PMN's and lymphocytes decreased to 50 and 85% of preinjection levels, respectively, 8 weeks after injection of either californium isotope (J. H. DOUGHERTY, 1971).

At the one low dose level, 0.28 to 0.3 μCi/kg, for which there is hematological data for ^{241}Am, ^{249}Cf, and ^{252}Cf, it appears that the californium isotopes were much more effective in depressing the leukocytes than was ^{241}Am.

2. Erythrocytes

a) Acutely lethal doses of ^{241}Am or ^{244}Cm. In dogs, at death 10 to 21 days after injection of 10 to 60 μCi/kg of ^{241}Am, red cell counts were only 50% of normal presumably because of massive internal bleeding (ZALIKIN et al., 1969). In rats, within 2 to 3 weeks after injection of 100 μCi/kg or more of either isotope, the number of red cells fell 20 to 40% below the control level, apparently also because of hemorrhage. Red cell counts remained low or continued to decline until death (LANGHAM and CARTER, 1951; ZALIKIN et al., 1968, 1969; SEMENOV, 1971a). Unusually large numbers of reticulocytes were observed in peripheral blood 2 to 4 weeks after injection of a large dose of ^{241}Am in rats (ZALIKIN et al., 1968).

b) Intermediate doses of ^{241}Am or ^{244}Cm. In rats the early decline in the red cell count was only 10 to 15% below the control level after injection of 10 to 60 μCi/kg of either isotope. Following an early recovery between 2 and 3 months after the injection, red cell counts declined again to 20 to 25% below normal between 3 to 6 months after injection. A long period of recovery was seen in the 6-month to 1-year survivors which was characterized by periodic wide fluctuations. In the 1-year survivors, red cell counts were within 10% of normal (LANGHAM and CARTER, 1951; ZALIKIN et al., 1968, 1969; SEMENOV, 1971a).

c) Low doses of ^{241}Am or ^{249}Cf. In beagles, 2.8 μCi/kg of ^{241}Am depressed the volume of packed red cells (VPRC) to about 15% below normal within 1 to 3 months after injection, where it remained for 7 to 9 months. After that time a

47 Californium-249 decays by the emission of α-particles with an average energy of 5.93 MeV. Californium-252 has two modes of decay; 97% by α-particle emission and 3.2% by spontaneous fission. Including the energy of the fission fragments, the average energy per disintegration of ^{252}Cf is 11.76 MeV (LLOYD et al., 1972b).

more severe anemia developed and was progressive until death at 13 to 15 months (J. H. DOUGHERTY, 1970). A comparable dose of ^{249}Cf affected the red cells more drastically. Both the VPRC and the hemoglobin fell to 55 to 60% of the preinjection levels by 4 weeks after injection; by 8 weeks they had recovered to 75 to 80% of preinjection levels. During the 4th week the number of reticulocytes in the peripheral blood rose sharply, reaching twice the normal level at 6 weeks, followed by a decline, so that by 8 weeks reticulocytes were present in normal numbers (J. H. DOUGHERTY, 1971).

In rats injected at doses lower than 10 μCi/kg of ^{241}Am, red cell counts fluctuated about the normal level for 3 to 5 months. By 9 months the red cell counts began to decline, and at 18 months they were 70% of the preinjection level. At dosages of 2.5 μCi/kg red cell counts were depressed after about the 6th month (ZALIKIN et al., 1968, 1969).

3. Thrombocytes

The effects of injected transplutonium elements on the numbers of circulating platelets has been studied in beagles. Injection of 2.8 μCi/kg of ^{241}Am reduced the platelets to 50% of the preinjection level within 2 to 4 weeks. By the third month after injection the platelets had recovered to 60% of normal. Eighteen months after injection of 0.9 μCi/kg of ^{241}Am, the platelet count was still only 60% of the preinjection level. No significant effects on the platelets were observed during 18 months after injection of 0.1 or 0.3 μCi/kg of ^{241}Am (J. H. DOUGHERTY, 1970).

Platelets were more severely affected by ^{249}Cf. Within 3 to 5 weeks after injection of 2.8 μCi/kg, the platelet count was less than 10% of the preinjection level and had recovered only to 40% at 8 weeks. The platelet count was depressed after injection of 0.28 μCi/kg of either ^{249}Cf or ^{252}Cf: It fell sharply after injection of ^{252}Cf and was 80% of the preinjection level within 1 to 2 weeks; the depression continued through 8 weeks. After injection of ^{249}Cf, the platelet count declined gradually and at 8 weeks was 80% of the preinjection value (J. H. DOUGHERTY, 1971).

4. Bone Marrow

Bone marrow biopsies of dogs injected with 10 to 60 μCi/kg of ^{241}Am showed fewer than 20% of the normal number of nucleated cells by the third day, and complete aplasia by 14 days (ZALIKIN et al., 1969). The marrow of four beagles was variable 127 to 1022 days after inhalation of an average ILB of 2.7 μCi/kg of an $^{241}AmO_2$ aerosol. Foci of complete acellularity, hypoplasia, and occasionally of hyperplasia, were found in the marrow taken from ribs, vertebrae, or long bones. The spleens of all four dogs contained foci of extramedullary hematopoiesis (R. G. THOMAS et al., 1972).

One year or more after injection of 7 μCi/kg of ^{241}Am into 7-week-old rats, foci of aplasia were seen in the marrow of the femoral metaphysis, and the marrow in the epiphysis was aplastic (BENSTED and TAYLOR, 1969). The marrow of Chinese hamsters was hypoplastic one year after injection of 5 μCi/kg of ^{241}Am (McKAY et al., 1972).

BENSTED et al. (1965) reported a case of myelogenous leukemia among 19 rats that survived more than one year after an injection of 2.5 μCi/kg of ^{241}Am. Two of 48 rats given 10.7 μCi/kg of $^{252}EsCl_3$ by intratracheal instillation developed myelogenous leukemia within 350 days (BALLOU and MORROW, 1972, and unpublished).

5. Hematological Effects of Inhaled ^{241}Am

Radiation dose rates in bone and liver are highest within a few hours after injection of an actinide isotope. Because of the delayed absorption after inhalation of these nuclides (Sec. IV.D), maximum radiation dose rates do not occur in bone and liver until several weeks after exposure. Therefore, the prompt depressions of leukocytes and platelets characteristic of large doses of injected actinides are not observed when comparable amounts are inhaled. R. G. THOMAS et al. (1972) have described the hematological effects in 4 beagles 127 to 1022 days after inhalation of an $^{241}AmO_2$ aerosol (average ILB was 2.7 μCi/kg). Lymphocytes were depressed to 40% of the preexposure level by 80 days, declined further to 20% by 512 days, and remained at that low level for the next 500 days. By 80 days the PMN's were reduced to 40% of the preexposure level. By 512 days the PMN's had recovered to 60% of the preexposure value; there was no further recovery. The VPRC gradually declined and was 85% of the preexposure level by 1022 days. Platelets decreased to 40% of the preexposure level by 256 days, but had recovered to 80 to 90% of normal by 1022 days.

BULDAKOV et al. (1972) reported leukopenia, depressed myelopoiesis, and reticulocytosis in dogs observed for up to two years after inhalation of 1.5 μCi/kg of an $^{241}Am(NO_3)_3$ aerosol.

Except for the absence of the early decreases, the effects of inhalation of 2.7 μCi/kg of $^{241}AmO_2$ on red cells, PMN's and platelets of beagles were generally similar to those seen after injection of a comparable dose of ^{241}Am citrate. After parenteral administration of ^{241}Am, lymphocytes are exposed to the radiation sources in liver and bone. Americium present in the lung constitutes an additional source of exposure. Later, loss of function of the tracheobronchial lymph nodes contributed further to the depressant effect.

C. Effect on Growth and Body Weight

Growth or weight gain ceased or was retarded in rats injected with ^{241}Am or ^{244}Cm at dosages greater than 5 μCi/kg or after intratracheal instillation of $^{253}EsCl_3$. After an acutely lethal dose (more than 50 μCi/kg of ^{241}Am or ^{244}Cm) the body weights of adult rats declined and were 85% of control weights at the time of death (ZALIKIN et al., 1968; SEMENOV, 1971a). A high percentage of rats injected at 35 days of age with 32 μCi/kg or more of ^{241}Am failed to gain weight (LANGHAM and CARTER, 1951). Weight gain of male rats was severely impaired after intratracheal instillation of $^{253}EsCl_3$; 150 days after instillation of 47.2 μCi/kg, body weights were 60 to 70% of age controls and 275 days after instillation of 10.7 μCi/kg body weights were 80% of the controls (BALLOU and MORROW, 1972, and unpublished).

D. Effects of Inhaled Transplutonium Elements in the Lung

Four beagles were exposed to an aerosol of $^{241}AmO_2$; the average ILB was 2.7 μCi/kg (R. G. THOMAS, 1970; R. G. THOMAS et al., 1972), and the dogs were killed 127, 256, 512, and 1022 days after exposure. Partial-body γ-ray counting and radiochemical analyses of lung tissue dissected at necropsy permitted estimation of the radiation doses absorbed in the lungs, which were 3000, 3200, 3800, and 5300 rad, respectively (the investigators stated that these were rough estimates). After 50 days the two dogs killed at 127 and 256 days exhibited increased respiratory rates and decreased tidal volumes.

Two dogs were euthanized because of their deteriorating physical condition. The 256-day dog exhibited progressive respiratory distress. The 512-day dog was killed because of general debility which included loss of weight and liver insufficiency. At necropsy the lungs of these two dogs showed striations of the pleural surface with alternating regions of greater radiation damage associated with the ribs (Barnes, 1970). Microscopic examination of the lungs of the 127-day dog revealed a few isolated areas of inflammation and some minor structural alterations in bronchiolar and alveolar duct epithelium. Microscopic findings in the other dogs were similar, but more pronounced. Lung pathology was most evident in the 256-day dog and included foci of dense fibrosis, which was mostly subpleural and was associated with thickening of the pleura. Small foci of mineralization were also present on the lungs of the dogs killed at 512 and 1022 days. Other pathology included local proliferation of alveolar cells, obliterative fibrosis of small arteries, and some dense parabronchiolar fibrosis. In all four dogs the tracheobronchial lymph nodes, which had the highest ^{241}Am tissue concentrations through 512 days, were small and showed medullary fibrosis and lymphoid depletion.

Two brief reports have appeared on a large series of dogs exposed to aerosols of soluble $^{241}Am(NO_3)_3$ or polymeric $^{239}Pu(NO_3)_4$ at the Institute of Biophysics in Moscow (Buldakov et al., 1972; Nifatov et al., 1972)[48]. Respiratory insufficiency was the stated cause of death in those dogs that succumbed within 6 months of the exposure to either nuclide. The dogs that inhaled the ^{239}Pu aerosol exhibited more pronounced clinical signs of altered lung function (increased density of lung fields in radiographs, higher respiratory rates, poorer O_2 saturation of arterial blood, and a greater difference in arterial-venous O_2 saturation) than the dogs exposed to an equal amount of the ^{241}Am aerosol.

Of those dogs exposed to an ^{241}Am aerosol, five died of a fibrinous purulent pneumonia 74 to 181 days after inhalation; the accumulated radiation doses in the lungs of these dogs were reported to be 655 to 4050 rad. Four dogs succumbed after 414 to 1108 days with signs of a moderate pulmonary sclerosis, but lung damage was apparently not the primary cause of death; the accumulated radiation dose in the lungs of these dogs was reported to be 2660 to 2990 rad.

Ballou and Morrow (1972, and unpublished) studied the effects of $^{253}EsCl_3$ instilled in the lungs of rats. Median survival time was 180 days after instillation of 47.2 μCi/kg, and was associated with respiratory distress. By 275 days there were only 28% (8/29) survivors at this dose level. The accumulated radiation dose to the lungs was reported to be 15000 rad at 275 days (decay of this 20.5-day nuclide is 99% complete by 150 days). By 275 days after exposure no deaths attributable to pulmonary difficulties had occurred in rats at the 10.7 μCi/kg level; their individual accumulated lung dose, 3000 rad.

E. Radiation Injury in the Skeleton

The skeletal effects of the transplutonium elements have not been thoroughly studied. However, based on the extensive experience with ^{239}Pu, one would expect gross damage to be dependent on the accumulated radiation dose in bone and

48 These reports are essentially abstracts; experimental details are scanty, and some of the information content may have been obscured in translation. The amounts of radioactivity inhaled by individual dogs, the ranges of IBB or ILB, and the numbers of dogs in the dose-level or nuclide groups were not reported.

the growth status of the skeleton. (For descriptions of the effects of ^{239}Pu in the skeleton, the interested reader is referred to: HELLER, 1948; LANGHAM and CARTER, 1951; FABRIKANT and SMITH, 1964; the appropriate chapters in T. F. DOUGHERTY et al., 1961; MAYS et al., 1967; STOVER and JEE, 1972; and Chap. 10 of this volume.) For ^{239}Pu, and presumably for the other surface-seeking α-emitters, gross structural abnormalities (stunting, fractures, altered radiographic densities and altered bone thicknesses) and microscopic changes (death of bone and cartilage cells, peritrabecular fibrosis, and obliteration of cortical vascular channels) would be expected to be more dramatic and to appear sooner in growing animals and in shortlived animals with more rapid bone turnover in adulthood.

1. Alterations of Growth and Structure of Bone

Cessation of growth of long bones and limb deformities were noted as early as 8 weeks after injection of 32 to 63 μCi/kg of ^{241}Am into 35-day-old male rats (LANGHAM and CARTER, 1951). Skeletal radiographs revealed fractures of the long bones, generalized osteoporosis, dense bars of bone along epiphyseal growth lines, and marked cortical thinning with occasional fibrocystic changes in the long bones. These skeletal abnormalities were similar to those seen in rats of the same age injected with ^{239}Pu at the same dose levels. Histological examination revealed extensive areas of bone necrosis. Skeletal anomalies similar to those just described were observed several months after injection of 5.8 to 7 μCi/kg of ^{241}Am into 7-week-old male rats (FABRIKANT and SMITH, 1964; D. M. TAYLOR and BENSTED, 1967; BENSTED and TAYLOR, 1969). Bone growth was reduced, and by 8 to 10 months after injection all long bones were stunted. Humeri and tibiae were 10% shorter and femora 15% shorter than in age controls. Radiographs at 10 months showed osteoporosis in the ends of the long bones and sclerosis in the femoral shaft. Bilateral symmetrical thinning of the cortex of the femoral metaphysis led to fractures of the knee. Thinning of the cortex was apparently not the result of an increase in the diameter of the marrow cavity, but rather the net result of continued periosteal resorption in the absence of the normal compensatory apposition of new bone on the endosteal surface. Gross damage has also been noted in the growth apparatus of the upper incisors (D. M. TAYLOR and BENSTED, 1967; BENSTED and TAYLOR, 1969). By 400 days post-injection the upper incisors had stopped growing; some were loosened and others were lost altogether, resulting in overgrowth of the lower incisors.

Growth retardation, spontaneous fractures, and loss of teeth have been observed within a few months after injection of 10 to 150 μCi/kg of $^{253}EsCl_3$ into either adult or 21-day-old rats. These effects were more pronounced in the rats injected as weanlings (MAHLUM and SIKOV, 1972).

Fairly extensive bone damage was observed in four beagles that inhaled an average ILB of 2.7 μCi/kg of an $^{241}AmO_2$ aerosol and were killed from 127 to 1022 days later (R. G. THOMAS et al., 1972). The effects in bone were reminiscent of those seen in beagles injected with ^{239}Pu. (See ^{239}Pu references cited above.) A series of radiographs (the first taken 1 year after exposure) of the skeletons of the 512-day and 1022 day dogs showed progressively decreasing bone density, coarsening and attenuation of trabecular patterns in vertebrae and the ends of long bones, and roughening of the endosteal surfaces of the long bones. Rib fractures were first noted in the 1022-day dog 550 days after exposure, and there were several unhealed fractures at the time of death. Microscopically, in the femora, ribs, and vertebrae, there was extensive endosteal fibrosis associated with focal osteoid thickening, focal resorption of spicules, irregular cortical thickening, and periosteal fibrosis. In the ribs, in addition to the ununited fractures, there were small areas of degeneration in the central cartilage of the costochondral junctions.

2. Induction of Bone Tumors

Tumors of bone, mostly osteogenic sarcomas, have been recognized for more than 40 years as a probable late consequence of radiation accumulated in bone from deposited ^{226}Ra (MARTLAND and HUMPHRIES, 1929). Osteogenic sarcomas have been observed in experimental animals injected with ^{239}Pu: mice (LISCO et al., 1947; FINKEL and BISKIS, 1962); growing and adult rats (LANGHAM and CARTER, 1951; FABRIKANT and SMITH, 1964; RUDNITSKAYA and MOSKALEV, 1970); and young adult beagles (references cited in Sec. XII. E). Osteogenic sarcomas have also been induced by ^{241}Am and ^{253}Es.

a) Estimation of radiation dose in rat bone. The range of radiation doses in the bones of ^{241}Am-injected rats was calculated from the dose rate (based on the µCi/kg injected) and the burden time (time of appearance of the first bone tumor and either the time of appearance of the last bone tumor or the time of death of the last rat in an experimental group). Radiation dose rate is given by the general equation

$$\text{Dose rate (rad/day)} = 51.2\,E\,C \qquad \text{Eq. (27)}$$

where E is the α-particle energy, 5.5 MeV, and C is the ^{241}Am concentration in bone which is given by the expression,

$$C\,(\mu\text{Ci/g}) = \frac{k\,(\mu\text{Ci/kg} \times 10^{-3})\ (\text{Body weight, g})}{(\text{Bone weight, g})}. \qquad \text{Eq. (28)}$$

For these approximate calculations, the fraction of injected ^{241}Am deposited in the skeleton, k, was assumed to be 0.3 of the administered dose on the average, and the weight of the rat skeleton was assumed to be 10% of the body weight. Substituting these values in Eq. (28) and combining terms, Eq. (28) reduces to

$$C\,(\mu\text{Ci/g}) = 3 \times 10^{-3}\,D, \qquad \text{Eq. (29)}$$

where D is the injected dose in µCi/kg of body weight. The accumulated radiation doses were calculated directly from the injected dose, D, and the time after injection, t days:

$$\text{Dose}_t\ (\text{rad}) = 0.845\,D\,t. \qquad \text{Eq. (30)}$$

b) Bone tumors induced in rats. In LANGHAM and CARTER's (1951) pioneer study of the biological effects of ^{241}Am, it was found that 19% (6/32) of rats that survived more than 290 days after injection of 32 to 63 µCi/kg of ^{241}Am at 35 days of age developed osteogenic sarcomas between 290 and 350 days. The accumulated radiation doses calculated from Eq. (30) ranged from 7800 to 16000 rad. In a series of experiments also using growing male rats (7 weeks old), but much lower injected doses of ^{241}Am: 16% (5/32) of rats injected with 2.5 µCi/kg developed bone tumors between 475 and 671 days after injection, estimated dose range, 1000 to 1400 rad (FABRIKANT and SMITH, 1964); 25% (5/20) of rats injected with 2.5 µCi/kg developed bone tumors 400 to 525 days after injection, estimated dose range, 800 to 1100 rad (BENSTED and TAYLOR, 1965; D. M. TAYLOR and BENSTED, 1967; BENSTED and TAYLOR, 1969); 17% (5/30) of rats injected with 5.8 µCi/kg developed bone tumors 365 to 510 days after injection, estimated dose range

1800 to 2500 rad (FABRIKANT and SMITH, 1964), and 47% (15/32) of rats injected with 7 μCi/kg developed bone tumors between 287 and 525 days after injection, estimated dose range, 1700 to 3100 rad (D. M. TAYLOR and BENSTED, 1967; BENSTED and TAYLOR, 1969).

In adult rats the incidence of osteogenic sarcomas has been reported to be 13.5% (7/52) among rats that lived longer than 300 days after injection of 2.5 to 10 μCi/kg of ^{241}Am (ZALIKIN et al., 1969; MOSKALEV and RUDNITSKAYA, 1970). Accumulated radiation doses in bone, estimated from Eq. (30), were 600 and 2500 rad 300 days after injection. They agreed reasonably well with the dose range calculated by the investigators, 500 to 5000 rad.

No bone tumors were found among 52 rats that were observed for longer than 500 days after injection of 0.37 μCi/kg of ^{241}Am at 110 days of age (DURBIN et al., 1967, 1972c). The bone doses estimated from Eq. (30) were 190 rad by 390 days post-injection and 250 rad by 800 days, the maximum survival time of this group of rats.

Although the results are preliminary, it is clear that osteogenic sarcomas in adult rats can also be induced by ^{253}Es. As of 280 to 350 days after intratracheal instillation of 10.7 μCi/kg of $^{253}EsCl_3$ into adult male rats, 4/48 rats had developed bone tumors. The accumulated radiation dose in bone was 300 rad at 20.5 days (one-half-life of the ^{253}Es isotope) and a maximum of 600 rad by 150 days after exposure (BALLOU and MORROW, 1972, and unpublished).

In the young rats bone tumors arose about equally in the axial and the appendicular skeleton. Combining all of the data from the studies of FABRIKANT and SMITH (1964), BENSTED et al. (1965), D. M. TAYLOR and BENSTED (1967), and BENSTED and TAYLOR (1969), bone tumors were distributed as follows: long bones, 16; vertebrae, 10; pelvis, 1; and jaw, 2. In the LANGHAM and CARTER (1951) study four tumors developed in long bones and one each in the spine and pelvis. The distribution of bone tumors induced in the young rats by ^{239}Pu was reported to be the same, but has not been reported for adult rats.

Much of the radiation dose absorbed in the skeletons of the growing rats from very high doses of ^{241}Am was wasted, perhaps by dissipation in bone that had already been killed by the continuous irradiation. The lowest dose associated with a bone tumor in the high-level injection study was 7800 rad. In the low-dose experiments the smallest tumorigenic dose was 800 rad, only one-ninth as large.

Although the range of doses required to elicit bone tumors was about the same in the growing and mature rats, tumors were induced more frequently in the growing rats. If all of the data in the low-dose studies are combined in such a fashion that equal numbers of rats are represented at the 2.5, 5.8, and 7 μCi/kg dose levels, the overall incidence of tumors is 29% (25/86) compared to an overall incidence in the adult rats (2.5 to 10 μCi/kg of ^{241}Am) of 13.5%. This finding is expected from the results of FABRIKANT and SMITH (1964) who found that after injection of 2.95 μCi/kg of ^{239}Pu the incidence of bone tumors in growing rats (7 weeks old at injection) and adult rats (7 months old at injection) was 25 and 5%, respectively. These investigators also indicated that the ^{239}Pu-induced tumors in the adult rats grew more slowly, were more confined, and had a lower tendency to invade surrounding tissue.

c) Bone tumors induced in dogs. Osteogenic sarcomas have been induced in beagles and mongrel dogs by ^{241}Am. Bone tumors developed in 10 mongrels[49] that survived 1044 to 1433 days after inhaling an $^{241}Am(NO_3)_3$ aerosol (BULDAKOV

49 Numbers of dogs and exact nuclide dosages were not reported.

et al., 1972; NIFATOV et al., 1972). Accumulated radiation doses in the skeleton were stated to be less than 2000 rad. Some dogs developed more than one tumor. Tumors were located in both the appendicular and axial skeleton, including the head; the most common location of tumors was in vertebrae.

Several groups of beagles have been injected with ^{241}Am, ^{249}Cf, or ^{252}Cf at various dose levels for long-term study at the University of Utah. Each dose level group will eventually contain 12 to 13 dogs, but they are introduced into the experiment at different times depending on the rate of production of the breeding colony. The first dogs were injected with ^{241}Am in June 1966, but only a few have been in the study long enough to achieve minimum tumorigenic doses. Of the 8 dogs injected with californium, none has been in the experiment long enough to develop a bone tumor. All of the results described below are as of March 1972 (LLOYD et al., 1972a; T. F. DOUGHERTY, 1972).

At the 0.9 μCi/kg level of ^{241}Am, 2 dogs developed bone tumors at 1132 and 1779 days after injection (accumulated radiation doses in bone were 1360 and 2220 rad, respectively). The accumulated radiation doses in the skeletons of 6 dogs in this group each exceeded 1360 rad. At the 0.3 μCi/kg level 2 dogs developed bone tumors at 1510 and 1917 days after injection (accumulated radiation doses in bone were 547 and 660 rad, respectively). The accumulated radiation doses in the skeletons of 4 dogs in this group exceeded 500 rad. At the 0.1 μCi/kg level one dog developed a bone tumor at 1847 days after injection (accumulated radiation dose in bone was 190 rad). The accumulated radiation doses in the skeletons of only 2 dogs in this group exceeded 190 rad.

The results obtained so far suggest that in beagles ^{241}Am will be about as effective for inducing bone tumors as is ^{239}Pu. The long-term studies in beagles of the higher dose levels of ^{239}Pu are complete. All of the dogs that accumulated more than 550 rad in their skeletons developed an osteogenic sarcoma. Osteogenic sarcomas were found in 10 of 12 dogs with accumulated bone doses of 230 to 460 rad, and in 9 of 13 dogs with accumulated bone doses of 170 to 220 rad (T. F. DOUGHERTY, 1972). At the highest dose levels (0.9, 0.3, and 0.1 μCi/kg) of ^{241}Am and ^{239}Pu for which there were long-term survivors, the minimum accumulated radiation doses in bone associated with bone tumor development were, respectively: for ^{241}Am 1360, 546, and 190 rad; for ^{239}Pu 1560, 550, and 230 rad.

In beagles, based on time from injection to death, ^{241}Am seems to be lethal sooner than ^{239}Pu, because of the greater amount of accumulated radiation damage in the soft tissues of ^{241}Am-injected animals. The 2 dogs injected with 2.8 μCi/kg of ^{241}Am died by 448 days of liver and kidney failure. At the same dose level of ^{239}Pu, 9 dogs died between 499 and 2059 days. Six deaths were related to bone only, two to soft tissue degeneration only, and one to a combination of both. At the 0.9 μCi/kg level of ^{241}Am, the first of 5 deaths occurred at 633 days, and all five deaths have involved radiation damage in one or more soft tissue; bone tumors also developed in two or them. At the 0.9 μCi/kg level of ^{239}Pu, all of the remaining dogs were euthanized between 1066 to 1724 days because of the development of bone tumors.

Although the Lawrence Berkeley Laboratory experiments with ^{241}Am in monkeys (see Sec. VII.D.3) are primarily metabolic, four ^{241}Am-injected monkeys have been observed for a time sufficient to permit them to accumulate significant skeletal radiation doses. All four were adults at the time of the injection (more than 6.5 years old with fully united epiphyses and fully erupted wisdom teeth). The accumulated radiation doses, based on the retention of ^{241}Am and the measured skeletal weight of each animal, range from 237 to 1023 rad accumulated during intervals of 735 to 2087 days after injection of an average dose of 0.65 μCi/kg of ^{241}Am. None of the four developed a bone tumor that could be discerned from skeletal radiographs or gross examination of defleshed bones.

F. Cytogenic Effects in Liver and Bone Marrow

The late effects of radiation, especially neoplastic changes, are believed to be the end result of the proliferation of genetically altered cells. The liver of the Chinese hamster, in which cell division has been stimulated by partial hepatectomy, has proven to be a useful model system in which to study the relationships between chromosome aberration frequencies and radiation dose or dose rate (McKay et al., 1970, 1972; Brooks et al., 1971, 1972).

Single doses of ^{241}Am citrate ranging from 0.0006 to 5 μCi/kg and of ^{252}Cf citrate ranging from 0.56 to 45 μCi/kg were injected into adult Chinese hamsters. Groups of hamsters were killed at intervals from 6 to 362 days after injection of ^{241}Am, and 6 to 42 days after injection of ^{252}Cf. Fifty-four hours before sacrifice a partial hepatectomy was performed to stimulate cell division, and four hours before sacrifice colchicine was administered to block cellular division at metaphase. Cells from livers or bone marrow of several animals were scored for chromosome aberrations (50 to 200 cells were scored at each possible postinjection time-dose level combination). The following types of chromosome changes were reported: isochromatid or chromatid deletions, chromosome exchanges, and dicentric, multicentric, or ring forms. Results were also expressed as total aberrations per cell. The accumulated radiation doses in the livers were calculated from tissue distribution data obtained in companion experiments.

After injection of ^{241}Am or ^{252}Cf the total number of metaphase chromosome aberrations in the liver was proportional to the accumulated radiation dose. The aberration frequency for ^{241}Am was 7.1×10^{-3} aberrations/cell/rad; for ^{252}Cf it was 3.3×10^{-3}, if radiation doses were calculated from the total energy per disintegration (fission fragments plus α-particles). If only the dose contributed by the α-particles was considered, the frequency of liver chromosome aberrations was the same as for ^{241}Am. (See footnote in Sec. XII.B.1.e). The investigators considered the following possibilities: (a) the energy of the fission fragments was so great that struck cells were probably killed outright and, therefore, did not divide and could not be scored; (b) the low frequency of emission of fission fragments (3.2% of decays) or their very short range in tissue might limit the number of cells hit, and these could have been overlooked by the scoring system, or (c) fission fragments had such a high linear energy transfer (LET) that for many biological effects much of their energy might have been wasted. They concluded that the fission fragments of ^{252}Cf were not contributing to the measured endpoint.

Based on accumulated radiation dose the frequency of liver cell aberrations after injection of the β-emitter, ^{144}Ce-^{144}Pr, was 3.1×10^{-4}. The effect was about 1/20 of that produced by ^{241}Am or ^{252}Cf. For the α-emitters, aberration frequency increased with total accumulated dose but was independent of the dose rate.

In the bone marrow studies, chromosome aberrations were scored only in the animals injected at levels greater than 5 μCi/kg of ^{252}Cf. Aberrations in chromosomes of marrow cells appeared to be related to dose rate. The average frequency of aberrations regardless of dose level or time after injection was 0.07 aberrations/cell, which was about four times the measured value in normal controls. The total scored aberration frequency was low; the frequency of chromatid and isochromatid deletions (which are usually reported as a function of dose rate) was high, as was the proportion of symmetrical to asymmetrical (dicentric) chromosome exchanges compared with that found in the liver. These results suggested to the investigators that certain chromosome aberrations were produced but not seen, because cell division in the rapidly dividing bone marrow cells placed a selective disadvantage on cells with certain types of chromosome aberrations, and the cells carrying them had been eliminated in successive cell divisions.

G. Effects on Reproduction and Offspring

A large study in rats of (a) the transfer of ^{237}Np, ^{239}Pu, and ^{241}Am via the placenta and the milk from injected mothers to their offspring, and (b) the effects of these nuclides on fertility, reproductive capacity, and development of offspring has been made at the Institute of Biophysics in Moscow. Only brief summaries of the results have been published (Moskalev et al., 1969; Ovcharenko, 1970, 1972).

In the studies on maternal transfer, female rats were injected intravenously with doses of 3 or 15 μCi/kg of ^{241}Am citrate at times from day 1 to day 19 of gestation and killed 48 hours later. As was observed by Hollins and Durakovic (1972) in lactating rats, the livers of the pregnant rats contained a greater fraction, and the skeletons a smaller fraction of the injected ^{241}Am, than was found in non-pregnant controls.

The amounts of ^{241}Am transferred from the mothers were greatest on the 19th day of gestation. Fetuses, placentas, and uterus contained 0.1, 0.9, and 0.27% of the mother's dose, respectively, when 3 μCi/kg was injected on the 13th day of gestation, and 0.45, 1.22, and 0.7%, respectively, when the dam was injected on the 19th day of gestation. The amounts of ^{241}Am transferred were less after injection of a larger dose, 15 μCi/kg. In a comparable experiment, the amounts of ^{239}Pu transferred to the fetuses, placentas, and uterus were three to six times greater than for ^{241}Am. A total of 1% of the mother's ^{241}Am dose was transferred to the offspring via the milk.

Radiation effects on the estrus cycle, embryonic deaths, and fertility of ^{241}Am-injected males and females were studied in rats injected with 2 to 96 μCi/kg of ^{241}Am at 1 to 180 days before mating. The number of implantations was normal (10.6 per pregnant female) when less than 4 μCi/kg of ^{241}Am was given 1 to 10 days before mating, but was only 6.6 per female at the 96 μCi/kg level. At the lower doses the numbers of embryonic deaths 14 or 17 days after conception (10% and 17%, respectively) were the same as in normal controls. At the highest dose level (96 μCi/kg of ^{241}Am, equal to the $LD_{50/30}$) embryonic deaths were 40 and 82% at 14 and 17 days of gestation, respectively. The intrauterine deaths were attributed to altered placental development and radiation stresses in the mothers (Moskalev et al., 1969).

In experiments in which ^{241}Am-injected males were mated to control females, intrauterine deaths were also increased.

Altered postnatal development of the offspring of ^{241}Am-injected mothers included poor survival, delayed development, variable weight, blood dyscrasias, altered radiosensitivity, and depressed reproductive capacity. The effects on the offspring were reported to depend on the dose administered to the mother.

Sikov and Mahlum (1972b) injected female rats with $^{253}EsCl_3$ on the 15th or 19th day of gestation, and the animals were killed 24 hours later. The concentrations of ^{253}Es in the fetuses were 0.002 and 0.007% of the mother's dose per gram of fetal tissue in the 16-day and 20-day fetuses, respectively. The ^{253}Es concentrations in placentas were about 100 times greater than in the fetuses. In the dams, ^{253}Es concentrations in liver, spleen, and femur ranged from 1.3 to 1.9%/g of wet tissue. Radiochemical analysis and autoradiography demonstrated that nearly all of the transferred ^{253}Es was in the fetal skeletons.

The average size of a litter was 9.5 fetuses, and the average weights of the fetuses were 0.5 and 4 g at 16 and 20 days, respectively (D. MAHLUM, personal communication). The total amounts of ^{253}Es transferred to the fetuses were 0.01 and 0.7% of the mother's dose at 16 and 20 days, respectively, in agreement with the extent of placental transfer of ^{241}Am given above.

Acknowledgments

The previously unpublished studies of ^{241}Am in rats and monkeys and of $^{249-252}$Cf in rats reported in this review cover a period of more than 12 years. They would never have been completed without the able assistance of my co-investigators MARILYN H. WILLIAMS and NYLAN JEUNG. Mrs. WILLIAMS cared for the animal colony, assisted with the injections, and performed most of the autopsies and the hundreds of radiochemical analyses. Miss JEUNG assisted with some of those aspects of the investigations and in addition prepared tissues for microscopic analysis, prepared the autoradiographs and photomicrographs, and analyzed the tissue sections and autoradiographs.

The author is indebted to the many investigators who participated in the symposium on the Biological Implications of the Transuranium Elements held at the Battelle Northwest Laboratory, Richland, September 1971, and who made preprints of their manuscripts available. Without their cooperation fully one-half of the material covered in this review would not have been available on time to be included.

My sincere thanks go also to my colleagues in other Laboratories who generously made available to me previously unpublished data. They are DAVID R. ATHERTON and RAY D. LLOYD of the Radiobiology Laboratory at the University of Utah, JOHN E. BALLOU and VICTOR H. SMITH of the Biology Department of the Battelle Northwest Laboratory, STEPHEN R. WRIGHT of the Lawrence Berkeley Laboratory, ANTONE L. BROOKS of the Fission Products Inhalation Group at the Lovelace Foundation, Albuquerque, and GEORGE A. SACHER of the Biology and Medicine Division of the Argonne National Laboratory.

Thanks go also to Mrs. MARY WILDENSTEN who carefully and critically edited the manuscript, to Mr. JOHN FLAMBARD for the line drawings, and to Mrs. GRACE WALPOLE who prepared the Tables.

Appendix A: Calculation of ILB in Intratracheal Intubation Studies with Rats

ZALIKIN et al. (1968) and TSEVELEVA and YEROKHIN (1969) reported the results of their intratracheal intubation studies with $^{241}AmCl_3$ or $^{241}Am(NO_3)_3$ as percent of amount introduced into the lungs, i.e., percent of administered dose. In both experiments 25 to 30% of the administered dose was missing from rats killed from $t=0$ to $t=1$ hr after administration—the ^{241}Am was not in the lungs nor had it been absorbed into the body. Excreta were not collected in either study, however, ZALIKIN et al. (1968) measured ^{241}Am in the upper respiratory and gastrointestinal tracts—they found 7% of the administered dose in the trachea and gastrointestinal contents 1 hr after administration. The ^{241}Am content of lungs and body changed very little in the succeeding 24 hr.

Clearance of a kinetic compartment in 1 hr requires a half-time of about 10 min. The best available lung model does *not* include a *pulmonary* clearance mechanism coupled to gastrointestinal tract elimination with a half-time shorter than 1 day (for man). However, the N-P and T-B compartments are cleared via the gastrointestinal tract with half-times of 4 to 10 min (ICRP, 1966). One must conclude that the missing 25 to 30% of the instilled ^{241}Am solutions was not actually administered, or remained in the upper airways and never reached the lung, or was drawn from the lung into the upper airways as the tracheal cannula was being removed. In the latter cases the ^{241}Am in the upper respiratory tract would be promptly cleared to the gastrointestinal tract. In these studies the ILB was assumed to be equal to the dose recovered at $t=0$ to $t=1$ hr less the ^{241}Am content of the upper respiratory and gastrointestinal tracts.

Hungate et al. (1972) expressed the results of their study of intratracheally instilled $^{253}EsCl_3$ in rats as percent of the administered dose. All of the administered ^{253}Es could be accounted for, if allowance were made for urinary excretion of the absorbed ^{253}Es. The ILB was assumed to be equal to the stated amount administered (IBB).

Appendix B: Calculation of the total amount of actinide that is transferred to rat liver after instillation into the lung, and calculation of the total amount of actinide absorbed from the rat lung

A large fraction of administered actinide is deposited in the liver of the rat whether the nuclide is introduced directly into the blood, or is absorbed from an intramuscular injection or from the lungs. The actinide deposited in the rat liver is promptly excreted in the feces. But, after administration by lung instillation some of the nuclide may at the same time be cleared from the lung into the gastrointestinal tract and excreted in feces. If the rate at which the nuclide is absorbed from the lung is not very different from the rate at which the nuclide deposited in liver is simultaneously excreted into the feces, the total transfer of nuclide from lung to liver cannot be measured directly.

The laws of radioactive decay provide an appropriate mathematical expression that describes the simultaneous growth and decay of a radioactive isotope (daughter) being formed from the decay of a parent isotope with a comparable half-life (Siri, 1949):

$$N_2 = \frac{T_2}{T_1 - T_2} N_0 \left(e^{-0.693\,t/T_1} - e^{-0.693\,t/T_2}\right). \qquad \text{Eq. A1}$$

N_2 is the amount of daughter present at time t ($N_2 = 0$ at $t=0$); N_0 is the amount of parent present at $t=0$; and T_1 and T_2 are the radioactive half-lives of the parent and daughter, respectively. When appropriate subscripts are inserted, Eq. A1 becomes

$$N_{li} = \frac{T_e}{T_a - T_e} f_{li} N_{lu} \left(e^{-0.693\,t/T_a} - e^{-0.693\,t/T_e}\right). \qquad \text{Eq. A2}$$

N_{li} is the amount of actinide in the liver at time t ($N_{li}=0$ at $t=0$); N_{lu} is the ILB, equal to 100%; f_{li} is the fraction of ILB absorbed and transferred to the liver; T_a is the half-time of absorption from the lung, and T_e is the half-time of transfer of actinide from liver to feces. It is assumed that f_{li}, T_a, and T_e are constant over the period of interest, and that T_a, the half-time of absorption from lung to liver, is the same as the half-time of transfer of nuclide from lung to bone.

As shown in Fig. 18.5, deposition of ^{241}Am in the skeleton reached an equilibrium value (Sk_{eq}) in about 10 days. This is an example of accumulation of a tracer in a long-lived phase, and its accumulation rate can be obtained by the method of COMAR (1955). The skeleton accumulation data from the three studies of inhalation or intratracheal instillation of ^{241}Am (ZALIKIN et al., 1968; TSEVELEVA and YEROKHIN, 1969; and NÉNOT et al., 1971a–c) were plotted linearly, and smooth curves (not shown) were drawn by eye. Several values for the amount of ^{241}Am in the skeleton at various times (Sk_t) were obtained for each experiment; from these it was possible to evaluate the expression $(1 - Sk_t/Sk_{eq})$ at each time t. Semilogarithmic plots (not shown) of $(1 - Sk_t/Sk_{eq})$ as functions of time indicated skeletal-uptake half-times of 3.2 days (ZALIKIN et al., 1968), 3.1 days (TSEVELEVA and YEROKHIN, 1969), and 3.4 days (NÉNOT et al., 1971a–c) with a mean of 3.2 ± 0.2 days.

It was assumed that during the first 30 days after exposure, the rate of elimination of ^{241}Am from the rat liver was the same as it would have been had the entire liver deposit been acquired on the first day (as it is after an intravenous or intraperitoneal injection). The results of several determinations of the half-time in the rat liver of parenterally administered ^{241}Am are given in Table 18.21; the mean half-time is 7.5 ± 0.5 days.

When appropriate substitutions are made, $T_a = 3.2$ days, $T_e = 7.5$ days, $N_{lu} = 100\%$, $\frac{T_e}{T_a - T_e} = -1.74$; and the terms are rearranged; an expression is obtained for f_{li}:

$$f_{li} = \frac{-N_{li}}{(1.74)(e^{-0.216t} - e^{-0.092t})}. \qquad \text{Eq. A3}$$

N_{li} and t were obtained from the results of the three experiments and f_{li} was evaluated for each: Five sets of data from 1 to 16 days after intratracheal instillation of $^{241}AmCl_3$ yielded a mean f_{li} of 59.6% of the ILB (ZALIKIN et al., 1968); five sets of data from 6 hr to 15 days after intratracheal instillation of $^{241}Am(NO_3)_3$ yielded a mean f_{li} of 52.2% of the ILB (TSEVELEVA and YEROKHIN, 1969); and two sets of data, 10 and 20 days after inhalation of an $^{241}Am(NO_3)_3$ aerosol yielded a mean f_{li} of 53.8% of the ILB (NÉNOT et al., 1971a–c). These calculated values of f_{li} compare favorably with total liver deposition of ^{241}Am administered parenterally (in the form of these same compounds) by these same investigators as shown in Table 18.18.

Absorption of ^{241}Am from the rat lung (shown in Table 18.9) was calculated at the time when the lung burden had been reduced to about 10% of the ILB (LB_{10}). Total absorption was assumed to equal the sum of the amounts of ^{241}Am (expressed as percent of the ILB) present at $t = LB_{10}$ in skeleton, liver, residual soft tissues, urine, and fecal excretion derived from the liver (Fe_{end}). Fe_{end} was calculated by difference,

$$Fe_{end}(\%) = f_{li} - L_t, \qquad \text{Eq. A4}$$

where L_t is the amount of ^{241}Am in the liver at $t = LB_{10}$. Urinary excretion in rats of absorbed ^{241}Am was estimated to be 27% of the maximum skeletal deposition based on the quantities of ^{241}Am in bone and urine 1 to 4 days after intramuscular injection of $^{241}AmCl_3$ (SCOTT et al., 1948a) or ^{241}Am citrate (DURBIN et al., 1967, 1972c).

Appen-

Tissue and organ weights and percent of body weight of standard man and laboratory elements[a]

Tissue	Standard man[b]		Cynomolgus monkey		Beagle dog	
	weight (g)	% of body wt.	weight (g)	% of body wt.	weight (g)	% of body wt.
Muscle	30000	43	1300	43	4300	43
Skin	6100	23	410	14	1510	15
Skeleton[a a]	10000	14	540	18	720	7.2
Blood[b]	5400	7.7	—	5.4	—	9.2
G.I. tract	3600	5.1	270	8.9	430	4.3
Liver	1700	2.4	74	2.5	400	4.0
Brain	1500	2.1	80	2.7	80	0.8
Lungs	1000	1.4	33	1.1	90	0.9
Kidneys	300	0.43	18	0.6	50	0.5
Heart	300	0.43	12	0.4	90	0.9
Spleen	150	0.21	5	0.17	20	0.2
Urinary bladder	150	0.21	—	—	15	0.15
Pancreas	70	0.10	3.8	0.12	20	0.2
Salivary glands	50	0.071	—	—	—	—
Testes	40	0.057	—	—	15	0.15
Eyes	30	0.043	7.2	0.024	—	—
Thyroid	20	0.029	0.5	0.016	0.9	0.01
Prostate	20	0.029	—	—	—	—
Adrenals	20	0.029	1.1	0.037	1.0	0.01
Thymus	10	0.014	—	—	—	—
Ovaries	(8)	(0.011)	0.34	0.011	1.0	0.01
Pituitary	0.6	0.0009	0.05	0.002	—	—
Uterus	—	—	5.5	0.18	—	—
Total body[a c]	70000	—	3000	—	10000	—

[a] Data sources: Standard Man (ICRP, 1959); cynomolgus monkey, female (P. W. Durbin and M. H. Williams, unpublished); beagle dog, male and female (Andersen and Johnson, 1962; Cuddihy et al., 1970; Atherton et al., 1969b); rat, Charles-River (Sprague-Dawley) female (Durbin et al., 1972e), male (R. N. Pierson and D. C. Price, unpublished); Chinese hamster, male, 100 days old (A. L. Brooks, unpublished), male more than 1 yr old, 39 g (G. A. Sacher, unpublished); mouse, LAF-1 female (P. W. Durbin, unpublished), BALB/c male (G. A. Sacher, unpublished).

Appendix D

Adult body weight (BW), glomerular filtration rate (GFR)[a], filtered volume (GFR $\times$ BW), plasma volume (V_p)[b], and total plasma volume ($\Sigma V_p = V_p \times$ BW) of several mammals.

	Body weight (kg)	GFR (ml/min/kg)	Filtered volume (ml/min)	V_p (ml/kg)	ΣV_p (ml)
Rat	0.25	6.0	1.5		
Cynomolgus monkey	3.3	(3.6)	11.9	36.4	120
Beagle	10	4.3	43	49.3	493
Man	70	1.7	120	42.8	3000
Cat	3.3	3.2	10.6	39	130
Rabbit	3.7	3.1	11.5	35.1	130

[a] References for GFR: rat, dog, cat, rabbit, man, H. W. Smith (1951); dog also Della Rosa et al. (1961). GFR for monkey estimated to be the mean for rabbit, cat and dog.
[b] References for V_p: rat, Everett et al. (1956); monkey, Gregerson et al. (1959); beagle, J. H. Dougherty and Seymour (1957); man, ICRP (1959); cat and rabbit (Teklad, Inc.).

dix C

animals that have been used in studies of the metabolism and toxicity of the transplutonium

Albino rat		Chinese hamster		Mouse (LAF-1)		Mouse (BALB/c)	
weight (g)	% of body wt.	weight (g)	% of body wt.	weight (g)	% of body wt.	weight (g)	% of body wt.
110	44	—	—	(10)[c]	(42)	—	—
30	12	—	—	3.4	14	—	—
23	8.2	—	—	2.0[ab]	8.3	—	—
—	5.1	—	—	—	5.0	—	—
22	9.0	2.6	9.6	2.6	9.6	—	—
7.6	3.1	1.4	5.2	1.6	6.7	1.5	5.5
2.0	0.8	0.66	2.4	0.44	1.8	0.46	1.7
1.8	0.7	0.33	1.2	0.3	1.2	—	—
1.6	0.65	0.33	1.2	0.32	1.3	0.35	1.3
1.2	0.5	0.15	0.56	0.13	0.54	0.13	0.47
0.6	0.24	0.33	1.2	0.14	0.55	0.071	0.26
—	—	—	—	0.025	0.1	—	—
—	—	—	—	0.19	0.76	—	—
—	—	—	—	0.17	0.7	—	—
(2.8)	(1.1)	0.16	0.59	—	—	0.16	0.58
—	—	—	—	0.047	0.19	—	—
0.025	0.009	—	—	0.0028	0.011	—	—
—	—	—	—	—	—	—	—
0.065	0.023	0.0057	0.021	0.012	0.049	0.0033	0.012
0.27	0.10	0.021	0.078	0.051	0.21	0.046	0.17
0.07	0.025	—	—	0.012	0.047	—	—
0.014	0.005	—	—	0.0023	0.0093	—	—
—	—	—	—	0.093	0.38	—	—
250	—	27	—	24.5	—	27	—

[b] Tissue weights of Standard Man are free of blood. Tissue weights of the animals include contained blood present after routine dissection.
[c] Pitts and Bullard (1968).
[aa] Skeleton includes cartilages, periosteum, and marrow.
[ab] Weight of mouse skeleton calculated to be 8.3% of the body weight (Parmley et al., 1961).
[ac] Fat can be calculated by difference of sum of all tissues from body weight.

References

Many of the references cited in this review appeared in books of conference proceedings or collections of individual papers. For those volumes containing more than one citation the reader is referred for publication details (date and place of conference, editors, publisher, and place of publication) to either the editors (of books cited as such) or to the first entry in the alphabetical listing.

Research and progress reports of several Laboratories have been rich sources of material. In order to conserve space these report citations have been abbreviated to the initials of the Laboratory followed by a report number.

ANL	Argonne National Laboratory.
BNWL	Battelle Northwest Laboratory.
CH	Metallurgical Laboratory (Manhattan Project) usually followed by a USAEC declassification number, MDDC- or AECD.
COO-119	University of Utah, Department of Anatomy, Radiobiology Project.
HW	Hanford Atomic Products Operation.
LA	Los Alamos Scientific Laboratory.
LF	Lovelace Foundation Fission Product Inhalation Program.
UCRL	University of California, Lawrence Radiation Laboratory.
USNRDL	U.S. Naval Radiological Defense Laboratory.

All of the numbered documents, progress reports, and English translations of French and Russian language publications are listed in the Cumulative Report Indexes of Nuclear Science Abstracts. The introductory pages of these Indexes include information on the availability for purchase from the USAEC Technical Information Center, Oak Ridge, Tennessee, and on the availability for loan from the several repository libraries in the United States; the National Lending Library, Boston Spa, England, the International Atomic Energy Agency, Vienna, Austria, and the Zentralstelle für Atomkernenergie-Dokumentation, Karlsruhe, W. Germany.

ADELSTEIN, S. J., VALLEE, B. L.: Copper. In: Mineral metabolism (C. L. COMAR and F. BRONNER, eds.), vol. 2B, p. 371–401. New York: Academic Press 1962.

ANDERSEN, A. C., JOHNSON, R. M.: Organ weights in the beagle. University of California, Davis, Radiobiology Project Annual Progress Report UCD-105, 187–189 (1962).

ARNOLD, E. D.: Economic and hazards potential of radioactive waste disposal. Presented at Amer. Nucl. Soc., annual meeting, Pittsburgh, Pennsylvania, 1957. Oak Ridge National Laboratory Document CF-57-7-31.

ARNOLD, E. D.: Radiation and environmental characteristics of the actinide isotopes. Trans. Amer. Nucl. Soc. 13, 219–220 (1970). (Abstr.).

ARNOLD, J. S.: An improved technic for liquid emulsion autoradiography. Proc. Soc. exp. Biol. (N.Y.) 85, 113–116 (1954).

ARNOLD, J. S., JEE, W. S. S.: Embedding and sectioning undecalcified bone and its application to radioautography. Stain Technol. 29, 225–229 (1954).

ARNOLD, J. S., JEE, W. S. S.: Bone growth and osteoclastic activity as indicated by radioautographic distribution of plutonium. Amer. J. Anat. 101, 367–418 (1957).

ARNOLD, J. S., JEE, W. S. S.: Autoradiography in the localization and radiation dosage of Ra^{226} and Pu^{239} in the bones of dogs. Lab. Invest. 8, 194–204 (1959).

ASLING, C. W., HAMILTON, J. G., AXELROD-HELLER, D., LOUIE, B. J.: The localization of certain alkaline and rare earth elements in the costochondral junction of the rat. Anat. Rec. 113, 285–300 (1952).

ATHERTON, D. R., LLOYD, R. D.: The distribution and retention of ^{249}Cf in beagle soft tissue. Hlth Phys. 22, 675–677 (1972).

ATHERTON, D. R., LLOYD, R. D., TAYLOR, G. N., STOVER, B. J., MAYS, C. W.: Distribution of ^{241}Am in the beagle. Radiat. Res. 35, 506–507 (1968). (Abstr.).

ATHERTON, D. R., STOVER, B. J., BRUENGER, F. W.: Fractional weight of the beagle skeleton. COO-119-240, 162–171 (1969b).

ATHERTON, D. R., STOVER, B. J., LLOYD, R. D., BRUENGER, F. W.: A comparison of the macrodistribution of ^{241}Am and ^{59}Fe in the young adult beagle. COO-119-240, 129–138 (1969a).

BAIR, W. J.: Plutonium inhalation studies. BNWL-1221, 7.1–7.25 (1970).

BALLOU, J. E., HESS, J. O.: Biliary plutonium excretion in the rat. Hlth Phys. 22, 369–372 (1972).

BALLOU, J. E., MORROW, W. G.: The disposition and long-term effect of intratracheally instilled ^{253}Es in rats. Hlth Phys. 23, 418 (1972). (Abstr.).

BARNES, J. E.: An autoradiographic study of the distribution of inhaled radionuclides in the respiratory tract of the beagle dog. LF-41, 88–93 (1970).

BAUER, G. C. H., CARLSSON, A., LINDQUIST, B.: Metabolism and homeostatic function of bone. In: Mineral metabolism, vol. 1B, p. 609–676, 1962. (See ADELSTEIN and VALLEE, 1962.)

BAYBARZ, R. D.: Dissociation constants of the transplutonium chelates with diethylenetriaminepentaacetic acid (DTPA) and the application of DTPA chelates to solvent extraction separations of transplutonium elements from the lanthanide elements. J. Inorg. Nucl. Chem. 27, 1831–1839 (1965).

BAYBARZ, R. D.: Dissociation constants of the transplutonium element chelates of 1,2,diaminocyclohexanetetraacetic acid. J. Inorg. Nucl. Chem. 28, 1055–1061 (1966).

BELYAEV, YU. A.: Americium-241 distribution in rats and the effect of complexing substances on its elimination. In: Radioaktivnye izotopy i organizm (YU. I. MOSKALEV, ed.). Moscow: Izdatel'stvo Meditsina 1969. English translation: AEC-tr-7195, p. 168–174.

BENSTED, J. P. M., TAYLOR, D. M.: A comparison of the carcinogenic effects of the bone seeking nuclides, ^{241}Am and ^{239}Pu. In: Symposium Ossium, London, England, 1969 (A. M. JELLIFFE and B. STRICKLAND, eds.), p. 211–214. Edinburgh-London: E. and S. Livingstone.

BENSTED, J. P. M., TAYLOR, D. M., SOWBY, F. D.: The carcinogenic effects of americium 241 and plutonium 239 in the rat. Brit. J. Radiol. 38, 920–925 (1965).

BERLINER, R. W.: Urine formation. In: The physiological basis of medical practice, 7th ed. (C. H. BEST and N. B. TAYLOR, eds.), p. 523–568. Baltimore: The Williams and Wilkins Co. 1961.

BLOOM, M. A.: Bone marrow. In: Histopathology of irradiation from external and internal sources (W. BLOOM, ed.). National Nuclear Energy Series, Div. IV, vol. 22I, p. 162–242. New York: McGraw-Hill Book Co. 1948.

BOROUGHS, H., TOWNSLEY, S. J., HIATT, R. W.: The metabolism of radionuclides by marine organisms. II. The uptake, accumulation and loss of yttrium^{91} by marine fish and the importance of short-lived radionuclides in the sea. Biol. Bull. **111**, 352–357 (1956).

BOTHWELL, T. H., FINCH, C. A.: Iron metabolism. Boston: Little, Brown and Co. 1962.

BOYD, G. A.: Autoradiography. New York: Academic Press 1955.

BRICK, I., RATH, C. E.: Evaluation of trisodium calcium diethylenetriamine penta-acetate in hemochromatosis and transfusion hemosiderosis. In: Iron metabolism. Proc. CIBA conference, Aix-en-Provence, France, 1963 (F. GROSS, ed.), p. 568–579. Berlin: Springer.

BRODSKY, A., WALD, N., HORM, I. S., VARZALY, B. J.: The removal of americium-241 from humans by DTPA. Hlth Phys. **17**, 379 (1969). (Abstr.)

BRODSKY, A., SAYEG, J. A., WALD, N., WECHSLER, R., CALDWELL, R.: The measurement and management of insoluble plutonium-americium inhalation in man. In: Radiation protection, Proc. Internat. Radiation Protection Assoc. Rome, Italy, 1966 (W. S. SYNDER, ed.), p. 1181–1190. Oxford: Pergamon Press.

BROOKS, A. L., JONES, R. K., HURNI, W. M., MCCLELLAN, R. O., STURBAUM, B.: Cytogenic and other effects of ^{144}Ce in Chinese hamsters. LF-41, 192–202 (1969).

BROOKS, A. L., MCCLELLAN, R. O.: Chromosome aberrations and other effects produced by ^{90}Sr-^{90}Y in Chinese hamsters. Int. J. Radiat. Biol. **16**, 545–561 (1969).

BROOKS, A. L., MCCLELLAN, R. O., MCKAY, L. R.: The relative biological effectiveness of alpha, beta and gamma irradiation in producing metaphase chromosome aberrations in vivo in Chinese hamster liver cells. Radiat. Res. **47**, 332–333 (1971). (Abstr.)

BROOKS, A. L., MEWHINNEY, J. A., MCCLELLAN, R. O.: The in vivo cytogenetic effects of ^{252}Cf on liver and bone marrow of the Chinese hamster. Hlth Phys. **22**, 701–706 (1972).

BRUENGER, F. W., ATHERTON, D. R., STEVENS, W.: Intracellular distribution of ^{249}Cf in canine liver. Hlth Phys. **22**, 685–689 (1972).

BRUENGER, F. W., ATHERTON, D. R., STEVENS, W., STOVER, B. J.: Interaction between blood constituents and some actinides. COO-119-244, 212–227 (1971).

BRUENGER, F. W., STEVENS, W., STOVER, B. J.: Americium-241 in the blood: In vivo and in vitro observations. Radiat. Res. **37**, 349–360 (1969a).

BRUENGER, F. W., STOVER, B. J., STEVENS, W.: On the binding of ^{241}Am in the canine liver. COO-119-240, 139–161 (1969b).

BULDAKOV, L. A., BUROV, N. I.: Behavior of Ce-144 in sheep of different age. Radiobiologiya **7**, 881–885 (1967). English translation: AEC-tr-6891.

BULDAKOV, L. A., KALMYKOVA, Z. I., NIFATOV, A. P., DOSHCHENKO, V. N., TSEVELEVA, I. A., MUSHKACHEVA, G. S., KUDASHEVA, N. P., PESTERNIKOV, V. M., MATVEEV, V. I., SURINA, A. G., KARPOVA, V. N.: Metabolism and biological effects of inhaled ^{241}Am and ^{239}Pu in dogs. Hlth Phys. **22**, 873–874 (1972).

BULDAKOV, L. A., LYUBCHANSKII, E. R., MOSKALEV. YU. I., NIFATOV, A. P.: Problemi toksikologii plutoniya, Moscow: Atomizdat 1969. English translation: LF-tr-41.

BUSHONG, S. C., PRASAD, N., BRINEY, S. A., OLIVER, G. D.: Radiocytogenic determination of the oxygen enhancement ratio of californium 252. Radiology **96**, 167–170 (1970).

CALDWELL, R.: Detection of insoluble alpha-emitters in the lung. Proc. 12th annual Bio-Assay and Analytical Chemistry Meeting, Gatlinburg, Tennessee, 1966. USAEC Document No. CONF-661018, p. 120–133.

CAMERON, J. R. (ed.): Bone Measurement, Proceedings of a conference, Chicago, Illinois, 1970. Oak Ridge: USAEC Information Division No. CONF-700515.

COCHRAN, K. W., DOULL, J., MAZUR, M., DUBOIS, K. P.: Acute toxicity of zirconium, columbium, strontium, lanthanum, cesium, tantalum, and yttrium. Arch. industr. Hyg. **1**, 637–650 (1950).

COLLINS, E. J., GARRETT, E. R., JOHNSON, R. L.: Effect of adrenal steroids on radio-calcium metabolism in dogs. Metabolism **11**, 716–726 (1962).

COMAR, C. L.: Radioisotopes in biology and agriculture, p. 18–25. New York: McGraw-Hill Book Co. 1955.

COPP, D. H., AXELROD, D. J., HAMILTON, J. G.: The deposition of radioactive metals in bone as a potential health hazard. Amer. J. Roentgenol. **58**, 10–16 (1947).

COPP, D. H., GREENBERG, D. M., HAMILTON, J. G., CHACE, M. J., MIDDLESWORTH, L. VAN, CUTHBERTSON, E. M., AXELROD, D. J.: The deposition of plutonium and certain fission products in bone as a decontamination problem. CH-3591 (1946). AECD-2483.

COPP, D. H., HAMILTON, J. G., JONES, D. C., THOMPSON, D. M., CRAMER, C.: The effect of age and low phosphorous rickets on calcification and the deposition of certain radioactive metals in bone. Trans. Third Conference on Metabolic Interrelations, 1951 (E. R. REIFENSTEIN, JR., ed.), p. 226–258. New York: Josiah Macy, Jr. Foundation.

Cotton, F. A., Wilkinson, G.: Advanced inorganic chemistry, 2nd ed. New York: Interscience Publishers 1966.

Crosby, W. H.: Regulation of iron metabolism. In: Regulation of hematopoiesis (A. S. Gordon, ed.), vol. I, p. 519–537. New York: Appleton-Century-Crofts 1970.

Cuddihy, R. G., Griffith, W. C., Boecker, B. B.: Tissue weight in adult beagle dogs. LF-43, 106–111 (1970).

Cummings, S. L., Bankert, L.: Uptake of ^{144}Ce, ^{147}Pm and ^{238}Pu by oat plants from soils. Radiol. Hlth Data Rep. **12**, No. 2, 83–85 (1971).

Davis, J. J., Hanson, W. C., Watson, D. G.: Some effects of environmental factors upon accumulation of worldwide fallout in natural populations. In: Radioecology, p. 35–38, 1961. (See Schultz and Klement, 1961.)

Davis, J. J., Perkins, R. W., Palmer, R. F., Hanson, W. C., Cline, J. F.: Radioactive materials in aquatic and terrestrial organisms exposed to reactor effluent water. Proc. Second Intern. Conf. Peaceful Uses Atomic Energy, Geneva, Switzerland **18**, 423–428 (1958).

Della Rosa, R. J., Smith, F. A., Stannard, J. N.: The renal excretion of strontium and calcium in dogs. Int. J. Radiat. Biol. **3**, 557–578 (1961).

Denham, D. H.: Health physics considerations in processing transplutonium elements. Hlth Phys. **16**, 475–487 (1969).

Dobson, E. L., Gofman, J. W., Jones, H. B., Kelly, L. S., Walker, L. A.: Studies with colloids containing radioisotopes of yttrium, zirconium, columbium and lanthanum. II. The controlled selective localization of radioisotopes of yttrium, zirconium and columbium in the bone marrow, liver, and spleen. J. Lab. clin. Med. **34**, 9–16 (1949).

Dobson, R. L.: Americium poisoning. In: Therapy of radioelement poisoning. ANL-5584, p. 28–35, 1955. (See Rosenthal, 1955.)

Dougherty, J. H.: Preliminary report on hematological effects of ^{241}Am in the beagle. COO-119-242, 354–362 (1970).

Dougherty, J. H.: Early hematological effects of californium in the beagle. COO-119-244, 117–125 (1971).

Dougherty, J. H.: ^{241}Am toxicity to leukocytes of beagles for the first year postinjection. Radiat. Res. **51**, 504–505 (1972a). (Abstr.)

Dougherty, J. H.: The hematological changes induced by ^{239}Pu in beagles. In: Radiobiology of plutonium, p. 75–86, 1972b. (See Stover and Jee, 1972.)

Dougherty, J. H., Seymour, K.: Hematology report. University of Utah Radiobiology Project Annual Report AECU-3522, 57–65 (1957).

Dougherty, T. F.: Injection tables. COO-119-246, 14–105 (1972).

Dougherty, T. F., Jee, W. S. S., Mays, C. W., Stover, B. J. (eds.): Some aspects of internal radiation. Proc. of a conference, Heber, Utah, 1961. Oxford: Pergamon Press.

Durbin, P. W.: Distribution of the transuranic elements in mammals. Hlth Phys. **8**, 665–671 (1962).

Durbin, P. W.: Plutonium in man. A new look at the old data. In: Radiobiology of plutonium, p. 469–530, 1972. (See Stover and Jee 1972.)

Durbin, P. W., Asling, C. W., Jeung, N., Williams, M. H., Post, J., Johnston, M. E., Hamilton, J. G.: The metabolism and toxicity of radium-223 in rats. UCRL-8189 (1958).

Durbin, P. W., Asling, C. W., Johnston, M. E., Hamilton, J. G., Williams, M. H.: The metabolism of the lanthanons in the rat. II. Time studies of the tissue deposition of intravenously administered radioisotopes. In: Rare earths in biochemical and medical research. Proc. of a conference, Oak Ridge, Tennessee, 1955 (G. C. Kyker and E. B. Anderson, eds.), p. 171–191. Oak Ridge Institute of Nuclear Studies Report ORINS-12.

Durbin, P. W., Horovitz, M. W., Close, E. R.: Plutonium deposition kinetics in the rat. Hlth Phys. **22**, 731–741 (1972a).

Durbin, P. W., Jeung, N., Williams, M. H.: Dynamics of ^{241}Am in the skeleton of the rat: A study of the relationship between behavior of bone-seeking elements and bone-growth status. In: Delayed effects of bone-seeking radionuclides, p. 137–156, 1967. (See Mays et al., 1967.)

Durbin, P. W., Jeung, N., Williams, M. H.: Distribution, retention and excretion of ^{241}Am in adult female rats and monkeys. Unpublished (1972c).

Durbin, P. W., Jeung, N., Williams, M. H.: Distribution, retention, and excretion of $^{249-252}$Cf in adult female rats. Unpublished (1972d).

Durbin, P. W., Jeung, N., Williams, M. H.: The growth and senescence of the female rat. Unpublished (1972e).

Durbin, P. W., Williams, M. H., Gee, M., Newman, R. H., Hamilton, J. G.: Metabolism of the lanthanons in the rat. Proc. Soc. exp. Biol. (N.Y.) **91**, 78–85 (1956). Original data contained in UCRL-3066.

DURBIN, P. W., WILLIAMS, M. H., JEUNG, N.: Metabolism of Americium^{241} in the monkey. Radiat. Res. **14**, 462 (1961). (Abstr.)

DURBIN, P. W., WILLIAMS, M. H., JEUNG, N.: Long-term excretion of 241-americium by the cynomolgus monkey. Hlth Phys. **23**, 417–418 (1972b). (Abstr.)

ERDMAN, H. E.: Effects of ingested Pu^{239} on fecundity, fertility, and life span of *Habrobracon* (Braconidae). Hlth Phys. **8**, 635–638 (1962).

EVERETT, N. B., SIMMONS, B., LASKER, E. P.: Distribution of blood (Fe^{59}) and plasma (I^{131}) volumes of rats determined by liquid nitrogen freezing. Circulat. Res. **4**, 419–424 (1956).

EYRING, L.: Progress in the science and technology of the rare earths, vol. 1. Oxford: Pergamon Press 1964.

FABRIKANT, J. J., SMITH, C. L. D.: Radiographic changes following the administration of bone-seeking radionuclides. Brit. J. Radiol. **37**, 53–62 (1964).

FASISKA, B. C., BOHNING, D. E., BRODSKY, A., HORM, J.: Urinary excretion of ^{241}Am under DTPA therapy. Hlth Phys. **21**, 523–529 (1971).

FAWWAZ, R. A., WINCHELL, H. S., POLLYCOVE, M., SARGENT, T., LAWRENCE, J. H.: Kinetics of oral iron absorption using iron-52, iron-55, and iron-59. J. nucl. Med. **7**, 349 (1966). (Abstr.)

FIGUEROA, W. G.: The enhancement of iron excretion in iron-storage diseases. In: Metal-binding in medicine. Proc. of a conference. Philadelphia, Pennsylvania, 1959 (M. J. SEVEN and L. A. JOHNSON, eds.), p. 146–153. Philadelphia: J. B. Lippincott Co.

FINKEL, M. P., BISKIS, B. O.: Toxicity of plutonium in mice. Hlth Phys. **8**, 565–579 (1962).

FINKLE, R. D., SNYDER, R. H., JACOBSON, L. O., KISIELESKI, W., LAWRENCE, B., SIMMONS, E. L.: The toxicity and metabolism of plutonium in laboratory animals. CH-3783 (1946). MDDC-1140.

FOREMAN, H.: Chelating experiments. UCRL-960, 27–32 (1950).

FOREMAN, H.: The pharmacology of some useful chelating agents. In: Metal-binding in medicine, p. 82–94, 1959. (See FIGUEROA, 1959.)

FOREMAN, H.: Studies on the mechanism of plutonium uptake by bone. Hlth. Phys. **8**, 713–716 ((1962a).

FOREMAN, H.: The effect of diethylenetriaminepentaacetic acid (DTPA) on acceleration of excretion of actinide elements. Hlth Phys. **8**, 735–738 (1962b).

FOREMAN, H., FINNEGAN, C.: Effect of ethylenediaminetetraacetic acid on deposition and excretion of certain rare earth elements. J. biol. Chem. **226**, 745–749 (1957).

FOREMAN, H., FUQUA, P. A., NORWOOD, W. D.: Experimental administration of ethylene-diaminetetraacetic acid in plutonium poisoning. Arch. industr. Hyg. **10**, 226–231 (1954).

FOREMAN, H., HAMILTON, J. G.: The use of chelating agents for accelerating excretion of radioelements. UCRL-1351 (1951).

FOREMAN, H., MOSS, W., EUSTLER, B. C.: Clinical experiences with radioactive materials. Amer. J. Roentgenol. **79**, 1071–1079 (1958).

FOREMAN, H., TRUJILLO, T. T., JOHNSON, O., FINNEGAN, C.: Ca EDTA and the excretion of plutonium. Proc. Soc. exp. Biol. (N.Y.) **89**, 339–342 (1955).

FRITZ, T. E., NORRIS, W. P., POOLE, C. M., REHFELD, C. E.: Cerium-144—toxic manifestations in beagle dogs. ANL-7278, 112–113 (1966).

FROST, H. M.: Dynamics of bone remodeling. In: Bone biodynamics. Proc. of a conference, Detroit, Michigan, 1963 (H. M. FROST, ed.), p. 315–333. Boston: Little, Brown and Co.

FUGER, J.: Ion exchange behavior and dissociation contants of americium, curium, and californium complexes with ethylenediaminetetraacetic acid. J. Inorg. Nucl. Chem. **5**, 332–338 (1958).

FUGER, J.: Actinide and lanthanide ion exchange separation studies. II. Separation by aminopolyacetic acids. J. Inorg. Nucl. Chem. **18**, 263–269 (1961).

FUJITA, D. K.: Some magnetic, spectroscopic, and crystallographic properties of berkelium, californium, and einsteinium. UCRL-19507 (1969).

GALLEGOS, A. F., BOYD, H. A., THOMAS, R. G., MCCLELLAN, R. O.: Retention and distribution of ^{244}Cm following inhalation of $^{244}CmCl_3$ and $^{244}CmO_{1.7}$ by beagle dogs. LF-43, 100–105 (1970).

GARCIA, J. F.: Radioiron time-distribution studies at various ages in the normal male rat. Amer. J. Physiol. **190**, 31–36 (1957).

GILLETTE, R.: Nuclear power in the USSR: American visitors find surprises. Science **173**, 1003–1006 (1971).

GLAZUNOV, V. V., PARCHEVSKII, V. P., FLEISHMAN, D. G.: The variation of the fission product content of *Cystoseira* in the Black Sea. Dokl. Akad. Nauk SSSR. **152**, 1222–1224 (1963). Nucl. Sci. Abstr. 18 : 4979.

GOFMAN, J. W., DE LALLA, O., JOHNSON, G., KOVICH, E. L., LOWE, O., MARTIN, W., PILUSO, D. L., TANDY, R. K., UPHAM, F., WEITZEL, R., WILBUR, D.: Chemical elements of the blood of man. UCRL-9897, 1–26 (1961).

GOLDSCHMIDT, V. M., BARTHE, T., LUNDE, G.: Geochemische Verteilungsgesetze der Elemente. V. Skr. norske Vidensk.-Akad. Oslo, I. Mat.-nat. Kl. No. 7 (1925).
GONG, J. K., SHIPMAN, W. H., WEISS, H. V., COHN, S. H.: Uptake of fission products and neutron-induced radionuclides by the clam. USNRDL-TR-119 (1956).
GRAN, F. C.: Studies on calcium and strontium-90 metabolism in rats. Norwegian Monographs on Medical Science. Oslo: University of Oslo Press 1960.
GREEN, R., CHARLTON, R., SEFTEL, H., BOTHWELL, T., MAYET, F., ADAMS, B., FINCH, C., LAYRISSE, M.: Body iron excretion in man. Amer. J. Med. **45**, 336–353 (1968).
GREENSDALE, A. E., BALLOU, N. E.: Physical state of fission product elements following their vaporization in distilled water and sea water. USNRDL-436 (1954).
GREGERSON, M. I., SEAR, H., RAWSON, R. A., CHIEN, S., SAIGER, G. L.: Cell volume, plasma volume, total blood volume and F_{cells} factor in the rhesus monkey. Amer. J. Physiol. **196**, 184–187 (1959).
GRIGORYAN, KH. V.: Build-up dynamics of chronically administered ^{144}Ce in the organism of animals. In: Raspredelenie i biologicheskoe deistvie radioaktivnykh izotopov (YU. I. MOSKALEV, ed.), p. 50–55. Moscow: Atomizdat 1966. English translation: USAEC-tr-6944 (Rev.).
HAMILTON, J. G.: The metabolism of the fission products and the heaviest elements. Radiology **49**, 325–343 (1947).
HAMILTON, J. G.: The metabolic properties of the fission products and actinide elements. Rev. Mod. Phys. **20**, 718–728 (1948).
HAMILTON, J. G., SCOTT, K. G.: Effect of calcium salt of versene upon metabolism of plutonium in the rat. Proc. Soc. exp. Biol. (N.Y.) **83**, 301–305 (1953).
HAMMARSTRÖM, L., NILSSON, A.: Radiopathology of americium 241. I. Distribution of americium in adult mice. Acta radiol. (Stockh.) **9**, 433–442 (1970a).
HAMMARSTRÖM, L., NILSSON, A.: Radiopathology of americium 241. II. Uptake in the developing teeth of rats. Acta radiol. (Stockh.) **9**, 609–617 (1970b).
HAMMOND, S. E., PUTZIER, E. A.: Observed effects of plutonium in wounds over a long period of time. Hlth Phys. **10**, 399–406 (1964).
HATCH, T. F., GROSS, P.: Pulmonary deposition and retention of inhaled aerosols. New York: Academic Press 1964.
HAWK, P. B., OSER, B. L., SUMMERSON, W. H.: Practical physiological chemistry, 12th ed., p. 370. Philadelphia: The Blakiston Co. 1947.
HEANEY, R. P., SKILLMAN, T. G.: Secretion and excretion of calcium by the human gastrointestinal tract. J. Lab. clin. Med. **64**, 29–41 (1964).
HELD, E. E.: Qualitative distribution of radionuclides at Rongelap atoll. In: Radioecology, p. 167–169, 1961. (See SCHULTZ and KLEMENT, 1961.)
HELLER, M.: Bone. In: Histopathology of irradiation from external and internal sources, p. 70–161, 1948. (See BLOOM, 1948.)
HERRING, G. M., VAUGHAN, J., WILLIAMSON, M.: Preliminary report on the site of localization and possible binding agent for yttrium, americium, and plutonium in cortical bone. Hlth Phys. 8, 717–724 (1962).
HOBBS, C. H., CUDDIHY, R. G., BENJAMIN, S. A., JONES, R. K., KANAPILLY, G. M., MAUDERLY, J. L., MCCLELLAN, R. O., PICKRELL, J. A.: Toxicity of inhaled ^{91}Y fused clay in beagle dogs. I. LF-43, 163–182 (1970).
HOLLINS, J. G., DURAKOVIC, A. B.: The metabolism of americium in lactating rats. Hlth Phys. **22**, 627–631 (1972).
HUNGATE, F. P. (ed.): Hanford symposium on radiation and terrestrial ecosystems. Hlth Phys. **11**, 1255–1676 (1965).
HUNGATE, F. P., BALLOU, J. E., MAHLUM, D. D., KASHIMA, M., SMITH, V. H., SANDERS, C. L., BAXTER, D. W., SIKOV, M. R., THOMPSON, R. C.: Preliminary data on ^{253}Es and ^{249}Bk metabolism in rats. Hlth Phys. **22**, 653–656 (1972).
HURME, V. O., VAN WAGENEN, G.: Basic data on the emergence of permanent teeth in the rhesus monkey (*Macaca mulatta*). Proc. Amer. phil. Soc. **105**, 105–140 (1961).
International Commission on Radiological Protection: Standard man. In: Report of committee II on permissible dose for internal radiation. Hlth Phys. **3**, 151 (1959).
International Commission on Radiological Protection. Deposition and retention models for internal dosimetry of the human respiratory tract, Task Group on Lung Dynamics (P. E. MORROW, chairman). Hlth Phys. **12**, 173–207 (1966).
IVANOV, V. N.: Effect of radioactive substances on the embryonic development of fish. In: Voprosy biookeangrafii. Materially II. Mezhdunarodnogo okeanografischeskogo kongressa, Moscow, USSR, 1966, p. 185–190. Kiev: Naukova Dumka. English translation: AEC-tr-6940.
JACOBSON, L., OVERSTREET, R.: The uptake by plants of plutonium and some products of nuclear fission absorbed on soil colloids. Soil Sci. **65**, 129–134 (1948).

JEANMAIRE, L., BALLADA, J.: Etude de deux cas de contamination par Am-241. In: Radiation protection problems relating to transuranium elements. Proc. of a seminar, Karlsruhe, W. Germany, 1970, p. 531–546. Luxembourg: Centre for Information and Documentation. English translation UCRL-trans-1459.

JEE, W. S. S.: Distribution and toxicity of ^{239}Pu in bone. Hlth Phys. **22**, 583–595 (1972).

JEE, W. S. S., ARNOLD, J. S.: Radioisotopes in the teeth of dogs. I. Arch. oral Biol. **2**, 215–238 (1960).

JOHNSON, G. T., KYKER, G. C.: Fission product and cerium uptake by bacteria, yeasts, and molds. J. Bact. **81**, 733–740 (1961).

JOHNSON, L. J.: Relative translocation and distribution of Pu and Am from experimental PuO_2 subcutaneous implants in beagles. Thesis. Colorado State University, Fort Collins 1969. USAEC Document COO-1787-6.

JOHNSON, L. J., BULL, E. H., LEBEL, J. L., WATTERS, R. L.: Kinetics of lymph node activity accumulation from subcutaneous PuO_2 implants. Hlth Phys. **18**, 416–418 (1970a).

JOHNSON, L. J., WATTERS, R. L., LAGERQUIST, C. R., HAMMOND, S. E.: Relative distribution of plutonium and americium following experimental PuO_2 implants. Hlth Phys. **19**, 743–749 (1970b).

JOHNSON, L. J., WATTERS, R. L., LEBEL, J. L., LAGERQUIST, C. R., HAMMOND, S. E.: The distribution of Pu and Am: Subcutaneous administration of PuO_2 and the effect of chelation therapy. In: Radiobiology of plutonium, p. 213–220, 1972. (See STOVER and JEE, 1972.)

KALLFELZ, F. A., COMAR, C. L., CASARETT, A. P., CRAIG, P. H.: Radiobiological effects of simulated nuclear power sources for artificial hearts. Trans. Amer. nucl. Soc. **13**, 499–500 (1970) (Abstr.)

KAMEDA, K.: Study on abundance of rare earth elements in marine organisms. J. Radiat. Res. (Tokyo) **3**, 89–103 (1962). In English.

KATZ, J. H.: Transferrin and its functions in the regulation of iron metabolism. In: Regulation of hematopoiesis (A. S. GORDON, ed.), vol. 1, p. 539–577. New York: Appleton-Century-Crofts 1970.

KAWIN, B., COPP, D. H., HAMILTON, J. G.: Studies of the metabolism of certain fission products and plutonium. UCRL-812 (1950).

KELLER, O. L., JR.: Transplutonium elements. Science **156**, 838–840 (1967).

KHMELEVA, N. N.: Accumulation of Sr, Ca, Ce, and Y radioisotopes by crustacea. Radiobiologiya **2**, 944–946 (1962). English translation: AEC-tr-5433.

KROLL, H.: Development of chelating agents potentially more effective than EDTA in radioelement removal. In: Therapy of radioelement poisoning, ANL-5584, p. 150–151 (1955). (See ROSENTHAL, 1955.)

KROLL, H., KORMAN, S., SIEGEL, E., HART, H. E., ROSOFF, B., SPENCER, H., LASZLO, D.: Excretion of yttrium and lanthanum chelates of cyclohexane, 1,2,transdiamine tetraacetic acid and diethylenetriaminepentaacetic acid in man. Nature (Lond.) **180**, 919–920 (1957).

KYKER, G.C.: Rare earths. In: Mineral metabolism, vol. 2B, p. 499–541, 1962. (See ADELSTEIN and VALLEE, 1962.)

LAFUMA, J., NÉNOT, J. C., MORIN, M.: Recherches expérimentales sur le traitement des contaminations par les transplutoniens. In: Radiation protection problems relating to transuranium elements, p. 249–258, 1970. (See JEANMAIRE and BALLADA, 1970.)

LAGERQUIST, C. R., PUTZIER, E. A., PILTINGSRUD, C. W.: Bioassay and body counter results for the first 2 years following an acute plutonium exposure. Hlth Phys. **13**, 965–972 (1967).

LANGHAM, W., CARTER, R. E.: The relative physiological and toxicological properties of americium and plutonium. LA-1309 (1951).

LANGHAM, W. H., BASSETT, S. H., HARRIS, P. S., CARTER, R. E.: Distribution and excretion of plutonium administered intravenously to man. LA-1151 (1950).

LANGHAM, W. H., LAWRENCE, J. N. P., MCCLELLAND, J., HEMPELMANN, L. H.: The Los Alamos Scientific Laboratory's experience with plutonium in man. Hlth Phys. **8**, 753–760 (1962).

LEWIN, R.: A study of metal complexes with proteins. Thesis. New York University 1954. Quoted by STERN (1955).

LINDENBAUM, A., ROSENTHAL, M. W.: Deposition patterns and toxicity of plutonium and americium in liver. Hlth Phys. **22**, 597–605 (1972).

LISCO, H., FINKEL, M. P., BRUES, A. M.: Carcinogenic properties of radioactive fission products and of plutonium. Radiology **49**, 361–363 (1947).

LLOYD, R. D., ATHERTON, D. R., GAUFIN, S. S., MAYS, C. W.: Distribution of injected ^{241}Am in the beagle skeleton. COO-119-244, 151–158 (1971).

LLOYD, R. D., ATHERTON, D. R., GAUFIN, S. S., MAYS, C. W., TAYLOR, G. N.: Am-241 skeletal distribution in beagles. COO-119-246, 263–265 (1972c).
LLOYD, R. D., ATHERTON, D. R., TAYLOR, G. N.: Tissue concentration of injected ^{241}Am in beagles. COO-119-246, 266–271 (1972d).
LLOYD, R. D., ATHERTON, D. R., TAYLOR, G. N., MAYS, C. W.: Soft tissue deposition of injected ^{241}Am in beagles. Hlth Phys. **15**, 175 (1968). (Abstr.)
LLOYD, R. D., JEE, W. S. S., ATHERTON, D. R., TAYLOR, G. N., MAYS, C. W.: Americium-241 in beagles: biological effects and skeletal distribution. In: Radiobiology of plutonium, p. 141–148, 1972a. (See STOVER and JEE, 1972.)
LLOYD, R. D., MAYS, C. W., ATHERTON, D. R., GAUFIN, S. S.: Californium retention in beagles during the first year after injection. COO-119-246, 289–298 (1972a).
LLOYD, R. D., MAYS, C. W., TAYLOR, G. N.: Strontium, radium, and americium metabolism in beagles. Hlth Phys. **17**, 384 (1969).
LLOYD, R. D., MAYS, C. W., TAYLOR, G. N., ATHERTON, D. R., SHABESTARI, L. R.: Retention of injected ^{241}Am in beagles. Hlth Phys. **13**, 938 (1967a). (Abstr.)
LLOYD, R. D., MAYS, C. W., TAYLOR, G. N., ATHERTON, D. R., SHABESTARI, L. R.: Americium metabolism in beagles. COO-119-236, 197–206 (1967b).
LLOYD, R. D., MAYS, C. W., TAYLOR, G. N., ATHERTON, D. R.: Americium-241 studies in beagles. Hlth Phys. **18**, 149–156 (1970).
LLOYD, R. D., MAYS, C. W., TAYLOR, G. N., WILLIAMS, J. L.: Californium excretion and retention by beagles injected with ^{249}Cf or ^{252}Cf. Hlth Phys. **22**, 667–673 (1972b).
LORENZ, E., CONGDON, C. C.: Radioactivity: biological effects of ionizing radiations. Ann. Rev. Med. **5**, 323–338 (1954).
MACKEY, J. L., POWELL, J. E., SPEDDING, F. H.: A calorimetric study of the reaction of rare-earth ions with EDTA in aqueous solution. J. Amer. chem. Soc. **84**, 2047–2050 (1962).
MAHLUM, D. D., SIKOV, M. R.: Metabolism and toxicity of ^{253}Es relative to age. Radiat. Res. **51**, 502–503 (1972). (Abstr.)
MALY, J., SIKKELAND, T., SILVA, R., GHIORSO, A.: Nobelium: Tracer chemistry of the divalent and trivalent ions. Science **160**, 1114–1115 (1968).
MARTELL, A. E., CALVIN, M.: The chemistry of the metal chelate compounds, p. 191–202. New York: Prentice-Hall 1952.
MARTLAND, H. S., HUMPHRIES, R. E.: Osteogenic sarcoma in dial painters using luminous paint. Arch. Path. **7**, 406–412 (1929).
MAUCHLINE, J., TAYLOR, A. M.: The accumulation of radionuclides by the thornback ray, *Raia clavata L.*, in the Irish Sea. Limnol. Oceanogr. **9**, 303–309 (1964).
MAY, H. A.: Preliminary report on the distribution of americium-241 following accidental inhalation. ANL-7489, 19–23 (1968).
MAYS, C. W., JEE, W. S. S., LLOYD, R. D., STOVER, B. J., DOUGHERTY, J. H., TAYLOR, G. N. (eds.): Delayed effects of bone-seeking radionuclides. Proc. of a conference, Sun Valley, Idaho, 1967. Salt Lake City: University of Utah Press. (Pub. 1969).
MCCLELLAN, R. O., BOYD, H. A., GALLEGOS, A. F., THOMAS, R. G.: Retention and distribution of ^{244}Cm following inhalation of $^{244}CmCl_3$ and $^{244}CmO_{1.73}$ by beagle dogs. Hlth Phys. **22**, 877–885 (1972).
MCCLELLAN, R. O., CASEY, H. W., BUSTAD, L. K.: Transfer of some transuranic elements to milk. Hlth Phys. **8**, 689–694 (1962b).
MCCLELLAN, R. O., CASEY, H. W., BUSTAD, L. K.: Transfer of some radionuclides to milk. HW-76000, 98–108 (1963).
MCCLELLAN, R. O., CASEY, H. W., CABLE, J. W., BUSTAD, L. K.: Transfer of heavy radionuclides to milk. HW-72500, 44–49 (1962a).
MCCLELLAN, R. O., RUPPRECHT, F. C. (eds.): The metabolism of inhaled $^{241}AmO_2$ in beagle dogs. LF-39, 148–150 (1968).
MCCLELLAN, R. O., RUPPRECHT, F. C. (eds.): Status of longevity and sacrifice experiments in beagles dogs. Appendix A. LF-43, 295–314 (1970).
MCKAY, L. R., BROOKS, A. L., MCCLELLAN, R. O.: Metabolism and toxicity of 241-Am in the Chinese hamster. LF-41, 203–308 (1969).
MCKAY, L. R., BROOKS, A. L., MCCLELLAN, R. O.: Retention, distribution, dose and cytogenetic effects of ^{241}Am citrate in the Chinese hamster. Hlth Phys. **22**, 633–640 (1972).
MCLEAN, F. C., HASTINGS, A. B.: Clinical estimation and significance of calcium-ion concentrations in the blood. Amer. J. med. Sci. **189**, 601–612 (1935).
MENZEL, R. G.: Soil-plant relationships of radioactive elements. Hlth Phys. **11**, 1325–1332 (1965).
MEWHINNEY, J. A., BROOKS, A. L., MCCLELLAN, R. O.: Comparison of the retention and distribution of injected ^{252}Cf in rats and Chinese hamsters. Hlth Phys. **22**, 695–700 (1972).

MEWHINNEY, J. A., HARRIS, M. D., III. Retention and distribution of ^{252}Cf: an interspecies comparison. Hlth Phys. **23**, 417 (1972). (Abstr.)
MEWHINNEY, J. A., ZIEMER, P. L., LANDHOLT, R. R.: Retention and distribution of californium-252 in the rat. Hlth Phys. **21**, 857–860 (1971).
MILLER, C. E.: The human spectrometer. ANL-5829, 144–166 (1957).
MITCHELL, H. H., CARD, L. E., HAMILTON, T. S.: A technical study of the growth of White Leghorn chickens. Illinois Agr. exp. Sta. Bull. No. 367, 81–139 (1931).
MOELLER, T., HSEU, T. M.: Observations on the rare earths—LXXVI. The stabilities of the *trans*-1,2,diaminocyclohexane-N,N′ tetraacetic acid chelates of the tripositive ions. J. Inorg. Nucl. Chem. **24**, 1635–1643 (1962).
MOELLER, T., THOMPSON, L. C.: Observations on the rare earths—LXXV. The stabilities of diethylenetriaminepentaacetic acid chelates. J. Inorg. Nucl. Chem. **24**, 499–519 (1962).
MOORE, C. V., DUBACH, R.: Iron. In: Mineral metabolism, vol. 2B, p. 287–348, 1962. (See ADELSTEIN and VALLEE, 1962.)
MORROW, P. E., GIBB, F. R., DAVIES, H., MITOLA, J., WOOD, D., WRAIGHT, N., CAMPBELL, H. S.: The retention and fate of inhaled plutonium dioxide in dogs. Hlth Phys. **13**, 113–133 (1967).
MOSKALEV, YU. I.: ^{239}Pu: Problems of its biological effect. Hlth Phys. **22**, 723–729 (1972).
MOSKALEV, YU. I., BULDAKOV, L. A., KOSHURNIKOVA, N. A., NIFATOV, A. P., RESHETOV, G. N.: Combined effect of ^{90}Sr, ^{144}Ce and ^{239}Pu on the rat organism (Reports 1 and 2). In: Raspredelenie i biologicheskoe deistvie radioaktivnykh izotopov, p. 346–365, 1966. (See GRIGORYAN, 1966.)
MOSKALEV, YU. I., BULDAKOV, L. A., LYAGINSKAYA, A. M., OVCHARENKO, E. P., EGOROVA, T. M.: Experimental study of radionuclide transfer through the placenta and their biological action on the fetus. In: Radiation biology of the fetal and juvenile mammal. Proc. of a conference, Richland, Washington, 1969 (M. R. SIKOV and D. D. MAHLUM, eds.), p. 153–160. AEC Symposium Series 17, Oak Ridge: USAEC Technical Information Division.
MOSKALEV, YU. I., RUDNITSKAYA, E. I.: Liver pathology under the action of americium-241. Radiobiologiya **10**, No. 5 (1970). (Abstr.) English translation: AEC-tr-7215, p. 240.
MRAZ, F. R., WRIGHT, P. L., FERGUSON, T. M., ANDERSON, D. L.: Fission product metabolism in hens and transference to eggs. Hlth Phys. **10**, 777–782 (1964).
National Research Council: Report of the panel on radioactivity in the marine environment of the National Academy of Science's Committee on Oceanography. Washington, D.C.: National Academy of Sciences (U.S.) 1971.
NELSON, D. J., EVANS, F. C. (eds.): Symposium on radioecology, Proc., Ann Arbor, Michigan, 1967. Oak Ridge: USAEC Technical Information Division No. CONF-670503.
NÉNOT, J. C., MASSE, R., MORIN, M., LAFUMA, J.: An experimental comparative study of the behavior of ^{237}Np, ^{238}Pu, ^{239}Pu, ^{241}Am, and ^{242}Cm in bone. Hlth Phys. **22**, 657–665 (1972).
NÉNOT, J. C., MORIN, M., LAFUMA, J.: Etude éxperimentale des contaminations par le curium-242 et de leur traitement. Hlth Phys. **18**, 613–622 (1970). English translation: AEC-tr-7178.
NÉNOT, J. C., MORIN, M., LAFUMA, J.: Etude métabolique et thérapeutique des contaminations respiratoires par certains actinides en solutions. Hlth Phys. **20**, 167–177 (1971a). English translation: ORNL-tr-2491.
NÉNOT, J. C., MORIN, M., LAFUMA, J.: Contaminations éxperimentales par l'américium en solution et leur traitement. Hlth Phys. **20**, 383–392 (1971b). English translation: LF-tr-61.
NÉNOT, J. C., MORIN, M., LAFUMA, J.: Etude éxperimentale de la décontamination du squelette après inhalation de nitrate d'américium. Hlth Phys. **21**, 395–400 (1971c). English translation: UCRL-tr-1477.
NEUMAN, W. F., NEUMAN, M. W.: The chemical dynamics of bone mineral. Chicago: Chicago University Press 1958.
NIFATOV, A. P., BULDAKOV, L. A., MATVEEV, V. I.: Some late effects after a single inhalation of ^{239}Pu and ^{241}Am in dogs. Hlth Phys. **22**, 875 (1972).
NORWOOD, W. D.: DTPA—effectiveness in removing internally deposited plutonium from humans. J. occup. Med. **2**, 371–376 (1960).
OLIVER, G. D., JR.: Aspects of radiation control for implants of californium-252. Hlth Phys. **21**, No. 6, Suppl. p. 27 (1971). (Abstr.)
OSANOV, D. P., KLYKOV, O. V., ERSHOV, E. B., RAKOVA, V. A.: Effective biological half life of radioactive isotopes in the skin. Radiobiologiya **11**, No. 1 (1971). (Abstr.) English translation: AEC-tr-7233, p. 236.
OVCHARENKO, E. P.: Distribution of transuranium elements in the bodies of pregnant rats. Radiobiologiya **10**, No. 2 (1970). (Abstr.) English translation: AEC-tr-7171, p. 219.
OVCHARENKO, E. P.: An experimental evaluation of the effects of transuranic elements on reproductive ability. Hlth Phys. **22**, 641 (1972).

Painter, E., Russell, E., Prosser, C. L., Swift, M. N., Kisieleski, W., Sacher, G.: Clinical physiology of dogs injected with plutonium. CH-3858 (1946). AECD-2042.

Palmer, H. E., Crook, G. H., Nelson, I. C.: The uptake, distribution and excretion of promethium in humans and the effect of DTPA on these parameters. BNWL-SA-1915 (1968a).

Palmer, H. E., Nelson, I. C., Crook, G. H.: The uptake, distribution and excretion of promethium isotope in humans. Hlth Phys. **15**, 187 (1968b). (Abstr.)

Parker, H. G., Low-Beer, A. de G., Isaac, E. L.: Comparison of retention and organ distribution of Am^{241} and Cf^{252} in mice: The effect of in vivo DTPA chelation. Hlth Phys. **8**, 679–684 (1962).

Parker, H. G., Wright, S. R., Low-Beer, A. de G., Yaeger, D. J.: The metabolism of ^{253}Es in mice. Hlth Phys. **22**, 647–651 (1972).

Parmley, W. W., Jensen, J. B., Mays, C. W.: Skeletal self-absorption of beta-particle energy. In: Some aspects of internal radiation, p. 437–451, 1961. (See Dougherty et al., 1961.)

Peterson, J. R.: The solution absorption spectrum of Bk^{+3} and the crystallography of berkelium dioxide, sesquioxide, trichloride, oxychloride and trifluoride. Thesis. University of California, Berkeley, 1967. UCRL-17875.

Pigford, T. H.: Protection of the public from radioactivity produced in nuclear reactors. Presented at IEEE, annual meeting, San Francisco, California, 1971.

Pitts, G. C., Bullard, T. R.: Some interspecific aspects of body composition in mammals. In: Body composition in animals and man, Proc. of a conference, Columbia, Missouri, 1967. Washington, D.C.: National Academy of Sciences (U.S.) Publication 1598.

Plummer, G. L.: Fallout radioisotopes in Georgia lichens. In: Symposium on radioecology, CONF-670503, p. 571–577, 1967. (See Nelson and Evans, 1967.)

Plutonium Project: Nuclei formed in fission: decay characteristics, fission yields, and chain relationships. J. Amer. chem. Soc. **68**, 2411–2442 (1946).

Pollycove, M.: Hemochromatosis. In: The metabolic basis of inherited disease, 2nd ed. (J. B. Stanbury, J. B. Wyngaarden, and D. S. Fredrickson, eds.), p. 780–810. New York: McGraw-Hill Book Co. 1966.

Popplewell, D. S., Boocock, G.: Distribution of some actinides in blood serum proteins. In: Diagnosis and treatment of deposited radionuclides. Proc. of a conference, Richland, Washington, 1967 (H. A. Kornberg and W. D. Norwood, eds.), p. 45–55. Amsterdam: Excerpta Medica Foundation.

Prosser, C. L., Pervinsek, W., Arnold, J., Svilha, G., Tompkins, P. C.: Accumulation and distribution of radioactive strontium, barium, lanthanum, fission mixture and sodium in goldfish. CH-3233 (1946). MDDC-496.

Rice, T. R.: The role of polyplankton in the cycling of radionuclides in the marine environment. In: Radioecology, p. 179–185. 1961. (See Schultz and Klement, 1961.)

Richards, O. W., Troutman, M. C.: Spectroscopic analysis of the mineral content of yeast grown on synthetic and natural media. J. Bact. **39**, 739–746 (1940).

Romney, E. M., Mork, H. M., Larson, K. H.: Persistence of plutonium in soil, plants, and small mammals. Hlth Phys. **19**, 487–491 (1970).

Rosen, J., Cohen, N., Wrenn, M. E.: In vivo measurement of americium-241 burdens in man. Hlth Phys. **21**, No. 6, Suppl. p. 19 (1971). (Abstr.)

Rosen, J. C., Cohen, N., Wrenn, M. E.: Short term metabolism of ^{241}Am in the adult baboon. Hlth Phys. **22**, 621–626 (1972).

Rosenthal, M. W.: Use of zirconium and other carriers in the removal of radioelements from the body; comments on the use of complexing agents. In: Therapy of radioelement poisoning. Proc. of a conference, Argonne National Laboratory, 1955 (M. W. Rosenthal, ed.), p. 100–113. ANL-5584.

Rosenthal, M. W.: Radioisotope absorption and elimination: Factors influencing elimination from the body. In: Radioisotopes in the biosphere. Proc. of a conference, Minneapolis, Minnesota, 1959 (R. S. Caldecott and L. A. Snyder, eds.), p. 541–563. Minneapolis: University of Minnesota Press.

Rowland, R. E., Miller, C. E., Marinelli, L. D.: An $americium^{241}$ accident. ANL-5518, 46–48 (1956).

Rudnitskaya, E. I., Moskalev, Yu. I.: Microdistribution and morphological changes in rats in the case of intravenous injection of americium-241. Radiobiologiya **10**, 570–574 (1970). English translation: AEC-tr-7205.

Rundo, J., Keane, A. T., May, H. A.: Measurement of americium-241 in two subjects. ANL-7760, Pt. II, 36–44 (1970).

Sacher, G. A.: Late effects of continuous irradiation: the relation of hematological injury to lethality. Laval méd. **34**, 163–168 (1963).

SANDERS, S. M., JR.: Excretion of ^{241}Am and ^{244}Cm by humans. Hlth Phys. **21**, No. 6, Suppl. p. 28 (1971). (Abstr.)

SCHOUR, I., MASSLER, M.: The teeth. In: The rat in laboratory investigation, 2nd. ed. (E. J. FARRIS and J. Q. GRIFFITH, JR. eds.), p. 104–165. Philadelphia: J. B. Lippincott Co. 1949.

SCHUBERT, J.: Treatment of plutonium poisoning by metal displacement. Science **105**, 389–390 (1947).

SCHUBERT, J.: An experimental study of the effect of zirconium and sodium citrate on the metabolism of plutonium and radioyttrium. J. Lab. clin. Med. **34**, 313–325 (1949).

SCHUBERT, J., FINKEL, M. P., WHITE, M. R., HIRSCH, G. M.: Plutonium and yttrium content of the blood, liver and skeleton of the rat at different times after intravenous administration. J. biol. Chem. **182**, 635–642 (1950).

SCHUBERT, J., FREID, J. F., ROSENTHAL, M. W., LINDENBAUM, A.: Tissue distribution of monomeric and polymeric plutonium as modified by a chelating agent. Radiat. Res. **15**, 220–226 (1961).

SCHULTZ, V., KLEMENT, A. W., JR. (eds.): Radioecology. Proc. of a conference, Fort Collins, Colorado, 1961. New York: Reinhold Publishing Co.

SCOTT, K., AXELROD, D. J., FISHER, H., CROWLEY, J. C., HAMILTON, J. G.: The metabolism of plutonium in rats following intramuscular injections. J. biol. Chem. **176**, 283–293 (1948b).

SCOTT, K. G., AXELROD, D. J., HAMILTON, J. G.: The metabolism of curium in the rat. J. biol. Chem. **177**, 325–335 (1949).

SCOTT, K. G., COPP, D. H., AXELROD, D. J., HAMILTON, J. G.: The metabolism of americium in the rat. J. biol. Chem. **175**, 691–703 (1948a).

SCOTT, K. G., OVERSTREET, R., JACOBSON, L., HAMILTON, J. G., FISHER, H., CROWLEY, J., CHAIKOFF, I. L., ENTENMAN, C., FISHLER, M., BARBER, A. J., LOOMIS, F.: The metabolism of carrier-free fission products in the rat. USAEC document MDDC-1275 (1947).

SEABORG, G. T.: Electronic structure of the heaviest elements. Metallurgical Laboratory memorandum MUC-GTS-858 (July 17, 1944). Reproduced as Appendix to paper 21.1, p. 1492–1524. In: The transuranium elements (G. T. SEABORG, J. J. KATZ, and W. M. MANNING, eds.). National Nuclear Energy Series Division IV, vol. 14B, Pt. 2. New York: McGraw-Hill Book Co. 1949.

SEABORG, G. T.: The chemical and radioactive properties of the heavy elements. Chem. Eng. News **23**, 2190–2193 (1945).

SEABORG, G. T.: The transuranium elements. Science **104**, 379–386 (1946).

SEABORG, G. T.: Mass production and practical applications of actinide elements. Isotopes Radiation Tech. **6**, 1–18 (1968).

SEABORG, G. T., JAMES, R. A., GHIORSO, A.: The new element curium (96). In: The transuranium elements, vol. 14B, Pt. 2, p. 1554–1571, 1949b. (See SEABORG, 1949).

SEABORG, G. T., JAMES, R. A., MORGAN, L. O.: The new element americium (atomic number 95). In: The transuranium elements, vol. 14B, Pt. 2, p. 1525–1553, 1949a. (See SEABORG, 1949.)

SEIDEL, A., VOLF, V.: Removal of internally deposited transuranium elements by Zn-DTPA. Hlth Phys. **22**, 779–783 (1972).

SEMENOV, A. I.: Biological effects of curium-244. Radiobiologiya **11**, 134–137 (1971a). English translation: AEC-tr-7233.

SEMENOV, A. I.: Kinetics of the exchange of curium-244. Radiobiologiya **11**, No. 1 (1971b). English translation: AEC-tr-7233; p. 237–238. (Abstr.)

SHELBY, C. A.: Radioactive fallout in *Sargassum* drift on Texas Gulf Coast beaches. Publ. Inst. Marine Sci., Univ. Texas **9**, 33–36 (1963).

SIKOV, M. R., MAHLUM, D. D.: Plutonium in the developing mammal. Hlth Phys. **22**, 707–712 (1972a).

SIKOV, M. R., MAHLUM, D. D.: Distribution of ^{253}Es in the fetoplacental unit of the rat. Radiat. Res. **51**, 543 (1972b). (Abstr.)

SILLEN, L. G., MARTELL, A. E.: Stability constants of metal-ion complexes. London: Chemical Society of London Special Publ. No. 17 (1964).

SIMEK, J. E., DAVIS, J. A., DAY, C. E., III, ANGINO, E. E.: Sorption of radioactive nuclides by *Sargassum fluitans* and *S. natans*. In: Symposium on radioecology, CONF-670503, p. 505–508, 1967. (See NELSON and EVANS, 1967.)

SIMPSON, M. E., ASLING, C. W., EVANS, H. M.: Some endocrine influences on skeletal growth and differentiation. Yale J. Biol. Med. **23**, 1–27 (1950).

SIRI, W. E.: Isotopic tracers and nuclear radiations, p. 163–171. New York: McGraw-Hill Book Co. 1949.

SMITH, H. W.: The kidney: Structure and function in health and disease. New York: Oxford University Press 1951.

SMITH, V. H.: Removal of internally deposited plutonium. HW-53500, 135–142 (1958a).
SMITH, V. H.: Removal of internally deposited plutonium. Nature (Lond.) **181**, 1792–1793 (1958b).
SMITH, V. H.: The removal of internally deposited plutonium. HW-59500, 63–72 (1959).
SMITH, V. H.: Therapeutic removal of internally deposited transuranium elements. Hlth Phys. **22**, 765–778 (1972a).
SMITH, V. H.: The biological disposition of $Es(NO_3)_3$ in rats after intravenous, intramuscular, subcutaneous and transcutaneous administration. BNWL-1650, vol. 1, Pt. 1, 279–283 (1972b).
SOWBY, F. D., TAYLOR, D. M.: Removal of internally deposited americium by chelating agents. Nature (Lond.) **187**, 612 (1960).
STEIDLE, H.: Seltene Erdmetalle. In: Handbuch der Experimentellen Pharmakologie (A. HEFFTER and W. HEUBNER, eds.), vol. 3, Pt. 4, p. 2189–2213. Berlin: Springer 1935.
STERN, K. G.: Interaction of rare-earth compounds with substances of biological interest. In: Rare earths in biochemical and medical research. ORINS-12, 143–161, 1955. (See DURBIN, et al., 1955.)
STEVENS, W., BRUENGER, F. W.: Interaction of ^{249}Cf and ^{252}Cf with constituents of dog and human blood. Hlth Phys. **22**, 679–683 (1972).
STEVENS, W., BRUENGER, F. W., ATHERTON, D. R., STOVER, B. J.: Subcellular distribution of ^{241}Am(III) and ^{239}Pu(IV) in livers studied serially. COO-119-244, 159–174 (1971).
STEVENS, W., BRUENGER, F. W., STOVER, B. J.: In vivo studies on the interaction of PuIV with blood constituents. Radiat. Res. **33**, 490–500 (1968a).
STEVENS, W., STOVER, B. J., BRUENGER, F. W.: Some observations on the deposition of ^{241}Am in the thyroid gland. COO-119-237, 133–134 (1968b).
STEVENS, W., STOVER, B. J., BRUENGER, F. W.: The distribution of ^{241}Am within selected soft tissues of beagle dogs. COO-119-240, 119–128 (1969a).
STEVENS, W., STOVER, B. J., BRUENGER, F. W., TAYLOR, G. N.: Some observations on the deposition of americium-241 in the thyroid gland of the beagle. Radiat. Res. **39**, 201–206 (1969b).
STONE, R. S.: Outline of health and biological research activities of the Metallurgical Laboratory, Spring 1944. In: Industrial medicine on the plutonium project. National Nuclear Energy Series Division IV, vol. 20, p. 8–9. New York: McGraw-Hill Book Co. 1949.
STOVER, B. J., ATHERTON, D. R., BRUENGER, F. W., BUSTER, D. S.: ^{239}PuIV: Its distribution in the beagle. In: Delayed effects of bone-seeking radionuclides, p. 109–123, 1967. (See MAYS et al., 1967.)
STOVER, B. J., ATHERTON, D. R., BUSTER, D. S.: Protracted hepatic, splenic, and renal retention of ^{239}Pu in the beagle. Hlth Phys. **20**, 369–374 (1971).
STOVER, B. J., ATHERTON, D. R., KELLER, N.: Metabolism of Pu^{239} in adult beagle dogs. Radiat. Res. **10**, 130–147 (1959).
STOVER, B. J., BRUENGER, F. W., STEVENS, W.: The reaction of PuIV with the iron transport system in human blood serum. Radiat. Res. **33**, 381–394 (1968).
STOVER, B. J., BRUENGER, F. W., STEVENS, W.: Association of americium with ferritin in the canine liver. Radiat. Res. **43**, 173–186 (1970).
STOVER, B. J., JEE, W. S. S. (eds.): Radiobiology of plutonium. Salt Lake City: The J. W. Press, University of Utah 1972.
STOVER, B. J., STEVENS, W., BRUENGER, F. W.: Chemical associations of ^{239}Pu(IV) and ^{241}Am(III) in blood, liver and thyroid. In: Radiobiology of plutonium, p. 129–140, 1972. (See STOVER and JEE, 1972.)
SWIFT, M. N., PROSSER, C. L.: The excretion, retention, distribution and clinical effects of $strontium^{89}$ in the dog. Part 1. CH-3843 (1947). MDDC-1388.
TAYLOR, D. M.: Some aspects of the comparative metabolism of plutonium and americium in rats. Hlth Phys. **8**, 673–677 (1962).
TAYLOR, D. M.: The effects of desferrioxamine on the retention of actinide elements in the rat. Hlth Phys. **13**, 135–140 (1967).
TAYLOR, D. M.: Interactions between transuranium elements and components of cells and tissues. Hlth Phys. **22**, 575–581 (1972).
TAYLOR, D. M., BENSTED, J. P. M.: Long-term biological damage from plutonium-239 and americium-241 in rats. In: Delayed effects of bone-seeking radionuclides, p. 357–370, 1967. (See MAYS, et al., 1967.)
TAYLOR, D. M., CHIPPERFIELD, A. R., JAMES, A. C.: The effects of tetracycline on the deposition of plutonium and related elements in rat bone. Hlth Phys. **21**, 197–204 (1971).
TAYLOR, D. M., SOWBY, F. D.: The removal of americium and plutonium from the rat by chelating agents. Phys. in Med. Biol. **7**, 83–91 (1962).
TAYLOR, D. M., SOWBY, F. D., KEMBER, N. F.: The metabolism of americium and plutonium in the rat. Phys. in Med. Biol. **6**, 73–86 (1961).

TAYLOR, G. N., JEE, W. S. S., DOCKUM, N. L., HROMYK, E.: Translocation of ^{239}Pu and ^{241}Am in beagle livers. Radiat. Res. **31**, 554 (1967). (Abstr.)

TAYLOR, G. N., JEE, W. S. S., DOCKUM, N., HROMYK, E.: Microscopic distribution of americium-241 in the beagle thyroid gland. Hlth Phys. **17**, 723–725 (1969b).

TAYLOR, G. N., JEE, W. S. S., MAYS, C. W., DELL, R. B., WILLIAMS, J. L., SHABESTARI, L.: Microscopic distribution of californium-249 and berkelium-249 in the soft tissues of beagles. Hlth Phys. **22**, 691–693 (1972).

TAYLOR, G. N., JEE, W. S. S., WILLIAMS, J. L., BURGGRAFF, B., ANGUS, W.: Microscopic distribution of ^{241}Am in the beagle. COO-119-240, 97–118 (1969a).

TEKLAD, INC.: Physiological data for common laboratory animals. Monmouth, Illinois. No date.

TEMPLETON, D. H., DAUBEN, C. H.: Lattice parameters of some rare earth compounds and a set of crystal radii. J. Amer. chem. Soc. **76**, 5237–5239 (1954).

THOMAS, R. G.: Estimation of ^{241}Am body burdens by analysis of whole-body scanning, excreta, and body weight data. Hlth Phys. **19**, 751–755 (1970).

THOMAS, R. G., MCCLELLAN, R. O., THOMAS, R. L., CHIFFELLE, T. L., HOBBS, C. H., JONES, R. K., MAUDERLY, J. L., PICKRELL, J. A.: Metabolism, dosimetry and biological effects of inhaled ^{241}Am in beagle dogs. Hlth Phys. **22**, 863–871 (1972).

THOMAS, R. G., THOMAS, R. L., MCCLELLAN, R. O.: The metabolism of ^{241}Am following inhalation by the beagle dog. A pilot study. Hlth Phys. **17**, 383 (1969). (Abstr.)

THOMAS, W. A., JACOBS, D. G.: Curium behavior in plants and soil. Soil Sci. **108**, 305–307 (1969).

THOMSON, D. M.: The effect of age and low phosphorus rickets on the metabolism of calcium^{45} in rats. Thesis. University of California, Berkeley, 1953. UCRL-2302.

TSEVELEVA, I. A., YEROKHIN, R. A.: Behavior of americium-241 in the body of rats under intraperitoneal and intratracheal administration. In: Radioaktivnye izotopy i organism, p. 161–168, 1969. (See BELYAEV, 1969.)

TURNER, G. A., TAYLOR, D. M.: The binding of plutonium to serum proteins in vitro. Radiat. Res. **36**, 22–30 (1968a).

TURNER, G. A., TAYLOR, D. M.: The transport of plutonium, americium, and curium in the blood of rats. Phys. in Med. Biol. **13**, 535–546 (1968b).

ULLBERG, S., NELSON, A., KRISTOFFERSSON, H., ENGSTRÖM, A.: Distribution of plutonium in mice: an autoradiographic study. Acta radiol. (Stockh.) **58**, 459–471 (1962).

U.S. Congress, Joint Committee on Atomic Energy: Hearings on the nature of radioactive fallout and its effects on man. May 27–29, June 3, 1957. Washington, D.C.: U.S. Government Printing Office.

U.S. Congress, Joint Committee on Atomic Energy: Hearings on fallout from nuclear weapons tests, May 5–8, 1959. Washington, D.C.: U.S.Government Printing Office.

United Nations: Proceedings of the Second International Conference on the Peaceful Uses of Atomic Energy, Geneva, Switzerland, 1958.

VALLEE, B. L.: Zinc. In: Mineral metabolism, vol. 2B, 443–482 (1962). (See ADELSTEIN and VALLEE, 1962.)

VAN WAGENEN, G., ASLING, C. W.: Roentgenographic estimation of bone age in the rhesus monkey (*Macaca mulatta*). Amer. J. Anat. **103**, 163–186 (1958).

VOLKOVA, G. A.: On the accumulation and excretion of the radioactive isotopes of seven chemical elements in dragonfly larvae. Zool. Zh. **42**, 138–139 (1963). English translation: OTS-63-21502.

WALLACE, A.: Increased uptake of ^{241}Am by plants caused by the chelating agent DTPA. Hlth Phys. **22**, 559–562 (1972).

WATTERS, R. L., LEBEL, J. J.: Progress in the beagle studies at Colorado State University. Hlth Phys. **22**, 811–814 (1972).

WELANDER, A. D.: Distribution of radionuclides in the environment of Eniwetok and Bikini atolls, August 1964. In: Symposium on radioecology, CONF-670503, p. 346–354, 1967. (See NELSON and EVANS, 1967.)

WILLIAMS, M. H., JEUNG, N., DURBIN, P. W.: Effect of a complexing agent and mode of administration on the distribution of curium-242 in the rat. UCRL-9617, 35–38 (1961).

WINCHELL, H. S., POLLYCOVE, M., KUSUBOV, N., FAWWAZ, R.: Kinetics of ^{131}I-Albumin and ^{131}I-transferrin in lymph. In: Symposium on the preparation and biomedical applications of labeled molecules. Proc. of a conference, Venice, Italy, 1964 (J. SIRCHIS, ed.), p. 273–286. Brussels: EURATOM.

WOLFE, D. A., SCHELSKE, C. L.: Accumulation of fallout radioisotopes by bivalve molluscs from the lower Trent and Neuse rivers. In: Symposium on radioecology, CONF-670503, p. 493–504, 1967. (See NELSON and EVANS, 1967.)

WRIGHT, S. R., L X-ray measurement of actinide metabolism using ^{243}Am in mice. Int. J. Appl. Radiat. Isotopes **23**, 423–429 (1972).

ZACHARIASEN, W. H.: Crystal chemistry of the 5f elements. In: The actinide elements (G. T. SEABORG and J. J. KATZ, eds.). National Nuclear Energy Series Division IV, vol. 14A, p. 769–796. New York: McGraw-Hill Book Co. 1954.

ZALIKIN, G. A., MOSKALEV, YU. I., PETROVICH, I. K.: Distribution and biological effects of americium-241. Radiobiologiya 8, 65–71 (1968). English translation AEC-tr-6950.

ZALIKIN, G. A., MOSKALEV, YU. I., PETROVICH, I. K., RUDNITSKAYA, E. I.: Biological effects of americium-241. Radiobiologiya **9**, 599–603 (1969). English translation: AEC-tr-7109.

ZESENKO, A. YA.: Distribution of radioactive substances in marine organisms and their associations. In: Vosprosy biookeangrafii, p. 210–218, 1966. (See IVANOV, 1966.)

ZLOBIN, V. S., MOKANER, O. V.: Mechanisms of the accumulation of ^{239}Pu and ^{210}Po by the brown alga *Ascophyllum nodosum* and marine phytoplankton. Radiobiologiya **10**, 584–589 (1970). English translation: AEC-tr-7205.

Chapter 19

Maximum Permissible Concentrations and Maximum Permissible Body Burdens for Transplutonic Elements

G. W. Dolphin

I. Introduction

The internationally recognised authority for setting permissible standards for radiation exposure is the International Commission on Radiological Protection (ICRP) which functions under the auspices of the International Congress on Radiology. The Commission was originally set up in 1928 to make recommendations on radiation safety standards for X-radiation. After the development of nuclear energy during the fifth decade of this century, the Commission widened its scope to include recommendations of safety standards for exposure to radionuclides which might become incorporated and at its meeting in London (ICRP, 1951) in 1950 set up several sub-committees to deal with the rapid growth of the use of radiation and radionuclides. One of these sub-committees was set up to deal with permissible doses for internal radiation, known as Committee 2.

In 1950, before Committee 2 was established, the Commission was not in a position to make firm recommendations for maximum permissible amounts and concentrations for a large number of radionuclides. However, it did deal with a few, namely radium-226, plutonium-239, strontium-89 and 90, polonium-210, hydrogen-3, carbon-14 as carbon dioxide in air, sodium-34, phosphorus-32, cobalt-60 and iodine-131. The Commission also drew attention to data on maximum permissible exposures to radionuclides for occupationally exposed workers in the U.S.A., Canada and the U.K. In 1953 the Sub-Committee on Permissible Internal Doses of the National Committee on Radiation Protection (NCRP, 1953) published its first report. Recommendations were made in this report for the maximum permissible body burden (MPBB) of americium-241 and curium-242 in the whole body and the maximum permissible concentrations in air and water, $(MPC)_a$ and $(MPC)_w$. The biological data on which the calculations of MPBB and MPC were based came from the animal work carried out by Hamilton and his colleagues (Hamilton, 1947, 1948; Scott et al., 1948, 1949), which is referred to in detail later.

Sub-Committee 2 published its first report in 1955 (ICRP, 1955). The only references to transplutonic radionuclides in this report were to americium-241 and curium-242. The MPBB and MPC values in this report were only marginally different from the values given in the earlier NCRP publication (NCRP, 1953) because both committees used the same basic methods of calculating the radiation standards from the biological data reported by Hamilton and his co-workers.

The current report of Sub-Committee 2 published in 1959 as ICRP Publication 2 (ICRP, 1959) gives a more comprehensive treatment of the transplutonic elements including two isotopes of americium, five of curium, two of californium and berkelium-249. Supplementary data were published by Committee 2 in 1962 (ICRP, 1962) for isotopes of americium, curium, berkelium and californium which were not previously considered and additional data on new elements einsteinium and fermium. However the biological basis for these recommended MPBB and MPC values was still the experiments of HAMILTON and his colleagues. An account of the methods used to derive the MPBB and MPC values given in Publication 2 is given in the next section.

During the last three years Sub-Committee 2 has been working on a new report and this should be available in 1973 or 1974. This new report will take into account more recent metabolic data from animal experiments and use new models for transfer of radionuclides from the gut and lungs to the blood and body tissues. Some comments on this forthcoming publication are given in the final section of this chapter.

II. Maximum Permissible Body Burdens (MPBB)

At present the basic standards in radiological protection of occupationally exposed workers to radionuclides are based on the radiation dose to the whole body or to organs within the body (ICRP, 1966). Briefly, these dose standards are 5 rem/year to the gonads and red bone marrow (and in the case of uniform irradiation, the whole body), 30 rem/year to skin, thyroid and bone, 75 rem/year to the hands, forearms, feet and ankles and 15 rem/year to all the other body organs. From these basic standards of annual dose the quantity of radionuclide permissible in the body may be calculated. This quantity, in μCi, is usually expressed in terms of the total body burden, q, as given by the equation:

$$q = \frac{2.8 \times 10^{-3}\, m\, R}{f_2\, E} \tag{1}$$

where m is the mass of the organ in grams, R is the permissible dose to the organ in rem per week, E is the effective absorbed energy in the organ per disintegration of the radionuclide in MeV, and f_2 is the fraction of the total amount of radionuclide in the body which is in the organ of reference.

A. Biological Parameters

In equation (1) there are two biological parameters; m, the mass of the organ of reference and f_2, the fraction in the organ of the total amount of radionuclide in the body. Values of m are given in a table of organ masses for a standard adult in ICRP Publication 2 (ICRP, 1959) and for example, the skeletal mass is 7 kg, the liver 1.7 kg and the kidney 300 g. Values of f_2 must be obtained by reference to animal experiments as there are no data for the metabolism and distribution of transplutonic radionuclides in man.

In ICRP Publication 2 (ICRP, 1959) values of f_2 for americium and curium have been obtained from the results of animal experiments as reported by HAMILTON and his colleagues (HAMILTON, 1947, 1948; SCOTT et al., 1948, 1949). Some results of their work on the initial and final distribution of americium-241 and curium-242 injected into rats are summarised in Table 19.1. From these data the initial transfer from blood to bone, liver and kidney were derived by ICRP Committee 2 and the chosen values are given in column 5 as a fraction, f_2' of the amount injected into the animal. It is noted that the values of f_2' in Table 19.2 for liver are lower than those measured after one day post injection of the animals.

Table 19.1. Summary of the distribution among body organs of rats following injection of americium-241 and curium-242 as chlorides and the values chosen for f_2' and f_2 in Publication 2 (ICRP, 1959)

Nuclide and compound	Organ of reference or excreta	Deposition in organs % injected amount per organ		Chosen value f_2', fraction from blood to organ of reference	f_2, fraction in organ of reference of that in total body for stable isotope
		1 day	256 days		
^{241}Am chloride	bone	23.2	24.8	0.25	0.9
	liver	54.8	0.68	0.35	0.06
	kidney	4.43	0.34	0.03	0.04
	urine	5.53	6.02		
	faeces	4.73	66.4		
^{241}Cm chloride	bone	25.1	21.9	0.3	0.9
	liver	59.6	1.1	0.4	0.05
	kidney	2.83	0.47	0.02	0.02
	urine	1.71	6.04		
	faeces	2.85	69.6		

No explanation for this choice by ICRP Committee 2 is available to the author. If the metabolism of americium and californium in man is the same as that in rats then for radiological protection purposes the liver is only important for short-lived radionuclides because it quickly loses its initial content of radionuclide by excretion, via the bile and into the faeces. More recent work with dogs (LLOYD et al., 1970) shows that americium is retained in the liver for many years and is not excreted in significant amounts. The values of f_2 for the stable isotopes, if they existed for the transplutonic elements, are obtained from the distribution of the radionuclide found at the end of the experiment, 256 days post injection, given in column 4 of Table 19.2. For both radionuclides about 90% of the retained activity is in the bone with a few percent in the liver and kidneys. Although the experiment lasted for only 256 days the retention curve at this time was changing very little and hence these values are taken to represent the distribution pattern following chronic intake of the hypothetical stable element.

In order to calculate the MPBB from equation (1) a value for f_2, the fraction of radionuclide in the organ relative to that in the total body is required. Generally this fraction is difficult to obtain from experimental data and some approximations have to be made but fortunately the fraction is not required in the calculation of MPC values. For americium and curium f_2 was obtained from the animal data in the following way. For example, the fraction, f_2, for bone may be defined as the ratio of the time integral of activity (μCi-days) in the bone to that in the total body. Following a unit injection of activity into the blood this fraction may be represented by the equation:

$$f_2 = \frac{T^{\text{bone}}}{T^{\text{body}}} f_2' \tag{2}$$

where T^{bone} and T^{body} are the effective half-lives of the radionuclide in the bone and total body respectively. This equation is an approximation in that the fraction of the injected activity rapidly excreted in the urine has been neglected but this was small in the animal experiments, as shown in Table 19.1. For stable elements equation (2) may be modified by substituting the biological half-lives for the effective half-lives:

$$f_2 = \frac{T_b^{\text{bone}}}{T_b^{\text{body}}} f_2' \tag{3}$$

Table 19.2. The permissible amount in body organs of reference and MPBB for americium, berkelium, californium, einsteinium and fermium. [Calculated using data from ICRP Publication 2 (ICRP, 1959) and Publication 6 (ICRP, 1962)]

Radio-nuclide	Organ of reference and mass (g) m	Fraction in organs of reference to that in total body (f_2)	Effective energy (MeV) E	Permissible amount in the organ of reference (f_2q) μCi	Maximum permissible body burden (q) μCi
^{241}Am	bone 7×10^3	0.71	280	0.042	0.059
^{242m}Am	bone 7×10^3	0.54	304	0.039	0.072
^{242}Am	bone 7×10^3	0.25	340	0.035	0.14
^{243}Am	bone 7×10^3	0.88	270	0.043	0.049
^{244}Am	bone 7×10^3	0.25	250	0.047	0.19
^{241}Am	liver 1.7×10^3	0.07	57	0.025	0.36
^{242m}Am	liver 1.7×10^3	0.078	57	0.025	0.32
^{242}Am	liver 1.7×10^3	0.35	63	0.023	0.066
^{243}Am	liver 1.7×10^3	0.06	54	0.026	0.43
^{244}Am	liver 1.7×10^3	0.35	20	0.071	0.12
^{241}Am	kidney 300	0.04	57	0.0044	0.10
^{242m}Am	kidney 300	0.037	60	0.0042	0.10
^{242}Am	kidney 300	0.03	65	0.0039	0.10
^{243}Am	kidney 300	0.04	54	0.0047	0.12
^{244}Am	kidney 300	0.03	44	0.0057	0.14
^{242}Cm	bone 7×10^3	0.3	400	0.029	0.096
^{243}Cm	bone 7×10^3	0.39	300	0.039	0.10
^{244}Cm	bone 7×10^3	0.35	300	0.039	0.11
^{245}Cm	bone 7×10^3	0.9	280	0.042	0.047
^{246}Cm	bone 7×10^3	0.9	280	0.042	0.047
^{247}Cm	bone 7×10^3	0.91	270	0.044	0.048
^{248}Cm	bone 7×10^3	0.91	2300	0.0051	0.0056
^{249}Cm	bone 7×10^3	0.3	27	0.44	1.4
^{242}Cm	liver 1.7×10^3	0.38	78	0.018	0.047
^{243}Cm	liver 1.7×10^3	0.12	60	0.024	0.20
^{244}Cm	liver 1.7×10^3	0.16	60	0.024	0.15
^{245}Cm	liver 1.7×10^3	0.05	56	0.025	0.50
^{246}Cm	liver 1.7×10^3	0.05	56	0.025	0.50
^{247}Cm	liver 1.7×10^3	0.05	55	0.026	0.52
^{248}Cm	liver 1.7×10^3	0.05	470	0.003	0.06
^{249}Cm	liver 1.7×10^3	—	—	—	—
^{242}Cm	kidney 300	0.02	78	0.0032	0.16
^{243}Cm	kidney 300	0.02	60	0.0042	0.21
^{244}Cm	kidney 300	0.02	60	0.0042	0.21
^{245}Cm	kidney 300	0.02	56	0.0045	0.22
^{246}Cm	kidney 300	0.02	56	0.0045	0.22
^{247}Cm	kidney 300	0.02	55	0.0045	0.22
^{248}Cm	kidney 300	0.02	470	0.00053	0.026
^{249}Cm	kidney 300	—	—	—	—
^{249}Bk	bone 7×10^3	0.8	20	0.59	0.73
^{250}Bk	bone 7×10^3	0.8	290	0.041	0.051
^{249}Cf	bone 7×10^3	0.8	300	0.039	0.049
^{250}Cf	bone 7×10^3	0.8	310	0.038	0.047
^{251}Cf	bone 7×10^3	0.88	290	0.041	0.047
^{252}Cf	bone 7×10^3	0.8	1100	0.011	0.014
^{253}Cf	bone 7×10^3	0.8	370	0.032	0.040
^{254}Cf	bone 7×10^3	0.8	18900	0.0006	0.0007

Table 19.2 (continued)

Radio-nuclide	Organ of reference and mass (g) m	Fraction in organs of reference to that in total body (f_2)	Effective energy (MeV) E	Permissible amount in the organ of reference (f_2q) μCi	Maximum permissible body burden (q) μCi
^{253}Es	bone 7×10^3	0.8	370	0.032	0.040
^{254m}Es	bone 7×10^3	0.8	660	0.018	0.022
^{254}Es	bone 7×10^3	0.8	620	0.019	0.023
^{255}Es	bone 7×10^3	0.8	380	0.031	0.039
^{254}Fm	bone 7×10^3	0.8	660	0.018	0.022
^{255}Fm	bone 7×10^3	0.8	380	0.031	0.039
^{256}Fm	bone 7×10^3	0.8	18000	0.0007	0.0009

where the subscript b indicates the biological half-lives of the parameters for bone and total body.

For all transplutonic elements the biological half-life for retention in bone is taken to be 3.7×10^4 days; in the absence of other information this was chosen to be the same as the value given by LANGHAM (1957) for plutonium retention in the human skeleton. By substitution of this value in equation (3) and values for f_2' and f_2 from Table 19.1 a value for retention of the element in the total body is obtained. This is transformed into the effective half-life of retention of the radioisotopes in the total body by substitution in equation (4) and this is used to calculate f_2 in equation (2).

It should be noted that in calculations for radiological protection purposes, given in Publication 2 (ICRP, 1959), the retention of a radionuclide in the body is assumed to be represented by a single exponential which is sometimes contrary to experimental data where the retention function may have several exponential terms. If the retention is multi-exponential then the fraction of the initial uptake to blood reaching an organ of reference is obtained by extrapolating back to the exponential component with the longest half-life, although this procedure does not appear to have been followed in Publication 2 for obtaining the liver retention from the experimental data of HAMILTON and his colleagues for americium-241 and curium-242.

As retention is assumed to be exponential there is a simple relationship between effective (T), biological (T_b) and radioactive (T_r) half-lives as given by the following equation:

$$\frac{1}{T} = \frac{1}{T_b} + \frac{1}{T_r}. \tag{4}$$

This equation is used to obtain the effective half-lives required in equation (2) from the biological half-life in the total body obtained from equation (3).

B. Effective Energy

The physical quantity required in evaluating equation (1) is the effective energy of disintegration. This is obtained by multiplying the energy per disintegration by some or all of the following five factors as appropriate.

1. The Relative Biological Effectiveness (RBE) of the Radiation

In Publication 2 (ICRP, 1959) the values assigned to RBE were 10 for alpha particles and 1 for beta and gamma rays. In more recent publications of ICRP (ICRP, 1962, 1966) the RBE has been replaced by a Quality Factor (QF) but there has been no change in the numerical values of the multiplying factors for the various types of radiation.

2. The Relative Damage Factor, n, for Radionuclides Deposited in Bone

This term was introduced for radiological protection purposes to account for the observation that, in animal experiments, some radionuclides deposited in bone produce more damage than radium for the same radiation dose, even when corrected for RBE. This is probably due to the different patterns of deposition in the bone, for radium follows the metabolic pathways of calcium and enters the inorganic bone crystal, whereas plutonium and the transplutonic elements are deposited on bone surfaces close to osteogenic cells from which osteosarcomas may arise. The relative damage factor has the value 5 for particulate radiation from all radionuclides other than radium isotopes.

3. Absorbed Fraction

A factor was introduced to take account of the fraction of the total energy released per disintegration which was absorbed in the organ containing the radionuclide. For particulate radiation this was taken as 100%, for gamma radiation in small organs, such as the kidney, most of the emitted radiation energy is not deposited inside the organ so that the fraction is small. Methods were given in Publication 2 (ICRP, 1959) for calculating approximate values of absorbed fraction. Recently Monte Carlo methods have been used and more exact values are available (Snyder et al., 1969; Ellett and Humes, 1971).

4. Fraction of Energy Released from Daughters

In the case of radionuclides having radioactive daughters with half-lives comparable with 50 years, the expected working life of a radiation worker, allowance is made for the energy absorbed in the reference organ from decay of these radioactive daughters. For example, curium-242 decays with a half-life of 162 days to plutonium-238 with a half-life of 3.3×10^4 days. The energies of alpha disintegration of curium-242 and plutonium-238 are 6.3 and 5.7 MeV, respectively. For purposes of calculating the MPBB from equation (1) it is assumed that the daughter product, plutonium-238, remains at the site of production in the bone and that 29% of the plutonium-238 activity produced by decay will decay during 50 years. Hence the effective decay energy of curium-242 is $6.3 + 5.7 \times 0.29 = 7.95$ MeV. To obtain the effective energy for curium-242 deposited in bone this value must be multiplied by an RBE factor of 10 and a relative damage factor of 5 making an effective energy of decay of 297.5 MeV.

5. Spontaneous Fission

Some isotopes of curium, californium, einsteinium and fermium undergo spontaneous fission and consequently the effective energy of disintegration used in the calculations takes into account the kinetic energy of the fission fragments, the instantaneous gamma-ray energy and the kinetic energy of fission neutrons. The total energy per disintegration is about 200 MeV which is considerably higher than for any other type of decay (ICRP, 1962).

III. MPBB and the Critical Organ

Values for the permissible amount in the body organs, $f_2 q$, may be calculated from equation (1) and these are given in column 5 of Table 19.2 and the corresponding values of MPBB are given in column 6. The biological data in column 3 for the americium and curium isotopes have been derived as described in a previous section from the experimental data given in Table 19.1. Only three body organs are considered as these were found to be the important sites of retention in the animal experiments. However, there were no appropriate biological data available for berkelium, californium, einsteinium and fermium when Publication 2 (ICRP, 1959) and Publication 6 (ICRP, 1962) were prepared so the value of f_2 given for the only organ of reference considered, bone, was obtained by comparison with plutonium-239.

For radiological protection purposes the critical organ is taken as that where radiation damage causes the greatest biological effect on the whole body. Although other factors such as essentialness of the organ to the body and the radio-sensitivity of an organ could in principle affect the choice of critical organ, the only criterion for the choice in practice is the highest concentration of radionuclide per gram of tissue. For the long-lived transplutonic radionuclides the critical organ is the bone as the MPBB calculated with the bone as the organ of reference is the most restrictive. This holds for all the radionuclides in Table 19.2 with the exception of americium-242 and curium-242. These radionuclides have short half-lives; americium-242 decays with a half-life of 0.67 days to curium-242 which has a half-life of 163 days. These short half-lives make the liver the organ associated with the lowest MPBB and hence the critical organ.

IV. Maximum Permissible Concentrations

The primary objective of radiological protection is to ensure that biological damage from exposure to ionising radiation is kept to an acceptable amount. To achieve this objective upper limits of environmental contamination are recommended at which a person may be exposed continuously without suffering or accumulating unacceptable amounts of damage. These upper limits are termed maximum permissible concentrations and are derived for air and water $(\mathrm{MPC})_a$ and $(\mathrm{MPC})_w$, respectively. Two values are derived for each radionuclide corresponding to two possible types of exposure, 40 hours per week for occupational exposure and 168 hours per week for continuous exposure. The MPC is derived in such a way that continuous exposure would eventually lead to the recommended dose limit, 5, 15, or 13 rem per year, depending on which organ is critical. For long-lived radionuclides in the bone the dose limit of 30 rem per year will only be reached after 50 years of continuous exposure.

The values of MPC are calculated on the basis that the elimination of radionuclides from a body organ both by biological pathways and by radioactive decay follows a simple exponential law. The amount of radionuclide in an organ which delivers the annual dose limit is $q f_2$ as defined in equation (1). The rate of elimination from the organ is determined by the effective half-life T and the corresponding elimination constant. If P, in microcuries per day, is the rate of entry of a radionuclide into the organ, then the rate of change of the amount of radionuclide in the organ is:

$$\frac{d}{dt}(q f_2) = p - \lambda q f_2. \tag{5}$$

The solution of this equation for the conditions that $qf_2=0$ at $t=0$ is:

$$qf_2=\frac{P}{\lambda}(1-e^{-\lambda t})$$

and

$$P=\frac{qf_2\lambda}{(1-e^{-\lambda t})}. \tag{6}$$

From equation (6) the value of P can be defined in terms of the permissible organ content, qf_2, the effective elimination constant λ for the organ, and t is taken as 18250 days which represent exposure over a working life, 50 years. The parameter P can be expressed in terms of the MPC for a radionuclide in air or drinking water by use of simple models to represent the fraction transferred to the organ of reference from inhaled air, f_a, and from ingested water, f_w. The model for the transfer of soluble radionuclides from the air inhaled into the lung is that 75% is deposited on lung surfaces and 25% exhaled immediately. Of the 75% deposited, one-third passes through the lung walls into the body fluids and two-thirds are transported up the ciliated epithelium of the larger airways to the throat and swallowed. Some of the 50% swallowed may be absorbed by the blood from the gut and this fraction is denoted by f_1. This model is modified for insoluble radionuclides and 12.5% is retained in the pulmonary region of the lung with a half-life of 120 days for all radionuclides except for plutonium where the half-life is taken as one year. In this model for insoluble radionuclides the ultimate fate of the material transferred from the lung to the body fluids is neglected for radiological protection purposes. From the inhalation model for soluble material the value of the fraction transferred from air to the organ of reference is given by the equation:

$$f_a=(0.25+0.5\,f_1)\,f_2' \tag{7}$$

where f_2' is the fraction transferred from the blood to the organ of reference.

For ingestion the fraction transferred to the organ from water is the product of the fraction crossing the gut wall into the blood, f_1, and the fraction transferred from the blood to the organ of reference, as in the equation below:

$$f_w=f_1f_2'. \tag{8}$$

For the intake by inhalation or ingestion of insoluble radionuclides the lung or the gut are the only possible critical organs as might be expected because, in principle, no insoluble material is transferred into the body. The model for retention of insoluble material in the lung has been referred to above and the model to represent passage of material through the gut is too complex for discussion in this chapter. The reader is referred to Publication 2 (ICRP, 1959), and papers by Eve (1966) and Dolphin and Eve (1966, 1968) for comprehensive reviews of the radiological protection aspects of passage of radionuclides through the gut.

Let the daily intake of air and water be D_a and D_w cm³ respectively, then the amount of radionuclide transferred per day from the air or water to the organ of reference is given:

$$P=f_aD_a\,(\mathrm{MPC})_a \tag{9}$$

or

$$P=f_wD_w\,(\mathrm{MPC})_w. \tag{10}$$

A standard adult inhales about 2×10^7 cm³ of air per day and it is assumed that half this amount is inhaled during the working day of eight hours. His water intake is taken to be 2200 cm³ per day and half this amount during the working period. In calculating the average amount inhaled per working day allowance is made for a five-day working week and a fifty-week working year, so that the daily intake averaged throughout the year is obtained by multiplying by the factors $^{50}/_{52}$ and $^{5}/_{7}$.

The above values of air and water consumption for a standard adult during working hours together with values for P from equations (9) and (10) are substituted into equation (6) as follows:

$$(\mathrm{MPC})_a = \frac{10^{-7}\, q f_2}{T\, f_a\, (1 - \mathrm{e}^{-\lambda t})}, \tag{11}$$

$$(\mathrm{MPC})_w = \frac{9.2 \times 10^{-4}\, q f_2}{T\, f_w\, (1 - e^{-\lambda t})}. \tag{12}$$

The values of MPC in μCi per cm³, calculated from equations (11) and (12) are for a forty-hour working week and must be divided by 2.92 to obtain the appropriate values for continuous exposure.

Values for $(\mathrm{MPC})_a$, in μCi per cm³, from Publication 2 (ICRP, 1959) and Publication 6 (ICRP, 1962) are given in Table 19.3 for occupational exposure to soluble and insoluble forms of the transplutonic elements. The most restrictive

Table 19.3. $(\mathrm{MPC})_a$ for 40 hours exposure per week to soluble and insoluble radioactive material as given in Publication 2 (ICRP, 1959) and Publication 6 (ICRP, 1962)

Radionuclide	$(\mathrm{MPC})_a$ for organs of reference (μCi/cm³)			
	soluble		insoluble	
	bone	liver	lung	gut
^{241}Am	6×10^{-12}	9×10^{-12}	1×10^{-10}	1×10^{-7}
^{242m}Am	6×10^{-12}	9×10^{-12}	3×10^{-10}	5×10^{-7}
^{242}Am	8×10^{-8}	4×10^{-8}	5×10^{-8}	7×10^{-7}
^{243}Am	6×10^{-12}	9×10^{-12}	1×10^{-10}	1×10^{-7}
^{242}Cm	2×10^{-10}	1×10^{-10}	2×10^{-10}	1×10^{-7}
^{243}Cm	6×10^{-12}	1×10^{-11}	1×10^{-10}	1×10^{-7}
^{244}Cm	9×10^{-12}	1×10^{-11}	1×10^{-10}	1×10^{-7}
^{245}Cm	5×10^{-12}	8×10^{-12}	1×10^{-10}	1×10^{-7}
^{246}Cm	5×10^{-12}	8×10^{-12}	1×10^{-10}	1×10^{-7}
^{247}Cm	5×10^{-12}	9×10^{-12}	1×10^{-10}	1×10^{-7}
^{248}Cm	6×10^{-13}	1×10^{-12}	1×10^{-11}	7×10^{-9}
^{249}Cm	1×10^{-5}	—	5×10^{-5}	1×10^{-5}
^{249}Bk	9×10^{-10}	—	1×10^{-7}	3×10^{-6}
^{250}Bk	1×10^{-7}	—	2×10^{-6}	1×10^{-6}
^{249}Cf	2×10^{-12}	—	1×10^{-10}	1×10^{-7}
^{250}Cf	5×10^{-12}	—	1×10^{-10}	1×10^{-7}
^{251}Cf	2×10^{-12}	—	1×10^{-10}	1×10^{-7}
^{252}Cf	6×10^{-12}	—	3×10^{-11}	4×10^{-8}
^{253}Cf	8×10^{-10}	—	8×10^{-10}	7×10^{-7}
^{254}Cf	5×10^{-12}	—	5×10^{-12}	6×10^{-10}
^{253}Es	8×10^{-10}	—	6×10^{-10}	1×10^{-7}
^{254m}Es	5×10^{-9}	—	6×10^{-9}	1×10^{-7}
^{254}Es	2×10^{-11}	—	1×10^{-10}	7×10^{-8}
^{255}Es	5×10^{-10}	—	4×10^{-10}	1×10^{-7}
^{254}Fm	6×10^{-8}	—	7×10^{-8}	6×10^{-7}
^{255}Fm	2×10^{-8}	—	1×10^{-8}	1×10^{-7}
^{256}Fm	3×10^{-9}	—	2×10^{-9}	5×10^{-9}

$(MPC)_a$ values are for bone as the organ of reference except for the short-lived radionuclides, curium-242 and americium-242, where the most restrictive $(MPC)_a$ is for the liver. Due to the lack of metabolic data on the higher radionuclides only values for bone as the organ of reference have been calculated.

For inhalation of insoluble compounds the lung is the critical organ in most cases and the $(MPC)_a$ is calculated from equation (1) and (11) using $m = 1000$ g, $T_b = 120$ days, $f_a = 0.125$ and $R = 0.3$ rem/week (15 rem per year). The gut is the critical organ for curium-249 and this is due to its very short half-life of 64 minutes.

There is little transfer of transplutonic elements across the gut wall; in Publication 2 (ICRP, 1959) f_1 is less than 10^{-4} for americium and curium as obtained by comparison with the rare earth elements and equal to 3×10^{-5} for berkelium, californium, einsteinium and fermium as obtained by comparison with plutonium. These low values indicate that large quantities of the element could pass through the gut without significant transfer to the body organs, and in particular, to the bone. Experimental work with animals fed large amounts of alpha emitting radionuclides show that little biological damage is caused by radiation to the intestinal walls (KATZ et al., 1955; WEEKS et al., 1956). This is probably due to the mucosal layer lining the gut (EVE, 1966) which is thick enough to absorb most of the alpha energy, and in Publication 2 (ICRP, 1959) only 1% of the alpha energy was taken to be effectively absorbed in the cells of the gut wall. As a consequence of this low value of f_1 and the low effective energy of alpha emitters, the $(MPC)_w$ values given in Publication 2 (ICRP, 1959) are relatively large for transplutonic elements mainly lying between 10^{-3} and 10^{-4} μCi/cm^3 of water for occupational exposure.

V. Future Developments

The values of MPBB and MPC are derived from the equations given in the previous sections and these equations are developed from the fundamental concepts of radiological protection. It is changes in these which may affect the presently accepted values and possible changes are discussed below.

A. Radiation Dose Standards

In radiological protection the radiation dose of interest has always been the average dose to the whole body, or to the body organs or to the body tissues, and currently the average doses are 5, 15, and 30 rem/year. These have evolved over many years from experience and observation of exposed individuals and populations, and also from a large number of animal experiments. However, at the present time there is some criticism of these values and they are under review (ICRP, 1969) and changes may occur in the future.

B. Dosimetric Models

The average radiation dose to the body or organs may not be the most appropriate dose in considering the consequences of irradiation from radionuclides deposited in the skeleton. Recently, attention has focused on the osteogenic cells (SPIERS, 1970) on bone surfaces; the cells from which osteosarcomas are thought to originate. Models for skeletal deposition and bone geometry are now being developed by a Task Group of ICRP (MARSHALL, 1970) and these may eventually lead to meaningful dose calculations to the osteogenic cells from alpha-particles emitted by transplutonic nuclides deposited in the skeleton.

C. Metabolic Models

It is clear from the earlier section in this chapter on biological parameters that more recent biological data on the translocation and deposition of americium and curium should be incorporated in the ICRP recommendations. For example, a long-term study of the retention and possible pathological effects of americium chloride injected into beagles has started at the University of Utah. Preliminary data have been reported by STEVENS et al. (1969) and LLOYD et al. (1970). At the Lovelace Foundation, Albuquerque, a study has started on the effects of inhaled americium particulate and preliminary data have been reported (THOMAS, 1970). These preliminary data show that liver retention in beagles is considerably higher than in rats and the initial deposition is also higher at about 50% compared with 40% in skeleton. Data on liver retention in rats are not applicable to humans for it is now known that rats rapidly clear some radionuclides into the bile and faeces. In the future, if the MPBB and MPC values are calculated using data from experiments with beagles, the liver will assume a greater importance and will certainly be the critical organ for the shorter-lived radionuclides, i.e. those radionuclides with a half-life less than about one hundred years.

D. Lung Model

The lung model used in Publication 2 (ICRP, 1959) is very simple; this is not always a disadvantage in radiological protection. Recently a Task Group of ICRP (TGLD, 1966) proposed a more complex model for deposition and retention of inhaled particulates. This model takes into account particle size distribution and its effect on deposition, and insoluble particulate is taken to be retained with a half-life of about 500 days rather than the lower value of 120 days used previously. These proposed changes will lead to more restrictive values of MPC.

E. Gut Model

A new model has been proposed for the dosimetry of the gut (DOLPHIN and EVE, 1966). This has not yet been introduced into ICRP recommendations but its use will lead to values very similar to those already in use as recommended in Publication 2 (ICRP, 1959).

At present there is every indication that the recommended values of MPC and MPBB will become more restrictive. However, if more restrictive values are necessary, it is hoped that they are introduced as a result of greater scientific understanding and not from pressure of uninformed people incited by misguided individual scientists. For we must never over-look the tremendous benefits which will fall to society from the careful exploitation of plutonium and other radionuclides as controlled power sources.

References

DOLPHIN, G. W., EVE, I. S.: Dosimetry of the gastrointestinal tract. Hlth Phys. **12**, 163–172 (1966).

DOLPHIN, G. W., EVE, I. S.: Some aspects of the radiological protection and dosimetry of the gastrointestinal tract. In: Gastrointestinal radiation injury, p. 465–477, ed. M. F. Sullivan. Amsterdam: Excerpta Medica Foundation 1968.

ELLET, W. H., HUMES, R. M.: Absorbed fractions for small volumes containing photon emitting radioactivity. J. nucl. Med., Suppl. **5**, 25–32 (1971).

EVE, I. S.: A review of the physiology of the gastrointestinal tract in relation to radiation dose from radiactive materials. Hlth Phys. **12**, 131–161 (1966).

Hamilton, J. G.: The metabolism of the fission products and the heaviest elements. Radiology **49**, 325–342 (1947).
Hamilton, J. G.: The metabolic properties of the fission and actinide elements. Rev. Mod. Phys. **20**, 718–728 (1948).
International Commission on Radiological Protection: International recommendations on radiological protection. Brit. J. Radiol. **24**, 95–102 (1951).
International Commission on Radiological Protection: Recommendations of the International Commission on Radiological Protection. Brit. J. Radiol., Suppl. **6** (1955).
International Commission on Radiological Protection: Publication 2: Report of Committee 2 on Permissible Dose for Internal Radiation. London: Pergamon Press 1959.
International Commission on Radiological Protection: Publication 6: Recommendations of the International Commission on Radiological Protection. London: Pergamon Press 1962.
International Commission on Radiological Protection: Publication 9: Recommendations of the International Commission on Radiological Protection. London: Pergamon Press 1966.
International Commission on Radiological Protection: Publication 14: Radiosensitivity and spatial distribution of dose. London: Pergamon Press 1969.
Katz, J., Kornberg, H. A., Parker, H. M.: Absorption of plutonium fed chronically to rats. I. Fraction deposited in skeleton and soft tissues following oral administration of solutions of very low mass concentrations. Amer. J. Roentgenol. **73**, 303–308 (1955).
Langham, W. H.: The application of excretion analyses to the determination of body burden of radioactive isotopes. Brit. J. Radiol., Suppl. **7**. 95–113 (1957).
Lloyd, R. D., Mays, C. W., Taylor, G. N., Atherton, D. R.: Americium-241 studies in beagles. Hlth Phys. **18**, 149–156 (1970).
Marshall, J. H.: A comprehensive model of alkaline earth metabolism in bone: A preliminary report. Argonne National Laboratory Radiological Physics Division Annual Report, ANL-7760, p. 95–109 (1970).
National Committee on Radiation Protection: Maximum permissible amounts of radioisotopes in the human body and maximum permissible concentrations in air and water. Handbook 52. Washington, D.C.: National Bureau of Standards 1953.
Scott, K. G., Axelrod, D. J., Hamilton, J. G.: The metabolism of curium in the rat. J. biol. Chem. **177**, 325–335 (1949).
Scott, K. G., Copp, D. H., Axelrod, D. J., Hamilton, J. G.: The metabolism of americium in the rat. J. biol. Chem. **175**, 691–703 (1948).
Spiers, F. W.: Absorbed dose to the cells of bone. In: Radiation dosimetry, p. 213–244, ed. I. Mirić. Belgrade: Boris Kidrič Institute of Nuclear Sciences 1971.
Stevens, W., Stover, B. J., Bruenger, F. W., Taylor, G. N.: Some observations on the deposition of americium-241 in the thyroid gland of the beagle. Radiat. Res. **39**, 201–206 (1969).
Synder, W. S., Fisher, H. L., Ford, M. R., Warner, G. G.: Estimates of absorbed fractions for monoenergetic photon sources uniformly distributed in various organs of a heterogeneous phantom. J. nucl. Med., Suppl. **3**, 7–52 (1969).
Task Group on Lung Dynamics: Deposition and retention models for internal dosimetry of the human respiratory tract. Hlth Phys. **12**, 173–208 (1966).
Thomas, R. G.: Estimation of ^{241}Am body burdens by analysis of whole-body scanning, excreta and body weight data. Hlth Phys. **19**, 751–755 (1970).
Weeks, M. H., Katz, J., Oakley, W. D., Ballou, J. E., George, L. A., Bustad, L. K., Thompson, R. C., Kornberg, H. A.: Further studies on the gastrointestinal absorption of plutonium. Radiat. Res. **4**, 339–345 (1956).

Chapter 20

Bioassay of Transplutonium Elements

Anne de G. Low-Beer

I. Introduction

Since the beginning of the atomic age the overriding concern in bioassay programs has been the detection of uranium and plutonium. These nuclides have presented the chief danger of internal contamination of workers in the atomic energy industry. The irradiation of uranium and plutonium produces transplutonium elements amounting to only a few percent of the target material. These elements have, therefore, constituted a relatively small hazard incidental to the far greater danger of contamination from uranium and plutonium. Indeed, concern for the transplutonium elements as potential hazards to workers was first engendered by reports of accidental contamination with americium by Foreman et al. (1958), and by Dobson (1955), and Rowland et al. (1955). The case reported by Foreman received his exposure by inhalation. He was treated with a chelating agent (EDTA) and retention of americium was evaluated by radiochemical analysis of excreta, which was carried on for several years. The case reported by Dobson and the Rowland group received his exposure by puncture wound, and is believed to be the first instance in which both radiochemical analysis of excreta and whole body counting were used to measure retention of the isotope.

Today the transplutonium elements are being produced in vastly increased amounts due to their potential usefulness (chiefly americium and californium) in research and medicine (see Chap. 21). Their prominence in bioassay programs of the future is therefore assured. The methodology of bioassay was developed as a means of evaluating human exposure to plutonium and uranium. The general considerations that govern methodology have been described in Chap. 13, and are applicable to the transplutonium elements. However, many of the isotopes of these elements have properties that impose a need for special precautions in monitoring, and in handling samples. Spontaneous fission gives rise to large amounts of high energy gamma radiation and occurs in many isotopes of the transplutonium series (Lederer et al., 1967). Californium-252 produces neutrons amounting to approximately 3.6 per fission (U.S. Atomic Energy Commission, 1968). This property, which makes ^{252}Cf useful as a therapeutic agent, constitutes an additional hazard to involved personnel. Adequate shielding is therefore necessary for safe handling of any material that contains any of these radionuclides. Furthermore, increasing specific activity of the transplutonium isotopes with increasing atomic number, adds to their toxicity as sources of internal contamination. This consideration should have bearing on frequency of sampling for potentially exposed personnel.

Chap. 13 includes a fairly detailed description of instruments used for the detection of radioactivity *in vivo* and *in vitro*. None of this material is repeated here. The present chapter is concerned with radiochemical and detection methods that have been designed to take advantage of the chemical and physical properties of the transplutonium elements that differ from those of plutonium. These relate to the predominantly trivalent state of the transplutonium elements and to the emission of abundant high energy gamma rays by many of them.

II. Chemical Bases for Separation of the Transplutonium Elements

The transplutonium elements were identified by PERLMAN and STREET (1954) as americium, atomic number 95; curium, 96; berkelium, 97; and californium, 98. The chemistry of these elements was elucidated in classical articles by these authors and by SEABORG (1954). Brief reviews of the radiochemistry of these elements are contained in two publications of the Nuclear Science Series of the National Academy of Sciences (U.S.A.E.C.) (1960). Discovery of additional elements in the actinide series has extended the designation of transplutonium elements to all the predominantly trivalent members of the series.

SEABORG et al. (1949) observed that the 5f electron shell of the actinide series was in the process of being filled, and correctly predicted increasing stability of the III state in the actinides which would culminate in a very stable state in elements 95 and 96. The analogy is at once apparent between the trivalent actinide and the lanthanide elements in which the 4f electron shell is being filled. The transplutonium elements in fact must be regarded as homologues of the lanthanides, with chemical properties more similar to each other and to the lanthanides than to the preceding members of the actinide series. For this reason the transplutonium elements are readily separated from more highly oxidized uranium, plutonium, and neptunium; oxidation-reduction reactions are not applicable in the separation of the transplutonium elements from each other[1]; separation of transplutonium elements from lanthanides is difficult.

Early work on separation of the transplutonium elements has been reviewed by THOMPSON et al. (1949), WERNER and PERLMAN (1949), and by HYDE (1954). A review by STARÝ (1966) summarizes more recent progress in this area. The chemical properties that have governed the development of radiochemical procedures for bioassay are reviewed briefly in the following sections. A more comprehensive treatment of this subject will be found in chap. 17 of this volume.

A. Precipitation

The transplutonium elements and plutonium in the reduced state (but not radium or uranium), undergo precipitation in the presence of suitable carriers, with phosphate, oxalate, and fluoride ions. LAWROSKI (1955) and HILL and COOPER (1958) used coprecipitation with bismuth phosphate in early separations of plutonium from uranium. The solubility of TP-phosphate is very low at molar H^+ concentration and is comparable to the solubility of rare earth phosphates. Coprecipitation with oxalate or fluoride ions must also be carried out at low pH to avoid precipitation of divalent ions, and requires a rare earth element, usually cerium or lanthanum, as carrier. In any precipitation method, therefore, complete

1 Americium, although predominantly trivalent in aqueous conditions, can be oxidized by very strong reagents. MOORE (1963) has shown that peroxidisulfate catalyzed by silver oxidizes americium to VI. In this state americium is not precipitated by fluoride, and this affords a means of separating it from other transplutonium elements, since later members of the series are not oxidizable by any means.

purification of the actinides requires separation from rare earth elements. THOMPSON et al. (1949) showed that in the presence of fluosilic acid precipitation of transplutonium elements with fluoride ion is inhibited, while precipitation of rare earth elements is nearly complete. By this means 90% group separation of transplutonium and lanthanide elements has been accomplished. SILL (1969) found that all of the actinide elements can be precipitated on barium sulfate, and that under suitable conditions sequential separation of individual members of the entire actinide series can be accomplished.

Precipitation methods, even when the carrier cation remains, result in preparations from which fission products and other interfering ions have been removed. They may be regarded as a first step in the purification of the transplutonium elements.

B. Ion Exchange Chromatography

Ion exchange is generally regarded as the most efficient means of separating transplutonium elements from lanthanides. Group and intragroup separations can be carried out on cation and anion exchange resins.

1. Cation Exchangers

Recovery of transplutonium elements from cation exchange resins involves first their adsorption by resin through exchange with a replaceable ion or group such as H^+, NH_4^+; and second, their selective desorption through complexation with an appropriate eluant. STREET et al. (1950) showed that transplutonium elements can be eluted as a group by 20% ethanol in 13.3 M hydrochloric acid. Since the transplutonium ions form more stable complexes with Cl^- than the lanthanides, they are eluted first, but differences in stability constants of the chloride complexes of transplutonium elements are not sufficient to effect separation of individual elements. DIAMOND et al. (1954) and SURLS and CHOPPIN (1957) have made group separations by using 1.8 M ammonium thiocyanate as elutant, though less efficiently than with alcoholic 13.3 M hydrochloric acid. Individual separation of transplutonium elements can be accomplished with organic eluting agents because of the marked differences in stability constants of their complexes. HARRIS and TOMKINS (1947), THOMPSON et al. (1949, 1954), and WERNER and PERLMAN (1949) successfully separated americium and curium by elution with 0.25 M citrate at pH 3.05. The slowness of elution by this method has led to its replacement by a number of other organic acids of which alpha-hydroxybutyric acid has been shown by CHOPPIN et al. 1962, GATTI et al. (1959), BRANDSTETR et al. (1964) to be most effective in separating transplutonium elements from each other. AUER-WELSBACH et al. (1967) have reported successful separation of americium and curium from cation exchange columns with alpha-hydroxybutyric acid. The disadvantage of the method is the fact that stability constants of the transplutonium complexes, while differing from one another, are too similar to the stability constants of some of the lanthanide complexes to afford a means of complete separation of transplutonium and lanthanide elements. VOJTĚCH and SPICIN (1961) have shown that this difficulty can be partly overcome by use of organic-aqueous solutions (e.g., dioxan-water, acetone-water), which alter the ratio of stability constants for lanthanide complexes, and thereby increase their separability. Inspection of stability constants of transplutonium elements with a number of complexing agents suggests that aminopolyacetic acids should afford good separation. FUGER (1958) has reported good separation of americium, curium, berkelium, and californium with dilute ethylene-diaminetetraacetic acid (EDTA) at low pH, and diethylenetriaminepentaacetic acid (DTPA) has been used at the Oak Ridge National Laboratory (1961) in the separation of americium from promethium.

2. Anion Exchangers

In the use of anion exchange resins the transplutonium elements are introduced to the column in the form of anionic complexes which are adsorbed on the resin by replacement, and are then eluted selectively by suitable complexing agents. Elements which form the weakest complexes are eluted first, and are followed in order of increasing stability. Since stability of chloride complexes increases with atomic number, elution with 13 M hydrochloric acid elutes americium and curium before berkelium and californium. THOMPSON et al. (1949), WERNER and PERLMAN (1949), HULET et al. (1961) have shown that this effect is increased if 10 M lithium chloride acidified to 0.1 M in hydrochloric acid is substituted for concentrated hydrochloric acid.

MARCUS and GIVAN (1962) have investigated lithium nitrate as an elutant and have obtained a high degree of separation of a mixture of light lanthanides and of the americium-curium pair, but no separation of transplutonium elements from lanthanides was possible. In this system elements are eluted in order of decreasing atomic number—the reverse of the order obtained with lithium chloride, but identical with the elution pattern produced by organic complexing agents in cation exchange resins.

SURLS and CHOPPIN (1957) have found that thiocyanate in anion exchange systems forms less stable complexes with lanthanides than with transplutonium elements. COLEMAN et al. (1956) have used this method to separate gram quantities of americium from large amounts of rare earth elements. The use of organic complexing agents in the elution of anion exchange resins has not been reported.

C. Solvent Extraction

Extraction chromatography affords another means by which differences in stability constants of metal complexes can be exploited to effect the separation of transplutonium elements. The procedure consists in mixing an aqueous solution of metal ions with an orgnaic solvent which contains a suitable complexing agent. The metal complex is thus extracted into the organic phase. This method has the advantage of speed as compared with ion exchange chromatography, and is therefore suitable for routine analysis. It is, however, less exact than ion exchange methods. A large number of organic solvents have been investigated in the development of this method. Only those that are widely used in analytical procedures will be mentioned.

1. Tributylphosphate

Tributylphosphate (TBP) is one of the oldest complexing agents, and was used originally for the purification of plutonium. Its usefulness in separation of transplutonium elements has been established by the work of many investigators. HESFORD et al. (1959) showed that the highest degree of extractability by TBP is obtained from nitric acid with high concentration of nitrate ion, and high concentration of TBP in the organic phase. Subsequent investigation by BEST et al. (1959) as well as studies by GRAY and THOMPSON (1952) and by BRANDSTETR et al. (1964) demonstrated that higher separation factors can be obtained by extracting from 12 M hydrochloric acid than from 13.1 M nitric acid. ISAAC et al. (1960) separated a mixture of americium and californium by this method. The most efficient organic solvent for TBP extractions is decane.

2. Thenoyltrifluoroacetone

Thenoyltrifluoroacetone (HTTA), a β-diketone, is used chiefly in separation of transplutonium elements from uranium, neptunium, plutonium, and fission products. MAGNUSON and ANDERSON (1954) have discussed its applicability to extraction of trivalent ions and have pointed out that extractability is inversely proportional to the third power of H^+ concentration in the aqueous phase. They have also shown that extraction constants of transplutonium elements are diminished in benzene and toluene, the most frequently used organic solvents, while VOJTĚCH and SPICIN (1961) have reported that oxygen-containing solvents increase extraction constants. SCHRECK (1967) has investigated the chelation and extraction of trivalent actinides with various 1–3 beta diketones and has calculated stability constants for these elements under different conditions of chelation.

3. Organophosphoric Compounds

Attention has been directed to the use of neutral organophosphoric compounds as extractants of transplutonium elements by GUREEV et al. (1964) who have investigated a number of such compounds and have concluded that their efficiency in effecting separations is influenced by the following factors: (1) the introduction of positive groups, such as an alkyl radical, enhances extractability—extractability is optimum with a 9 carbon radical chain—beyond this steric hindrance acts to reduce extractability; (2) introduction of a negative group, such as an aromatic group, reduces extractability; (3) extractability decreases in proportion to the number of C-O-P bonds. In the systems investigated by these authors, using decane as the organic solvent, highest extraction constants were found for both americium and curium with the compound $(C_9H_{17})_3PO$. Recently dialkylphosphoric and dialkylphosphonic acids have come into widespread use for group separation of transplutonium elements from lanthanides and for individual separation of transplutonium elements. The compounds that are most extensively employed are di(2 ethylhexyl)phosphoric acid (HDEHP) which is used in a system of toluene-1 M hydrochloric acid for transplutonium separations; and 2-ethylhexylphenylphosphonic acid (HEHØP) in a system of diethylbenzene-2 M hydrochloric acid. The advantages of these compounds as stated by STARÝ (1966) are their high degree of selectivity for the separation of transplutonium elements, and the fact that they can be used at low pH so that difficulties attributable to hydrolysis, which are experienced in other extraction systems, are avoided. PEPPARD et al. (1959) have described the separation of californium, einsteinium, and fermium from americium and curium with HDEHP, and BAYBARZ (1962) reported a separation factor of 100 for californium and curium with HEHØP. BUTLER and HALL (1970) have demonstrated the superior effectiveness of dibutyl N-N-diethylcarbamylphosphonate (DDCP) for the separation of trivalent actinides. Unlike many other extractants, this compound is not pH dependent, and according to these authors its use results in higher recovery and greater precision (see Methods). Separation of individual transplutonium elements by extraction chromatography using HEDHP as the stationary phase has been reported by KOOI et al. (1964), HULET (1964), PIERCE and PECK (1962), SICKIERSKI and SOCHOCKA (1964), BOSHOLM and GOSSE-RUYKEN (1964), BUTLER (1965) HOROWITZ et al. (1969). In such systems the extractant is adsorbed in Celite or diatomaceous earth, and dilute mineral acid is used to effect sequential elution of transplutonium elements.

D. Surface Adsorption

The phenomenon of adsorption of cations on the walls of laboratory glassware is well known. This surface adsorption is greatest at near neutral pH, and while it occurs to some extent with most cations, it is most marked in the case of lanthanide and actinide elements. The phenomenon of surface adsorption has been regarded as a problem to be taken into account in the storage and processing of solutions in glassware. It is mentioned here because it has been successfully exploited by Eakins and Gomm (1968) as a method for the determination of actinide elements in biological materials (*vide infra*). From the work of Lieser (1965), Mellish et al. (1966), and of Benes and Garba (1966), it has been determined: (1) that surface adsorption is independent of concentration up to concentrations of about 10^{-3} M, and that above this adsorption increases with concentration; (2) that adsorption is pH dependent, increasing markedly with increasing pH. These facts suggest that surface adsorption is in part due to an ion exchange mechanism dependent on dissociation of the $\equiv$SiOH groups on the surface of glass, and to the hydrolytic behavior of the adsorbate with increasing pH.

III. Procedures Used in Bioassay

The chemical methods which have been discussed in the preceding section were developed for the most part for relatively large scale separations, and for the detection of high levels of radioactivity. In bioassay the situation is quite different. Chemical procedures require the separation of microgram or submicrogram amounts from very large amounts of other materials; the final preparation must meet the criteria for acceptable counting efficiency for very low levels of alpha activity.

All of the radiochemical procedures employed in bioassay programs depend on utilization of one or more of the principles which have been cited. Most of them, however, involve modifications of method designed to overcome the problems encountered in dealing with biological materials. Thus the radiochemical aspect of bioassay is an eclectic art. Many laboratories have developed procedures satisfactory for their purposes, but which are not used uniformly throughout the laboratories of any one country, much less of the world. Moreover, many of the procedures that are in use have not reached the world literature. For these reasons a questionnaire was sent to a large number of laboratories in preparation for the compilation of this review. The response to this questionnaire was most generous, and the kindness of the many respondents is gratefully acknowledged. Examination of the responses indicates that at this time the transplutonium elements command very little attention in routine bioassay programs. Nevertheless a number of methods for determination of transplutonium elements have been submitted. The following section presents a prototype method illustrating each of the principles embodied in Sec. II. A list of all laboratories that have contributed information pertinent to the transplutonium elements, and a brief description of methods involved, will be found in appendix 1. It will be noted that the methods that are described do not provide for individual separation of transplutonium elements. The need for such individual separation is largely obviated by the capability for identification of radioisotopes by alpha pulse height analysis.

A. Gross Alpha Methods

1. Precipitation (Schubert et al., 1951)[2]

a) Dissolve the cooled pure white ash (see Chap. 13) as completely as possible by adding 5 ml of concentrated HNO_3 followed by 30 ml of distilled water. Rinse the speedy vap and the sides of the beaker with distilled water. Bring the contents of the beaker to boiling and, after cooling decant into a 90-ml pyrex centrifuge tube. If a large residue of undissolved material remains in the beaker, repeat this step and add the product to the centrifuge tube. Otherwise, rinse the beaker thoroughly with 2 N HNO_3 and add the rinsings to the centrifuge tube. Centrifuge for 10 minutes at 2000 rpm.

b) Prepare a 250-ml beaker by placing in it a teflon magnetic stirrer and 2 ml of a saturated solution of SO_2 (6% sulfurous acid). This will reduce plutonium state. Carefully decant into the beaker the supernatant from Step 1, using extreme care to see that no solids are transferred. Use one extra beaker for the preparation of a reagent blank.

c) To the residue remaining in the centrifuge tube from Step 2, add 30 ml of 2 N HNO_3. Heat gently until fumes appear. Centrifuge for 5 minutes at 2000 rpm. Add this supernatant carefully to the contents of the beaker.

d) Add 6 ml of 2 N HNO_3 to the residue remaining in the centrifuge tube and heat to boiling. Centrifuge for 5 minutes at 2000 rpm and add the supernatant to the material collected from Steps 2 and 3. Discard the precipitate.

e) Titrate the combined supernatants, with stirring, to pH 1.7 with 50% NaOH, taking care to adjust the temperature gauge of the pH meter to reflect the actual temperature of the solution. Take the final pH reading when the temperature is 20 to 23° C. This titration should be carried out in an ice water bath.

f) Decant the titrated solution into two 90-ml pyrex centrifuge tubes, keeping the volumes of the two tubes equal. Rinse the beaker with distilled water and add these rinses to the tubes.

g) Immerse the tubes in a constant temperature water bath at 80 to 85° C and equip each tube with a mechanical stirrer. Equilibrate the temperature for 10 to 15 minutes with stirring, and then add 1 ml of concentrated H_3PO_4 followed by 1 ml of $Bi(NO_3)_3$ (100 mg Bi^{+++}/ml, dissolved in 6 N HNO_3). Stir for one hour.

h) Remove the tubes from the bath, rinse the stirrers into the tubes with distilled water, and centrifuge for 5 minutes at 2000 rpm. Discard the supernatant.

i) Transfer the precipitate from each tube into a 40-ml centrifuge tube. This is accomplished by breaking up the precipitate with a fine stream of distilled water and thoroughly rinsing the tube. Do not attempt to transfer the precipitate of more than one tube at a time. Save the large tube from which the transfer was made. Centrifuge the tube containing the precipitate for 5 minutes at 2000 rpm and discard the supernatant. Repeat these steps for the second large tube so that the precipitate from both of the large tubes will be combined in one 40-ml tube.

j) To each of the large tubes in which the $BiPO_4$ precipitate was formed, add 4 ml of 6 N HCl. Swirl the tubes so that the acid is in contact with the walls of the tube, and then transfer into the tube containing the precipitate, taking care to see that any precipitate which may adhere to the walls of the smaller tube is washed down. Using a fine glass stirring rod, dissolve the precipitate completely in the acid. If any turbidity persists, add a few drops of concentrated HCl.

2 This method as developed by the original authors has been modified as to minor details by many laboratories. The protocol given here describes the practice at Lawrence Berkeley Laboratory, Berkeley, California. Thorium, plutonium, actinium, neptunium and the transplutonium elements are recovered in this procedure.

k) To each tube add 0.3 ml of $La(NO_3)_3$[3] (containing 1.0 mg La^{+++}/ml), and 1 ml of concentrated hydrofluoric acid. Mix by gently swirling the tubes. Allow to stand for 8 to 10 minutes at room temperature. Centrifuge for 5 minutes at 2000 rpm and discard the supernatant.

l) Wash the precipitate once with about 10 ml of dilute hydrofluoric acid (1 ml concentrated HF/100 ml distilled water containing 2 ml 6 N HCl). Centrifuge immediately for 5 minutes at 2000 rpm and discard the supernatant.

m) Invert the tubes over several thicknesses of absorbent paper, and allow to drain for 15 to 20 minutes.

n) Add a few drops of distilled water to each tube and, using a capillary pipette equipped with a rubber bulb, break up the precipitate. Pipette the resulting slurry onto planchets of platinum or stainless steel ($1^1/_2''$ inner diameter, 2″ outer diameter, depth $^1/_8''$). Take care to distribute the slurry evenly over the surface of the planchet.

o) Dry the samples by means of an infrared lamp, adding water as necessary until all material in the tube has been transferred. When the sample is completely dry, flame to cherry red over a Fisher burner.

Recovery of americium is $95 \pm 5\%$, curium $85 \pm 5\%$.

2. Surface Adsorption (Eakins and Gomm, 1968)

a) Dissolve the white phosphate residue from alkaline phosphate precipitation (see Chap. 13) in 100 ml of 1 M nitric acid and heat on a hot plate for 40 min (see Note A).

b) Remove the beaker from the hot plate and add 5 ml of 1 M sodium sulphite solution (see Note B). Allow to stand for 5 min.

c) Transfer the solution to a 600 ml beaker and dilute to 500 ml with water. Adjust the solution to pH 5.0 with ammonia (see Note C).

d) Pour the solution through a glass fibre filter paper previously prepared (see Note D).

e) Wash the filter paper twice with 200 ml of water.

f) Elute the alpha activity with 50 ml of 6 M hydrochloric acid and two 10 ml water washes into a 250 ml beaker.

g) Add 2 glass beads to the solution in the beaker and boil down to a few ml.

h) Evaporate the solution to dryness on a platinum counting tray, using several water washes to ensure quantitative transfer to the tray. Flame the source and count in a suitably calibrated alpha counter.

Note A. The hot plate should be on "high" and up to temperature before the beaker is placed on it. Otherwise ensure that the nitric acid actually boils for 20 min.

Note B. If the presence of neptunium is suspected in the sample, 5 ml of 7 M hydroxylamine hydrochloride solution should be added instead of sodium sulphite and the solution boiled for a further 10 min.

Note C. Use ammonia solution (Sp. gr. 0.88) until the pH is between 2 and 3, then use 1 M ammonia solution added dropwise for the final pH adjustment.

Note D. A Whatman 9.0 cm GF/A glass fibre filter paper is placed on a Buchner funnel and washed with 100 ml of 6 M hydrochloric acid. It is then washed twice with about 10 ml of water and finally with 200 ml of water before use.

Recovery of americium and curium is in excess of 80%.

---

3 Lanthanum used in this procedure must be purified to avoid spurious counts. The method of purification is given in Appendix 2.

3. Anion Exchange (Henley, 1965)

a) Column preparation. The resin column is made by sealing a platinum tube, 0.5 mm inner diameter and approximately 1 inch long into the stem of a $2^1/_2$ inch pyrex funnel. Insert a small glass wool plug to hold the resin and fill the stem of the funnel with Dowex 2×10, 50–100 mesh. Condition the resin with 10 M hydrochloric acid 0.01 M in nitric acid.

b) Column operation:

(1) The prepared sample (alkaline phosphate precipitation) is dissolved in 15 ml of 10 M hydrochloric acid 0.01 M in nitric acid[4]. If necessary use heat to bring about complete solution.

(2) Pass the solution over the prepared resin column. The flow rate is approximately 0.5 ml/min.

(3) Collect the effluent in a 100 ml beaker. Wash the column with 25 ml of 10 M hydrochloric acid and collect washing in the same beaker.

(4) Evaporate the effluent and wash to dryness. This fraction contains the trivalent actinide elements, lanthanides, the strontium-yttrium pair, and zinc.

(5) The dry residue from Step 4 may then be processed for alpha counting by one of the other gross alpha methods, or the transplutonium elements may be purified by cation exchange or solvent extraction. Recovery of californium by this method is reported as better than 90%.

(6) Plutonium, neptunium, protactinium, and uranium may be removed sequentially from the anion exchange column by suitable eluting agents.

B. Specific Actinide Determination (Low-Beer and Story, 1962)

For identification of activity detected in the gross alpha procedure or in Fraction 1 of Procedure II.B., the following chromatographic method is used[5].

1. Dissolve the lanthanum fluoride precipitate of Procedure II.A. in 0.5 ml of 0.1 N HCl saturated with H_3BO_3, and introduce onto a 3 mm × 8 cm column of resin, Dowex 50×4, 200–400 mesh in the hydrogen form.

2. Wash the column with 0.5 ml of 0.05 N HCl, followed by 0.1 ml of 2 N HCl. Discard the washings.

3. Elute by passing 0.1 ml of 2 N HCl through the column and follow with 1.5 ml of 20% ethanol saturated with HCl. This is made by bubbling HCl gas into a solution of 80 parts concentrated HCl and 20 parts absolute ethanol. Collect the entire eluate in one tube and bring the volume to 22 ml with distilled water.

4. Place the eluate on a 3 mm × 5 cm column of resin, Dowex 1×8, 200–400 mesh in the chloride form. Wash the column three times with 0.5 ml of 2 N HCl. Collect the effluent and all washings. Evaporate to dryness in a hot oil bath.

5. Dissolve the residue in 0.5 ml of 0.05-N HCl and introduce onto a column of resin, 3 mm × 5 cm, Dowex 50×4, 200–400 mesh in the hydrogen form, and operated at a constant temperature of 87.0° C. To maintain this temperature, the column is equipped with a glass jacket which is connected to a flask in which trichloroethylene is boiled under reflux. For best results the resin and all solutions applied to the column should be warmed just prior to use.

---

4 If analytical grade resin is used, nitric acid can be omitted from this solution. This is recommended for recovery of americium.

5 It is, of course, possible to identify ^{253}Es by determining its rate of decay. Where two or more elements may be present in combination, studies of decay rates do not provide a satisfactory means of identification.

6. Wash the column with 0.65 ml of 2 N HCl and discard the washings.

7. Elute with 6 N HCl, collect drops 5 through 30 in a 10-ml beaker and evaporate to incipient dryness. Retain for electrodeposition.

C. Solvent Extraction with Bidentate Organophosphorus (Butler and Hall, 1970)

Residue of the wet ashing procedure for preparation of urine for analysis is dissolved in 8 N HCl and extracted with 10% triisooctylamin-exylene (TIOA). The organic phase will contain plutonium, neptunium and uranium, while the trivalent actinides do not extract under these conditions. The aqueous phase which contains the trivalent actinides is evaporated to dryness. Forty ml of 8 N HNO_3 is added to the residue and evaporated to dryness.

The beaker is allowed to cool to room temperature and 20 ml of distilled water is added, followed by 10 ml of concentrated HNO_3. The beaker is then heated to near boiling, and 20 ml of concentrated HNO_3 is added. At this point solution will be complete, and H^+ concentration will be 11–12 N.

The solution is transferred to a 125 ml separatory funnel and the beaker is rinsed with two 10 ml volumes of concentrated HNO_3. The rinses are added to the funnel. The solution is allowed to cool to room temperature and 1 ml of dibutyl N, N diethylcarbamyl phosphonate (DDCP) is added. The solutions are mixed vigorously for 10 seconds and the layers are allowed to separate for 30 minutes. The aqueous phase is drained from the funnel and discarded.

The organic phase is washed with 10 ml 12 N HNO_3 by shaking vigorously for one second, and allowing the phases to separate. The aqueous phase is drained and discarded.

Five ml toluene is added to the organic phase. This reduces the affinity of the actinides for DDCP and reduces the transfer of nonoxidizable phosphate to the strip solution.

Twenty ml of 2 N HNO_3 is added to the funnel and shaken vigorously for 10 seconds. After separation of the phases, the aqueous layer is drained into a 100 ml beaker. This step is repeated with a second 20 ml of 2 N HNO_3.

The combined 2 N HNO_3 strip solutions are evaporated and the residue is either mounted on a platinum or stainless steel planchet for total alpha counting or electrodeposited for alpha pulse height analysis. Overall sensitivity of the method is 0.02 ± 0.01 dpm/sample.

Note. This procedure may be carried out on a 250 ml aliquot of a 24-hour urine sample.

IV. Detection of Radioactivity

When the identity of a radionuclide that may be present in a biological sample is known, or when only one radionuclide is known to be present, a count of total or "gross" alpha a ctivity is adequate. For this purpose, the end product of the chemical procedure that has been employed to isolate the activity may be counted directly. When it is necessary to identify a radionuclide or to establish the relative activities due to two or more elements or isotopes, spectrometry must be invoked. Many alpha emitting nuclides have abundant gamma emissions of characteristic energy, and all have low energy photon spectra due to X-rays. These qualities, as well as the characteristic alpha energies of the transplutonium elements, may be used for their detection and identification.

A. Alpha Counting

Methods of alpha counting described in Chap. 13 apply to the transplutonium elements and are not repeated here.

1. Total Alpha Counting

Total or "gross" alpha counts are made on pulse or scintillation counters. Total alpha activity of the samples is calculated by:

$$\frac{(\text{net cpm of sample} \pm \sigma) \times 100}{\text{C.E.} \times \text{C.R.} \times \text{S.A.} \times \text{aliquot}} = (\text{dpm} \pm \sigma) \text{ whole sample.}$$

Nuclides which decay with a short half-life such as ^{253}Es, require that the observed count be corrected for decay:

$$\text{dpm}_t = \text{dpm}_0 \times e^{-\frac{k}{t_{1/2}} t}$$

dpm_t = count on day t
dpm_0 = count on day of excretion or ingestion
$k = 0.693$
t = interval from time 0 to time of count
$t_{1/2}$ = radioactive half-life of the nuclide.

2. Alpha Spectrometry

The transplutonium elements emit alpha particles with energies in the range of about 5 to 8 MeV. The alpha pulse height system described in Chap. 13 is used to identify these nuclides when they are present either alone or in combination. Pulse height analysis requires that the sample be electrodeposited on platinum or stainless steel.

a) Electrodeposition

Several methods for electrodeposition of transplutonium elements have been reported by a number of investigators. These methods apply as well to plutonium that has been reduced to the trivalent state. Yakovlev et al. (1957) developed one of the earliest procedures for electrodeposition of americium and curium; Fuger (1962) used aqueous solutions of EDTA as the plating solution: Donnan and Dukes (1964) used a carrier technique. The method that has gained widest acceptance was reported by Mitchell (1960). This procedure, which involves modifications of a number of earlier techniques, may be summarized briefly as follows:

1. The activity isolated by a chemical procedure such as ion exchange chromatography or solvent extraction is evaporated to dryness several times with hydrochloric acid. The residue is taken up in 1 ml of HCl and transferred to an electrodeposition cell with one or two washes of distilled water.

2. Volume is brought to 4 to 5 ml with ammonium chloride containing 0.1 to 0.2 grams of chloride per ml. A drop of methyl red is added, and the solution is made alkaline with concentrated ammonium hydroxide; 2 N HCl is added dropwise with stirring until the solution is just on the acid side.

3. The anode is placed so that its tip is within 5 mm of the cathode, and electrodeposition is carried on for 15 minutes with a current of 0.6 to 1 ampere. One minute before turning off the power, a few drops of concentrated ammonium hydroxide are added to the cell.

4. The cell is disassembled, and the cathode is rinsed with distilled water, acetone, and dried in air. It is flamed to cherry red, and is then ready for counting. Yields for americium and curium are 95 to 100% by this method.

5. Electrodeposition cells consist typically of a metal frame which holds the cathode (platinum or stainless steel), and a glass or lucite tower held in place by a gasket. The gasket size determines the area to be electroplated, which may be as small as 3 mm^2 or as large as 3 cm^2. With large area samples it is customary to use a Frisch grid in making spectrometric analyses as this provides collimation and greatly improves the resolution for such preparations. The anode is a platinum-iridium wire which in Mitchell's system is operated as a stirrer.

Sementsova (1962) investigated the use of nitric acid as a plating solution. A modification of the method suggested by Alloway (1971) has been found very satisfactory at the Lawrence Berkeley Laboratory. The purified activity is evaporated to dryness several times in dilute nitric acid. It is taken up in 4 to 5 ml of 0.05 N HNO_3 and transferred to an electrodeposition cell with a plating area of 3 sq. mm. Plating is carried out for 50 minutes at 30 volts. The current does not exceed 120 milliamperes. At the end of the plating time 2 or 3 ml of concentrated ammonium hydroxide are added, and plating is continued for one minute. The cathode is removed, rinsed, and flamed as described above. Yields of 97 to 102% have been obtained for americium, curium, californium and einsteinium.

b) *Monomolecular Layers*

A procedure employing an alternative technique to electrodeposition has been developed by Deal and Chanda (1949). This consists in extracting the purified nuclide of interest into a 0.25 M solution of TTA in benzene. The nuclide is extracted from an aqueous solution of suitable pH (determined by consulting the extraction coefficients of Poskanzer and Foreman (1961). Extraction is carried out by shaking the two phases in a 10 ml screw cap vial. The vial is then centrifuged to separate the phases and the upper organic layer is transferred by micropipette in small increments to a platinum planchet in which a small depression has been stamped. The planchet is dried by infrared lamp after each addition, and is finally flamed over a burner. The method results in rapid production of very satisfactory point sources suitable for high resolution alpha particle pulse height analysis. Its disadvantage lies in the fact that the transplutonium elements have different extraction coefficients and, therefore, a different pH adjustment is required for each of them. This limits the capability of the system to dealing with one nuclide at a time, whereas all the transplutonium elements behave similarly in the electrodeposition methods that have been described above.

B. Gamma Spectrometry

Because of the long range of gamma rays, many radionuclides can be detected by *in vivo* counting. Whole body counting has developed as a method of taking advantage of this phenomenon. The technique is applicable to nuclides that have abundant gamma emissions in the energy range above approximately 0.1 MeV. Some of the transplutonium elements are readily detected by whole body counting while others, including plutonium, are conspicuously lacking in the properties that make this feasible. Technological advances in recent years have revolutionized *in vivo* counting by opening the way to detection and identification of all radionuclides by means of their low energy photons. The entire subject of gamma spectrometry is presented in admirable detail by Crouthamel (1970).

1. Whole Body Counting

Development of a capability for *in vivo* gamma spectrometry was an early feature of the atomic age. One of the first applications of whole body counting to evaluation of human contamination by radionuclides was made at the Argonne National Laboratory by MILLER and MARINELLI (1950) who measured the radium content of residents of the Chicago area who had, at an earlier time, received radium for therapeutic purposes. The ensuing years were marked by great progress in the development of this technique, and whole body counting has become an essential component of a complete program for the protection of human beings from radiation hazard. The early history of this development has been traced by SPIERS (1962). Whole body counters are coordinated systems including a scintillator—plastic, liquid, or crystal; a photomultiplier tube; and a multichannel analyzer. According to a survey by MEHL and RUNDO (1963), crystal scintillators have been used predominantly throughout the world, although liquid scintillators have been preferred by some laboratories. The latter have been described by VAN DILLA and ANDERSON (1962). The typical crystal scintillator is NaI activated by Tl. Its design and method of operation have been described by MAY and MARINELLI (1962). Counting efficiency is proportional to the dimensions of the crystal, and increases up to values of 12 and 4 inches for diameter and thickness respectively. The three major problems in the operation of whole body counters are background, geometry, and correction of observed counting rates for absorption. Background counts can be minimized by excluding all extraneous contributions by extensive shielding with lead-lined steel. However, background counts are also contributed by the crystal and these are proportional to its thickness. Constant geometry can be achieved by positioning all subjects in an identical manner. This is usually accomplished by using a reclining chair in which the trunk and torso of the subject form an arc, and the relationship of the body to the detector is the same for all subjects. Calibration with phantoms establishes the effect on count rate of absorption of a radionuclide in bone or soft tissue. When these conditions are observed, reproducible counts can be made, and peaks due to energies that are separated by at least 100 keV are readily distinguishable in 15 to 30 minute periods of counting.

Whole body counting has been used to measure the natural radioactivity in the body due to ^{40}K, and that which is due to ^{137}Cs—a radionuclide that has been distributed ubiquitously as a result of weapons testing. For rapid and direct evaluation of internal contamination by radionuclides that emit abundant gamma rays of high energy, whole body counting is the method of choice. Table 20.1. shows the detection limits of *in vivo* gamma spectrometry for some of the transplutonium isotopes.

Table 20.1. HEID et al. (1971)

Isotope	Maximum permissible lung burden (μCi)	Gamma photons (keV)	Minimum detectable amount in chest area (μCi)
Plutonium-239	1.6×10^{-2}	39 and 384	4×10^{-1}
Americium-241	1.5×10^{-2}	60	2×10^{-4}
Americium-243	1.6×10^{-2}	75	2×10^{-4}
Curium-244	1.4×10^{-2}	43	4×10^{-1}
Berkelium-249	12.0	320	1×10^{2}
Californium-252	4×10^{-3}	100	5×10^{-1}
Einsteinium-254	1.2×10^{-2}	63	4×10^{-3}

It will be seen that except for ^{241}Am, ^{243}Am, and ^{254}Es, detection limits exceed the maximum permissible burden. WRIGHT and PARKER (1971) have found that ^{252}Cf can be measured by gamma spectrometry if the gamma emission of the daughter fission products are utilized. These are in equilibrium with ^{252}Cf, and have energies in the range 100 to 800 keV, and abundance of approximately 34%. For such isotopes as ^{239}Pu, ^{244}Cm, ^{249}Bk, and ^{253}Es, however, the limitations of conventional whole body counting leave the bioassayist entirely dependent upon the indirect method of determining alpha activity in excreta.

The subject of whole body counting cannot be concluded without mention of the scintillator camera developed by ANGER (1966). PARKER and WRIGHT (1971) have demonstrated good agreement in detection limits and accurate localization for pure sources of ^{241}Am and ^{252}Cf counted in the whole body counter and by the Anger camera. Increasing distribution of the camera throughout the world should make it a useful alternative to the whole body counter for evaluation of appropriate radionuclides.

C. Low Energy Photon Spectrometry

Impetus for the use of low energy photon counting for *in vivo* counting was engendered by the limitations of conventional whole body counting for detection of ^{239}Pu. The principles of low energy photon counting are equally applicable in the study of transplutonium elements. HEID *et al.* (1971) have shown that detection limits for many of these elements are markedly decreased when energies in the L X-ray range are measured with thin NaI scintillators.

Table 20.2. HEID et al. (1971)

Isotope	X-rays (keV)	Minimum detectable amount in chest area (μCi)
Plutonium-239	17	1×10^{-2}
Americium-241	17	5×10^{-4}
Americium-243	17	3×10^{-3}
Curium-244	17	3×10^{-3}
Berkelium-249	~18	1×10^{5}
Californium-252	~19	3×10^{-3}
Einsteinium-254	~20	2×10^{-4}

PARKER et al. (1971) have used a counter consisting of two 5″ by 3 mm NaI crystals horizontally opposed and mounted in a small animal counter to follow excretion and retention of ^{253}Es in mice. Counts were made of the 10 to 39 keV energy range, and although the mice had received doses of only 25 nCi, their retention and excretion were readily followed. After sacrifice, counts were made of skeletons, carcasses, and organs of interest, and these were found to be in good agreement with alpha determinations which were made on the same tissues and on the excreta. It is probable that the attenuation of counting rate by absorption in bone would preclude the use of this technique for *in vivo* counting of ^{253}Es in animals larger than the mouse. However, the effectiveness of the method for determining activity in excreta and isolated tissues has been clearly demonstrated.

V. Interpretation of Bioassay Data

The interpretation of bioassay results involving the transplutonium elements depends on models that have been developed to describe the metabolic behavior of plutonium (Chap. 13).

The great question that remains unanswered at this time concerns the applicability of the plutonium models to evaluation of contamination by transplutonium elements. Data from recent investigations (see Chap. 18) indicate that the transplutonium elements have greater mobility, and are excreted more rapidly than plutonium. The oxides of these elements appear to be more soluble in body fluids than those of plutonium, and the urinary to fecal ratios of excretion appear to increase as one goes toward the right of the actinide series. It is too soon to generalize from these observations, and the crying need of the moment is for more animal experiments, and for systematic study of any human beings who become contaminated. SNYDER (1971) has analyzed some of the recent data from animal experiments with americium and californium, and has concluded that while there are significant differences in excretion rates as compared with plutonium, these differences may be factors of only two to five and not orders of magnitude. Since the excretion rates of the transplutonium elements are in all cases higher than those of plutonium, some comfort may be found in the fact that interpretation based on the plutonium models will err on the side of caution when the models are applied to transplutonium metabolism.

Appendix 1

Laboratories reporting methods of determining transplutonium elements in biological materials

Laboratory	Elements of Concern	Method	Applications	References
Argonne National Laboratory	Am, Cm, Cf	sequential separation by solvent extraction with TIOA and HDEHP	urine, feces, blood sputum, nose blows	SEDLET (unpublished)
Atomic Energy Research Establishment, Harwell, United Kingdom	Am, Cm	glass adsorption (see text)	urine, feces, blood nose blows	EAKINS and GOMM (1968)
Dow Chemical Co., Rocky Flats, Golden, Colo.	Am	separation on cation exchange resin	urine, feces, sputum, blood	BOKOWSKI and ALLEN (1970)
E. I. du Pont de Nemours, Savannah, River Plant, Aiken, S. C.	Am-Cm, Cf, Es	solvent extraction with TIOA and DDCP (see text)	urine, feces	BUTLER (1965), BUTLER and HALL (1970)
Kernforschungszentrum Karlsruhe, Germany	Am, Cm, Cf	solvent extraction with HDEHP	all biological materials	SCHIEFER-DECKER (1968)
Lawrence Berkeley Laboratory, University of California	Am, Cm, Cf Es	coprecipitation with LaF_3 followed by group separation by ion exchange (see text)	all biological materials	SCHUBERT et al. (1951), LOWBEER and STORY (1962)
Lawrence Livermore Laboratory, University of California	Cm	coprecipitation with LaF_3-cation exchange separation	urine and feces	DUPZYK and BIGGS (1960)
Los Alamos Scientific Laboratory	Am	separation on anion exchange resin	urine and feces	CAMPBELL and MOSS (1965)
Oak Ridge National Laboratory	trivalent actinides	separation of trivalent actinides on anion exchange resin followed by $BiPO_4$ and LaF_3 precipitation (see text)	urine and feces	SCHUBERT et al. (1951), HENLEY (1965)
Sandia Laboratory Albuquerque, N.M.	actinides other than U or Pu	electrodeposition from acid solution of ashed residue	urine	MITCHELL (1960)
Trapelo West, Richmond, Calif.	Am	extraction with HDEHP using Cm tracer. Final purification on ion exchange resin	urine	WEISSMAN et al. (1968)
U.S.A.E.C. Health and Safety Laboratory, New York, N.Y.	Am	LaF_3 coprecipitation-oxidation of Am by peroxidisulfate	urine, feces, tissues	MOORE (1963)
U.S.A.E.C. Health Services Laboratory, Nuclear Reactor Station, Idaho Falls, Idaho	Am, Cm, Cf	sequential extraction from $Al(NO_3)_3$ with aliquot (quarternary amine)	urine and feces	SILL (in preparation)
United States Testing Co., Richland, Washington	Am, Cm	solvent extraction with HDEHP	urine	BUTLER (1965)
University of Pittsburgh, Radiation Health Dept., Graduate School of Public Health	Am	solvent, extraction with HDEHP	urine, feces, blood, tissues	HORM (in press)

Appendix 2

Lanthanum Nitrate Purification

1. Place 10 g of lanthanum nitrate [$La(NO_3)_3 \cdot 6H_2O$–AR] in a 400 ml beaker and dissolve in 200 ml of distilled water.

2. Add 50 ml of ammonium hydroxide (NH_4OH 28%, sp. gr. 0.90) to the beaker and stir.

3. Transfer the solution to four 90 ml centrifuge tubes. Rinse the beaker with distilled water and add the washings to the tubes to give equal volumes. Centrifuge at 2000 rpm for 5 minutes. Discard the supernatant fluid.

4. Add conc. nitric acid, with stirring, to Tube 1 until the precipitate is just dissolved. Transfer the clear solution to Tube 2. Rinse Tube 1 with 2 to 3 ml of conc. nitric acid and add to Tube 2. Stir and add conc. nitric acid, if necessary, until the precipitate is just dissolved. Transfer the solution to Tubes 3 and 4 in the same manner. Transfer the combined solutions in Tube 4 to four clean 90 ml centrifuge tubes. Divide the solution to give equal volumes.

5. Add 10 ml of ammonium hydroxide to each tube and stir. Centrifuge at 2000 rpm for 5 minutes and discard the supernatant.

6. Repeat Steps 4 and 5.

7. As in Step 4, add conc. nitric acid with stirring to Tube 1 until the precipitate is just dissolved. Transfer the clear solution to Tube 2. Rinse Tube 1 with 2 to 3 ml of conc. nitric acid and add to Tube 2. Stir and add conc. nitric acid, if necessary, until the precipitate just dissolves. Transfer to Tubes 3 and 4 in the same manner. Transfer the clear solution to a 250 ml graduated cylinder. Rinse the centrifuge tubes with 2 to 3 ml distilled water and add to the solution in the cylinder.

8. Add distilled water to the solution to give a total volume of 160 ml. Stir and transfer to a glass-stoppered bottle. This gives a solution containing 20 mg of La^{+3} per milliliter. Dilute stock solution with 6 N HNO_3 to give a solution containing 1 mg La^{+3} per ml.

TIOA Purification

1. Filter tri-isooctylamine (TIOA) through glass wool before use to remove solids.

2. Add 50 ml TIOA to 50 ml xylene in a 250 ml separatory funnel. Add 100 ml 8 N HCl, shake gently, vent, shake 1 minute, vent (there will be some heat evolution). Let phases separate (about 5 min). Organic phase will turn red. Draw off the lower acidic (aqueous) layer and discard.

3. Add 50 ml 8 N HCl, add 1 ml 2 M $NaNO_2$ (sodium nitrite), shake for 1 minute. (Organic layer turns dark red-brown.) Let phases separate, draw off and discard lower layer.

4. Add 50 ml 8 N HCl, add 5 ml 30% H_2O_2 (hydrogen peroxide), shake gently, vent, shake 3 minutes, vent after each minute. Let phases separate (very cloudy), draw off and discard lower layer. [Heat and (Cl_2) chlorine evolution.] Repeat Step 4.

5. Add 50 ml 8 N HCl, shake 1 minute, let phases separate, drain and discard lower layer.

6. Add 50 ml 0.1 N HCl, shake 1 minute, let phases separate, drain and discard lower layer, repeat Step 6.

7. Add 50 ml deionized water, shake for 1 minute, let phases separate, drain and discard lower layer.

8. Add 50 ml 8 N HCl, shake for 1 minute, let phases separate, draw off and discard lower layer, repeat Step 8.

9. Drain organic phase into a 600 ml beaker, add 372 ml xylene (to dilute to 10% (w/w) TIOA-xylene). Let stand covered overnight for cloudiness to clear and layers to separate. Decant into a 500 ml glass stoppered storage bottle.

References

ALLAWAY, D.: Personal communication 1971.

ANGER, H. O.: Sensitivity, resolution and linearity of the scintillation camera. Trans. Nucl. Sci 13, 380–392 (1966).

AUER-WELSBACH, H., BILDSTEIN, H., MÜLNER, P.: Chromatographische Trennung von Curium-242 und Americium-241 und die spektroskopische Überprüfung ihrer α-Aktivität. Mh. Chem. 98, 61–67 (1967).

BAYBARZ, D. D.: Separation of transplutonium elements by phosphonate extraction. Oak Ridge National Laboratory Report ORNL-3273 (1962).

BENES, P., GARBA, A.: A radiotracer study of manganese adsorption on glass from very dilute aqueous solutions. Radiochem. Acta 5, 99–104 (1966).

BEST, G. F., HESFORD, E., MCKAY, H. A. C.: Tri-n-butyl phosphate as an extracting agent for inorganic nitrates, VII. J. Inorg. Anal. Chem. 12, 136–140 (1959).

BOSHOLM, J., GOSSE-RUYKEN, H.: J. prakt. Chem. 26, 83–89 (1964).

BRANDSTETR, J., WANG, T., GABRILOV, K. A., GWOZDZ, E., MALÝ, J., TAUBE, M.: Extraction properties of fermium and mendelevium. Radiokhimiya 6, 22–30 (1964).

BROWN, K. B.: Extraction of lanthanide and transplutonium elements by dialkyl phosphoric acid and amines. Chemical Technology Division, Chemical Development Section C, Progress Report on Separation Chemistry and Separation Research, Jan.–June, 1963, ORNL-3496.

BUTLER, F. E.: Determination of uranium and americium-curium in urine by liquid ion exchange. Analyt. Chem. 37, 340–342 (1965).

BUTLER, F. E., HALL, R. M.: Determination of actinides in biological samples with bidentate organophosphorus compounds. Analyt. Chem. 42, 1073–1076 (1970).

CHOPPIN, G. R., MAY, D., HOLM, L. W.: Ion exchange studies of alpha-hydroxy carboxylic acid—lanthanide and actinide systems. In: Radioisotopes in the physical sciences and industry, vol. III, p. 283–299. Vienna: IAEA 1962.

COLEMAN, J. S., PENNEMAN, R. A., KEENAN, T. K., LAMAR, L. E., ARMSTRONG, D. E., ASPREY, L. B.: An ion exchanger process for gram scale separation of americium from rare earths. J. Inorg. Nucl. Chem. 3, 327–328 (1956).

CROUTHAMEL, E. E.: Applied gamma ray spectrometry, 2nd ed. (ADAMS and DAM, eds.). New York-Oxford: Pergamon Press 1970.

DEAL, R. A., CHANDA, R. N.: A rapid source preparation technique for high resolution alpha particle spectroscopy. Nucl. Instr. Meth. 69, 88–92 (1969).

DIAMOND, R. M., STREET, K., JR., SEABORG, G. T.: An ion exchange study of possible hybridized 5f bonding in the actinides. J. Amer. chem. Soc. 76, 1461–1469 (1954).

DOBSON, R. L.: Americium poisoning—therapy of radioelement poisoning. ANL-5584, p. 28–35 (1955).

DONNAN, M. Y., DUKES, E. K.: Carrier technique for quantitative electrodeposition of actinides. Analyt. Chem. 36, 392–394 (1964).

DREXLER, G., PERZL, F.: Spectrometry of low-energy γ and x-rays with Ge(Li) detectors. Nucl. Instr. Meth. 48, 332–334 (1967).

EAKINS, J. D., GOMM, P. J.: The determination of gross alpha activity in urine by adsorption on glass filter paper. Hlth Phys. 14, 461–472 (1968).

FINE, S., HENDEE, C. F.: X-ray critical-absorption and emission energies in keV. In: Radiological health handbook, Rev. ed., p. 162. U.S. Department of Health, Education and Welfare, Public Health Service, 1970.

FOREMAN, H., MOSS, W., ENSETTER, B. C.: Clinical experiences with radioactive materials. Amer. J. Roentgenol. 79, 1071–1079 (1958).

FUGER, J.: Ion exchange behavior and dissociation constants of Am, Cm, and Cf complexes with EDTA. J. Inorg. Nucl. Chem. 5, 332–338 (1958).

FUGER, J.: Preparation of thin electrodeposits of actinides from aqueous solutions of EDTA. Bull. Soc. chim. Belg. 71, 134–141 (1962).

GATTI, R. C., PHILLIPS, L., SIKKELAND, T., MUGA, M. L., THOMPSON, S. G.: Ion exchange behavior of mendelavium. J. Inorg. Nucl. Chem. 11, 251–253 (1959).

GRAY, P. R., THOMPSON, S. G.: Lawrence Radiation Laboratory, Berkeley, Chemistry Division Quarterly Report, Sept.–Nov., 1952, UCRL-2069.
GUREEV, E. S., DEDOV, V. B., KARPACHAVA, S. M., LEBEDEV, I. A., SHVETSOV, I. D., YAKOLEV, M., RHYZOV, N., TRUKCHLAYEV, P. S.: Proc. Third Int. Conf. Peaceful Uses of Atomic Energy, p. 553–560. Geneva: IAEA, 1964.
HARLEY, J. H.: Sampling and analysis for assessment of body burdens. In: Assessment of radioactivity in man, vol. 1, p. 155–167. Vienna: IAEA 1964.
HARRIS, D. H., TOMPKINS, E. R.: Separations of several rare earths of the curium group. J. Amer. chem. Soc. **69**, 3792–3800 (1947).
HENLEY, L. C.: Urinalysis by ion exchange. Proc. 11th Annual Bioassay and Analytical Chemistry Meeting, Albuquerque, N. M., 1965, Conf.-651008, Biology and Medicine, USAEC TID-4500, p. 1–19.
HESFORD, E., JACKSON, E. E., MCKAY, H. A. C.: Tri-n-butyl phosphateas an extracting agent for inorganic nitrates—VI. J. Inorg. Nucl. Chem. **9**, 279–289 (1959).
HILL, O. F., COOPER, V. R.: Scale-up problems in the Pu separations program. Ind. Eng. Chem. **50**, 599–602 (1958).
HOROWITZ, E. P., BLOOMQUIST, C. A. A., HENDERSON, J. D.: The extraction chromatography of californium, einsteinium and fermium with di(2-ethylhexyl)-orthophosphoric acid. J. Inorg. Nucl. Chem. **31**, 1149–1166 (1969).
HULET, E. K.: An investigation of the extraction chromatography of Am (VI) and Bk (IV). J. Inorg. Nucl. Chem. **26**, 1721–1726 (1964).
HYDE, E. K.: Radiochemical separations of the actinide elements. In: The actinide elements. National Nuclear Energy Series Div. IV., vol. **14**A, p. 542–595 (SEABORG and KATZ, eds.). New York: McGraw-Hill 1954.
ISAAC, N. M., WILKINS, J. W., FIELD, P. R.: Separation of transcurium elements from large quanities of Cm by liquid-liquid extraction. J. Inorg. Nucl. Chem. **15**, 151–157 (1960).
KETELLE, B. H., BOYD, G. E.: Exchange adsorption of ions from aqueous solutions by organic zeolites. IV. Separation of the Y group rare earths. J. Amer. chem. Soc. **69**, 2800–2812 (1947).
KOOI, J., BODEN, R., WIKSTRA, J.: Separation of Am(Cm), Bk, and Cf by extraction chromatography. J. Inorg. Nucl. Chem. **26**, 2300–2302 (1964).
LAWROSKI, S.: Survey of separations process other than solvent recovery. Chem. Eng. Prog. **51**, 461–466 (1955).
LEDERER, C. M., HOLLANDER, J. M., PERLMAN, I.: Table of isotopes, 6th ed. New York: John Wiley & Sons, Inc. 1967.
LIESER, K. H.: Kationenaustausch und Adsorption von Radionukliden an Glasoberflächen. Radiochim. Acta **4**, 225–227 (1965).
LOW-BEER, A. DE G., STORY, T. L., JR.: A method for determining low level alpha activity from Am, Cm and Cf in biological materials. Proc. 8th Annual Bioassay and Analytical Chemistry Meeting, DP-831, p. 58–63 (1962).
MAGNUSON, L. B., ANDERSON, M. L.: Chelation of the +3 ions of elements 95 through 100 with thenoyltrifluoroacetone. J. Amer. chem. Soc. **76**, 6207 (1954).
MARCUS, Y., GIVON, M.: Anion exchange of metal complexes XIII the actinide (III)—nitrate system. IAR (1962).
MAY, H. A., MARINELLI, L. D.: Sodium iodide systems: Optimum crystal dimensions and origin of background. In: Whole body counting, p. 15–36. Vienna: IAEA 1962.
MEHL, H. G., RUNDO, J.: Preliminary results of a world survey of whole body monitors. Hlth Phys. **9**, 607–614 (1963).
MELLISH, C. E., HILL, M. R., GLANVILLE, D. E.: UKAEA Report AERE-R-5323 (1966).
MILLER, C. E., MARINELLI, L. D.: ANL-4531 (1950) 77 pages.
MITCHELL, R. F.: Electrodeposition of actinide elements at tracer concentrations. Analyt. Chem. **32**, 326–328 (1960).
MOORE, F. L.: Separation of Am from other elements: Application to the purification and radiochemical determination of Am. Analyt. Chem. **35**, 715–719 (1963).
MUELLER, W., FUGER, J., DUYCKAERTS, G.: Complex formation in tertiary amine extraction of trivalent metals. 8th Int. Conf. on Coordination Chemistry, Vienna (1964), Euratom 2169e.
National Council on Radiation Protection and Measurements Reports No 33 (1968) and No 39 (1971), Washinton, D.C.
National Research Council, National Academy of Science: The radiochemistry of americium and curium. NAS-NS-3006 (1960).
National Research Council, National Academy of Science: The radiochemistry of the transcurium elements. NAS-NS-3031 (1960).
PARKER, H. G., WRIGHT, S. R.: Whole-body counting of ^{241}Am and ^{252}Cf with the Anger camera. Health Phys., **22**, 891–898 (1972).

PARKER, H. G., WRIGHT, S. R., LOW-BEER, A. DE G., YAEGER, D. J.: The metabolism of ^{253}Es in mice. Health Phys., **22**, 641–652 (1972).

PEPPARD, D. F., MASON, G. W., DRISCOLL, W. J., MCCARTY, S.: Application of phosphoric acid esters to the isolation of certain transplutonides by liquid-liquid extraction. J. Inorg. Nucl. Chem. **12**, 141–149 (1959).

PERLMAN, I., STREET, K., JR.: The chemistry of the transplutonium elements. In: The actinide elements, National Nuclear Energy Series Div. IV, vol. 14A, p. 507–541 (SEABORG and KATZ, eds.). New York: McGraw-Hill 1954.

PIERCE, T. B., PECK, P. F.: Use of di(2-ethylhexyl) orthophosphoric acid for the separation of the rare earths by reverse phase partition chromatography. Nature (Lond.) **195**, 597 (1962).

POSKANZER, A. M., FOREMAN, B. M., JR.: A summary of TTA extraction coefficients. J. Inorg. Nucl. Chem. **16**, 323–336 (1961).

ROWLAND, R. E., MILLER, E. C., MARINELLI, L. D.: An americium accident. ANL-5518, p. 46–48 (1956).

SCHIEFERDECKER, H.: A method for separation of Am, Cm and Cf. KFK-80 (1968).

SCHRECK, H.: Complexing and extraction of trivalent actinium, americium, curium and californium with acetyl acetone and several of its derivatives. Kernforschungszentrum, Karlsruhe KFK-672, 83 pages (1967).

SCHUBERT, J., MYERS, L. S., JR., JACKSON, J. A.: Analytical procedures of the Bioassay Group at the Argonne National Laboratory. ANL-4509 (1951).

SEABORG, G. T.: Correlation of properties of actinide transition series. In: The actinide elements. National Nuclear Energy Series Div. IV, vol. 14A, p. 733–768 (SEABORG and KATZ, eds.). New York: McGraw-Hill 1954.

SEABORG, G. T., JAMES, R. A., GHIORSO, A.: The new element curium (atomic number 96). In: Transuranium elements. National Nuclear Energy Series Div. IV, vol. 14B, p. 1554–1585 (SEABORG and KATZ, eds.). New York: McGraw-Hill 1949a.

SEABORG, G. T., JAMES, R. A., MORGAN, L. O.: The new element americium (atomic number 95). In: Transuranium elements. National Nuclear Energy Series Div. IV, vol. 14B, p. 1525–1553 (SEABORG and KATZ, eds.). New York: McGraw-Hill 1949b.

SEMENTSOVA, A. K.: Electrolytic precipitation of Am and Cm from their aqueous solutions onto a Pt cathode. Radiokhimiya **4**, 669–671 (1962).

SIEKIERSKI, S., SOCHACKA, R. J.: Reversed phase partition chromatography with HDEHP as the stationary phase. Part I. Separation of rare earths. J. Chromatogr. **16**, 376–384 (1964).

SILL, C. W.: Separation and radiochemical determination of uranium and the transuranium elements using barium sulfate. Hlth Phys. **17**, 89–107 (1969).

SNYDER, W. S.: Mathematical models for internal dose-americium and californium. Presented at Annual Health Physics Society Meeting, July 1971, New York.

SPIERS, F. W.: Whole-body counting: An introductory review. In: Whole body counting, p. 3–12. Vienna: IAEA 1962.

STARÝ, J.: Separation of transplutonium elements. Talanta **13**, 421–437 (1966).

STREET, K., JR., HIGGINS, G. H., SEABORG, G. T.: The separation of americium and curium from the rare earth elements. J. Amer. chem. Soc. **72**, 2790–2797 (1950).

SURLS, J. P., CHOPPIN, G. R.: Equilibrium absorption of lanthanides, americium and curium on Dowex 50 resin. J. Amer. chem. Soc. **79**, 855–859 (1957).

THOMPSON, S. G., HARVEY, B. H., CHOPPIN, G. R., SEABORG, G. T.: Chemical properties of elements 99 and 100. J. Amer. chem. Soc. **76**, 6229–6236 (1954).

THOMPSON, S. G., MORGAN, L. O., JAMES, R. A., PERLMAN, I.: The tracer chemistry of americium and curium in aqueous solutions. In: The transuranium elements. National Nuclear Energy Series Div. IV, vol. 14B, p. 1339–1362 (SEABORG and KATZ, eds.). New York: McGraw-Hill 1949.

U. S. Atomic Energy Commission: Californium-252. Proc. of a Symposium American Nuclear Society, New York, 1968. Conf. 681032, p. 41.

VAN DILLA, M. A., ANDERSON, E. C.: Human counters using liquid scintillators. In: Whole body counting, p. 41–58. Vienna: IAEA 1962.

VOJTECH, O., SPICIN, V. I.: Dokl. Akad. Nauk. SSSR **136**, 339–341 (1961).

WERNER, L. B., PERLMAN, I.: The preparation and isolation of curium. In: Transuranium elements. National Nuclear Energy Series Div. IV, vol. 14B, p. 1586–1654 (SEABORG and KATZ, eds.). New York: McGraw-Hill 1949.

WRIGHT, S. R., PARKER, H. G.: Unpublished material (1971).

YAKOVLEV, G. N., CHULKOV, P. M., DEDOV, V. M., KOSYSKOV, V. N., SOBELEV, Y. P.: The preparation of thin films of Pu, Am, and Cm by an electrolytic method. J. Nucl. Energy **5**, No 1, 159–161 (1957).

Chapter 21

Medical Uses: Americium-241; Californium-252

GLENN T. SEABORG

With 9 Figures

I. Medical Uses

Radioactive isotopes in general have become an essential part of clinical medicine. In diagnostic procedures, the rate of appearance, the localization and the disappearance of small amounts of internally administered radionuclides are used to determine such factors as organ size and function. In therapeutic uses, isotope sources deliver large amounts of ionizing radiation to selected areas of the body.

The use of radioisotopes of the transplutonium series in the field of medicine is severely restricted for a number of reasons. From a medical viewpoint, internal administration of elements of this series for either diagnostic or therapeutic purposes is precluded by their extreme toxicity. Hence, medical use of the transplutonium elements is limited to encapsulated sources for radiotherapy or selected medical applications. In the following pages, the use of americium-241 and californium-252 radiation sources will be discussed.

II. Americium-241

Americium was discovered in 1945 by SEABORG, JAMES, MORGAN, and GHIORSO (1949, 1950) at the Metallurgical Laboratory of the University of Chicago. These investigators found that plutonium was the source of a long-lived alpha particle emitter which could be separated chemically from the parent. The daughter isotope was shown to have the mass number 241 and atomic number 95, resulting from beta particle decay of plutonium-241. Americium-241 was also produced by decay from plutonium-241 formed by the bombardment of uranium-238 with alpha particles.

Americium-241 decays essentially entirely by alpha particle emission with a half-life of 434 years. The decay scheme is complex, with many gamma transitions. The 59.57 KeV transition heavily predominates. One gamma photon of this energy is emitted for every 2.5 alpha particles. The 59.57 KeV gamma ray is used in nuclear medicine; other gamma rays and copious L X-rays are either too low in abundance to interfere or can be filtered out.

Kilogram quantities of americium-241 resulting from the decay of plutonium-241 have been recovered routinely from plutonium produced in military production reactors or power reactors. Gamma ray sources of many sizes and shapes are available from commercial suppliers.

A. Fluorescent Scanning of the Thyroid

Fluorescent scanning of the thyroid is a new technique for imaging of the thyroid without administering radioactive isotopes to the patient. The basic concept of this technique is illustrated in Fig. 21.1. In this procedure, K-shell fluorescence of the stable iodine within the thyroid gland is induced by exposure to an external americium-241 source. The fluorescent radiation produced is detected by means of a high resolution lithium-drifted silicon detector.

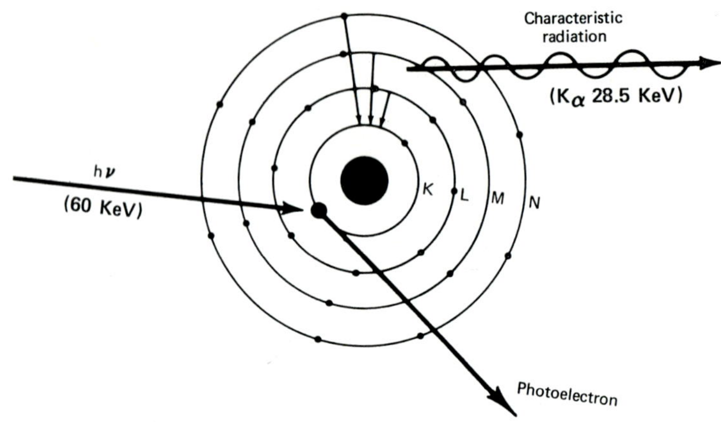

Fig. 21.1. Production of characteristic radiation. The 60 KeV gamma ray (hν) of ^{241}Am displaces a K shell electron (photoelectron) from the target iodine atom. Replacement of this shell vacancy from an outer ring results in production of a 28.5 KeV characteristic X-ray (characteristic radiation). (Reproduced with permission from P. B. Hoffer and A. Gottschalk, Radiology **99**, 117–124, 1971)

The energy of the K-shell fluorescent X-rays of each element is unique and the number of characteristic X-rays induced is proportional to the concentration of the particular element. For elements with low atomic numbers, the excited fluorescent X-rays are of low energy and readily absorbed by overlying tissue. In the case of iodine in the thyroid, however, the energy of the fluorescent X-rays (28.5 KeV with a half-value layer of tissue of approximately 2 cm) is sufficient to be detected outside the body.

Application of fluorescent scanning is limited by a number of factors. The specific element to be excited must be present in sufficient concentration for the characteristic X-ray fluorescence to be detected above the background. Also, the energy of the exciting radiation must be above the K-shell edge of the element to be scanned. A high resolution detector allowing a large percentage of the characteristic radiation to be detected and excluding scatter from other sources is essential.

In the system described by Hoffer et al. (1969a), the gamma ray source and the detector are mounted on the probe of a commercial scanner and moved in a rectilinear raster over the area to be scanned. The americium-241 source is in the form of encapsulated pellets of the sesquioxide and is shielded and collimated. The detector, a lithium-drifted silicon crystal, must be kept at liquid nitrogen temperature. This is accomplished by a built-in insulated reservoir (Fig. 21.2) for the liquid nitrogen.

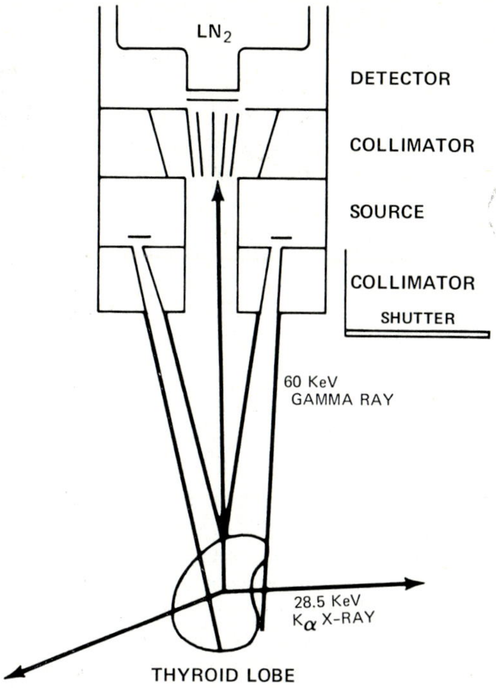

Fig. 21.2. Schematic drawing of fluorescent scanner. A beam of 60 KeV gamma rays from americium-241 (Source) irradiates a portion of thyroid. Some of the characteristic radiation produced is seen by the lithium-drifted silicon detector (Detector) which is housed at the base of a liquid nitrogen cooling reservoir. (Reproduced with permission from P. B. Hoffer and A. Gottschalk, Radiology **99**, 117–124, 1971)

Under normal physiological conditions, the thyroid gland contains approximately 0.4% iodine by weight (McGavack, 1951), an amount sufficient to act as target for fluorescent scanning[1].

A great advantage of fluorescent scanning is that only the area of tissue in the field of view of the excitor-detector system is being irradiated during the scanning procedure. The total exposed area is restricted to the area that is scanned and the radiation dose to that area can be kept reasonably low. Currently, the radiation dose from fluorescent scans is in the order of 15 millirads. This is much lower than the radiation dose received from thyroid scans with iodine-131 (50 to 150 rads) or iodine-125 (40–80 rads) (Maynard, 1969).

The information obtained from the fluorescent scans is unique (Fig. 21.3) in the sense that it describes the distribution of iodine within the gland and, since the number of X-rays produced are proportional to the concentration of iodine in the tissue, it allows a subjective estimate to be made of the amount of iodine present. In contrast, radioactive iodine studies give direct information about the functional metabolic activity of the thyroid and only indirect information with respect to the distribution of iodine within the gland. Combination of the two examinations, i.e., radioisotope and fluorescent scans, should give a more complete picture of the thyroid function *in vivo*.

1 The applicability of fluorescent scanning of organs other than the thyroid has been explored by Hoffer et al. (1969b) but has not been evaluated clinically.

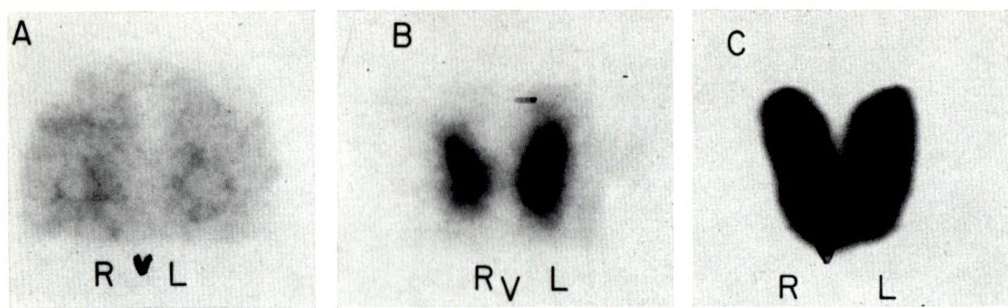

Fig. 21.3. These three cases demonstrate the subjective assessment of thyroidal iodine concentration discussed in the text. *Case A*. Low iodine content in a hypothyroid patient. The gland is difficult to perceive. Note the tracheal air column. *Case B*. Normal iodine content in euthyroid patient. *Case C*. Increased iodine content in a patient with a binding defect. (Reproduced with permission from P. B. Hoffer and A. Gottschalk, Radiology **99**, 117–124, 1971)

In a recent study, Hoffer and Gottschalk (1971) evaluated the use of fluorescent thyroid scanning in a series of 76 patients. In the majority of these patients, a conventional radioisotope scan and a 24-hour iodine-131 uptake study were done in conjunction with the fluorescent scan. It is interesting that in comparing the technical quality of the scans, the fluorescent scans, in general, were judged to be of equal or superior quality. Poor quality of scans was attributed to technical imperfections of the equipment used at the time of the study.

From the results of these studies, Hoffer and Gottschalk concluded that:

(1) Fluorescent scanning is possible in a variety of situations when radionuclide scanning is either impossible or undesirable. These include pregnancy, thyroid replacement therapy, and low nuclide uptake (e.g., flooded iodine pool).

(2) In this series, the clinically hypothyroid patients have fluorescent scans that always demonstrate diminished iodine content. We have not examined anyone with secondary hypothyroidism with this method; the fact that thyroids "at rest" from replacement doses of thyroxine show normal iodine concentrations (i.e., normal fluorescent scans) suggests that the technique could also be used to separate primary from secondary hypothyroidism.

(3) In active Graves' disease, the fluorescent scan demonstrates a normal to large gland with *normal* or *increased* iodine content.

(4) Some nodules that are "cold" on routine radionuclide scans contain enough iodoprotein to show "structure" on fluorescent scans. Our series is still too small to determine if etiologic characterization will be possible in these patients.

(5) Autonomous nodules may show either increased or decreased iodine concentration. It is possible to see thyroid tissue suppressed by toxic nodules directly by fluorescent scanning without the necessity of TSH stimulation.

(6) No consistent appearance is demonstrated in thyroiditis.

B. Americium-241 Transmission Scanning

Anger et al. (1966) described the use of americium-241 sources for whole-body transmission scanning in a so-called Hybrid Scanner. The principal use of this scanning device is to determine the localization of administered radionuclides in various parts of the body, such as iron-52 and iron-59.

The patient lies on a table (Fig. 21.4) that slowly moves over the scintillation counters. A single americium-241 source is located about 3 meters above this bank of counters. Pulse height selection is used to separate pulses from americium-241 gamma rays from the pulses originating from the radionuclide administered to the patient.

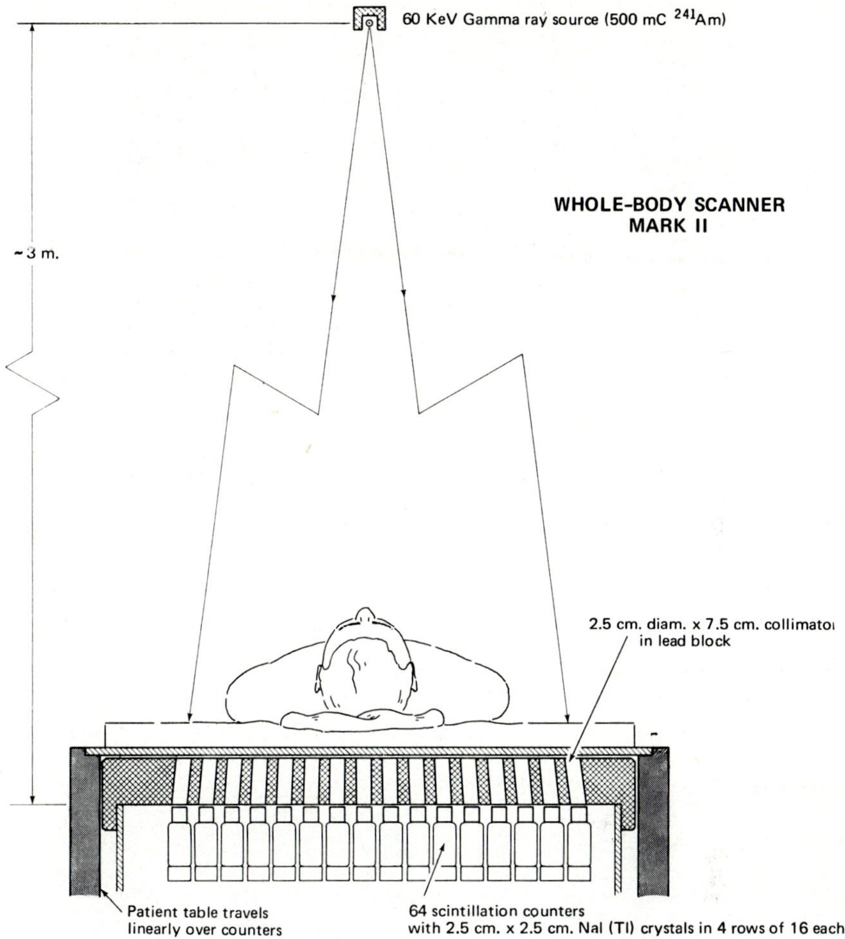

Fig. 21.4. ANGER Whole-Body Scanner Mark II. Patient traverses over fixed array of 64 scintillation counters each with 2.5 cm-dia X 2.5 cm-thick NaI(Tl) crystal. Each counter has single-channel collimator aimed upward towards ^{241}Am source located 3 m above. (Reproduced with permission from P. RONAI, H. S. WINCHELL, H. D. ANGER and J. H. LAWRENCE, Journal of Nuclear Medicine **10**, 469–474, 1969)

The two separate images are placed on a photographic film in such a way that the gamma ray image of the distribution of the isotope is superimposed upon the gamma ray transmission image which gives an accurate outline of the patient's body.

The clinical uses of this instrument were recently reported by RONAI et al. (1969) from which Fig. 21.5 and 21.6, showing whole-body transmission scans, were obtained. It is clear that whole-body scans obtained in this manner result in accurate visualization of the localization of administered radionuclides in the body.

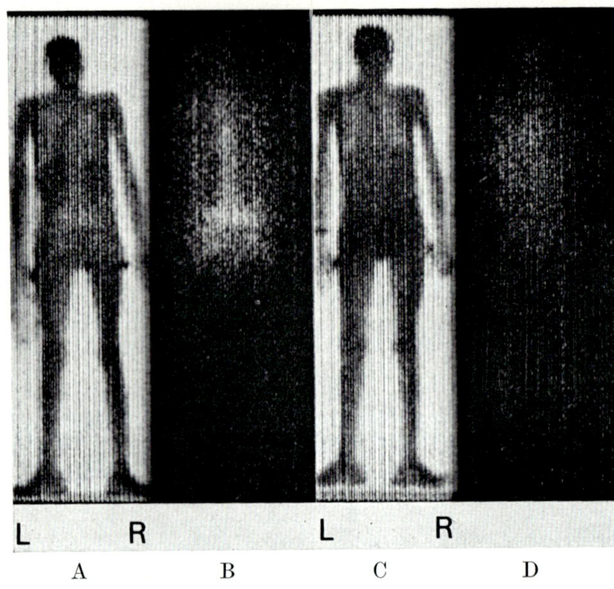

Fig. 21.5 A—D. Whole-body scans (posterior views) at 24 hours and 2 weeks after radioiron administration in patient with endogenous hemochromatosis after depletion of iron stores by phlebotomy. Distribution of radionuclides (B and D) is essentially normal except for minimal splenic red-cell sequestration. A and C are ^{241}Am transmission pictures to show body outline. (Reproduced with permission from P. Ronai, H. S. Winchell, H. D. Anger and J. H. Lawrence, Journal of Nuclear Medicine **10**, 469–474, 1969)

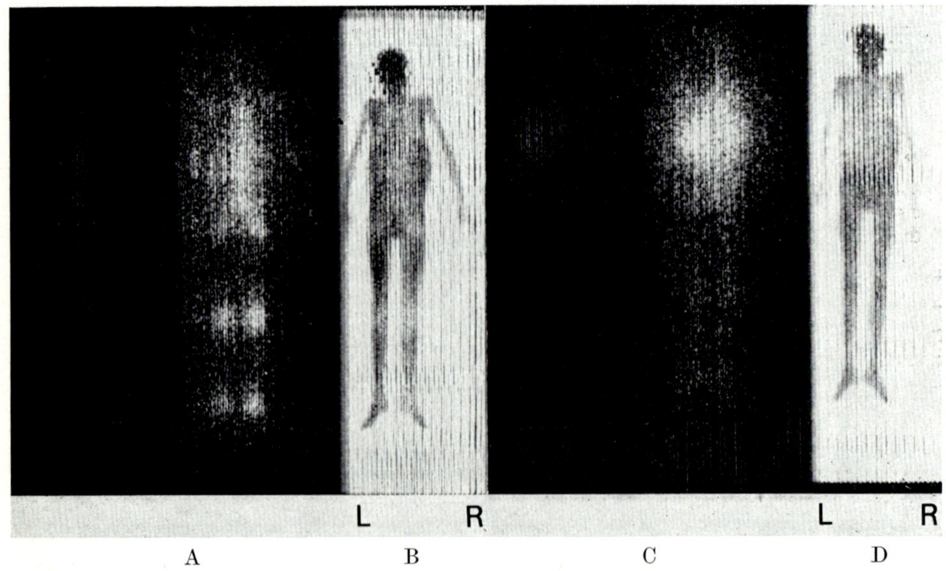

Fig. 21.6 A—D. Whole-body scans (posterior views) to show peripheral extension of erythropoietic marrow and splenic sequestration of red cells in 14-year old girl with cystic fibrosis (A and C). B and D represent ^{241}Am body outline. (Reproduced with permission from P. Ronai, H. S. Winchell, H. D. Anger and J. H. Lawrence, Journal of Nuclear Medicine **10**, 469–474, 1969)

C. Use of Americium-241 for Bone Density Measurements

An improved system for measuring bone mineral content *in vivo* using photon absorption techniques has been developed by CAMERON and co-workers at the University of Wisconsin (1963–1967). This method requires an essentially monochromatic photon source; americium-241 whose predominant photon emission has an energy of 59.6 KeV is one of the sources that has been used for this purpose. The collimated source is located below and a collimated scintillation detector above the limb to be measured as shown in Fig. 21.7.

The amount of bone mineral in the path traversed by the photon beam is given by the equation:

$$M_B = \frac{\varrho_B \ln\left(\frac{I_0^*}{I}\right)}{(\mu_B)(\varrho_B) - (\mu_S)(\varrho_S)}$$

M_B = bone mineral mass per unit area in radiation beam path (gm/cm²); I_0^* = beam intensity after passing through soft tissue; μ_B = mass absorption coefficient of bone mineral; μ_S = mass absorption coefficient of soft tissue; ϱ_B = density of bone mineral; ϱ_S = density of soft tissue; I = beam intensity after passing through bone mineral and soft tissue.

For this relationship to hold true, the thickness of soft tissue or soft tissue plus bone must be constant. In practice this is accomplished by surrounding the limb to be measured with a moldable soft-tissue-equivalent plastic.

The source-detector combination is passed transversely over the limb to be measured. Photon intensity measurements are recorded, yielding a plot similar to that shown in Fig. 21.8 where the cross-hatched area is proportional to the bone mineral mass per unit length.

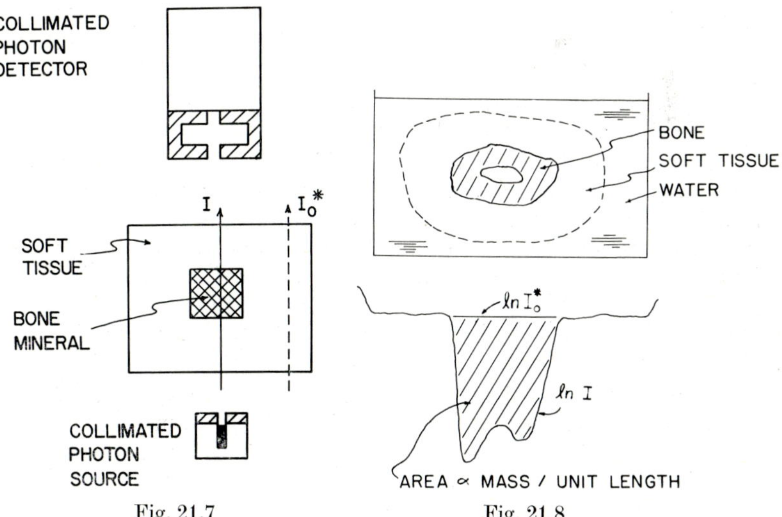

Fig. 21.7 Fig. 21.8

Fig. 21.7. Schematic drawing of bone densitometer as developed by CAMERON. (Reproduced with permission from J. A. SORENSON and J. R. CAMERON, Journal of Bone and Joint Surgery **49** A, 481–497, 1967)

Fig. 21.8. Relation of photon beam absorption to bone mineral content as indicated by the scan area. (Reproduced with permission from J. A. SORENSON and J. R. CAMERON, Journal of Bone and Joint Surgery **49** A, 481–498, 1967)

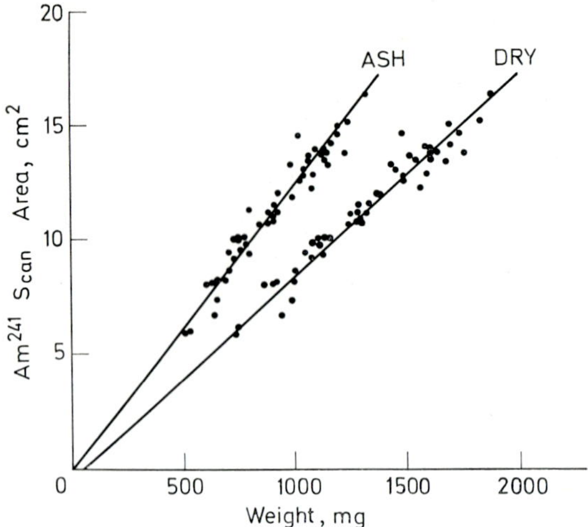

Fig. 21.9. Relationship of scanning area to weight for human bone. (Reproduced with permission from J. R. CAMERON, R. B. MAZESS and J. A. SORENSON, Investigative Radiology 3, 141, 1968)

In Fig. 21.9 the relationship of the observed scanning area to the dry, fat-free weight of ash weight for sections of human bone is shown. Instruments employing this principle are now commercially available and are being used to assess bone mineral content in healthy or diseased human subjects more accurately and at a much lower radiation dose to the subject than was possible with earlier densitometric determination of bone mineral using X-rays.

III. Californium-252

Californium was discovered in 1950 by THOMPSON, STREET, GHIORSO, and SEABORG (1950) at Berkeley. Small amounts of curium-242 were bombarded with 35 MeV helium ions in the 60-inch cyclotron, producing small amounts of californium-245. Subsequently, californium-252 has been produced by bombarding plutonium-239 with neutrons. It results from the successive capture of 13 neutrons. The availability of very high flux reactors makes it practical to produce up to milligram quantities of californium-252 for individual radiation sources and total quantities up to grams per year.

The property of californium-252 which distinguishes it from other readily available man-made isotopes is that it emits a large number of neutrons by spontaneous fission. Californium-252 decays by two modes: by emission of an alpha particle with a half-life of 2.6 years and by spontaneous fission with a half-life of 85.5 years. The spontaneous fission results in mixed neutron-gamma ray emission which makes the nuclide of interest for medical purposes. About 37% of the absorbed dose in tissue results from the gamma component, which is in the 0.5 to 1.0 MeV range. The remainder of the dose is contributed by the fission spectrum neutrons having an average energy of 2.3 MeV and a modal energy of about 1 MeV. The radiological properties are summarized in Table 21.1.

Table 21.1. Nuclear properties of ^{252}Cf

Effective half life	2.646 ± 0.004 years (METTA et al., 1965)
Alpha decay half-life	2.731 ± 0.007 years (METTA et al., 1965)
Spontaneous fission half-life	85.5 ± 0.5 years (METTA et al., 1965)
Average neutron energy	2.348 MeV (MEADOWS, 1967)
Neutron energy spectrum maximum	1.0 MeV (REINING, 1968)
Gamma emission rate	1.3×10^{13} photons per second per gram (STODDARD, 1965)
Neutrons per spontaneous fission	3.76 (FILLMORE, 1968)
Specific neutron activity	4.4×10^{9} neutrons per second per curie
Neutron emission rate	2.34×10^{12} neutrons per second per gram

A. Use of Californium-252 in Cancer Therapy

There are a number of theoretical reasons why neutrons should be more effective in cancer therapy than gamma or X-radiation. While tumors are well oxygenated at the periphery, toward the center the oxygen concentration decreases rapidly. At the same time, it is known that the sensitivity of cells to ionizing radiations increases with the oxygen concentration within the cell. One rationale for the use of californium-252 is based on the expectation that the oxygen enhancement ratio (OER[2]) for californium-252 neutrons is significantly smaller than for gamma rays. This is important since it is felt that one of the reasons for the failure to cure tumors by conventional radiation therapy may be due to the relative radioresistance of the more centrally placed hypoxic cells. On the other hand, with densely ionizing radiation (LET[3] > 20 KeV/μ), each "hit" or event may be lethal to a cell, with no apparent recovery following single, or between multiple doses (BARENDSEN et al., 1960; BARENDSEN, 1962). With high LET, the presence of oxygen becomes less important to the biological event. Several varieties of high LET radiations are available but the most practical one at present for clinical purposes is fast neutron radiation.

Generator- or cyclotron-produced neutrons have a wide range of LET, ranging from 10 to 1000 KeV/μ, but are mainly between 10–30 KeV/μ. Fission spectrum neutrons are more attractive in that their range is about 50–60 KeV/μ. There are, however, a number of problems associated with obtaining a suitable flux of neutrons for clinical purposes from generators, cyclotrons and reactors, such as cost, equipment maintenance, and accessibility. Most of the disadvantages could be overcome by using californium-252 as the fission neutron source in radiotherapy, as first suggested by SCHLEA and STODDARD (1965).

The neutrons from californium-252 have relatively poor penetration in tissue; hence the use of this isotope may be restricted to intracavitary and interstitial applications. The considerable flexibility in source design will allow treatment of any sites that could be treated with radium or similar sources.

Since the idea of the dependence of radiosensitivity on oxygenation is central to the rationale for the use of californium-252 neutrons, it is essential that a number of radiobiological measurements be made before proceeding with the clinical application of californium.

Several medical centers are presently participating in an evaluation program for the medical uses of californium. A number of studies are in progress or have been completed concerning the oxygen enhancement ratio (OER), relative bio-

2 OER: The ratio of doses which give the same biological effect in the absence and presence of oxygen.

3 LET: Linear Energy Transfer, i.e., the rate of energy loss along the path of the individual ionizing radiation (KeV/μ).

Table 21.2. OER for ^{252}Cf

Test system	Dose/rate	OER	Reference
V. faba roots	16 rads/h	1.66	FILLMORE (1968)
Chinese hamster ovary cells *in vitro*	60 rads/h	1.75	BUSHONG et al. (1970)
^{388}P Leukemia cells	195 rads/h	1.3:1	BERRY et al. (1970)
HeLa cells	16 rads/h	1.2	FAIRCHILD et al. (1970)

logical effect (RBE[4]), tolerance dose, dose-rate effect, and the dose distribution. Results to date indicate that the interstitial or intracavitary use of californium-252 may be quite effective in the therapy of certain types of cancer tumors. Some of the studies are summarized below.

Investigations using several test systems show that the OER for californium-252 is significantly lower than for X- or gamma-rays. Additionally, it approximates that obtained for neutrons from sources such as generators, cyclotrons, and reactors (BRENNAN, 1969). They OER's reported to date are summarized in Table 21.2.

The OER for seedlings of *Vicia faba* exposed at a rate of 16.02 rads per hour was 1.66 (HALL and FAIRCHILD, 1970), which is identical to that previously reported for acute exposures to 14 MeV neutrons measured with the same system (HALL and CAVANAGH, 1967). This compares with an OER of 2.02 for radium and 2.58 for X-rays using *V. faba* seedlings. Based on the cytogenetic effects in Chinese hamster ovary cells an OER of 1.75 was obtained (BUSHONG et al., 1970), compared with an OER of 2.5 using radium (FOWLER, 1967). The reproductive survival of ^{388}P leukemia cells assessed *in vivo* by their ability to produce tumors in isologous recipient mice after irradiation *in vitro* indicated a therapeutic gain factor of 1.3:1 for californium-252 compared with radium-226 (BERRY et al., 1970). For protracted irradiations of human cancer cells (HeLa cells) in culture, an OER of about 1.2 was obtained (FAIRCHILD et al., 1970).

In the early forties, a series of patients were treated for cancer with fast neutrons (STONE, 1948). Damage to normal tissues occurred in these patients, and subsequently the use of fast neutrons for clinical purposes was discouraged for many years. It is believed now that the late damage that occurred was due to an overexposure based on the fact that the RBE for neutrons was actually higher than was assumed in calculating the exposure doses. A number of studies are now under way to determine the RBE for fast neutrons from californium-252 sources and are summarized in Table 21.3.

Over the range of dose rates tested and depending on the system used, the RBE of californium-252 neutrons varies from about 2.5 to 10. In initial human studies, a dose rate of about 15 to 20 rads per hour will probably be used. However, the total time in which this dose would be given is still undetermined. Studies with swine skin (ATKINS et al., 1970) show a skin tolerance level of 1725 rads (end point was moist desquamation) compared to 8400 rads with radium exposures, for an RBE of 4.9. They also concluded that any increase in dose rate of californium-252 would have little effect on the skin tolerance level. Cellular repopulation and recovery appear to be minimal during the irradiation period, so changes in the RBE's may depend mostly upon the dose rate of the gamma or X-ray sources with which californium-252 is being compared.

On the basis of these preliminary studies, the effectiveness of californium-252 cancer radiotherapy and undesired consequences to normal tissues has been

4 RBE: The inverse ratio of the amount of absorbed radiation required to produce a given effect, to a standard radiation required to produce the same effect.

Table 21.3. RBE-^{252}Cf

Test system	^{252}Cf dose/rate	Comparison radiation dose/rate	RBE[a]	Reference
V. faba roots	16 rads/h	Radium: 50 rads/h	6.5	HALL (1970)
V. faba roots	82 rads/h	Radium: 50 rads/h	5.3	HALL (1970)
HeLa cells	16.1 rads/h	Radium: 31 rads/h	2.86	FAIRCHILD et al. (1969)
HeLa cells	16.1 rads/h	250 KVP X-ray 111 rads/min	2.34	FAIRCHILD et al. (1969)
Micro colony assay mouse small int.	34 rads/h	Radium: 34 rads/h	5–10	WITHERS et al. (1970)
Pig skin	12, 15, 23 rads/h	Radium: 35, 68 rads/h ^{137}Cs: 192 rads/h	2.5–5	ATKINS et al. (1970)

[a] The RBE's determined include the gamma component of the ^{252}Cf; if based solely on the neutron component, they would be higher.

Table 21.4. Californium-252 implants in patients[a]

Patient number	Region implanted	Total quantity (μg)	Treatment (time/hr)	Treatment (dose rate) (rad/hr) (neutron, gamma)	Biological (dose/rate[b]) (rem/hr)	Type implant
1 (11–13–69)	Inguinal node	18	109	7.0, 3.5	56	Single plane
2 (2–18–70)	Vaginal	23	32	10.0, 5.0	80	Volume
3 (3–17–70)	Inguinal node	10	100	5.0, 2.5	40	Single plane
4 (4–30–70)	Submandibular node	15	40	7.0, 4.0	56	Double plane
5 (5–4–70)	Vaginal	21	24	5.0, 2.8	40	Volume
6 (6–23–70)	Lip	10	30	8.0, 4.0	64	Single plane
7 (7–9–70)	Skin of cheek	13.5	67	8.0, 3.5	64	Single plane
8 (7–14–70)	Skin metastasis	25	70.5	10.0, 10.0	100	Single plane
9 (7–28–70)	Lip	11.5	26	10.0, 4.5	79	Single plane
10 (8–25–70)	Vaginal	16	51	3.0, 2.0	25	Volume
11 (9–1–70)	Vaginal	23.7	25	8.0, 4.0	64	Volume
12 (9–28–70)	Urethal	10.5	69	5.0, 3.0	40	Volume
13 (10–1–70)	Neck	23	65.7	8.0, 4.0	64	Curved plane
14 (10–6–70)	Tongue	12	70.3	7.0, 4.0	57	Double plane
15 (11–10–70)	Tongue	18.5	67	5.0, 3.0	41	Volume

[a] This series of patients is from the University of Texas, M. D. Anderson Hospital and Tumor Institute. Table taken from *Californium-252 Progress*, Number 6, page 13, January 1971, USAEC.

[b] Biological dose rate derived by weighting the physical dose rate according to the formula $7.5 D_{neutron} + 1.0\ D_{gamma}$.

studied in carefully selected patients with advanced cancers (Table 21.4). The results to date permit cautious optimism and merit further studies in humans in this phase of the program. Following the initial evaluation of these limited studies, a determination will be made as to the justification of controlled and comparative clinical trials at various medical centers.

References

ANGER, H. O.: Whole-body scanner Mark II. J. nucl. Med. **7**, 311 (1966).

ATKINS, H. L., FAIRCHILD, R. G., ROBERTSON, J. S., LEHMAN, W. P.: Comparison of irradiation by californium-252 and radium on the skin of swine. Radiology **96**, 161–165 (1970).

BARENDSEN, G. W.: Dose-survival curves of human cells in tissue culture irradiated with alpha-, beta, 20 KV X- and 200 KV X-radiation. Nature (Lond.) **193**, 1153–1155 (1962).

Barendsen, G. W., Bensker, T. L. J., Vergroesen, A. J., Budke, L.: Effects of different ionizing radiations on human cells in tissue culture. II. Biological experiments. Radiat. Res. **13**, 841–849 (1960).

Berry, R. J., Brennan, J. T., Mettler, F. A., Werner, J. L.: Survival of murine leukemia cells exposed to ^{252}Cf neutrons under conditions simulating implantation of these radiation sources with tissue in radiotherapy. Abstract Cd-7, 18th Annual Meeting, Radiation Research Society. Radiat. Res. **43**, 215–216 (1970).

Brennan, J. T.: Fast neutrons for radiation therapy. Radiol. Clin. N. Amer. **7**, No 2, 365–374 (1969).

Bushong, S. C., Prasad, N., Briney, S. A., Oliver, G. D.: The oxygen enhancement ratio of californium-252, Abstract Cd-5, 18th Annual Meeting, Radiation Research Society. Radiat. Res. **43**, 214–215 (1970).

Cameron, J. R., Sorenson, J.: Measurement of bone mineral *in vivo*: An improved method. Science **142**, 230–232 (1963).

Fairchild, R. G., Drew, R. M., Atkins, H. L.: The relative biological effect of ^{252}Cf radiation on HeLa cells in culture. Radiology **93**, 1187–1189 (1969).

Fairchild, R. G., Drew, R. M., Atkins, H. L.: The oxygen enhancement ratio for protracted irradiation with ^{252}Cf. Radiology **96**, 661–665 (1970).

Fillmore, F. L.: Recommended values for the number of neutrons per fission. J. nucl. Energy **22**, 79–97 (1968).

Fowler, J.: In: Modern trends in radiotherapy (T. J. Deeley and C. A. Woods, eds.). London: Butterworth 1967.

Ghiorso, A., James, R. A., Morgan, L. O., Seaborg, G. T.: Preparation of transplutonium isotopes by neutron irradiation. Phys. Rev. **78**, 472 (1950).

Hall, E. J.: RBE and OER measurements for the mixed neutron and gamma-ray emission from californium-252, Abstract Cd-4, 18th Annual Meeting, Radiation Research Society. Radiat. Res. **43**, 214 (1970).

Hall, E. J., Cavanagh, J.: The oxygen effect for acute and protracted radiation exposures measured with seedlings of *vicia faba*. Brit. J. Radiol. **40**, 128–133 (1967).

Hall, E. J., Fairchild, R. G.: Radiobiological measurements with californium-252. Brit. J. Radiol. **43**, 263–266 (1970).

Hoffer, P. B., Charleston, D. B., Beck, R. N., Gottschalk, A.: Fluorescent scanning; a new investigative technique. In: Medical radioisotope scintigraphy, vol. 1, p. 261–271. Vienna: International Atomic Energy Agency 1969.

Hoffer, P. B., Gottschalk, A.: Fluorescent thyroid scanning. Radiology **99**, 117–124 (1971).

Hoffer, P. B., Polcyn, R. E., Moody, R., Lowe, H. J., Gottschalk, A.: Fluorescent detection: Application to the study of cerebral blood flow. J. nucl. Med. **10**, 651–653 (1969).

Maynard, C. D.: In: Clinical nuclear medicine. Philadelphia: Lea and Febiger 1969.

Meadows, J. W.: ^{252}Cf fission neutron spectrum from 0.003 to 15.0 MeV. Phys. Rev. **157**, 1076–1082 (1967).

McGavack, B. A.: In: The thyroid. St. Louis: C. V. Mosby 1951.

Metta, D., Diamond, H., Barnes, R. F., Milsted, J., Gray, J., Jr., Henderson, D. J., Stevens, C. M.: Nuclear constants of nine transplutonium nuclides. J. inorg. nucl. Chem. **27**, 33–39 (1965).

Reining, W. C.: Advantages and applications of ^{252}Cf as a neutron source. Nucl. appl. **5**, 24–25 (1968).

Ronai, P., Winchell, H. S., Anger, H. O., Lawrence, J. H.: Whole-body scanning of ^{59}Fe for evaluating body distribution of erythropoietic marrow, splenic sequestration of red cells and hepatic disposition of iron. J. nucl. Med. **10**, 469–474 (1969).

Schlea, C. S., Stoddard, D. H.: Californium isotopes proposed for intracavitary and interstitial radiation therapy with neutrons. Nature (Lond.) **206**, 1058–1059 (1965).

Seaborg, G. T., James, R. A., Morgan, L. O.: The new element americium (atomic number 95). In: The transuranium elements, vol. IV–14B, Paper 22.1, Nat. nucl. ener. Series, p. 1525–1553. New York: McGraw-Hill Book Co. 1949. See also Ghiorso et al., 1950.

Sorenson, J. A., Cameron, J. R.: A reliable *in vivo* measurement of bone mineral content. J. Bone Jt Surg. A **49**, 481–497 (1967).

Stoddard, D. H.: Radiation properties of californium-252. USAEC Report DP-986, E. I. du Pont de Nemours and Co., Savannah River Lab., Aiken, S. C. (1965).

Stone, R. S.: Neutron therapy and specific ionization. Janeway Memorial Lecture. Amer. J. Roentgenol. **59**, 771–785 (1948).

Thompson, S. G., Street, K., Jr., Ghiorso, A., Seaborg, G. T.: The new element californium (atomic number 98). Phys. Rev. **80**, 790–796 (1950).

Withers, H. R., Oliver, G., Glenn, D. W.: Response of mouse intestine to irradiation by californium-252 neutrons. Abstract Cd-8, 18th Annual Meeting, Radiation Research Society. Radiat. Res. **43**, 216 (1970).

Author Index

Page numbers in *italics* refer to the bibliography

Subject Index

Prepared by Anne de G. Low-Beer

Handbuch der experimentellen Pharmakologie / Handbook of Experimental Pharmacology

Heffter – Heubner. New Series

Vol. IV: **General Pharmacology**
ISBN 3-540-04845-6 DM 76,00; US $34.30

Vol. X: **Die Pharmakologie anorganischer Anionen**
ISBN 3-540-01465-9 DM 205,00; US $92.40

Vol. XI: **Lobelin und Lobeliaalkaloide**
ISBN 3-540-01910-3 DM 16,00; US $7.30

Vol. XII: **Morphin und morphinähnlich wirkende Verbindungen**
ISBN 3-540-02158-2 DM 79,00; US $35.60

Vol. XIII: **The Alkali Metal Ions in Biology.** In preparation

Vol. XIV:

Part 1 **The Adrenocortical Hormones I**
ISBN 3-540-02830-7 DM 260,00; US $117.10

Part 2 **The Adrenocortical Hormones II**
ISBN 3-540-03146-4 DM 75,00; US $33.80

Part 3 **The Adrenocortical Hormones III**
ISBN 3-540-04147-8 DM 152,00; US $68.50

Vol. XV: **Cholinesterases and Anticholinesterase Agents**
ISBN 3-540-02988-5 DM 328,00; US $147.80

Vol. XVI: **Erzeugung von Krankheitszuständen durch das Experiment**

Part 1 **Blut / Blood.** In preparation

Part 2 **Atemwege**
ISBN 3-540-04517-1 DM 124,00; US $55.90

Part 3 **Herz und Kreislauf / Heart and Circulation.** In preparation

Part 4 **Niere, Nierenbecken, Blase**
ISBN 3-540-03305-X DM 152,00; US $68.50

Part 5 **Leber.** In preparation

Part 6 **Schilddrüse.** In preparation

Part 7 **Zentralnervensystem**
ISBN 3-540-02831-5 DM 123,00; US $55.40

Part 8 **Stütz- und Hartgewebe**
ISBN 3-540-04518-X DM 98,00; US $44.20

Part 9 **Infektionen I**
ISBN 3-540-03147-2 DM 152,00; US $68.50

Part 10 **Infektionen II**
ISBN 3-540-03531-1 DM 169,00; US $76.20

Part 11 A **Infektionen III**
ISBN 3-540-03840-X DM 150,00; US $67.60

Part 11 B **Infektionen IV.** In preparation

Part 12 **Tumoren I**
ISBN 3-540-03532-X DM 150,00; US $67.60

Part 13 **Tumoren II**
ISBN 3-540-03533-8 DM 90,00; US $40.60

Part 14 **Tumoren III.** In preparation

Part 15 **Kohlenhydratstoffwechsel, Fieber / Carbohydrate Metabolism, Fever**
ISBN 3-540-03534-6 DM 150,00; US $67.60

Vol. XVII:

Part 1 **Ions, alcalino-terreux I. Systèmes isolés**
ISBN 3-540-02989-3 DM 162,00; US $73.00

Part 2 **Ions alcalino-terreux II. Organismes entiers**
ISBN 3-540-03148-0 DM 218,00; US $98.20

Vol. XVIII:			
Part 1	**Histamine**		
	ISBN 3-540-03535-4	DM 202,00;	US $91.00
Part 2	**Anti-Histaminics.** In preparation		
Vol. XIX:	**5-Hydroxytryptamine and Related Indolealkylamines**		
	ISBN 3-540-03536-2	DM 202,00;	US $91.00
Vol. XX:			
Part 1	**Pharmacology of Fluorides I**		
	ISBN 3-540-03537-0	DM 158,00;	US $71.20
Part 2	**Pharmacology of Fluorides II**		
	ISBN 3-540-04846-4	DM 146,00;	US $65.80
Vol. XXI:	**Beryllium**		
	ISBN 3-540-03538-9	DM 40,00;	US $18.10
Vol. XXII:			
Part 1	**Die Gestagene I**		
	ISBN 3-540-04148-6	DM 320,00;	US $144.20
Part 2	**Die Gestagene II**		
	ISBN 3-540-04519-8	DM 320,00;	US $144.20
Vol. XXIII:	**Neurohypophysial Hormones and Similar Polypeptides**		
	ISBN 3-540-04149-4	DM 192,00;	US $86.50
Vol. XXIV:	**Diuretica**		
	ISBN 3-540-04520-1	DM 248,00;	US $111.70
Vol. XXV:	**Bradykinin, Kallidin and Kallikrein**		
	ISBN 3-540-04847-2	DM 248,00;	US $111.70
Vol. XXVI:	**Vergleichende Pharmakologie von Überträgersubstanzen in tiersystematischer Darstellung**		
	ISBN 3-540-05132-5	DM 148,00;	US $66.70
Vol. XXVII:	**Anticoagulantien**		
	ISBN 3-540-05133-3	DM 198,00;	US $89.20
Vol. XXVIII:			
Part 1	**Concepts in Biochemical Pharmacology I**		
	ISBN 3-540-05134-1	DM 174,00;	US $78.40
Part 2	**Concepts in Biochemical Pharmacology II**		
	ISBN 3-540-05389-1	DM 248,00;	US $111.70
Part 3	**Concepts in Biochemical Pharmacology III.** In preparation		
Vol. XXIX:	**Oral wirksame Antidiabetika**		
	ISBN 3-540-05554-1	DM 296,00;	US $133.40
Vol. XXX:	**Modern Inhalation Anesthetics**		
	ISBN 3-540-05135-X	DM 198,00;	US $89.20
Vol. XXXI:	**Antianginal Drugs**		
	ISBN 3-540-05365-4	DM 168,00;	US $75.70
Vol. XXXII:			
Part 1	**Insulin I**		
	ISBN 3-540-05470-7	DM 248,00;	US $111.70
Part 2	**Insulin II.** In preparation		
Vol. XXXIII:	**Catecholamines**		
	ISBN 3-540-05517-7	DM 396,00;	US $178.40
Vol. XXXIV:	**Secretin, Cholecystokinin-Pancreozymin and Gastrin**		
	ISBN 3-540-05952-0	DM 165,00;	US $74.40
Vol. XXXV:			
Part 1	**Androgene I**		
	ISBN 3-540-05706-4	DM 298,00;	US $134.30
Part 2	**Androgene II und Antiandrogene.** In preparation		
Vol. XXXVI:	**Uranium-Plutonium-Transplutonic Elements**		
	ISBN 3-540-06168-1	DM 374,00;	US $168.50
Vol. XXXVII:	**Angiotensin**		
	ISBN 3-540-06276-9	DM 224,00;	US $100,90
Vol. XXXVIII:	**Antineoplastic and Immunosuppressive Agents.**		
	ISBN 3-530-06402-8	In preparation